Statistics

UNLOCKING THE POWER OF DATA

SECOND EDITION

Robin H. Lock
St. Lawrence University

Patti Frazer Lock
St. Lawrence University

Kari Lock Morgan
Pennsylvania State University

Eric F. Lock
University of Minnesota

Dennis F. Lock
Miami Dolphins

WILEY

VICE PRESIDENT AND DIRECTOR	Laurie Rosatone
SENIOR ACQUISITIONS EDITOR	Joanna Dingle
DEVELOPMENTAL EDITOR	Adria Giattino
FREELANCE DEVELOPMENTAL EDITOR	Anne Scanlan-Rohrer
EDITORIAL ASSISTANT	Giana Milazzo
SENIOR CONTENT MANAGER	Valerie Zaborski
SENIOR PRODUCTION EDITOR	Laura Abrams
MARKETING MANAGER	John LaVacca
SENIOR PRODUCT DESIGNER	Tom Kulesa
DESIGNER	Thomas Nery
PHOTO EDITOR	Billy Ray
COVER DESIGN	Thomas Nery
COVER PHOTO	© Simone Brandt/Alamy Limited

This book was set in 10/12 TimesTen by SPi Global, and printed and bound by Quad/Graphics. The cover was printed by Quad/Graphics.

This book is printed on acid free paper. ∞

Founded in 1807, John Wiley & Sons, Inc. has been a valued source of knowledge and understanding for more than 200 years, helping people around the world meet their needs and fulfill their aspirations. Our company is built on a foundation of principles that include responsibility to the communities we serve and where we live and work. In 2008, we launched a Corporate Citizenship Initiative, a global effort to address the environmental, social, economic, and ethical challenges we face in our business. Among the issues we are addressing are carbon impact, paper specifications and procurement, ethical conduct within our business and among our vendors, and community and charitable support. For more information, please visit our website: www.wiley.com/go/citizenship.

Evaluation copies are provided to qualified academics and professionals for review purposes only, for use in their courses during the next academic year. These copies are licensed and may not be sold or transferred to a third party. Upon completion of the review period, please return the evaluation copy to Wiley. Return instructions and a free of charge return mailing label are available at www.wiley.com/go/returnlabel. If you have chosen to adopt this textbook for use in your course, please accept this book as your complimentary desk copy. Outside of the United States, please contact your local sales representative.

The inside back cover will contain printing identification and country of origin if omitted from this page. In addition, if the ISBN on the back cover differs from the ISBN on this page, the one on the back cover is correct.

ISBN: 978-1-119-30884-3

Printed in the United States of America

V10018677_052220

StatKey to accompany *Statistics: Unlocking the Power of Data*
by Lock, Lock, Lock, Lock, and Lock

Descriptive Statistics and Graphs	Bootstrap Confidence Intervals	Randomization Hypothesis Tests
One Quantitative Variable	CI For Single Mean, Median, St.Dev.	Test For Single Mean
One Categorical Variable	CI For Single Proportion	Test for Single Proportion
One Quantitative and One Categorical Variable	CI For Difference In Means	Test For Difference in Means
Two Categorical Variables	CI For Difference In Proportions	Test For Difference In Proportions
Two Quantitative Variables	CI For Slope, Correlation	Test For Slope, Correlation

Sampling Distributions	Mean		Proportion	
Theoretical Distributions	Normal	t	χ^2	F
More Advanced Randomization Tests	χ^2 Goodness-of-fit	χ^2 Test for Association	ANOVA for Difference in Means	ANOVA for Regression

StatKey Randomization Test for a Difference in Means

Leniency and Smiles ▾   Show Data Table   Edit Data   Upload File   Change Column(s)

Randomization method   Reallocate Groups ▾

Generate 1 Sample   Generate 10 Samples   Generate 100 Samples   Generate 1000 Samples   Reset Plot

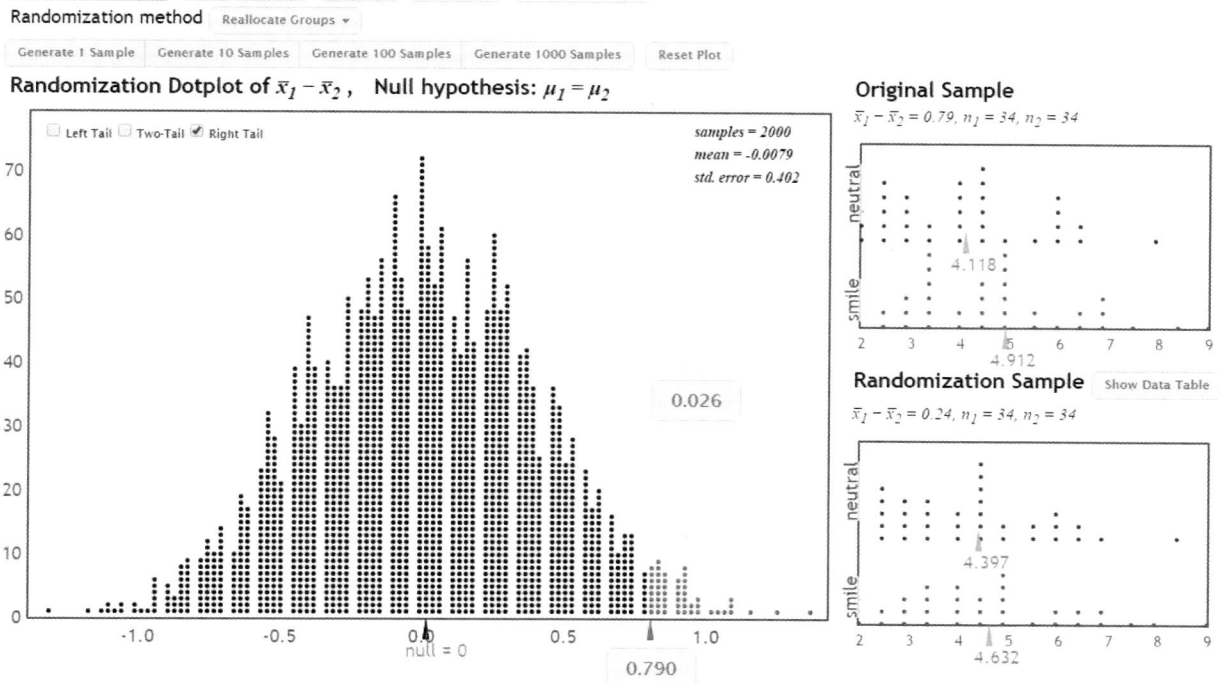

Randomization Dotplot of $\bar{x}_1 - \bar{x}_2$, Null hypothesis: $\mu_1 = \mu_2$

samples = 2000
mean = -0.0079
std. error = 0.402

0.026

null = 0

0.790

Original Sample
$\bar{x}_1 - \bar{x}_2 = 0.79, n_1 = 34, n_2 = 34$

4.118
4.912

Randomization Sample   Show Data Table
$\bar{x}_1 - \bar{x}_2 = 0.24, n_1 = 34, n_2 = 34$

4.397
4.632

You will find *StatKey* and many additional resources (including short, helpful videos for all examples and all learning goals; electronic copies of all datasets; and technology help for a wide variety of platforms) at

www.wiley.com/college/lock

You can also find *StatKey* at

www.lock5stat.com/statkey

A message from the Locks

Data are everywhere—in vast quantities, spanning almost every topic. Being able to make sense of all this information is becoming both a coveted and necessary skill. This book will help you learn how to effectively collect and analyze data, enabling you to investigate any questions you wish to ask. The goal of this book is to help you unlock the power of data!

An essential component of statistics is **randomness**. Rather than viewing randomness as a confused jumble of numbers (as the random number table on the front cover might appear), you will learn how to use randomness to your advantage, and come to view it as one of the most powerful tools available for making new discoveries and bringing clarity to the world.

CONTENTS

"Statistical thinking will one day be as necessary a qualification for efficient citizenship as the ability to read and write."

–H.G. Wells

Why We Wrote this Book

Helping students make sense of data will serve them well in life and in any field they might choose. Our goal in writing this book is to help students understand, appreciate, and use the power of statistics and to help instructors teach an outstanding course in statistics.

The text is designed for use in an introductory statistics course. The focus throughout is on data analysis and the primary goal is to enable students to effectively collect data, analyze data, and interpret conclusions drawn from data. The text is driven by real data and real applications. Although the only prerequisite is a minimal working knowledge of algebra, students completing the course should be able to accurately interpret statistical results and to analyze straightforward datasets. The text is designed to give students a sense of the power of data analysis; our hope is that many students learning from this book will want to continue developing their statistical knowledge.

Students who learn from this text should finish with

- A solid conceptual understanding of the key concepts of statistical inference: estimation with intervals and testing for significance.
- The ability to do straightforward data analysis, including summarizing data, visualizing data, and inference using either traditional methods or modern resampling methods.
- Experience using technology to perform a variety of different statistical procedures.
- An understanding of the importance of data collection, the ability to recognize limitations in data collection methods, and an awareness of the role that data collection plays in determining the scope of inference.
- The knowledge of which statistical methods to use in which situations and the ability to interpret the results effectively and in context.
- An awareness of the power of data.

Building Conceptual Understanding with Simulation Methods

This book takes a unique approach of utilizing computer simulation methods to introduce students to the key ideas of statistical inference. Methods such as bootstrap intervals and randomization tests are very intuitive to novice students and capitalize on visual learning skills students bring to the classroom. With proper use of computer support, they are accessible at very early stages of a course with little formal background. Our text introduces statistical inference through these resampling and randomization methods, not only because these methods are becoming increasingly important for statisticians in their own right but also because they are outstanding in building students' conceptual understanding of the key ideas.

Our text includes the more traditional methods such as t-tests, chi-square tests, etc., but only after students have developed a strong intuitive understanding of inference through randomization methods. At this point students have a conceptual understanding and appreciation for the results they can then compute using the more traditional methods. We believe that this approach helps students realize that although the formulas may take different forms for different types of data, the conceptual framework underlying most statistical methods remains the same. Our experience has been that after using the intuitive simulation-based methods to introduce the core ideas, students understand and can move quickly through most of the traditional techniques.

Sir R.A. Fisher, widely considered the father of modern statistics, said of simulation and permutation methods in 1936:

"Actually, the statistician does not carry out this very simple and very tedious process, but his conclusions have no justification beyond the fact that they agree with those which could have been arrived at by this elementary method."

Modern technology has made these methods, too 'tedious' to apply in 1936, now readily accessible. As George Cobb wrote in 2007:

"... despite broad acceptance and rapid growth in enrollments, the consensus curriculum is still an unwitting prisoner of history. What we teach is largely the technical machinery of numerical approximations based on the normal distribution and its many subsidiary cogs. This machinery was once necessary, because the conceptually simpler alternative based on permutations was computationally beyond our reach. Before computers statisticians had no choice. These days we have no excuse. Randomization-based inference makes a direct connection between data production and the logic of inference that deserves to be at the core of every introductory course."

Building Understanding and Proficiency with Technology

Technology is an integral part of modern statistics, but this text does not require any specific software. We have developed a user-friendly set of online interactive dynamic tools, *StatKey*, to illustrate key ideas and analyze data with modern simulation-based methods. *StatKey* is freely available with data from the text integrated. We also provide **Companion Manuals**, tied directly to the text, for other popular technology options. The text uses many real datasets which are electronically available in multiple formats.

Building a Framework for the Big Picture: Essential Synthesis

One of the drawbacks of many current texts is the fragmentation of ideas into disjoint pieces. While the segmentation helps students understand the individual pieces, we believe that integration of the parts into a coherent whole is also essential. To address this we have sections called Essential Synthesis at the end of each unit, in which students are asked to take a step back and look at the big picture. We hope that these sections, which include case studies, will help to prepare students for the kind of statistical thinking they will encounter after finishing the course.

Building Student Interest with Engaging Examples and Exercises

This text contains over 300 fully worked-out examples and over 1800 exercises, which are the heart of this text and the key to learning statistics. One of the great things about statistics is that it is relevant in so many fields. We have tried to find studies and datasets that will capture the interest of students—and instructors! We hope all users of this text find many fun and useful tidbits of information from the datasets, examples, and exercises, above and beyond the statistical knowledge gained.

The exercise sets at the end of every section assess computation, interpretation, and understanding using a variety of problem types. Some features of the exercise sets include:

- *Skill Builders.* After every section, the exercise set starts with skill-building exercises, designed to be straightforward and to ensure that students have the basic skills and confidence to tackle the more involved problems with real data.

- *Lots of real data.* After the opening skill builders, the vast majority of the exercises in a section involve real data from a wide variety of disciplines. These allow students to practice the ideas of the section and to see how statistics is used in actual practice in addition to illustrating the power and wide applicability of statistics. These exercises call for interpretations of the statistical findings in the context of a real situation.

- *Exercises using technology.* While many exercises provide summary statistics, some problems in each exercise set invite students to use technology to analyze raw data. All datasets, and software-specific companion manuals, are available electronically.

- *Essential synthesis and review.* Exercises at the end of each unit let students choose from among a wider assortment of possible approaches, without the guiding cues associated with section-specific exercise sets. These exercises help students see the big picture and prepare them for determining appropriate analysis methods.

Building Confidence with Robust Student and Instructor Resources

This text has many additional resources designed to facilitate and enhance its use in teaching and learning statistics. The following are all readily accessible and organized to make them easy to find and easy to use. Almost all were written exclusively by the authors.

Resources for students and instructors:
- *StatKey*; online interactive dynamic tools (*www.lock5stat.com/statkey*)
- Software-specific companion manuals (*www.wiley.com/college/lock*)
- All datasets in multiple formats (*www.wiley.com/college/lock*)
- Video solutions for all examples and video tutorials for all learning goals (*www.wiley.com/college/lock*)
- WileyPLUS—an innovative, research-based online environment for effective teaching and learning
- Student solution manual with fully worked solutions to odd-numbered exercises

Resources for instructors
- Complete instructors manual with sample syllabi, teaching tips and recommended class examples, class activities, homework assignments, and student project assignments
- Short videos with teaching tips for instructors, for every section
- Detailed interactive class activities with handouts, for every section

- PowerPoint slides, for every section, with or without integrated clicker questions
- In-class example worksheets ready to go, for every section
- Clicker questions, for every section
- A variety of different types of student projects, for every unit
- Fully worked out solutions to all exercises
- Test bank with a wide variety of question types
- The full WileyPLUS learning management system at your disposal

Content and Organization

UNIT A: Data

The first unit deals with data—how to obtain data (Chapter 1) and how to summarize and visualize the information in data (Chapter 2). We explore how the method of data collection influences the types of conclusions that can be drawn and how the type of data (categorical or quantitative) helps determine the most appropriate numerical and/or graphical technique for describing a single variable or investigating a relationship between two variables. We end the unit discussing multiple variables and exploring a variety of additional ways to display data.

UNIT B: Understanding Inference

In Unit B we develop the key ideas of statistical inference—estimation and testing—using simulation methods to build understanding and to carry out the analysis. Chapter 3 introduces the idea of using information from a single sample to provide an estimate for a population, and uses a bootstrap distribution to determine the uncertainty in the estimate. In Chapter 4 we illustrate the important ideas for testing statistical hypotheses, again using simple simulations that mimic the random processes of data production.

UNIT C: Inference with Normal and t-Distributions

In Unit C we see how theoretical distributions, such as the classic, bell-shaped normal curve, can be used to approximate the distributions of sample statistics that we encounter in Unit B. Chapter 5 shows, in general, how the normal curve can be used to facilitate constructing confidence intervals and conducting hypothesis tests. In Chapter 6 we see how to estimate standard errors with formulas and use the normal or t-distributions in situations involving means, proportions, differences in means, and differences in proportions. Since the main ideas of inference have already been covered in Unit B, Chapter 6 has many very short sections that can be combined and covered in almost any order.

UNIT D: Inference for Multiple Parameters

In Unit D we consider statistical inference for situations with multiple parameters: testing categorical variables with more than two categories (chi-square tests in Chapter 7), comparing means between more than two groups (ANOVA in Chapter 8), making inferences using the slope and intercept of a regression model (simple linear regression in Chapter 9), and building regression models with more than one explanatory variable (multiple regression in Chapter 10).

The Big Picture: Essential Synthesis

This section gives a quick overview of all of the units and asks students to put the pieces together with questions related to a case study on speed dating that draws on ideas from throughout the text.

This is an optional chapter covering basic ideas of formal probability theory. The material in this chapter is independent of the other chapters and can be covered at any point in a course or omitted entirely.

Changes in the Second Edition

- **Many New Exercises.** The Second Edition includes over 300 completely new exercises, almost all of which use real data. As always, our goal has been to continue to try to find datasets and studies of interest to students and instructors.

- **Many Updated Exercises.** In addition to the many new exercises, this edition also includes over 100 exercises that have been updated with new data.

- **Multiple Variables and Data Visualization.** Chapter 2 has a new section *2.7: Multiple Variables and Data Visualization*, in which we consider a variety of creative and effective ways to visualize data with additional variables and/or data that include geographic or time variables, and illustrate the value of including additional variables.

- **Chapter 4 Reorganized.** Chapter 4 on hypothesis tests has been reorganized to focus more explicitly on one key idea at a time (hypotheses, randomization distributions, p-values, significance) before discussing additional considerations about testing.

- **Chapter 5 Rewritten.** Chapter 5 has been rewritten to improve the transition from Unit B (using simulation methods) to Unit C (using normal and t-based inference).

- **Chapter 6 Sections Re-labeled.** The sections of Chapter 6 have been re-numbered, and some of the sections have been made more concise, to further emphasize the fact that the sections are short and designed to be combined and covered in almost any order.

- **Probability Chapter Renamed.** The Probability chapter has been renamed Chapter P to further emphasize the fact that the chapter is independent of the rest of the material and can be omitted or covered at any point in the course. In addition, a new section has been added to the chapter, on density curves and the normal distribution.

- **StatKey Enhanced.** The online interactive dynamic software *StatKey* has been enhanced, including adding the option to upload whole datasets with multiple variables.

- **And Much More!** We have also made many additional edits to the text to improve the flow and clarity, keep it current, and respond to student and user feedback.

Tips for Students

- **Do the Exercises!** The key to learning statistics is to try lots of the exercises. We hope you find them interesting!

- **Videos** To aid student learning, we have created video solutions for all examples and short video tutorials for all learning goals. These are available through WileyPLUS or the Student Companion Site. Check them out!

- **Partial Answers** Partial answers to the odd-numbered problems are included in the back of the book. These are *partial* answers, not full solutions or even complete answers. In particular, many exercises expect you to interpret or explain or show details, and you should do so! (Full solutions to the odd-numbered problems are included with WileyPLUS or the Student Solutions Manual.)

- **Exercises Referencing Exercises** Many of the datasets and studies included in this book are introduced and then referenced again later. When this happens, we include the earlier reference for your information, but *you should not need to refer back to the earlier reference.* All necessary information will be included in the later problem. The reference is there in case you get curious and want more information or the source citation.

- **Accuracy** Statistics is not an exact science. Just about everything is an approximation, with very few exactly correct values. Don't worry if your answer is slightly different from your friend's answer, or slightly different from the answer in the back of the book. Especially with the simulation methods of Chapters 3 and 4, a certain amount of variability in answers is a natural and inevitable part of the process.

Acknowledgments

The team at John Wiley & Sons, including Joanna Dingle, Anne Scanlan-Rohrer, Tom Kulesa, John LaVacca, Laura Abrams, Adria Giattino, Giana Milazzo, Tom Nery, Billy Ray, Valerie Zaborski, and Laurie Rosatone, has provided wonderful support and guidance, while also being great fun to work with. We especially appreciate the fact that they have shared our great enthusiasm for this project throughout our work together.

Ed Harcourt, Rich Sharp, Kevin Angstadt, and Yuxi Zhang are the programmers behind *StatKey*. We are incredibly appreciative of all of their efforts to bring our shared vision of these tools into a working reality and the many helpful suggestions for enhancements. Thanks also to John Lock for hosting the *lock5stat.com* website.

Ann Cannon did a fabulous job of the monumental task of accuracy checking the entire book, all exercise solutions, and making many helpful suggestions along the way.

Many people helped us collect lots of interesting datasets and applications that are so vital to a modern statistics text. These people include Edith Frazer, Zan Armstrong, Adam Pearce, Jim Vallandingham, Rick Cleary, Ivan Ramler, Tom DeRosa, Judy Graham, Bruce Frazer, Serge Onyper, Pamela Thatcher, Brad Baldwin, Laura Fonken, Jeremy Groves, Michael Frazer, Linda Casserly, Paul Doty, and Ellen Langer. We appreciate that more and more researchers are willing to share their data to help students see the relevance of statistics to so many fields.

We appreciate many valuable discussions with colleagues and friends in the statistics education community who have been active in developing this new approach to teaching statistics, including Beth Chance, Laura Chihara, George Cobb, Bob del Mas, Michelle Everson, Joan Garfield, Jeff Hamrick, Tim Hesterberg, John Holcomb, Rebekah Isaak, Laura Le, Allan Rossman, Andrew Zieffler, and Laura Ziegler. Thanks also to Jeff Tecosky-Feldman and Dan Flath for their early support and encouragement for this project. Special thanks to Jessica Chapman for her excellent work on the Test Bank.

We thank Roxy Peck for the use of her home in Los Osos, CA for a sabbatical in the spring of 2011. Many of the words in this text were originally penned while appreciating her wonderful view of Morro Bay.

We thank our students who provided much valuable feedback on all drafts of the book, and who continue to help us improve the text and help us find interesting datasets. We particularly thank Adrian Recinos, who read early drafts of Chapters 3 and 4 to help check that they would be accessible to students with no previous background in statistics.

We thank the many reviewers (listed at the end of this section) for their helpful comments, suggestions, and insights. They have helped make this text significantly better, and have helped give us confidence that this approach can work in a wide variety of settings.

We owe our love of both teaching and statistics at least in part to Ron Frazer, who was teaching innovative statistics classes as far back as the 60's and 70's. He read early versions of the book and was full of excitement and enthusiasm for the project. He spent 88 years enjoying life and laughing, and his constant smile and optimism are greatly missed.

The Second Edition of this book was written amidst many wonderful additions to our growing family. We are especially grateful to Eugene Morgan, Amy Lock, and Nidhi Kohli for their love and support of this family and this project. All three share our love of statistics and have patiently put up with countless family conversations about the book. We are also very grateful for the love and joy brought into our lives by the next generation of Locks, all of whom have been born since the First Edition of this book: Axel, Cal, and Daisy Lock Morgan, and Jocelyn Lock, who we hope will also come to share our love of statistics!

Suggestions?

Our goal is to design materials that enable instructors to teach an excellent course and help students learn, enjoy, and appreciate the subject of statistics. If you have suggestions for ways in which we can improve the text and/or the available resources, making them more helpful, accurate, clear, or interesting, please let us know. We would love to hear from you! Our contact information is at *www.lock5stat.com*.

We hope you enjoy the journey!

Robin H. Lock   Patti Frazer Lock

Kari Lock Morgan

Eric F. Lock   Dennis F. Lock

Reviewers for the Second Edition

Alyssa Armstrong	*Wittenberg University*	Barb Barnet	*University of Wisconsin - Platteville*
Kevin Beard	*University of Vermont*	Barbara Bennie	*University of Wisconsin - La Crosse*
Karen Benway	*University of Vermont*	Dale Bowman	*University of Memphis*
Patricia Buchanan	*Penn State University*	Coskun Cetin	*California State University - Sacramento*
Jill A. Cochran	*Berry College*	Rafael Diaz	*California State University - Sacramento*
Kathryn Dobeck	*Lorain County Community College*	Brandi Falley	*Texas Women's University*
John Fieberg	*University of Minnesota*	Elizabeth Brondos Fry	*University of Minnesota*
Jennifer Galovich	*College of St. Benedict and St. John's University*	Steven T. Garren	*James Madison University*
Mohinder Grewal	*Memorial University of Newfoundland*	Robert Hauss	*Mt. Hood Community College*
James E. Helmreich	*Marist College*	Carla L. Hill	*Marist College*
Martin Jones	*College of Charleston*	Jeff Kollath	*Oregon State University*
Paul Kvam	*University of Richmond*	David Laffie	*(formerly) California State University - East Bay*
Bernadette Lanciaux	*Rochester Institute of Technology*	Anne Marie S. Marshall	*Berry College*
Gregory Mathews	*Loyola University Chicago*	Scott Maxwell	*University of Notre Dame*
Kathleen McLaughlin	*University of Connecticut*	Sarah A Mustillo	*University of Notre Dame*
Elaine T. Newman	*Sonoma State University*	Rachel Rader	*Ohio Northern University*
David M. Reineke	*University of Wisconsin - La Crosse*	Rachel Roe-Dale	*Skidmore College*
Kimberly A. Roth	*Juniata College*	Dinesh Sharma	*James Madison University*
Alla Sikorskii	*Michigan State University*	Karen Starin	*Columbus State Community College*
Paul Stephenson	*Grand Valley State University*	Asokan Variyath	*Memorial University*
Lissa J. Yogan	*Valparaiso University*	Laura Ziegler	*Iowa State University*

Reviewers and Class Testers for the First Edition

Wendy Ahrensen	*South Dakota St.*	Diane Benner	*Harrisburg Comm. College*
Steven Bogart	*Shoreline Comm. College*	Mark Bulmer	*University of Queensland*
Ken Butler	*Toronto-Scarborough*	Ann Cannon	*Cornell College*
John "Butch" Chapelle	*Brookstone School*	George Cobb	*Mt. Holyoke University*
Steven Condly	*U.S. Military Academy*	Salil Das	*Prince Georges Comm. College*
Jackie Dietz	*Meredith College*	Carolyn Dobler	*Gustavus Adolphus*
Robert Dobrow	*Carleton College*	Christiana Drake	*Univ. of Calif. - Davis*
Katherine Earles	*Wichita State*	Laura Estersohn	*Scarsdale High School*
Karen A. Estes	*St. Petersburg College*	Soheila Fardanesh	*Towson University*
Diane Fisher	*Louisiana - Lafayette*	Brad Fulton	*Duke University*
Steven Garren	*James Madison Univ.*	Mark Glickman	*Boston University*
Brenda Gunderson	*Univ. of Michigan*	Aaron Gutknecht	*Tarrant County C.C.*
Ian Harris	*Southern Methodist Univ.*	John Holliday	*North Georgia College*
Pat Humphrey	*Georgia Southern Univ.*	Robert Huotari	*Glendale Comm.College*
Debra Hydorn	*Univ. of Mary Washington*	Kelly Jackson	*Camden County College*
Brian Jersky	*Macquarie University*	Matthew Jones	*Austin Peay St. Univ.*
James Lang	*Valencia Comm. College*	Lisa Lendway	*University of Minnesota*
Stephen Lee	*University of Idaho*	Christopher Malone	*Winona State*
Catherine Matos	*Clayton State*	Billie May	*Clayton State*
Monnie McGee	*Southern Methodist Univ.*	William Meisel	*Florida St.-Jacksonville*
Matthew Mitchell	*Florida St.-Jacksonville*	Lori Murray	*Western University*
Perpetua Lynne Nielsen	*Brigham Young University*	Julia Norton	*Cal. State - East Bay*
Nabendu Pal	*Louisiana - Lafayette*	Alison Paradise	*Univ. of Puget Sound*
Iwan Praton	*Franklin & Marshall College*	Guoqi Qian	*Univ. of Melbourne*
Christian Rau	*Monash University*	Jerome Reiter	*Duke University*
Thomas Roe	*South Dakota State*	Rachel Roe-Dale	*Skidmore College*
Yuliya Romanyuk	*King's University College*	Charles Scheim	*Hartwick College*
Edith Seier	*East Tennessee State*	Therese Shelton	*Southwestern Univ.*
Benjamin Sherwood	*University of Minnesota*	Sean Simpson	*Westchester Comm. College*
Dalene Stangl	*Duke University*	Robert Stephenson	*Iowa State*
Sheila Weaver	*Univ. of Vermont*	John Weber	*Georgia Perimeter College*
Alison Weir	*Toronto-Mississauaga*	Ian Weir	*Univ. of the West of England*
Rebecca Wong	*West Valley College*	Laura Ziegler	*University of Minnesota*

WileyPLUS with ORION

A personalized, adaptive learning experience.

WileyPLUS with ORION delivers easy-to-use analytics that help educators and students see strengths and weaknesses to give learners the best chance of succeeding in the course.

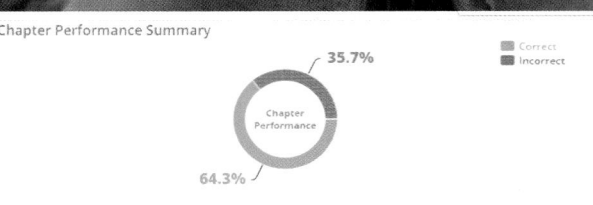
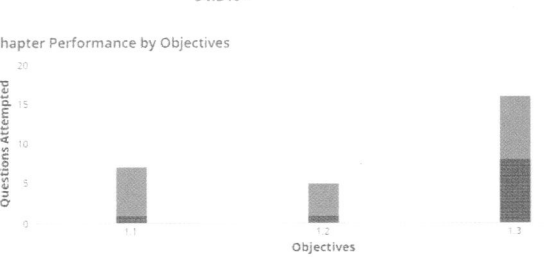

Identify which students are struggling early in the semester.

Educators assess the real-time engagement and performance of each student to inform teaching decisions. Students always know what they need to work on.

Help students organize their learning and get the practice they need.

With ORION's adaptive practice, students quickly understand what they know and don't know. They can then decide to study or practice based on their proficiency.

Measure outcomes to promote continuous improvement.

With visual reports, it's easy for both students and educators to gauge problem areas and act on what's most important.

www.ORION.wileyplus.com

Data

"For Today's Graduate, Just One Word: Statistics"

New York Times headline, August 5, 2009

In this unit, we learn how to collect and describe data. We explore how data collection influences the types of conclusions that can be drawn, and discover ways to summarize and visualize data.

CHAPTER 1

Collecting
Data

"You can't fix by analysis what you bungled by design."

Richard Light, Judith Singer, and John Willett in *By Design*

Questions and Issues

Here are some of the questions and issues we will discuss in this chapter:

- Is there a "sprinting gene"?
- Does tagging penguins for identification purposes harm them?
- Do humans subconsciously give off chemical signals (pheromones)?
- What proportion of people using a public restroom wash their hands?
- If parents could turn back time, would they still choose to have children?
- Why do adolescent spiders engage in foreplay?
- How broadly do experiences of parents affect their future children?
- What percent of college professors consider themselves "above-average" teachers?
- Does giving lots of high fives to teammates help sports teams win?
- Which is better for peak performance: a short mild warm-up or a long intense warm-up?
- Does the color red increase the attractiveness of women to men?
- Are city dwellers more likely than country dwellers to have mood and anxiety disorders?
- Is there truth to the saying "beauty sleep"?
- What percent of young adults in the US move back in with their parents?
- Does turning up the music in a bar cause people to drink more beer?
- Is your nose getting bigger?
- Does watching cat videos improve mood?
- Does sleep deprivation hurt one's ability to interpret facial expressions?
- Do artificial sweeteners cause weight gain?
- Does late night eating impair concentration?

1.1 THE STRUCTURE OF DATA

We are being inundated with data. It is estimated that the amount of new technical information is doubling every two years, and that over 7.2 zettabytes (that's 7.2×10^{21} bytes) of unique new information will be generated this year.[1] That is more than was generated during the entire 5000-year period before you were born. An incredible amount of data is readily available to us on the Internet and elsewhere. The people who are able to analyze this information are going to have great jobs and are going to be valuable in virtually every field. One of the wonderful things about statistics is that it is relevant in so many areas. Whatever your focus and your future career plans, it is likely that you will need statistical knowledge to make smart decisions in your field and in everyday life. As we will see in this text, effective collection and analysis of data can lead to very powerful results.

Statistics *is the science of collecting, describing, and analyzing data.* In this chapter, we discuss effective ways to *collect* data. In Chapter 2, we discuss methods to *describe* data. The rest of the chapters are devoted to ways of *analyzing* data to make effective conclusions and to uncover hidden phenomena.

DATA 1.1	A Student Survey

For several years, a first-day survey has been administered to students in an introductory statistics class at one university. Some of the data for a few of the students are displayed in Table 1.1. A more complete table with data for 362 students and 17 variables can be found in the file **StudentSurvey**.[2] ■

Cases and Variables

The subjects/objects that we obtain information about are called the *cases* or *units* in a dataset. In the **StudentSurvey** dataset, the cases are the students who completed the survey. Each row of the dataset corresponds to a different case.

A *variable* is any characteristic that is recorded for each case. Each column of our dataset corresponds to a different variable. The data in Table 1.1 show eight variables (in addition to the ID column), each describing a different characteristic of the students taking the survey.

Table 1.1 *Partial results from a student survey*

ID	Gender	Smoke	Award	Exercise	TV	GPA	Pulse	Birth
1	M	No	Olympic	10	1	3.13	54	4
2	F	Yes	Academy	4	7	2.5	66	2
3	M	No	Nobel	14	5	2.55	130	1
4	M	No	Nobel	3	1	3.1	78	1
5	F	No	Nobel	3	3	2.7	40	1
6	F	No	Nobel	5	4	3.2	80	2
7	F	No	Olympic	10	10	2.77	94	1
8	M	No	Olympic	13	8	3.3	77	1
9	F	No	Nobel	3	6	2.8	60	2
10	F	No	Nobel	12	1	3.7	94	8

[1]*http://www.emc.com/leadership/programs/digital-universe.htm.* Accessed January 2015.
[2]Most datasets used in this text, and descriptions, are available electronically. They can be found at *www.wiley.com/college/lock.* See the Preface for more information. Descriptions of many datasets can also be found in Appendix B.

Cases and Variables

We obtain information about **cases** or **units** in a dataset, and generally record the information for each case in a row of a data table.

A **variable** is any characteristic that is recorded for each case. The variables generally correspond to the columns in a data table.

In any dataset, it is important to understand exactly what each variable is measuring and how the values are coded. For the data in Table 1.1, the first column is ID, to provide an identifier for each of the individuals in the study. In addition, we have:

Gender	M for male and F for female
Smoke	Does the student smoke: yes or no
Award	Award the student prefers to win: Academy Award, Olympic gold medal, or Nobel Prize
Exercise	Number of hours spent exercising per week
TV	Number of hours spent watching television per week
GPA	Current grade point average on a 4-point scale
Pulse	Pulse rate in number of beats per minute at the time of the survey
Birth	Birth order: 1 for first/oldest, 2 for second born, etc.

Example 1.1

Explain what each variable tells us about the student with ID 1 in the first row of Table 1.1.

Solution Student 1 is a male who does not smoke and who would prefer to win an Olympic gold medal over an Academy Award or a Nobel Prize. He says that he exercises 10 hours a week, watches television one hour a week, and that his grade point average is 3.13. His pulse rate was 54 beats per minute at the time of the survey, and he is the fourth oldest child in his family.

Categorical and Quantitative Variables

In determining the most appropriate ways to summarize or analyze data, it is useful to classify variables as either *categorical* or *quantitative*.

Categorical and Quantitative Variables

A **categorical variable** divides the cases into groups, placing each case into exactly one of two or more categories.

A **quantitative variable** measures or records a numerical quantity for each case. Numerical operations like adding and averaging make sense for quantitative variables.

We may use numbers to code the categories of a categorical variable, but this does not make the variable quantitative unless the numbers have a quantitative meaning. For example, "gender" is categorical even if we choose to record the results as 1 for male and 2 for female, since we are more likely to be interested in how many are in each category rather than an average numerical value. In other situations, we might choose to convert a quantitative variable into categorical groups. For example, "household income" is quantitative if we record the specific values but is categorical if we instead record only an income category ("low," "medium," "high") for each household.

Example 1.2

Classify each of the variables in the student survey data in Table 1.1 as either categorical or quantitative.

Solution

Note that the ID column is neither a quantitative nor a categorical variable. A dataset often has a column with names or ID numbers that are for reference only.

- *Gender* is categorical since it classifies students into the two categories of male and female.
- *Smoke* is categorical since it classifies students as smokers or nonsmokers.
- *Award* is categorical since students are classified depending on which award is preferred.
- *Exercise*, *TV*, *GPA*, and *Pulse* are all quantitative since each measures a numerical characteristic of each student. It makes sense to compute an average for each variable, such as an average number of hours of exercise a week.
- *Birth* is a somewhat ambiguous variable, as it could be considered either quantitative or categorical depending on how we use it. If we want to find an average birth order, we consider the variable quantitative. However, if we are more interested in knowing how many first-borns, how many second-borns, and so on, are in the data, we consider the variable categorical. Either answer is acceptable.

Investigating Variables and Relationships between Variables

In this book, we discuss ways to describe and analyze a single variable and to describe and analyze relationships between two or more variables. For example, in the student survey data, we might be interested in the following questions, each about a single variable:

- What percentage of students smoke?
- What is the average number of hours a week spent exercising?
- Are there students with unusually high or low pulse rates?
- Which award is the most desired?
- How does the average GPA of students in the survey compare to the average GPA of all students at this university?

Often the most interesting questions arise as we look at relationships between variables. In the student survey data, for example, we might ask the following questions about relationships between variables:

- Who smokes more, males or females?
- Do students who exercise more tend to prefer an Olympic gold medal? Do students who watch lots of television tend to prefer an Academy Award?
- Do males or females watch more television?
- Do students who exercise more tend to have lower pulse rates?
- Do first-borns generally have higher grade point averages?

These examples show that relationships might be between two categorical variables, two quantitative variables, or a quantitative and a categorical variable. In the following chapters, we examine statistical techniques for exploring the nature of relationships in each of these situations.

DATA 1.2

Data on Countries

As of this writing, there are 215 countries listed by the World Bank.[3] A great deal of information about these countries (such as energy use, birth rate, life expectancy) is in the full dataset under the name **AllCountries**. ∎

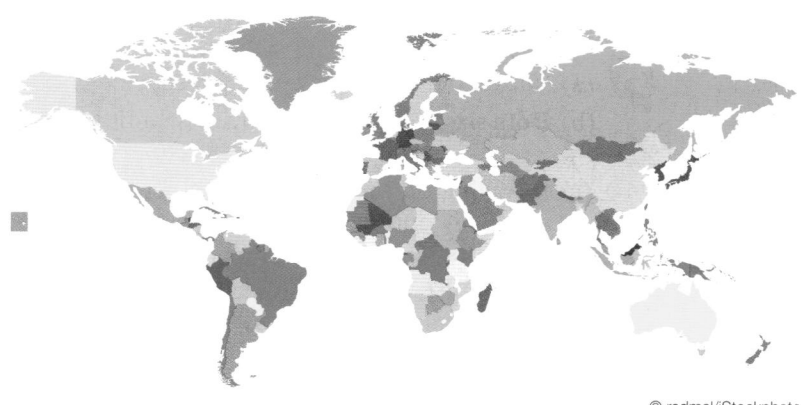

© redmal/iStockphoto

Countries of the world

Example 1.3

The dataset **AllCountries** includes information on the percent of people in each country with access to the Internet.

(a) Data from Iceland were used to determine that 96.5% of Icelanders have access to the Internet, the highest rate of any country. What are the cases in the data from Iceland? What variable is used? Is it categorical or quantitative?

(b) In the **AllCountries** dataset, we record the percent of people with access to the Internet for each country. What are the cases in that dataset? What is the relevant variable? Is it categorical or quantitative?

Solution

(a) For determining the rate of Internet usage in Iceland, the cases are people in Iceland, and the relevant variable is whether or not each person has access to the Internet. This is a categorical variable.

(b) In the **AllCountries** dataset, the cases are the countries of the world. The variable is the proportion with access to the Internet. For each country, we record a numerical value. These values range from a low of 0.9% in Eritrea to the high of 96.5% in Iceland, and the average is 43.02%. This is a quantitative variable.

As we see in the previous example, we need to think carefully about what the cases are and what is being recorded in each case in order to determine whether a variable is categorical or quantitative.

---

[3]*http://data.worldbank.org/indicator/IT.NET.USER.P2.* Data include information on both countries and economies, accessed May 2015.

Example 1.4

In later chapters, we examine some of the following issues using the data in **AllCountries**. Indicate whether each question is about a single variable or a relationship between variables. Also indicate whether the variables are quantitative or categorical.

(a) How much energy does the average country use in a year?

(b) Do countries larger in area tend to have a more rural population?

(c) What is the relationship, if any, between a country's government spending on the military and on health care?

(d) Is the birth rate higher in developed or undeveloped countries?

(e) Which country has the highest percent of elderly people?

Solution

(a) The amount of energy used is a single quantitative variable.

(b) Both size and percent rural are quantitative variables, so this is a question about a relationship between two quantitative variables.

(c) Spending on the military and spending on health care are both quantitative, so this is another question about the relationship between two quantitative variables.

(d) Birth rate is a quantitative variable and whether or not a country is developed is a categorical variable, so this is asking about a relationship between a quantitative variable and a categorical variable.

(e) Because the cases are countries, percent elderly is a single quantitative variable.

Using Data to Answer a Question

The **StudentSurvey** and **AllCountries** datasets contain lots of information and we can use that information to learn more about students and countries. Increasingly, in this data-driven world, we have large amounts of data and we want to "mine" it for valuable information. Often, however, the order is reversed: We have a question of interest and we need to collect data that will help us answer that question.

© Pete Saloutos/iStockphoto

Is there a "sprinting gene"?

Example 1.5

Is There a "Sprinting Gene"?

A gene called *ACTN3* encodes a protein which functions in fast-twitch muscles. Some people have a variant of this gene that cannot yield this protein. (So we might call the gene variant a possible *non*-sprinting gene.) To address the question of whether this gene is associated with sprinting ability, geneticists tested people from three different groups: world-class sprinters, world-class marathon runners, and a control group of non-athletes. In the samples tested, 6% of the sprinters had the gene variant, compared with 18% of the non-athletes and 24% of the marathon runners. This study[4] suggests that sprinters are less likely than non-sprinters to have the gene variant.

(a) What are the cases and variables in this study? Indicate whether each variable is categorical or quantitative.

(b) What might a table of data look like for this study? Give a table with a possible first two cases filled in.

Solution

(a) The cases are the people included in the study. One variable is whether the individual has the gene variant or not. Since we record simply "yes" or "no," this is a categorical variable. The second variable keeps track of the group to which the individual belongs. This is also a categorical variable, with three possible categories (sprinter, marathon runner, or non-athlete). We are interested in the relationship between these two categorical variables.

(b) The table of data must record answers for each of these variables and may or may not have an identifier column. Table 1.2 shows a possible first two rows for this dataset.

Table 1.2 *Possible table to investigate whether there is a sprinter's gene*

Name	Gene Variant	Group
Allan	Yes	Marathon runner
Beth	No	Sprinter
. . .	. . .	. . .

Example 1.6

What's a Habanero?

A habanero chili is an extremely spicy pepper (roughly 500 times hotter than a jalapeño) that is used to create fiery food. The vice president of product development and marketing for the Carl's Jr. restaurant chain[5] is considering adding a habanero burger to the menu. In developing an advertising campaign, one of the issues he must deal with is whether people even know what the term "habanero" means. He identifies three specific questions of interest and plans to survey customers who visit the chain's restaurants in various parts of the country.

• What proportion of customers know and understand what "habanero" means?

• What proportion of customers are interested in trying a habanero burger?

• How do these proportions change for different regions of the country?

---

[4]Yang, N., et al., "ACTN3 genotype is associated with human elite athletic performance," *American Journal of Human Genetics*, September 2003; 73: 627–631.

[5]With thanks to Bruce Frazer.

(a) Identify the cases for the data he collects.

(b) Describe three variables that should be included in the dataset.

Solution (a) The cases in the habanero marketing study are the individual customers that respond to the survey request.

(b) Here are three variables that should be included in the data:

- *Know* = yes or no, depending on whether the customer knows the term "habanero"

- *Try* = yes or no, to indicate the customer's willingness to try a habanero burger

- *Region* = area in the country where the customer lives

All three variables are categorical.

　　Notice that for each case (customer), we record a value for each of the variables. Each customer would be represented in the dataset by a different row in the data table.

© Keith Szafranski/iStockphoto

Does tagging penguins harm them?

DATA　1.3　　**Tagging Penguins**

Do metal tags on penguins harm them? Scientists trying to tell penguins apart have several ways to tag the birds. One method involves wrapping metal strips with ID numbers around the penguin's flipper, while another involves electronic tags. Neither tag seems to physically harm the penguins. However, since tagged penguins are used to study *all* penguins, scientists wanted to determine whether the metal tags have any significant effect on the penguins. Data were collected over a 10-year time span from a sample of 100 penguins that were randomly given either metal or electronic tags. This included information on number of chicks, survival over the decade, and length of time on foraging trips.[6] ■

Example 1.7

In the study on penguin tags:

(a) What are the cases? What are the variables? Identify each variable as categorical or quantitative.

[6]Saraux, C., et al., "Reliability of flipper-banded penguins as indicators of climate change," *Nature*, January 2011; 469: 203–206.

(b) What information do the scientists hope to gain from the data? Describe at least one question in which they might be interested.

Solution

(a) The cases are the tagged penguins. The variables are the type of tag (categorical), number of chicks (quantitative), survival or not (categorical), and length of time on foraging trips (quantitative).

(b) The scientists want to determine whether there is a relationship between the type of tag and any of the other variables. For example, they might be interested in knowing whether survival rates differ between the penguins with the metal tags and penguins with the electronic tags.

In Example 1.7, we are investigating whether one of the variables (the type of tag) helps us explain or predict values of the other variables. In this situation, we call the type of tag the *explanatory variable* and the other variables the *response variables*. One way to remember these names is the explanatory variable helps explain the response variable, and the response variable responds to the explanatory variable.

Explanatory and Response Variables

If we are using one variable to help us understand or predict values of another variable, we call the former the **explanatory variable** and the latter the **response variable**.

Example 1.8

In Example 1.4, we considered the following three questions about relationships between variables in the **AllCountries** dataset. Identify the explanatory variable and the response variable if it makes sense to do so.

(a) Do countries larger in area tend to have a more rural population?

(b) Is the birth rate higher in developed or undeveloped countries?

(c) What is the relationship, if any, between a country's government spending on the military and on health care?

Solution

(a) The question indicates that we think area might influence the percent of a country that is rural, so we call area the explanatory variable and percent rural the response variable.

(b) The question indicates that we think whether or not a country is developed might influence the birth rate, so the explanatory variable is whether the country is developed or undeveloped and the response variable is birth rate.

(c) There is no indication in this situation of why we might identify either of the two variables (spending on military and spending on health care) as explanatory or as response. In a relationship between two variables, we don't always identify one as the explanatory variable and one as the response variable.

Different Ways to Answer a Broad Question

A pheromone is a chemical signal given off by one member of a species that influences other members of the species. Many studies (involving data, of course!) provide evidence that different types of animals give off pheromones. It is currently under debate whether humans also communicate subconsciously through pheromones. How might we collect data to answer the question of whether there are human pheromones? We might start by narrowing the question to one that is not so general. For example, are there pheromones in female tears that affect the behavior of males?

Several studies[7] suggest that the scent of tears from a crying woman may reduce sexual interest in men. However, to determine whether this effect is caused subconsciously by pheromones (rather than by obvious social influences), we need to think carefully about how to collect data. How might you collect data to answer this question? Three different methods were used in the studies. See what you think of the three methods.

- In one study, 24 men in their twenties had a pad attached to their upper lip that contained either tears collected from women who watched sad films or a salt solution that had been trickled down the same women's faces. Neither substance had a perceptible odor. The men who had tears on the upper lip rated female faces as less sexually alluring than the men who had salt solution on the upper lip.

- In a second study, 50 men who sniffed women's tears showed reduced levels of testosterone relative to levels after sniffing a salt solution.

- In a third study involving 16 men, those who sniffed female tears displayed significantly reduced brain-cell activity in areas that had reacted strongly to an erotic movie, whereas those who sniffed a salt solution did not show the same reduced activity.

Example 1.9

For each of the three studies on women's tears, state the explanatory and response variables.

Solution   In all three studies, the explanatory variable is whether tears or a salt solution is used. In the first study, the response variable is how sexually alluring males rated female faces, in the second study it is testosterone levels, and in the third study it is brain cell activity.

All three of these studies describe data collected in a careful way to answer a question. How to collect data in a way that helps us understand real phenomena is the subject of the rest of this chapter.

We have described several datasets, studies, and questions in this section, involving students, countries, sprinter genes, habanero burgers, penguins, and pheromones. If you are intrigued by any of these questions, keep reading! We examine all of them in more detail in the pages ahead.

SECTION LEARNING GOALS

You should now have the understanding and skills to:

- Recognize that a dataset consists of cases and variables
- Identify variables as either categorical or quantitative
- Determine explanatory and response variables where appropriate
- Describe how data might be used to address a specific question
- Recognize that understanding statistics allows us to investigate a wide variety of interesting phenomena

---

[7]Gelstein, S., et al., "Human Tears Contain a Chemosignal," *Science*, January 2011; 331(6014): 226–230.

Exercises for Section **1.1**

SKILL BUILDER 1

For the situations described in Exercises 1.1 to 1.6:

(a) What are the cases?

(b) What is the variable and is it quantitative or categorical?

1.1 People in a city are asked if they support a new recycling law.

1.2 Record the percentage change in the price of a stock for 100 stocks publicly traded on Wall Street.

1.3 Collect data from a sample of teenagers with a question that asks "Do you eat at least five servings a day of fruits and vegetables?"

1.4 Measure the shelf life of bunches of bananas (the number of days until the bananas go bad) for a large sample.

1.5 Estimate the bending strength of beams by bending 10 beams until they break and recording the force at which the beams broke.

1.6 Record whether or not the literacy rate is over 75% for each country in the world.

SKILL BUILDER 2

In Exercises 1.7 to 1.10, a relationship between two variables is described. In each case, we can think of one variable as helping to explain the other. Identify the explanatory variable and the response variable.

1.7 Lung capacity and number of years smoking cigarettes

1.8 Amount of fertilizer used and the yield of a crop

1.9 Blood alcohol content (BAC) and number of alcoholic drinks consumed

1.10 Year and the world record time in a marathon

1.11 Student Survey Variables Data 1.1 introduced the dataset **StudentSurvey**, and Example 1.2 identified seven of the variables in that dataset as categorical or quantitative. The remaining variables are:

Year	First Year, Sophomore, Junior, Senior
Height	In inches
Weight	In pounds
Siblings	Number of siblings the person has
VerbalSAT	Score on the Verbal section of the SAT exam
MathSAT	Score on the Math section of the SAT exam
SAT	Sum of the scores on the Verbal and Math sections of the SAT exam
HigherSAT	Which is higher, Math SAT score or Verbal SAT score?

(a) Indicate whether each variable is quantitative or categorical.

(b) List at least two questions we might ask about any one of these individual variables.

(c) List at least two questions we might ask about relationships between any two (or more) of these variables.

1.12 Countries of the World Information about the world's countries is given in **AllCountries**, introduced in Data 1.2 on page 7. You can find a description of the variables in Appendix B. For the full dataset:

(a) Indicate which of the variables are quantitative and which are categorical.

(b) List at least two questions we might ask about any one of these individual variables.

(c) List at least two questions we might ask about relationships between any two (or more) of these variables.

1.13 Spider Sex Play Spiders regularly engage in spider foreplay that does not culminate in mating. Male spiders mature faster than female spiders and often practice the mating routine on not-yet-mature females. Since male spiders run the risk of getting eaten by female spiders, biologists wondered why spiders engage in this behavior. In one study,[8] some spiders were allowed to participate in these near-matings, while other maturing spiders were isolated. When the spiders were fully mature, the scientists observed real matings. They discovered that if either partner had participated at least once in mock sex, the pair reached the point of real mating significantly faster than inexperienced spiders did. (Mating faster is, apparently, a real advantage in the spider world.) Describe the variables, indicate whether each variable is quantitative or categorical, and indicate the explanatory and response variables.

[8]Pruitt, J., paper presented at the Society for Integrative and Comparative Biology Annual Meeting, January 2011, and reported in "For spiders, sex play has its pluses," *Science News*, January 29, 2011.

1.14 Hormones and Fish Fertility When women take birth control pills, some of the hormones found in the pills eventually make their way into lakes and waterways. In one study, a water sample was taken from various lakes. The data indicate that as the concentration of estrogen in the lake water goes up, the fertility level of fish in the lake goes down. The estrogen level is measured in parts per trillion (ppt) and the fertility level is recorded as the percent of eggs fertilized. What are the cases in this study? What are the variables? Classify each variable as either categorical or quantitative.

1.15 Fast-Twitch Muscles and Race Example 1.5 studied a variant of the gene *ACTN3* which inhibits fast-twitch muscles and seems to be less prevalent in sprinters. A separate study[9] indicated ethnic differences: Approximately 20% of a sample of Caucasians, approximately 25% of a sample of Asians, and approximately 1% of a sample of Africans had the gene variant. What are the variables in this study? Classify each as categorical or quantitative.

1.16 Rowing Solo Across the Atlantic Ocean On January 14, 2012, Andrew Brown of Great Britain set the world record time (40 days) for rowing solo across the northern Atlantic Ocean. On March 14, 2010, Katie Spotz of the United States became the youngest person to ever row solo across the Atlantic when she completed it in 70 days at the age of 22 years old. Table 1.3 shows times for males and females who rowed solo across the Atlantic Ocean in the last few years.[10]

(a) How many cases are there in this dataset? How many variables are there and what are they? Is each categorical or quantitative?

(b) Display the information in Table 1.3 as a dataset with cases as rows and variables as columns.

Table 1.3 *Number of days to row alone across the Atlantic Ocean*

Male times:	40, 87, 78, 106, 67
Female times:	70, 153, 81

1.17 Largest Cities in the World Seven of the ten largest cities in the world are in the Eastern Hemisphere (including the largest: Tokyo, Japan) and three are in the Western Hemisphere.[11] Table 1.4

shows the populations, in millions of people, for these cities.

(a) How many cases are there in this dataset? How many variables are there and what are they? Is each categorical or quantitative?

(b) Display the information in Table 1.4 as a dataset with cases as rows and variables as columns.

Table 1.4 *Population, in millions, of the world's largest cities*

Eastern hemisphere:	37, 26, 23, 22, 21, 21, 21
Western hemisphere:	21, 20, 19

1.18 Trans-Generational Effects of Diet Can experiences of parents affect future children? New studies[12] suggest that they can: Early life experiences of parents appear to cause permanent changes in sperm and eggs. In one study, some male rats were fed a high-fat diet with 43% of calories from fat (a typical American diet), while others were fed a normal healthy rat diet. Not surprisingly, the rats fed the high-fat diet were far more likely than the normal-diet rats to develop metabolic syndrome (characterized by such things as excess weight, excess fat, insulin resistance, and glucose intolerance.) What surprised the scientists was that the daughters of these rats were also far more likely to develop metabolic syndrome than the daughters of rats fed healthy diets. None of the daughters and none of the mothers ate a high-fat diet and the fathers did not have any contact with the daughters. The high-fat diet of the fathers appeared to cause negative effects for their daughters. What are the two main variables in this study? Is each categorical or quantitative? Identify the explanatory and response variables.

1.19 Trans-Generational Effects of Environment In Exercise 1.18, we ask whether experiences of parents can affect future children, and describe a study that suggests the answer is yes. A second study, described in the same reference, shows similar effects. Young female mice were assigned to either live for two weeks in an enriched environment or not. Matching what has been seen in other similar experiments, the adult mice who had been exposed to an enriched environment were smarter (in the sense that they learned how to navigate mazes faster) than the mice that did not have that experience. The other interesting result, however, was that the offspring of the mice exposed to the enriched environment were also smarter than the

[9]North, K., et al., "A common nonsense mutation results in α-actinin-3 deficiency in the general population," *Nature Genetics*, April 1999; 21(4): 353–354.

[10]http://www.oceanrowing.com/statistics/ocean_rowing_records2.htm.

[11]http://www.worldatlas.com/citypops.htm. Accessed June 2015.

[12]Begley, S., "Sins of the Grandfathers," *Newsweek*, November 8, 2010; 48–50.

offspring of the other mice, even though none of the offspring were exposed to an enriched environment themselves. What are the two main variables in this study? Is each categorical or quantitative? Identify explanatory and response variables.

1.20 Hookahs and Health Hookahs are waterpipes used for smoking flavored tobacco. One study[13] of 3770 university students in North Carolina found that 40% had smoked a hookah at least once, with many claiming that the hookah smoke is safer than cigarette smoke. However, a second study observed people at a hookah bar and recorded the length of the session, the frequency of puffing, and the depth of inhalation. An average session lasted one hour and the smoke inhaled from an average session was equal to the smoke in more than 100 cigarettes. Finally, a third study measured the amount of tar, nicotine, and heavy metals in samples of hookah smoke, finding that the water in a hookah filters out only a very small percentage of these chemicals. Based on these studies and others, many states are introducing laws to ban or limit hookah bars. In each of the three studies, identify the individual cases, the variables, and whether each variable is quantitative or categorical.

1.21 Is Your Nose Getting Bigger? Next time you see an elderly man, check out his nose and ears! While most parts of the human body stop growing as we reach adulthood, studies show that noses and ears continue to grow larger throughout our lifetime. In one study[14] examining noses, researchers report "Age significantly influenced all analyzed measurements:" including volume, surface area, height, and width of noses. The gender of the 859 participants in the study was also recorded, and the study reports that "male increments in nasal dimensions were larger than female ones."

(a) How many variables are mentioned in this description?

(b) How many of the variables are categorical? How many are quantitative?

(c) If we create a dataset of the information with cases as rows and variables as columns, how many rows and how many columns would the dataset have?

1.22 Don't Text While Studying! For the 2015 Intel Science Fair, two brothers in high school recruited 47 of their classmates to take part in a two-stage study. Participants had to read two different passages and then answer questions on them, and each person's score was recorded for each of the two tests. There were no distractions for one of the passages, but participants received text messages while they read the other passage. Participants scored significantly worse when distracted by incoming texts. Participants were also asked if they thought they were good at multitasking (yes or no) but "even students who were confident of their abilities did just as poorly on the test while texting."[15]

(a) What are the cases?

(b) What are the variables? Is each variable categorical or quantitative?

(c) If we create a dataset of the information with cases as rows and variables as columns, how many rows and how many columns would the dataset have?

1.23 Help for Insomniacs A recent study shows that just one session of cognitive behavioral therapy can help people with insomnia.[16] In the study, forty people who had been diagnosed with insomnia were randomly divided into two groups of 20 each. People in one group received a one-hour cognitive behavioral therapy session while those in the other group received no treatment. Three months later, 14 of those in the therapy group reported sleep improvements while only 3 people in the other group reported improvements.

(a) What are the cases in this study?

(b) What are the relevant variables? Identify each as categorical or quantitative.

(c) If we create a dataset of the information with cases as rows and variables as columns, how many rows and how many columns would the dataset have?

1.24 How Are Age and Income Related? An economist collects data from many people to determine how age and income are related. How the data is collected determines whether the variables are quantitative or categorical. Describe how the information might be recorded if we regard both variables as quantitative. Then describe a different way

[13]Quenqua, D., "Putting a Crimp in the Hookah," *New York Times*, May 31, 2011, p A1.

[14]Sforza, C., Grandi, G., De Menezes, M., Tartaglia, G.M., and Ferrario, V.F., "Age- and sex-related changes in the normal human external nose," *Forensic Science International*, January 30, 2011; 204(1–3): 205.e1–9.

[15]Perkins, S., "Studying? Don't answer that text!" *Science News*, July 23, 2015.

[16]Ellis, J.G., Cushing, T., and Germain, A., "Treating acute insomnia: a randomized controlled trial of a 'single-shot' of cognitive behavioral therapy for insomnia," *SLEEP*, 2015; 38(6): 971–978.

to record information about these two variables that would make the variables categorical.

1.25 Political Party and Voter Turnout Suppose that we want to investigate the question "Does voter turnout differ by political party?" How might we collect data to answer this question? What would the cases be? What would the variable(s) be?

1.26 Wealth and Happiness Are richer people happier? How might we collect data to answer this question? What would the cases be? What would the variable(s) be?

1.27 Choose Your Own Question Come up with your own question you would like to be able to answer. What is the question? How might you collect data to answer this question? What would the cases be? What would the variable(s) be?

1.2 SAMPLING FROM A POPULATION

While most of this textbook is devoted to analyzing data, the way in which data are *collected* is critical. Data collected well can yield powerful insights and discoveries. Data collected poorly can yield very misleading results. Being able to think critically about the method of data collection is crucial for making or interpreting data-based claims. In the rest of this chapter, we address some of the most important issues that need to be considered when collecting data.

Samples from Populations

The US Census is conducted every 10 years and attempts to gather data about all people living in the US. For example, the census shows that, for people living in the US who are at least 25 years old, 84.6% have at least a high school degree and 27.5% have at least a college bachelor's degree.[17] The cases in the census dataset are all residents of the US, and there are many variables measured on these cases. The US census attempts to gather information from an entire *population*. In **AllCountries**, introduced as Data 1.2 on page 7, the cases are countries. This is another example of a dataset on an entire population because we have data on every country.

Usually, it is not feasible to gather data for an entire population. If we want to estimate the percent of people who wash their hands after using a public restroom, it is certainly not possible to observe all people all the time. If we want to try out a new drug (with possible side effects) to treat cancer, it is not safe to immediately give it to all patients and sit back to observe what happens. If we want to estimate what percentage of people will react positively to a new advertising campaign, it is not feasible to show the ads to everyone and then track their responses. In most circumstances, we can only work with a *sample* from what might be a very large population.

> **Samples from Populations**
>
> A **population** includes all individuals or objects of interest.
>
> Data are collected from a **sample**, which is a subset of the population.

Example 1.10

To estimate what percent of people in the US wash their hands after using a public restroom, researchers pretended to comb their hair while observing 6000 people in public restrooms throughout the United States. They found that 85% of the people who were observed washed their hands after going to the bathroom.[18] What is the sample in this study? What is a reasonable population to which we might generalize?

---

[17] *http://factfinder.census.gov.*
[18] Zezima, K., "For many, 'Washroom' seems to be just a name," *New York Times*, September 14, 2010, p. A14.

Solution The sample is the 6000 people who were observed. A reasonable population to generalize to would be all people in the US. There are other reasonable answers to give for the population, such as all people in the US who use public restrooms or all people in the US who use public restrooms in the cities in which the study was conducted. Also, people might behave differently when alone than when there is someone else in the restroom with them, so we might want to restrict the population to people in a restroom with someone else.

We denote the size of the sample with the letter n. In Example 1.10, $n = 6000$ because there are 6000 people in the sample. Usually, the sample size, n, is much smaller than the size of the entire population.

Since we rarely have data on the entire population, a key question is how to use the information in a sample to make reliable statements about the population. This is called *statistical inference*.

Statistical Inference

Statistical inference is the process of using data from a sample to gain information about the population.

Figure 1.1 diagrams the process of selecting a sample from a population, and then using that sample to make inferences about the population. Much of the data analysis discussed in this text focuses on the latter step, statistical inference. However, the first step, selecting a sample from the population, is critical because the process used to collect the sample determines whether valid inference is even possible.

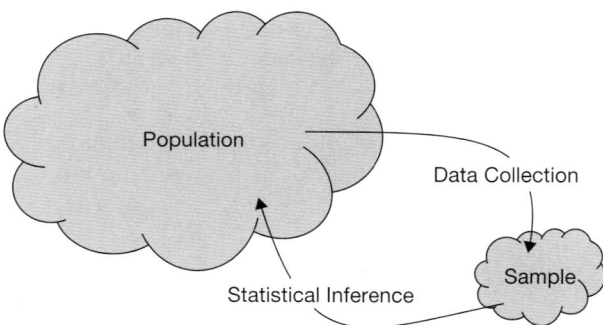

Figure 1.1 *From population to sample and from sample to population*

Sampling Bias

Example 1.11

Dewey Defeats Truman

The day after the 1948 presidential election, the *Chicago Tribune* ran the headline "Dewey Defeats Truman." However, Harry S Truman defeated Thomas E. Dewey to become the 33rd president of the United States. The newspaper went to press before all the results had come in, and the headline was based partly on the results of a large telephone poll which showed Dewey sweeping Truman.

(a) What is the sample and what is the population?

(b) What did the pollsters want to infer about the population based on the sample?

(c) Why do you think the telephone poll yielded such inaccurate results?

Solution   (a) The sample is all the people who participated in the telephone poll. The population is all voting Americans.

(b) The pollsters wanted to estimate the percentage of all voting Americans who would vote for each candidate.

(c) One reason the telephone poll may have yielded inaccurate results is that people with telephones in 1948 were not representative of all American voters. People with telephones tended to be wealthier and prefer Dewey while people without phones tended to prefer Truman.

Photo by Underwood Archives/Getty Images

A triumphant Harry S Truman holds the Chicago Tribune published with the incorrect headline "Dewey defeats Truman"

The previous example illustrates *sampling bias*, because the method of selecting the sample biased the results by selecting only people with telephones.

> **Sampling Bias**
>
> **Sampling bias** occurs when the method of selecting a sample causes the sample to differ from the population in some relevant way. If sampling bias exists, then we cannot trust generalizations from the sample to the population.

Example 1.12

After a flight, one of the authors received an email from the airline asking her to fill out a survey regarding her satisfaction with the travel experience. The airline analyzes the data from all responses to such emails.

(a) What is the sample and in what population is the airline interested?

(b) Do you expect these survey results to accurately portray customer satisfaction?

Solution   (a) The sample is all people who choose to fill out the survey and the population is all people who fly this airline.

(b) The survey results will probably not accurately portray customer satisfaction. Many people won't bother to fill out the survey if the flight was uneventful, while people with a particularly bad or good experience are more likely to fill out the survey.

 A sample comprised of volunteers (like the airline survey) often creates sampling bias in opinion surveys, because the people who choose to participate (the sample) often have more extreme opinions than the population.

To avoid sampling bias, we try to obtain a sample that is *representative* of the population. A representative sample resembles the population, only in smaller numbers. The telephone survey in 1948 reached only people wealthy enough to own a telephone, causing the sample to be wealthier than the population, so it was not a representative sample. The more representative a sample is, the more valuable the sample is for making inferences about the population.

Example 1.13

An online poll conducted in April 2014 on *biblegateway.com* asked, "How often do you talk about the Bible in your normal course of conversation?" Over 5000 people answered the question, and 78% of respondents chose the most frequent option: Multiple times a week. Can we infer that 78% of people talk about the bible multiple times a week? Why or why not?

Solution

 No. People who visit the website for *Bible Gateway* and choose to take the poll are probably more likely than the general public to talk about the bible. This sample is not representative of the population of all people, so the results cannot be generalized to all people.

Simple Random Sample

Since a representative sample is essential for drawing valid inference to the population, you are probably wondering how to select such a sample! The key is *random sampling*. We can imagine putting the names of all the cases in the population into a hat and drawing out names to be in our sample. Random sampling avoids sampling bias.

> ### Simple Random Sample
>
> When choosing a **simple random sample** of n units, all groups of size n in the population have the same chance of becoming the sample. As a result, in a simple random sample, each unit of the population has an equal chance of being selected, regardless of the other units chosen for the sample.
>
> Taking a simple random sample avoids sampling bias.

Part of the power of statistics lies in this amazing fact: A simple random sample tends to produce a good representative sample of the population. At the time of writing this book, the population of the United States is more than 300 million people. Although the census collects some data on the entire population, for many questions of interest we are forced to rely on a small sample of individuals. Amazingly, if a simple random sample is selected, even a small sample can yield valid inferences for all 300 million Americans!

Example 1.14

Election Polling

Right before the 2012 presidential election, Google Consumer Surveys[19] randomly sampled and collected data on $n = 3252$ Americans. The sample showed Barack Obama ahead of Mitt Romney in the national popular vote by 2.3 percentage points. Can we generalize these results to the entire population of 126 million voters in order to estimate the popular vote in the election?

Solution

Yes! Because the poll included a random sample of voters, the results from the sample should generalize to the population. In the actual election, Obama won the national popular vote by 2.6 percentage points. Of course, the sample data do not perfectly match the population data (in Chapter 3 we will learn how closely we expect sample results to match population results), but the fact that we can get such an accurate guess from sampling only a very small fraction of the population is quite astonishing!

Analogy to Soup

Patti McConville/GettyImages, Inc.

Sampling the soup

Do we need to eat an entire large pot of soup to know what the soup tastes like? No! As long as the soup is well mixed, a few spoonfuls will give us a pretty good idea of the taste of the soup. This is the idea behind sampling: Just sampling a small part of the soup (or population) can still give us a good sense of the whole. However, if the soup is not mixed, so that the broth is at the top, the meat at the bottom, and so on, then a few spoonfuls will *not* give us a good sense of the taste of the soup. This is analogous to taking a biased sample. Mixing the soup randomizes the sample, so that the small part we taste is representative of the entire large pot.

How Do We Select a Random Sample?

You may think that you are capable of "randomly" selecting samples on your own, but you are wrong! Deborah Nolan, a statistics professor, has half of her students flip a coin and write the resulting sequence of heads and tails on the board

---

[19]Data from *http://www.fivethirtyeight.com*, "Which Polls Fared Best (and Worst) in the 2012 Presidential Race," November 10, 2012, and Google Consumer Surveys, November 5, 2012, *http://www.google .com/insights/consumersurveys/view?survey=6iwb56wuu5vc6&question=2&filter=&rw=1.*

(flipping a coin is truly random), and the other half of her students generate their own sequence of heads and tails without actually flipping the coin, trying to fake randomness. She can always tell the difference.[20] How can she tell? Because students (and their professors and all people!) are very bad at actual randomness.

Similarly, you may think you can select representative samples better than randomness can, but again, you are most likely wrong! Just as humans are surprisingly bad at faking randomness, humans are surprisingly bad at selecting representative samples. We tend to oversample some types of units and undersample less obvious types, even when we are explicitly trying hard not to do so. Luckily, randomness is surprisingly good at selecting a representative sample.

If we can't do randomness ourselves, how *do* we select a random sample? As mentioned, one way is to draw names out of a hat. A more common way is to use technology. Some forms of technology can automatically draw a sample randomly from a list of the entire population, mimicking the process of drawing names from a hat. Other forms produce random numbers, in which case we give a number to each unit in the population, and our sample becomes the units corresponding to the selected numbers.

Example 1.15

The dataset **AllCountries** contains data on 215 countries or economies. Select a random sample of 5 of these.

Solution

There are many ways to do this, depending on the technology or method available. One way is to number the countries from 1 to 215 and then use a random number generator to select five of the numbers between 1 and 215. Suppose we do this and get the numbers

46   85   152   161   49

As we see in the dataset, the corresponding countries are Costa Rica (46), Hungary (85), Peru (152), Samoa (161), and Cuba (49). These five countries are a random sample from the population of all countries. Of course, the numbers are randomly generated, so each sample generated this way is likely to be different. We talk more about the variability of random samples in Chapter 3.

Random Sampling Caution

In statistics, random is NOT the same as haphazard! We cannot obtain a random sample by haphazardly picking a sample on our own. We must use a formal random sampling method such as technology or drawing names out of a hat.

Realities of Random Sampling

While a random sample is ideal, often it may not be achievable. A list of the entire population may not exist, it may be impossible to contact some members of the population, or it may be too expensive or time consuming to do so. Often we must make do with whatever sample is convenient. The study can still be worth doing, but we have to be very careful when drawing inferences to the population and should at least try to avoid obvious sampling bias as much as possible.

---

[20]Gelman, A. and Nolan, D., *Teaching Statistics: A Bag of Tricks*, Oxford University Press, New York, 2002.

Example 1.16

If the *Chicago Tribune* had wanted to more accurately predict the outcome of the 1948 presidential election, what should they have done? Why do you think they didn't do this?

Solution   To more accurately predict the 1948 presidential election, they should have selected a random sample from the list of all registered voters and then asked the selected people who they would vote for. There are several possible reasons they did not select a random sample. They may not have had a list of all registered voters available from which to sample randomly. Also, collecting data from a random sample of voters might have required traveling to homes all over the country, which would have been time consuming and expensive. Sampling only people with telephones was cheaper and more convenient.

When it is difficult to take a random sample from the population of interest, we may have to redefine the population to which we generalize.

Example 1.17

What Proportion of People Are Vegetarian?

To determine what proportion of people are vegetarian, we would need to take a random sample of all people, which would be extremely difficult or impossible. How might we redefine our population and question so that it is possible to obtain an accurate estimate?

Solution   One option is to narrow our population to those living in Boston and ask, "What proportion of Bostonians are vegetarian?" It would be possible to take a random sample of telephone numbers from a Boston phone book and call and ask whether they eat meat. In this case our population would only include people with land line phone numbers listed in the Boston phone book so would not include people who rely only on cell phones or who have no phone at all.

For simplicity we only describe a simple random sample in detail, but there are other types of random samples. If we want to know the average weight of a population and want to ensure that the proportion of males and females in our sample matches that of the population, we may take two simple random samples, one within males and one within females. For a study on high school students, it is hard to take a simple random sample. We might first take a simple random sample of schools and then, within each of the sampled schools, take a simple random sample of students. These random sampling schemes are more complicated than the simple random sample but can still yield valid inferences.

Other Sources of Bias

Sampling bias is not the only form of bias that can occur when collecting data. Particularly when collecting data on humans, even if we have a good random sample, there are other issues that might bias the results.

Bias

Bias exists when the method of collecting data causes the sample data to inaccurately reflect the population.

Bias can occur when people we have selected to be in our sample choose not to participate. If the people who choose to respond would answer differently than the people who choose not to respond, results will be biased.

Example 1.18

In 1997 in Somerset (a county in England), a study was conducted on lifestyle choices associated with health.[21] A random sample of 6009 residents of Somerset were mailed a questionnaire that they were asked to fill out and return, and 57.6% of the people in the sample returned the questionnaire. Do you think health-related behavior such as exercise and smoking are accurately portrayed by the data collected?

Solution

 Probably not. People who returned the questionnaire may have been more proud of their responses, or may have been more organized and motivated in general, so more likely to lead a healthy lifestyle.

The researchers followed up with phone interviews for a random sample of those who had not responded. As suspected, the non-responders were quite different regarding health behaviors. For example, only 35.9% of initial responders reported getting no moderate or vigorous physical activity, while this percentage was almost doubled, 69.6%, for non-responders. Using only the data from the initial responders is very misleading.

 The way questions are worded can also bias the results. In 1941 Daniel Rugg[22] asked people the same question in two different ways. When asked "Do you think that the United States should allow public speeches against democracy?" 21% said the speeches should be allowed. However, when asked "Do you think that the United States should forbid public speeches against democracy?" 39% said the speeches should not be forbidden. Merely changing the wording of the question nearly doubled the percentage of people in favor of allowing (not forbidding) public speeches against democracy.

© Carmen Martínez Banús/iStockphoto

Would you have children?

[21]Hill, A., Roberts, J., Ewings, P., and Gunnell, D., "Non-response bias in a lifestyle survey," *Journal of Public Health Medicine*, June 1997; 19(2): 203–207.
[22]Rugg, D., "Experiments in wording questions," *Public Opinion Quarterly*, 1941; 5: 91–92.

> **DATA 1.4**   **"If You Had It to Do Over Again, Would You Have Children?"**
>
> In 1976, a young couple wrote to the popular columnist Ann Landers, asking for advice on whether or not to have children.[23] Ann ran the letter from the young couple (which included various reasons not to have kids, but no positive aspects of parenting) and asked her readers to respond to the question "If you had it to do over again, would you have children?" Her request for data yielded over 10,000 responses, and to her surprise, only 30% of readers answered "Yes." She later published these results in *Good Housekeeping*, writing that she was "stunned, disturbed, and just plain flummoxed" by the survey results. She again asked readers to answer the exact same question, and this time 95% of responders said "Yes."   ■

Example 1.19

In Data 1.4, why do you think the two percentages, 30% and 95%, are so drastically different?

Solution

The initial request for data was in a column with a letter stating many reasons not to have kids, which may have brought these issues to the minds of the responders. The second request was in an article mentioning Ann Landers' dismay at parents answering no, which may have influenced responses. The context in which the question is asked can bias answers one way or another.

Sampling bias is also present, since readers of her column in the newspaper and readers of *Good Housekeeping* and readers who choose to respond to each request for data are probably not representative of the population and probably differ from each other. For the first request, people with more negative experiences with children may have been encouraged to respond, while the opposite may have been true in the second case. You may be able to think of additional reasons for the discrepancy in the sample results.

Example 1.20

Suppose you are considering having children and would really like to know whether more parents are happy about having kids or regret their decision. Which percentages in Data 1.4 can you trust? How would you go about collecting data you can trust?

Solution

Since both of these samples only include people who decided to write in (volunteer samples) instead of taking a random sample, both almost definitely contain sampling bias, so neither should be trusted. To collect data you can trust, you should take a random sample of all parents (or perhaps take a random sample of all parents of your nationality).

Newsday took a random sample of all parents in the US, asking the same question as in Data 1.4. In this random sample, 91% said "Yes," they would have children again if given the choice. This doesn't mean that exactly 91% of parents are happy they had kids, but because it was a random sample, it does mean that the true percentage is close to 91%. In Chapter 3 we'll learn how to assess exactly how close we expect it to be. (Notice that the initial sample result of 30% is extremely misleading!)

Bias may also be introduced if people do not answer truthfully. If the sample data cannot be trusted, neither can generalizations from the sample to the population.

[23]*http://www.stats.uwo.ca/faculty/bellhouse/stat353annlanders.pdf.*

Example 1.21

Illicit Drug Use

The 2009 National Survey on Drug Use and Health[24] selected a random sample of US college students and asked them about illicit drug use, among other things. In the sample, 22.7% of the students reported using illicit drugs in the past year. Do you think this is an accurate portrayal of the percentage of all college students using illicit drugs?

Solution

This may be an underestimate. Even if the survey is anonymous, students may be reluctant to report illicit drug use on an official survey and thus may not answer truthfully.

Bias in data collection can result in many other ways not discussed here. The most important message is to always think critically about the way data are collected and to recognize that not all methods of data collection lead to valid inferences. Recognizing sources of bias is often simply common sense, and you will instantly become a more statistically literate individual if, each time you are presented with a statistic, you just stop, inquire, and think about how the data were collected.

SECTION LEARNING GOALS

You should now have the understanding and skills to:

- Distinguish between a sample and a population
- Recognize when it is appropriate to use sample data to infer information about the population
- Critically examine the way a sample is selected, identifying possible sources of sampling bias
- Recognize that random sampling is a powerful way to avoid sampling bias
- Identify other potential sources of bias that may arise in studies on humans

Exercises for Section **1.2**

SKILL BUILDER 1
In Exercises 1.28 to 1.31, state whether the data are best described as a population or a sample.

1.28 To estimate size of trout in a lake, an angler records the weight of 12 trout he catches over a weekend.

1.29 A subscription-based music website tracks its total number of active users.

1.30 The US Department of Transportation announces that of the 250 million registered passenger vehicles in the US, 2.1% are electro-gas hybrids.

1.31 A questionnaire to understand athletic participation on a college campus is emailed to 50 college students, and all of them respond.

SKILL BUILDER 2
In Exercises 1.32 to 1.35, describe the sample and describe a reasonable population.

1.32 A sociologist conducting a survey at a mall interviews 120 people about their cell phone use.

[24]Substance Abuse and Mental Health Services Administration, Results from the 2009 National Survey on Drug Use and Health: Volume I. Summary of National Findings (Office of Applied Studies, NSDUH Series H-38A, HHS Publication No. SMA 10-4856Findings), Rockville, MD, 2010, *https://nsduhweb .rti.org/*.

1.33 Five hundred Canadian adults are asked if they are proficient on a musical instrument.

1.34 A cell phone carrier sends a satisfaction survey to 100 randomly selected customers.

1.35 The Nielsen Corporation attaches databoxes to televisions in 1000 households throughout the US to monitor what shows are being watched and produce the Nielsen Ratings for television.

SKILL BUILDER 3

In Exercises 1.36 to 1.39, a biased sampling situation is described. In each case, give:

(a) The sample

(b) The population of interest

(c) A population we can generalize to given the sample

1.36 To estimate the proportion of Americans who support changing the drinking age from 21 to 18, a random sample of 100 college students are asked the question "Would you support a measure to lower the drinking age from 21 to 18?"

1.37 To estimate the average number of tweets from all twitter accounts in 2015, one of the authors randomly selected 10 of his followers and counted their tweets.

1.38 To investigate interest across all residents of the US in a new type of ice skate, a random sample of 1500 people in Minnesota are asked about their interest in the product.

1.39 To determine the height distribution of female high school students, the rosters are collected from 20 randomly selected high school girls basketball teams.

SKILL BUILDER 4

In Exercises 1.40 to 1.45, state whether or not the sampling method described produces a random sample from the given population.

1.40 The population is incoming students at a particular university. The name of each incoming student is thrown into a hat, the names are mixed, and 20 names (each corresponding to a different student) are drawn from the hat.

1.41 The population is the approximately 25,000 protein-coding genes in human DNA. Each gene is assigned a number (from 1 to 25,000), and computer software is used to randomly select 100 of these numbers yielding a sample of 100 genes.

1.42 The population is all employees at a company. All employees are emailed a link to a survey.

1.43 The population is adults between the ages of 18 and 22. A sample of 100 students is collected from a local university, and each student at the university had an equal chance of being selected for the sample.

1.44 The population is all trees in a forest. We walk through the forest and pick out trees that appear to be representative of all the trees in the forest.

1.45 The population is all people who visit the website *CNN.com*. All visitors to the website are invited to take part in the daily online poll.

IS IT BIASED?

In Exercises 1.46 to 1.50, indicate whether we should trust the results of the study. Is the method of data collection biased? If it is, explain why.

1.46 Ask a random sample of students at the library on a Friday night "How many hours a week do you study?" to collect data to estimate the average number of hours a week that all college students study.

1.47 Ask a random sample of people in a given school district, "Excellent teachers are essential to the well-being of children in this community, and teachers truly deserve a salary raise this year. Do you agree?" Use the results to estimate the proportion of all people in the school district who support giving teachers a raise.

1.48 Take 10 apples off the top of a truckload of apples and measure the amount of bruising on those apples to estimate how much bruising there is, on average, in the whole truckload.

1.49 Take a random sample of one type of printer and test each printer to see how many pages of text each will print before the ink runs out. Use the average from the sample to estimate how many pages, on average, all printers of this type will last before the ink runs out.

1.50 Send an email to a random sample of students at a university asking them to reply to the question: "Do you think this university should fund an ultimate frisbee team?" A small number of students reply. Use the replies to estimate the proportion of all students at the university who support this use of funds.

1.51 Do Parents Regret Having Children? In Data 1.4 on page 24, we describe the results of a question asked by a national newspaper columnist: "If you had it to do over again, would you have children?" In addition to those results and a follow-up national survey, the *Kansas City Star* selected a random sample of parents from Kansas City and asked them the same question. In this sample, 94%

said "Yes." To what population can this statistic be generalized?

1.52 Wearing a Uniform to Work The website *fox6now.com* held an online poll in June 2015 asking "What do you think about the concept of having an everyday uniform for work, like Steve Jobs did?" Of the people who answered the question, 24% said they loved the idea, 58% said they hated the idea, and 18% said that they already wore a uniform to work.

(a) Are the people who answered the poll likely to be representative of all adult workers? Why or why not?

(b) Is it reasonable to generalize this result and estimate that 24% of all adult workers would like to wear a uniform to work?

1.53 Canadians Stream Music In a random sample of 3500 Canadian consumers, about 71% report that they regularly stream music.[25]

(a) Is the sample likely to be representative of all Canadian consumers? Why or why not?

(b) Is it reasonable to generalize this result and estimate that about 71% of all Canadian consumers regularly stream music?

1.54 Climate Change In July 2015, a poll asked a random sample of 1,236 registered voters in Iowa whether they agree or disagree that the world needs to do more to combat climate change.[26] The results show that 65% agree, while 25% disagree and 10% don't know.

(a) What is the sample? What is the intended population?

(b) Is it reasonable to generalize this result and estimate that 65% of all registered voters in Iowa agree that the world needs to do more to combat climate change?

1.55 How Many People Wash Their Hands After Using the Washroom? In Example 1.10 on page 16, we introduce a study by researchers from Harris Interactive who were interested in determining what percent of people wash their hands after using the washroom. They collected data by standing in public restrooms and pretending to comb their hair or put on make-up as they observed patrons' behavior.[27]

Public restrooms were observed at Turner's Field in Atlanta, Penn Station and Grand Central Station in New York, the Museum of Science and Industry and the Shedd Aquarium in Chicago, and the Ferry Terminal Farmers Market in San Francisco. Of the over 6000 people whose behavior was observed, 85% washed their hands. Women were more likely to wash their hands: 93% of women washed, while only 77% of men did. The Museum of Science and Industry in Chicago had the highest hand-washing rate, while men at Turner's Field in Atlanta had the lowest.

(a) What are the cases? What are the variables? Classify each variable as quantitative or categorical.

(b) In a separate telephone survey of more than 1000 adults, more than 96% said they always wash their hands after using a public restroom. Why do you think there is such a discrepancy in the percent from the telephone survey compared to the percent observed?

1.56 Teaching Ability In a sample survey of professors at the University of Nebraska, 94% of them described themselves as "above average" teachers.[28]

(a) What is the sample? What is the population?

(b) Based on the information provided, can we conclude that the study suffers from sampling bias?

(c) Is 94% a good estimate for the percentage of above-average teachers at the University of Nebraska? If not, why not?

1.57 Effects of Alcohol and Marijuana In 1986 the Federal Office of Road Safety in Australia conducted an experiment to assess the effects of alcohol and marijuana on mood and performance.[29] Participants were volunteers who responded to advertisements for the study on two rock radio stations in Sydney. Each volunteer was given a randomly determined combination of the two drugs, then tested and observed. Is the sample likely representative of all Australians? Why or why not?

1.58 What Percent of Young Adults Move Back in with Their Parents? The Pew Research Center polled a random sample of $n = 808$ US residents between the ages of 18 and 34. Of those in the sample, 24% had moved back in with their parents for

[25]"What Moves Today's Teenage Canadian Music Fan?," *http://www.nielsen.com/ca/en/insights/news/2015/what-moves-todays-teenage-canadian-music-fan.html*, Neilsen, Media and Entertainment, June 2, 2015.

[26]*http://www.quinnipiac.edu/news-and-events/quinnipiac-university-poll/2016-presidential-swing-state-polls*. Accessed July 2015.

[27]Bakalar, "Study: More people washing hands after using bathroom," *Salem News*, September 14, 2010.

[28]Cross, P., "Not can, but *will* college teaching be improved?," *New Directions for Higher Education*, 1977; 17: 115.

[29]Chesher, G., Dauncey, H., Crawford, J., and Horn, K., "The Interaction between Alcohol and Marijuana: A Dose Dependent Study on the Effects on Human Moods and Performance Skills," Report No. C40, Federal Office of Road Safety, Federal Department of Transport, Australia, 1986.

economic reasons after living on their own.[30] Do you think that this sample of 808 people is a representative sample of all US residents between the ages of 18 and 34? Why or why not?

1.59 Do Cat Videos Improve Mood? As part of an "internet cat videos/photos" study, Dr. Jessica Gall Myrick posted an on-line survey to Facebook and Twitter asking a series of questions regarding how individuals felt before and after the last time they watched a cat video on the Internet.[31] One of the goals of the study was to determine how watching cat videos affects an individual's energy and emotional state. People were asked to share the link, and everyone who clicked the link and completed the survey was included in the sample. More than 6,000 individuals completed the survey, and the study found that after watching a cat video people generally reported more energy, fewer negative emotions, and more positive emotions.

(a) Would this be considered a simple random sample from a target population? Why or why not?

(b) Ignoring sampling bias, what other ways could bias have been introduced into this study?

1.60 Diet Cola and Weight Gain in Rats A study[32] fed one group of rats a diet that included yogurt sweetened with sugar, and another group of rats a diet that included yogurt sweetened with a zero-calorie artificial sweetener commonly found in diet cola. The rats that were fed a zero-calorie sweetener gained more weight and more body fat compared to the rats that were fed sugar. After the study was published, many news articles discussed the implication that people who drink diet soda gain more weight. Explain why we cannot conclude that this is necessarily true.

1.61 Armoring Military Planes During the Second World War, the U.S. military collected data on bullet holes found in B-24 bombers that returned from flight missions. The data showed that most bullet holes were found in the wings and tail of the aircraft. Therefore, the military reasoned that more armor should be added to these regions, as they are more likely to be shot. Abraham Wold, a famous

statistician of the era, is reported to have argued against this reasoning. In fact, he argued that based on these data more armor should be added to the center of the plane, and NOT the wings and tail. What was Wald's argument?

1.62 Employment Surveys Employment statistics in the US are often based on two nationwide monthly surveys: the Current Population Survey (CPS) and the Current Employment Statistics (CES) survey. The CPS samples approximately 60,000 US households and collects the employment status, job type, and demographic information of each resident in the household. The CES survey samples 140,000 nonfarm businesses and government agencies and collects the number of payroll jobs, pay rates, and related information for each firm.

(a) What is the population in the CPS survey?

(b) What is the population in the CES survey?

(c) For each of the following statistical questions, state whether the results from the CPS or CES survey would be more relevant.

  i. Do larger companies tend to have higher salaries?

  ii. What percentage of Americans are self-employed?

  iii. Are married men more or less likely to be employed than single men?

1.63 National Health Statistics The Centers for Disease Control and Prevention (CDC) administers a large number of survey programs for monitoring the status of health and health care in the US. One of these programs is the National Health and Nutrition Examination Survey (NHANES), which interviews and examines a random sample of about 5000 people in the US each year. The survey includes questions about health, nutrition, and behavior, while the examination includes physical measurements and lab tests. Another program is the National Hospital Ambulatory Medical Care Survey (NHAMCS), which includes information from hospital records for a random sample of individuals treated in hospital emergency rooms around the country.

(a) To what population can we reasonably generalize findings from the NHANES?

(b) To what population can we reasonably generalize findings from the NHAMCS?

(c) For each of the questions below, indicate which survey, NHANES or NHAMCS, would probably be more appropriate to address the issue.

  i. Are overweight people more likely to develop diabetes?

---

[30]Parker, K., "The Boomerang Generation: Feeling OK about Living with Mom and Dad," Pew Research Center, March 15, 2012.

[31]Gall Myrick, J., "Emotion regulation, procrastination, and watching cat videos online: Who watches Internet cats, why, and to what effect?," *Computers in Human Behavior*, June 12, 2015.

[32]Swithers, S.E., Sample, C.H., and Davidson, T.L., "Adverse effects of high-intensity sweeteners on energy intake and weight control in male and obesity-prone female rats." *Behavioral neuroscience*, 2013; 127(2), 262.

ii. What proportion of emergency room visits in the US involve sports-related injuries?

iii. Is there a difference in the average waiting time to be seen by an emergency room physician between male and female patients?

iv. What proportion of US residents have visited an emergency room within the past year?

1.64 Interviewing the Film Crew on Hollywood Movies There were 970 movies made in Hollywood between 2007 and 2013. Suppose that, for a documentary about Hollywood film crews, a random sample of 5 of these movies will be selected for in-depth interviews with the crew members. Assuming the movies are numbered 1 to 970, use a random number generator or table to select a random sample of five movies by number. Indicate which numbers were selected. (If you want to know which movies you selected, check out the dataset **HollywoodMovies**.)

1.65 Sampling Some Hardee's Restaurants The Hardee's Restaurant chain has about 1900 quick-serve restaurants in 30 US states and 9 countries.[33] Suppose that a member of the Hardee's administration wishes to visit six of these restaurants, randomly selected, to gather some first-hand data. Suppose the restaurants are numbered 1 to 1900. Use a random-number generator or table to select the numbers for 6 of the restaurants to be in the sample.

---

[33] hardees.com/company/franchise.

1.3 EXPERIMENTS AND OBSERVATIONAL STUDIES

Association and Causation

Three neighbors in a small town in northern New York State enjoy living in a climate that has four distinct seasons: warm summers, cold winters, and moderate temperatures in the spring and fall. They also share an interest in using data to help make decisions about questions they encounter at home and at work.

- Living in the first house is a professor at the local college. She's been looking at recent heating bills and comparing them to data on average outside temperature. Not surprisingly, when the temperature is lower, her heating bills tend to be much higher. She wonders, "It's going to be an especially cold winter; should I budget for higher heating costs?"

- Her neighbor is the plant manager for a large manufacturing plant. He's also been looking at heating data and has noticed that when the building's heating plant is used, there are more employees missing work due to back pain or colds and flu. He wonders, "Could emissions from the heating system be having adverse health effects on the workers?"

- The third neighbor is the local highway superintendent. He is looking at data on the amount of salt spread on the local roads and the number of auto accidents. (In northern climates, salt is spread on roads to help melt snow and ice and improve traction.) The data clearly show that weeks when lots of salt is used also tend to have more accidents. He wonders, "Should we cut down on the amount of salt we spread on the roads so that we have fewer accidents?"

Each of these situations involves a relationship between two variables. In each scenario, variations in one of the variables tend to occur in some regular way with changes in the other variable: lower temperatures go along with higher heating costs, more employees have health issues when there is more activity at the heating plant, and more salt goes with more accidents. When this occurs, we say there is an *association* between the two variables.

Association

Two variables are **associated** if values of one variable tend to be related to the values of the other variable.

The three neighbors share a desirable habit of using data to help make decisions, but they are not all doing so wisely. While colder outside temperatures probably force the professor's furnace to burn more fuel, do you think that using less salt on icy roads will make them safer? The key point is that an association between two variables, even a very strong one, does not imply that there is a *cause and effect* relationship between the two variables.

> **Causation**
>
> Two variables are **causally associated** if changing the value of one variable influences the value of the other variable.

The distinction between association and causation is subtle, but important. In a causal relationship, manipulating one of the variables tends to cause a change in the other. For example, we put more pressure on the gas pedal and a car goes faster. When an association is not causal, changing one of the variables will not produce a predictable change in the other. Causation often implies a particular direction, so colder outside temperatures might cause a furnace to use more fuel to keep the professor's house warm, but if she increases her heating costs by buying more expensive fuel, we should not expect the outdoor temperatures to fall!

Recall from Section 1.1 that values of an explanatory variable might help predict values of a response variable. These terms help us make the direction of a causal relationship more clear: We say changing the explanatory variable tends to cause the response variable to change. A causal statement (or any association statement) means that the relationship holds as an overall trend—not necessarily in every case.

Example 1.22

For each sentence discussing two variables, state whether the sentence implies no association between the variables, association without implying causation, or association with causation. If there is causation, indicate which variable is the explanatory variable and which is the response variable.

(a) Studies show that taking a practice exam increases your score on an exam.

(b) Families with many cars tend to also own many television sets.

(c) Sales are the same even with different levels of spending on advertising.

(d) Taking a low-dose aspirin a day reduces the risk of heart attacks.

(e) Goldfish who live in large ponds are usually larger than goldfish who live in small ponds.

(f) Putting a goldfish into a larger pond will cause it to grow larger.

Solution (a) This sentence implies that, in general, taking a practice exam *causes* an increase in the exam grade. This is association with causation. The explanatory variable is whether or not a practice exam was taken and the response variable is the score on the exam.

(b) This sentence implies association, since we are told that one variable (number of TVs) tends to be higher when the other (number of cars) is higher. However, it does not imply causation since we do not expect that buying another television set will somehow cause us to own more cars, or that buying another car will somehow cause us to own more television sets! This is association without causation.

(c) Because sales don't vary in any systematic way as advertising varies, there is no association.

(d) This sentence indicates association with causation. In this case, the sentence makes clear that a daily low-dose aspirin *causes* heart attack risk to go down. The explanatory variable is taking aspirin and the response variable is heart attack risk.

(e) This sentence implies association, but it only states that larger fish tend to be in larger ponds, so it does not imply causation.

(f) This sentence implies association with causation. The explanatory variable is the size of the pond and the response variable is the size of the goldfish.

Contrast the sentences in Example 1.22 parts (e) and (f). Both sentences are correct, but one implies causation (moving to a larger pond makes the fish grow bigger) and one does not (bigger fish just happen to reside in larger ponds). Recognizing the distinction is important, since implying causation incorrectly is one of the most common mistakes in statistics. Try to get in the habit of noticing when a sentence implies causation and when it doesn't.

Many decisions are made based on whether or not an association is causal. For example, in the 1950s, people began to recognize that there was an association between smoking and lung cancer, but there was a debate that lasted for decades over whether smoking *causes* lung cancer. It is now generally agreed that smoking causes lung cancer, and this has led to a substantial decline in smoking rates in the US. The fact that smoking causes lung cancer does not mean that everyone who smokes will get lung cancer, but it does mean that people who smoke are more likely to get it (in fact, 10 to 20 times more likely[34]). Other important causal questions, such as whether cell phones cause cancer or whether global warming is causing an increase in extreme weather events, remain topics of research and debate. One of the goals of this section is to help you determine when a study can, and cannot, establish causality.

Confounding Variables

Why are some variables associated even when they have no cause and effect relationship in either direction? As the next example illustrates, the reason is often the effect of other variables.

DATA 1.5   **Vehicles and Life Expectancy**

The US government collects data from many sources on a yearly basis. For example, Table 1.5 shows the number of vehicles (in millions) registered in the US[35] and the average life expectancy (in years) of babies born[36] in the US every four years from 1970 to 2010. A more complete dataset with values for each of the years from 1970 through 2013 is stored in **LifeExpectancyVehicles**. If we plot the points in Table 1.5, we obtain the graph in Figure 1.2. (This graph is an example of a scatterplot, which we discuss in Chapter 2.) As we see in the table and the graph, these two variables are very strongly associated; the more vehicles that are registered, the longer people are expected to live.   ■

---

[34] *http://www.cdc.gov/cancer/lung/basic_info/risk_factors.htm#1.*
[35] Vehicle registrations from US Census Bureau, *http://www.census.gov/compendia/statab/cats/transportation.html.*
[36] Centers for Disease Control and Prevention, National Center for Health Statistics, Health Data Interactive, *www.cdc.gov/nchs/hdi.htm*, Accessed June 2015.

Table 1.5 *Vehicle registrations (millions) and life expectancy*

Year	Vehicles	Life Expectancy
1970	108.4	70.8
1974	129.9	72.0
1978	148.4	73.5
1982	159.6	74.5
1986	175.7	74.7
1990	188.8	75.4
1994	198.0	75.7
1998	211.6	76.7
2002	229.6	77.3
2006	244.2	77.7
2010	242.1	78.7

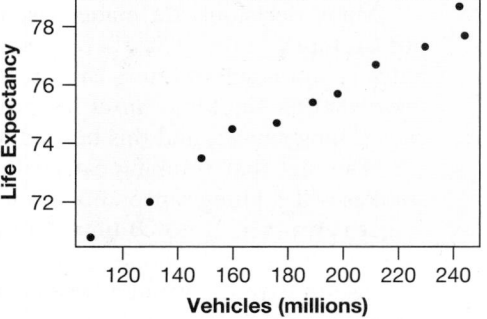

Figure 1.2 *A strong association between vehicles and life expectancy*

There is a clear association between vehicle registrations and life expectancy. Is this a *causal* association? If so, which way might it go? Do people live longer because they have a car to drive? When people live longer, do they have time to buy more vehicles? Or is there something else driving this association?

> **Confounding Variable**
>
> A **confounding variable**, also known as a **confounding factor** or **lurking variable**,[37] is a third variable that is associated with both the explanatory variable and the response variable. A confounding variable can offer a plausible explanation for an association between two variables of interest.

Example 1.23

Describe a possible confounding variable in Data 1.5 about vehicle registrations and life expectancy.

Solution

One confounding variable is the year. As time goes along, the population grows so more vehicles are registered and improvements in medical care help people live longer. Both variables naturally tend to increase as the year increases and may have little direct relationship with each other. The years are an explanation for the association between vehicle registrations and life expectancy.

[37]Some statisticians distinguish between confounding variables and lurking variables. However, for simplicity in this book we treat them as synonymous.

When faced with a strong association such as that between vehicles and life expectancy, it can be tempting to immediately jump to conclusions of causality. However, it is important to stop and think about whether there are confounding variables which could be explaining the association instead.

Example 1.24

In 2008, the *Los Angeles Times* published a headline[38] that included "Hospitals… Riskier than a Casino in Event of Cardiac Arrest." The article, based on a study published in the *New England Journal of Medicine*,[39] states that the widespread availability of defibrillators and bystanders in public places like casinos leads to a higher survival rate than hospitals in the case of cardiac arrest.

(a) What are the primary variables of interest in this study? Which is the explanatory variable and which is the response variable?

(b) Give an example of one potential confounding variable in this study.

(c) If you are having a heart attack, would you go to a hospital or a casino?

Solution

(a) The two primary variables of interest are the place of cardiac arrest (explanatory) and whether or not the person survives (response).

(b) A confounding variable is the health of the person at the time of cardiac arrest. Older, frailer, sicker people are more likely to be in the hospital and also less likely to survive (not because they are in a hospital, but just because they are weaker to begin with). Someone in a casino is much more likely to be in better physical shape, and thus better able to survive a heart attack. Notice that the confounding variable (health of the person) influences *both* of the variables of interest: where the person might be and whether the person is likely to survive.

(c) If you are having a heart attack, you should go to a hospital! Even though casinos have a higher survival rate, this can be explained by the confounding variable, and we cannot conclude that being in a casino *causes* a higher survival rate. For a person of a given health status, it is probably safer to be in a hospital under the care of professionals.

Many seemingly surprising claims in the media (such as that hospitals are riskier than casinos) can be explained simply by the presence of a confounding variable. Knowing how and when to be on the lookout for confounding variables is essential for statistical literacy and for assessing any data-based claims.

Observational Studies vs Experiments

How can we establish (statistically) when an association represents a causal relationship? The key is in how the data are collected. If we want to study how the explanatory variable influences the response variable, we have to be able to control or specify the values of the explanatory variable to make sure it is not associated with any potential confounding variables.

Note that in data such as **LifeExpectancyVehicles** or the study of cardiac arrest we merely collect available data after the fact. We call data collected in this way, with no effort or ability to manipulate the variables of interest, an *observational study*.

---

[38]Maugh, T., "Study finds hospitals slow to defibrillate: Researchers say they're riskier than a casino in event of cardiac arrest," *Los Angeles Times*, January 3, 2008.

[39]Chan, P., Krumholz, H., Nichol, G., and Nallamothu, B., American Heart Association National Registry of Cardiopulmonary Resuscitation Investigators, "Delayed Time to Defibrillation after In-Hospital Cardiac Arrest," *New England Journal of Medicine*, 2008; 358: 9–17.

With observational data we can never be certain that an apparent association is not due to some confounding variable, and thus the association is not evidence of a causal relationship.

The alternative is to intentionally control one or more of the explanatory variables when producing the data to see how the response variable changes. We call this method of data collection a *statistical experiment*. With a well-designed experiment, we can make conclusions about causation when we see a strong association, since the method for assigning the values of the explanatory variable(s) are not influenced by any confounding variables.

> **Observational Studies and Experiments**
>
> An **experiment** is a study in which the researcher actively controls one or more of the explanatory variables.
>
> An **observational study** is a study in which the researcher does not actively control the value of any variable but simply observes the values as they naturally exist.

Example 1.25

Both studies below are designed to examine the effect of fertilizer on the yield of an apple orchard. Indicate whether each method of collecting the data is an experiment or an observational study.

(a) Researchers find several different apple orchards and record the amount of fertilizer that had been used and the yield of the orchards.

(b) Researchers find several different apple orchards and assign different amounts of fertilizer to each orchard. They record the resulting yield from each.

Solution

(a) This is an observational study, since data were recorded after the fact and no variables were actively manipulated. Notice that there are many possible confounding variables that might be associated with both the amount of fertilizer used and the yield, such as the quality of soil.

(b) This is an experiment since the amount of fertilizer was *assigned* to different orchards. The researchers actively manipulated the assignment of the fertilizer variable, in order to determine the effect on the yield variable.

Do high fives help teams win?

Example 1.26

Basketball High Fives

In the 2011 NBA (National Basketball Association) finals, the Dallas Mavericks defeated the Miami Heat. One headline on NBC sports[40] stated, "Miami's real problem this series: Not enough high fives," citing a study[41] that found that teams exhibiting more "touching," such as high fives, early in the season had better performance later in the season. Is this study an experiment or an observational study? Does the study provide evidence that additional high fiving improves basketball performance?

Solution

The study is an observational study, because researchers did not manipulate or assign the number of high fives. The word "improves" implies causality, but because this was an observational study, confounding variables are likely and causality cannot be established. This study does not provide evidence that giving high fives improves basketball performance.

One possible confounding variable in Example 1.26 is how well a team gets along, which is likely to be associated both with the number of high fives and a team's performance. While we consider methods to account for some confounding variables later in this text, additional confounding variables may still exist. In an observational study, there is no way of guaranteeing that we haven't missed one.

Causation Caution

It is difficult to avoid confounding variables in observational studies. For this reason, observational studies can almost never be used to establish causality.

Randomized Experiments

In an experiment, the researcher controls the assignment of one or more variables. This power can allow the researcher to avoid confounding variables and identify causal relationships, if used correctly. But how can the researcher possibly avoid all potential confounding variables? The key is surprisingly simple: a *randomized experiment*. Just as randomness solved the problem of sampling bias, randomness can also solve the problem of confounding variables.

Randomized Experiment

In a **randomized experiment** the value of the explanatory variable for each unit is determined randomly, before the response variable is measured.

If a randomized experiment yields an association between the two variables, we can establish a causal relationship from the explanatory to the response variable.

---

[40]*http://probasketballtalk.nbcsports.com/2011/06/09/miami's-real-problem-this-series-not-enough-high-fives/.*
[41]Kraus, M., Huang, C., and Keltner, D., "Tactile communication, cooperation, and performance: An ethological study of the NBA," *Emotion*, 2010; 10(5): 745–749.

In a randomized experiment, we don't expect the explanatory variable to be associated with any other variables at the onset of the study, because its values were determined simply by random chance. If nothing is associated with the explanatory variable, then confounding variables do not exist! For this reason, if an association is found between the explanatory and response variables in a randomized experiment, we can conclude that it indeed is a causal association.

Recall from Section 1.2 that "random" does not mean haphazard. To ensure that the explanatory variable is determined by random chance alone, a formal randomization method (such as flipping a coin, dealing cards, drawing names out of a hat, or using technology) must be used.

Example 1.27

A college professor writes two final exams for her class of 50 students and would like to know if the two exams are similar in difficulty. On exam day, she gives Exam A to the first 25 students to enter the classroom and Exam B to the remaining 25 students.

(a) What is the explanatory variable? What is the response variable?

(b) Is this a randomized experiment? What might be a confounding variable?

Solution

(a) The explanatory variable is the exam the student took (A or B); the response variable is the exam score.

(b) No, this is not a randomized experiment. The exam students take is determined by when they enter the room, which is not random. Students that arrive especially early may be more motivated, and those that straggle in late may be less likely to perform well; time of arrival is a confounding variable.

Example 1.28

The following year the professor decides to do a truly randomized experiment. She prints the name of each of her students on an index card, shuffles the cards, and deals them into two piles. On exam day, she gives Exam A to the students with names dealt into one pile, and Exam B to the other pile. After grading, she observes that the students taking Exam B had much higher scores than those who took Exam A. Can we conclude that Exam B was easier?

Solution

Yes! This experiment provides evidence that Exam B was easier. Only random chance determined which student got which exam, so there is no reason to suspect confounding variables.

The key idea of Section 1.2 was that results from a sample can only be generalized to the population if the sampling units were selected *randomly* from the population. The key idea of this section is that causality can only be established if the values of the explanatory variable are *randomly* assigned to the units. Randomness is the essential ingredient in both cases, but the type of randomness should not be confused. In the first case we are randomly determining which units will be a part of the study. In the second case we are randomly determining which value of the explanatory variable will be assigned to each of the units already selected to be in our sample. Lack of randomness in either stage drastically influences the types of conclusions that can be made: Lack of randomness in sampling prevents generalizing to the population, lack of randomness in assigning the values of the explanatory variable prevents making causal conclusions. See Figure 1.3.

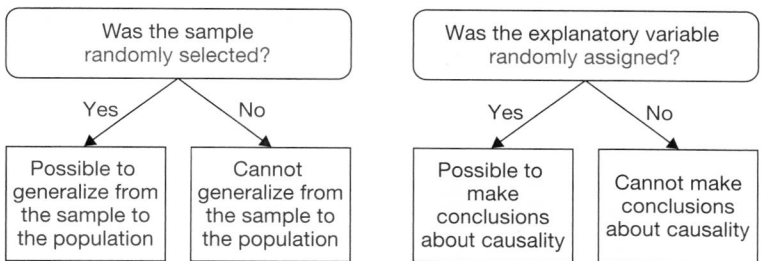

Figure 1.3 *Two fundamental questions about data collection*

DATA 1.6

Physicians' Health Study

Does anyone you know regularly take a low-dose aspirin? If so, it may be because of a randomized experiment conducted in the 1980s, the Physicians' Health Study.[42] The study recruited 22,071 male physicians and randomly assigned half of them to take an aspirin every other day for about five years and the other half to take a fake aspirin pill instead. They found that the physicians who took the real aspirin had 44% fewer heart attacks than those taking the fake aspirin. ■

The study in Data 1.6 is a randomized experiment because the researchers randomly determined which physicians received the real aspirin. The physicians themselves had no choice and in fact did not even know which pill they were taking. Because the physicians were split into two groups randomly, the only difference between the groups should be the aspirin. Therefore, we can conclude that the difference in heart attack rates must be *caused* by the aspirin. From this study we can conclude that regularly taking a low-dose aspirin reduces the risk of heart attack.

Many ideas of experimental design were originally developed for medical studies (such as Data 1.6) or agricultural experiments (like the fertilizer example of Example 1.25). For this reason, we often refer to values of the explanatory variable which the researcher controls as *treatments*. In Data 1.6, the treatments are the real aspirin and the fake aspirin.

Example 1.29

Warming Up

Warming up is a regular part of almost every athlete's pre-game routine, but the optimal amount to warm up is not always known. Cyclists typically have a very intense warm-up, and a study[43] in 2011 tests whether a shorter, less intense warm-up is better. Ten cyclists were recruited from the Calgary Track Cycling League and completed both a traditional intense warm-up and a shorter, less physically demanding, experimental warm-up. Each cyclist completed each warm-up at different times, and the order in which the warm-ups were performed was randomized. After each warm-up, performance was measured. The study found performance to be better after the shorter warm-up.

(a) What are the treatments?

(b) What conclusion can we draw from this study?

---

[42]The Steering Committee of the Physicians' Health Study Research Group, "Final report on the aspirin component of the ongoing Physicians' Health Study," *New England Journal of Medicine*, 1989; 321: 129–135.

[43]Tomaras, E. and Macintosh, B., "Less is More: Standard Warm-up Causes Fatigue and Less Warm-up Permits Greater Cycling Power Output," *Journal of Applied Physiology*, May 5, 2011.

Solution (a) There are two treatments to be compared: the more intense traditional warm-up and the shorter, less intense, experimental warm-up.

(b) Because the order of the warm-ups was *randomized*, causal conclusions can be made. The shorter warm-up causes the cyclists to perform better.

In Example 1.28 and Data 1.6, it was randomly determined which units got which treatment. In Example 1.29, every unit got both treatments, but the *order* of the treatments was randomly determined. Both ways of randomization yield valid randomized experiments. The former is known as a *randomized comparative experiment* because two groups of units are compared. The latter is known as a *matched pairs experiment*, because each unit forms a pair of data values (one under each treatment), and comparisons are made within each pair. These are only two of many different ways to incorporate randomization into an experiment.

> **Two Types of Randomized Experiments**
>
> In a **randomized comparative experiment**, we randomly assign cases to different treatment groups and then compare results on the response variable(s).
>
> In a **matched pairs experiment**, each case gets both treatments in random order (or cases get paired up in some other obvious way), and we examine individual differences in the response variable between the two treatments.

Example 1.30

Is the Dominant Hand Stronger?

We wish to run an experiment using 30 right-handed people to determine whether gripping strength in the dominant hand is greater than gripping strength in the other hand.

(a) Describe the experiment if we use a randomized comparative design.

(b) Describe the experiment if we use a matched pairs design.

(c) Which design makes more sense in this case?

Solution (a) Using a randomized comparative design, we randomly divide the 30 people into two groups of 15 each. We measure gripping strength in the right hand for one of the groups and in the left hand for the other group, and compare results.

(b) In a matched pairs experiment, we measure the gripping strength in both hands for each of the 30 people. The data are "paired" because we compare the right- and left-handed gripping strength for each person, and examine the difference between the two values. We randomize the order in which participants use the hands: some (randomly determined) doing the right hand first and some the left hand first. Notice that all participants are doing both, unlike in the experiment described in part (a) with two distinct groups each assigned a different treatment.

(c) A matched pairs experiment makes sense here because hand-gripping strength can vary a great deal between different people and it makes sense to compare a person's right-hand strength to his or her own left-hand strength.

Control Groups, Placebos, and Blinding

The Physicians' Health Study illustrates many aspects of a well-designed experiment. The participants who did not take an aspirin pill are an example of a *control group*. Nothing was done to this group that might directly influence the response variable. The control group provides a good comparison for the group that actually got the treatment of interest. Not all good experiments need a control group. There is no control, for example, in Example 1.28 when testing to see if one exam is more difficult than the other. In all cases, however, procedures for handling the groups should match as closely as possible so that effective comparisons can be made.

If people believe they are getting an effective treatment, they may experience the desired effect regardless of whether the treatment is any good. This phenomenon is called the *placebo effect*. Although perhaps not intuitive, the placebo effect has been studied extensively and can be very powerful. A *placebo* is a fake pill or treatment, and placebos are often used to control for the placebo effect in experiments. The fake aspirin pill given to participants in the control group of the Physicians' Health Study is an example of a placebo.

Using a placebo is not helpful, however, if participants know they are not getting the real treatment. This is one of the reasons that *blinding* is so important. In a *single-blind* experiment, the participants are not told which group they are in. In a *double-blind* experiment, the participants are not told which group they are in *and* the people interacting with the participants and recording the results of the response variable also do not know who is in which group. The Physicians' Health Study was double-blind: The people taking the pills did not know whether they were taking an aspirin or a placebo and the doctors treating them and determining who had heart attacks also did not know.

DATA 1.7

Sham Knee Surgery

For people suffering from arthritis of the knee, arthroscopic surgery has been one possible treatment. In the mid-1990s, a study[44] was conducted in which 10 men with arthritic knees were scheduled for surgery. They were all treated exactly the same except for one key difference: only some of them actually had the surgery! Once each patient was in the operating room and anesthetized, the surgeon looked at a randomly generated code indicating whether he should do the full surgery or just make three small incisions in the knee and stitch up the patient to leave a scar. All patients received the same post-operative care, rehabilitation, and were later evaluated by staff who didn't know which treatment they had. The result? The men who got the sham knee surgery and the men who got the real knee surgery showed similar and indistinguishable levels of improvement. ∎

Example 1.31

Discuss the experiment in Data 1.7. How is randomization used? Is there a placebo? Is the study double-blind? Why did the doctors make incisions in the knees of the men not getting the surgery?

Solution   Randomization was used to divide the men into groups, determining who got the real surgery and who didn't. The placebo was the fake surgery. Because the placebo surgery should match the real surgery as much as possible, those in the placebo group still received incisions and stitches. The men needed similar scars so that both the patients and the staff giving follow-up care were blind as to who actually had surgery done inside their knee. This made the study double-blind.

[44]Talbot, M., "The Placebo Prescription," *The New York Times*, January 9, 2000.

You may wonder whether data from only 10 patients are sufficient to make strong conclusions about the best treatment plan for arthritic knees. That would be a valid concern. In general, we would like to *replicate* each treatment on as many experimental units as is feasible. In many situations a small pilot study, such as the one described in Data 1.7, is used for initial guidance before undertaking a larger, more expensive experiment. In the case of the placebo knee surgery, a follow-up study with 180 patients produced similar results[45] – indicating that full knee surgery may not be needed for patients with this condition.

Example 1.32

Does an injection of caffeine help rats learn a maze faster? Design an experiment to investigate this question. Incorporate elements of a well-designed experiment.

Solution

We take the rats that are available for the study and *randomly* divide them into two groups. One group will get a shot of caffeine, while the other group will get a shot of saline solution (placebo). We have the rats run the maze and record their times. Don't tell the rats which group they are in! Ideally, all people who come in contact with the rats (the people giving the shots, the people recording the maze times, and so on) should not know which rats are in which group. This makes the study double-blind. Only the statistician analyzing the data will know which rats are in which group. (We describe here a randomized comparative experiment. A matched pairs experiment would also work, and in that case we would also use a placebo and blinding.)

Realities of Randomized Experiments

Randomization should always be used in designing an experiment. Blinding and the use of a placebo treatment should be used when appropriate and possible. However, there are often ethical considerations that preclude the use of an experiment in any form. For example, imagine designing an experiment to determine whether cell phones cause cancer or whether air pollution leads to adverse health consequences. It would not be appropriate to require people to wear a cell phone on their head for large amounts of time to see if they have higher cancer rates! Similarly, it would not be appropriate to require some people to live in areas with more polluted air. In situations such as these, observational studies can at least help us determine associations.

SECTION LEARNING GOALS

You should now have the understanding and skills to:

- Recognize that not every association implies causation
- Identify potential confounding variables in a study
- Distinguish between an observational study and a randomized experiment
- Recognize that only randomized experiments can lead to claims of causation
- Explain how and why placebos and blinding are used in experiments
- Distinguish between a randomized comparative experiment and a matched pairs experiment
- Design and implement a randomized experiment

[45]Moseley, J., et al., "A Controlled Trial of Arthroscopic Surgery for Osteoarthritis of the Knee," *The New England Journal of Medicine*, 2002; 347: 81–88.

Exercises for Section **1.3**

SKILL BUILDER 1

In Exercises 1.66 to 1.71, we give a headline that recently appeared online or in print. State whether the claim is one of association and causation, association only, or neither association nor causation.

1.66 Daily exercise improves mental performance.

1.67 Among college students, no link found between number of friends on social networking websites and size of the university.

1.68 Cell phone radiation leads to deaths in honey bees.

1.69 Wealthy people are more likely than other folks to lie, cheat, and steal.

1.70 Cat owners tend to be more educated than dog owners.

1.71 Want to lose weight? Eat more fiber!

SKILL BUILDER 2

Exercises 1.72 to 1.77 describe an association between two variables. Give a confounding variable that may help to account for this association.

1.72 More ice cream sales have been linked to more deaths by drowning.

1.73 The total amount of beef consumed and the total amount of pork consumed worldwide are closely related over the past 100 years.

1.74 People who own a yacht are more likely to buy a sports car.

1.75 Sales of toboggans tend to be higher when sales of mittens are higher.

1.76 Air pollution is higher in places with a higher proportion of paved ground relative to grassy ground.

1.77 People with shorter hair tend to be taller.

SKILL BUILDER 3

In Exercises 1.78 to 1.81, we describe data collection methods to answer a question of interest. Are we describing an experiment or an observational study?

1.78 To examine whether eating brown rice affects metabolism, we ask a random sample of people whether they eat brown rice and we also measure their metabolism rate.

1.79 To examine whether playing music in a store increases the amount customers spend, we randomly assign some stores to play music and some to stay silent and compare the average amount spent by customers.

1.80 To examine whether planting trees reduces air pollution, we find a sample of city blocks with similar levels of air pollution and we then plant trees in half of the blocks in the sample. After waiting an appropriate amount of time, we measure air pollution levels.

1.81 To examine whether farm-grown salmon contain more omega-3 oils if water is more acidic, we collect samples of salmon and water from multiple fish farms to see if the two variables are related.

REVISITING QUESTIONS FROM SECTION 1.1

Exercises 1.82 to 1.84 refer to questions of interest asked in Section 1.1 in which we describe data collection methods. Indicate whether the data come from an experiment or an observational study.

1.82 "Is there a sprinting gene?" Introduced in Example 1.5 on page 9.

1.83 "Do metal tags on penguins harm them?" Introduced in Data 1.3 on page 10.

1.84 "Are there human pheromones?" Introduced on page 11. Three studies are described; indicate whether each of them is an experiment or an observational study.

1.85 Salt on Roads and Accidents Three situations are described at the start of this section, on page 29. In the third bullet, we describe an association between the amount of salt spread on the roads and the number of accidents. Describe a possible confounding variable and explain how it fits the definition of a confounding variable.

1.86 Height and Reading Ability In elementary school (grades 1 to 6), there is a strong association between a child's height and the child's reading ability. Taller children tend to be able to read at a higher level. However, there is a very significant confounding variable that is influencing both height and reading ability. What is it?

1.87 Music Volume and Beer Consumption In 2008, a study[46] was conducted measuring the impact that music volume has on beer consumption. The researchers went into bars, controlled the music

[46]Gueguen, N., Jacob, C., Le Guellec, H., Morineau, T., and Lourel, M., "Sound Level of Environmental Music and Drinking Behavior: A Field Experiment with Beer Drinkers," *Alcoholism: Clinical and Experimental Research*, 2008; 32: 1795–1798.

volume, and measured how much beer was consumed. The article states that "the sound level of the environmental music was manipulated according to a randomization scheme." It was found that louder music corresponds to more beer consumption. Does this provide evidence that louder music causes people to drink more beer? Why or why not?

1.88 Nuts and Cholesterol Several studies have been performed to examine the relationship between nut consumption and cholesterol levels. Here we consider two such studies. In Study 1,[47] participants were assigned into two groups: one group was given nuts to eat each day, and the other group was told to consume a diet without nuts. In Study 2,[48] participants were free to follow their own diet, and reported how many nuts they consumed. Cholesterol levels were measured for all participants, and both studies found that nut consumption was associated with lower levels of LDL ("bad") cholesterol. Based on the information above, which study do you think provides better evidence that nut consumption reduces LDL cholesterol? Explain your answer.

1.89 Antibiotics in Infancy and Obesity in Adults "Antibiotics in infancy may cause obesity in adults," claims a recent headline.[49] A study in mice randomly assigned infant mice to either be given antibiotics or not, and the mice given antibiotics were more likely to be obese as adults. A separate study in humans found that children who had been given antibiotics before they were a year old (for example, for an ear infection) were more likely to be obese as adults. (Researchers believe the effect may be due to changes in the gut microbiome.) Based on these studies, is the headline an appropriate conclusion to make:

(a) For mice?

(b) For humans?

1.90 Do Online Cat Videos Improve Mood? Exercise 1.59 on page 28 introduced a study on cat videos, in which people who clicked on the link were asked questions regarding their mood before and after the most recent time they watched a cat video. Overall, participants reported that after watching a cat video they had significantly more energy, fewer negative emotions, and more positive emotions. Can we conclude from this study that watching cat videos increases energy and improves emotional state?

1.91 Green Spaces Make Kids Smarter A recent article[50] claims that "Green Spaces Make Kids Smarter." The study described in the article involved 2,623 schoolchildren in Barcelona. The researchers measured the amount of greenery around the children's schools, and then measured the children's working memories and attention spans. The children who had more vegetation around their schools did better on the memory and attention tests.

(a) What are the cases in this study?

(b) What is the explanatory variable?

(c) What is the response variable?

(d) Does the headline imply causation?

(e) Is the study an experiment or an observational study?

(f) Is it appropriate to conclude causation in this case?

(g) Suggest a possible confounding variable, and explain why it meets the requirements of a confounding variable.

1.92 Infections Can Lower IQ A headline in June 2015 proclaims "Infections can lower IQ."[51] The headline is based on a study in which scientists gave an IQ test to Danish men at age 19. They also analyzed the hospital records of the men and found that 35% of them had been in a hospital with an infection such as an STI or a urinary tract infection. The average IQ score was lower for the men who had an infection than for the men who hadn't.

(a) What are the cases in this study?

(b) What is the explanatory variable? Is it categorical or quantitative?

(c) What is the response variable? Is it categorical or quantitative?

(d) Does the headline imply causation?

(e) Is the study an experiment or an observational study?

(f) Is it appropriate to conclude causation in this case?

1.93 Sitting Is the New Smoking A 2014 headline reads "Sitting Is the New Smoking: Ways a

[47]Morgan, W.A., and Clayshulte, B.J., "Pecans lower low density lipoprotein cholesterol in people with normal lipid levels." *Journal of the American Dietetic Association*, 200; 100(3), 312–318.
[48]Li, T.Y., Brennan, A.M., Wedick, N.M., Mantzoros, C., Rifai, N., and Hu, F.B. "Regular consumption of nuts is associated with a lower risk of cardiovascular disease in women with type 2 diabetes." *The Journal of Nutrition*, 2009; 139(7), 1333–1338.
[49]Saey, T.H., "Antibiotics in infancy may cause obesity in adults," *Science News*, September 20, 2014.
[50]Khazan, O., "Green Spaces Make Kids Smarter," *The Atlantic*, June 16, 2016.
[51]"Infections can lower IQ," *The Week*, June 12, 2015, p. 18.

Sedentary Lifestyle is Killing You,"[52] and explains the mounting evidence for ways in which sitting is bad for you. A more recent large 2015 study[53] contributed to this evidence by following 69,260 men and 77,462 women and finding that for women, those who spent more leisure time sitting were significantly more likely to get cancer.

(a) What are the explanatory and response variables for the 2015 study?

(b) Is the 2015 study an observational study or a randomized experiment?

(c) Can we conclude from the 2015 study that spending more leisure time sitting causes cancer in women? Why or why not?

(d) Can we conclude from the 2015 study that spending more leisure time sitting does not cause cancer in women?

1.94 Late Night Eating It is well-known that lack of sleep impairs concentration and alertness, and this might be due partly to late night food consumption. A 2015 study[54] took 44 people aged 21 to 50 and gave them unlimited access to food and drink during the day, but allowed them only 4 hours of sleep per night for three consecutive nights. On the fourth night, all participants again had to stay up until 4 am, but this time participants were randomized into two groups; one group was only given access to water from 10 pm until their bedtime at 4 am while the other group still had unlimited access to food and drink for all hours. The group forced to fast from 10 pm on performed significantly better on tests of reaction time and had fewer attention lapses than the group with access to late night food.

(a) What are the explanatory and response variables?

(b) Is this an observational study or a randomized experiment?

(c) Can we conclude that eating late at night worsens some of the typical effects of sleep deprivation (reaction time and attention lapses)?

(d) Are there likely to be confounding variables? Why or why not?

1.95 To Spoon or Not to Spoon? Does cuddling after sex help boost sexual and relationship satisfaction? A study[55] involving 335 participants involved in romantic relationships found that people who reported more time spent on cuddling and affection after sex were more satisfied with their sex lives and relationships. This fact held true for both men and women. The average amount of time spent cuddling after sex was 15 minutes, and time spent on after sex affection was more strongly associated with sexual and relationship satisfaction than time spent on either foreplay or sex itself.

(a) Is this an observational study or a randomized experiment?

(b) Can we conclude that spending more time on affection after sex increases sexual and relationship satisfaction?

(c) A headline for an article[56] describing this study was titled "To Spoon or Not to Spoon? After-Sex Affection Boosts Sexual and Relationship Satisfaction." Does the study support this title?

(d) The title of the scientific article in which the study was originally published is "Post sex affectionate exchanges promote sexual and relationship satisfaction." Does the study support this title?

1.96 Does Early Language Reduce Tantrums? A recent headline reads "Early Language Skills Reduce Preschool Tantrums, Study Finds,"[57] and the article offers a potential explanation for this: "Verbalizing their frustrations may help little ones cope." The article refers to a study that recorded the language skill level and the number of tantrums of a sample of preschoolers.

(a) Is this an observational study or a randomized experiment?

(b) Can we conclude that "Early Language Skills Reduce Preschool Tantrums"? Why or why not?

(c) Give a potential confounding variable.

[52]"Sitting Is the New Smoking: Ways a Sedentary Lifestyle Is Killing You," *http://www.huffingtonpost.com/the-active-times/ sitting-is-the-new-smokin_b_5890006.html*, September 29, 2014, Accessed July 17, 2015.

[53]Patel, A.V., et al., "Leisure-time spent sitting and site-specific cancer incidence in a large US cohort," *Cancer Epidemiology, Biomarkers & Prevention*, June 30, 2015, doi:10.1158/1055-9965.EPI-15-0237.

[54]University of Pennsylvania School of Medicine. "Eating less during late night hours may stave off some effects of sleep deprivation." *ScienceDaily*, June 4, 2015 *www.sciencedaily.com/ releases/2015/06/150604141905.htm*.

[55]Muise, A., Giang, E., and Impett, E.A., "Post sex affectionate exchanges promote sexual and relationship satisfaction," *Archives of Sexual Behavior,* October 2014; 43(7): 1391–1402.

[56]"To Spoon or Not to Spoon? After-Sex Affection Boosts Sexual and Relationship Satisfaction," *http://www.scienceofrelationships .com/home/2014/5/16/to-spoon-or-not-to-spoon-after-sex-affec tion-boosts-sexual-a.html*, Accessed July 17, 2015.

[57]"Early Language Skills Reduce Preschool Tantrums, Study Finds," *US News and World Report, http://health.usnews.com/ health-news/news/articles/2012/12/20/early-language-skills-redu ce-preschool-tantrums-study-finds*, 20 December 2012, Accessed July 17, 2015.

1.97 Sleep and Recognition of Facial Expressions The ability to recognize and interpret facial expressions is key to successful human interaction. Could this ability be compromised by sleep deprivation? A 2015 study[58] took 18 healthy young adult volunteers and exposed them to 70 images of facial expressions, ranging from friendly to threatening. They were each shown images both after a full night of sleep and after sleep deprivation (24 hours of being awake), and whether each individual got a full night of sleep or was kept awake first was randomly determined. The study found that people were much worse at recognizing facial expressions after they had been kept awake.

(a) What are the explanatory and response variables?

(b) Is this an observational study or a randomized experiment? If it is a randomized experiment, is it a randomized comparative experiment or a matched pairs experiment?

(c) Can we conclude that missing a night of sleep hinders the ability to recognize facial expressions? Why or why not?

(d) In addition, for the people who had slept, the study found a strong positive association between quality of Rapid Eye Movement (REM) sleep and ability to recognize facial expressions. Can we conclude that better quality of REM sleep improves ability to recognize facial expressions? Why or why not? (*Hint:* What is the explanatory variable in this case? Was it randomly assigned?)

1.98 Diet Cola and Weight Gain in Humans A study[59] found that American senior citizens who report drinking diet soda regularly experience a greater increase in weight and waist circumference than those who do not drink diet soda regularly.

(a) From these results, can we conclude that drinking diet soda causes weight gain? Explain why or why not.

(b) Consider the results of this study on senior citizens, and the randomized experiment on rats introduced in Exercise 1.60 on page 28, which showed a similar association. Discuss what these two studies together might imply about the likelihood that diet cola causes weight gain in humans.

1.99 Does Red Increase Men's Attraction to Women? A study[60] examined the impact of the color red on how attractive men perceive women to be. In the study, men were randomly divided into two groups and were asked to rate the attractiveness of women on a scale of 1 (not at all attractive) to 9 (extremely attractive). One group of men were shown pictures of women on a white background and the other group were shown the same pictures of women on a red background. The men who saw women on the red background rated them as more attractive. All participants and those showing the pictures and collecting the data were not aware of the purpose of the study.

(a) Is this an experiment or an observational study? Explain.

(b) What is the explanatory variable and what is the response variable? Identify each as categorical or quantitative.

(c) How was randomization used in this experiment? How was blinding used?

(d) Can we conclude that using a red background color instead of white increases men's attractiveness rating of women's pictures?

1.100 Urban Brains and Rural Brains A study published in 2010 showed that city dwellers have a 21% higher risk of developing anxiety disorders and a 39% higher risk of developing mood disorders than those who live in the country. A follow-up study published in 2011 used brain scans of city dwellers and country dwellers as they took a difficult math test.[61] To increase the stress of the participants, those conducting the study tried to humiliate the participants by telling them how poorly they were doing on the test. The brain scans showed very different levels of activity in stress centers of the brain, with the urban dwellers having greater brain activity than rural dwellers in areas that react to stress.

(a) Is the 2010 study an experiment or an observational study?

(b) Can we conclude from the 2010 study that living in a city increases a person's likelihood of developing an anxiety disorder or mood disorder?

[58]Goldstein-Piekarski, A., et al., "Sleep Deprivation Impairs the Human Central and Peripheral Nervous System Discrimination of Social Threat," *The Journal of Neuroscience*, July 15, 2015; 35(28): 10135–10145; doi: 10.1523/JNEUROSCI.5254-14.2015

[59]Fowler, S.P., Williams, K., and Hazuda, H.P. "Diet Soda Intake Is Associated with LongTerm Increases in Waist Circumference in a Bioethnic Cohort of Older Adults: The San Antonio Longitudinal Study of Aging." *Journal of the American Geriatrics Society*, 2015; 63(4), 708–715.

[60]Elliot, A. and Niesta, D., "Romantic Red: Red Enhances Men's Attraction to Women," *Journal of Personality and Social Psychology*, 2008; 95(5): 1150–1164.

[61]"A New York state of mind," *The Economist*, June 25, 2011, p. 94.

(c) Is the 2011 study an experiment or an observational study?

(d) In the 2011 study, what is the explanatory variable and what is the response variable? Indicate whether each is categorical or quantitative.

(e) Can we conclude from the 2011 study that living in a city increases activity in stress centers of the brain when a person is under stress?

1.101 Split the Bill? When the time comes for a group of people eating together at a restaurant to pay their bill, sometimes they might agree to split the costs equally and other times will pay individually. If this decision were made in advance, would it affect what they order? Suppose that you'd like to do an experiment to address this question. The variables you will record are the type of *payment* (split or individual), *sex* of each person, number of *items* ordered, and the *cost* of each person's order. Identify which of these variables should be treated as *explanatory* and which as *response*. For each explanatory variable, indicate whether or not it should be randomly assigned.

1.102 Be Sure to Get Your Beauty Sleep! New research[62] supports the idea that people who get a good night's sleep look more attractive. In the study, 23 subjects ages 18 to 31 were photographed twice, once after a good night's sleep and once after being kept awake for 31 hours. Hair, make-up, clothing, and lighting were the same for both photographs. Observers then rated the photographs for attractiveness, and the average rating under the two conditions was compared. The researchers report in the *British Medical Journal* that "Our findings show that sleep-deprived people appear less attractive compared with when they are well rested."

(a) What is the explanatory variable? What is the response variable?

(b) Is this an experiment or an observational study? If it is an experiment, is it a randomized comparative design or a matched pairs design?

(c) Can we conclude that sleep deprivation *causes* people to look less attractive? Why or why not?

1.103 Do Antidepressants Work? Following the steps given, design a randomized comparative experiment to test whether fluoxetine (the active ingredient in Prozac pills) is effective at reducing depression. The participants are 50 people suffering from depression and the response variable is the change on a standard questionnaire measuring level of depression.

(a) Describe how randomization will be used in the design.

(b) Describe how a placebo will be used.

(c) Describe how to make the experiment double-blind.

1.104 Do Children Need Sleep to Grow? About 60% of a child's growth hormone is secreted during sleep, so it is believed that a lack of sleep in children might stunt growth.[63]

(a) What is the explanatory variable and what is the response variable in this association?

(b) Describe a randomized comparative experiment to test this association.

(c) Explain why it is difficult (and unethical) to get objective verification of this possible causal relationship.

1.105 Carbo Loading It is commonly accepted that athletes should "carbo load," that is, eat lots of carbohydrates, the day before an event requiring physical endurance. Is there any truth to this? Suppose you want to design an experiment to find out for yourself: "Does carbo loading actually improve athletic performance the following day?" You recruit 50 athletes to participate in your study.

(a) How would you design a randomized comparative experiment?

(b) How would you design a matched pairs experiment?

(c) Which design do you think is better for this situation? Why?

1.106 Alcohol and Reaction Time Does alcohol increase reaction time? Design a randomized experiment to address this question using the method described in each case. Assume the participants are 40 college seniors and the response variable is time to react to an image on a screen after drinking either alcohol or water. Be sure to explain how randomization is used in each case.

(a) A randomized comparative experiment with two groups getting two separate treatments

(b) A matched pairs experiment

1.107 Causation and Confounding Causation does not necessarily mean that there is no confounding variable. Give an example of an association between two variables that have a causal relationship AND have a confounding variable.

[62]Stein, R., "Beauty sleep no myth, study finds," *Washington Post*, *washingtonpost.com*, Accessed December 15, 2010.

[63]Rochman, B., "Please, Please, Go to Sleep," *Time Magazine*, March 26, 2012, p. 46.

CHAPTER 2

Describing Data

"*Technology [has] allowed us to collect vast amounts of data in almost every business. The people who are able to in a sophisticated and practical way analyze that data are going to have terrific jobs.*"

Chrystia Freeland, Managing Editor, *Financial Times**

*Speaking on CNN *Your Money*, November 29, 2009.
Top left: Image Source/Getty Images, Inc., Top right: Erik Von Weber/Getty Images, Inc., Bottom right: ©Dumrong Khajaroen/iStockphoto

Questions and Issues

Here are some of the questions and issues we will discuss in this chapter:

- Can dogs smell cancer in humans?
- What percent of college students say stress negatively affects their grades?
- How big is the home field advantage in soccer?
- Does electrical stimulation of the brain help with problem solving?
- Should males with a laptop computer worry about lowering their sperm count?
- What percent of people smoke?
- Can cricket chirps be used to predict the temperature?
- Which coffee has more caffeine: light roast or dark roast?
- How much do restaurant customers tip?
- How heavily does economic growth depend on the inflation rate?
- How does a person's body posture affect levels of stress?
- Do movies with larger budgets get higher audience ratings?
- Does it pay to get a college degree?
- What percent of college students have been in an emotionally abusive relationship?
- Does sexual frustration increase the desire for alcohol?
- What percent of NBA basketball players *never* attempt a 3-point shot?
- Are there "commitment genes"? Are there "cheating genes"?
- Do antibiotics in infancy affect the tendency to be overweight?
- Do online sites charge different amounts depending on the browser you use?
- What is the impact of "social jetlag" in which weekend sleep time is different from weekday sleep time?

2.1 CATEGORICAL VARIABLES

In Chapter 1, we learned that there are two types of variables, categorical and quantitative. In this chapter, we see how to describe both types of variables and the relationships between them. In each case, the description takes two parts: We see how to visualize the data using graphs and we see how to summarize key aspects of the data using numerical quantities, called *summary statistics*. We start by investigating categorical variables in this section.

© Denis Zbukarev/iStockphoto

Does each person have one true love?

Do you believe that there is only one true love for each person? What proportion of people do you think share your opinion? A recent survey addressed this question.

DATA 2.1

Is there one true love for each person?

A nationwide US telephone survey conducted by the Pew Foundation[1] in October 2010 asked 2625 adults ages 18 and older, "Some people say there is only one true love for each person. Do you agree or disagree?" In addition to finding out the proportion who agree with the statement, the Pew Foundation also wanted to find out if the proportion who agree is different between males and females, and whether the proportion who agree is different based on level of education (no college, some college, or college degree). The survey participants were selected randomly, by landlines and cell phones. ■

Example 2.1

What is the sample? What is the population? Do you believe the method of sampling introduces any bias? Can the sample results be generalized to the population?

Solution The sample is the 2625 people who were surveyed. The population is all US adults ages 18 or older who have a landline telephone or cell phone. Since the sampling was random, there is no sampling bias. There are no obvious other forms of bias, so the sample results can probably generalize to the population.

---

[1] Adapted from "The Decline of Marriage and Rise of New Families," Social and Demographic Trends, Pew Research Center, *http://www.pewresearch.org*, released November 18, 2010.

Example 2.2

What are the cases in the survey about one true love? What are the variables? Are the variables categorical or quantitative? How many rows and how many columns would the data table have?

Solution

The cases are the adults who answered the survey questions. The description indicates that there are at least three variables. One variable is whether the responder agrees or disagrees with the statement that each person has only one true love. A second variable is gender and a third is level of education. All three variables are categorical. The data table will have one row for each person who was surveyed, so there will be 2625 rows. There is a column for each variable, and there are at least three variables so there will be at least three columns.

One Categorical Variable

Of the $n = 2625$ people who responded to the survey, 735 agree with the statement that there is only one true love for each person, while 1812 disagree and 78 say they don't know. Table 2.1 displays these results in a *frequency table*, which gives the counts in each category of a categorical variable.

Table 2.1 *A frequency table: Is there one true love for each person?*

Response	Frequency
Agree	735
Disagree	1812
Don't know	78
Total	2625

What proportion of the responders agree with the statement that we all have exactly one true love? We have

$$\text{Proportion who agree} = \frac{\text{Number who agree}}{\text{Total number}} = \frac{735}{2625} = 0.28$$

This proportion is a summary statistic that helps describe the categorical variable for this group of adults. We see that the proportion who agree that there is one true love is 0.28 or 28%.[2]

Proportion

The **proportion** in some category is found by

$$\text{Proportion in a category} = \frac{\text{Number in that category}}{\text{Total number}}$$

---

[2]The two values 0.28 and 28% are equivalent and we use them interchangeably.

Proportions are also called *relative frequencies*, and we can display them in a *relative frequency table*. The proportions in a relative frequency table will add to 1 (or approximately 1 if there is some round-off error). Relative frequencies allow us to make comparisons without referring to the sample size.

Example 2.3

Find the proportion of responders who disagree in the one-true-love survey and the proportion who don't know and display all the proportions in a relative frequency table.

Solution   The proportion of responders who disagree is $1812/2625 = 0.69$ and the proportion who responded that they didn't know is $78/2625 = 0.03$. A frequency table or relative frequency table includes all possible categories for a categorical variable, so the relative frequency table includes the three possible answers of "Agree," "Disagree," and "Don't know," with the corresponding proportions. See Table 2.2. The proportions add to 1, as we expect.

Table 2.2 *A relative frequency table: Is there one true love for each person?*

Response	Relative Frequency
Agree	0.28
Disagree	0.69
Don't know	0.03
Total	1.00

Visualizing the Data in One Categorical Variable

Figure 2.1(a) displays a *bar chart* of the results in Table 2.1. The vertical axis gives the frequency (or count), and a bar of the appropriate height is shown for each category. Notice that if we used relative frequencies or percentages instead of frequencies, the bar chart would be identical except for the scale on the vertical axis. The categories can be displayed in any order on the horizontal axis. Another way to display proportions for a categorical variable, common in the popular media, is with a *pie chart*, as in Figure 2.1(b), in which the proportions correspond to the areas of sectors of a circle.

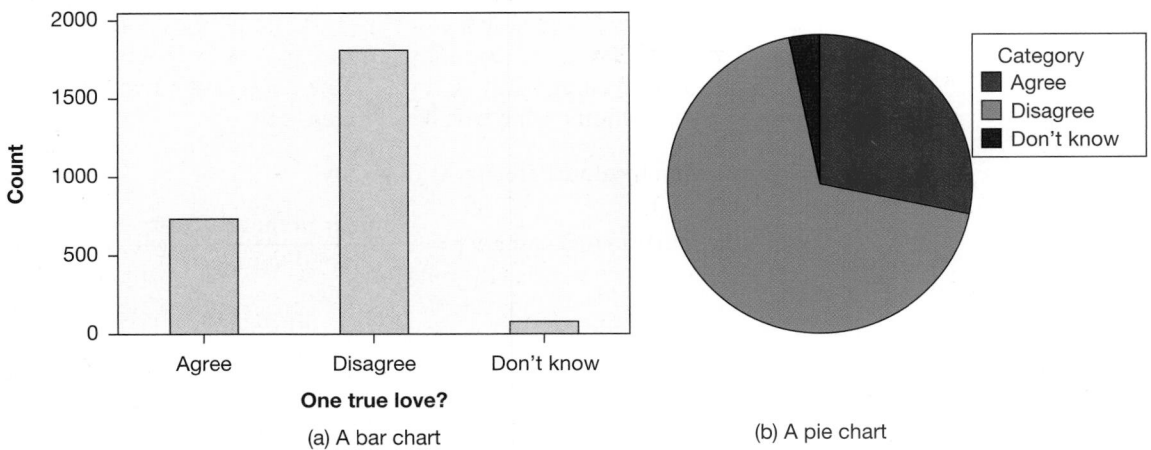

(a) A bar chart

(b) A pie chart

Figure 2.1 *Is there one true love for each person?*

Notation for a Proportion

As we saw in Chapter 1, it is important to distinguish between a population and a sample. For this reason, we often use different notation to indicate whether a quantity such as a proportion comes from a sample or an entire population.

> **Notation for a Proportion**
>
> The proportion for a *sample* is denoted $\hat{p}$ and read "*p*-hat."
> The proportion for a *population* is denoted p.

Example 2.4

In each of the situations below,[3] find the proportion of people who identify themselves as Hispanic or Latino. Use the correct notation with each answer.

(a) The 2010 US Census shows a US population of 308,745,538 people, and 50,325,523 of these people identify themselves as Hispanic or Latino.

(b) A random sample of 300 US citizens in Colorado shows that 62 of them identify themselves as Hispanic or Latino.

Solution

(a) The US Census gives information about essentially all residents of the US. This is a population, so we use the notation p for the proportion and we have

$$p = \frac{50,325,523}{308,745,538} = 0.163$$

(b) This is a sample of the population of Colorado, so we use the notation $\hat{p}$ for the sample proportion. We have

$$\hat{p} = \frac{62}{300} = 0.207$$

Two Categorical Variables: Two-Way Tables

Does the proportion of people who agree that there is exactly one true love for each person differ between males and females? Does it differ based on the education level of the responders? Both questions are asking about a *relationship* between two categorical variables. We investigate the question about gender here and investigate the effect of education level in Exercise 2.17.

To investigate a possible relationship between two categorical variables we use a *two-way table*. In a two-way table, we add a second dimension to a frequency table to account for the second categorical variable. Table 2.3 shows a two-way table for the responses to the question of one true love by gender.

> **Two-Way Table**
>
> A **two-way table** is used to show the relationship between two categorical variables. The categories for one variable are listed down the side (rows) and the categories for the second variable are listed across the top (columns). Each cell of the table contains the count of the number of data cases that are in both the row and column categories.

---

[3] *http://www.2010.census.gov/2010census.*

Table 2.3 *Two-way table: Is there one true love for each person?*

	Male	Female
Agree	372	363
Disagree	807	1005
Don't know	34	44

It is often helpful to also include the totals (both for rows and columns) in the margins of a two-way table, as in Table 2.4. Notice the column labeled "Total" corresponds exactly to the frequency table in Table 2.1.

Table 2.4 *Two-way table with row and column totals*

	Male	Female	Total
Agree	372	363	735
Disagree	807	1005	1812
Don't know	34	44	78
Total	1213	1412	2625

So, are men or women more hopelessly romantic? The two-way table can help us decide.

Example 2.5

Use Table 2.4 to answer the following questions.

(a) What proportion of females agree?

(b) What proportion of the people who agree are female?

(c) What proportion of males agree?

(d) Are females or males more likely to believe in one true love?

(e) What proportion of survey responders are female?

Solution

(a) To determine what proportion of females agree, we are interested only in the females, so we use only that column. We divide the number of females who agree (363) by the total number of females (1412):

$$\text{Proportion of females who agree} = \frac{\text{Number of females who agree}}{\text{Number of females}}$$
$$= \frac{363}{1412} = 0.26$$

(b) To determine what proportion of the people who agree are female, we are interested only in the people who agree, so we use only that row. We have

$$\text{Proportion who agree that are female} = \frac{\text{Number of females who agree}}{\text{Number who agree}}$$
$$= \frac{363}{735} = 0.49$$

Notice that the answers for parts (a) and (b) are NOT the same! The proportion in part (a) is probably more useful. More females than males happened to be included in the survey, and this affects the proportion in part (b), but not in part (a).

(c) To determine what proportion of males agree, we have

$$\text{Proportion of males who agree} = \frac{\text{Number of males who agree}}{\text{Number of males}}$$
$$= \frac{372}{1213} = 0.31$$

(d) We see in part (c) that 31% of the males in the survey agree that there is one true love for each person, while we see in part (a) that only 26% of the females agree with that statement. In this sample, males are more likely than females to believe in one true love.

(e) To determine what proportion of all the survey responders are female, we use the total row. We have

$$\text{Proportion of females} = \frac{\text{Number of females}}{\text{Total number}} = \frac{1412}{2625} = 0.54$$

We see that 54% of the survey respondents are female and the other 46% are male.

Be sure to read questions carefully when using a two-way table. The questions "What proportion of females agree?" and "What proportion of people who agree are female?" in Example 2.5(a) and (b) sound similar but are asking different questions. Think about the proportion of US senators who are male and the proportion of males who are US senators; clearly, these proportions are not the same!

Example 2.6

In the **StudentSurvey** dataset, students are asked which award they would prefer to win: an Academy Award, a Nobel Prize, or an Olympic gold medal. The data show that 20 of the 31 students who prefer an Academy Award are female, 76 of the 149 students who prefer a Nobel Prize are female, and 73 of the 182 who prefer an Olympic gold medal are female.

(a) Create a two-way table for these variables.

(b) Which award is the most popular with these students? What proportion of all students selected this award?

Solution

(a) The relevant variables are gender and which award is preferred. Table 2.5 shows a two-way table with three columns for award and two rows for gender. It doesn't matter which variable corresponds to rows and which to columns, but we need to be sure that all categories are listed for each variable. The numbers given in the problem are shown in bold, and the rest of the numbers can be calculated accordingly.

Table 2.5 *Two-way table of gender and preferred award*

	Academy	Nobel	Olympic	Total
Female	**20**	**76**	**73**	169
Male	11	73	109	193
Total	**31**	**149**	**182**	362

(b) More students selected an Olympic gold medal than either of the others, so that award is the most popular. We have

$$\text{Proportion selecting Olympic} = \frac{\text{Number selecting Olympic}}{\text{Total number}} = \frac{182}{362} = 0.503$$

We see that 50.3%, or about half, of the students prefer an Olympic gold medal to the other options.

Difference in Proportions

Example 2.7

In Example 2.6, we see that the most popular award is the Olympic gold medal. Is preferring an Olympic gold medal associated with gender? Use Table 2.5 to determine the *difference* between the proportion of males who prefer an Olympic gold medal and the proportion of females who prefer an Olympic gold medal.

Solution 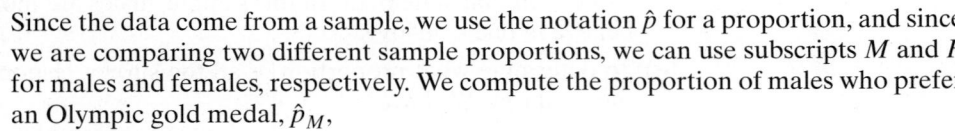 Since the data come from a sample, we use the notation $\hat{p}$ for a proportion, and since we are comparing two different sample proportions, we can use subscripts M and F for males and females, respectively. We compute the proportion of males who prefer an Olympic gold medal, $\hat{p}_M$,

$$\hat{p}_M = \frac{109}{193} = 0.565$$

and the proportion of females who prefer an Olympic gold medal, $\hat{p}_F$,

$$\hat{p}_F = \frac{73}{169} = 0.432$$

The difference in proportions is

$$\hat{p}_M - \hat{p}_F = 0.565 - 0.432 = 0.133$$

Males in the sample are much more likely to prefer an Olympic gold medal, so there appears to be an association between gender and preferring an Olympic gold medal.

As in Example 2.7, we often use subscripts to denote specific sample proportions and take the difference between two proportions. Computing a difference in proportions is a useful measure of association between two categorical variables, and in later chapters we develop methods to determine if this association is likely to be present in the entire population.

Visualizing a Relationship between Two Categorical Variables

There are several different types of graphs to use to visualize a relationship between two categorical variables. One is a *segmented bar chart*, shown in Figure 2.2(a), which gives a graphical display of the results in Table 2.5. In a

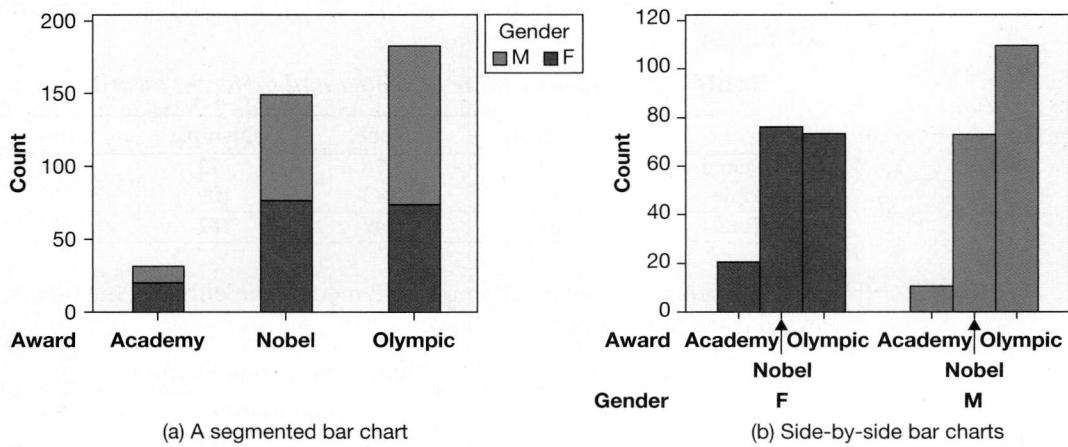

(a) A segmented bar chart   (b) Side-by-side bar charts

Figure 2.2 *Displaying a relationship between gender and preferred award*

segmented bar chart, the height of each bar represents the number of students selecting each award, while the color (in this case, red for females and green for males) indicates how many of the students who preferred each type were male and how many were female.

Example 2.8

Use Figure 2.2(a) to determine which award was most preferred overall and which award was most preferred by females. Explain.

Solution

From the height of the bars, we see that more students in general preferred an Olympic gold medal. From the colors, we see that the preferred award for females was a Nobel prize, by a small margin over an Olympic gold medal.

This same information can instead be displayed in *side-by-side bar charts*, in which separate bar charts are given for each group in one of the categorical variables. In Figure 2.2(b), we show bar charts for each gender, males and females. We could have also decided to show gender bar charts for each award. The graph we choose to display depends on what information we hope to convey about the data. Graphs such as a segmented bar chart or side-by-side bar charts are called *comparative plots* since they allow us to compare groups in a categorical variable.

SECTION LEARNING GOALS

You should now have the understanding and skills to:

- Display information from a categorical variable in a table or graph
- Use information about a categorical variable to find a proportion, with correct notation
- Display information about a relationship between two categorical variables in a two-way table
- Use a two-way table to find proportions
- Interpret graphs involving two categorical variables

Exercises for Section **2.1**

SKILL BUILDER 1
Exercises 2.1 to 2.4 provide information about data in **StudentSurvey**. Find the sample proportion $\hat{p}$.

2.1 The survey students consisted of 169 females and 193 males. Find $\hat{p}$, the proportion who are female.

2.2 The survey included 43 students who smoke and 319 who don't. Find $\hat{p}$, the proportion who smoke.

2.3 Of the students who identified their class year in the survey, 94 were first years, 195 were sophomores, 35 were juniors, and 36 were seniors. Find $\hat{p}$, the proportion who are upperclass students (juniors or seniors.)

2.4 The math SAT score is higher than the verbal SAT score for 205 of the 355 students who answered the questions about SAT scores. Find $\hat{p}$, the proportion for whom the math SAT score is higher.

SKILL BUILDER 2
In Exercises 2.5 to 2.8, give the relevant proportion using correct notation.

2.5 In the 2010 US Census, we see that 37,342,870 people, or 12.4% of all residents, are foreign-born.[4]

[4]*http://www.census.gov.*

2.6 A recent headline states that "45% Think Children of Illegal Immigrants Should Be Able to Attend Public School." The report gives the results of a survey of 1000 randomly selected likely voters in the US.[5]

2.7 A survey conducted of 1060 randomly selected US teens aged 13 to 17 found that 605 of them say they have made a new friend online.[6]

2.8 Of all 1,672,395 members of the high school class of 2014 who took the SAT (Scholastic Aptitude Test), 793,986 were minority students.[7]

SKILL BUILDER 3

In Exercises 2.9 and 2.10, data from the **StudentSurvey** dataset are given. Construct a relative frequency table of the data using the categories given. Give the relative frequencies rounded to three decimal places.

2.9 Of the 362 students who answered the question about what award they would prefer, 31 preferred an Academy Award, 149 preferred a Nobel Prize, and 182 preferred an Olympic gold medal.

2.10 Of the 361 students who answered the question about the number of piercings they had in their body, 188 had no piercings, 82 had one or two piercings, and the rest had more than two.

SKILL BUILDER 4

In Exercises 2.11 and 2.12, a two-way table is shown for two groups, 1 and 2, and two possible outcomes, A and B. In each case,

(a) What proportion of all cases had Outcome A?

(b) What proportion of all cases are in Group 1?

(c) What proportion of cases in Group 1 had Outcome B?

(d) What proportion of cases who had Outcome A were in Group 2?

2.11

	Outcome A	Outcome B	Total
Group 1	20	80	100
Group 2	60	40	100
Total	80	120	200

2.12

	Outcome A	Outcome B	Total
Group 1	40	10	50
Group 2	30	20	50
Total	70	30	100

2.13 Home Field Advantage in Soccer In the book *Scorecasting*,[8] we learn that "Across 43 professional soccer leagues in 24 different countries spanning Europe, South America, Asia, Africa, Australia, and the United States (covering more than 66,000 games), the home field advantage [percent of games won by the home team] in soccer worldwide is 62.4%." Is this a population or a sample? What are the cases and approximately how many are there? What is the variable and is it categorical or quantitative? What is the relevant statistic, including correct notation?

2.14 Airborne Antibiotics A recent study shows that antibiotics added to animal feed to accelerate growth can become airborne. Some of these drugs can be toxic if inhaled and may increase the evolution of antibiotic-resistant bacteria. Scientists[9] analyzed 20 samples of dust particles from animal farms. Tylosin, an antibiotic used in animal feed that is chemically related to erythromycin, showed up in 16 of the samples.

(a) What is the variable in this study? What are the individual cases?

(b) Display the results in a frequency table.

(c) Make a bar chart of the data.

(d) Give a relative frequency table of the data.

2.15 Rock-Paper-Scissors Rock-Paper-Scissors, also called Roshambo, is a popular two-player game often used to quickly determine a winner and loser. In the game, each player puts out a fist (rock), a flat hand (paper), or a hand with two fingers extended (scissors). In the game, rock beats scissors which beats paper which beats rock. The question is: Are the three options selected equally often by players? Knowing the relative frequencies with which the options are selected would give a player a significant advantage. A study[10] observed 119 people playing Rock-Paper-Scissors. Their choices are shown in Table 2.6.

(a) What is the sample in this case? What is the population? What does the variable measure?

(b) Construct a relative frequency table of the results.

[5]Rassmussen Reports, October 10, 2013.

[6]Lenhart, A., "Teens, Technology, and Friendships," Pew Research Center, *pewresearch.org*, August 6, 2015.

[7]*https://www.collegeboard.org/program-results/2014/sat.*

[8]Moskowitz, T. and Wertheim, L., *Scorecasting*, Crown Archetype, New York, 2011, p. 113.

[9]Hamscher, G., et al., "Antibiotics in Dust Originating from a Pig-Fattening Farm: A New Source of Health Hazard for Farmers?" *Environmental Health Perspectives*, October 2003; 111(13): 1590–1594.

[10]Eyler, D., Shalla, Z., Doumaux, A., and McDevitt, T., "Winning at Rock-Paper-Scissors," *College Mathematics Journal*, March 2009.

(c) If we assume that the sample relative frequencies from part (b) are similar for the entire population, which option should you play if you want the odds in your favor?

(d) The same study determined that, in repeated plays, a player is more likely to repeat the option just picked than to switch to a different option. If your opponent just played paper, which option should you pick for the next round?

Table 2.6 *Frequencies in Rock-Paper-Scissors*

Option Selected	Frequency
Rock	66
Paper	39
Scissors	14
Total	119

2.16 Sports-Related Concussions in Kids Researchers examined all sports-related concussions reported to an emergency room for children ages 5 to 18 in the United States over the course of one year.[11] Table 2.7 displays the number of concussions in each of the major activity categories.

(a) Are these results from a population or a sample?

(b) What proportion of concussions came from playing football?

(c) What proportion of concussions came from riding bicycles?

(d) Can we conclude that, at least in terms of concussions, riding bicycles is more dangerous to children in the US than playing football? Why or why not?

Table 2.7 *Concussion frequency by activity in children ages 5–18*

Activity	Frequency
Bicycles	23,405
Football	20,293
Basketball	11,507
Playground	10,414
Soccer	7,667
Baseball	7,433
All-Terrain Vehicle	5,220
Hockey	4,111
Skateboarding	4,408
Swimming	3,846
Horseback Riding	2,648
Total	100,952

[11]Valasek, A. and Walter, K.D., "Pediatric and Adolescent Concussion," *Springer*, New York, 2012.

2.17 Does Belief in One True Love Differ by Education Level? In Data 2.1 on page 48, we introduce a study in which people were asked whether they agreed or disagreed with the statement that there is only one true love for each person. Is the level of a person's education related to the answer given, and if so, how? Table 2.8 gives a two-way table showing the results for these two variables. A person's education is categorized as HS (high school degree or less), Some (some college), or College (college graduate or higher).

(a) Create a new two-way table with row and column totals added.

(b) Find the percent who agree that there is only one true love, for each education level. Does there seem to be an association between education level and agreement with the statement? If so, in what direction?

(c) What percent of people participating in the survey have a college degree or higher?

(d) What percent of the people who disagree with the statement have a high school degree or less?

Table 2.8 *Education level and belief in one true love*

	HS	Some	College
Agree	363	176	196
Disagree	557	466	789
Don't know	20	26	32

2.18 Who Smokes More: Males or Females The **StudentSurvey** dataset includes variables on gender and on whether or not the student smokes. Who smokes more: males or females? Table 2.9 shows a two-way table of these two variables.

(a) Which gender has a higher percentage of smokers: males or females?

(b) What is the proportion of smokers for the entire sample?

(c) What proportion of the smokers in the sample are female?

Table 2.9 *Smoking habits by gender*

	Female	Male	Total
Don't smoke	153	166	319
Smoke	16	27	43
Total	169	193	362

2.19 Antibiotics in Infancy and Weight A Canadian longitudinal study[12] examines whether giving antibiotics in infancy increases the likelihood that the child will be overweight later in life. The children were classified as having received antibiotics or not during the first year of life and then being overweight or not at 9 years old. The study included 616 children, and the results are shown in Table 2.10.

(a) What proportion of all children in the study were given antibiotics during the first year of life?

(b) What proportion of all children in the study were classified as overweight at age 9?

(c) What proportion of those receiving antibiotics were classified as overweight at age 9?

(d) What proportion of those not receiving antibiotics were classified as overweight at age 9?

(e) If we use $\hat{p}_A$ to denote the proportion from part (c) and $\hat{p}_N$ to denote the proportion from part (d), calculate the difference in proportion being overweight, $\hat{p}_A - \hat{p}_N$, between those who were exposed to antibiotics and those who weren't.

(f) What proportion of all children classified as overweight were given antibiotics during the first year of life?

Table 2.10 *Does antibiotic exposure early in life affect weight later in life?*

	Overweight	Not overweight	Total
Antibiotics	144	294	438
No antibiotics	37	141	178
Total	181	435	616

2.20 Culture and Mental Illness A recent study[13] examining the link between schizophrenia and culture interviewed 60 people who had been diagnosed with schizophrenia and who heard voices in their heads. The participants were evenly split between the US, India, and Ghana, and each was interviewed to determine whether the voices were mostly negative, mostly neutral, or mostly positive. The results are shown in Table 2.11. "Learned cultural expectations about the nature of mind and self" appear to influence how the voices are perceived.

(a) What proportion of all the participants felt that the voices are mostly negative?

(b) What proportion of all US participants felt that the voices are mostly negative?

(c) What proportion of non-US participants felt that the voices are mostly negative?

(d) What proportion of participants hearing positive voices are from the US?

(e) Does culture appear to be associated with how voices are perceived by people with schizophrenia?

Table 2.11 *How do people with schizophrenia perceive the voices?*

	US	India	Ghana	Total
Negative	14	4	2	20
Neutral	6	3	2	11
Positive	0	13	16	29
Total	20	20	20	60

2.21 Does It Pay to Get a College Degree? The Bureau of Labor Statistics[14] in the US tells us that, in 2010, the unemployment rate for high school graduates with no college degree is 9.7% while the unemployment rate for college graduates with a bachelor's degree is only 5.2%. Find the difference in proportions of those unemployed between these two groups and give the correct notation for the difference, with a minus sign. Since the data come from the census, you can assume that the values are from a population rather than a sample. Use the correct notation for population proportions, and use subscripts on the proportions to identify the two groups.

2.22 Can Dogs Smell Cancer? Scientists are working to train dogs to smell cancer, including early stage cancer that might not be detected with other means. In previous studies, dogs have been able to distinguish the smell of bladder cancer, lung cancer, and breast cancer. Now, it appears that a dog in Japan has been trained to smell bowel cancer.[15] Researchers collected breath and stool samples from patients with bowel cancer as well as from healthy people. The dog was given five samples in each test, one from a patient with cancer and four from healthy volunteers. The dog correctly selected the cancer sample in 33 out of 36 breath tests and in 37 out of 38 stool tests.

[12]Azad, M.B., Bridgman, S.L., Becker, A.B. and Kozyrskyj, A.L., "Infant antibiotic exposure and the development of childhood overweight and central adiposity," *International Journal of Obesity* (2014) 38, 1290–1298.

[13]Bower, B., "Hallucinated voices' attitudes vary with culture," *Science News*, December 10, 2014.

[14]Thompson, D., "What's More Expensive than College? Not Going to College," *The Atlantic*, March 27, 2012.

[15]"Dog Detects Bowel Cancer," CNN Health Online, January 31, 2011.

(a) The cases in this study are the individual tests. What are the variables?

(b) Make a two-way table displaying the results of the study. Include the totals.

(c) What proportion of the breath samples did the dog get correct? What proportion of the stool samples did the dog get correct?

(d) Of all the tests the dog got correct, what proportion were stool tests?

2.23 Is There a Genetic Marker for Dyslexia? A disruption of a gene called *DYXC1* on chromosome 15 for humans may be related to an increased risk of developing dyslexia. Researchers[16] studied the gene in 109 people diagnosed with dyslexia and in a control group of 195 others who had no learning disorder. The *DYXC1* break occurred in 10 of those with dyslexia and in 5 of those in the control group.

(a) Is this an experiment or an observational study? What are the variables?

(b) How many rows and how many columns will the data table have? Assume rows are the cases and columns are the variables. (There might be an extra column for identification purposes; do not count this column in your total.)

(c) Display the results of the study in a two-way table.

(d) To see if there appears to be a substantial difference between the group with dyslexia and the control group, compare the proportion of each group who have the break on the *DYXC1* gene.

(e) Does there appear to be an association between this genetic marker and dyslexia for the people in this sample? (We will see in Chapter 4 whether we can generalize this result to the entire population.)

(f) If the association appears to be strong, can we assume that the gene disruption causes dyslexia? Why or why not?

2.24 Help for Insomniacs In Exercise 1.23, we learned of a study to determine whether just one session of cognitive behavioral therapy can help people with insomnia. In the study, forty people who had been diagnosed with insomnia were randomly divided into two groups of 20 each. People in one group received a one-hour cognitive behavioral therapy session while those in the other group received no treatment. Three months later, 14 of those in the therapy group reported sleep improvements while only 3 people in the other group reported improvements.

[16]*Science News*, August 30, 2003, p. 131.

(a) Create a two-way table of the data. Include totals across and down.

(b) How many of the 40 people in the study reported sleep improvement?

(c) Of the people receiving the therapy session, what proportion reported sleep improvements?

(d) What proportion of people who did not receive therapy reported sleep improvements?

(e) If we use $\hat{p}_T$ to denote the proportion from part (c) and use $\hat{p}_N$ to denote the proportion from part (d), calculate the difference in proportion reporting sleep improvements, $\hat{p}_T - \hat{p}_N$, between those getting therapy and those not getting therapy.

2.25 Electrical Stimulation for Fresh Insight? If we have learned to solve problems by one method, we often have difficulty bringing new insight to similar problems. However, electrical stimulation of the brain appears to help subjects come up with fresh insight. In a recent experiment[17] conducted at the University of Sydney in Australia, 40 participants were trained to solve problems in a certain way and then asked to solve an unfamiliar problem that required fresh insight. Half of the participants were randomly assigned to receive non-invasive electrical stimulation of the brain while the other half (control group) received sham stimulation as a placebo. The participants did not know which group they were in. In the control group, 20% of the participants successfully solved the problem while 60% of the participants who received brain stimulation solved the problem.

(a) Is this an experiment or an observational study? Explain.

(b) From the description, does it appear that the study is double-blind, single-blind, or not blind?

(c) What are the variables? Indicate whether each is categorical or quantitative.

(d) Make a two-way table of the data.

(e) What percent of the people who correctly solved the problem had the electrical stimulation?

(f) Give values for $\hat{p}_E$, the proportion of people in the electrical stimulation group to solve the problem, and $\hat{p}_S$, the proportion of people in the sham stimulation group to solve the problem. What is the difference in proportions $\hat{p}_E - \hat{p}_S$?

(g) Does electrical stimulation of the brain appear to help insight?

[17]Chi, R. and Snyder, A., "Facilitate Insight by Non-Invasive Brain Stimulation," *PLoS ONE*, 2011; 6(2).

NATIONAL COLLEGE HEALTH ASSESSMENT SURVEY

Exercises 2.26 to 2.29 use data on college students collected from the American College Health Association–National College Health Assessment survey[18] conducted in Fall 2011. The survey was administered at 44 colleges and universities representing a broad assortment of types of schools and representing all major regions of the country. At each school, the survey was administered to either all students or a random sample of students, and more than 27,000 students participated in the survey.

2.26 Emotionally Abusive Relationships Students in the ACHA–NCHA survey were asked, "Within the last 12 months, have you been in a relationship (meaning an intimate/coupled/partnered relationship) that was emotionally abusive?" The results are given in Table 2.12.

(a) What percent of all respondents have been in an emotionally abusive relationship?

(b) What percent of the people who have been in an emotionally abusive relationship are male?

(c) What percent of males have been in an emotionally abusive relationship?

(d) What percent of females have been in an emotionally abusive relationship?

Table 2.12 *Have you been in an emotionally abusive relationship?*

	Male	Female	Total
No	8352	16,276	24,628
Yes	593	2034	2627
Total	8945	18,310	27,255

[18]*http://www.acha-ncha.org/docs/ACHA-NCHA-II_Reference Group_DataReport_Fall2011.pdf.*

2.27 Binge Drinking Students in the ACHA–NCHA survey were asked, "Within the last two weeks, how many times have you had five or more drinks of alcohol at a sitting?" The results are given in Table 2.13.

Table 2.13 *In the last two weeks, how many times have you had five or more drinks of alcohol?*

	Male	Female	Total
0	5402	13,310	18,712
1–2	2147	3678	5825
3–4	912	966	1878
5 +	495	358	853
Total	8956	18,312	27,268

(a) What percent of all respondents answered zero?

(b) Of the students who answered five or more days, what percent are male?

(c) What percent of males report having five or more drinks at a sitting on three or more days in the last two weeks?

(d) What percent of females report having five or more drinks at a sitting on three or more days in the last two weeks?

2.28 How Accurate Are Student Perceptions? Students in the ACHA–NCHA survey were asked two questions about alcohol use, one about their own personal consumption of alcohol and one about their perception of other students' consumption of alcohol. Figure 2.3(a) shows side-by-side bar charts for responses to the question "The last time you 'partied'/socialized, how many drinks of alcohol did you have?" while Figure 2.3(b) shows side-by-side bar charts for responses to the question "How many drinks of alcohol do you think the typical student at your school had the last time he/she 'partied'/socialized?"

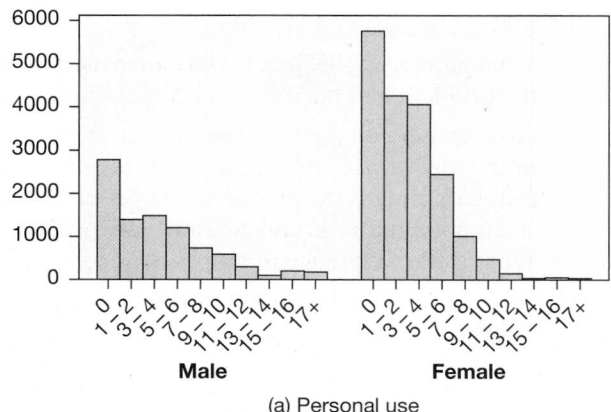
(a) Personal use

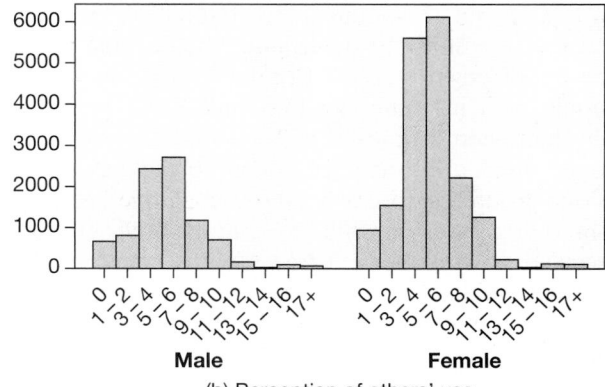
(b) Perception of others' use

Figure 2.3 *How many drinks of alcohol?*

(a) What is the most likely response for both males and females when asked about their own personal alcohol use?

(b) What is the most likely response for both males and females when asked about alcohol use of a "typical student"?

(c) Do students' perceptions of "typical" alcohol use match reality? (This phenomenon extends what we learned about the inability of students to select unbiased samples in Chapter 1. In this case, students tend to notice heavy drinkers disproportionately.)

2.29 Does Stress Affect Academic Performance? Students in the ACHA–NCHA survey were asked, "Within the last 12 months, has stress negatively affected your academics?" Figure 2.4(a) shows a segmented bar chart for response frequencies, while Figure 2.4(b) shows a segmented bar chart for response relative frequencies as percents. Possible responses were "I haven't had any stress," shown in red, "I've had stress but it hasn't hurt my grades," shown in green, or "I've had stress and it has hurt my grades," shown in blue.

(a) Did more males or more females answer the survey or did approximately equal numbers of males and females participate? Is graph (a) or (b) more helpful to answer this question?

(b) Did a greater *number* of males or females say they had no stress or is it approximately equal between males and females? Is graph (a) or (b) more helpful to answer this question?

(c) Did a greater *percent* of males or females say they had no stress or is it approximately equal between males and females? Is graph (a) or (b) more helpful to answer this question?

(d) Did a greater percent of males or females say that stress affected their grades or is it approximately equal between males and females? Is graph (a) or (b) more helpful to answer this question?

FINANCIAL INCENTIVES TO QUIT SMOKING

Exercises 2.30 to 2.32 deal with an experiment to study the effects of financial incentives to quit smoking.[19] Smokers at a company were invited to participate in a smoking cessation program and randomly assigned to one of two groups. Those in the *Reward* group would get a cash award if they stopped smoking for six months. Those in the *Deposit* group were asked to deposit some money which they would get back along with a substantial bonus if they stopped smoking.

2.30 Success at Quitting Smoking After six months, 156 of the 914 smokers who accepted the invitation to be in the reward-only program stopped smoking, while 78 of the 146 smokers who paid a deposit quit. Set up a two-way table and compare the success rates between participants who entered the two programs.

2.31 Agreeing to Participate The random assignment at the start of the experiment put 1017 smokers in the *Reward* group and 914 of them agreed to participate. However, only 146 of the 1053 smokers assigned to the *Deposit* group agreed to participate (since they had to risk some of their own money). Set up a two-way table and compare the participation rates between subjects assigned to the two treatment groups.

2.32 Accounting for Participation Rates Exercises 2.30 and 2.31 show that smokers who agreed

[19]Halpern, S., et al., "Randomized Trial of Four Financial-Incentive Programs for Smoking Cessation," *New England Journal of Medicine*, 2015, 372:2108–2117

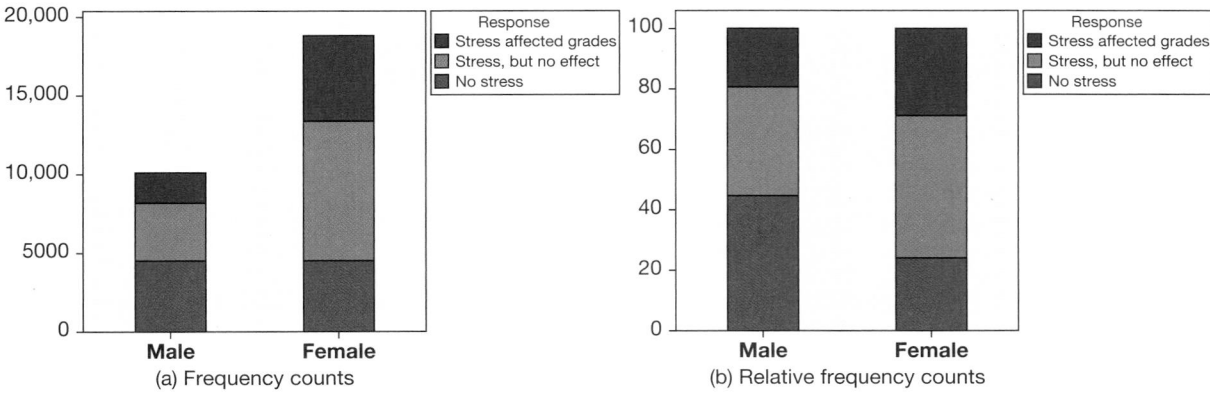

Figure 2.4 *Has stress hurt your grades?*

to be in the *Deposit* group (having their own money at risk) were much more likely to quit smoking than those enrolled in the *Reward* group, but many fewer of the original subjects assigned to that group agreed to participate. The researchers also found that 30 of the smokers originally in the *Deposit* group, who declined to participate, ended up stopping smoking on their own, while only 3 of the decliners in the *Reward* group stopped. Use this information and the data in the two exercises to find and compare the proportion of the subjects originally assigned to each group who were successful in quitting smoking for six months. For reference, another group of subjects at this company who got no financial incentive had about a 6.0% success rate in quitting.

2.33 Association or No Association? Suppose an experiment will randomly divide 40 cases between two possible treatments, *A* and *B*, and will then record two possible outcomes, Successful or Not successful. The outline of a two-way table is shown in Table 2.14. In each case below, fill in the table with possible values to show:

(a) A clear association between treatment and outcome.

(b) No association at all between treatment and outcome.

Table 2.14 *Fill in the blanks to show (a) Association or (b) No association*

	Successful	Not successful	Total
Treatment A			20
Treatment B			20
Total			40

2.34 Vaccine for Malaria In order for a vaccine to be effective, it should reduce a person's chance of acquiring a disease. Consider a hypothetical vaccine for malaria—a tropical disease that kills between 1.5 and 2.7 million people every year.[20] Suppose the vaccine is tested with 500 volunteers in a village who are malaria free at the beginning of the trial. Two hundred of the volunteers will get the experimental vaccine and the rest will not be vaccinated. Suppose that the chance of contracting malaria is 10% for those who are not vaccinated. Construct a two-way table to show the results of the experiment if:

(a) The vaccine has no effect.

(b) The vaccine cuts the risk of contracting malaria in half.

[20]World Health Organization.

2.35 Class Year in Statistics The **StudentSurvey** data file contains information from a survey done the first day of an introductory statistics course. The *Year* variable shows the class year (FirstYear, Sophomore, Junior, or Senior) for each student.

(a) Use technology to construct a frequency table showing the number of students in each class year. Ignore any cases where the year is missing.

(b) Find the relative frequency for the class year with the largest number of students taking this course.

2.36 Bistro Servers The **RestaurantTips** data file comes from a sample of 157 bills at a small bistro/wine bar. Three different servers (labeled A, B, and C) worked on the nights the bills were collected. Use technology to find the relative frequency for the server who had the most bills.

2.37 Class Year by Gender Exercise 2.35 deals with the distribution of class *Year* for students in an introductory statistics course. The **StudentSurvey** data also has information on the *Gender* for each student. Use technology to produce a two-way table showing the gender distribution within each class year. Comment on any interesting features in the relationship.

2.38 Credit Card by Server The **RestaurantTips** data in Exercise 2.36 also has information on whether each bill was paid with a credit card or cash (*Credit* = y or n). Use technology to produce a two-way table showing the credit/cash distribution for each server. Comment on any interesting features in the relationship.

2.39 Graph Class Year by Gender Use technology and the data in **StudentSurvey** to construct a graph of the relationship between class *Year* and *Gender* for the situation in Exercise 2.37.

2.40 Graph Credit Card by Server Use technology and the data in **RestaurantTips** to construct a graph of the relationship between *Server* and *Credit* for the situation in Exercise 2.38.

2.41 Which of These Things Is Not Like the Other? Four students were working together on a project and one of the parts involved making a graph to display the relationship in a two-way table of data with two categorical variables: college accept/reject decision and type of high school (public, private, parochial). The graphs submitted by each student are shown in Figure 2.5. Three are from the same data, but one is inconsistent with the other three. Which is the bogus graph? Explain.

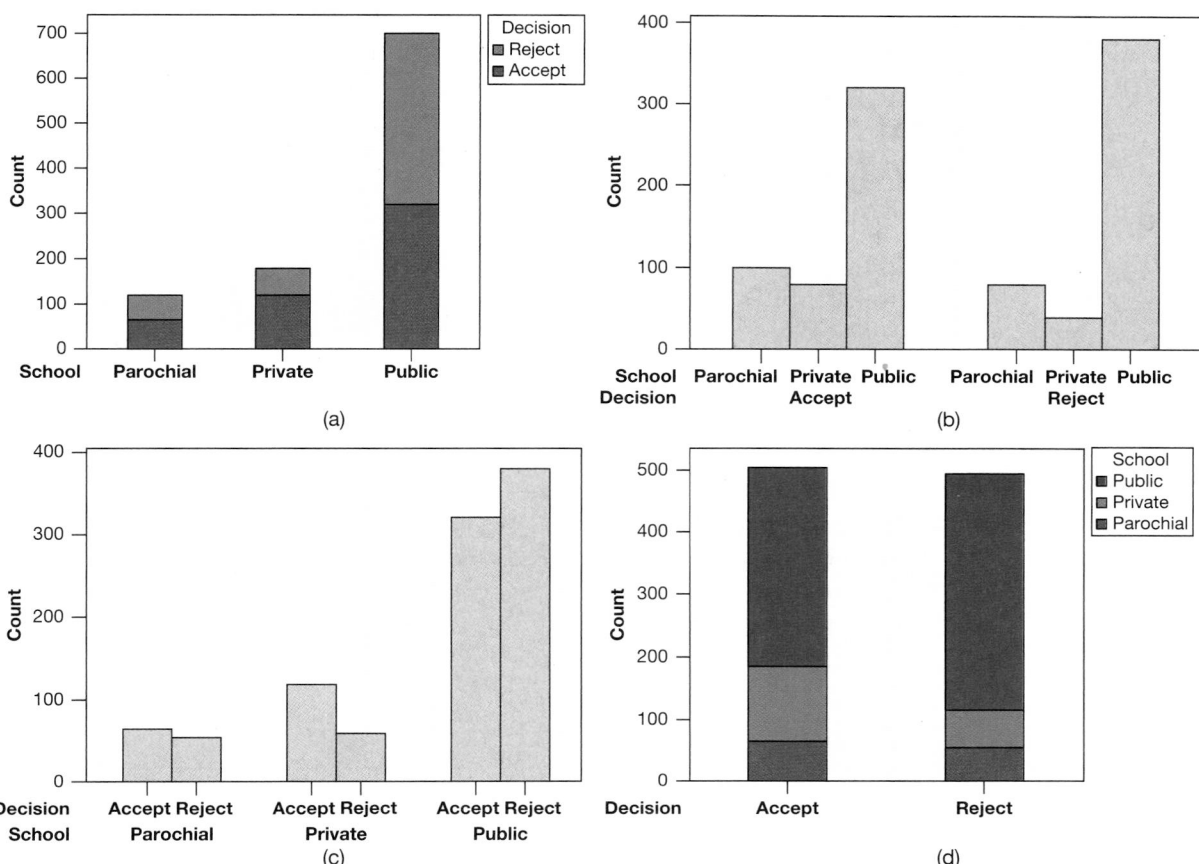

Figure 2.5 *Views of the same two-way table—with one error*

2.2 ONE QUANTITATIVE VARIABLE: SHAPE AND CENTER

In Section 2.1, we see how to describe categorical variables. In this section, we begin to investigate quantitative variables. In describing a single quantitative variable, we generally consider the following three questions:

- What is the general *shape* of the data?
- Where are the data values *centered*?
- How do the data *vary*?

These are all aspects of what we call the *distribution* of the data. In this section, we focus on the first two questions and leave the third question, on variability, to Section 2.3.

The Shape of a Distribution

We begin by looking at graphical displays as a way of understanding the shape of a distribution.

Dotplots

A common way to visualize the shape of a moderately sized dataset is a *dotplot*. We create a dotplot by using an axis with a scale appropriate for the numbers in the dataset and placing a dot over the axis for each case in the dataset. If there are multiple data values that are the same, we stack the dots vertically. To illustrate a dotplot, we look at some data on the typical lifespan for several mammals.

© Britta Kasholm-Tengve/iStockphoto

How long does an elephant live?

DATA 2.2	Longevity of Mammals

The dataset **MammalLongevity** includes information on longevity (typical lifespan), in years, for 40 species of mammals as well as information on length of gestation for these same mammals.[21] The longevity data are given in Table 2.15.   ■

Table 2.15 *Longevity of mammals*

Species	Longevity	Species	Longevity	Species	Longevity
Baboon	20	Elephant	40	Mouse	3
Black bear	18	Elk	15	Opossum	1
Grizzly bear	25	Fox	7	Pig	10
Polar bear	20	Giraffe	10	Puma	12
Beaver	5	Goat	8	Rabbit	5
Buffalo	15	Gorilla	20	Rhinoceros	15
Camel	12	Guinea pig	4	Sea lion	12
Cat	12	Hippopotamus	25	Sheep	12
Chimpanzee	20	Horse	20	Squirrel	10
Chipmunk	6	Kangaroo	7	Tiger	16
Cow	15	Leopard	12	Wolf	5
Deer	8	Lion	15	Zebra	15
Dog	12	Monkey	15		
Donkey	12	Moose	12		

A dotplot of the longevity data is shown in Figure 2.6. We see a horizontal scale from 0 to 40 to accommodate the range of lifespans. Quite a few mammals have lifespans of 12, 15, and 20 years. All but one typically live between 1 and 25 years, while the elephant's lifespan of 40 years is much higher than the rest. The value of 40 years appears to be an *outlier* for longevity in this group of mammals.

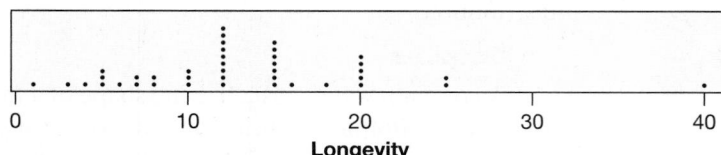

Figure 2.6 *Dotplot of longevity of mammals*

[21] *2010 World Almanac*, p. 292.

Outliers

An **outlier** is an observed value that is notably distinct from the other values in a dataset. Usually, an outlier is much larger or much smaller than the rest of the data values.

Histograms

An alternative graph for displaying a distribution of data is a *histogram*. If we group the longevity data into five-year intervals (1–5 years, 6–10 years, and so on), we obtain the frequency table in Table 2.16. We see that, for example, six of the mammals in the sample have longevity between 1 and 5 years.

The histogram for this dataset is shown in Figure 2.7. The frequency count of 6 for values between 1 and 5 in Table 2.16 corresponds to a vertical bar of height 6 over the interval from 1 to 5 in Figure 2.7. Similarly, we draw vertical bars of heights corresponding to all the frequencies in Table 2.16. Histograms are similar to bar charts for a categorical variable, except that a histogram always has a numerical scale on the horizontal axis. The histogram of mammal longevities in Figure 2.7 shows the relatively symmetric nature of most of the data, with an outlier (the elephant) in the class from 36 to 40.

Table 2.16 *Frequency counts for longevity of mammals*

Longevity (years)	Frequency Count
1–5	6
6–10	8
11–15	16
16–20	7
21–25	2
26–30	0
31–35	0
36–40	1
Total	40

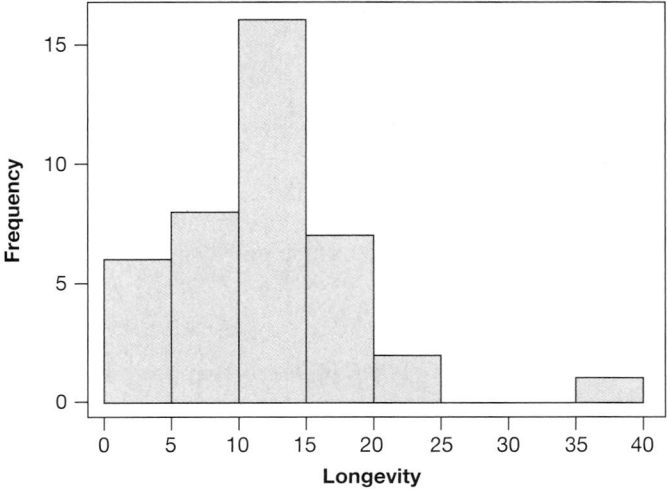

Figure 2.7 *Histogram of longevity of mammals*

Symmetric and Skewed Distributions

We are often interested in the general "big picture" shape of a distribution. A distribution is considered *symmetric* if we can fold the plot (either a histogram or dotplot) over a vertical center line and the two sides match closely. When we consider the shape of a dataset, we ask: Is it approximately symmetric? If not, is the data piled up on one side? If so, which side? Are there outliers? These are all questions that a histogram or dotplot can help us answer.

Example 2.9

The **StudentSurvey** dataset introduced in Data 1.1 on page 4 contains results for 362 students and many variables. Figure 2.8 shows histograms for three of the quantitative variables: *Pulse* (pulse rate in number of beats per minute), *Exercise* (number of hours of exercise per week), and *Piercings* (number of body piercings). Describe each histogram.

Solution

(a) In the histogram for *Pulse*, we see that almost all pulse rates are between about 35 beats per minute and about 100 beats per minute, with two possible outliers at about 120 and 130. Other than the outliers, this histogram is quite symmetric.

(b) In the histogram for *Exercise*, the number of hours spent exercising appears to range from about 0 hours per week to about 30 hours per week, with a possible outlier at 40. This histogram is not very symmetric, since the data stretch out more to the right than to the left.

(c) The histogram for *Piercings* is even more asymmetric than the one for *Exercise*. It does not stretch out at all to the left and stretches out quite a bit to the right. Notice the peak at 0, for all the people with no piercings, and the peak at 2, likely due to students who have pierced ears and no other piercings.

The histogram in Figure 2.8(a) is called *symmetric* and *bell-shaped*. The sort of non-symmetric distributions we see in Figures 2.8(b) and (c) are called *skewed*. The direction of skewness is determined by the longer tail. In both cases, we see that the right tail of the distribution is longer than the left tail, so we say that these distributions are *skewed to the right*.

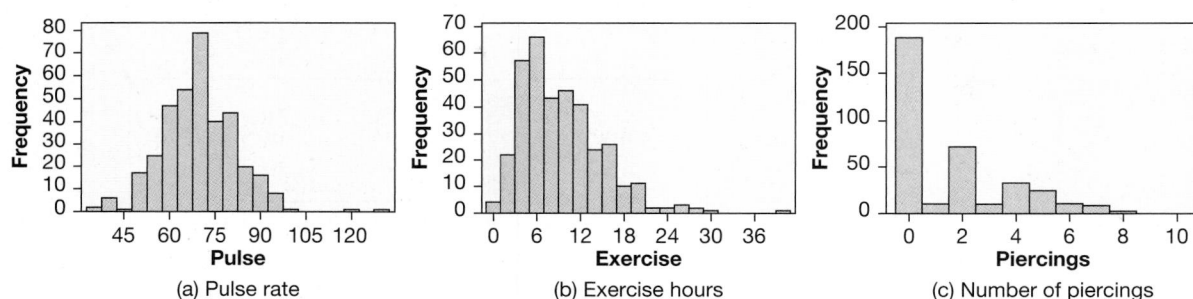

Figure 2.8 *Three histograms for the student survey data*

Using a Curve to Represent the Shape of a Histogram

We often draw smooth curves to illustrate the general shape of a distribution. Smoothing a histogram into a curve helps us to see the shape of the distribution with less jagged edges at the corners. When we describe a histogram with a smooth curve, we don't try to match every bump and dip seen in a particular sample. Rather we find a relatively simple curve that follows the general pattern in the data. Figure 2.9 gives examples of curves showing several common shapes for distributions.

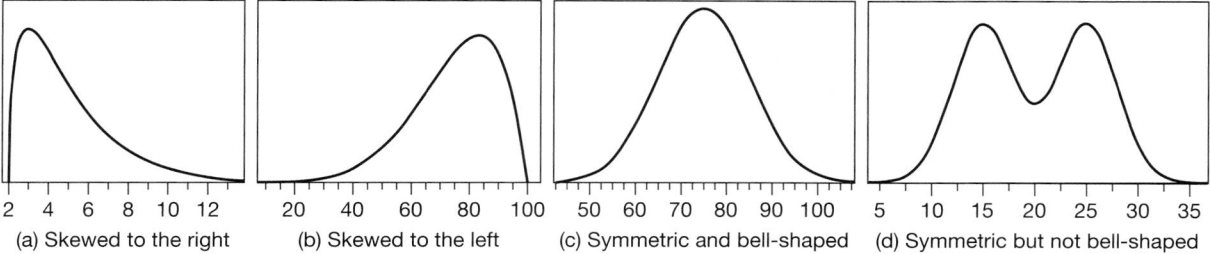

Figure 2.9 *Common shapes for distributions*

Common Shapes for Distributions

A distribution shown in a histogram or dotplot is called:

- **Symmetric** if the two sides approximately match when folded on a vertical center line
- **Skewed to the right** if the data are piled up on the left and the tail extends relatively far out to the right
- **Skewed to the left** if the data are piled up on the right and the tail extends relatively far out to the left
- **Bell-shaped** if the data are symmetric and, in addition, have the shape shown in Figure 2.9(c)

Of course, many other shapes are also possible.

The Center of a Distribution

A graph is useful to help us visualize the shape of a distribution. We can also summarize important features of a distribution numerically. Two summary statistics that describe the center or location of a distribution for a single quantitative variable are the *mean* and the *median*.

Mean

The mean for a single quantitative variable is the numerical average of the data values:

$$\text{Mean} = \frac{\text{Sum of all data values}}{\text{Number of data values}}$$

To express the calculation of the mean in a mathematical formula, we let n represent the number of data cases in a dataset and $x_1, x_2, \ldots, x_n$ represent the numerical values for the quantitative variable of interest.

Mean

The **mean** of the data values for a single quantitative variable is given by

$$\text{Mean} = \frac{x_1 + x_2 + \cdots + x_n}{n} = \frac{\Sigma x}{n}$$

The Greek letter Σ is used as a shorthand for adding all of the x values. For example, the longevity data in Table 2.15 yield a sum of $\Sigma x = 526$ years and thus the mean longevity for this sample of 40 mammals is $\frac{526}{40} = 13.15$ years.

Notation for a Mean

As with a proportion, we use different notation to indicate whether a mean summarizes the data from a sample or a population.

> **Notation for a Mean**
>
> The mean of a *sample* is denoted $\bar{x}$ and read "x-bar."
>
> The mean of a *population* is denoted μ, which is the Greek letter "mu."

Example 2.10

Give the notation for the mean in each case.

(a) For a random sample of 50 seniors from a large high school, the average SAT (Scholastic Aptitude Test) score was 582 on the Math portion of the test.

(b) About 1.67 million students in the class of 2014 took the SAT,[22] and the average score overall on the Math portion was 513.

Solution

(a) The mean of 582 represents the mean of a sample, so we use the notation $\bar{x}$ for the mean, and we have $\bar{x} = 582$.

(b) The mean of 513 represents the mean for everyone who took the exam in the class of 2014, so we use the notation μ for the population mean, and we have $\mu = 513$.

Median

The *median* is another statistic used to summarize the center of a set of numbers. If the numbers in a dataset are arranged in order from smallest to largest, the median is the middle value in the list. If there are an even number of values in the dataset, then there is not a unique middle value and we use the average of the two middle values.

> **Median**
>
> The **median** of a set of data values for a single quantitative variable, denoted m, is
>
> - the middle entry if an ordered list of the data values contains an odd number of entries, or
> - the average of the middle two values if an ordered list contains an even number of entries.
>
> The median splits the data in half.[23]

---

[22] *http://www.sat.collegeboard.org/scores.*
[23] If there are duplicate values at the median, we may not have exactly half on either side.

For example, the middle two values of the 40 mammal lifespans are both 12, so the median lifespan is 12 years. Notice that the dotplot in Figure 2.6 shows that roughly half of the species live less than 12 years, and the other half live more.

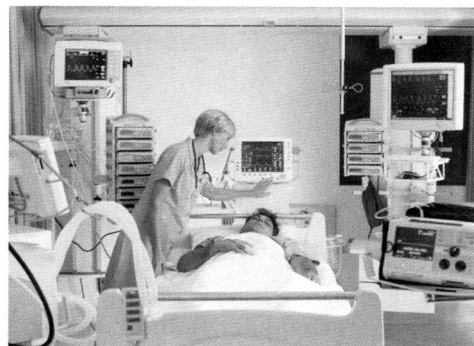

Martin Barraud/Getty Images, Inc.

An intensive care unit

DATA 2.3	**Admissions to an Intensive Care Unit**

The dataset **ICUAdmissions**[24] includes information on 200 patients admitted to the Intensive Care Unit at a hospital. Twenty variables are given for the patients being admitted, including age, gender, race, heart rate, systolic blood pressure, whether or not CPR was administered, and whether or not the patient lived or died. ∎

Example 2.11

Find the median and the mean for the heart rates, in beats per minute, of 20-year-old patients and 55-year-old patients from the **ICUAdmissions** study.

(a) 20-year-old patients: 108,   68,   80,   83,   72
(b) 55-year-old patients: 86,   86,   92,   100,   112,   116,   136,   140

Solution

(a) To find the median for the 20-year-old patients, we put the heart rates in order from smallest to largest:

$$68 \quad 72 \quad 80 \quad 83 \quad 108$$

The middle number is the third largest value, so the median heart rate for these patients is 80 beats per minute.

There are five numbers in the dataset for the 20-year-old patients, so the mean is the sum of the values divided by 5. These values are from a sample, so we use the notation $\bar{x}$ and we have

$$\text{Mean} = \bar{x} = \frac{108 + 68 + 80 + 83 + 72}{5} = \frac{411}{5} = 82.2 \text{ beats per minute}$$

For this dataset the mean and the median are close but not the same.

(b) For the 55-year-old patients, note that the data are already listed in numerical order. Since there are an even number ($n = 8$) of values in this dataset, there is

[24]DASL dataset downloaded from *http://lib.stat.cmu.edu/DASL/Datafiles/ICU.html.*

no single number in the middle. The median is the average of the two middle numbers:

$$m = \frac{100 + 112}{2} = 106 \text{ beats per minute}$$

The mean of the heart rates for the 55-year-old patients is

$$\bar{x} = \frac{868}{8} = 108.5 \text{ beats per minute}$$

In this case, also, the mean and median are relatively close but not equal.

Resistance

The term *resistance* is related to the impact of outliers on a statistic. We examine the effect of an outlier on the mean and the median.

Example 2.12

In Example 2.11(a), we saw that the mean and the median heart rate for $n = 5$ ICU patients in their twenties are given by

$$\bar{x} = 82.2 \text{ bpm} \qquad \text{and} \qquad m = 80 \text{ bpm}$$

Suppose that the patient with a heart rate of 108 bpm instead had an extremely high heart rate of 200 bpm. How does this change affect the mean and median?

Solution

The median doesn't change at all, since 80 is still the middle value. The effect on the mean is substantial: We see that with the change the mean increases to $\bar{x} = 100.6$ beats per minute. The extreme value of 200 has a large effect on the mean but little effect on the median.

> **Resistance**
>
> In general, we say that a statistic is **resistant** if it is relatively unaffected by extreme values. The median is resistant, while the mean is not.

The mean and the median both provide valuable measures of the center of a dataset. Knowing that outliers have a substantial effect on the mean but not the median can help determine which is more meaningful in different situations.

Example 2.13

As in most professional sports, some star players in the National Football League (NFL) in the US are paid much more than most other players. In particular, four players (all quarterbacks) were paid salaries greater than $20 million in 2014. Two measures of the center of the player salary distribution for the 2014–2015 NFL season are

$$\$872{,}000 \qquad \text{and} \qquad \$2.14 \text{ million}$$

(a) One of the two values is the mean and the other is the median. Which is which? Explain your reasoning.

(b) In salary negotiations, which measure (the mean or the median) are the owners more likely to find relevant? Which are the players more likely to find relevant? Explain.

Solution   (a) There are some high outliers in the data, representing the players who make a very high salary. These high outliers will pull the mean up above the median. The mean is $2.14 million and the median is $872,000.

(b) The owners will find the mean more relevant, since they are concerned about the total payroll, which is the mean times the number of players. The players are likely to find the median more relevant, since half of the players make less than the median and half make more. The high outliers influence the mean but are irrelevant to the salaries of most players. Both measures give an appropriate measure of center for the distribution of player salaries, but they give significantly different values. This is one of the reasons that salary negotiations can often be so difficult.

Visualizing the Mean and the Median on a Graph

The mean is the "balancing point" of a dotplot or histogram in the sense that it is the point on the horizontal axis that balances the graph. In contrast, the median splits the dots of a dotplot, or area in the boxes of a histogram, into two equal halves.

© Dennis Hallinan/Alamy Limited

Fishing in a Florida lake

DATA 2.4   **Florida Lakes**

The **FloridaLakes** dataset[25] describes characteristics of water samples taken at $n = 53$ Florida lakes. Alkalinity (concentration of calcium carbonate in mg/L) and acidity (pH) are given for each lake. In addition, the average mercury level is recorded for a sample of fish (largemouth bass) from each lake. A standardized mercury level is obtained by adjusting the mercury averages to account for the age of the fish in each sample. Notice that the cases are the 53 lakes and that all four variables are quantitative.   ■

Example 2.14

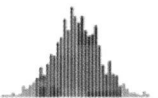

Using the *Alkalinity* values for the $n = 53$ lakes in the **FloridaLakes** dataset:

(a) Use technology to create a histogram of the alkalinity values. What is the shape of the histogram?

(b) Which do you expect to be larger for this sample of alkalinity values, the mean or the median? Why?

[25]Lange, T., Royals, H., and Connor, L., "Mercury accumulation in largemouth bass (*Micropterus salmoides*) in a Florida Lake," *Archives of Environmental Contamination and Toxicology*, 2004; 27(4): 466–471.

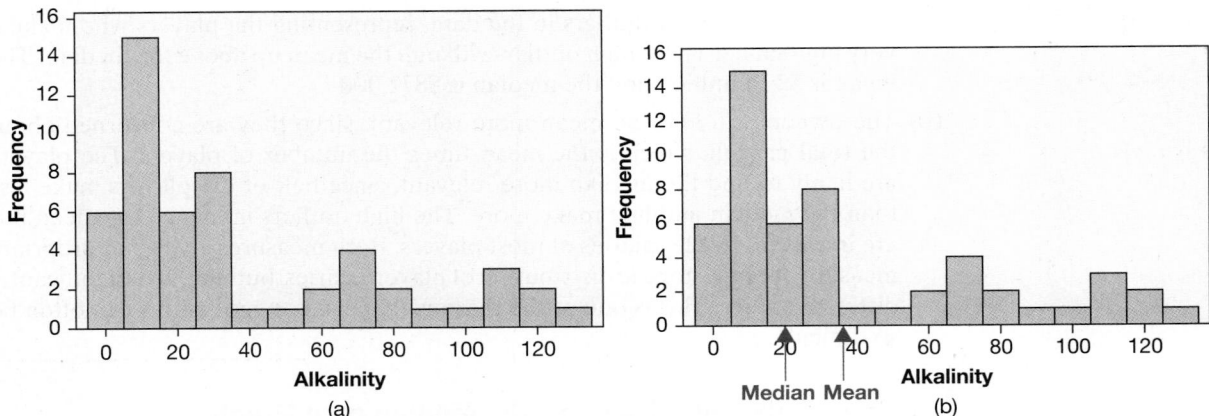

Figure 2.10 *Alkalinity in Florida lakes*

(c) Use technology to compute the mean and the median alkalinity for this sample.

(d) Locate the mean and the median on the histogram.

Solution

(a) We use technology to create the histogram of alkalinity values shown in Figure 2.10(a). There are many alkalinity values between 0 and 40, with a few large values extending out as far as 130 or so. The data are clearly skewed to the right.

(b) The few very large values on the right will pull up the mean and won't affect the median very much, so we expect the mean to be larger than the median.

(c) Using technology, we compute the mean to be $\bar{x} = 37.5$ mg/L and the median to be $m = 19.6$ mg/L. The median splits the data in half: There are 26 values above the median and 26 values below it.

(d) See Figure 2.10(b). The mean is the balance point for the histogram, and the median splits the data in half. The mean is substantially larger than the median, and almost two-thirds of the lakes (35 out of 53) have alkalinity levels below the mean. The data are skewed to the right, and the values out in the extended right tail pull the mean up quite a bit.

Since the median cuts a histogram in half, if a histogram is symmetric, the median is right in the middle and approximately equal to the mean. If the data are skewed to the right, as we see in Figure 2.10, the values in the extended right tail pull the mean up but have little effect on the median. In this case, the mean is bigger than the median. Similarly, if data are skewed to the left, the mean is less than the median. See Figure 2.11.

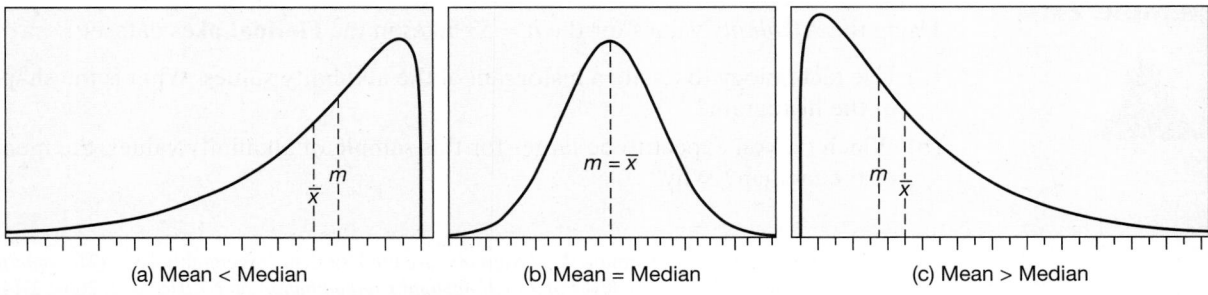

Figure 2.11 *Mean and median for different shaped distributions*

SECTION LEARNING GOALS

You should now have the understanding and skills to:

▶ • Use a dotplot or histogram to describe the shape of a distribution

▶ • Find the mean and the median for a set of data values, with appropriate notation

▶ • Identify the approximate locations of the mean and the median on a dotplot or histogram

▶ • Explain how outliers and skewness affect the values for the mean and median

Exercises for Section 2.2

SKILL BUILDER 1

Exercises 2.42 to 2.48 refer to histograms A through H in Figure 2.12.

2.42 Which histograms are skewed to the left?

2.43 Which histograms are skewed to the right?

2.44 Which histograms are approximately symmetric?

2.45 Which histograms are approximately symmetric and bell-shaped?

2.46 For each of the four histograms A, B, C, and D, state whether the mean is likely to be larger than the median, smaller than the median, or approximately equal to the median.

2.47 For each of the four histograms E, F, G, and H, state whether the mean is likely to be larger than the median, smaller than the median, or approximately equal to the median.

2.48 Which of the distributions is likely to have the largest mean? The smallest mean?

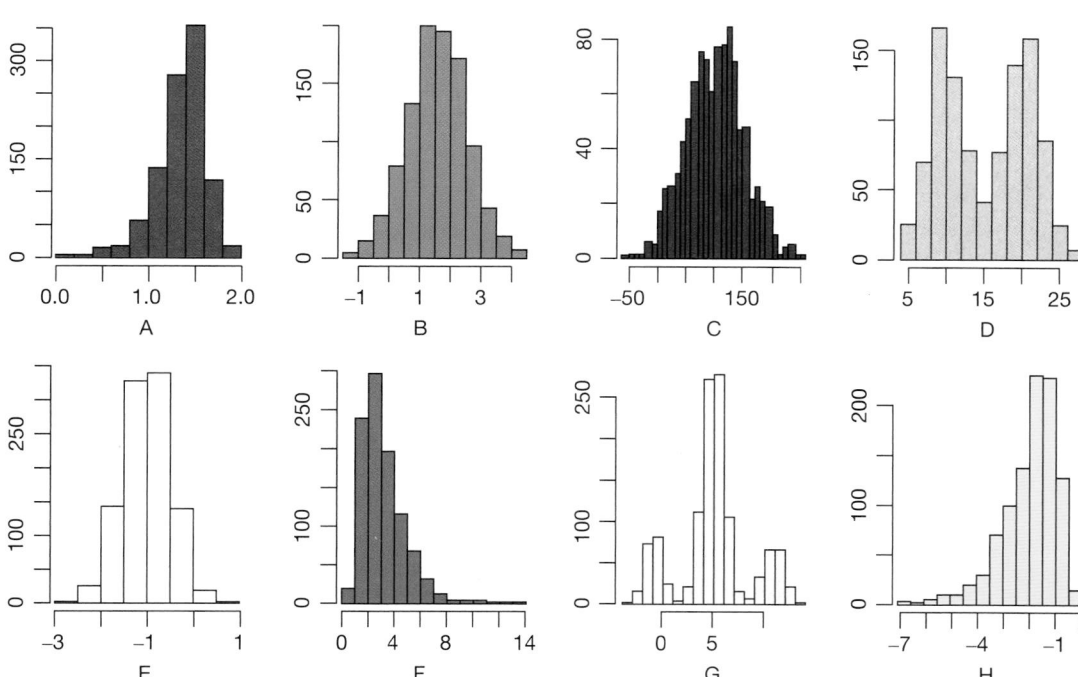

Figure 2.12 *Eight histograms*

SKILL BUILDER 2

In Exercises 2.49 to 2.52, draw any dotplot to show a dataset that is

2.49 Clearly skewed to the left.

2.50 Approximately symmetric and bell-shaped.

2.51 Approximately symmetric but not bell-shaped.

2.52 Clearly skewed to the right.

SKILL BUILDER 3

For each set of data in Exercises 2.53 to 2.56:

(a) Find the mean $\bar{x}$.

(b) Find the median m.

(c) Indicate whether there appear to be any outliers. If so, what are they?

2.53 8,   12,   3,   18,   15

2.54 41,   53,   38,   32,   115,   47,   50

2.55 15,   22,   12,   28,   58,   18,   25,   18

2.56 110,   112,   118,   119,   122,   125,   129, 135,   138,   140

SKILL BUILDER 4

In Exercises 2.57 to 2.60, give the correct notation for the mean.

2.57 The average number of calories eaten in one day is 2386 calories for a sample of 100 participants.

2.58 The average number of text messages sent in a day was 67, in a sample of US smartphone users ages 18–24, according to a survey conducted by Experian.[26]

2.59 The average number of yards per punt for all punts in the National Football League is 41.5 yards.[27]

2.60 The average number of television sets owned per household for all households in the US is 2.6.[28]

2.61 Arsenic in Toenails Arsenic is toxic to humans, and people can be exposed to it through contaminated drinking water, food, dust, and soil. Scientists have devised an interesting new way to measure a person's level of arsenic poisoning: by examining toenail clippings. In a recent study,[29]

scientists measured the level of arsenic (in mg/kg) in toenail clippings of eight people who lived near a former arsenic mine in Great Britain. The following levels were recorded:

0.8   1.9   2.7   3.4   3.9   7.1   11.9   26.0

(a) Do you expect the mean or the median of these toenail arsenic levels to be larger? Why?

(b) Calculate the mean and the median.

2.62 Fiber in the Diet The number of grams of fiber eaten in one day for a sample of ten people are

10   11   11   14   15   17   21   24   28   115

(a) Find the mean and the median for these data.

(b) The value of 115 appears to be an obvious outlier. Compute the mean and the median for the nine numbers with the outlier excluded.

(c) Comment on the effect of the outlier on the mean and on the median.

2.63 Online Cat Videos In Exercise 1.59 on page 28, we introduced a study looking at the effect of watching cat videos on mood and energy. The authors asked participants how many cats they currently own and report 2.39 as the measure of center for this right-skewed distribution.

(a) Is this measure of center the mean number of cats or the median number of cats? (*Hint:* Think about how the two numbers are calculated.)

(b) Would we expect the mean number of cats to be greater than or less than the median?

2.64 Insect Weights Consider a dataset giving the adult weight of species of insects. Most species of insects weigh less than 5 grams, but there are a few species that weigh a great deal, including the largest insect known: the rare and endangered Giant Weta from New Zealand, which can weigh as much as 71 grams. Describe the shape of the distribution of weights of insects. Is it symmetric or skewed? If it is skewed, is it skewed to the left or skewed to the right? Which will be larger, the mean or the median?

2.65 Population of States in the US The dataset **USStates** has a great deal of information about the 50 states, including population. Figure 2.13 shows a histogram of the population, in millions, of the 50 states in the US.

(a) Do these values represent a population or a sample?

(b) Describe the shape of the distribution: Is it approximately symmetric, skewed to the right, skewed to the left, or none of these? Are there any outliers?

[26]Cocotas, A., "Chart of the Day: Kids Send a Mind Boggling Number of Texts Every Month," *Business Insider*, March 22, 2013.

[27]Moskowitz, T. and Wertheim, L., *Scorecasting*, Crown Archetype, New York, 2011, p. 119.

[28]http://www.census.gov.

[29]Button, M., Jenkin, G., Harrington, C., and Watts, M., "Human toenails as a biomarker of exposure to elevated environment arsenic," *Journal of Environmental Monitoring*, 2009; 11(3): 610–617. Data are reproduced from summary statistics and are approximate.

(c) Estimate the median population.

(d) Estimate the mean population.

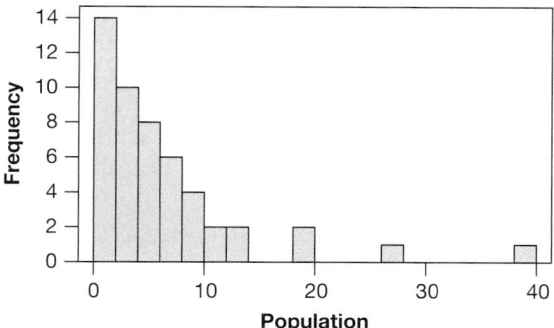

Figure 2.13 *Population, in millions, of the 50 states*

2.66 Is Language Biased toward Happiness? "Are natural languages neutrally, positively, or negatively biased?" That is the question a recent study[30] set out to answer. They found the top 5000 words used in English in each of four different places: Twitter, books on the Google Book Project, *The New York Times*, and music lyrics. The resulting complete list was 10,222 unique words in the English language. Each word was then evaluated independently by 50 different people, each giving a rating on how the word made them feel on a 1 to 9 scale, where 1 = least happy, 5 = neutral, and 9 = most happy. (The highest rated word was "laughter," while the lowest was "terrorist.") The distributions of the ratings for all 10,222 words for each of the four media sources were surprisingly similar, and all had approximately the shape shown in Figure 2.14.

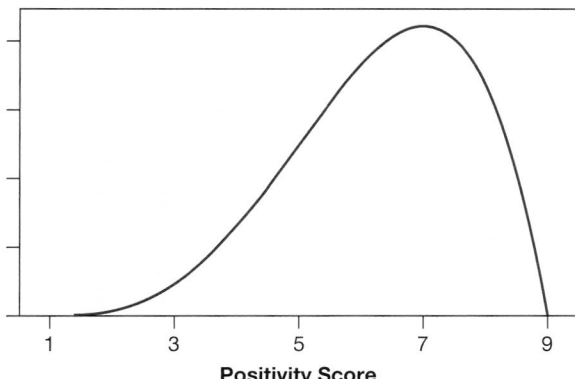

Figure 2.14 *Distribution of ratings of words where 9 = most positive*

[30]Kloumann, I., Danforth, C., Harris, K., Bliss, C., and Dodds, P., "Positivity of the English Language," *PLoS ONE*, 2012; 7(1).

(a) Describe the shape of the distribution.

(b) Which of the following values is closest to the median of the distribution:

3.5   5   6.5   7   7.5   8

(c) Will the mean be smaller or larger than the value you gave for the median in part (b)?

2.67 Life Expectancy Life expectancy for all the different countries in the world ranges from a low of only 45.6 years (in Sierra Leone) to a high of 83.8 years (in Hong Kong). Life expectancies are clustered at the high end, with about half of all the countries having a life expectancy between about 74 and the maximum of 83.8. A few countries, such as Sierra Leone, have a very low life expectancy. The full dataset is in **AllCountries**.

(a) What is the shape of the distribution of life expectancies for all countries?

(b) From the information given, estimate the median of the life expectancies.

(c) Will the mean be larger or smaller than the median?

2.68 Making Friends Online A survey conducted in March 2015 asked 1060 teens to estimate, on average, the number of friends they had made online. While 43% had not made any friends online, a small number of the teens had made many friends online.

(a) Do you expect the distribution of number of friends made online to be symmetric, skewed to the right, or skewed to the left?

(b) Two measures of center for this distribution are 1 friend and 5.3 friends.[31] Which is most likely to be the mean and which is most likely to be the median? Explain your reasoning.

2.69 Donating Blood to Grandma? Can young blood help old brains? Several studies[32] in mice indicate that it might. In the studies, old mice (equivalent to about a 70-year-old person) were randomly assigned to receive blood plasma either from a young mouse (equivalent to about a 25-year-old person) or another old mouse. The mice receiving the young blood showed multiple signs of a reversal of brain aging. One of the studies[33] measured exercise

[31]Lenhart, A., "Teens, Technology, and Friendships," Pew Research Center, *http://www.pewresearch.org*, August 6, 2015. Value for the mean is estimated from information given.

[32]Sanders, L., "Young blood proven good for old brain," *Science News*, 185(11), May 31, 2014.

[33]Manisha, S., et al., "Restoring systemic GDF11 Levels Reverses Age-Related Dysfunction in Mouse Skeletal Muscle," *Science*, May 9, 2014. Values are estimated from summary statistics and graphs.

endurance using maximum treadmill runtime in a 90-minute window. The number of minutes of runtime are given in Table 2.17 for the 17 mice receiving plasma from young mice and the 13 mice receiving plasma from old mice. The data are also available in **YoungBlood**.

Table 2.17 *Number of minutes on a treadmill*

Young	27	28	31	35	39	40	45
	46	55	56	59	68	76	90
	90	90	90				
Old	19	21	22	25	28	29	29
	31	36	42	50	51	68	

(a) Calculate $\bar{x}_Y$, the mean number of minutes on the treadmill for those mice receiving young blood.

(b) Calculate $\bar{x}_O$, the mean number of minutes on the treadmill for those mice receiving old blood.

(c) To measure the effect size of the young blood, we are interested in the difference in means $\bar{x}_Y - \bar{x}_O$. What is this difference? Interpret the result in terms of minutes on a treadmill.

(d) Does this data come from an experiment or an observational study?

(e) If the difference is found to be significant, can we conclude that young blood increases exercise endurance in old mice? (Researchers are just beginning to start similar studies on humans.)

2.70 Price Differentiating E-commerce websites "alter results depending on whether consumers use smartphones or particular web browsers,"[34] reports a new study. The researchers created clean accounts without cookies or browser history and then searched for specific items at different websites using different devices and browsers. On one travel site, for example, prices given for hotels were cheaper when using Safari on an iPhone than when using Chrome on an Android. At Home Depot, the average price of 20 items when searching from a smartphone was $230, while the average price when searching from a desktop was $120. For the Home Depot data:

(a) Give notation for the two mean prices given, using subscripts to distinguish them.

(b) Find the difference in means, and give notation for the result.

2.71 Does It Pay to Get a College Degree? In Exercise 2.21 on page 58, we saw that those with

a college degree were much more likely to be employed. The same article also gives statistics on earnings in the US in 2009 by education level. The median weekly earnings for high school graduates with no college degree was $626, while the median weekly earnings for college graduates with a bachelor's degree was $1025. Give correct notation for and find the difference in medians, using the notation for a median, subscripts to identify the two groups, and a minus sign.

2.72 Does Sexual Frustration Increase the Desire for Alcohol? Apparently, sexual frustration increases the desire for alcohol, at least in fruit flies. Scientists[35] randomly put 24 fruit flies into one of two situations. The 12 fruit flies in the "mating" group were allowed to mate freely with many available females eager to mate. The 12 in the "rejected" group were put with females that had already mated and thus rejected any courtship advances. After four days of either freely mating or constant rejection, the fruit flies spent three days with unlimited access to both normal fruit fly food and the same food soaked in alcohol. The percent of time each fly chose the alcoholic food was measured. The fruit flies that had freely mated chose the two types of food about equally often, choosing the alcohol variety on average 47% of the time. The rejected males, however, showed a strong preference for the food soaked in alcohol, selecting it on average 73% of the time. (The study was designed to study a chemical in the brain called neuropeptide that might play a role in addiction.)

(a) Is this an experiment or an observational study?

(b) What are the cases in this study? What are the variables? Which is the explanatory variable and which is the response variable?

(c) We are interested in the difference in means, where the means measure the average percent preference for alcohol (0.47 and 0.73 in this case). Find the difference in means and give the correct notation for your answer, using the correct notation for a mean, subscripts to identify groups, and a minus sign.

(d) Can we conclude that rejection increases a male fruit fly's desire for alcohol? Explain.

2.73 Create a Dataset Give any set of five numbers satisfying the condition that:

(a) The mean of the numbers is substantially less than the median.

---

[34]Ehrenberg, R., "E-commerce sites personalize search results to maximize profits," *Science News*, November 29, 2014.

[35]Shohat-Ophir, G., Kaun, K., Azanchi, R., and Heberlein, U., "Sexual Deprivation Increases Ethanol Intake in *Drosophila*," *Science*, March 16, 2012; 335(6074): 1351–1355.

(b) The mean of the numbers is substantially more than the median.

(c) The mean and the median are equal.

2.74 Describe a Variable Describe one quantitative variable that you believe will give data that are skewed to the right, and explain your reasoning. Do not use a variable that has already been discussed.

2.75 Distribution of Birth Rate The *BirthRate* variable in the **AllCountries** dataset gives the birth rate per 1000 people for all the countries in the world. Use technology to create a histogram for this variable, and describe the shape of the distribution: is it relatively symmetric, mildly skewed to the left or right, or strongly skewed to the left or right?

2.76 Distribution of Death Rate The *DeathRate* variable in the **AllCountries** dataset gives the death rate per 1000 people for all the countries in the world. Use technology to create a histogram for this variable, and describe the shape of the distribution: is it relatively symmetric, mildly skewed to the left or right, or strongly skewed to the left or right?

2.77 Number of Children Table 2.18 shows the number of women (per 1000) between 15 and 50 years of age who have been married grouped by the number of children they have had. Table 2.19 gives the same information for women who have never been married.[36]

[36]*http://www.census.gov/hhes/fertility/data/cps/2014.html*, Table 1, June 2014.

(a) Without doing any calculations, which of the two samples appears to have the highest mean number of children? Which of the distributions appears to have the mean most different from the median? Why?

(b) Find the median for each dataset.

Table 2.18 *Women who have been married*

Number of Children	Women per 1000
0	162
1	213
2	344
3	182
4	69
5+	32

Table 2.19 *Women who have never been married*

Number of Children	Women per 1000
0	752
1	117
2	72
3	35
4	15
5+	10

2.3 ONE QUANTITATIVE VARIABLE: MEASURES OF SPREAD

So far, we have looked at two important summary statistics for a single quantitative variable: the mean and the median. Although there are important differences between them, both of these measurements tell us something about the "middle" or "center" of a dataset. When we give a statistical summary of the values in a dataset, we are interested in not just the center of the data but also how spread out the data are. Knowing that the average high temperature in Des Moines, Iowa, in April is 62°F is helpful, but it is also helpful to know that the historical range is between 8°F and 97°F! In this section, we examine additional measures of location and measures of spread.

Using Technology to Compute Summary Statistics

In practice, we generally use technology to compute the summary statistics for a dataset. For assistance in using a wide variety of different types of technology and software, see the available supplementary resources.

Example 2.15

Des Moines vs San Francisco Temperatures

Average temperature on April 14th for the 21 years ending in 2015 is given in Table 2.20 for Des Moines, Iowa, and San Francisco, California.[37] Use technology and the data in **April14Temps** to find the mean and the median temperature on April 14th for each city.

Table 2.20 *Temperature on April 14th*

Des Moines	56.0	37.5	37.2	56.0	54.3	63.3	54.7	60.6
	70.6	53.7	52.9	74.9	44.4	40.3	44.4	71.0
	56.8	59.2	53.3	35.7	56.2			
San Francisco	51.0	55.3	55.7	48.7	56.2	57.2	49.5	61.0
	51.4	55.8	53.0	58.1	54.2	53.4	49.9	53.8
	51.4	52.3	52.1	57.3	55.8			

Solution

Computer output from one statistics package for the Des Moines temperatures is shown in Figure 2.15 and an image of the descriptive statistics as they appear on a TI-83 graphing calculator or on StatKey is shown in Figure 2.16. We see from any of these options that for Des Moines the mean temperature is 53.95°F and the median is 54.7°F. Similar output shows that for San Francisco the mean temperature is 53.96°F and the median is 53.8°F.

Figure 2.15 *Output from Minitab with summary statistics*

Descriptive Statistics: DesMoines

Variable	N	Mean	SE Mean	StDev	Minimum	Q1	Median	Q3	Maximum
DesMoines	21	53.95	2.41	11.05	35.70	44.40	54.70	59.90	74.90

Figure 2.16 *Output from TI-83 or StatKey*

1-Var Stats

$\bar{x} = 53.95238095$
$\Sigma x = 1133$
$\Sigma x^2 = 63570.9$
$Sx = 11.05181519$
$\sigma x = 10.78546725$
$\downarrow n = 21$
$minX = 35.7$
$Q_1 = 44.4$
$Med = 54.7$
$Q_3 = 59.9$
$maxX = 74.9$

(a) *TI-83*

Summary Statistics

Statistics	Value
Sample Size	21
Mean	53.952
Standard Deviation	11.052
Minimum	35.7
Q_1	44.400
Median	54.700
Q_3	59.200
Maximum	74.9

(b) *StatKey*

Standard Deviation

We see in Example 2.15 that the means and medians for temperatures in Des Moines and San Francisco are almost identical. However, the dotplots in Figure 2.17 show that, while the centers may be similar, the distributions are very different. The temperatures in San Francisco are clustered closely around the center, while the temperatures in Des Moines are very spread out. We say that the temperatures in Des

---

[37] *http://academic.udayton.edu/kissock/http/Weather/citylistUS.htm* Accessed June 2015.

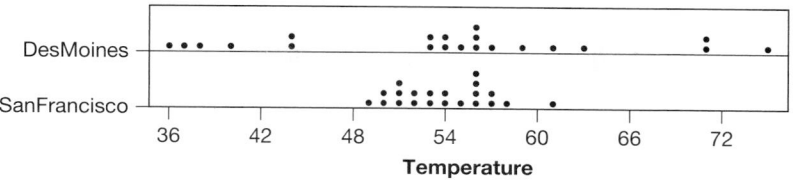

Figure 2.17 *Which city has greater variability in temperature?*

Moines have greater *variability* or greater *spread*. The *standard deviation* is a statistic that measures how much variability there is in the data.

In a sample, the *deviation* of an individual data value x from the mean $\bar{x}$ is simply the difference $x - \bar{x}$. We see in Example 2.15 that the mean April 14th temperature in Des Moines is $\bar{x} = 53.95$, so the deviation for the first data value listed, 56.0, is $56.0 - 53.95 = 2.05$. Since values can fall above and below the mean, some deviations are negative and some are positive. In fact, if you add up all of the deviations, the sum will always be zero.

The standard deviation is computed by squaring each deviation (to make them all positive), adding up these squared deviations, dividing by $n - 1$ (to take an approximate average), and taking the square root of the result (to undo the earlier squaring). It is not necessary to fully understand the details of this computation.[38] However, interpreting the standard deviation correctly is essential: A larger standard deviation means the data values are more spread out and have more variability.

Definition of Standard Deviation

The **standard deviation** for a quantitative variable measures the spread of the data in a sample:

$$\text{Standard deviation} = \sqrt{\frac{\Sigma(x - \bar{x})^2}{n - 1}}$$

The standard deviation gives a rough estimate of the typical distance of a data value from the mean. The larger the standard deviation, the more variability there is in the data and the more spread out the data are.

In practice, we generally compute the standard deviation using a calculator or computer software. The units for the standard deviation are the same as the units for the original data.

Notation for the Standard Deviation

The standard deviation of a *sample* is denoted s, and measures how spread out the data are from the sample mean $\bar{x}$.

The standard deviation of a *population*[39] is denoted σ, which is the Greek letter "sigma," and measures how spread out the data are from the population mean μ.

---

[38]Two natural questions here are (1) Why square everything and then take a square root? and (2) Why divide by $n - 1$ instead of n (like a mean)? Both have justifications but are beyond the scope of this textbook.
[39]The formula is modified slightly for a population.

Example 2.16

Temperatures on April 14th in Des Moines and San Francisco are given in Table 2.20 and shown in Figure 2.17.

(a) Which dataset do we expect to have a larger standard deviation? Why?

(b) Use technology to find the standard deviation for each dataset and compare your answers.

Solution

(a) The Des Moines temperatures are more spread out, so we expect this dataset to have a larger standard deviation.

(b) We use technology to find the standard deviation. In Figure 2.15, standard deviation is denoted "StDev." In Figure 2.16, the standard deviation is given by "S_x" for the TI-83, and is labeled as "Standard Deviation" in *StatKey*. In all three cases, we see that the standard deviation for the sample of Des Moines temperatures is about $s = 11.05°F$. Similar output for the San Francisco temperatures shows that the standard deviation for those 21 values is $s = 3.15°F$. As we expect, the standard deviation is larger for the Des Moines temperatures than for the San Francisco temperatures.

Interpreting the Standard Deviation

Since the standard deviation is computed using the deviations from the mean, we get a rough sense of the magnitude of s by considering the typical distance of a data value from the mean. The following rule of thumb is helpful for interpreting the standard deviation for distributions that are symmetric and bell-shaped.

> **Using the Standard Deviation: The 95% Rule**
>
> If a distribution of data is approximately symmetric and bell-shaped, about 95% of the data should fall within two standard deviations of the mean. This means that about 95% of the data in a sample from a bell-shaped distribution should fall in the interval from $\bar{x} - 2s$ to $\bar{x} + 2s$. See Figure 2.18.

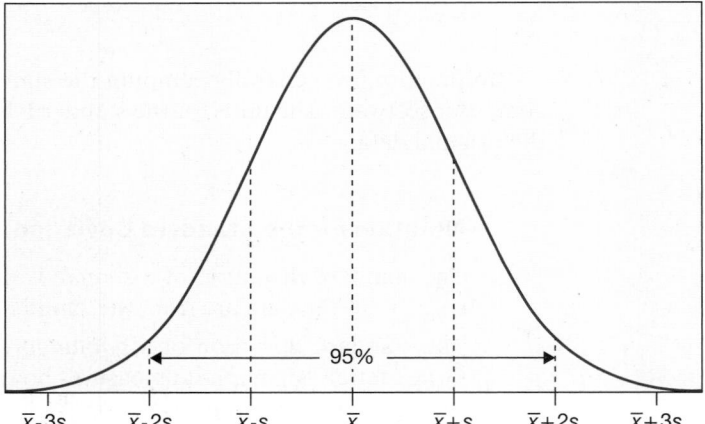

Figure 2.18 *Most data are within two standard deviations of the mean*

Example 2.17

We see in Example 2.9 on page 66 that the distribution for pulse rates from the **StudentSurvey** data is symmetric and approximately bell-shaped. Use the fact that the mean of the pulse rates is $\bar{x} = 69.6$ and the standard deviation is $s = 12.2$ to give an interval that is likely to contain about 95% of the pulse rates for students.

Solution    To identify pulse rates within two standard deviations of the mean, we compute

$$\bar{x} \quad \pm \quad 2s$$
$$69.6 \quad \pm \quad 2(12.2)$$
$$69.6 \quad \pm \quad 24.4$$
$$69.6 - 24.4 = 45.2 \quad \text{and} \quad 69.6 + 24.4 = 94.0.$$

Roughly 95% of the pulse rates are between 45.2 and 94.0 beats per minute.

neil denham/Alamy Stock Photo

How fast can a car travel a quarter-mile?

DATA 2.5

2015 New Car Models

The dataset **Cars2015** contains information for a sample of 110 new car models[40] in 2015. There are many variables given for these cars, including model, price, miles per gallon, and weight. One of the variables, *QtrMile*, shows the time (in seconds) needed for a car to travel one-quarter mile from a standing start. ■

Example 2.18

A histogram of the quarter-mile times is shown in Figure 2.19. Is the distribution approximately symmetric and bell-shaped? Use the histogram to give a rough estimate of the mean and standard deviation of quarter-mile times.

Figure 2.19 *Estimate the mean and the standard deviation*

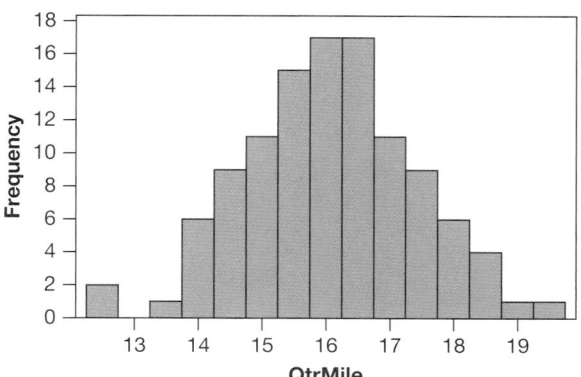

---

[40]Data from *Consumer Reports New Car Buying Guide 2015.*

Solution The histogram is relatively symmetric and bell-shaped. The mean appears to be approximately 16 seconds. To estimate the standard deviation, we estimate an interval centered at 16 that contains approximately 95% of the data. The interval from 13 to 19 appears to contain almost all the data. Since 13 is 3 units below the mean of 16 and 19 is 3 units above the mean, by the 95% rule we estimate that 2 times the standard deviation is 3, so the standard deviation appears to be approximately 1.5 seconds. Note that we can only get a rough approximation from the histogram. To find the exact values of the mean and standard deviation, we would use technology and all the values in the dataset. For the *QtrMile* variable in this example, we find $\bar{x} = 16.1$ seconds and $s = 1.31$ seconds, so the rough approximation worked reasonably well.

Understanding *z*-Scores

A doctor finds that a patient has a a systolic blood pressure of 200 mmHg. Just how unusual is this? Sometimes a single data value is meaningless without knowing the center and spread of a distribution. A common way to determine how unusual a single data value is, that is independent of the units used, is to count how many standard deviations it is away from the mean. This quantity is known as the *z-score*.

> **Number of Standard Deviations from the Mean: *z*-Scores**
>
> The *z*-score for a data value, x, from a sample with mean $\bar{x}$ and standard deviation s is defined to be
>
> $$z\text{-score} = \frac{x - \bar{x}}{s}$$
>
> For a population, $\bar{x}$ is replaced with μ and s is replaced with σ.
>
> The *z*-score tells how many standard deviations the value is from the mean, and is independent of the unit of measurement.

If the data have a distribution that is symmetric and bell-shaped, we know from the 95% rule that about 95% of the data will fall within two standard deviations of the mean. This means that only about 5% of the data values will have *z*-scores beyond ±2.

Example 2.19

One of the patients (ID#772) in the ICU study (Data 2.3 on page 69) had a high systolic blood pressure of 204 mmHg and a low pulse rate of 52 bpm. Which of these values is more unusual relative to the other patients in the sample? The summary statistics for systolic blood pressure show a mean of 132.3 and standard deviation of 32.95, while the heart rates have a mean of 98.9 and standard deviation of 26.83.

Solution We compute the *z*-scores for this patient's blood pressure and heart rate:

$$\text{Blood pressure:} \quad z = \frac{x - \bar{x}}{s} = \frac{204 - 132.3}{32.95} = 2.18$$

This patient's blood pressure is slightly more than two standard deviations above the sample mean.

$$\text{Heart rate:} \quad z = \frac{x - \bar{x}}{s} = \frac{52 - 98.9}{26.83} = -1.75$$

This patient's heart rate is less than two standard deviations below the sample mean heart rate. The high blood pressure is somewhat more unusual than the low heart rate.

Percentiles

We turn now to an alternate way to give information about a distribution. The Scholastic Aptitude Test (SAT) is given several times a year to secondary school students and is often used in admissions decisions by colleges and universities. The SAT has three parts: Critical Reading, Mathematics, and Writing. Each is scored on a scale of 200 to 800. The SAT aims to have the average score close to 500 in each part. For students in the graduating class of 2014, the averages were as follows:

Critical Reading: 497   Mathematics: 513   Writing: 487

When students receive their score reports, they see their score as well as their *percentile*. For example, for the class of 2014, a score of 620 in Critical Reading is the 84th percentile. This means that 84% of the students in the class of 2014 who took the exam scored less than or equal to 620. The 30th percentile in Mathematics is a score of 450, which means that 30% of the students scored less than 450.

> **Percentiles**
>
> The Pth **percentile** is the value of a quantitative variable which is greater than P percent of the data.[41]

Example 2.20

Standard & Poor's maintains one of the most widely followed indices of large-cap American stocks: the S&P 500. The index includes stocks of 500 companies in industries in the US economy. A histogram of the daily volume (in millions of shares) for the S&P 500 stock index for every day in 2014 is shown in Figure 2.20. The data are stored in **SandP500**. Use the histogram to roughly estimate and interpret the 25th percentile and the 90th percentile.

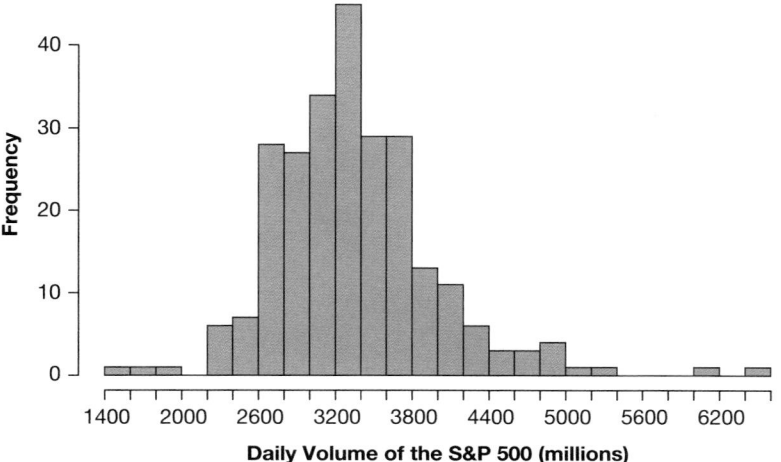

Figure 2.20 *Daily volume for the S&P 500 index in 2014*

[41]Different software packages may give slightly different answers for percentiles. Some sources, for example, define the Pth percentile as the value which is greater than *or equal to* P percent of the values. For large datasets, however, the numbers will generally be very similar.

Solution The 25th percentile is the value with a quarter of the values below or equal to it. This is the value where 25% of the area of the histogram lies to the left. This appears to be about 3000 million. We do not expect you to compute this exactly, but simply be able to give an estimate. A volume of about 4200 million has roughly 10% of the data values above it (and 90% below), so the 90th percentile is about 4200 million.

Five Number Summary

The *minimum* and *maximum* in a dataset identify the extremes of the distribution: the smallest and largest values, respectively. The median is the 50th percentile, since it divides the data into two equal halves. If we divide each of those halves again, we obtain two additional statistics known as the *first (Q_1)* and *third (Q_3) quartiles*, which are the 25th and 75th percentiles. Together these five numbers provide a good summary of important characteristics of the distribution and are known as the *five number summary*.

> **Five Number Summary**
>
> We define
>
> **Five Number Summary** = (minimum, Q_1, median, Q_3, maximum)
>
> where
>
> $$Q_1 = \text{First quartile} = \text{25th percentile}$$
> $$Q_3 = \text{Third quartile} = \text{75th percentile}$$
>
> The five number summary divides the dataset into fourths: about 25% of the data fall between any two consecutive numbers in the five number summary.

© kristian sekulic/iStockphoto

How many hours a week do students exercise?

Example 2.21

The five number summary for the number of hours spent exercising a week for the **StudentSurvey** sample is (0, 5, 8, 12, 40). Explain what this tells us about the amount of exercise for this group of students.

Solution All of the students exercise between 0 and 40 hours per week. The 25% of students who exercise the least exercise between 0 and 5 hours a week, and the 25% of students who exercise the most exercise between 12 and 40 hours a week. The middle 50% of students exercise between 5 and 12 hours a week, with half exercising less than 8 hours per week and half exercising more than 8 hours per week.

Range and Interquartile Range

The five number summary provides two additional opportunities for summarizing the amount of spread in the data, the *range* and the *interquartile range*.

> **Range and Interquartile Range**
>
> From the five number summary, we can compute the following two statistics:
>
> $$\textbf{Range} = \text{Maximum} - \text{Minimum}$$
> $$\textbf{Interquartile range (IQR)} = Q_3 - Q_1$$

Example 2.22

The five number summary for the mammal longevity data in Table 2.15 on page 64 is $(1, 8, 12, 16, 40)$. Find the range and interquartile range for this dataset.

Solution From the five number summary $(1, 8, 12, 16, 40)$, we see that the minimum longevity is 1 and the maximum is 40, so the range is $40 - 1 = 39$ years. The first quartile is 8 and the third quartile is 16, so the interquartile range is $IQR = 16 - 8 = 8$ years.

Note that the range and interquartile range calculated in Example 2.22 (39 and 8, respectively) are numerical values not intervals. Also notice that if the elephant (longevity $= 40$) were omitted from the sample, the range would be reduced to $25 - 1 = 24$ years while the IQR would go down by just one year, $15 - 8 = 7$. In general, although the range is a very easy statistic to compute, the IQR is a more resistant measure of spread.

Example 2.23

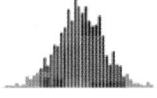

Using the temperature data for Des Moines and San Francisco given in Table 2.20, find the five number summary for the temperatures in each city. Find the range and the IQR for each dataset and compare the results for the two cities.

Solution We use technology to find the five number summaries. From the output in Figure 2.15 or Figure 2.16 on page 78, we see that the five number summary for Des Moines temperatures is $(35.7, 44.4, 54.7, 59.9, 74.9)$. (Notice that the value for Q_3 is slightly different between the three outputs. You may get slightly different values for the quartiles depending on which technology you use.) The five number summary for San Francisco temperatures is $(48.7, 51.4, 53.8, 56.0, 61.0)$. We have

$$\text{Range for Des Moines} = 74.9 - 35.7 = 39.2°F$$
$$\text{Range for San Francisco} = 61.0 - 48.7 = 12.3°F$$

$$\text{IQR for Des Moines} = 59.9 - 44.4 = 15.5°F$$
$$\text{IQR for San Francisco} = 56.0 - 51.4 = 4.6°F$$

The range and IQR are much larger for the Des Moines data than the San Francisco data. Temperatures are much more variable in central Iowa than they are on the California coast!

Choosing Measures of Center and Spread

Because the standard deviation measures how much the data values deviate from the mean, it makes sense to use the standard deviation as a measure of variability when the mean is used as a measure of center. Both the mean and standard deviation have the advantage that they use all of the data values. However, they are not resistant to outliers. The median and IQR are resistant to outliers. Furthermore, if there are outliers or the data are heavily skewed, the five number summary can give more information (such as direction of skewness) than the mean and standard deviation.

Example 2.24

Example 2.13 on page 70 describes salaries in the US National Football League, in which some star players are paid much more than most other players.

(a) We see in that example that players prefer to use the median ($872,000 in 2014) as a measure of center since they don't want the results heavily influenced by a few huge outlier salaries. What should they use as a measure of spread?

(b) We also see that the owners of the teams prefer to use the mean ($2.14 million in 2014) as a measure of center since they want to use a measure that includes all the salaries. What should they use as a measure of spread?

Solution

(a) The interquartile range (IQR) should be used with the median as a measure of spread. Both come from the five number summary, and both the median and the IQR are resistant to outliers.

(b) The standard deviation should be used with the mean as a measure of spread. Both the mean and the standard deviation use all the data values in their computation.

SECTION LEARNING GOALS

You should now have the understanding and skills to:

- Use technology to compute summary statistics for a quantitative variable
- Recognize the uses and meaning of the standard deviation
- Compute and interpret a *z*-score
- Interpret a five number summary or percentiles
- Use the range, the interquartile range, and the standard deviation as measures of spread
- Describe the advantages and disadvantages of the different measures of center and spread

Exercises for Section **2.3**

SKILL BUILDER 1

For the datasets in Exercises 2.78 to 2.83, use technology to find the following values:

(a) The mean and the standard deviation.

(b) The five number summary.

2.78 10, 11, 13, 14, 14, 17, 18, 20, 21, 25, 28

2.79 1, 3, 4, 5, 7, 10, 18, 20, 25, 31, 42

2.80 4, 5, 8, 4, 11, 8, 18, 12, 5, 15, 22, 7, 14, 11, 12

2.81 25, 72, 77, 31, 80, 80, 64, 39, 75, 58, 43, 67, 54, 71, 60

2.82 The variable *Exercise*, number of hours spent exercising per week, in the **StudentSurvey** dataset.

2.83 The variable *TV*, number of hours spent watching television per week, in the **StudentSurvey** dataset.

SKILL BUILDER 2

In Exercises 2.84 and 2.85, match the standard deviations with the histograms.

2.84 Match the three standard deviations $s = 1$, $s = 3$, and $s = 5$ with the three histograms in Figure 2.21.

2.85 Match each standard deviation with one of the histograms in Figure 2.22.

(a) $s = 0.5$

(b) $s = 10$

(c) $s = 50$

(d) $s = 1$

(e) $s = 1000$

(f) $s = 0.29$

SKILL BUILDER 3

In Exercises 2.86 and 2.87, match each five number summary with the corresponding histogram.

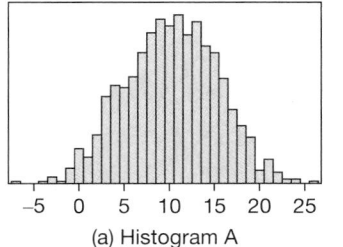 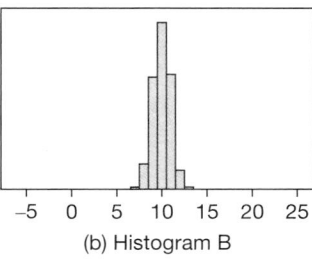 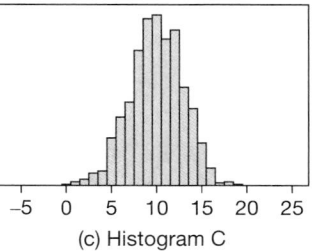

Figure 2.21 *Three histograms for Exercise 2.84*

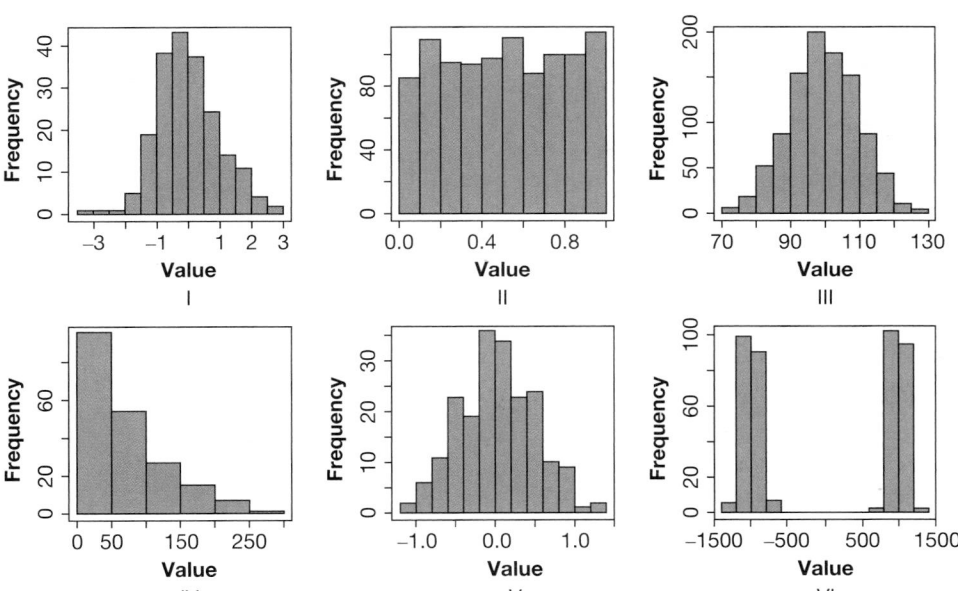

Figure 2.22 *Histograms for Exercises 2.85 and 2.86*

2.86 Match each five number summary with one of the histograms in Figure 2.22.

(a) (0, 0.25, 0.5, 0.75, 1)

(b) (−1.08, −0.30, 0.01, 0.35, 1.27)

(c) (0.64, 27.25, 53.16, 100, 275.7)

(d) (−3.5, −0.63, −0.11, 0.59, 2.66)

(e) (71.45, 92.77, 99.41, 106.60, 129.70)

(f) (−1296, −1005, −705, 998, 1312)

2.87 Match each five number summary with one of the histograms in Figure 2.23. The scale is the same on all four histograms.

(a) (1, 3, 5, 7, 9)

(b) (1, 4, 5, 6, 9)

(c) (1, 5, 7, 8, 9)

(d) (1, 1, 2, 4, 9)

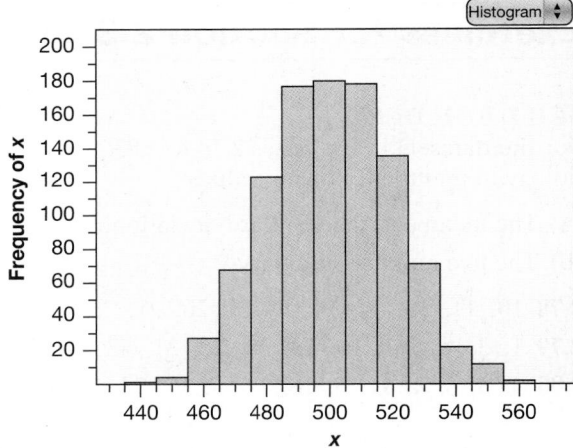

Figure 2.24 *Histogram for Exercises 2.88 to 2.90*

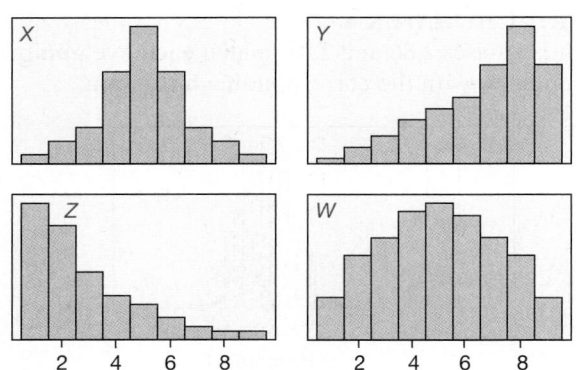

Figure 2.23 *Match five number summaries in Exercise 2.87*

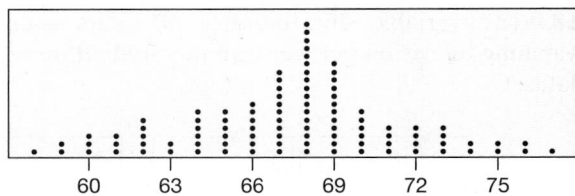

Figure 2.25 *Dotplot with n = 100 for Exercises 2.91 to 2.93*

SKILL BUILDER 4

In Exercises 2.88 to 2.93, estimate the summary statistics requested, using the histogram in Figure 2.24 for Exercises 2.88 to 2.90 and the dotplot in Figure 2.25 for Exercises 2.91 to 2.93. There are $n = 100$ data points included in the dotplot.

2.88 Estimate the mean and the standard deviation for the data in the histogram in Figure 2.24.

2.89 Estimate values at the 10th percentile and the 90th percentile for the data in Figure 2.24.

2.90 Estimate the five number summary for the data in Figure 2.24.

2.91 Estimate the mean and the standard deviation for the data in the dotplot in Figure 2.25.

2.92 Estimate values at the 10th percentile and the 90th percentile for the data in Figure 2.25.

2.93 Estimate the five number summary for the data in Figure 2.25.

SKILL BUILDER 5

In Exercises 2.94 to 2.97, indicate whether the five number summary corresponds most likely to a distribution that is skewed to the left, skewed to the right, or symmetric.

2.94 (15, 25, 30, 35, 45)

2.95 (100, 110, 115, 160, 220)

2.96 (0, 15, 22, 24, 27)

2.97 (22.4, 30.1, 36.3, 42.5, 50.7)

SKILL BUILDER 6: Z-SCORES

In Exercises 2.98 to 2.101, find and interpret the z-score for the data value given.

2.98 The value 243 in a dataset with mean 200 and standard deviation 25.

2.99 The value 88 in a dataset with mean 96 and standard deviation 10.

2.100 The value 5.2 in a dataset with mean 12 and standard deviation 2.3.

2.101 The value 8.1 in a dataset with mean 5 and standard deviation 2.

SKILL BUILDER 7: THE 95% RULE

In Exercises 2.102 to 2.105, use the 95% rule and the fact that the summary statistics come from a distribution that is symmetric and bell-shaped to find an interval that is expected to contain about 95% of the data values.

2.102 A bell-shaped distribution with mean 200 and standard deviation 25.

2.103 A bell-shaped distribution with mean 10 and standard deviation 3.

2.104 A bell-shaped distribution with mean 1000 and standard deviation 10.

2.105 A bell-shaped distribution with mean 1500 and standard deviation 300.

2.106 Estimating Summary Statistics For the dataset

$$45, 46, 48, 49, 49, 50, 50, 52, 52, 54, 57, 57, 58, 58, 60, 61$$

(a) Without doing any calculations, estimate which of the following numbers is closest to the mean:

$$60, \quad 53, \quad 47, \quad 58$$

(b) Without doing any calculations, estimate which of the following numbers is closest to the standard deviation:

$$52, \quad 5, \quad 1, \quad 10, \quad 55$$

(c) Use technology to find the mean and the standard deviation for this dataset.

2.107 Percent Obese by State Computer output giving descriptive statistics for the percent of the population that is obese for each of the 50 US states, from the **USStates** dataset, is given in Figure 2.26. Since all 50 US states are included, this is a population not a sample.

(a) What are the mean and the standard deviation? Include appropriate notation with your answers.

(b) Calculate the z-score for the largest value and interpret it in terms of standard deviations. Do the same for the smallest value.

(c) This distribution is relatively symmetric and bell-shaped. Give an interval that is likely to contain about 95% of the data values.

2.108 Five Number Summary for Percent Obese by State Computer output giving descriptive statistics for the percent of the population that is obese for each of the 50 US states, from the **USStates** dataset, is given in Figure 2.26.

(a) What is the five number summary?

(b) Give the range and the IQR.

(c) What can we conclude from the five number summary about the location of the 15th percentile? The 60th percentile?

2.109 Public Expenditure on Education Figure 2.27 shows the public expenditure on education as percentage of Gross Domestic Product (GDP) for all countries.[42] The mean expenditure is $\mu = 4.7\%$ and the standard deviation of the expenditures is $\sigma = 2\%$. The data are stored in **EducationLiteracy**.

(a) The United States spends 5.2% of it's GDP on education. *Without doing any calculations yet*, will the z-score for the US be positive, negative, or zero? Why?

(b) Calculate the z-score for the US.

(c) There are two high outliers; Lesotho (a small country completely surrounded by South Africa) spends 13% of it's GDP on education and Cuba spends 12.8%. Equatorial Guinea spends the lowest percentage on education at only 0.8%. Calculate the range.

(d) The five number summary for this data set is (0.8, 3.2, 4.6, 5.6, 13). Calculate the IQR.

2.110 How Many Hot Dogs Can You Eat in Ten Minutes? Every Fourth of July, Nathan's Famous in New York City holds a hot dog eating contest, in

[42]Data from The World Bank, the most recent data available for each country as of July 2015 obtained from *http://www.knoema.com*.

Descriptive Statistics: Obese

Variable	N	N*	Mean	SE Mean	StDev	Minimum	Q1	Median	Q3	Maximum
Obese	50	0	28.766	0.476	3.369	21.300	26.375	29.400	31.150	35.100

Figure 2.26 *Percent of the population that is obese by state*

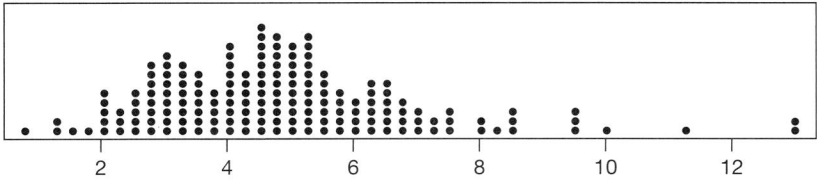

Figure 2.27 *Public expenditure on education as percentage of GDP*

which contestants try to eat as many hot dogs as possible in 10 minutes.[43] In 2015, thousands of people watched the event live on Coney Island, and it was broadcast live to many more on ESPN. The winning number of hot dogs along with the year is shown in Table 2.21 and is available in the dataset **HotDogs**.

(a) Use technology to find the mean and the standard deviation of the 14 numbers.

(b) How many of the 14 values are above the mean? How many are above the mean for the five values in the earlier five years (2002–2006)? How many are above the mean for the remaining nine values (2007–2015)?

Table 2.21 *Winning number of hot dogs in the hot dog eating contest*

Year	Hot Dogs
2015	62
2014	61
2013	69
2012	68
2011	62
2010	54
2009	68
2008	59
2007	66
2006	54
2005	49
2004	54
2003	45
2002	50

2.111 The Hot Dog Eating Rivalry: Matched Pairs In Exercise 2.110, we introduce Nathan's Famous Hot Dog Eating Contest. Every year from 2002 to 2014, either Joey Chestnut of California or Takeru Kobayashi of Japan won the contest. In five of those years, both men competed and the results of the rivalry are shown in Table 2.22. (After the tie in 2008, Joey Chestnut won in an overtime.) Because the conditions of the year matter, this is a *matched pairs* situation, with the two men going against each other each year. In a matched-pairs situation, we use the summary statistics of the *differences* between the pairs of values.

(a) For each of the five years, find the difference in number of hot dogs eaten between Joey and Takeru. For example, in 2009, the difference is 68 − 64 = 4. Since it is important to always subtract the same way (in this case, Joey's value minus Takeru's value), some of the differences will be negative.

(b) Use technology to find the mean and the standard deviation of the differences.

2.112 Grade Point Averages A histogram of the $n = 345$ grade point averages reported by students in the **StudentSurvey** dataset is shown in Figure 2.28.

(a) Estimate and interpret the 10th percentile and the 75th percentile.

(b) Estimate the range.

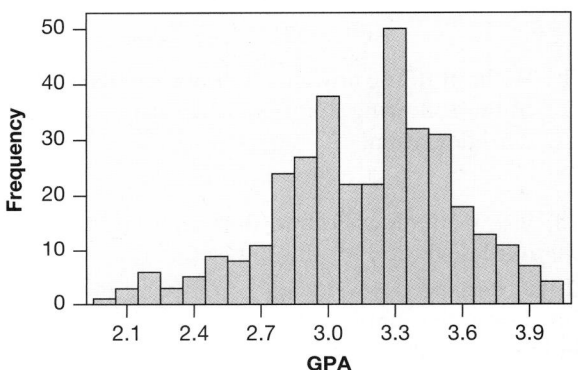

Figure 2.28 *Estimate the 10th percentile and 75th percentile*

2.113 Laptop Computers and Sperm Count Studies have shown that heating the scrotum by just 1°C can reduce sperm count and sperm quality, so men concerned about fertility are cautioned to avoid too much time in the hot tub or sauna. A new study[44] suggests that men also keep their laptop computers off their laps. The study measured scrotal temperature in 29 healthy male volunteers as they sat with legs together and a laptop computer on the lap. Temperature increase in the left scrotum over a 60-minute session is given as 2.31 ± 0.96 and a note tells us that "Temperatures are given as °C; values are shown as mean ± SD." The abbreviation SD stands for standard deviation. (Men who sit with

Table 2.22 *Hot dog eating rivalry*

Year	Joey Chestnut	Takeru Kobayashi
2009	68	64
2008	59	59
2007	66	63
2006	52	54
2005	32	49

[43]*nathansfamous.com.*

[44]Sheynkin, Y., et al., "Protection from scrotal hyperthermia in laptop computer users," *Fertility and Sterility*, February 2011; 92(2): 647–651.

their legs together without a laptop computer do not show an increase in temperature.)

(a) If we assume that the distribution of the temperature increases for the 29 men is symmetric and bell-shaped, find an interval that we expect to contain about 95% of the temperature increases.

(b) Find and interpret the z-score for one of the men, who had a temperature increase of 4.9°.

2.114 Estimating Monthly Retail Sales A histogram of total US monthly retail sales, in billions of dollars, for the 136 months starting with January 2000 is shown in Figure 2.29. Is the distribution approximately symmetric and bell-shaped? Use the histogram to give a rough estimate of the mean and standard deviation of monthly US retail sales totals.

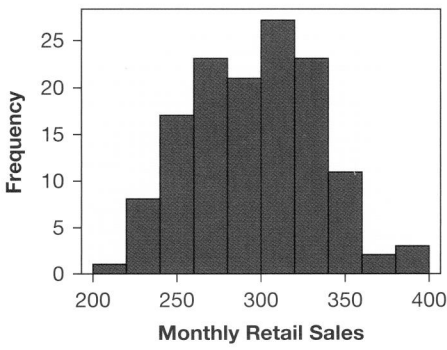

Figure 2.29 *Estimate the mean and the standard deviation*

2.115 Summarizing Monthly Retail Sales US monthly retail sales, in billions of dollars, for the 136 months starting with January 2000 is given in the **RetailSales** dataset, in the variable *Sales*, and shown in Figure 2.29. Use technology to find the mean and the standard deviation for this sample of US monthly retail sales. Use the 95% Rule to compute an interval that is likely to contain about 95% of the data.

STATISTICS FOR NBA PLAYERS IN 2014–2015 Exercises 2.116 to 2.118 refer to the dataset **NBAPlayers2015**, which contains information on many variables for players in the NBA (National Basketball Association) during the 2014–2015 season. The dataset includes information for all players who averaged more than 24 minutes per game, and includes $n = 182$ players and 25 variables.

2.116 Distribution of Three-Point Attempts in the NBA In basketball, a basket is awarded three points (rather than the usual two) if it is shot from farther away. Some players attempt lots of three-point shots and quite a few attempt none, as we see in

the distribution of number of three-point attempts by players in the NBA in Figure 2.30. The data are available in **NBAPlayers2015** under the variable name *FG3Attempt*. Is it appropriate to use the 95% rule with this dataset? Why or why not?

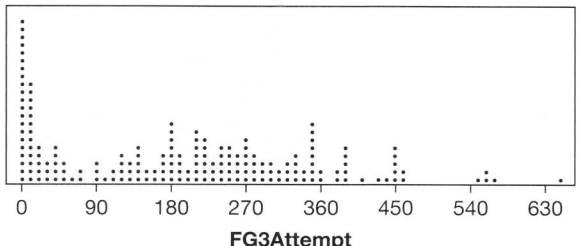

Figure 2.30 *Number of three-point shot attempts in the NBA, by player*

2.117 Distribution of Blocked Shots in the NBA The variable *Blocks* in the dataset **NBAPlayers2015** includes information on the number of blocked shots during the season for each of the 182 players in the dataset.

(a) Use technology to find the mean and the standard deviation of the number of blocked shots.

(b) Use technology to find the five number summary for the same variable.

(c) Which set of summary statistics, those from part (a) or part (b), is more resistant to outliers and more appropriate if the data are heavily skewed?

(d) Use technology to create a graph of the data in *Blocks* and describe the shape of the distribution.

(e) Is it appropriate to use the 95% rule with these data? Why or why not?

2.118 Which Accomplishment of LeBron James Is Most Impressive? Table 2.23 shows the means and standard deviations for four of the variables in the **NBAPlayers2015** dataset. *FGPct* is the field goal percentage, *Points* is total number of points scored during the season, *Assists* is total number of assists during the season, and *Steals* is total number of steals during the season. LeBron James had a field goal percentage of 0.488, scored 1743 points, had 511 assists, and had 109 steals. Find the z-score for each of LeBron's statistics. Use the z-scores to determine, relative to the other players in the NBA that season, which statistic of LeBron's is the most impressive. Which is the least impressive?

Table 2.23 *Summary statistics on NBA players*

Variable	Mean	Standard Deviation
FGPct	0.453	0.054
Points	915	357
Assists	205	149
Steals	67.5	33.6

2.119 SAT Scores Stanley, a recent high school student, took the SAT exam in 2014 and got a 600 in all three components (Critical Reading, Math, and Writing). He was interested in how well he did compared to the rest of his peers. Table 2.24 shows the summary statistics for all students in 2014.[45]

(a) Calculate *z*-scores for all three of Stanley's scores using the summary statistics in Table 2.24.

(b) Which of Stanley's three scores is the most unusual relative to his peers? Which is the least unusual?

(c) In which component did Stanley perform best relative to his peers?

Table 2.24 *Summary statistics for SAT scores*

Component	Mean	Standard Deviation
Critical Reading	497	115
Math	513	120
Writing	487	115

2.120 Comparing Global Internet Connections The Nielsen Company measured connection speeds on home computers in nine different countries and wanted to determine whether connection speed affects the amount of time consumers spend online.[46] Table 2.25 shows the percent of Internet users with a "fast" connection (defined as 2Mb or faster) and the average amount of time spent online, defined as total hours connected to the web from a home computer during the month of February 2011. The data are also available in the dataset **GlobalInternet**.

(a) Use technology to find the mean and standard deviation of the nine values for percent with a fast connection.

(b) Use technology to find the mean and standard deviation of the nine values for time online.

(c) Does there seem to be a relationship between the two variables? Explain. (We examine this relationship further in Section 2.5.)

[45] *https://secure-media.collegeboard.org/digitalServices/pdf/sat/sat-percentile-ranks-crit-reading-math-writing-2014.pdf*
[46] "Swiss Lead in Speed: Comparing Global Internet Connections," NielsenWire, April 1, 2011.

Table 2.25 *Internet connection speed and hours online*

Country	Percent Fast Connection	Hours Online
Switzerland	88	20.18
United States	70	26.26
Germany	72	28.04
Australia	64	23.02
United Kingdom	75	28.48
France	70	27.49
Spain	69	26.97
Italy	64	23.59
Brazil	21	31.58

2.121 Daily Calorie Consumption The five number summary for daily calorie consumption for the $n = 315$ participants in the **NutritionStudy** is (445, 1334, 1667, 2106, 6662).

(a) Give the range and the IQR.

(b) Which of the following numbers is most likely to be the mean of this dataset? Explain.

$$1550 \quad 1667 \quad 1796 \quad 3605$$

(c) Which of the following numbers is most likely to be the standard deviation of this dataset? Explain.

$$5.72 \quad 158 \quad 680 \quad 1897 \quad 5315$$

2.122 Largest and Smallest Standard Deviation Using only the whole numbers 1 through 9 as possible data values, create a dataset with $n = 6$ and $\bar{x} = 5$ and with:

(a) Standard deviation as small as possible.

(b) Standard deviation as large as possible.

USING THE 95% RULE TO DRAW SMOOTH BELL-SHAPED CURVES
In Exercises 2.123 to 2.126, sketch a curve showing a distribution that is symmetric and bell-shaped and has approximately the given mean and standard deviation. In each case, draw the curve on a horizontal axis with scale 0 to 10.

2.123 Mean 3 and standard deviation 1.

2.124 Mean 7 and standard deviation 1.

2.125 Mean 5 and standard deviation 2.

2.126 Mean 5 and standard deviation 0.5.

2.127 Rotten Tomatoes Movie Ratings The variable *RottenTomatoes* in the **HollywoodMovies** dataset gives the critics' rating on the *Rotten Tomatoes* website of movies that came out of Hollywood between

2007 and 2013. Use technology to find the mean, the standard deviation, and the five number summary for the data in this variable.

2.128 Audience Movie Ratings The variable *AudienceScore* in the **HollywoodMovies** dataset gives the audience rating on the *Rotten Tomatoes* website of movies that came out of Hollywood between 2007 and 2013. Use technology to find the mean, the standard deviation, and the five number summary for the data in this variable.

2.129 Using the Five Number Summary to Visualize Shape of a Distribution Draw a histogram or a smooth curve illustrating the shape of a distribution with the properties that:

(a) The range is 100 and the interquartile range is 10.

(b) The range is 50 and the interquartile range is 40.

2.130 Rough Rule of Thumb for the Standard Deviation According to the 95% rule, the largest value in a sample from a distribution that is approximately symmetric and bell-shaped should be between 2 and 3 standard deviations above the mean, while the smallest value should be between 2 and 3 standard

deviations below the mean. Thus the range should be roughly 4 to 6 times the standard deviation. As a rough rule of thumb, we can get a quick estimate of the standard deviation for a bell-shaped distribution by dividing the range by 5. Check how well this quick estimate works in the following situations.

(a) Pulse rates from the **StudentSurvey** dataset discussed in Example 2.17 on page 80. The five number summary of pulse rates is $(35, 62, 70, 78, 130)$ and the standard deviation is $s = 12.2$ bpm. Find the rough estimate using all the data, and then excluding the two outliers at 120 and 130, which leaves the maximum at 96.

(b) Number of hours a week spent exercising from the **StudentSurvey** dataset discussed in Example 2.21 on page 84. The five number summary of this dataset is $(0, 5, 8, 12, 40)$ and the standard deviation is $s = 5.741$ hours.

(c) Longevity of mammals from the **Mammal-Longevity** dataset discussed in Example 2.22 on page 85. The five number summary of the longevity values is $(1, 8, 12, 16, 40)$ and the standard deviation is $s = 7.24$ years.

2.4 BOXPLOTS AND QUANTITATIVE/CATEGORICAL RELATIONSHIPS

In this section, we examine a relationship between a quantitative variable and a categorical variable by examining both comparative summary statistics and graphical displays. Before we get there, however, we look at one more graphical display for a single quantitative variable that is particularly useful for comparing groups.

Boxplots

A *boxplot* is a graphical display of the five number summary for a single quantitative variable. It shows the general shape of the distribution, identifies the middle 50% of the data, and highlights any outliers.

Boxplots

A **boxplot** includes:

- A numerical scale appropriate for the data values.
- A box stretching from Q_1 to Q_3.
- A line that divides the box drawn at the median.
- A line from each quartile to the most extreme data value that is *not* an outlier.
- Each outlier plotted individually with a symbol such as an asterisk or a dot.

Example 2.25

Draw a boxplot for the data in **MammalLongevity**, with five number summary $(1, 8, 12, 16, 40)$.

Solution

The boxplot for mammal longevities is shown in Figure 2.31. The box covers the interval from the first quartile of 8 years to the third quartile of 16 years and is divided at the median of 12 years. The line to the left of the lower quartile goes all the way to the minimum longevity at 1 year, since there were no small outliers. The line to the right stops at the largest data value (25, grizzly bear and hippopotamus) that is not an outlier. The only clear outlier is the elephant longevity of 40 years, so this value is plotted with an individual symbol at the maximum of 40 years.

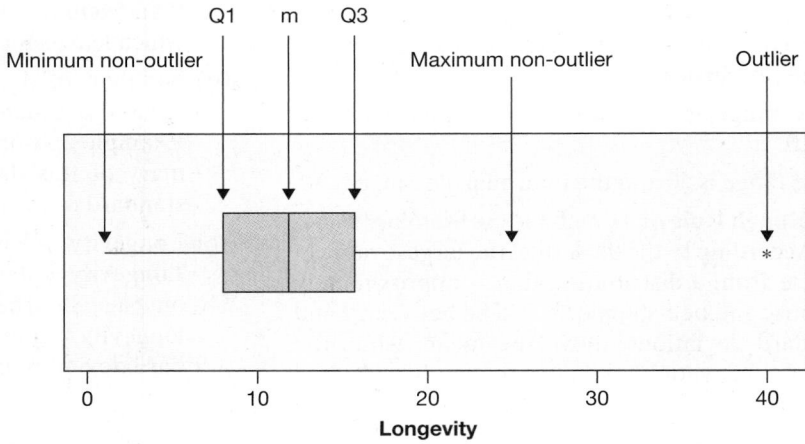

Figure 2.31 *Boxplot of longevity of mammals*

DATA 2.6

US States

The dataset **USStates** includes many variables measured for each of the 50 states in the US. Some of the variables included for each state are average household income, percent to graduate high school, health statistics such as consumption of fruits and vegetables, percent obese, percent of smokers, and some results from the 2012 US presidential election.[47] ■

Example 2.26

One of the quantitative variables in the **USStates** dataset is *Vegetables*, which gives the percentage of the population that eats at least one serving of vegetables per day for each of the states. Figure 2.32 shows a boxplot of the percent for all 50 states.

(a) Discuss what the boxplot tells us about the distribution of this variable.

(b) Estimate the five number summary.

Solution

(a) The distribution of percentages is relatively symmetric, with one low outlier, and centered around 77. The percent of people who eat vegetables appears to range from about 67% to about 84%. The middle 50% of percentages is between about 75% and 80%, with a median value at about 77%. The low outlier is about 67% and represents the state of Louisiana.

(b) The five number summary appears to be approximately $(67, 75, 77, 80, 84)$.

---

[47]Data from a variety of sources, mostly *http://www.census.gov*.

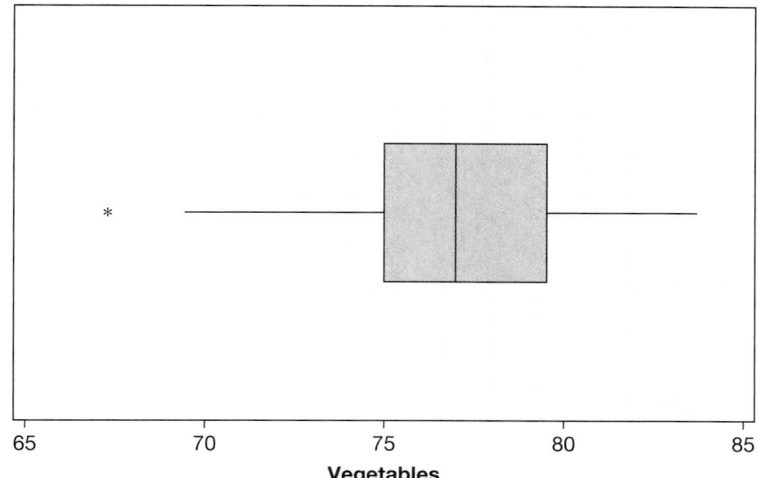

Figure 2.32 *Percent of people who eat at least one serving of vegetables per day, by state*

© Ben Molyneux/Alamy Stock Photo

How much did it cost to make this movie?

DATA 2.7   **Hollywood Movies**

Over 900 movies came out of Hollywood between 2007 and 2013 and the dataset **HollywoodMovies** contains lots of information on these movies, such as studio, genre, budget, audience ratings, box office average opening weekend, world gross, and others.[48] ■

Example 2.27

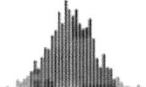

One of the quantitative variables in the **HollywoodMovies** dataset is *Budget*, which gives the budget, in millions of dollars, to make each movie. Figure 2.33 shows the boxplot for the budget of all Hollywood movies that came out in 2013.

(a) Discuss what the boxplot tells us about the distribution of this variable.

(b) What movies do the two largest outliers correspond to?

(c) What was the budget to make *Frozen*? Is it an outlier?

[48]McCandless, D., "Most Profitable Hollywood Movies," from "Information is Beautiful," *http://www.davidmccandless.com*, Accessed July 2015.

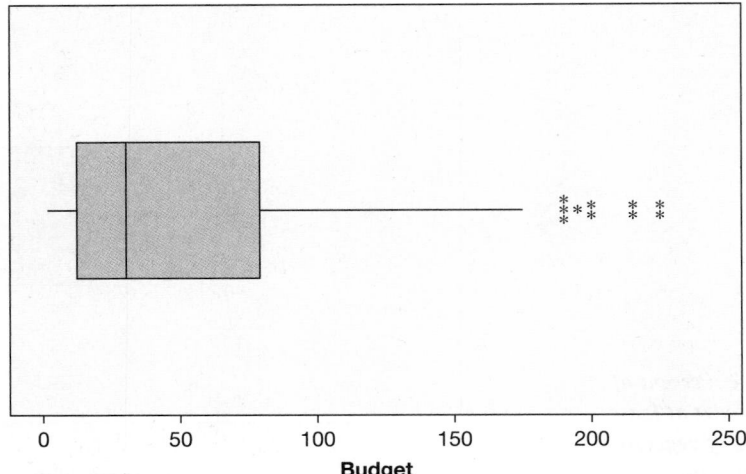

Figure 2.33 *Budget, in millions of dollars, of Hollywood movies*

Solution

(a) Because the minimum, first quartile, and median are so close together, we see that half the data values are packed in a small interval, then the other half extend out quite far to the right. These data are skewed to the right, with many large outliers.

(b) The two largest outliers represent movies with budgets of about 225 million dollars. We see from the dataset **HollywoodMovies** that the two movies are *Man of Steel* and *The Desolation of Smaug*.

(c) We see from the dataset that the budget for *Frozen* was 150 million dollars. We see in the boxplot that this is not an outlier.

Detection of Outliers

Consider again the data on mammal longevity in Data 2.2 on page 64. Our intuition suggests that the longevity of 40 years for the elephant is an unusually high value compared to the other lifespans in this sample. How do we determine objectively when such a value is an outlier? The criteria should depend on some measure of location for "typical" values and a measure of spread to help us judge when a data point is "far" from those typical cases. One approach, typically used for identifying outliers for boxplots, uses the quartiles and interquartile range. As a rule, most data values will fall within about $1.5(IQR)$'s of the quartiles.[49]

> #### *IQR* Method for Detecting Outliers
>
> For boxplots, we call a data value an **outlier** if it is
>
> Smaller than $\quad Q_1 - 1.5(IQR) \quad$ or $\quad$ Larger than $\quad Q_3 + 1.5(IQR)$

Example 2.28

According to the IQR method, is the elephant an outlier for the mammal longevities in the dataset **MammalLongevity**? Are any other mammals outliers in that dataset?

---

[49]In practice, determining outliers requires judgment and understanding of the context. We present a specific method here, but no single method is universally used for determining outliers.

Solution The five number summary for mammal longevities is $(1, 8, 12, 16, 40)$. We have $Q_1 = 8$ and $Q_3 = 16$ so the interquartile range is $IQR = 16 - 8 = 8$. We compute

$$Q_1 - 1.5(IQR) = 8 - 1.5(8) = 8 - 12 = -4$$

and

$$Q_3 + 1.5(IQR) = 16 + 1.5(8) = 16 + 12 = 28$$

Clearly, there are no mammals with negative lifetimes, so there can be no outliers below the lower value of -4. On the upper side, the elephant, as expected, is clearly an outlier beyond the value of 28 years. No other mammal in this sample exceeds that value, so the elephant is the only outlier in the longevity data.

One Quantitative and One Categorical Variable

Do men or women tend to watch more television? Is survival time for cancer patients related to genetic variations? How do April temperatures in Des Moines compare to those in San Francisco? These questions each involve a relationship between a quantitative variable (amount of TV, survival time, temperature) and a categorical variable (gender, type of gene, city). One of the best ways to visualize such relationships is to use *side-by-side graphs*. Showing graphs with the same axis facilitates the comparison of the distributions.

> **Visualizing a Relationship between Quantitative and Categorical Variables**
>
> **Side-by-side graphs** are used to visualize the relationship between a quantitative variable and a categorical variable. The side-by-side graph includes a graph for the quantitative variable (such as a boxplot, histogram, or dotplot) for each group in the categorical variable, all using a common numeric axis.

Erik Von Weber/Getty Images, Inc.

Who watches more TV, males or females?

Example 2.29

Who watches more TV, males or females? The data in **StudentSurvey** contains the categorical variable *Gender* as well as the quantitative variable *TV* for the number of hours spent watching television per week. For these students, is there an association between gender and number of hours spent watching television? Use the side-by-side graphs in Figure 2.34 showing the distribution of hours spent watching television for males and females to discuss how the distributions compare.

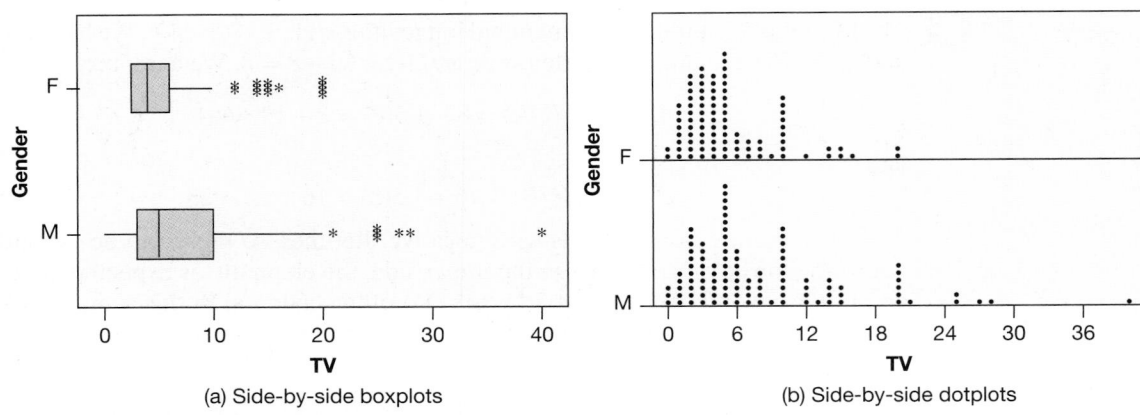

(a) Side-by-side boxplots   (b) Side-by-side dotplots

Figure 2.34 *Who watches more TV, males or females?*

Solution Both distributions are skewed to the right and have many outliers. There appears to be an association: In this group of students, males tend to watch more television. In fact, we see in Figure 2.34(a) that the females who watch about 15 hours of TV a week are considered outliers, whereas males who watch the same amount of television are not so unusual. The minimum, first quartile, and median are relatively similar for the two genders, but the third quartile for males is much higher than the third quartile for females and the maximum for males is also much higher. While the medians are similar, the distribution for males is more highly skewed to the right, so the mean for males will be higher than the mean for females.

DATA 2.8 **Genetics and Cancer Survival**

Genetic profiles may help oncologists predict the survival time for cancer patients. In a recent study,[50] scientists looked for variations in two genes that encode proteins governing DNA repair in 103 advanced lung cancer patients. Variations of both genes were found in 13 of the patients, variations on just one of the genes were found in 64 of the patients, and 26 of the patients had neither variation. The scientists compared the survival time on chemotherapy for the patients in each of the three groups. (The study lasted 60 months.) ■

Example 2.30

In Data 2.8, we are interested in whether the genetic differences can help us predict survival time.

(a) What is the explanatory variable and what is the response variable? Indicate whether each is categorical or quantitative.

(b) Figure 2.35 shows side-by-side boxplots for the three groups. Discuss what the graph shows. What conjectures might we make about these genetic variations and survival time?

(c) Can we conclude that having one or both of the gene variations reduces survival time? Why or why not?

[50] Adapted from Gurubhagavatula, S., "XPD and XRCC1 Genetic Polymorphisms are Associated with Overall Survival (OS) in Advanced Non-Small Cell Lung Cancer (NSCLC) Patients Treated with Platinum Chemotherapy," Abstract 491, paper presented at the Annual Meeting of the American Society of Clinical Oncology, June 2003.

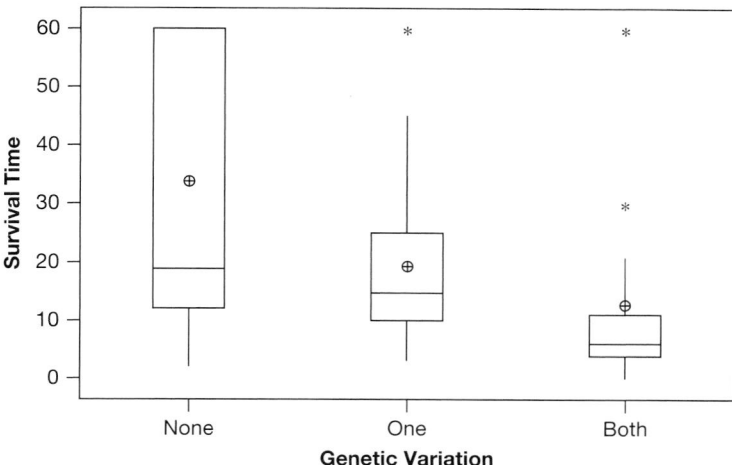

Figure 2.35 *Lung cancer survival times*

Solution   (a) The explanatory variable is which of the three groups the patient is in, which is categorical. The response variable is survival time, which is quantitative.

(b) The side-by-side boxplots show that survival time is substantially shorter for patients with variations on both genes, and tends to be longest for patients with neither variation. Based on this study, we might hypothesize that survival time for patients undergoing chemotherapy for advanced lung cancer is reduced when variations of either gene are present and more seriously shortened when both variations are present.

(c) No. Although there appears to be a strong association between these two variables, the data come from an observational study and we cannot conclude that there is a cause and effect relationship. As we learned in Chapter 1, only when we have conducted a randomized experiment can we conclude that one variable is causing another to change.

What about the added circles we see in the boxplots in Figure 2.35? Often, we use circles such as these to indicate the means in each case. Notice in Figure 2.35 that the mean survival time is highest for those with neither gene variation and lowest for those with both variations. The large outliers for those with both gene variations are so extreme that the mean is even larger than the third quartile.

Comparative Summary Statistics

Most statistical software packages will give summary statistics for a quantitative variable by categorical groups. This gives us a numerical way to compare the two variables in addition to the graphical way provided by side-by-side graphs.

Example 2.31

Use the computer output provided in Figure 2.36, based on the **StudentSurvey** dataset, to compare the mean and standard deviation for number of hours spent watching television per week broken down by gender. Find the difference in means, using appropriate notation, and interpret it in terms of television viewing habits.

Solution   The output in Figure 2.36 shows that there were 169 females who filled out the survey, and these females watched TV for a mean of 5.237 hours per week with a standard deviation of 4.1 hours per week. There were 192 males in the survey and these males

Descriptive Statistics: TV

Variable	Gender	N	Mean	SE Mean	StDev	Minimum	Q1	Median	Q3	Maximum
TV	F	169	5.237	0.315	4.100	0.000	2.500	4.000	6.000	20.000
	M	192	7.620	0.464	6.427	0.000	3.000	5.000	10.000	40.000

Figure 2.36 *Output comparing TV watching by gender*

had a mean of 7.620 hours spent watching TV per week with a standard deviation of 6.427. Both the mean and the standard deviation are larger for the males, which matches what we see in the graphs in Figure 2.34.

Using the notation $\bar{x}_m$ for the male mean and $\bar{x}_f$ for the female mean, the difference in means is

$$\bar{x}_m - \bar{x}_f = 7.620 - 5.237 = 2.383$$

In this sample, on average the males watched an additional 2.383 hours of television per week.

SECTION LEARNING GOALS

You should now have the understanding and skills to:

- Identify outliers in a dataset based on the *IQR* method
- Use a boxplot to describe data for a single quantitative variable
- Use a side-by-side graph to visualize a relationship between quantitative and categorical variables
- Examine a relationship between quantitative and categorical variables using comparative summary statistics

---

Exercises for Section 2.4

SKILL BUILDER 1

In Exercises 2.131 and 2.132, match the five number summaries with the boxplots.

2.131 Match each five number summary with one of the boxplots in Figure 2.37.

(a) (2, 12, 14, 17, 25)

(b) (5, 15, 18, 20, 23)

(c) (10, 12, 13, 18, 25)

(d) (12, 12, 15, 20, 24)

2.132 Match each five number summary with one of the boxplots in Figure 2.38.

(a) (1, 18, 20, 22, 25)

(b) (1, 10, 15, 20, 25)

(c) (1, 3, 5, 10, 25)

(d) (1, 1, 10, 15, 25)

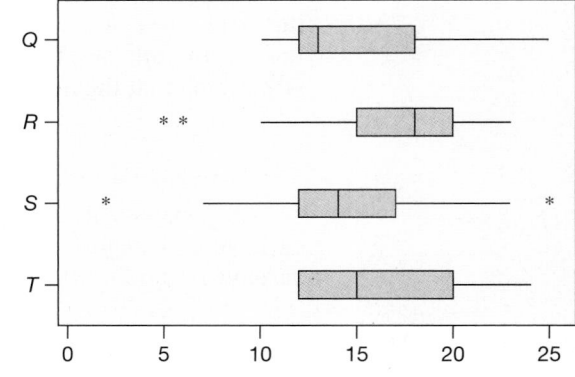

Figure 2.37 *Match five number summaries in Exercise 2.131*

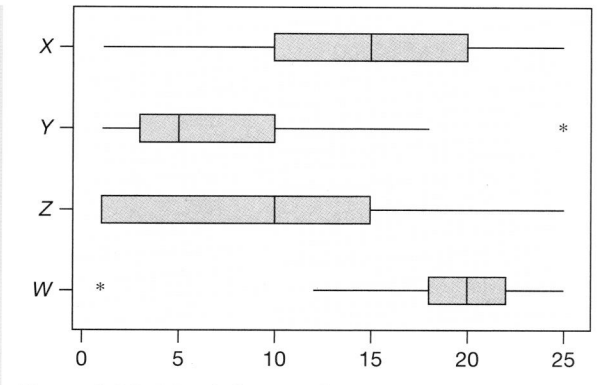

Figure 2.38 *Match five number summaries in Exercise 2.132*

SKILL BUILDER 2

Exercises 2.133 to 2.136 show a boxplot for a set of data. In each case:

(a) Indicate whether the distribution of the data appears to be skewed to the left, skewed to the right, approximately symmetric, or none of these.

(b) Are there any outliers? If so, how many and are they high outliers or low outliers?

(c) Give a rough approximation for the mean of the dataset.

2.133

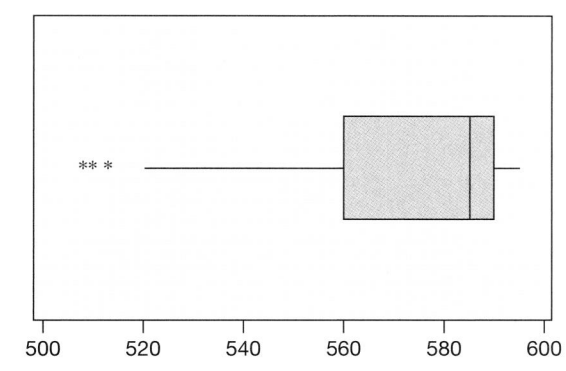

2.134

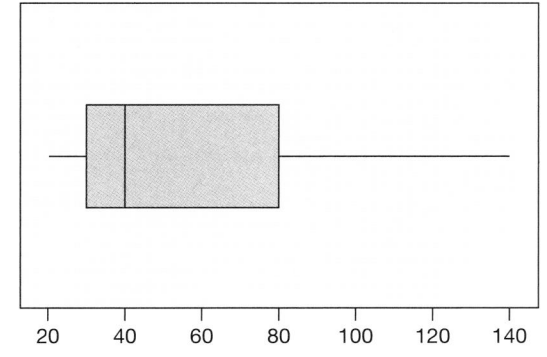

2.135

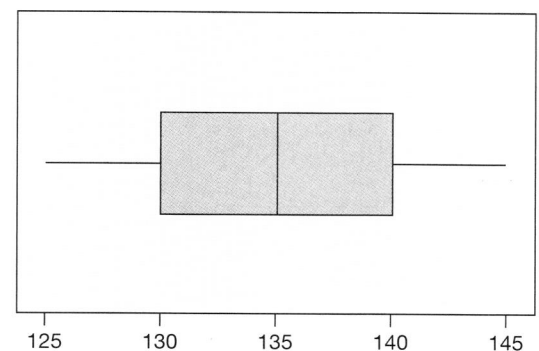

2.136

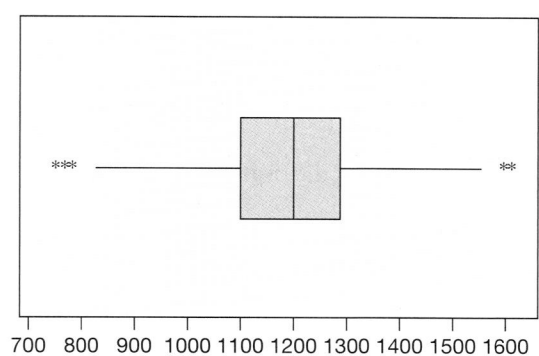

SKILL BUILDER 3

Exercises 2.137 to 2.140 each describe a sample. The information given includes the five number summary, the sample size, and the largest and smallest data values in the tails of the distribution. In each case:

(a) Clearly identify any outliers, using the IQR method.

(b) Draw a boxplot.

2.137 Five number summary: (210, 260, 270, 300, 320); $n = 500$.
Tails: 210, 215, 217, 221, 225, …, 318, 319, 319, 319, 320, 320.

2.138 Five number summary: (15, 42, 52, 56, 71); $n = 120$.
Tails: 15, 20, 28, 30, 31, …, 64, 65, 65, 66, 71.

2.139 Five number summary: (42, 72, 78, 80, 99); $n = 120$.
Tails: 42, 63, 65, 67, 68, …, 88, 89, 95, 96, 99.

2.140 Five number summary: (5, 10, 12, 16, 30); $n = 40$.
Tails: 5, 5, 6, 6, 6, …, 22, 22, 23, 28, 30.

2.141 Literacy Rate Figure 2.39 gives a boxplot showing the literacy rate of the countries of the world.[51]

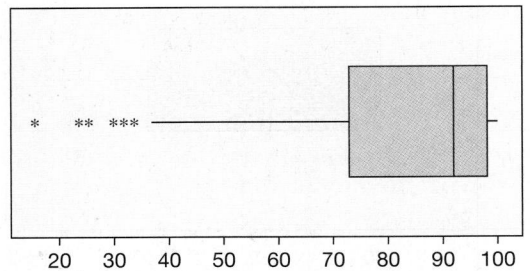

Figure 2.39 *Literacy rate for countries of the world*

(a) Describe the shape of the distribution.

(b) The middle half of all literacy rates are between approximately what two values?

(c) Approximate the five-number summary from the boxplot.

(d) Some countries did not have data on literacy rate available from the World Bank. Do you suspect the true median is higher or lower than that shown on the boxplot from the available data? Why?

2.142 Young Blood Helps Old Brains Exercise 2.69 on page 75 introduces a study in which old mice were randomly assigned to receive transfusions of blood from either young mice or old mice. Researchers then measured, among other things, the number of minutes each mouse was able to run on a treadmill. The results are given in the side-by-side boxplots in Figure 2.40.

(a) Estimate the median runtime for the mice receiving old blood.

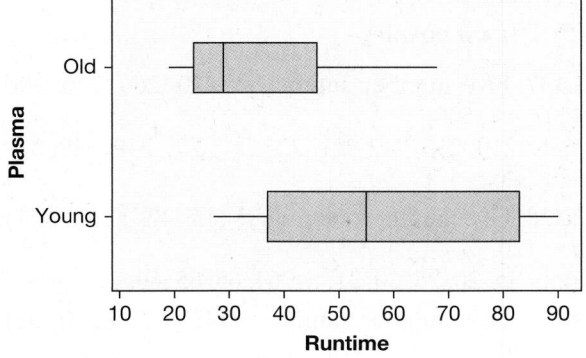

Figure 2.40 *Minutes on a treadmill after receiving blood from old or young mice*

[51]Data from Education Statistics (World Bank), May 2015. Most recent data available obtained from *http://www.knoema.com.*

(b) Do we expect the mean runtime for the mice receiving old blood to be larger than, smaller than, or about the same as the median for these mice?

(c) Which group of mice, those receiving old blood or those receiving young blood, appear to be able to run for more minutes?

(d) Does there appear to be an association between runtime and whether or not the mouse received young blood or old blood?

(e) Are there any outliers in either group? If so, which one(s)?

2.143 Football and Brain Size A recent study[52] examined the relationship of football and concussions on hippocampus volume in the brain. The study included three groups with $n = 25$ in each group: heathy controls who had never played football, football players with no history of concussions, and football players with a history of concussions. Figure 2.41 shows side-by-side boxplots for total hippocampus volume, in μL, for the three groups.

(a) Is the explanatory variable categorical or quantitative? Is the response variable categorical or quantitative?

(b) Which group has the largest hippocampal volume? Which group has the smallest?

(c) Are there any outliers in any of the groups? If so, which one(s)?

(d) Estimate the third quartile for the football players without a history of concussion.

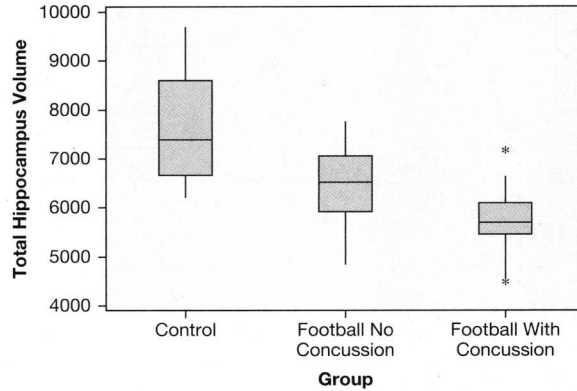

Figure 2.41 *Relationship of football and concussions on brain hippocampus size*

[52]Singh, R., et al., "Relationship of Collegiate Football Experience and Concussion with Hippocampal Volume and Cognitive Outcomes," *JAMA*, 311(18), 2014. Data values are estimated from information in the paper.

(e) Does there appear to be an association between football experience and hippocampus size?

(f) Can we conclude that playing football decreases hippocampus volume? Why or why not?

2.144 Ronda Rousey Fight Times Perhaps the most popular fighter since the turn of the decade, Ronda Rousey is famous for defeating her opponents quickly. The five number summary for the times of her first 12 UFC (Ultimate Fighting Championship) fights, in seconds, is (14, 25, 44, 64, 289).

(a) Only three of her fights have lasted more than a minute, at 289, 267, and 66 seconds, respectively. Use the *IQR* method to see which, if any, of these values are high outliers.

(b) Are there any low outliers in these data, according to the *IQR* method?

(c) Draw the boxplot for Ronda Rousey's fight times.

(d) Based on the boxplot or five number summary, would we expect Ronda's mean fight time to be greater than or less than her median?

INVESTIGATING HOLLYWOOD MOVIES

In Exercises 2.145 to 2.148, we use data from **HollywoodMovies** introduced in Data 2.7 on page 95. The dataset includes information on all movies to come out of Hollywood between 2007 and 2013.

2.145 How Profitable Are Hollywood Movies? One of the variables in the **HollywoodMovies** dataset is *Profitability*, which measures the proportion of the budget recovered in revenue from the movie. A profitability less than 100 means the movie did not make enough money to cover the budget, while a profitability greater than 100 means it made a profit. A boxplot of the profitability ratings of all 896 movies is shown in Figure 2.42. (The largest outlier is the movie *The Devil Inside*, which had a very small budget and relatively high gross revenue.)

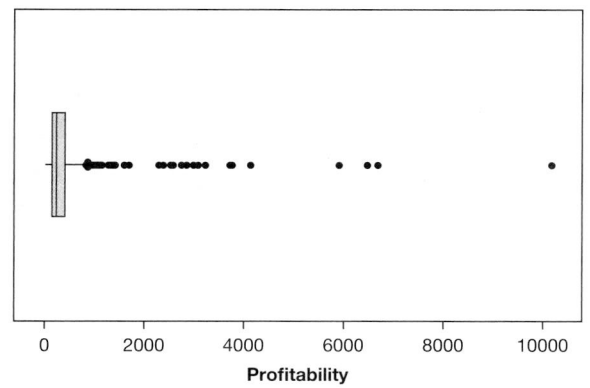

Figure 2.42 *Profitability of Hollywood movies*

(a) Describe the shape of the distribution.

(b) Estimate the range.

(c) Estimate the median. Interpret it in terms of profitability of movies.

(d) Do we expect the mean to be greater than or less than the median?

2.146 Audience Scores on Rotten Tomatoes The variable *AudienceScore* in the dataset **HollywoodMovies** gives audience scores (on a scale from 1 to 100) from the Rotten Tomatoes website. The five number summary of these scores is (19, 49, 61, 74, 96). Are there any outliers in these scores, according to the *IQR* method? How bad would an average audience score rating have to be on Rotten Tomatoes to qualify as a low outlier?

2.147 Do Movie Budgets Differ Based on the Genre of the Movie? The dataset **HollywoodMovies** includes a quantitative variable on the *Budget* of the movie, in millions of dollars, as well as a categorical variable classifying each movie by its *Genre*. Figure 2.43 shows side-by-side boxplots investigating a relationship between these two variables for movies made in 2011, using four of the possible genres.

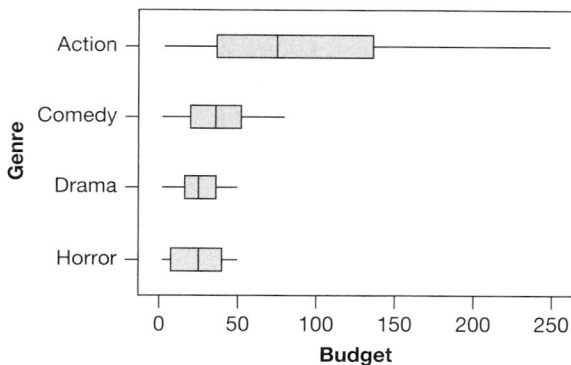

Figure 2.43 *Movie budgets (in millions of dollars) based on genre*

(a) Which genre appears to have the largest budgets? Which appears to have the smallest?

(b) Which genre has the biggest spread in its budgets? Which has the smallest spread?

(c) Does there appear to be an association between genre of a movie and size of the budget? Explain.

2.148 Do Audience Ratings Differ Based on the Genre of the Movie? The dataset **HollywoodMovies** includes a quantitative variable on the *AudienceScore* of the movie as well as a categorical variable

Variable	Genre	N	Mean	StDev	Minimum	Q1	Median	Q3	Maximum
Audience Score	Action	32	58.63	18.39	32.00	44.50	51.00	78.00	93.00
	Comedy	27	59.11	15.68	31.00	48.00	58.00	71.00	93.00
	Drama	21	72.10	14.55	46.00	59.00	72.00	84.50	91.00
	Horror	17	48.65	15.88	25.00	34.00	52.00	60.50	78.00

classifying each movie by its *Genre*. The computer output above gives summary statistics for audience ratings based on genre for a sample of the movies, using four of the possible genres.

(a) Which genre has the highest mean audience score? The lowest mean audience score?

(b) Which genre has the highest median score? The lowest median score?

(c) In which genre is the lowest score, and what is that score? In which genre is the highest score, and what is that score?

(d) Which genre has the largest number of movies in that category?

(e) Calculate the difference in mean score between comedies and horror movies, and give notation with your answer, using $\bar{x}_C$ for the mean comedy score and $\bar{x}_H$ for the mean horror score.

2.149 Physical Activity by Region of the Country in the US The variables in **USStates** include the percent of the people in each state who say they engage in at least 150 minutes of physical activity each week as well as the region of the country in which the state is found (Midwest, Northeast, South, or West). One of these variables is quantitative and one is categorical, and Figure 2.44 allows us to visualize the relationship between the two variables.

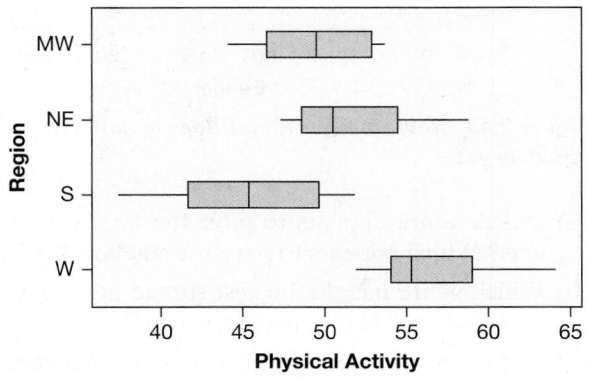

Figure 2.44 *Physical activity in US states by region of the country*

(a) Which region shows the lowest level of physical activity? Which region shows the highest?

(b) Are there any outliers? If so, where?

(c) Does there appear to be an association between amount of physical activity and region of the country?

2.150 Hits in Baseball Major League Baseball is split into two leagues, the National League (NL) and the American League (AL). The main difference between the two leagues is that pitchers take at bats in the National League but not in the American League. Are total team hits different between the two leagues? Figure 2.45 shows side-by-side boxplots for the two leagues. The data are stored in **BaseballHits**.

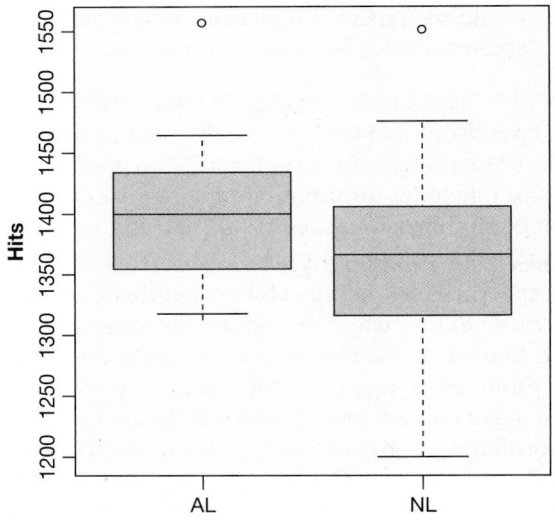

Figure 2.45 *Side by side boxplots for hits by league*

(a) Estimate the median number of hits for each league, and estimate the difference in median hits between the two leagues. Which league appears to get more hits?

(b) What is the other obvious difference (apparent in Figure 2.45) between the two leagues?

2.151 Concentration of Retinol by Vitamin Use Figure 2.46 displays the relationship between vitamin use and the concentration of retinol (a micronutrient) in the blood for a sample of $n = 315$ individuals. (The full dataset, with more variables, is available in **NutritionStudy**.) Does there seem to be an association between these two variables?

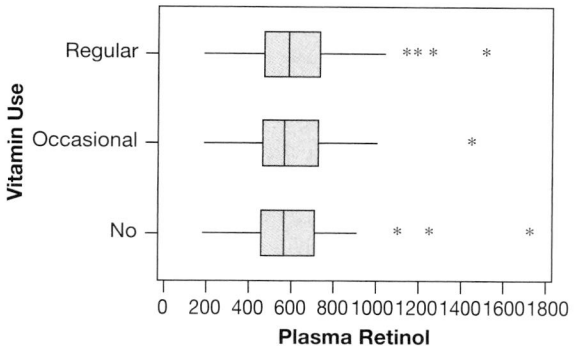

Figure 2.46 *Concentration of retinol by vitamin use*

2.152 How Do Honeybees Communicate Quality? When honeybees are looking for a new home, they send out scouts to explore options. When a scout returns, she does a "waggle dance" with multiple circuit repetitions to tell the swarm about the option she found.[53] The bees then decide between the options and pick the best one. Scientists wanted to find out how honeybees decide which is the best option, so they took a swarm of honeybees to an island with only two possible options for new homes: one of very high honeybee quality and one of low quality. They then kept track of the scouts who visited each option and counted the number of waggle dance circuits each scout bee did when describing the option.[54] Comparative dotplots of number of circuits performed by the 41 bees that visited the high-quality option and the 37 bees that visited the low-quality option are shown in Figure 2.47. The data are available in **HoneybeeCircuits**.

(a) Does there appear to be an association between number of circuits in the waggle dance and the quality of the site? If so, describe the association.

(b) The five number summary for the number of circuits for those describing the high-quality site is $(0, 7.5, 80, 122.5, 440)$, while the five number

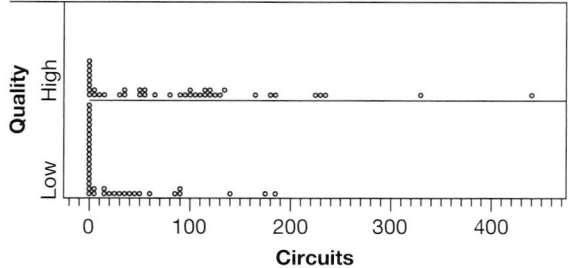

Figure 2.47 *Number of circuits completed in the honeybee waggle dance*

[53]Check out a honeybee waggle dance on YouTube!
[54]Seeley, T., *Honeybee Democracy*, Princeton University Press, Princeton, NJ, 2010, p. 128.

summary for those describing the low-quality site is $(0, 0, 0, 42.5, 185)$. Use the *IQR* method to identify any outliers in either group. Justify your answer.

(c) The mean for the high-quality group is $\bar{x}_H = 90.5$ with a standard deviation of 94.6, while the mean for the low-quality group is $\bar{x}_L = 30.0$ with a standard deviation of 49.4. What is the difference in means, $\bar{x}_H - \bar{x}_L$?

(d) Find the *z*-score for the largest value in the high-quality group and the *z*-score for the largest value in the low-quality group. Which is larger relative to its group?

(e) Is it appropriate to use the 95% rule with either set of data?

2.153 Split the Bill? When the time comes for a group of people eating together at a restaurant to pay their bill, sometimes they might agree to split the costs equally and other times will pay individually. If this decision was made in advance, would it affect what they order? Some reseachers[55] conducted an experiment where 48 Israeli college students were randomly assigned to eat in groups of six (three females and three males). Before ordering, half of the groups were told they would be responsible for paying for meals individually and half were told to split the cost equally among the six participants. The number of items ordered (*Items*) and cost of each person's order (*Cost*) in new Israeli shekels (ILS) was recorded and appears in the dataset **SplitBill**.

(a) Provide a numerical summary to compare costs of orders between subjects paying individual bills and those splitting the bill.

(b) Create and comment on a graphical display to compare costs between these two groups.

2.154 Restaurant Bill by Sex Exercise 2.153 describes a study on the cost of meals when groups pay individually or split the bill. One of the variables in **SplitBill** also records the *Sex* of each subject.

(a) Provide a numerical summary to compare costs of orders between females and males.

(b) Create and comment on a graphical display to compare costs by *Sex*.

2.155 Better Traffic Flow Have you ever driven along a street where it seems that every traffic light is red when you get there? Some engineers in Dresden, Germany, are looking at ways to improve traffic flow by enabling traffic lights to communicate information about traffic flow with nearby traffic

[55]Gneezy, U., Haruvy, E., and Yafe, H. "The Inefficiency of Splitting the Bill," *The Economic Journal*, 2004; 114, 265–280.

lights. The data in **TrafficFlow** show results of one experiment[56] that simulated buses moving along a street and recorded the delay time (in seconds) for both a fixed time and a flexible system of lights. The simulation was repeated under both conditions for a total of 24 trials.

(a) What is the explanatory variable? What is the response variable? Is each categorical or quantitative?

(b) Use technology to find the mean and the standard deviation for the delay times under each of the two conditions (*Timed* and *Flexible*). Does the flexible system seem to reduce delay time?

(c) The data in **TrafficFlow** are paired since we have two values, timed and flexible, for each simulation run. For paired data we generally compute the *difference* for each pair. In this example, the dataset includes a variable called *Difference* that stores the difference *Timed − Flexible* for each simulation run. Use technology to find the mean and standard deviation of these differences.

(d) Use technology to draw a boxplot of the differences. Are there any outliers?

DRAW THESE SIDE-BY-SIDE BOXPLOTS

Exercises 2.156 and 2.157 examine issues of location and spread for boxplots. In each case, draw side-by-side boxplots of the datasets on the same scale. There are many possible answers.

[56]Lammer, S. and Helbing, D., "Self-Stabilizing Decentralized Signal Control of Realistic, Saturated Network Traffic," Santa Fe Institute, Santa Fe, NM, working paper No. 10-09-019, September 2010.

2.156 One dataset has median 25, interquartile range 20, and range 30. The other dataset has median 75, interquartile range 20, and range 30.

2.157 One dataset has median 50, interquartile range 20, and range 40. A second dataset has median 50, interquartile range 50, and range 100. A third dataset has median 50, interquartile range 50, and range 60.

2.158   Examine a Relationship in StudentSurvey From the **StudentSurvey** dataset, select any categorical variable and select any quantitative variable. Use technology to create side-by-side boxplots to examine the relationship between the variables. State which two variables you are using and describe what you see in the boxplots. In addition, use technology to compute comparative summary statistics and compare means and standard deviations for the different groups.

2.159   Examine a Relationship in USStates Exercise 2.149 examines the relationship between region of the country and level of physical activity of the population of US states. From the **USStates** dataset, examine a different relationship between a categorical variable and a quantitative variable. Select one of each type of variable and use technology to create side-by-side boxplots to examine the relationship between the variables. State which two variables you are using and describe what you see in the boxplots. In addition, use technology to compute comparative summary statistics and compare means and standard deviations for the different groups.

2.5 TWO QUANTITATIVE VARIABLES: SCATTERPLOT AND CORRELATION

In Section 2.1, we look at relationships between two categorical variables, and in Section 2.4, we investigate relationships between a categorical and a quantitative variable. In this section, we look at relationships between two quantitative variables.

DATA 2.9

Presidential Approval Ratings and Re-election

When a US president runs for re-election, how strong is the relationship between the president's approval rating and the outcome of the election? Table 2.26 includes all the presidential elections since 1940 in which an incumbent was running and shows the presidential approval rating at the time of the election and the margin of victory or defeat for the president in the election.[57] The data are available in **ElectionMargin**.   ∎

[57]Data obtained from *http://www.fivethirtyeight.com* and *http://www.realclearpolitics.com*.

Table 2.26 *Presidential approval rating and margin of victory or defeat*

Year	Candidate	Approval	Margin	Result
1940	Roosevelt	62	10.0	Won
1948	Truman	50	4.5	Won
1956	Eisenhower	70	15.4	Won
1964	Johnson	67	22.6	Won
1972	Nixon	57	23.2	Won
1976	Ford	48	−2.1	Lost
1980	Carter	31	−9.7	Lost
1984	Reagan	57	18.2	Won
1992	G. H. W. Bush	39	−5.5	Lost
1996	Clinton	55	8.5	Won
2004	G. W. Bush	49	2.4	Won
2012	Obama	50	3.9	Won

Example 2.32

(a) What was the highest approval rating for any of the losing presidents? What was the lowest approval rating for any of the winning presidents? Make a conjecture about the approval rating needed by a sitting president in order to win re-election.

(b) Approval rating and margin of victory are both quantitative variables. Does there seem to be an association between the two variables?

Solution

(a) Three presidents lost, and the highest approval rating among them is 48%. Nine presidents won, and the lowest approval rating among them is 49%. It appears that a president needs an approval rating of 49% or higher to win re-election.

(b) In general, it appears that a higher approval rating corresponds to a larger margin of victory, although the association is not perfect.

Visualizing a Relationship between Two Quantitative Variables: Scatterplots

The standard way to display the relationship between two quantitative variables is to extend the notion of a dotplot for a single quantitative variable to a two-dimensional graph known as a *scatterplot*. To examine a relationship between two quantitative variables, we have *paired* data, where each data case has values for both of the quantitative variables.

Scatterplot

A **scatterplot** is a graph of the relationship between two quantitative variables.

A scatterplot includes a pair of axes with appropriate numerical scales, one for each variable. The paired data for each case are plotted as a point on the scatterplot. If there are explanatory and response variables, we put the explanatory variable on the horizontal axis and the response variable on the vertical axis.

Example 2.33

Solution

Draw a scatterplot for the data on approval rating and margin of victory in Table 2.26.

We believe approval ratings may help us predict the margin of victory, so the explanatory variable is approval rating and the response variable is margin of victory. We put approval rating on the horizontal axis and margin of victory on the vertical axis. The 12 data pairs are plotted as 12 points in the scatterplot of Figure 2.48. The point corresponding to Roosevelt in 1940, with an approval rating of 62% and a margin of victory of 10 points, is indicated. We notice from the upward trend of the points that the margin of victory does tend to increase as the approval rating increases.

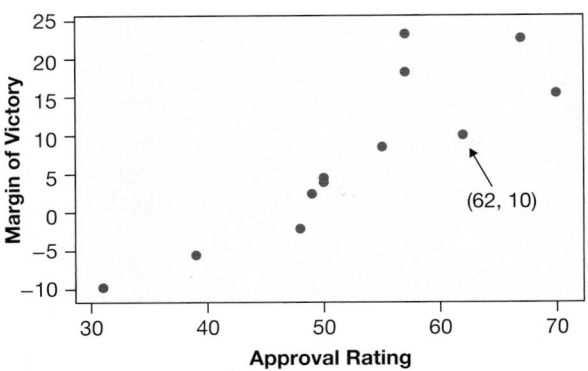

Figure 2.48 *Approval rating and margin of victory*

Interpreting a Scatterplot

When looking at a scatterplot, we often address the following questions:

- Do the points form a clear trend with a particular direction, are they more scattered about a general trend, or is there no obvious pattern?
- If there is a trend, is it generally upward or generally downward as we look from left to right? A general upward trend is called a *positive* association while a general downward trend is called a *negative* association.
- If there is a trend, does it seem to follow a straight line, which we call a *linear association*, or some other curve or pattern?
- Are there any outlier points that are clearly distinct from a general pattern in the data?

For the presidential re-election data in Figure 2.48, we see a positive association since there is an upward trend in margin of victory as approval increases. While the points certainly do not all fall exactly on a straight line, we can imagine drawing a line to match the general trend of the data. There is a general linear trend, and it is a relatively strong association.

Example 2.34

Scatterplots Using Data from Florida Lakes

Four scatterplots are shown in Figure 2.49 using data from the **FloridaLakes** dataset, introduced in Data 2.4 on page 71. For each pair of variables, discuss the information contained in the scatterplot. If there appears to be a positive or negative association, discuss what that means in the specific context.

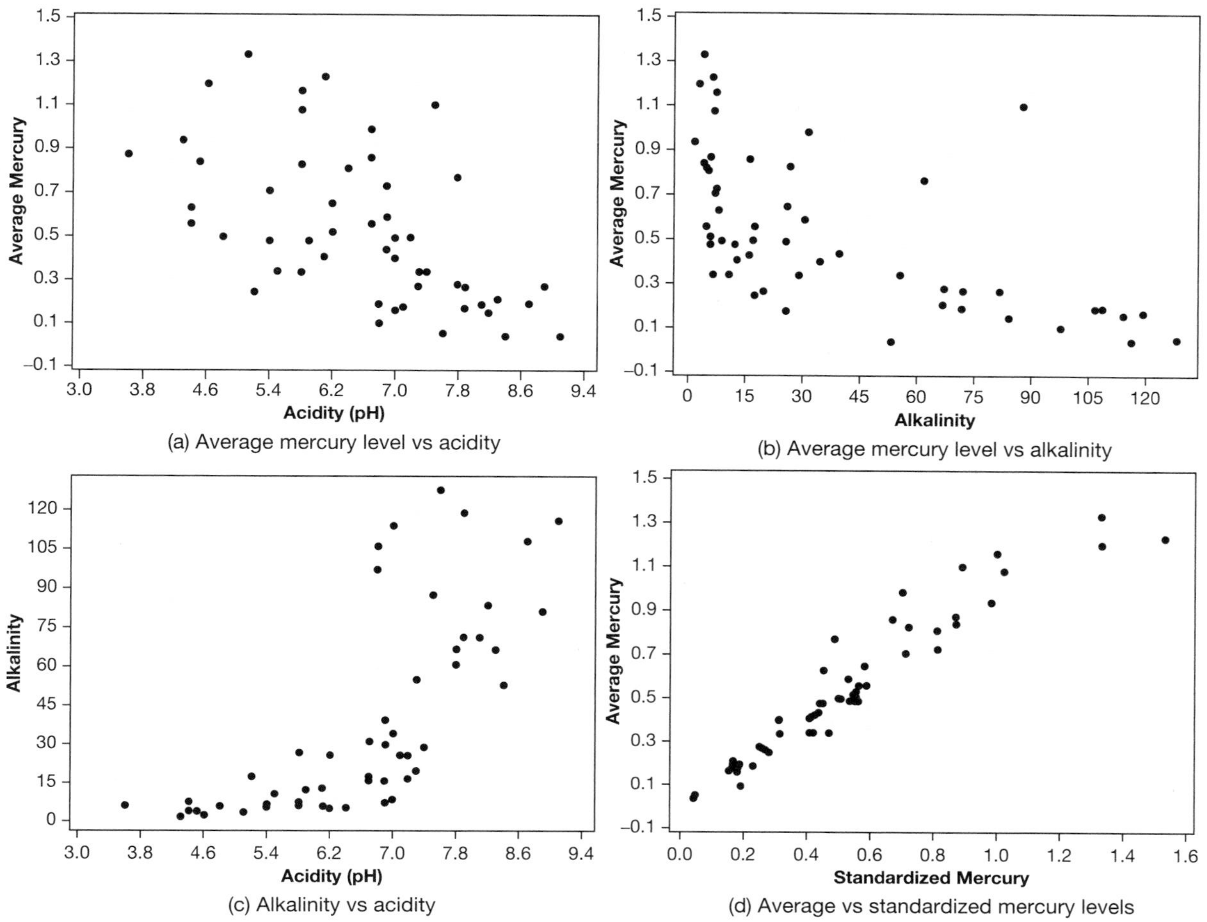

Figure 2.49 *Scatterplots of data from Florida lakes*

Solution

(a) Acidity appears to have a negative linear association with average mercury level, but not a strong one as the points are scattered widely around any straight line. Since the association is negative, larger values of acidity tend to be associated with smaller levels of mercury.

(b) Alkalinity also is negatively associated with average mercury level, with a slightly stronger association along a more curved trend. One lake with a high average mercury level around 1.1 ppm also has a high alkalinity at almost 90 mg/L and is clearly away from the general trend of the data. Note that neither of the values for this lake would be considered outliers for the individual variables, but the data pair stands out in the scatterplot so it is considered an outlier. Since the association is negative, larger values of alkalinity tend to be associated with smaller levels of mercury.

(c) There is a positive association since the acidity increases with alkalinity along a curved pattern. Since the association is positive, larger values of acidity tend to be associated with larger values of alkalinity.

(d) The average mercury levels show a strong positive association with the standardized mercury levels that fit fairly closely to a straight line. Since the association is positive, larger levels of standardized mercury tend to be associated with larger levels of mercury.

Summarizing a Relationship between Two Quantitative Variables: Correlation

Just as the mean or median summarizes the center and the standard deviation or interquartile range measures the spread of the distribution for a single quantitative variable, we need a numerical statistic to measure the strength and direction of association between two quantitative variables. One such statistic is the *correlation*.

> **Correlation**
>
> The **correlation** is a measure of the strength and direction of linear association between two quantitative variables.

As with previous summary statistics, we generally use technology to compute a correlation. Also as with some of our previous summary statistics, the notation we use for correlation depends on whether we are referring to a sample or a population.

> **Notation for the Correlation**
>
> The correlation between two quantitative variables of a *sample* is denoted r.
>
> The correlation between two quantitative variables of a *population* is denoted ρ, which is the Greek letter "rho".

Properties of the Correlation

Table 2.27 shows correlations for each of the pairs of variables that have been displayed in scatterplots earlier in this section.

Table 2.27 *Compare these correlations to their scatterplots*

Variable 1	Variable 2	Correlation
Margin of victory	Approval rating	0.86
Average mercury	Acidity	−0.58
Average mercury	Alkalinity	−0.59
Alkalinity	Acidity	0.72
Average mercury	Standardized mercury	0.96

Notice that all the correlations in the table are between −1 and +1. We see that a positive correlation corresponds to a positive association and a negative correlation corresponds to a negative association. Notice also that correlation values closer to 1 or −1 correspond to stronger linear associations. We make these observations more precise in the following list of properties.

> **Properties of the Correlation**
>
> The sample correlation r has the following properties:
> - Correlation is always between −1 and 1, inclusive: $-1 \leq r \leq 1$.
> - The sign of r (positive or negative) indicates the direction of association.
> - Values of r close to +1 or −1 show a strong linear relationship, while values of r close to 0 show no linear relationship.

- The correlation r has no units and is independent of the scale of either variable.
- The correlation is symmetric: The correlation between variables x and y is the same as the correlation between y and x.

The population correlation ρ also satisfies these properties.

© Dumrong Khajaroen/iStockphoto

Is the chirp rate of crickets associated with the temperature?

DATA 2.10   **Cricket Chirps and Temperature**

Common folk wisdom claims that one can determine the temperature on a summer evening by counting how fast the crickets are chirping. Is there really an association between chirp rate and temperature? The data in Table 2.28 were collected by E. A. Bessey and C. A. Bessey,[58] who measured chirp rates for crickets and temperatures during the summer of 1898. The data are also stored in **CricketChirps**. ■

Table 2.28 *Cricket chirps and temperature*

Temperature (° F)	54.5	59.5	63.5	67.5	72.0	78.5	83.0
Chirps (per minute)	81	97	103	123	150	182	195

Example 2.35

A scatterplot of the data in Table 2.28 is given in Figure 2.50.

(a) Use the scatterplot to estimate the correlation between chirp rate and temperature. Explain your reasoning.

(b) Use technology to find the correlation and use correct notation.

(c) Are chirp rate and temperature associated?

Solution   (a) Figure 2.50 shows a very strong positive linear trend in the data, so we expect the correlation to be close to +1. Since the points do not all lie exactly on a line, the correlation will be slightly less than 1.

[58]Bessey, E. A. and Bessey, C. A., "Further Notes on Thermometer Crickets," *American Naturalist*, 1898; 32: 263–264.

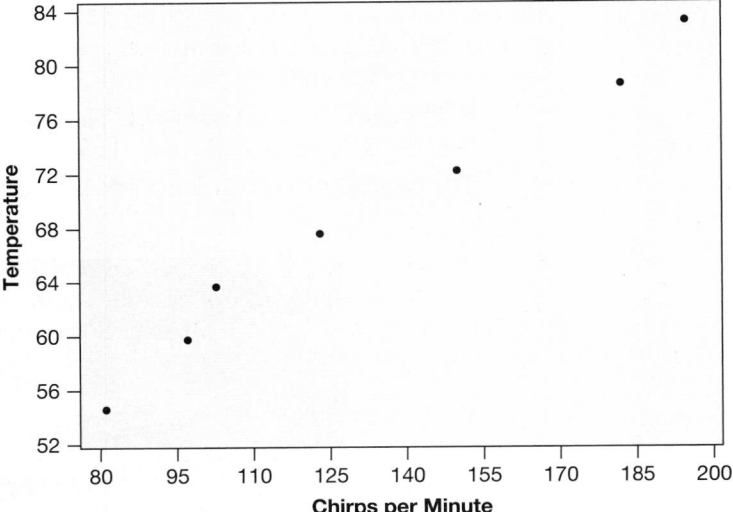

Figure 2.50 *Scatterplot of chirp rate and temperature*

(b) We use the notation r for this sample correlation. Using technology, we see that $r = 0.99$, matching the very strong positive linear relationship we see in the scatterplot.

(c) Yes, cricket chirp rates and air temperature are strongly associated!

Correlation Cautions

Example 2.36

Figure 2.51 shows the estimated average life expectancy[59] (in years) for a sample of 40 countries against the average amount of fat[60] (measured in grams per capita per day) in the food supply for each country. The scatterplot shows a clear positive

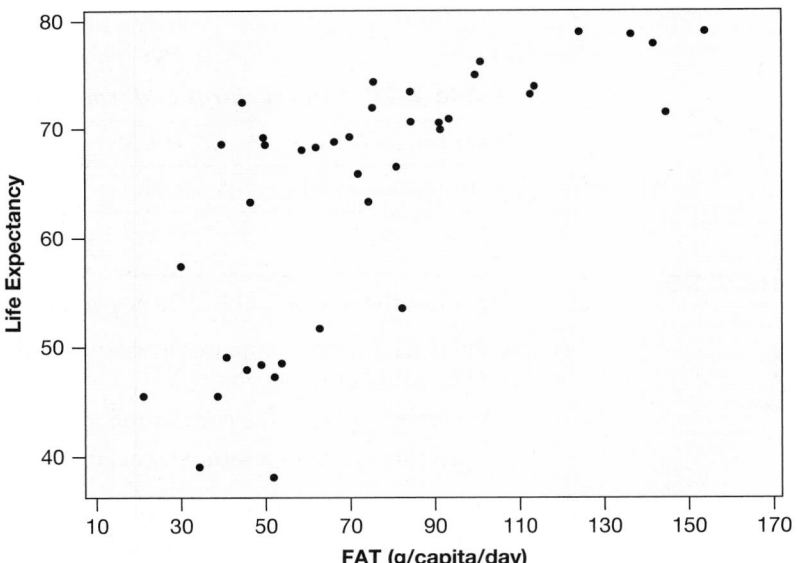

Figure 2.51 *Life expectancy vs grams of fat in daily diet for 40 countries*

[59]United Nations Development Program, *Human Development Report 2003.*
[60]Food and Agriculture Organization of the United Nations.

association $(r = 0.70)$ between these two variables. The countries with short life expectancies all have below-average fat consumption, while the countries consuming more than 100 grams of fat on average all have life expectancies over 70 years. Does this mean that we should eat more fat in order to live longer?

Solution   No! Just because there is a strong association between these two variables, it would be inappropriate to conclude that changing one of them (for example, increasing fat in the diet) would *cause* a corresponding change in the other variable (lifetime). An observational study was used to collect these data, so we cannot conclude that there is a causal relationship. One likely confounding variable is the wealth of the country, which is associated with both life expectancy and fat consumption.

A strong correlation does not necessarily imply a causal association! As we saw in Chapter 1, we need to be aware of confounding variables and we need to pay attention to whether the data come from an experiment or an observational study.

 Correlation Caution #1

A strong positive or negative correlation does not (necessarily) imply a cause and effect relationship between the two variables.

Example 2.37

Core body temperature for an individual person tends to fluctuate during the day according to a regular circadian rhythm. Suppose that body temperatures for an adult woman are recorded every hour of the day, starting at 6 am. The results are shown in Figure 2.52. Does there appear to be an association between the time of day and body temperature? Estimate the correlation between the hour of the day and the woman's body temperature.

Solution   There is a regular pattern with temperatures rising in the morning, staying fairly constant throughout the day, and then falling at night, so these variables are associated. Despite this association, the correlation between these two variables will be

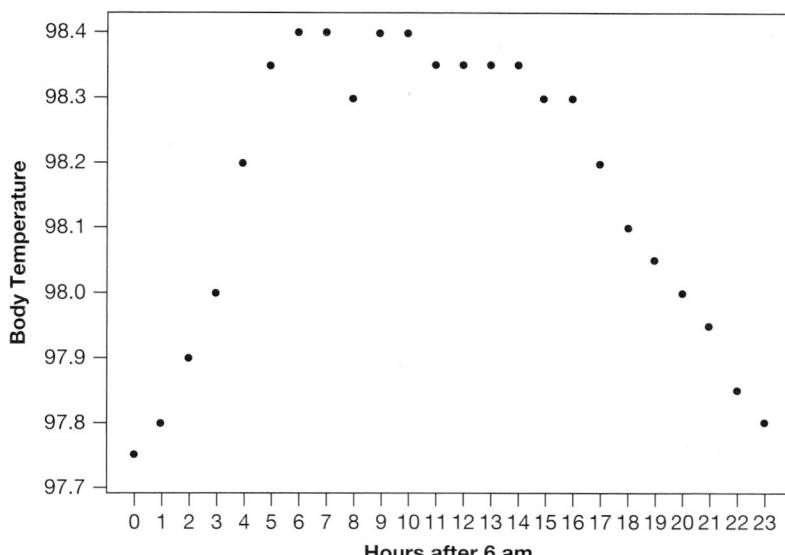

Figure 2.52 *Hourly body temperatures*

near zero. (For Figure 2.52 the actual correlation is $r = -0.08$.) The beginning hours appear to have a positive association but the trend is negative for the later hours. Remember that correlation measures the strength of a *linear* relationship between two variables.

Correlation Caution #2

A correlation near zero does not (necessarily) mean that the two variables are not associated, since the correlation measures only the strength of a *linear* relationship.

DATA 2.11 **Effects of Diet on Levels of Retinol and Beta-carotene**

In a study on the association between diet and levels of retinol and beta-carotene in the blood stream, researchers recorded a variety of dietary and demographic variables for the subjects. Variables include alcohol consumption, average daily calories, age, gender, multivitamin use, fat grams per day, fiber grams per day, smoking habits, etc. The data are available in **NutritionStudy**. ■

Example 2.38

Figure 2.53 shows the alcohol consumption (drinks per week) and average daily caloric intake for 91 subjects who are at least 60 years old, from the data in **NutritionStudy**. Notice the distinct outlier who claims to down 203 drinks per week as part of a 6662 calorie diet! This is almost certainly an incorrect observation. The second plot in Figure 2.53 shows these same data with the outlier omitted. How do you think the correlation between calories and alcohol consumption changes when the outlier is deleted?

Solution

The correlation between alcohol consumption and daily calories is $r = 0.72$ with the outlier present, but only $r = 0.15$ when that data point is omitted. What initially might look like a strong association between alcohol consumption and daily calories turns out to be much weaker when the extreme outlier is removed.

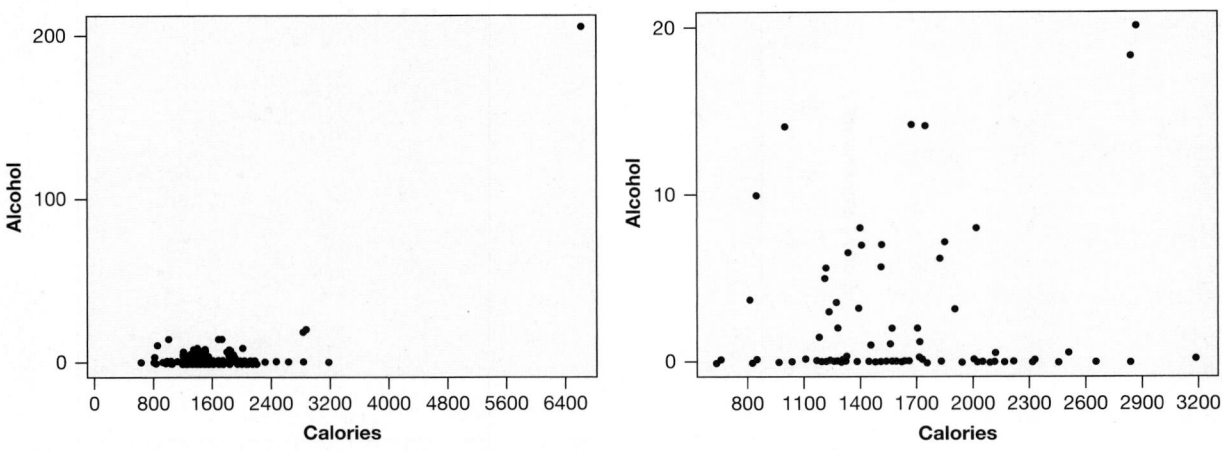

Figure 2.53 *Alcohol consumption vs calories (with and without an outlier)*

Correlation Caution #3

Correlation can be heavily influenced by outliers. Always plot your data!

A Formula for Correlation

We routinely rely on technology to compute correlations, but you may be wondering how such computations are done. While computing a correlation "by hand" is tedious and often not very informative, a formula, such as the one shown below, can be helpful in understanding how the correlation works:

$$r = \frac{1}{n-1} \sum \left(\frac{x - \bar{x}}{s_x} \right) \left(\frac{y - \bar{y}}{s_y} \right)$$

Essentially this involves converting all values for both variables to z-scores, which puts the correlation on a fixed -1 to $+1$ scale and makes it independent of the scale of measurement. For a positive association, large values for x tend to occur with large values of y (both z-scores are positive) and small values (with negative z-scores) tend to occur together. In either case the products are positive, which leads to a positive sum. For a negative association, the z-scores tend to have opposite signs (small x with large y and vice versa) so the products tend to be negative.

SECTION LEARNING GOALS

You should now have the understanding and skills to:

- Describe an association displayed in a scatterplot
- Explain what a positive or negative association means between two variables
- Interpret a correlation
- Use technology to find a correlation
- Recognize that correlation does not imply cause and effect
- Recognize that you should always plot your data in addition to interpreting numerical summaries

Exercises for Section **2.5**

SKILL BUILDER 1
Match the scatterplots in Figure 2.54 with the correlation values in Exercises 2.160 to 2.163.

2.160 $r = -1$

2.161 $r = 0$

2.162 $r = 0.8$

2.163 $r = 1$

SKILL BUILDER 2
Match the scatterplots in Figure 2.55 with the correlation values in Exercises 2.164 to 2.167.

2.164 $r = 0.09$

2.165 $r = -0.38$

2.166 $r = 0.89$

2.167 $r = -0.81$

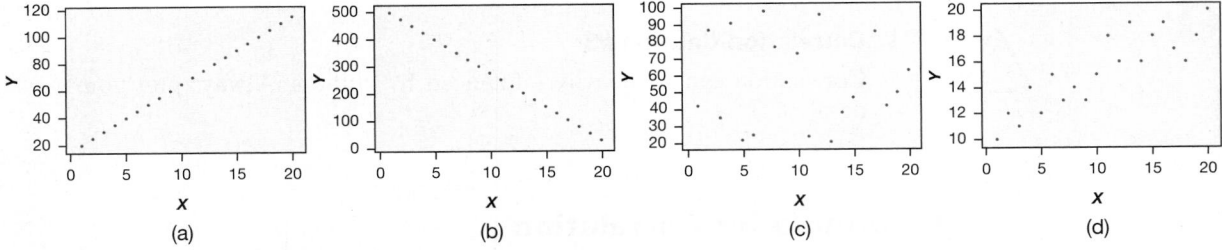

Figure 2.54 *Match the correlations to the scatterplots*

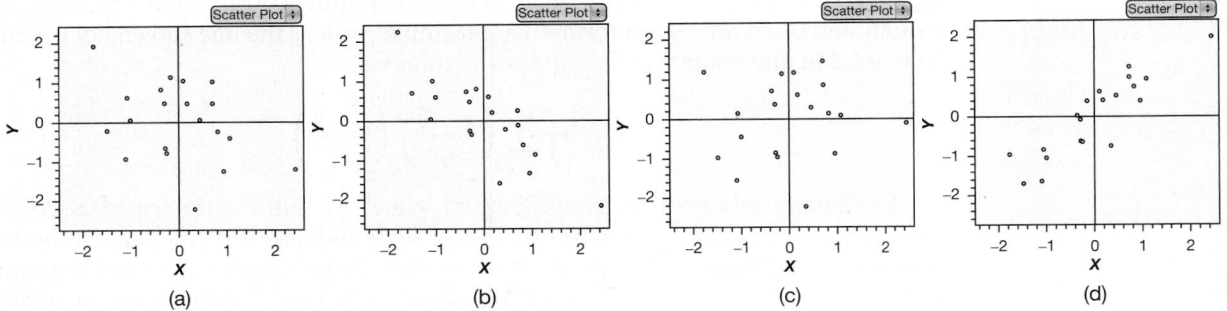

Figure 2.55 *Match the correlations to the scatterplots*

SKILL BUILDER 3

In Exercises 2.168 to 2.173, two quantitative variables are described. Do you expect a positive or negative association between the two variables? Explain your choice.

2.168 Size of a house *and* Cost to heat the house.

2.169 Distance driven since the last fill-up of the gas tank *and* Amount of gas left in the tank.

2.170 Outside temperature *and* Amount of clothes worn.

2.171 Number of text messages sent on a cell phone *and* Number of text messages received on the phone.

2.172 Number of people in a square mile *and* Number of trees in the square mile.

2.173 Amount of time spent studying *and* Grade on the exam.

SKILL BUILDER 4

In Exercises 2.174 and 2.175, make a scatterplot of the data. Put the *X* variable on the horizontal axis and the *Y* variable on the vertical axis.

2.174

X	3	5	2	7	6
Y	1	2	1.5	3	2.5

2.175

X	15	20	25	30	35	40	45	50
Y	532	466	478	320	303	349	275	221

SKILL BUILDER 5

In Exercises 2.176 and 2.177, use technology to find the correlation for the data indicated.

2.176 The data in Exercise 2.174.

2.177 The data in Exercise 2.175.

2.178 Light Roast or Dark Roast for Your Coffee? A somewhat surprising fact about coffee is that the longer it is roasted, the less caffeine it has. Thus an "extra bold" dark roast coffee actually has less caffeine than a light roast coffee. What is the explanatory variable and what is the response variable? Do the two variables have a negative association or a positive association?

2.179 Mother's Love, Hippocampus, and Resiliency Multiple studies[61] in both animals and humans show the importance of a mother's love (or the unconditional love of any close person to a child) in a child's brain development. A recent study shows that children with nurturing mothers had a substantially larger area of the brain called the hippocampus than children with less nurturing mothers. This is important because other studies have shown that the size of the hippocampus matters: People with large hippocampus area are more resilient and are more likely to be able to weather the stresses and strains of daily life. These observations come from experiments in animals and observational studies in humans.

[61] Raison, C., "Love key to brain development in children," cnn.com, *The Chart*, March 12, 2012.

(a) Is the amount of maternal nurturing one receives as a child positively or negatively associated with hippocampus size?

(b) Is hippocampus size positively or negatively associated with resiliency and the ability to weather the stresses of life?

(c) How might a randomized experiment be designed to test the effect described in part (a) in humans? Would such an experiment be ethical?

(d) Can we conclude that maternal nurturing in humans causes the hippocampus to grow larger? Can we conclude that maternal nurturing in animals (such as mice, who were used in many of the experiments) causes the hippocampus to grow larger? Explain.

2.180 Commitment Genes and Cheating Genes In earlier studies, scientists reported finding a "commitment gene" in men, in which men with a certain gene variant were much less likely to commit to a monogamous relationship.[62] That study involved only men (and we return to it later in this text), but a new study, involving birds this time rather than humans, shows that female infidelity may be inherited.[63] Scientists recorded who mated with or rebuffed whom for five generations of captive zebra finches, for a total of 800 males and 754 females. Zebra finches are believed to be a monogamous species, but the study found that mothers who cheat with multiple partners often had daughters who also cheat with multiple partners. To identify whether the effect was genetic or environmental, the scientists switched many of the chicks from their original nests. More cheating by the biological mother was strongly associated with more cheating by the daughter. Is this a positive or negative association?

[62]Timmer, J., "Men with genetic variant struggle with commitment," *http://www.arstechnica.com*, reporting on a study in *Proceedings of the National Academy of Science*, 2009.
[63]Millus, S., "Female infidelity may be inherited," *Science News*, July 16, 2011, p. 10.

2.181 Social Jetlag Social jetlag refers to the difference between circadian and social clocks, and is measured as the difference in sleep and wake times between work days and free days. For example, if you sleep between 11 pm and 7 am on weekdays but from 2 am to 10 am on weekends, then your social jetlag is three hours, or equivalent to flying from the West Coast of the US to the East every Friday and back every Sunday. Numerous studies have shown that social jetlag is detrimental to health. One recent study[64] measured the self-reported social jetlag of 145 healthy participants, and found that increased social jetlag was associated with a higher BMI (body mass index), higher cortisol (stress hormone) levels, higher scores on a depression scale, fewer hours of sleep during the week, less physical activity, and a higher resting heart rate.

(a) Indicate whether social jetlag has a positive or negative correlation with each variable listed: BMI, cortisol level, depression score, weekday hours of sleep, physical activity, heart rate.

(b) Can we conclude that social jetlag causes the adverse effects described in the study?

2.182 Football, Brain Size, and Cognitive Scores Exercise 2.143 on page 102 introduces a study that examines the association between playing football, brain size as measured by left hippocampal volume (in μL), and percentile on a cognitive reaction test. Figure 2.56 gives two scatterplots. Both have number of years playing football as the explanatory variable while Graph (a) has cognitive percentile as the response variable and Graph (b) has hippocampal volume as the response variable.

(a) The two corresponding correlations are −0.465 and −0.366. Which correlation goes with which scatterplot?

[64]Rutters, F., et al., "Is social jetlag associated with an adverse endocrine, behavioral, and cardiovascular risk profile?" *J Biol Rhythms*, October 2014; 29(5):377–383.

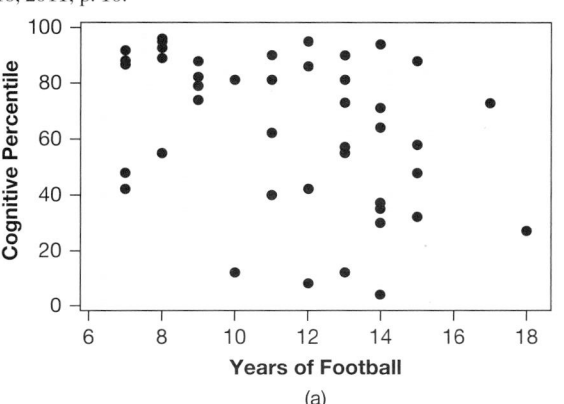 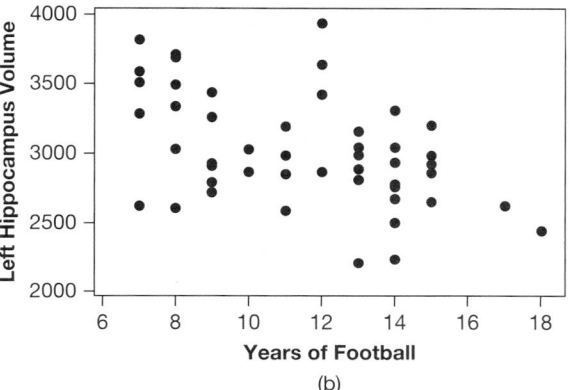

Figure 2.56 *Football, cognitive percentile, and brain size*

(b) Both correlations are negative. Interpret what this means in terms of football, brain size, and cognitive percentile.

2.183 NFL Pre-Season Does pre-season success indicate regular season success in the US National Football League? We looked at the number of pre-season wins and regular season wins for all 32 NFL teams over a 10-year span.

(a) What would a positive association imply about the relationship between pre-season and regular season success in the NFL? What would a negative association imply?

(b) The correlation between these two variables is $r = 0.067$. What does this correlation tell you about the strength of a linear relationship between these two variables?

2.184 What's Wrong with the Statement? A researcher claims to have evidence of a strong positive correlation ($r = 0.88$) between a person's blood alcohol content (BAC) and the type of alcoholic drink consumed (beer, wine, or hard liquor). Explain, statistically, why this claim makes no sense.

2.185 Help for Insomniacs In Exercise 1.23, we learned of a study in which participants were randomly assigned to receive or not receive cognitive behavioral therapy (CBT), and then reported whether or not they experienced any sleep improvement. One news magazine reporting this study said "Sleep improvements were strongly correlated with CBT." Why is this an incorrect use of the statistics word *correlation*?

2.186 Genetic Diversity and Distance from Africa It is hypothesized that humans originated in East Africa, and migrated from there. We compute a measure of genetic diversity for different populations,[65] and the geographic distance of each population from East Africa (Addis Ababa, Ethiopia), as one would travel over the surface of the earth by land (migration long ago is thought to have happened by land). The relationship between these two variables is shown in Figure 2.57 and the data are given in **GeneticDiversity**.

(a) Describe the relationship between genetic diversity and distance from East Africa. Does there appear to be an association? If so, it is positive or negative? Strong or weak? Linear or nonlinear?

[65]Calculated from data from Ramachandran, S., Deshpande, O., Roseman, C.C., Rosenberg, N.A., Feldman, M.W., Cavalli-Sforza, L.L. "Support from the relationship of genetic and geographic distance in human populations for a serial founder effect originating in Africa," *Proceedings of the National Academy of Sciences*, 2005, 102: 15942–15947.

(b) Which of the following values gives the correlation between these two variables: $r = -1.22$, $r = -0.83$, $r = -0.14$, $r = 0.14$, $r = 0.83$, or $r = 1.22$?

(c) On which continent is the population with the lowest genetic diversity? On which continent is the population that is farthest from East Africa (by land)?

(d) Populations with greater genetic diversity are thought to be better able to adapt to changing environments, because more genetic diversity provides more opportunities for natural selection. Based only on this information and Figure 2.57, do populations closer or farther from East Africa appear to be better suited to adapt to change?

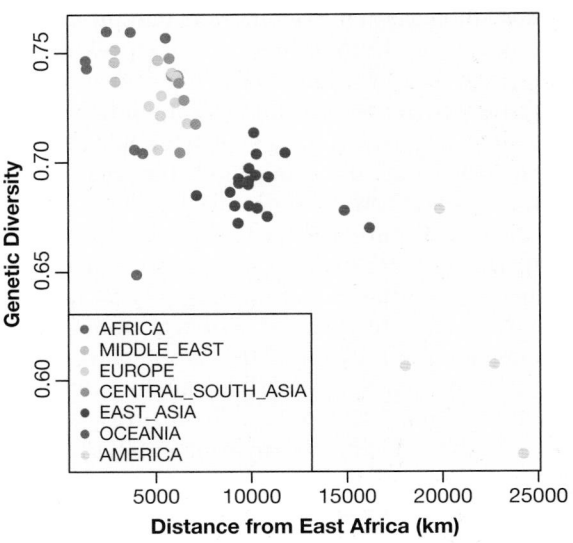

Figure 2.57 *Genetic diversity of populations by distance from East Africa*

2.187 The Happy Planet Index The website *TED.com* offers free short presentations, called TED Talks, on a variety of interesting subjects. One of the talks is called "The Happy Planet Index," by Nic Marks.[66] Marks comments that we regularly measure and report economic data on countries, such as Gross National Product, when we really ought to be measuring the well-being of the people in the countries. He calls this measure *Happiness*, with larger numbers indicating greater happiness, health, and well-being. In addition, he believes we ought to be measuring the ecological footprint, per capita, of the country, with larger numbers

[66]Marks, N., "The Happy Planet Index," *http://www.TED.com/talks*, August 29, 2010.

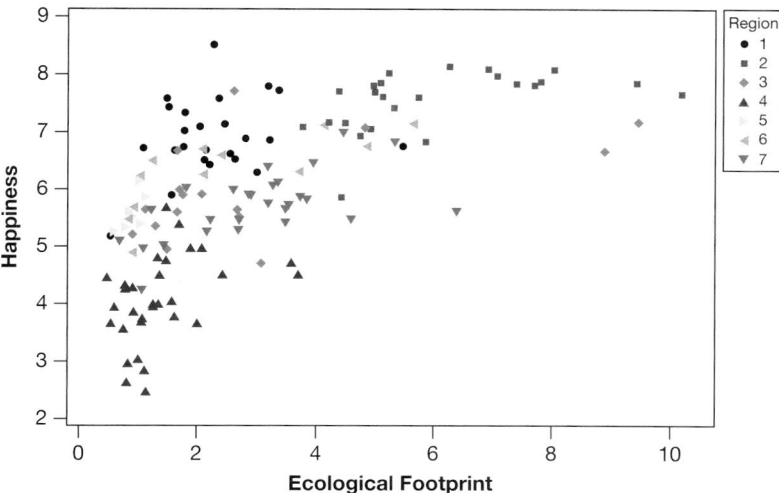

Figure 2.58 *Happiness and ecological footprint*

indicating greater use of resources (such as gas and electricity) and more damage to the planet. Figure 2.58 shows a scatterplot of these two quantitative variables. The data are given in **HappyPlanetIndex**.

(a) Does there appear to be a mostly positive or mostly negative association between these two variables? What does that mean for these two variables?

(b) Describe the happiness and ecological footprint of a country in the bottom left of the graph.

(c) Costa Rica has the highest *Happiness* index. Find it on the graph and estimate its ecological footprint score.

(d) For ecological footprints between 0 and 6, does a larger ecological footprint tend to be associated with more happiness? What about for ecological footprints between 6 and 10? Discuss this result in context.

(e) Marks believes we should be working to move all countries to the top left of the graph, closer to Costa Rica. What attributes does a country in the upper left of the graph possess?

(f) This graph shows a third variable as well: region of the world. One way to depict a categorical variable on a scatterplot is using different colors or shapes for different categories. The code is given in the top right, and is categorized as follows: 1 = Latin America, 2 = Western nations, 3 = Middle East, 4 = Sub-Saharan Africa, 5 = South Asia, 6 = East Asia, 7 = former Communist countries. Discuss one observation of an association between region and the two quantitative variables.

(g) If the goal is to move all countries to the top left, how should efforts be directed for those in the bottom left? How should efforts be directed for those in the top right?

2.188 Ages of Husbands and Wives Suppose we record the husband's age and the wife's age for many randomly selected couples.

(a) What would it mean about ages of couples if these two variables had a negative relationship?

(b) What would it mean about ages of couples if these two variables had a positive relationship?

(c) Which do you think is more likely, a negative or a positive relationship?

(d) Do you expect a strong or a weak relationship in the data? Why?

(e) Would a strong correlation imply there is an association between husband age and wife age?

2.189 Is Your Body Language Closed or Open? A closed body posture includes sitting hunched over or standing with arms crossed rather than sitting or standing up straight and having the arms more open. According to a recent study, people who were rated as having a more closed body posture "had higher levels of stress hormones and said they felt less powerful than those who had a more open pose."[67]

(a) What are the variables in this study? Is each variable categorical or quantitative? Assume participants had body language rated on a numerical scale from low values representing more closed to larger values representing more

[67]"Don't Slouch!" *Consumer Reports OnHealth*, February 2011; 23(2):3.

open. Assume also that participants were rated on a numerical scale indicating whether each felt less powerful (low values) or more powerful (higher values).

(b) Do the results of the study indicate a positive or negative relationship between the body language scores and levels of stress hormones? Would your answer be different if the scale had been reversed for the body language scores?

(c) Do the results of the study indicate a positive or negative relationship between the body language scores and the scores on the feelings of power? Would your answer be different if both scales were reversed? Would your answer be different if only one of the scales had been reversed?

2.190 SAT Scores: Math vs Verbal The **Student-Survey** dataset includes scores on the Math and Verbal portions of the SAT exam.

(a) What would a positive relationship between these two variables imply about SAT scores? What would a negative relationship imply?

(b) Figure 2.59 shows a scatterplot of these two variables. For each corner of the scatterplot (top left, top right, bottom left, bottom right), describe a student whose SAT scores place him or her in that corner.

(c) Does there appear to be a strong linear relationship between these two variables? What does that tell you about SAT scores?

(d) Which of the following is most likely to be the correlation between these two variables?

−0.941, −0.605, −0.235, 0.445, 0.751, 0.955

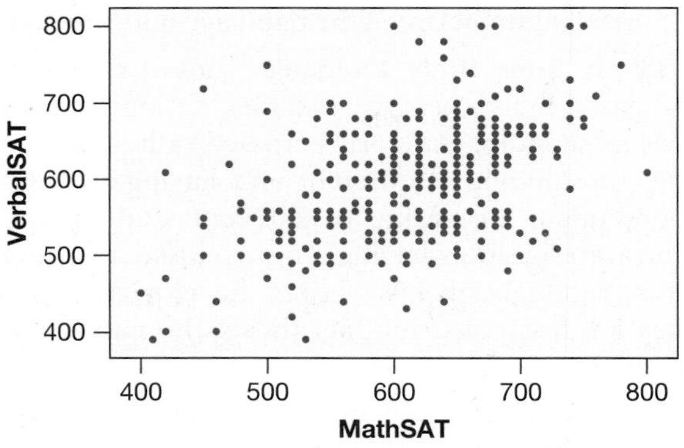

Figure 2.59 *MathSAT score and VerbalSAT score*

2.191 Exercising or Watching TV? The **Student-Survey** dataset includes information on the number of hours a week students say they exercise and the number of hours a week students say they watch television.

(a) What would a positive relationship between these two variables imply about the way students spend their time? What would a negative relationship imply?

(b) For each corner of the scatterplot of these two variables shown in Figure 2.60 (top left, top right, bottom left, bottom right), describe a student whose daily habits place him or her in that corner.

(c) There are two outliers in this scatterplot. Describe the student corresponding to the outlier on the right. Describe the student corresponding to the outlier on the top.

(d) The correlation between these two variables is $r = 0.01$. What does this correlation tell you about the strength of a linear relationship between these two variables?

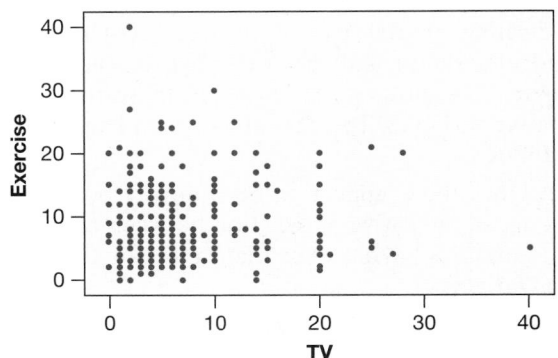

Figure 2.60 *Number of hours a week of exercise and of television watching*

2.192 Females Rating Males on OKCupid The OKCupid dating site provides lots of very interesting data.[68] Figure 2.61 shows a scatterplot of the age of males that females find most attractive, based on the age of the females doing the rating. The X-variable is the age of heterosexual females using the OKCupid site. For each age, the Y-variable gives the age of males that are rated most attractive by women at that age. So, for example, the dot that is farthest left shows that 20-year-old women find men who are age 23 the most attractive. The $Y = X$ line is also shown on the graph, for reference. (The comparable graph for males is given in Exercise 2.193.)

(a) At what age(s) do women find men the same age as themselves the most attractive?

(b) What age men do women in their early 20s find the most attractive: younger men, older men, or men the same age as themselves?

[68]Matlin, C., "Matchmaker, Matchmaker, Make Me a Spreadsheet," *http://fivethirtyeight.com*, September 9, 2014. Based on data from *http://blog.okcupid.com*.

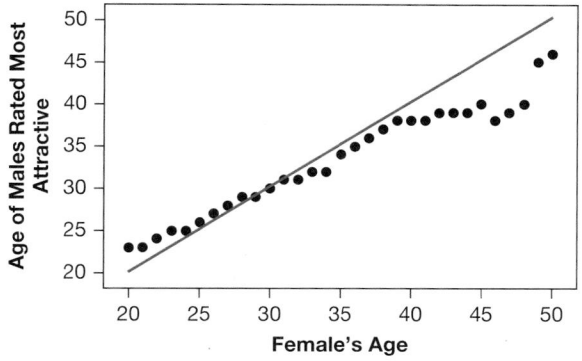

Figure 2.61 *Women rating men*

(c) What age men do women in their 40s find the most attractive: younger men, older men, or men the same age as themselves?

(d) Which of the following is likely to be closest to the correlation between these two variables?

0.9, 0, −0.9

2.193 Males Rating Females on OKCupid Exercise 2.192 introduced data showing the age of males that females find most attractive, based on the age of the females doing the rating. Here we examine the ratings males give for females. Figure 2.62 shows a scatterplot of the age of females that males find most attractive, based on the age of the males doing the rating. The *X*-variable is the age of heterosexual males using the OKCupid site. For each age, the *Y*-variable gives the age of females that are rated most attractive by men at that age. So, for example, the dot that is farthest left shows that 20-year-old men find women who are also age 20 the most attractive. The *Y* = *X* line is shown on the graph, for reference.

(a) At what age(s) do men find women the same age as themselves the most attractive?

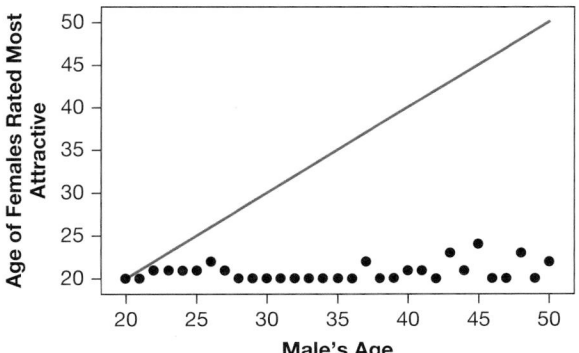

Figure 2.62 *Men rating women*

(b) What age range for women do *all* ages of men find most attractive?

(c) Which of the following is likely to be closest to the correlation between these two variables?

0.9, 0, −0.9

2.194 Comparing Global Internet Connections In Exercise 2.120 on page 92, we discuss a study in which the Nielsen Company measured connection speeds on home computers in nine different countries in order to determine whether connection speed affects the amount of time consumers spend online.[69] Table 2.29 shows the percent of Internet users with a "fast" connection (defined as 2Mb or faster) and the average amount of time spent online, defined as total hours connected to the Web from a home computer during the month of February 2011. The data are also available in the dataset **GlobalInternet**.

(a) What would a positive association mean between these two variables? Explain why a positive relationship might make sense in this context.

(b) What would a negative association mean between these two variables? Explain why a negative relationship might make sense in this context.

Table 2.29 *Internet connection speed and hours online*

Country	Percent Fast Connection	Hours Online
Switzerland	88	20.18
United States	70	26.26
Germany	72	28.04
Australia	64	23.02
United Kingdom	75	28.48
France	70	27.49
Spain	69	26.97
Italy	64	23.59
Brazil	21	31.58

(c) Make a scatterplot of the data, using connection speed as the explanatory variable and time online as the response variable. Is there a positive or negative relationship? Are there any outliers? If so, indicate the country associated with each outlier and describe the characteristics that make it an outlier for the scatterplot.

[69]"Swiss Lead in Speed: Comparing Global Internet Connections," *http://www.nielsen.com/us/en/insights/news/2011/swiss-lead-in-speed-comparing-global-internet-connections.html*, April 1, 2011.

(d) If we eliminate any outliers from the scatterplot, does it appear that the remaining countries have a positive or negative relationship between these two variables?

(e) Use technology to compute the correlation. Is the correlation affected by the outliers?

(f) Can we conclude that a faster connection speed causes people to spend more time online?

2.195 Iris Petals Allometry is the area of biology that studies how different parts of a body grow in relation to other parts. Figure 2.63 shows a scatterplot[70] comparing the length and width of petals of irises.

(a) Does there appear to be a positive or negative association between petal width and petal length? Explain what this tells us about petals.

(b) Discuss the strength of a linear relationship between these two variables.

(c) Estimate the correlation.

(d) Are there any clear outliers in the data?

(e) Estimate the width of the petal that has a length of 30 mm.

(f) There are at least two different types of irises included in the study. Explain how the scatterplot helps illustrate this, and name one difference between the types that the scatterplot makes obvious.

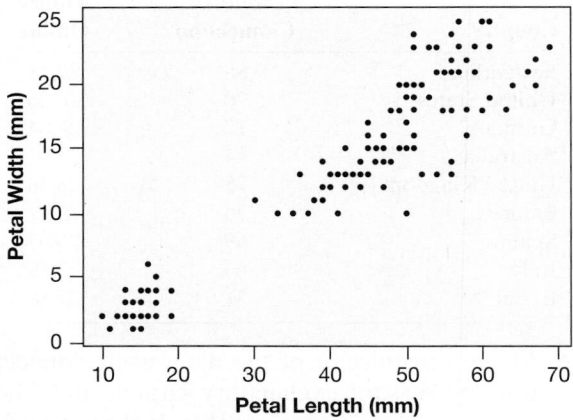

Figure 2.63 *Iris petals*

2.196 Create a Scatterplot Draw any scatterplot satisfying the following conditions:

(a) $n = 10$ and $r = 1$

(b) $n = 8$ and $r = -1$

(c) $n = 5$ and $r = 0$

2.197 Offensive Rebounds vs Defensive Rebounds The dataset **NBAPlayers2015** is introduced on page 91, and includes many variables about players in the National Basketball Association in 2014–2015.

(a) Use technology to create a scatterplot for the relationship between the number of offensive rebounds in the season and the number of defensive rebounds. (Put offensive rebounds on the horizontal axis.)

(b) Does the relationship appear to be positive or negative? What does that mean for these two variables? How strong is the relationship?

(c) There appear to be two outliers in the top right. Who are they?

(d) Use technology to find the correlation between these two variables.

2.198 Do Movies with Larger Budgets Get Higher Audience Ratings? The dataset **HollywoodMovies** is introduced on page 95, and includes many variables for movies, including *Budget* and *AudienceScore*.

(a) Use technology to create a scatterplot to show the relationship between the budget of a movie, in millions of dollars, and the audience score. We want to see if the budget has an effect on the audience score.

(b) There is an outlier with a very large budget. What is the audience rating for this movie and what movie is it? For the data value with the highest audience rating, what is the budget and what movie is it?

(c) Use technology to find the correlation between these two variables.

2.199 Pick a Relationship to Examine Choose one of the following datasets: **USStates, HollywoodMovies, AllCountries,** or **NBAPlayers2015,** and then select any two quantitative variables that we have not yet analyzed. Use technology to graph a scatterplot of the two variables and discuss what you see. Is there a linear relationship? If so, is the association positive or negative? How strong is the trend? Are there any outliers? If so, identify them by name. In addition, use technology to find the correlation. Does the correlation match what you see in the scatterplot? Be sure to state the dataset and variables you use.

---

[70]R.A. Fisher's's iris data downloaded from *http://lib.stat.cmu .edu/DASL/Datafiles/Fisher'sIrises.html.*

2.6 TWO QUANTITATIVE VARIABLES: LINEAR REGRESSION

In Section 2.5, we investigate the relationship between two quantitative variables. In this section, we discuss how to use one of the variables to predict the other when there is a linear trend.

Image Source/Getty Images, Inc.

Can we predict the size of a tip?

DATA 2.12

Restaurant Tips

The owner[71] of a bistro called *First Crush* in Potsdam, New York, is interested in studying the tipping patterns of its patrons. He collected restaurant bills over a two-week period that he believes provide a good sample of his customers. The data from 157 bills are stored in **RestaurantTips** and include the amount of the bill, size of the tip, percentage tip, number of customers in the group, whether or not a credit card was used, day of the week, and a coded identity of the server. ■

For the restaurant tips data, we want to use the bill amount to predict the tip amount, so the explanatory variable is the amount of the bill and the response variable is the amount of the tip. A scatterplot of this relationship is shown in Figure 2.64.

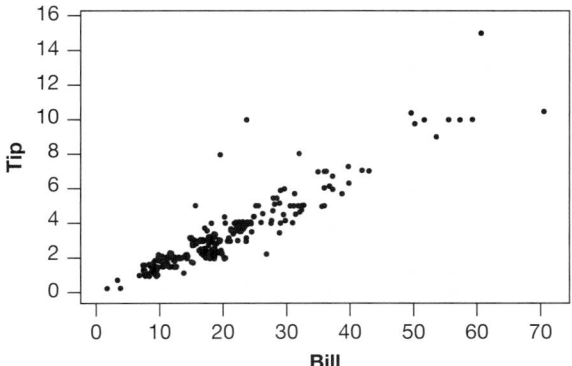

Figure 2.64 *Tip vs Bill for a sample of First Crush customers*

[71]Thanks to Tom DeRosa for providing the tipping data.

Example 2.39

(a) Use Figure 2.64 to describe the relationship between the bill amount and the tip amount at this restaurant.

(b) Use technology to find the correlation between these two variables.

(c) Draw a line on the scatterplot that seems to fit the data well.

Solution

(a) Figure 2.64 shows a strong positive linear relationship in the data, with a few outliers (big tippers!) above the main pattern.

(b) Using technology, we see that the correlation is $r = 0.915$, reinforcing the fact that the data have a strong positive linear relationship.

(c) There are many lines we could draw that fit the data reasonably well. Try drawing some! Which of the lines you drew do you think fits the data the best? One line that fits the data particularly well is shown in Figure 2.65.

The Regression Line

The process of fitting a line to a set of data is called *linear regression* and the line of best fit is called the *regression line*. The regression line for the restaurant tips data is shown in Figure 2.65 and we see that it seems to fit the data very well. The regression line provides a model of a linear association between two variables, and we can use the regression line on a scatterplot to give a predicted value of the response variable, based on a given value of the explanatory variable.

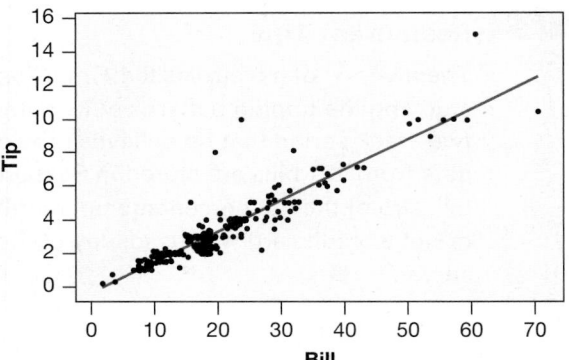

Figure 2.65 *How well does this line fit the data?*

Example 2.40

Use the regression line in Figure 2.65 to estimate the predicted tip amount on a $60 bill.

Solution

The predicted tip amount for a $60 bill is about $10, because the point on the regression line above *Bill* = 60 is at a height of about $10 on the vertical *Tip* scale.

Usually, rather than estimating predictions using a graph, we use the equation of the regression line. Recall that the equation for a line is given by $y = a + bx$ where the constant a represents the y-intercept and the coefficient b represents the slope of the line.[72] Finding the regression line, then, means finding values for the slope and

---

[72] You may have learned the equation for a line as $y = mx + b$. Statisticians prefer to use $y = a + bx$. In either case, the coefficient of x is the slope and the constant term is the vertical intercept.

intercept of the line that best describes the linear trend of the data. This can be done on many calculators and computer programs.

To help distinguish between the predicted and observed values of the response variable, we often add a "hat" to the response variable name to denote the predicted value. Thus if our data pairs are (x, y) with x as the explanatory variable and y as the response variable, the regression line is given by

$$\hat{y} = a + bx$$

Explanatory and Response Variables

The regression line to predict y from x is NOT the same as the regression line to predict x from y. Be sure to always pay attention to which is the explanatory variable and which is the response variable! A regression line is always in the form

$$\widehat{Response} = a + b \cdot Explanatory$$

For the restaurant tips data, the equation of the regression line shown in Figure 2.65 is

$$\widehat{Tip} = -0.292 + 0.182 \cdot Bill$$

The y-intercept of this line is -0.292 and the slope is 0.182.

Using the Equation of the Regression Line to Make Predictions

The equation of the regression line is often also called a *prediction equation* because we can use it to make predictions. We substitute the value of the explanatory variable into the prediction equation to calculate the predicted response.

Example 2.41

Three different bill amounts from the **RestaurantTips** dataset are given. In each case, use the regression line $\widehat{Tip} = -0.292 + 0.182 \cdot Bill$ to predict the tip.

(a) A bill of \$59.33
(b) A bill of \$9.52
(c) A bill of \$23.70

Solution   (a) If the bill is \$59.33, we have

$$
\begin{aligned}
\widehat{Tip} &= -0.292 + 0.182 \cdot Bill \\
&= -0.292 + 0.182(59.33) \\
&= 10.506
\end{aligned}
$$

The predicted size of the tip is 10.506 or about \$10.51.

(b) For a bill of \$9.52, we have $\widehat{Tip} = -0.292 + 0.182(9.52) = 1.441 \approx \$1.44$.

(c) For a bill of \$23.70, we have $\widehat{Tip} = -0.292 + 0.182(23.70) = 4.021 \approx \$4.02$.

The predicted value is an estimate of the average response value for that particular value of the explanatory variable. We expect actual values to be above or below this amount.

Residuals

In Example 2.41, we found the predicted tip for three of the bills in the restaurant tips dataset. We can look in the dataset to see how close these predictions are to the actual tip amount for those bills. The *residual* is the difference between the observed value and the predicted value. On a scatterplot, the predicted value is the height of the regression line for a given *Bill* amount and the observed value is the height of the particular data point with that *Bill* amount, so the residual is the vertical distance from the point to the line. The residual for one data value is shown in Figure 2.66.

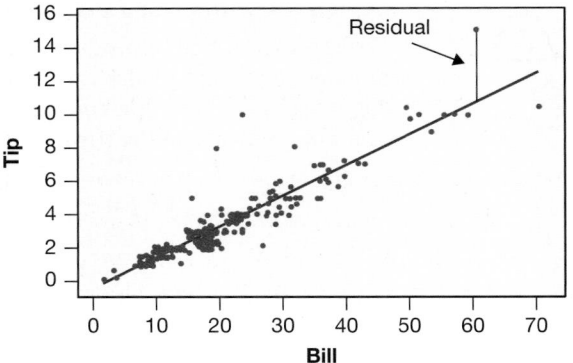

Figure 2.66 *A residual is the vertical deviation from a point to the line*

Residuals

The **residual** at a data value is the difference between the observed and predicted values of the response variable:

$$Residual = Observed - Predicted = y - \widehat{y}$$

On a scatterplot, the residual represents the vertical deviation from the line to a data point. Points above the line will have positive residuals and points below the line will have negative residuals. If the predicted values closely match the observed data values, the residuals will be small.

Example 2.42

In Example 2.41, we find the predicted tip amount for three different bills in the **RestaurantTips** dataset. The actual tips left by each of these customers are shown below. Use this information to calculate the residuals for each of these sample points.

(a) The tip left on a bill of $59.33 was $10.00.

(b) The tip left on a bill of $9.52 was $1.00.

(c) The tip left on a bill of $23.70 was $10.00.

Solution

(a) The observed tip left on the bill of $59.33 is $10.00 and we see in Example 2.41(a) that the predicted tip is $10.51. The observed tip is a bit less than the predicted tip. We have

$$Residual = Observed - Predicted = 10.00 - 10.51 = -0.51$$

(b) The observed tip left on the bill of $9.52 is just $1.00, and we see in Example 2.41(b) that the predicted tip for a bill this size is $1.44, so

$$Residual = Observed - Predicted = 1.00 - 1.44 = -0.44$$

(c) The observed tip left on a bill of $23.70 (the first case in the dataset) is $10.00 and we see in Example 2.41(c) that the predicted tip is only $4.02. The observed tip is quite a bit larger than the predicted tip and we have

$$Residual = Observed - Predicted = 10.00 - 4.02 = 5.98$$

This is one of the largest residuals. The server would be quite happy to receive this extra large tip!

Example 2.43

Data 2.9 on page 106 introduced data that show the approval rating of a president running for re-election and the resulting margin of victory or defeat for the president in the election. The data are in **ElectionMargin**.

(a) The regression line for these 12 data points is

$$\widehat{Margin} = -36.8 + 0.839(Approval)$$

Calculate the predicted values and the residuals for all the data points.

(b) Show the residuals as distances on a scatterplot with the regression line.

(c) Which residual is the largest? For this largest residual, is the observed margin higher or lower than the margin predicted by the regression line? To which president and year does this residual correspond?

Solution

(a) We use the regression line to find the predicted value for each data point, and then subtract to find the residuals. The results are given in Table 2.30. Some of the residuals are positive and some are negative, reflecting the fact that some of the data points lie above the regression line and some lie below.

(b) See Figure 2.67. At a given approval rating, such as 62, the observed margin (10) corresponds to the height of the data point, while the predicted value (15.23) corresponds to the height of the line at an approval rating of 62. Notice that in this case the line lies above the data point, and the difference between the observed value and the predicted value is the length of the vertical line joining the point to the line.

(c) The largest residual is 12.16. The observed margin of victory is 23.2, high above the predicted value of 11.04. We see in Figure 2.67 that this is the point with the greatest vertical deviation from the line. Looking back at Table 2.26 on page 107, we see that this residual corresponds to President Nixon in 1972.

Table 2.30 *Predicted margin and residuals for presidential incumbents*

Approval	Actual Margin	Predicted Margin	Residual
62	10.0	15.23	−5.23
50	4.5	5.17	−0.67
70	15.4	21.94	−6.54
67	22.6	19.43	3.17
57	23.2	11.04	12.16
48	−2.1	3.49	−5.59
31	−9.7	−10.76	1.06
57	18.2	11.04	7.16
39	−5.5	−4.05	−1.45
55	8.5	9.36	−0.86
49	2.4	4.33	−1.93
50	3.9	5.17	−1.27

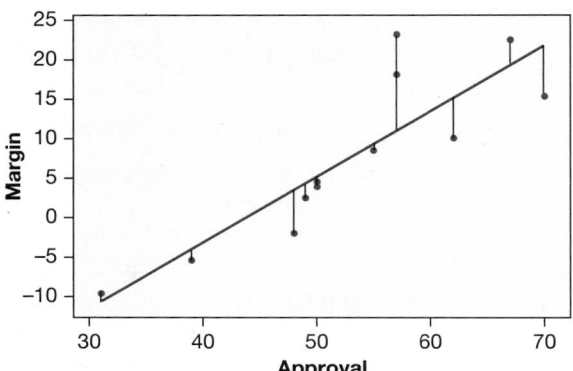

Figure 2.67 *Residuals measure vertical deviations from the line to the points*

What Does "Line of Best Fit" Mean?

How can we determine which line is the best fit for a set of data? And what do we even mean by "best fit"? Our goal is to find the line that provides the best predictions for the observed values of the response variable. The line that fits the data best should then be one where the residuals are close to zero. In particular, we usually try to make the squares of the residuals, $(y - \hat{y})^2$, small. The *least squares line* is the line with the slope and intercept that makes the sum of the squared residuals as small as it can possibly be.

> **Least Squares Line**
>
> The **least squares line**, also called the **line of best fit**, is the line that minimizes the sum of the squared residuals.

Throughout this text, we use the terms *regression line* and *least squares line* interchangeably.

We should expect observed values to fall both above and below the line of best fit, so residuals are both positive and negative. This is one reason why we square them. In fact, if we add up all of the residuals from the regression line, the sum will always be zero.

Interpreting the Slope and Intercept of the Regression Line

Recall that the regression line for the **RestaurantTips** data is

$$\widehat{Tip} = -0.292 + 0.182 \cdot Bill$$

How can we interpret the slope 0.182 and intercept −0.292?

Recall that for a general line $y = a + bx$, the slope represents the change in y over the change in x. If the change in x is 1, then the slope represents the change in y. The intercept represents the value of y when x is zero.

> **Interpreting the Slope and Intercept of the Regression Line**
>
> For the regression line $\hat{y} = a + bx$,
>
> - The slope b represents the predicted change in the response variable y given a one unit increase in the explanatory variable x.
>
> - The intercept a represents the predicted value of the response variable y if the explanatory variable x is zero. The interpretation may be nonsensical since it is often not reasonable for the explanatory variable to be zero.

Example 2.44

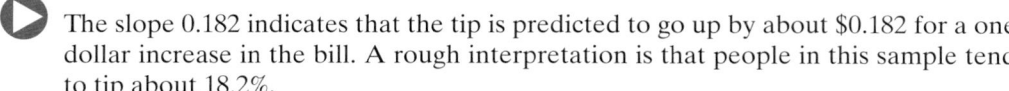

For the **RestaurantTips** regression line $\widehat{Tip} = -0.292 + 0.182 \cdot Bill$, interpret the slope and the intercept in context.

Solution

The slope 0.182 indicates that the tip is predicted to go up by about $0.182 for a one dollar increase in the bill. A rough interpretation is that people in this sample tend to tip about 18.2%.

The intercept −0.292 indicates that the tip will be −$0.292 if the bill is $0. Since a bill is rarely zero dollars and a tip cannot be negative, this makes little sense.

Example 2.45

In Example 2.34 on page 108, we consider some scatterplots from the dataset **FloridaLakes** showing relationships between acidity, alkalinity, and fish mercury levels in $n = 53$ Florida lakes. We wish to predict a quantity that is difficult to measure (mercury level of fish) using a value that is more easily obtained from a water sample (acidity). We see in Example 2.34 that there appears to be a negative linear association between these two variables, so a regression line is appropriate.

(a) Use technology to find the regression line to predict *Mercury* from *pH*, and plot it on a scatterplot of the data.

(b) Interpret the slope of the regression line in the context of Florida lakes.

(c) Put an arrow on the scatterplot pointing to the data for Puzzle Lake, which has an acidity of 7.5 and an average mercury level of 1.10 ppm. Calculate the predicted mercury level for Puzzle Lake and compare it to the observed mercury level. Calculate the residual.

Solution

(a) We use technology to find the regression line:

$$\widehat{Mercury} = 1.53 - 0.1523 \cdot pH$$

For the scatterplot, since we are predicting mercury level from pH, the pH variable goes on the horizontal axis and the mercury variable goes on the vertical axis. The line is plotted with the data in Figure 2.68.

(b) The slope in the prediction equation represents the expected change in the response variable for a one unit increase in the explanatory variable. Since the slope in this case is −0.1523, we expect the average mercury level in fish to decrease by about 0.1523 ppm for each increase of 1 in the pH of the lake water.

(c) See the arrow in Figure 2.68. The predicted value for Puzzle Lake is $\widehat{Mercury} = 1.53 - 0.1523 \cdot (7.5) = 0.388$ ppm. The observed value of 1.10 is quite a bit higher than the predicted value for this lake. The residual is $1.10 - 0.388 = 0.712$, the largest residual of all 53 lakes.

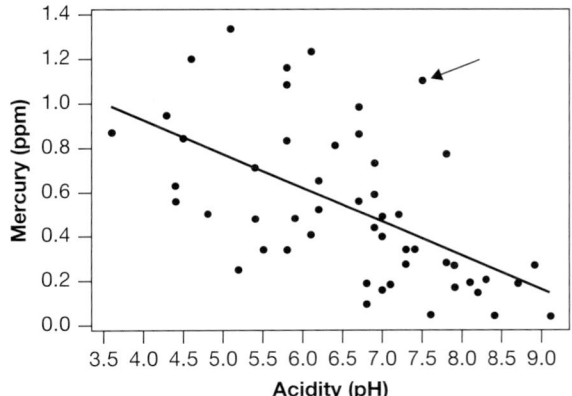

Figure 2.68 *Using acidity to predict average mercury level in fish*

Notation for the Slope

We have seen that we use the notation b for the slope of a regression line that comes from a sample. What about the regression line for a population? The dataset on presidential elections used to create the regression line $\widehat{Margin} = -36.8 + 0.839 \cdot Approval$ in Example 2.43 represents the population of all relevant US presidential elections since 1940. As we have seen with other quantities, the notation we use for the slope of the regression line of a population is different than the notation we use for the slope of the regression line of a sample. For the slope of a regression line for a population, we use the Greek letter β (beta).

Regression Cautions

In the solution to Example 2.44, we see that predicting the tip for a bill of $0 does not make any sense. Since the bill amounts in that dataset range from $1.66 to $70.51, it also would not make sense to use the regression line to predict the tip on a bill of $1000. In general, it is not appropriate to use regression lines to make predictions using values of the explanatory variable that are far from the range of values used to create the line. This is called *extrapolating* too far from the original data.

> **Regression Caution #1**
>
> Avoid trying to apply a regression line to predict values far from those that were used to create it.

Example 2.46

In Example 2.45 on page 129, we used the acidity (pH) of Florida lakes to predict mercury levels in fish. Suppose that, instead of mercury, we use acidity to predict the calcium concentration (mg/l) in Florida lakes. Figure 2.69 shows a scatterplot of these data with the regression line $\widehat{Calcium} = -51.4 + 11.17 \cdot pH$ for the 53 lakes in our sample. Give an interpretation for the slope in this situation. Does the intercept make sense? Comment on how well the linear prediction equation describes the relationship between these two variables.

Solution

The slope of 11.17 in the prediction equation indicates that the calcium concentration in lake water increases by about 11.17 mg/l when the pH goes up by one. The intercept does not have a physical interpretation since there are no lakes with a pH of zero and a negative calcium concentration makes no sense. Although there is clearly a positive association between acidity and calcium concentration, the relationship is not a linear one. The pattern in the scatterplot indicates a curved pattern that

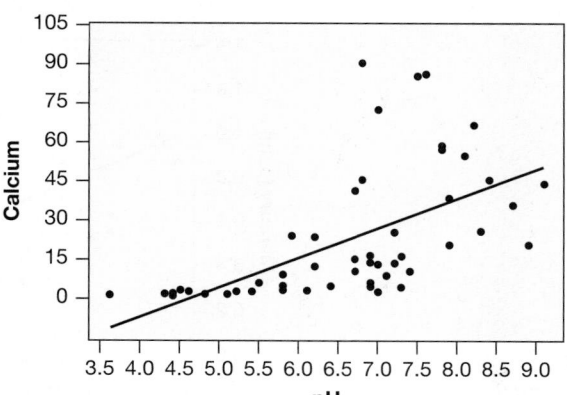

Figure 2.69 *Using acidity to predict calcium concentration*

increases more steeply as pH increases. The least squares line predicts negative calcium concentrations (which are impossible) for pH levels as large as 4.5, which are within the domain of lakes in this sample.

The correlation between acidity and average mercury levels in Figure 2.68 is −0.575 while acidity and calcium concentration in Figure 2.69 have a correlation of 0.577. Although these correlations are close in magnitude, linear regression is a more appropriate model for the first situation than it is for the second. It is always important to plot the data and look for patterns that may or may not follow a linear trend.

Regression Caution #2

Plot the data! Although the regression line can be calculated for *any* set of paired quantitative variables, it is only appropriate to use a regression line when there is a linear trend in the data.

Finally, when we plot the data, we also look for outliers that may exert a strong influence on the regression line, similar to what we see for correlation in Figure 2.53 on page 114.

Regression Caution #3

Outliers can have a strong influence on the regression line, just as we saw for correlation. In particular, data points for which the explanatory value is an outlier are often called *influential points* because they exert an overly strong effect on the regression line.

SECTION LEARNING GOALS

You should now have the understanding and skills to:

- Use technology to find the regression line for a dataset with two quantitative variables
- Calculate predicted values from a regression line
- Interpret the slope (and intercept, when appropriate) of a regression line in context
- Calculate residuals and visualize residuals on a scatterplot
- Beware of extrapolating too far out when making predictions
- Recognize the importance of plotting your data

Exercises for Section **2.6**

SKILL BUILDER 1

In Exercises 2.200 to 2.203, two variables are defined, a regression equation is given, and one data point is given.

(a) Find the predicted value for the data point and compute the residual.

(b) Interpret the slope in context.

(c) Interpret the intercept in context, and if the intercept makes no sense in this context, explain why.

2.200 Hgt = height in inches, Age = age in years of a child.
$\widehat{Hgt} = 24.3 + 2.74(Age)$; data point is a child 12 years old who is 60 inches tall.

2.201 BAC = blood alcohol content (% of alcohol in the blood), $Drinks$ = number of alcoholic drinks.
$\widehat{BAC} = -0.0127 + 0.018(Drinks)$; data point is an individual who consumed 3 drinks and had a BAC of 0.08.

2.202 $Weight$ = maximum weight capable of bench pressing (pounds), $Training$ = number of hours spent lifting weights a week.
$\widehat{Weight} = 95 + 11.7(Training)$; data point is an individual who trains 5 hours a week and can bench 150 pounds.

2.203 $Study$ = number of hours spent studying for an exam, $Grade$ = grade on the exam.
$\widehat{Grade} = 41.0 + 3.8(Study)$; data point is a student who studied 10 hours and received an 81 on the exam.

SKILL BUILDER 2
Use technology to find the regression line to predict Y from X in Exercises 2.204 to 2.207.

2.204

X	3	5	2	7	6
Y	1	2	1.5	3	2.5

2.205

X	2	4	6	8	10	12
Y	50	58	55	61	69	68

2.206

X	10	20	30	40	50	60
Y	112	85	92	71	64	70

2.207

X	15	20	25	30	35	40	45	50
Y	532	466	478	320	303	349	275	221

2.208 Concentration of CO_2 in the Atmosphere
Levels of carbon dioxide (CO_2) in the atmosphere are rising rapidly, far above any levels ever before recorded. Levels were around 278 parts per million in 1800, before the Industrial Age, and had never, in the hundreds of thousands of years before that, gone above 300 ppm. Levels are now over 400 ppm. Table 2.31 shows the rapid rise of CO_2 concentrations over the 50 years from 1960–2010, also available in **CarbonDioxide**.[73] We can use this information to predict CO_2 levels in different years.

(a) What is the explanatory variable? What is the response variable?

(b) Draw a scatterplot of the data. Does there appear to be a linear relationship in the data?

(c) Use technology to find the correlation between year and CO_2 levels. Does the value of the correlation support your answer to part (b)?

(d) Use technology to calculate the regression line to predict CO_2 from year.

(e) Interpret the slope of the regression line, in terms of carbon dioxide concentrations.

(f) What is the intercept of the line? Does it make sense in context? Why or why not?

(g) Use the regression line to predict the CO_2 level in 2003. In 2020.

(h) Find the residual for 2010.

Table 2.31 *Concentration of carbon dioxide in the atmosphere*

Year	CO_2
1960	316.91
1965	320.04
1970	325.68
1975	331.08
1980	338.68
1985	345.87
1990	354.16
1995	360.62
2000	369.40
2005	379.76
2010	389.78

2.209 The Honeybee Waggle Dance When honeybee scouts find a food source or a nice site for a new home, they communicate the location to the rest of the swarm by doing a "waggle dance."[74] They point in the direction of the site and dance longer for sites farther away. The rest of the bees use the duration of the dance to predict distance to the site.

[73]Dr. Pieter Tans, NOAA/ESRL, *http://www.esrl.noaa.gov/gmd/ccgg/trends/*. Values recorded at the Mauna Loa Observatory in Hawaii.
[74]Check out a honeybee waggle dance on YouTube!

Table 2.32 *Duration of a honeybee waggle dance to indicate distance to the source*

Distance	Duration
200	0.40
250	0.45
500	0.95
950	1.30
1950	2.00
3500	3.10
4300	4.10

Table 2.32 shows the distance, in meters, and the duration of the dance, in seconds, for seven honeybee scouts.[75] This information is also given in **HoneybeeWaggle**.

(a) Which is the explanatory variable? Which is the response variable?

(b) Figure 2.70 shows a scatterplot of the data. Does there appear to be a linear trend in the data? If so, is it positive or negative?

(c) Use technology to find the correlation between the two variables.

(d) Use technology to find the regression line to predict distance from duration.

(e) Interpret the slope of the line in context.

(f) Predict the distance to the site if a honeybee does a waggle dance lasting 1 second. Lasting 3 seconds.

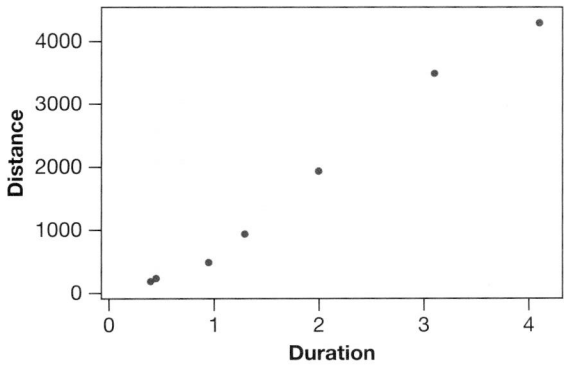

Figure 2.70 *Using dance duration to predict distance to source*

2.210 Is It Getting Harder to Win a Hot Dog Eating Contest? Every Fourth of July, Nathan's Famous in New York City holds a hot dog eating contest, which

we discuss in Exercise 2.110. Table 2.21 on page 89 shows the winning number of hot dogs eaten every year from 2002 to 2015, and the data are also available in **HotDogs**. Figure 2.71 shows the scatterplot with the regression line.

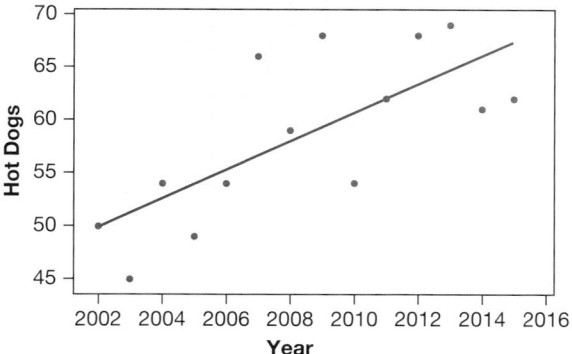

Figure 2.71 *Winning number of hot dogs*

(a) Is the trend in the data mostly positive or negative?

(b) Using Figure 2.71, is the residual larger in 2007 or 2008? Is the residual positive or negative in 2010?

(c) Use technology to find the correlation.

(d) Use technology to find the regression line to predict the winning number of hot dogs from the year.

(e) Interpret the slope of the regression line.

(f) Predict the winning number of hot dogs in 2016. (*Bonus*: Find the actual winning number in 2016 and compute the residual.)

(g) Why would it not be appropriate to use this regression line to predict the winning number of hot dogs in 2025?

2.211 Oxygen and Lung Cancer A recent study[76] has found an association between elevation and lung cancer incidence. Incidence of cancer appears to be lower at higher elevations, possibly because of lower oxygen levels in the air. We are told that "for every one km rise in elevation, lung cancer incidence decreased by 7.23" where cancer incidence is given in cases per 100,000 individuals.

(a) Is this a positive or negative association?

[75]Seeley, T., *Honeybee Democracy*, Princeton University Press, Princeton, NJ, 2010, p. 128.

[76]Simeonov, K.P., and Himmelstein, D.S., "Lung cancer incidence decreases with elevation: Evidence for oxygen as an inhaled carcinogen," *Peer Journal*, 2015; 3:e705.

(b) Which of the following quantities is given in the sentence in quotes: correlation, slope of regression line, intercept of regression line, or none of these?

(c) What is the explanatory variable? What is the response variable?

2.212 Football and Cognitive Percentile Exercise 2.143 on page 102 introduces a study that examines several variables on collegiate football players, including the variable *Years*, which is number of years playing football, and the variable *Cognition*, which gives percentile on a cognitive reaction test. Exercise 2.182 shows a scatterplot for these two variables and gives the correlation as −0.366. The regression line for predicting *Cognition* from *Years* is:

$$\widehat{Cognition} = 102 - 3.34 \cdot Years$$

(a) Predict the cognitive percentile for someone who has played football for 8 years and for someone who has played football for 14 years.

(b) Interpret the slope in terms of football and cognitive percentile.

(c) All the participants had played between 7 and 18 years of football. Is it reasonable to interpret the intercept in context? Why or why not?

2.213 Football and Brain Size Exercise 2.143 on page 102 introduces a study that examines several variables on collegiate football players, including the variable *Years*, which is number of years playing football, and the variable *BrainSize*, which is volume of the left hippocampus in the brain measured in μL. Figure 2.72 shows a scatterplot of these two variables along with the regression line. For each of the following cases, estimate from the graph the number of years of football, the predicted brain size, the actual brain size, and the residual.

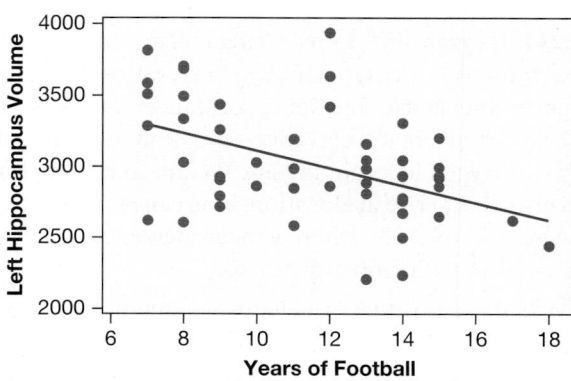

Figure 2.72 *Relationship of football experience and brain hippocampus size*

(a) The case with 18 years of football experience.

(b) The case with the largest positive residual.

(c) The case with the largest negative residual.

2.214 Runs and Wins in Baseball In Exercise 2.150 on page 104, we looked at the relationship between total hits by team in the 2014 season and division (NL or AL) in baseball. Two other variables in the **BaseballHits** dataset are the number of wins and the number of runs scored during the season. The dataset consists of values for each variable from all 30 MLB teams. From these data we calculate the regression line:

$$\widehat{Wins} = 34.85 + 0.070(Runs)$$

(a) Which is the explanatory and which is the response variable in this regression line?

(b) Interpret the intercept and slope in context.

(c) The San Francisco Giants won 88 games while scoring 665 runs in 2014. Predict the number of games won by San Francisco using the regression line. Calculate the residual. Were the Giants efficient at winning games with 665 runs?

2.215 Presidential Elections In Example 2.43 on page 127, we used the approval rating of a president running for re-election to predict the margin of victory or defeat in the election. We saw that the least squares line is $\widehat{Margin} = -36.76 + 0.839\,(Approval)$. Interpret the slope and the intercept of the line in context.

2.216 Height and Weight Using the data in the **StudentSurvey** dataset, we use technology to find that a regression line to predict weight (in pounds) from height (in inches) is

$$\widehat{Weight} = -170 + 4.82(Height)$$

(a) What weight does the line predict for a person who is 5 feet tall (60 inches)? What weight is predicted for someone 6 feet tall (72 inches)?

(b) What is the slope of the line? Interpret it in context.

(c) What is the intercept of the line? If it is reasonable to do so, interpret it in context. If it is not reasonable, explain why not.

(d) What weight does the regression line predict for a baby who is 20 inches long? Why is it not appropriate to use the regression line in this case?

2.217 NFL Pre-Season Using 10 years of National Football League (NFL) data, we calculate the following regression line to predict regular season wins

(*Wins*) by number of wins in the 4 pre-season games (*PreSeason*):

$$\widehat{Wins} = 7.5 + 0.2(PreSeason)$$

(a) Which is the explanatory variable, and which is the response variable in this regression line?

(b) How many wins does the regression line predict for a team that won 2 games in pre-season?

(c) What is the slope of the line? Interpret it in context.

(d) What is the intercept of the line? If it is reasonable to do so, interpret it in context. If it is not reasonable, explain why not.

(e) How many regular season wins does the regression line predict for a team that wins 100 pre-season games? Why is it not appropriate to use the regression line in this case?

2.218 Is the Honeybee Population Shrinking? The **Honeybee** dataset contains data collected from the USDA on the estimated number of honeybee colonies (in thousands) for the years 1995 through 2012.[77] We use technology to find that a regression line to predict number of (thousand) colonies from year (in calendar year) is

$$\widehat{Colonies} = 19,291,511 - 8.358(Year)$$

(a) Interpret the slope of the line in context.

(b) Often researchers will adjust a year explanatory variable such that it represents years since the first year data were colleected. Why might they do this? (*Hint:* Consider interpreting the y-intercept in this regression line.)

(c) Predict the bee population in 2100. Is this prediction appropriate (why or why not)?

PREDICTING PERCENT BODY FAT
Exercises 2.219 to 2.221 use the dataset **BodyFat**, which gives the percent of weight made up of body fat for 100 men as well as other variables such as *Age*, *Weight* (in pounds), *Height* (in inches), and circumference (in cm) measurements for the *Neck*, *Chest*, *Abdomen*, *Ankle*, *Biceps*, and *Wrist*.[78]

2.219 Using Weight to Predict Body Fat Figure 2.73 shows the data and regression line for using weight to predict body fat percentage. For the case

[77]USDA National Agriculture and Statistical Services, *http://usda.mannlib.cornell.edu/MannUsda/viewDocumentInfo.do?documentID=1191*. Accessed September 2015.

[78]A sample taken from data provided by R. Johnson in "Fitting Percentage of Body Fat to Simple Body Measurements," *Journal of Statistics Education*, 1996, *http://www.amstat.org/publications/jse/v4n1/datasets.johnson.html*.

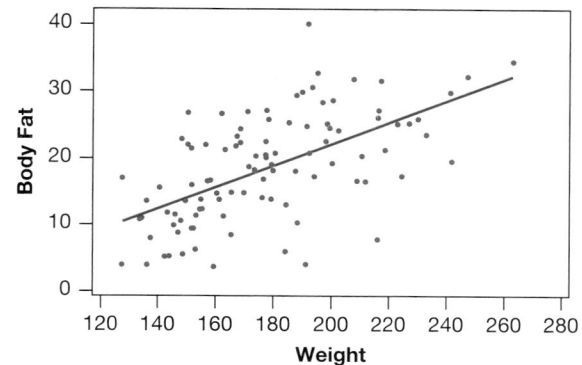

Figure 2.73 *Using weight to predict percent body fat*

with the largest positive residual, estimate the values of both variables. In addition, estimate the predicted body fat percent and the residual for that point.

2.220 Using Abdomen Circumference to Predict Body Fat Figure 2.74 shows the data and regression line for using abdomen circumference to predict body fat percentage.

(a) Which scatterplot, the one using *Weight* in Figure 2.73 or the one using *Abdomen* in Figure 2.74, appears to contain data with a larger correlation?

(b) In Figure 2.74, one person has a very large abdomen circumference of about 127 cm. Estimate the actual body fat percent for this person as well as the predicted body fat percent.

(c) Use Figure 2.74 to estimate the abdomen circumference for the person with about 40% body fat. In addition, estimate the residual for this person.

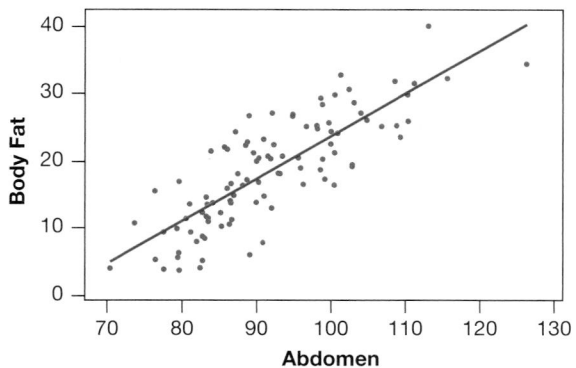

Figure 2.74 *Using abdomen circumference to predict percent body fat*

2.221 Using Neck Circumference to Predict Body Fat The regression line for predicting body fat percent using neck circumference is

$$\widehat{BodyFat} = -47.9 + 1.75 \cdot Neck$$

(a) What body fat percent does the line predict for a person with a neck circumference of 35 cm? Of 40 cm?

(b) Interpret the slope of the line in context.

(c) One of the men in the study had a neck circumference of 38.7 cm and a body fat percent of 11.3. Find the residual for this man.

2.222 Predicting World Gross Revenue for a Movie from Its Opening Weekend Use the data in **HollywoodMovies** to use revenue from a movie's opening weekend (*OpeningWeekend*) to predict total world gross revenues by the end of the year (*WorldGross*). Both variables are in millions of dollars.

(a) Use technology to create a scatterplot for this relationship. Describe the scatterplot: Is there a linear trend? How strong is it? Is it positive or negative? Does it look as if revenue from a movie's opening weekend is a good predictor of its future total earnings?

(b) The scatterplot contains an outlier with an unusually high world gross. Use the dataset to identify this movie.

(c) Use technology to find the correlation between these variables.

(d) Use technology to find the regression line.

(e) Use the regression line to predict world gross revenues for a movie that makes 50 million dollars in its opening weekend.

2.223 Using Life Expectancy to Predict Happiness In Exercise 2.187 on page 118, we introduce the dataset **HappyPlanetIndex**, which includes information for 143 countries to produce a "happiness" rating as a score of the health and well-being of the country's citizens, as well as information on the ecological footprint of the country. One of the variables used to create the happiness rating is life expectancy in years. We explore here how well this variable, *LifeExpectancy*, predicts the happiness rating, *Happiness*.

(a) Using technology and the data in **HappyPlanetIndex**, create a scatterplot to use *LifeExpectancy* to predict *Happiness*. Is there enough of a linear trend so that it is reasonable to construct a regression line?

(b) Find a formula for the regression line and display the line on the scatterplot.

(c) Interpret the slope of the regression line in context.

2.224 Pick a Relationship to Examine Choose one of the following datasets: **USStates**, **StudentSurvey**, **AllCountries**, or **NBAPlayers2015**, and then select any two quantitative variables that we have not yet analyzed. Use technology to create a scatterplot of the two variables with the regression line on it and discuss what you see. If there is a reasonable linear relationship, find a formula for the regression line. If not, find two other quantitative variables that do have a reasonable linear relationship and find the regression line for them. Indicate whether there are any outliers in the dataset that might be influential points or have large residuals. Be sure to state the dataset and variables you use.

2.225 The Impact of Strong Economic Growth In 2011, the Congressional Budget Office predicted that the US economy would grow by 2.8% per year on average over the decade from 2011 to 2021. At this rate, in 2021, the ratio of national debt to GDP (gross domestic product) is predicted to be 76% and the federal deficit is predicted to be $861 billion. Both predictions depend heavily on the growth rate. If the growth rate is 3.3% over the same decade, for example, the predicted 2021 debt-to-GDP ratio is 66% and the predicted 2021 deficit is $521 billion. If the growth rate is even stronger, at 3.9%, the predicted 2021 debt-to-GDP ratio is 55% and the predicted 2021 deficit is $113 billion.[79]

(a) There are only three individual cases given (for three different economic scenarios), and for each we are given values of three variables. What are the variables?

(b) Use technology and the three cases given to find the regression line for predicting 2021 debt-to-GDP ratio from the average growth rate over the decade 2011 to 2021.

(c) Interpret the slope and intercept of the line from part (b) in context.

(d) What 2021 debt-to-GDP ratio does the model in part (b) predict if growth is 2%? If it is 4%?

(e) Studies indicate that a country's economic growth slows if the debt-to-GDP ratio hits 90%. Using the model from part (b), at what growth rate would we expect the ratio in the US to hit 90% in 2021?

[79]Gandel, S., "Higher growth could mean our debt worries are all for nothing," *Time Magazine*, March 7, 2011, p. 20.

(f) Use technology and the three cases given to find the regression line for predicting the deficit (in billions of dollars) in 2021 from the average growth rate over the decade 2011 to 2021.

(g) Interpret the slope and intercept of the line from part (f) in context.

(h) What 2021 deficit does the model in part (f) predict if growth is 2%? If it is 4%?

(i) The deficit in 2011 was $1.4 trillion. What growth rate would leave the deficit at this level in 2021?

2.7 DATA VISUALIZATION AND MULTIPLE VARIABLES

In Sections 2.1 through 2.6 we consider basic graphs that can be used to visualize the distribution of a single variable (such as a histogram or a barchart), or a relationship between two variables (such as a scatterplot or side-by-side boxplot). Often we may wish to extend these basic methods or create an entirely new type of graph to convey more information. For example, we may wish to display more than two variables, to incorporate geographic information, to track data over time, or to connect our graph to the specific applied context of our dataset in other ways. Our only guiding principle is to facilitate quick and accurate interpretation of data, and this allows plenty of room for creativity. Good data visualization ("data viz") can involve elements of statistics, computer graphics, and artistic design.

Here we explore some additional, more advanced, techniques to visualize data, including ways to visually explore relationships between multiple variables.

Augmented Scatterplots for More than Two Variables

The scatterplot, introduced in Section 2.5, displays a relationship between two quantitative variables by letting the horizontal axis correspond to one variable, the vertical axis correspond to the other variable, and plotting a point for each pair of values. For a basic scatterplot the type of "point" (for example, a black circle) does not change. However, we can incorporate other variables by letting the size, shape, or color of the points vary. A scatterplot in which the size of the point depends on another variable is sometimes called a *bubble chart* or *bubble plot*.

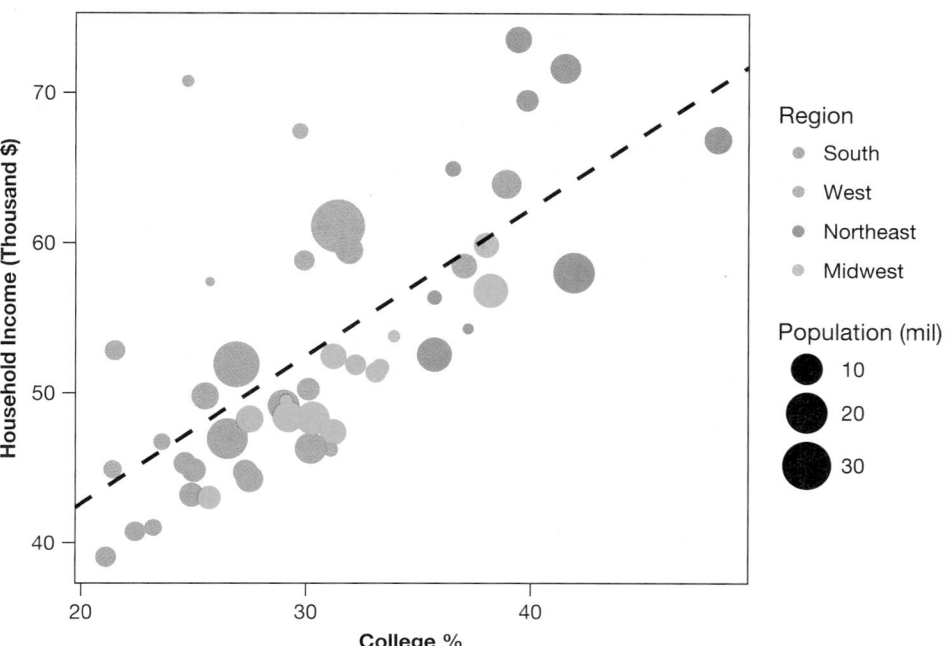

Figure 2.75 *Augmented scatterplot showing various information for all US states*

For example, Figure 2.75 gives a scatterplot of household income vs percent with a college degree for all US States, using the **USStates** dataset. The size of each point is proportional to the population of the state, and the color of each point indicates the region of the state (South, West, Northeast, or Midwest). In this single figure we are showing four variables: three quantitative variables (*HouseholdIncome*, *College*, *Population*) and one categorical variable (*Region*). The regression line for predicting *HouseholdIncome* from *College* is also shown. This augmented scatterplot can be used to answer questions about any combination of these four variables.

Example 2.47

Figure 2.75 shows income, education, population, and region of US states.

(a) What is the approximate population of the state with the highest household income, and in what region is it located?

(b) On average, which region of the US has the highest rate of college education?

(c) Is there a region of the US that appears to have high household income relative to college education rate?

Solution

(a) Comparing the size of the point with highest income with the legend on the right, we see that the state's population is slightly smaller than 10 million (somewhere between 5 and 10 million), and by its color we see that it is in the northeast region. The state is Maryland.

(b) The northeast states (green points) tend to be toward the right of the plot, indicating that these states generally have higher college education rates.

(c) Western states (purple points) generally appear to have high household income relative to their college education rate. There are several purple points that are far above the regression line.

If we are not restricted to the printed page, scatterplots and other graphs can be augmented further by making them *interactive* or *dynamic*. An interactive graph can give additional information based on user input. For example, an interactive version of Figure 2.75 may give the name of the corresponding state if your mouse curser hovers over a particular point. A dynamic graph can change over time, like a movie.

A great example of augmented scatterplots that are both dynamic and interactive is the Gapminder software (*https://www.gapminder.org/tools*). This online applet can display worldwide health, economic, and environmental data aggregated by country over time. There are several variables to choose for the horizontal and vertical axes of a scatterplot, and the applet allows us to see how the scatterplot has changed historically year-by-year. By default, the color of a given point gives the world region of the country, and the size of the points give the population of the country.

Example 2.48

Gapminder: Child Mortality vs Income

In the Gapminder applet, set the vertical axis to *child mortality rate* (deaths under age 5 per 1000 births) and the horizontal axis to *income per person* (GDP in current US dollars, per person). Click the play icon and observe how the scatterplot

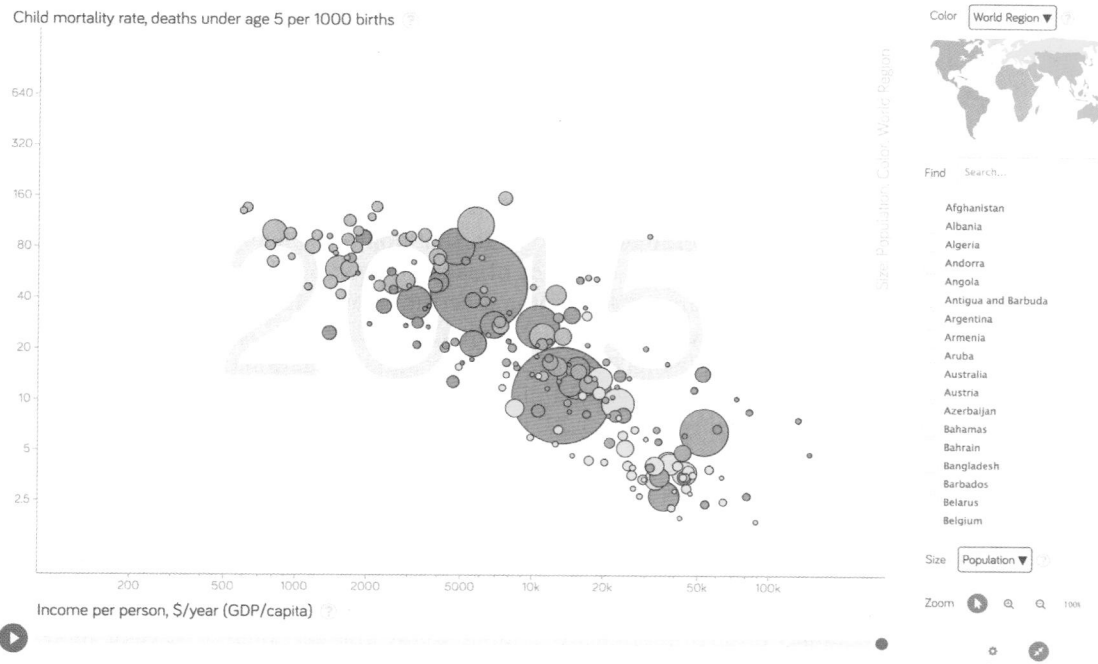

Figure 2.76 *A screenshot of the Gapminder software**

changes over time.[80] This dynamic scatterplot allows us to answer the following questions:

(a) In 2015, do child mortality rate and income per person have a positive or negative association?

(b) In 1800, which country has the largest income per person? In 2000?

(c) In 1800, what are the two largest countries by population? In 2015?

(d) In general, what is the worldwide trend in child mortality rate and income per person from 1800 to 2015?

Solution

(a) Pause at the year 2015. Child mortality rate and income per person have a clear negative association (see Figure 2.76).

(b) When we start at the first frame (1800) and hover over the rightmost point, we see that the country with the largest income per person is the Netherlands. When we pause at the year 2000, we see that the country with the highest income per person is Qatar.

(c) When we start at 1800 and hover over the two largest points, we see that they are China and India. When we do the same in 2015, we see that the two largest countries by population are still China and India.

(d) Play the graph from start to finish. In general, with each passing year we see child mortality decrease and income per person increase.

---

[80]For a very exciting commentary on this dynamic scatterplot, see the TED talk by Hans Rosling, a founder of Gapminder: *https://www.ted.com/talks/hans_rosling_shows_the_best_stats_you_ve_ever_seen*.

Displaying Geographic Data

Data are popularly referred to as "numbers + context," and it is nice when visualization of data connects to the context, in addition to simply showing the numbers. If the data correspond to geographic entities, one way to connect to their context is to display data or summary statistics on a map. A common approach is to color each geographic unit (for example: states, countries, or cities) by the variable of interest. If the variable is categorical we can use a different color for each category, analogous to using different colors for different points in an augmented scatterplot. If the variable is quantitative, the color of each unit can be given by a color-scale, in which values are associated with a color spectrum (for example: light to dark, or green to blue); this is sometimes called a *heatmap*.

As an example of a heatmap, we consider population density from the **AllCountries** dataset. Here our geographic units are countries, and we can show all countries on a world map. In Figure 2.77, the color of each country is determined by matching that country's population density in people per square kilometer with a color on the spectrum from light yellow (low density) to dark red (high density).

Example 2.49

Population Density Heatmap

Using the population density in Figure 2.77, rank the following countries from lowest to highest population density: United States, India, Canada, Mexico, and China.

Solution Looking at their colors, we see that Canada (light yellow) is the least dense, then the United States (yellow), then Mexico (orange), then China (light red), then India (red).

Population Density (Million per 1000 sq.km)

Figure 2.77 *A map of world countries, colored by population density.*

Displaying Data Over Time

In many situations, data are measured over multiple points in time. In such cases we often plot *time* on the horizontal axis and a quantitative variable of interest on the vertical axis to produce a *time series plot*. Since we typically have just one data value at each time period, we usually connect the points for adjacent time periods with line segments to help visualize trends and pattern in the data over time (and may omit plotting the points themselves.) Although there may be lots of short term fluctuations in a time series plot, we try to look for long term general trends, possible striking departures from trends, and potential seasonal patterns that are repeated at regular intervals.

Example 2.50

Quarterly Housing Starts (2000–2015)

A common measure of economic activity is the number of new homes being constructed. Data in **HouseStarts** include the number (in thousands) of new residential houses started in the US[81] each quarter from 2000 to 2015. Figure 2.78(a) shows a time series plot of the data over all 16 years and Figure 2.78(b) focuses on just the five year period from 2011 to 2015 to better see the quarterly seasonal pattern. Discuss what these plots tell us about trends in housing starts over this time period.

Solution

In Figure 2.78(a) we see that housing starts in the US generally increased until about 2006 when a substantial decline started that lasted until around 2009 (corresponding to the worldwide economic slowdown often called the Great Recession). Housing starts began to recover after 2009 and show modest increases through 2015, but still lag well below the levels of the early 2000's.

In Figure 2.78(b) we examine the recovery more closely and notice a seasonal pattern that occurs throughout the time series. Housing starts tend to be low in the winter (Q1=January-March), rise rapidly to a high level in the spring (Q2=April-June), drop slightly, but stay pretty high in the summer (Q3=July-September), and then fall quite a bit in the fall (Q4=October-December). This pattern, a clear reflection of the influence of favorable building weather, repeats over the five-year period, even as the overall trend shows a steady increase.

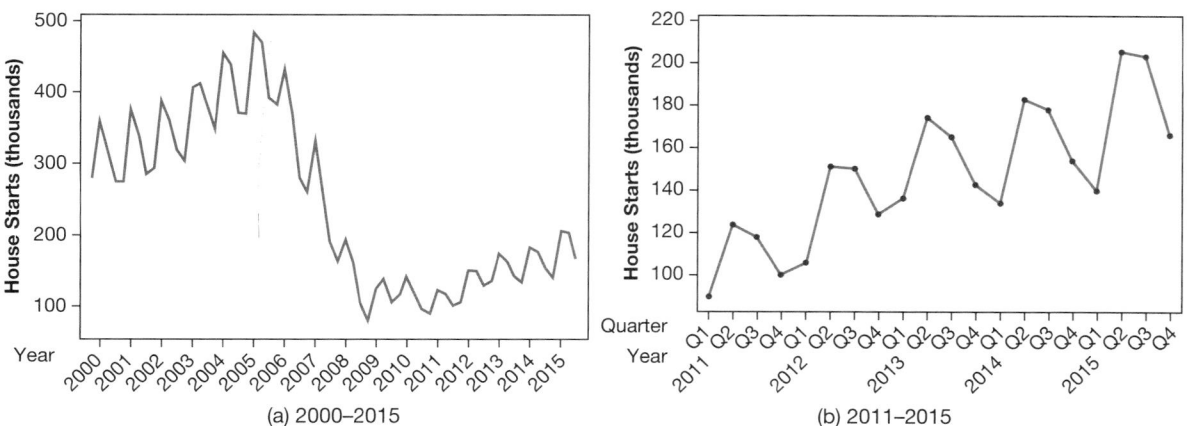

Figure 2.78 *Quarterly housing starts in the United States*

[81] Data on housing starts downloaded from the US Census Bureau website at *https://www.census.gov/econ/currentdata/*

DATA 2.13

Berkeley Growth Study

In the 1940s and 1950s, the heights of 39 boys and 54 girls, in centimeters, were measured at 30 different time points between the ages of 1 and 18 years as part of the University of California Berkeley growth study.[82] The data are available in **HeightData**. ∎

The Berkeley growth data involves many repeated measures over time (*age*), so it makes sense to plot the data in a chart where *age* is given on the horizontal axis, and the *height* of the children is given on the vertical axis. To display the data for a single child, we could simply make a time series plot of *height* that shows the measurements at each of the 30 different *age* times for that child. To show the data for all children at once we can use a *spaghetti plot*. In a spaghetti plot, we plot all available measurements and connect the points within each subject (in this case, each child) with a line, like many strands of spaghetti.

This is a spaghetti pot, NOT a spaghetti plot

In Figure 2.79 a line shows the growth trajectory for each child in the Berkeley growth study. We again make use of color, letting red lines correspond to girls and blue lines correspond to boys.

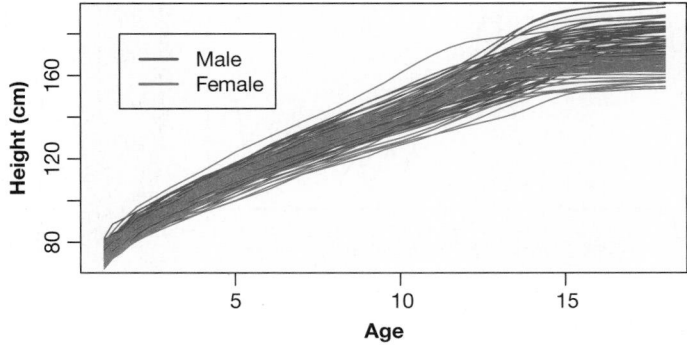

Figure 2.79 *Spaghetti plot of heights in the Berkeley growth study, colored by gender*

[82]Tuddenham, R.D., and Snyder, M.M., (1954). "Physical growth of California boys and girls from birth to age 18", *University of California Publications in Child Development*, 1, 183–364.

Example 2.51

Using Figure 2.79, showing childhood growth:

(a) Generally when do girls tend to stop growing in height? When do boys stop growing in height?

(b) Was the tallest participant in this study from ages 2 to 12 a boy or a girl?

Solution

(a) For girls, growth appears to slow substantially between ages 13 and 14 (after this, the slope of the red lines are very small). For boys, growth appears to slow between the ages of 15 and 16.

(b) A red line is above all others between age 2 and age 12, so the tallest individual between these ages is a girl.

Breaking it Down

Many of the basic plots that we've learned in previous sections can easily be extended just by breaking it down by a categorical variable. In other words, instead of looking at one plot containing all the cases, we break it down into several plots, where each plot includes only the cases within a certain subgroup. Often, breaking the data down into subgroups can reveal new and important insights.

DATA 2.14

Discrimination among the Developmentally Disabled?

The California Department of Developmental Services (DDS) allocates funds to support developmentally disabled California residents (such as those with autism, cerebral palsy, or intellectual disabilities) and their families. We'll refer to those supported by DDS as DDS consumers. An allegation of discrimination was made when it was found that average annual expenditures were substantially lower for Hispanic consumers than for white non-Hispanic consumers (who, for simplicity below, we refer to simply as white consumers.) The dataset **DDS** includes data on annual expenditure (in $), ethnicity, age, and gender for 1000 DDS consumers.[83] Do these data provide evidence of discrimination?   ■

Example 2.52

Expenditure by Ethnicity

Compare annual expenditure for Hispanic consumers versus white consumers.

Solution

Figure 2.80 provides a visual comparison of annual DDS expenditure values for Hispanic consumers versus white consumers. From the graph, it is immediately apparent that expenditures tend to be much higher for white consumers than for Hispanic consumers. Computing means for each group, we find that the average annual expenditure is $11,066 for Hispanic consumers, as opposed to $24,698 for white consumers. Based on this analysis of two variables, it appears that these data provide very strong evidence of discrimination (we could formalize this with techniques we'll learn in Chapter 4 or 6; this difference is *extremely* significant), with expenditures much higher, on average, for white non-Hispanics than for Hispanics.

[83] Taylor, S.A. and Mickel, A.E. (2014). "Simpson's Paradox: A Data Set and Discrimination Case Study Exercise," *Journal of Statistics Education*, **22**(1). The dataset has been altered slightly for privacy reasons, but is based on actual DDS consumers.

Figure 2.80 *Annual expenditure for Hispanic consumers vs white consumers*

An initial bivariate exploration of expenditure by ethnicity finds very strong evidence of discrimination. What happens if we "break it down" by age? To make the data easier to visualize in separate groups, we'll work with age group categories rather than age itself.

Example 2.53

Expenditure by Ethnicity, Broken Down by Age Group

Compare annual expenditure for Hispanic consumers versus white consumers, broken down by age group.

Solution   Figure 2.81 provides a visual comparison of annual expenditure values for Hispanic residents versus white residents, broken down by age group. Each set of side-by-side boxplots can be interpreted in the usual way, except that now we have a separate set of side-by-side boxplots for each age group (with age groups shown along the top). Take a close look at this plot, and in particular, look at the comparison of Hispanic to white expenditures *within each age group*. We now see that, within *each* age group, expenditures are actually *higher* for Hispanics than for whites!

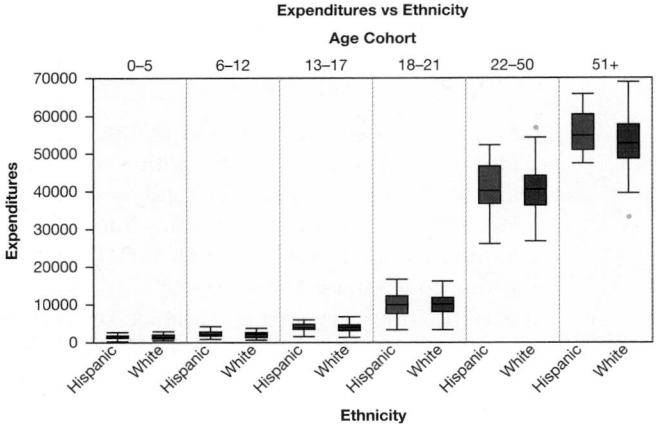

Figure 2.81 *Expenditure for Hispanic vs white, by age group*

In Example 2.53 we find that within age groups, expenditures are consistently higher for Hispanics than for whites, yet in Example 2.52 we find that expenditures overall are higher for whites than for Hispanics. How is this possible?? The explanation lies in one additional data visualization.

Example 2.54

Ethnicity by Age Group

Compare ethnicity counts by age group.

Solution

Figure 2.82 shows a bar chart displaying the number of people within each ethnic category, broken down by age group. For example, the first red bar goes up to 44 on the y-axis, showing that there are 44 Hispanic consumers within the 0-5 years old age group. This plot shows that the sample contains many more Hispanic children than white children, and many more white adults than Hispanic adults.

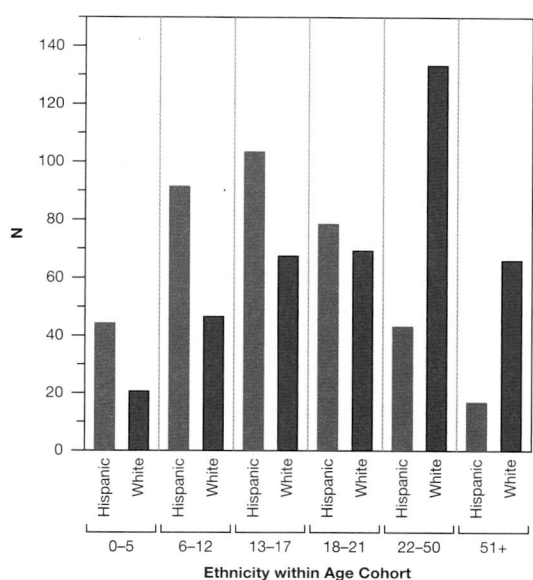

Figure 2.82 *Ethnicity counts by age group*

Combining the information from Figure 2.81 and Figure 2.82, we see that white consumers tend to be older than Hispanic consumers (Figure 2.82), and older consumers tend to receive higher expenditures than younger consumers (Figure 2.81). This explains why the white consumers receive higher expenditures overall: not because of discrimination, but just because they tend to be older. These visualizations are shown together in Figure 2.83, illustrating this point. Here age is a confounding variable, as introduced in Section 1.3: age is associated with both ethnicity and expenditure, confounding the relationship. Failing to account for the

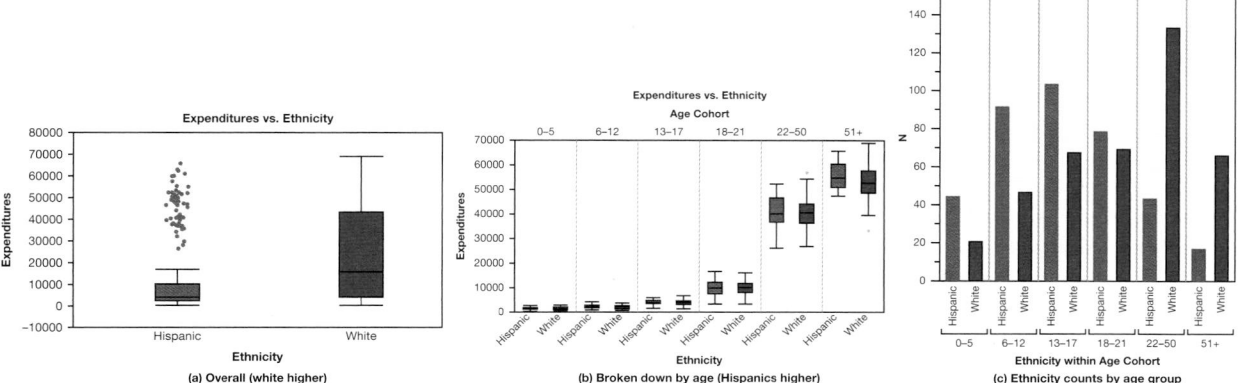

Figure 2.83 *Comparing the visualizations of DDS expenditure by ethnicity, with age as a confounding variable*

confounding variable can paint a very misleading picture. Luckily, DDS called in a statistician, and was not falsely charged with discrimination.

This is an example of *Simpson's Paradox*, which occurs when the relationship between two variables switches when the data are broken down by a third variable. Although a true reversal is rare, it is often true that the relationship between two variables differs when a third variable is taken into account. Incorporating more than just two variables can be important for revealing the true story the data have to tell.

Other Visualization Examples

The internet has many creative, interactive, and fascinating visualizations of data. Here, we link to a small sample of great visualization examples that we encourage you to check out:

- This interactive visual from the *NY Times* allows you to guess the relationship between income and percent of children attending college, then compares your guess with the guesses of others and the reality: *http://www.nytimes.com/interactive/ 2015/05/28/upshot/you-draw-it-how-family-income-affects-childrens-college-chances.html.*

- This interactive visual from *fivethirtyeight.com* is an interactive spaghetti plot of the performance history of every team in the national basketball association: *http://projects.fivethirtyeight.com/complete-history-of-the-nba/.*

- This visual, by Mauro Martino (*mamartino.com*), shows the increase in political polarization in the United States Congress since 1948: *http://www.mamartino.com/ projects/rise_of_partisanship/.*

- This creative dynamic visualization by Nathan Yau (*flowingdata.com*) shows a day in the life of 1,000 people : *http://flowingdata.com/2015/12/15/a-day-in-the-life-of-americans/.*

SECTION LEARNING GOALS

You should now have the understanding and skills to:

- Interpret information from a variety of data visualizations
- Recognize ways to include multiple variables, and the value of including additional variables, in a display
- Recognize ways to include geographic data in a display
- Recognize ways to display time-dependent data
- Recognize that there are many effective and creative ways to display data

Exercises for Section **2.7**

There are many links included in these exercises, and some will probably break during the life of this text. We apologize in advance for any inconvenience.

2.226 Considering Direction of Association

(a) How many variables are included in the scatterplot in Figure 2.84(a)? Identify each as categorical or quantitative. Estimate the range for Variable1 and for Variable2.

(b) In Figure 2.84(a), does the association between the variables appear to be positive or negative?

(c) Figure 2.84(b) shows the same scatterplot with regression line added. Which variable is the

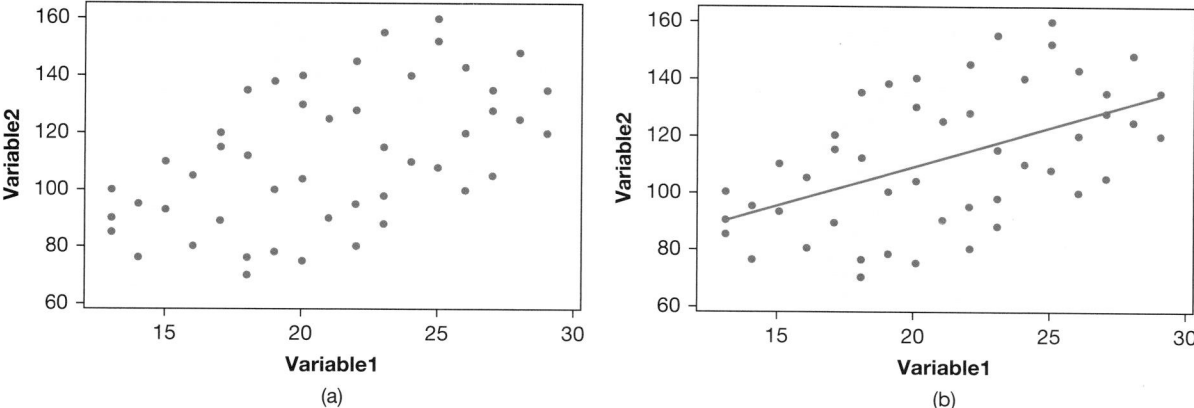

Figure 2.84 *Describe the association between these variables*

response variable? Does the line show a positive or negative association?

(d) Figure 2.85(a) shows the same scatterplot with a third variable included. Is the new variable categorical or quantitative? If categorical, how many categories? If quantitative, estimate the range.

(e) In Figure 2.85(a), if we only consider cases in Group A, does the association between Variable1 and Variable2 appear to be positive or negative? How about in Group B? Group C? Group D?

(f) Figure 2.85(b) shows the same scatterplot as Figure 2.85(a) with regression lines added within each of the four groups. Does the regression line for Group A show a positive or negative association? How about Group B? Group C? Group D?

(g) What happens to the direction of association shown in Figure 2.84 when we add the additional information contained in Variable3 as in

Figure 2.85? (This is an example of Simpson's Paradox for quantitative variables.)

2.227 Visualizing the Happy Planet Index Figure 2.86 shows a scatterplot illustrating three different variables from the dataset **HappyPlanetIndex**, introduced in Exercise 2.187. The variable *Happiness* is a measure of the well-being of a country, with larger numbers indicating greater happiness, health, and well-being. The variable *Footprint* is a per capita measure of the ecological impact of a country on the environment, with larger numbers indicating greater use of resources (such as gas and electricity) and more damage to the planet. A third variable, *Region*, is given by the code shown in the top right, and is categorized as follows: 1 = Latin America, 2 = Western nations, 3 = Middle East, 4 = Sub-Saharan Africa, 5 = South Asia, 6 = East Asia, 7 = former Communist countries.

(a) Classify each of the three variables as categorical or quantitative.

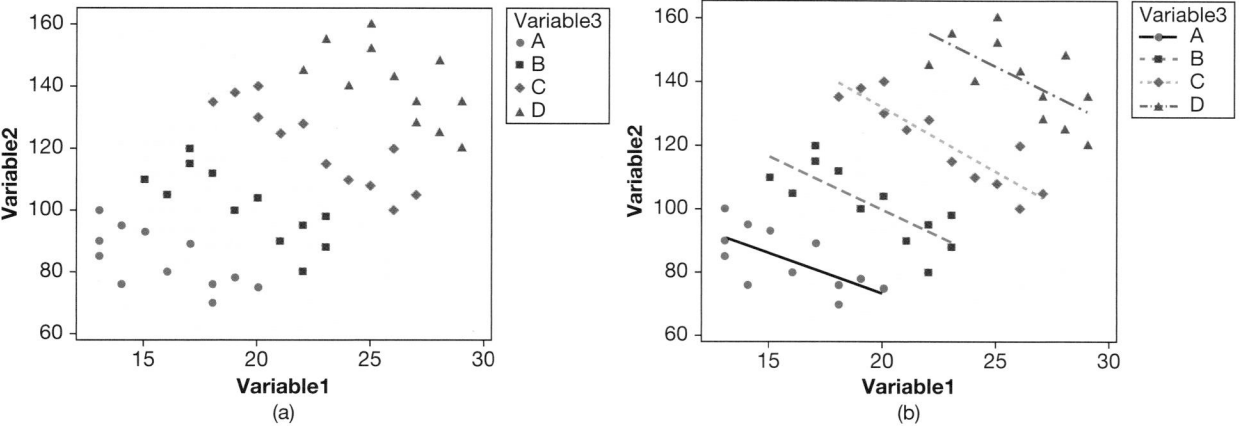

Figure 2.85 *Describe the association in each group*

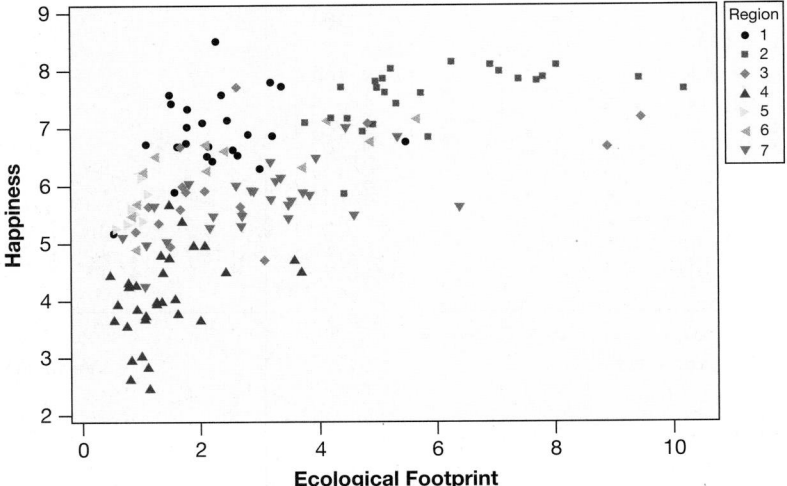

Figure 2.86 *Happiness, ecological footprint, and region*

(b) Which two regions seem to have the greatest happiness score? Which region seems to have the greatest ecological footprint?

(c) Which region seems to have the lowest happiness score? Does the ecological footprint tend to be high or low in that area?

(d) Looking at the scatterplot overall and ignoring region, does there appear to be a positive relationship between happiness score and ecological footprint?

(e) Considering only region 2 (Western nations), does there appear to be a positive relationship between happiness score and ecological footprint?

(f) The country with the highest happiness score is Costa Rica. Is it in the top left, top right, bottom left, or bottom right of the scatterplot?

(g) According to Nic Marks, the developer of the Happy Planet Index, we should be trying to move more countries to the top left of the scatterplot. To do this for countries in region 4 (Sub-Saharan Africa), which variable should we focus on and should we be trying to increase or decrease this variable for these countries? For countries in region 2 (Western nations), which variable should we focus on and should we be trying to increase or decrease this variable for these countries?

2.228 Height, Weight, and BodyFat Figure 2.87 shows a bubble plot of height, weight, and body fat percentage for a sample of 100 men, from the dataset **BodyFat**, introduced in Exercise 2.219. The body fat percentage is indicated by the size of the bubble for each case.

(a) How many variables are shown in the scatterplot? Identify each as categorical or quantitative.

(b) Ignoring bubble size, does there appear to be a positive or negative relationship between height and weight?

(c) Do the bubbles tend to be larger on the top half of the scatterplot or the bottom half? Interpret this in context and in terms of the relevant variables.

(d) Body fat percentage depends on more than just height and weight. There are two cases who are about 66 inches tall, one weighing about 125 pounds and the other about 140 pounds. Which has the larger body fat percentage?

(e) There are two cases weighing about 125 pounds, one about 66 inches tall and the other about 67 inches tall. Which has a larger body fat percentage?

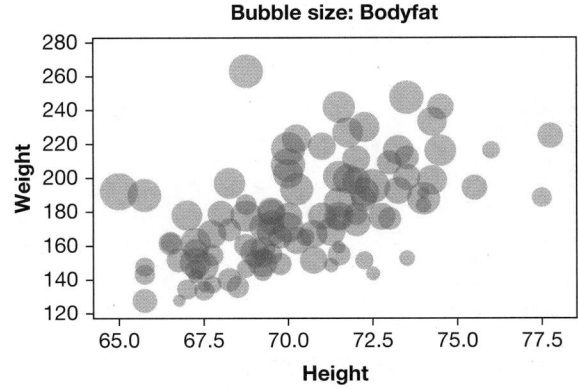

Figure 2.87 *Height, weight, and body fat*

(f) Does the person with the largest weight have a body fat percentage that is relatively large, relatively small, or pretty average?

(g) Does the person with the largest height have a body fat percentage that is relatively large, relatively small, or pretty average? How about the person with the third largest height?

(h) This particular sample included only males, but if we created a similar graph for a dataset including both males and females, indicate one way in which we could incorporate the fourth variable of gender in the graph.

2.229 Visualizing Football and Brain Size Exercise 2.143 introduces a study in which the number of years playing football and the size of the hippocampus in the brain were recorded for each person in the study. There were three different groups in the study: football players who had been diagnosed with at least one concussion, football players who had never been diagnosed with a concussion, and a control group of people who had never played football. Figure 2.88(a) shows a graph that incorporates all three of these variables.

(a) Identify each variable as quantitative or categorical.

(b) Why are all the blue dots stacked up on the left?

(c) Overall, does there appear to be a positive or negative association (or no association) between years playing football and hippocampus size?

(d) Figure 2.88(b) shows the same graph with regression lines for the two groups of football players. Which of the groups has the line that is lower on the graph? What does this tell us in the context of the three variables?

(e) Which of the groups has the line with the steeper slope?

2.230 Carbon Dioxide Levels Over Time Scientists are concerned about global warming and the effect of carbon dioxide emissions on the atmosphere. Figure 2.89 shows the concentration of carbon dioxide (CO_2) in the atmosphere, in parts per million (ppm), over two different time intervals. Often, plots of data over time can look very different depending on the time interval selected for the graph. Figure 2.89(a) shows the concentration[84] of CO_2 during the period from 1959 to 2015, while Figure 2.89(b) shows the concentration[85] over a very different window: the last 400,000 years!!

(a) Using Figure 2.89(a), estimate the CO_2 concentration in 1960. Estimate the CO_2 concentration in 2015.

(b) Is CO_2 concentration primarily increasing, decreasing, or oscillating up and down during the period from 1959 to 2015?

(c) In the period of time shown in Figure 2.89(b), is CO_2 concentration primarily increasing, decreasing, or oscillating up and down?

(d) In Figure 2.89(b), locate the portion of data that is shown in Figure 2.89(a). What does the curve look like on that piece?

(e) What was the highest concentration of CO_2 ever in the 400,000 year history of the Earth, before 1950?

(f) For more data visualization on this subject, watch the 3-minute animated video "CO2 Movie" at *http://www.esrl.noaa.gov/gmd/ccgg/trends/history.html*.

[84]Dr. Pieter Tans, NOAA/ESRL (*www.esrl.noaa.gov/gmd/ccgg/trends/*) and Dr. Ralph Keeling, Scripps Institution of Oceanography (*scrippsco2.ucsd.edu/*).
[85]Data from the Vostok Ice Core project, Barnola, J.M., Raynaud, D., Lorius, C., and Barkov, N.I., *http://cdiac.ornl.gov/ftp/trends/co2/vostok.icecore.co2*

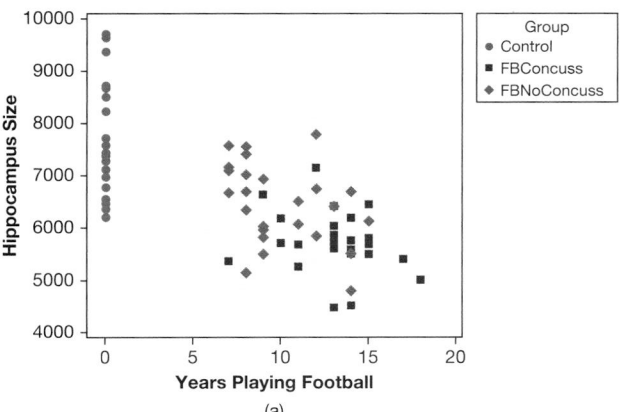
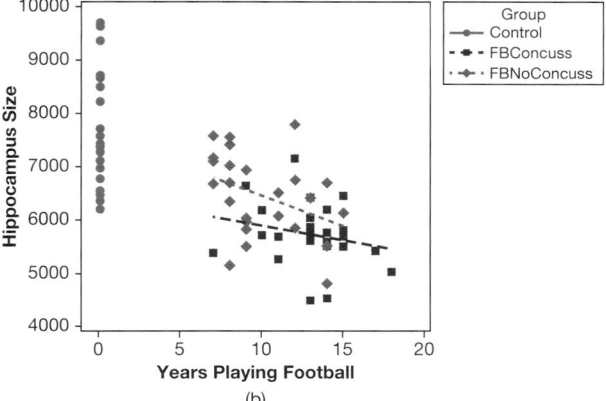

Figure 2.88 *Brain size, football experience, and concussions*

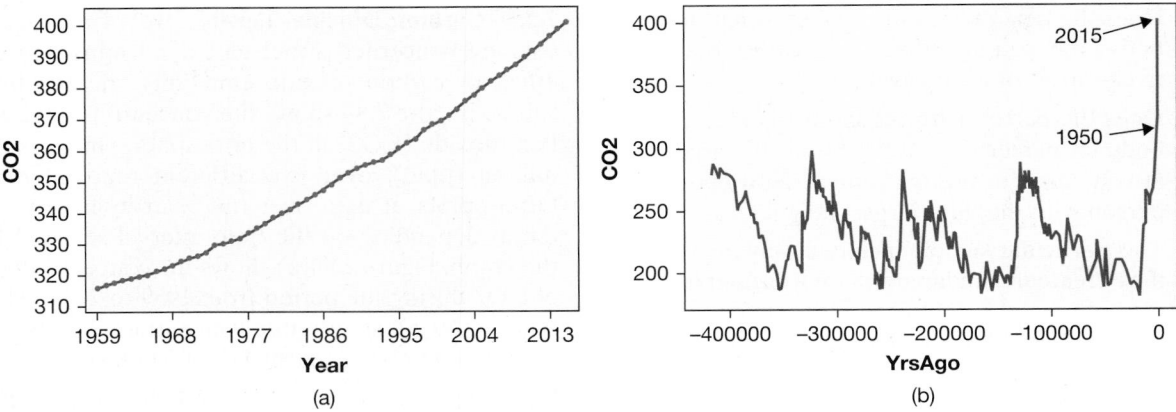

Figure 2.89 *Carbon dioxide levels over time, in two very different windows*

2.231 Forty-Yard Dash at the NFL Combine Every year the National Football League invites 335 draft eligible college football players to a scouting combine where they participate in a variety of drills and exercises. One of the more popular drills, called the 40-yard dash, is the time it takes each player to run 40-yards. We computed the average 40-yard dash time every season from 1990 to 2016 for all players at two positions: wide receiver (WR), who receive passes from the quarterback, and defensive cornerback (DC), who try to stop the wide receivers from catching the ball. These data are presented as a spaghetti plot in Figure 2.90.

(a) Describe the general trend of both positions. Are the players getting faster or slower?

(b) In 2016 which position had a faster average 40 time?

(c) Does one position appear to be consistently faster than the other?

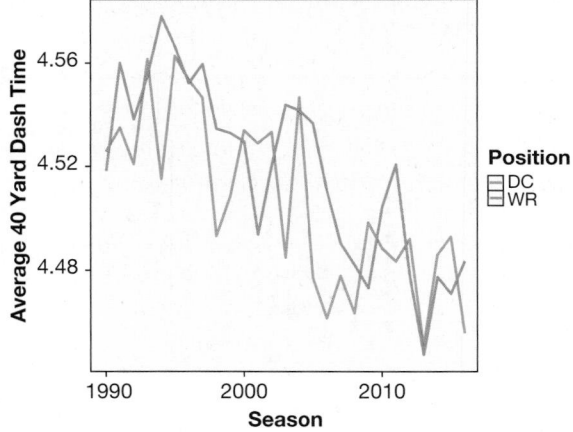

Figure 2.90 *Average 40 yard dash time by season for WR and DC*

2.232 Golden State Warriors: One Game During a record breaking season the Golden State Warriors of the National Basketball Association (NBA) won 24 straight games to start the 2015–2016 season. Adam Pearce plotted the point differential, Golden State points − Opponent points, each minute of the first 16 games of the streak.[86] One of those games, a 119 to 104 victory on November 6th, is plotted in Figure 2.91(a).

(a) Were the Warriors ever losing in this game (point differential below 0)?

(b) The game is split into quarters, demonstrated by the minutes remaining where 48 to 36 is 1st quarter, 36 to 24 is 2nd quarter, 24 to 12 is 3rd quarter, 12 to 0 is 4th quarter. In which quarter did the Warriors have their largest lead?

2.233 Golden State Warriors: First Half of Season Exercise 2.232 plotted the Golden State Warriors point differential, Golden State points − Opponent points, each minute of one game during their record breaking 2015–2016 season. Adam Pearce[87] of *road-tolarissa.com* also plotted the game state (winning or losing or tied) every minute of every game over the first half of the Warriors season in Figure 2.91(b). Within this plot each orange dot above 0 indicates a game where they were winning at that minute (point differential above 0), and each purple dot below 0 indicates a game where they were losing (point differential below 0) at that minute. The darkness of the dot indicates how far the point differential was above or below 0.

---

[86] Used with permission from *http://roadtolarissa.com/gsw-streak/*. Check out more of Adam Pearce's work at *roadtolarissa.com*.

[87] Used with permission. Check out more of Adam Pearce's work at *roadtolarissa.com*.

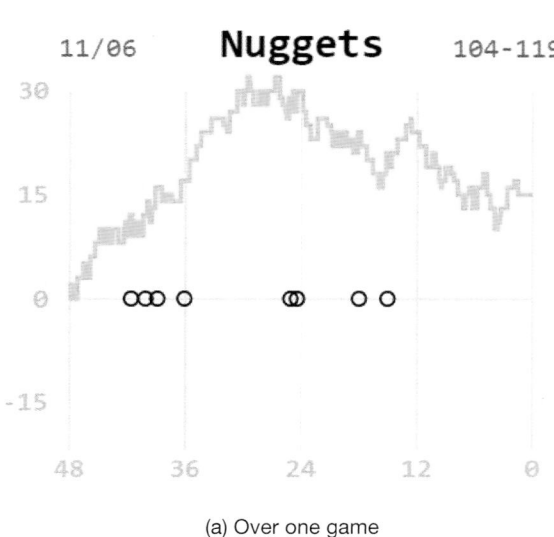

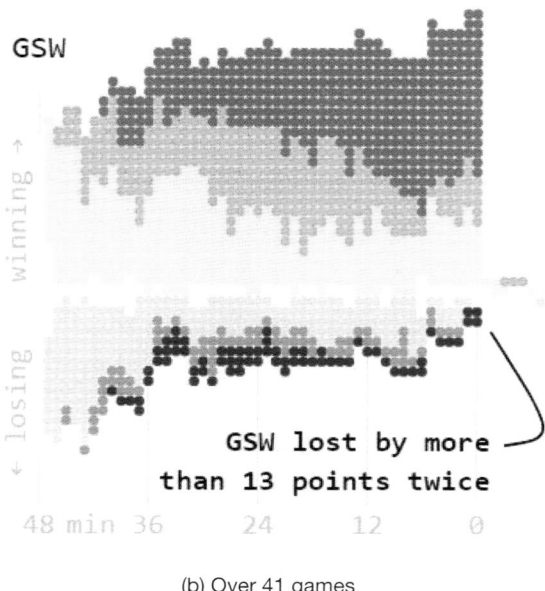

(a) Over one game

(b) Over 41 games

Figure 2.91 *Score difference by minutes*

(a) How many games were the Warriors losing at the 24-minute mark (halftime)?

(b) How many games were the Warriors losing at the 0-minute mark (the end of the game)?

2.234 Marriage Age vs Number of Children Using the Gapminder software (*https://www.gapminder.org/tools*), set the vertical axis to *Age at 1st marriage (women)* and the horizontal axis to *Babies per woman*. This scatterplot shows the mean age at which woman marry, and the mean number of children they have, for various countries. Click the play icon and observe how the scatterplot changes over time, then answer the following questions:

(a) Overall, is there a positive or negative association between *Age at 1st marriage* and *Babies per woman*?

(b) Describe what happens to the number of babies per woman, and age at 1st marriage, between 1941 and 1943 in Russia (at the height of World War II).

(c) Describe what happens to the number of babies per woman, and the age at 1st marriage, in Libya from 1973 to 2005.

2.235 Income Distribution by Country A *mountain chart* is a creative way to display the distribution of a quantitative variable over different categories. The overall distribution is shown as a smoothed histogram, and the area underneath is colored according to different categories. The distribution of the variable for each category corresponds to the size of its colored area. The Gapminder software includes a mountain chart that shows the distribution of income broken down by world region and country (*http://www.gapminder.org/tools/mountain*). Click the play icon to see how the mountain chart changes from 1800 to present day, then answer the following questions:

(a) Extreme poverty is defined as living on less than $2 a day. In 1970, the majority of people living in extreme poverty came from which world region?

(b) In 2015, the majority of people living in extreme poverty came from which world region?

(c) Describe the shape of the worldwide distribution of income in 1970. In 2015.

2.236 A Day in the Life This creative dynamic visualization by Nathan Yau (*flowingdata.com*) shows a day in the life of 1000 different representative Americans based on survey data: *http://flowingdata.com/2015/12/15/a-day-in-the-life-of-americans/*. Watch the dynamic visualization over the entire day, then answer the following questions:

(a) At 6 am, what are the majority of Americans doing?

(b) At 10 am, what is the most common activity for Americans?

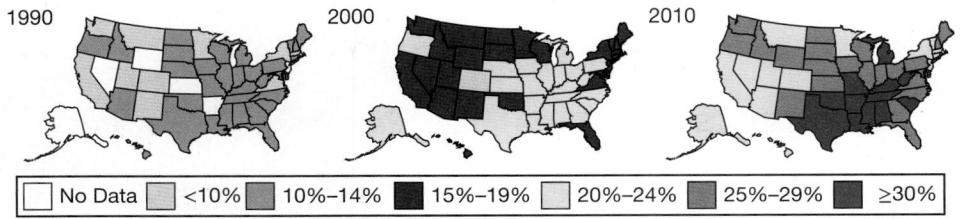

1990 2000 2010

No Data | <10% | 10%–14% | 15%–19% | 20%–24% | 25%–29% | ≥30%

Figure 2.92 *Obesity rates in the US over time*

(c) Are more Americans eating & drinking at 6 pm or 7 pm?

(d) Describe one additional fact from this visualization that you find particularly interesting or surprising.

2.237 US Obesity Levels by State and over Time These questions refer to the graphs found at *http://stateofobesity.org/adult-obesity/* which show a sequence of maps of US states, colored by the proportion of the adult population classified as obese, for many different years between 1990 and 2014. Three of the maps, from 1990, 2000, and 2010, as well as a key for the color coding of percent obese, are shown in Figure 2.92.

(a) The cases are US states and one variable is year. Another variable is percent obese in that state, which could be a quantitative variable but here is classified into categories, making it a categorical variable. Using the legend given in Figure 2.92 (and ignoring the category of "No Data"), how many different categories are shown? (There is one more category already in the online graphs as this book goes to print and maybe another one by the time you are reading this!)

(b) Using Figure 2.92, what appears to be the highest category needed for any state in 1990? In 2000? In 2010?

2.238 US Obesity Levels by State over Many Years Exercise 2.237 deals with some graphs showing information about the distribution of obesity rates in states over three different years. The website *http://stateofobesity.org/adult-obesity/* shows similar graphs for a wider selection of years. Use the graphs at the website to answer the questions below.

(a) What is the first year recorded in which the 15-19% category was needed, and how many states are in that category in that year? What is the first year the 20-24% category was needed? The 25-29% category? The 30-34% category? The 35%+ category?

(b) If you are in the US right now as you read this, what state are you in? In what obesity rate category did that state fall in 1990? In what category is it in now? If you are not in the US right now as you read this, find out the current percent obese of the country you are in. Name a state (and year, if needed) which roughly matches that value.

2.239 Spaghetti Plots of US Obesity Levels by State over Many Years Exercises 2.237 and 2.238 look at geographic plots of obesity rates in different years. The website *http://stateofobesity.org/adult-obesity/* also shows a spaghetti plot (on the right) which tracks the obesity rate of each state for the years from 1990. Hovering over any strand highlights that state (click to select it or click on the state in the map) and then you can point along the strand to see the obesity rate for that year. Use this graph to answer the questions below.

(a) Between 1990 and 2014, did the percent obese more than double: for every state or no states or just some states?

(b) Is there more variability in obesity rates between states in 1990 or in 2014?

(c) Identify the state with the largest percent obese in 1990, and give the state name and the percent obese at that time. In addition, identify the state with the smallest percent obese in 2014, and give the state name and the percent obese at that time.

2.240 What Do You Call a Sweetened Carbonated Beverage? If you reach for a sweetened carbonated beverage, do you refer to it as soda, pop, coke, or a soft drink? Different regions of the United States use different terms, as shown in this heat map: *discovermagazine.com/galleries/2013/june/regional-us-language-dialect*.[88] If you live in the United States, specify where you live and which term is

[88]"Soda or Pop? Maps Show Americans' Colorful Dialect Differences," *Discover Magazine*, 6/7/13. *discovermagazine.com/galleries/2013/june/regional-us-language-dialect*. Visualization by Joshua Katz (NC State University), Data from Bert Vaux (Cambridge University).

predominantly used there. If you do not live in the United States, choose a location in the US and specify the location and which term is predominantly used there.

2.241 The Dude Map Go to *http://qz.com/316906/the-dude-map-how-american-men-refer-to-their-bros/* to see "The dude map: How Americans refer to their bros,"[89] a heat map of the United States displaying how common the words "dude," "bro," "buddy," "fella," and "pal" are across the US.

(a) In which region of the country is "bro" most commonly used?

(b) In which region of the country is "buddy" most commonly used?

2.242 Cotton Pricing Although the abundance and availability of data have increased rapidly due to technology advances, and computers make fancier and fancier visualizations, data visualization itself is not new. This link *http://www.handsomeatlas.com/us-census-statistical-atlas-1880/manufactures-specific-cotton-goods* provides a data visualization of the consumption and price of cotton from 1880. Zoom in on the graph to better see certain features.

(a) In 1880, which state spent the most money (per capita) on cotton? Which spent the least?

(b) In which year (1825 to 1880) was cotton most expensive?

2.243 A Map of All Americans! Visualization can often be an effective way to make sense of very large datasets. The Census Dot Map at *https://demographics.virginia.edu/DotMap/* displays the race and location of *every* American recorded by the 2010 Census; that's over 300 million data points displayed simultaneously on one map! If you were counted in the US Census in 2010, you can find your dot on the map!

---

[89]Sonnad, N., "The dude map: How Americans refer to their bros," *Quartz*, 12/23/14.

(a) Zoom out to see the entire United States. Which half of the country is more heavily populated, the East or the West?

(b) Zoom in to see the town/city your school is located in.[90] (Click on "Add map labels" to help you find it.) Are there any noticeable racial/ethnic patterns, with any areas predominantly populated by a particular racial/ethnic group? If so, describe one of these noticeable characteristics. If not, describe a noticeable characteristic about the population density in your town.

2.244 Names over Time The website *http://www.visualcinnamon.com/babynamesus* gives a spaghetti plot showing the popularity of the top 10 baby names for each year 1880 to 2014 (use the window scroller at the bottom to select the timespan shown). By default, girl names are shown. What was the top baby girl name in 2014? In 1880?

FISHER'S IRIS DATA

Exercises 2.245 to 2.247 refer to the data in **FisherIris**, from a paper published in 1936 by Sir R.A. Fisher, widely considered the father of modern statistics.[91] The cases are 150 irises and there are five variables: Type of iris is categorical, while petal width, petal length, sepal width, and sepal length (all in millimeters) are quantitative. Sepals are the green leaves underneath the petals, providing support for the petals.

2.245 Petal Length and Petal Width Figure 2.93(a) shows a scatterplot of the two quantitative variables petal length and petal width.

(a) Explain how the scatterplot appears to show at least two different types of irises.

---

[90]If outside the US, pick a city to look at and specify which city you pick.

[91]Fisher, R.A., "The use of multiple measurements in taxonomic problems," *Annals of Eugenics*, 7(2), 1936, 179–188.

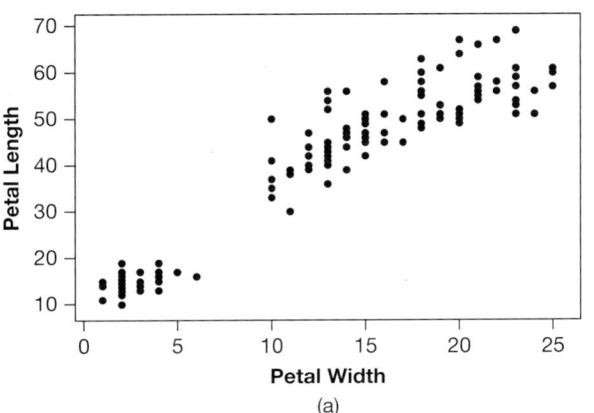

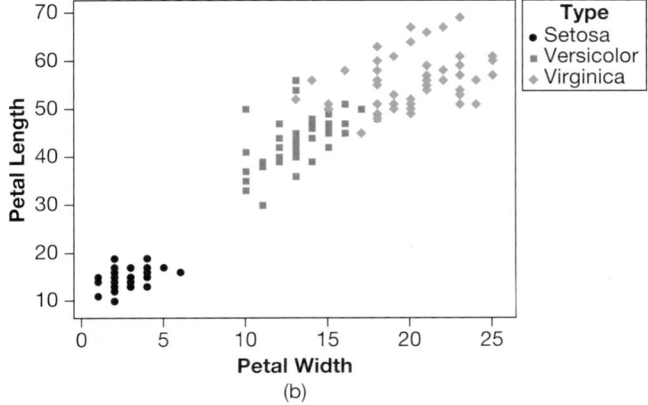

Figure 2.93 *Petal length and petal width*

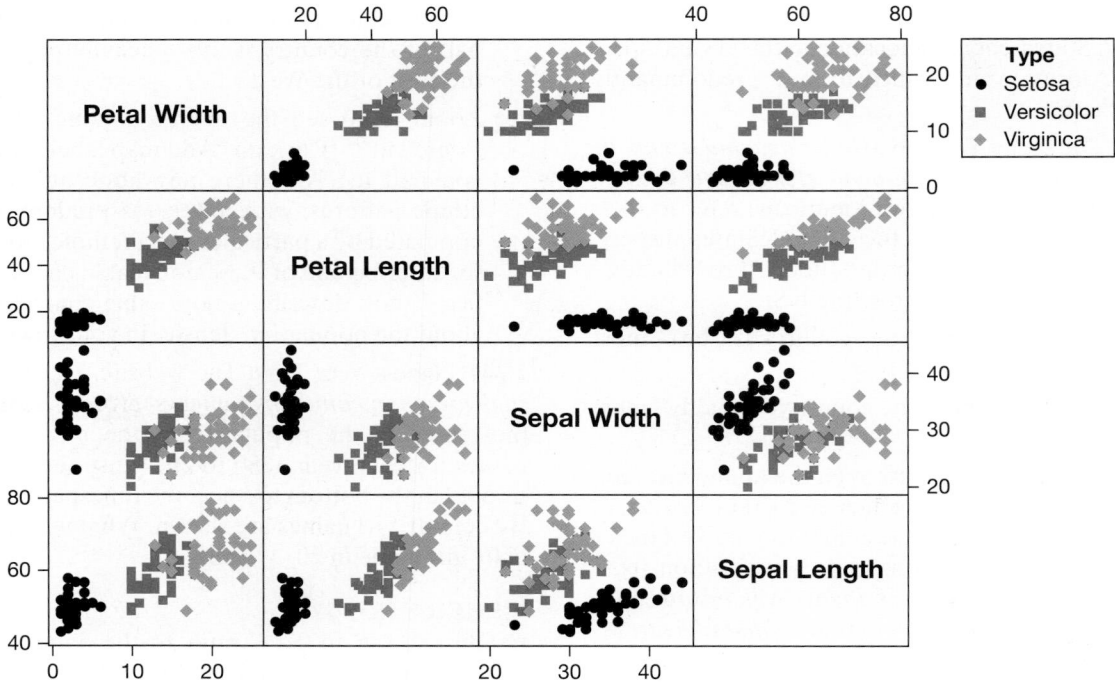

Figure 2.94 *A scatterplot matrix*

(b) Figure 2.93(b) shows the same scatterplot with the *Type* categorical variable included. We can now see the that there are three different types included: Setosa, Versacolor, and Virginica. Which type has the smallest petals? Which type has the largest petals?

2.246 A Scatterplot Matrix There are five variables in the dataset, and all are included in the *scatterplot matrix* shown in Figure 2.94. The graph shows scatterplots for each pair of quantitative variables, with the *Type* categorical variable included on each. For example, the second one down on the left is the same scatterplot as in Figure 2.93(b), with petal width on the horizontal axis and petal length on the vertical axis.

(a) Which shows a clearer distinction between Versicolor and Virginica: a scatterplot of petal length and petal width or a scatterplot of sepal length and sepal width?

(b) Considering only the Setosa type of iris, does the association between sepal width and petal length appear to be positive, negative, or neither?

2.247 A 3-Dimensional Scatterplot We have seen that we can use a bubble plot to show a third quantitative variable on a scatterplot. Another way to show three quantitative variables together is to use

a *3-dimensional scatterplot*, such as the one showing petal length, petal width, and sepal length in Figure 2.95. In this case, which variable is on the vertical axis? Which color dots are highest up (meaning they have the largest values for that variable)?

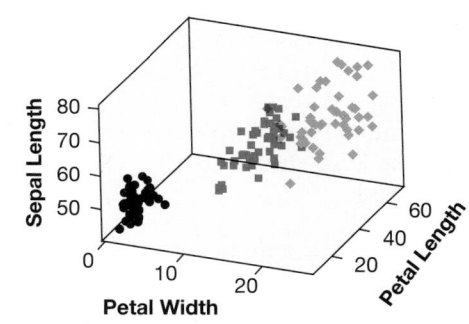

Figure 2.95 *A 3-dimensional scatterplot*

VISUALIZING CHANGES IN THE US POPULATION OVER TIME

The website *http://www.pewresearch.org/next-america/#Two-Dramas-in-Slow-Motion* shows two data visualizations of the the US population changing from 1950 or 1960 to 2060 (projected). The first visualization is broken down by both sex and age, and the second visualization is broken down by race and ethnicity. Exercises 2.248 and 2.249 pertain to this article.

2.248 Look at the dynamic visualization titled "U.S. Age Pyramid Becomes a Rectangle" (the first visualization on the page). The data visualization is essentially histograms of age, but the histograms are turned sideways and broken down by sex, comparing males and females.

(a) Go to View Data for 1950 (the baby boomer band will be at the bottom). Describe the distribution of ages (for either sex) in 1950.

(b) In 1950, is the age distribution left-skewed, right-skewed, or symmetric? (Think of what the histogram would look like on a regular number line with 0 on the left.)

(c) Click the Animation Control button to watch the distribution change over time. How does the (projected) age distribution in 2060 differ from the age distribution in 1950? (Note the age distribution in 2060 is close to what is referred to as a "uniform" distribution, a histogram that is essentially flat.)

(d) In 2060, are there projected to be more males or females in the 85+ range?

2.249 Look at the visualization titled "Changing Face of America" (the second visualization on the page). This is a new kind of visualization in which the total for each year is scaled to 100%, and the colors are shaded according to the percentage of the population comprised of each racial/ethnic group.

(a) Which racial/ethnic group is decreasing the most in terms of percentage of the US population?

(b) Which racial/ethnic group is increasing the most in terms of percentage of the US population?

(c) Which racial/ethnic group is staying the most constant in terms of percentage of the US population?

2.250 The Wind Map The website *hint.fm/wind/* shows the current wind patterns across the US. In order to generate this map, what two variables are being recorded at weather stations across the US?

2.251 Are Carbon or Steel Bikes Faster? Dr. Jeremy Groves was interested in whether his carbon bike or his steel bike led to a shorter commute time. To answer this, he flipped a coin each day to randomly decide whether to ride his 20.9 lb (9.5 kg) carbon bike or his 29.75 lb (13.5 kg) steel bike for his 27 mile round trip commute. His data for 56 days are stored in **BikeCommute**. (We're not sure why distance wasn't always the same, but apparently it wasn't.) Upon inspection of the data, he finds that the commute took an average of 107.8 minutes on the steel bike and 108.3 minutes on the carbon bike,

suggesting the steel bike is faster, but he also finds that the average speed on the steel bike was 15.05 mph and the average speed on the carbon bike was 15.19 mph, suggesting that the carbon bike is faster. What's going on?!?

(a) Using Figure 2.96, what is the most obvious difference between commutes on the steel and the carbon bike?

(b) Use your answer to part (a) to explain why the carbon bike is slightly faster in terms of average speed, but yields a longer commute time, on average.

(c) In this study investigating whether the steel or the carbon bike yields a shorter commute time, a confounding variable is present. What is the confounding variable?

(d) What advice would you give to Dr. Groves to minimize his commute time?

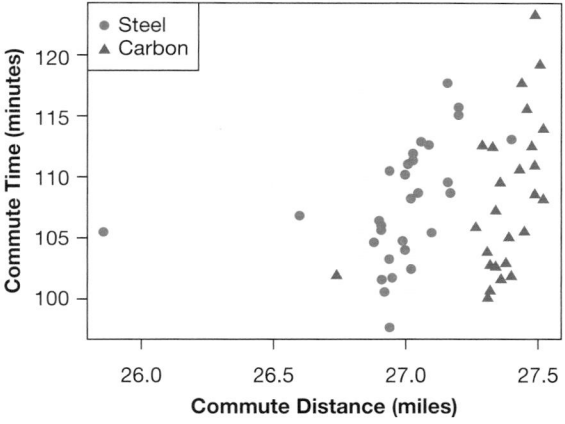

Figure 2.96 *Commute distance and time, by type of bike*

2.252 Political Polarization Pew Research Center collected data on the same 10 political value questions from 1994 to 2014 and combines these responses to place each person on a scale ranging from consistently liberal to consistently conservative.[92] Visit *http://www.people-press.org/2014/06/12/section-1-growing-ideological-consistency/#interactive* to see a visualization of responses (in the form of a smoothed histogram), broken down by political party, changing over time. Click "Animate data from 1994–2014" to dynamically watch the distribution of responses changing over time.

[92]"Political Polarization in the American Public," *Pew Research Center*, June 12, 2014, *http://www.people-press.org/2014/06/12/section-1-growing-ideological-consistency/#interactive*

(a) Describe what happened to the median Democrat value in the years 1994–2014.

(b) Describe what happened to the median Republican value in the years 1994–2014.

(c) Is there more political polarization (less overlap between parties) in 1994 or 2014?

(d) In what year did the two medians start moving rapidly away from each other?

(e) By default, the general population results are shown. Click "POLITICALLY ACTIVE" just above the visualization to see results only for the third of the public who are most politically active. In 2014, are the politically active people more or less politically polarized than the general population?

(f) Read the first two paragraphs of the article. Do you think you learn more from reading the text or from looking at the data visualization? (*Note*: This is a matter of opinion and there is no right or wrong answer, but it is worth thinking about.)

2.253 Regional Support for Same Sex Marriage The website *http://www.pewresearch.org/ fact-tank/2014/10/15/gay-marriage-arrives-in-the -south-where-the-public-is-less-enthused/*[93] shows the changing views on gay marriage from 2003 to 2014, by region. Scroll down to the visualization titled "Regional Support for Same-Sex Marriage".

(a) In 2014, which region of the country had the highest support for same-sex marriage?

[93]Lipka, M., "Gay marriage arrives in the South, where the public is less enthused" *Pew Research Center*, October 15, 2014.

(b) In 2014, which region of the country had the lowest support for same-sex marriage?

(c) Although regions have different starting levels of support, the *increase* in support for same-sex marriage is remarkably consistent across many of the different regions. Which region of the country displayed the smallest increase? The largest?

(d) The author (or the statistician / data scientist) decided to display these data with a separate time series plot for each region. Name two other ways these data could have been visualized.

2.254 Kidney Stone Treatment A study[94] collected data comparing treatments for kidney stones. Two of the treatments studied were open surgery and percutaneous nephrolithotomy. Treatment was deemed "successful" if, after three months, the kidney stones were either eliminated or less than 2 mm. The latter treatment (nephrolithotomy) is cheaper and less invasive, but is it as successful? Results are shown in Figure 2.97, first overall and then broken down by stone size. (If the answers are not obvious visually, in each case you can calculate the proportion of successes using the numbers shown on the graph.)

(a) When all stone sizes are considered, which treatment is more successful?

(b) When only small kidney stones are considered, which treatment is more successful?

(c) When only large kidney stones are considered, which treatment is more successful?

[94]Charig, R., Webb, D.R., Payne, S. (1986). "Comparison of treatment of renal calculi by open surgery, percutaneous nephrolithotomy, and extracorporeal shockwave lithotripsy". *British Medical Journal* (Clinical Residents Edition), 292(6524): 879–882.

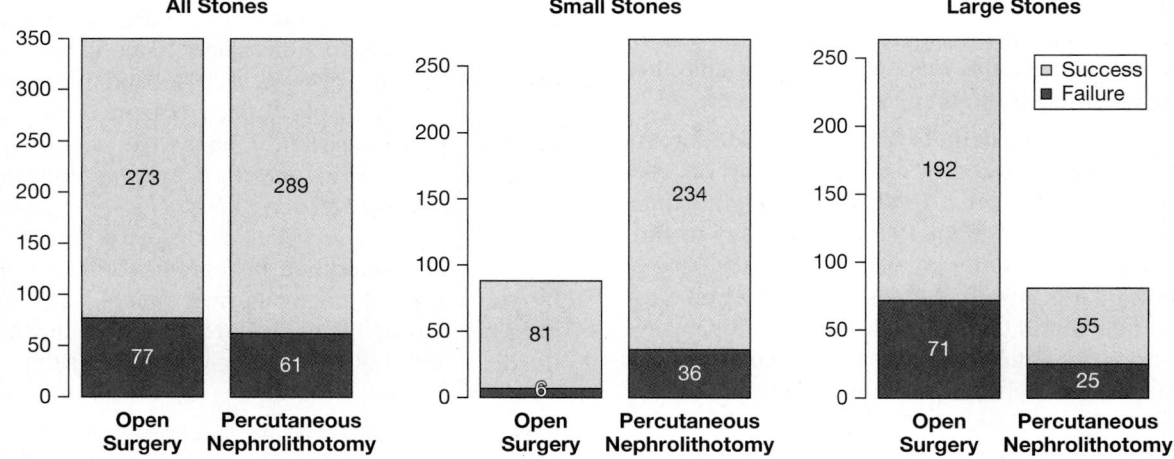

Figure 2.97 *Success rates, first for all stones, and then broken down by stone size*

(d) Which stone size results in higher success rates, regardless of treatment type?

(e) Which treatment is more commonly used for small stones?

(f) Which treatment is more commonly used for large stones?

(g) This is an example of Simpson's Paradox. Use your answers to parts (d) to (f) to explain how one treatment can be better for both small AND large stones, yet the other treatment appears to be better overall.

(h) Do you think this was a randomized experiment, with treatment randomly assigned? Why or why not?

2.255 Draw a Graph for Income vs College! The *New York Times* created an interesting interactive graph on income versus percent of children who attend college[95]. They ask you to first draw what you think the graph might look like, based on your intuition, and then compare your guess to the actual graph. The site also then shows a visualization of everyone's guesses. Go to *http://www.nytimes .com/interactive/2015/05/28/upshot/you-draw-it-how -family-income-affects-childrens-college-chances .html*, make your guess by drawing a line on the plot, and then click "I'm Done." How did you do? (The title above the plot showing your guess and reality tells you how you did — just reproduce the title phrase for your answer.)

2.256 What's Really Warming the World? Scientists at NASA collected data to study which forces, including both natural and human factors, are responsible for the increase in observed temperature in the last two centuries.[96] Go to *http://www .bloomberg.com/graphics/2015-whats-warming-the- world/* to see their resulting data visualization, an animated spaghetti plot (click the down arrow or scroll down with your mouse to see each new line appear). According to this data visualization, what is warming the world?

2.257 The Racial Divide The website *http://vallandin gham.me/racial_divide/#pt* uses data from the US Census to visualize where whites and blacks live in different cities. Figure 2.98 gives a heat map of all the census tracks in St. Louis, with each track colored according to the racial composition (white to black).

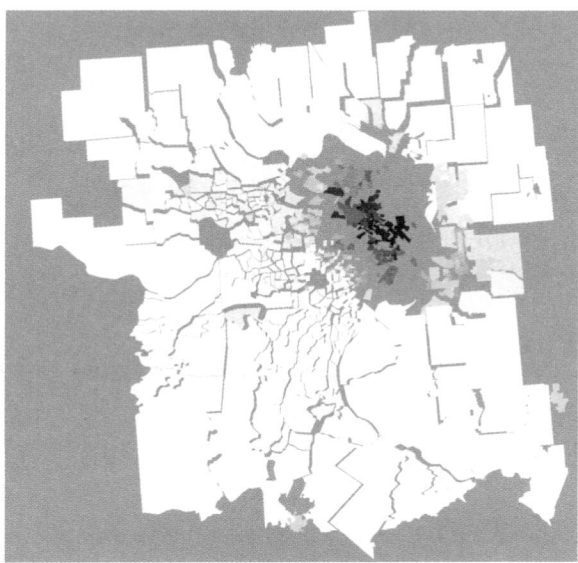

Figure 2.98 *The census tracks of St. Louis, colored by racial composition*

Also, the space between tracks is shown proportional to the change in racial composition between neighboring tracks. Comment on what you see.

2.258 Cloud Cover in San Francisco Often, the same dataset can be visualized in many different ways. Figure 2.99 shows two different visualizations of San Francisco's typical cloud cover (as a percent of the sky) for each day of the year and time of day, based on the same data from the last 30 years.[97] Figure 2.99(a) shows a spaghetti plot with each hour of the day depicted with a separate strand, and Figure 2.99(b) displays the data with a different curve for every day of the year (the center of the circle is 0% cloud cover, the outer circle is 100%). Both visualizations are created using the same data, and both convey the same information; you may use whichever you find more intuitive to answer the following questions.

(a) Do mornings or evenings typically have more cloud cover, in general?

(b) Which season (winter, spring, summer, or fall) typically shows the most variability in cloud cover throughout the day?

(c) Which visualization do you find easier to interpret? (The answer may depend on the question of interest and there is no right or wrong answer.)

[95] Aisch, G., Cox, A., and Quealy, K., "You Draw It: How Family Income Predicts Children's college Chances," *New York Times*, May 28, 2015.

[96] Roston, E. and Migliozzi, B., "What's Really Warming the World?," *Bloomberg*, June 24, 2015.

[97] Visualizations created by Zan Armstrong and used with permission. Check out more of her work at *zanarmstrong.com*.

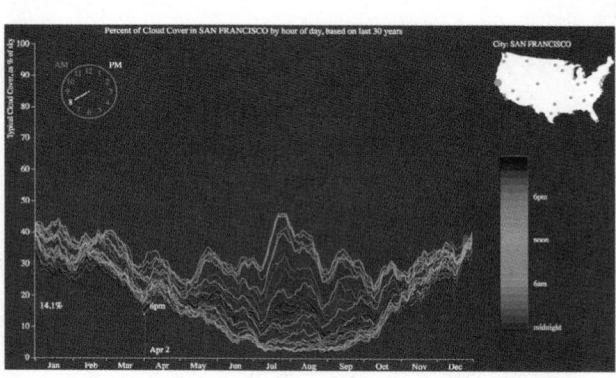

(a) Spaghetti plot by days

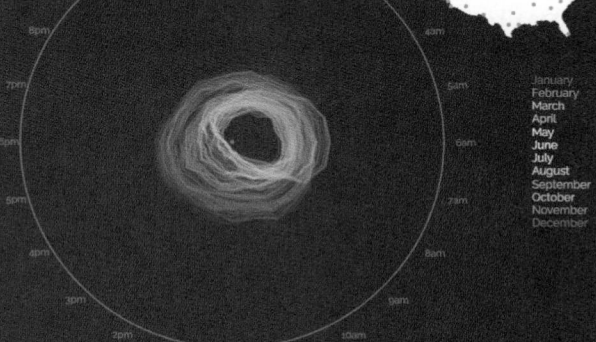

(b) Time of day in a circle

Figure 2.99 *Percent of cloud cover for San Francisco by time and day*

2.259 Cloud Cover in San Francisco–Online The plots in Exercise 2.258 on cloud cover in San Francisco can be found online at

weatherlines.zanarmstrong.com

if you prefer Figure 2.99(a) or

weather.zanarmstrong.com

if you prefer Figure 2.99(b). In the interactive display you can hover over points to get more information. You can also click on the map to change the city or the drop down menu to change the weather statistic that is plotted. Use the interactive plots at this website to answer the questions below.

(a) In San Francisco, approximately what time of day has the highest percent cloud cover in August?

(b) Which season tends to be the least windy for Chicago (the "Windy City")?

SCIENCE DATA STORIES VIDEOS
Data visualizations can also include video. In 2016, Science magazine hosted a Data Stories competition, in which participants upload a short video visualizing and telling a story with data. The winners can be viewed at *www.sciencemag.org/projects/data-stories/winners*. Exercises 2.260 to 2.263 pertain to the winners and finalists in this competition.

2.260 The Corporate Winner and People's Choice Winner was Daniel Gallagher from NASA's Scientific Visualization Studio, for his video "Martian Atmosphere Loss Explained by NASA." Watch this video and briefly describe the main message of the video.

2.261 The Professional Winner was RJ Andrews from Info We Trust, for the video "Are Gazelles Endangered?"

(a) Watch this video. What data are this video conveying?

(b) You can interact with the data and learn about other animals at this site: *http://www.infowetrust.com/endangeredsafari/*. Go to this site and hover over an animal on the interactive visualization. Indicate what animal you chose, whether its population is increasing or decreasing, and the endangered status of the animal. (These details appear at the bottom when you hover over an animal.)

2.262 The Student Winner was Ulf Aslak Jensen, for the video "How People Gather: An Interactive Visualization Approach." Watch this video, and answer the following questions:

(a) What data are this video displaying?

(b) You can explore the data shown in the video on your own at *https://ulfaslak.com/portfolio/Visualisation/*. Interact with the data, and report one of your findings.

2.263 The finalists can be viewed at *http://www.sciencemag.org/projects/data-stories/finalists*. Pick a video that interests you, watch it, and answer the following questions:

(a) Give a link to the chosen video.

(b) What data are being displayed in the video?

(c) What did you learn from the video?

GOOGLE PUBLIC DATA
The link *http://www.google.com/publicdata/directory* brings up some data visualizations created from public online data. Hovering your mouse over the little circles below the visualization brings you to a different data visualization (or you can just wait for the image to change). Many of the visualizations are dynamic, and by clicking them and then pressing the play button, you can watch them change over time. Also, when you click on the image, you can then edit it, including changing the variables and cases (often countries) displayed, or even the way the data are visualized. You can also hover over the points to see case labels and get more accurate information. Exercises 2.264 to 2.266 pertain to this website.

2.264 Find an example of an augmented scatterplot and click on the image. You can answer the following questions using either the default variables and cases, or else use the menu on the left to select variables and cases you are more interested in.

(a) Take a screenshot of the visualization (use the most recent year if multiple years are available) and include it.

(b) Which variable is displayed on the x-axis? The y-axis? The color of the points? The size of the points?

(c) Describe what you see in this static visualization. (You don't have to describe everything, just choose a few of the most obvious or interesting features.)

(d) Choose one point and hover over it to see which country (or case) it corresponds to. Give the values of each variable for this country (or case).

(e) If this is a dynamic graph, press play to watch the trend over time. Comment on what is happening over time.

(f) If this is a dynamic graph, choose one case, identify its name, and explain how this case changes over time.

2.265 Find an example of plot displaying geographic data and click on the image. You can answer the following questions using either the default variables and cases, or else use the menu on the left to select variables and cases you are more interested in.

(a) Take a screenshot of the visualization (use the most recent year if multiple years are available) and include it.

(b) Which variable is displayed by color? By point size (if the plot has points)?

(c) Describe what you see in this static visualization. (You don't have to describe everything, just choose a few of the most obvious or interesting features.)

(d) Choose one point and hover over it to see which country (or case) it corresponds to. Give the variable value(s) for this country (or case).

(e) If this is a dynamic graph, press play to watch the trend over time. Comment on what is happening over time.

(f) If this is a dynamic graph, choose one case, identify it's name, and explain how this case changes over time.

2.266 Find an example of a spaghetti plot and click on the image. You can answer the following questions using either the default variables and cases, or else use the menu on the left to select variables and cases you are more interested in.

(a) Take a screenshot of the visualization and include it.

(b) What is this plot displaying?

(c) Describe the overall trend in this visualization.

(d) Choose one case, identify it, and describe the trend for this particular case.

2.267 Find your own! Find your own data visualization online.[98]

(a) Include a screenshot of the visualization.

(b) What data are being displayed?

(c) Describe the story told by the visualization.

2.268 Monthly City Temperatures The data file **CityTemps** contains the average monthly temperature (in °C) for the cities of Moscow (Russia), Melbourne (Australia), and San Francisco (United States) in each of the years 2014 and 2015.

(a) Use time series plots and/or spaghetti plots to compare monthly temperatures between Moscow and San Francisco.

(b) Use time series plots and/or spaghetti plots to compare monthly temperatures between Melbourne and San Francisco.

2.269 Create Your Own: Augmented Scatterplot Using any of the datasets that come with this text that include at least two quantitative variables and at least one categorical variable (or any other

[98]If you need inspiration, check out *flowingdata.com, driven-by-data.net/*, or search for "New York Times Data Visualization" in Google images.

dataset that you find interesting and that meets these conditions), use statistical software to create an augmented scatterplot that identifies the dots by the category that they are in. Indicate the dataset, the cases, and the variables that you use. Comment (in context) about any interesting features revealed in your plot.

2.270 Create Your Own: Bubble Plot Using any of the datasets that come with this text that include at least three quantitative variables (or any other

dataset that you find interesting and that meets this condition), use statistical software to create a bubble plot of the data. Indicate the dataset, the cases, and the variables that you use. Specify which variable represents the size of the bubble. Comment (in context) about any interesting features revealed in your plot.

2.271 Create Your Own: Be Creative!! Create your own data visualization, and describe it. Be creative!!

Summary: Collecting Data

In Chapter 1, we learn about appropriate ways to *collect* data. A dataset consists of values for one or more variables that record or measure information for each of the cases in a sample or population. A variable is generally classified as either categorical, if it divides the data cases into groups, or quantitative, if it measures some numerical quantity.

What we can infer about a population based on the data in a sample depends on the method of data collection. We try to collect a sample that is representative of the population and that avoids sampling bias. The most effective way to avoid sampling bias is to select a random sample. Also, we try to avoid other possible sources of bias by considering things like the wording of a question. The key is to always think carefully about whether the method used to collect data might introduce any bias.

Data collected to analyze a relationship between variables can come from an observational study or a randomized experiment. In an observational study, we need to be wary of confounding variables. A randomized experiment allows us to avoid confounding variables by actively manipulating one of the variables. The handling of different treatment groups in an experiment should be as similar as possible, with the use of blinding (double-blind or single-blind) and a placebo treatment when appropriate.

The only way to infer a causal association between variables statistically is through data obtained from a randomized experiment. One of the most common and serious mistakes in all of statistics comes from a failure to appreciate the importance of this statement.

Summary: Describing Data

In Chapter 2, we learn about methods to *display* and *summarize* data. We use statistical graphs to display information about the variables, and summary statistics to quantify aspects of that information. The type of graph or statistic we use often depends on the types of variables (quantitative or categorical), as summarized below and in Table A.1.

Describing a Single Variable

- Categorical variable
  - Graphical display: *bar chart, pie chart*
  - Summary statistics: *frequency table, relative frequency table, proportion*
- Quantitative variable
  - Graphical display: *dotplot, histogram, boxplot*
  - Summary statistics:
    * Center: *mean, median*
    * Other locations: *maximum, minimum, first quartile, third quartile*
    * Spread: *standard deviation, interquartile range, range*

Describing a Relationship between Two Variables

- Categorical vs Categorical
  - Graphical display: *segmented or side-by-side bar chart*
  - Summary statistics: *two-way table, difference in proportions*
- Categorical vs Quantitative
  - Graphical display: *side-by-side boxplots (or other side-by-side graphs)*
  - Summary statistics: *statistics for the quantitative variable within each category of the categorical variable, difference in means*
- Quantitative vs Quantitative
  - Graphical display: *scatterplot*
  - Summary statistics: *correlation, regression line*

Some statistics (such as the median and interquartile range) are resistant to the effects of outliers, while others (such as the mean, standard deviation, correlation, and regression line) can be strongly influenced by extreme values. We discuss two methods for identifying possible outliers in quantitative data: using z-scores or using $1.5 \cdot IQR$.

The techniques discussed in Chapter 2 allow us to look for patterns, find anomalies, and suggest relationships within a given set of data. Many of the conclusions we draw are fairly informal. We might see that the sample mean for one group is larger than that of another group, but we are not ready yet to determine whether that difference might extend to the entire population or whether it is likely due just to random chance. We return to these ideas in Chapter 3 when we study more formal techniques for using the information in sample data to make inferences about the nature of a given population.

Table A.1 *Appropriate graphical displays and summary statistics by variable type*

Variable(s)	Graphical Displays	Summary Statistics
One categorical	Bar chart Pie chart	Proportion Frequency table Relative frequency table
One quantitative	Histogram Dotplot Boxplot	Mean Median Standard deviation Five-number summary IQR, Range
Two categorical	Side-by-side bar chart Segmented bar chart	Two-way table Difference in proportions
One quantitative and one categorical	Side-by-side boxplots Side-by-side histograms Side-by-side dotplots	Any quantitative statistic broken down by groups Difference in means
Two quantitative	Scatterplot	Correlation Regression line

Case Study: Sleep, Alcohol, Depression, and Cognition

UpperCut Images/Getty Images, Inc.

Is this an early morning class?

DATA A.1   **Sleep Study with College Students**

A recent study[81] examines the relationship between sleep habits, alcohol use, academic performance, measures of depression and stress, and other variables in US college students. The data were obtained from a sample of 253 students who did skills tests to measure cognitive function, completed a survey that asked many questions about attitudes and habits, and kept a sleep diary to record time and quality of sleep over a two-week period. Some data from this study, including 27 different variables, are available in **SleepStudy**.   ■

Example A.1

What are the cases in this study? What is the sample size n? How many rows and how many columns will the dataset have, if we use cases as rows and variables as columns? To what population do we want to generalize?

Solution   The cases are the students who participated in the study, and the sample size is $n = 253$. The dataset will have 253 rows and 27 columns. The population is US college students.

[81] Onyper, S., Thacher, P., Gilbert, J., and Gradess, S., "Class Start Times, Sleep, and Academic Performance in College: A Path Analysis," *Chronobiology International*, April 2012; 29(3): 318–335. Thanks to Serge Onyper and Pamela Thacher for sharing the data with us.

Example A.2

Six variables from the study are described below. Which of these variables are categorical and which are quantitative?

> *DASScore*   Amount of depression, anxiety, and stress, on a
>   0 to 100 scale, with higher values indicating
>   more depression, anxiety, and/or stress
> *Stress*   Amount of stress coded as Normal or High
> *LarkOwl*   Circadian preference, identified as early riser (Lark),
>   night owl (Owl), or Neither
> *AlcoholUse*   Self-reported as Abstain, Light, Moderate, or Heavy
> *PoorSleepQuality* A measure of sleep quality on a 0 to 20 scale, with higher
>   values indicating poorer sleep quality
> *CognitionZScore* Z-score based on several cognitive skills tests

Solution We see that *Stress*, *LarkOwl*, and *AlcoholUse* are categorical, while *DASScore*, *PoorSleepQuality*, and *CognitionZScore* are quantitative.

Example A.3

Does this study describe an experiment or an observational study? Can we infer causation from the results? What do we have to assume in order to generalize from the sample to the broader population?

Solution This study describes an observational study, since no explanatory variables were manipulated. Since the data come from an observational study, we cannot infer causation from the results. In order to generalize from the sample to the population, we need to assume that the sample is representative of the population and is not biased.

Example A.4

Using the variables described in Example A.2, indicate how we might display the relevant data graphically and what summary statistics we might use to examine the variable or relationship in each case below.

(a) The distribution of *DAS* scores

(b) The relationship between stress and circadian preference

(c) The relationship between alcohol use and cognitive skills

(d) The relationship between sleep quality and *DAS* score

Solution (a) *DAS* scores are quantitative, so we might display the results using a dotplot, a histogram, or a boxplot. Relevant summary statistics are the mean and standard deviation and/or the five number summary.

(b) Both variables are categorical so we would most likely display the data in a two-way table. We might compare the proportion that have high stress in each of the three circadian preference groups.

(c) Alcohol use is a categorical variable and cognition scores are a quantitative variable so we might use side-by-side boxplots to visualize this relationship. It would make sense to compare the mean cognitive score for each group, or to look at the difference in mean cognitive score between the groups.

(d) Sleep quality and *DAS* score are both quantitative so we would use a scatterplot to visualize the relationship. The relevant summary statistics are correlation and possibly a regression line.

In the examples that follow, we explore all the variables and relationships described in Example A.4. Use technology to verify the graphs and summary statistics described in the examples!

Example A.5

Are You a Lark or an Owl?

Past studies[82] indicate that about 10% of us are morning people (Larks) while 20% are evening people (Owls) and the rest aren't specifically classified as either. Studies also indicate that this circadian preference may not be settled until 22 years of age or later. Table A.2 shows the number in each category for the 253 college students in the study described in Data A.1. In this sample of college students, what proportion are Larks? Include correct notation with your answer.

Table A.2 *Circadian preference: Are you a lark or an owl?*

Type	Frequency
Lark	41
Neither	163
Owl	49
Total	253

Solution   The proportion of Larks in the sample is

$$\hat{p} = \frac{41}{253} = 0.162.$$

Surprisingly, there are more early risers in this sample of college students than expected.

Example A.6

Depression/Anxiety/Stress Scores

A histogram of *DAS* scores is given in Figure A.1.

(a) Describe the shape of the distribution.

(b) Is the mean or the median likely to be larger?

(c) For these scores, the mean is 20.04 and the standard deviation is 16.54. The highest *DAS* score is 82. Find and interpret the z-score for this value.

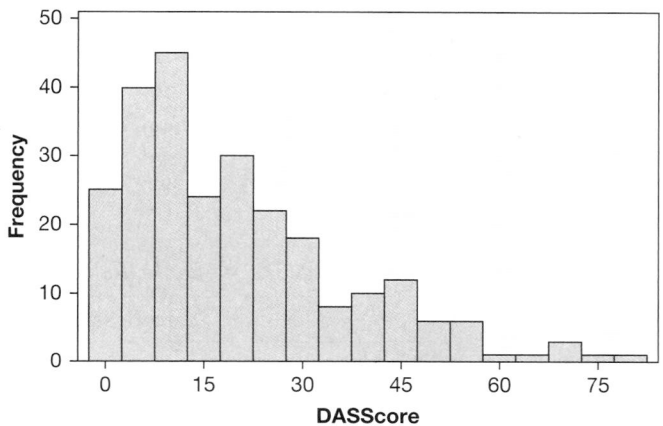

Figure A.1 *Describe the distribution of DAS scores*

---

[82] *http://www.nasw.org/users/llamberg/larkowl.htm.*

Solution (a) The distribution of *DAS* scores is skewed to the right. There are many scores between 0 and 20, and then a few much higher scores.

(b) The large values in the right tail will pull the mean up, so the mean will be larger.

(c) The *z*-score for the highest score is

$$Z\text{-score} = \frac{82 - 20.04}{16.54} = 3.746.$$

This student's *DAS* score is 3.746 standard deviations above the mean.

Example A.7

Stress and Circadian Preference

Are stress levels of college students affected by circadian preference? Table A.3 shows a two-way table for these two variables.

(a) What proportion of all the students in the sample report a high level of stress?

(b) Give the proportion reporting a high level of stress in each of the three circadian preference groups. Use subscripts to identify the three proportions. Which group has the lowest proportion reporting a high level of stress?

(c) What proportion of those with high stress are Larks?

Table A.3 *Stress levels and circadian preference*

	Lark	Neither	Owl	Total
Normal	31	125	41	197
High	10	38	8	56
Total	41	163	49	253

Solution (a) We see that the proportion reporting a high level of stress is $\hat{p} = 56/253 = 0.221$.

(b) Using subscripts *L*, *N*, and *O* for Lark, Neither, and Owl, respectively, we calculate the proportion reporting a high level of stress in each group:

$$\hat{p}_L = \frac{10}{41} = 0.244, \qquad \hat{p}_N = \frac{38}{163} = 0.233, \qquad \hat{p}_O = \frac{8}{49} = 0.163.$$

We see that the night owls have the lowest proportion reporting a high level of stress.

(c) The proportion of those with high stress that are Larks is $10/56 = 0.179$.

Example A.8

Are Cognitive Skills and Alcohol Use Related?

Is there a relationship between cognitive skills and alcohol use? The variable *CognitionZscore* is quantitative while the variable *AlcoholUse* is categorical. Summary statistics are given in Table A.4 and side-by-side boxplots are shown in Figure A.2.

Table A.4 *Cognitive skills and alcohol use*

	Sample Size	Mean	St.Dev.
Abstain	34	0.0688	0.7157
Light	83	0.1302	0.7482
Moderate	120	−0.0785	0.6714
Heavy	16	−0.2338	0.6469
Overall	253	0.000	0.7068

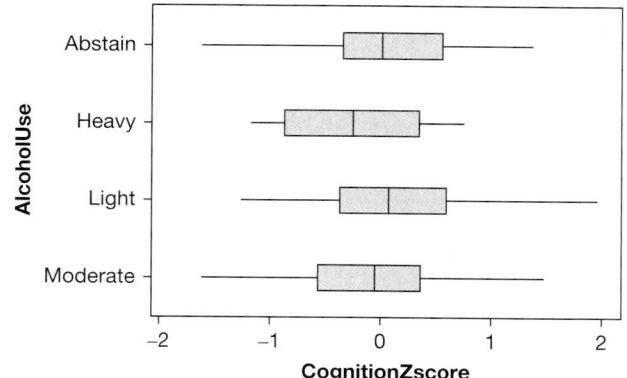

Figure A.2 *Is there a difference in cognitive skills based on alcohol use?*

(a) Which alcohol use group has the highest mean cognitive score? Which group has the lowest? Compute the difference in means between these two groups.

(b) Can we conclude that heavy drinking causes a reduction in cognitive skills? Why or why not?

(c) Do the side-by-side boxplots appear to show a strong association between the two variables?

Solution

(a) We see that the light drinkers have the highest mean cognitive score while the heavy drinkers have the lowest. The difference in mean cognitive score between the two groups is $0.1302 - (-0.2338) = 0.364$.

(b) No, we cannot make any causation conclusions from the study since it was not an experiment.

(c) While it appears that cognition scores tend to be lower in heavy drinkers, there is a great deal of overlap in all four boxplots so the association does not appear to be very strong.

Example A.9

Sleep Quality and DAS Score

We are interested in the effect of the DAS score on sleep quality. The scatterplot of these two variables along with the regression line is given in Figure A.3. Recall that for the variable *PoorSleepQuality*, higher values indicate poorer sleep

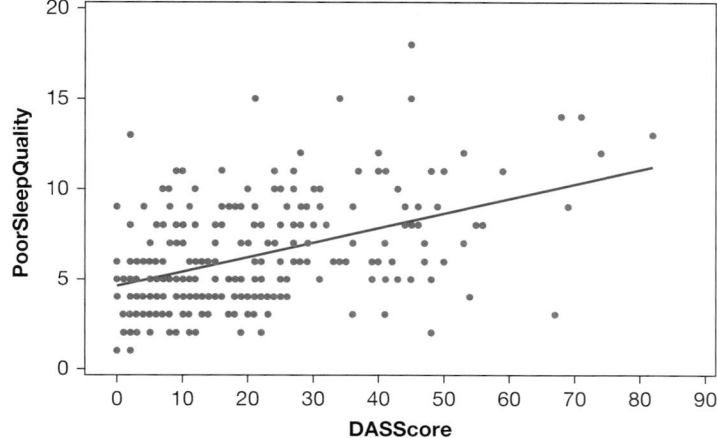

Figure A.3 *Sleep quality and depression scores*

quality, and for the variable *DASScore*, higher values indicate more depression, anxiety, and/or stress.

(a) For the point with the largest residual, use the graph to estimate the *DAS* score, the sleep quality score, and the predicted sleep quality score.

(b) In the context of these variables, describe a person whose values on these two variables places them in the lower right corner of the scatterplot.

(c) Which of the following values is the best estimate of the correlation between these two variables:

$$-0.9 \quad -0.5 \quad 0.05 \quad 0.5 \quad 0.9 \quad 2.0$$

(d) Use technology to find the correlation.

(e) Explain what a positive association means in this context.

Solution

(a) The point with the largest residual on the scatterplot appears to have a *DASScore* of about 45, a *PoorSleepQuality* score of about 18, and a predicted *PoorSleepQuality* score of about 7.5.

(b) A point in the lower right of the scatterplot would represent a person with a relatively high *DASScore* and a relatively low *PoorSleepQuality* score. This indicates a person with high levels of depression, stress, and/or anxiety but who sleeps very well. There aren't many points like this on the scatterplot!

(c) The scatterplot shows a clear positive trend, but the linear relationship is not that strong. The best answer is a correlation of about 0.5.

(d) Using technology, we see that $r = 0.457$.

(e) A positive association means that as *DASScore* increases, we expect *PoorSleepQuality* to increase. In context, this means that people with higher levels of depression, anxiety, and/or stress tend to have poorer sleep quality.

Example A.10

Predicting Sleep Quality

The regression line for predicting sleep quality from *DAS* score is:

$$\widehat{PoorSleepQuality} = 4.64 + 0.0806 \cdot DASScore.$$

(a) Find the predicted value for *PoorSleepQuality* and calculate the residual for the person who has a *DAS* score of 50 and a *PoorSleepQuality* index of 6.

(b) Interpret the slope in context.

(c) Interpret the intercept in context if it makes sense to do so.

Solution

(a) When $DASScore = 50$, we have $\widehat{PoorSleepQuality} = 4.64 + 0.0806(50) = 8.67$. The predicted sleep quality index is 8.67. The actual sleep quality index for this person is 6, so the residual is $6 - 8.67 = -2.67$.

(b) The slope is 0.0806. As the *DASScore* goes up by 1 unit, the predicted *PoorSleepQuality* score goes up by 0.0806.

(c) The intercept is 4.64. The predicted *PoorSleepQuality* value is 4.64 for a person whose *DASScore* is zero. This is a reasonable interpretation, since it is possible to have a *DASScore* of 0 (corresponding to no or very low levels of depression, anxiety, and stress) and we see in the scatterplot of the data that several people in the study did have a value of 0 for the *DASScore*.

Exercises for UNIT A: Essential Synthesis

RANDOM SAMPLING AND RANDOM ASSIGNMENT

We have seen that random sampling allows us to generalize to a broader population and that random assignment to groups allows us to conclude causation. A study may include either one of these, or both, or neither. For the each of the questions in Exercises A.1 to A.6:

(a) Does the study appear to use random sampling? (Yes or No)

(b) Does the study appear to use random assignment? (Yes or No)

A.1 To predict the outcome of an election, 2500 likely voters are randomly selected to take a survey.

A.2 The students in an Introductory Statistics class are asked to complete a survey about study habits.

A.3 In the oncology unit of a hospital, half the breast cancer patients are randomly selected to receive a new drug while the rest will receive a placebo.

A.4 At a large university, 100 students are randomly selected to take part in a marketing study. Half of these students will be randomly assigned to view one advertising campaign while the other half will watch a different advertising campaign.

A.5 Students in a Psychology class are shown a scary movie while eating popcorn, to examine the impact of fear on appetite.

A.6 In a study examining the reaction of mice to alcohol, all 50 mice in a lab will be randomly assigned to either have alcohol mixed in with their water or to have non-alcoholic liquid of similar taste and caloric content mixed in with their water.

WHICH GRAPH AND STATISTIC?

For the each of the questions in Exercises A.7 to A.16:

(a) From the following list, choose the type(s) of variable(s) that the question pertains to.

> *One categorical variable*
> *One quantitative variable*
> *One categorical variable and one quantitative variable*
> *Two categorical variables*
> *Two quantitative variables*

(b) From the following list, choose the appropriate type(s) of graph(s) that could be used to visualize data corresponding to the question.

> *Segmented or side-by-side bar charts*
> *Side-by-side boxplots, dotplots, or histograms*
> *Bar chart or pie chart*
> *Scatterplot*
> *Histogram, dotplot, or boxplot*

(c) From the following list, choose the appropriate type(s) of statistic(s) that could be used to summarize data corresponding to the question.

> *Correlation or slope from regression*
> *Two-way table or difference in proportions*

Mean, median, standard deviation, range, IQR
Statistics by group or difference in means
Frequency or relative frequency table, proportion

A.7 What is the current public opinion on capital punishment? Do more people support or oppose it?

A.8 Is there an association between the diameter of the plate used and how much food is consumed?

A.9 Do people who take a multivitamin live longer than those who don't?

A.10 How many hours do college students sleep each night?

A.11 Are males or females more likely to be homosexual?

A.12 Do college graduates who take a statistics course in college earn more at age 40 than those who don't take statistics?

A.13 How far away are stars in the Milky Way Galaxy?

A.14 What percentage of first dates yield second dates?

A.15 Is there an association between how long a child is breastfed and the weight of the child at age 2?

A.16 Is there an association between the color of a car and whether that car has been pulled over for speeding?

A.17 Does Eye Black Work for Athletes? Athletes routinely swipe black grease under their eyes to help cut down on glare on sunny days. Recently, some athletes have switched from grease to patches of black tape. Does either method work? Which is best? A study[83] helped to answer these questions. A sample of 46 subjects were tested using the Pelli-Robson contrast chart, which gives a numerical rating for ability to discern contrast against a sunlit background. Subjects were then randomly assigned to one of three groups and tested again. One group used black grease, one used black tape patches, and one used clear petroleum jelly. The group wearing the black grease was the only group to show significant improvement in discerning contrast in sunlight.

(a) Is this an experiment or an observational study? Explain.

(b) Why is this study not double-blind (or even single-blind)?

(c) What is the sample in this study? Give a reasonable intended population.

(d) What are the variables in the study? Identify each as either categorical or quantitative.

(e) What sort of graph would you use to display the results of the study?

A.18 Penguin Tags May Do Harm In Data 1.3 on page 10, we describe a 10-year study in which scientists investigated the effect of tagging penguins with either a metal strip or an electronic tag. In the study, a sample of 100 penguins were randomly assigned to one of the two groups and then followed for 10 years. The study found that, overall, penguins banded using a metal strip had fewer chicks, had a lower survival rate (percent to survive over the decade), and on average took significantly longer on foraging trips than penguins who were tagged with an electronic tag.[84]

(a) What are the cases in this study? What are the variables? Identify each variable as categorical or quantitative.

(b) The description above indicates that the scientists found a strong association between the type of tag and various measures, with the metal-tagged penguins having less success. Can we conclude that the metal tag is causing the problems?

(c) To investigate a relationship between each of the following two variables, what graph or table might we use? What statistics might we compare or use?

  i. Type of tag and number of chicks

  ii. Type of tag and survival

  iii. Type of tag and foraging time

  iv. Foraging time and number of chicks

  v. Foraging time and survival

A.19 What Webpages Do Students Visit during Class? In a study[85] investigating how students use their laptop computers in class, researchers recruited 45 students at one university in the

[83]DeBroff, B. and Pahk, P., "The Ability of Periorbitally Applied Antiglare Products to Improve Contrast Sensitivity in Conditions of Sunlight Exposure," *Archives of Ophthalmology*, July 2003; 121: 997–1001.

[84]Saraux, C., et. al., "Reliability of flipper-banded penguins as indicators of climate change," *Nature*, 13 January 2011; 469: 203–206.

[85]Kraushaar, J. and Novak, D., "Examining the Affects of Student Multitasking with Laptops during the Lecture," *Journal of Information Systems Education*, 2010; 21(2): 241–251.

Northeast who regularly take their laptops to class. Software was installed on each of their computers that logged information on the applications the computer was running, including how long each was open and which was the primary focus on the monitor. Logs were kept over multiple lectures. On average, the students cycled through 65 active windows per lecture, with one student averaging 174 active windows per lecture! The researchers developed a rubric to distinguish productive class-related applications from distractive ones, such as email and social networking sites. For each student, they recorded the percent of active windows that were distractive and the percent of time spent on distractive windows. They found that, on average, 62% of the windows students open in class are completely unrelated to the class, and students had distracting windows open and active 42% of the time, on average. Finally, the study included a measure of how each student performed on a test of the relevant material. Not surprisingly, the study finds that the students who spent more time on distracting websites generally had lower test scores.

(a) What are the cases in this dataset? What is the sample size? Is the sample a random sample?

(b) Is this an experiment or an observational study? Explain.

(c) From the description given, what variables are recorded for each case? Identify each as categorical or quantitative.

(d) What graph(s) might we use to display the data about the number of active windows opened per lecture? What graph is most appropriate if we want to quickly determine whether the maximum value (174) is an outlier?

(e) The last sentence of the paragraph describes an association. What graph might be used to display this association? What statistic might be used to quantify it? Is it a positive or negative association?

(f) From the information given, can we conclude that students who allocate their cognitive resources to distracting sites during class get lower grades because of it? Why or why not?

(g) For the association described in part (e), what is the explanatory variable? What is the response variable?

(h) Describe the design of a study that might allow us to make the conclusion in part (f). Comment on the feasibility of conducting such a study.

A.20 Laptop Computers and Sperm Count In Exercise 2.113 on page 90, we discuss a study about the effect of heat from laptop computers on scrotum temperature in men. Heating the scrotum by just 1°C can reduce sperm count and quality, and repeated increases in temperature can have a long-term effect. In a new study,[86] temperature increases in the right scrotum over one hour were measured in °C while men sat with a laptop computer on their lap. Three different conditions were tested. In Session 1, men sat with legs close together. In Session 2, the legs were kept close together and a lap pad was used to separate the laptop computer from the legs. In Session 3, no lap pad was used but the legs were spread farther apart. The sessions were conducted on three different days with the same volunteers. (Sitting with legs together without a laptop does not increase temperature.) A histogram of the values from Session 2 is shown in Figure A.4, and summary statistics from computer output for all three sessions are shown in Figure A.5.

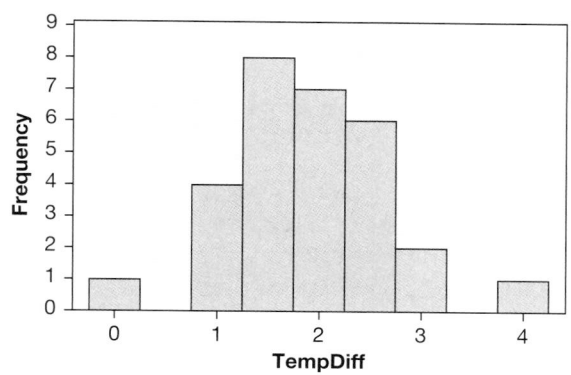

Figure A.4 _Histogram of Session 2_

[86]Data are approximated from summary statistics given in: Sheynkin, Y., et. al., "Protection from scrotal hyperthermia in laptop computer users," _Fertility and Sterility_, February 2011; 95(2): 647–651.

Variable	Session	N	Mean	StDev	Minimum	Q1	Median	Q3	Maximum
TempDiff	1	29	2.325	1.058	−0.076	1.571	2.328	2.806	4.915
	2	29	1.911	0.747	0.158	1.398	1.936	2.475	3.806
	3	29	1.494	0.617	0.534	1.005	1.439	1.898	3.141

Figure A.5 _Output with summary statistics on temperature increases_

(a) Describe the shape of the histogram of the temperature changes from Session 2.

(b) How many men participated in each session?

(c) Give the summary statistics from Session 1. What are the mean and standard deviation? What is the five number summary?

(d) Find the z-score for the smallest value in Session 3. Is the smallest value more than two standard deviations from the mean?

(e) If the histogram in Figure A.4 makes it appropriate to do so, use the mean and standard deviation from Session 2 to find an interval that is likely to contain 95% of the values. If the shape of the histogram makes this rule inappropriate, say so.

(f) Use the IQR for Session 1 to determine if the largest value in Session 1 is an outlier. Show your work.

(g) Side-by-side boxplots for the three sessions are shown in Figure A.6. Describe what you see. How many outliers are there in each session? Which situation produced the largest temperature increase? Which is more effective at reducing the negative effects of the laptop: using a lap pad or sitting with legs farther apart?

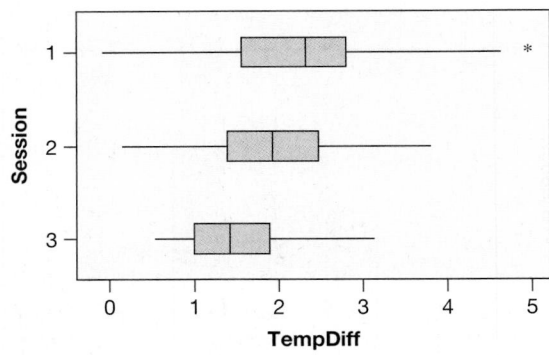

Figure A.6 *Comparing temperature increase for the three sessions*

A.21 A New Drug for Bladder Cancer Eighty-six patients with bladder cancer participated in a study in which all tumors were removed and then the subjects were monitored to see if the tumors returned. Patients were randomly assigned to one of two treatment groups: one group received a placebo and one group received the drug thiotepa. The study was double-blind. Tumors returned in 29 out of 48 patients in the placebo group and in 18 out of 38 patients in the thiotepa group.[87]

(a) What is the sample in this study? What is the intended population?

(b) What are the variables in this study? Classify each as categorical or quantitative.

(c) Is this an experiment or an observational study? What does it mean to say that the study was "double-blind"?

(d) What kind of graph or table might be used to display the data? Display the data in this way.

(e) Compute relevant statistics to compare the success rate of the two groups. Does the drug appear to be more effective than the placebo?

A.22 Genetic Diversity and Distance from Africa Exercise 2.186 on page 118 describes a data set exploring the relationship between the genetic diversity of a population and geographic distance of each population from East Africa. These data are displayed in Figure A.7 along with the regression line. The regression equation is $\widehat{diversity} = 0.76 - 0.0000067 distance$.

(a) Is the association between genetic diversity and distance from East Africa approximately linear?

(b) Interpret the slope in context.

(c) Interpret the intercept in context.

(d) The Mayan population in Mexico is 19,847 km from East Africa (by land). Calculate the predicted genetic diversity for the Mayans.

(e) The actual genetic diversity calculated for the Mayan population is 0.678. Will the point on the graph that corresponds to Maya be above or below the line?

(f) Calculate the residual for Maya.

A.23 How Much Do People Tip in Restaurants? Data 2.12 on page 123 introduces a dataset containing information on customers' tipping patterns in a restaurant. The data are available in **RestaurantTips**.

(a) What are the cases? What is the sample size?

(b) What are the variables? Identify each variable as quantitative or categorical.

[87]Wei, L., Lin, D., and Weissfeld, L., "Regression-analysis of multivariate incomplete failure time data by modeling marginal distributions," *Journal of the American Statistical Association*, 1989; 84: 1065–73.

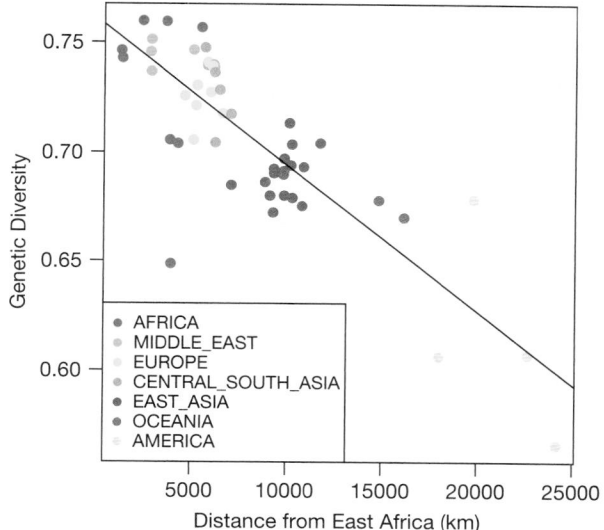

Figure A.7 *Genetic diversity of populations by distance from East Africa.*

(c) The variable *PctTip* includes information on the tip as a percent of the bill. Use technology to find the mean tip percentage, the standard deviation, and the five number summary. How large or small does a tip percentage have to be to qualify as an outlier relative to this dataset?

(d) Use technology to create a histogram of *PctTip* and describe its shape.

(e) Use technology to create a two-way table of *Credit* (yes or no depending on whether the bill was paid with a credit card) and *Day* (the day of the week). Compute the proportion of bills paid with a credit card on Thursday; do the same for Friday. Does there appear to be an association between whether it is Thursday or Friday and whether a person pays with a credit card or cash? Why do you think this might be so? (*Hint:* For many people, Friday is payday.)

(f) We might be interested in how the tip percentage, in the *PctTip* variable, varies for different servers (*Server*)? What graph should be used to examine a relationship between these two variables? Use technology to create such a graph and comment on the relationship. Which server appears to make the highest percent tips?

(g) Does the size of the bill (*Bill*) influence the tip percentage (*PctTip*)? In addressing this question, what is the explanatory variable? What is the response variable? Use technology to draw a scatterplot. Are there any outliers in the

scatterplot? Ignoring the outliers, does there appear to be a positive, negative, or no relationship between these two variables?

(h) Use technology to find the correlation between *Bill* and *PctTip*.

A.24 Analyzing Data from US States The dataset **USStates** has a great deal of information on the 50 states in the US, including two categorical variables and 14 quantitative variables. In the questions that follow, we ask you to use technology to do some analysis of this dataset.

(a) Choose one of the two categorical variables and use technology to create a frequency table and a relative frequency table of the values.

(b) Choose one of the quantitative variables and use technology to create a histogram. Describe the shape of the histogram. For the same variable, create a boxplot. Are there any outliers? Finally, for the same variable, give summary statistics: mean, standard deviation, and the five number summary.

(c) Choose any quantitative variable and any categorical variable and use technology to create a side-by-side boxplot. Describe what you see in the graph, and discuss any association that might exist between the variables, as evidenced by the graph.

(d) Create a two-way table of the two categorical variables. Find appropriate proportions to help you determine if there is an association between the two variables, and explain your reasoning.

(e) Choose any two quantitative variables and use technology to create a scatterplot. Describe the scatterplot: Is there an obvious positive or negative linear trend? Are there any outliers? Use technology to find the correlation and the least squares line to predict one variable from the other. Interpret the slope of the line in context.

Review Exercises for UNIT A

A.25 Student Survey: Sample or Population? The results of a student survey were introduced in Data 1.1 on page 4. Is the dataset from a sample or a population? If it is from a sample, describe a relevant population to which we might make inferences.

A.26 Intensive Care Unit Admissions Data 2.3 on page 69 introduced the dataset **ICUAdmissions**, which contains 20 different variables recorded for 200 patients admitted to the Intensive Care Unit at a hospital.

(a) What is the sample? What is a reasonable population?

(b) Indicate which variables are quantitative.

(c) List at least two questions we might ask about any one of these individual variables.

(d) List at least two questions we might ask about relationships between any two (or more) of these variables.

A.27 Exercise, Protein, and Muscle Mass A Dutch study implies that exercising before eating protein might help the body convert more of the protein into muscle. In the study, 48 men were randomly assigned to either exercise or rest for 30 minutes. At the end of the 30 minutes, all drank a protein shake and had their muscle-protein synthesis measured. Regardless of age, exerciser's bodies converted more of the protein to muscle than the resting bodies.[88]

(a) What is the sample? What is a reasonable population?

(b) What are the variables? Include all variables mentioned in the description.

(c) Identify all variables as either categorical or quantitative.

A.28 Diet and Retinol and Beta-Carotene Levels The data from a study[89] examining the association between diet and plasma retinol and plasma beta-carotene levels are given in **NutritionStudy**. The dataset has 315 cases (people who have the measurements taken) and 16 columns that are described in Appendix B.

(a) Indicate which of the variables are quantitative and which are categorical.

(b) Discuss one possible relationship of interest in this dataset between two categorical variables. Between two quantitative variables. Between one categorical and one quantitative variable.

A.29 Psychological and Physiological Effects of Meditation Forty-one employees of a biotechnology company participated in a study[90] that examines the immunological and psychological effects of meditation. Twenty-five of the participants, chosen at random, completed an 8-week meditation program while the other sixteen employees did no meditation. Brain wave activity across the front of the left hemisphere was measured for all participants before, immediately following, and four months after the program. (Studies have suggested that increased activity in this part of the brain is associated with decreases in negative emotions and increases in positive emotions.) All 41 people received an influenza vaccination at the end of the program and their immune response to the vaccine was measured through blood samples taken one month and two months later. All participants also completed surveys designed to measure negative and positive emotions before and after the course. The surveys produced two numerical scores (one for positive emotions and one for negative emotions) in both situations.

Meditators showed an increase in brain wave activity, a decrease in reported negative feelings, and no change in reported positive feelings. Non-meditators showed no significant change in any of these areas. Meditators had a stronger antibody response to the vaccine than the non-meditators.

(a) What are the cases in this study? How many cases are there?

[88] Published online, *American Journal of Clinical Nutrition*, November 27, 2010, reported in *Consumer Reports OnHealth*, March 2011; 23(3): 3.

[89] Nierenberg, D., et. al., "Determinants of plasma levels of beta-carotene and retinol," *American Journal of Epidemiology*, 1989 Sep; 130(3): 511–21.

[90] Davidson, R., et. al., "Alterations in brain and immune function produced by mindfulness meditation," *Psychosomatic Medicine*, July/August, 2003, 65:564–570.

(b) What are the variables? Which are categorical and which are quantitative?

(c) Which variable is the explanatory variable?

(d) How many rows and how many columns will the dataset contain if we assume that each data case is a row and each variable is a column?

A.30 Special Shakes A large restaurant chain (see Example 1.6) periodically offers special milk shake flavors for a limited time. Suppose that the contenders for the next special flavor are Green Mint, Orange Crush, Egg Nog, and Piña Colada. The chain plans to collect data from customers on these flavors, and there are several ways they might solicit responses. For each of the options below, state the number of variables needed to code the information in a dataset, whether the variable(s) is/are categorical or quantitative, and what sort of values should be recorded.

(a) "Which of the four flavors is most appealing to you?"

(b) "Put a check next to any of the four flavors you find appealing."

(c) "Please rank the four flavors with 1=most appealing and 4=least appealing."

(d) "Rate each of the four flavors on a 1 to 10 scale with 10=extremely appealing and 1=very unappealing."

A.31 Sample and Population for Fishing A fishing boat captain examines one day's catch of fish to see if the average weight of fish in that area is large enough to make fishing there profitable. Describe the sample and describe a reasonable population.

A.32 Sample and Population for Chips Ahoy! A hungry yet diligent snacker eats an entire package of Chips Ahoy! cookies while counting and recording the number of chocolate chips in each cookie. Describe the sample and describe a reasonable population.

A.33 Does Physical Beauty Matter? One of the daily polls on *CNN.com* asked "Does Physical Beauty Matter to You?" Of 38,485 people responding, 79% said yes and 21% said no. Can we conclude that about 79% of all people think physical beauty matters? Why or why not? In making such a conclusion, what are we considering the sample? What are we considering the population? Is there any bias in the sampling method?

A.34 Do Tanning Salons Mislead Their Customers? Investigators posing as fair-skinned teenage girls contacted 300 tanning salons nationwide, including at least three randomly selected in each state. The investigators report that 90% of the salons stated that indoor tanning did not pose a health risk and over half (51%) of the salons denied that indoor tanning would increase a fair-skinned teenager's risk of developing skin cancer. Going even further, 78% of the tanning salons even claimed that indoor tanning is beneficial to health.[91] (In fact, many studies have shown that tanning is dangerous, especially for teenagers, and that tanning raises the risk of melanoma, the deadliest type of skin cancer, by 74%.)

(a) What is the sample?

(b) Do you think the sample is representative of all tanning salons in the US?

(c) Although the sample is random, discuss why the results do not paint an accurate picture of the dangers of tanning.

(d) Do you think the study accurately portrays the messages tanning salons give to teenage girls?

A.35 Strawberry Fields A strawberry farmer has planted 100 rows of plants, each 12 inches apart, and there are about 300 plants in each row. He would like to select a random sample of 30 plants to estimate the average number and weight of berries per plant.

(a) Explain how he might choose the specific plants to include in the sample.

(b) Carry out your procedure from (a) to identify the first three plants selected for the sample.

A.36 Fish Consumption and Intelligence In 2000, a study[92] was conducted on 4000 Swedish 15-year-old males. The boys were surveyed and asked, among other things, how often they consume fish each week. Three years later, these answers were linked to the boys' scores as 18-year-olds on an intelligence test. The study found that boys who consume fish at least once a week scored higher on the intelligence test.

(a) Is this an experiment or an observational study? Explain.

(b) What are the explanatory and response variables?

[91] "Congressional Report Exposes Tanning Industry's Misleading Messaging to Teens," http://www.skincancer.org/news/tanning/tanningreport, a report released by the House Committee on Energy and Commerce, February 1, 2012.

[92] Aberg, M., et. al., "Fish intake of Swedish male adolescents is a predictor of cognitive performance," *Acta Paediatrica*, 2009; 98 (3):555.

(c) Give an example of a potential confounding factor.

(d) Does this study provide evidence that eating fish once a week improves cognitive ability?

A.37 First Quiz Easy or Hard? In an introductory statistics class in which regular quizzes are given, should the first quiz be easy (to give students confidence) or hard (to convince students to work harder)? The response variable will be grades on a later exam that is common to all students.

(a) Describe an observational study to answer this question.

(b) Describe a confounding variable that is likely to impact the results of the observational study.

(c) Describe a randomized experiment designed to answer this question.

A.38 Outwit the Grim Reaper by Walking Faster! The title of this exercise was a recent headline.[93] The article goes on to describe a study in which men's walking speeds at age 70 were measured and then the men were followed over several years. In the study, men who walked slowly were more likely to die. The article concludes that "Men can elude the Grim Reaper by walking at speeds of at least 3 miles per hour." What common mistake is this article making? What is a possible confounding variable?

A.39 Shoveling Snow Three situations are described at the start of Section 1.3 on page 29. In the second bullet, we describe an association between activity at a building's heating plant and more employees missing work due to back pain. A confounding variable in this case is amount of snow. Describe how snowfall meets the definition of a confounding variable by describing how it might be associated with both variables of interest.

A.40 Exercise and Alzheimer's Disease A headline at *MSNBC.com*[94] stated "One way to ward off Alzheimer's: Take a hike. Study: Walking at least one mile a day reduces risk of cognitive impairment by half." The article reports on a study[95] showing that elderly people who walked a lot tended to have more brain mass after nine years and were less likely to develop dementia that can lead to Alzheimer's

disease than subjects who walked less. At the start of the study the researchers measured the walking habits of the elderly subjects and then followed up with measures of brain volume nine years later. Assuming that active walkers really did have more brain mass and fewer dementia symptoms, is the headline justified?

A.41 Single-Sex Dorms and Hooking Up The president of a large university recently announced[96] that the school would be switching to dorms that are all single-sex, because, he says, research shows that single-sex dorms reduce the number of student hook-ups for casual sex. He cites studies showing that, in universities that offer both same-sex and co-ed housing, students in co-ed dorms report hooking up for casual sex more often.

(a) What are the cases in the studies cited by the university president? What are the two variables being discussed? Identify each as categorical or quantitative.

(b) Which is the explanatory variable and which is the response variable?

(c) According to the second sentence, does there appear to be an association between the variables?

(d) Use the first sentence to determine whether the university president is assuming a causal relationship between the variables.

(e) Use the second sentence to determine whether the cited studies appear to be observational studies or experiments?

(f) Name a confounding variable that might be influencing the association. (*Hint*: Students usually request one type of dorm or the other.)

(g) Can we conclude from the information in the studies that single-sex dorms reduce the number of student hook-ups?

(h) What common mistake is the university president making?

A.42 Percent of College Graduates by Region of the US The dataset **USStates** includes information on the percent of the population to graduate from college (of those age 25 and older) for each US state. Figure A.8 shows side-by-side boxplots for percent of college graduates by region of the country (Midwest, Northeast, South, and West.)

[93]"Outwit the Grim Reaper by Walking Faster," medicalexpress.com, posted December 16, 2011.

[94]http://www.msnbc.msn.com/id/39657391/ns/health-alzheimer's_disease.

[95]Erickson, K., et. al., "Physical activity predicts gray matter volume in late adulthood: The Cardiovascular Health Study," *Neurology*, published online October 13, 2010.

[96]Stepp, L., "Single-sex dorms won't stop drinking or 'hooking-up'," www.cnn.com, June 16, 2011.

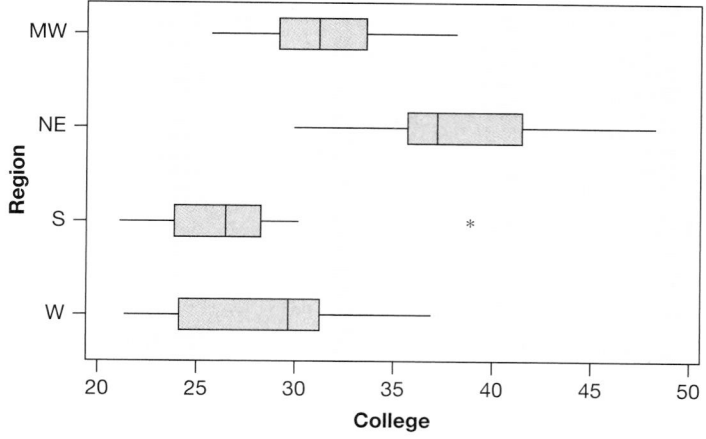

Figure A.8 *Percent of college graduates by region of the US*

(a) What are the variables and is each categorical or quantitative?

(b) Describe the results seen in the graph. Which region has the highest percent of college graduates? Which has the lowest? Are there any outliers and, if so, where?

(c) Does there appear to be an association between these two variables?

(d) Can we conclude that there is causation between the two variables: that one of the variables is causing changes in the other?

A.43 Moving in Sync Increases Feelings of Connection If you want people to agree with you, get them to join you in a line dance or to march in lock-step with you. A recent study[97] shows that we feel more emotionally connected to one another when we're moving in sync. In the study, researchers asked 70 college students to walk behind an accomplice either matching stride for stride, walking completely out of sync, or walking at any comfortable pace. The students were randomly divided between the three options. After following the accomplice around campus, the students were asked to rate how close they felt to the accomplice, how much they liked the accomplice, and how similar they felt to the accomplice. Ratings were on a 7-point scale (which we will treat as quantitative) with 7 representing highest levels of closeness, liking, and similarity. On all three questions, those who had been forced to walk in sync gave substantially higher ratings than either of the other two groups.

(a) What are the cases? What is the sample size?

(b) What are the variables?

(c) Is this an experiment or an observational study?

(d) Use the information given to draw a rough sketch of possible side-by-side boxplots comparing the three groups on the similarity rankings. Be sure the sketch shows the association described.

(e) In a second part of the experiment, the students were encouraged by the accomplice to funnel live pill bugs into a grinder labeled an "extermination machine." Those who had marched in step with the accomplice followed orders and "killed" the most pill bugs. (The pill bugs were actually secretly funneled to safety.) What graph would we use to look at a relationship of number of pill bugs killed by which treatment group the student was in? What graph would we use to look at the association of number of pill bugs killed with the rating given on the liking accomplice scale?

A.44 Pricing and Social Responsibility An experiment on pricing and social responsibility was conducted using a popular ride at a large amusement park, where digital photos are taken of the riders and offered for sale at the end of the ride.[98] The experiment was designed to determine the effect of pricing strategy under four conditions: the normal pricing strategy used by the ride; the effect when customers are allowed to pay whatever they want; the effect when customers are told that half the revenue is donated to charity; and the effect when customers can pay what they want *and* half the money is donated to charity. The experimenter had the amusement park try all four pricing

[97]Carroll, L., "Moving in sync makes people think alike, study finds," *The Body Odd*, msnbc.com, posted January 18, 2012.

[98]Nelson, L., Pricing Strategy and Corporate Social Responsibility, Research News from the Haas School of Business, October 13, 2010.

Table A.5 *Which pricing strategy is best?*

Pricing Strategy	Proportion Buying	Mean Amount Paid
(a) Standard	0.005	$12.95
(b) Pay what you want	0.08	$ 0.92
(c) Half to charity	0.006	$12.95
(d) Both (b) and (c)	0.04	$ 5.50

strategies, and the proportion of riders buying the photos and the mean price paid are given for each of the pricing scenarios in Table A.5. The ride has 15,000 customers per day, and photos normally cost $12.95. Compute the daily total revenue for the company under each of the scenarios. (The total number of customers buying the photo is the proportion buying times the 15,000 customers. The total revenue is the number buying photos times the mean price paid. Also don't forget to donate half the money to charity when required!) What pricing strategy should the managers of the business use if they are only concerned about maximizing revenue? What pricing strategy should they use if they want high revenue combined with social responsibility?

A.45 Sampling Some Frazer Computing Customers Frazer Computing, a company that leases software, has over 16,000 used car dealer customers.[99] The company wants to contact 10 of these car dealers, randomly selected, to conduct in-depth interviews on how the software is meeting their needs. Suppose the car dealers are numbered 1 to 16,000. Use a random number generator or table to select the numbers for the 10 dealers to be in the sample.

A.46 Driving with a Pet on Your Lap Over 30,000 people participated in an online poll on cnn.com[100] asking "Have you ever driven with a pet on your lap?" The results show that 34% answered yes and 66% answered no. Can we conclude that 34% of all people have driven with a pet on their lap? Can we conclude that 34% of all people who visit cnn.com have driven with a pet on their lap? Explain.

A.47 Does Smiling Increase Positive Emotions Scientists came up with a clever way to test whether the physical act of smiling increases positive emotions.[101] They randomly divided participants into two groups of 24 each. The "smiling" group was asked to hold a pencil between their teeth (which forces the face into a smile), while the "non-smiling" group was asked to hold a pencil

between their lips (which does not). Participants were not told the purpose of the experiment. They then rated video clips on a scale from −9 (very negative) to +9 (very positive). The ratings of the two groups did not differ on the negative clips, but the mean for the smiling group on a positive clip (from a *Tom & Jerry* cartoon) was 7.8 while it was 5.9 for the non-smiling group.

(a) Is this an experiment or an observational study?

(b) Why is it important that participants were not told the purpose of the study?

(c) Find the difference in means for the ratings on the positive clip, and use notation, with subscripts and a minus sign, for your answer.

(d) If the difference in means from part (c) is considered substantial, can we conclude that forcing the facial muscles into a smile in this way increases positive emotions?

A.48 What Type of Cell Phone? A 2012 survey[102] examined cell phone ownership by US adults. The results of the survey are shown in Table A.6.

Table A.6 *Frequencies in cell phone ownership*

Cell Phone Owned	Frequency
Android smartphone	458
iPhone smartphone	437
Blackberry smartphone	141
Cell phone not smartphone	924
No cell phone	293
Total	2253

(a) Make a relative frequency table of the data. Give results to three decimal places.

(b) What percent of the survey respondents do not own a cell phone? What percent own a cell phone but not a smartphone? What percent own a smartphone?

A.49 Near-Death Experiences People who have a brush with death occasionally report experiencing a near-death experience, which includes the sensation of seeing a bright light or feeling separated from one's body or sensing time speeding up or slowing down. Researchers[103] interviewed 1595 people admitted to a hospital cardiac care unit during a recent 30-month period. Patients were classified as cardiac arrest patients (in which the heart briefly stops after beating unusually quickly)

[99]Thanks to Michael J. Frazer, President of Frazer Computing, for the information.

[100]cnn.com, poll results from April 12, 2012.

[101]Soossignan, R., "Duchenne smile, emotional experience, and automatic reactivity: A test of the facial feedback hypothesis," *Emotion*, March 2002; 2(1): 52–74.

[102]"Nearly Half of American Adults are Smartphone Owners," Pew Research Center, pewresearch.org, March 1, 2012.

[103]Greyson, B., "Incidence and correlates of near-death experiences on a cardiac care unit," *General Hospital Psychiatry*, July/August 2003; 25:269–276.

or patients suffering other serious heart problems (such as heart attacks). The study found that 27 individuals reported having had a near-death experience, including 11 of the 116 cardiac arrest patients. Make a two-way table of the data. Compute the appropriate percentages to compare the rate of near-death experiences between the two groups. Describe the results.

A.50 Painkillers and Miscarriage A recent study[104] examined the link between miscarriage and the use of painkillers during pregnancy. Scientists interviewed 1009 women soon after they got positive results from pregnancy tests about their use of painkillers around the time of conception or in the early weeks of pregnancy. The researchers then recorded which of the pregnancies were successfully carried to term. The results are in Table A.7.

Table A.7 *Does the use of painkillers increase the risk of miscarriage?*

	Miscarriage	Total
Aspirin	5	22
Ibuprofen	13	53
Acetaminophen	24	172
No painkiller	103	762
Total	145	1009

(a) What percent of the pregnancies ended in miscarriage?

(b) Compute the percent of miscarriages for each of the four groups. Discuss the results.

(c) Is this an experiment or an observational study? Describe how confounding variables might affect the results.

(d) Aspirin and ibuprofen belong to a class of medications called nonsteroidal anti-inflammatory drugs, or NSAIDs. What percent of women taking NSAIDs miscarried? Does the use of NSAIDs appear to increase the risk of miscarrying? Does the use of acetaminophen appear to increase the risk? What advice would you give pregnant women?

(e) Is Table A.7 a two-way table? If not, construct one for these data.

(f) What percent of all women who miscarried had taken no painkillers?

A.51 Smoking and Pregnancy Rate Studies have concluded that smoking while pregnant can have negative consequences, but could smoking also negatively affect one's ability to become pregnant? A study collected data on 678 women who had gone off birth control with the intention of becoming pregnant.[105] Smokers were defined as those who smoked at least one cigarette a day prior to pregnancy. We are interested in the pregnancy rate during the first cycle off birth control. The results are summarized in Table A.8.

Table A.8 *Smoking and pregnancy rate*

	Smoker	Non-smoker	Total
Pregnant	38	206	244
Not pregnant	97	337	434
Total	135	543	678

(a) Is this an experiment or an observational study? Can we use the data to determine whether smoking influences one's ability to get pregnant? Why or why not?

(b) What is the population of interest?

(c) What is the proportion of women successfully pregnant after their first cycle ($\hat{p}$)? Proportion of smokers successful ($\hat{p}_s$)? Proportion of non-smokers successful ($\hat{p}_{ns}$)?

(d) Find and interpret ($\hat{p}_{ns} - \hat{p}_s$) the difference in proportion of success between non-smokers and smokers.

A.52 Age of Patients with Back Pain Figure A.9 shows a histogram of the ages of $n = 279$ patients

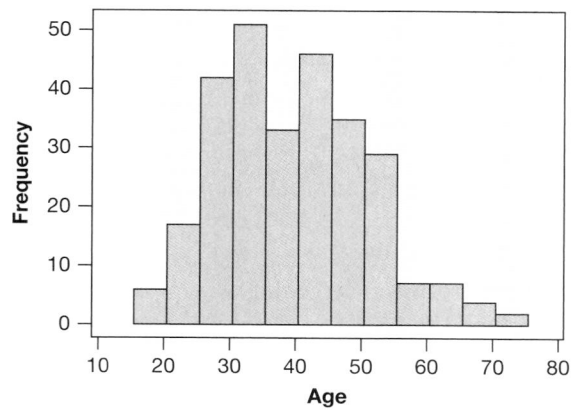

Figure A.9 *Age of patients with back pain*

[104]Li, D-K., et al., "Use of NSAIDs in pregnancy increases risk of miscarriage," *British Medical Journal*, August 16, 2003; 327(7411): 1.

[105]Baird, D. and Wilcox, A., "Cigarette Smoking Associated With Delayed Conception," *Journal of the American Medical Association*, June 2011; 305(23): 2379–2484.

being treated for back pain.[106] Estimate the mean and standard deviation of the ages of back pain patients.

A.53 Lie Detection Is lie detection software accurate? A recent study was conducted in order to test the accuracy of a commonly used method of lie detection.[107] The researchers are specifically interested in how lie detectors perform when an individual is stressed. A sample of 48 participants were gathered and attached to the lie detection device. They were asked to read deceptive (lying) material out loud while receiving an electric shock (to add stress). The lie detector failed to report deception in 17 of the 48 participants.

(a) What is the sample in this study? What is the population of interest? What does the variable measure?

(b) What proportion of time does the lie detector fail to report deception?

(c) If you were a prosecutor, would you recommend this lie detector to reveal deception?

A.54 Lie Detection of Truthful Material Exercise A.53 describes a study in which lie detector accuracy is checked by having participants read deceptive material. In addition to deceptive material, the individuals were also asked to read truthful material. The electric shock was again included to add stress. The lie detector accurately reported no deception in 21 of the 48 participants.

(a) What proportion of the time does the lie detector incorrectly report deception?

(b) If you were on a jury, would you trust results from this device?

A.55 Clutch Sizes of Birds A naturalist counts the number of baby birds, or *clutch size*, in a sample of 130 different nests. A histogram of her results is shown in Figure A.10. Is the distribution approximately symmetric and bell-shaped? Estimate the mean of the clutch sizes. Estimate the standard deviation of the clutch sizes.

A.56 PSA Cancer Screening A sample of 30 men were given the PSA (prostate specific antigen) test to screen for prostate cancer. For the 30 values obtained, the median score was $m = 3$ and the mean

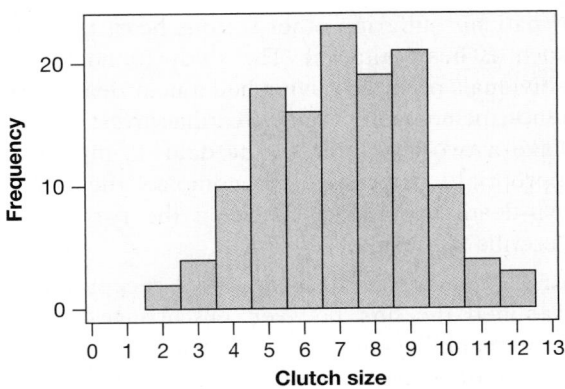

Figure A.10 *Estimate the mean and the standard deviation*

was $\bar{x} = 11$. Explain how it is possible for the mean and the median to be so different. What is likely to be true about the shape of the distribution of PSA scores?

A.57 The Growing Season The number of consecutive frost-free days in a year is called the growing season. A farmer considering moving to a new region finds that the median growing season for the area for the last 50 years is 275 days while the mean growing season is 240 days.

(a) Explain how it is possible for the mean to be so much lower than the median, and describe the distribution of the growing season lengths in this area for the last 50 years.

(b) Draw a possible curve for the shape of this distribution. Label the mean and median on the horizontal axis.

(c) Describe the likely shape of the distribution.

A.58 Normal Body Temperature It is commonly believed that normal human body temperature is 98.6°F (or 37°C). In fact, "normal" temperature can vary from person to person, and for a given person it can vary over the course of a day. Table A.9 gives a set of temperature readings of a healthy woman taken over a two-day period.

Table A.9 *Body temperature during the day*

97.2	97.6	98.4	98.5	98.3	97.7
97.3	97.7	98.5	98.5	98.4	97.9

(a) Make a dotplot of the data.

(b) Compute the mean of the data and locate it on the dotplot as the balance point.

(c) Compute the median of the data and locate it on the dotplot as the midway point.

[106]Sample dataset from Student Version 12 of Minitab Statistical Software.
[107]Hollien, H., Harnsberger, J., Martin, C., and Hollien, K., "Evaluation of the NITV CVSA," *Journal of Forensic Sciences*, January 2010; 53(1):183–193.

A.59 Beta-Carotene Levels in the Blood The plasma beta-carotene level (concentration of beta-carotene in the blood), in ng/ml, was measured for a sample of $n = 315$ individuals, and the results[108] are shown in the histogram in Figure A.11.

(a) Describe the shape of this distribution. Is it symmetric or skewed? Are there any obvious outliers?

(b) Estimate the median of this sample.

(c) Estimate the mean of this sample.

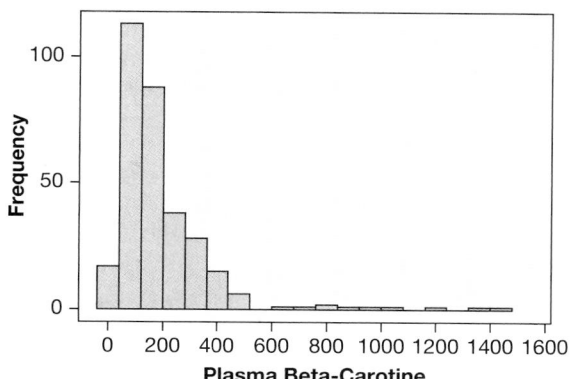

Figure A.11 *Concentration of beta-carotene in the blood*

A.60 Time Spent Exercising, between Males and Females Often we are interested not just in a single mean but in a difference in means between two groups. In the **StudentSurvey** data, there are 36 seniors: 26 males and 10 females. Table A.10 gives the number of hours per week that each said he or she spent exercising.

Table A.10 *Number of hours spent exercising a week*

Females	4	2	5	6	12	15	10
	5	0	5				

Males	10	10	6	5	7	8	4
	12	12	4	15	10	5	5
	2	2	7	3	5	15	6
	6	5	0	8	5		

(a) Calculate $\bar{x}_f$, the mean number of hours spent exercising by the females.

(b) Calculate $\bar{x}_m$, the mean number of hours spent exercising by the males.

[108]http://lib.stat.cmu.edu/datasets/Plasma_Retinol, accessed November 24, 2003.

(c) Compute the difference, $\bar{x}_m - \bar{x}_f$, and interpret it in context.

A.61 Mean or Median Calculate the mean and the median for the numbers

$$1, \quad 1, \quad 1, \quad 1, \quad 1, \quad 1, \quad 2, \quad 5, \quad 7, \quad 12$$

Which do you think is a better measure of center for this set of values? Why? (There is no right answer, but think about which you would use.)

A.62 Time in Days to Row Solo across the Atlantic Ocean Exercise 1.16 on page 14 gives a sample of eight times, in days, to row solo across the Atlantic Ocean. The times are

$$40, \quad 87, \quad 78, \quad 106, \quad 67, \quad 70, \quad 153, \quad 81$$

(a) Use technology to find the mean and standard deviation of the eight times.

(b) Find and interpret the z-scores for the longest time and shortest time in the sample.

A.63 Arsenic in Toenails Exercise 2.61 on page 74 discusses the use of toenail clippings to measure the level of arsenic exposure of individuals in Great Britain. A similar study was conducted in the US. Table A.11 gives toenail arsenic concentrations (in ppm) for 19 individuals with private wells in New Hampshire, and the data are also available in **ToenailArsenic**. Such concentrations prove to be an effective indicator of ingestion of arsenic-containing water.[109]

Table A.11 *Arsenic concentration in toenail clippings*

0.119	0.118	0.099	0.118	0.275	0.358	0.080
0.158	0.310	0.105	0.073	0.832	0.517	0.851
0.269	0.433	0.141	0.135	0.175		

(a) Use technology to find the mean and standard deviation.

(b) Compute the z-score for the largest concentration and interpret it.

(c) Use technology to find the five number summary.

(d) What is the range? What is the IQR?

A.64 A Dotplot of Arsenic in Toenails Figure A.12 shows a dotplot of the arsenic concentrations in Table A.11.

[109]Adapted from Karagas, M., et. al., "Toenail samples as an indicator of drinking water arsenic exposure," *Cancer Epidemiology, Biomarkers and Prevention*, 1996; 5: 849–852.

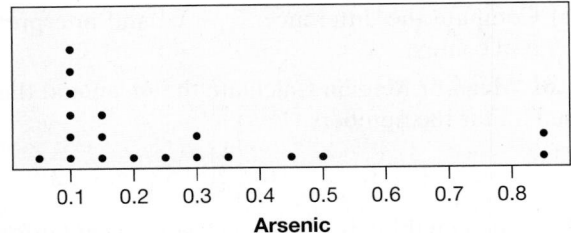

Figure A.12 *Dotplot of arsenic concentration in toenails*

(a) Which measures of center and spread are most appropriate for this distribution: the mean and standard deviation or the five number summary? Explain.

(b) Is it appropriate to use the 95% rule about having 95% of the data within two standard deviations for this distribution? Why or why not?

A.65 Jogging Times Consider the jogging times from a set of 5-mile runs by two different runners in Table A.12.

Table A.12 *Jogging times*

Jogger 1	Jogger 2
44	48
45	49
43	38
48	40
45	50

(a) Which runner is faster on average?

(b) What is the main difference in the jogging times of joggers 1 and 2?

A.66 Mammal Longevities Table 2.15 on page 64 shows longevity (typical lifespan) in years for 40 species of mammals, and the data are also available in **MammalLongevity**.

(a) Use technology to find the mean and standard deviation of the 40 values.

(b) The elephant's longevity of 40 years appears to be an outlier in the dotplot in Figure 2.6 on page 64. Find and interpret the z-score for the elephant.

A.67 Diabetes Drug Benefits Heart Patients Rosiglitazone is normally prescribed to control blood sugar in people with Type II diabetes, but it may also provide a benefit to heart patients. A study[110] identified 95 people with Type II diabetes who were undergoing angioplasty to open coronary

arteries. For six months after the angioplasty, about half the patients received daily oral doses of rosiglitazone, while the rest received a placebo. The groups were randomly assigned and the study was double-blind. Each patient was then tested to see if blood vessel blockage was greater or less than 50%. Since the goal is to limit the propensity of blood vessels to close again after angioplasty, an outcome of less than 50% is desirable. The results are shown in Table A.13.

Table A.13 *Is the drug effective at keeping blockage less than 50%?*

	Greater than 50%	Less than 50%
Rosiglitazone	5	42
Placebo	21	27

(a) How many patients received the drug? How many received a placebo?

(b) What percent of all patients in the study had less than 50% blockage within 6 months of the angioplasty?

(c) Of the patients with greater than 50% blockage, what percent were on rosiglitazone?

(d) What percent of the patients given a placebo had less than 50% blockage?

(e) We wish to compare the effectiveness of rosiglitazone to a placebo at keeping blockage to less than 50%. Calculate the relevant percentages to make this comparison and discuss the results.

(f) Does rosiglitazone appear to be effective?

A.68 Number of Cell Phone Calls per Day A survey of $n = 1917$ cell phone users in May 2010 asked "On an average day, about how many phone calls do you make and receive on your cell phone?" The results[111] are displayed in Table A.14.

Table A.14 *Number of cell phone calls made or received per day*

Number of Calls	Percentage
0	5%
1–5	44%
6–10	22%
11–20	14%
21–30	5%
More than 30	6%
Don't know	3%

[110]"Double Duty: Diabetes drug protects reopened heart vessels," *Science News*, June 21, 2003; 163(25):389–390.

[111]Princeton Survey Research Results, "Spring Change Assessment Survey," June 4, 2010.

(a) What is the sample? What is the intended population?

(b) Is this a frequency table or a relative frequency table?

(c) We can't draw an accurate histogram from the data in the table since the first category has a single value and the last is unbounded (even if we ignore the nonresponses). However, you should still be able to discuss the general shape of the distribution. For example, are the data skewed or relatively symmetric? Explain your reasoning.

(d) The article lists two statistics for the "center" of these data: 5.00 and 13.10. Which is the mean and which is the median? Explain.

A.69 Prostate Cancer and a Drug for Baldness The drug finasteride is marketed as Propecia to help protect against male pattern baldness, and it also may protect against prostate cancer. A large sample[112] of healthy men over age 55 were randomly assigned to receive either a daily finasteride pill or a placebo. The study lasted seven years and the men had annual check-ups and a biopsy at the end of the study. Prostate cancer was found in 804 of 4368 men taking finasteride and in 1145 of the 4692 men taking a placebo.

(a) Is this an experiment or an observational study? The study was double-blind; what does that mean?

(b) What are the variables in the study?

(c) Make a two-way table to display the results. Include the row and column totals.

(d) What percent of men in the study received finasteride?

(e) What percent of the men with prostate cancer were in the placebo group?

(f) Compare the percent of men getting prostate cancer between the two groups. Does finasteride appear to offer some protection against prostate cancer?

A.70 Length of Calls on a Cell Phone

(a) Do you expect the distribution of the lengths of all phone calls made on a cell phone during one month to be symmetric, skewed to the right, or skewed to the left?

(b) Two measures of center for this distribution are 2.5 minutes and 13.7 minutes. Which is most likely to be the mean and which is most likely to be the median? Explain your reasoning.

A.71 Fighting Insomnia Studies have shown that behavioral changes and prescription medication can help older people fight insomnia. Researchers[113] followed the sleep patterns of 72 people whose average age was 65. Seventeen people took a medication called temazepam an hour before bedtime, 18 people received training in techniques to improve sleep, 19 did both, and 18 took a placebo and received no training. The results are summarized in Table A.15. Find the following proportions for this sample.

Table A.15 *Treating insomnia*

Improvement	Medication	Training	Both	Neither	Total
Much	5	7	10	0	22
Some	4	5	6	1	16
None	8	6	3	17	34
Total	17	18	19	18	72

(a) The proportion who experienced much improved sleep quality

(b) The proportion of those who took medication (with or without training) who experienced much improvement

(c) The proportion of those with no improvement who used temazepam

(d) The proportion of those who did not have training who experienced some or much improvement

A.72 Comparing Two Drugs in Dialysis Patients Many kidney dialysis patients get vitamin D injections to correct for a lack of calcium. Two forms of vitamin D injections are used: calcitriol and paricalcitol. In the first study[114] to compare survival rates of patients getting one drug or the other, the records of 67,000 dialysis patients were examined. Half received one drug; the other half the other drug. After three years, 58.7% of those getting paricalcitol had survived, while only 51.5% of those getting calcitriol had survived. What percent of the survivors had received paricalcitol? Construct an approximate two-way table of the data (due to rounding of the percentages we can't recover the exact frequency counts).

[112]Thompson, I., et. al., "The influence of finasteride on the development of prostate cancer," *New England Journal of Medicine*, July 17, 2003; 349(3): 215–24.

[113]Information adapted from *Science News*, April 3, 1999.

[114]Teng, M., et. al., "Survival of patients undergoing hemodialysis with paricalcitol or calcitriol Therapy," *New England Journal of Medicine*, July 31, 2003; 349(5): 446–456.

A.73 Birth Rate Is the birth rate different in developed and undeveloped countries? In the dataset **AllCountries**, we have information on the birth rate of all 215 countries as well as an indicator for whether the country is considered a developed or undeveloped nation.[115] Use the five number summaries for each group of countries in Table A.16 to answer the following questions.

Table A.16 *Five number summaries for birth rate in developed and undeveloped nations*

	Min	1st Quartile	Median	3rd Quartile	Max
Developed	7.9	9.5	11.4	14.5	22.7
Undeveloped	9.5	18.8	23.7	31.5	44.1

(a) Does the birth rate distribution appear to be different in developed and undeveloped countries?

(b) Would any of the undeveloped countries be outliers if they were considered developed? What about developed countries if they were considered undeveloped?

(c) Libya is on the border between being considered developed or undeveloped. The birth rate in Libya is 23.3. Is this an outlier for undeveloped countries? Is it an outlier for developed countries?

(d) Using the five number summaries, make a rough sketch of side-by-side boxplots for birth rate (ignoring outliers).

A.74 Draw a Boxplot

(a) Draw a boxplot for data that illustrate a distribution skewed to the right.

(b) Draw a boxplot for data that illustrate a distribution skewed to the left.

(c) Draw a boxplot for data that illustrate a symmetric distribution.

A.75 Variability by Age in Systolic Blood Pressure How does the variability in systolic blood pressure compare between ICU patients in their teens and those in their eighties for the patients in the dataset **ICUAdmissions**? The values for each group are given in Table A.17. Use technology to find

the five number summary, the range and IQR, and the standard deviation in each case and compare the measures of spread for the two groups.

Table A.17 *Systolic blood pressure of ICU patients*

Teens	100	100	104	104	112	130		
	130	136	140	140	142	146	156	
Eighties	80	100	100	110	110	122	130	
	135	136	138	140	141	162	190	190

A.76 Examining Blood Pressure by Age Are any of the systolic blood pressures in Exercise A.75 for patients in their teens or eighties outliers within their group? Justify your answer.

A.77 Infection in Dialysis Patients Table A.18 gives data showing the time to infection, at the point of insertion of the catheter, for kidney patients using portable dialysis equipment. There are 38 patients, and the data give the first observation for each patient.[116] The five number summary for the data is (2, 15, 46, 149, 536).

Table A.18 *Time to infection for dialysis patients*

2	5	6	7	7	8	12	13
15	15	17	22	22	23	24	27
30	34	39	53	54	63	96	113
119	130	132	141	149	152	152	185
190	292	402	447	511	536		

(a) Identify any outliers in the data. Justify your answer.

(b) Draw the boxplot.

EFFECT OF DIET ON NUTRIENTS IN THE BLOOD
Exercises A.78 and A.79 use data from **NutritionStudy** on dietary variables and concentrations of micronutrients in the blood for a sample of $n = 315$ individuals.

A.78 Daily Calorie Consumption The five number summary for daily calorie consumption is (445, 1334, 1667, 2106, 6662).

(a) The ten largest data values are given below. Which (if any) of these is an outlier?

3185   3228   3258   3328   3450   3457
3511   3711   4374   6662

[115]In this exercise nations are considered undeveloped if the average electricity used per person is less then 2500 kWh a year (coded with "1" in the *Developed* variable of **AllCountries**). We combined the other two categories into a single category. Seventy-eight countries were excluded due to missing data.

[116]McGilchrist, C. and Aisbett, C., "Regression with frailty in survival analysis," *Biometrics*, 1991; 47: 461–466.

(b) Determine whether there are any low outliers. Show your work.

(c) Draw the boxplot for the calorie data.

A.79 Daily Calories by Gender Figure A.13 shows side-by-side boxplots comparing calorie consumption by gender.

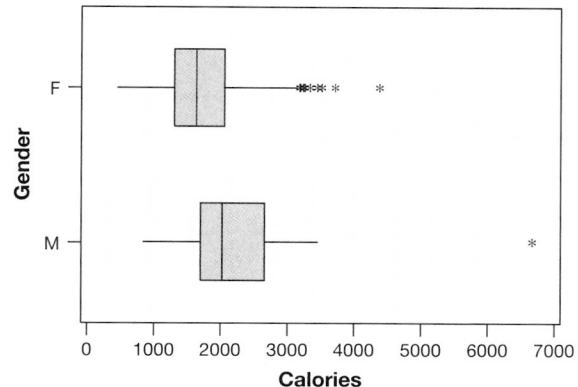

Figure A.13 *Calorie consumption by gender*

(a) Which gender has the largest median daily calorie consumption? Which gender has the largest outlier? Which gender has the most outliers?

(b) Does there seem to be an association between gender and calorie consumption? Explain.

A.80 Systolic Blood Pressure: Boxplot Figure A.14 shows the boxplot for the systolic blood pressures for all 200 patients in the ICU study in **ICUAdmissions**. Discuss what information this graph gives about the distribution of blood pressures in this sample of patients. What is the five-number summary?

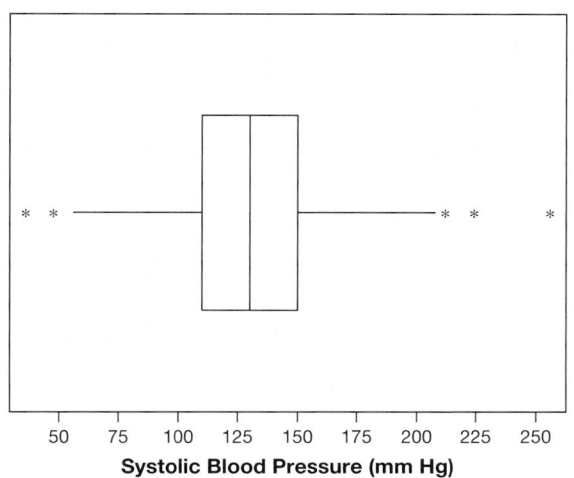

Figure A.14 *Systolic blood pressure of ICU patients*

A.81 Systolic Blood Pressure and Survival The data in **ICUAdmissions** contains a categorical variable *Status* indicating whether each patient lived (0) or died (1). Is there a relationship between the status (lived/died) and the systolic blood pressures? Use the side-by-side boxplots showing the systolic blood pressures for these two groups of patients in Figure A.15 to discuss how the distributions compare.

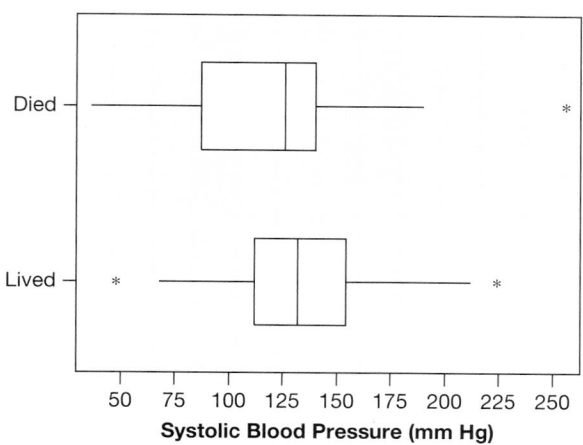

Figure A.15 *Systolic blood pressures of patients who lived or died*

A.82 Effect of Calcium on Fish In a study[117] to determine how the calcium level of water affects respiration rate of fish, 360 fish in a sample were randomly divided into three tanks with different levels of calcium: low, medium, and high. The respiration rate of the fish, in beats per minute, was then measured. The dataset is in **FishGills3** and the two variables are *Calcium* and *GillRate*.

(a) Use technology to create side-by-side boxplots for gill rate in the three different calcium conditions. Describe the relationship between the two variables.

(b) Use technology to obtain comparative summary statistics for gill rate in the three different calcium conditions and give the mean and the standard deviation for the gill rates in each of the three calcium conditions.

(c) Is this study an experiment or an observational study?

A.83 Systolic Blood Pressure: Histogram Figure A.16 shows a histogram of the systolic blood pressure (in mm Hg) for all 200 patients admitted to the Intensive Care Unit, from the **ICUAdmissions**

[117]Thanks to Professor Brad Baldwin of St. Lawrence University for this dataset.

dataset. The mean and standard deviation of these 200 numbers are $\bar{x} = 132.28$ and $s = 32.95$.

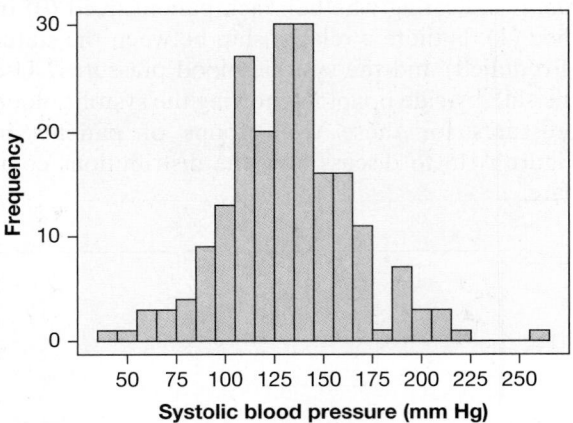

Figure A.16 *Systolic blood pressure*

(a) Is it appropriate to use the 95% rule with these data? Why or why not?

(b) Use the 95% rule to give an interval that is likely to contain about 95% of the data values.

(c) Use the data in the **ICUAdmissions** dataset to count the actual number of data values, and find the percent of data values, lying within the interval given in part (b).

(d) Is the result for the sample close to the result predicted by the 95% rule?

A.84 Heart Rates Figure A.17 shows a histogram of the heart rate data, in beats per minute, from **ICUAdmissions**. The values come from $n = 200$ patients being admitted to the Intensive Care Unit at a hospital.

(a) Estimate the mean and the standard deviation.

(b) Estimate the 10th percentile and interpret it.

(c) Estimate the range.

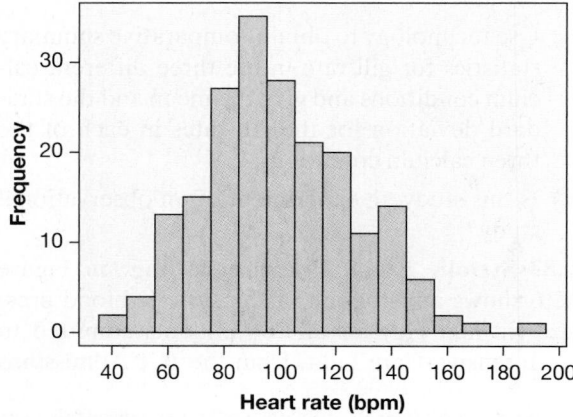

Figure A.17 *Heart rate in beats per minute*

A.85 A Small Sample of SAT Scores A random sample of seven statistics students were taken from the **StudentSurvey** dataset. The Math and Verbal SAT scores for the seven students are shown in Table A.19. We are interested in predicting scores on the verbal section using scores on the math section.

(a) Use technology to plot the data with a regression line to predict verbal scores based on math scores.

(b) Use technology to find the correlation between these seven verbal and math scores.

(c) Based on this sample, is it reasonable to use a regression line to predict verbal scores using math scores? Explain your answer using what you have found in (a) and (b).

Table A.19 *SAT scores*

Math	Verbal
720	450
650	530
670	680
660	680
550	560
620	600
680	670

A.86 Positive or Negative Association? In each case, decide whether the relationship between the variables is more likely to be positive or negative. In the cases where it makes sense to view one variable as an explanatory variable and the other as the response variable, identify which is which.

(a) Number of years spent smoking cigarettes and lung capacity

(b) Height and weight

(c) Systolic blood pressure and diastolic blood pressure

A.87 Effect of Outliers For the five data points in Table A.20:

(a) Make a scatterplot of the data.

(b) Use technology to find the correlation.

(c) Make a new scatterplot showing these five data points together with an additional data point at $(10, 10)$.

(d) Use technology to find the correlation for this larger dataset with six points.

Table A.20 *What is the correlation?*

x	2	1	4	5	3
y	5	3	4	3	4

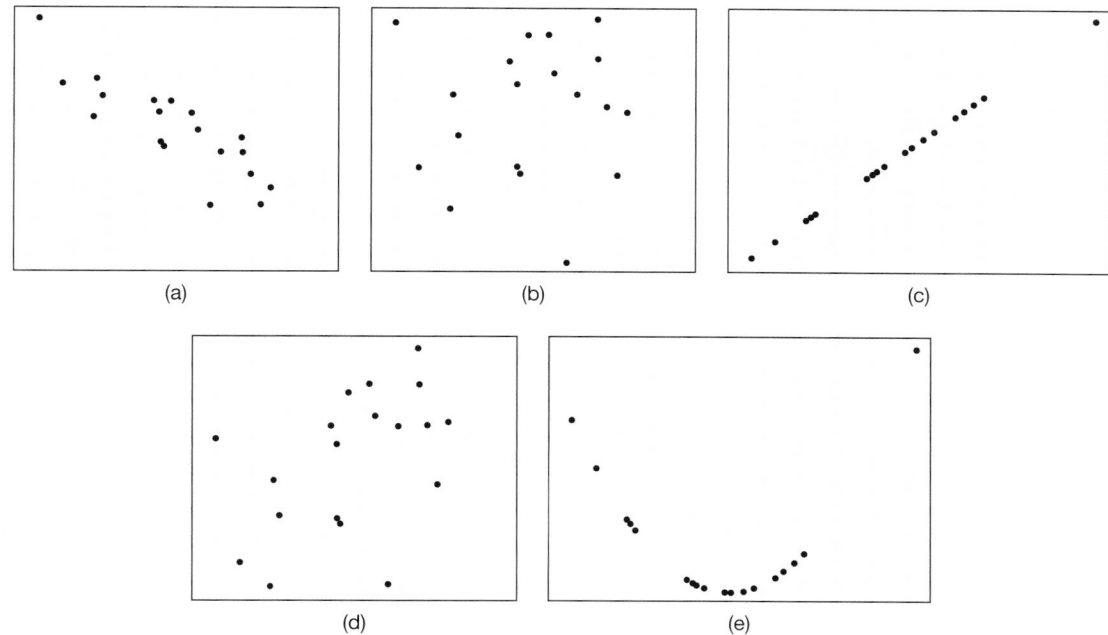

Figure A.18 *Estimate the correlation*

(e) Discuss the effect of an outlier on the correlation.

A.88 Estimate the Correlation Give a rough estimate of the correlation *r* for each of the scatterplots shown in Figure A.18.

A.89 A Sample of Height and Weight Figure A.19 shows a scatterplot of height and weight for a new sample of 105 college students.

(a) Does there appear to be a positive or a negative relationship in the data?

(b) Describe the body shape of the individuals whose points are labeled by A, B, C, and D.

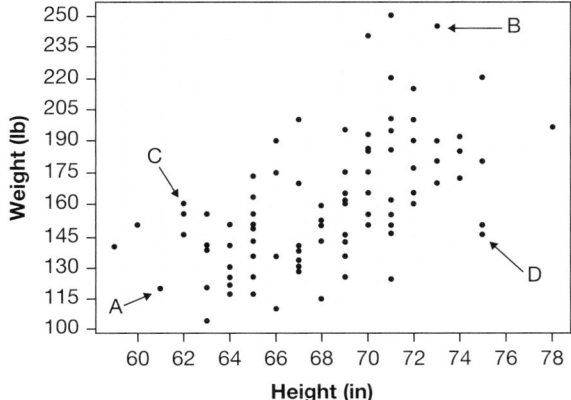

Figure A.19 *Scatterplot of height and weight with n = 105*

A.90 Fat and Fiber and Calories The dataset **NutritionStudy** contains information on daily calorie consumption, fat consumption, and fiber consumption for 315 healthy individuals. Figure A.20 shows a scatterplot of fat vs calories and a scatterplot of fiber vs calories. (In these figures, we have omitted one extreme outlier.)

(a) Does there appear to be a positive or negative correlation between calories and fat? Between calories and fiber?

(b) Judging from the scatterplots, which correlation appears to be larger: between calories and fat or between calories and fiber?

(c) One person in the study consumed over 4000 calories daily. Approximately what was the fat consumption for this person? The fiber consumption? Is the value an extreme value for either fat or fiber as an individual variable?

A.91 Height and Weight The quantitative variables *Height* (in inches) and *Weight* (in pounds) are included in the **StudentSurvey** dataset.

(a) What would a positive association mean for these two variables? What would a negative association mean? Which do you expect is more likely?

(b) Figure A.21 shows a scatterplot of the data. Does there appear to be a positive or negative relationship between height and weight? How strong does the trend appear to be? Does it appear to be approximately a linear trend?

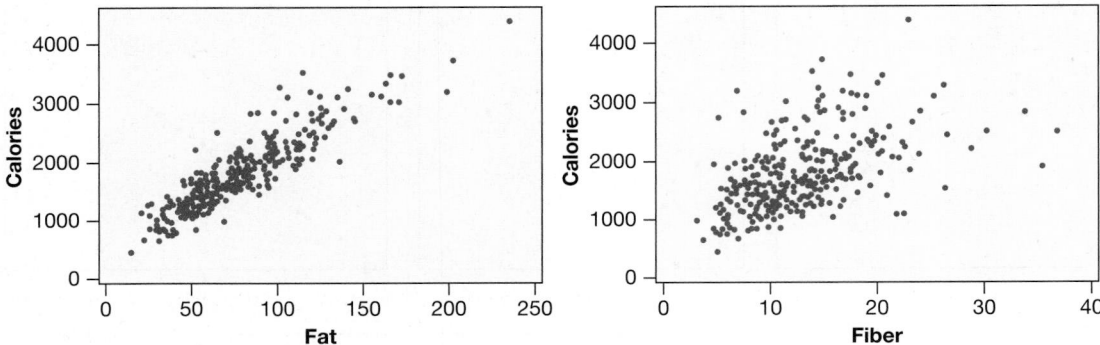

Figure A.20 *Calorie consumption vs fat or fiber consumption*

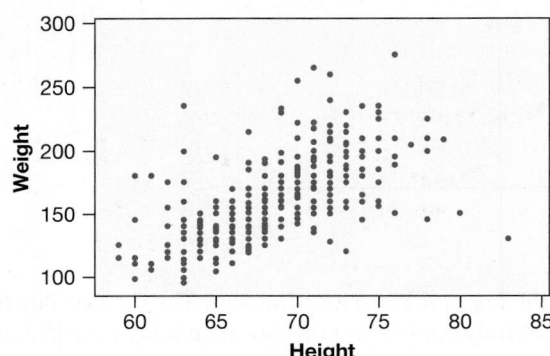

Figure A.21 *Student survey: height and weight*

(c) Describe the person represented by the outlier in the lower right corner.

A.92 Blood Pressure and Heart Rate In Example 2.19 on page 82 we computed z-scores for patient #772 in the **ICUAdmissions** dataset, who had a high systolic blood pressure reading of 204 but a low pulse rate of 52 bpm.

(a) Find the point corresponding to patient #772 on the scatterplot of blood pressure vs heart rate shown in Figure A.22.

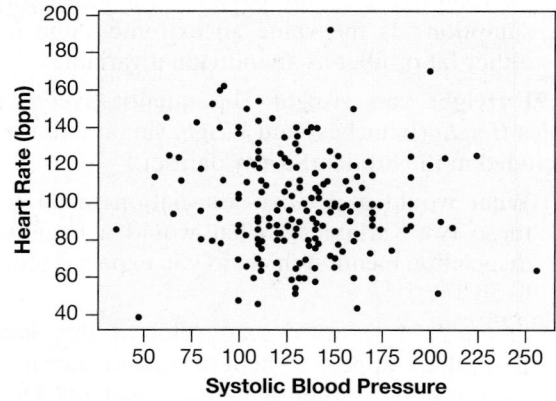

Figure A.22 *Blood pressure vs heart rate for ICU patients*

(b) Patient #772 has a high blood pressure reading but a low pulse rate. Does the scatterplot in Figure A.22 support a conjecture that these two variables have a negative association?

A.93 An Outlier in Jogging Times Table A.21 gives the times for five races in which two joggers participated.

Table A.21 *Jogging times*

Jogger A	Jogger B
44	48
45	49
43	38
48	40
45	50

(a) Use technology to construct a scatterplot of the race times.

(b) Use technology to find the correlation.

(c) A sixth race is held on a very windy day, and jogger A takes 50 minutes while jogger B takes a whole hour to complete the race. Recalculate the correlation with this point added.

(d) Compare correlations from parts (b) and (c). Did adding the results from the windy day have an effect on the relationship between the two joggers?

A.94 Cricket Chirps and Temperature In the **CricketChirp** dataset given in Table 2.28 on page 111, we learn that the chirp rate of crickets is related to the temperature of the air.

(a) Figure A.23 shows the seven points together with the regression line. Does there appear to be a linear relationship between these two variables? How strong is it? Is it positive or negative?

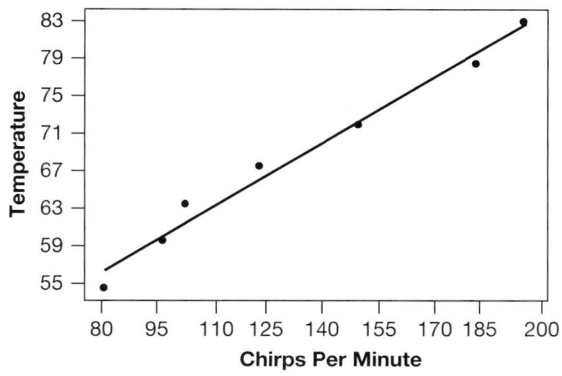

Figure A.23 *How big are these residuals?*

(b) Use technology to find the formula for the regression line for the seven data points.

(c) Calculate the predicted values and the residuals for all seven data points.

A.95 Land Area and Rural Population Two variables in the dataset **AllCountries** are the size of the country (in 1000 sq km) and the percent of the population living in rural areas. We are interested in using the size of the country (*LandArea*) to predict the percent rural (*Rural*). The values of these variables for a random sample of ten countries is shown, with the 3-letter country codes, in Table A.22, and is also available in **TenCountries**. Figure A.24 shows a scatterplot of the data.

(a) What is the explanatory variable? What is the response variable?

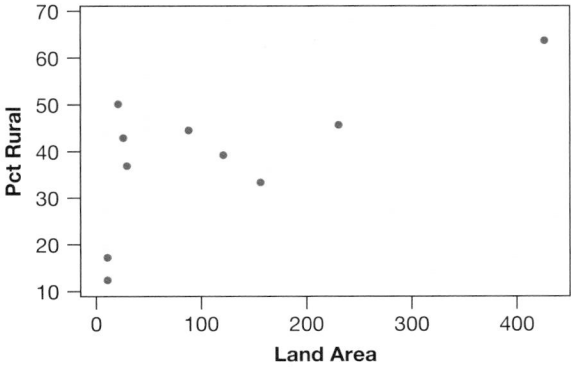

Figure A.24 *Scatterplot of land area and percent rural*

(b) Without doing any calculations, which do you think is the most likely correlation between the two variables?

$$0.00, \quad 0.60, \quad -0.60, \quad 60$$

(c) Use technology to find the regression line to predict percent rural from land area, and interpret the slope.

(d) Does the intercept make sense in this situation?

(e) Which country is the most influential on this regression line (use the 3 letter code)?

(f) Use the regression line to predict the percent of the US population living in rural areas, given that the area of the US is 9147.4 thousand sq km. Does the prediction seem reasonable? Explain why it is not appropriate to use this regression line to predict the percent rural for the US.

A.96 Adding One Point to Land Area and Rural Population In Exercise A.95, we use a random sample of 10 countries to use the size of a country to predict the percent of the population living in rural areas. We now see how results change if we add the United States (Land Area: 9147.4, Rural: 18.7%) to the sample.

(a) Use technology to find the new regression line using the 11 data points.

(b) The slope of the regression line using the original 10 points in Exercise A.95 is about 0.08. Compare the slope with US added to the slope without US. Does adding US have a strong effect on the slope? Why or why not? (*Hint*: Plot the data!)

(c) Predict the percent rural for US with the new regression line. Is this prediction better than the prediction given in Exercise A.95 (which was 752%)?

A.97 The Effect of a Hyper-Aggressive Male If a male wants mating success, he should not hang out with hyper-aggressive males. They tend to scare away all the females. At least, that is the message from a study[118] on water striders. Water striders are

[118]Sih, A. and Watters, J., "The mix matters: behavioural types and group dynamics in water striders," *Behaviour*, 2005; 142 (9–10):1423.

Table A.22 *Land area (in 1000 sq km) and percent living in rural areas*

Country	SRB	BHS	SVN	UZB	TUN	ARM	ROU	MKD	LBN	PRK
Land Area	87.46	10.01	20.14	425.40	155.36	28.47	230.02	25.22	10.23	120.41
Rural	44.6	17.3	50.2	63.8	33.5	37.0	45.8	43.0	12.5	39.4

common bugs that skate across the surface of water. Water striders have different personalities and some of the males are hyper-aggressive, meaning they jump on and wrestle with any other water strider near them. Individually, because hyper-aggressive males are much more active, they tend to have better mating success than more inactive striders. This study examined the effect they have on a group. Four males and three females were put in each of ten pools of water. Half of the groups had a hyper-aggressive male as one of the males and half did not. The proportion of time females are in hiding was measured for each of the 10 groups, and a measure of mean mating activity was also measured with higher numbers meaning more mating. Results are shown in Table A.23 and are available in **WaterStriders**.

Table A.23 *Effect of a hyper-aggressive male on water striders*

Aggressive Male?	Females Hiding	Mating Activity
No	0	0.48
No	0	0.48
No	0	0.45
No	0.09	0.30
No	0.13	0.49
Yes	0.16	0.41
Yes	0.17	0.57
Yes	0.25	0.36
Yes	0.55	0.45
Yes	0.82	0.11

(a) For the five groups with no hyper-aggressive male, find the mean and standard deviation of the proportion of time females stay in hiding. Also find the same summary statistics for the five groups with a hyper-aggressive male. Does there seem to be a difference in the proportion of time females are in hiding between the two groups?

(b) Using all 10 data points, make a scatterplot of the proportion of time females are hiding and the mean mating activity. We consider the proportion of time in hiding to be the explanatory variable.

(c) Using all 10 points, find the regression line to predict mean mating activity from the proportion of time females spend hiding.

(d) For each of the two means found in part (a), find the predicted mating activity for that proportion of time in hiding. What is the predicted mean mating activity if there is not a hyper-aggressive male in the group? What is the predicted mean mating activity if there is a hyper-aggressive male in the group?

(e) What advice would you give to a male water strider that wants to mate?

A.98 Predicting Percent of College Graduates Using High School Graduation Rates Exercise A.42 on page 176 uses data in the **USStates** dataset to examine the percent of adults to graduate college in US states by region. The dataset also includes information on the percent to graduate high school in each state. We use the percent to graduate high school to predict the percent to graduate college. A scatterplot with regression line for all 50 states is shown in Figure A.25.

(a) The formula for the regression line is $\widehat{College} = -53.2 + 0.94 \cdot HighSchool$. Interpret the slope of the line in context.

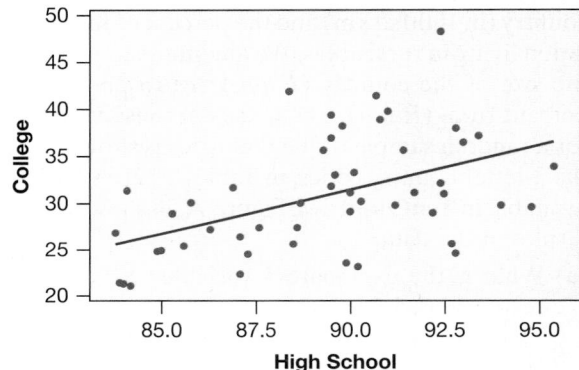

Figure A.25 *Predicting percent to graduate college using percent to graduate high school*

(b) What is the predicted percent to graduate college if 85% graduate high school? What is it if 90% graduate high school?

(c) Massachusetts appears to have a particularly large positive residual. Massachusetts has 92.4% graduating high school and 48.3% graduating college. Compute the residual for Massachusetts.

A.99 Predicting Percent of College Graduates Using Income In Exercise A.98, we used the percent of the population graduating high school to predict the percent to graduate college, using data in **USStates**. It is likely that the mean household income(in thousands) in the state might also be a reasonable predictor. Figure A.26 shows a scatterplot with regression line for these two variables.

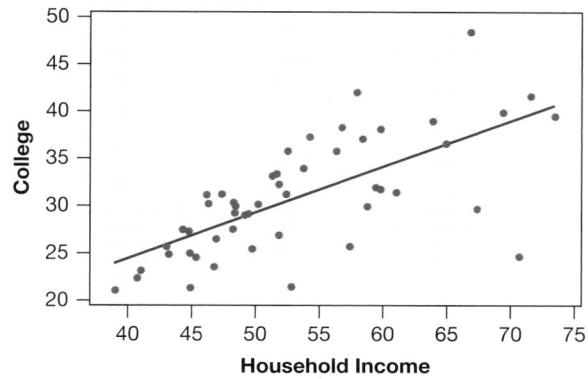

Figure A.26 *Predicting college graduation rate from household income*

(a) Describe the scatterplot in Figure A.26. Is there a linear trend? Is it positive or negative? Are there any really obvious outliers?

(b) Use Figures A.25 and A.26 to decide which variable, percent to graduate high school or household income, is more strongly correlated with percent to graduate college.

(c) For the state with the largest positive residual, estimate from the graph the household income, the percent graduating college, and the predicted percent to graduate college.

(d) For the state with the largest negative residual, estimate from the graph the household income, the percent graduating college, and the predicted percent to graduate college.

A.100 Investigating NBA Statistics The dataset **NBAPlayers2015** is introduced on page 91, and includes many variables about players in the National Basketball Association in 2014–2015. In this exercise, we'll use *FTPct*, the percent of free throws made, to predict *FGPct*, the percent of field goals made.

(a) Make a scatterplot of this relationship. Is there a linear trend? If so, is it positive or negative?

Indicate what positive/negative means in this situation.

(b) Are there any outliers on the scatterplot? If so, describe them. Identify the players by name.

(c) Use technology to find the correlation between the two variables.

(d) Use technology to find a formula for the regression line and also to plot it on the scatterplot.

(e) Find the predicted field goal percentage for a player who has a 0.70 free throw percentage.

A.101 Investigating the Happy Planet Index In Exercise 2.187 on page 118, we introduced the dataset **HappyPlanetIndex**. This exercise asks you to use technology to investigate more variables and relationships in that dataset.

(a) Use technology to create a frequency table for the number of countries in each of the different regions. (The code for each region is given in Exercise 2.187 on page 118).

(b) Use technology to create both a histogram and a boxplot for any one of the quantitative variables. Describe the shape of the distribution and indicate whether there are any outliers. In addition, give summary statistics (mean, standard deviation, five number summary) for the variable.

(c) For a different quantitative variable than the one you used in part (b), create a side-by-side boxplot for your variable by region. Discuss what you see in the graph.

(d) Pick two quantitative variables where you believe one might be useful to predict the other. Create a scatterplot and discuss what you see. Find the correlation between the variables and find the regression line. Use the regression line to make at least one prediction.

Understanding Inference

"The sexy job in the next 10 years will be statisticians."

Hal Varian, Chief Economist at Google*

UNIT OUTLINE

3 Confidence Intervals
4 Hypothesis Tests
Essential Synthesis

In this unit, we develop the key ideas of statistical inference–estimation and testing–using simulation methods to build understanding and to carry out the analysis.

*New York Times, August 6, 2009, p. A1

CHAPTER 3

Confidence Intervals

"Knowing what to measure and how to measure it makes a complicated world less so. If you learn to look at data in the right way, you can explain riddles that otherwise might have seemed impossible. Because there is nothing like the sheer power of numbers to scrub away layers of confusion and contradiction."

Levitt and Dubner*

* *Freakonomics*, HarperCollins, NY, 2005, p. 13.
Top left: ©Pavel Losevsky/iStockphoto, Top right: ©Mikkel William Nielsen/iStockphoto, Bottom right: Aspen Photo/Shutterstock

Questions and Issues

Here are some of the questions and issues we will discuss in this chapter:

- Do they really **pay** you to go to graduate school in statistics? (Yes)
- What proportion of US residents have a college degree?
- What is the average number of calls per day by cell phone users?
- When studying for a test, is it better to mix up the topics or study one topic at a time?
- What proportion of adults send text messages? What proportion of teens do?
- What proportion of those inducted into the Rock and Roll Hall of Fame are performers?
- What proportion of young adults in the US have ever been arrested?
- If a person overeats for a month and then loses the weight, are there long-term effects?
- How much BPA (the chemical bisphenol A) is in your canned soup?
- Does playing action video games improve a person's ability to make accurate quick decisions?
- Does adding indifferent people to a group make it more democratic?
- Are rats compassionate? Are female rats more compassionate than male rats?
- Does drinking tea help the immune system?
- How often, on average, do people laugh in a day?
- Do people order more when splitting the bill?
- Do antibacterial products in soap actually increase infections?
- What is the effect of eating organic for one week?
- Are consumers willing to pay more for socially responsible products?
- Are fears learned as teenagers harder to unlearn?
- Do people find solitude distressing?

3.1 SAMPLING DISTRIBUTIONS

In Chapter 1 we discuss data collection: methods for obtaining sample data from a population of interest. In this chapter we begin the process of going in the other direction: using the information in the sample to understand what might be true about the entire population. If all we see are the data in the sample, what conclusions can we draw about the population? How sure are we about the accuracy of those conclusions? Recall from Chapter 1 that this process is known as *statistical inference*.

> **Statistical Inference**
>
> **Statistical inference** is the process of drawing conclusions about the entire population based on the information in a sample.

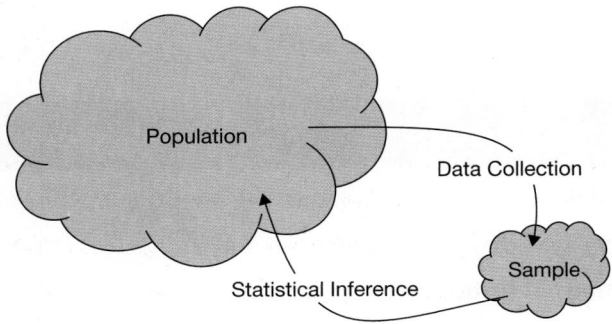

Statistical inference uses sample data to understand a population

Population Parameters and Sample Statistics

To help identify whether we are working with the entire population or just a sample, we use the term *parameter* to identify a quantity measured for the population and *statistic* for a quantity measured for a sample.

> **Parameters vs Statistics**
>
> A **parameter** is a number that describes some aspect of a population.
>
> A **statistic** is a number that is computed from the data in a sample.

As we saw in Chapter 2, although the name (such as "mean" or "proportion") for a statistic and parameter is generally the same, we often use different notation to distinguish the two. For example, we use μ (mu) as a parameter to denote the mean for a population and $\bar{x}$ as a statistic for the mean of a sample. Table 3.1 summarizes

Table 3.1 *Notation for common parameters and statistics*

	Population Parameter	**Sample Statistic**
Mean	μ	$\bar{x}$
Standard deviation	σ	s
Proportion	p	$\hat{p}$
Correlation	ρ	r
Slope (regression)	β	b

common notation for some population parameters and corresponding sample statistics. The notation for each should look familiar from Chapter 2.

Example 3.1

Proportion of College Graduates

The US Census states that 27.5% of US adults who are at least 25 years old have a college bachelor's degree or higher. Suppose that in a random sample of $n = 200$ US residents who are 25 or older, 58 of them have a college bachelor's degree or higher. What is the population parameter? What is the sample statistic? Use correct notation for each answer.

Solution

The population parameter is the proportion with a bachelor's degree for *all* US adults who are at least 25 years old; it is $p = 0.275$. The sample statistic is the proportion with a bachelor's degree for all people in the sample; it is $\hat{p} = 58/200 = 0.29$.

Sample Statistics as Estimates of Population Parameters

On April 29, 2011, Prince William married Kate Middleton (now Duchess Catherine) in London. The Pew Research Center reports that 34% of US adults watched some or all of the royal wedding.[1] How do we know that 34% of all US adults watched? Did anyone ask *you* if you watched it? In order to know for sure what proportion of US adults watched the wedding, we would need to ask *all* US adults whether or not they watched. This would be very difficult to do. As we will see, however, we can estimate the population parameter quite accurately with a sample statistic, as long as we use a random sample (as discussed in Chapter 1). In the case of the royal wedding, the estimate is based on a poll using a random sample of 1006 US adults.

In general, to answer a question about a population parameter *exactly*, we need to collect data from every individual in the population and then compute the quantity of interest. That is not feasible in most settings. Instead, we can select a sample from the population, calculate the quantity of interest for the sample, and use this sample statistic to estimate the value for the whole population.

> **Best Estimate**
>
> If we only have one sample and we don't know the value of the population parameter, the sample statistic is our **best estimate** of the true value of the population parameter.

Example 3.2

Fuel economy information for cars is determined by the EPA (Environmental Protection Agency) by testing a sample of cars.[2] Based on a sample of $n = 12$ Toyota Prius cars in 2012, the average fuel economy was 48.3 mpg. State the population and parameter of interest. Use the information from the sample to give the best estimate of the population parameter.

---

[1]Pew Research Center, "Too Much Coverage: Birth Certificate, Royal Wedding," *http://www.pewresearch.org*, May 3, 2011.
[2]*http://www.epa.gov/fueleconomy/data.htm*.

Solution The population is all Toyota Prius cars manufactured in 2012. The population parameter of interest is μ, the mean fuel economy (mpg) for all 2012 Toyota Prius cars. For this sample, $\bar{x} = 48.3$. Unless we have additional information, the best estimate of the population parameter is the sample statistic of 48.3. Notice that to find μ exactly, we would have to obtain information on the fuel economy for *every* 2012 Toyota Prius.

Example 3.3

For each of the questions below, identify the population parameter(s) of interest and the sample statistic we might use to estimate the parameter.

(a) What is the mean commute time for workers in a particular city?

(b) What is the correlation between the size of dinner bills and the size of tips at a restaurant?

(c) How much difference is there in the proportion of 30 to 39-year-old US residents who have only a cell phone (no land line phone) compared to 50 to 59-year-olds in the US?

Solution (a) The relevant parameter is μ, the mean commute time for all people who work in the city. We estimate it using $\bar{x}$, the mean from a random sample of people who work in the city.

(b) The relevant parameter is ρ, the correlation between the bill amount and tip size for all dinner checks at the restaurant. We estimate it using r, the correlation from a random sample of dinner checks.

(c) The relevant quantity is $p_1 - p_2$, the difference in the proportion of all 30 to 39-year-old US residents with only a cell phone (p_1) and the proportion with the same property among all 50 to 59-year-olds (p_2). We estimate it using $\hat{p}_1 - \hat{p}_2$, the difference in sample proportions computed from random samples taken in each age group.

Variability of Sample Statistics

We usually think of a parameter as a fixed value[3] while the sample statistic varies from sample to sample, depending on which cases are selected to be in the sample. We would like to know the value of the population parameter, but this usually cannot be computed directly because it is often very difficult or impossible to collect data from every member of the population. The sample statistic might vary depending on the sample, but at least we can compute its value.

In Example 3.3, we describe several situations where we might use a sample statistic to estimate a population parameter. How accurate can we expect the estimates to be? That is one of the fundamental questions of statistical inference. Because the value of the parameter is usually fixed but unknown, while the value of the statistic is known but varies depending on the sample, the key to addressing this question is to understand how the value of the sample statistic varies from sample to sample.

Consider the average fuel economy for 2012 Toyota Prius cars in Example 3.2. The average observed in the sample is $\bar{x} = 48.3$. Now suppose we were to take a new random sample of $n = 12$ cars and calculate the sample average. A new sample

[3]In reality, a population may not be static and the value of a parameter might change slightly, for example, if a new person moves into a city. We assume that such changes are negligible when measuring a quantity for the entire population.

average of $\bar{x} = 48.2$ (very close to 48.3!) would suggest low variability in the statistic from sample to sample, suggesting the original estimate of 48.3 is pretty accurate. However, a new sample average of 56.8 (pretty far from 48.3) would suggest high variability in the statistic from sample to sample, giving a large amount of uncertainty surrounding the original estimate.

Of course, it's hard to judge variability accurately from just two sample means. To get a better estimate of the variability in the means we should consider many more samples. One way to do this is to use computer simulations of samples from a known population, as illustrated in the following examples.

Aspen Photo/Shutterstock

How much do baseball players get paid?

DATA 3.1

Major League Baseball Salaries

The dataset **BaseballSalaries2015** shows the salary (in millions of dollars) for all 868 players who were on the active roster for Major League Baseball teams at the start of the 2015 season.[4] The file also shows the team and primary position for each player. We will treat this as the population of all MLB players at that time. ∎

Example 3.4

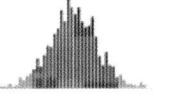

What is the average salary for MLB players on opening day in 2015? Use the correct notation for your answer.

Solution

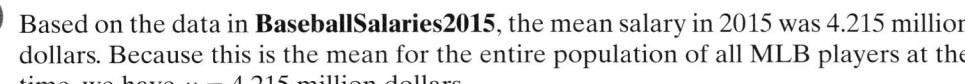

Based on the data in **BaseballSalaries2015**, the mean salary in 2015 was 4.215 million dollars. Because this is the mean for the entire population of all MLB players at the time, we have $\mu = 4.215$ million dollars.

[4]Downloaded from USA Today's Salary Database at *http://www.usatoday.com/sports/mlb/salaries*.

Example 3.5

Solution

Select a random sample of 30 players from the file **BaseballSalaries2015** and compute the mean salary for the players in your sample. Use the correct notation for your answer. Compare your answer to the population mean.

We use *StatKey*, other technology, or a random number table to select the sample. One sample is given in Table 3.2. We compute the mean salary for this sample to be $\bar{x} = 4.550$ million. The sample mean is similar to the population mean of $\mu = 4.215$ million but is not exactly the same.

Table 3.2 *A sample of 30 MLB salaries in 2015*

Player	Salary	Player	Salary	Player	Salary
Kevin Kiermaier	0.514	Will Smith	0.513	Kelly Johnson	1.500
Joe Mauer	23.000	Logan Verrett	0.508	Jean Segura	0.534
Buddy Carlyle	0.725	Carson Smith	0.510	Xavier Cedeno	0.517
Jumbo Diaz	0.510	Dan Jennings	0.523	Asdrubal Cabrera	7.500
Jason Garcia	0.508	Jeremy Affeldt	6.000	Cory Luebke	5.375
Nick Franklin	0.511	Aaron Harang	5.000	Welington Castillo	2.100
Antonio Bastardo	3.100	Chad Billingsley	1.500	Will Harris	0.514
Shelby Miller	0.535	Jake Arrieta	3.630	Kurt Suzuki	6.000
Justin Verlander	28.000	Aramis Ramirez	14.000	Sean Doolittle	0.780
Michael Bourn	13.500	Edinson Volquez	7.500	Michael McKenry	1.088

That the sample statistic in Example 3.5 does not exactly match the population parameter is not surprising: We don't expect to get exactly the mean of the entire population from every sample we choose, but we hope that our sample mean is close to the population mean.

A different random sample of salaries for 30 MLB players is shown in Table 3.3. The mean salary for this sample is $\bar{x} = 3.715$ million. Again, the sample mean is somewhat similar to the population mean of $\mu = 4.215$ million but is not exactly the same. Note that this sample mean is also different from the sample mean found from the sample in Table 3.2.

If everyone in your statistics class selects a random sample of size 30 from the population of baseball salaries and computes the sample mean, there will be many

Table 3.3 *Another sample of 30 baseball player salaries in 2015*

Player	Salary	Player	Salary	Player	Salary
Matt Szczur	0.509	Corey Hart	2.500	Jordan Pacheco	0.900
Gerald Laird	1.000	Anthony Rizzo	5.286	Dustin Pedroia	12.142
Jarrod Parker	0.850	Luke Hochevar	4.000	Christian Bethancourt	0.508
Odrisamer Despaigne	0.517	Kevin Pillar	0.512	Ben Revere	4.100
Hernan Perez	0.509	David Hernandez	2.000	Eduardo Escobar	0.533
Pablo Sandoval	17.600	Torii Hunter	10.500	Alejandro De Aza	5.000
Odubel Herrera	0.508	Oscar Hernandez	0.508	Nick Franklin	0.511
Chris Davis	12.000	Logan Verrett	0.508	Gregor Blanco	3.600
Jose Alvarez	0.510	Jake McGee	3.550	Hector Santiago	2.290
Carlos Beltran	15.000	Corey Kluber	1.000	Justin Turner	2.500

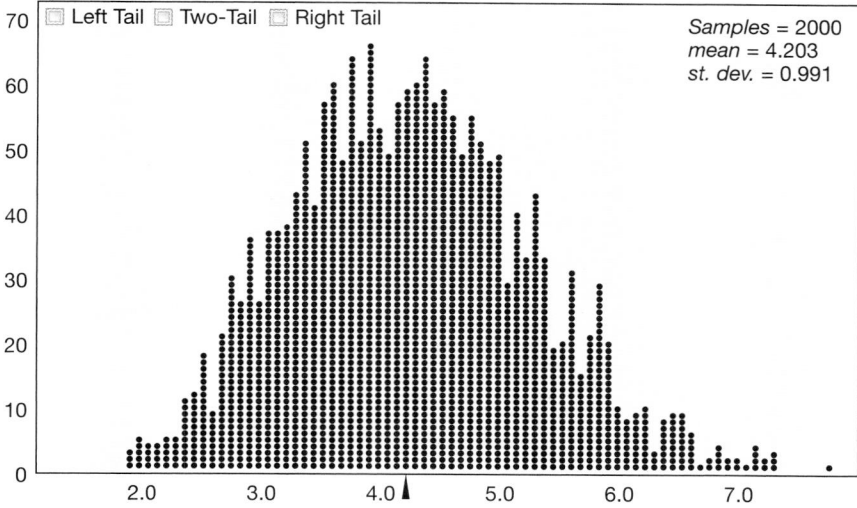

Figure 3.1 *2,000 means for samples of size n = 30 from BaseballSalaries2015*

different results. Try it! (In fact, from a population of size 868, there are 3.25×10^{55} different samples of size 30 that can be selected!) We expect these sample means to be clustered around the true population mean of $\mu = 4.215$. To see that this is so, we use *StatKey* or other technology to take 2000 random samples of size $n = 30$ from our population and compute the sample mean in each case. A dotplot of the results is shown in Figure 3.1. The sample means of $\bar{x} = 4.550$ and $\bar{x} = 3.715$ from the two random samples above correspond to two of the dots in this dotplot.

Notice in Figure 3.1 that we do indeed have many different values for the sample means, but the distribution of sample means is fairly symmetric (despite the heavily right-skewed distribution of salaries in the population) and centered approximately at the population mean of 4.215. From Figure 3.1 we see that most sample means for samples of size 30 fall between about 2 and 7 million dollars. We will see that the bell-shaped curve seen in this distribution is very predictable. The distribution of sample statistics for many samples, such as those illustrated in Figure 3.1, is called a *sampling distribution*.

Sampling Distribution

A **sampling distribution** is the distribution of sample statistics computed for different samples of the same size from the same population.

A sampling distribution shows us how the sample statistic varies from sample to sample.

Figure 3.1 illustrates the sampling distribution for sample means based on samples of size 30 from the population of salaries of all MLB players in 2015. Of course, we don't show the means for all 3.25×10^{55} possible samples. However, the approximation based on 2000 samples should be sufficient to give us a good sense of the shape, center, and spread of the sampling distribution.

Sampling distributions apply to every statistic that we saw in Chapter 2 and lots more! We look next at a sampling distribution for a proportion.

Example 3.6

In Example 3.1 on page 197 we see that 27.5% of US adults at least 25 years old have a college bachelor's degree or higher. Investigate the behavior of sample proportions from this population by using *StatKey* or other technology to simulate lots of random samples of size $n = 200$ when the population proportion is $p = 0.275$. Describe the shape, center, and spread of the distribution of sample proportions.

Solution

Figure 3.2 illustrates the sampling distribution of proportions for 1000 samples, each of size $n = 200$ when $p = 0.275$. We see that the sampling distribution of simulated $\hat{p}$ values is relatively symmetric, centered around the population proportion of $p = 0.275$, ranges from about 0.175 to 0.38, and again has the shape of a bell-shaped curve. Note that the sample statistic $\hat{p} = 0.29$ mentioned in Example 3.1 is just one of the dots in this dotplot.

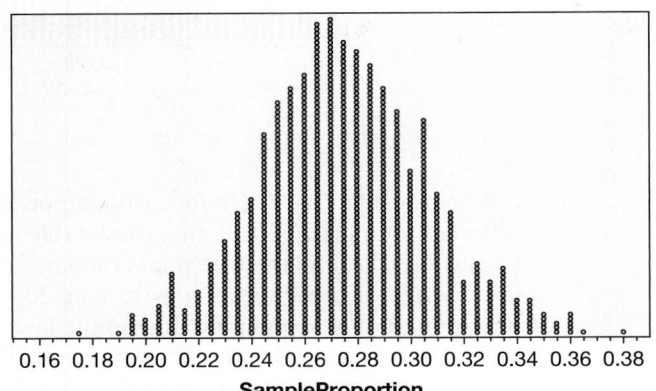

Figure 3.2 *Sample proportions when $n = 200$ and $p = 0.275$*

The distributions of sample proportions in Figure 3.2 and sample means in Figure 3.1 have a similar shape. Both are symmetric, bell-shaped curves centered at the population parameter. As we will see, this is a very common pattern and can often be predicted with statistical theory. If samples are randomly selected and the sample size is large enough, the corresponding sample statistics will often have a symmetric, bell-shaped distribution centered at the value of the parameter. In later chapters we formalize the idea of a bell-shaped distribution and elaborate on how large a sample size is "large enough."

> **Shape and Center of a Sampling Distribution**
>
> For most of the parameters we consider, if samples are randomly selected and the sample size is large enough, the sampling distribution will be symmetric and bell-shaped and centered at the value of the population parameter.

Measuring Sampling Variability: The Standard Error

What we really care about is the *spread* of the sampling distribution (the variability of the statistic from sample to sample). Knowing how much a statistic varies from sample to sample is key in helping us know how accurate an estimate is.

One measure of variability associated with the sample statistic can be found by computing the standard deviation of the sample statistics in a sampling distribution.

Although this is no different from the standard deviation of sample values we saw in Chapter 2, the standard deviation of a sample statistic is so important that it gets its own name: the *standard error* of the statistic. The different name helps to distinguish between the variability in the sample statistics and the variability among the values within a particular sample. We think of the standard error as a "typical" distance between the sample statistics and the population parameter.

> **Standard Error**
>
> The **standard error** of a statistic, denoted *SE*, is the standard deviation of the sample statistic.

In situations such as the mean baseball salary in Example 3.5 and the proportion of college graduates in Example 3.6 where we can simulate values of a statistic for many samples from a population, we can estimate the standard error of the statistic by finding the standard deviation of the simulated statistics.

Example 3.7

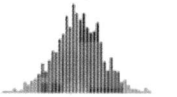

Use *StatKey* or other technology to estimate the standard error for the sampling distributions of the following:

(a) Mean salary for baseball players in samples of size 30 (as in Example 3.5)
(b) Proportion of college graduates in samples of size 200 (as in Example 3.6)

Solution

The standard error is the standard deviation of all the simulated sample statistics. In *StatKey*, this standard deviation is given in the upper right corner of the box containing the sampling distribution (see Figure 3.1). With other technology, once we have the sampling distribution we find the standard deviation of the values in the same way as in Chapter 2.

(a) For the 2000 means for simulated samples of $n = 30$ baseball salaries shown in Figure 3.1, we find the standard deviation of the 2000 means to be 0.991 so we have $SE = 0.991$.

(b) For the 1000 proportions of college graduates in simulated samples of size 200 shown in Figure 3.2, we find the standard deviation of the 1000 proportions to be 0.03, so we have $SE = 0.03$.

Since these standard errors are estimated from a set of random simulations, the values might change slightly from one simulation to another.

Recall from Section 2.3 that when distributions are relatively symmetric and bell-shaped, the 95% rule tells us that approximately 95% of the data values fall within two standard deviations of the mean. Applying the 95% rule to sampling distributions, we see that about 95% of the sample statistics will fall within two standard errors of the mean. This allows us to get a rough estimate of the standard error directly from the dotplot of the sampling distribution, even if we don't have the individual values for each dot.

Example 3.8

Use the 95% rule to estimate the standard error for the following sampling distributions:

(a) Mean salary for baseball players in samples of size 30 (from Figure 3.1)
(b) Proportion of college graduates in samples of size 200 (from Figure 3.2)

Solution   (a) In Figure 3.1, we see that the middle 95% of sample means appear to range from about 2.5 to about 6.5. This should span about two standard errors below the mean and two standard errors above the mean. We estimate the standard error to be about $(6.5 - 2.5)/4 = 1.0$.

  (b) In Figure 3.2, we see that the middle 95% of sample proportions appear to range from about 0.21 to 0.34, or about 0.065 above and below the mean of $p = 0.275$. We estimate the standard error to be about $0.065/2 = 0.0325$.

These rough estimates from the graphs match what we calculated in Example 3.7.

A low standard error means statistics vary little from sample to sample, so we can be more certain that our sample statistic is a reasonable estimate. In Section 3.2, we will learn more about how to use the standard error to quantify the uncertainty in an estimate.

The Importance of Sample Size

Example 3.9

In Example 3.1, we learn that the population proportion of college graduates in the US is $p = 0.275$, and Figure 3.2 on page 202 shows the sampling distribution for the sample proportion of college graduates when repeatedly taking samples of size $n = 200$ from the population. How does this distribution change for other sample sizes? Figure 3.3 shows the distributions of sample proportions for many (simulated) random samples of size $n = 50$, $n = 200$, and $n = 1000$. Discuss the effect of sample size on the center and variability of the distributions.

Solution   The center appears to be close to the population proportion of $p = 0.275$ in all three distributions, but the variability is quite different. As the sample size increases, the variability decreases and a sample proportion is likely to be closer to the population proportion. In other words, as the sample size increases, the standard error decreases.

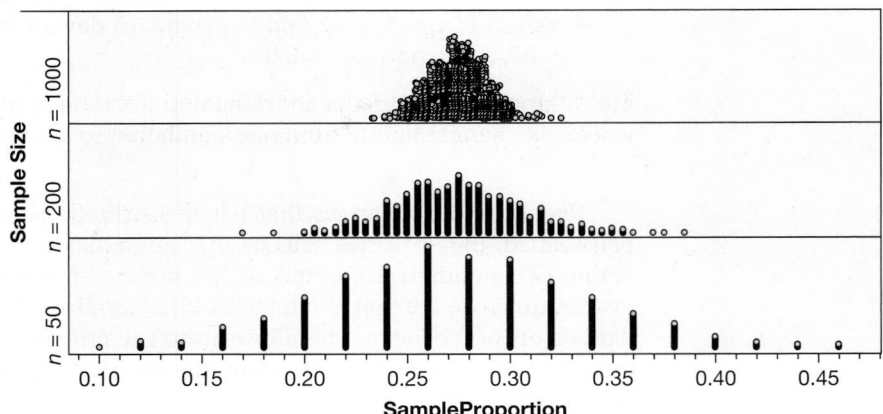

Figure 3.3 *What effect does sample size have on the distributions?*

We see in Example 3.9 that the larger the sample size the lower the variability in the sampling distribution, so the smaller the standard error of the sample statistic. This makes sense: A larger sample allows us to collect more information and estimate a population parameter more precisely. If the sample were the entire population, then the sample statistic would match the population parameter exactly and the sampling distribution would be one stack of dots over a single value!

> **Sample Size Matters!**
>
> As the sample size increases, the variability of sample statistics tends to decrease and sample statistics tend to be closer to the true value of the population parameter.

Example 3.10

Here are five possible standard errors for proportions of college graduates using different size samples:

$$SE = 0.005 \qquad SE = 0.014 \qquad SE = 0.032 \qquad SE = 0.063 \qquad SE = 0.120$$

For each of the three sample sizes shown in Figure 3.3, use the 95% rule to choose the most appropriate standard error from the five options listed.

Solution Since each of the distributions is centered near $p = 0.275$, we consider the interval $0.275 \pm 2 \cdot SE$ and see which standard error gives an interval that contains about 95% of the distribution of simulated $\hat{p}$'s shown in Figure 3.3.

n=1000: It appears that $SE = 0.014$ is the best choice, since the interval on either side of $p = 0.275$ would go from $0.275 \pm 2(0.014)$ which is 0.247 to 0.303. This looks like a reasonable interval to contain the middle 95% of the values in the dotplot shown in the top panel of Figure 3.3, when the sample size is $n = 1000$.

n=200: It appears that $SE = 0.032$ is the best choice, since the interval on either side of $p = 0.275$ would go from $0.275 \pm 2(0.032)$ which is 0.211 to 0.339. This looks like a reasonable interval to contain the middle 95% of the values in the dotplot shown in the middle panel of Figure 3.3, when the sample size is $n = 200$.

n=50: It appears that $SE = 0.063$ is the best choice, since the interval on either side of $p = 0.275$ would go from $0.275 \pm 2(0.063)$ which is 0.149 to 0.401. This looks like a reasonable interval to contain the middle 95% of the values in the dotplot shown in the bottom panel of Figure 3.3, when the sample size is $n = 50$.

The standard error of $SE = 0.005$ is too small for any of these plots, and $SE = 0.120$ would give an interval that is too large.

We see again in Example 3.10 that as the sample size increases, the standard error decreases, so the sample statistic generally becomes a better estimate of the population parameter.

Importance of Random Sampling

So far, the sampling distributions we have looked at have all been centered around the population parameter. It is important that samples were selected at random in each of these cases. Too often this is overlooked. Random sampling will generally yield a sampling distribution centered around the population parameter, but, as we learned in Section 1.2, non-random samples may be biased, in which case the sampling distribution may be centered around the wrong value.

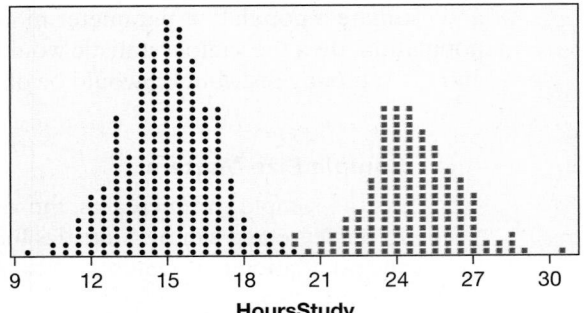

Figure 3.4 *Sample means: Which color shows a biased sampling method?*

Example 3.11

Suppose that students at one college study, on average, 15 hours a week. Two different students, Judy and Mark, are curious about sampling. They each sample $n = 50$ students many times, ask each student the number of hours they study a week, and record the mean of each sample. Judy takes many random samples of 50 students from the entire student body, while Mark takes many samples of 50 students by asking students in the library. The sampling distributions generated by Mark and Judy are shown with different colors in Figure 3.4. Which set of sample means (red or black) were produced by Judy? Why did Mark and Judy get such different results?

Solution

Judy was utilizing random sampling, so we expect her sample means to be centered around the true average of 15 hours a week. Therefore, we can conclude that her sample means correspond to the black dots. Mark chose to take a convenient sampling approach, rather than take a random sample. Due to this fact his samples are not representative of the population (students sampled in the library are likely to study more), so his sample means are biased to overestimate the average number of hours students study.

Inference Caution

Statistical inference is built on the assumption that samples are drawn randomly from a population. Collecting the sample data in a way that biases the results can lead to false conclusions about the population.

In this section, we've learned that statistics vary from sample to sample, and that a sample statistic can be used as an estimate for an unknown fixed population parameter. However, a sample statistic will usually not match the population parameter exactly, and a key question is how accurate we expect our estimate to be. We explore this by looking at many statistics computed from many samples of the same size from the same population, which together form a sampling distribution. The standard deviation of the sampling distribution, called the standard error, is a common way of measuring the variability of a statistic. Knowing how much a statistic varies from sample to sample will allow us to determine the uncertainty in our estimate, a concept we will explore in more detail in the next section.

SECTION LEARNING GOALS

You should now have the understanding and skills to:

▷ • Distinguish between a population parameter and a sample statistic, recognizing that a parameter is fixed while a statistic varies from sample to sample

▷ • Compute an estimate for a parameter using an appropriate statistic from a sample

▷ • Recognize that a sampling distribution shows how sample statistics tend to vary

▷ • Recognize that statistics from random samples tend to be centered at the population parameter

▷ • Estimate the standard error of a statistic from its sampling distribution

▷ • Explain how sample size affects a sampling distribution

Exercises for Section 3.1

SKILL BUILDER 1

In Exercises 3.1 to 3.5, state whether the quantity described is a parameter or a statistic and give the correct notation.

3.1 Average household income for all houses in the US, using data from the US Census.

3.2 Correlation between height and weight for players on the 2014 Brazil World Cup Team, using data from all 23 players on the roster.

3.3 Proportion of people who use an electric toothbrush, using data from a sample of 300 adults.

3.4 Proportion of registered voters in a county who voted in the last election, using data from the county voting records.

3.5 Average number of television sets per household in North Carolina, using data from a sample of 1000 households.

SKILL BUILDER 2

In Exercises 3.6 to 3.11, give the correct notation for the quantity described and give its value.

3.6 Proportion of families in the US who were homeless in 2010. The number of homeless families[5] in 2010 was about 170,000 while the total number of families is given in the 2010 Census as 78 million.

3.7 Average enrollment in charter schools in Illinois. In 2014, there were 148 charter schools in the state of Illinois[6] and the total number of students attending the charter schools was 59,388.

3.8 Proportion of US adults who own a cell phone. In a survey of 1006 US adults in 2014, 90% said they had a cell phone.[7]

3.9 Correlation between age and heart rate for patients admitted to an Intensive Care Unit. Data from the 200 patients included in the file **ICUAdmissions** gives a correlation of 0.037.

3.10 Mean number of cell phone calls made or received per day by cell phone users. In a survey of 1917 cell phone users, the mean was 13.10 phone calls a day.[8]

3.11 Correlation between points and penalty minutes for all 24 players with at least 10 games played for the 2014–2015 Ottawa Senators[9] NHL hockey team. The data are given in Table 3.4 and the full data are available in the file **OttawaSenators**.

[5]Luo, M, "Number of Families in Shelters Rises," *New York Times*, September 12, 2010.

[6]Data obtained from *www.incschools.org*, October 2015.

[7]Pew Research Center's Mobile Technology Fact Sheet, accessed October 2015 at *http://www.pewinternet.org/fact-sheets/mobile-technology-fact-sheet/*.

[8]"Spring Change Assessment Survey 2010," Princeton Survey Research Associates International, 6/4/10, accessed via "Cell Phones and American Adults," Amanda Lenhart, Pew Research Center's Internet and American Life Project, accessed at *http://pewinternet.org/Reports/2010/Cell-Phones-and-American-Adults/Overview.aspx*.

[9]Data obtained from *http://www.hockey-reference.com/teams/OTT/2015.html*.

Table 3.4 *Points and penalty minutes for the 2014–2015 Ottawa Senators NHL team*

Points	66	64	64	54	48	46	36	34	27	26	23	21
Pen mins	42	14	36	24	14	20	36	33	32	67	30	6
Points	19	15	13	12	11	11	9	7	3	3	3	1
Pen mins	9	14	28	97	107	18	45	78	8	18	14	29

SKILL BUILDER 3

Exercises 3.12 to 3.15 refer to the sampling distributions given in Figure 3.5. In each case, estimate the value of the population parameter and estimate the standard error for the sample statistic.

3.12 Figure 3.5(a) shows sample proportions from samples of size $n = 40$ from a population.

3.13 Figure 3.5(b) shows sample means from samples of size $n = 30$ from a population.

3.14 Figure 3.5(c) shows sample means from samples of size $n = 100$ from a population.

3.15 Figure 3.5(d) shows sample proportions from samples of size $n = 180$ from a population.

SKILL BUILDER 4

Exercises 3.16 to 3.19 refer to the sampling distributions given in Figure 3.5. Several possible values are given for a sample statistic. In each case, indicate whether each value is (i) reasonably likely to occur from a sample of this size, (ii) unusual but might occur occasionally, or (iii) extremely unlikely to ever occur.

3.16 Using the sampling distribution shown in Figure 3.5(a), how likely are these sample proportions:

$$\text{(a) } \hat{p} = 0.1 \qquad \text{(b) } \hat{p} = 0.35 \qquad \text{(c) } \hat{p} = 0.6$$

3.17 Using the sampling distribution shown in Figure 3.5(b), how likely are these sample means:

$$\text{(a) } \overline{x} = 70 \qquad \text{(b) } \overline{x} = 100 \qquad \text{(c) } \overline{x} = 140$$

3.18 Using the sampling distribution shown in Figure 3.5(c), how likely are these sample means:

$$\text{(a) } \overline{x} = 250 \qquad \text{(b) } \overline{x} = 305 \qquad \text{(c) } \overline{x} = 315$$

3.19 Using the sampling distribution shown in Figure 3.5(d), how likely are these sample proportions:

$$\text{(a) } \hat{p} = 0.72 \qquad \text{(b) } \hat{p} = 0.88 \qquad \text{(c) } \hat{p} = 0.95$$

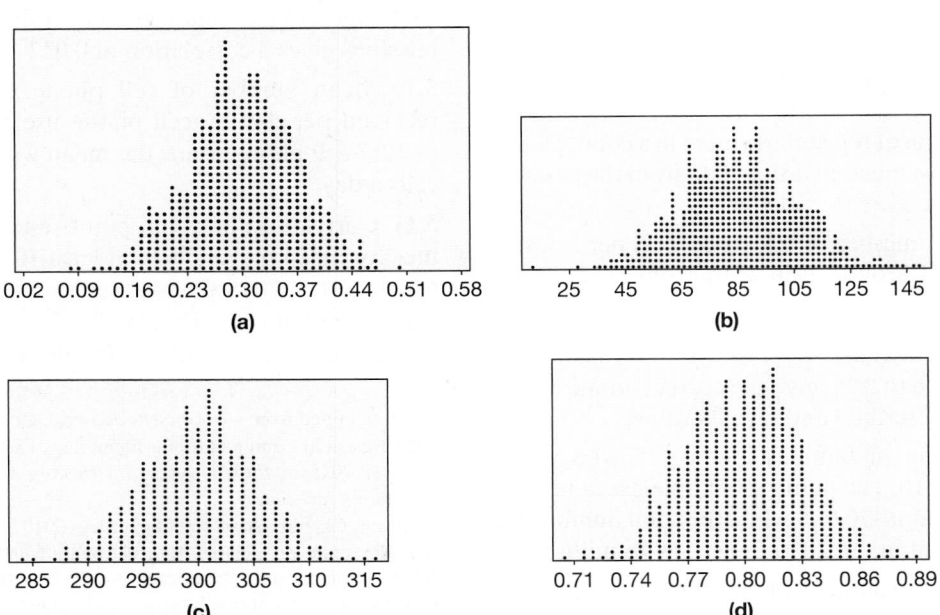

Figure 3.5 *Four sampling distributions*

3.20 Downloading Apps for Your Smartphone A random sample of $n = 461$ smartphone users in the US in January 2015 found that 355 of them have downloaded an app.[10]

(a) Give notation for the parameter of interest, and define the parameter in this context.

(b) Give notation for the quantity that gives the best estimate and give its value.

(c) What would we have to do to calculate the parameter exactly?

3.21 How Many Apps for Your Smartphone? Exercise 3.20 describes a study about smartphone users in the US downloading apps for their smartphone. Of the $n = 355$ smartphone users who had downloaded an app, the average number of apps downloaded was 19.7.

(a) Give notation for the parameter of interest, and define the parameter in this context.

(b) Give notation for the quantity that gives the best estimate and give its value.

(c) What would we have to do to calculate the parameter exactly?

3.22 Socially Conscious Consumers In March 2015, a Nielsen global online survey "found that consumers are increasingly willing to pay more for socially responsible products."[11] Over 30,000 people in 60 countries were polled about their purchasing habits, and 66% of respondents said that they were willing to pay more for products and services from companies who are committed to positive social and environmental impact. We are interested in estimating the proportion of all consumers willing to pay more. Give notation for the quantity we are estimating, notation for the quantity we are using to make the estimate, and the value of the best estimate. Be sure to clearly define any parameters in the context of this situation.

3.23 Florida Lakes Florida has over 7700 lakes.[12] We wish to estimate the correlation between the pH levels of all Florida lakes and the mercury levels of fish in the lakes. We see in Data 2.4 on page 71 that the correlation between these two variables for a sample of $n = 53$ of the lakes is -0.575.

(a) Give notation for the quantity we are estimating, notation for the quantity we use to make the estimate, and the value of the best estimate.

(b) Why is an estimate necessary here? What would we have to do to calculate the exact value of the quantity we are estimating?

3.24 Topical Painkiller Ointment The use of topical painkiller ointment or gel rather than pills for pain relief was approved just within the last few years in the US for prescription use only.[13] Insurance records show that the average copayment for a month's supply of topical painkiller ointment for regular users is $30. A sample of $n = 75$ regular users found a sample mean copayment of $27.90.

(a) Identify each of 30 and 27.90 as a parameter or a statistic and give the appropriate notation for each.

(b) If we take 1000 samples of size $n = 75$ from the population of all copayments for a month's supply of topical painkiller ointment for regular users and plot the sample means on a dotplot, describe the shape you would expect to see in the plot and where it would be centered.

(c) How many dots will be on the dotplot you described in part (b)? What will each dot represent?

3.25 Average Household Size The latest US Census lists the average household size for all households in the US as 2.61. (A household is all people occupying a housing unit as their primary place of residence.) Figure 3.6 shows possible distributions of means for 1000 samples of household sizes. The scale on the horizontal axis is the same in all four cases.

(a) Assume that two of the distributions show results from 1000 random samples, while two others show distributions from a sampling method that is biased. Which two dotplots appear to show samples produced using a biased

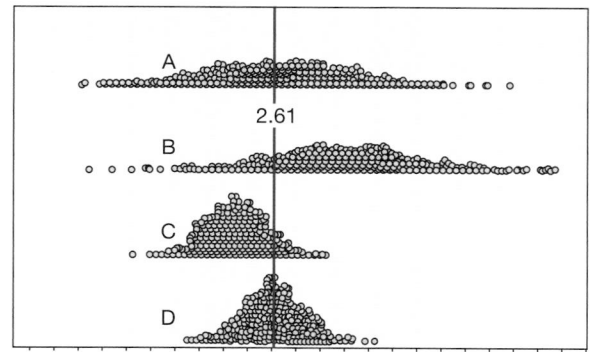

Figure 3.6 *Sets of 1000 sample means*

[10]Olmstead, K. and Atkinson, M., "Apps Permissions in the Google Play Store," *pewresearch.org*, November 10, 2015.

[11]"Sustainable Selections: How Socially Responsible Companies Are Turning a Profit," *www.nielsen.com*, October 12, 2015.

[12]*www.stateofflorida.com/florquicfac.html*.

[13]Tarkan, L., "Topical Gel Catches up with Pills for Relief," *The New York Times*, September 6, 2010.

sampling method? Explain your reasoning. Pick one of the distributions that you listed as biased and describe a sampling method that might produce this bias.

(b) For the two distributions that appear to show results from random samples, suppose that one comes from 1000 samples of size $n = 100$ and one comes from 1000 samples of size $n = 500$. Which distribution goes with which sample size? Explain.

3.26 Proportion of US Residents Less than 25 Years Old The US Census indicates that 35% of US residents are less than 25 years old. Figure 3.7 shows possible sampling distributions for the proportion of a sample less than 25 years old, for samples of size $n = 20$, $n = 100$, and $n = 500$.

(a) Which distribution goes with which sample size?

(b) If we use a proportion $\hat{p}$, based on a sample of size $n = 20$, to estimate the population parameter $p = 0.35$, would it be very surprising to get an estimate that is off by more than 0.10 (that is, the sample proportion is less than 0.25 or greater than 0.45)? How about with a sample of size $n = 100$? How about with a sample of size $n = 500$?

(c) Repeat part (b) if we ask about the sample proportion being off by just 0.05 or more.

(d) Using parts (b) and (c), comment on the effect that sample size has on the accuracy of an estimate.

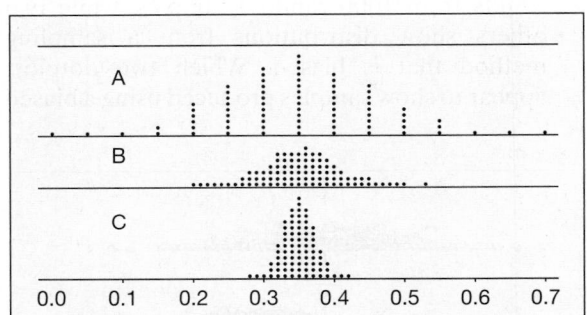

Figure 3.7 *Match the dotplots with the sample size*

3.27 Mix It Up for Better Learning In preparing for a test on a set of material, is it better to study one topic at a time or to study topics mixed together? In one study,[14] a sample of fourth graders were

taught four equations. Half of the children learned by studying repeated examples of one equation at a time, while the other half studied mixed problem sets that included examples of all four types of calculations grouped together. A day later, all the students were given a test on the material. The students in the mixed practice group had an average grade of 77, while the students in the one-at-a-time group had an average grade of 38. What is the best estimate for the difference in the average grade between fourth-grade students who study mixed problems and those who study each equation independently? Give notation (as a difference with a minus sign) for the quantity we are trying to estimate, notation for the quantity that gives the best estimate, and the value of the best estimate. Be sure to clearly define any parameters in the context of this situation.

3.28 What Proportion of Adults and Teens Text Message? A study of $n = 2252$ adults age 18 or older found that 72% of the cell phone users send and receive text messages.[15] A study of $n = 800$ teens age 12 to 17 found that 87% of the teen cell phone users send and receive text messages. What is the best estimate for the difference in the proportion of cell phone users who use text messages, between adults (defined as 18 and over) and teens? Give notation (as a difference with a minus sign) for the quantity we are trying to estimate, notation for the quantity that gives the best estimate, and the value of the best estimate. Be sure to clearly define any parameters in the context of this situation.

3.29 Hollywood Movies Data 2.7 on page 95 introduces the dataset **HollywoodMovies**, which contains information on more than 900 movies that came out of Hollywood between 2007 and 2013.[16] One of the variables is the budget (in millions of dollars) to make the movie. Figure 3.8 shows two boxplots. One represents the budget data for one random sample of size $n = 30$. The other represents the values in a sampling distribution of 1000 means of budget data from samples of size 30.

(a) Which is which? Explain.

(b) From the boxplot showing the data from one random sample, what does one value in the sample represent? How many values are included

[14]Rohrer, D. and Taylor, K., "The Effects of Interleaved Practice," *Applied Cognitive Psychology*, 2010;24(6):#837–848.

[15]Lenhart A., "Cell Phones and American Adults," Pew Research Center's Internet and American Life Project, accessed at *http://pewinternet.org/Reports/2010/Cell-Phones-and-American-Adults/Overview.aspx*.

[16]McCandless, D., "Most Profitable Hollywood Movies," "Information is Beautiful," *davidmccandless.com*, accessed July 2015.

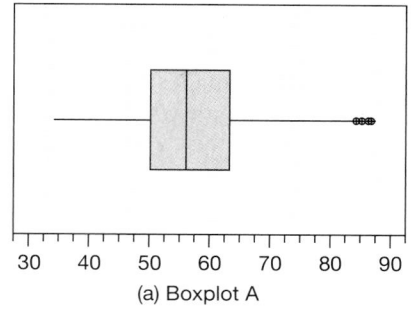 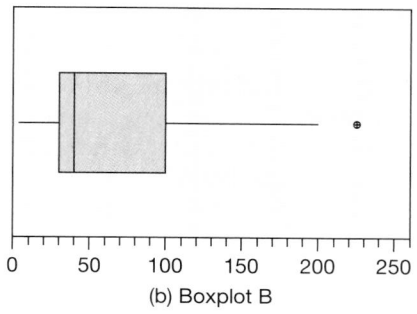

(a) Boxplot A   (b) Boxplot B

Figure 3.8 *One sample and one sampling distribution: Which is which?*

in the data to make the boxplot? Estimate the minimum and maximum values. Give a rough estimate of the mean of the values and use appropriate notation for your answer.

(c) From the boxplot showing the data from a sampling distribution, what does one value in the sampling distribution represent? How many values are included in the data to make the boxplot? Estimate the minimum and maximum values. Give a rough estimate of the value of the population parameter and use appropriate notation for your answer.

3.30 What Percent of the US Population Are Senior Citizens? People 65 years and older are the fastest growing segment of the US population, and constituted 13% of the population in 2010.[17] Figure 3.9 shows sample proportions from two sampling distributions of the proportion 65 years and older in

the US: One shows samples of size 100, and the other shows samples of size 1000.

(a) What is the center of both distributions?

(b) What is the approximate minimum and maximum of each distribution?

(c) Give a rough estimate of the standard error in each case.

(d) Suppose you take one more sample in each case. Would a sample proportion of 0.17 (that is, 17% senior citizens in the sample) be surprising to see from a sample of size 100? Would it be surprising from a sample of size 1000?

3.31 Graduate Programs in Statistics! One of the many wonderful things about studying statistics is that graduate programs in statistics often pay their graduate students, which means that many graduate students in statistics are able to attend graduate school tuition free with an assistantship or fellowship. In 2009, there were 82 US statistics or biostatistics doctoral programs for which enrollment data were available.[18] The dataset **StatisticsPhD** lists all these schools together with the total enrollment of full-time graduate students in each program in 2009.

(a) Use *StatKey* or other technology to select a random sample of 10 of the 82 enrollment values. Indicate which values you've selected and compute the sample mean.

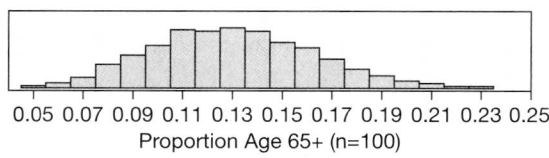

Proportion Age 65+ (n=100)

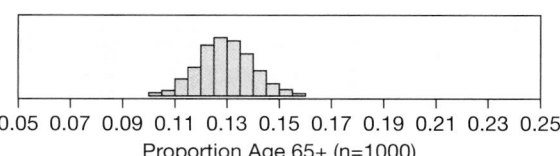

0.05 0.07 0.09 0.11 0.13 0.15 0.17 0.19 0.21 0.23 0.25
Proportion Age 65+ (n=1000)

Figure 3.9 *Sampling distributions for n = 100 and n = 1000*

---

[17]*www.census.gov.*

[18]Full list of the 82 Group IV departments was obtained at *http://www.ams.org/profession/data/annual-survey/group_iv.* Data on enrollment obtained primarily from *Assistantships and Graduate Fellowships in the Mathematical Sciences*, 2009, American Mathematical Society. The list does not include combined departments of mathematics and statistics and does not include departments that did not reply to the AMS survey.

(b) Repeat part (a) by taking a second sample and calculating the mean.

(c) Find the mean enrollment for the entire population of these 82 graduate programs. Use correct notation for your answer. Comment on the accuracy of using the sample means found in parts (a) and (b) to estimate the population mean.

(d) Give a rough sketch of the sampling distribution if we calculate many sample means taking samples of size $n = 10$ from this population of enrollment values. What shape will it have and where will it be centered?

3.32 Average Salary of NFL Players The dataset **NFLContracts2015** contains the yearly salary (in millions of dollars) from the contracts of all players on a National Football League (NFL) roster at the start of the 2015 season.[19]

(a) Use *StatKey* or other technology to select a random sample of 5 NFL contract *YearlySalary* values. Indicate which players you've selected and compute the sample mean.

(b) Repeat part (a) by taking a second sample of 5 values, again indicating which players you selected and computing the sample mean.

(c) Find the mean for the entire population of players. Include notation for this mean. Comment on the accuracy of using the sample means found in parts (a) and (b) to estimate the population mean.

3.33 A Sampling Distribution for Statistics Graduate Programs Exercise 3.31 introduced the dataset **StatisticsPhD**, which gives enrollment for all 82 graduate statistics programs in the US in 2009. Use *StatKey* or other technology to generate a sampling distribution of sample means using a sample size of $n = 10$ from the values in this dataset. What shape does the distribution have? Approximately where is it centered? What is the standard error (in other words, what is the standard deviation of the sample means)?

3.34 A Sampling Distribution for Average Salary of NFL Players Use *StatKey* or other technology to generate a sampling distribution of sample means using a sample of size $n = 5$ from the *YearlySalary* values in the dataset **NFLContracts2015**, which gives the total and yearly money values from the contracts of all NFL players in 2015.

[19]2015 NFL Contract information was collected on 9/16/2015 from *OverTheCap.com*.

(a) What are the smallest and largest *YearlySalary* values in the population?

(b) What are the smallest and largest sample means in the sampling distribution?

(c) What is the standard error (that is, the standard deviation of the sample means) for the sampling distribution for samples of size $n = 5$?

(d) Generate a new sampling distribution with samples of size $n = 50$. What is the standard error for this sampling distribution?

3.35 What Is an Average Budget for a Hollywood Movie? Data 2.7 on page 95 introduces the dataset **HollywoodMovies**, which contains information on more than 900 movies that came out of Hollywood between 2007 and 2013. We will consider this the population of all movies produced in Hollywood during this time period.

(a) Find the mean and standard deviation for the budgets (in millions of dollars) of all Hollywood movies between 2007 and 2013. Use the correct notation with your answer.

(b) Use *StatKey* or other technology to generate a sampling distribution for the sample mean of budgets of Hollywood movies during this period using a sample size of $n = 20$. Give the shape and center of the sampling distribution and give the standard error.

3.36 College Graduates In Example 3.1 on page 197, we see that 27.5% of US adults are college graduates.

(a) Use *StatKey* or other technology to generate a sampling distribution for the sample proportion of college graduates using a sample size of $n = 50$. Generate at least 1000 sample proportions. Give the shape and center of the sampling distribution and give the standard error.

(b) Repeat part (a) using a sample size of $n = 500$.

3.37 Gender in the Rock and Roll Hall of Fame From its founding through 2015, the Rock and Roll Hall of Fame has inducted 303 groups or individuals. Forty-seven of the inductees have been female or have included female members.[20] The full dataset is available in **RockandRoll**.

(a) What proportion of inductees have been female or have included female members? Use the correct notation with your answer.

(b) If we took many samples of size 50 from the population of all inductees and recorded the

[20]Rock and Roll Hall of Fame website: *rockhall.com/inductees*.

proportion female or with female members for each sample, what shape do we expect the distribution of sample proportions to have? Where do we expect it to be centered?

3.38 Performers in the Rock and Roll Hall of Fame From its founding through 2015, the Rock and Roll Hall of Fame has inducted 303 groups or individuals, and 206 of the inductees have been performers while the rest have been related to the world of music in some way other than as a performer. The full dataset is available in **RockandRoll**.

(a) What proportion of inductees have been performers? Use the correct notation with your answer.

(b) If we took many samples of size 50 from the population of all inductees and recorded the proportion who were performers for each sample, what shape do we expect the distribution of sample proportions to have? Where do we expect it to be centered?

3.39 A Sampling Distribution for Gender in the Rock and Roll Hall of Fame Exercise 3.37 tells us that 47 of the 303 inductees to the Rock and Roll Hall of Fame have been female or have included female members. The data are given in **RockandRoll**. Using all inductees as your population:

(a) Use *StatKey* or other technology to take many random samples of size $n = 10$ and compute the sample proportion that are female or with female members. What is the standard error for

these sample proportions? What is the value of the sample proportion farthest from the population proportion of $p = 0.155$? How far away is it?

(b) Repeat part (a) using samples of size $n = 20$.

(c) Repeat part (a) using samples of size $n = 50$.

(d) Use your answers to parts (a), (b), and (c) to comment on the effect of increasing the sample size on the accuracy of using a sample proportion to estimate the population proportion.

3.40 A Sampling Distribution for Performers in the Rock and Roll Hall of Fame Exercise 3.38 tells us that 206 of the 303 inductees to the Rock and Roll Hall of Fame have been performers. The data are given in **RockandRoll**. Using all inductees as your population:

(a) Use *StatKey* or other technology to take many random samples of size $n = 10$ and compute the sample proportion that are performers. What is the standard error of the sample proportions? What is the value of the sample proportion farthest from the population proportion of $p = 0.68$? How far away is it?

(b) Repeat part (a) using samples of size $n = 20$.

(c) Repeat part (a) using samples of size $n = 50$.

(d) Use your answers to parts (a), (b), and (c) to comment on the effect of increasing the sample size on the accuracy of using a sample proportion to estimate the population proportion.

3.2 UNDERSTANDING AND INTERPRETING CONFIDENCE INTERVALS

We can use a sample statistic to estimate a population parameter, but we know that sample statistics vary from sample to sample. Because of this variation, we usually give a range of plausible values for the population parameter rather than just a single best estimate. Whether we are using a single best estimate or a range of plausible values, however, the first step is to correctly identify the parameter that is being estimated.

Identifying the Parameter of Interest

In Chapter 1, we see that variables can be classified as categorical or quantitative, and in Chapter 2, we explore summary statistics and graphs for each type of variable individually and for each possible relationship between two variables. The parameters that we focus on in this chapter correspond to these same variables and relationships. Table 3.5 summarizes the different situations, and reviews the notation for the population parameter being estimated and the sample statistic that gives the best estimate.

Table 3.5 *Identifying the correct parameter and statistic*

Variable(s)	Estimating:	Parameter	Statistic
Single categorical variable	Proportion	p	$\hat{p}$
Single quantitative variable	Mean	μ	$\bar{x}$
Two categorical variables	Difference in proportions	$p_1 - p_2$	$\hat{p}_1 - \hat{p}_2$
One categorical and one quantitative	Difference in means	$\mu_1 - \mu_2$	$\bar{x}_1 - \bar{x}_2$
Two quantitative variables	Correlation	ρ	r

Estimating with an Interval: Margin of Error

We use the value of a statistic computed from a sample to give an estimate for a parameter of a population. However, since statistics vary from sample to sample, a single estimate is often not sufficient. We need some measure of accuracy associated with our estimate. Thankfully, we can use knowledge about how sample statistics vary to find a *margin of error* for the estimate. This allows us to construct an *interval* that gives a range of plausible values for the population parameter.

Example 3.12

Is a Television Set a Necessity?

The percent of Americans saying that a television set is a necessity has dropped dramatically in recent years. In a nationwide survey of 1484 people ages 18 and older living in the continental United States, only 42% say that a television set is a necessity rather than a luxury.[21] The article goes on to say "the margin of sampling error is plus or minus 3.0 percentage points." Use the information from this article to find an interval that gives a range of plausible values for the proportion of people 18 and older living in the continental United States who believe that a television set is a necessity.

Solution We are estimating a proportion so the notation for the thing we are estimating is the population proportion p. The proportion who believe a television set is a necessity in the sample is $\hat{p} = 0.42$. The phrase "margin of sampling error is plus or minus 3.0 percentage points" indicates that the true proportion for the entire population of all American adults is probably within 3% (or 0.03) on either side of this estimate. Thus an interval giving plausible values for the population proportion is

$$0.42 \pm 0.03$$

Since $0.42 - 0.03 = 0.39$ and $0.42 + 0.03 = 0.45$, the interval is 0.39 to 0.45, or from 39% to 45%.

Let's take a minute to think about the information in Example 3.12. There are about 240,000,000 people age 18 and older living in the continental United States, and we only asked 1484 of them the question. It is remarkable that we can be relatively confident that our estimate will be off by at most ±0.03 even though we only asked a very small portion of the entire population. This is part of the amazing power of statistics!

[21]Taylor, P. and Wang, W., "The Fading Glory of the Television and Telephone," Pew Research Center, August 19, 2010, *http://pewsocialtrends.org/pubs/762/fading-glory-television-telephone-luxury-necessity#prc-jump.*

We often use *interval notation* to denote an interval. For example, for the interval from 0.39 to 0.45 in Example 3.12, we write (0.39, 0.45).

Using a Margin of Error to Give a Range of Plausible Values

Recall that the best estimate for a population parameter is the relevant sample statistic. We can expand this idea to give a range of plausible values for the population parameter using:

Sample statistic ± margin of error

where the **margin of error** is a number that reflects the precision of the sample statistic as an estimate for this parameter.

Example 3.13

Suppose the results of an election poll show the proportion supporting a particular candidate is $\hat{p} = 0.54$. We would like to know how close the true p is to $\hat{p}$. Two possible margins of error are shown. In each case, indicate whether we can be reasonably sure that this candidate will win the majority of votes and win the election.

(a) Margin of error is 0.02

(b) Margin of error is 0.10

Solution

(a) If the margin of error is 0.02, then our interval giving plausible values for p is 0.54 ± 0.02, which gives an interval of 0.52 to 0.56. All plausible values of the true proportion are greater than one-half, so we can be reasonably sure that this candidate will win the election.

(b) If the margin of error is 0.10, then our interval is 0.54 ± 0.10, which gives an interval of 0.44 to 0.64. Since this interval contains values of p that are less than one-half, we would be less certain about the result of the election.

Confidence Intervals

The "range of plausible values" interpretation for an interval estimate can be refined with the notion of a *confidence interval*. A confidence interval is an interval, computed from a sample, that has a predetermined chance of capturing the value of the population parameter. Remember that the parameter is a fixed value; it is the sample that is prone to variability. The method used to construct a confidence interval should capture the parameter for a predetermined proportion of all possible samples. Some (hopefully most) samples will give intervals that contain the parameter and some (hopefully just a few) will give intervals that miss the target.

Confidence Interval

A **confidence interval** for a parameter is an interval computed from sample data by a method that will capture the parameter for a specified proportion of all samples.

The success rate (proportion of all samples whose intervals contain the parameter) is known as the **confidence level**.

Recall that for a symmetric, bell-shaped distribution, roughly 95% of the values fall within two standard deviations of the center. Therefore, we can assume that the sample statistic will be within two standard errors of the parameter about 95% of the time. Thus the interval *Statistic* $\pm 2 \cdot SE$, where *SE* stands for the standard error, will contain the population parameter for about 95% of all samples. If we have a way to estimate the standard error (*SE*), and if the sampling distribution is relatively symmetric and bell-shaped, this is one way to construct an approximate 95% confidence interval.

> **95% Confidence Interval Using the Standard Error**
>
> If we can estimate the standard error *SE* and if the sampling distribution is relatively symmetric and bell-shaped, a 95% confidence interval can be estimated using
>
> $$Statistic \pm 2 \cdot SE$$

Example 3.14

The salaries for a random sample of 30 Major League Baseball players at the start of the 2015 season are given in Table 3.2 on page 200. For this sample, we have $n = 30$ with $\bar{x} = 4.55$ million and $s = 6.81$ million.

In Example 3.7 on page 203, we estimate that the standard error for means based on samples of size $n = 30$ from this population is about 0.99.

(a) Use the information in this one sample and the estimated standard error of 0.99 to find a 95% confidence interval for the average salary in 2015 for Major League Baseball players. Also give the best estimate, the margin of error, and give notation for the parameter we are estimating.

(b) For these data, we know that the true population parameter is $\mu = 4.215$. Does the confidence interval generated from this one sample contain the true value of the parameter?

(c) The standard deviation of 6.81 and the standard error of 0.99 are quite different. Explain the distinction between them.

Solution

(a) We are estimating a mean μ, where μ represents the average salary for all baseball players in 2015. The best estimate for μ using this one sample is $\bar{x}$, so a 95% confidence interval is given by

$$
\begin{array}{rcl}
Statistic & \pm & 2 \cdot SE \\
\bar{x} & \pm & 2 \cdot SE \\
4.55 & \pm & 2(0.99) \\
4.55 & \pm & 1.98 \\
2.57 & \text{to} & 6.53
\end{array}
$$

A 95% confidence interval for the mean baseball salary goes from 2.57 to 6.53 million. Since the confidence interval is 4.55 ± 1.98, the best estimate using this one sample is 4.55 million dollars and the margin of error is 1.98.

(b) The population mean 4.215 falls within the interval from 2.57 to 6.53, so in this case, the confidence interval generated from one sample of 30 values does contain the population parameter.

(c) The standard deviation of 6.81 is the standard deviation of the 30 individual salaries in our sample. We see in Table 3.2 that the individual salaries are quite spread out. The standard error of 0.99 is the standard deviation of the sample *means* if we sampled 30 player salaries at a time and computed the sample means over and over (as shown in the sampling distribution in Figure 3.1). These means are much less spread out than the individual values.

Margin of error, standard error, and *standard deviation of a sample* are all different! Be careful to distinguish correctly between them. The margin of error is the amount added and subtracted in a confidence interval. The standard error is the standard deviation of the sample statistics if we could take many samples of the same size. The standard deviation of a sample is the standard deviation of the individual values in that one sample.

Understanding Confidence Intervals

In Section 3.1, we see that the proportion of US adults with a college degree is 0.275. Figure 3.10 shows the sampling distribution (the same one as in Figure 3.2 on page 202) of the proportion of adults with a college degree for 1000 samples of size 200. Each of the dots in Figure 3.10 represents the proportion with a college degree for a different possible random sample of size $n = 200$ from a population with parameter $p = 0.275$. Any one of those dots represents a sample statistic we might actually see, and we could find a separate confidence interval for each of the dots in that sampling distribution. How many of these intervals will contain the parameter value of $p = 0.275$?

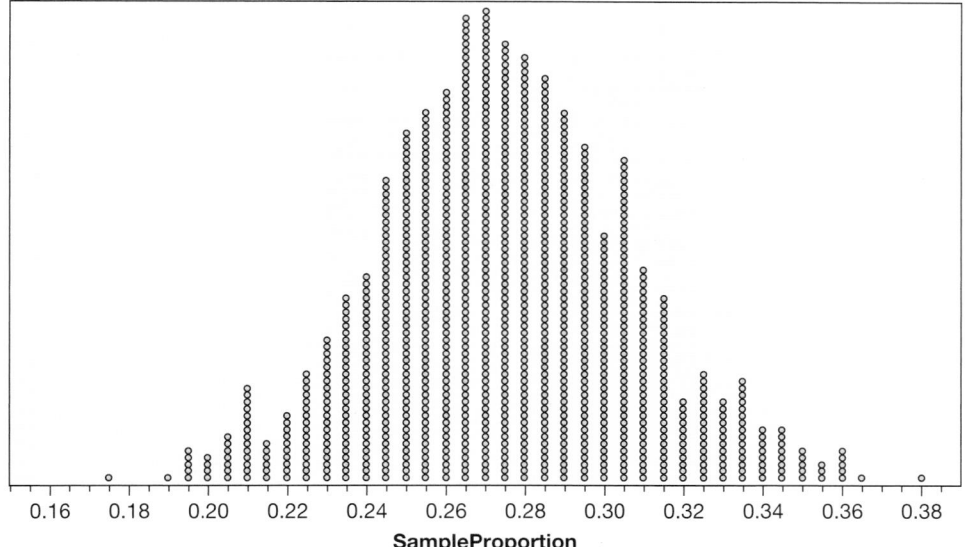

Figure 3.10 *Sample proportions for samples of size n = 200 when p = 0.275*

Example 3.15

Each of the three values listed below is one of the sample proportions shown in the dotplot in Figure 3.10. Find a 95% confidence interval using the sample proportion and the fact that the standard error is approximately 0.03 ($SE = 0.03$). In each case, also locate the sample proportion on the sampling distribution and indicate whether the 95% confidence interval captures the true population proportion.

(a) $\hat{p} = 0.26$

(b) $\hat{p} = 0.32$

(c) $\hat{p} = 0.20$

Solution

We use the sample proportion $\hat{p}$ as our best estimate for the population proportion, so in each case, we find a 95% confidence interval using

$$\hat{p} \pm 2 \cdot SE$$

(a) For $\hat{p} = 0.26$ the interval is $0.26 \pm 2 \cdot 0.03 = (0.20, 0.32)$. We see in Figure 3.10 that a sample proportion of $\hat{p} = 0.26$ is close to the center of the sampling distribution. The confidence interval $(0.20, 0.32)$ does include the population proportion of 0.275.

(b) For $\hat{p} = 0.32$ the interval is $0.32 \pm 2 \cdot 0.03 = (0.26, 0.38)$. We see in Figure 3.10 that a sample proportion of $\hat{p} = 0.32$ is farther from the center of the sampling distribution, but not way out in one of the tails. The confidence interval $(0.26, 0.38)$ does include the population proportion of 0.275.

(c) For $\hat{p} = 0.20$ the interval is $0.20 \pm 2 \cdot 0.03 = (0.14, 0.26)$. We see in Figure 3.10 that a sample proportion of $\hat{p} = 0.20$ is quite far out in the left tail of the sampling distribution and is not very close to the center. In fact, this sample proportion is outside of the middle 95% of values, so it is more than $2 \cdot SE$ away from the center. The confidence interval $(0.14, 0.26)$ does *not* include the population proportion of 0.275.

Note that two of the confidence intervals found in Example 3.15 successfully capture the population parameter of $p = 0.275$, but the third interval, the one generated from $\hat{p} = 0.20$, fails to contain $p = 0.275$. Remember that a 95% confidence interval should work only about 95% of the time. The sample proportion $\hat{p} = 0.20$ is in a pretty unusual place of the sampling distribution in Figure 3.10. Any of the (rare) samples that fall this far away will produce intervals that miss the parameter. This will happen about 5% of the time — precisely those samples that fall in the most extreme tails of the sampling distribution.

Figure 3.11 shows the sampling distribution for the proportion of college graduates in samples of size $n = 200$ (it should look very similar to Figure 3.10), although now the dots (statistics) are colored according to how far they are from the true population proportion of $p = 0.275$. Notice that all the statistics colored blue are within two standard errors of $p = 0.275$, and these comprise about 95% of all statistics.

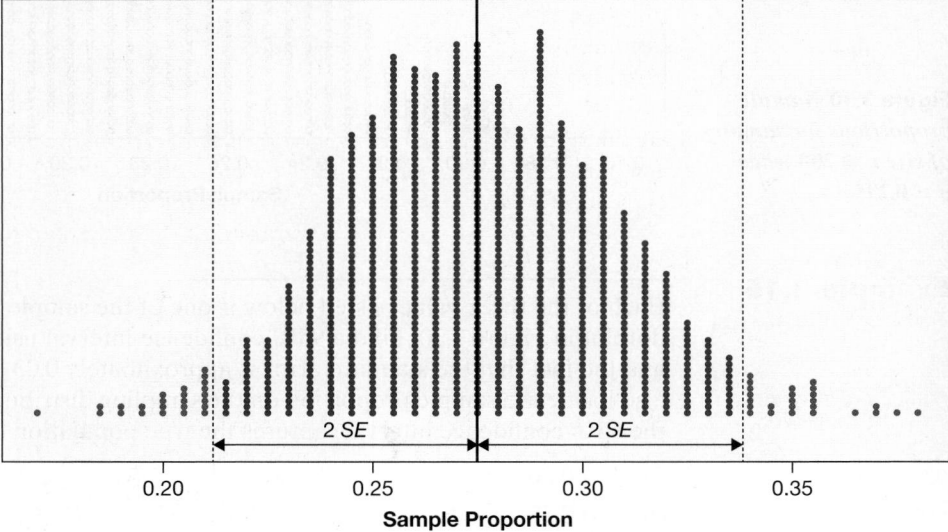

Figure 3.11 *Sampling distribution for proportion of college graduates in samples of size $n = 200$. Statistics within $2 \times SE$ of the true proportion $p = 0.275$ are colored blue, and statistics not within this range are colored red.*

The more extreme statistics, those farther than two standard errors from $p = 0.275$, are colored red and comprise about 5% of all statistics. The blue statistics, those closer to the true parameter, will lead to confidence intervals that contain the truth, $p = 0.275$, while the red statistics, those farther from $p = 0.275$, will lead to confidence intervals that miss the true parameter.

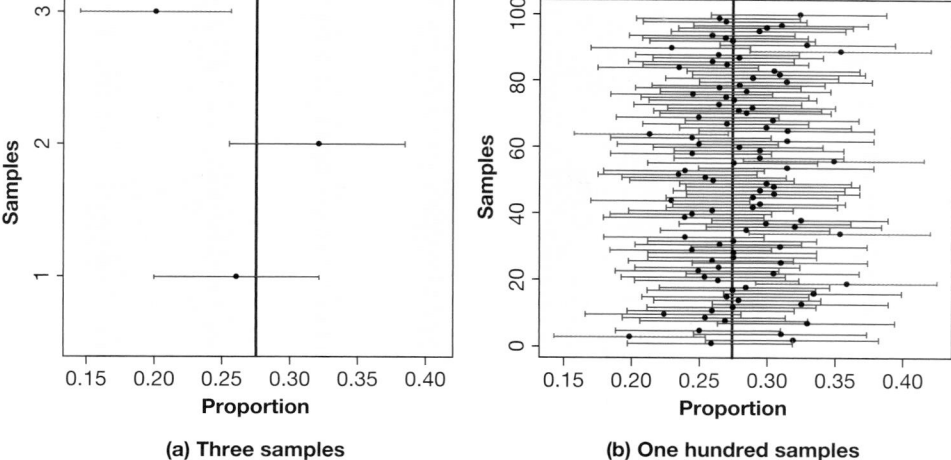

Figure 3.12 *Plots showing 95% confidence intervals for many samples*

(a) Three samples

(b) One hundred samples

Figure 3.12(a) on the left shows the three 95% confidence intervals from Example 3.15, with samples 1, 2, and 3 corresponding to those computed in parts (a), (b), and (c), respectively. Each horizontal line represents one of the confidence intervals (with a black dot showing the sample proportion), while the vertical line represents the population parameter of $p = 0.275$. The confidence interval that fails to capture the parameter value is shown in red.

Extending this idea to more sample proportions, Figure 3.12(b) shows confidence intervals for 100 of the sample proportions from Figure 3.11. Again, each horizontal line represents a confidence interval and we can see which intervals (shown in red) fail to capture the parameter value of $p = 0.275$. In this instance 6 of the 100 intervals miss the mark, while the other 94 are successful. Notice that the intervals are changing from sample to sample, while the parameter value ($p = 0.275$) stays fixed. Over the long run, for many such intervals, about 95% will successfully contain the parameter, while about 5% will miss it. That is what we mean by "95% confidence."

The parallelism in colors between Figures 3.11 and 3.12 is not a coincidence; any of the 5% of statistics colored red in the tails of the sampling distribution, those farther than $2 \times SE$ from the parameter, will lead to a confidence interval colored red that misses the true parameter. Likewise, any of the 95% of statistics colored blue in the middle of the sampling distribution will lead to a confidence interval colored blue that captures the true parameter.

Interpreting Confidence Intervals

We usually only have one sample and we do not know what the population parameter is. Referring to Figure 3.12(b), we don't know if our sample is one of the ones producing a "blue" confidence interval (successfully capturing the true population parameter) or one of the few producing a "red" confidence interval (and missing the mark). That is why we use words such as "95% confident" or "95% sure" when we interpret a confidence interval.

> **Interpreting Confidence Level**
>
> The confidence level indicates how sure we are that our interval contains the population parameter. For example, we interpret a 95% confidence interval by saying we are 95% sure that the interval contains the population parameter.

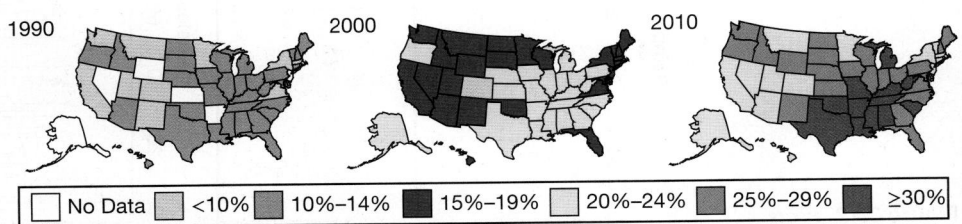

| | No Data | <10% | 10%–14% | 15%–19% | 20%–24% | 25%–29% | ≥30% |

Obesity rates in the US over time

DATA 3.2 | **Obesity in America**

Obesity is one of the most serious public health concerns of the 21st century, and has become one of the leading preventable causes of death worldwide.[22] Besides being a health issue, obesity is an important economic concern. It is estimated that in 2008, medical care costs due to obesity totaled about $147 billion.[23] Moreover, obesity rates are increasing at an alarming rate (see animation at *http://www.cdc.gov/obesity/data/trends.html* which includes the figures above). The Centers for Disease Control and Prevention (CDC) annually conducts a large national survey, based on a random sample of adults living in US states and territories, called the Behavioral Risk Factor Surveillance System.[24] Height and weight, from which we calculate body mass index (used to categorize "obese"), are among hundreds of variables collected on over 450,000 people in 2010. ■

Example 3.16

Average BMI

Body mass index (BMI) is calculated as $\frac{\text{weight in pounds}}{(\text{height in inches})^2} \times 703$. For the sample described in Data 3.2, the sample mean BMI for those surveyed is $\bar{x} = 27.655$, with a standard error of $SE = 0.009$.

(a) Give a 95% confidence interval for the average BMI for all adults living in the US, and interpret this interval.

(b) A BMI ≥ 25 is classified as overweight. Is it plausible that America's overall average BMI would *not* be classified as overweight?

[22]Barness, L.A., Opitz, J.M., and Gilbert-Barness, E. "Obesity: Genetic, Molecular, and Environmental Aspects," *Am. J. Med. Genet. A* 2007; 143(24): 3016–34.

[23]Finkelstein, E.A.; Trogdon, J.G., Cohen, J.W., and Dietz, W., "Annual Medical Spending Attributable to Obesity: Payer- and Service-Specific Estimates," *Health Affairs* 2009; 28(5): w822–w831.

[24]Centers for Disease Control and Prevention (CDC), *Behavioral Risk Factor Surveillance System Survey Data*, U.S. Department of Health and Human Services, Centers for Disease Control and Prevention, Atlanta, 2010.

Solution (a) Let μ represent the mean BMI for all adults living in the US. A 95% confidence interval is given by

$$
\begin{aligned}
\textit{Statistic} \;&\pm\; 2 \cdot SE \\
\bar{x} \;&\pm\; 2 \cdot SE \\
27.655 \;&\pm\; 2(0.009) \\
27.655 \;&\pm\; 0.018 \\
27.637 \;&\text{to}\; 27.673
\end{aligned}
$$

A 95% confidence interval for μ is (27.637, 27.673). We are 95% sure that the *mean* BMI for *all* adults living in the US in 2010 is between 27.637 and 27.673.

(b) Because the entire interval is above 25, we are (unfortunately) confident that America's average BMI in 2010 is considered overweight.

Notice that the interval in Example 3.16 is extremely narrow. This is because of the extremely large sample size ($n = 451,075$). Remember, the larger the sample size, the lower the variability in the sample statistics, so the smaller the standard error. The fact that we can be 95% sure that the average BMI of all 250 million plus American adults is between 27.63 and 27.67 is quite remarkable! Sampling only a small fraction of the population (chances are no one asked you your height and weight for a national survey), we can get a *very* accurate estimate of the average for the entire population.

Example 3.17

BMI and Exercise

The survey in Data 3.2 also asked people whether they had exercised in the past 30 days, to which $\hat{p} = 0.726$ of the people, with $SE = 0.0007$, responded yes. We expect that people who exercise have lower BMIs, but by how much? The sample difference in mean BMI between those who said they had exercised in the past 30 days and those who said they hadn't is $\bar{x}_E - \bar{x}_N = -1.915$, with a standard error of 0.016. Give and interpret a 95% confidence interval for the difference in mean BMI between people who exercise and those who don't for the entire US adult population. Is it plausible that there is no difference in average BMI between exercisers and non-exercisers?

Solution We are estimating a difference in means, so notation for the thing we are estimating (using meaningful subscripts) is $\mu_E - \mu_N$. Let $\mu_E - \mu_N$ be the mean BMI in 2010 for all adults living in the US who had exercised in the past 30 days minus the mean BMI in 2010 for all adults living in the US who had not exercised in the past 30 days. The relevant statistic is $\bar{x}_E - \bar{x}_N$. A 95% confidence interval for $\mu_E - \mu_N$ is

$$
\begin{aligned}
\textit{Statistic} \;&\pm\; 2 \cdot SE \\
(\bar{x}_E - \bar{x}_N) \;&\pm\; 2 \cdot SE \\
-1.915 \;&\pm\; 2(0.016) \\
-1.915 \;&\pm\; 0.032 \\
-1.947 \;&\text{to}\; -1.883
\end{aligned}
$$

We are 95% confident that people living in the US in 2010 who exercise (at least once in 30 days) have mean BMI between 1.883 and 1.947 lower than people living in the US in 2010 who do not exercise. Because this interval does not contain 0, we can be confident that there *is* a difference in average BMI between exercisers and non-exercisers.

Example 3.18

Obesity Prevalence

The CDC classifies an adult as "obese" if BMI $\geq$ 30. Based on the data from the 2010 survey, a 95% confidence interval for the proportion of all adults living in the US that were obese in 2010, p_{2010}, is $(0.276, 0.278)$. Based on the data from the 2000 survey, a 95% confidence interval for the proportion of all adults living in the US that were obese in 2000, p_{2000}, is $(0.195, 0.199)$.

(a) Interpret both confidence intervals.

(b) Do you think the sample size for the 2000 survey was smaller or larger than the sample size for the 2010 survey? Why?

(c) The confidence interval for the difference in proportions, $p_{2010} - p_{2000}$, based on the two sets of survey data is $(0.049, 0.111)$. Interpret this interval.

Solution

(a) We are 95% confident that the proportion of all adults living in the US who were obese in 2010 is between 0.276 and 0.278. We are also 95% confident that the proportion of all adults living in the US who were obese in 2000 is between 0.195 and 0.199.

(b) The sample size for the 2000 survey was smaller, because the margin of error for the 2000 interval is larger, meaning the standard error is larger, and smaller sample sizes correspond to larger standard errors. (In fact, the 2000 sample size was 184,450.)

(c) We are 95% sure that the proportion of adults living in the US that were obese increased by anywhere from 0.049 to 0.111 from 2000 to 2010. (This is a very large increase in the obesity rate over the span of just 10 years!)

Common Misinterpretations of Confidence Intervals

Misinterpretation 1 *A 95% confidence interval contains 95% of the data in the population.*

This is a common mistake! The 95% confidence interval for the mean BMI of US adults computed in Example 3.16 goes from 27.637 to 27.673. It certainly isn't true that 95% of all adults living in the US have a BMI within the narrow range of 27.637 and 27.673! Remember that the confidence interval is built to try and capture a *parameter* (in this case the mean) of the *population*, not individual cases. The correct statement is that we are 95% confident that the *population mean* is in the interval.

Misinterpretation 2 *I am 95% sure that the mean of a sample will fall within a 95% confidence interval for the mean.*

Again, this is a false statement. The correct statement is that we are 95% sure that the mean of the *population* will fall within a 95% confidence interval for the mean. In fact, we are 100% sure that the mean of our *sample* falls within the confidence interval since we constructed the confidence interval around the sample mean. A confidence interval gives us information about the population parameter, not about means of other samples. Remember that the interval is making a statement about where the population parameter is likely to be found.

Misinterpretation 3 *The probability that the population parameter is in this particular 95% confidence interval is 0.95.*

False! Remember that what varies are the statistics from sample to sample, not the population parameter. Once we have constructed an interval such as 27.637 to 27.673 for the mean BMI, it either does or does not contain the true average BMI for all US adults. The actual mean doesn't bounce around and fall within that interval some proportion of the time. That is why we use language such as *"we are 95% confident that ..."* or *"we are 95% sure that ..."* rather than making an explicit probability statement. This helps distinguish between the *method* of constructing the interval working 95% of the time (or for 95% of all samples), rather than a success rate for a particular interval.

But How Can We Construct a Confidence Interval?

We now know how to interpret a confidence interval and how to construct a 95% confidence interval *if we know the standard error*. But how can we find the standard error if we only have one sample? And how do we construct intervals for confidence levels other than 95%? If we had a sampling distribution, we could get a good idea of how far the sample statistic tends to vary from the parameter. But in reality we only have the data in our original sample to work with! We do not know the true value of the parameter, and do not have the funds to sample thousands of times from the population to generate a sampling distribution. How can we know how accurate our estimate is?

You may have detected this subtle impracticality in our approach so far: in order to understand the variability of the sample statistic, we need to know the population parameter or be able to take thousands of real samples from the population. However, this would defeat the point of using a sample statistic to estimate a population parameter! In most real situations we only have the information from just *one* sample, just one of the dots in the sampling distribution dotplots shown. How can we possibly determine how much a statistic varies from sample to sample if we only have one sample?!?

Amazingly, it *is* possible to assess variability of a sample statistic using just the data from one sample, and we will see how to do it in the next section.

SECTION LEARNING GOALS

You should now have the understanding and skills to:

- ▶ • Identify the parameter being estimated and the sample statistic that gives the best estimate
- ▶ • Construct a confidence interval for a parameter based on a sample statistic and a margin of error
- ▶ • Use a confidence interval to recognize plausible values of the population parameter
- ▶ • Construct a confidence interval for a parameter given a sample statistic and an estimate of the standard error
- ▶ • Interpret (in context) what a confidence interval says about a population parameter

---

Exercises for Section **3.2**

SKILL BUILDER 1

In Exercises 3.41 to 3.44, data from a sample is being used to estimate something about a population. In each case:

(a) Give notation for the quantity that is being estimated.

(b) Give notation for the quantity that gives the best estimate.

3.41 A random sample of registered voters in the US is used to estimate the proportion of all US registered voters who voted in the last election.

3.42 A random sample of maple trees in a forest is used to estimate the mean base circumference of all maple trees in the forest.

3.43 Random samples of organic eggs and eggs that are not organic are used to estimate the difference in mean protein level between the two types of eggs.

3.44 Random samples of people in Canada and people in Sweden are used to estimate the difference between the two countries in the proportion of people who have seen a hockey game (at any level) in the past year.

SKILL BUILDER 2

In Exercises 3.45 to 3.48, construct an interval giving a range of plausible values for the given parameter using the given sample statistic and margin of error.

3.45 For μ, using $\bar{x} = 25$ with margin of error 3.

3.46 For p, using $\hat{p} = 0.37$ with margin of error 0.02.

3.47 For ρ, using $r = 0.62$ with margin of error 0.05.

3.48 For $\mu_1 - \mu_2$, using $\bar{x}_1 - \bar{x}_2 = 5$ with margin of error 8.

SKILL BUILDER 3

In Exercises 3.49 and 3.50, a 95% confidence interval is given, followed by possible values of the population parameter. Indicate which of the values are plausible values for the parameter and which are not.

3.49 A 95% confidence interval for a mean is 112.1 to 128.2. Is the value given a plausible value of μ?

(a) $\mu = 121$

(b) $\mu = 113.4$

(c) $\mu = 105.3$

3.50 A 95% confidence interval for a proportion is 0.72 to 0.79. Is the value given a plausible value of p?

(a) $p = 0.85$

(b) $p = 0.75$

(c) $p = 0.07$

SKILL BUILDER 4

In Exercises 3.51 to 3.56, information about a sample is given. Assuming that the sampling distribution is symmetric and bell-shaped, use the information to give a 95% confidence interval, and indicate the parameter being estimated.

3.51 $\hat{p} = 0.32$ and the standard error is 0.04.

3.52 $\bar{x} = 55$ and the standard error is 1.5.

3.53 $r = 0.34$ and the standard error is 0.02.

3.54 $r = -0.46$ and the margin of error for 95% confidence is 0.05.

3.55 $\bar{x}_1 - \bar{x}_2 = 3.0$ and the margin of error for 95% confidence is 1.2.

3.56 $\hat{p}_1 - \hat{p}_2 = 0.08$ and the margin of error for 95% confidence is $\pm 3\%$.

3.57 Adolescent Brains Are Different Researchers continue to find evidence that brains of adolescents behave quite differently than either brains of adults or brains of children. In particular, adolescents seem to hold on more strongly to fear associations than either children or adults, suggesting that frightening connections made during the teen years are particularly hard to unlearn. In one study,[25] participants first learned to associate fear with a particular sound. In the second part of the study, participants heard the sound without the fear-causing mechanism, and their ability to "unlearn" the connection was measured. A physiological measure of fear was used, and larger numbers indicate less fear. We are estimating the difference in mean response between adults and teenagers. The mean response for adults in the study was 0.225 and the mean response for teenagers in the study was 0.059. We are told that the standard error of the estimate is 0.091.

(a) Give notation for the quantity being estimated.

(b) Give notation for the quantity that gives the best estimate, and give its value.

(c) Give a 95% confidence interval for the quantity being estimated.

(d) Is this an experiment or an observational study?

3.58 Do You Find Solitude Distressing? "For many people, being left alone with their thoughts is a most undesirable activity," says a psychologist involved in a study examining reactions to solitude.[26] In the study, 146 college students were asked to hand over their cell phones and sit alone, thinking, for about 10 minutes. Afterward, 76 of the participants rated the experience as unpleasant. Use this information to estimate the proportion of all college students who would find it unpleasant to sit alone with their thoughts. (This reaction is not limited to college students: in a follow-up study involving adults ages 18 to 77, a similar outcome was reported.)

(a) Give notation for the quantity being estimated, and define any parameters used.

(b) Give notation for the quantity that gives the best estimate, and give its value.

[25] Sanders, L., "Adolescent brains open to change," *Science News*, October 31, 2015.
[26] Bower, B., "People will take pain over being left alone with their thoughts," *Science News*, 186(3), August 9, 2014, p. 12.

(c) Give a 95% confidence interval for the quantity being estimated, given that the margin of error for the estimate is 8%.

3.59 Do You Prefer Pain over Solitude? Exercise 3.58 describes a study in which college students found it unpleasant to sit alone and think. The same article describes a second study in which college students appear to prefer receiving an electric shock to sitting in solitude. The article states that "when asked to spend 15 minutes in solitary thought, 12 of 18 men and 6 of 24 women voluntarily gave themselves at least one electric shock." Use this information to estimate the difference between men and women in the proportion preferring pain over solitude. The standard error of the estimate is 0.154.

(a) Give notation for the quantity being estimated, and define any parameters used.

(b) Give notation for the quantity that gives the best estimate, and give its value.

(c) Give a 95% confidence interval for the quantity being estimated.

(d) Is "no difference" between males and females a plausible value for the difference in proportions?

3.60 Moose Drool Makes Grass More Appetizing Different species can interact in interesting ways. One type of grass produces the toxin ergovaline at levels about 1.0 part per million in order to keep grazing animals away. However, a recent study[27] has found that the saliva from a moose counteracts these toxins and makes the grass more appetizing (for the moose). Scientists estimate that, after treatment with moose drool, mean level of the toxin ergovaline (in ppm) on the grass is 0.183. The standard error for this estimate is 0.016.

(a) Give notation for the quantity being estimated, and define any parameters used.

(b) Give notation for the quantity that gives the best estimate, and give its value.

(c) Give a 95% confidence interval for the quantity being estimated. Interpret the interval in context.

3.61 Have You Ever Been Arrested? According to a recent study of 7335 young people in the US, 30% had been arrested[28] for a crime other than a

traffic violation by the age of 23. Crimes included such things as vandalism, underage drinking, drunken driving, shoplifting, and drug possession.

(a) Is the 30% a parameter or a statistic? Use the correct notation.

(b) Use the information given to estimate a parameter, and clearly define the parameter being estimated.

(c) The margin of error for the estimate in part (b) is 0.01. Use this information to give a range of plausible values for the parameter.

(d) Given the margin of error in part (c), if we asked *all* young people in the US if they have ever been arrested, is it likely that the actual proportion is less than 25%?

3.62 Employer-Based Health Insurance A report from a Gallup poll[29] in 2011 started by saying, "Forty-five percent of American adults reported getting their health insurance from an employer...." Later in the article we find information on the sampling method, "a random sample of 147,291 adults, aged 18 and over, living in the US," and a sentence about the accuracy of the results, "the maximum margin of sampling error is ±1 percentage point."

(a) What is the population? What is the sample? What is the population parameter of interest? What is the relevant statistic?

(b) Use the margin of error[30] to give an interval showing plausible values for the parameter of interest. Interpret it in terms of getting health insurance from an employer.

3.63 Is a Car a Necessity? A random sample of $n = 1483$ adults in the US were asked whether they consider a car a necessity or a luxury,[31] and we find that a 95% confidence interval for the proportion saying that it is a necessity is 0.83 to 0.89. Explain the meaning of this confidence interval in the appropriate context.

3.64 Number of Text Messages a Day A random sample of $n = 755$ US cell phone users age 18 and older in May 2011 found that the average number

---

[27]Tanentzap, A.J., Vicari, M., Bazely, D.R., "Ungulate saliva inhibits a grass-endophyte mutualism," *Biology Letters*, 10(7), July 2014.

[28]From a study in *USA Today*, quoted in *The Week*, 2012; 11: 547–548.

[29]*http://www.gallup.com/poll/148079/Employer-Based-Health-Insurance-Declines-Further.aspx.*

[30]Actually, the margin of error is significantly less than ±1% for this sample, but the Gallup Poll rounded up to the nearest whole number.

[31]Taylor, P. and Wang, W., "The Fading Glory of the Television and Telephone," Pew Research Center, *http://pewsocial trends.org/pubs/762/fading-glory-television-telephone-luxury-necessity#prc-jump*, accessed August 19, 2010.

of text messages sent or received per day is 41.5 messages,[32] with standard error about 6.1.

(a) State the population and parameter of interest. Use the information from the sample to give the best estimate of the population parameter.

(b) Find and interpret a 95% confidence interval for the mean number of text messages.

3.65 What Proportion Believe in One True Love? In Data 2.1 on page 48, we describe a study in which a random sample of 2625 US adults were asked whether they agree or disagree that there is "only one true love for each person." The study tells us that 735 of those polled said they agree with the statement. The standard error for this sample proportion is 0.009. Define the parameter being estimated, give the best estimate, the margin of error, and find and interpret a 95% confidence interval.

3.66 Males vs Females and One True Love In Data 2.1 on page 48, we describe a study in which a random sample of 2625 US adults were asked whether they agree or disagree that there is "only one true love for each person." The response and gender of the participants is shown in Table 3.6. Use the information in the table to construct and interpret a 95% confidence interval for the difference in the proportion who agree, between males and females, using the fact that the standard error for the difference is 0.018. Is it plausible that there is no difference between males and females in the proportion who agree that each person has only one true love?

Table 3.6 *Is there one true love for each person?*

	Male	Female	Total
Agree	372	363	735
Disagree	807	1005	1812
Don't know	34	44	78
Total	1213	1412	2625

3.67 Playing Video Games A new study provides some evidence that playing action video games strengthens a person's ability to translate sensory information quickly into accurate decisions. Researchers had 23 male volunteers with an average age of 20 look at moving arrays on a computer screen and indicate the direction in which the dots were moving.[33] Half of the volunteers (11 men) reported playing action video games at least five times a week for the previous year, while the other 12 reported no video game playing in the previous year. The *response time* and the *accuracy score* were both measured. A 95% confidence interval for the mean response time for game players minus the mean response time for non-players is −1.8 to −1.2 seconds, while a 95% confidence interval for mean accuracy score for game players minus mean accuracy score for non-players is −4.2 to +5.8.

(a) Interpret the meaning of the 95% confidence interval for difference in mean response time.

(b) Is it plausible that game players and non-game players are basically the same in response time? Why or why not? If not, which group is faster (with a smaller response time)?

(c) Interpret the meaning of the 95% confidence interval for difference in mean accuracy score.

(d) Is it plausible that game players and non-game players are basically the same in accuracy? Why or why not? If not, which group is more accurate?

3.68 Bisphenol A in Your Soup Cans Bisphenol A (BPA) is in the lining of most canned goods, and recent studies have shown a positive association between BPA exposure and behavior and health problems. How much does canned soup consumption increase urinary BPA concentration? That was the question addressed in a recent study[34] in which consumption of canned soup over five days was associated with a more than 1000% increase in urinary BPA. In the study, 75 participants ate either canned soup or fresh soup for lunch for five days. On the fifth day, urinary BPA levels were measured. After a two-day break, the participants switched groups and repeated the process. The difference in BPA levels between the two treatments was measured for each participant. The study reports that a 95% confidence interval for the difference in means (canned minus fresh) is 19.6 to 25.5 μg/L.

(a) Is this a randomized comparative experiment or a matched pairs experiment? Why might this type of experiment have been used?

(b) What parameter are we estimating?

(c) Interpret the confidence interval in terms of BPA concentrations.

[32]Smith, A., "Americans and Text Messaging," Pew Research Center, *http://www.pewinternet.org/Reports/2011/Cell-Phone-Texting-2011/Main-Report/How-Americans-Use-Text-Messaging.aspx*, accessed September 19, 2011.

[33]Green, et al., "Improved probabilistic inference as a general learning mechanism with action video games," *Current Biology*, 2010; 20(September 14): 1.

[34]Carwile, J., Ye, X., Zhou, X., Calafat, A., and Michels, K., "Canned Soup Consumption and Urinary Bisphenol A: A Randomized Crossover Trial," *Journal of the American Medical Association*, 2011; 306(20): 2218–2220.

(d) If the study had included 500 participants instead of 75, would you expect the confidence interval to be *wider* or *narrower*?

3.69 Predicting Election Results Throughout the US presidential election of 2016, polls gave regular updates on the sample proportion supporting each candidate and the margin of error for the estimates. This attempt to predict the outcome of an election is a common use of polls. In each case below, the proportion of voters who intend to vote for each of two candidates is given as well as a margin of error for the estimates. Indicate whether we can be relatively confident that candidate A would win if the election were held at the time of the poll. (Assume the candidate who gets more than 50% of the vote wins.)

(a) Candidate A: 54%   Candidate B: 46%
    Margin of error: ±5%

(b) Candidate A: 52%   Candidate B: 48%
    Margin of error: ±1%

(c) Candidate A: 53%   Candidate B: 47%
    Margin of error: ±2%

(d) Candidate A: 58%   Candidate B: 42%
    Margin of error: ±10%

3.70 Effect of Overeating for One Month: Average Long-Term Weight Gain Overeating for just four weeks can increase fat mass and weight over two years later, a Swedish study shows.[35] Researchers recruited 18 healthy and normal-weight people with an average age of 26. For a four-week period, participants increased calorie intake by 70% (mostly by eating fast food) and limited daily activity to a maximum of 5000 steps per day (considered sedentary). Not surprisingly, weight and body fat of the participants went up significantly during the study and then decreased after the study ended. Participants are believed to have returned to the diet and lifestyle they had before the experiment. However, two and a half years after the experiment, the mean weight gain for participants was 6.8 lbs with a standard error of 1.2 lbs. A control group that did not binge had no change in weight.

(a) What is the relevant parameter?

(b) How could we find the actual exact value of the parameter?

(c) Give a 95% confidence interval for the parameter and interpret it.

(d) Give the margin of error and interpret it.

FISH DEMOCRACIES
Exercises 3.71 to 3.73 consider the question (using fish) of whether uncommitted members of a group make it more democratic. It has been argued that individuals with weak preferences are particularly vulnerable to a vocal opinionated minority. However, recent studies, including computer simulations, observational studies with humans, and experiments with fish, all suggest that adding uncommitted members to a group might make for more democratic decisions by taking control away from an opinionated minority.[36] In the experiment with fish, golden shiners (small freshwater fish who have a very strong tendency to stick together in schools) were trained to swim toward either yellow or blue marks to receive a treat. Those swimming toward the yellow mark were trained more to develop stronger preferences and became the fish version of individuals with strong opinions. When a minority of five opinionated fish (wanting to aim for the yellow mark) were mixed with a majority of six less opinionated fish (wanting to aim for the blue mark), the group swam toward the minority yellow mark almost all the time. When some untrained fish with no prior preferences were added, however, the majority opinion prevailed most of the time.[37] Exercises 3.71 to 3.73 elaborate on this study.

3.71 Training Fish to Pick a Color Fish can be trained quite easily. With just seven days of training, golden shiner fish learn to pick a color (yellow or blue) to receive a treat, and the fish will swim to that color immediately. On the first day of training, however, it takes them some time. In the study described under *Fish Democracies* above, the mean time for the fish in the study to reach the yellow mark is $\bar{x} = 51$ seconds with a standard error for this statistic of 2.4. Find and interpret a 95% confidence interval for the mean time it takes a golden shiner fish to reach the yellow mark. Is it plausible that the average time it takes fish to find the mark is 60 seconds? Is it plausible that it is 55 seconds?

3.72 How Often Does the Fish Majority Win? In a school of fish with a minority of strongly opinionated fish wanting to aim for the yellow mark and a majority of less passionate fish wanting to aim for the blue mark, as described under *Fish Democracies* above, a 95% confidence interval for the proportion of times the majority wins (they go to the blue mark) is 0.09 to 0.26. Interpret this confidence interval. Is it

[35]Ernersson, A., Nystrom, F., and Linsstrrom, T., "Long-term Increase of Fat Mass after a Four Week Intervention with a Fast-food Hyper-alimentation and Limitation of Physical Activity," *Nutrition & Metabolism*, 2010; 7: 68. Some of the data is estimated from available information.

[36]Milius, S., "Uncommitted Newbies Can Foil Forceful Few," *Science News*, 2012: 181(1); 18.
[37]Couzin, I., et al., "Uninformed Individuals Promote Democratic Consensus in Animal Groups," *Science*, 2011; 334(6062): 1578–80.

plausible that fish in this situation are equally likely to go for either of the two options?

3.73 What Is the Effect of Including Some Indifferent Fish? In the experiment described above under *Fish Democracies*, the schools of fish in the study with an opinionated minority and a less passionate majority picked the majority option only about 17% of the time. However, when groups also included 10 fish with no opinion, the schools of fish picked the majority option 61% of the time. We want to estimate the effect of adding the fish with no opinion to the group, which means we want to estimate the difference in the two proportions. We learn from the study that the standard error for estimating this difference is about 0.14. Define the parameter we are estimating, give the best point estimate, and find and interpret a 95% confidence interval. Is it plausible that adding indifferent fish really has no effect on the outcome?

3.74 Student Misinterpretations Suppose that a student is working on a statistics project using data on pulse rates collected from a random sample of 100 students from her college. She finds a 95% confidence interval for mean pulse rate to be (65.5, 71.8).

Discuss how each of the statements below would indicate an *improper* interpretation of this interval.

(a) I am 95% sure that all students will have pulse rates between 65.5 and 71.8 beats per minute.

(b) I am 95% sure that the mean pulse rate for this sample of students will fall between 65.5 and 71.8 beats per minute.

(c) I am 95% sure that the confidence interval for the average pulse rate of all students at this college goes from 65.5 to 71.8 beats per minute.

(d) I am sure that 95% of all students at this college will have pulse rates between 65.5 and 71.8 beats per minute.

(e) I am 95% sure that the mean pulse rate for all US college students is between 65.5 and 71.8 beats per minute.

(f) Of the mean pulse rates for students at this college, 95% will fall between 65.5 and 71.8 beats per minute.

(g) Of random samples of this size taken from students at this college, 95% will have mean pulse rates between 65.5 and 71.8 beats per minute.

3.3 CONSTRUCTING BOOTSTRAP CONFIDENCE INTERVALS

The distributions of sample statistics considered so far in this chapter require us to either already know the value of the population parameter or to have the resources to take thousands of different samples. In most situations, neither of these is an option.

In this section we introduce a method for estimating the variability of a statistic that uses only the data in the original sample. This clever technique, called *bootstrapping*,[38] allows us to approximate a sampling distribution and estimate a standard error using just the information in that one sample.

© Pavel Losevsky/iStockphoto

What is the average commute time in Atlanta?

[38]The term *bootstrap* was coined by Brad Efron and Robert Tibsharani to reflect the phrase "pulling oneself up by one's own bootstraps."

| **DATA 3.3** | ### Commuting in Atlanta |

What is the average commuting time for people who live and work in the Atlanta metropolitan area? It's not very feasible to contact *all* Atlanta residents and ask about their commutes, but the US Census Bureau regularly collects data from carefully selected samples of residents in many areas. One such data source is the American Housing Survey (AHS), which contains information about housing and living conditions for samples from the country as a whole and certain metropolitan areas. The data in **CommuteAtlanta** includes cases where the respondent worked somewhere other than home in the Atlanta metropolitan area.[39] Among the questions asked were the time (in minutes) and distance (in miles) that respondents typically traveled on their commute to work each day. ■

The commute times for this sample of 500 Atlantans are shown in the dotplot of Figure 3.13. The sample mean is $\bar{x} = 29.11$ minutes and the standard deviation in the sample is $s = 20.7$ minutes. The distribution of commute times is somewhat right skewed with clusters at regular intervals that reflect many respondents rounding their estimates to the nearest 5 or 10 minutes. Based on this sample we have a best estimate of 29.11 minutes for μ, the mean commute time for all workers in metropolitan Atlanta. How accurate is that estimate likely to be?

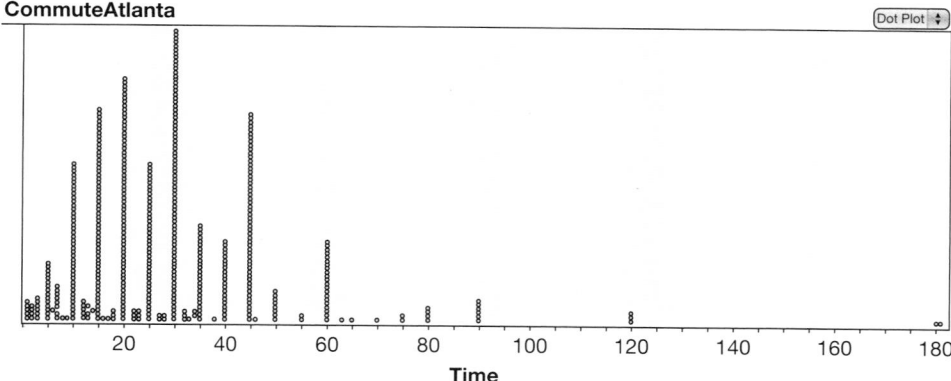

Figure 3.13 *Sample of 500 Atlanta commute times*

To get a range of plausible values for the mean commute time of all Atlantans it would help to see a sampling distribution of means for many samples of size 500. However, we don't have data for the population of all Atlanta commuters, and if we did we could easily find the population mean exactly! We only have the commuting times for the single sample of 500 workers. How can we use the information in that sample to assess how much the means for other samples of 500 Atlantans might vary?

Bootstrap Samples

Ideally, we'd like to sample repeatedly from the population to create a sampling distribution. How can we make the sample data look like data from the entire population? The key idea is to assume for the moment that the population of all commuters in Atlanta is basically just many, many copies of the commuters in our

---

[39]Sample chosen using DataFerret at *http://www.thedataweb.org/index.html.*

original sample. See Figure 3.14, which illustrates this concept for a very small sample of six stick figures, in which we assume the population is just many copies of the sample. If we make lots of copies of the sample and then sample repeatedly from this hypothetical "population," we are coming as close as we can to mimicking the process of sampling repeatedly from the population.

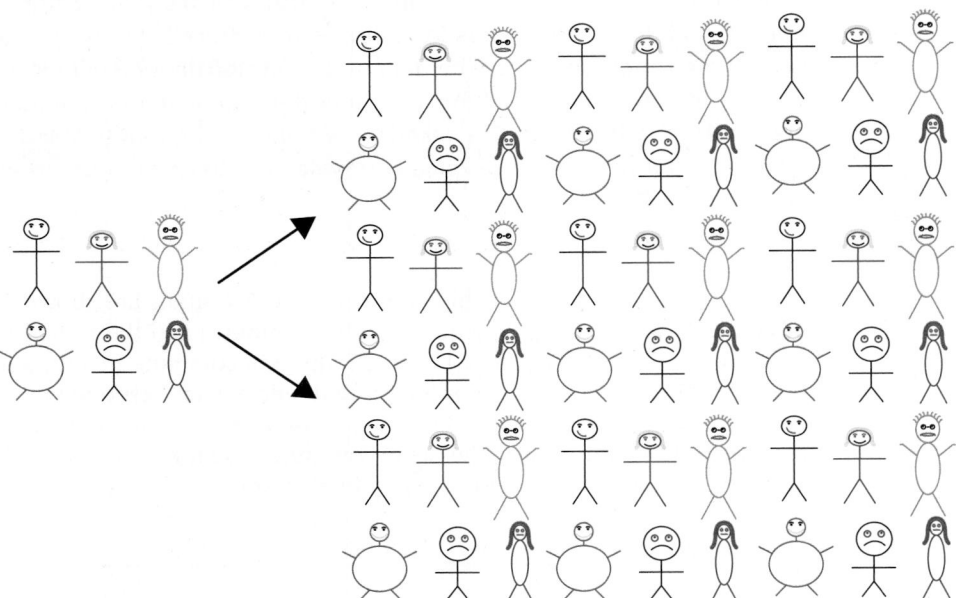

Figure 3.14 *Using a sample to represent a population*

In practice, instead of actually making many copies of the sample and sampling from that, we use a sampling technique that is equivalent: we sample *with replacement* from the original sample. Sampling with replacement means that once a commuter has been selected for the sample, he or she is still available to be selected again. This is because we're assuming that each commuter in the original sample actually represents many fellow Atlantans with a similar commute time. Each sample selected in this way, with replacement from the original sample, is called a *bootstrap sample*.

Recall from Section 3.1 that the variability of a sample statistic depends on the size of the sample. Because we are trying to uncover the variability of the sample statistic, it is important that each bootstrap sample is the same size as the original sample. For the Atlanta commuters, each bootstrap sample will be of size $n = 500$.

For each bootstrap sample, we compute the statistic of interest, giving us a *bootstrap statistic*. For the Atlanta commuters, we compute a bootstrap statistic as the sample mean commute time for a bootstrap sample. If we take many bootstrap samples and compute a bootstrap statistic from each, the distribution of these bootstrap statistics will help us understand the distribution of the sample statistic. Table 3.7 shows the sample means for 10 different bootstrap samples of size 500 taken with replacement from the original commute times in **CommuteAtlanta**.

Table 3.7 *Mean commute times for 10 bootstrap samples of n = 500 Atlantans*

28.06	29.21	28.43	28.97	29.95	28.67	30.57	29.22	27.78	29.58

Bootstrap Distribution

Based on just the 10 bootstrap statistics in Table 3.7, we can begin to get some feel for how accurately we can estimate the mean commute time based on a sample of size 500. Note that, for the hypothetical population we simulate when sampling with replacement from the original sample, we *know* that the "population" mean is the sample mean, 29.11 minutes. Thus the bootstrap sample means give us a good idea of how close means for samples of size 500 should be to a "true" mean. For the 10 samples in Table 3.7 the biggest discrepancy is the seventh sample mean (30.57), which is still within 1.46 minutes of 29.11.

Of course, with computer technology, we aren't limited to just 10 bootstrap samples. We can get a much better picture of the variability in the means for samples of size 500 by generating many such samples and collecting the sample means. Figure 3.15 shows a dotplot of the sample means for 1000 samples of size 500, taken with replacement, from the original sample of Atlanta commute times. This gives a good representation of the *bootstrap distribution* for mean Atlanta commute times. We see that the distribution is relatively symmetric, bell-shaped, and centered near the original sample mean of 29.11.

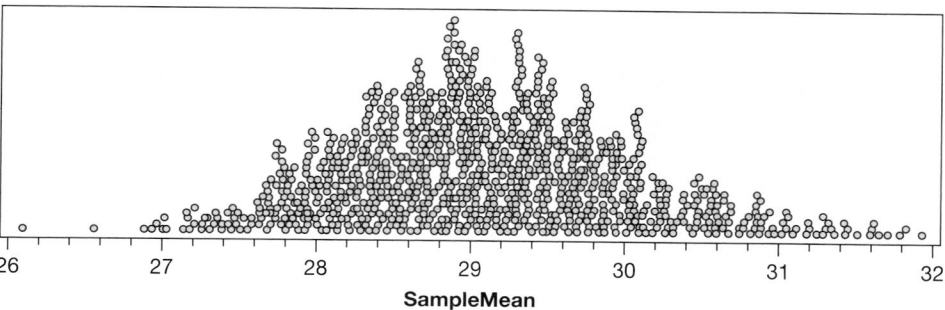

Figure 3.15 *Commuting time means for 1000 bootstrap samples of size n = 500*

Generating a Bootstrap Distribution

To generate a bootstrap distribution, we:

- Generate **bootstrap samples** by sampling with replacement from the original sample, using the same sample size.
- Compute the statistic of interest, called a **bootstrap statistic**, for each of the bootstrap samples.
- Collect the statistics for many bootstrap samples to create a **bootstrap distribution**.

This process is illustrated in Figure 3.16.

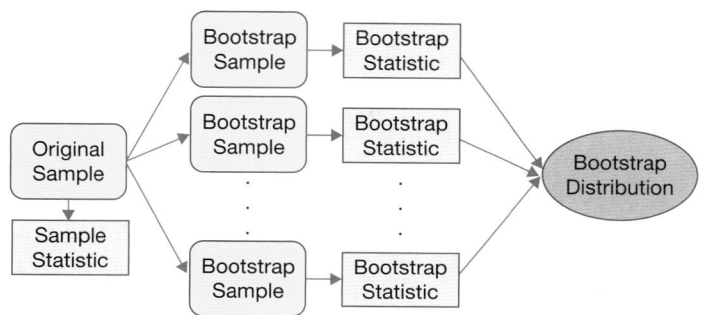

Figure 3.16 *Generating a bootstrap distribution*

Assuming the original sample is chosen randomly from the population, the bootstrap distribution generally gives a good approximation to a sampling distribution that we might see if we were able to collect lots of samples from the entire population, but is centered around the sample statistic rather than the population parameter. This allows us to get a good idea of how variable our sample statistic is, and how close we can expect it to be to the population parameter. In Figure 3.15 we see that none of the 1000 sample means are more than three minutes away from the center of the bootstrap distribution. Thus, we are quite confident that a sample of 500 Atlanta commuters will give an estimate that is within three minutes of the mean commute time for the entire population.

Example 3.19

Mixed Nuts with Peanuts

Containers of mixed nuts often contain peanuts as well as cashews, pecans, almonds, and other nuts. For one brand, we want to estimate the proportion of mixed nuts that are peanuts. We get a jar of the nuts and assume that the nuts in that container represent a random sample of all the mixed nuts sold by that company. We open the jar and count 100 nuts of which 52 are peanuts. The estimated proportion of peanuts is $\hat{p} = 52/100 = 0.52$.

(a) How could we physically use the jar of nuts to construct one bootstrap sample? What would we record to find the bootstrap statistic?

(b) If we create a bootstrap distribution by collecting many bootstrap statistics, describe the center and anticipated shape of the distribution.

(c) Use *StatKey* or other technology to create a bootstrap distribution.

Solution

(a) To find a bootstrap sample we need to select 100 nuts from the original sample with replacement. To accomplish this we could shake the nuts in the jar, reach in and pick one at random, record whether or not it is a peanut, and *put it back in the jar*. (This is what sampling with replacement means.) Repeat this process 99 more times to simulate a new sample of 100 nuts. The bootstrap statistic is the proportion of peanuts among the 100 nuts selected.

(b) Since the bootstrap statistics come from the original sample with a sample proportion of 0.52, we expect the bootstrap distribution to be centered at 0.52. Since we are simulating a sampling distribution, we think it is likely that the distribution will be bell-shaped.

(c) While it would be time consuming to repeat the physical sampling process described in part (a) many times, it is relatively easy to use *StatKey* or other technology to simulate the process automatically. Figure 3.17 shows a dotplot of the bootstrap distribution of sample proportions for 1000 samples of size 100, simulated from the original sample with 52 peanuts out of 100. As expected, we see a symmetric, bell shape distribution, centered near the value of the statistic in the original sample (0.52).

Example 3.20

Laughter in Adults

How often do you laugh? Estimates vary greatly in how often, on average, adults laugh in a typical day. (Different sources indicate that the average is 10, or 15, or 40, depending on the source, although all studies conclude that adults laugh significantly

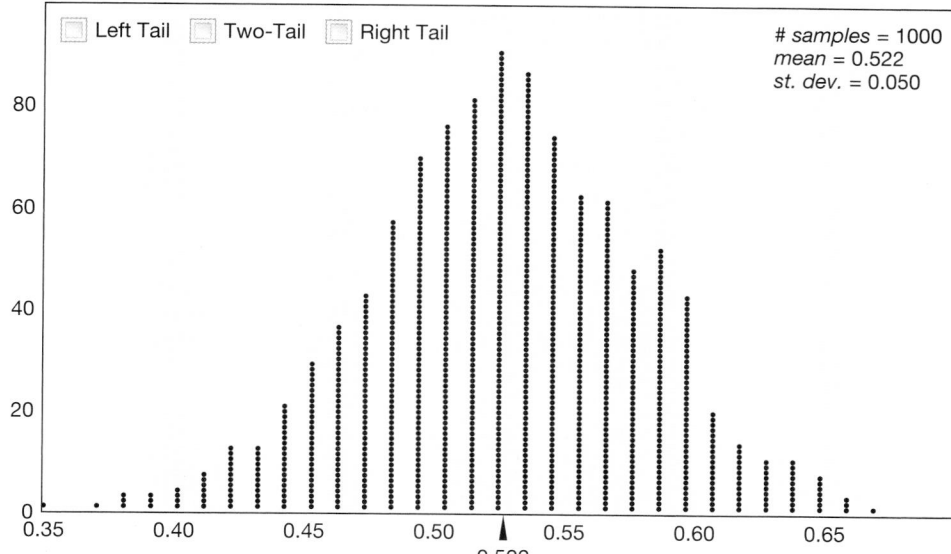

Figure 3.17 *Bootstrap proportions for 1000 samples simulated from a sample with $\hat{p} = 0.52$ and $n = 100$*

less than children.) Suppose that one study randomly selects six adults and records how often these adults laugh in a day, with the results given in Table 3.8.

(a) Define the parameter we are estimating and find the best point estimate from this sample.

(b) Describe how to use cards to generate one bootstrap sample. What statistic would we record for this sample?

(c) Generate several bootstrap samples this way, and compute the mean for each.

(d) If we generated many such bootstrap statistics, where will the bootstrap distribution be centered?

Table 3.8 *Number of laughs in a day*

16	22	9	31	6	42

Solution

(a) We are estimating μ, the average number of laughs in a typical day for all adults. The best point estimate is the mean from our sample, which we calculate to be $\bar{x} = 21.0$.

(b) Since there are six values in the sample, we use six cards and put the six values on the cards. We then mix them up, pick one, and write down the value. (Since there are six values, we could also roll a six-sided die to randomly select one of the numbers.) Then we put the card back (since we are sampling with replacement), mix the cards up, and draw out another. We do this six times to obtain a bootstrap sample of size 6. Since we are interested in the mean, the statistic we record is the mean of the six values.

(c) Several bootstrap samples are shown in Table 3.9. Answers will vary, but all bootstrap samples will have the same sample size, $n = 6$, and will only include values already present in the original sample.

(d) If we calculated many bootstrap statistics to generate a bootstrap distribution, it would be centered at the value of the original sample statistic, which is $\bar{x} = 21.0$.

Table 3.9 *Three bootstrap samples*

Bootstrap Sample 1:	16	31	9	16	6	42	Mean = 20.0
Bootstrap Sample 2:	31	16	16	6	31	22	Mean = 20.33
Bootstrap Sample 3:	42	31	42	9	42	22	Mean = 31.33

Estimating Standard Error from a Bootstrap Distribution

The variability of bootstrap statistics is similar to the variability of sample statistics if we were to sample repeatedly from the population, so we can use the standard deviation of the bootstrap distribution to estimate the standard error of the sample statistic.

> ### Standard Error from a Bootstrap Distribution
> The standard deviation of the bootstrap statistics in a bootstrap distribution gives a good approximation of the standard error of the statistic.

Example 3.21

Use the information in Figure 3.17 to find the standard error of the sample proportion when estimating the proportion of peanuts in mixed nuts with a sample of size 100.

Solution

The information in the upper corner of Figure 3.17 indicates that the standard deviation of those 1000 bootstrap proportions is 0.050, so we use that value as an estimate of the standard error for the proportion.

The 1000 bootstrap means for Atlanta commute times in Figure 3.15 have a standard deviation of 0.915 minutes, so we have $SE = 0.915$ for the sample mean commute time based on samples of size $n = 500$. The standard error depends on the size and variability of the original sample, but not on the number of bootstrap samples (provided we use enough bootstrap samples to obtain a reasonable estimate).

Because the estimated SE is based on simulated bootstrap samples, it will vary slightly from simulation to simulation. A different set of 1000 commute bootstrap means produced a standard error estimate of 0.932 (similar to the previous estimate of 0.915), and 1000 new simulated mixed nut samples gave an estimated standard error of 0.048 (also similar to the previous estimate of $SE = 0.050$.) In practice, these subtle differences are almost always negligible. However, a more accurate estimate can easily be achieved by simulating more bootstrap samples: The more bootstrap samples, the more accurate the estimated SE will be. In this text we often use 1000 bootstrap samples so that the individual bootstrap statistics are visible in plots, but 10,000 or more bootstrap samples are more often used in practice.[40] If we create 100,000 bootstrap samples for the Atlanta commute times, the SE is 0.927 in one simulation, 0.928 in another simulation, and 0.927 in a third simulation: We are now estimating within 1 one-thousandth of a minute.

[40]The number of bootstrap samples you find may depend on the speed of your technology.

95% Confidence Interval Based on a Bootstrap Standard Error

Recall from Section 3.2 that we can use the standard error to construct a 95% confidence interval by going two standard errors on either side of the original statistic. Now, we can use this idea more practically by using the bootstrap distribution to estimate the standard error.

A 95% Confidence Interval Using a Bootstrap Standard Error

When a bootstrap distribution for a sample statistic is symmetric and bell-shaped, we estimate a 95% confidence interval using

$$Statistic \pm 2 \cdot SE$$

where SE denotes the standard error of the statistic estimated from the bootstrap distribution.

Example 3.22

Use the standard errors found in previous examples to find and interpret 95% confidence intervals for

(a) the mean Atlanta commute time, and

(b) the proportion of peanuts in mixed nuts.

In addition, give the margin of error for both intervals.

Solution (a) The sample mean from the original sample of 500 Atlanta commuters is $\bar{x} = 29.11$ minutes and the estimated standard error for this mean from the bootstrap distribution in Figure 3.15 is 0.915. Going two standard errors on either side of the sample statistic gives

$$\bar{x} \pm 2 \cdot SE$$
$$29.11 \pm 2(0.915)$$
$$29.11 \pm 1.83$$

or an interval from $29.11 - 1.83 = 27.28$ minutes to $29.11 + 1.83 = 30.94$ minutes. The margin of error is 1.83 minutes, and we are 95% confident that the mean commute time for all Atlanta commuters is between 27.28 minutes and 30.94 minutes.

(b) The original sample has a proportion of $\hat{p} = 0.52$ peanuts, and the estimated standard error for this proportion from Example 3.21 is 0.050. Going two standard errors on either side of the estimate gives

$$\hat{p} \pm 2 \cdot SE$$
$$0.52 \pm 2(0.050)$$
$$0.52 \pm 0.10$$

or an interval from $0.52 - 0.10 = 0.42$ to $0.52 + 0.10 = 0.62$. The margin of error is 0.10, and we are 95% confident that between 42% and 62% of all mixed nuts from this company are peanuts.

We now have a very powerful technique for constructing confidence intervals for a wide variety of parameters. As long as we can do the following:

- Find a sample statistic to serve as a point estimate for the parameter.
- Compute bootstrap statistics for many samples with replacement from the original sample.
- Estimate the standard error from the bootstrap distribution.
- Check that the bootstrap distribution is reasonably symmetric and bell-shaped.

Then we can use $statistic \pm 2 \cdot SE$ to estimate a 95% confidence interval for the parameter.

But what about other confidence levels, like 90% or 99%? We explore an alternate method for obtaining a confidence interval from a bootstrap distribution in the next section, which will address this question and provide even more general results.

SECTION LEARNING GOALS

You should now have the understanding and skills to:

- ● Describe how to select a bootstrap sample to compute a bootstrap statistic
- ● Recognize that a bootstrap distribution tends to be centered at the value of the original statistic
- ● Use technology to create a bootstrap distribution
- ● Estimate the standard error of a statistic from a bootstrap distribution
- ● Construct a 95% confidence interval for a parameter based on a sample statistic and the standard error from a bootstrap distribution

---

Exercises for Section 3.3

SKILL BUILDER 1

In Exercises 3.75 and 3.76, a sample is given. Indicate whether each option is a possible bootstrap sample from this original sample.

3.75 Original sample: 17, 10, 15, 21, 13, 18.
Do the values given constitute a possible bootstrap sample from the original sample?

(a) 10, 12, 17, 18, 20, 21

(b) 10, 15, 17

(c) 10, 13, 15, 17, 18, 21

(d) 18, 13, 21, 17, 15, 13, 10

(e) 13, 10, 21, 10, 18, 17

3.76 Original sample: 85, 72, 79, 97, 88.
Do the values given constitute a possible bootstrap sample from the original sample?

(a) 79, 79, 97, 85, 88

(b) 72, 79, 85, 88, 97

(c) 85, 88, 97, 72

(d) 88, 97, 81, 78, 85

(e) 97, 85, 79, 85, 97

(f) 72, 72, 79, 72, 79

SKILL BUILDER 2

In Exercises 3.77 to 3.80, use the bootstrap distributions in Figure 3.18 to estimate the value of the sample statistic and the standard error, and then use this information to give a 95% confidence interval. In addition, give notation for the parameter being estimated.

3.77 The bootstrap distribution in Figure 3.18(a), generated for a sample proportion

3.78 The bootstrap distribution in Figure 3.18(b), generated for a sample mean

3.79 The bootstrap distribution in Figure 3.18(c), generated for a sample correlation

3.80 The bootstrap distribution in Figure 3.18(d), generated for a difference in sample means

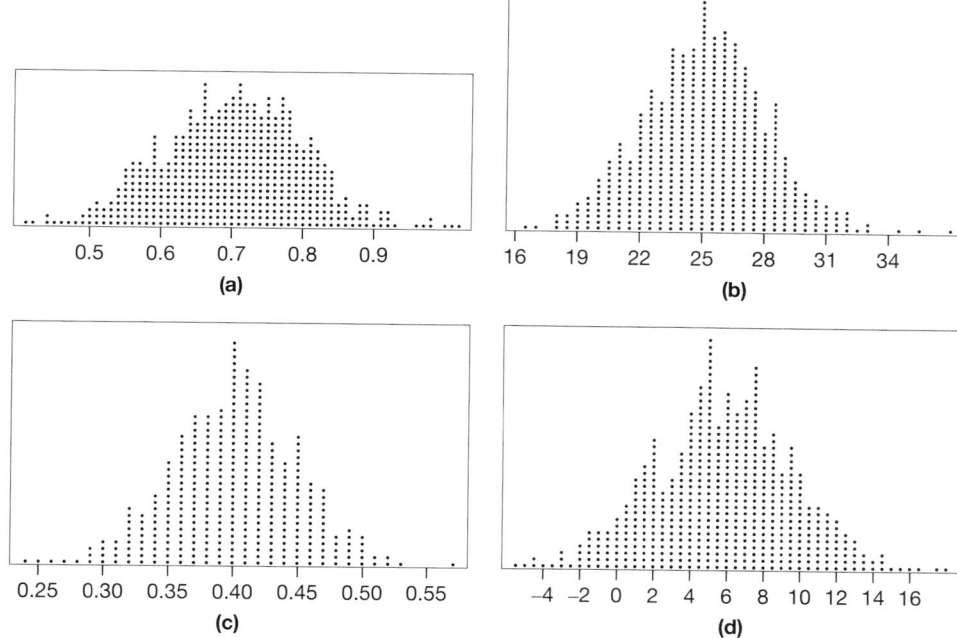

Figure 3.18 *Four bootstrap distributions*

SKILL BUILDER 3

Exercises 3.81 to 3.84 give information about the proportion of a sample that agrees with a certain statement. Use *StatKey* or other technology to estimate the standard error from a bootstrap distribution generated from the sample. Then use the standard error to give a 95% confidence interval for the proportion of the population to agree with the statement. *StatKey* tip: Use "CI for Single Proportion" and then "Edit Data" to enter the sample information.

3.81   In a random sample of 100 people, 35 agree.

3.82   In a random sample of 250 people, 180 agree.

3.83   In a random sample of 400 people, 112 agree and 288 disagree.

3.84   In a random sample of 1000 people, 382 people agree, 578 disagree, and 40 are undecided.

3.85   Hitchhiker Snails A type of small snail is very widespread in Japan, and colonies of the snails that are genetically similar have been found very far apart. Scientists wondered how the snails could travel such long distances. A recent study[41] provides the answer. Biologist Shinichiro Wada fed 174 live snails to birds and found that 26 of the snails were excreted live out the other end. The snails

apparently are able to seal their shells shut to keep the digestive fluids from getting in.

(a) What is the best estimate for the proportion of all snails of this type to live after being eaten by a bird?

(b) Figure 3.19 shows a bootstrap distribution based on this sample. Estimate the standard error.

(c) Use the standard error from part (b) to find and interpret a 95% confidence interval for the

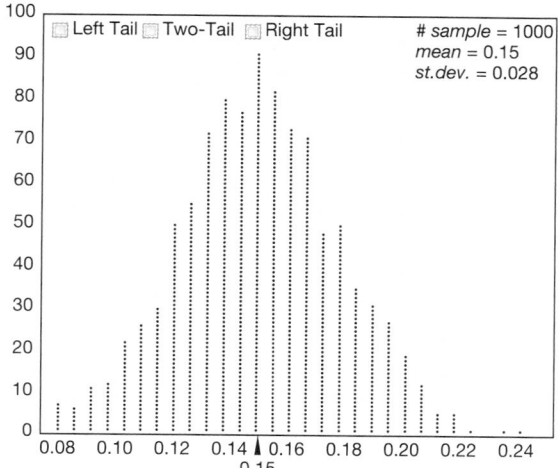

Figure 3.19 *Bootstrap distribution of the sample proportion of snails that live*

[41]Yong, E., "The Scatological Hitchhiker Snail," *Discover*, October 2011, 13.

proportion of all snails of this type to live after being eaten by a bird.

(d) Using your answer to part (c), is it plausible that 20% of all snails of this type live after being eaten by a bird?

3.86   Ants on a Sandwich How many ants will climb on a piece of a peanut butter sandwich left on the ground near an ant hill? To study this, a student in Australia left a piece of a sandwich for several minutes, then covered it with a jar and counted the number of ants. He did this eight times, and the results are shown in Table 3.10. (In fact, he also conducted an experiment to see if there is a difference in number of ants based on the sandwich filling. The details of that experiment are given in Chapter 8, and the full dataset is in **SandwichAnts**.)[42]

Table 3.10 *Number of ants on a sandwich*

Number of ants	43	59	22	25	36	47	19	21

(a) Find the mean and standard deviation of the sample.

(b) Describe how we could use eight slips of paper to create one bootstrap statistic. Be specific.

(c) What do we expect to be the shape and center of the bootstrap distribution?

(d) What is the population parameter of interest? What is the best estimate for that parameter?

(e) A bootstrap distribution of 5000 bootstrap statistics gives a standard error of 4.85. Use the standard error to find and interpret a 95% confidence interval for the parameter defined in part (d).

3.87   Brain Hippocampal Size Exercise 2.143 introduces a study examining the relationship between football playing and hippocampal volume, in μL,

in the brain. We use the $n = 25$ participants in the control group to estimate average brain hippocampus volume for all non-football playing people. Use the dotplots in Figure 3.20 to answer the following questions.

(a) Two dotplots are shown: one shows the original data and one shows a bootstrap distribution from the data. Which is which?

(b) Estimate the sample mean hippocampal volume based on the graphs. Give the correct notation and the value.

(c) Use the graphs to roughly estimate the standard error if we use this sample mean to estimate hippocampal volume.

(d) Does the standard deviation of the original data appear to be larger or smaller than the standard error?

(e) Use the estimated mean and the estimated standard error to find and interpret a rough 95% confidence interval for the mean hippocampal volume.

3.88   Rats with Compassion The phrase "You dirty rat" does rats a disservice. In a recent study,[43] rats showed compassion that surprised scientists. Twenty-three of the 30 rats in the study freed another trapped rat in their cage, even when chocolate served as a distraction and even when the rats would then have to share the chocolate with their freed companion. (Rats, it turns out, love chocolate.) Rats did not open the cage when it was empty or when there was a stuffed animal inside, only when a fellow rat was trapped. We wish to use the sample to estimate the proportion of rats that would show empathy in this way. The data are available in the dataset **CompassionateRats**.

(a) Give the relevant parameter and its best estimate.

(b) Describe how to use 30 slips of paper to create one bootstrap statistic. Be specific.

[42]Mackisack, M., "Favourite Experiments: An Addendum to What Is the Use of Experiments Conducted by Statistics Students?," *Journal of Statistics Education*, 1994, *http://www.amstat.org/publications/jse/v2n1/mackisack.supp.html*.

[43]Bartal, I.B., Decety, J., and Mason, P., "Empathy and Pro-Social Behavior in Rats," *Science*, 2011; 224(6061):1427–30.

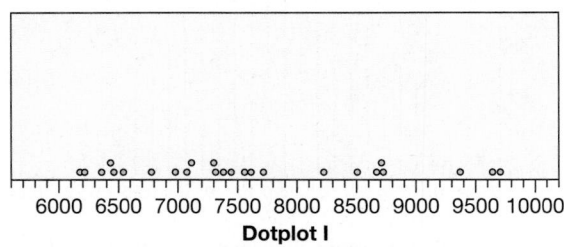

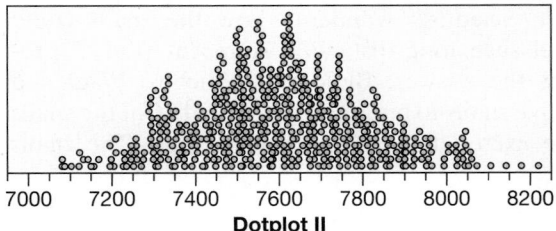

Figure 3.20 *Which is the bootstrap distribution?*

(c) Use *StatKey* or other technology to create a bootstrap distribution. Describe the shape and center of the bootstrap distribution. What is the standard error?

(d) Use the standard error to find and interpret a 95% confidence interval for the proportion of rats likely to show empathy.

3.89 Are Female Rats More Compassionate Than Male Rats? Exercise 3.88 describes a study in which rats showed compassion by freeing a trapped rat. In the study, all six of the six female rats showed compassion by freeing the trapped rat while 17 of the 24 male rats did so. Use the results of this study to give a best estimate for the difference in proportion of rats showing compassion, between female rats and male rats. Then use *StatKey* or other technology to estimate the standard error[44] and use it to compute a 95% confidence interval for the difference in proportions. Use the interval to determine whether it is plausible that male and female rats are equally compassionate (i.e., that the difference in proportions is zero). The data are available in the dataset **CompassionateRats**.

3.90 Teens Are More Likely to Send Text Messages Exercise 3.28 on page 210 compares studies which measure the proportions of adult and teen cell phone users that send/receive text messages. The summary statistics are repeated below:

Group	Sample Size	Proportion
Teen	$n_t = 800$	$\hat{p}_t = 0.87$
Adult	$n_a = 2252$	$\hat{p}_a = 0.72$

Figure 3.21 shows a distribution for the differences in sample proportions ($\hat{p}_t - \hat{p}_a$) for 5000 bootstrap

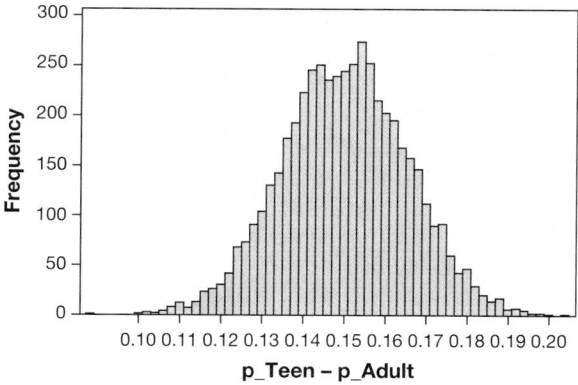

Figure 3.21 *Bootstrap difference in sample proportions of teen and adult cell phone users who text*

samples (taking 800 values with replacement from the original teen sample and 2252 from the adults).

(a) Based on the bootstrap distribution, which is the most reasonable estimate of the standard error for the difference in proportions: $SE = 0.015$, 0.030, 0.050, 0.10, or 0.15? Explain the reason for your choice.

(b) Using your choice for the SE estimate in part (a), find and interpret a 95% confidence interval for the difference in proportion of teen and adult cell phone users who send/receive text messages.

3.91 Tea, Coffee, and Your Immune System Researchers suspect that drinking tea might enhance the production of interferon gamma, a molecule that helps the immune system fight bacteria, viruses, and tumors. A recent study[45] involved 21 healthy people who did not normally drink tea or coffee. Eleven of the participants were randomly assigned to drink five or six cups of tea a day, while 10 were asked to drink the same amount of coffee. After two weeks, blood samples were exposed to an antigen and production of interferon gamma was measured.[46] The results are shown in Table 3.11 and are available in **ImmuneTea**. We are interested in estimating the effect size, the increase in average interferon gamma production for drinking tea when compared to coffee. Use *StatKey* or other technology to estimate the difference in mean production for tea drinkers minus coffee drinkers. Give the standard error for the difference and a 95% confidence interval. Interpret the result in context.

Table 3.11 *Immune system response in tea and coffee drinkers*

Tea	5	11	13	18	20	47
	48	52	55	56	58	
Coffee	0	0	3	11	15	16
	21	21	38	52		

3.92 Better Traffic Flow Exercise 2.155 on page 105 introduces the dataset **TrafficFlow**, which gives delay time in seconds for 24 simulation runs

[45] Adapted from Kamath et. al., "Antigens in Tea-Beverage Prime Human Vγ2Vδ2 T Cells *in vitro* and *in vivo* for Memory and Non-memory Antibacterial Cytokine Responses," Proceedings of the National Academy of Sciences, May 13, 2003.
[46] To be specific, peripheral blood mononuclear cells were cultured with the antigen alkylamine ethylalamine in an enzyme linked immunospot assay to the frequency of interferon-gamma-producing cells.

[44] In practice we should raise a caution here, since the proportion for female rats will be $\hat{p} = 1$ for every bootstrap sample.

in Dresden, Germany, comparing the current timed traffic light system on each run to a proposed flexible traffic light system in which lights communicate traffic flow information to neighboring lights. On average, public transportation was delayed 105 seconds under the timed system and 44 seconds under the flexible system. Since this is a matched pairs experiment, we are interested in the difference in times between the two methods for each of the 24 simulations. For the $n = 24$ differences D, we saw in Exercise 2.155 that $\overline{x}_D = 61$ seconds with $s_D = 15.19$ seconds. We wish to estimate the average time savings for public transportation on this stretch of road if the city of Dresden moves to the new system.

(a) What parameter are we estimating? Give correct notation.

(b) Suppose that we write the 24 differences on 24 slips of paper. Describe how to physically use the paper slips to create a bootstrap sample.

(c) What statistic do we record for this one bootstrap sample?

(d) If we create a bootstrap distribution using many of these bootstrap statistics, what shape do we expect it to have and where do we expect it to be centered?

(e) How can we use the values in the bootstrap distribution to find the standard error?

(f) The standard error is 3.1 for one set of 10,000 bootstrap samples. Find and interpret a 95% confidence interval for the average time savings.

3.93 Automobile Depreciation For a random sample of 20 automobile models, we record the value of the model as a new car and the value after the car has been purchased and driven 10 miles.[47] The difference between these two values is a measure of the depreciation on the car just by driving it off the lot. *Depreciation* values from our sample of 20 automobile models can be found in the dataset **CarDepreciation**.

(a) Find the mean and standard deviation of the *Depreciation* amounts in **CarDepreciation**.

(b) Use *StatKey* or other technology to create a bootstrap distribution of the sample mean of depreciations. Describe the shape, center, and spread of this distribution.

(c) Use the standard error obtained in your bootstrap distribution to find and interpret a 95%

confidence interval for the mean amount a new car depreciates by driving it off the lot.

3.94 Correlation between Price and Depreciation in Automobiles The data in **CarDepreciation** given in Exercise 3.93 contains information on both *New* price and *Depreciation* for a sample of 20 automobile models.

(a) Find the correlation between *New* price and *Depreciation* from the original sample of 20 automobiles.

(b) Use *StatKey* or other technology to create a bootstrap distribution of correlations and report the standard error.

(c) Using the standard error, create and interpret a 95% confidence interval for the correlation between *New* price and *Depreciation* of automobile models.

3.95 Headaches and Handedness A study was conducted to investigate the relationship between severe headaches and being left- or right-handed.[48] (Incidentally, Lisa Kudrow, who played Phoebe Buffay on the hit sitcom "Friends," is an author on this study.) Of 273 participants with cluster headaches, 24 were left-handed. Of 477 participants with migraine headaches, 42 were left-handed.

(a) Give an estimate for the proportion of cluster headache sufferers who are left-handed.

(b) Use *StatKey* or other technology to construct and interpret a 95% confidence interval for the proportion of cluster headache sufferers who are left-handed.

(c) Give an estimate for the proportion of migraine sufferers who are left-handed.

(d) Use *StatKey* or other technology to construct and interpret a 95% confidence interval for the proportion of migraine sufferers who are left-handed.

(e) Compare your confidence intervals in parts (b) and (d). Which is more narrow? Explain why.

SYNCHRONIZED DANCING, ANYONE?

Exercises 3.96 to 3.101 use data from a study designed to examine the effect of doing synchronized movements (such as marching in step or doing synchronized dance steps) and the effect of exertion on many different variables, such as pain tolerance and attitudes toward others. In the study, 264 high school students in Brazil were randomly assigned to

---

[47]New and used automobile costs were determined using 2015 models on *kellybluebook.com*.

[48]Messinger, H.B., Messinger, M.I., Kudrow, L., and Kudrow, L.V. "Handedness and headache." *Cephalalgia*, 1994; 14:64–67

one of four groups reflecting whether or not movements were synchronized (*Synch*= yes or no) and level of activity (*Exertion*= high or low).[49] Participants rated how close they felt to others in their group both before (*CloseBefore*) and after (*CloseAfter*) the activity, using a 7-point scale (1=least close to 7=most close). Participants also had their pain tolerance measured using pressure from a blood pressure cuff, by indicating when the pressure became too uncomfortable (up to a maximum pressure of 300 mmHg). Higher numbers for this *PainTolerance* measure indicate higher pain tolerance. The full dataset is available in **Synchronized-Movement**. For each of the following problems:

(a) Give notation for the quantity we are estimating, and define any relevant parameters.

(b) Use *StatKey* or other technology to find the value of the sample statistic. Give the correct notation with your answer.

(c) Use *StatKey* or other technology to find the standard error for the estimate.

(d) Use the standard error to give a 95% confidence interval for the quantity we are estimating.

(e) Interpret the confidence interval in context.

3.96 How Close Do You Feel to Others? Use the closeness ratings before the activity (*CloseBefore*) to estimate the mean closeness rating one person would assign to others in a group.

3.97 Does Synchronization Increase Feelings of Closeness? Use the closeness ratings given after the activity (*CloseAfter*) to estimate the difference in mean rating of closeness between those who have just done a synchronized activity and those who do a non-synchronized activity.

3.98 Does Synchronization Boost Pain Tolerance? Use the pain tolerance ratings (*PainTolerance*) after the activity to estimate the difference in mean pain tolerance between those who just completed a synchronized activity and those who did a non-synchronized activity.

3.99 Does Exertion Boost Pain Tolerance? Use the pain tolerance ratings after the activity to estimate the difference in mean pain tolerance between those who just completed a high exertion activity and those who completed a low exertion activity.

3.100 What Proportion Go to Maximum Pressure? We see that 75 of the 264 people in the study allowed the pressure to reach its maximum level of 300 mmHg, without ever saying that the pain was

too much (*MaxPressure*=yes). Use this information to estimate the proportion of people who would allow the pressure to reach its maximum level.

3.101 Are Males or Females More Likely to Go to Maximum Pressure? The study recorded whether participants were female or male (*Sex*= F or M), and we see that 33 of the 165 females and 42 of the 99 males allowed the pressure to reach its maximum level of 300 mmHg after treatment, without ever saying that the pain was too much. Use this information to estimate the difference in proportion of people who would allow the pressure to reach its maximum level after treatment, between females and males.

3.102 NHL Penalty Minutes Table 3.4 on page 208 shows the number of points scored and penalty minutes for 24 ice hockey players on the Ottawa Senators NHL team for the 2014–2015 season. The data are also stored in **OttawaSenators**. Assume that we consider these players to be a sample of all NHL players.

(a) Create a dotplot of the distribution of penalty minutes (*PenMin*) for the original sample of 24 players. Comment on the shape, paying particular attention to skewness and possible outliers.

(b) Find the mean and standard deviation of the penalty minute values for the original sample.

(c) Use *StatKey* or other technology to construct a bootstrap distribution for the mean penalty minutes for samples of size $n = 24$ NHL players. Comment on the shape of this distribution, especially compared to the shape of the original sample.

(d) Compute the standard deviation of the bootstrap means using the distribution in part (c). Compare this value to the standard deviation of the penalty minutes in the original sample.

(e) Construct a 95% confidence interval for the mean penalty minutes of NHL players.

(f) Give a reason why it might *not* be reasonable to use the players on one team as a sample of all players in a league.

3.103 Standard Deviation of NHL Penalty Minutes Exercise 3.102 describes data on the number of penalty minutes for Ottawa Senators NHL players. The sample has a fairly large standard deviation, $s = 27.3$ minutes. Use *StatKey* or other technology to create a bootstrap distribution, estimate the standard error, and give a 95% confidence interval for the *standard deviation* of penalty minutes for NHL players. Assume that the data in **OttawaSenators** can be viewed as a reasonable sample of all NHL players.

[49]Tarr, B., Launay, J., Cohen, E., and Dunbar, R., "Synchrony and exertion during dance independently raise pain threshold and encourage social bonding," *Biology Letters*, 11(10), October 2015.

3.4 BOOTSTRAP CONFIDENCE INTERVALS USING PERCENTILES

Confidence Intervals Based on Bootstrap Percentiles

If we were only concerned with 95% confidence intervals and always had a symmetric, bell-shaped bootstrap distribution, the rough *Statistic* $\pm 2 \cdot SE$ interval we computed in Section 3.3 would probably be all that we need. But we might have a bootstrap distribution that is symmetric but subtly flatter (or steeper) so that more (or less) than 95% of bootstrap statistics are within two standard errors of the center. Or we might want more (say 99%) or less (perhaps 90%) confidence that the method will produce a successful interval.

Fortunately, the bootstrap distribution provides a method to address both of these concerns. Rather than using the 95% rule and $\pm 2 \cdot SE$ as a yardstick to estimate the middle 95% of the bootstrap statistics, we can use the percentiles of the bootstrap distribution to locate the actual middle 95%! If we want the middle 95% of the bootstrap distribution (the values that are most likely to be close to the center), we can just chop off the lowest 2.5% and highest 2.5% of the bootstrap statistics to produce an interval.

Example 3.23

Figure 3.15 on page 231 shows a bootstrap distribution of sample means based on a sample of commute times (in minutes) for 500 residents of metropolitan Atlanta. That figure is reproduced in Figure 3.22 where we also indicate the boundaries for the middle 95% of the data, leaving 2.5% of the values in each tail. Use these boundaries to find and interpret a 95% confidence interval for Atlanta commute times.

Solution

The 2.5%-tile of the bootstrap distribution is at 27.43 minutes and the 97.5%-tile is at 31.05 minutes. Thus the 95% confidence interval for mean commute time in Atlanta, based on the original sample, goes from 27.43 to 31.05 minutes. We are 95% sure that the mean commute time for all Atlanta commuters is between 27.43 and 31.05 minutes.

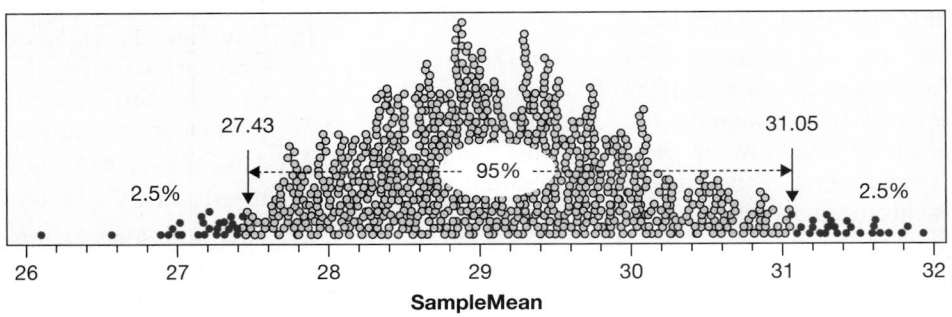

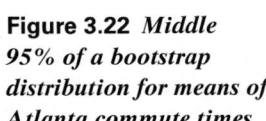

Figure 3.22 *Middle 95% of a bootstrap distribution for means of Atlanta commute times*

If we could actually poll every single commuter in Atlanta and find the commute times and calculate the population mean, the resulting value would either lie within the 95% confidence interval of 27.43 to 31.05 or it would not. Remember that when we say we are "95% sure," we just mean that 95% of intervals constructed using this method will contain the population parameter.

The 95% confidence interval calculated based on percentiles in Example 3.23 is similar to the 95% confidence interval based on two standard error bounds calculated in Example 3.22 that went from 27.28 to 30.94 minutes. If the bootstrap distribution is symmetric and bell-shaped, the two methods give approximately the same results for a 95% confidence interval.

Example 3.24

Use the bootstrap distribution in Figure 3.22 to estimate 99% and 90% confidence intervals for the mean Atlanta commute time.

Solution

Since the bootstrap distribution used 1000 samples, the middle 99% of the values would include 990 bootstrap means, leaving just five values in each of the tails. In Figure 3.22 this would put boundaries near 27.0 and 31.6. For a 90% confidence interval we need the 5%-tile and 95%-tile, leaving roughly 50 values in each tail. This gives a 90% confidence interval for mean commute times between about 27.7 and 30.7 minutes. More precise values for the percentiles found with computer software are shown in Figure 3.23.

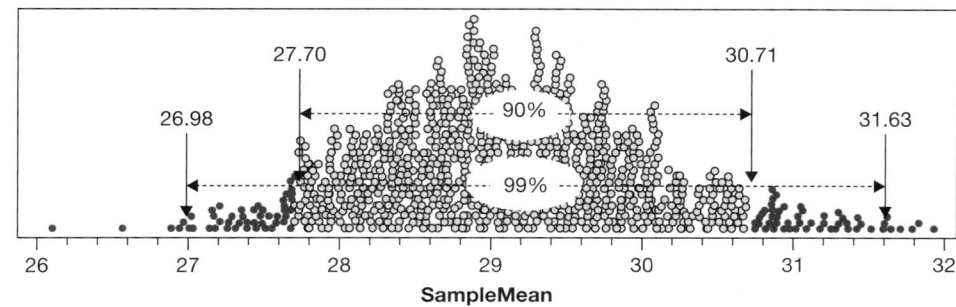

Figure 3.23 *99% and 90% confidence intervals for mean commute time in Atlanta*

Figures 3.22 and 3.23 make it clear that to get more confidence that our interval contains the true mean, we need to use a wider interval. This is generally the case for any particular sample.

Why don't we look for a 100% confidence interval? We might be 100% sure that the mean commute time in Atlanta is between 0 and 200 minutes, but is that interval of any practical use? In general we need to balance a reasonable chance of capturing the parameter of interest with a need to narrow in on where the parameter might be. That is why we commonly use confidence levels like 90%, 95%, or 99% depending on the trade-off we are willing to make between a precise, narrow interval and a good chance that it succeeds.

Constructing a Confidence Interval from the Percentiles of a Bootstrap Distribution

If the bootstrap distribution is approximately symmetric, we construct a confidence interval by finding the percentiles in the bootstrap distribution so that the proportion of bootstrap statistics between the percentiles matches the desired confidence level.

Finding Confidence Intervals for Many Different Parameters

These procedures for finding bootstrap confidence intervals are quite flexible and can be applied in a wide variety of different situations and with many different parameters. The basic procedure is very straightforward. As long as we can construct a reasonable bootstrap distribution, we can use it to estimate a confidence interval. The tools in *StatKey* automate this process, as do many statistical software packages.

The process of creating bootstrap samples can require a bit more thought when the sampling process is more involved. We should always create bootstrap statistics as similar as possible to the relevant statistic from the original data, as illustrated in the next two examples.

Example 3.25

Who Exercises More: Males or Females?

Fifty students were asked how many hours a week they exercise, and the results are included in the dataset **ExerciseHours**. Figure 3.24 shows comparative boxplots of the number of hours spent exercising, and we compute the summary statistics to be $\bar{x}_M = 12.4$ and $s_M = 8.80$ with $n_M = 20$ for the males and $\bar{x}_F = 9.4$ and $s_F = 7.41$ with $n_F = 30$ for the females. How big might the difference in mean hours spent exercising be, between males and females?

(a) Use the sample to give a best estimate for the difference in mean hours spent exercising between males and females.

(b) Describe the process we would use to compute one bootstrap statistic from the sample.

(c) Use *StatKey* or other technology to find and interpret a 95% confidence interval for the difference in mean number of hours spent exercising.

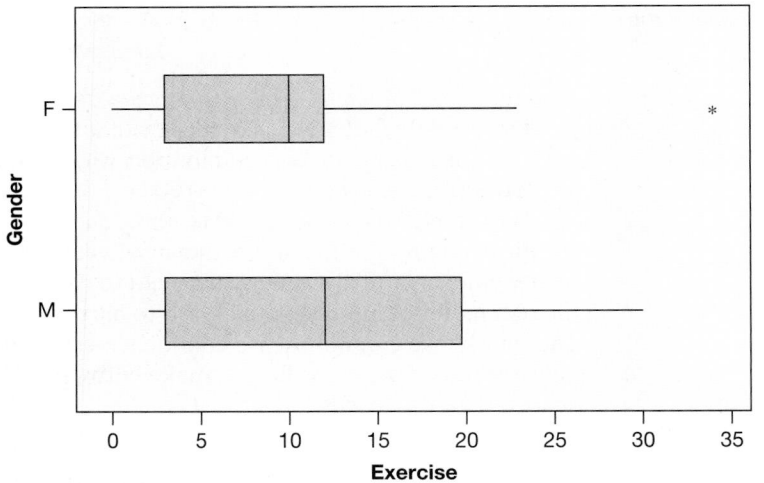

Figure 3.24 *Number of hours a week spent exercising: males and females*

Solution

(a) We estimate the difference in mean exercise times between males and females ($\mu_M - \mu_F$) with the difference in the sample means, $\bar{x}_M - \bar{x}_F = 12.4 - 9.4 = 3.0$ hours per week. In other words, we estimate that males spend, on average, three more hours a week exercising than females spend.

(b) To match the original data as closely as possible, for each bootstrap sample we take 20 male times with replacement from the original 20 male values and 30 female times with replacement from the original 30 female values. To compute the bootstrap statistic, we compute the sample means for males and females, and find the difference in the two means, mimicking the statistic found in the original sample.

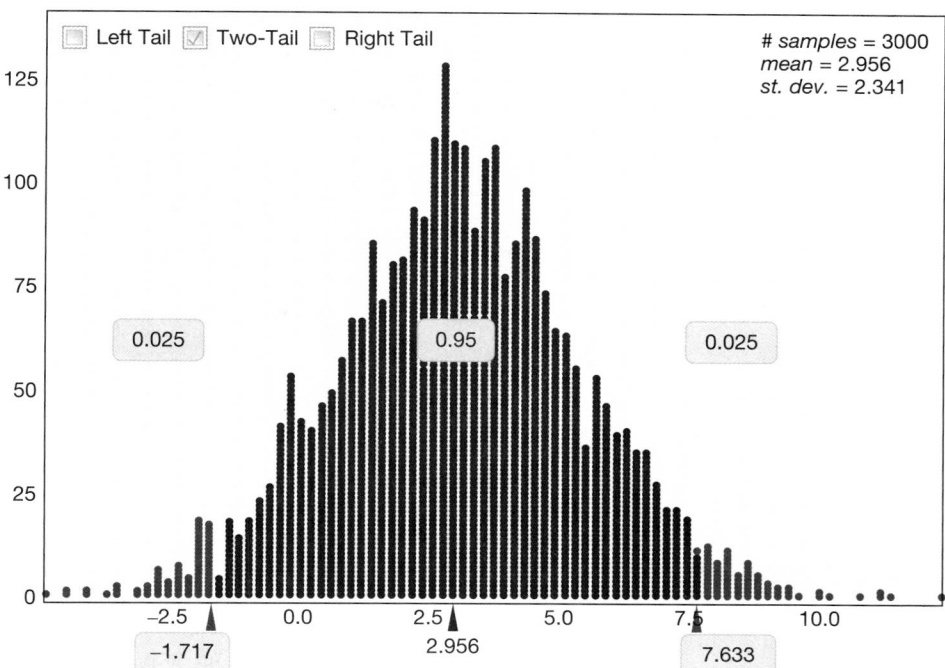

Figure 3.25 *Bootstrap distribution for difference in mean time spent exercising*

(c) Figure 3.25 displays a dotplot from *StatKey* showing these differences in means based on 3000 bootstrap samples.

Using the 2.5%-tile and 97.5%-tiles from the bootstrap distribution in Figure 3.25 we get a 95% confidence interval for the difference in mean exercise hours between men and women from −1.717 hours to 7.633 hours.

Since the bootstrap distribution is relatively symmetric and bell-shaped, we can also (or instead) use its standard error to find a 95% confidence interval. The standard deviation for the bootstrap statistics, found in the upper corner of Figure 3.25, is 2.341, so we estimate the standard error of the statistic $\bar{x}_M - \bar{x}_F$ to be $SE = 2.341$. We find an interval estimate for the difference in the population means with

$$Statistic \pm 2 \cdot SE$$
$$(\bar{x}_M - \bar{x}_F) \pm 2 \cdot SE$$
$$(12.4 - 9.4) \pm 2 \cdot (2.341)$$
$$3.0 \pm 4.68$$
$$-1.68 \text{ to } 7.68$$

While this is not exactly the same as the interval we obtained from the percentiles, there is not much practical difference between them.

To interpret the percentile interval, we are 95% sure that the difference in mean time spent exercising between males and females is between −1.72 and 7.63 hours per week. To make the direction of the difference more explicit, we might revise the interpretation to say that we are 95% sure that the mean exercise time for males is between 1.72 hours less and 7.63 hours more than mean exercise time for females. Since 0 is within this interval and thus a plausible value for $\mu_M - \mu_F$, it is plausible that there is no difference in mean exercise times between males and females.

© Mikkel William Nielsen/iStockphoto

How do price and mileage correlate for used Mustangs?

DATA 3.4	**Mustang Prices**

A statistics student, Gabe McBride, was interested in prices for used Mustang cars being offered for sale on an Internet site. He sampled 25 cars from the website and recorded the age (in years), mileage (in thousands of miles), and asking price (in $1000s) for each car in his sample. The data are stored in **MustangPrice** and the scatterplot in Figure 3.26 shows the relationship between the *Miles* on each car and the *Price*. Not surprisingly, we see a strong negative association showing the price of a used Mustang tends to be lower if it has been driven for more miles. The correlation between *Price* and *Miles* for this sample is $r = -0.825$. ■

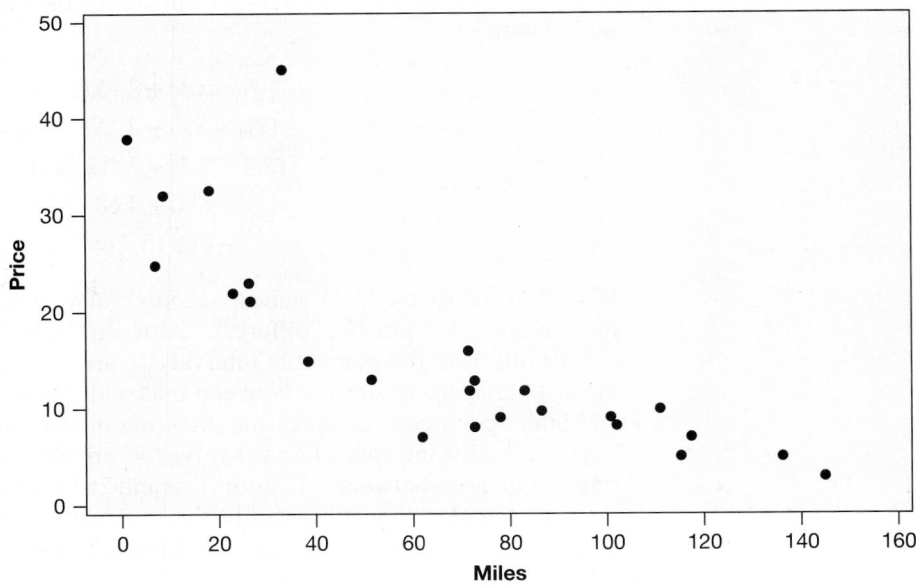

Figure 3.26 *Price (in $1000s) and mileage (in 1000s) for a sample of 25 used Mustang cars*

Example 3.26

(a) Describe how we might create one bootstrap statistic for the correlation between price and mileage of used Mustang cars, using the data described in Data 3.4.

(b) Figure 3.27 shows a dotplot of the correlations between *Price* and *Miles* for each of 5000 bootstrap samples from the **MustangPrice** data, and Table 3.12 gives some percentiles from this bootstrap distribution. Use this information to create a 98% confidence interval for the correlation between *Price* and *Miles* for the population of all Mustangs for sale at this website. Interpret the interval in context.

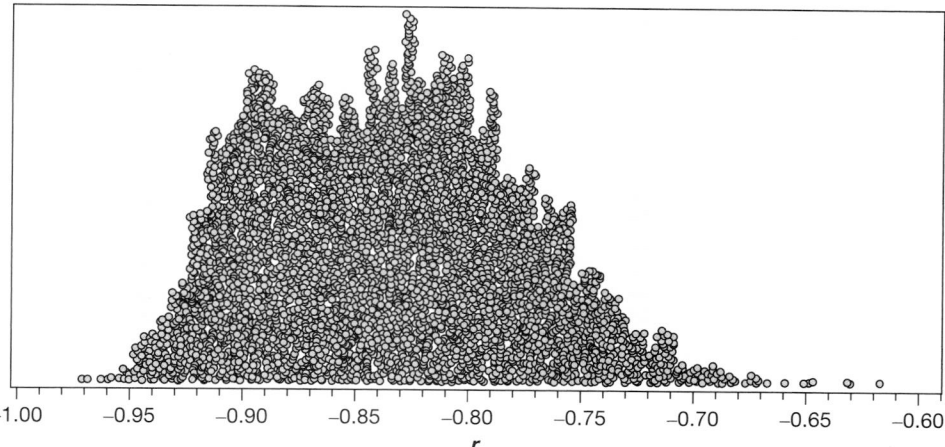

Figure 3.27 *Bootstrap correlations between Price and Miles for 5000 samples of size 25*

Table 3.12 *Percentiles from a bootstrap distribution of Mustang correlations*

	0.5%	1.0%	2.0%	2.5%	5.0%	95.0%	97.5%	98.0%	99.0%	99.5%
Percentile	−0.945	−0.940	−0.931	−0.928	−0.919	−0.741	−0.723	−0.717	−0.705	−0.689

Solution

(a) Because the correlation is based on ordered pairs of data (price and mileage), we compute a bootstrap statistic by sampling (with replacement) ordered pairs from the original sample. We select 25 ordered pairs in this way (to match the original sample size) and compute the correlation of the 25 ordered pairs for one bootstrap statistic.

(b) For a 98% confidence interval we need to take 1% from each tail of the bootstrap distribution, so we use the 1%-tile and 99%-tile from Table 3.12. This gives us an interval from −0.940 to −0.705. Based on this sample of 25 Mustangs, we are 98% sure that the correlation between price and mileage for all used Mustangs for sale at this Internet site is somewhere between −0.940 and −0.705.

Another Look at the Effect of Sample Size

In Example 3.22, we calculated a 95% confidence interval for the proportion of peanuts in mixed nuts based on a sample proportion of $\hat{p} = 0.52$ and a sample size of $n = 100$. The next example investigates how the result changes if we have the same sample proportion but a larger sample size.

Example 3.27

Suppose a sample of size $n = 400$ mixed nuts contains 208 peanuts, so the proportion of peanuts is $\hat{p} = 0.52$. Use this sample data to compute a 95% confidence interval for the proportion of peanuts. Compare your answer to the 95% confidence interval of 0.42 to 0.62 based on a sample of size $n = 100$ given in Example 3.22.

Solution

Figure 3.28 shows a dotplot of the bootstrap proportions for 1000 simulated samples of size 400. We see that a 95% confidence interval for the proportion of peanuts goes from 0.472 to 0.568. This confidence interval for a sample size of 400 is considerably narrower than the interval based on a sample size of 100; in fact, it is about half the width. The margin of error has gone from about 0.10 to about 0.05.

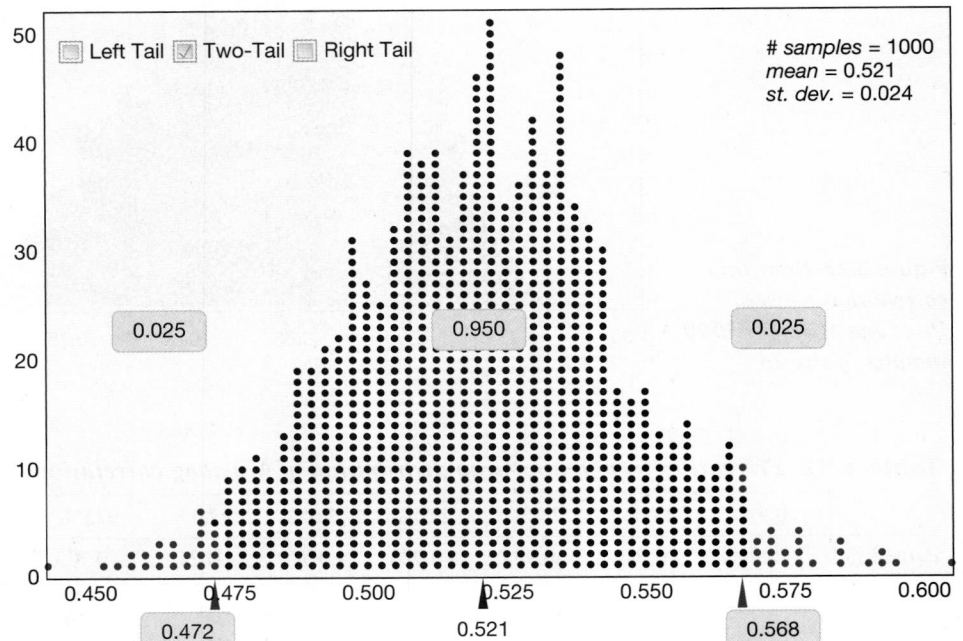

Figure 3.28 *Bootstrap proportions for 1000 samples simulated from a sample with $\hat{p} = 0.52$ and $n = 400$*

At first glance the bootstrap distribution for proportions based on samples of size $n = 400$ might look similar to Figure 3.17, which used samples of size $n = 100$. However, pay close attention to the scale for the horizontal axis. As we saw with the sampling distribution, when the sample size is larger, the bootstrap proportions tend to be closer to the center proportion of 0.52. This is consistent with the fact that the estimated standard error of the proportion based on the larger samples is $SE = 0.024$, about half of the standard error when $n = 100$. We improve the accuracy of our estimate, and reduce the width of our interval, by taking a larger sample.

Larger Sample Size Increases Accuracy

A larger sample size tends to increase the accuracy of the estimate, giving a smaller standard error and reducing the width of a confidence interval.

One Caution on Constructing Bootstrap Confidence Intervals

Example 3.28

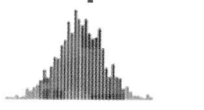

Bootstrap Intervals Don't Always Work Well

Use *StatKey* or other technology to create a bootstrap distribution for the *median* price of Mustangs using the sample of 25 cars in **MustangPrice**. Explain why it would not be appropriate to use the bootstrap distribution to construct a 95% confidence interval for the median price of mustangs.

Solution

We create 5000 bootstrap samples, each with 25 cars selected from the original sample, and find the median price for each sample. Figure 3.29 shows a dotplot of the 5000 bootstrap medians. While the mechanics of constructing a confidence interval from this bootstrap distribution appear very straightforward, it is important to always pause first and take a good look at the bootstrap distribution. This plot looks quite different from the bootstrap distributions we have seen in other examples. Notice that the median for 25 data points is always one of the data values, so the choices for bootstrap medians are limited to the original 25 prices. For example, a percentile can be at prices of 16 or 21, but never in between. When using the percentiles of the bootstrap distribution or using the $\pm 2 \cdot SE$ method, we need to make sure that the bootstrap distribution is reasonably symmetric around the original statistic and reasonably bell-shaped. In this case, it is not appropriate to use this bootstrap distribution to find a confidence interval.

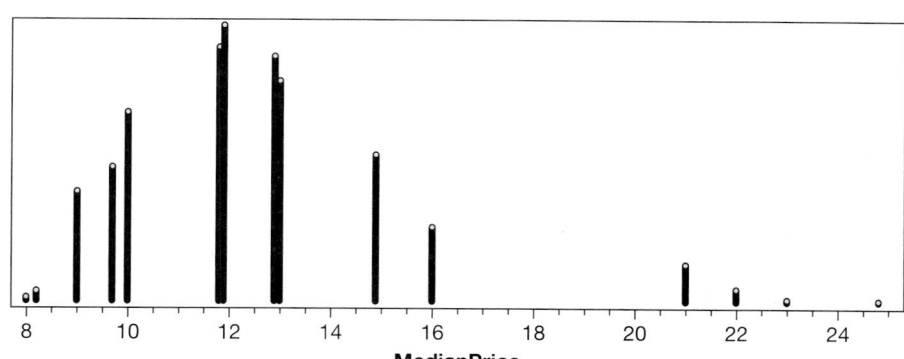

Figure 3.29 *Bootstrap medians for Mustang prices (n = 25)*

You should always look at a plot of the bootstrap distribution. If the plot is poorly behaved (for example, heavily skewed or isolated clumps of values), you should not have much confidence in the intervals it produces. Fortunately, for most of the statistics we consider, the bootstrap distributions work well.

SECTION LEARNING GOALS

You should now have the understanding and skills to:

- Construct a confidence interval based on the percentiles of a bootstrap distribution

- Describe the process of constructing a bootstrap statistic for many different parameters

- Explain how the width of an interval is affected by the desired level of confidence and the sample size

- Recognize when it is appropriate to construct a bootstrap confidence interval using percentiles or the standard error

Exercises for Section 3.4

SKILL BUILDER 1

3.104 To create a confidence interval from a bootstrap distribution using percentiles, we keep the middle values and chop off a certain percent from each tail. Indicate what percent of values must be chopped off from each tail for each confidence level given.

(a) 95%

(b) 90%

(c) 98%

(d) 99%

SKILL BUILDER 2

3.105 To create a confidence interval from a bootstrap distribution using percentiles, we keep the middle values and chop off some number of the lowest values and the highest values. If our bootstrap distribution contains values for 1000 bootstrap samples, indicate how many we chop off at each end for each confidence level given.

(a) 95%

(b) 90%

(c) 98%

(d) 99%

SKILL BUILDER 3

In estimating the mean score on a fitness exam, we use an original sample of size $n = 30$ and a bootstrap distribution containing 5000 bootstrap samples to obtain a 95% confidence interval of 67 to 73. In Exercises 3.106 to 3.111, a change in this process is described. If all else stays the same, which of the following confidence intervals (A, B, or C) is the most likely result after the change:

    A. 66 to 74 B. 67 to 73 C. 67.5 to 72.5

3.106 Using the data to find a 99% confidence interval.

3.107 Using the data to find a 90% confidence interval.

3.108 Using an original sample of size $n = 45$.

3.109 Using an original sample of size $n = 16$.

3.110 Using 10,000 bootstrap samples for the distribution.

3.111 Using 1000 bootstrap samples for the distribution.

SKILL BUILDER 4

Exercises 3.112 to 3.115 give information about the proportion of a sample that agree with a certain statement. Use *StatKey* or other technology to find a confidence interval at the given confidence level for the proportion of the population to agree, using percentiles from a bootstrap distribution. *StatKey* tip: Use "CI for Single Proportion" and then "Edit Data" to enter the sample information.

3.112 Find a 95% confidence interval if 35 agree in a random sample of 100 people.

3.113 Find a 95% confidence interval if 180 agree in a random sample of 250 people.

3.114 Find a 90% confidence interval if 112 agree and 288 disagree in a random sample of 400 people.

3.115 Find a 99% confidence interval if, in a random sample of 1000 people, 382 agree, 578 disagree, and 40 can't decide.

3.116 IQ Scores A sample of 10 IQ scores was used to create the bootstrap distribution of sample means in Figure 3.30.

(a) Estimate the mean of the original sample of IQ scores.

(b) The distribution was created using 1000 bootstrap statistics. Use the distribution to estimate a 99% confidence interval for the mean IQ score for the population. Explain your reasoning.

3.117 Average Penalty Minutes in the NHL In Exercise 3.102 on page 241, we construct a 95% confidence interval for mean penalty minutes given

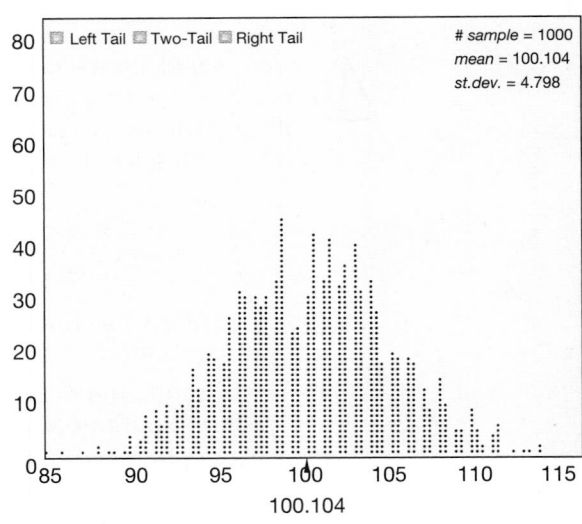

Figure 3.30 *Bootstrap distribution of sample means of IQ scores*

Table 3.13 *Percentiles for a bootstrap distribution of penalty minutes*

	0.5%	1.0%	2.0%	2.5%	5.0%	95.0%	97.5%	98.0%	99.0%	99.5%
Percentile	21.8	22.8	23.9	24.4	25.8	43.5	45.5	46.1	47.7	49.5

to NHL players in a season using data from players on the Ottawa Senators as our sample. Some percentiles from a bootstrap distribution of 5000 sample means are shown in Table 3.13. Use this information to find and interpret a 98% confidence interval for the mean penalty minutes of NHL players. Assume that the players on this team are a reasonable sample from the population of all players.

3.118 How Important Is Regular Exercise? In a recent poll[50] of 1000 American adults, the number saying that exercise is an important part of daily life was 753. Use *StatKey* or other technology to find and interpret a 90% confidence interval for the proportion of American adults who think exercise is an important part of daily life.

3.119 Many Europeans Don't Recognize Signs of Stroke or Heart Attack Across nine European countries in a large-scale survey, people had a hard time identifying signs of a stroke or heart attack. The survey[51] included 10,228 inhabitants of Austria, France, Germany, Italy, the Netherlands, Poland, Russia, Spain, and the United Kingdom. Participants ages ranged from 14 to 98. Of those surveyed, less than half (4910) linked arm or shoulder

pain to heart attacks. Use *StatKey* to find and interpret a 99% confidence interval for the proportion of Europeans (from these nine countries) who can identify arm or shoulder pain as a symptom of a heart attack. Can we be 99% confident that the proportion is less than half?

THE ORGANIC EFFECT

Exercises 3.120 to 3.123 pertain to a 2015 study[52] which took a Swedish family that ate a conventional diet (non-organic), and then had them eat only organic for two weeks. Pesticide concentrations for several different pesticides were measured in μg/g creatinine by testing morning urine. Multiple measurements[53] were taken for each person before the switch to organic foods, and then again after participants had been eating organic for at least one week. The results are pretty compelling, and are summarized in a short video[54] which as of this writing has had over 30 million views online. The data are visualized in Figure 3.31 for eight different detected pesticides,

[50]*Rasmussen Reports*, "75% Say Exercise is Important in Daily Life," March 26, 2011.

[51]Mata, J., Frank, R., and Gigerenza, G., "Symptom Recognition of Heart Attack and Stroke in Nine European Countries: A Representative Aurvey", *Health Expectations*, 2012; doi: 10.1111/j.1369-7625.2011.00764.x.

[52]Magner, J., Wallberg, P., Sandberg, J., and Cousins, A.P. (2015). "Human exposure to pesticides from food: A pilot study," *IVL Swedish Environmental Research Institute. https://www.coop.se/PageFiles/429812/Coop%20Ekoeffekten_Report%20ENG.pdf*, January 2015.

[53]For illustrative purposes we will assume the measurements were far enough apart to be unrelated.

[54]*www.youtube.com/watch?v=oB6fUqmyKC8.*

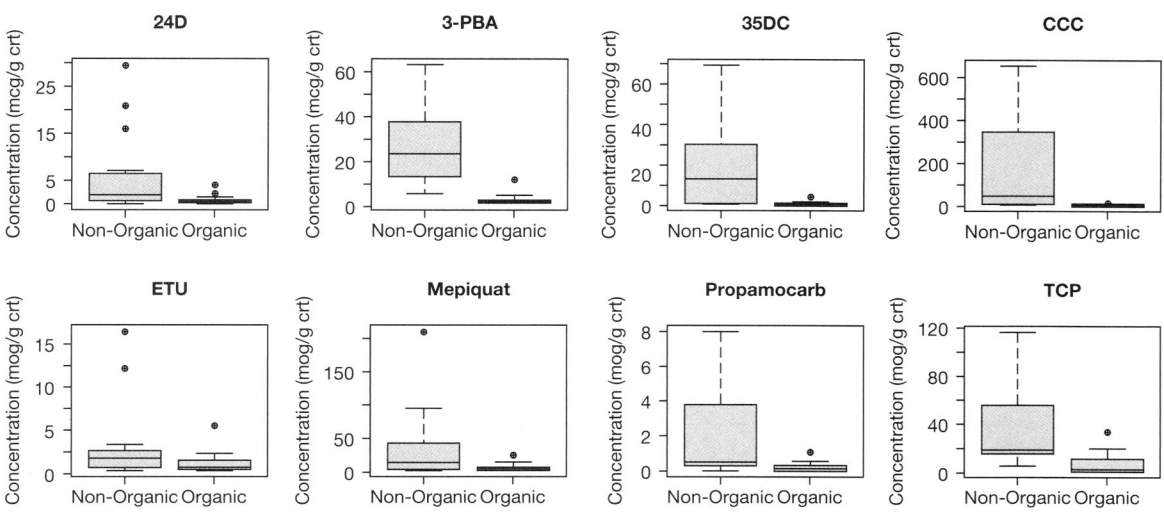

Figure 3.31 *Concentrations of eight different pesticides while eating non-organic versus eating organic*

and can be found in **OrganicEffect**. How do pesticide levels in the body differ after eating organic versus non-organic?

3.120 Eating Organic and 3-PBA Levels We first study 3-PBA, a commonly used insecticide found in grains, fruits, and vegetables. How much higher are 3-PBA concentrations while not eating organic versus eating organic? A bootstrap distribution based on 1000 samples of the mean concentration before the switch minus the mean concentration after switching to organic is shown in Figure 3.32.

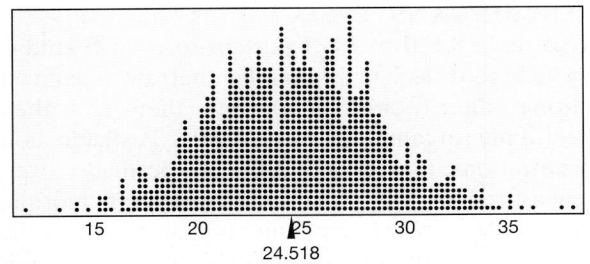

Figure 3.32 *Bootstrap distribution based on 1000 simulations for Exercise 3.120*

(a) Define a relevant parameter of interest, using correct notation.

(b) Use the bootstrap distribution to estimate the sample statistic, including correct notation.

(c) Use the bootstrap distribution to estimate a 99% confidence interval.

(d) Interpret this interval in context.

(e) Does this interval provide evidence that concentrations of the pesticide 3-PBA are lower while eating organic? Why or why not?

(f) Can we conclude that eating organic *causes* lower concentrations of 3-PBA? Why or why not?

3.121 Using the Paired Nature of the Data This is actually a paired data situation (since each person was measured before and after eating organic), so we create a new variable, the *differences* in pesticide concentration, by taking pesticide concentration for a particular pesticide before eating organic (on say, the first day of the week) minus pesticide

concentration after eating organic. Because this is now a single quantitative variable to analyze, we are interested in inference for just a single mean, the average difference in pesticide concentration. A bootstrap distribution for the mean of the differences for the pesticide 3-PBA was created, and some of the percentiles for this distribution are given in Table 3.14.

(a) Give the 99% confidence interval.

(b) Interpret this interval in context.

3.122 Investigating the Width of a Confidence Interval Comparing Exercise 3.120 to Exercise 3.121, you should have found that the confidence interval when utilizing the paired structure of the data was narrower than the confidence interval ignoring this structure (this will generally be the case, and is the primary reason for pairing). How else could we change the width of the confidence interval? More specifically, for each of the following changes, would the width of the confidence interval likely *increase*, *decrease*, or *remain the same*?

(a) Increase the sample size.

(b) Simulate more bootstrap samples.

(c) Decrease the confidence level from 99% to 95%.

3.123 What Proportion Have Pesticides Detected? In addition to the quantitative variable pesticide concentration, the researchers also report whether or not the pesticide was detected in the urine (at standard detection levels). Before the participants started eating organic, 111 of the 240 measurements (combining all pesticides and people) yielded a positive pesticide detection. While eating organic, only 24 of the 240 measurements resulted in a positive pesticide detection.

(a) Calculate the sample difference in proportions: proportion of measurements resulting in pesticide detection while eating non-organic minus proportion of measurements resulting in pesticide detection while eating organic.

(b) Figure 3.33 gives a bootstrap distribution for the difference in proportions, based on 1000 simulated bootstrap samples. Approximate a 98% confidence interval.

(c) Interpret this interval in context.

Table 3.14 *Table of percentiles for the bootstrap distribution for Exercise 3.121*

	0.5%	1.0%	2.5%	5.0%	10%	...	90%	95%	97.5%	99%	99.5%
Percentile	15.8	16.6	17.5	18.7	19.9	...	28.9	30.0	30.9	31.8	32.4

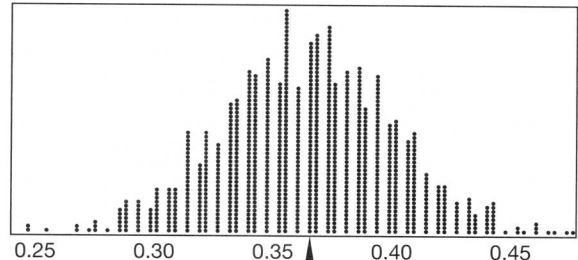

Figure 3.33 *Bootstrap distribution for the difference in proportions*

3.124 Donating Blood to Grandma? There is some evidence that "young blood" might improve the health, both physically and cognitively, of elderly people (or mice). Exercise 2.69 on page 75 introduces one study in which old mice were randomly assigned to receive transfusions of blood from either young mice or old mice. Researchers then measured the number of minutes each of the old mice was able to run on a treadmill. The data are stored in **YoungBlood**. We wish to estimate the difference in the mean length of time on the treadmill, between those mice getting young blood and those mice getting old blood. Use *StatKey* or other technology to find and interpret a 90% confidence interval for the difference in means.

3.125 Does a Common Ingredient in Soap Increase Staph Infections? Triclosan is a compound often added to products such as soaps, lotions, and toothpaste. It is antimicrobial, so we expect it to lower one's chance of having a staph infection. However, the opposite was found in a recent study.[55] Microbiologists swabbed the noses of 90 people, and recorded which had detectable levels of triclosan and which had evidence of carrying the staph bacteria, which greatly increases one's chance of having a serious staph infection. The results are shown in Table 3.15. Use the results of this study to estimate the difference in the proportion of people with staph bacteria, between those with triclosan in their

Table 3.15 *Does triclosan increase staph infections?*

	Staph	No Staph	Total
Triclosan	24	13	37
No Triclosan	15	38	53
Total	39	51	90

[55]Mole, B., "Triclosan aids nasal invasions by staph," *Science News*, April 15, 2014.

system and those without. (As a result of this study and others, the US Food and Drug Administration is investigating whether adding triclosan to personal care products is safe and effective.)

(a) Give notation for the parameter we are estimating.

(b) Give notation for the best estimate and give its value.

(c) Use *StatKey* or other technology to find a 99% confidence interval.

(d) Interpret this interval in context. Can we conclude that people with triclosan in their systems are more likely to have staph infections? Can we conclude that triclosan causes the increase in staph infections?

3.126 Average Tip for a Waitress Data 2.12 on page 123 describes information from a sample of 157 restaurant bills collected at the *First Crush* bistro. The data is available in **RestaurantTips**. Create a bootstrap distribution using this data and find and interpret a 95% confidence interval for the average tip left at this restaurant. Find the confidence interval two ways: using the standard error and using percentiles. Compare your results.

3.127 Daily Tip Revenue for a Waitress Data 2.12 on page 123 describes information from a sample of 157 restaurant bills collected at the *First Crush* bistro. The data is available in **RestaurantTips**. Two intervals are given below for the average tip left at a restaurant; one is a 90% confidence interval and one is a 99% confidence interval.

Interval A: 3.55 to 4.15   Interval B: 3.35 to 4.35

(a) Which one is the 90% confidence interval? Which one is the 99% confidence interval?

(b) One waitress generally waits on 20 tables in an average shift. Give a range for her expected daily tip revenue, using both 90% and 99% confidence. Interpret your results.

3.128 Who Smokes More: Male Students or Female Students? Data 1.1 on page 4 includes lots of information on a sample of 362 college students. The complete dataset is available at **StudentSurvey**. We see that 27 of the 193 males in the sample smoke while 16 of the 169 females in the sample smoke.

(a) What is the best point estimate for the difference in the proportion of smokers, using male proportion minus female proportion? Which gender smokes more in the sample?

(b) Find and interpret a 99% confidence interval for the difference in proportions.

3.129 Home Field Advantage Is there a home field advantage in soccer? We are specifically interested in the Football Association (FA) premier league, a football (soccer) league in Great Britain known for having especially passionate fans. We took a sample of 120 matches (excluding all ties) and found that the home team was victorious in 70 cases.[56]

(a) What is the population of interest? What is the specific population parameter of interest?

(b) Estimate the population parameter using the sample.

(c) Using *StatKey* or other technology, construct and interpret a 90% confidence interval.

(d) Using *StatKey* or other technology, construct and interpret a 99% confidence interval.

(e) Based on this sample and the results in parts (c) and (d), are we 90% confident a home field advantage exists? Are we 99% confident?

3.130 Using Percentiles to Estimate Tea vs Coffee Immune Response In Exercise 3.91, we introduce a study to estimate the difference in mean immune response (as measured in the study) between tea drinkers and coffee drinkers. The data are given in Table 3.11 on page 239 and are available in **ImmuneTea**.

(a) Give a best estimate for the difference in means: tea drinkers mean immune response minus coffee drinkers mean immune response.

(b) What quantity are we estimating? Give the correct notation.

(c) Using *StatKey* or other technology, construct and interpret a 90% confidence interval.

(d) Using *StatKey* or other technology, construct and interpret a 99% confidence interval.

(e) Based on this sample and the results in parts (c) and (d), are we 90% confident that tea drinkers have a stronger immune response? Are we 99% confident?

3.131 Effect of Splitting the Bill Exercise 2.153 on page 105 describes a study on the cost of restaurant meals when groups pay individually or split the bill equally. In the experiment 24 subjects ordered meals in groups which paid individually and 24 were in groups which agreed to split the costs. The data in **SplitBill** includes the cost of what each person ordered (in Israeli shekels) and the type of payment (*Individual* or *Split*). Use this information to construct a bootstrap distribution and find a 95%

confidence interval for the difference in means (*Individual* minus *Split*) between the two situations.

3.132 St. Louis vs Atlanta Commute Times The datafile **CommuteAtlanta** contains a sample of commute times for 500 workers in the Atlanta area as described in Data 3.3 on page 229. The data in **CommuteStLouis** has similar information on the commuting habits of a random sample of 500 residents from metropolitan St. Louis. Figure 3.34 shows comparative boxplots of the commute times for the two samples. We wish to estimate the difference in mean commute time between Atlanta and St. Louis.

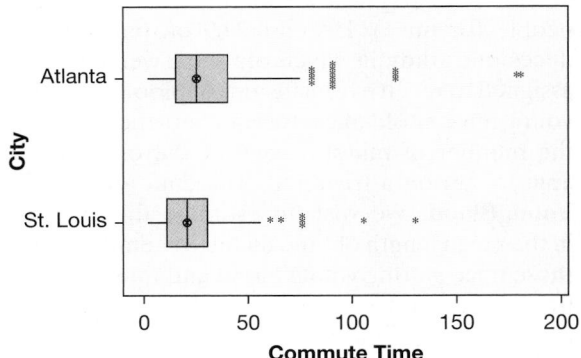

Figure 3.34 *Commute times for samples in Atlanta and St. Louis*

(a) Discuss and compare the boxplots in Figure 3.34. Which city appears to have the longer average commute time?

(b) Give notation for the parameter we are estimating and give the best estimate from the data.

(c) Describe how to compute one bootstrap statistic from this data.

(d) Use *StatKey* or other technology to create a bootstrap distribution for the difference in mean commute times between the two cities and use the standard error to find and interpret a 95% confidence interval.

3.133 Effect of Overeating for One Month: Correlation between Short-Term and Long-Term Weight Gain In Exercise 3.70 on page 227, we describe a study in which participants ate significantly more and exercised significantly less for a month. Two and a half years later, participants weighed an average of 6.8 pounds more than at the start of the experiment (while the weights of a control group had not

[56]*http://www.premierleague.com/page/Home/0,,12306,00.html*

changed). Is the amount of weight gained over the following 2.5 years directly related to how much weight was gained during the one-month period? For the 18 participants, the correlation between increase of body weight during the one-month intervention and increase of body weight after 30 months is $r = 0.21$. We want to estimate, for the population of all adults, the correlation between weight gain over one month of bingeing and the effect of that month on a person's weight 2.5 years later.

(a) What is the population parameter of interest? What is the best estimate for that parameter?

(b) To find the sample correlation $r = 0.21$, we used a dataset containing 18 ordered pairs (weight gain over the one month and weight gain 2.5 years later for each individual in the study). Describe how to use this data to obtain one bootstrap sample.

(c) What statistic is recorded for the bootstrap sample?

(d) Suppose that we use technology to calculate the relevant statistic for 1000 bootstrap samples. Describe how to find the standard error using those bootstrap statistics.

(e) The standard error for one set of bootstrap statistics is 0.14. Calculate a 95% confidence interval for the correlation.

(f) Use the confidence interval from part (e) to indicate whether you are confident that there is a positive correlation between amount of weight gain during the one-month intervention and amount of weight gained over the next 2.5 years, or whether it is plausible that

there is no correlation at all. Explain your reasoning.

(g) Will a 90% confidence interval most likely be *wider* or *narrower* than the 95% confidence interval found in part (e)?

3.134 Mustang Prices and Car Sales Figure 3.35 shows bootstrap distributions for the *standard deviation* of two different datasets. In each case, if appropriate, use the bootstrap distribution to estimate and interpret a 95% confidence interval for the population standard deviation. If not appropriate, explain why not.

(a) Standard deviation of prices of used Mustang cars (in thousands of dollars), introduced in Data 3.4 on page 246, with bootstrap distribution in Figure 3.35(a).

(b) Standard deviation of monthly sales of new cars, using the following sales figures for a sample of five months: 658, 456, 830, 696, 385, with bootstrap distribution in Figure 3.35(b).

3.135 Small Sample Size and Outliers As we have seen, bootstrap distributions are generally symmetric and bell-shaped and centered at the value of the original sample statistic. However, strange things can happen when the sample size is small and there is an outlier present. Use *StatKey* or other technology to create a bootstrap distribution for the *standard deviation* based on the following data:

| 8 | 10 | 7 | 12 | 13 | 8 | 10 | 50 |

Describe the shape of the distribution. Is it appropriate to construct a confidence interval from this distribution? Explain why the distribution might have the shape it does.

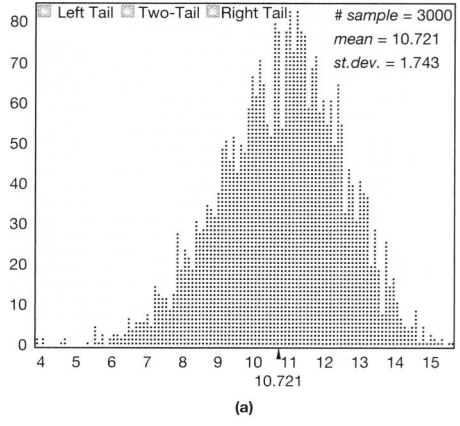

(a)

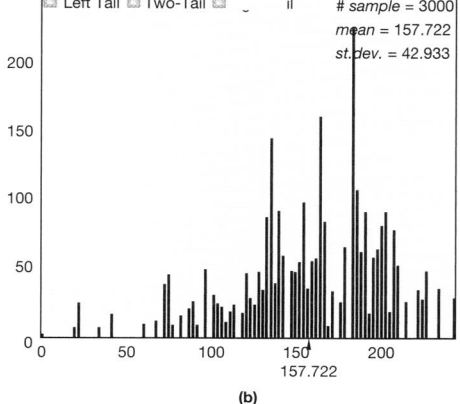

(b)

Figure 3.35 *Bootstrap distributions for standard deviation*

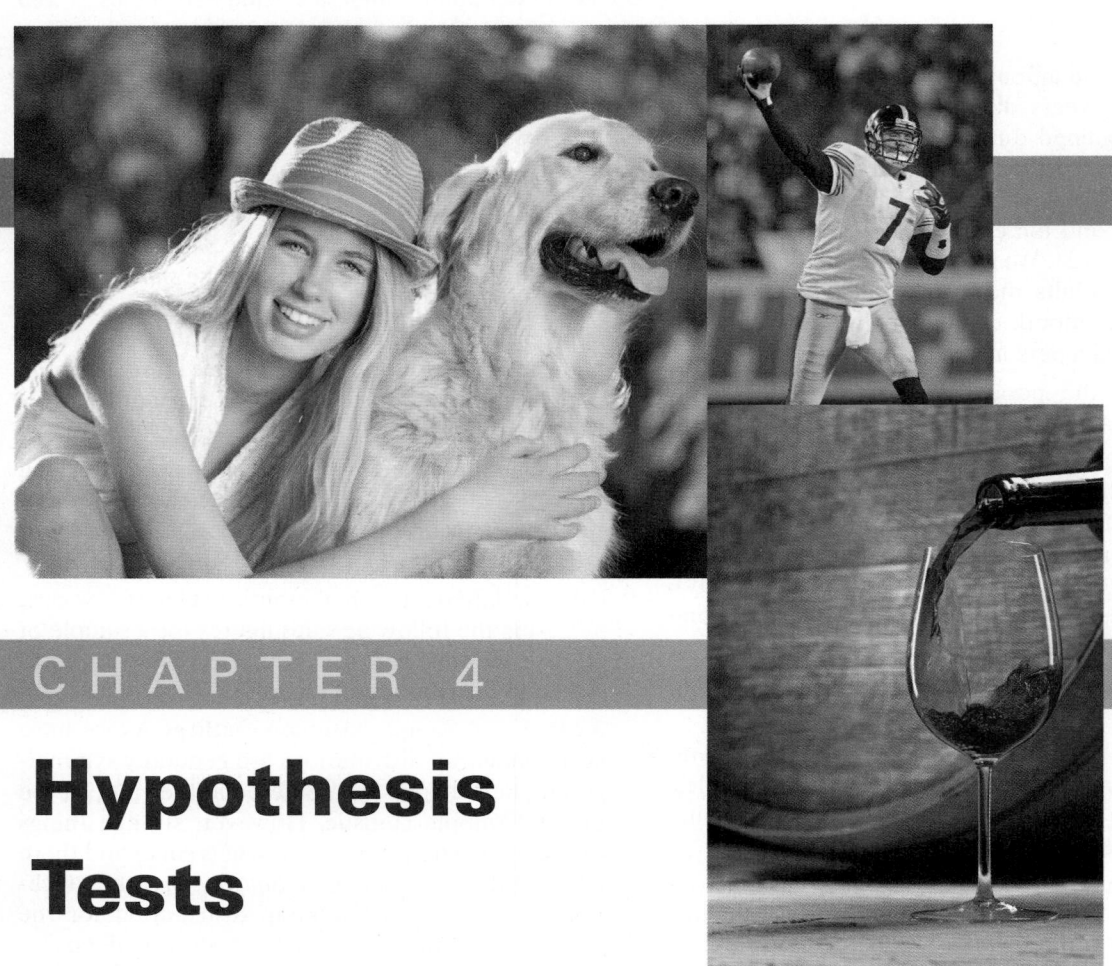

CHAPTER 4

Hypothesis Tests

"Few things mislead us more than failing to grasp simple statistical principles. Understanding what counts as evidence should trump memorizing the structural formulas for alkanes."

Sharon Begley, Science Editor for *Newsweek Magazine*[1]

[1]Begley, S., "Wanted: BS Detectors, What Science Ed Should Really Teach," *Newsweek*, November 8, 2010, p. 26.
Top left: ©Kevin Klöpper/iStockphoto, Top right: Peter G. Aiken/Getty Images, Inc., Bottom right: ©Mark Swallow/iStockphoto

Questions and Issues

Here are some of the questions and issues we will discuss in this chapter:

- Does leaving a light on at night affect weight gain?

- Does just imagining moving a muscle strengthen it?

- Do males or females play more video games?

- Does playing football affect brain size?

- Is there a relationship between owning a cat and mental illness?

- Does a high fat diet affect memory in children?

- Do sports teams wearing aggressive-looking uniforms tend to get more penalties?

- Is ADHD more likely if pesticide exposure is high?

- Are mosquitoes more attracted to someone who has been drinking beer?

- If you want to remember something, should you take a nap or have some caffeine?

- If you get called before a disciplinary panel, should you smile or maintain a serious expression?

- We know exercise is good for the body. Is it also good for the brain?

- Does the price you pay for something impact your sense of how effective it is?

- Does massage help muscles recover from exercise stress?

- Are lions more likely to attack after a full moon?

- What percent of couples say that they are "In a relationship" on Facebook?

- Do people read faster using a printed book or a Kindle or iPad?

4.1 INTRODUCING HYPOTHESIS TESTS

In statistical inference, we use data from a sample to make conclusions about a population. The two main areas of statistical inference are *estimation* and *testing*. In Chapter 3, we see how to estimate with a confidence interval. In this chapter, we see how to use a *hypothesis test* to answer questions such as:

- Does radiation from cell phones affect brain activity?
- Are mosquitoes more attracted to someone who has been drinking beer?
- If you want to remember something, should you take a nap or have some caffeine?
- Does leaving a light on at night affect weight gain?

In every case, we use data from a sample to answer a specific question about a population.

© Johannes Norpoth/iStockphoto

Does light at night make mice fatter?

DATA 4.1	**Does Light at Night Affect Weight Gain?**

Numerous studies have shown that exposure to light at night is harmful to human health. A recent study[2] examines the effect of light at night on body mass gain in mice. Eighteen mice were randomly assigned to one of two groups: the *Dark* group had a normal light/dark cycle with darkness at night and the *Light* group had a dim light on at night, equivalent to having a television set on in a room.[3] The body mass gain (*BMGain*), in grams, was recorded after three weeks, and the results are given in **LightatNight** and in Table 4.1. Do the data in this study provide convincing evidence that having a light on at night increases weight gain in mice? ∎

Table 4.1 *Body mass gain with Light or Dark at night*

Light	9.17	6.94	4.99	1.71	5.43	10.26	4.67	11.67	7.15	5.33
Dark	2.83	4.60	6.52	2.27	5.95	4.21	4.00	2.53		

[2]Fonken, L., et. al., "Light at night increases body mass by shifting time of food intake," *Proceedings of the National Academy of Sciences*, October 26, 2010; 107(43): 18664–18669.
[3]Additional results from this study are given in the datasets **LightatNight4Weeks** and **LightatNight8Weeks**. Both include three groups (with the third group having a bright light on at night), and many additional variables.

Example 4.1

In Data 4.1:

(a) Is the study an experiment or an observational study? What are the cases? What are the variables?

(b) Use appropriate summary statistics and a graph to compare the two groups.

Solution

(a) Since the mice were randomly assigned to the two groups, this is an experiment. The cases are the 18 mice. There are two variables: whether the mouse is assigned to the Light or Dark group (which is categorical) and the body mass gain for that mouse (which is quantitative).

(b) The sample means are

$$\bar{x}_L = 6.732 \quad \text{and} \quad \bar{x}_D = 4.114,$$

where $\bar{x}_L$ is mean body mass gain (in grams) for the mice with a light on at night and $\bar{x}_D$ is mean body mass gain for the mice with darkness at night. Figure 4.1 shows side-by-side dotplots.

Figure 4.1 *Body mass gain with Dark or Light at night*

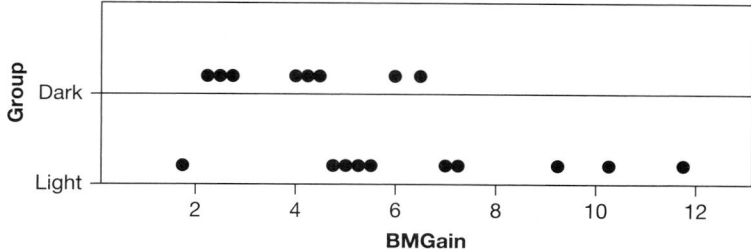

We see that, in the sample, the mice in the light group gained more weight, on average, than the mice in the dark group. The question of interest is whether this effect holds in the population. To address this question, let's think about what might happen if light condition has no effect on weight gain in mice.

We know from Chapter 3 that sample statistics vary, so we know that there will be random variation in the sample means. Even if light condition doesn't matter, we don't expect the two sample means to be exactly identical. Thus, there are two possible reasons for the difference in sample means: one is that the difference really is due to the effect of light at night and the other is that the difference is just due to random variation. Which is correct? How extreme does the difference in sample means have to be in order to argue against random chance? These are the types of questions we'll be discussing in this chapter.

In Data 4.1, we're using data from the sample ($\bar{x}_L = 6.732$ compared to $\bar{x}_D = 4.114$) to assess a claim about a population (light at night increases weight gain in mice). This is the essence of all statistical tests: using data from a sample to assess a claim about a population.

Statistical Tests

A **statistical test** is used to determine whether results from a sample are convincing enough to allow us to conclude something about the population.

Null and Alternative Hypotheses

In Chapter 3, we use data from a sample to create a confidence interval for a population parameter. In this chapter, we use data from a sample to help us decide between two competing *hypotheses* about a population. In Data 4.1, one hypothesis is that light at night really does increase weight gain in mice, and the competing hypothesis is that light condition does not affect weight gain in mice. We make these hypotheses more concrete by specifying them in terms of the population parameter(s) of interest.

In this case, we are interested in the difference in population means μ_L and μ_D, where μ_L is mean body mass gain for mice with a light on at night and μ_D is mean body mass gain for mice with darkness at night. If light condition doesn't affect weight gain, we have $\mu_L = \mu_D$. However, if a light at night does increase weight gain, we have $\mu_L > \mu_D$. Which is correct: $\mu_L = \mu_D$ or $\mu_L > \mu_D$? We use the data in the sample to try to answer this question.

We refer to the competing claims about the population as the *null hypothesis*, denoted by H_0, and the *alternative hypothesis*, denoted by H_a. The roles of these two hypotheses are *not* interchangeable. The claim for which we seek significant evidence ($\mu_L > \mu_D$ in the light at night example) is assigned to the alternative hypothesis. Usually, the null hypothesis is a claim that there really is "no effect" or "no difference." In testing whether light at night affects weight gain in mice, the hypotheses are

$$\text{Null Hypothesis } (H_0): \quad \mu_L = \mu_D$$
$$\text{Alternative Hypothesis } (H_a): \quad \mu_L > \mu_D$$

In many cases, the null hypothesis represents the status quo or that nothing interesting is happening. The alternative is usually what the experimenter or researcher wants to establish or find evidence for. A hypothesis test is designed to measure evidence against the null hypothesis, and determine whether this evidence is sufficient to conclude in favor of the alternative hypothesis.

Null and Alternative Hypotheses

Null Hypothesis (H_0): Claim that there is no effect or no difference.
Alternative Hypothesis (H_a): Claim for which we seek significant evidence.

In a hypothesis test, we examine whether sample data provide enough evidence to refute the null hypothesis and support the alternative hypothesis.

Note that the hypotheses are written in terms of the population parameters μ_L and μ_D, not in terms of the sample statistics $\bar{x}_L$ and $\bar{x}_D$. We *know* that, in the sample, mean weight gain is larger for the mice with light at night. The key question is whether the sample provides convincing evidence that mean weight gain for all mice with a light on at night is larger than mean weight gain for all mice with darkness at night.

In Data 4.1, we describe a hypothesis test comparing two means. Just as we discuss confidence intervals for any population parameter in Chapter 3, statistical tests can apply to any population parameter. (This would be a good time to review the summary of different parameters given in Table 3.5 on page 214.) In the next two examples, we consider a hypothesis test for a single proportion and a hypothesis test for a difference in proportions.

Example 4.2

Have you ever forgotten a to-go box?

A server at a restaurant recently told one of the authors that, when diners at a restaurant request a to-go box to take leftover food home, about 30% of them forget the box and leave it sitting on the table when they leave. Suppose the restaurant plans to take a random sample of 120 diners requesting a to-go box, and count how many of them forget to take their leftover food. State the null and alternative hypotheses in a test to determine whether the sample provides enough evidence to refute the server's estimate of 30%, defining any parameters used.

Solution

 This is a hypothesis test for a proportion, and the relevant parameter is p = the proportion of all diners requesting a to-go box at this restaurant who forget it when they leave. We are seeing if there is evidence that p is different from 30%, so the hypotheses are:

$$H_0: p = 0.30$$
$$H_a: p \neq 0.30$$

Note that the sample statistic, $\hat{p}$, will probably not equal 0.3 exactly, but the question is whether $\hat{p}$ will be far enough away from 0.3 to conclude that the *population parameter*, p, does not equal 0.3.

Example 4.3

Breaking the Cycle of Cocaine Addiction

Cocaine addiction is very hard to break. Even among addicts trying hard to break the addiction, relapse (going back to using cocaine) is common. Researchers try different treatment drugs to see if the relapse rate can be reduced.[4] One randomized double-blind experiment compares the proportion that relapse between those taking the drug being tested and those in a control group taking a placebo. The question of interest is whether the drug reduces the proportion of addicts that relapse. Define the parameters of interest and state the null and alternative hypotheses.

Solution

 This is a hypothesis test for a difference in proportions, and we are comparing the two proportions p_D and p_C where

p_D = the proportion of addicts taking the treatment drug who relapse, and

p_C = the proportion of addicts taking a control placebo who relapse.

The test is determining whether there is evidence that the proportion who relapse is lower in people taking the treatment drug, so the null and alternative hypotheses are:

$$H_0: p_D = p_C$$
$$H_a: p_D < p_C$$

We see again that the hypotheses are stated in terms of the relevant population parameters, not the sample statistics.

Notice that, in general, the null hypothesis is a statement of equality, while the alternative hypothesis contains a range of values, using notation indicating greater than, not equal to, or less than. This is because in a hypothesis test, we measure evidence against the null hypothesis, and it is relatively straightforward to assess evidence against a statement of equality.

---

[4]We will examine one of these studies later in this chapter.

While the null hypothesis of "no difference" is the same in each case, the alternative hypothesis depends on the question of interest. In general, the question of interest, and therefore the null and alternative hypotheses, should be determined before any data are examined.

In Data 4.1 about the effect of light at night on weight gain, we could interpret the null hypothesis as simply "Light at night has no effect on weight gain" rather than the more specific claim that the means are equal. For the sake of simplicity in this book, we will generally choose to express hypotheses in terms of parameters, even when the hypothesis is actually more general, such as "no effect."

Stating Null and Alternative Hypotheses

- H_0 and H_a are claims about *population parameters*, not sample statistics.

- In general, the null hypothesis H_0 is a statement of equality (=), while the alternative hypothesis uses notation indicating greater than (>), not equal to (≠), or less than (<), depending on the question of interest.

Evidence Against the Null Hypothesis (H_0) and in Support of the Alternative Hypothesis (H_a)

In a hypothesis test, the alternative hypothesis is the claim for which we seek evidence. The goal of the test is to determine whether the sample data provide enough evidence to refute the null hypothesis in favor of the alternative hypothesis.

Example 4.4

Example 4.2 gives null and alternative hypotheses for a test for the proportion forgetting a to-go box in a restaurant. The sample involved 120 people who requested a to-go box. In each situation below, discuss the strength of evidence against the null hypothesis and in support of the alternative hypothesis. Which provides the strongest evidence? Which provides the least evidence?

(a) 36 forget out of 120   (b) 37 forget out of 120
(c) 90 forget out of 120   (d) 27 forget out of 120

Solution 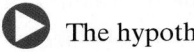 The hypotheses are

$$H_0: p = 0.30$$
$$H_a: p \neq 0.30$$

Since the sample size of $n = 120$ is the same in every case, the sample proportion furthest from 0.30 will provide the strongest evidence against H_0 and in favor of H_a.

(a) The sample proportion is $\hat{p} = 36/120 = 0.30$. This provides no evidence at all against H_0 since it is exactly 0.30.

(b) The sample proportion in this case, $\hat{p} = 37/120 = 0.308$, isn't exactly 0.30, but it is unlikely to provide convincing evidence against H_0.

(c) The sample proportion is $\hat{p} = 90/120 = 0.75$. This is very far from 0.30, and likely to provide convincing evidence against H_0 and in support of H_a.

(d) The sample proportion is $\hat{p} = 27/120 = 0.225$. The conclusion here is not obvious. We need to read the next two sections to learn how to formally determine whether such data provide convincing evidence against H_0.

The sample results in part (c) provide the strongest evidence for H_a while the sample results in part (a) provide no evidence at all for H_a.

Garrett Ellwood/Getty Images, Inc.   Tom Hauck/Getty Images, Inc.

Most and least malevolent NFL team logos

DATA 4.2

Do Teams with Malevolent Uniforms Get More Penalties?

Frank and Gilovich[5] describe a study of relationships between the type of uniforms worn by professional sports teams and the aggressiveness of the team. They consider teams from the National Football League (NFL) and National Hockey League (NHL). Participants with no knowledge of the teams rated the jerseys on characteristics such as timid/aggressive, nice/mean, and good/bad. The averages of these responses produced a "malevolence" index with higher scores signifying impressions of more malevolent (evil-looking) uniforms. To measure aggressiveness, the authors used the amount of penalties (yards for football and minutes for hockey) converted to z-scores and averaged for each team over the seasons from 1970 to 1986. The data are shown in Table 4.2 and stored in **MalevolentUniformsNFL** and **MalevolentUniformsNHL**. We are interested in whether teams with more malevolent uniforms get more penalties. ∎

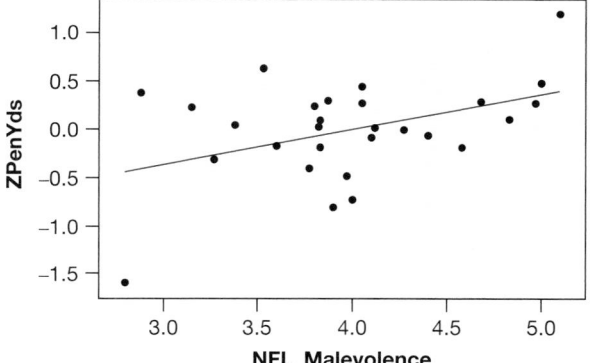

Figure 4.2 *Relationship between penalties and uniform malevolence for NFL teams*

Figure 4.2 shows a scatterplot with regression line of the malevolence ratings vs z-scores of the penalty yardage for the $n = 28$ NFL teams in this dataset. The graph shows a somewhat positive association: Teams with more malevolent uniforms tend to have more penalty yards. In fact the most penalized team (LA Raiders, now in Oakland) had the most malevolent uniform, and the least penalized team (Miami Dolphins) had the least malevolent uniform. The sample correlation between malevolence and penalties for the 28 teams is $r = 0.43$. Does this provide evidence to conclude that the true correlation is really positive?

[5]Frank, M.G., and Gilovich, T., "The Dark Side of Self- and Social Perception: Black Uniforms and Aggression in Professional Sports," *Journal of Personality and Social Psychology*, 1988; 54(1):74–85.

Table 4.2 *Malevolence rating of uniforms and z-scores for penalties*

NFLTeam	Malevolence	ZPenYds	NHLTeam	Malevolence	ZPenMin
LA Raiders	5.10	1.19	Vancouver	5.33	0.88
Pittsburgh	5.00	0.48	Philadelphia	5.17	2.01
Cincinnati	4.97	0.27	Boston	5.13	0.42
New Orleans	4.83	0.10	New Jersey	4.45	−0.78
Chicago	4.68	0.29	Pittsburgh	4.27	0.64
Kansas City	4.58	−0.19	Chicago	4.18	−0.02
Washington	4.40	−0.07	Montreal	4.18	−0.70
St. Louis	4.27	−0.01	Detroit	4.15	0.44
NY Jets	4.12	0.01	Edmonton	4.15	0.58
LA Rams	4.10	−0.09	Calgary	4.13	−0.40
Cleveland	4.05	0.44	LA Kings	4.05	−0.20
San Diego	4.05	0.27	Buffalo	4.00	−0.68
Green Bay	4.00	−0.73	Minnesota	4.00	−0.11
Philadelphia	3.97	−0.49	NY Rangers	3.90	−0.31
Minnesota	3.90	−0.81	NY Islanders	3.80	−0.35
Atlanta	3.87	0.30	Winnipeg	3.78	−0.30
Indianapolis	3.83	−0.19	St. Louis	3.75	−0.09
San Francisco	3.83	0.09	Washington	3.73	−0.07
Seattle	3.82	0.02	Toronto	3.58	0.34
Denver	3.80	0.24	Quebec	3.33	0.41
Tampa Bay	3.77	−0.41	Hartford	3.32	−0.34
New England	3.60	−0.18			
Buffalo	3.53	0.63			
Detroit	3.38	0.04			
NY Giants	3.27	−0.32			
Dallas	3.15	0.23			
Houston	2.88	0.38			
Miami	2.80	−1.60			

Example 4.5

Define the parameter of interest and state the null and alternative hypotheses.

Solution

The parameter of interest is the correlation ρ between malevolence of uniforms and number of penalty yards. We are testing to see if the correlation is positive, so the hypotheses are

$$H_0 : \rho = 0$$
$$H_a : \rho > 0$$

Even if there were no relationship between the types of jerseys and penalties for the teams, we would not expect the correlation for any sample of teams and seasons to be *exactly* zero. Once again, the key question is whether the statistic for this sample (in this case the sample correlation r) is farther away from zero than we would reasonably expect to see by random chance alone. In other words, is it unusual to see a sample correlation as high as $r = 0.43$ if the null hypothesis of $\rho = 0$ is really true?

Example 4.6

If the sample correlation of $r = 0.43$ provides convincing evidence against H_0 and in support of H_a, what does that mean in the context of malevolent uniforms?

Solution

If the sample data provide convincing evidence against H_0 and for H_a, then we have convincing evidence that the true correlation is positive, indicating that teams with more malevolent uniforms tend to be more heavily penalized. It also means that we are unlikely to get a sample correlation as high as $r = 0.43$ just by random chance if the true correlation ρ is really zero.

© Lauri Patterson/iStockphoto

How much arsenic is in this chicken?

DATA 4.3

Arsenic Levels in Chicken Meat

Arsenic-based additives in chicken feed have been banned by the European Union but are mixed in the diet of about 70% of the 9 billion broiler chickens produced annually in the US.[6] Many restaurant and supermarket chains are working to reduce the amount of arsenic in the chicken they sell. To accomplish this, one chain plans to measure, for each supplier, the amount of arsenic in a random sample of chickens. The chain will cancel its relationship with a supplier if the sample provides sufficient evidence that the average amount of arsenic in chicken provided by that supplier is greater than 80 ppb (parts per billion).   ■

Example 4.7

For the situation in Data 4.3, define the population parameter(s) and state the null and alternative hypotheses.

Solution

The parameter of interest is μ, the mean arsenic level in all chickens from a supplier. We are testing to see if the mean is greater than 80, so the hypotheses are

$$H_0: \mu = 80$$
$$H_a: \mu > 80$$

---

[6]"Arsenic in Chicken Production," *Chemical and Engineering News: Government and Policy*, 2007; 85(15):34–35.

Since we are testing to see if there is evidence that the mean is greater than 80, it is clear that the alternative hypothesis is $H_a : \mu > 80$. For the null hypothesis, writing $H_0 : \mu \leq 80$ makes intuitive sense, as any arsenic level less than 80 is satisfactory. However, it is easier to assess the extremity of our data for a single, specific value ($H_0 : \mu = 80$). This is a conservative choice; if the sample mean is large enough to provide evidence against H_0 when $\mu = 80$, it would provide even more evidence when compared to $\mu = 78$ or $\mu = 75$. Thus, for convenience, we generally choose to write the null hypothesis as an equality.

Example 4.8

Suppose the chain measures arsenic levels in chickens sampled randomly from three different suppliers, with data given in Figure 4.3.

(a) Which of the samples shows the strongest evidence for the alternative hypothesis?

(b) Which of the samples shows no evidence in support of the alternative hypothesis?

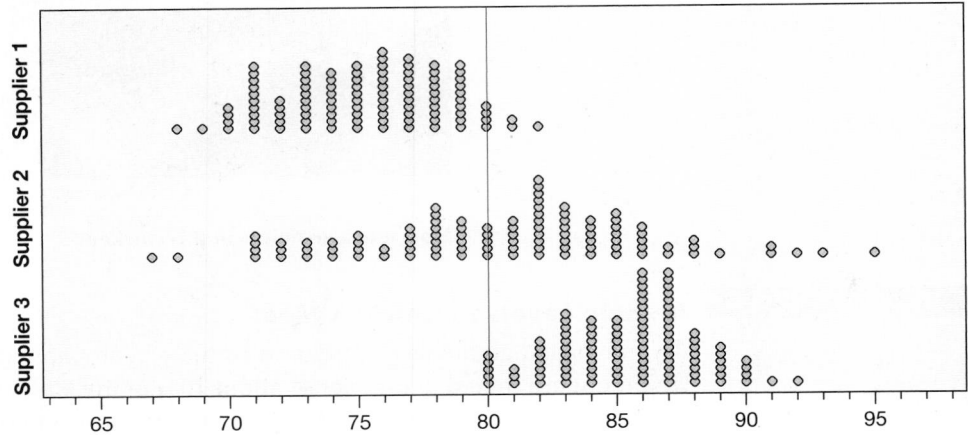

Figure 4.3 *Arsenic levels in chicken samples from three different suppliers*

Solution (a) The sample from Supplier 3 shows the strongest evidence of an average arsenic amount greater than 80, because it has the highest sample mean and all of the sampled chickens have arsenic levels at least 80.

(b) The sample from Supplier 1 shows no evidence of an average arsenic amount greater than 80, since the mean of that sample is less than 80.

In this section, we've learned that evidence for a claim about a population can be assessed using data from a sample. If the sample data are unlikely to occur just by random chance when the null hypothesis (usually "no effect") is true, then we have evidence that there is some effect and that the alternative hypothesis is true. We understand that you don't yet know how to determine what is "likely" to occur by random chance when the null hypothesis is true, and that you are probably eager to learn. That is the topic of the next section. By the end of the chapter, we'll return to the examples in this section as well as the situations described in the exercises and find out which of them provide convincing evidence for the question of interest and which do not.

SECTION LEARNING GOALS

You should now have the understanding and skills to:

▷ • Recognize when and why statistical tests are needed
▷ • Specify null and alternative hypotheses based on a question of interest, defining relevant parameters
▷ • Recognize that hypothesis tests examine whether sample data provide sufficient evidence to refute the null hypothesis and support the alternative hypothesis
▷ • Compare the strength of evidence that different samples have about the same hypotheses

Exercises for Section **4.1**

SKILL BUILDER 1

In Exercises 4.1 to 4.4, a situation is described for a statistical test and some hypothetical sample results are given. In each case:

(a) State which of the possible sample results provides the most evidence for the claim.

(b) State which (if any) of the possible results provide no evidence for the claim.

4.1 Testing to see if there is evidence that the population mean for mathematics placement exam scores is greater than 25. Use Figure 4.4.

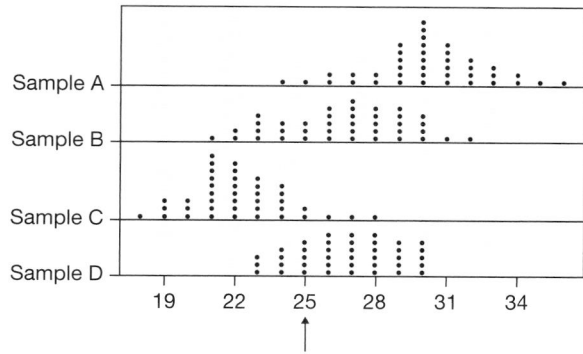

Figure 4.4 *Samples for Exercise 4.1*

4.2 Testing to see if there is evidence that the mean service time at Restaurant #1 is less than the mean service time at Restaurant #2. Use Figure 4.5 and assume that the sample sizes are all the same. Sample means are shown with circles on the boxplots.

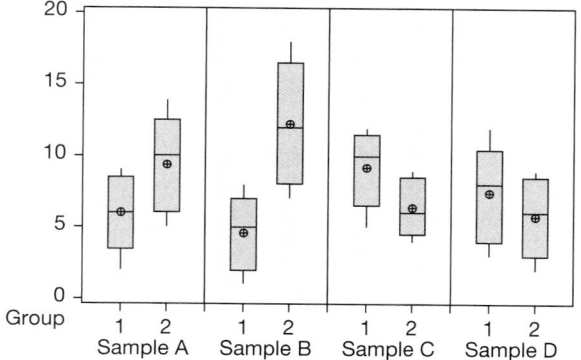

Figure 4.5 *Samples for Exercise 4.2*

4.3 Testing to see if there is evidence that the correlation between exam grades and hours playing video games is negative for a population of students. Use Figure 4.6.

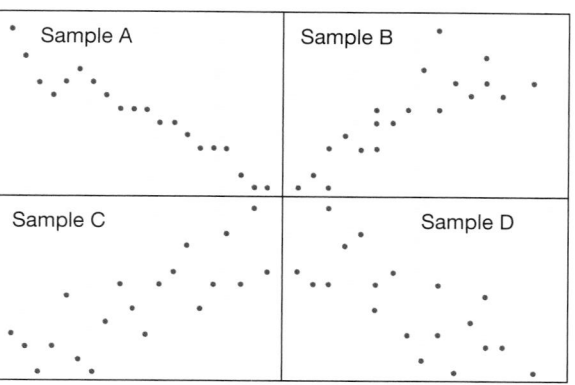

Figure 4.6 *Samples for Exercise 4.3*

4.4 Testing to see if there is evidence that the proportion of US citizens who can name the capital city of Canada is greater than 0.75. Use the following possible sample results:

Sample A:	31 successes out of 40
Sample B:	34 successes out of 40
Sample C:	27 successes out of 40
Sample D:	38 successes out of 40

SKILL BUILDER 2

In Exercises 4.5 to 4.8, state the null and alternative hypotheses for the statistical test described.

4.5 Testing to see if there is evidence that the mean of group A is not the same as the mean of group B.

4.6 Testing to see if there is evidence that a proportion is greater than 0.3.

4.7 Testing to see if there is evidence that a mean is less than 50.

4.8 Testing to see if there is evidence that the correlation between two variables is negative.

SKILL BUILDER 3

In Exercises 4.9 to 4.13, a situation is described for a statistical test. In each case, define the relevant parameter(s) and state the null and alternative hypotheses.

4.9 Testing to see if there is evidence that the proportion of people who smoke is greater for males than for females.

4.10 Testing to see if there is evidence that a correlation between height and salary is significant (that is, different than zero).

4.11 Testing to see if there is evidence that the percentage of a population who watch the Home Shopping Network is less than 20%.

4.12 Testing to see if average sales are higher in stores where customers are approached by salespeople than in stores where they aren't.

4.13 Testing to see if there is evidence that the mean time spent studying per week is different between first-year students and upperclass students.

SKILL BUILDER 4

In Exercises 4.14 and 4.15, determine whether the sets of hypotheses given are valid hypotheses.

4.14 State whether each set of hypotheses is valid for a statistical test. If not valid, explain why not.

(a) $H_0 : \mu = 15$ vs $H_a : \mu \neq 15$
(b) $H_0 : p \neq 0.5$ vs $H_a : p = 0.5$

(c) $H_0 : p_1 < p_2$ vs $H_a : p_1 > p_2$
(d) $H_0 : \bar{x}_1 = \bar{x}_2$ vs $H_a : \bar{x}_1 \neq \bar{x}_2$

4.15 State whether each set of hypotheses is valid for a statistical test. If not valid, explain why not.

(a) $H_0 : \rho = 0$ vs $H_a : \rho < 0$
(b) $H_0 : \hat{p} = 0.3$ vs $H_a : \hat{p} \neq 0.3$
(c) $H_0 : \mu_1 \neq \mu_2$ vs $H_a : \mu_1 = \mu_2$
(d) $H_0 : p = 25$ vs $H_a : p \neq 25$

4.16 Pesticides and ADHD Are children with higher exposure to pesticides more likely to develop ADHD (attention-deficit/hyperactivity disorder)? In one study, authors measured levels of urinary dialkyl phosphate (DAP, a common pesticide) concentrations and ascertained ADHD diagnostic status (Yes/No) for 1139 children who were representative of the general US population.[7] The subjects were divided into two groups based on high or low pesticide concentrations, and we compare the proportion with ADHD in each group.

(a) Define the relevant parameter(s) and state the null and alternative hypotheses.

(b) In the sample, children with high pesticide levels were more likely to be diagnosed with ADHD. Can we necessarily conclude that, in the population, children with high pesticide levels are more likely to be diagnosed with ADHD? (Whether or not we can make this generalization is, in fact, the statistical question of interest.)

(c) In the study, evidence was found to support the alternative hypothesis. Explain what that means in the context of pesticide exposure and ADHD?

4.17 Beer and Mosquitoes Does consuming beer attract mosquitoes? A study done in Burkino Faso, Africa, about the spread of malaria investigated the connection between beer consumption and mosquito attraction.[8] In the experiment, 25 volunteers consumed a liter of beer while 18 volunteers consumed a liter of water. The volunteers were assigned to the two groups randomly. The attractiveness to mosquitoes of each volunteer was tested twice: before the beer or water and after. Mosquitoes were released and caught in traps as they approached the volunteers. For the beer group, the total number of mosquitoes caught in the

[7]Bouchard, M., Bellinger, D., Wright, R., and Weisskopf, M., "Attention-Deficit/Hyperactivity Disorder and Urinary Metabolites of Organophosphate Pesticides," *Pediatrics*, 2010; 125: e1270–e1277.
[8]Lefvre, T., et al. "Beer Consumption Increases Human Attractiveness to Malaria Mosquitoes," 2010; 5(3): e9546.

traps before consumption was 434 and the total was 590 after consumption. For the water group, the total was 337 before and 345 after.

(a) Define the relevant parameter(s) and state the null and alternative hypotheses for a test to see if, after consumption, the average number of mosquitoes is higher for people who drink beer.

(b) Compute the average number of mosquitoes per volunteer before consumption for each group and compare the results. Are the two sample means different? Do you expect that this difference is just the result of random chance?

(c) Compute the average number of mosquitoes per volunteer after consumption for each group and compare the results. Are the two sample means different? Do you expect that this difference is just the result of random chance?

(d) If the difference in part (c) provides convincing evidence for the alternative hypothesis, what can we conclude about beer consumption and mosquitoes?

(e) If the difference in part (c) provides convincing evidence for the alternative hypothesis, do we have evidence that beer consumption increases mosquito attraction? Why or why not?

4.18 Guilty Verdicts in Court Cases A reporter on *cnn.com* stated in July 2010 that 95% of all court cases that go to trial result in a guilty verdict. To test the accuracy of this claim, we collect a random sample of 2000 court cases that went to trial and record the proportion that resulted in a guilty verdict.

(a) What is/are the relevant parameter(s)? What sample statistic(s) is/are used to conduct the test?

(b) State the null and alternative hypotheses.

(c) We assess evidence by considering how likely our sample results are *when H_0 is true*. What does that mean in this case?

4.19 Exercise and the Brain It is well established that exercise is beneficial for our bodies. Recent studies appear to indicate that exercise can also do wonders for our brains, or, at least, the brains of mice. In a randomized experiment, one group of mice was given access to a running wheel while a second group of mice was kept sedentary. According to an article describing the study, "The brains of mice and rats that were allowed to run on wheels

pulsed with vigorous, newly born neurons, and those animals then breezed through mazes and other tests of rodent IQ"[9] compared to the sedentary mice. Studies are examining the reasons for these beneficial effects of exercise on rodent (and perhaps human) intelligence. High levels of BMP (bone-morphogenetic protein) in the brain seem to make stem cells less active, which makes the brain slower and less nimble. Exercise seems to reduce the level of BMP in the brain. Additionally, exercise increases a brain protein called noggin, which improves the brain's ability. Indeed, large doses of noggin turned mice into "little mouse geniuses," according to Dr. Kessler, one of the lead authors of the study. While research is ongoing in determining how strong the effects are, all evidence points to the fact that exercise is good for the brain. Several tests involving these studies are described. In each case, define the relevant parameters and state the null and alternative hypotheses.

(a) Testing to see if there is evidence that mice allowed to exercise have lower levels of BMP in the brain on average than sedentary mice.

(b) Testing to see if there is evidence that mice allowed to exercise have higher levels of noggin in the brain on average than sedentary mice.

(c) Testing to see if there is evidence of a negative correlation between the level of BMP and the level of noggin in the brains of mice.

4.20 Taste Test A taste test is conducted between two brands of diet cola, Brand A and Brand B, to determine if there is evidence that more people prefer Brand A. A total of 100 people participate in the taste test.

(a) Define the relevant parameter(s) and state the null and alternative hypotheses.

(b) Give an example of possible sample results that would provide strong evidence that more people prefer Brand A. (Give your results as number choosing Brand A and number choosing Brand B.)

(c) Give an example of possible sample results that would provide no evidence to support the claim that more people prefer Brand A.

(d) Give an example of possible sample results for which the results would be inconclusive: the sample provides some evidence that Brand A is preferred but the evidence is not strong.

[9]Reynolds, G., "Phys Ed: Your Brain on Exercise," *The New York Times*, July 7, 2010.

INTENSIVE CARE UNIT (ICU) ADMISSIONS
Exercises 4.21 to 4.25 describe tests we might conduct based on Data 2.3, introduced on page 69. This dataset, stored in **ICUAdmissions**, contains information about a sample of patients admitted to a hospital Intensive Care Unit (ICU). For each of the research questions below, define any relevant parameters and state the appropriate null and alternative hypotheses.

4.21 Is there evidence that mean heart rate is higher in male ICU patients than in female ICU patients?

4.22 Is there a difference in the proportion who receive CPR based on whether the patient's race is white or black?

4.23 Is there a positive linear association between systolic blood pressure and heart rate?

4.24 Is either gender over-represented in patients in the ICU or is the gender breakdown about equal?

4.25 Is the average age of ICU patients at this hospital greater than 50?

4.26 Income East and West of the Mississippi For a random sample of households in the US, we record annual household income, whether the location is east or west of the Mississippi River, and number of children. We are interested in determining whether there is a difference in average household income between those east of the Mississippi and those west of the Mississippi.

(a) Define the relevant parameter(s) and state the null and alternative hypotheses.

(b) What statistic(s) from the sample would we use to estimate the difference?

4.27 Relationship between Income and Number of Children Exercise 4.26 discusses a sample of households in the US. We are interested in determining whether or not there is a linear relationship between household income and number of children.

(a) Define the relevant parameter(s) and state the null and alternative hypotheses.

(b) Which sample correlation shows more evidence of a relationship, $r = 0.25$ or $r = 0.75$?

(c) Which sample correlation shows more evidence of a relationship, $r = 0.50$ or $r = -0.50$?

4.28 Red Wine and Weight Loss Resveratrol, a compound in grapes and red wine, has been shown to promote weight loss in rodents and now in a primate.[10] Lemurs fed a resveratrol supplement for four weeks had decreased food intake, increased metabolic rate, and a reduction in seasonal body mass gain compared to a control group. Suppose a hypothetical study is done for a different primate species, with one group given a resveratrol supplement and the other group given a placebo. We wish to see if there is evidence that resveratrol increases the mean metabolism rate for this species. (This exercise presents hypothetical data. We will see the results from the actual study later in this chapter.)

(a) Define the relevant parameter(s) and state the null and alternative hypotheses.

(b) Possible sample results for species A are shown in Figure 4.7(a) with the mean indicated by a circle on the boxplots. In the sample, is the mean greater for the resveratrol group? Can we necessarily conclude that resveratrol increases the metabolism rate for this species?

(c) Possible sample results for species B are shown in Figure 4.7(b) and the sample sizes are the same as for species A. For which of the two species, A or B, is the evidence stronger that resveratrol increases the metabolism rate for this species? Explain your reasoning.

---

[10]BioMed Central, "Lemurs Lose Weight with 'Life-Extending' Supplement Resveratrol," *ScienceDaily*, June 22, 2010.

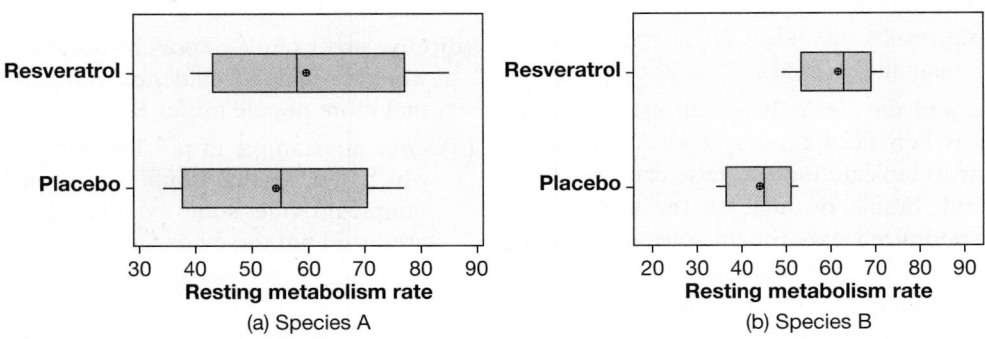

Figure 4.7 *Does red wine boost metabolism rates?*

4.29 The Lady Tasting Tea By some accounts, the first formal hypothesis test to use statistics involved the claim of a lady tasting tea.[11] In the 1920's Muriel Bristol-Roach, a British biological scientist, was at a tea party where she claimed to be able to tell whether milk was poured into a cup before or after the tea. R.A. Fisher, an eminent statistician, was also attending the party. As a natural skeptic, Fisher assumed that Muriel had no ability to distinguish whether the milk or tea was poured first, and decided to test her claim. An experiment was designed in which Muriel would be presented with some cups of tea with the milk poured first, and some cups with the tea poured first.

(a) In plain English (no symbols), describe the null and alternative hypotheses for this scenario.

(b) Let p be the true proportion of times Muriel can guess correctly. State the null and alternative hypothesis in terms of p.

4.30 Flaxseed and Omega-3 Studies have shown that omega-3 fatty acids have a wide variety of health benefits. Omega-3 oils can be found in foods such as fish, walnuts, and flaxseed. A company selling milled flaxseed advertises that one tablespoon of the product contains, on average, at least 3800 mg of ALNA, the primary omega-3.

(a) The company plans to conduct a test to ensure that there is sufficient evidence that its claim is correct. To be safe, the company wants to make sure that evidence shows the average is higher than 3800 mg. What are the null and alternative hypotheses?

(b) Suppose, instead, that a consumer organization plans to conduct a test to see if there is evidence *against* the claim that the product contains an average of 3800 mg per tablespoon. The consumer organization will only take action if it finds evidence that the claim made by the company is false and that the actual average amount of omega-3 is less than 3800 mg. What are the null and alternative hypotheses?

STATISTICAL TESTS?

In Exercises 4.31 to 4.37, indicate whether the analysis involves a statistical test. If it does involve a statistical test, state the population parameter(s) of interest and the null and alternative hypotheses.

[11]Salzburg, D. (2002) *The Lady Tasting Tea: How Statistics Revolutionized Science in the Twentieth Century*, New York: W.H. Freeman.

4.31 Polling 1000 people in a large community to determine the average number of hours a day people watch television.

4.32 Polling 1000 people in a large community to determine if there is evidence for the claim that the percentage of people in the community living in a mobile home is greater then 10%.

4.33 Utilizing the census of a community, which includes information about all residents of the community, to determine if there is evidence for the claim that the percentage of people in the community living in a mobile home is greater than 10%.

4.34 Testing 100 right-handed participants on the reaction time of their left and right hands to determine if there is evidence for the claim that the right hand reacts faster than the left.

4.35 Testing 50 people in a driving simulator to find the average reaction time to hit the brakes when an object is seen in the view ahead.

4.36 Giving a Coke/Pepsi taste test to random people in New York City to determine if there is evidence for the claim that Pepsi is preferred.

4.37 Using the complete voting records of a county to see if there is evidence that more than 50% of the eligible voters in the county voted in the last election.

4.38 Influencing Voters When getting voters to support a candidate in an election, is there a difference between a recorded phone call from the candidate or a flyer about the candidate sent through the mail? A sample of 500 voters is randomly divided into two groups of 250 each, with one group getting the phone call and one group getting the flyer. The voters are then contacted to see if they plan to vote for the candidate in question. We wish to see if there is evidence that the proportions of support are different between the two methods of campaigning.

(a) Define the relevant parameter(s) and state the null and alternative hypotheses.

(b) Possible sample results are shown in Table 4.3. Compute the two sample proportions: $\hat{p}_c$, the

Table 4.3 *Sample A: Is a phone call or a flyer more effective?*

Sample A	Will Vote for Candidate	Will Not Vote for Candidate
Phone call	152	98
Flyer	145	105

proportion of voters getting the phone call who say they will vote for the candidate, and $\hat{p}_f$, the proportion of voters getting the flyer who say they will vote for the candidate. Is there a difference in the sample proportions?

(c) A different set of possible sample results are shown in Table 4.4. Compute the same two sample proportions for this table.

Table 4.4 *Sample B: Is a phone call or a flyer more effective?*

Sample B	Will Vote for Candidate	Will Not Vote for Candidate
Phone call	188	62
Flyer	120	130

(d) Which of the two samples seems to offer stronger evidence of a difference in effectiveness between the two campaign methods? Explain your reasoning.

4.39 Influencing Voters: Is a Phone Call More Effective? Suppose, as in Exercise 4.38, that we wish to compare methods of influencing voters to support a particular candidate, but in this case we are specifically interested in testing whether a phone call is more effective than a flyer. Suppose also that our random sample consists of only 200 voters, with 100 chosen at random to get the flyer and the rest getting a phone call.

(a) State the null and alternative hypotheses in this situation.

(b) Display in a two-way table possible sample results that would offer clear evidence that the phone call is more effective.

(c) Display in a two-way table possible sample results that offer no evidence at all that the phone call is more effective.

(d) Display in a two-way table possible sample results for which the outcome is not clear: there is some evidence in the sample that the phone call is more effective but it is possibly only due to random chance and likely not strong enough to generalize to the population.

4.2 MEASURING EVIDENCE WITH P-VALUES

In Section 4.1, we see that a hypothesis test examines whether data from a sample provide enough evidence to refute the null hypothesis and support the alternative hypothesis. In this section, we learn how to measure this evidence.

Let's return to the situation described in Data 4.1 on page 258 comparing weight gain between mice with and without a light on at night. The hypotheses are:

$$H_0: \mu_L = \mu_D$$
$$H_a: \mu_L > \mu_D$$

where μ_L and μ_D represent the mean weight gain for all mice in the light and dark conditions, respectively.

The data from the experiment are stored in **LightatNight** and shown again in Table 4.5, and we see that the difference in this sample is $\bar{x}_L - \bar{x}_D = 6.732 - 4.114 = 2.618$ grams. In the sample, mice in the group with light at night gained more weight, on average, than mice with darkness at night. There are two possible explanations for this: it might be that light at night is affecting weight gain, or it might be that the difference is just due to random chance and has nothing to do with light condition. How can we determine whether there is enough evidence *against* the random chance option to be convincing that light at night matters? We need to have a sense of what is likely to occur just by random chance. Again, understanding random variation is key!

Table 4.5 *Body mass gain with Light or Dark at night*

Light	9.17	6.94	4.99	1.71	5.43	10.26	4.67	11.67	7.15	5.33	$\bar{x}_L = 6.732$
Dark	2.83	4.60	6.52	2.27	5.95	4.21	4.00	2.53			$\bar{x}_D = 4.114$

Randomization Distribution

In Chapter 3, we used a bootstrap distribution to understand how statistics randomly vary from sample to sample. Here, we want to understand how statistics randomly vary from sample to sample, *if the null hypothesis is true*. Thus, when testing hypotheses, we simulate samples in a way that is consistent with the null hypothesis. We call these *randomization samples*.

Just as we did with bootstrap samples, for each simulated randomization sample, we calculate the statistic of interest. We collect the value of the statistic for many randomization samples to generate a *randomization distribution*. This distribution approximates a sampling distribution of the statistic when the null hypothesis is true.

Randomization Distribution

Simulate many samples assuming the null hypothesis is true. Collect the values of a sample statistic for each simulated sample to create a **randomization distribution**.

The randomization distribution will be centered at the value indicated by the null hypothesis and shows what values of the sample statistic are likely to occur by random chance, if the null hypothesis is true.

Figure 4.8 shows a randomization distribution for 3000 values of $\overline{x}_L - \overline{x}_D$ generated by assuming the null hypothesis of no difference in the average weight gain, $\mu_L = \mu_D$. Note that this randomization distribution is centered near zero since, assuming H_0 is true, we have $\mu_L - \mu_D = 0$. The values in the dotplot show what we might expect to see for sample mean differences if the light and dark conditions really have no effect on weight gain in mice.

To generate randomization samples under the null hypothesis from the original mice data, we assume that a mouse's weight gain would be the same whether it was

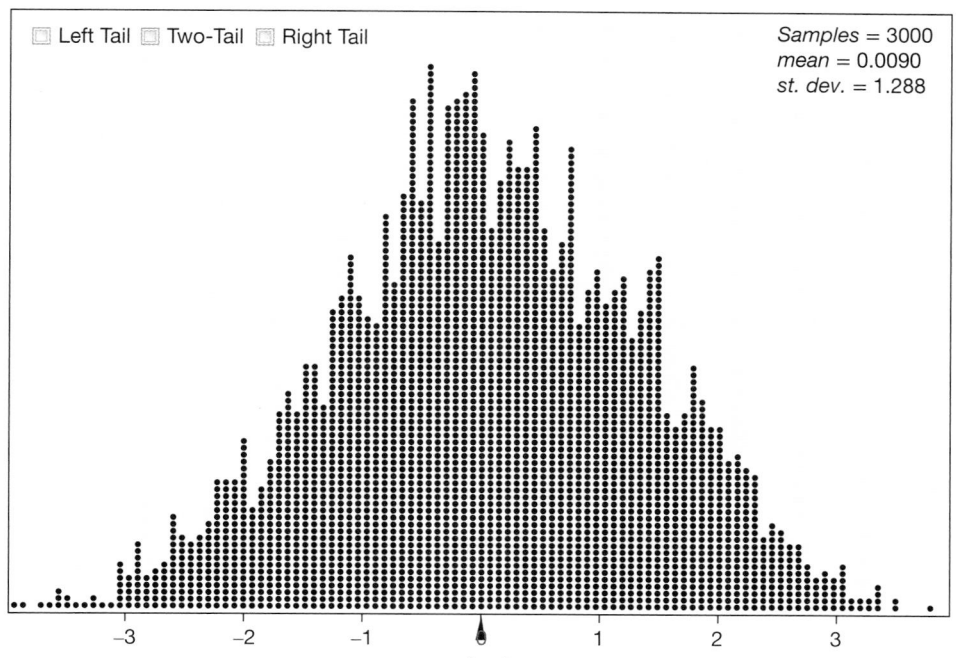

Figure 4.8 *Distribution of differences in means for 3,000 randomizations*

assigned to the Light group or the Dark group. Under the null hypothesis, any of the values in the Light group could just as easily have come from the Dark group and vice versa if a different random assignment had been made at the start of the experiment. Thus, to simulate what would happen just by random chance, we randomly reassign the 18 observed weight gains to two groups so that 10 values are put in the Light group and the remaining eight are assigned to the Dark group.

For example, the data in Table 4.6 show the same 18 body mass gains after a new random assignment into the two groups. We see that the largest body mass gain of 11.67 grams was in the Light group and got randomly assigned to stay in the Light group, but that the next two highest values, 10.26 and 9.17, were both in the Light group but got randomly assigned to the Dark group in this simulated assignment.

Table 4.6 *Simulated randomization of body mass gain to groups*

Light	5.43	7.15	2.27	1.71	6.52	6.94	4.00	11.67	2.83	5.33	$\overline{x}_L = 5.385$
Dark	5.95	10.26	4.60	4.21	2.53	9.17	4.67	4.99			$\overline{x}_D = 5.798$

The relevant sample statistic is the difference in sample means, $\overline{x}_L - \overline{x}_D$, and we compute this difference for each such randomly generated assignment. For the simulated assignment shown in Table 4.6, the difference in the sample means is $\overline{x}_L - \overline{x}_D = 5.385 - 5.798 = -0.413$. Note that this statistic, by itself, is not very meaningful. It is just the value obtained from one simulated randomization, just one possible value we might see, if the null hypothesis were true, and just one of the dots shown in the randomization distribution shown in Figure 4.8.

Imagine putting all 18 sample values on index cards, shuffling the deck and repeatedly dealing them at random into two piles, with 10 cards in the "Light" group and 8 in the "Dark" group. Each such random deal represents a different random assignment of mice to the two experimental groups and, if the null hypothesis (no effect due to light at night) is true, gives a plausible value for the difference in the two sample means. If we repeat this process many times, we obtain a randomization distribution, such as the one shown in Figure 4.8, of plausible differences that might happen by random chance if the null hypothesis is true.

There are different ways to create randomization distributions, based on the parameters involved and the way the data were collected. We discuss some of these methods in more detail in Section 4.5. In every case, however, the randomization distribution shows what statistics are likely to occur by random chance, if the null hypothesis is true. We use technology, such as the online tools at *StatKey*, to automate the creation of a randomization distribution.[12]

Example 4.9

Use the randomization distribution in Figure 4.8 to determine whether each of the following differences in sample means is likely to occur by random chance, if light at night has no effect on weight gain in mice.

$$\text{(a)} \quad \overline{x}_L - \overline{x}_D = 0.75 \qquad \text{(b)} \quad \overline{x}_L - \overline{x}_D = 4.0$$

Solution

(a) We see that a difference in means of 0.75 happened frequently in our simulated samples, so this difference in sample means is likely to occur by random chance if light at night has no effect.

(b) We see in Figure 4.8 that a difference in means as extreme as 4.0 never happened in 3000 simulated samples, so this difference in sample means is very unlikely to occur just by random chance if light at night has no effect.

[12]Supplementary materials with instructions for creating randomization distributions using various statistical software packages are available online.

We see in Example 4.9 that a difference in means of 0.75 is relatively likely to occur just by random chance, while a difference in means of 4.0 is very unlikely to occur just by random chance. What about the actual difference in means we saw in the experiment, $\overline{x}_L - \overline{x}_D = 6.732 - 4.114 = 2.618$? How likely is this sample result, if light at night has no effect? We need a more formal way to answer this question.

Measuring Strength of Evidence with a P-value

The randomization distribution gives us a way to see what kinds of statistics would occur, just by random chance, if the null hypothesis is true. The next natural step is to find the observed statistic in this randomization distribution: how unusual would our observed sample results be, if the null hypothesis were true? This leads us to one of the most important ideas of statistical inference: the *p-value*. The p-value measures how extreme sample results would be, if the null hypothesis were true.

> ### The P-value
>
> The **p-value** is the proportion of samples, when the null hypothesis is true, that would give a statistic as extreme as (or more extreme than) the observed sample.

There are various ways to calculate p-values. In this chapter we'll take an approach similar to the bootstrapping procedures of Chapter 3 and calculate the p-value by seeing where the sample statistic lies in a randomization distribution. Since the randomization distribution shows what is likely to occur by random chance if the null hypothesis is true, we find the p-value by determining what proportion of the simulated statistics are as extreme as the observed statistic.

Example 4.10

Explain, using the definition of a p-value, how we can find the p-value for the light at night experiment from the randomization distribution in Figure 4.8.

Solution The randomization distribution in Figure 4.8 shows 3000 simulated samples generated by assuming the null hypothesis is true. To find the p-value, we find the proportion of these simulated samples that have statistics as extreme as the statistic observed in the original sample, $\overline{x}_L - \overline{x}_D = 2.618$.

Figure 4.9 shows in red the simulated statistics in the randomization distribution that are at or beyond 2.618. We see that only 54 of the 3000 simulated samples have differences in means that are so large. Thus, we have

$$\text{p-value} = 54/3000 = 0.018.$$

If light at night does not affect weight gain, there is only about a 0.018 chance of getting a difference in means as extreme as the observed 2.618.

Recall that there were two possible reasons for the sample difference in means of 2.618: one possibility was that light at night really does tend to increase weight gain (H_a), and the other possibility was that the difference is just due to random variation and light at night does not affect weight gain in mice (H_0). Because a statistic this extreme is relatively unlikely to occur if the null hypothesis is true (only a 0.018 chance), we have fairly strong evidence against the null hypothesis and for the alternative hypothesis that light at night really does increase weight gain in mice.

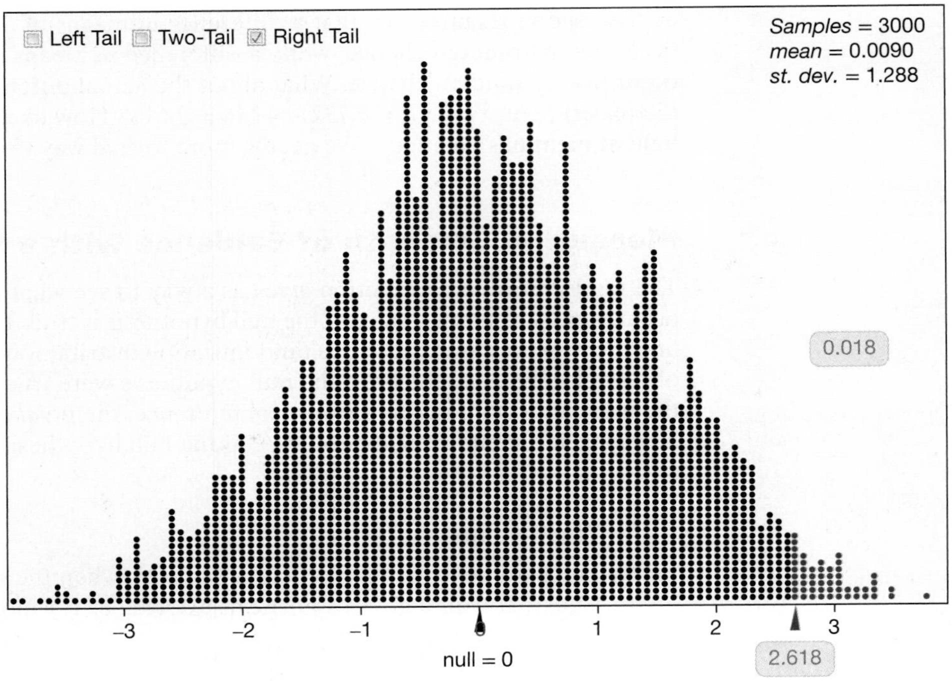

Figure 4.9 *P-value: Finding the proportion as extreme as the observed sample statistic*

Example 4.11

Use Figure 4.8 or Figure 4.9 to determine which would have a smaller p-value: a sample difference in means of $\bar{x}_L - \bar{x}_D = 2$ or a sample difference in means of $\bar{x}_L - \bar{x}_D = 3$?

Solution

We see that there are many more dots (simulated statistics) beyond 2 than beyond 3, so, because the p-value is the proportion of dots beyond the observed statistic, the p-value is greater for $\bar{x}_L - \bar{x}_D = 2$ than for $\bar{x}_L - \bar{x}_D = 3$.

P-values from Randomization Distributions

To find the p-value from a randomization distribution:

- We find the observed statistic in the randomization distribution.
- The p-value is the proportion of simulated statistics that are as extreme as the observed statistic.

Sample statistics farther out in the tail of the randomization distribution give smaller p-values.

P-values and the Alternative Hypothesis

As we have seen, we create the randomization distribution by assuming the null hypothesis is true. Does the alternative hypothesis play any role in making a randomization distribution or computing a p-value? For creating a randomization distribution, the answer is no. The randomization samples depend on the null hypothesis, but not the alternative. However, the alternative hypothesis is important in determining the p-value because it determines which tail(s) to use to calculate the p-value. In some tests, the alternative hypothesis specifies a particular direction (greater than or less than). We refer to these as *right-tailed* or *left-tailed* tests, depending on whether the alternative hypothesis is greater than or less than, respectively.[13] In other cases, we are only looking to see if there is a difference without specifying in advance in which direction it might lie. These are called *two-tailed* tests. The definition of "more extreme" when computing a p-value depends on whether the alternative hypothesis yields a test that is right-, left-, or two-tailed.

In the light at night example, the alternative hypothesis is $H_a : \mu_L > \mu_D$. We are looking to see if the sample data support this claim that $\mu_L - \mu_D > 0$, which means we are looking at the right side of the randomization distribution where values of $\bar{x}_L - \bar{x}_D$ are large. In this case, "more extreme than" means larger than, and we see in Figure 4.9 that the p-value is the proportion of simulated samples to the right of the observed statistic. This is an example of a right-tail test. The next two examples illustrate left-tail and two-tail tests.

Example 4.12

Why do students hate zeros so much?

In a student survey, 250 students were asked to write down a 4 digit random number. Only 12 of the 250 numbers ended with a zero. Does this provide evidence that students do not like zeros in the last digit (and are not very good at making up random numbers)? If the numbers were truly random, the values in the last digit would be evenly spread out between the 10 digits, so the proportion of zeros would be $p = 0.1$.

(a) State the null and alternative hypotheses.

(b) Where will the randomization distribution be centered?

(c) Is this a right-tail, left-tail, or two-tailed test?

(d) Give the value and notation for the sample statistic.

(e) Use *StatKey* or other technology to generate a randomization distribution, and find the p-value.

Solution

(a) We are looking to see if p, the proportion of zeros that students put in the last digit of random numbers, is less than the expected value of 0.1, so the hypotheses are:
$$H_0 : \quad p = 0.1$$
$$H_a : \quad p < 0.1$$

(b) The randomization distribution is created assuming the null hypothesis of $p = 0.1$ is true. Thus, the randomization distribution will be centered at 0.1.

(c) Since we are seeing if there is evidence that $p < 0.1$, we are interested in values to the left of 0.1 on the number line, so this is a left-tail test.

(d) The statistic for the original sample of students is $\hat{p} = 12/250 = 0.048$.

(e) Figure 4.10 shows one randomization distribution for 1000 simulated samples. We see that the dotplot is centered at 0.1, as expected. In this case, "more extreme than" means less than, and we look for dots (in red) to the left of the sample statistic of 0.048. In fact, there is only one dot out of 1000 simulated samples that is as extreme as the original data from the students, so we see that
$$\text{p-value} = \frac{1}{1000} = 0.001.$$

If students are just as likely to choose zeros as any other number, the chance of seeing a sample proportion as low as 0.048 for this size sample is only about 0.001.

[13] We may also refer to one-tailed tests as *upper* or *lower* tail tests, depending on which side of the randomization distribution gives evidence for H_a.

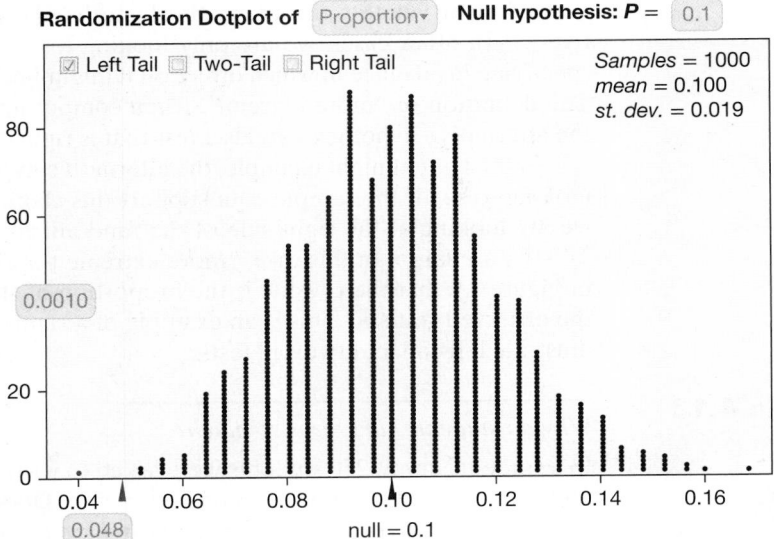

Figure 4.10 *P-value in a left-tail test*

Nebojsa Bobic/Shutterstock

Do men or women play more video games?

DATA 4.4	**Do Men or Women Play More Video Games?**

A Pew Research study[14] conducted in July 2015 asked 2001 US adults whether they play "video games on a computer, TV, gaming console, or portable device like a cell phone." In the sample, 501 of the 1000 males and 477 of the 1001 females said that they played video games. Does this provide evidence of a difference in the proportion playing video games, between males and females? ■

Example 4.13

In Data 4.4, state the null and alternative hypotheses and define any parameters used. Where will the randomization distribution be centered? Is the test a left-tail, right-tail, or two-tail test? Give notation and value for the relevant sample statistic.

[14]Duggan, M., "Gaming and Gamers," *pewresearch.org*, December 15, 2015.

Solution   This is a hypothesis test for a difference in proportions. The hypotheses are:

$$H_0: \quad p_M = p_F$$
$$H_a: \quad p_M \neq p_F$$

where p_M is the proportion of US adult males who say they play video games and p_F is the proportion of US adult females who say they play video games. Since the null hypothesis is $p_M = p_F$, the randomization distribution of differences will be centered at zero. Since the question is asking whether there is a difference (not whether one specific proportion is larger than the other), the alternative hypothesis is $p_M \neq p_F$. We are interested in extreme values on either side, so this is a two-tail test. Since $\hat{p}_M = 501/1000 = 0.501$ and $\hat{p}_F = 477/1001 = 0.477$, the relevant sample statistic is

$$\hat{p}_M - \hat{p}_F = 0.501 - 0.477 = 0.024.$$

Example 4.14

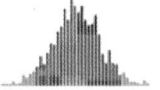

Use *StatKey* or other technology to generate a randomization distribution and find the p-value for the statistic in Example 4.13.

Solution   Figure 4.11 shows a randomization distribution for 2000 simulated samples. As expected the randomization distribution is centered at zero. The sample statistic of 0.024 is to the right of zero on the number line and we see that the proportion of simulated samples more extreme on the right is 0.141. Since this is a two-tail test, however, we care about values more extreme *on either side*, so we multiply the proportion on the right by 2:

$$\text{p-value} = 2(0.141) = 0.282.$$

If there is no difference in proportion playing video games between males and females, there is a 0.282 chance of seeing a difference this extreme for samples this size.

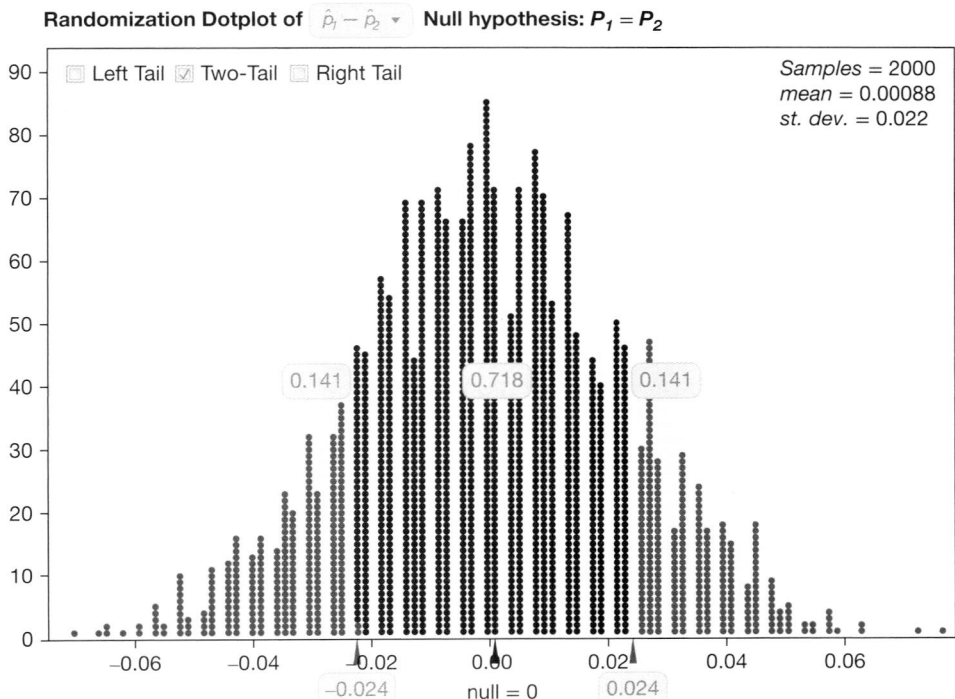

Figure 4.11 *P-value in a two-tailed test*

> ## P-value and the Alternative Hypothesis
>
> *Right-tailed test*: If H_a contains >, the p-value is the proportion of simulated statistics greater than or equal to the observed statistic, in the right tail.
>
> *Left-tailed test*: If H_a contains <, the p-value is the proportion of simulated statistics less than or equal to the observed statistic, in the left tail.
>
> *Two-tailed test*: If H_a contains ≠, we find the proportion of simulated statistics in the *smaller* tail at or beyond the observed statistic, and then to find the p-value, we double this proportion to account for the other tail.

Take care when applying this method for a two-tailed test to always use the proportion in the smaller tail. Otherwise, we might get a p-value larger than 1.0, which is impossible!

Order Matters! The order in which we subtract to find the sample statistic should match the alternative hypothesis. In the experiment on light at night, with μ_L and μ_D representing mean body mass gain for mice with light or dark at night, respectively, the alternative hypothesis is $\mu_L > \mu_D$ which is the same as $\mu_L - \mu_D > 0$. To match this alternative hypothesis, we find the difference in sample means the same way: $\overline{x}_L - \overline{x}_D = 2.618$. We could instead have used $\mu_D < \mu_L$ for the alternative hypothesis (since seeing if the mean is less for those in darkness is the same as seeing if the mean is greater for those in light) but then we would need to use $\overline{x}_D - \overline{x}_L = -2.618$ for the sample statistic. We should always be sure that the order in which we give the parameters in the alternative hypothesis matches the order in which we subtract in the observed statistic. This is particularly important in one-tailed tests.

Just as different bootstrap distributions gave slightly different confidence intervals for the same sample, different randomization distributions will give slightly different p-values. Different simulations yield slightly different counts and p-value estimates which are similar, but not identical. Our goal in constructing the randomization distribution is to get an idea of whether the sample data is unusual if the null hypothesis is true, and variation in the third digit of the p-value is generally not something to worry about. However, just as with confidence intervals, if we do care about accuracy even in the third decimal place, we can simply increase the number of simulated randomizations.

A randomization distribution allows us to see what kinds of statistics we would observe, just by random chance, if the null hypothesis is true. The p-value then measures how extreme the observed statistic would be, if the null hypothesis were true, by seeing how extreme the observed statistic is in the randomization distribution. See Figure 4.12.

The p-value allows us to assess whether the null hypothesis is plausible (the observed statistic could easily arise, just by random chance, if the null hypothesis is true), or not plausible (the observed statistic would rarely be seen, just by random chance, if the null hypothesis is true). How small does the p-value need to be for us to decide the sample provides enough evidence to reject the null hypothesis in favor of the alternative hypothesis? That is the topic we consider in the next section.

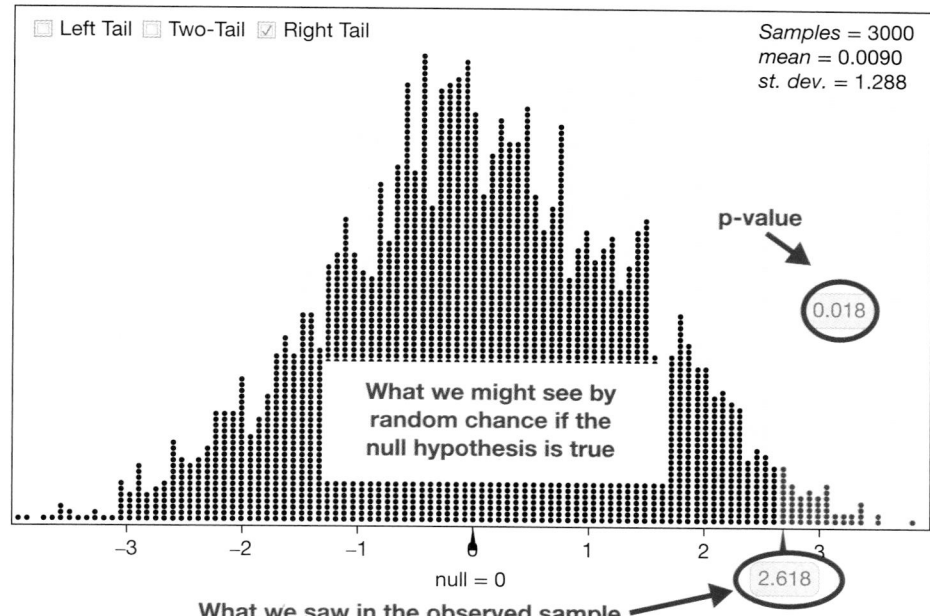

Figure 4.12 *Summary: P-value from a randomization distribution*

SECTION LEARNING GOALS

You should now have the understanding and skills to:

- ◉ • Recognize that a randomization distribution shows what is likely to happen by random chance if the null hypothesis is true
- ◉ • Use technology to create a randomization distribution
- ◉ • Interpret a p-value as the proportion of samples that would give a statistic as extreme as the observed sample, if the null hypothesis is true
- ◉ • Distinguish between one-tailed and two-tailed tests in finding p-values
- ◉ • Find a p-value from a randomization distribution

Exercises for Section 4.2

SKILL BUILDER 1

In Exercises 4.40 to 4.44, null and alternative hypotheses for a test are given. Give the notation ($\bar{x}$, for example) for a sample statistic we might record for each simulated sample to create the randomization distribution.

4.40 $H_0 : p = 0.5$ vs $H_a : p \neq 0.5$

4.41 $H_0 : \mu = 15$ vs $H_a : \mu < 15$

4.42 $H_0 : \rho = 0$ vs $H_a : \rho \neq 0$

4.43 $H_0 : \mu_1 = \mu_2$ vs $H_a : \mu_1 > \mu_2$

4.44 $H_0 : p_1 = p_2$ vs $H_a : p_1 \neq p_2$

SKILL BUILDER 2

In Exercises 4.45 to 4.49, the null and alternative hypotheses for a test are given as well as some information about the actual sample(s) and the statistic that is computed for each randomization sample. Indicate where the randomization distribution will be centered. In addition, indicate whether the test is a left-tail test, a right-tail test, or a two-tailed test.

4.45 Hypotheses: $H_0 : p = 0.5$ vs $H_a : p < 0.5$
Sample: $\hat{p} = 0.4, n = 30$
Randomization statistic = $\hat{p}$

4.46 Hypotheses: $H_0 : \mu = 10$ vs $H_a : \mu > 10$
Sample: $\bar{x} = 12, s = 3.8, n = 40$
Randomization statistic $= \bar{x}$

4.47 Hypotheses: $H_0 : \rho = 0$ vs $H_a : \rho \neq 0$
Sample: $r = -0.29, n = 50$
Randomization statistic $= r$

4.48 Hypotheses: $H_0 : \mu_1 = \mu_2$ vs $H_a : \mu_1 \neq \mu_2$
Sample: $\bar{x}_1 = 2.7$ and $\bar{x}_2 = 2.1$
Randomization statistic $= \bar{x}_1 - \bar{x}_2$

4.49 Hypotheses: $H_0 : p_1 = p_2$ vs $H_a : p_1 > p_2$
Sample: $\hat{p}_1 = 0.3, n_1 = 20$ and $\hat{p}_2 = 0.167, n_2 = 12$
Randomization statistic $= \hat{p}_1 - \hat{p}_2$

SKILL BUILDER 3

In Exercises 4.50 and 4.51, a randomization distribution is given for a hypothesis test, and shows what values of the sample statistic are likely to occur if the null hypothesis is true. Several possible values are given for a sample statistic. In each case, indicate whether seeing a sample statistic as extreme as the value given is (i) reasonably likely to occur when the null hypothesis is true, (ii) unusual but might occur occasionally when the null hypothesis is true, or (iii) extremely unlikely to ever occur when the null hypothesis is true.

4.50 Figure 4.13(a) shows a randomization distribution for a hypothesis test with $H_0 : p = 0.30$. Answer the question for these possible sample proportions:

(a) $\hat{p} = 0.1$ (b) $\hat{p} = 0.35$ (c) $\hat{p} = 0.6$

4.51 Figure 4.13(b) shows a randomization distribution for a hypothesis test with $H_0 : \mu = 300$. Answer the question for these possible sample means:

(a) $\bar{x} = 250$ (b) $\bar{x} = 305$ (c) $\bar{x} = 315$

SKILL BUILDER 4

In Exercises 4.52 to 4.54, a randomization distribution based on 1000 simulated samples is given along with the relevant null and alternative hypotheses. Which p-value most closely matches the observed statistic?

4.52 Figure 4.14 shows a randomization distribution for testing $H_0 : \mu = 50$ vs $H_a : \mu > 50$. In each case, use the distribution to decide which value is closer to the p-value for the observed sample mean.

(a) The p-value for $\bar{x} = 68$ is closest to: 0.01 or 0.25?

(b) The p-value for $\bar{x} = 54$ is closest to: 0.10 or 0.30?

(c) The p-value for $\bar{x} = 63$ is closest to: 0.05 or 0.50?

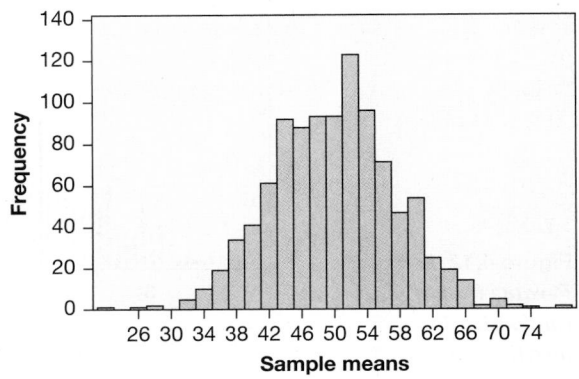

Figure 4.14 *Randomization distribution for Exercise 4.52*

4.53 Figure 4.15 shows a randomization distribution for testing $H_0 : p = 0.3$ vs $H_a : p < 0.3$. In each case, use the distribution to decide which value is closer to the p-value for the observed sample proportion.

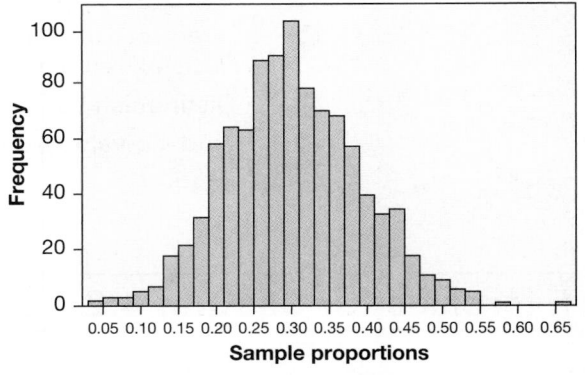

Figure 4.15 *Randomization distribution for Exercise 4.53*

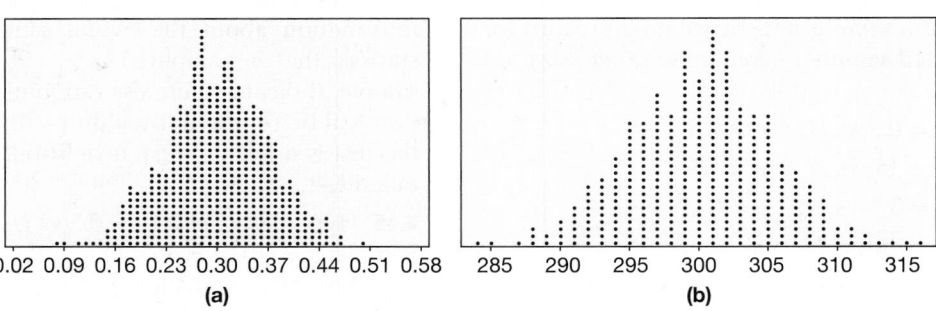

Figure 4.13 *Randomization distributions for Skill Builder 3*

(a) The p-value for $\hat{p} = 0.25$ is closest to: 0.001 or 0.30?

(b) The p-value for $\hat{p} = 0.15$ is closest to: 0.04 or 0.40?

(c) The p-value for $\hat{p} = 0.35$ is closest to: 0.30 or 0.70?

4.54 Figure 4.16 shows a randomization distribution for testing $H_0 : \mu_1 = \mu_2$ versus $H_a : \mu_1 \neq \mu_2$. The statistic used for each sample is $D = \bar{x}_1 - \bar{x}_2$. In each case, use the distribution to decide which value is closer to the p-value for the observed difference in sample means.

(a) The p-value for $D = \bar{x}_1 - \bar{x}_2 = -2.9$ is closest to: 0.01 or 0.25?

(b) The p-value for $D = \bar{x}_1 - \bar{x}_2 = 1.2$ is closest to: 0.30 or 0.60?

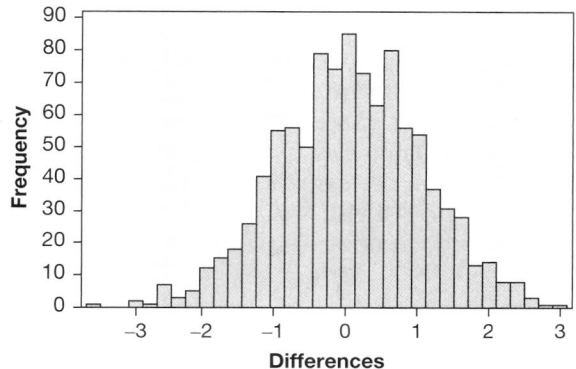

Figure 4.16 *Randomization distribution for Exercises 4.54 to 4.58*

SKILL BUILDER 5
Exercises 4.55 to 4.58 also refer to Figure 4.16, which shows a randomization distribution for hypotheses $H_0 : \mu_1 = \mu_2$ vs $H_a : \mu_1 \neq \mu_2$. The statistic used for each sample is $D = \bar{x}_1 - \bar{x}_2$. Answer parts (a) and (b) using the two possible sample results given in each exercise.

(a) For each D-value, sketch a smooth curve to roughly approximate the distribution in Figure 4.16, mark the D-value on the horizontal axis, and shade in the proportion of area corresponding to the p-value.

(b) Which sample provides the strongest evidence against H_0? Why?

4.55 $D = 2.8$ or $D = 1.3$

4.56 $D = 0.7$ or $D = -1.3$

4.57 $\bar{x}_1 = 17.3, \bar{x}_2 = 18.7$ or $\bar{x}_1 = 19.0, \bar{x}_2 = 15.4$

4.58 $\bar{x}_1 = 95.7, \bar{x}_2 = 93.5$ or $\bar{x}_1 = 94.1, \bar{x}_2 = 96.3$

SKILL BUILDER 6
Exercises 4.59 to 4.64 give null and alternative hypotheses for a population proportion, as well as sample results. Use *StatKey* or other technology to generate a randomization distribution and calculate a p-value. *StatKey* tip: Use "Test for a Single Proportion" and then "Edit Data" to enter the sample information.

4.59 Hypotheses: $H_0 : p = 0.5$ vs $H_a : p > 0.5$
Sample data: $\hat{p} = 30/50 = 0.60$ with $n = 50$

4.60 Hypotheses: $H_0 : p = 0.5$ vs $H_a : p < 0.5$
Sample data: $\hat{p} = 38/100 = 0.38$ with $n = 100$

4.61 Hypotheses: $H_0 : p = 0.7$ vs $H_a : p < 0.7$
Sample data: $\hat{p} = 125/200 = 0.625$ with $n = 200$

4.62 Hypotheses: $H_0 : p = 0.6$ vs $H_a : p > 0.6$
Sample data: $\hat{p} = 52/80 = 0.65$ with $n = 80$

4.63 Hypotheses: $H_0 : p = 0.5$ vs $H_a : p \neq 0.5$
Sample data: $\hat{p} = 42/100 = 0.42$ with $n = 100$

4.64 Hypotheses: $H_0 : p = 0.5$ vs $H_a : p \neq 0.5$
Sample data: $\hat{p} = 28/40 = 0.70$ with $n = 40$

4.65 Finger Tapping and Caffeine The effects of caffeine on the body have been extensively studied. In one experiment,[15] researchers examined whether caffeine increases the rate at which people are able to tap their fingers. Twenty students were randomly divided into two groups of 10 students each, with one group receiving caffeinated coffee and one group receiving decaffeinated coffee. The study was double-blind, and after a 2-hour period, each student was tested to measure finger tapping rate (taps per minute). The goal of the experiment was to determine whether caffeine produces an increase in the average tap rate. The finger-tapping rates measured in this experiment are stored in **CaffeineTaps**. A distribution of differences in means, $\bar{x}_c - \bar{x}_n$, for randomization samples from this experiment is given in Figure 4.17.

(a) State the null and alternative hypotheses.

(b) Sketch a smooth curve that roughly approximates the distribution in Figure 4.17 and shade in the proportion of area corresponding to the p-value for a difference in average sample tap rates of $\bar{x}_c - \bar{x}_n = 1.6$. Which of the following values is closest to the p-value: 0.60, 0.45, 0.11, or 0.03?

(c) On another sketch of the distribution in Figure 4.17, shade in the proportion of area corresponding to the p-value for a difference in

[15]Hand, A.J., Daly, F., Lund, A.D., McConway, K.J. and Ostrowski, E., *Handbook of Small Data Sets*, Chapman and Hall, London, 1994, p. 40,

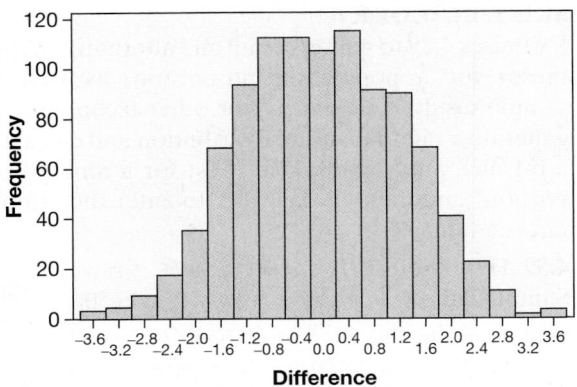

Figure 4.17 *Distribution of differences in means for 1000 randomizations when* $\mu_c = \mu_n$

average sample tap rates of $\bar{x}_c - \bar{x}_n = 2.4$. Which of the following values is closest to the p-value: 0.60, 0.45, 0.11, or 0.03?

(d) Which of the results given in parts (b) and (c) (a difference of 1.6 or a difference of 2.4) is likely to provide stronger evidence that caffeine increases average finger-tapping rate?

4.66 Influencing Voters: Is a Phone Call Better Than a Flyer? Exercise 4.39 on page 272 describes a study to investigate whether a recorded phone call is more effective than a flyer in persuading voters to vote for a particular candidate. The response variable is the proportion of voters planning to vote for the candidate, with p_c and p_f representing the proportions for the two methods (receiving a phone call and receiving a flyer, respectively). The sample statistic of interest is $\hat{p}_c - \hat{p}_f$. We are testing $H_0 : p_c = p_f$ vs $H_a : p_c > p_f$. A randomization distribution of differences in proportions, $\hat{p}_c - \hat{p}_f$, for this test is shown in Figure 4.18.

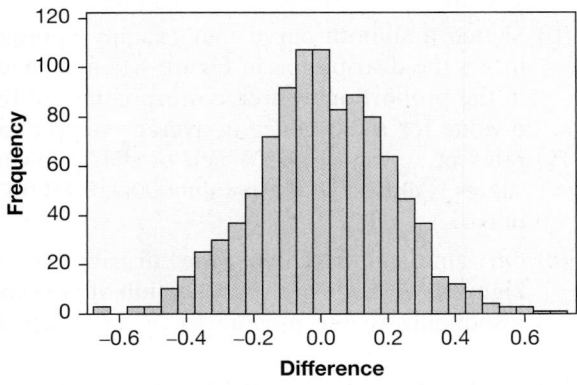

Figure 4.18 *Randomization distribution using n = 1000 for testing* $H_0 : p_c = p_f$

(a) Sketch a smooth curve that roughly approximates the distribution in Figure 4.18 and shade in the proportion of the area corresponding to the p-value for the sample statistic $\hat{p}_c - \hat{p}_f = 0.3$.

(b) Four possible sample statistics are given, along with four possible p-values. Match the statistics with the p-values.

$$\hat{p}_c - \hat{p}_f : 0.1, \ 0.3, \ 0.5, \ 0.65$$
$$\text{p-values} : 0.012, \ 0.001, \ 0.365, \ 0.085$$

(c) Interpret the p-value 0.001 in terms of the probability of the results happening by random chance.

(d) Of the four statistics given in part (b), which is likely to provide the strongest evidence that a phone call is more effective?

4.67 Influencing Voters: Is There a Difference in Effectiveness between a Phone Call and a Flyer? Exercise 4.38 on page 271 describes a study to investigate which method, a recorded phone call or a flyer, is more effective in persuading voters to vote for a particular candidate. Since in this case, the alternative hypothesis is not specified in a particular direction, the hypotheses are: $H_0 : p_c = p_f$ vs $H_a : p_c \neq p_f$. All else is as in Exercise 4.66, including the randomization distribution for $\hat{p}_c - \hat{p}_f$ shown in Figure 4.18.

(a) Sketch smooth curves that roughly approximate the distribution in Figure 4.18 and shade in the proportion of the area corresponding to the p-value for each of $\hat{p}_c - \hat{p}_f = 0.2$ and $\hat{p}_c - \hat{p}_f = -0.4$.

(b) Two possible sample statistics are given below, along with several possible p-values. Select the most accurate p-value for each sample statistic.

$$\hat{p}_c - \hat{p}_f : 0.2, -0.4$$
$$\text{p-values} : 0.008, 0.066, 0.150, 0.392, 0.842$$

(c) Of the two statistics given in part (b), which is likely to provide the strongest evidence that the methods are not equally effective?

4.68 If You Break an Arm, Use Your Brain! Multiple studies are helping us understand the power of the mind on the body. One such study[16] examines the effect of imagery on muscle strength. In the study, 29 healthy individuals underwent four weeks of wrist immobilization to induce weakness. Half

[16]Clark, B.C., et al., "The power of the mind: The cortex as a critical determinant of muscle strength/weakness," *J Neurophysiol*, 112: 3219–26, October, 2014.

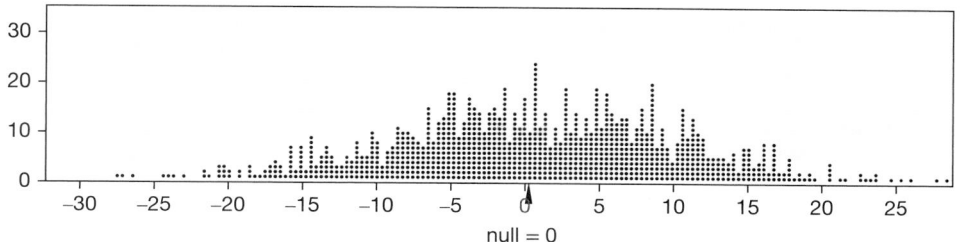

Figure 4.19 *Does imagining motion reduce muscle loss?*

of the participants were randomly assigned to mentally imagine strong muscle contractions in that arm (without actually moving the arm.) At the end of the four weeks, the percentage decrease in muscle strength was measured for all the participants, and we are testing to see if mean decrease is greater for people who do not use mental imaging when muscles are immobilized than for people who do. For the 15 participants not using any mental imagery, the mean percent decrease in muscle strength was 51.2 while for the 14 participants using imagery, it was 24.5. A randomization distribution of differences in means (mean decrease with no imagery minus mean decrease with imagery) using 1000 simulated samples is shown in Figure 4.19.

(a) State the null and alternative hypotheses.

(b) Give the notation and value for the sample statistic.

(c) Use the randomization distribution to estimate the p-value.

4.69 Colonoscopy, Anyone? A colonoscopy is a screening test for colon cancer, recommended as a routine test for adults over age 50. One study[17] provides some evidence that this test saves lives. The proportion of people with colon polyps expected to die from colon cancer is 0.01. A sample of 2602 people who had polyps removed during a colonoscopy were followed for 20 years, and 12 of them died from colon cancer. We want to assess the strength of evidence that the proportion of people who die from colon cancer after having polyps removed in a colonoscopy is less than the expected proportion (without a colonoscopy) of 0.01.

(a) What are the null and alternative hypotheses?

(b) What is the sample proportion?

(c) Figure 4.20 shows a randomization distribution of proportions for this test. Use the fact that there are 1000 dots in the distribution to find the p-value. Explain your reasoning.

[17]Zauber, et al., "Colonoscopic Polypectomy and Long-Term Prevention of Colorectal-Cancer Deaths," *New England Journal of Medicine*, 2012; 366: 687–96.

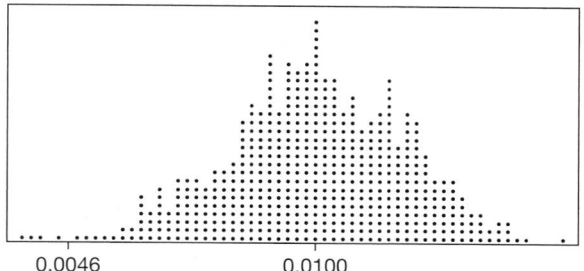

Figure 4.20 *Randomization distribution for 1000 samples testing effectiveness of colonoscopies*

4.70 Red or Pink Display of Fertility Female primates visibly display their fertile window, often with red or pink coloration. Do humans also do this? A study[18] looked at whether human females are more likely to wear red or pink during their fertile window (days 6–14 of their cycle). They collected data on 24 female undergraduates at the University of British Columbia, and asked each how many days it had been since her last period, and observed the color of her shirt. Of the 10 females in their fertile window, 4 were wearing red or pink shirts. Of the 14 females not in their fertile window, only 1 was wearing a red or pink shirt.

(a) State the null and alternative hypotheses.

(b) Calculate the relevant sample statistic, $\hat{p}_f - \hat{p}_{nf}$, for the difference in proportion wearing a pink or red shirt between the fertile and not fertile groups.

(c) For the 1000 statistics obtained from the simulated randomization samples, only 6 different values of the statistic $\hat{p}_f - \hat{p}_{nf}$ are possible. Table 4.7 shows the number of times each difference occurred among the 1000 randomizations. Calculate the p-value.

[18]Beall, A.T. and Tracy, J.L. (2013). "Women Are More Likely to Wear Red or Pink at Peak Fertility," *Psychological Science*, **24**(9): 1837–1841. doi: 10.1177/0956797613476045

Table 4.7 *Randomization distribution for difference in proportion wearing red or pink in 1000 samples*

$\hat{p}_f - \hat{p}_{nf}$	−0.357	−0.186	−0.014	0.157	0.329	0.500
Count	39	223	401	264	68	5

4.71 Football Experience and Brain Size Exercise 2.143 on page 102 introduces a study examining the relationship of football and concussions on hippocampus volume in the brain. The study included three groups with $n = 25$ in each group: heathy controls who had never played football, football players with no history of concussions, and football players with a history of concussions. In this exercise, we use the first two groups (*Control* vs *FBNoConcuss*) to test to see if there is evidence that the average brain size in people who have never played football is larger than average brain size for football players with no history of concussions. The data for this exercise are in **FootballBrain** with hippocampus brain size (in μL) stored in a variable called *Hipp*.

(a) What are the null and alternative hypotheses?

(b) Use *StatKey* or other technology to find the value of the relevant sample statistic.

(c) Use *StatKey* or other technology to find the p-value.

(d) Does it appear that this difference in brain size is just due to random chance?

4.72 Concussions and Brain Size Exercise 4.71 discusses a study comparing average brain size between three groups of subjects. In this exercise, we test for evidence that average brain size is larger in football players who have never had a concussion (*FBNoConcuss*) than in football players with a history of concussions (*FBConcuss*). The data are in **FootballBrain** where the variable *Hipp* measures brain size as the volume of the hippocampus (in μL) for each subject.

(a) What are the null and alternative hypotheses?

(b) Use *StatKey* or other technology to find the value of the relevant sample statistic.

(c) Use *StatKey* or other technology to find the p-value.

(d) Does it appear that this difference in brain size is just due to random chance?

4.73 Do You Own a Smartphone? A study[19] conducted in July 2015 examines smartphone owner-ship by US adults. A random sample of 2001 people were surveyed, and the study shows that 688 of the 989 men own a smartphone and 671 of the 1012 women own a smartphone. We want to test whether the survey results provide evidence of a difference in the proportion owning a smartphone between men and women.

(a) State the null and alternative hypotheses, and define the parameters.

(b) Give the notation and value of the sample statistic. In the sample, which group has higher smartphone ownership: men or women?

(c) Use *StatKey* or other technology to find the p-value.

4.74 Do You Own a Tablet? A study[20] conducted in June 2015 examines ownership of tablet computers by US adults. A random sample of 959 people were surveyed, and we are told that 197 of the 455 men own a tablet and 235 of the 504 women own a tablet. We want to test whether the survey results provide evidence of a difference in the proportion owning a tablet between men and women.

(a) State the null and alternative hypotheses, and define the parameters.

(b) Give the notation and value of the sample statistic. In the sample, which group has higher tablet ownership: men or women?

(c) Use *StatKey* or other technology to find the p-value.

4.75 Are Antimicrobial Ingredients Having the Opposite Effect? Exercise 3.125 on page 253 introduced a study examining the effect of triclosan on staph infections. Triclosan is found in soap and many other products and is antimicrobial, so we expect it to lower one's chance of having a staph infection. However, the opposite was found in the study. Microbiologists swabbed the noses of 90 people, and recorded which had detectable levels of triclosan and which had evidence of carrying the staph bacteria. The results are shown in Table 4.8.

Table 4.8 *Does triclosan increase staph infections?*

	Staph	No Staph	Total
Triclosan	24	13	37
No Triclosan	15	38	53
Total	39	51	90

[19]Anderson, M., "The Demographics of Device Ownership," *pewresearch.org*, October 29, 2015.

[20]Anderson, M., "The Demographics of Device Ownership," *pewresearch.org*, October 29, 2015.

(a) What proportion of those with triclosan have the staph bacteria? What proportion of those without triclosan have the staph bacteria? What is the sample difference in proportions?

(b) We wish to test to see if the effect found in the sample generalizes to the population. In testing to see if there is evidence that people with triclosan are more likely to have staph bacteria, what are the null and alternative hypotheses?

(c) Use *StatKey* or other technology to find the p-value for the test.

4.76 Testing the Lady Tasting Tea Exercise 4.29 on page 271 introduces a historical scenario in which a British woman, Muriel Bristol-Roach, claimed to be able to tell whether milk had been poured into a cup before or after the tea. An experiment was conducted in which Muriel was presented with 8 cups of tea, and for each cup she correctly guessed whether the milk was added first or the tea added first. Let's assume that Muriel did not know beforehand how many of the 8 cups had tea first and how many had milk first. Let p represent the true proportion of times Muriel can guess correctly. Our hypotheses are $H_0 : p = 0.5$ (random guessing) and $H_a : p > 0.5$.

(a) Give the value and notation for the sample statistic.

(b) Use *StatKey* or other technology to generate a randomization distribution, and find the p-value.

4.77 Definition of a P-value Using the definition of a p-value, explain why the area in the tail of a randomization distribution is used to compute a p-value.

4.78 Rolling Dice You roll a die 60 times and record the sample proportion of 5's, and you want to test whether the die is biased to give more 5's than a fair die would ordinarily give. To find the p-value for your sample data, you create a randomization distribution of proportions of 5's in many simulated samples of size 60 with a fair die.

(a) State the null and alternative hypotheses.

(b) Where will the center of the distribution be? Why?

(c) Give an example of a sample proportion for which the number of 5's obtained is *less* than what you would expect in a fair die.

(d) Will your answer to part (c) lie on the left or the right of the center of the randomization distribution?

(e) To find the p-value for your answer to part (c), would you look at the left, right, or both tails?

(f) For your answer in part (c), can you say anything about the size of the p-value?

4.79 What Is Your Lucky Number? Thirty students are asked to choose a random number between 0 and 9, inclusive, to create a data set of $n = 30$ digits. If the numbers are truly random, we would expect about three 0's, three 1's, three 2's, and so on. If the dataset includes eight 7's, how unusual is that? If we look exclusively at the number of 7's, we expect the proportion of 7's to be 0.1 (since there are 10 possible numbers) and the number of 7's to be 3 in a sample of size 30. We are testing $H_0 : p = 0.1$ vs $H_a : p \neq 0.1$, where p is the proportion of 7's. We can generate the randomization distribution by generating 1000 sets of 30 random digits and recording $X =$ the number of 7's in each simulated sample. See Figure 4.21.

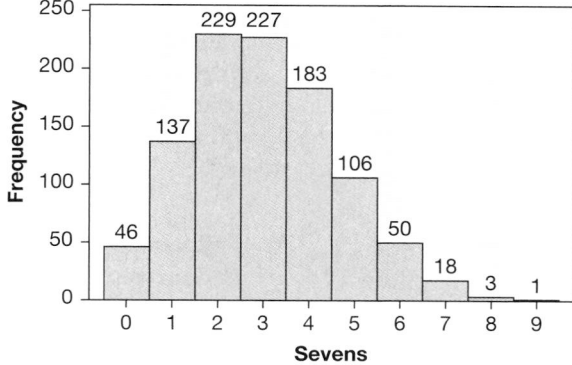

Figure 4.21 *Randomization distribution for 1000 samples of number of 7's in 30 digits when $H_0 : p = 0.1$*

(a) Notice that this randomization distribution is *not* symmetric. This is a two-tailed test, so we need to consider both "tails." How far is $X = 8$ from the expected value of 3? What number would be equally far out on the other side? Explain why it is better in this situation to double the observed one-tailed p-value rather than to add the exact values on both sides.

(b) What is the p-value for the observed statistic of $X = 8$ sevens when doing the two-tailed test?

(c) The randomization distribution in Figure 4.21 would apply to any digit (not just 7's) if the null hypothesis is $H_0 : p = 0.1$. Suppose we want to test if students tend to avoid choosing zero when picking a random digit. If we now let p be the proportion of 0's all students choose, the alternative would be $H_a : p < 0.1$. What is the smallest p-value we could get using the randomization distribution in Figure 4.21? What would have to happen in the sample of digits from 30 students for this p-value to occur?

4.3 DETERMINING STATISTICAL SIGNIFICANCE

In Section 4.1, we see that a hypothesis test involves determining whether sample data provide enough evidence to refute a null hypothesis and support an alternative hypothesis. In Section 4.2, we see that a p-value measures how unlikely a sample statistic is, if the null hypothesis is true. In this section, we connect these two ideas.

Example 4.15

Do more than 30% of restaurant patrons forget a to-go box?

In Example 4.2 on page 261, we examine a claim that about 30% of people in a restaurant who request a to-go box forget it when they leave. If we want to see if there is evidence that the true proportion is larger than 30%, then the null and alternative hypotheses are

$$H_0 : \quad p = 0.30$$
$$H_a : \quad p > 0.30$$

where p is the proportion of diners at this restaurant who forget their to-go box. For each sample proportion, find the p-value and interpret it in terms of evidence for the alternative hypothesis. Which p-value provides stronger evidence for the alternative hypothesis?

(a) $\hat{p} = \frac{39}{120} = 0.325$ (b) $\hat{p} = \frac{51}{120} = 0.425$

Solution

In both cases, the sample proportion is greater than 0.3. Is it greater just because of random variation or because the true population proportion really is greater than 0.3?

(a) Figure 4.22(a) shows a randomization distribution for this test and we see that the sample proportion $\hat{p} = 0.325$ is in a likely part of the distribution, with a p-value of 0.283. Lots of the simulated samples were more extreme than this one, and we see that this sample result is not very unusual just by random chance if the proportion is really 0.3. This sample result doesn't provide much evidence to refute the null hypothesis.

(b) Figure 4.22(b) shows the same randomization distribution and we see that the sample proportion $\hat{p} = 0.425$ is way out in the tail, with a p-value of 0.002. This sample proportion is very unlikely to happen just by random chance if the true proportion is really 0.3. This sample appears to provide evidence that the true proportion is greater than 0.3. We have evidence to refute the null hypothesis and support the alternative hypothesis!

The p-value of 0.002 provides stronger evidence for the alternative hypothesis than the p-value of 0.283.

Interpreting a P-value as Strength of Evidence

The smaller the p-value, the stronger the statistical evidence is against the null hypothesis and in support of the alternative hypothesis.

Statistical Significance

If the p-value is small enough, the sample results are more extreme than we would reasonably expect to see by random chance if the null hypothesis were true. If this is the case, we say the data are *statistically significant*. Statistically significant data

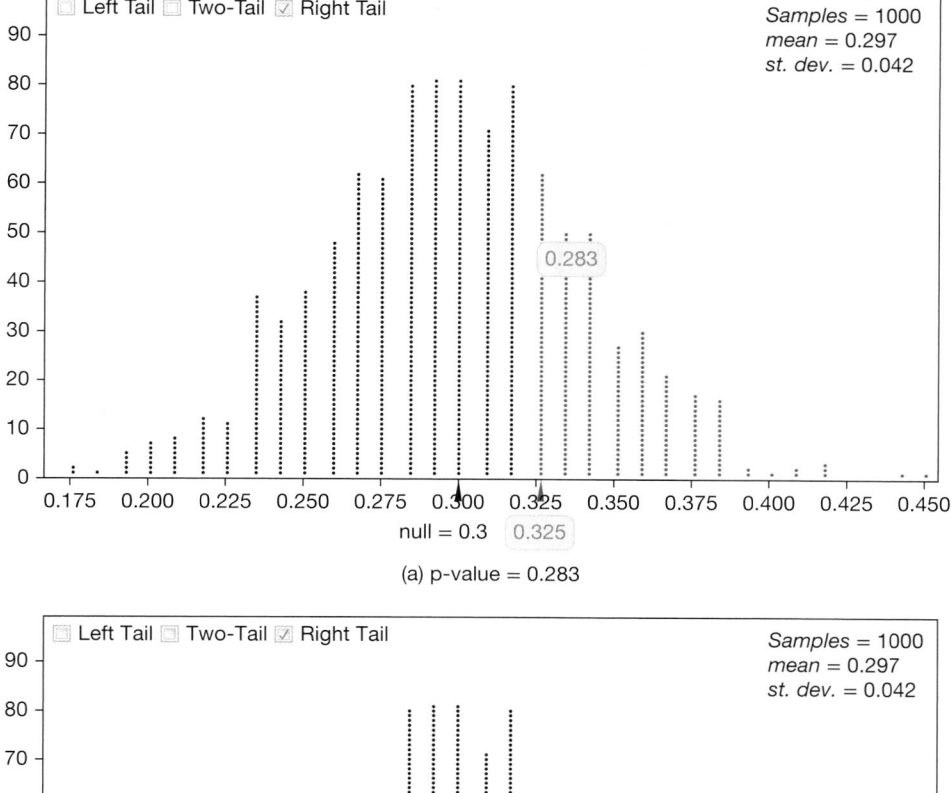

Figure 4.22 *Which p-value provides more evidence for the alternative hypothesis?*

provide convincing evidence against the null hypothesis in favor of the alternative, and allow us to use our sample results to support that claim about the population.

Statistical Significance

If the p-value is small enough, then results as extreme as the observed sample statistic are unlikely to occur by random chance alone (assuming the null hypothesis is true), and we say the sample results are **statistically significant**.

If our sample is statistically significant, we have convincing evidence against H_0 and in favor of H_a.

Example 4.16

In Data 4.3 on page 265, a company is testing whether chicken meat from a supplier has an average arsenic level higher than 80 ppb. The hypotheses are

$$H_0 : \quad \mu = 80$$
$$H_a : \quad \mu > 80$$

where μ is the mean arsenic level in chicken from this supplier.

(a) If sample results are statistically significant, what can the company conclude?

(b) If sample results are not statistically significant, what can the company conclude?

Solution

(a) If results are statistically significant, the company has found evidence that the average level of arsenic in chickens from that supplier is greater than 80, and the company should stop buying chicken from that supplier.

(b) If results are not statistically significant, the company cannot conclude anything specific about the average level of arsenic in chickens from that supplier. The company would not have sufficient evidence to cancel its relationship with the supplier, since the mean arsenic level may or may not be greater than 80 ppb.

You're probably wondering, *how small does a p-value have to be for us to achieve statistical significance?* If we agree that a p-value of 0.0001 is clearly statistically significant and a p-value of 0.50 is not, there must be some point between 0.0001 and 0.50 where we cross the threshold between statistical significance and random chance. That point, measuring when something becomes rare enough to be called "unusual," might vary a lot from person to person. We should agree in advance on a reasonable cutoff point. Statisticians call this cutoff point the *significance level* of a test and usually denote it with the Greek letter α (alpha). For example, if $\alpha = 0.05$ we say we are doing a 5% test and will call the results statistically significant if the p-value for the sample is smaller than 0.05. Often, short hand notation such as $P < 0.05$ is used to indicate that the p-value is less than 0.05, which means the results are significant at a 5% level.

> **Significance Level**
>
> The **significance level**, α, for a test of hypotheses is a boundary below which we conclude that a p-value shows statistically significant evidence against the null hypothesis.
>
> Common significance levels are $\alpha = 0.05$, $\alpha = 0.01$, or $\alpha = 0.10$. If a significance level is not specified, we use $\alpha = 0.05$.

Statistical Decisions

A small p-value means that the sample result is unlikely to occur by random chance alone and provides evidence against the null hypothesis, H_0, in favor of the alternative, H_a. If the evidence is strong enough against the null hypothesis, we can *reject the null hypothesis* in favor of the alternative. On the other hand, if the data are reasonably likely to occur when the null hypothesis is true, we *do not reject the null hypothesis*. Given a specific significance level, α, the formal decision in a statistical test, based on comparing the p-value from a sample to α, is very straightforward.

Formal Statistical Decisions

Given a significance level α and the p-value from a sample:

If the p-value $< \alpha$:   *Reject H_0*
This means the results are significant and we have convincing evidence that H_a is true.

If the p-value $\geq \alpha$:   *Do Not Reject H_0*
This means the results are not significant and we do not have convincing evidence that H_a is true.

Example 4.17

Students hate zeros!

In Example 4.12 on page 277, we are testing $H_0 : p = 0.1$ vs $H_a : p < 0.1$ where p represents the proportion of zeros that students put in the last digit of "random" numbers. We see in that example that the p-value is 0.001. If the significance level is 0.05, what decision do we make? Does the decision change if we use $\alpha = 0.10$ or $\alpha = 0.01$?

Solution

Since the p-value is less than a significance level of 0.05, we reject H_0 and find evidence that the proportion of zeros that students put in the last digit of their random numbers is less than the expected value of 0.1 for true random numbers. Since the p-value is also less than both 0.10 and 0.01, the decision and interpretation are the same for those significance levels. Students are not very good at generating their own random numbers!

In a formal hypothesis test, after making the decision of whether to reject or not reject the null hypothesis, we always follow up with a conclusion in context, stated in terms of the alternative hypothesis and referring back to the question of interest.

Example 4.18

Light at Night: The Conclusion!

In Data 4.1, we are testing to see if mean body mass gain for mice with a light on at night, μ_L, is greater than mean body mass gain for mice with darkness at night, μ_D, with hypotheses $H_0 : \mu_L = \mu_D$ vs $H_a : \mu_L > \mu_D$. In Example 4.10 on page 275, we see that the p-value for the test is

$$\text{p-value} = 0.018.$$

Using a 5% significance level,

(a) Are the results statistically significant?

(b) Give the formal decision of the test.

(c) State the conclusion in context.

(d) Can we conclude that having a light on at night *causes* increased weight gain in mice?

Solution

(a) Yes, the results are statistically significant because the p-value of 0.018 is less than the significance level of $\alpha = 0.05$.

(b) Since the p-value of 0.018 is less than the significance level of 0.05, the formal conclusion is to "Reject H_0."

(c) Since we reject H_0, we have evidence for the alternative hypothesis, $\mu_L > \mu_D$. We have convincing evidence that average weight gain is greater for mice with a light on at night than for mice with darkness at night.

(d) Yes. The results are statistically significant *and* the data come from an experiment, so we can conclude that there is causation.

Example 4.19

Who plays more video games?

In Example 4.14 on page 279, we are testing to see if there is a difference in the proportion who play video games, between males and females, using $H_a: p_M \neq p_F$. The p-value of the test is 0.282. Give the generic decision and the conclusion in context.

Solution   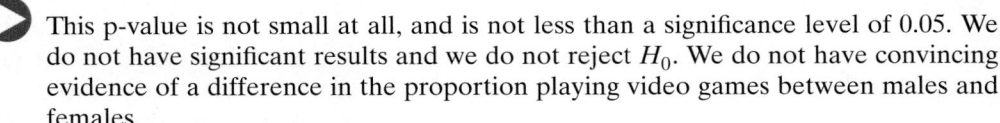 This p-value is not small at all, and is not less than a significance level of 0.05. We do not have significant results and we do not reject H_0. We do not have convincing evidence of a difference in the proportion playing video games between males and females.

Notice that the formal decision is generally made in terms of whether or not we reject the null hypothesis: Reject H_0 or do not reject H_0. If the data are significant, we reject H_0. If the data are not significant, we do not reject H_0. When the sample is not significant, as in Example 4.19, we do not say that we "accept H_0." Finding a lack of convincing evidence against the null hypothesis should never be confused with finding strong evidence *for* the null hypothesis. In fact, in a hypothesis test, we can find evidence for the alternative hypothesis, but the conclusion is never that we have found evidence for the null hypothesis. The next example illustrates this point.

Example 4.20

Walking Elephants

Suppose that we have a mystery animal named X and consider the hypotheses

$$H_0: \quad \text{X is an elephant}$$
$$H_a: \quad \text{X is not an elephant}$$

What conclusion would you draw from each of the following pieces of evidence?

(a) X has four legs.

(b) X walks on two legs.

Solution   (a) It is not at all unusual for an elephant to have four legs, so that evidence would certainly not lead to rejecting this null hypothesis. However, we do not "Accept H_0" and we do not conclude that X *must* be an elephant or even that we have strong evidence X is an elephant. Rather we say that the data do not provide significant evidence against H_0 and we cannot determine whether X is or is not an elephant.

(b) While it is not impossible for an elephant to walk on two legs (for example, you might think of trained circus elephants), it is certainly very uncommon. So "walking on two legs" would be sufficient evidence to reject H_0 and conclude X is probably not an elephant.

This reinforces the distinction between the null hypothesis and the alternative hypothesis as described in Section 4.1: a hypothesis test is designed only to find evidence *for* the alternative hypothesis (usually the claim for which you seek evidence) and *against* the null hypothesis.

Michael Edwards/Stone/Getty Images, Inc.

An elephant standing on two legs

Hypothesis Tests

A formal hypothesis test includes the following components:

- State the null and alternative hypotheses (defining parameters when necessary)
- Determine the value of the observed sample statistic
- Find the p-value
- Make a generic decision about H_0: Reject H_0 or do not reject H_0
- Write a sentence explaining the conclusion of the test in context, indicating whether or not we have convincing evidence for H_a and referring back to the question of interest.

DATA 4.5

Smiles and Leniency

Can a simple smile have an effect on punishment assigned following an infraction? Hecht and LeFrance conducted a study examining the effect of a smile on the leniency of disciplinary action for wrongdoers.[21] Participants in the experiment took on the role of members of a college disciplinary panel judging students accused of cheating. For each suspect, along with a description of the offense, a picture was provided with either a smile or neutral facial expression. A leniency score was calculated based on the disciplinary decisions made by the participants. The full data can be found in **Smiles**. The experimenters are testing to see if the average leniency score is different for smiling students than it is for students with a neutral facial expression.   ■

[21]LeFrance, M., and Hecht, M.A., "Why Smiles Generate Leniency," *Personality and Social Psychology Bulletin*, 1995; 21:207–214.

© Cameron Whitman/iStockphoto

A neutral expression and a smiling expression: Which student gets the harsher punishment?

Example 4.21

In the experiment described in Data 4.5, there were 34 participants in each group who made disciplinary decisions that were interpreted to give a leniency score (on a 10 point scale) for each case. The mean leniency score for the smiling group was 4.91 while the mean leniency score for the neutral group was 4.12. Conduct a hypothesis test to see whether smiling has an effect on leniency. Use a 5% significance level.

Solution

We are comparing two means in this test, so the relevant parameters are μ_s, the true mean score for smiling students, and μ_n, the true mean score for students with a neutral expression. We are testing to see if there is evidence of a difference in average leniency score, so the hypotheses are:

$$H_0: \quad \mu_s = \mu_n$$
$$H_a: \quad \mu_s \neq \mu_n$$

We are told that $\bar{x}_s = 4.91$ and $\bar{x}_n = 4.12$, so the observed difference in sample means is

$$\bar{x}_s - \bar{x}_n = 4.91 - 4.12 = 0.79.$$

The randomization distribution in Figure 4.23 shows the results of the differences in sample means for 1000 simulations where the 34 "smile" and "neutral" labels were

Figure 4.23
Randomization distribution of differences in leniency means, $\bar{x}_s - \bar{x}_n$

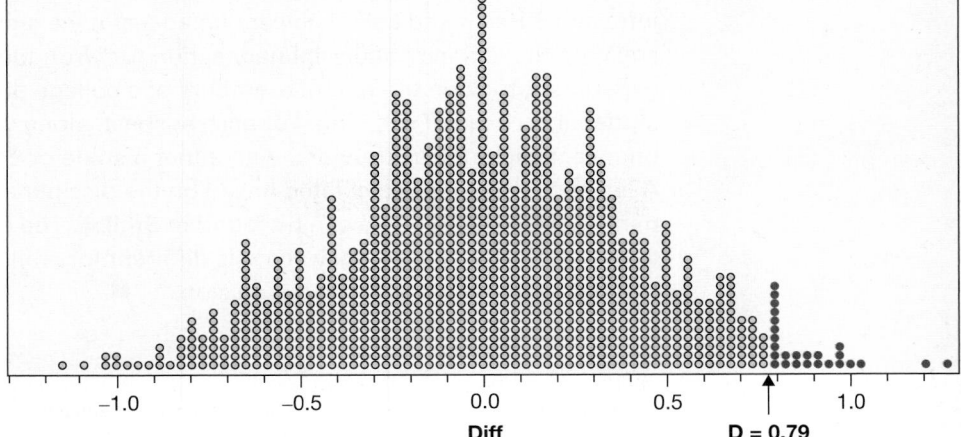

randomly assigned to the 68 leniency scores. There are 23 values in the upper tail of the 1000 simulations that are larger than the observed sample difference of 0.79. Since we are doing a two-tailed test, we have

$$\text{p-value} = 2 \cdot \frac{23}{1000} = 0.046.$$

This p-value is less than the significance level of $\alpha = 0.05$ so the formal decision is to reject H_0. We have found evidence that smiling makes a difference and we expect more leniency, on average, to be awarded to smiling suspects. If you go before a disciplinary panel, you should smile!

Example 4.22

If we change the score for just one of the participants in the smiling and leniency experiment by a single point, either less lenient for someone in the smile group or more lenient for someone in the neutral group, the difference in means becomes $\bar{x}_s - \bar{x}_n = 0.76$ and four *new* points in the randomization distribution would exceed this difference. Find the p-value for this new difference in means, and use a 5% significance level to give the conclusion of the test.

Solution

The randomization distribution has $23 + 4 = 27$ cases above $\bar{x}_s - \bar{x}_n = 0.76$ which produces a p-value of $2 \cdot 27/1000 = 0.054$. This p-value is not less than 5% so we do not reject H_0 and thus conclude that we do not have sufficient evidence to show that smiling makes a difference in the amount of leniency. If you go before a disciplinary panel, it may not matter whether you smile or maintain a neutral expression.

Less Formal Statistical Decisions

Notice in Example 4.22 that changing just one person's score by a single point dramatically changes the conclusion of the test. One of the drawbacks of the classical approach to hypothesis testing is that it forces us to make very black-white decisions. The formal decision is either "Reject H_0" or "Do not reject H_0" depending on whether or not the p-value is less than the desired significance level. This scenario for a 5% test is illustrated in Figure 4.24(a).

In some situations we might feel more comfortable with a less prescriptive decision. We might be "pretty sure" that H_0 should be rejected or find some, but not entirely convincing, evidence against it. For this reason we sometimes interpret a p-value less formally by merely indicating the strength of evidence it shows against the null hypothesis. For example, the p-values of 0.046 and 0.054 in Examples 4.21 and 4.22 might both be interpreted as showing moderate but not very strong evidence that smiling helps increase leniency.

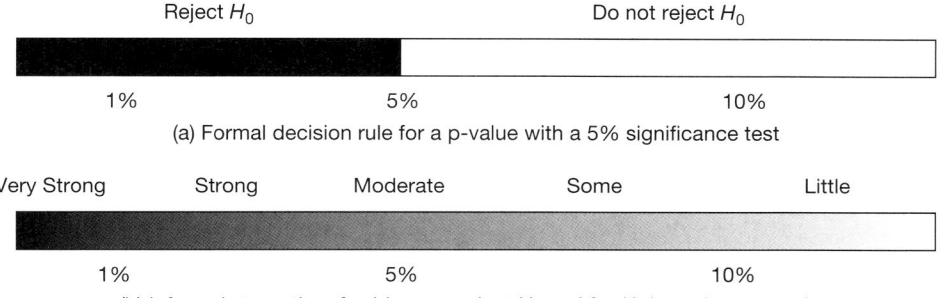

Figure 4.24 *Formal vs Informal statistical decisions*

(a) Formal decision rule for a p-value with a 5% significance test

(b) Informal strengths of evidence against H_0 and for H_a based on a p-value

Figure 4.24(b) gives a schematic representation of a less formal way to interpret p-values as strength of evidence against a null hypothesis. Contrast this with the formal decision rule shown in Figure 4.24(a). Which way is right? They both have their merits. As we continue studying significance testing, keep both approaches in mind so that you can make a concrete decision for a given significance level but also interpret any p-value as a measure of strength of evidence.

SECTION LEARNING GOALS

You should now have the understanding and skills to:

- Recognize that smaller p-values give stronger evidence in support of the alternative hypothesis
- Demonstrate an understanding of the concept of statistical significance
- Make a formal decision in a hypothesis test by comparing a p-value to a given significance level
- State the conclusion to a hypothesis test in context
- Make a less formal decision that reflects the strength of evidence in a p-value
- Conduct a hypothesis test for a variety of situations

Exercises for Section 4.3

SKILL BUILDER 1
In Exercises 4.80 to 4.83, two p-values are given. Which one provides the strongest evidence against H_0?

4.80 p-value = 0.90 or p-value = 0.08

4.81 p-value = 0.04 or p-value = 0.62

4.82 p-value = 0.007 or p-value = 0.13

4.83 p-value = 0.02 or p-value = 0.0008

SKILL BUILDER 2
Exercises 4.84 to 4.87 give a p-value. State the conclusion of the test based on this p-value in terms of "Reject H_0" or "Do not reject H_0", if we use a 5% significance level.

4.84 p-value = 0.0007

4.85 p-value = 0.0320

4.86 p-value = 0.2531

4.87 p-value = 0.1145

SKILL BUILDER 3
In Exercises 4.88 to 4.91, using the p-value given, are the results significant at a 10% level? At a 5% level? At a 1% level?

4.88 p-value = 0.0320

4.89 p-value = 0.2800

4.90 p-value = 0.008

4.91 p-value = 0.0621

SKILL BUILDER 4
In Exercises 4.92 and 4.93, match the four p-values with the appropriate conclusion:

(a) The evidence against the null hypothesis is significant, but only at the 10% level.

(b) The evidence against the null and in favor of the alternative is very strong.

(c) There is not enough evidence to reject the null hypothesis, even at the 10% level.

(d) The result is significant at a 5% level but not at a 1% level.

4.92   I. 0.0875   II. 0.5457
       III. 0.0217   IV. 0.00003

4.93   I. 0.00008   II. 0.0571
       III. 0.0368   IV. 0.1753

4.94 **Significance Levels** Test A is described in a journal article as being significant with "$P < .01$";

Test B in the same article is described as being significant with "$P < .10$." Using only this information, which test would you suspect provides stronger evidence for its alternative hypothesis?

4.95 High Fat Diet Affects Young Memory? Numerous studies have shown that a high fat diet can have a negative effect on a child's health. A new study[22] suggests that a high fat diet early in life might also have a significant effect on memory and spatial ability. In the double-blind study, young rats were randomly assigned to either a high-fat diet group or to a control group. After 12 weeks on the diets, the rats were given tests of their spatial memory. The article states that "spatial memory was significantly impaired" for the high-fat diet rats, and also tells us that "there were no significant differences in amount of time exploring objects" between the two groups. The p-values for the two tests are 0.0001 and 0.7.

(a) Which p-value goes with the test of spatial memory? Which p-value goes with the test of time exploring objects?

(b) The title of the article describing the study states "A high-fat diet causes impairment" in spatial memory. Is the wording in the title justified (for rats)? Why or why not?

4.96 Multiple Sclerosis and Sunlight It is believed that sunlight offers some protection against multiple sclerosis (MS) since the disease is rare near the equator and more prevalent at high latitudes. What is it about sunlight that offers this protection? To find out, researchers[23] injected mice with proteins that induce a condition in mice comparable to MS in humans. The control mice got only the injection, while a second group of mice were exposed to UV light before and after the injection, and a third group of mice received vitamin D supplements before and after the injection. In the test comparing UV light to the control group, evidence was found that the mice exposed to UV suppressed the MS-like disease significantly better than the control mice. In the test comparing mice getting vitamin D supplements to the control group, the mice given the vitamin D did not fare significantly better than the control group. If the p-values for the two tests are 0.472 and 0.002, which p-value goes with which test?

4.97 Arsenic in Chicken Data 4.3 on page 265 discusses a test to determine if the mean level of arsenic

in chicken meat is above 80 ppb. If a restaurant chain finds significant evidence that the mean arsenic level is above 80, the chain will stop using that supplier of chicken meat. The hypotheses are

$$H_0 : \quad \mu = 80$$
$$H_a : \quad \mu > 80$$

where μ represents the mean arsenic level in all chicken meat from that supplier. Samples from two different suppliers are analyzed, and the resulting p-values are given:

Sample from Supplier A: p-value is 0.0003
Sample from Supplier B: p-value is 0.3500

(a) Interpret each p-value in terms of the probability of the results happening by random chance.

(b) Which p-value shows stronger evidence for the alternative hypothesis? What does this mean in terms of arsenic and chickens?

(c) Which supplier, A or B, should the chain get chickens from in order to avoid too high a level of arsenic?

IS YOUR NOSE GETTING BIGGER?
Exercise 1.21 on page 15 introduces a study examining whether or not noses continue to grow throughout a person's lifetime. The study included many measurements (including size of the nose as measured by total volume) and included multiple tests. Exercises 4.98 to 4.100 each describe one of the tests along with information about the p-value for that test. In each case:

(a) State the hypotheses.

(b) Give the formal decision using a 5% significance level, and interpret the conclusion in context.

4.98 In a test to see whether males, on average, have bigger noses than females, the study indicates that "$p < 0.01$."

4.99 In a test to see whether there is a difference between males and females in average nasal tip angle, the study indicates that "$p > 0.05$."

4.100 In a test to see whether there is a positive linear relationship between age and nose size, the study indicates that "$p < 0.001$."

4.101 Antibiotics in Infancy and Weight Exercise 2.19 on page 58 introduces a study examining whether giving antibiotics in infancy increases the likelihood that the child will be overweight. Prescription records were examined to determine whether or not antibiotics were prescribed during the first year of a child's life, and each child was classified as overweight or not at age 12. (Exercise 2.19 looked at the results for age 9.) The researchers

[22]Underwood, E.L. and Thompson, L.T., "A high-fat diet causes impairment in hippocampal memory and sex-dependent alterations in peripheral metabolism," *Neural Plasticity*, August 2015.
[23]Seppa, N., "Sunlight may cut MS risk by itself," *Science News*, April 24, 2010, p. 9, reporting on a study in the *Proceedings of the National Academy of Science*, March 22, 2010.

compared the proportion overweight in each group. The study concludes that: "Infants receiving antibiotics in the first year of life were more likely to be overweight later in childhood compared with those who were unexposed (32.4% versus 18.2% at age 12, $P = 0.002$)."

(a) What is the explanatory variable? What is the response variable? Classify each as categorical or quantitative.

(b) Is this an experiment or an observational study?

(c) State the null and alternative hypotheses and define the parameters.

(d) Give notation and the value of the relevant sample statistic.

(e) Use the p-value to give the formal conclusion of the test (Reject H_0 or Do not reject H_0) and to give an indication of the strength of evidence for the result.

(f) Can we conclude that whether or not children receive antibiotics in infancy causes the difference in proportion classified as overweight?

4.102 Sleep or Caffeine for Memory? The consumption of caffeine to benefit alertness is a common activity practiced by 90% of adults in North America. Often caffeine is used in order to replace the need for sleep. One study[24] compares students' ability to recall memorized information after either the consumption of caffeine or a brief sleep. A random sample of 35 adults (between the ages of 18 and 39) were randomly divided into three groups and verbally given a list of 24 words to memorize. During a break, one of the groups takes a nap for an hour and a half, another group is kept awake and then given a caffeine pill an hour prior to testing, and the third group is given a placebo. The response variable of interest is the number of words participants are able to recall following the break. The summary statistics for the three groups are in Table 4.9. We are interested in testing whether there is evidence of a difference in average recall ability between any

Table 4.9 *Effect of sleep and caffeine on memory*

Group	Sample size	Mean	Standard Deviation
Sleep	12	15.25	3.3
Caffeine	12	12.25	3.5
Placebo	11	13.70	3.0

[24]Mednick, S., Cai, D., Kanady, J., and Drummond, S., "Comparing the benefits of caffeine, naps and placebo on verbal, motor and perceptual memory," *Behavioural Brain Research*, 2008; 193: 79–86.

two of the treatments. Thus we have three possible tests between different pairs of groups: Sleep vs Caffeine, Sleep vs Placebo, and Caffeine vs Placebo.

(a) In the test comparing the sleep group to the caffeine group, the p-value is 0.003. What is the conclusion of the test? In the sample, which group had better recall ability? According to the test results, do you think sleep is really better than caffeine for recall ability?

(b) In the test comparing the sleep group to the placebo group, the p-value is 0.06. What is the conclusion of the test using a 5% significance level? If we use a 10% significance level? How strong is the evidence of a difference in mean recall ability between these two treatments?

(c) In the test comparing the caffeine group to the placebo group, the p-value is 0.22. What is the conclusion of the test? In the sample, which group had better recall ability? According to the test results, would we be justified in concluding that caffeine impairs recall ability?

(d) According to this study, what should you do before an exam that asks you to recall information?

4.103 Price and Marketing How influenced are consumers by price and marketing? If something costs more, do our expectations lead us to believe it is better? Because expectations play such a large role in reality, can a product that costs more (but is in reality identical) actually be more effective? Baba Shiv, a neuroeconomist at Stanford, conducted a study[25] involving 204 undergraduates. In the study, all students consumed a popular energy drink which claims on its packaging to increase mental acuity. The students were then asked to solve a series of puzzles. The students were charged either regular price ($1.89) for the drink or a discount price ($0.89). The students receiving the discount price were told that they were able to buy the drink at a discount since the drinks had been purchased in bulk. The authors of the study describe the results: "the number of puzzles solved was lower in the reduced-price condition ($M = 4.2$) than in the regular-price condition ($M = 5.8$) ... $p < .0001$."

(a) What can you conclude from the study? How strong is the evidence for the conclusion?

(b) These results have been replicated in many similar studies. As Jonah Lehrer tells us: "According to Shiv, a kind of placebo effect is at work.

[25]Shiv, B., Carmon, Z., and Ariely, D., "Placebo Effects of Marketing Actions: Consumers May Get What They Pay For," *Journal of Marketing Research*, 2005; 42:383–93.

Since we expect cheaper goods to be less effective, they generally are less effective, even if they are identical to more expensive products. This is why brand-name aspirin works better than generic aspirin and why Coke tastes better than cheaper colas, even if most consumers can't tell the difference in blind taste tests."[26] Discuss the implications of this research in marketing and pricing.

4.104 Mercury Levels in Fish Figure 4.25 shows a scatterplot of the acidity (pH) for a sample of $n = 53$ Florida lakes vs the average mercury level (ppm) found in fish taken from each lake. The full dataset is introduced in Data 2.4 on page 71 and is available in **FloridaLakes**. There appears to be a negative trend in the scatterplot, and we wish to test whether there is significant evidence of a negative association between pH and mercury levels.

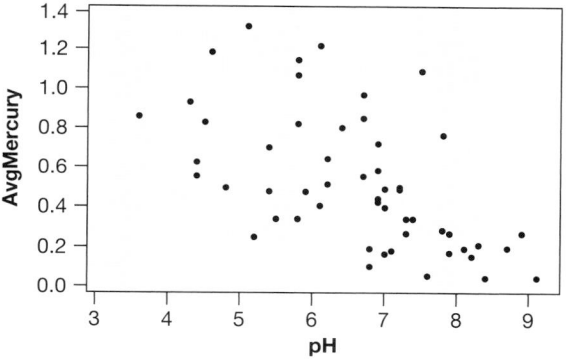

Figure 4.25 *Water pH vs mercury levels of fish in Florida lakes*

(a) What are the null and alternative hypotheses?

(b) For these data, a statistical software package produces the following output:

$$r = -0.575 \qquad p\text{-}value = 0.000017$$

Use the p-value to give the conclusion of the test. Include an assessment of the strength of the evidence and state your result in terms of rejecting or failing to reject H_0 *and* in terms of pH and mercury.

(c) Is this convincing evidence that low pH *causes* the average mercury level in fish to increase? Why or why not?

4.105 Penalty Shots in Soccer An article noted that it may be possible to accurately predict which way a

penalty-shot kicker in soccer will direct his shot.[27] The study finds that certain types of body language by a soccer player – called "tells" – can be accurately read to predict whether the ball will go left or right. For a given body movement leading up to the kick, the question is whether there is strong evidence that the proportion of kicks that go right is significantly different from one-half.

(a) What are the null and alternative hypotheses in this situation?

(b) If sample results for one type of body movement give a p-value of 0.3184, what is the conclusion of the test? Should a goalie learn to distinguish this movement?

(c) If sample results for a different type of body movement give a p-value of 0.0006, what is the conclusion of the test? Should a goalie learn to distinguish this movement?

4.106 ADHD and Pesticides In Exercise 4.16 on page 268, we describe an observational study investigating a possible relationship between exposure to organophosphate pesticides as measured in urinary metabolites (DAP) and diagnosis of ADHD (attention-deficit/hyperactivity disorder). In reporting the results of this study, the authors[28] make the following statements:

• "The threshold for statistical significance was set at $P < .05$."

• "The odds of meeting the . . . criteria for ADHD increased with the urinary concentrations of total DAP metabolites."

• "The association was statistically significant."

(a) What can we conclude about the p-value obtained in analyzing the data?

(b) Based on these statements, can we distinguish whether the evidence of association is very strong vs moderately strong? Why or why not?

(c) Can we conclude that exposure to pesticides is related to the likelihood of an ADHD diagnosis?

(d) Can we conclude that exposure to pesticides *causes* more cases of ADHD? Why or why not?

4.107 Diabetes and Pollution Diabetes tends to be more prevalent in urban populations, but why this

[26]Lehrer, J., "Grape expectations: What wine can tell us about the nature of reality," *The Boston Globe*, February 28, 2008.

[27]"A Penalty Kicker's Cues," *The Week*, July 16, 2010, p. 21.
[28]Bouchard, M., Bellinger, D., Wright, R., and Weisskopf, M., "Attention-Deficit/Hyperactivity Disorder and Urinary Metabolites of Organophosphate Pesticides," *Pediatrics*, 2010; 125: e1270–e1277.

is so is not fully understood. A study[29] on mice was designed to investigate the link between diabetes and air pollution. The study involved 28 mice, with 14 randomly selected to have filtered air pumped into their cage while the other 14 breathed particulate matter that simulated common air pollution. The response variable is the amount of insulin resistance each mouse had after 24 weeks. Higher insulin resistance indicates a greater risk for developing diabetes.

(a) Is this an observational study or randomized experiment?

(b) If we are interested in whether there is a difference in mean insulin resistance between the two groups, what are the null and alternative hypotheses?

(c) The difference in sample means from the original sample is $D = \bar{x}_{FA} - \bar{x}_{PM} = 1.8 - 6.2 = -4.4$. Figure 4.26 shows 1000 random assignments of insulin resistant scores from the original sample to each of the two groups. Is it likely we will reject the null hypothesis?

(d) What is the p-value?

(e) State the conclusion of the test in context.

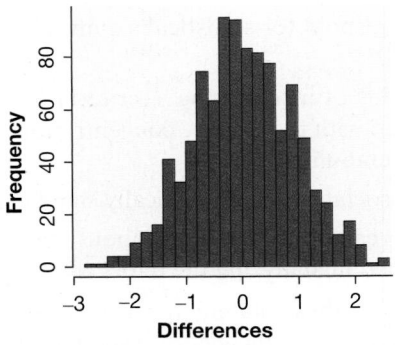

Figure 4.26 *Randomization distribution for 1000 simulations with* $H_0 : \mu_{FA} = \mu_{PM}$

4.108 Beer and Mosquitoes Does consuming beer attract mosquitoes? Exercise 4.17 on page 268 discusses an experiment done in Africa testing possible ways to reduce the spread of malaria by mosquitoes. In the experiment, 43 volunteers were randomly assigned to consume either a liter of beer or a liter of water, and the attractiveness to mosquitoes of each

volunteer was measured. The experiment was designed to test whether beer consumption increases mosquito attraction. The report[30] states that "Beer consumption, as opposed to water consumption, significantly increased the activation...of *An. gambiae* [mosquitoes]...($P < 0.001$)."

(a) Is this convincing evidence that consuming beer is associated with higher mosquito attraction? Why or why not?

(b) How strong is the evidence for the result? Explain.

(c) Based on these results, it is reasonable to conclude that consuming beer *causes* an increase in mosquito attraction? Why or why not?

4.109 Exercise and the Brain Exercise 4.19 on page 269 describes a study investigating the effects of exercise on cognitive function.[31] Separate groups of mice were exposed to running wheels for 0, 2, 4, 7, or 10 days. Cognitive function was measured by Y-maze performance. The study was testing whether exercise improves brain function, whether exercise reduces levels of BMP (a protein which makes the brain slower and less nimble), and whether exercise increases the levels of noggin (which improves the brain's ability). For each of the results quoted in parts (a), (b), and (c), interpret the information about the p-value in terms of evidence for the effect.

(a) "Exercise improved Y-maze performance in most mice by the 7th day of exposure, with further increases after 10 days for all mice tested ($p < .01$)."

(b) "After only two days of running, BMP...was reduced...and it remained decreased for all subsequent time-points ($p < .01$)."

(c) "Levels of noggin...did not change until 4 days, but had increased 1.5-fold by 7–10 days of exercise ($p < .001$)."

(d) Which of the tests appears to show the strongest statistical effect?

(e) What (if anything) can we conclude about the effects of exercise on mice?

4.110 Watch Out for Lions after a Full Moon Scientists studying lion attacks on humans in Tanzania[32] found that 95 lion attacks happened

---

[29]Data recreated from information in Sun, et al. "Ambient Air Pollution Exaggerates Adipose Inflammation and Insulin Resistance in a Mouse Model of Diet-Induced Obesity," *Journal of the American Heart Association*, 2009; 119 (4): 538–46.

[30]Lefvre, T., et al. "Beer Consumption Increases Human Attractiveness to Malaria Mosquitoes." *PLoS ONE*, 2010; 5(3): e9546.

[31]Gobeske, K., et al. "BMP Signaling Mediates Effects of Exercise on Hippocampal Neurogenesis and Cognition in Mice," *PLoS One*, 2009; 4(10); e7506.

[32]Packer, C., Swanson, A., Ikanda, D., and Kushnir, H., "Fear of Darkness, the Full Moon and the Nocturnal Ecology of African Lions," *PLoS ONE* 2011; 6(7): e22285.

between 6 pm and 10 pm within either five days before a full moon or five days after a full moon. Of these, 71 happened during the five days after the full moon while the other 24 happened during the five days before the full moon. Does this sample of lion attacks provide evidence that attacks are more likely after a full moon? In other words, is there evidence that attacks are not equally split between the two five-day periods? Use *StatKey* or other technology to find the p-value, and be sure to show all details of the test. (Note that this is a test for a single proportion since the data come from one sample.)

4.111 Electrical Stimulation for Fresh Insight? Exercise 2.25 on page 59 introduces a study in which 40 participants are trained to solve problems in a certain way and then asked to solve an unfamiliar problem that requires fresh insight. Half of the participants were randomly assigned to receive electrical stimulation of the brain while the other half (control group) received sham stimulation as a placebo. The results are shown in Table 4.10.

Table 4.10 *Does electrical brain stimulation bring fresh insight to a problem?*

Treatment	Solved	Not Solved
Sham	4	16
Electrical	12	8

(a) Use *StatKey* or other technology to create a randomization distribution to test whether the proportion able to solve the problem was significantly higher in the group receiving electrical stimulation of the brain. Be sure to state the hypotheses, give the p-value, and clearly state the conclusion in context.

(b) Can we conclude that electrical stimulation of the brain helps people solve a new problem that needs fresh insight?

4.112 Does Massage Help Heal Muscles Strained by Exercise? After exercise, massage is often used to relieve pain, and a recent study[33] shows that it also may relieve inflammation and help muscles heal. In the study, 11 male participants who had just strenuously exercised had 10 minutes of massage on one quadricep and no treatment on the other, with treatment randomly assigned. After 2.5 hours, muscle biopsies were taken and production of the

inflammatory cytokine interleukin-6 was measured relative to the resting level. The differences (control minus massage) are given in Table 4.11.

Table 4.11 *Inflammation in muscle: control minus massage*

0.6	4.7	3.8	0.4	1.5	−1.2	2.8	−0.4	1.4	3.5	−2.8

(a) Is this an experiment or an observational study? Why is it not double blind?

(b) What is the sample mean difference in inflammation between no massage and massage?

(c) We want to test to see if the population mean difference μ_D is greater than zero, meaning muscle with no treatment has more inflammation than muscle that has been massaged. State the null and alternative hypotheses.

(d) Use *StatKey* or other technology to find the p-value from a randomization distribution.

(e) Are the results significant at a 5% level? At a 1% level? State the conclusion of the test if we assume a 5% significance level (as the authors of the study did).

4.113 The Ignorance Survey: United States The Ignorance Surveys were conducted in 2013 using random sampling methods in four different countries under the leadership of Hans Rosling, a Swedish statistician and international health advocate. The survey questions were designed to assess the ignorance of the public to global population trends. The survey was not just designed to measure ignorance (no information), but if preconceived notions can lead to more wrong answers than would be expected by random guessing. One question asked, "In the last 20 years the proportion of the world population living in extreme poverty has...," and three choices were provided: 1) "almost doubled" 2) "remained more or less the same," and 3) "almost halved." Of 1005 US respondents, just 5% gave the correct answer: "almost halved."[34] We would like to test if the percent of correct choices is significantly different than what would be expected if the participants were just randomly guessing between the three choices.

(a) What are the null and alternative hypotheses?

(b) Using *StatKey* or other technology, construct a randomization distribution and compute the p-value.

(c) State the conclusion in context.

---

[33]Data approximated from summary statistics in Crane, J., et al., "Massage Therapy Attenuates Inflammatory Signaling After Exercise-Induced Muscle Damage," *Science Translational Medicine*, February 1, 2012.

[34]*http://www.gapminder.org/GapminderMedia/wp-uploads/ Results-from-the-Ignorance-Survey-in-the-US..pdf*

4.114 The Ignorance Survey: United Kingdon Exercise 4.113 refers to a survey used to assess the ignorance of the public to global population trends. A similar survey was conducted in the United Kingdom, where respondents were asked if they had a university degree. One question asked, "In the last 20 years the proportion of the world population living in extreme poverty has...," and three choices were provided: 1)"increased" 2) "remained more or less the same," and 3) "decreased." Of 373 university degree holders, 45 responded with the correct answer: decreased; of 639 non-degree respondents, 57 responded with the correct answer.[35] We would like to test if the percent of correct answers is significantly different between degree holders and non-degree holders.

(a) What are the null and alternative hypotheses?

(b) Using *StatKey* or other technology, construct a randomization distribution and compute the p-value.

(c) State the conclusion in context.

4.115 Flying Home for the Holidays Does the airline you choose affect when you'll arrive at your destination? The dataset **DecemberFlights** contains the difference between actual and scheduled arrival time from 1000 randomly sampled December flights for two of the major North American airlines, *Delta Air Lines* and *United Air Lines*. A negative difference indicates a flight arrived early. We are interested in testing whether the average difference between actual and scheduled arrival time is different between the two airlines.

(a) Define any relevant parameter(s) and state the null and alternative hypotheses.

(b) Find the sample mean of each group, and calculate the difference in sample means.

(c) Use *StatKey* or other technology to create a randomization distribution and find the p-value.

(d) At a significance level of $\alpha = 0.01$, what is the conclusion of the test? Interpret the conclusion in context.

LIZARDS AND INVASIVE FIRE ANTS

Exercises 4.116 and 4.117 address lizard behavior in response to fire ants. The red imported fire ant, *Solenopsis invicta*, is native to South America, but has an expansive invasive range, including much of the southern United States (invasion of this ant is predicted to go global). In the United States, these ants occupy similar habitats as fence lizards. The ants eat the lizards and the lizards eat the ants, and in either scenario the venom from the fire ant can be fatal to the lizard. A study[36] explored the question of whether lizards learn to adapt their behavior if their environment has been invaded by fire ants. The researchers selected lizards from an uninvaded habitat (eastern Arkansas) and lizards from an invaded habitat (southern Alabama, which has been invaded for more than 70 years) and exposed them to fire ants. They measured how long it takes each lizard to flee and the number of twitches each lizard does. The data are stored in **FireAnts**.

4.116 If lizards adapt their behavior to the fire ants, then lizards from the invaded habitats should flee from the fire ants faster than lizards from the uninvaded habitats. Test this hypothesis. The variable *Flee* gives time to flee, measured in seconds, and lizards taking more than a minute to flee have recorded responses of 61 seconds.

(a) State the null and alternative hypotheses.

(b) Use technology to calculate the p-value.

(c) What (if anything) does this p-value tell you about lizards and fire ants?

(d) Can we conclude that living in a habitat invaded by fire ants causes lizards to adapt their behavior and flee faster when exposed to fire ants? Why or why not?

4.117 If lizards adapt their behavior to the fire ants, then lizards from the invaded habitats should twitch more than lizards from uninvaded habitats when exposed to red imported fire ants (twitching helps to repel the ants). Test this hypothesis. The variable *Twitches* is the number of twitches exhibited by each lizard in the first minute after exposure.

(a) State the null and alternative hypotheses.

(b) Use technology to calculate the p-value.

(c) What (if anything) does this p-value tell you about lizards and fire ants?

4.118 Split the Bill? Exercise 2.153 on page 105 describes a study to compare the cost of restaurant meals when people pay individually versus splitting the bill as a group. In the experiment 48 subjects were randomly assigned to eight groups of six diners each. Half of the people were told that they would each be responsible for individual meal costs and the other half were told to split the cost equally among the six people at the table. The data in **Split-Bill** includes the cost of what each person ordered

---

[35]Counts approximated from the percentages reported at *http://www.gapminder.org/news/highlights-from-ignorance-survey-in-the-uk/*

[36]Langkilde, T. (2009). "Invasive fire ants alter behavior and morphology of native lizards," *Ecology*, 90(1): 208–217. Thanks to Dr. Langkilde for providing the data.

(in Israeli shekels) and the payment method (*Individual* or *Split*). Use StatKey or other technology to construct a randomization distribution using these data to test whether there is sufficient evidence to conclude that the mean cost is lower when diners are paying individually than when they split the bill equally.

4.119 Cat Ownership and Schizophrenia Could owning a cat as a child be related to mental illness later in life? Toxoplasmosis is a disease transmitted primarily through contact with cat feces, and has recently been linked with schizophrenia and other mental illnesses. Also, people infected with Toxoplasmosis tend to like cats more and are 2.5 times more likely to get in a car accident, due to delayed reaction times. The CDC estimates that about 22.5% of Americans are infected with Toxoplasmosis (most have no symptoms), and this prevalence can be as high as 95% in other parts of the world. A study[37] randomly selected 262 people registered with the National Alliance for the Mentally Ill (NAMI), almost all of whom had schizophrenia, and for each person selected, chose two people from families without mental illness who were the same age, sex, and socioeconomic status as the person selected from NAMI. Each participant was asked whether or not they owned a cat as a child. The results showed that 136 of the 262 people in the mentally ill group had owned a cat, while 220 of the 522 people in the not mentally ill group had owned a cat.

(a) This is known as a case-control study, where *cases* are selected as people with a specific disease or trait, and controls are chosen to be people without the disease or trait being studied.

[37]Torrey, E.F., Simmons, W., Yolken, R.H. (2015). "Is childhood cat ownership a risk factor for schizophrenia later in life?," *Schizophrenia Research*, June 2015, 165(1):1–2.

Both cases and controls are then asked about some variable from their past being studied as a potential risk factor. This is particularly useful for studying rare diseases (such as schizophrenia), because the design ensures a sufficient sample size of people with the disease. Can case-control studies such as this be used to infer a causal relationship between the hypothesized risk factor (e.g., cat ownership) and the disease (e.g., schizophrenia)? Why or why not?

(b) In case-control studies, controls are usually chosen to be similar to the cases. For example, in this study each control was chosen to be the same age, sex, and socioeconomic status as the corresponding case. Why choose controls who are similar to the cases?

(c) For this study, calculate the relevant difference in proportions; proportion of cases (those with schizophrenia) who owned a cat as a child minus proportion of controls (no mental illness) who owned a cat as a child.

(d) For testing the hypothesis that the proportion of cat owners is higher in the schizophrenic group than the control group, use technology to generate a randomization distribution and calculate the p-value.

(e) Do you think this provides evidence that there is an association between owning a cat as a child and developing schizophrenia?[38] Why or why not?

[38]Even if you owned a cat as a child, you probably do not have to worry. Schizophrenia has a strong genetic component, and most cat owners do not go on to develop schizophrenia. However, because of this study, you may want to think twice before letting children (or pregnant women) come into contact with cat feces. If you love cats but are worried by this study, cats always kept indoors are thought to be safe.

4.4 A CLOSER LOOK AT TESTING

Hypothesis testing is very powerful, as it helps shed light on whether an observed effect is real or just due to random chance. However, statistical significance is not foolproof, and it is possible to make the wrong decision, rejecting a true null hypothesis or not rejecting a false null hypothesis. This section discusses common pitfalls associated with formal hypothesis testing, along with factors that influence the chances of these errors occurring.

Type I and Type II Errors

Formal hypothesis testing produces one of two possible generic decisions: "reject H_0" or "do not reject H_0." In reality, the claims about the population described by H_0 and H_a might be either true or false. Perhaps light at night really does increase weight gain (H_a), or maybe this phenomenon doesn't exist at all (H_0) and sample differences

Table 4.12 *Possible errors in a formal statistical decision*

		Decision:	
		Reject H_0	Do not reject H_0
Reality:	H_0 is true	Type I error	No error
	H_0 is false	No error	Type II error

just reflect random variation. When we make a formal decision to "reject H_0," we generally are accepting some risk that H_0 might actually be true. For example, we may have been unlucky and stumbled upon one of those "1 in a 1000" samples that are very rare to see when H_0 holds but still are not impossible. This is an example of what we call a *Type I error*: rejecting a true H_0. The other possible error to make in a statistical test is to fail to reject H_0 when it is false and the alternative H_a is actually true. We call this a *Type II error*: failing to reject a false H_0. See Table 4.12.

In medical terms we often think of a Type I error as a "false positive" – a test that indicates a patient has an illness when actually none is present, and a Type II error as a "false negative" – a test that fails to detect an actual illness.

Example 4.23

Describe the consequences of making Type I and Type II errors in each case.

(a) In the light at night experiment where we test $H_0 : \mu_L = \mu_D$ vs $H_a : \mu_L > \mu_D$

(b) In Example 4.20 where we have a mystery animal named X and test $H_0 :$ X is an elephant vs $H_a :$ X is not an elephant

Solution

(a) A Type I error is to reject a true H_0. In the light at night study, a Type I error is to conclude that light at night increases weight gain when actually there is no effect.

A Type II error is to fail to reject a false H_0. In this case, a Type II error means the test based on our sample data does not convince us that light increases weight gain when it actually does.

(b) If we see evidence (perhaps that X walks on two legs) that is so rare we conclude that X is not an elephant and it turns out that X is an elephant (perhaps trained in a circus), we have made a Type I error.

For a Type II error, we might find evidence (perhaps having four legs) that is not unusual for an elephant, so we do not reject H_0 and then discover that X is actually a giraffe.

If our results are significant and we reject H_0, there is usually no way of knowing whether we are correct or whether we have made a Type I error. If our results are insignificant and we fail to reject H_0, we could be correct or we could have made a Type II error. While we can never rule out these possibilities entirely, we do have some control over the chance of making these errors.

Significance Level and Errors

While we wish to avoid both types of errors, in practice we have to accept some trade-off between them. If we make it very hard to reject H_0, we could reduce the chance of making a Type I error, but then we would make Type II errors more often. On the other hand, making it easier to reject H_0 would reduce the chance of making a Type II error, but increase the chance of making a Type I error and we would end up rejecting too many H_0's that were actually true. This balance is set by how easy or hard it is to reject H_0, which is exactly determined by the significance level! To decrease the chance of making a Type I error, we make it harder to reject H_0 by using a lower significance level. To decrease the chance of making a Type II error, we make it easier to reject H_0 by using a higher significance level.

If we use $\alpha = 0.05$ in a right-tail test, we decide to reject H_0 for all sample values in the 5% of area in the upper tail of a randomization distribution. If instead we use $\alpha = 0.01$, we only reject H_0 for sample values in the extreme upper 1% of area. If we make α smaller, fewer samples would be that extreme, meaning we would reject H_0 less often.

Because 5% of statistics will be in the most extreme 5% of the randomization distribution, these 5% of samples will yield p-values less than 0.05. Because a randomization distribution is created assuming the null hypothesis is true, this means 5% of samples will lead to rejecting H_0 at $\alpha = 0.05$, even when H_0 is true. This idea generalizes beyond 0.05: if the null hypothesis is true, α is the probability of making a Type I error.

Understanding a Significance Level

The significance level, α, represents the tolerable probability of making a Type I error.

If the consequences of a Type I error are severe (for example, approving a new drug that is potentially dangerous) we might use a very small α (perhaps even $\alpha = 0.005$). If the consequences of a Type II error are severe (for example, failing to diagnose a treatable disease), we would want to make it easier to reject H_0, so might use a relatively large α. However, remember that there is always a trade-off between the two types of errors, so we usually use the common significance levels of 5%, 10% or 1%.

Example 4.24

Analogy to Law

It is often helpful to think of significance tests as similar to cases in a court of law. For each italicized word or phrase below, give the analogy in a statistical test.

(a) A person is *innocent* until proven *guilty*.

(b) The *evidence* provided must indicate the suspect's guilt beyond a *reasonable doubt*.

(c) There are two types of errors a jury can make:
   • *Releasing a guilty person*
   • *Convicting an innocent person*

Solution   (a) "Innocent" is the null hypothesis, H_0 (the status quo that we assume to be the situation until we see convincing evidence to the contrary). "Guilty" represents the alternative hypothesis, H_a (the claim that instigates the trial).

(b) The "evidence" is the data from the sample and its p-value. The "reasonable doubt" corresponds to the significance level, α. We reject the claim of innocence (H_0) and determine the suspect is guilty (H_a) when the evidence (p-value) is very unlikely (less than α) to occur if the suspect is really innocent.

(c) "Releasing a guilty person" corresponds to a Type II error, since we fail to find evidence to reject a false H_0. "Convicting an innocent person" corresponds to a Type I error, since we (incorrectly) find sufficient evidence in the data to reject a true H_0. As in our legal system, we are usually more worried about a Type I error (convicting an innocent person) than about a Type II error (releasing a guilty person). Also as in our legal system, there is a trade-off between the two kinds of errors when we test hypotheses.

The Problem of Multiple Testing

We just learned that if the null hypothesis is true, then 5% of hypothesis tests using $\alpha = 0.05$ will incorrectly reject the null hypothesis. This issue becomes even more important when doing multiple hypothesis tests. Of all hypothesis tests conducted for a true null hypothesis, using $\alpha = 0.05$, 5% of the tests will lead to rejecting the null hypothesis! In other words, if you do 100 hypothesis tests, all testing for an effect that doesn't exist (the null is true), about 5% of them will incorrectly reject the null.

Example 4.25

Opening an Umbrella Indoors

Is it really bad luck to open an umbrella indoors? Suppose researchers all over the world set out to actually test this idea, each randomizing people to either open an umbrella indoors or to open an umbrella outdoors, and somehow measure "luck" afterwards. If there are 100 people all testing this phenomenon at $\alpha = 0.05$, and if opening an umbrella indoors does *not* bring bad luck, then about how many people do you expect to get statistically significant results?

Solution

If the null hypothesis is true (opening an umbrella indoors has no effect on luck), then about 5% of the hypothesis tests will get p-values less than 0.05 just by random chance, so about 5 of the 100 people testing this phenomena will get statistically significant results.

If multiple hypothesis tests are conducted for an effect that doesn't exist, some of them may get significant results just by chance. The more hypothesis tests being conducted, the higher the chance that at least one of those tests will make a Type I error. This problem is known as the problem of *multiple testing*.

The Problem of Multiple Testing

When multiple tests are conducted, if the null hypotheses are all true, the proportion of all the tests that will yield statistically significant results just by random chance is about α, the significance level.

This issue is made even worse by the fact that usually only significant results are published. This problem is known as *publication bias*. If only significant results are published, then no one knows of all the studies producing insignificant results. Consider the umbrella example. If the five statistically significant studies are all published, and we do not know about the 95 insignificant studies, we might take this as convincing evidence that opening an umbrella indoors really does cause bad luck. Unfortunately this is a very real problem with scientific research.

Publication bias

Often, only significant results are published. If many tests are conducted, some of them will be significant just by chance, and it may be only these studies that we hear about.

The problem of multiple testing can also occur when one researcher is testing multiple hypotheses.

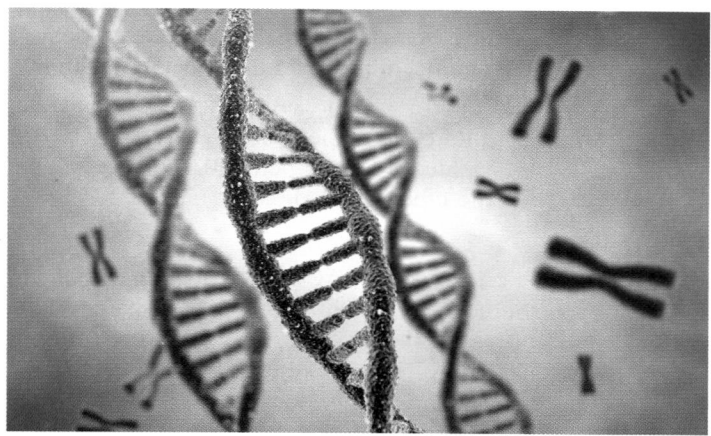

nobeastsofierce/Shutterstock

Are there genes related to leukemia?

DATA 4.6   ## Genes and Leukemia

Genome association studies, tests for whether genes are associated with certain diseases or other traits, are currently widely used in medical research, particularly in cancer research. Typically, DNA is collected from a group of people, some of whom have the disease in question, and some of whom don't. These DNA data are made up of values for thousands of different genes, and each gene is tested to see if there is a difference between the diseased patients and the healthy patients. Results can then be useful in risk assessment, diagnosis, and the search for a cure. One of the most famous genome association studies tested for genetic differences between patients with two different types of leukemia (acute myeloid leukemia and acute lymphoblastic leukemia).[39] In this study, scientists collected data on 7129 different genes for 38 patients with leukemia.   ■

Example 4.26

Genes and Leukemia

Data 4.6 refers to a study in which data included information on 7129 genes, and each gene was tested for a difference between the two types of leukemia.

(a) If all tests used a significance level of $\alpha = 0.01$, and if there are no genetic differences between the two types of leukemia, about how many of the genes would be found to be significantly different between the two groups?

(b) Do we have reason to believe that all of the genes found to be statistically significant are actually associated with the type of leukemia?

(c) In the actual study, 11% of tests for each gene yielded p-values less than 0.01. Do we have reason to believe that there is some association between genes and the type of leukemia?

Solution   (a) If there are no genetic differences between the two types of leukemia, then we would expect about 0.01 of the tests to yield statistically significant results just by random chance. We expect about $0.01 \times 7129 \approx 71$ of the genes to be found

[39] Golub, T.R., et al. "Molecular Classification of Cancer: Class Discovery and Class Prediction by Gene Expression Monitoring," *Science*, 1999; 286:531–537.

to be significantly different between the two groups, even if no differences actually exist.

(b) Because we expect 71 genes to be found significant just by random chance even if no associations exist, we should not believe that all genes found to be statistically significant are actually associated with the type of leukemia.

(c) If there were no association between genes and leukemia, we would only expect about 1% of the tests to yield p-values less than 0.01. Because 11% of the genes yielded p-values below 0.01, some of them are probably truly associated with the type of leukemia.

There are many ways of dealing with the problem of multiple testing,[40] but those methods are outside the scope of this text. The most important thing is to be aware of the problem, and to realize that when doing multiple hypothesis tests, some are likely to be significant just by random chance.

Replicating Results

If results are statistically significant we unfortunately have no way of knowing (unless we can measure the entire population) whether the alternative hypothesis is actually true or whether a Type I error occurred. In some situations, we may be extra suspicious of Type I errors, such as when multiple tests were performed or when a result is surprising or goes against current scientific theory. In other situations, making a Type I error can be very problematic, such as when the study results have important implications that will change current practice. For example, it would be very bad to start prescribing a new drug with potential side effects to millions of people if the drug does not work. In situations such as these, studies yielding statistical significance should be *replicated* or *reproduced* with another study.

Example 4.27

Clinical Trials

Clinical trials are studies on people investigating a new drug or medical treatment, and in the United States these are conducted in four phases. During Phase I, the new treatment is tested for safety on a small group of people. In Phase II, an experiment is conducted to test the effectiveness of the new treatment (usually against a placebo). In Phase III, if the Phase II experiment yielded statistical significance, another experiment is conducted (usually with a larger sample size and comparing to an existing treatment). Phase IV consists of data collection on users after the drug has gone to market, which occurs only if both Phase II and Phase III yield significant results. Explain why two separate randomized experiments are required for clinical trials.

Solution

If the Phase II experiment yields statistically significant results, we can't tell for sure whether the treatment effect is real or whether a Type I error occurred. Because we want to be sure the treatment really works before approving it for widespread use, we perform another experiment in an attempt to replicate the results. If the Phase III experiment also yields statistically significant results, we can be much more certain that the treatment is actually effective. If the Phase III experiment does *not* yield statistically significant results, perhaps the Phase II significance was just the result of a Type I error.

---

[40] One common way, known as Bonferroni's correction, is to divide the significance level by the number of tests. For $\alpha = 0.05$ and 100 tests, a p-value would have to be less than $0.05/100 = 0.0005$ to be deemed statistically significant.

Replication is an important safeguard in clinical trials for two reasons. One is that many pharmaceutical companies are testing new drugs and treatments all the time, so we have a multiple testing problem. The second reason is that it could be very harmful to society to conclude a new drug or medical treatment is effective when it is not: people could suffer side effects for no reason, or they could choose this treatment over another option that actually works.

Replication with Another Study

By attempting to *replicate* significant results with another study, this second study can either:

(a) reject H_0, providing further confirmation that H_a is true
(b) fail to reject H_0, suggesting that the first study may have yielded a Type I error

Because replication of results is an important part of science, it is good practice to conduct your study in a way that is fully *reproducible*, making each step of the data collection and analysis transparent to other researchers.

Practical vs Statistical Significance

Suppose that a company offers an online tutorial course to help high school students improve their scores when retaking a standardized exam such as the Scholastic Aptitude Test (SAT). Does the online course improve scores? We might use a hypothesis test to determine if there is an improvement in scores and a confidence interval to determine the size of the improvement.

Suppose we set up an experiment to measure the change in SAT score by randomly assigning students to either take the course or just study on their own before retaking the SAT. We let μ_c be the mean change in SAT scores for those taking the online course and μ_{nc} be the mean change for those who just prepare on their own with no course. This gives the hypotheses

$$H_0: \quad \mu_c = \mu_{nc}$$
$$H_a: \quad \mu_c > \mu_{nc}$$

Suppose that we randomly assign 2000 students to take the online course and another 2000 students to a "no course" group. Figure 4.27 shows possible histograms of the score changes for both groups. Although some students in both groups do worse (i.e., have a negative change) when they retake the exam, in general students tend to do better the second time. The mean change for the sample of students taking the online course is $\bar{x}_c = 42.7$ points improvement and for the other group without the course the sample mean change is $\bar{x}_{nc} = 38.5$ points. The difference is $\bar{x}_c - \bar{x}_{nc} = 42.7 - 38.5 = 4.2$ points and a randomization distribution shows the upper tail p-value is about 0.0038. For any reasonable significance level this is a small p-value so we have very strong evidence to reject H_0 and conclude that the mean improvement in SAT scores is higher for students who use the online course.

We not only care about significance but also want to know how much higher the average improvement is for students who use the online course. For this, we compute an interval estimate. A 95% confidence interval for difference in mean improvement in SAT scores for students who use the online course minus students who don't is (1.04, 7.36) points. Is an average improvement between 1 and 7 points worth it?

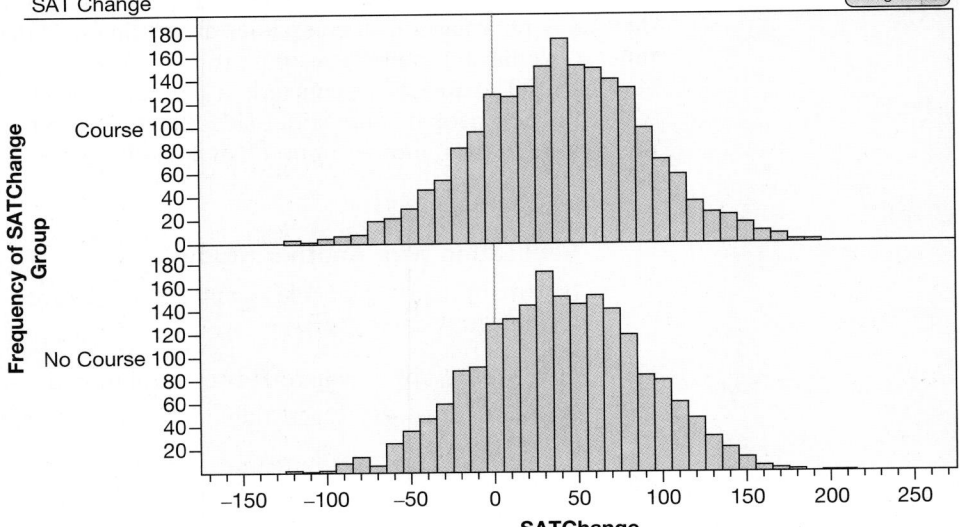

Figure 4.27
Hypothetical SAT score changes for groups of 2000 students with/without an online course

Now suppose that the online prep course costs $3000 and takes more than 50 hours to complete. Would you be willing to spend that time and money to earn (on average) roughly 4 more points than you might get by preparing on your own (on an exam that is scored out of 800 points)? Would that magnitude of a score change really make much difference in how your SAT score is viewed by a college admissions officer?

Example 4.28

In testing whether an online prep course for the SAT test improves scores, we saw that the average increase is 4.2 points and the p-value for the test is 0.0038. Are the results statistically significant? Are the results practically significant?

Solution Since the p-value is very low, at 0.0038, the results are definitely statistically significant. Since the average improvement is only 4.2 points, however, the results are probably not practically significant. It is probably not worth taking the online course for such a small change.

This hypothetical example demonstrates that a difference that is *statistically* significant might not have much *practical* significance. Especially when the sample sizes are large, a rather small difference (such as 4 points on an 800-point SAT exam) might turn out to be statistically significant. That does not necessarily mean that the difference is going to be particularly important to individuals making a decision (such as whether or not to take the online course). While some small differences may be important and large samples can help reveal the true effects, we should not make the mistake of automatically assuming that anything that is statistically significant is practically significant.

The Effect of Sample Size

We see in Section 3.1 that as the sample size increases, the variability of sample statistics tends to decrease and sample statistics tend to be closer to the true value of the population parameter. This idea applies here as well. As the sample size increases, there will be less spread in the kinds of statistics we will see just by random chance, when the null hypothesis is true, and statistics in the randomization distribution will be more closely concentrated around the null value.

Example 4.29

Let's return to Example 4.15 examining the proportion, p, of customers who forget to take their to-go box, with $H_0 : p = 0.3$ and $H_a : p > 0.3$. In the following situations, the sample statistic is the same but the sample size is different. In each case, find the p-value, make a formal decision (using $\alpha = 0.05$), and give a conclusion in context.

(a) $\hat{p} = 0.35$ with $n = 100$

(b) $\hat{p} = 0.35$ with $n = 1000$

Solution

(a) The randomization distribution for samples of size 100 is shown in Figure 4.28(a), and yields a p-value of 0.166. This is greater than $\alpha = 0.05$, so we do not reject H_0. We would not have convincing evidence, based on the sample of 100 customers with $\hat{p} = 0.35$, that more than 30% of people forget their to-go box at this restaurant.

(b) The randomization distribution for samples of size 1000 is shown in Figure 4.28(b), and yields a p-value of 0.001. This is less than $\alpha = 0.05$, so we reject H_0. We would have strong evidence from a sample of 1000 customers with $\hat{p} = 0.35$ that more than 30% of people forget their to-go box at this restaurant.

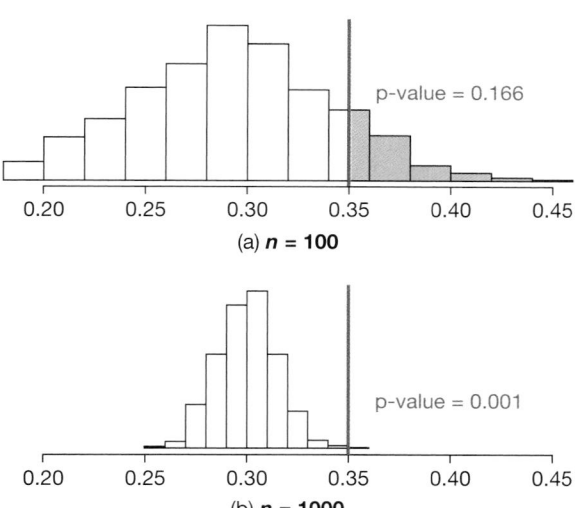

Figure 4.28
Randomization distributions for
$H_0 : p = 0.3$

Notice in Example 4.29 that the hypotheses and sample statistic are the same in both cases, and the only difference is the sample size. In Figure 4.28, the randomization distribution with $n = 100$ has more variability than the randomization distribution with $n = 1000$. As a result, if the proportion really is 0.3, it is quite likely to see a statistic as high as $\hat{p} = 0.35$ if $n = 100$, but very unlikely to see a statistic that large if $n = 1000$. Because a larger sample size decreases the spread of values we might see just by random chance, a larger sample size makes it easier to reject H_0 when the alternative hypothesis is true.

Sample Size and Significance

A larger sample size makes it easier to find significant results, if H_a is true.

This has several important practical implications. The first is that to maximize your chances of finding significant results, you should always use the largest sample size possible. A larger sample size decreases the chance of making a Type II error. In this case there is no trade-off with Type I errors, because the chance of a Type I error if H_0 is true is simply α, and is not affected by sample size. A larger sample size is always better!

The second implication is that a small sample size can make it hard to find significant results, because your observed statistic would have to be very far from the null value to constitute convincing evidence. This means with a small sample size we may often fail to reject H_0, even if the alternative is true.

Finally, in this age of "big data," we often have samples consisting of millions of credit card records or billions of mouse clicks, where almost every test will yield significant results. This makes it important to avoid confusing statistical significance with practical significance.

Effects of Sample Size

- With a small sample size, it may be hard to find significant results, even when the alternative hypothesis is true.
- With a large sample size, it is easier to find significant results when the alternative hypothesis is true, but we should be especially careful to distinguish between statistical significance and practical significance.

In this section we have learned about some of the ways in which formal hypothesis testing can go wrong. If we reject H_0, it is always possible that the null hypothesis is actually true, and that we made a Type I error. If we fail to reject H_0, it is always possible that the alternative hypothesis is true, and that we made a Type II error. Although we can never fully prevent errors, we should know how to control the chances of making errors (such as by choosing the significance level or increasing the sample size), be able to recognize common situations that are prone to errors (Type I errors are more common when multiple tests are performed and Type II errors are more common with small sample sizes), and recognize the importance of replication as a tool to reveal or dispel Type I errors.

In general, when performing statistical inference, it is important to remember that intervals will not always capture the truth, results can be deemed statistically significant even when the null hypothesis is in fact true, and failing to reject the null hypothesis does not mean the null hypothesis is true.

SECTION LEARNING GOALS

You should now have the understanding and skills to:

- Interpret Type I and Type II errors in hypothesis tests
- Recognize a significance level as measuring the tolerable chance of making a Type I error
- Explain the potential problem with significant results when doing multiple tests
- Recognize the value of replicating a study that shows significant results
- Recognize that statistical significance is not always the same as practical significance
- Recognize that larger sample sizes make it easier to achieve statistical significance if the alternative hypothesis is true

Exercises for Section 4.4

SKILL BUILDER 1

In Exercises 4.120 to 4.123, the same sample statistic is used to test a hypothesis, using different sample sizes. In each case, use *StatKey* or other technology to find the p-value and indicate whether the results are significant at a 5% level. Which sample size provides the strongest evidence for the alternative hypothesis?

4.120 Testing $H_0 : p = 0.5$ vs $H_a : p > 0.5$ using $\hat{p} = 0.55$ with each of the following sample sizes:

(a) $\hat{p} = 55/100 = 0.55$

(b) $\hat{p} = 275/500 = 0.55$

(c) $\hat{p} = 550/1000 = 0.55$

4.121 Testing $H_0 : p = 0.5$ vs $H_a : p > 0.5$ using $\hat{p} = 0.58$ with each of the following sample sizes:

(a) $\hat{p} = 29/50 = 0.58$

(b) $\hat{p} = 290/500 = 0.58$

4.122 Testing $H_0 : p_1 = p_2$ vs $H_a : p_1 > p_2$ using $\hat{p}_1 - \hat{p}_2 = 0.8 - 0.7 = 0.10$ with each of the following sample sizes:

(a) $\hat{p}_1 = 24/30 = 0.8$ and $\hat{p}_2 = 14/20 = 0.7$

(b) $\hat{p}_1 = 240/300 = 0.8$ and $\hat{p}_2 = 140/200 = 0.7$

4.123 Testing $H_0 : p_1 = p_2$ vs $H_a : p_1 > p_2$ using $\hat{p}_1 - \hat{p}_2 = 0.45 - 0.30 = 0.15$ with each of the following sample sizes:

(a) $\hat{p}_1 = 9/20 = 0.45$ and $\hat{p}_2 = 6/20 = 0.30$

(b) $\hat{p}_1 = 90/200 = 0.45$ and $\hat{p}_2 = 60/200 = 0.30$

(c) $\hat{p}_1 = 900/2000 = 0.45$ and $\hat{p}_2 = 600/2000 = 0.30$

SKILL BUILDER 2

In Exercises 4.124 to 4.127, we are conducting many hypothesis tests to test a claim. In every case, *assume that the null hypothesis is true*. Approximately how many of the tests will incorrectly find significance?

4.124 100 tests conducted using a significance level of 5%.

4.125 300 tests using a significance level of 1%.

4.126 40 tests using a significance level of 10%.

4.127 800 tests using a significance level of 5%.

4.128 Interpreting a P-value In each case, indicate whether the statement is a proper interpretation of what a p-value measures.

(a) The probability the null hypothesis H_0 is true.

(b) The probability that the alternative hypothesis H_a is true.

(c) The probability of seeing data as extreme as the sample, when the null hypothesis H_0 is true.

(d) The probability of making a Type I error if the null hypothesis H_0 is true.

(e) The probability of making a Type II error if the alternative hypothesis H_a is true.

4.129 Translating Information to Other Significance Levels Suppose in a two-tailed test of $H_0 : \rho = 0$ vs $H_a : \rho \neq 0$, we reject H_0 when using a 5% significance level. Which of the conclusions below (if any) would also definitely be valid for the same data? Explain your reasoning in each case.

(a) Reject $H_0 : \rho = 0$ in favor of $H_a : \rho \neq 0$ at a 1% significance level.

(b) Reject $H_0 : \rho = 0$ in favor of $H_a : \rho \neq 0$ at a 10% significance level.

(c) Reject $H_0 : \rho = 0$ in favor of the one-tail alternative, $H_a : \rho > 0$, at a 5% significance level, assuming the sample correlation is positive.

4.130 Euchre One of the authors and some statistician friends have an ongoing series of Euchre games that will stop when one of the two teams is deemed to be *statistically significantly* better than the other team. Euchre is a card game and each game results in a win for one team and a loss for the other. Only two teams are competing in this series, which we'll call team A and team B.

(a) Define the parameter(s) of interest.

(b) What are the null and alternative hypotheses if the goal is to determine if either team is statistically significantly better than the other at winning Euchre?

(c) What sample statistic(s) would they need to measure as the games go on?

(d) Could the winner be determined after one or two games? Why or why not?

(e) Which significance level, 5% or 1%, will make the game last longer?

4.131 Flying Home for the Holidays, On Time In Exercise 4.115 on page 302, we compared the average difference between actual and scheduled arrival times for December flights on two major airlines: *Delta* and *United*. Suppose now that we are only interested in the proportion of flights arriving more than 30 minutes after the scheduled time. Of the 1,000 *Delta* flights, 67 arrived more than 30 minutes late, and of the 1,000 *United* flights, 160 arrived more than 30 minutes late. We are testing to see if this provides evidence to conclude that the proportion of flights that are over 30 minutes late is different between flying *United* or *Delta*.

(a) State the null and alternative hypothesis.

(b) What statistic will be recorded for each of the simulated samples to create the randomization distribution? What is the value of that statistic for the observed sample?

(c) Use *StatKey* or other technology to create a randomization distribution. Estimate the p-value for the observed statistic found in part (b).

(d) At a significance level of $\alpha = 0.01$, what is the conclusion of the test? Interpret in context.

(e) Now assume we had only collected samples of size 75, but got essentially the same proportions (5/75 late flights for *Delta* and 12/75 late flights for *United*). Repeating steps (b) through (d) on these smaller samples, do you come to the same conclusion?

4.132 Flaxseed and Omega-3 Exercise 4.30 on page 271 describes a company that advertises that its milled flaxseed contains, on average, at least 3800 mg of ALNA, the primary omega-3 fatty acid in flaxseed, per tablespoon. In each case below, which of the standard significance levels, 1% or 5% or 10%, makes the most sense for that situation?

(a) The company plans to conduct a test just to double-check that its claim is correct. The company is eager to find evidence that the average amount per tablespoon is greater than 3800 (their alternative hypothesis), and is not really worried about making a mistake. The test is internal to the company and there are unlikely to be any real consequences either way.

(b) Suppose, instead, that a consumer organization plans to conduct a test to see if there is evidence *against* the claim that the product contains at least 3800 mg per tablespoon. If the organization finds evidence that the advertising claim is false, it will file a lawsuit against the flaxseed company. The organization wants to be very sure that the evidence is strong, since if the company is sued incorrectly, there could be very serious consequences.

SELECTING A SIGNIFICANCE LEVEL

For each situation described in Exercises 4.133 to 4.138, indicate whether it makes more sense to use a relatively large significance level (such as $\alpha = 0.10$) or a relatively small significance level (such as $\alpha = 0.01$).

4.133 Testing a new drug with potentially dangerous side effects to see if it is significantly better than the drug currently in use. If it is found to be more effective, it will be prescribed to millions of people.

4.134 Using your statistics class as a sample to see if there is evidence of a difference between male and female students in how many hours are spent watching television per week.

4.135 Using a sample of 10 games each to see if your average score at Wii bowling is significantly more than your friend's average score.

4.136 Testing to see if a well-known company is lying in its advertising. If there is evidence that the company is lying, the Federal Trade Commission will file a lawsuit against them.

4.137 Testing to see whether taking a vitamin supplement each day has significant health benefits. There are no (known) harmful side effects of the supplement.

4.138 A pharmaceutical company is testing to see whether its new drug is significantly better than the existing drug on the market. It is more expensive than the existing drug. Which significance level would the company prefer? Which significance level would the consumer prefer?

TYPE I AND TYPE II ERRORS

For each situation given in Exercises 4.139 to 4.143, describe what it means in that context to make a Type I and Type II error. Personally, which do you feel is a worse error to make in the given situation?

4.139 The situation described in Exercise 4.133.

4.140 The situation described in Exercise 4.134.

4.141 The situation described in Exercise 4.135.

4.142 The situation described in Exercise 4.136.

4.143 The situation described in Exercise 4.137.

4.144 Influencing Voters Exercise 4.39 on page 272 describes a possible study to see if there is evidence that a recorded phone call is more effective than a mailed flyer in getting voters to support a certain candidate. The study assumes a significance level of $\alpha = 0.05$.

(a) What is the conclusion in the context of this study if the p-value for the test is 0.027?

(b) In the conclusion in part (a), which type of error are we possibly making: Type I or Type II? Describe what that type of error means in this situation.

(c) What is the conclusion if the p-value for the test is 0.18?

(d) In the conclusion in part (c), which type of error are we possibly making: Type I or Type II? Describe what that type of error means in this situation.

4.145 Significant and Insignificant Results

(a) If we are conducting a statistical test and determine that our sample shows significant results, there are two possible realities: We are right in our conclusion or we are wrong. In each case, describe the situation in terms of hypotheses and/or errors.

(b) If we are conducting a statistical test and determine that our sample shows insignificant results, there are two possible realities: We are right in our conclusion or we are wrong. In each case, describe the situation in terms of hypotheses and/or errors.

(c) Explain why we generally won't ever know which of the realities (in either case) is correct.

4.146 Weight Loss Program Suppose that a weight loss company advertises that people using its program lose an average of 8 pounds the first month, and that the Federal Trade Commission (the main government agency responsible for truth in advertising) is gathering evidence to see if this advertising claim is accurate. If the FTC finds evidence that the average is less than 8 pounds, the agency will file a lawsuit against the company for false advertising.

(a) What are the null and alternative hypotheses the FTC should use?

(b) Suppose that the FTC gathers information from a very large random sample of patrons and finds that the average weight loss during the first month in the program is $\bar{x} = 7.9$ pounds with a p-value for this result of 0.006. What is the conclusion of the test? Are the results statistically significant?

(c) Do you think the results of the test are practically significant? In other words, do you think patrons of the weight loss program will care that the average is 7.9 pounds lost rather than 8.0 pounds lost? Discuss the difference between practical significance and statistical significance in this context.

4.147 Do iPads Help Kindergartners Learn: A Subtest The Auburn, Maine, school district conducted an early literacy experiment in the fall of 2011. In September, half of the kindergarten classes were randomly assigned iPads (the intervention group) while the other half of the classes got them in December (the control group.) Kids were tested in September and December and the study measures the average difference in score gains between the control

and intervention group.[41] The experimenters tested whether the mean score for the intervention group was higher on the HRSIW subtest (Hearing and Recording Sounds in Words) than the mean score for the control group.

(a) State the null and alternative hypotheses of the test and define any relevant parameters.

(b) The p-value for the test is 0.02. State the conclusion of the test in context. Are the results statistically significant at the 5% level?

(c) The effect size was about two points, which means the mean score for the intervention group was approximately two points higher than the mean score for the control group on this subtest. A school board member argues, "While these results might be statistically significant, they may not be practically significant." What does she mean by this in this context?

4.148 Do iPads Help Kindergartners Learn: A Series of Tests Exercise 4.147 introduces a study in which half of the kindergarten classes in a school district are randomly assigned to receive iPads. We learn that the results are significant at the 5% level (the mean for the iPad group is significantly higher than for the control group) for the results on the HRSIW subtest. In fact, the HRSIW subtest was one of 10 subtests and the results were not significant for the other 9 tests. Explain, using the problem of multiple tests, why we might want to hesitate before we run out to buy iPads for all kindergartners based on the results of this study.

4.149 Eating Breakfast Cereal and Conceiving Boys *Newscientist.com* ran the headline "Breakfast Cereals Boost Chances of Conceiving Boys," based on an article which found that women who eat breakfast cereal before becoming pregnant are significantly more likely to conceive boys.[42] The study used a significance level of $\alpha = 0.01$. The researchers kept track of 133 foods and, for each food, tested whether there was a difference in the proportion conceiving boys between women who ate the food and women who didn't. Of all the foods, only breakfast cereal showed a significant difference.

(a) If none of the 133 foods actually have an effect on the gender of a conceived child, how many (if any) of the individual tests would you expect to

[41] Reich, J., "Are iPads Making a Significant Difference? Findings from Auburn, Maine," *Ed Tech Researcher*, February 17, 2012.

[42] Mathews, F., Johnson, P.J., and Neil, A., "You are what your mother eats: Evidence for maternal preconception diet influencing foetal sex in humans." *Proceedings of the Royal Society B: Biological Sciences*. 2008; 275: 1643,1661–68.

show a significant result just by random chance? Explain. (*Hint*: Pay attention to the significance level.)

(b) Do you think the researchers made a Type I error? Why or why not?

(c) Even if you could somehow ascertain that the researchers did not make a Type I error, that is, women who eat breakfast cereals are actually more likely to give birth to boys, should you believe the headline "Breakfast Cereals Boost Chances of Conceiving Boys"? Why or why not?

4.150 Approval from the FDA for Antidepressants The FDA (US Food and Drug Administration) is responsible for approving all new drugs sold in the US. In order to approve a new drug for use as an antidepressant, the FDA requires two results from randomized double-blind experiments showing the drug is more effective than a placebo at a 5% level. The FDA does not put a limit on the number of times a drug company can try such experiments. Explain, using the problem of multiple tests, why the FDA might want to rethink its guidelines.

4.151 Does Massage Really Help Reduce Inflammation in Muscles? In Exercise 4.112 on page 301, we learn that massage helps reduce levels of the inflammatory cytokine interleukin-6 in muscles when muscle tissue is tested 2.5 hours after massage. The results were significant at the 5% level. However, the authors of the study actually performed 42 different tests: They tested for significance with 21 different compounds in muscles and at two different times (right after the massage and 2.5 hours after).

(a) Given this new information, should we have less confidence in the one result described in the earlier exercise? Why?

(b) Sixteen of the tests done by the authors involved measuring the effects of massage on muscle metabolites. None of these tests were significant. Do you think massage affects muscle metabolites?

(c) Eight of the tests done by the authors (including the one described in the earlier exercise) involved measuring the effects of massage on inflammation in the muscle. Four of these tests were significant. Do you think it is safe to conclude that massage really does reduce inflammation?

4.152 Cat Ownership and Schizophrenia Exercise 4.119 on page 303 revealed an association between owning a cat as a child and developing schizophrenia later in life. Many people enjoy cats as pets, so this conclusion has profound implications and could change pet ownership habits substantially. However, because of the chance for false positives (Type I errors) and potential problems with generalizability, good scientific conclusions rarely rest on a foundation of just one study. Because of this, significant results often require *replication* with follow up studies before they are truly trusted. If study results can be replicated, especially in a slightly different setting, they become more trustworthy, and if results can not be replicated, suspicions of a Type I error (significant results by random chance) or a lack of generalizability from the setting of the initial study may arise. In fact, the paper[43] cited in Exercise 4.119 actually provided three different datasets, all from different years (1982, 1992, and 1997) and with different choices for choosing the control group. The sample proportions for each dataset, with the sample sizes in the denominator, are given in Table 4.13.

(a) As we know, statistics vary from sample to sample naturally, so it is not surprising that the sample proportions differ slightly from year to year. However, does the relative consistency of the sample proportions affect the credibility of any single dataset?

(b) Use technology to calculate the p-value for each dataset, testing the alternative hypothesis that the proportion of cat owners is higher among schizophrenics.

(c) Do all datasets yield significant results? Should this increase or decrease potential suspicions that the significance of any single study may have been just a Type I error?

[43]Torrey, E.F., Simmons, W., and Yolken, R.H. (2015). "Is childhood cat ownership a risk factor for schizophrenia later in life," *Schizophrenia Research*, June 2015, 165(1):1–2.

Table 4.13 *Cat Ownership and Schizophrenia*

Year	Proportion of schizophrenics who owned cats as children	Proportion of controls who owned cats as children
1982 Data (Analyzed in 2015)	$1075/2125 = 0.506$	$2065/4847 = 0.426$
1992 Data	$84/165 = 0.509$	$65/165 = 0.394$
1997 Data	$136/262 = 0.519$	$220/522 = 0.421$

(d) Why is the p-value lowest for the 1982 data, even though this dataset yields the smallest difference in proportions? Similarly, why is the p-value highest for the 1992 data, even though this data yielded the largest difference in proportions?

4.153 Mating Choice and Offspring Fitness Does the ability to choose a mate improve offspring fitness in fruit flies? Researchers have studied this by taking female fruit flies and randomly dividing them into two groups; one group is put into a cage with a large number of males and able to freely choose who to mate with, while flies in the other group are each put into individual vials, each with only one male, giving no choice in who to mate with. Females are then put into egg laying chambers, and a certain number of larvae collected. Do the larvae from the mate choice group exhibit higher survival rates? A study[44] published in *Nature* found that mate choice *does* increase offspring fitness in fruit flies (with p-value < 0.02), yet this result went against conventional wisdom in genetics and was quite controversial. Researchers attempted to replicate this result with a series of related experiments,[45] with data provided in **MateChoice**.

(a) In the first replication experiment, using the same species of fruit fly as the original *Nature* study, 6067 of the 10000 larvae from the mate choice group survived and 5976 of the 10000 larvae from the no mate choice group survived. Calculate the p-value.

(b) Using a significance level of $\alpha = 0.05$ and p-value from (a), state the conclusion in context.

(c) Actually, the 10,000 larvae in each group came from a series of 50 different runs of the experiment, with 200 larvae in each group for each run. The researchers believe that conditions differ from run to run, and thus it makes sense to treat each *run* as a case (rather than each fly). In this analysis, we are looking at paired data, and the response variable would be the *difference* in the number of larvae surviving between the choice group and the no choice group, for each of the 50 runs. The counts (*Choice* and *NoChoice* and difference (*Choice − NoChoice*) in number of surviving larva are stored in **MateChoice**. Using the single variable of differences,

[44]Patridge, L. (1980). "Mate choice increases a component of offspring fitness in fruit flies," *Nature*, 283:290–291, 1/17/80.
[45]Schaeffer, S.W., Brown, C.J., and Anderson, W.W. (1984). "Does mate choice affect fitness?" *Genetics*, 107:s94. Thank to Stephen Schaeffer for providing the data. The data is included here with permission from Wyatt W. Anderson (University of Georgia), Celeste J. Brown (University of Idaho), and Stephen W. Schaeffer (Penn State).

calculate the p-value for testing whether the average difference is greater than 0. (*Hint*: this is a single quantitative variable, so the corresponding test would be for a single mean.)

(d) Using a significance level of $\alpha = 0.05$ and the p-value from (c), state the conclusion in context.

(e) The experiment being tested in parts (a)–(d) was designed to mimic the experiment from the original study, yet the original study yielded significant results while this study did not. If mate choice really does improve offspring fitness in fruit flies, did the follow-up study being analyzed in parts (a)–(d) make a Type I, Type II, or no error?

(f) If mate choice really does not improve offspring fitness in fruit flies, did the original *Nature* study make a Type I, Type II, or no error?[46]

4.154 Mating Choice and Offspring Fitness: Mini-Experiments Exercise 4.153 explores the question of whether mate choice improves offspring fitness in fruit flies, and describes two seemingly identical experiments yielding conflicting results (one significant, one insignificant). In fact, the second source was actually a series of three different experiments, and each full experiment was comprised of 50 different mini-experiments (runs), 10 each on five different days.

(a) Suppose each of the 50 mini-experiments from the first study were analyzed individually. If mating choice has no impact on offspring fitness, about how many of these 50 p-values would you expect to yield significant results at $\alpha = 0.05$?

(b) The 50 p-values, testing the alternative $H_a : p_C > p_{NC}$ (proportion of flies surviving is higher in the mate choice group) are given below:

Day 1: 0.96 0.85 0.14 0.54 0.76 0.98 0.33 0.84 0.21 0.89
Day 2: 0.89 0.66 0.67 0.88 1.00 0.01 1.00 0.77 0.95 0.27
Day 3: 0.58 0.11 0.02 0.00 0.62 0.01 0.79 0.08 0.96 0.00
Day 4: 0.89 0.13 0.34 0.18 0.11 0.66 0.01 0.31 0.69 0.19
Day 5: 0.42 0.06 0.31 0.24 0.24 0.16 0.17 0.03 0.02 0.11

How many are actually significant using $\alpha = 0.05$?

(c) You may notice that two p-values (the fourth and last run on day 3) are 0.00 when rounded to two decimal places. The second of these is actually

[46]Subsequent studies suggest that this is more likely the true scenario, at least for populations in genetic equilibrium.

0.0001 if we report more decimal places. This is very significant! Would it be appropriate and/or ethical to just report this one run, yielding highly statistically significant evidence that mate choice improves offspring fitness? Explain.

(d) You may also notice that two of the p-values on day 2 are 1 (rounded to two decimal places). If we had been testing the *opposite* alternative, H_a : $p_C < p_{NC}$ (proportion surviving is *lower* in the mate choice group), would these two runs yield significant results that *preventing* mate choice actually improves offspring fitness? Explain.

(e) Using the problem of multiple testing, explain why it is possible to get such contradictory results.

(f) In fact, the replication studies involved three different sets of experiments, each with a different type of mutant fly introduced for competition, and with two different species of fruit flies being studied. While the original *Nature* study yielded statistically significant results in favor of mate choice, and the first replication study yielded insignificant results, the latter two replication studies yielded statistically significant results in favor of no mate choice. Explain why replication of studies (the replication of a study in a different setting or by a different researcher) is an important part of science.

4.5 MAKING CONNECTIONS

In this section, we connect some of the ideas of the previous sections together. We start by connecting the bootstrap distributions of Chapter 3 with the randomization distributions of this chapter. We then examine the connection between confidence intervals and hypothesis tests. Finally, we connect the definition of randomization distributions with how randomization distributions are actually created.

Connecting Randomization and Bootstrap Distributions

In Chapter 3 we examine methods to construct confidence intervals for population parameters. We sample (with replacement) from the original sample to create a *bootstrap distribution* of possible values for a sample statistic. Based on this distribution, we produce a range of plausible values for the parameter so that we have some degree of certainty that the interval will capture the actual parameter value for the population.

In this chapter we develop methods to test claims about populations. After specifying null and alternative hypotheses, we assess the evidence in a sample by constructing a *randomization distribution* of possible sample statistics that we might see by random chance, if the null hypothesis were true. If the original sample statistic falls in an unlikely location of the randomization distribution, we have evidence to reject the null hypothesis in favor of the alternative.

You have probably noticed similarities in these two approaches. Both use some sort of random process to simulate many samples and then collect values of a sample statistic for each of those samples to form a distribution. In both cases we are generally concerned with distinguishing between "typical" values in the middle of a distribution and "unusual" values in one or both tails. Assuming that the values in a bootstrap or randomization distribution reflect what we might expect to see if we could generate many sets of sample data, we use the information based on our original sample to make some inference about what actually might be true about a population, parameter, or relationship.

> ## Sampling Distribution, Bootstrap Distribution, and Randomization Distribution
>
> A *sampling distribution* shows the distribution of sample statistics from a population, and is generally centered at the true value of the population parameter.
>
> A *bootstrap distribution* simulates a distribution of sample statistics for the population, but is generally centered at the value of the original sample statistic.
>
> A *randomization distribution* simulates a distribution of sample statistics for a population in which the null hypothesis is true, and is generally centered at the value of the null parameter.

We take a closer look at the connection between a bootstrap distribution and a randomization distribution at the end of this section (page 330).

Connecting Confidence Intervals and Hypothesis Tests

In Chapter 3, we see that a confidence interval shows us the plausible values of the population parameter. In Chapter 4, we use a hypothesis test to determine whether a given parameter in a null hypothesis is plausible or not. Thus, we can use a confidence interval to make a decision in a hypothesis test, and we can use a hypothesis test to determine whether a given value will be inside a confidence interval!

> ## Connection between Confidence Intervals and Hypothesis Tests
>
> The formal decision to a two-tailed hypothesis test is related to whether or not the null parameter falls within a confidence interval:
>
> - When the parameter value given in H_0 falls *outside* of a 95% confidence interval, then it is not a plausible value for the parameter and we should reject H_0 at a 5% level in a two-tailed test.
> - When the parameter value specified by H_0 falls *inside* of a 95% confidence interval, then it is a plausible value for the parameter and we should not reject H_0 at a 5% level in a two-tailed test.

This relationship is very flexible: It can be applied to different parameters and we can use different significance levels by adjusting the confidence level accordingly. For example, a 1% test would correspond to seeing if the hypothesized value is within a 99% confidence interval and a significance level of 10% would use a 90% confidence interval.

Example 4.30

CAOS Exam

The Comprehensive Assessment of Outcomes in Statistics[47] (CAOS) exam is a standardized test for assessing students' knowledge of statistical concepts. The questions on this exam have been tested extensively to establish benchmarks for how well students do when answering them. One of the tougher questions, dealing with misinterpretations of a confidence interval, is answered correctly by about 42% of all statistics students. A statistics instructor gets the results for 30 students in a class and finds that 17 of the students ($\hat{p} = 17/30 = 0.567$) answered the confidence interval question correctly. Based on these sample results a 95% confidence interval for the proportion of students with this instructor who get the question correct goes from 0.39 to 0.75. We assume that the 30 students who answered the question are a random sample of this instructor's students.

(a) Based on this confidence interval, is the instructor justified in saying the proportion of his students who get the question correct is different from the baseline national proportion of $p = 0.42$?

(b) This question is in a multiple-choice format with four possible answers, only one of which is correct. Can the instructor conclude that his students are not just guessing on this question?

Solution

(a) If the hypotheses are $H_0 : p = 0.42$ and $H_a : p \neq 0.42$, we see that the null proportion is within the 95% confidence interval, $(0.39, 0.75)$, so using a 5% significance level we would not reject H_0. The instructor would not have sufficient evidence to conclude that the proportion correct for his students is different than 0.42.

(b) If students are just guessing, the proportion correct for a question with four choices is $p = 0.25$. Since 0.25 is not within the 95% confidence interval, we reject H_0 and the instructor can conclude (using a 5% significance level) that the proportion of correct answers for this question is different from 0.25. The students are doing better than merely guessing at random.

**About 59% of Americans favored a ban
on smoking in restaurants**

[47] *https://app.gen.umn.edu/artist/caos.html.*

Example 4.31

Smoking in Restaurants

Attitudes on banning smoking in restaurants have evolved over time and may impact public policy. In a Gallup poll of American adults in August 2010, 59% of the respondents favored a total ban on smoking in restaurants.[48] In a similar survey a decade earlier the proportion who favored such a ban was only 40%. We use these two samples to construct a 95% confidence interval for the difference in proportion of support for a smoking ban in restaurants between these two years, $p_2 - p_1$, where p_2 is the proportion in 2010 and p_1 is the proportion in 2000. The confidence interval for the difference in proportions is 0.147 to 0.233.

(a) Does this confidence interval provide sufficient evidence at a 5% level that the proportion of Americans supporting a ban on smoking in restaurants was different in 2010 than it was in 2000?

(b) What conclusions (if any) could we draw if the significance level was 10% or 1%?

Solution

(a) When testing $H_0 : p_2 = p_1$, the null difference in proportions is $p_2 - p_1 = 0$. Since the 95% confidence interval for $p_2 - p_1$ does not include zero, we have sufficient evidence (at a 5% level) to reject H_0 and conclude that the proportion of Americans favoring the smoking ban changed over the decade.

Since the confidence interval includes only positive differences, we can go even further and conclude that the proportion supporting such a ban was *higher* in 2010 than it was in 2000. This conclusion may seem more appropriate for a one-tailed test, but note that a sample statistic which is far enough in the tail to reject H_0 for a two-tailed test will also reject H_0 for a one-tailed test in that direction.

(b) Since part (a) indicates that we should reject H_0 at a 5% significance level, we know we would also reject H_0 at the larger 10% level and draw the same conclusion. However, we cannot reach a firm decision for a 1% test based only on the results of the 95% confidence interval for the difference in proportions. Since that is a stricter significance level, we would need to either construct a 99% confidence interval for the difference in proportions or carry out the actual details of the hypothesis test to make a decision at the 1% level.

Confidence Intervals and Hypothesis Tests Are Both Still Important!

Since we can use a confidence interval to make a conclusion in a hypothesis test, you might be wondering why we bother with significance tests at all. Couldn't we just always compute a confidence interval and then check whether or not it includes some hypothesized value for a parameter? If we adopted this approach, we could make a reject/not reject decision, but we lose information about the strength of evidence. For example, when actually doing a hypothesis test for the situation in Example 4.31, the p-value is less than 0.0001, indicating very strong evidence that the proportion of Americans who support a total ban on smoking in restaurants has increased over the decade from 2000 to 2010. On the other hand, the question of interest is often "how big is the difference?" not does a difference exist at all. In that case the confidence interval for the difference in proportions, (0.147, 0.233), is more useful than just knowing that the p-value is very small. Confidence intervals and hypothesis tests are both important inference procedures, and which is most relevant in a particular situation depends on the question of interest.

[48]*http://www.gallup.com/poll/141809/Americans-Smoking-Off-Menu-Restaurants.aspx.*

Example 4.32

For each question, indicate whether it is best assessed using a confidence interval or a hypothesis test, or whether statistical inference is not relevant to answer it.

(a) Estimate the proportion of infant car seats that are installed correctly, using a sample of new parents leaving the hospital.

(b) Use the sample from part (a) to determine whether the proportion of car seats installed correctly differs between two of the leading infant car seat brands.

(c) Look in the US Congressional Record to see what proportion of times a senator was absent for a vote during the previous year.

Solution

(a) We are estimating a population proportion, so a confidence interval is most appropriate.

(b) We want to answer the question: Is the proportion different between the two brands? A hypothesis test is most appropriate.

(c) Inference is not relevant in this case since we have access to the entire population of all votes taken during the year. Remember that we use inference procedures when we want to generalize beyond a sample.

Creating Randomization Distributions

Although we introduce randomization distributions in Section 4.2, our main focus in that section is on understanding a p-value. In this section, we focus more explicitly on how randomization distributions are created. As we saw in Section 4.2, we select randomization samples by focusing on two important goals.

> **Generating Randomization Samples**
>
> The main criteria to consider when creating randomization samples for a statistical test are:
>
> • Be consistent with the null hypothesis.
> • Use the data in the original sample.
>
> Ideally, we also try to reflect the way the original data were collected.

Several examples of randomization distributions are given in Section 4.2. For the difference in means test for the light at night experiment, the mice were randomly assigned to one of two treatment groups (Light or Dark) and their body mass gain was measured after treatment. The null hypothesis states that the treatment has no effect on the mean body mass gain ($H_0 : \mu_L = \mu_D$). To generate randomization samples, we reassign the treatment labels at random to body mass gains (to satisfy H_0 and use the original sample).

Our goal in that situation is to generate lots of samples that mimic what we see in the original sample but use a random process (random assignment of mice to groups) that is consistent with both the null hypothesis and the way the actual sample was generated. By doing so we can assess how unlikely the original sample results might be when H_0 is true.

In this section, we describe four additional examples: randomization tests for a difference in two proportions, a correlation, a single proportion, and a single mean. Before reading the details for each example, you might think for a minute about how you might generate randomization samples for each situation that satisfy the criteria in the box above. For example, you might put the original data on cards to shuffle, flip a coin, or use the data to create a new population that satisfies the null hypothesis. One goal in this section is to understand the basic principles behind creating a randomization distribution.

Randomization Test for a Difference in Proportions: Cocaine Addiction

DATA 4.7

Cocaine Addiction

Cocaine addiction is very hard to break. Even among addicts trying hard to break the addiction, relapse is common. (A relapse is when a person trying to break out of the addiction fails and uses cocaine again.) One experiment[49] investigates the effectiveness of the two drugs desipramine and lithium in the treatment of cocaine addiction. The subjects in the study were cocaine addicts seeking treatment, and the study lasted six weeks. The 72 subjects were randomly assigned to one of three groups (desipramine, lithium, or a placebo, with 24 subjects in each group) and the study was double blind. The results of the study are summarized in Table 4.14. ∎

Table 4.14 *Treatment for cocaine addiction*

	Relapse	No relapse
Desipramine	10	14
Lithium	18	6
Placebo	20	4

For now, we focus on comparing the data for those in the lithium group with those taking the placebo. (Exercise 4.180 asks you to consider desipramine versus a placebo.) The question of interest is whether lithium is more effective at preventing relapse than taking an inert pill. We define parameters p_l and p_n, the proportion who relapse while taking lithium and the placebo, respectively, and test

$$H_0 : \quad p_l = p_n$$
$$H_a : \quad p_l < p_n$$

In the sample data we see that the proportion of subjects using lithium who relapsed ($\hat{p}_l = 18/24 = 0.75$) is smaller than the proportion who relapsed with the placebo ($\hat{p}_n = 20/24 = 0.83$). That result is in the direction of H_a, but is that difference statistically significant? We construct a randomization distribution and use it to address this question.

Example 4.33

Explain how to use cards to generate one randomization sample for the test to see if lithium is more effective than a placebo. What statistic is recorded for the sample?

Solution   Since this was a designed experiment and the treatments (lithium or placebo) were assigned at random, our procedure for generating randomization samples should also be based on random assignments. Suppose that the null hypothesis is true and lithium is no more effective than the placebo, so all participants would have had the same

---

[49]Gawin, F., et. al., "Desipramine Facilitation of Initial Cocaine Abstinence", *Archives of General Psychiatry*, 1989; 46(2):117–121.

response (relapse or no relapse) if they had been assigned to the other group. If we construct a deck of 48 cards with 38 "relapse" cards and 10 "no relapse" cards, we could shuffle the deck and deal the cards into two piles of 24 to simulate the random assignments into the lithium and placebo groups. Randomly dealing the cards into these two piles of 24 gives us one randomization sample. What statistic should we record for each of the randomization samples? Since we are conducting a test for a difference in proportions, an obvious choice is the difference in sample proportions, $\hat{p}_l - \hat{p}_n$.

While dealing cards may help us understand what is going on in a randomization distribution, in practice, of course, we use technology to generate a randomization distribution and to compute a p-value.

Example 4.34

Use *StatKey* or other technology to generate a randomization distribution for the difference in proportions between the lithium group and the placebo group. Use the sample data to find a p-value. What is the conclusion of the test?

Solution

Figure 4.29 shows a dotplot of the difference in proportions for 5000 randomization samples. As expected, the distribution is centered approximately at zero. The original sample difference in proportions is $\hat{p}_l - \hat{p}_n = 0.75 - 0.83 = -0.08$. Since the alternative hypothesis is $H_a : p_l < p_n$, this is a left-tail test. We see in Figure 4.29 that the area to the left of -0.08 is 0.36. The p-value is 0.36. Since this p-value is not less than any reasonable significance level, we do not reject H_0. We do not have sufficient evidence that lithium works better than a placebo when treating cocaine addiction.

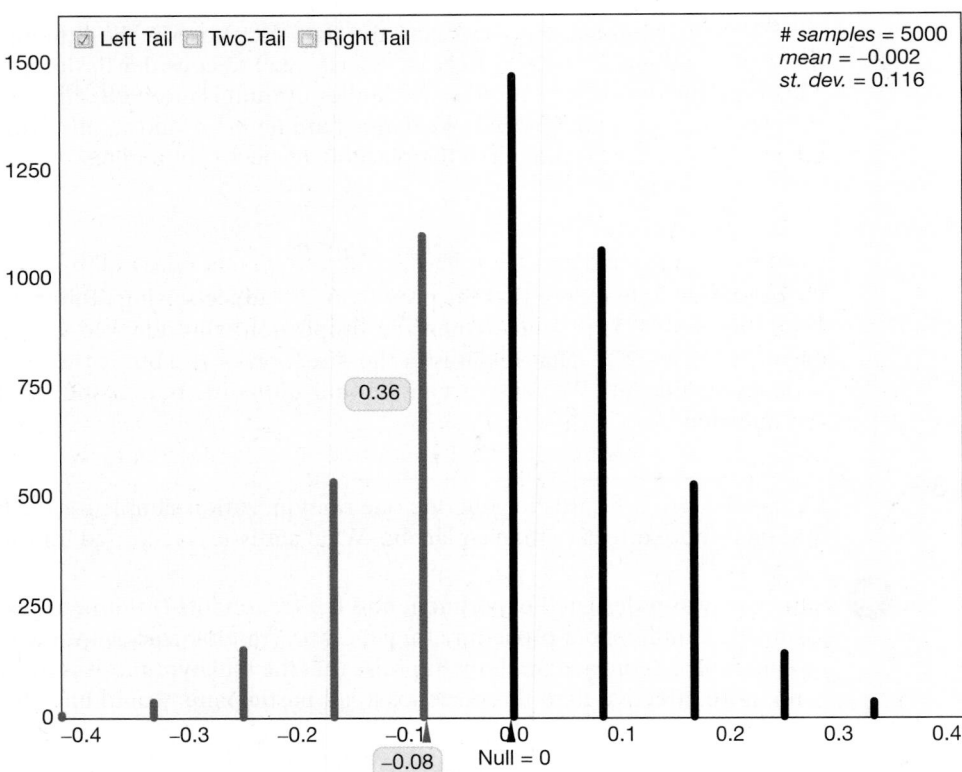

Figure 4.29 *Randomization distribution of difference in proportions*

Using Other Statistics

In Example 4.33, we dealt cards into two piles and computed the difference in proportion of relapses between the two piles. Now imagine that you were actually shuffling and dealing the cards over and over again. You might soon realize that you don't really need to divide each relapse count by 24, since comparing the difference in the counts is equally effective. After a few more randomizations you might also realize that you just need to count the relapse cards in the lithium pile, since the total number of relapse cards is always the same (38). All three of these ways to record a sample statistic are effective, since each offers an equivalent way to measure how extreme the original sample results are. We could have used any of these three statistics in constructing a randomization distribution.

We often have this sort of flexibility in choosing a sample statistic. One of the powerful aspects of the randomization approach is that we can apply it to whatever statistic makes sense for a sample. As long as the statistic we use consistently measures which samples are more extreme than the original data, the results (i.e., the p-values) from the randomization distributions will be the same.

Randomization Test for a Correlation: Malevolent Uniforms and Penalties

In Data 4.2 on page 263 we consider the question of whether the perceived malevolence score of NFL team jerseys (*NFL_Malevolence*) is related to the aggressiveness of the team as measured by a standardized score for number of penalty yards (*ZPenYds*). If we let ρ be the population correlation for all teams in all years, we want to see if malevolence is positively associated with penalty yards. We have

$$H_0: \quad \rho = 0$$
$$H_a: \quad \rho > 0$$

The data for the study are stored in the **MalevolentUniformsNFL** file. For the sample of $n = 28$ NFL teams in the years of the original study, the correlation between *NFL_Malevolence* and *ZPenYds* is $r = 0.43$. We create a randomization distribution of sample correlations to assess the strength of evidence that the original sample correlation, $r = 0.43$, has in this situation and make a conclusion about a possible relationship between the perceived malevolence of uniforms and penalty yards.

Example 4.35

To construct the randomization distribution, we assume the null hypothesis is true. What does that mean in this case? How might we construct a randomization sample using the original data while also assuming that the null hypothesis is true? What statistic do we record for each randomization sample?

Solution The null hypothesis is $\rho = 0$, which means that *NFL_Malevolence* and *ZPenYds* are really unrelated. This would mean that there is no connection between the two data columns, and that any number in one column could just as easily be matched with any number in the other column. One way to simulate this physically is to put the 28 *ZPenYds* values on index cards, shuffle, and randomly deal the cards to the 28 teams. This way the *ZPenYds* value that gets paired with each *NFL_Malevolence* value happens by random chance. In this method, we use the data that we have while also forcing the null hypothesis to be true. For each such randomization sample, we compute the sample correlation.

After computing one randomization statistic as in Example 4.35, we shuffle the cards again and deal out another set of *ZPenYds* assignments. Using this process of creating the randomization samples, we ensure no association between *ZPenYds*

Table 4.15 *Original ZPenYds and four random assignments*

NFL Team	Malevolence	ZPenYds	ZPenYds_1	ZPenYds_2	ZPenYds_3	ZPenYds_4
LA Raiders	5.10	1.19	−0.19	0.02	−0.41	1.19
Pittsburgh	5.00	0.48	0.02	0.10	0.27	−0.19
Cincinnati	4.97	0.27	0.38	0.23	−0.01	0.24
New Orleans	4.83	0.10	−0.49	−0.07	−0.73	−0.01
Chicago	4.68	0.29	0.10	1.19	0.38	−0.07
Kansas City	4.58	−0.19	−0.01	0.48	−0.18	0.27
Washington	4.40	−0.07	1.19	0.27	−0.49	0.23
St. Louis	4.27	−0.01	−0.32	0.24	−0.81	−0.81
NY Jets	4.12	0.01	0.01	−0.32	0.02	0.48
LA Rams	4.10	−0.09	0.23	−0.49	−1.60	0.10
Cleveland	4.05	0.44	−0.73	−0.01	0.63	0.38
San Diego	4.05	0.27	0.48	0.29	−0.07	−0.73
Green Bay	4.00	−0.73	−0.18	0.04	0.29	0.29
Philadelphia	3.97	−0.49	0.29	0.09	0.44	−1.60
Minnesota	3.90	−0.81	−0.09	−0.41	−0.19	0.44
Atlanta	3.87	0.30	−0.19	−0.19	0.27	−0.18
Indianapolis	3.83	−0.19	−0.07	0.38	0.09	0.02
San Francisco	3.83	0.09	−1.60	0.01	−0.32	−0.32
Seattle	3.82	0.02	0.09	−0.18	1.19	−0.41
Denver	3.80	0.24	0.27	−1.60	0.30	0.27
Tampa Bay	3.77	−0.41	0.30	−0.09	0.01	0.01
New England	3.60	−0.18	−0.41	0.27	0.04	−0.19
Buffalo	3.53	0.63	0.27	−0.81	0.10	−0.09
Detroit	3.38	0.04	0.44	0.44	−0.09	−0.49
NY Giants	3.27	−0.32	0.04	0.30	0.48	0.09
Dallas	3.15	0.23	−0.81	−0.19	−0.19	0.04
Houston	2.88	0.38	0.63	0.63	0.24	0.30
Miami	2.80	−1.60	0.24	−0.73	0.23	0.63
Correlation		**0.43**	**−0.02**	**0.23**	**−0.26**	**0.08**

and *NFL_Malevolence* values, as required by the null hypothesis. Thus we can use the original data to build up a distribution of typical sample correlations under the assumption that the null hypothesis, $\rho = 0$, holds. Table 4.15 shows the original malevolent uniform data and four random assignments of the *ZPenYds* values along with the sample correlation of each assignment.

Of course, in practice we use technology to simulate this process and generate the randomization distribution of sample correlations more efficiently, as in the next example.

Example 4.36

Use *StatKey* or other technology to create a randomization distribution and find a p-value for the malevolent uniform data. What conclusion can we make about the relationship (if any) between the perceived malevolence of uniforms and penalty yards for NFL teams?

Solution

Figure 4.30 shows a randomization distribution for these data. Since $H_a : \rho > 0$, this is a right-tail test. The sample correlation is $r = 0.43$ and we are interested in what proportion of the randomization statistics are more extreme than $r = 0.43$. We see in Figure 4.30 that

$$\text{p-value} = 0.011$$

This small p-value gives fairly strong evidence to conclude that there is a positive association between the malevolence score of NFL uniforms and the number of penalty yards a team receives.

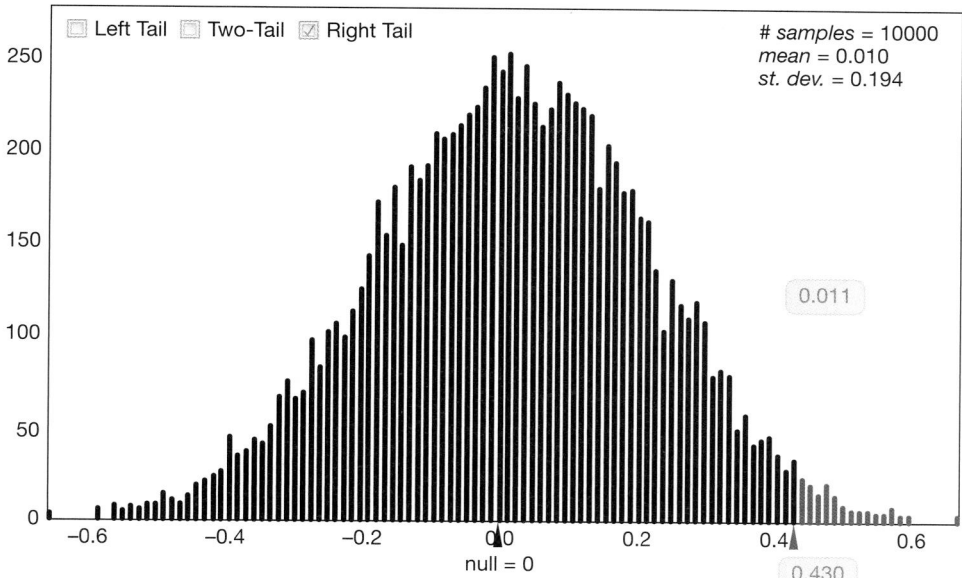

Figure 4.30 *Randomization distribution of NFL_Malevolence vs ZPenYds correlations when $\rho = 0$*

Take care when interpreting this conclusion to avoid assuming a cause-effect relationship since these data come from an observational study and not an experiment. It may be true that referees deal more harshly with malevolent-looking players or that donning such a uniform might instill a more aggressive attitude. However, it might also be the case that owners or coaching staffs who value aggressiveness might select players with such attitudes and also choose a more malevolent team uniform.

Randomization Test for a Proportion: CAOS Exam

Example 4.30 on page 320 describes a situation where a statistics instructor is interested in evaluating the performance of a sample of 30 students on a particular question of a multiple choice exam. Each question had four possible choices, so the proportion correct would be $p = 0.25$ if students are just randomly guessing, but higher if the students know something about the substance of the question.

Example 4.37

Describe how we might use cards to create a randomization sample of 30 values that satisfies $H_0 : p = 0.25$. Also, explain how to compute a randomization statistic for that sample.

Solution

Since the multiple choice question has four choices, we can use four cards, one labeled "Right" and the other three labeled "Wrong." To simulate an answer for a student who is just guessing, we just select a card at random and record the result. To get a randomization sample for 30 such students, we repeat this process 30 times, returning the chosen card and reshuffling after each selection to keep $p = 0.25$. Thus we sample with replacement from the four cards until we have 30 answers, then find the sample statistic as the proportion of right answers in the randomization sample.

We can easily generalize this process to test any null proportion. The CAOS example also included a test of $H_0 : p = 0.42$ vs $H_a : p > 0,42$ to see if the proportion correct for the professor's students was clearly higher than a benchmark proportion for this question. We could match this null hypothesis by sampling with replacement from 100 cards which had 42 labeled "Right." Of course, in practice we use technology such as *StatKey* to simulate this process much more efficiently.

Randomization Test for a Mean: Body Temperature

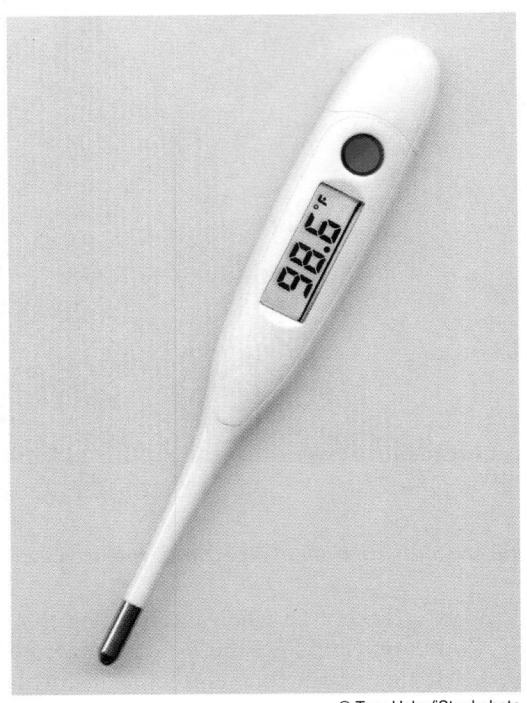

© Tom Hahn/iStockphoto

Is the average body temperature 98.6°F?

DATA 4.8

Body Temperature

What is the average body temperature for healthy humans? Most people using a Fahrenheit scale will say 98.6°F. This "fact" has been well established for many years. But is it possible that the average body temperature has changed over time? Perhaps people are getting warmer in response to global warming of the environment or have slightly lower temperatures than they had in past centuries. Allen Shoemaker[50] presented some data derived from a study of healthy adults which contains body temperature, gender, and pulse rate for each subject. The data in **BodyTemp50** contains a sample of $n = 50$ cases from that dataset with information for 25 males and 25 females. Figure 4.31 shows a dotplot of the body temperature readings in the variable *BodyTemp*. The mean in this sample is $\bar{x} = 98.26$ and the standard deviation is $s = 0.765$. Do these data provide significant evidence (at a 5% level) that the average body temperature is really different from the standard 98.6°F?   ■

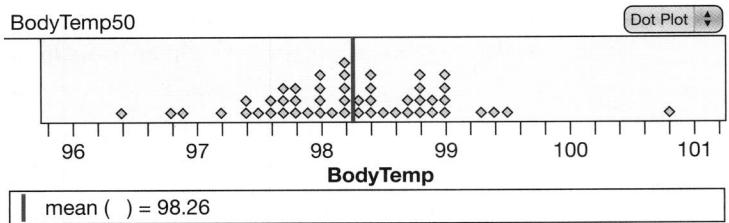

Figure 4.31 *Sample of body temperatures for 50 people*

---

[50]Shoemaker, A., "What's Normal? – Temperature, Gender and Heartrate," *Journal of Statistics Education*, 1996; 4(2).

Of course, we don't expect every sample of 50 people to have a mean body temperature of exactly 98.6. There will be some variation from sample to sample. Once again, the important question is whether a sample mean of 98.26 is farther away than we would expect to see by random chance alone if the true mean body temperature, μ, is really 98.6. The relevant hypotheses for a two-tailed test are

$$H_0 : \quad \mu = 98.6$$
$$H_a : \quad \mu \neq 98.6$$

In order to construct a randomization distribution to assess the evidence in this sample, we need to find a way to generate new samples that are consistent with the null hypothesis that the population mean is 98.6. We also want the simulated samples to reflect the structure in the population that is represented by our original sample of 50 values. In Chapter 3 we saw that sampling with replacement from the original sample (bootstrapping) is a good way to simulate samples from a population that is similar to our sample. The same idea applies here, except that we now have the additional constraint that we need to require that "population" to have a specific mean (98.6).

Example 4.38

Describe how to construct one randomization sample using the original data and assuming the null hypothesis is true. What statistic do we record for that sample?

Solution

One easy way to use the original data while simultaneously ensuring that the null hypothesis ($\mu = 98.6$) is satisfied is to add a constant amount to every value in our original sample. Since the sample mean, $\bar{x} = 98.26$, is 0.34° below the hypothesized mean, $\mu = 98.6$, we can add 0.34 to each temperature reading to produce a new set of temperatures with a mean exactly equal to 98.6. This has the advantage of preserving the general structure of the original sample data, while shifting the mean to be consistent with the null hypothesis. To generate a randomization distribution of sample means assuming the null hypothesis is true, we select samples of 50 temperatures at a time (with replacement) from the shifted data values and compute the mean for each sample. A set of sample means generated by this process will be a randomization distribution of values produced at random under the null hypothesis that $\mu = 98.6$.

Example 4.39

Use *StatKey* or other technology to generate a randomization distribution for the test about body temperature. Find and interpret the p-value.

Solution

Figure 4.32 shows a dotplot of the sample means for 10,000 samples generated by the randomization process described in Example 4.38. As expected, the distribution is centered at the null value of 98.6. As with any randomization distribution, the key question is where does the original sample mean ($\bar{x} = 98.26$) fall in the distribution? In this instance, we see in Figure 4.32 that only 8 of the 10,000 simulated means are as small (or smaller) than 98.26. This is a two-tail test since the alternative hypothesis is $H_a : \mu \neq 98.6$. Doubling to account for both tails gives

$$\text{p-value} = 2 \cdot (8/10000) = 0.0016$$

This very small p-value (well below $\alpha = 0.05$) gives strong evidence against the null hypothesis that the average body temperature is 98.6°F and indicates that the mean human body temperature is probably lower. Note that, even though the sample leads to convincingly rejecting the null hypothesis, the practical effect of assuming the average body temperature may be closer to 98.3°F than 98.6°F is pretty minimal.

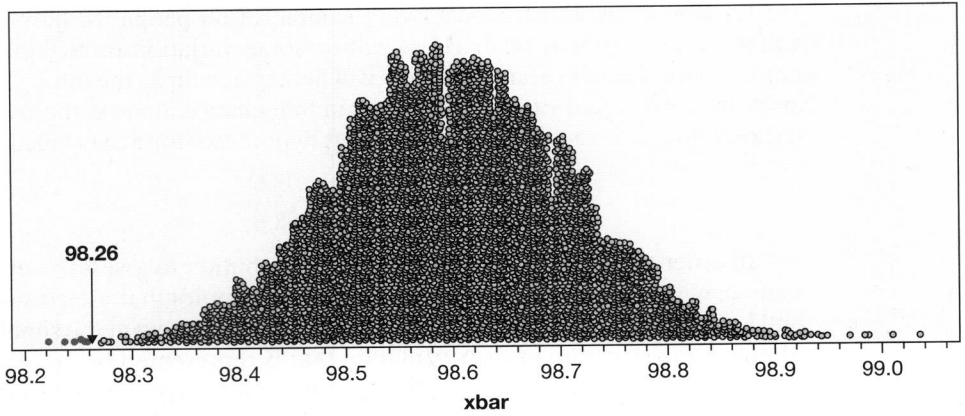

Figure 4.32 *Randomization distribution of body temperature means when μ = 98.6*

Choosing a Randomization Method

For a specific hypothesis test we may have several valid options for generating randomization samples. Recall that the key requirements are to be consistent with the null hypothesis and use the original data. When testing for a relationship between two variables, such as uniform malevolence and penalty yards (page 325) or drug treatment and cocaine relapse (page 323), the null hypothesis is often "no relationship." This can be simulated in randomization samples by scrambling the values of one of the variables so that we know the results are unrelated to the other variable. In other situations, such as testing the proportion of correct answers on the CAOS exam (page 327) or mean body temperature (page 329) we created a new population that matched the null hypothesis and then took new randomization samples with replacement from that null population. These are just some ways to simulate randomization samples, but not the only possibilities. Exercises 4.186 and 4.187 on page 339 illustrate several different randomization methods we could use to test the same hypotheses about a difference in means.

A randomization distribution shows what kinds of statistics we would observe by random chance if the null hypothesis is true. So far, we've focused on the "null hypothesis is true" part of the definition, but a more subtle point is that "random chance" can depend on the type of randomness in the data collection. For example, we might have random sampling of cases in an observational study or random assignment of subjects to treatment groups in an experiment. In simulating randomization samples for the body temperatures in Example 4.38, we randomly resample from a "population" with a mean of 98.6, reflecting the random sampling of participants in that study. In Example 4.33 on cocaine relapse, we randomly reassign the values to groups, reflecting the random assignment to groups in that experiment. Although reflecting the way the original data were collected like this is desirable, it usually does not make much of a difference in practice, and as long as we generate randomization samples based on the sample data and satisfying the null hypothesis, we should get an accurate p-value.

Another Look at Connecting Intervals and Tests

We take a more explicit look here at the connection between randomization and bootstrap distributions, and between intervals and tests. In Data 4.8, we consider measurements of body temperature for a sample of $n = 50$ individuals to test $H_0 : \mu = 98.6$ vs $H_a : \mu \neq 98.6$, where μ is the average body temperature. The mean in the sample is $\bar{x} = 98.26$, so we construct a randomization distribution by adding the difference, 0.34, to each of the sample values, creating a "population" that matches the null mean of 98.6, and then sampling with replacement from that new sample. The original sample mean (98.26) is well out in the tail of this randomization distribution (estimated p-value = 0.0016). This shows significant evidence in the sample to reject H_0 and conclude that the average body temperature is probably less than 98.6°F.

Now suppose that we use the original data to find a 95% confidence interval for the average body temperature, μ, by constructing a bootstrap distribution. This involves sampling (with replacement) from the original sample and computing the mean for each sample. How does this differ from the randomization distribution we use in the test? The procedures are exactly the same, except that one set of values has been shifted by 0.34°F. The two distributions are displayed in Figure 4.33. Note that any of the bootstrap samples might have been selected as a sample in the randomization distribution, with the only difference being that each of the values would be 0.34°F larger in the randomization sample to account for the adjustment to a null mean of 98.6°F.

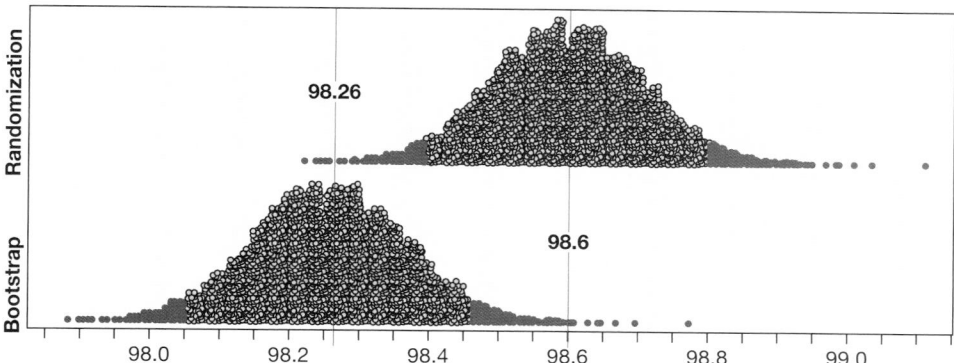

Figure 4.33 *Bootstrap and randomization distributions for body temperatures with* $H_0 : \mu = 98.6$

To find a 95% confidence interval from the bootstrap distribution of Figure 4.33 we need to find values with just 2.5% of the samples beyond them in each tail. This interval goes from 98.05 to 98.47. Thus, based on this sample, we are relatively sure that mean body temperature for the population is somewhere between 98.05°F and 98.47°F.

Note that, looking at the bootstrap confidence interval, the hypothesized value, $\mu = 98.6$, is *not* within the 95% confidence interval and, looking at the randomization distribution for the test, the mean of the sample, $\bar{x} = 98.26$, falls in the extreme tail of the distribution. This is not a coincidence! If 98.6°F is not a plausible value for the population mean, we should see this with both the confidence interval and the hypothesis test. The values in the lower and upper 2.5% tails of the randomization distribution (including the original sample mean of $\bar{x} = 98.26$) are values of sample means that would be extreme if H_0 were true and thus would lead to rejecting $H_0 : \mu = 98.6$ at a 5% level. The values in the lower and upper 2.5% tails of the bootstrap distribution (including the null mean of $\mu = 98.6$) are values of means that would be outside of the 95% confidence bounds and thus are considered unlikely candidates to be the actual mean for the population.

Example 4.40

Suppose we observe the same data (so $\bar{x} = 98.26$) but are instead testing $H_0 : \mu = 98.4$ versus $H_a : \mu \neq 98.4$. How would Figure 4.33 change? Would the confidence interval contain the null value of $\mu = 98.4$? Would we reject the null hypothesis when $\bar{x} = 98.26$?

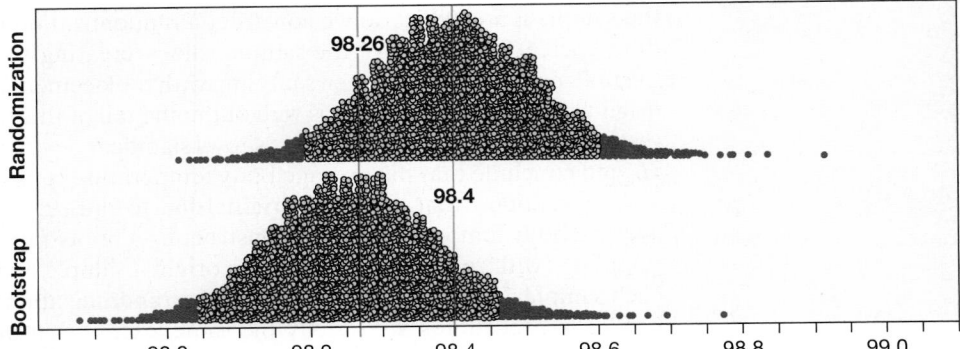

Figure 4.34 *Bootstrap and randomization distributions for body temperatures with* $H_0 : \mu = 98.4$

Solution

Since the bootstrap distribution and corresponding confidence interval don't depend on the hypotheses, they would remain unchanged. When testing $H_0 : \mu = 98.4$ the randomization samples would only be shifted to the right by 0.14 to be centered at 98.4 as shown in Figure 4.34. Now we see that the hypothesized value, $\mu = 98.4$ is contained within the 95% confidence interval and the sample mean, $\bar{x} = 98.26$, falls in the "typical" region of the randomization distribution, so the null hypothesis would not be rejected at a 5% level.

In general, we see that a sample statistic lies in the tail of the randomization distribution when the null hypothesized parameter lies in the tail of the bootstrap distribution, and that the sample statistic lies in the typical part of the randomization distribution when the null hypothesized parameter lies in the typical part of the bootstrap distribution (i.e., in the confidence interval). While this relationship is precise for a mean, the idea extends (somewhat more loosely) to any parameter, leading to the box earlier in this section. Note that, especially when doing confidence intervals and tests using simulation methods, the correspondence is not exact. For example, the precise boundaries for the 2.5%-points in the tails of either a randomization or a bootstrap distribution will vary slightly depending on the particular batch of simulated samples.

SECTION LEARNING GOALS

You should now have the understanding and skills to:

- Interpret a confidence interval as the plausible values of a parameter that would not be rejected in a two-tailed hypothesis test

- Determine the decision for a two-tailed hypothesis test from an appropriately constructed confidence interval

- Recognize whether inference is applicable in a given situation, and if so, whether a confidence interval and/or hypothesis test is most appropriate

- For a given sample and null hypothesis, describe the process of creating a randomization distribution

Exercises for Section **4.5**

SKILL BUILDER 1

For each question in Exercises 4.155 and 4.156, indicate whether it is best assessed by using a confidence interval or a hypothesis test or whether statistical inference is not relevant to answer it.

4.155 (a) What percent of US voters support instituting a national kindergarten through 12th grade math curriculum?

(b) Do basketball players hit a higher proportion of free throws when they are playing at home than when they are playing away?

(c) Do a majority of adults riding a bicycle wear a helmet?

(d) On average, were the 23 players on the 2014 Canadian Olympic hockey team older than the 23 players on the 2014 US Olympic hockey team?

4.156 (a) What proportion of people using a public restroom wash their hands after going to the bathroom?

(b) On average, how much more do adults who played sports in high school exercise than adults who did not play sports in high school?

(c) In 2010, what percent of the US Senate voted to confirm Elena Kagan as a member of the Supreme Court?

(d) What is the average daily calorie intake of 20-year-old males?

SKILL BUILDER 2

In Exercises 4.157 to 4.160, hypotheses for a statistical test are given, followed by several possible confidence intervals for different samples. In each case, use the confidence interval to state a conclusion of the test for that sample and give the significance level used.

4.157 Hypotheses: $H_0 : \mu = 15$ vs $H_a : \mu \neq 15$

(a) 95% confidence interval for μ: 13.9 to 16.2

(b) 95% confidence interval for μ: 12.7 to 14.8

(c) 90% confidence interval for μ: 13.5 to 16.5

4.158 Hypotheses: $H_0 : p = 0.5$ vs $H_a : p \neq 0.5$

(a) 95% confidence interval for p: 0.53 to 0.57

(b) 95% confidence interval for p: 0.41 to 0.52

(c) 99% confidence interval for p: 0.35 to 0.55

4.159 Hypotheses: $H_0 : \rho = 0$ vs $H_a : \rho \neq 0$. In addition, in each case for which the results are significant, give the sign of the correlation.

(a) 95% confidence interval for ρ: 0.07 to 0.15.

(b) 90% confidence interval for ρ: -0.39 to -0.78.

(c) 99% confidence interval for ρ: -0.06 to 0.03.

4.160 Hypotheses: $H_0 : \mu_1 = \mu_2$ vs $H_a : \mu_1 \neq \mu_2$. In addition, in each case for which the results are significant, state which group (1 or 2) has the larger mean.

(a) 95% confidence interval for $\mu_1 - \mu_2$: 0.12 to 0.54

(b) 99% confidence interval for $\mu_1 - \mu_2$: -2.1 to 5.4

(c) 90% confidence interval for $\mu_1 - \mu_2$: -10.8 to -3.7

SKILL BUILDER 3

In Exercises 4.161 to 4.163, a confidence interval for a sample is given, followed by several hypotheses to test using that sample. In each case, use the confidence interval to give a conclusion of the test (if possible) and also state the significance level you are using.

4.161 A 95% confidence interval for p: 0.48 to 0.57

(a) $H_0 : p = 0.5$ vs $H_a : p \neq 0.5$

(b) $H_0 : p = 0.75$ vs $H_a : p \neq 0.75$

(c) $H_0 : p = 0.4$ vs $H_a : p \neq 0.4$

4.162 A 99% confidence interval for μ: 134 to 161

(a) $H_0 : \mu = 100$ vs $H_a : \mu \neq 100$

(b) $H_0 : \mu = 150$ vs $H_a : \mu \neq 150$

(c) $H_0 : \mu = 200$ vs $H_a : \mu \neq 200$

4.163 A 90% confidence interval for $p_1 - p_2$: 0.07 to 0.18

(a) $H_0 : p_1 = p_2$ vs $H_a : p_1 \neq p_2$

(b) $H_0 : p_1 = p_2$ vs $H_a : p_1 > p_2$

(c) $H_0 : p_1 = p_2$ vs $H_a : p_1 < p_2$

4.164 Approval Rating for Congress In a Gallup poll[51] conducted in December 2015, a random sample of $n = 824$ American adults were asked "Do you

---

[51]*http://www.gallup.com/file/poll/187853/Congress-Approval-151217.pdf.*

approve or disapprove of the way Congress is handling its job?" The proportion who said they approve is $\hat{p} = 0.13$, and a 95% confidence interval for Congressional job approval is 0.107 to 0.153. If we use a 5% significance level, what is the conclusion if we are:

(a) Testing to see if there is evidence that the job approval rating is different than 14%. (This happens to be the average sample approval rating from the six months prior to this poll.)

(b) Testing to see if there is evidence that the job approval rating is different than 9%. (This happens to be the lowest sample Congressional approval rating Gallup ever recorded through the time of the poll.)

4.165 Car Window Skin Cancer? A study suggests that exposure to UV rays through the car window may increase the risk of skin cancer.[52] The study reviewed the records of all 1,050 skin cancer patients referred to the St. Louis University Cancer Center in 2004. Of the 42 patients with melanoma, the cancer occurred on the left side of the body in 31 patients and on the right side in the other 11.

(a) Is this an experiment or an observational study?

(b) Of the patients with melanoma, what proportion had the cancer on the left side?

(c) A bootstrap 95% confidence interval for the proportion of melanomas occurring on the left is 0.579 to 0.861. Clearly interpret the confidence interval in the context of the problem.

(d) Suppose the question of interest is whether melanomas are more likely to occur on the left side than on the right. State the null and alternative hypotheses.

(e) Is this a one-tailed or two-tailed test?

(f) Use the confidence interval given in part (c) to predict the results of the hypothesis test in part (d). Explain your reasoning.

(g) A randomization distribution gives the p-value as 0.003 for testing the hypotheses given in part (d). What is the conclusion of the test in the context of this study?

(h) The authors hypothesize that skin cancers are more prevalent on the left because of the sunlight coming in through car windows. (Windows protect against UVB rays but not UVA rays.) Do the data in this study support a conclusion that more melanomas occur on the left side because of increased exposure to sunlight on that side for drivers?

4.166 Print vs E-books Suppose you want to find out if reading speed is any different between a print book and an e-book.

(a) Clearly describe how you might set up an experiment to test this. Give details.

(b) Why is a hypothesis test valuable here? What additional information does a hypothesis test give us beyond the descriptive statistics we discuss in Chapter 2?

(c) Why is a confidence interval valuable here? What additional information does a confidence interval give us beyond the descriptive statistics of Chapter 2 and the results of a hypothesis test described in part (b)?

(d) A similar study[53] has been conducted, and reports that "the difference between Kindle and the book was significant at the $p < .01$ level, and the difference between the iPad and the book was marginally significant at $p = .06$." The report also stated that "the iPad measured at 6.2% slower reading speed than the printed book, whereas the Kindle measured at 10.7% slower than print. However, the difference between the two devices [iPad and Kindle] was not statistically significant because of the data's fairly high variability." Can you tell from the first quotation which method of reading (print or e-book) was faster in the sample or do you need the second quotation for that? Explain the results in your own words.

4.167 Are You "In a Relationship"? A study[54] shows that relationship status on Facebook matters to couples. The study included 58 college-age heterosexual couples who had been in a relationship for an average of 19 months. In 45 of the 58 couples, both partners reported being in a relationship on Facebook. In 31 of the 58 couples, both partners showed their dating partner in their Facebook profile picture. Men were somewhat more likely to include their partner in the picture than vice versa. However, the study states: "Females' indication that they are in a relationship was not as important to their male partners compared with how females felt about male partners indicating they are in a relationship."

[52]"Surprising Skin Cancer Risk: Too Much Driving," *LiveScience.com*, May 7, 2010, reporting on Butler, S., and Fosko, S., "Increased Prevalence of Left-sided Skin Cancers", *Journal of the American Academy of Dermatology*, published online, March 12, 2010.

[53]Neilsen, J., "iPad and Kindle Reading Speeds," *www.useit.com/alertbox/ipad-kindle-reading.html*, accessed July 2010.
[54]Roan, S., "The True Meaning of Facebook's 'in a Relationship'," *Los Angeles Times*, February 23, 2012, reporting on a study in *Cyberpsychology, Behavior, and Social Networking*.

Using a population of college-age heterosexual couples who have been in a relationship for an average of 19 months:

(a) A 95% confidence interval for the proportion with both partners reporting being in a relationship on Facebook is about 0.66 to 0.88. What is the conclusion in a hypothesis test to see if the proportion is different from 0.5? What significance level is being used?

(b) A 95% confidence interval for the proportion with both partners showing their dating partner in their Facebook profile picture is about 0.40 to 0.66. What is the conclusion in a hypothesis test to see if the proportion is different from 0.5? What significance level is being used?

4.168 Testing for a Gender Difference in Compassionate Rats In Exercise 3.89 on page 239, we found a 95% confidence interval for the difference in proportion of rats showing compassion, using the proportion of female rats minus the proportion of male rats, to be 0.104 to 0.480. In testing whether there is a difference in these two proportions:

(a) What are the null and alternative hypotheses?

(b) Using the confidence interval, what is the conclusion of the test? Include an indication of the significance level.

(c) Based on this study would you say that female rats or male rats are more likely to show compassion (or are the results inconclusive)?

4.169 Testing for a Home Field Advantage in Soccer In Exercise 3.129 on page 254, we see that the home team was victorious in 70 games out of a sample of 120 games in the FA premier league, a football (soccer) league in Great Britain. We wish to investigate the proportion p of all games won by the home team in this league.

(a) Use *StatKey* or other technology to find and interpret a 90% confidence interval for the proportion of games won by the home team.

(b) State the null and alternative hypotheses for a test to see if there is evidence that the proportion is different from 0.5.

(c) Use the confidence interval from part (a) to make a conclusion in the test from part (b). State the confidence level used.

(d) Use *StatKey* or other technology to create a randomization distribution and find the p-value for the test in part (b).

(e) Clearly interpret the result of the test using the p-value and using a 10% significance level. Does your answer match your answer from part (c)?

(f) What information does the confidence interval give that the p-value doesn't? What information does the p-value give that the confidence interval doesn't?

(g) What's the main difference between the bootstrap distribution of part (a) and the randomization distribution of part (d)?

4.170 Change in Stock Prices Standard & Poor's maintains one of the most widely followed indices of large-cap American stocks: the S&P 500. The index includes stocks of 500 companies in industries in the US economy. A random sample of 50 of these companies was selected, and the change in the price of the stock (in dollars) over the 5-day period from August 2 to 6, 2010 was recorded for each company in the sample. The data are available in **StockChanges**.

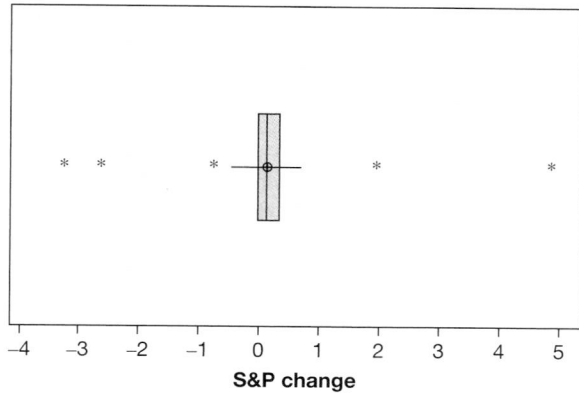

Figure 4.35 *Changes in stock prices on the S&P 500 over a 5-day period*

(a) Is this an experiment or an observational study? How was randomization used in the study, if at all? Do you believe the method of data collection introduced any bias?

(b) Describe one way to select a random sample of size 50 from a population of 500 stocks.

(c) Figure 4.35 shows a boxplot of the data. Describe what this plot shows about the distribution of stock price changes in this sample.

(d) Give relevant summary statistics to describe the distribution of stock price changes numerically.

(e) Use *StatKey* or other technology to calculate a 95% confidence interval for the mean change in all S&P stock prices. Clearly interpret the result in context.

(f) Use the confidence interval from part (e) to predict the results of a hypothesis test to see if the mean change for all S&P 500 stocks over this period is different from zero. State the hypotheses and significance level you use and state the conclusion.

(g) Now give the null and alternative hypotheses in a test to see if the average five-day change is positive. Use *StatKey* or other technology to find a p-value of the test and clearly state the conclusion.

(h) If you made an error in your decision in part (g), would it be a Type I error or a Type II error? Can you think of a way to actually find out if this error occurred?

4.171 How Long Do Mammals Live? Data 2.2 on page 64 includes information on longevity (typical lifespan), in years, for 40 species of mammals.

(a) Use the data, available in **MammalLongevity**, and *StatKey* or other technology to test to see if average lifespan of mammal species is different from 10 years. Include all details of the test: the hypotheses, the p-value, and the conclusion in context.

(b) Use the result of the test to determine whether $\mu = 10$ would be included as a plausible value in a 95% confidence interval of average mammal lifespan. Explain.

4.172 How Long Are Mammals Pregnant? Data 2.2 on page 64 includes information on length of gestation (length of pregnancy in days) for 40 species of mammals.

(a) Use the data, available in **MammalLongevity**, and *StatKey* or other technology to test to see if average gestation of mammals is different from 200 days. Include all details of the test: the hypotheses, the p-value, and the conclusion in context.

(b) Use the result of the test to indicate whether $\mu = 200$ would be included as a plausible value in a 95% confidence interval of average mammal gestation time. Explain.

4.173 Finger Tapping and Caffeine In Exercise 4.65 on page 283 we look at finger-tapping rates to see if ingesting caffeine increases average tap rate. The sample data for the 20 subjects (10 randomly getting caffeine and 10 with no caffeine) are given in Table 4.16. To create a randomization distribution

for this test, we assume the null hypothesis $\mu_c = \mu_n$ is true, that is, there is no difference in average tap rate between the caffeine and no caffeine groups.

(a) Create one randomization sample by randomly separating the 20 data values into two groups. (One way to do this is to write the 20 tap rate values on index cards, shuffle, and deal them into two groups of 10.)

(b) Find the sample mean of each group and calculate the difference, $\bar{x}_c - \bar{x}_n$, in the simulated sample means.

(c) The difference in sample means found in part (b) is one data point in a randomization distribution. Make a rough sketch of the randomization distribution shown in Figure 4.17 on page 284 and locate your randomization statistic on the sketch.

4.174 Arsenic in Chicken Data 4.3 on page 265 introduces a situation in which a restaurant chain is measuring the levels of arsenic in chicken from its suppliers. The question is whether there is evidence that the mean level of arsenic is greater than 80 ppb, so we are testing $H_0 : \mu = 80$ vs $H_a : \mu > 80$, where μ represents the average level of arsenic in all chicken from a certain supplier. It takes money and time to test for arsenic, so samples are often small. Suppose $n = 6$ chickens from one supplier are tested, and the levels of arsenic (in ppb) are:

$$68, \quad 75, \quad 81, \quad 93, \quad 95, \quad 134$$

(a) What is the sample mean for the data?

(b) Translate the original sample data by the appropriate amount to create a new dataset in which the null hypothesis is true. How do the sample size and standard deviation of this new dataset compare to the sample size and standard deviation of the original dataset?

(c) Write the six new data values from part (b) on six cards. Sample from these cards with replacement to generate one randomization sample. (Select a card at random, record the value, put it back, select another at random, until you have a sample of size 6, to match the original sample size.) List the values in the sample and give the sample mean.

(d) Generate 9 more simulated samples, for a total of 10 samples for a randomization distribution.

Table 4.16 *Finger tap rates for subjects with and without caffeine*

Caffeine	246	248	250	252	248	250	246	248	245	250	$\bar{x}_c = 248.3$
No caffeine	242	245	244	248	247	248	242	244	246	242	$\bar{x}_n = 244.8$

Give the sample mean in each case and create a small dotplot. Use an arrow to locate the original sample mean on your dotplot.

4.175 A Randomization Distribution for Arsenic in Chicken For the study in Exercise 4.174, use *StatKey* or other technology to create the randomization distribution for this test. Find the p-value and give a conclusion for the test. Should the restaurant chain stop ordering chickens from that supplier?

4.176 Effect of Sleep and Caffeine on Memory Exercise 4.102 on page 298 describes a study in which a random sample of 24 adults are divided equally into two groups and given a list of 24 words to memorize. During a break, one group takes a 90-minute nap while another group is given a caffeine pill. The response variable of interest is the number of words participants are able to recall following the break. We are testing to see if there is a difference in the average number of words a person can recall depending on whether the person slept or ingested caffeine. The data[55] are shown in Table 4.17 and are available in **SleepCaffeine**.

(a) Define any relevant parameter(s) and state the null and alternative hypotheses.

(b) What assumption do we make in creating the randomization distribution?

(c) What statistic will we record for each of the simulated samples to create the randomization distribution? What is the value of that statistic for the observed sample?

(d) Where will the randomization distribution be centered?

(e) Find one point on the randomization distribution by randomly dividing the 24 data values into two groups. Describe how you divide the data into two groups and show the values in each group for the simulated sample. Compute the sample mean in each group and compute the difference in the sample means for this simulated result.

(f) Use *StatKey* or other technology to create a randomization distribution. Estimate the p-value for the observed difference in means given in part (c).

[55]These data are recreated from the published summary statistics and are estimates of the actual data.

(g) At a significance level of $\alpha = 0.01$, what is the conclusion of the test? Interpret the result in context.

4.177 Hurricane Rate The data in **Hurricanes** contains the number of hurricanes that made landfall on the eastern coast of the United States over the 101 years from 1914 to 2014. Suppose we are interested in testing whether the number of hurricanes is increasing over time.

(a) State the null and alternative hypotheses for testing whether the correlation between year and number of hurricanes is positive, which would indicate the number of hurricanes is increasing.

(b) Describe in detail how you would create a randomization distribution to test this claim (if you had many more hours to do this exercise and no access to technology).

4.178 Finding the Hurricane Rate P-value Use *StatKey* or other technology to create a randomization distribution for the correlation described in Exercise 4.177. What is the p-value for this test? Is this convincing evidence that the number of hurricanes that make landfall on the eastern coast of the United States is increasing over time?

4.179 Hockey Malevolence Data 4.2 on page 263 describes a study of a possible relationship between the perceived malevolence of a team's uniforms and penalties called against the team. In Example 4.36 on page 326 we construct a randomization distribution to test whether there is evidence of a positive correlation between these two variables for NFL teams. The data in **MalevolentUniformsNHL** has information on uniform malevolence and penalty minutes (standardized as *z*-scores) for National Hockey League (NHL) teams. Use *StatKey* or other technology to perform a test similar to the one in Example 4.36 using the NHL hockey data. Use a 5% significance level and be sure to show all details of the test.

4.180 Desipramine vs Placebo in Cocaine Addiction In this exercise, we see that it is possible to use counts instead of proportions in testing a categorical variable. Data 4.7 describes an experiment to investigate the effectiveness of the two drugs desipramine and lithium in the treatment of cocaine addiction. The results of the study are summarized in Table 4.14 on page 323. The comparison of lithium

Table 4.17 *Number of words recalled*

Sleep	14	18	11	13	18	17	21	9	16	17	14	15	Mean = 15.25
Caffeine	12	12	14	13	6	18	14	16	10	7	15	10	Mean = 12.25

to the placebo is the subject of Example 4.34. In this exercise, we test the success of desipramine against a placebo using a different statistic than that used in Example 4.34. Let p_d and p_c be the proportion of patients who relapse in the desipramine group and the control group, respectively. We are testing whether desipramine has a lower relapse rate then a placebo.

(a) What are the null and alternative hypotheses?

(b) From Table 4.14 we see that 20 of the 24 placebo patients relapsed, while 10 of the 24 desipramine patients relapsed. The observed difference in relapses for our sample is

$$D = \text{desipramine relapses} - \text{placebo relapses}$$
$$= 10 - 20 = -10$$

If we use this difference in number of relapses as our sample statistic, where will the randomization distribution be centered? Why?

(c) If the null hypothesis is true (and desipramine has no effect beyond a placebo), we imagine that the 48 patients have the same relapse behavior regardless of which group they are in. We create the randomization distribution by simulating lots of random assignments of patients to the two groups and computing the difference in number of desipramine minus placebo relapses for each assignment. Describe how you could use index cards to create one simulated sample. How many cards do you need? What will you put on them? What will you do with them?

4.181 Testing Desipramine vs Placebo in Cocaine Addiction

(a) For the study in Exercise 4.180, use *StatKey* or other technology to create a randomization distribution for these data. Use the distribution to calculate a p-value. Interpret the results.

(b) In Example 4.34 on page 324, we saw that the p-value was 0.36 for testing whether lithium is better than a placebo in battling cocaine addiction. Using this p-value and your result from part (a), which drug shows stronger evidence that it is better than a placebo? Explain.

4.182 The Lady Tasting Tea Exercises 4.29 on page 271 and 4.76 on page 287 describe a historical scenario in which a British woman, Muriel Bristol-Roach, claimed to be able to tell whether milk had been poured into a cup before or after the tea. An experiment was conducted in which Muriel was presented with 8 cups of tea, and asked to guess whether the milk or tea was poured first. Our null hypothesis (H_0) is that Muriel has no ability to tell whether the milk was poured first. We would like to create a randomization distribution for $\hat{p}$, the proportion of cups out of 8 that Muriel guesses correctly under H_0. Describe a possible approach to generate randomization samples for each of the following scenarios:

(a) Muriel does not know beforehand how many cups have milk poured first.

(b) Muriel knows that 4 cups will have milk poured first and 4 will have tea poured first.

4.183 Quiz vs Lecture Pulse Rates Do you think that students undergo physiological changes when in potentially stressful situations such as taking a quiz or exam? A sample of statistics students were interrupted in the middle of a quiz and asked to record their pulse rates (beats for a 1-minute period). Ten of the students had also measured their pulse rate while sitting in class listening to a lecture, and these values were matched with their quiz pulse rates. The data appear in Table 4.18 and are stored in **QuizPulse10**. Note that this is paired data since we have two values, a quiz and a lecture pulse rate, for each student in the sample. The question of interest is whether quiz pulse rates tend to be higher, on average, than lecture pulse rates. (*Hint*: Since this is paired data, we work with the differences in pulse rate for each student between quiz and lecture. If the differences are $D =$ quiz pulse rate minus lecture pulse rate, the question of interest is whether μ_D is greater than zero.)

Table 4.18 *Quiz and Lecture pulse rates for 10 students*

Student	1	2	3	4	5	6	7	8	9	10
Quiz	75	52	52	80	56	90	76	71	70	66
Lecture	73	53	47	88	55	70	61	75	61	78

(a) Define the parameter(s) of interest and state the null and alternative hypotheses.

(b) Determine an appropriate statistic to measure and compute its value for the original sample.

(c) Describe a method to generate randomization samples that is consistent with the null hypothesis and reflects the paired nature of the data. There are several viable methods. You might use shuffled index cards, a coin, or some other randomization procedure.

(d) Carry out your procedure to generate one randomization sample and compute the statistic you chose in part (b) for this sample.

(e) Is the statistic for your randomization sample more extreme (in the direction of the alternative) than the original sample?

4.184 Testing Quiz vs Lecture Pulse Rates Use *StatKey* or other technology to create a randomization distribution for the paired data in the quiz-lecture pulse test described in Exercise 4.183. Find the p-value for the original sample and determine if there is sufficient evidence to conclude that the mean pulse rate during a quiz is larger than the mean pulse rate during lecture. (*Hint*: As described in Exercise 4.183, be sure to pay attention to the paired nature of the data. In particular, you will need to use the differences [$D = $ Quiz pulse$-$Lecture pulse] for each person as your data and conduct a test for a mean to determine whether the average difference is larger than zero.)

4.185 Clicker Questions A statistics instructor would like to ask "clicker" questions that about 80% of her students in a large lecture class will get correct. A higher proportion would be too easy and a lower proportion might discourage students. Suppose that she tries a sample of questions and receives 76 correct answers and 24 incorrect answers among 100 responses. The hypotheses of interest are $H_0 : p = 0.80$ vs $H_a : p \neq 0.80$. Discuss whether or not the methods described below would be appropriate ways to generate randomization samples in this setting. Explain your reasoning in each case.

(a) Sample 100 answers (with replacement) from the original student responses. Count the number of correct responses.

(b) Sample 100 answers (with replacement) from a set consisting of 8 correct responses and 2 incorrect responses. Count the number of correct responses.

4.186 Exercise Hours Introductory statistics students fill out a survey on the first day of class. One of the questions asked is "How many hours of exercise do you typically get each week?" Responses for a sample of 50 students are introduced in Example 3.25 on page 244 and stored in the file **ExerciseHours**. The summary statistics are shown in the computer output

below. The mean hours of exercise for the combined sample of 50 students is 10.6 hours per week and the standard deviation is 8.04. We are interested in whether these sample data provide evidence that the mean number of hours of exercise per week is different between male and female statistics students.

Variable	Gender	N	Mean	StDev	Minimum	Maximum
Exercise	F	30	9.40	7.41	0.00	34.00
	M	20	12.40	8.80	2.00	30.00

Discuss whether or not the methods described below would be appropriate ways to generate randomization samples that are consistent with $H_0 : \mu_F = \mu_M$ vs $H_a : \mu_F \neq \mu_M$. Explain your reasoning in each case.

(a) Randomly label 30 of the actual exercise values with "F" for the female group and the remaining 20 exercise values with "M" for the males. Compute the difference in the sample means, $\bar{x}_F - \bar{x}_M$.

(b) Add 1.2 to every female exercise value to give a new mean of 10.6 and subtract 1.8 from each male exercise value to move their mean to 10.6 (and match the females). Sample 30 values (with replacement) from the shifted female values and 20 values (with replacement) from the shifted male values. Compute the difference in the sample means, $\bar{x}_F - \bar{x}_M$.

(c) Combine all 50 sample values into one set of data having a mean amount of 10.6 hours. Select 30 values (with replacement) to represent a sample of female exercise hours and 20 values (also with replacement) for a sample of male exercise values. Compute the difference in the sample means, $\bar{x}_F - \bar{x}_M$.

4.187 Different Randomization Distributions for Exercise Hours Use *StatKey* or other technology and the data in **ExerciseHours** to carry out any two of the three randomization procedures as described in parts (a) to (c) in Exercise 4.186 comparing mean hours of exercise per week by gender. Are the results relatively consistent or are they quite different? What conclusion would you draw about the relationship (if any) between gender and amount of exercise?

Unit B introduces the key ideas of *statistical inference*. Statistical inference enables us to use information in a sample to understand properties of a population. Statistical inference can be very powerful. As we have seen, data from just a small subset of a population can often be used to give very accurate estimates and make very specific conclusions about the entire population. We can use the data in a sample to *estimate* one or more population parameter(s), create a *confidence interval* for the parameter(s), and *test* a hypothesis about the parameter(s).

For any of the methods discussed in Chapters 3 and 4, it is important to remember the lessons of Chapter 1: For statistical inference to be valid, the data must be collected in a way that does not introduce bias. If the data are collected in an appropriate way, we can learn remarkable things from just one sample.

Summary: Confidence Intervals

We estimate a population parameter using a sample statistic. Since such statistics vary from sample to sample, we need to get some sense of the accuracy of the statistic, for example, with a *margin of error*. This leads to the concept of an *interval estimate* as a range of plausible values for the population parameter. When we construct this interval using a method that has some predetermined chance of capturing the true parameter, we get a *confidence interval*. The correct interpretation of a confidence interval is important:

We have some level of confidence that the population parameter is contained within the confidence interval.

We describe two methods to compute a confidence interval. Both use a bootstrap distribution, created using the key idea that if the sample is representative of the population, then the population can be approximated by many, many copies of the sample. To construct a bootstrap distribution we:

- Generate bootstrap samples, with replacement, from the original sample, using the same sample size
- Compute the statistic of interest for each of the bootstrap samples
- Collect the statistics from many (usually at least 1000) bootstrap samples into a bootstrap distribution

Once we have a bootstrap distribution, we have two methods to construct an interval estimate:

Method 1: Estimate *SE*, the standard error of the statistic, as the standard deviation of the bootstrap distribution. The 95% confidence interval is then

$$\text{Sample statistic} \quad \pm \quad 2 \cdot SE$$

Method 2: Use percentiles of the bootstrap distribution to chop off the tails of the bootstrap distribution and keep a specified percentage (determined by the confidence level) of the values in the middle.

Both methods apply to a wide variety of parameters and situations, and can be used whenever the bootstrap distribution is approximately symmetric. They each have strengths in helping us understand the ideas behind interval estimation. For 95% confidence, the two methods usually give very similar answers. In later chapters we will learn other methods for constructing confidence intervals for specific parameters.

Summary: Hypothesis Tests

Hypothesis tests are used to investigate claims about population parameters. We use the question of interest to determine the two competing *hypotheses*: The null hypothesis is generally that there is no effect or no difference while the alternative hypothesis is the claim for which we seek evidence. The null hypothesis is the default assumption; we only conclude in favor of the alternative hypothesis if the evidence in the sample supports the alternative hypothesis and provides strong evidence against the null hypothesis. If the evidence is inconclusive, we stick with the null hypothesis.

We measure the strength of evidence against the null hypothesis using a *p-value*. A p-value is the probability of obtaining a sample statistic as extreme as (or more extreme than) the observed sample statistic, when the null hypothesis is true. A small p-value means that the observed sample results would be unlikely to happen just by random chance, if the null hypothesis were true, and thus provides evidence against the null hypothesis. *The smaller the p-value, the stronger the evidence against the null hypothesis and in support of the alternative hypothesis.*

When making specific decisions based on the p-value, we use a pre-specified *significance level*. If the p-value is less than the significance level, we reject H_0, conclude that there is evidence to support the alternative hypothesis, and say the results are statistically significant. If the p-value is not less than the significance level, we do not reject H_0, we have an inconclusive test, and we say the results are not statistically significant at that level. The conclusion should always be given in context to answer the question of interest.

We calculate a p-value by constructing a randomization distribution of possible sample statistics that we might see by random chance, if the null hypothesis were true. A randomization distribution is constructed by simulating many samples in a way that:

- Assumes the null hypothesis is true
- Uses the original sample data

The p-value is the proportion of the randomization distribution that is as extreme as, or more extreme than, the observed sample statistic. If the original sample falls out in the tails of the randomization distribution, then a result this extreme is unlikely to occur if the null hypothesis is true, and we have evidence against the null hypothesis in favor of the alternative.

Connecting Confidence Intervals and Hypothesis Tests

The two processes of interval estimation and significance testing are related, and, in many circumstances, each one can tell us something about the other. If the null parameter in a test falls outside the corresponding confidence interval for the same data, we are likely to reject that null hypothesis. On the other hand, if the null parameter in a test falls inside the confidence interval, we will likely not have sufficient evidence to reject the null hypothesis. The two processes are designed to give different information, but both are based on understanding how far the sample statistic might be from an unknown population parameter (in interval estimation) or a hypothesized population parameter (in testing). Creating a bootstrap distribution or randomization distribution helps us visualize and estimate this variability.

Case Study: Restaurant Tips

The exercises at the end of this section include several case studies that ask you to tie together the pieces of statistical inference learned so far. In addition to connecting the ideas that we have already discussed, you now have the power to extend these ideas to new situations. The methods we have covered have few conditions and great flexibility. To illustrate that flexibility, we ask you, in the examples that follow, to extend these ideas to a new parameter: the slope of a regression line.

Data 2.12 on page 123 describes information from a sample of 157 restaurant bills collected at the *First Crush* bistro. The relevant data file is **RestaurantTips**. In Chapter 2 we calculated a regression line with these data to investigate how the tip amount is related to the bill size. However, in Chapter 2 our analysis was limited to just the sample data. Now, with our newfound ability to perform statistical inference, we can extend the results from the sample to make conclusions about the population!

Example B.1

Data Collection

What population can we draw inferences about from the data in **RestaurantTips**? The sample was generated by collecting all bills over several nights at the restaurant. Is this a problem for making inferences?

Solution Because the data are all from one restaurant, the population of interest is all bills and tips at this restaurant. The original sample was not a random sample of all bills, but rather the data were collected from all bills in a certain time frame. That might introduce bias if the days for the bills in the sample are different in some substantial way (for example, over a holiday season or only weekends). However, the owner indicates that the days for which bills were sampled are representative of the business at his restaurant. As data analysts we might alert the owner to the possibility of bias when reporting our findings, but we proceed for now with the assumption that the sample is representative of all bills at this restaurant.

Example B.2

Interval or Test?

There are many questions we could ask about the **RestaurantTips** dataset. For each question below, indicate whether it is best assessed by using a confidence interval, a hypothesis test, or whether statistical inference is not relevant to answer it. Assume the population is all bills given to customers of the *First Crush* bistro.

(a) Estimate the size of an average bill at this restaurant.

(b) Is there evidence that customers at this restaurant leave an average tip greater than 15% of the bill?

(c) For what proportion of the 157 bills included in the dataset did the customer leave a tip greater than 20% of the bill?

(d) Is there a significant difference in the average tip left between waitress A and waitress B?

(e) What proportion of customers at the restaurant have a bill greater than $30?

Solution 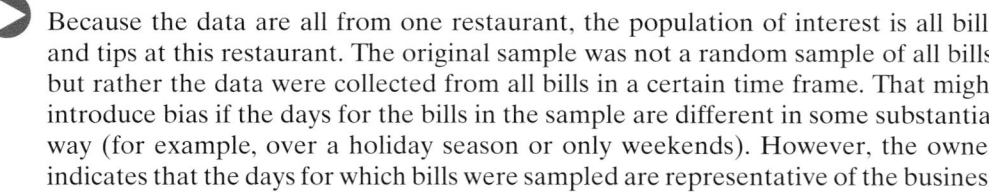 (a) We want to find an estimate and are not given any specific claim to test, so a confidence interval is most appropriate.

(b) Since we are specifically testing a claim about the average tip percentage, we use a hypothesis test to address this claim.

(c) This is a question about the 157 values in the dataset, *not* about the population. Statistical inference is not appropriate here, since we can find the proportion exactly from the dataset.

(d) Since we are testing a claim about a difference in means, we use a hypothesis test.

(e) We are estimating a proportion and are not given any specific claim to test, so a confidence interval is most appropriate.

Example B.3

Find the Regression Line

Find the equation of the least squares line for predicting the *Tip* amount based on the *Bill*. Interpret the slope of that line in context and include a plot to show the relationship.

Solution

Using statistical software with the data in **RestaurantTips** gives the prediction equation

$$\widehat{Tip} = -0.292 + 0.182 \cdot Bill$$

The slope of 0.182 indicates that for every extra dollar in the restaurant bill the tip will increase, on average, by about $0.18. This means the typical tip rate at this restaurant is roughly 18% of the total bill.

Figure B.1 shows a scatterplot of the relationship between *Tip* and *Bill* with the regression line drawn on it. We see a fairly strong, positive, linear association.

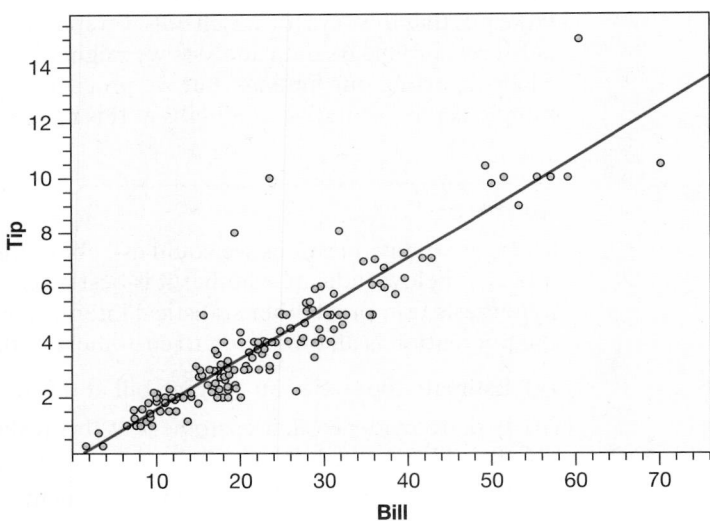

Figure B.1 *Tip vs Bill for n = 157 restaurant customers*

Example B.4

Confidence Interval for a Slope

(a) Describe how to use this sample to construct a bootstrap distribution for the slope of the regression line of *Tip* on *Bill*.

(b) A dotplot for one such bootstrap distribution from 100 bootstrap samples is shown in Figure B.2. Use the plot to estimate a 90% confidence interval for the slope of this regression line. Be sure to include an interpretation (in context) of the interval.

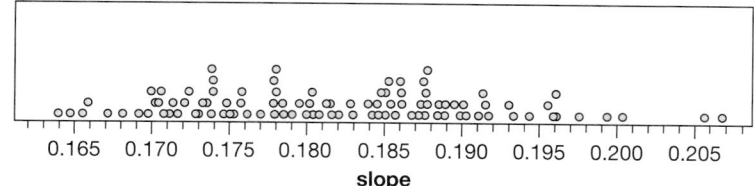

Figure B.2 *Bootstrap distribution of 100 sample slopes*

▶ (a) To construct a bootstrap distribution for the sample slope, we select samples of size $n = 157$, with replacement, from the cases in the **RestaurantTips** dataset. For each sample, we run the regression model, compute the sample slope, and save it to form the bootstrap distribution.

(b) Because the dotplot in Figure B.2 is based on the slopes from 100 bootstrap samples, we need to find the cutoffs for the upper and lower 5% in each tail to get the boundaries for a 90% confidence interval. Removing the smallest 5 and largest 5 bootstrap slopes leaves values ranging from about 0.168 to 0.197. Thus we are roughly 90% sure that the slope (or average tip rate) for the population of customers at this restaurant is somewhere between 0.168 and 0.197. Note that 100 bootstrap samples is a convenient number for finding the boundaries by eye from a dotplot, but in practice we should use a larger number of simulated samples and rely on technology to help with the counting.

Example B.5

Test for Slope Using a Confidence Interval

(a) If the amount of tip is unrelated to the size of the bill, the population slope for this relationship would be zero. On the other hand, we generally suspect that the *Tip* tends to increase as the *Bill* increases. What are the null and alternative hypotheses for testing whether the sample provides evidence that the slope of the regression relationship between these two variables is positive? (*Hint:* Use the Greek letter β [beta] to represent the slope for the population.)

(b) Can we make a decision about the outcome of the test (assuming a 10% significance level) based solely on the confidence interval for the slope found in Example B.4? If so, explain the decision. If not, explain why we cannot reach a conclusion.

▶ (a) If we let β denote the population slope for the relationship between amount of tips and size of bills at this restaurant, the hypotheses to test whether or not there is a positive slope are

$$H_0 : \quad \beta = 0$$
$$H_a : \quad \beta > 0$$

(b) The 90% confidence interval for the slope, found in Example B.4, is (0.168, 0.197). It does not include the null value of zero, so we reject H_0 in favor of a two-tailed alternative at a 10% level. In fact, since the confidence interval includes only positive values, we can be fairly sure that the true slope is above zero. Thus we have evidence that there is some positive slope for the relationship between the amount of a tip and the size of a bill at this restaurant.

What about Tip Percentage?

The data in **RestaurantTips** also include a variable showing the tip amount expressed as a percentage of the bill (*PctTip*). Most people use a fairly regular percentage (which may vary from person to person) of the total bill when deciding how big a tip to leave. Some economists[49] have theorized that people tend to reduce that percentage when the bill gets large, but larger groups of customers might be more generous as a group due to peer pressure. We can use the **RestaurantTip** data to see if there is evidence to support either theory—or perhaps there is no consistent relationship between the size of the bill and percent tip. Figure B.3 shows a scatterplot with regression line for this relationship. The sample slope in the regression line, $\widehat{PctTip} = 15.5 + 0.0488 \cdot Bill$, is positive, but looks pretty close to zero. Just looking at this scatterplot, it is hard to tell whether this slope is significantly different from zero. We need to conduct a hypothesis test.

Example B.6

Another Test for Slope, Using a Randomization Distribution

Perform a hypothesis test based on a randomization distribution to see if there is sufficient evidence to conclude that the slope of the relationship between *PctTip* and *Bill* is different from zero.

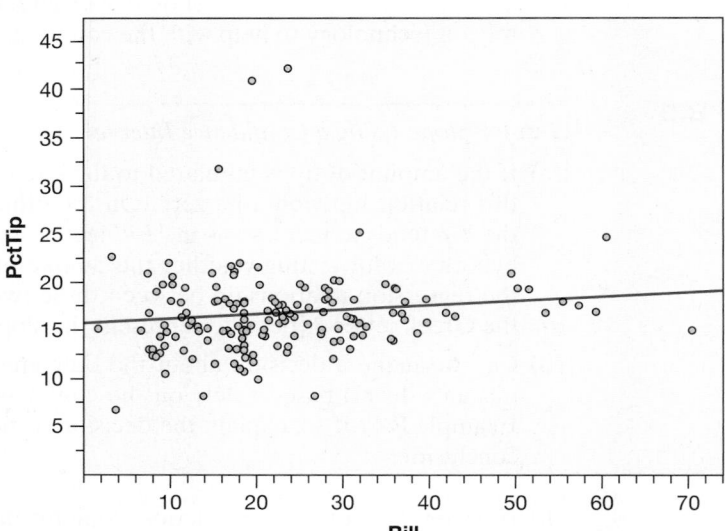

Figure B.3 *Tip percentage vs size of restaurant bill*

Solution

If we now let β denote the slope for predicting tip percentage based on bills at this restaurant, the relevant hypotheses are

$$H_0: \quad \beta = 0$$
$$H_a: \quad \beta \neq 0$$

Borrowing an idea from the randomization test for correlation in Example 4.35 on page 325, we can simulate data under the null hypothesis of no relationship ($\beta = 0$)

[49]Loewenstein, G. and Prelec, D. (1992). *Anomalies in Intertemporal Choice: Evidence and an Interpretation*, *Quarterly Journal of Economics*, 1992;107:573–97.

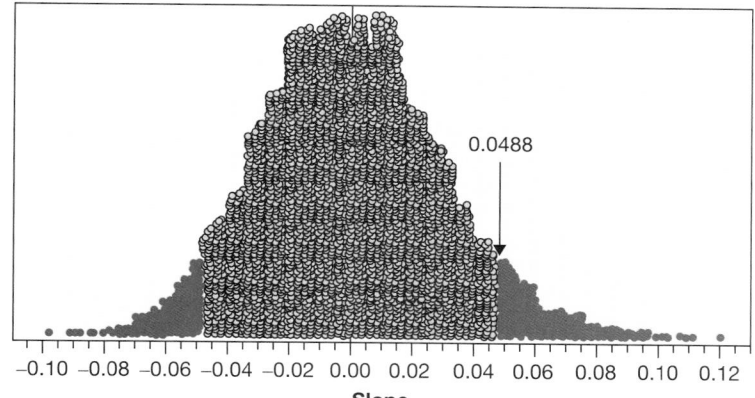

Figure B.4
Randomization distribution for slopes of PctTip vs Bill under
$H_0 : \beta = 0$

by randomly assigning the tip percentages in the dataset to the bill amounts. For each rearrangement, we compute the new regression line and save the sample slope to create a randomization distribution. Figure B.4 shows one such distribution with slopes for 10,000 simulated samples.

The location of the slope from our original sample, $b = 0.0488$, is indicated on the randomization distribution. It turns out that 545 of the 10,000 samples simulated under the null hypothesis of zero slope produced sample slopes above 0.0488. Doubling to account for a two-tail test gives a p-value of $2 \cdot 545/10,000 = 0.109$, which is not very small. The sample does not have enough evidence to conclude that the slope between *PctTip* and *Bill* is different from zero. Since the slope in our sample is positive, there is certainly no evidence to support the economists' claim of a negative relationship.

Example B.7

What About the Outliers?

Figures B.1 and B.3 both show a few possible outliers from the pattern of the rest of the data. Three very generous customers left tips that were more than 30% of the bill. Do those points have a large effect on the conclusions of the slopes for either of these relationships (*Tip* vs *Bill* or *PctTip* vs *Bill*)? One way to investigate this question is to omit those cases from the data and re-run the analysis without them.

Solution

After dropping the three generous data points, the new least squares lines with the remaining 154 cases for both relationships are shown in Figure B.5. The outliers have a negligible effect on the slope of the relationship between *Tip* and *Bill*. It barely changes from 0.182 with the outliers to 0.183 without them. A 90% confidence interval for this slope, based on the data without outliers, goes from 0.173 to 0.193, which is a bit narrower, but otherwise similar to the 90% interval, (0.168, 0.197), from the full data.

The regression equation for predicting percentage tip, $\widehat{PctTip} = 14.9 + 0.056 \cdot Bill$, is a bit steeper when the outliers are removed. When testing $H_0 : \beta = 0$ vs $H_a : \beta \neq 0$ for this new slope, the p-value for one set of 10,000 randomizations turns out to be 0.006. This p-value is quite small, showing that, when we remove the outlier big tippers, there is a significant positive association with the percentage of the tip tending to increase with larger bills.

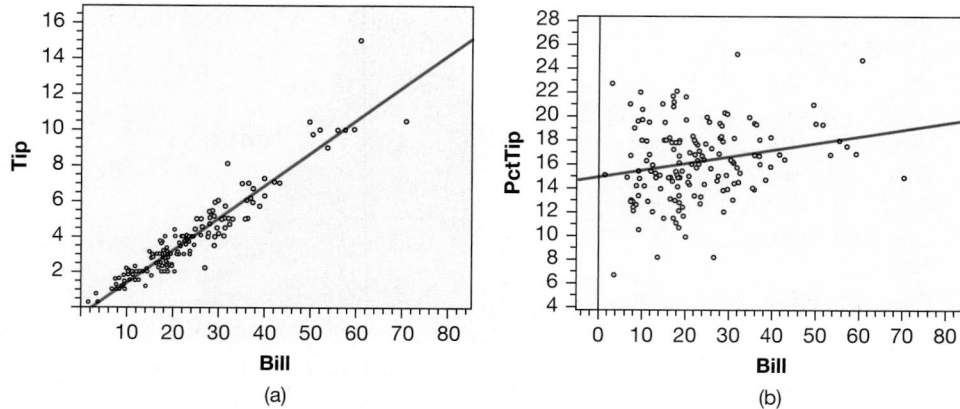

Figure B.5 *Regressions based on Bill with three outliers removed*

SECTION LEARNING GOALS

You should now have the understanding and skills to:

▶ • Demonstrate an understanding of estimation and testing and how they fit together

▶ • Distinguish whether an interval or a test is more appropriate for addressing a particular question

▶ • Apply the concepts of estimation and testing to answer questions using real data

---

Exercises for UNIT B: Essential Synthesis

B.1 Does Vitamin C Cure the Common Cold? A study conducted on a college campus tested to see whether students with colds who are given large doses of vitamin C recover faster than students who are not given vitamin C. The p-value for the test is 0.003.

(a) Given the p-value, what is the conclusion of the test: Reject H_0 or do not reject H_0?

(b) Results of statistical inference are only as good as the data used to obtain the results. No matter how low a p-value is, it has no relevance (and we can't trust conclusions from it) if the data were collected in a way that biases the results. Describe an *inappropriate* method of collecting the data for this study that would bias the results so much that a conclusion based on the p-value is very unreliable.

(c) Describe a method of collecting the data that would allow us to interpret the p-value

appropriately and to extend the results to the broader student population.

(d) Assuming the data were collected as you describe in part (c), use the p-value to make a conclusion about vitamin C as a treatment for students with a common cold.

B.2 Can Dogs Smell Cancer? Can dogs provide an easy noninvasive way to detect cancer? Several methods have been used to test this. In a recent study,[50] five dogs were trained over a three-week period to smell cancer in breath samples. To collect the data, cancer patients who had just been diagnosed and had not yet started treatment were asked to breathe into a tube. Breath samples were also collected from healthy control adults. Dogs were

---

[50]McCulloch, M., et al., "Diagnostic Accuracy of Canine Scent Detection in Early- and Late-Stage Lung and Breast Cancers," *Integrative Cancer Therapies*, 2006;5(1):30–39.

trained to sit if a breath sample came from a cancer patient. After training, the dogs were presented with breath samples from new subjects, with the samples randomly presented in a double-blind environment. The study was done for patients who were in different stages of cancer and for lung and breast cancer patients. The results for all groups were similar. The data for early-stage breast cancer are presented in Table B.1.

(a) Discuss the data collection for this study. Why is it important that the samples are from new subjects whose samples the dogs have not encountered before? That they are randomly presented? That the study is double-blind? That patients have not yet started treatment? Do you think this experiment was well designed?

(b) In the study for lung cancer, the experimenters had to account for the effect of smoking. Why?

(c) The question of interest is whether dogs are more likely to sit if the subject has cancer than if the subject does not have cancer. State the null and alternative hypotheses and give the relevant sample statistics.

(d) Without doing any computations, just looking at the data, do you expect the p-value to be relatively large or relatively small? Why? How strong is the evidence? (Sometimes, in extreme cases such as this one, we see results so obvious that a formal test may not be necessary. Unless results are exceptionally obvious, however, you should confirm your intuition with statistical inference.)

Table B.1 *Can dogs smell cancer?*

	Control	Cancer
Dog doesn't sit	99	3
Dog sits	6	45

B.3   Diet Cola and Calcium A study[51] examined the effect of diet cola consumption on calcium levels in women. A sample of 16 healthy women aged 18 to 40 were randomly assigned to drink 24 ounces of either diet cola or water. Their urine was collected for three hours after ingestion of the beverage and calcium excretion (in mg) was measured. The researchers were investigating whether

[51]Larson, N.S., et al., "Effect of Diet Cola on Urine Calcium Excretion," *Endocrine Reviews*, 2010;31(3):S1070. These data are recreated from the published summary statistics and are estimates of the actual data.

diet cola leaches calcium out of the system, which would increase the amount of calcium in the urine for diet cola drinkers. The data are given in Table B.2 and stored in **ColaCalcium**.

(a) Using *StatKey* or other technology, carry out an appropriate inference procedure to address the question of whether or not the mean amount of calcium lost for women who drink diet cola is more than for women who drink water.

(b) If the analysis in part (a) indicates that the results are significant, construct a 95% confidence interval to estimate the size of the effect. If the results in part (a) are not significant, combine the data into one sample of 16 values and use it to construct a 95% confidence interval for the average amount of calcium excreted. Be sure to interpret whichever interval you construct.

Table B.2 *Do diet cola drinkers excrete more calcium?*

Diet cola	50	62	48	55	58	61	58	56
Water	48	46	54	45	53	46	53	48

B.4   NFL Overtime The National Football League (NFL) has used a sudden death overtime period since 1974 to help decide a winner in games that are tied at the end of regulation time. Before the overtime starts, a coin flip is used to determine which team gets the ball first. Some critics of the system complain that the team that wins the coin flip has an unfair advantage. In the 445 overtime NFL games in the period between 1974 and 2009, the team winning the coin toss has won 240 times and lost 188 times and 17 games have ended in a tie when neither team scored during the overtime. When considering the impact of overtime policy for future games, we'll consider these games as a sample of all possible NFL games.

(a) Discarding the 17 tie games, we see that the winner of the coin flip has gone on to win 240 of the 428 games where a winner is determined in overtime. Does this provide sufficient evidence to show that the team winning the coin flip has an advantage? Use *StatKey* or other technology and assume that the league uses a 5% significance level.

(b) The NFL changed a rule before the 1994 season (moving the kickoff line back 5 yards) that might affect this analysis. For 188 games (again ignoring ties) from 1974 to 1993, the winner of the

Table B.3 *Videogames and GPA*

Student Brought Videogame	Roommate Brought Videogame	Sample Size	Mean GPA	Std. Dev.
No	No	88	3.128	0.590
Yes	No	44	3.039	0.689
No	Yes	38	2.932	0.699
Yes	Yes	40	2.754	0.639

coin flip won 94 times and lost 94 times. In 240 games played between 1994 and 2009 (after the rule change) the winner of the coin flip won 146 games and lost 94. Organize this information in a two-way table and discuss any statistical evidence for a difference in the advantage (if any exists at all) for the team winning the coin flip under the new and old rules.

B.5 Impact of College Roommates on Grades How much of an effect does your roommate have on your grades? In particular, does it matter whether your roommate brings a videogame to college? A study[52] examining this question looked at $n = 210$ students entering Berea College as first-year students in the Fall of 2001 who were randomly assigned a roommate. The explanatory variable is whether or not the roommate brought a videogame to college and the response variable is grade point average (GPA) for the first semester.

(a) Discuss one of the important strengths of the method of data collection.

(b) In conducting a test to see whether GPA is lower on average for students whose roommate brings a videogame to campus, define the parameter(s) of interest and state the null and alternative hypotheses.

(c) The p-value for the test in part (b) is 0.036. What is the conclusion for a 5% significance level?

(d) We are interested in seeing how large the roommate effect is on GPA. A 90% confidence interval for $\mu_v - \mu_n$ is (−0.315, −0.015), where μ_v is the average GPA for first-year students whose roommate brings a videogame to college and μ_n is the average GPA for first-year students whose roommate does not bring a videogame to college. Explain how you can tell just from the confidence interval which group has a higher average GPA. Interpret the confidence interval in terms of roommates, videogames, and GPA.

(e) The researchers also record whether the student him- or herself brought a videogame to college. We conduct the same test as in part (b), to see if having a roommate bring a videogame to college hurts GPA, for each of these groups separately. For the test for students who do not themselves bring a videogame to college, the p-value is 0.068. What is the conclusion, using a 5% significance level?

(f) For the test for students who themselves bring a videogame to campus, the p-value for the test is 0.026. What is the conclusion, again using a 5% significance level?

(g) Using the p-values in parts (e) and (f), for which group of students (those who bring a videogame or those who do not) does having a roommate bring a videogame have a larger effect on GPA? Does this match what you would expect? Explain.

(h) For students who bring a videogame to campus, a 90% confidence interval for $\mu_v - \mu_n$ is (−0.526, −0.044). Interpret this confidence interval in context and compare the effect size to that found for all students in part (d).

(i) The summary statistics are in Table B.3. Comment on the effect on GPA of videogames at college in general.

(j) Describe at least one additional test you might conduct using the data summarized in Table B.3.

B.6 Husbands Older Than Wives? A sample of marriage licenses in St. Lawrence County[53] in Northern New York State gives the ages of husbands and wives at the time of marriage for 105 newly married couples. The data are stored in **MarriageAges** and the first few cases from this file are shown in Table B.4. The question of interest is whether or not husbands tend to be older than their wives. Use *StatKey* or other technology and statistical inference to address this issue based on the questions

[52]Stinebrickner, R. and Stinebrickner, T., "The Causal Effect of Studying on Academic Performance," *The B.E. Journal of Economic Analysis & Policy*, 2008; 8(1) (Frontiers), Article 14.

[53]Thanks to Linda Casserly at the County Clerk's office for the data.

in parts (a) and (b). In all cases be sure to interpret your findings in the context of this problem, stating to what population (if any) your findings apply.

(a) When getting married, is the average age for husbands greater than the average age for wives? (*Hint*: The data are paired.)

(b) Is the proportion of couples for which the husband is older greater than 50%?

(c) For any significant results in parts (a) and (b), construct and interpret an interval for the size of the effect.

Table B.4 *First ten cases in MarriageAges, giving ages from marriage licenses*

Husband	53	38	46	30	31	26	29	48	65	29	...
Wife	50	34	44	36	23	31	25	51	46	26	...

B.7   Correlation between Ages of Husbands and Wives Exercise B.6 describes data on ages of husbands and wives at the time of marriage.

(a) Do you expect the correlation between the ages at marriage of husbands and wives to be positive, negative, or near zero? Explain.

(b) Using the data in **MarriageAges**, find the sample correlation and display the data in a scatterplot. Describe what you observe about the data in the scatterplot.

(c) Use *StatKey* or other technology to construct and interpret a 95% confidence interval for the correlation between the ages of husbands and wives when they get married.

(d) Does the correlation between ages help address the question in the previous exercise about whether husbands tend to be older than their wives?

Review Exercises for UNIT B

B.8   Customized Home Pages A random sample of $n = 1675$ Internet users in the US in January 2010 found that 469 of them have customized their web browser's home page to include news from sources and on topics that particularly interest them.[54] State the population and parameter of interest. Use the information from the sample to give the best estimate of the population parameter. What would we have to do to calculate the value of the parameter exactly?

B.9   Laptop Computers A survey conducted in May of 2010 asked 2252 adults in the US "Do you own a laptop computer?" The number saying yes was 1238. What is the best estimate for the proportion of US adults owning a laptop computer? Give notation for the quantity we are estimating, notation for the quantity we are using to make the estimate, and the value of the best estimate. Be sure to clearly define any parameters in the context of this situation.

B.10   Average GRE Scores The GRE (Graduate Record Exam) is like the SAT exam except it is used for application to graduate school instead of college. The mean GRE scores[55] for all examinees

tested between July 1, 2006, and June 20, 2009, are as follows: Verbal 456, Quantitative 590, Analytic Writing 3.8. If we consider the population to be all people who took the test during this time period, are these parameters or statistics? What notation would be appropriate for each of them? Suppose we take 1000 different random samples, each of size $n = 50$, from each of the three exam types and record the mean score for each sample. Where would the distribution of sample means be centered for each type of exam?

B.11   Do Violent Movies Lead to Violence in Society? A national telephone survey[56] reports that 57% of those surveyed think violent movies lead to more violence in society. The survey included a random sample of 1000 American adults and reports: "The margin of sampling error is +/− 3 percentage points with a 95% level of confidence."

(a) Define the relevant population and parameter. Based on the data given, what is the best point estimate for this parameter?

(b) Find and interpret a 95% confidence interval for the parameter defined in part (a).

B.12   Carbon Stored in Forest Biomass Scientists hoping to curb deforestation estimate that the carbon stored in tropical forests in Latin America,

[54]Purcell, K., Rainie, L., Mitchell, A., Rosenthal, T., and Olmstead, K., "Understanding the Participatory News Consumer," Pew Research Center, March 1, 2010, *http://www .pewinternet.org/Reports/2010/Online-News.aspx*.
[55]*http://ets.org/s/gre/pdf/gre_interpreting_scores.pdf*.
[56]"57% Believe Violence in Movies Leads to Violence in Society," *Rasmussen Reports*, February 14, 2012.

sub-Saharan Africa, and southeast Asia has a total biomass of 247 gigatons spread over 2.5 billion hectares. The scientists[57] measured carbon levels at 4079 inventory plots and also used information from satellite images. A 95% confidence interval for the mean amount of carbon per square kilometer in tropical forests in Latin America, sub-Saharan Africa, and southeast Asia is 9600 to 13,600 tons. Interpret this confidence interval. In addition, explain what the scientists would have to do to calculate the estimated quantity exactly.

B.13 What Proportion of Hollywood Movies Are Comedies? Data 2.7 on page 95 introduces the dataset **HollywoodMovies**, which contains information on over 900 movies to come out of Hollywood between 2007 and 2013. Of the 691 movies that have a genre listed, 177 are classified as comedies.

(a) What proportion of Hollywood movies with genres listed in this period were comedies? Use the correct notation with your answer.

(b) If we took many samples of size 50 from the population of all Hollywood movies with genres listed from this period and recorded the proportion of comedies for each sample, what shape do we expect the distribution of sample proportions to have? Where do we expect it to be centered?

B.14 Sampling Distributions for Proportion of Hollywood Movies That Are Comedies The dataset **HollywoodMovies** contains information on more than 900 movies to come out of Hollywood between 2007 and 2013. We see in Exercise B.13 that 177 of the 691 movies that have genres listed were comedies. If we take 1000 samples of size $n = 50$ from the population of movies in this period with genres listed and record the proportion of movies in the sample that are comedies, we get the sampling distribution shown in Figure B.6. In each case below, fill in the blank with the best option provided.

(a) The standard error of this sampling distribution is approximately _____.

   0.02   0.07   0.13   0.17   0.20

(b) If we create a new sampling distribution using samples of size $n = 100$, we expect the center of the new distribution to be _____ the center of the distribution shown in Figure B.6.

   smaller than   about the same as   larger than

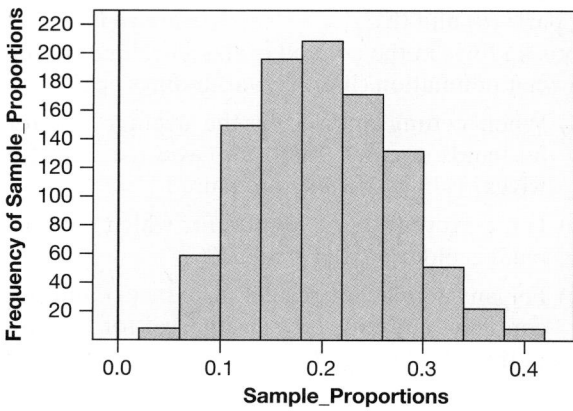

Figure B.6 *Sampling distribution using 1000 samples of size $n = 50$*

(c) If we create a new sampling distribution using samples of size $n = 100$, we expect the standard error of the new distribution to be _____ the standard error of the distribution shown in Figure B.6.

   smaller than   about the same as   larger than

(d) If we create a new sampling distribution using 5000 samples of size $n = 50$, we expect the center of the new distribution to be _____ the center of the distribution shown in Figure B.6.

   smaller than   about the same as   larger than

(e) If we create a new sampling distribution using 5000 samples of size $n = 50$, we expect the standard error of the new distribution to be _____ the standard error of the distribution shown in Figure B.6.

   smaller than   about the same as   larger than

B.15 Defective Screws Suppose that 5% of the screws a company sells are defective. Figure B.7 shows sample proportions from two sampling distributions: One shows samples of size 100, and the other shows samples of size 1000.

(a) What is the center of both distributions?

(b) What is the approximate minimum and maximum of each distribution?

(c) Give a rough estimate of the standard error in each case.

(d) Suppose you take one more sample in each case. Would a sample proportion of 0.08 (that is, 8% defective in the sample) be reasonably likely from a sample of size 100? Would it be reasonably likely from a sample of size 1000?

[57] Saatchi, S.S., et al., "Benchmark Map of Forest Carbon Stocks in Tropical Regions Across Three Continents," *Proceedings of the National Academy of Sciences*, May 31, 2011.

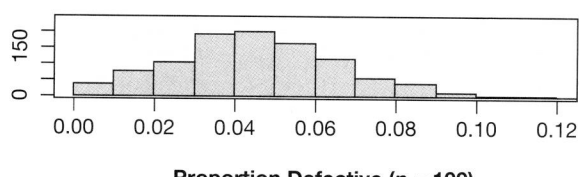

Proportion Defective (n = 100)

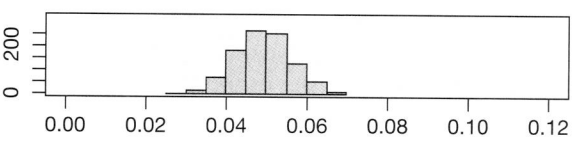

Proportion Defective (n = 1000)

Figure B.7 *Sampling distributions for n = 100 and n = 1000 screws*

B.16   Number of Screws in a Box A company that sells boxes of screws claims that a box of its screws contains on average 50 screws ($\mu = 50$). Figure B.8 shows a distribution of sample means collected from many simulated random samples of size 10 boxes.

(a) For a random sample of 10 boxes, is it unlikely that the sample mean will be more than 2 screws different from μ? What about more than 5? 10?

(b) If you bought a random sample of 10 boxes at the hardware store and the mean number of screws per box was 42, would you conclude that the company's claim ($\mu = 50$) is likely to be incorrect?

(c) If you bought a random box at the hardware store and it only contained 42 screws, would you conclude that the company's claim is likely to be incorrect?

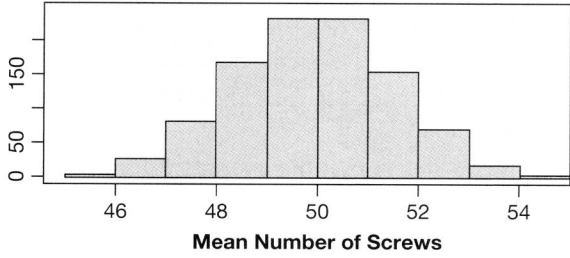

Mean Number of Screws

Figure B.8 *Sampling distribution for average number of screws in a box*

B.17   Average Points for a Hockey Player Table 3.4 on page 208 gives the number of points for all 24 players on the Ottawa Senators NHL hockey team, also available in the dataset **OttawaSenators**.

(a) Use *StatKey*, other technology, or a random number table to select a random sample of 5 of the 24 *Points* values. Indicate which values you've selected and compute the sample mean.

(b) Repeat part (a) by taking a second sample and calculating the mean.

(c) Find the mean for the entire population of these 24 players. Use correct notation for your answer. Comment on the accuracy of using the sample means found in parts (a) and (b) to estimate the population mean.

(d) Give a rough sketch of the sampling distribution if we calculate many sample means taking samples of size $n = 5$ from this population of *Points* values. What shape will it have and where will it be centered?

B.18   Time to Finish in 2012 Olympic Men's Marathon In the 2012 Olympic Men's Marathon, 85 athletes finished the race. Their times are stored in the file **OlympicMarathon**. Use the times stored in the *Minutes* column.

(a) Use *StatKey*, other technology, or a random number table to randomly select 10 values. Indicate which values you've selected and compute the sample mean.

(b) Repeat part (a) by taking a second sample and calculating the mean. Make a mini-dotplot by plotting the two sample means on a dotplot.

(c) Find the mean for the entire population of 76 race times. Use correct notation for your answer. Comment on the accuracy of using the sample means found in parts (a) and (b) to estimate the population mean.

(d) Suppose we take many samples of size $n = 10$ from this population of values and plot the mean for each sample on a dotplot. Describe the shape and center of the result. Draw a rough sketch of a possible distribution for these means.

B.19   A Sampling Distribution for Average Points for a Hockey Player Use *StatKey* or other technology to generate a sampling distribution of sample means using a sample size of $n = 5$ from the *Points* values in Table 3.4 on page 208, which gives the number of points for all 24 players on the Ottawa Senators NHL hockey team, also available in the dataset **OttawaSenators**. What are the smallest and largest sample means in the distribution? What is the standard deviation of the sample means (in other words, what is the standard error?)

B.20   A Sampling Distribution for Time to Finish in 2012 Olympic Men's Marathon Use *StatKey* or other technology to generate a sampling distribution of sample means using a sample size of $n = 10$ from the population of all times to finish the 2012 Olympic Men's Marathon, available in the *Minutes* column of the file **OlympicMarathon**. What are the smallest and largest sample means in the distribution? What is the standard deviation of the sample means (in other words, what is the standard error?)

B.21 Cell Phones in the Classroom Many professors do not like having cell phones ring during class. A recent study[58] appears to justify this reaction, by showing that a ringing cell phone can adversely affect student learning. In the experiment, students in a college classroom were exposed to a ringing cell phone during a psychology lecture. In the first part of the experiment, performance on a quiz revealed significantly lower accuracy rates on material presented while the phone was ringing. In a second part of the experiment, proximity of the students to the ringing phone was measured and results showed that the location of the ringing phone within the classroom was not associated with performance. The p-values for the two tests were 0.93 and 0.0004. Which p-value goes with which test? For the significant result, describe the strength of the evidence in context.

B.22 What Proportion of US Adults Exercise? In Example 3.17 on page 221, we learn that, in a random sample of over 450,000 US adults, the proportion of people who say they exercised at some point in the last 30 days is $\hat{p} = 0.726$ with a standard error of $SE = 0.0007$. Find and interpret a 95% confidence interval for the proportion of US adults who have exercised in the last 30 days.

B.23 Do Ovulating Women Affect Men's Speech? Studies suggest that when young men interact with a woman who is in the fertile period of her menstrual cycle, they pick up subconsciously on subtle changes in her skin tone, voice, and scent. A new study[59] suggests that they may even change their speech patterns. The experiment included 123 male and 5 female college students, all of them heterosexual. The men were randomly divided into two groups with one group paired with a woman in the fertile phase of her cycle and the other group with a woman in a different stage of her cycle. The women

were used equally in the two different stages. For the men paired with a less fertile woman, 38 of the 61 men copied their partner's sentence construction in a task to describe an object. For the men paired with a woman at peak fertility, 30 of the 62 men copied their partner's sentence construction. The experimenters hypothesized that men might be less likely to copy their partner during peak fertility in a (subconscious) attempt to attract more attention to themselves. Use *StatKey* or other technology to create a randomization distribution and conduct a hypothesis test to see if the proportion copying sentence structure is less when the woman is at peak fertility. Include all details of the test. Are the results significant at a 5% level? Are they significant at a 10% level?

B.24 Estimating Pizza Girl's Tips A pizza delivery person was interested in knowing how she spends her time and what her actual hourly earnings are, so she recorded all of her deliveries and tips and how she spent every minute of her time over three shifts, on one Friday night and two Saturday nights. She discusses the results, and much more, on "Diary of a Pizza Girl" on the Slice website.[60] Some of these data are available in **PizzaGirl**. The average tip for pizza deliveries on the nights sampled is $\bar{x} = \$3.04$. If we want to use this sample mean to estimate the average tip for all deliveries, the margin of error is 0.86. Find an interval estimate for the average tip for all pizza deliveries she makes. What do we have to assume about the sample in order for this point estimate and interval estimate to be valid?

B.25 Price of Textbooks We select a random sample of $n = 10$ textbooks at a university bookstore and are testing to see if there is evidence that the average price of textbooks at that store is greater than $100. Give an example of possible sets of 10 prices that would provide:

(a) Strong evidence that the average price of the store's textbooks is greater than $100.

(b) No evidence at all that the average price is greater than $100.

(c) Some evidence that the average price is greater than $100 but not strong evidence.

B.26 Most Americans Don't Go Out to Movies According to a survey,[61] most Americans prefer to watch a movie in the comfort of their own home rather than going out to a theater. In the telephone

---

[58]Shelton, J., Elliott, E., Eaves, S., and Exner, A., "The Distracting Effects of a Ringing Cell Phone: An Investigation of the Laboratory and the Classroom Setting," *Journal of Environmental Psychology*, 2009.

[59]Data approximated from information given in "How Ovulating Women Affect Men's Speech," the chart, *CNNHealth.com*, February 8, 2012.

[60]http://slice.seriouseats.com/archives/2010/04/statistical-analysis-of-a-pizza-delivery-shift-20100429.html.

[61]"56% Rarely Go To Movies," *Rasmussen Reports*, February 7, 2012.

Table B.5 *Prices of skateboards for sale online*

19.95	24.99	39.99	34.99	30.99	92.50	84.99	119.99	19.99	114.99
44.99	50.00	84.99	29.91	159.99	61.99	25.00	27.50	84.99	199.00

survey of 1000 randomly selected American adults, 56% say they rarely or never go out to the movies, while 32% go "occasionally" and 12% go at least once a month. We wish to estimate the proportion of American adults that rarely or never go out to the movies, and the report tells us: "The margin of sampling error is +/− 3 percentage points with a 95% level of confidence." Find and interpret a 95% confidence interval for the proportion rarely or never going out. Can we be relatively sure that the percent rarely or never going out to the movies is greater than 50%?

B.27 Skateboard Prices A sample of prices of skateboards for sale online[62] is shown in Table B.5 and is available in the dataset **SkateboardPrices**.

(a) What are the mean and standard deviation of the 20 skateboard prices?

(b) Describe how to use the data to select one bootstrap sample. What statistic is recorded from the sample?

(c) What shape and center do we expect the bootstrap distribution to have?

(d) One bootstrap distribution gives a standard error of 10.9. Find and interpret a 95% confidence interval.

B.28 Comparing Methods for Having Dogs Identify Cancer in People Exercise 2.22 on page 58 describes a study in which scientists train dogs to smell cancer. Researchers collected breath and stool samples from patients with cancer as well as from healthy people. A trained dog was given five samples, randomly displayed, in each test, one from a patient with cancer and four from healthy volunteers. The results are displayed in Table B.6. Use *StatKey* or other technology to use a bootstrap distribution to find and interpret a 90% confidence

interval for the difference in the proportion of times the dog correctly picks out the cancer sample, between the two types of samples. Is it plausible that there is no difference in the effectiveness in the two types of methods (breath or stool)?

Table B.6 *Can dogs smell cancer?*

	Breath Test	Stool Test	Total
Dog selects cancer	33	37	70
Dog does not select cancer	3	1	4
Total	36	38	74

B.29 Standard Deviation of Penalty Minutes in the NHL Exercise 3.103 on page 241 asked you to use the standard error to construct a 95% confidence interval for the standard deviation of penalty minutes for NHL players.

(a) Assuming the data in **OttawaSenators** can be viewed as a reasonable sample of all NHL players, use *StatKey* or other technology and percentiles of a bootstrap distribution to find and interpret a 95% confidence interval for the standard deviation of NHL penalty minutes for players in a season.

(b) What is the standard deviation for the original sample? Is the standard deviation for the original sample exactly in the middle of the confidence interval found in part (a)?

B.30 Average Land Area in Countries of the World Table B.7 shows land area (in 1000 sq km) and percent living in rural areas for a random sample of 10 countries selected from the **AllCountries** dataset. The data for this sample is stored in **TenCountries**. Use *StatKey* or other technology and this sample to find and interpret a 99% confidence interval for the average country size, in 1000 sq km.

[62]Random sample taken from all skateboards available for sale on eBay on February 12, 2012.

Table B.7 *Land area (in 1000 sq km) and percent living in rural areas*

Country	SRB	BHS	SVN	UZB	TUN	ARM	ROU	MKD	LBN	PRK
Land Area:	87.46	10.01	20.14	425.40	155.36	28.47	230.02	25.22	10.23	120.41
Rural:	44.6	17.3	50.2	63.8	33.5	37.0	45.8	43.0	12.5	39.4

B.31 Land Area and Percent Rural in Countries of the World Table B.7 shows land area (in 1000 sq km) and percent living in rural areas for a random sample of 10 countries from the **AllCountries** dataset that are stored in **TenCountries**.

(a) Using the data in the sample, find the slope of the regression line to predict the percent of the population living in rural areas using the land area (in 1000 sq km).

(b) Using *StatKey* or other technology and percentiles from a bootstrap distribution of this sample, find a 95% confidence interval to estimate the true slope (for all 213 countries) for predicting percent rural using the land area.

(c) The actual population slope is essentially 0. Does your 95% confidence interval from part (b) succeed in capturing the true slope from all 213 countries?

B.32 Effect of Smoking on Pregnancy Rate Exercise A.51 on page 179 introduces a study of 678 women who had gone off birth control with the intention of becoming pregnant. Table B.8 includes information on whether or not a woman was a smoker and whether or not the woman became pregnant during the first cycle. We wish to estimate the difference in the proportion who successfully get pregnant, between smokers and non-smokers.

(a) Find the best point estimate for the difference in proportions.

(b) Use *StatKey* or other technology to find and interpret a 90% confidence interval for the difference in proportions. Is it plausible that smoking has no effect on pregnancy rate?

Table B.8 *Smoking and pregnancy rate*

	Smoker	Non-smoker	Total
Pregnant	38	206	244
Not pregnant	97	337	434
Total	135	543	678

B.33 Testing the Effect of Smoking on Pregnancy Rate Exercise B.32 discusses a study to see if smoking might be negatively related to a woman's ability to become pregnant. The study looks at the proportion of successful pregnancies in two groups, smokers and non-smokers, and the results are summarized in Table B.8. In this exercise, we are interested in conducting a hypothesis test to determine if there is evidence that the proportion of

successful pregnancies is lower among smokers than non-smokers.

(a) Is this a one- or two-tailed test?

(b) What are the null and alternative hypotheses?

(c) If the null hypothesis is true (and smoking has no effect on pregnancy rate), we expect the 678 women to have the same pregnancy success rate regardless of smoking habits. We might create a randomization distribution by simulating many random assignments of women to the two groups. How many women would you randomly assign to the smoking group? The nonsmoking group?

(d) Figure B.9 shows the counts for successful pregnancies in the smoking group from 1000 such simulations. Using the same statistic (count for successful pregnancies in the smoking group) from the original sample and using this figure, which value would best approximate the p-value for our test?

$$0.0001 \quad 0.04 \quad 0.38 \quad 0.50 \quad 0.95$$

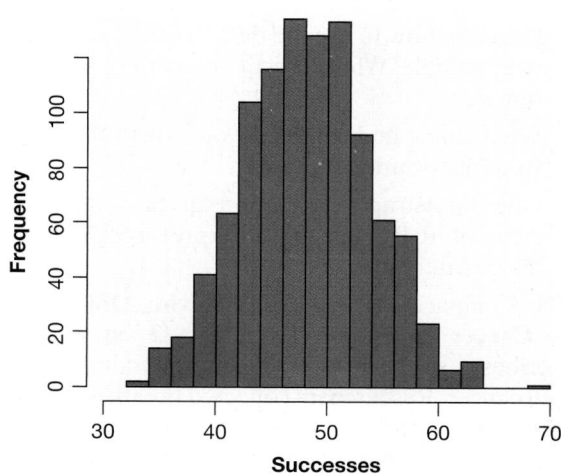

Figure B.9 *Randomization distribution for 1000 simulations* $H_0 : p_s = p_{ns}$

B.34 Taxes and Soda Consumption: Dotplots of Samples The average American drinks approximately 50 gallons of soda (pop) a year, delivering approximately 50,000 calories and no nutrition.[63] Some legislators are recommending instituting a sales tax on soda to raise revenue and fight obesity. Will a sales tax impact consumption? Suppose that

---

[63]Kiviat, B., "Tax and Sip," *Time Magazine*, July 12, 2010, p. 51–52.

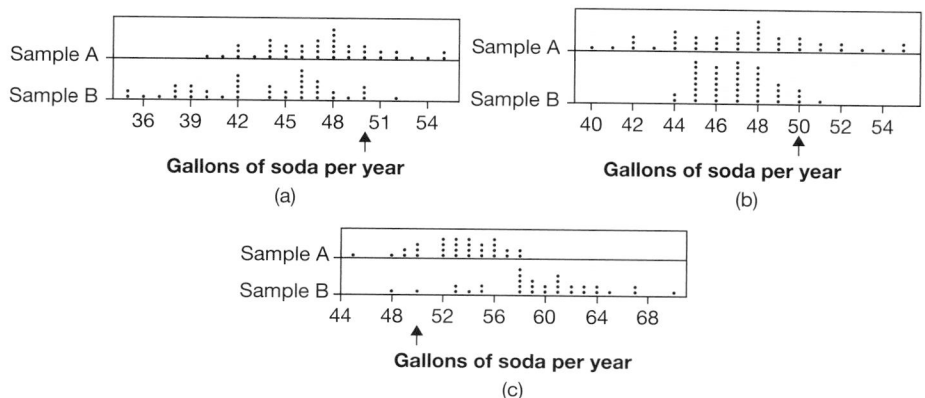

Figure B.10 *Samples for Exercise B.34*

a sales tax on soda will be added in a random sample of communities to measure the impact on soda consumption. We wish to determine whether average per-capita consumption of taxed soda is significantly less than 50 gallons a year. Figure B.10 shows dotplots of three pairs of possible sample results. In each case, indicate whether the results of Sample A or Sample B show stronger evidence that average consumption of taxed soda is below 50, or state that neither sample shows evidence that the mean is below 50. Explain your reasoning in each case.

B.35 Taxes and Soda Consumption: Boxplots of Samples We extend the situation described in Exercise B.34 to each of the pairs of boxplots in

Figure B.11. In each case, indicate whether the results of Sample A or Sample B show stronger evidence that average consumption of taxed soda is below 50, or state that neither sample shows evidence of this. Notice that sample sizes are shown on the side of the boxplots.

B.36 Taxes on Soda: Interpreting P-values Exercises B.34 and B.35 describe a study to determine whether a sales tax on soda will reduce consumption of soda in the US below the current per-capita level of about 50 gallons of soda per year. The hypotheses for the test are $H_0 : \mu = 50$ vs $H_a : \mu < 50$, where μ represents the average annual consumption of soda in communities where the sales tax is implemented.

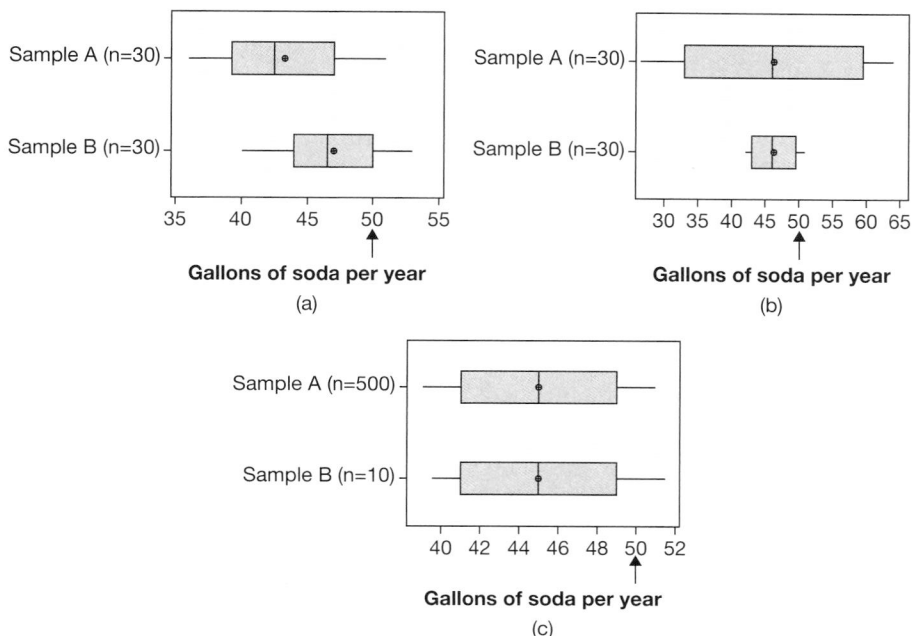

Figure B.11 *Samples for Exercise B.35*

(a) Suppose sample results give a p-value of 0.02. Interpret this p-value in terms of random chance and in the context of taxes and soda consumption.

(b) Now suppose sample results give a p-value of 0.41. Interpret this p-value in terms of random chance and in the context of taxes and soda consumption.

(c) Which p-value, 0.02 or 0.41, gives stronger evidence that a sales tax will reduce soda consumption?

(d) Which p-value, 0.02 or 0.41, is more statistically significant?

B.37 Standard Error for Proportion of Hollywood Movies That Are Action Movies Data 2.7 on page 95 introduced the dataset **HollywoodMovies**, which contains information on more than 900 movies that came out of Hollywood between 2007 and 2013. Of the 691 movies in this dataset that have a genre listed, 166 were classified as action movies.

(a) What proportion of Hollywood movies in the population of movies in this period that have a genre listed in the dataset were action movies? Use the correct notation with your answer.

(b) Use *StatKey* or other technology to generate a sampling distribution for the proportion of action movies for sample proportions of size $n = 40$. Give the shape and center of the sampling distribution and give the standard error.

B.38 Average Size of a Performing Group in the Rock and Roll Hall of Fame From its founding through 2015, the Rock and Roll Hall of Fame has inducted 303 groups or individuals, and 206 of the inductees have been performers while the rest have been related to the world of music in some way other than as a performer. The full dataset is available at **RockandRoll**. Some of the 206 performer inductees have been solo artists while some are groups with a large number of members. We are interested in the average number of members across all groups or individuals inducted as performers.

(a) What is the mean size of the performer inductee groups (including individuals)? Use the correct notation with your answer.

(b) Use technology to create a graph of all 206 values. Describe the shape, and identify the two groups with the largest number of people.

(c) Use technology to generate a sampling distribution for the mean size of the group using samples of size $n = 10$. Give the shape and center of the sampling distribution and give the standard error.

(d) What does one dot on the sampling distribution represent?

B.39 Sampling Distributions vs Bootstrap Distributions Given a specific sample to estimate a specific parameter from a population, what are the expected similarities and differences in the corresponding sampling distribution (using the given sample size) and bootstrap distribution (using the given sample)? In particular, for each aspect of a distribution listed below, indicate whether the values for the two distributions (sampling distribution and bootstrap distribution) are expected to be approximately the *same* or *different*. If they are different, explain how.

(a) The shape of the distribution

(b) The center of the distribution

(c) The spread of the distribution

(d) What one value (or dot) in the distribution represents

(e) The information needed in order to create the distribution

B.40 Bootstrap Distributions for Intervals vs Randomization Distributions for Tests What is the expected center of a bootstrap distribution generated to find a confidence interval? What is the expected center of a randomization distribution generated to test a hypothesis?

B.41 Paul the Octopus In the 2010 World Cup, Paul the Octopus (in a German aquarium) became famous for being correct in all eight of the predictions it made, including predicting Spain over Germany in a semifinal match. Before each game, two containers of food (mussels) were lowered into the octopus's tank. The containers were identical, except for country flags of the opposing teams, one on each container. Whichever container Paul opened was deemed his predicted winner.[64] Does Paul have psychic powers? In other words, is an 8-for-8 record significantly better than just guessing?

(a) State the null and alternative hypotheses.

(b) Simulate one point in the randomization distribution by flipping a coin eight times and counting the number of heads. Do this five times. Did you get any results as extreme as Paul the Octopus?

(c) Why is flipping a coin consistent with assuming the null hypothesis is true?

---

[64]For video of Paul go to *http://www.cnn.com/2010/SPORT/ football/07/08/germany.octopus.explainer/index.html*.

B.42 How Unlikely Is Paul the Octopus's Success? For the Paul the Octopus data in Exercise B.41, use *StatKey* or other technology to create a randomization distribution. Calculate a p-value. How unlikely is his success rate if Paul the Octopus is really not psychic?

B.43 Flipping Coins We flip a coin 150 times and get 90 heads, so the sample proportion of heads is $\hat{p} = 90/150 = 0.6$. To test whether this provides evidence that the coin is biased, we create a randomization distribution. Where will the distribution be centered? Why?

B.44 Exercise Hours: Males vs Females In Example 3.25 on page 244, we compare the mean hours of exercise per week between male and female college students. The sample results are included in the dataset **ExerciseHours**, where we see that the 20 men in the sample exercised for an average of 12.4 hours per week and the 30 women in the sample exercised for an average of 9.4 hours per week. Using the standard error for a bootstrap distribution, we find a 95% confidence interval for the difference in means $(\mu_m - \mu_f)$ to go from -1.75 hours to 7.75 hours. Use *StatKey* or other technology and a bootstrap distribution to find and interpret a 90% confidence interval for the difference in mean hours of exercise between males and females. How does your interval compare to the 95% confidence interval based on the standard error?

B.45 What Proportion Watch the Super Bowl? The Super Bowl is the final championship game in the National Football League in the US, and is one of the most watched television events of the year. In January 2016, just before Super Bowl 50, a random sample[65] of 7293 American adults were asked if they plan to watch the Super Bowl. A 95% confidence interval for the proportion planning to watch is 0.573 to 0.597.

(a) What is the population? What is the sample?

(b) Interpret the confidence interval in context.

(c) What is the best estimate for the proportion watching Super Bowl 50 and the margin of error for that estimate?

B.46 A Possible Fast-Acting Antidepressant Traditional antidepressants often take weeks or months to improve symptoms. A new study[66] may provide

a faster acting option. The anesthetic ketamine is very dangerous and can be deadly at high doses. However, low doses appear to have a rapid effect on levels of a brain compound linked to depression. In the study, mice receiving a single injection of ketamine showed fewer signs of depression within 30 minutes and the results lasted almost a week. One standard test of depression in mice is called the forced-swim test: Mice who are not depressed will struggle longer to stay afloat rather than giving up and sinking. The quantity measured is seconds that the mouse is immobilized, and lower numbers mean less depression. In a sample of 10 depressed mice 30 minutes after receiving a shot of ketamine, the mean number of seconds immobile was 135 with a standard error for the estimate of 6.

(a) Describe carefully how to use slips of paper containing the sample data to generate one bootstrap statistic. In particular, how many slips of paper are needed and what is on them? What do we do with them to obtain a bootstrap sample? What statistic do we then record?

(b) Find and interpret a 95% confidence interval for the mean time immobile in a forced-swim test for mice receiving a shot of ketamine.

(c) Researchers report that the average immobile time for depressed mice is about 160 seconds. Based on the interval in part (b), is 160 a plausible value for the mean immobile time for mice treated with ketamine?

B.47 A Test for a Possible Fast-Acting Antidepressant Exercise B.46 describes a study on the use of ketamine in treating depression in mice. Ten depressed mice given the drug had a mean score of 135 seconds on a forced-swim test used to measure depression (lower scores indicate less depression). The usual mean for depressed mice on this test is about 160 seconds.

(a) Using the parameter μ to denote the mean score on this test for depressed mice after treatment with ketamine, what are the null and alternative hypotheses for seeing if there is evidence that the mean score is lower than 160?

(b) Describe carefully how to use slips of paper to generate one randomization statistic for this test. In particular, how many slips of paper are needed and what do we write on them? What do we do with them to obtain a randomization sample? What statistic do we then record?

B.48 Proportion of a Country's Population with Access to the Internet One of the variables in the **AllCountries** dataset gives the percent of the

[65]"An Estimated 189 Million Americans Expected to Watch Super Bowl 50," National Retail Federation, January 28, 2016, *https://nrf.com/media/press-releases/estimated-189-million-americans-expected-watch-super-bowl-50*

[66]Autry, A., et al., "NMDA Receptor Blockade at Rest Triggers Rapid Behavioural Antidepressant Responses," *Nature*, online, June 15, 2011.

population of each country with access to the Internet. This information is available for all 203 countries (ignoring a few with missing values). We are interested in the average percent with Internet access.

(a) What is the mean percent with Internet access across all countries? What is the standard deviation of the values? Use the correct notation with your answers.

(b) Which country has the highest Internet access rate, and what is that percent? Which country has the lowest Internet access rate, and what is that percent? What is the Internet access rate for *your* country?

(c) Use *StatKey* or other technology to generate a sampling distribution for the mean Internet access rate using samples of size $n = 10$. Give the shape and center of the sampling distribution and give the standard error.

B.49 Cell Phones and Cancer Does heavy cell phone use increase the incidence of brain tumors? A study of cell phone use among 10,000 participants found that "the 10% who used their phones most often and for the longest time had a 40% higher risk of developing some form of brain cancer than those who didn't use a mobile phone."[67] Nonetheless, the results were not statistically significant. Epidemiologists Saracci and Samet write that the results "tell us that the question of whether mobile-phone use increases risks for brain cancers remains open." Based on this study, describe whether each statement below is plausible for this population:

(a) Heavy cell phone use has no effect on developing brain cancer.

(b) Heavy cell phone use is associated with an increased risk of brain cancer.

(c) Heavy cell phone use causes an increased risk of brain cancer.

B.50 Infections in Childbirth The Centers for Disease Control and Prevention (CDC) conducted a randomized trial in South Africa designed to test the effectiveness of an inexpensive wipe to be used during childbirth to prevent infections.[68] Half of the mothers were randomly assigned to have their birth canal wiped with a wipe treated with a drug called chlorhexidine before giving birth, and the other half to get wiped with a sterile wipe (a placebo). The response variable is whether or not the newborns develop an infection. The CDC hopes to find out whether there is evidence that babies delivered by the women getting the treated wipe are less likely to develop an infection.

(a) Define the relevant parameter(s) and state the null and alternative hypotheses.

(b) What is/are the sample statistic(s) to be used to test this claim?

(c) If the results are statistically significant, what would that imply about the wipes and infections?

(d) If the results are not statistically significant, what would that imply about the wipes and infections?

B.51 Mice and Pain Can you tell if a mouse is in pain by looking at its facial expression? One study believes you can. The study[69] created a "mouse grimace scale" and tested to see if there was a positive correlation between scores on that scale and the degree and duration of pain (based on injections of a weak and mildly painful solution). The study's authors believe that if the scale applies to other mammals as well, it could help veterinarians test how well painkillers and other medications work in animals.

(a) Define the relevant parameter(s) and state the null and alternative hypotheses.

(b) Since the study authors report that you can tell if a mouse is in pain by looking at its facial expression, do you think the data were found to be statistically significant? Explain.

(c) If another study were conducted testing the correlation between scores on the "mouse grimace scale" and a placebo (non-painful) solution, should we expect to see a sample correlation as extreme as that found in the original study? Explain. (For simplicity, assume we use a placebo that has no effect on the facial expressions of mice. Of course, in real life, you can never automatically assume that a placebo has no effect!)

(d) How would your answer to part (c) change if the original study results showed no evidence of a relationship between mouse grimaces and pain?

[67]Walsh, B., "A Study on Cell Phones and Cancer," *Time Magazine*, May 31, 2010, p. 15, reporting on a study in the *International Journal of Epidemiology*, May 17, 2010.

[68]Eriksen, N., Sweeten, K., and Blanco, J., "Chlorhexidine Vs. Sterile Vaginal Wash During Labor to Prevent Neonatal Infection," *Infectious Disease in Obstetrics and Gynecology*, 1997; 5(4): 286–290.

[69]"Of Mice and Pain," *The Week*, May 28, 2010, p. 21.

B.52 Measuring the Impact of Great Teachers An education study in Tennessee in the 1980s (known as Project Star) randomly assigned 12,000 students to kindergarten classes, with the result that all classes had fairly similar socioeconomic mixes of students.[70] The students are now about 35 years old, and the study is ongoing. In each case below, assume that we are conducting a test to compare performance of students taught by outstanding kindergarten teachers with performance of students taught by mediocre kindergarten teachers. What does the quoted information tell us about whether the p-value is relatively *large* or relatively *small* in a test for the indicated effect?

(a) On the tests at the end of the kindergarten school year, "some classes did far better than others. The differences were too big to be explained by randomness."

(b) By junior high and high school, the effect appears to be gone: "Children who had excellent early schooling do little better on tests than similar children who did not."

(c) Results reported in 2010 by economist Chetty, show that the effects seem to re-emerge in adulthood. The students who were in a classroom that made significant gains in kindergarten were significantly "more likely to go to college,…less likely to become single parents,…more likely to be saving for retirement,…Perhaps most striking, they were earning more." (Economists Chetty and Saez estimate that a standout kindergarten teacher is worth about $320,000 a year

[70]Leonhardt, D., "The Case for $320,000 Kindergarten Teachers," *The New York Times*, July 27, 2010, reporting on a study by R. Chetty, a Harvard economist, and his colleagues.

in increased future earnings of one class of students. If you had an outstanding grade-school teacher, consider sending a thank you note!)

B.53 Smiles and Leniency Data 4.2 on page 293 describes an experiment to study the effects of smiling on leniency in judging students accused of cheating. The full data are in **Smiles**. In Example 4.21 we consider hypotheses $H_0 : \mu_s = \mu_n$ vs $H_a : \mu_s > \mu_n$ to test if the data provide evidence that average leniency score is higher for smiling students (μ_s) than for students with a neutral expression (μ_n). A dotplot for the difference in sample means based on 1000 random assignments of leniency scores from the original sample to smile and neutral groups is shown in Figure B.12.

(a) The difference in sample means for the original sample is $D = \bar{x}_s - \bar{x}_n = 4.91 - 4.12 = 0.79$ (as shown in Figure B.12). What is the p-value for the one-tailed test? (*Hint:* There are 27 dots in the tail beyond 0.79.)

(b) In Example 4.22 on page 295 we consider the test with a two-tailed alternative, $H_0 : \mu_s = \mu_n$ vs $H_a : \mu_s \neq \mu_n$, where we make no assumption in advance on whether smiling helps or discourages leniency. How would the randomization distribution in Figure B.12 change for this test? How would the p-value change?

B.54 Estimating the Proportion with Employer-Based Health Insurance In Exercise 3.62 on page 225, we discuss a Gallup poll stating that the proportion of American adults getting health insurance from an employer is estimated to be 0.45. We are also told that, with 95% confidence, "the maximum margin of sampling error is ±1 percentage point" for this estimate. In fact, the Gallup

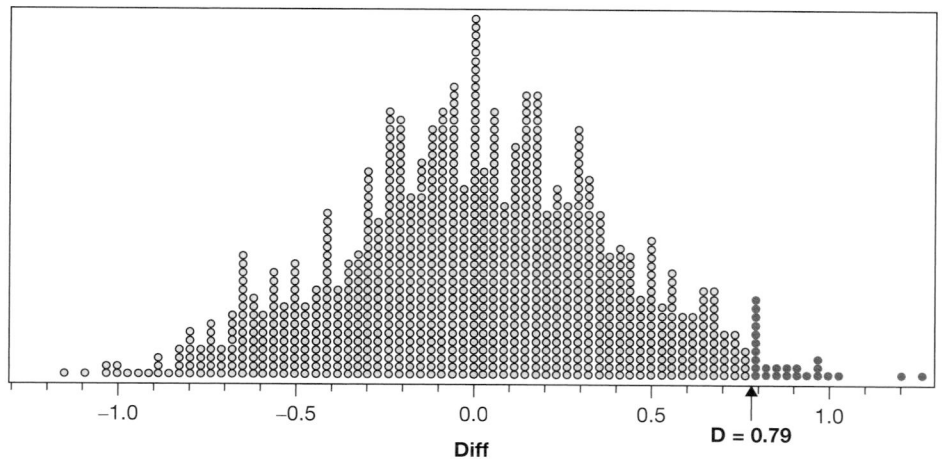

Figure B.12 *Randomization distribution for 1000 samples testing $H_0 : \mu_s = \mu_n$ using Smiles data*

organization rounded up the margin of error to the nearest whole number and the actual margin of error is quite a bit less. Figure B.13 shows a bootstrap distribution based on the sample results. Use the bootstrap distribution to estimate the standard error and find and interpret a 95% confidence interval.

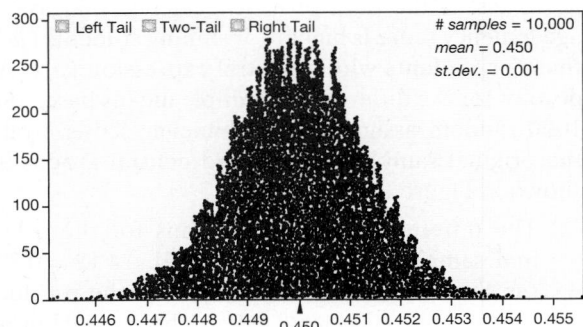

Figure B.13 *Bootstrap proportions for 10,000 samples based on a Gallup poll with n = 147,291 and $\hat{p}$ = 0.45*

B.55 False Positives in Lie Detection Is lie detection software accurate? Exercise A.23 on page 147 describes a study in which 48 individuals read a truthful passage while under stress and while connected to a lie detector. The lie detection software inaccurately reported deception in 57% of the cases. A bootstrap distribution shows an estimated standard error of 0.07.

(a) Give a best estimate for the population parameter of interest.

(b) Give a 95% confidence interval for this population parameter.

(c) Comment on the accuracy of this lie detector. Do you think results from this lie detector should hold up in court?

B.56 How Common are False Positives in Lie Detection? In Exercise B.55, we learn that when 48 stressed individuals read a truthful passage while being hooked up to a lie detector, the lie detection software inaccurately reported deception by 27 of them. Does this sample provide evidence that lie detection software will give inaccurate results more than half the time when used in situations such as this? State the null and alternative hypotheses. Use *StatKey* or other technology to create a randomization distribution, find a p-value, and give a clear conclusion in context.

B.57 Red Wine and Weight Loss Resveratrol, an ingredient in red wine and grapes, has been shown to promote weight loss in rodents. One study[71] investigates whether the same phenomenon holds true in primates. The grey mouse lemur, a primate, demonstrates seasonal spontaneous obesity in preparation for winter, doubling its body mass. A sample of six lemurs had their resting metabolic rate, body mass gain, food intake, and locomotor activity measured for one week prior to resveratrol supplementation (to serve as a baseline) and then the four indicators were measured again after treatment with a resveratrol supplement for four weeks. Some p-values for tests comparing the mean differences in these variables (before vs after treatment) are given below. In parts (a) to (d), state the conclusion of the test using a 5% significance level, and interpret the conclusion in context.

(a) In a test to see if mean resting metabolic rate is higher after treatment, $p = 0.013$.

(b) In a test to see if mean body mass gain is lower after treatment, $p = 0.007$.

(c) In a test to see if mean food intake is affected by the treatment, $p = 0.035$.

(d) In a test to see if mean locomotor activity is affected by the treatment, $p = 0.980$.

(e) In which test is the strongest evidence found? The weakest?

(f) How do your answers to parts (a) to (d) change if the researchers make their conclusions using a stricter 1% significance level?

(g) For each p-value, give an informal conclusion in the context of the problem describing the level of evidence for the result.

(h) The sample only included six lemurs. Do you think that we can generalize to the population of all lemurs that body mass gain is lower on average after four weeks of a resveratrol supplement? Why or why not?

B.58 Radiation from Cell Phones and Brain Activity Does heavy cell phone use affect brain activity? There is some concern about possible negative effects of radiofrequency signals delivered to the brain. In a randomized matched-pairs study,[72] 47 healthy participants had cell phones placed on the left and right ears. Brain glucose metabolism (a measure of brain activity) was measured for all

[71]BioMed Central. "Lemurs Lose Weight with 'Life-Extending' Supplement Resveratrol," *ScienceDaily*, June 22, 2010.

[72]Volkow, et al., "Effects of Cell Phone Radiofrequency Signal Exposure on Brain Glucose Metabolism," *Journal of the American Medical Association*, 2011; 305(8): 808–13.

participants under two conditions: with one cell phone turned on for 50 minutes (the "on" condition) and with both cell phones off (the "off" condition). The amplitude of radiofrequency waves emitted by the cell phones during the "on" condition was also measured.

(a) Is this an experiment or an observational study? Explain what it means to say that this was a "matched-pairs" study.

(b) How was randomization likely used in the study? Why did participants have cell phones on their ears during the "off" condition?

(c) The investigators were interested in seeing whether average brain glucose metabolism was different based on whether the cell phones were turned on or off. State the null and alternative hypotheses for this test.

(d) The p-value for the test in part (c) is 0.004. State the conclusion of this test in context.

(e) The investigators were also interested in seeing if brain glucose metabolism was significantly correlated with the amplitude of the radiofrequency waves. What graph might we use to visualize this relationship?

(f) State the null and alternative hypotheses for the test in part (e).

(g) The article states that the p-value for the test in part (e) satisfies $p < 0.001$. State the conclusion of this test in context.

B.59 Genetic Component of Autism It is estimated that in the general population, about 9 out of every 1000, or 0.009, children are diagnosed with autism. One study[73] included 92 six-month-old babies who had a sibling with autism. Twenty-eight of these babies were later diagnosed with autism. Use *StatKey* or other technology to find a 99% confidence interval for the proportion of siblings of autistic children likely to have autism. (In the study, brain scans taken at six-months old predicted almost perfectly which children would later be diagnosed with autism, providing the earliest known method for diagnosing the disease.)

B.60 Commuting Distances in Atlanta In addition to the commute *time* (in minutes), the **CommuteAtlanta** dataset gives the *distance* for the commutes (in miles) for 500 workers sampled from the Atlanta metropolitan area.

(a) Find the mean and standard deviation of the commute distances in **CommuteAtlanta**.

[73]*CBS Evening News*, February 17, 2012.

(b) Use *StatKey* or other technology to create a bootstrap distribution of the sample means of the distances. Describe the shape and center of the distribution.

(c) Use the bootstrap distribution to estimate the standard error for mean commute distance when using samples of size 500.

(d) Use the standard error to find and interpret a 95% confidence interval for the mean commute distance of Atlanta workers.

B.61 Mean of Atlanta Commuting Distances Exercise B.60 describes the variable *Distance* for the Atlanta commuter sample stored in **CommuteAtlanta**, giving the distance of each commute (in miles). Use *StatKey* or other technology to create a distribution with the mean distances for 5000 bootstrap samples and use it to find and interpret a 90% confidence interval for the mean commute distance in metropolitan Atlanta.

B.62 Correlation between Distance and Time for Atlanta Commutes The data in **CommuteAtlanta** contains information on both the *Distance* (in miles) and *Time* (in minutes) for a sample of 500 Atlanta commutes. We expect the correlation between these two variables to be positive, since longer distances tend to take more time.

(a) Find the correlation between *Distance* and *Time* for the original sample of 500 Atlanta commutes.

(b) Use *StatKey* or other technology to create a bootstrap distribution of correlations of *Distance* vs *Time* for at least 1000 bootstrap samples using the Atlanta commuting data. Give a rough sketch of the bootstrap distribution, find its center, and describe the shape.

(c) Estimate the standard error of the bootstrap correlations and use it to find a 95% confidence interval for the correlation between distance and time of Atlanta commutes.

(d) Mark where the interval estimate lies on your plot in part (b).

B.63 Confidence Intervals for Correlation using Atlanta Commutes In Exercise B.62 we use the standard error to construct a 95% confidence interval for the correlation between *Distance* (in miles) and *Time* (in minutes) for Atlanta commuters, based on the sample of size $n = 500$ in **CommuteAtlanta**.

(a) Describe a process that could be used to generate *one* value in the bootstrap distribution for this situation.

(b) Use *StatKey* or other technology to generate 1000 bootstrap correlations to find and interpret a 99% confidence interval for the correlation in this setting.

(c) Repeat (b) for 95% and 90% confidence levels. You do not need to repeat the interpretation.

(d) Describe how the interval changes as the confidence level decreases.

B.64 Classroom Games Two professors[74] at the University of Arizona were interested in whether having students actually play a game would help them analyze theoretical properties of the game. The professors performed an experiment in which students played one of two games before coming to a class where both games were discussed. Students were randomly assigned to which of the two games they played, which we'll call Game 1 and Game 2. On a later exam, students were asked to solve problems involving both games, with Question 1 referring to Game 1 and Question 2 referring to Game 2. When comparing the performance of the two groups on the exam question related to Game 1, they suspected that the mean for students who had played Game 1 (μ_1) would be higher than the mean for the other students, μ_2, so they considered the hypotheses $H_0 : \mu_1 = \mu_2$ vs $H_a : \mu_1 > \mu_2$.

(a) The paper states: "*test of difference in means results in a p-value of 0.7619.*" Do you think this provides sufficient evidence to conclude that playing Game 1 helped student performance on that exam question? Explain.

(b) If they were to repeat this experiment 1000 times, and there really is no effect from playing the game, roughly how many times would you expect the results to be as extreme as those observed in the actual study?

(c) When testing a difference in mean performance between the two groups on exam Question 2 related to Game 2 (so now the alternative is reversed to be $H_a : \mu_1 < \mu_2$ where μ_1 and μ_2 represent the mean on Question 2 for the respective groups), they computed a p-value of 0.5490. Explain what it means (in the context of this problem) for both p-values to be greater than 0.5.

B.65 Classroom Games: Is One Question Harder? Exercise B.64 describes an experiment involving playing games in class. One concern in the

experiment is that the exam question related to Game 1 might be a lot easier or harder than the question for Game 2. In fact, when they compared the mean performance of all students on Question 1 to Question 2 (using a two-tailed test for a difference in means), they report a p-value equal to 0.0012.

(a) If you were to repeat this experiment 1000 times, and there really is no difference in the difficulty of the questions, how often would you expect the means to be as different as observed in the actual study?

(b) Do you think this p-value indicates that there is a difference in the average difficulty of the two questions? Why or why not?

(c) Based on the information given, can you tell which (if either) of the two questions is easier?

B.66 Classroom Games Exercise B.64 describes a situation in which game theory students are randomly assigned to play either Game 1 or Game 2, and then are given an exam containing questions on both games. Two one-tailed tests were conducted: one testing whether students who played Game 1 did better than students who played Game 2 on the question about Game 1, and one testing whether students who played Game 2 did better than students who played Game 1 on the question about Game 2. The p-values were 0.762 and 0.549, respectively. The p-values greater than 0.5 mean that, in the sample, the students who played the *opposite* game did better on each question. What does this study tell us about possible effects of actually playing a game and answering a theoretical question about it? Explain.

B.67 Mercury Levels in Fish and pH in Lake Water Data 2.4 on page 71 introduces the dataset **FloridaLakes** and discusses the correlation between the acidity (pH) for a sample of $n = 53$ Florida lakes and the average mercury level (ppm) found in fish taken from each lake. We saw in Chapter 2 that there appears to be a negative trend in the scatterplot between the two variables. We wish to test whether there is significant evidence of a negative correlation between pH and mercury levels. A randomization distribution based on the data is shown in Figure B.14. The sample statistic of interest is the sample correlation.

(a) What are the null and alternative hypotheses?

(b) Use Figure B.14 to give a very rough estimate of the sample correlation corresponding to a p-value of 0.30. Explain your reasoning.

[74]Dufewenberg, M. and Swarthout, J.T., "Play to Learn? An Experiment," from a working paper, at *http://econ.arizona.edu/docs/Working_Papers/2009/Econ-WP-09-03.pdf*.

(c) Use Figure B.14 to give a very rough estimate of the sample correlation corresponding to a p-value of 0.01. Explain your reasoning.

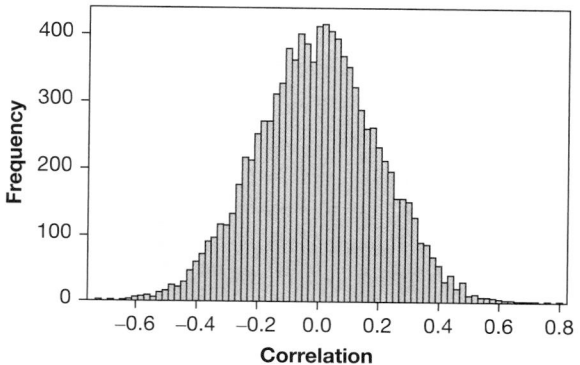

Figure B.14 *Randomization distribution of correlations for 10,000 samples using $H_0 : \rho = 0$*

B.68 Arsenic in Chicken In Data 4.3 on page 265 we describe a situation in which a restaurant chain will test for arsenic levels in a sample of chickens from a supplier. If there is evidence that the average level of arsenic is over 80 ppb, the chain will permanently cancel its relationship with the supplier. The null and alternative hypotheses are $H_0 : \mu = 80$ vs $H_a : \mu > 80$.

(a) What would it mean for analysts at the restaurant chain to make a Type I error in the context of this situation?

(b) What would it mean to make a Type II error in this situation?

(c) Does the word "error" mean that the person doing the test did something wrong (perhaps by sampling in a way that biased the results, making a mistake in data entry, or an arithmetic error)? Explain.

B.69 Medicinal Marijuana in HIV Treatment In 1980, the active ingredient in marijuana was shown to outperform a placebo in reducing nausea in chemotherapy patients, with a p-value of 0.0004. Many studies[75] are now underway to see if the drug has additional medicinal uses. In one controlled, randomized trial, 55 patients with HIV were randomly assigned to two groups, with one group getting cannabis (marijuana) and the other getting

[75]Seppa, N., "Not Just a High: Scientists Test Medicinal Marijuana Against MS, Inflammation, and Cancer," *ScienceNews*, June 19, 2010.

Table B.9 *Is marijuana effective at relieving pain in HIV patients?*

	Pain Reduced	Pain Not Reduced
Cannabis	14	13
Placebo	7	21

a placebo. All of the patients had severe neuropathic pain, and the response variable is whether or not pain was reduced by 30% or more (a standard benchmark in pain measurement). The results are shown in Table B.9. The question of interest is whether marijuana is more effective than a placebo in relieving pain.

(a) What are the null and alternative hypotheses?

(b) What are the sample proportions of patients with reduced pain in each group? Are the sample results in the direction of the alternative hypothesis?

(c) The US Food and Drug Administration (FDA) is reluctant to approve the medicinal use of cannabis unless the evidence supporting it is very strong because the drug has significant side effects. Do you expect the FDA to use a relatively small or relatively large significance level in making a conclusion from this test?

(d) What assumption do we make in creating the randomization distribution? If we use the difference in sample proportions, $D = \hat{p}_1 - \hat{p}_2$, as our sample statistic, where will the distribution be centered? Give a rough sketch of the general shape of the randomization distribution, showing the shape and where it is centered.

(e) What is the observed statistic from the sample? If the p-value for this test is 0.02, locate the observed statistic on your rough sketch of the randomization distribution.

(f) Use the p-value given in part (e) to give an informal conclusion to the test by describing the strength of evidence for the result.

(g) Combining your answers to parts (c) and (f), what is the likely formal conclusion of the test?

B.70 Finding a P-value for Marijuana for HIV Patients For the study in Exercise B.69, use *StatKey* or other technology to create the randomization distribution for this data. Use the distribution to calculate a p-value for the test, and compare this p-value to the one given in Exercise B.69(e). Use the p-value obtained in this exercise to assess the strength of evidence against the null hypothesis, in context.

B.71 Election Poll In October before the 2008 US presidential election, *ABC News* and the *Washington Post* jointly conducted a poll of "a random national sample" and asked people who they intended to vote for in the 2008 presidential election.[76] Of the 1057 sampled people who answered either Barack Obama or John McCain, 55.2% indicated that they would vote for Obama while 44.8% indicated that they would vote for McCain. While we now know the outcome of the election, at the time of the poll many people were very curious as to whether this significantly predicts a winner for the election. (While a candidate needs a majority of the electoral college vote to win an election, we'll simplify things and simply test whether the percentage of the popular vote for Obama is greater than 50%.)

(a) State the null and alternative hypotheses for testing whether more people would vote for Obama than McCain. (*Hint:* This is a test for a single proportion since there is a single variable with two possible outcomes.)

(b) Describe in detail how you could create a randomization distribution to test this (if you had many more hours to do this homework and no access to technology).

B.72 Finding the P-value for the Election Poll Use *StatKey* or other technology to create a randomization distribution for the poll described in Exercise B.71. What is the p-value for the test? Would this have been convincing evidence that Obama would win the election (at least the popular vote)? Now, knowing the true outcome of the election, does the test avoid making a Type I or Type II error?

B.73 Possible Errors in Testing Infections in Childbirth Exercise B.50 on page 360 describes a randomized trial in South Africa to test whether the proportion of babies born with infections is smaller if women in labor are treated with a wipe containing chlorohexidine rather than a sterile wipe (the placebo). A sample of $n = 481$ pregnant women were randomly split into the two groups. One goal of the study is to test $H_0 : p_c = p_w$ vs $H_a : p_c < p_w$, where p_c and p_w are the proportion of babies who develop infections during childbirth with the respective treatments.

(a) What does it mean to make a Type I error in this situation?

[76]http://www.washingtonpost.com/wp-srv/politics/polls/postpoll_101308.html.

(b) What does it mean to make a Type II error in this situation?

(c) In which of the following two situations should we select a smaller significance level:
- The drug chlorohexidine is very safe and known to have very few side effects.
- The drug chlorohexidine is relatively new and may have potentially harmful side effects for the mother and newborn child.

(d) The p-value for the data in this study is 0.32. What is the conclusion of the test?

(e) Does this conclusion mean that the treated wipes do not help prevent infections? Explain.

B.74 Mercury Levels in Fish The dataset **FloridaLakes** is introduced in Data 2.4 on page 71, and we return to the dataset in Exercise B.67 on page 364. To see if there is evidence of a negative correlation between pH in lakes and mercury levels in fish, the hypotheses are $H_0 : \rho = 0$ vs $H_a : \rho < 0$. For the observed sample, with $n = 53$, we have $r = -0.575$.

(a) What assumption do we make in creating the randomization distribution?

(b) Where will the randomization distribution be centered?

(c) Describe how you could use index cards to create one simulated sample. How many cards do you need? What will you put on them? What will you do with them? Once you have used the cards to create a simulated sample, what statistic will you calculate from it to use in a randomization distribution? (You don't have to actually create a simulated sample, just give a description of the process you would follow.)

B.75 A Randomization Distribution for Mercury Levels in Fish Use *StatKey* or other technology to create the randomization distribution for the situation in Exercise B.74. Use the distribution to calculate a p-value. Using $\alpha = 0.05$, state the conclusion in context.

B.76 Heart Rates and Blood Pressure Table B.10 shows the heart rates and systolic blood pressure for eight 55-year-old patients from the Intensive Care Unit data introduced in Data 2.3 on page 69 and available at **ICUAdmissions**. We are testing to see if the data provide evidence of a positive correlation between these two variables for 55-year-old ICU patients.

(a) Define any relevant parameter(s) and state the null and alternative hypotheses.

(b) What assumption do we make in creating the randomization distribution?

(c) What statistic will we record for each of the simulated samples to create the randomization distribution? What is the value of that statistic for the observed sample?

(d) Where will the randomization distribution be centered?

(e) Explain how we can create randomization samples to be consistent with the null hypothesis.

(f) Find one point on the randomization distribution by carrying out the procedure in part (e). Show the resulting values for the variables, and compute the sample correlation for the randomization sample.

(g) Find a second randomization sample and record its sample correlation.

Table B.10 *Are heart rate and systolic blood pressure positively correlated?*

Heart Rate	Systolic BP
86	110
86	188
92	128
100	122
112	132
116	140
136	190
140	138

B.77 Randomization Distribution for Heart Rate and Blood Pressure Use *StatKey* or other technology to create the randomization distribution for the data in Exercise B.76. Use the distribution to estimate the p-value for the test. Are the results statistically significant?

RANDOMIZATION SAMPLES

In Exercises B.78 to B.82, a situation is described for a statistical test. In Section 4.1 you were asked to state the null and alternative hypotheses (Exercises 4.9 to 4.13). Here, for each situation, describe how you might physically create one randomization sample and compute one randomization statistic (without using any technology) from a given sample. Be explicit enough that a classmate could follow your instructions (even if it might take a very long time).

B.78 Testing to see if there is evidence that the proportion of people who smoke is greater for males than for females.

B.79 Testing to see if there is evidence that a correlation between height and salary is significant (that is, different than zero).

B.80 Testing to see if there is evidence that the percentage of a population who watch the Home Shopping Network is less than 20%.

B.81 Testing to see if average sales are higher in stores where customers are approached by salespeople than in stores where they aren't.

B.82 Testing to see if there is evidence that the mean time spent studying per week is different between first-year students and upperclass students.

Inference with Normal and t-Distributions

"Statistics is now the sexiest subject around. "

Hans Rosling, Professor of International Health*

UNIT OUTLINE

5 Approximating with a Distribution

6 Inference for Means and Proportions

Essential Synthesis

In this unit, we use the normal and t-distributions, together with formulas for standard errors, to create confidence intervals and conduct hypothesis tests involving means and proportions.

*"When the Data Struts its Stuff," *The New York Times*, April 3, 2011, p. B3

CHAPTER 5

Approximating with a Distribution

*"The
normal
law of error
stands out in the
experience of man-
kind as one of the broad-
est generalizations of natural
philosophy. It serves as the guiding
instrument in researches in the physical
and social sciences and in medicine, agriculture and
engineering. It is an indispensable tool for the analysis and the
interpretation of the basic idea obtained by observation and experiment."*

–W. J. Youden*

Questions and Issues

Here are some of the questions and issues we will discuss in this chapter:

- How strong are the benefits of eating organic food?
- What proportion of US adults have used online dating or a dating app?
- Is there a home field advantage in American football?
- Do dogs notice whether someone is reliable or unreliable?
- What incentives are most effective in getting people to exercise?
- What incentives are most effective at helping people quit smoking?
- What is the most common sleep position?
- What proportion of adults say television is their main source of news?
- How often will a soccer goalie correctly guess the direction of a penalty kick?
- How helpful is it to use self-quizzes when studying?
- What proportion of teenagers have some hearing loss?
- What percent of air travelers prefer a window seat?
- Has smoke-free legislation had an effect on asthma rates?
- What percent of people rarely use cash?
- What proportion of cell phone users have downloaded an app?

5.1 HYPOTHESIS TESTS USING NORMAL DISTRIBUTIONS

If you skim through Chapters 3 and 4, you will notice that many of the graphs of sampling, bootstrap, and randomization distributions (for example, those reproduced in Figure 5.1) have a similar shape. This is not a coincidence. Under fairly general circumstances, the distribution of many common statistics will follow this same bell-shaped pattern. The formal name for this shape is a *normal* distribution. In this section we exploit this common shape to find p-values for hypothesis tests, and in the next section we use it to find confidence intervals.

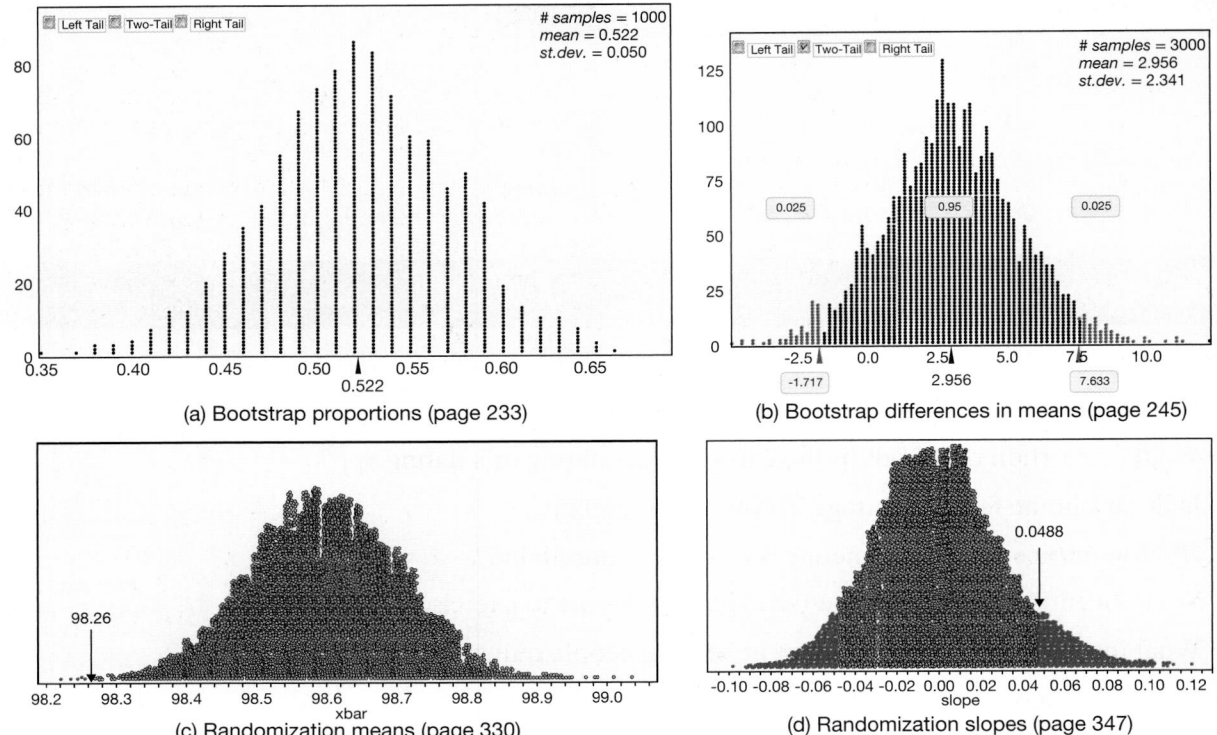

Figure 5.1 *Some bootstrap and randomization distributions*

Normal Distributions

A normal distribution has the shape of a bell-shaped curve. Every normal distribution has the same shape, but we get different normal distributions by changing either the mean (center) or the standard deviation (spread). The mean, μ, determines the center of the distribution, and the standard deviation, σ, determines the spread. Knowing just the mean and standard deviation of a normal distribution tells us what the entire distribution looks like.

Normal Distribution

A **normal distribution** follows a bell-shaped curve. We use the two parameters *mean*, μ, and *standard deviation*, σ, to distinguish one normal curve from another.

For shorthand we often use the notation $N(\mu, \sigma)$ to specify that a distribution is normal (N) with some mean (μ) and standard deviation (σ).

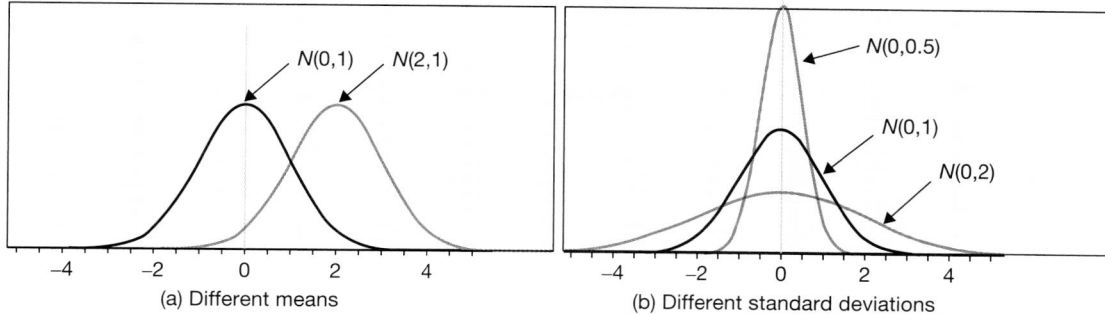

Figure 5.2 *Comparing normal curves*

Figure 5.2 shows how the normal distribution changes as the mean μ is shifted to move the curve horizontally or the standard deviation σ is changed to stretch or shrink the curve.

In a bootstrap distribution or randomization distribution, we are often interested in finding the proportion of statistics to the right or left of a certain value, with the total proportion of all the statistics being 1. When we use a normal distribution, this corresponds to finding an area to the right or left of a certain value, given that the total area under the curve is 1. Many technology options exist for finding the proportion of a normal distribution that falls beyond a specified endpoint.[1]

Example 5.1

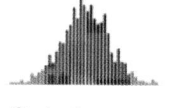

Find the area to the right of 95 in a normal distribution with mean 80 and standard deviation 10.

Solution

There are many different methods that can be used to find this area, and this is a good time to get comfortable with the method that you will use. Figure 5.3 shows how we find the area using *StatKey*. We see that the area to the right of 95 in the $N(80, 10)$ distribution is 0.067.

The Standard Normal Distribution

Because all the normal distributions look the same except for the horizontal scale, another common way to use normal distributions is to convert everything to

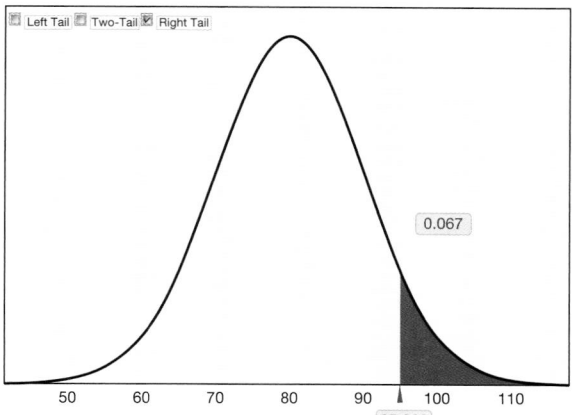

Figure 5.3 *Finding an area in the normal distribution N(80, 10)*

[1]For more information on specific options, see the technology manuals available in the online resources for this book.

one specific *standard normal* scale. The standard normal, $N(0, 1)$, has a mean of 0 and a standard deviation of 1. We often use the letter Z to denote a standard normal distribution.

To convert a value from a $N(\mu, \sigma)$ scale to a standard normal scale, we subtract the mean μ to shift the center to zero, then divide the result by the standard deviation σ to stretch (or shrink) the difference to match a standard deviation of one. If X is a value on the $N(\mu, \sigma)$ scale, then $Z = (X - \mu)/\sigma$ is the corresponding point on the $N(0, 1)$ scale.[2] You should recognize this as the *z-score* from page 82, because the standardized value just measures how many standard deviations a value is above or below the mean.

> ## Standard Normal Distribution
>
> The **standard normal** distribution has mean zero and standard deviation equal to one, $N(0, 1)$.
>
> To convert from an X value on a $N(\mu, \sigma)$ scale to a Z value on a $N(0, 1)$ scale, we standardize values with the z-score:
>
> $$Z = \frac{X - \mu}{\sigma}$$

Example 5.2

In Example 5.1, we find the area to the right of 95 in a normal distribution with mean 80 and standard deviation 10. Find the *z-score* for the endpoint of 95 and use the standard normal distribution to find the area to the right of that point.

Solution

We see that the *z*-score for 95 in a $N(80, 10)$ distribution is given by

$$z = \frac{X - \mu}{\sigma} = \frac{95 - 80}{10} = 1.5.$$

We see in Figure 5.4 that the area above 1.5 in the standard normal distribution is 0.067. This matches the area we found in Example 5.1, since it is the same area but converted to a standardized scale.

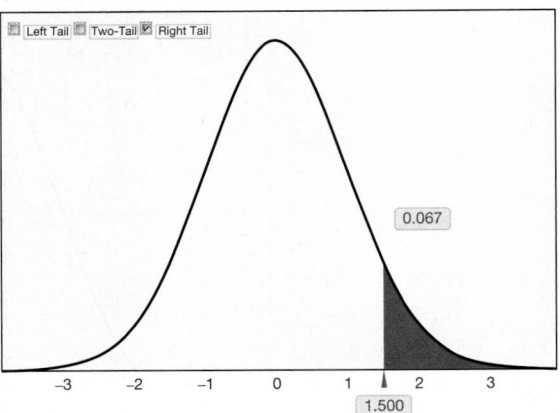

Figure 5.4 *Finding an area in a standard normal distribution*

[2] When technology is not available, a printed table with probabilities for certain standard normal endpoints can be used.

The Central Limit Theorem

The fact that so many sampling, bootstrap, and randomization distributions follow a normal distribution is not a coincidence. Statistical theory confirms that, for a large enough sample size, the distribution of many common sample statistics, such as means, proportions, differences in means, and differences in proportions, will all follow the pattern of a normal distribution. These results follow from one of the most important results in all of statistics: the *Central Limit Theorem*.

Central Limit Theorem (General)

For random samples with a sufficiently large sample size, the distribution of many common sample statistics can be approximated with a normal distribution.

What do we mean by "sufficiently large" sample size? We return to specific versions of the Central Limit Theorem for different parameters (such as means, proportions and differences) in Chapter 6 and address "sufficiently large" in each situation, as well as how to determine the mean and standard deviation for the approximating normal distribution. It is worth noting now that, in general, as the sample size *n* gets larger, the distribution of sample statistics tends to more closely resemble a normal distribution. In cases where a bootstrap distribution or randomization distribution follows the pattern of a normal distribution we can readily compute a confidence interval or p-value using that normal distribution.

Aleksandra Zaitseva / Shutterstock

Eating organic: Does it offer health benefits?

DATA 5.1

Eating Organic

Many people pay more to eat organic when possible, but does eating organic food actually offer health benefits? Spurred on by her parents debating this question, a 16-year-old girl decided to investigate it with a science project. She randomly divided fruit flies into two groups, and fed one group conventional food (potatoes, raisins, bananas, or soybeans) and the other group the same type of food, but the organic version. She then followed the flies and measured variables like longevity, fertility, stress resistance, and activity. Her results earned her honors in a national science competition, and publication in a prestigious journal.[3] Here we examine the survival of fruit flies eating potatoes. After 13 days, 318 of the 500 flies fed organic potatoes were still alive, as compared to 285 of the 500 flies fed conventional potatoes.[4] ◼

Example 5.3

State the null and alternative hypotheses for a test to investigate whether eating organic potatoes improves survival of fruit flies. Define the parameters and give the notation and value of the observed sample statistic.

Solution

Define p_o and p_c to be the proportion of flies alive after 13 days of eating organic and conventional potatoes, respectively. We are interested in whether the proportion of flies alive is higher for the flies that eat organic than for the flies that do not eat organic, so the hypotheses are

$$H_0: p_o = p_c$$
$$H_a: p_o > p_c$$

The observed sample statistic, the difference in proportions, is

$$\hat{p}_o - \hat{p}_c = \frac{318}{500} - \frac{285}{500} = 0.636 - 0.570 = 0.066.$$

The next step in the hypothesis test is to find the p-value and we illustrate below three different ways of accomplishing this goal.

Review: P-value from a Randomization Distribution

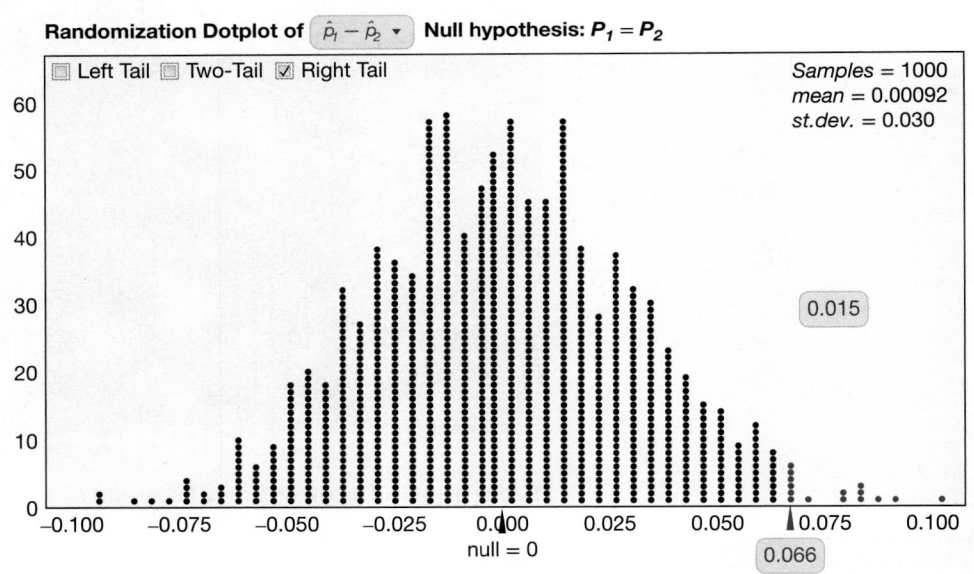

Figure 5.5
Randomization distribution and p-value

Example 5.4

Use a randomization distribution to find the p-value and complete the test about the effect of organic eating on fruit fly survival started in Example 5.3.

[3]Chhabra, R., Kolli, S., Bauer, J.H. (2013). "Organically Grown Food Provides Health Benefits to Drosophila melanogaster," *PLoS ONE*, 8(1): e52988. doi:10.1371/journal.pone.0052988
[4]Data approximated from paper.

Solution

We use technology to generate a randomization distribution and find the proportion of randomization values at or above the sample difference of $\hat{p}_o - \hat{p}_c = 0.066$. Figure 5.5 shows this p-value to be 0.015. Using $\alpha = 0.05$, this is low enough to reject H_0, and thus we have convincing evidence that eating organic as opposed to conventional potatoes causes fruit flies to live longer. We can make conclusions about causality because this was a randomized experiment.[5]

P-value from a Normal Distribution

When a randomization distribution displays the bell-shaped pattern of a normal distribution, as in Figure 5.5, we can use a normal distribution to compute a p-value in a hypothesis test. The process of choosing a null and alternative hypothesis in a particular situation is the same as in Chapter 4, and our interpretation of the outcome of the test remains the same: The smaller the p-value, the stronger the evidence against the null hypothesis. The only difference in the test is that we use a smooth normal distribution (rather than the empirical randomization distribution) to compute the p-value.

Example 5.5

What normal distribution best approximates the randomization distribution shown in Figure 5.5?

Solution

To choose a normal distribution, we need to specify both its mean and standard deviation. The randomization distribution in Figure 5.5 is centered at the null value[6] of 0, so the mean of the normal distribution would be 0. The standard deviation of the randomization distribution is the standard error, shown to be 0.03. This randomization distribution would be best approximated by a $N(0, 0.03)$ distribution, shown by the blue curve in Figure 5.6.

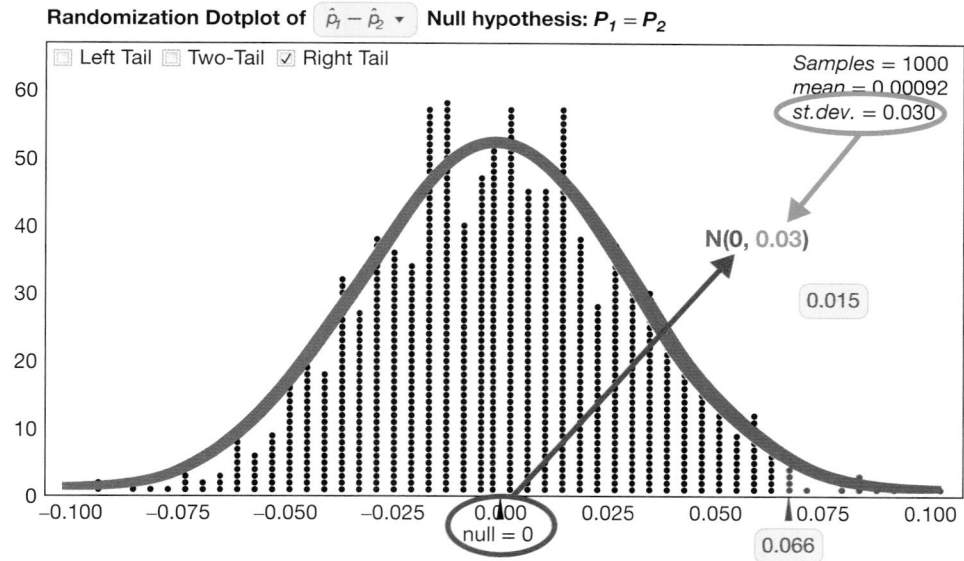

Figure 5.6
Randomization distribution with corresponding normal distribution

[5]The results are just starting to be significant after 13 days, but after 3 weeks the p-value is less than one in a trillion! For the flies eating soybeans, after 10 days *none* of the flies eating conventional soybeans were alive, as opposed to 60% of the flies eating organic soybeans (p-value $\approx$ 0). These experiments provide *very* strong evidence that eating organic has health benefits.
[6]The actual mean of the 2000 randomizations shown in the *StatKey* output is −0.00092, but this is just random fluctuation. If we use many, many randomizations, a bell-shaped randomization distribution should always be centered at the null value.

Once we have the normal distribution that approximates the randomization distribution, we no longer need all the dots under the curve, and can work directly with just the normal distribution. Exactly as we did with the randomization distribution, we find the p-value as the proportion of the distribution beyond the observed sample statistic.

Example 5.6

Solution

Use the $N(0, 0.03)$ distribution from Example 5.5 to find the p-value for the test described in Example 5.3.

We use the $N(0, 0.03)$ distribution to find the proportion beyond the observed sample statistic of 0.066, as shown in Figure 5.7. Note that the p-value of 0.014 obtained from the normal distribution is very close to the p-value of 0.015 found from the randomization distribution (which might vary slightly for a different set of randomizations). The interpretation of the p-value is exactly the same as in Example 5.4. The only thing different is that, in this example, we use a theoretical normal curve instead of a randomization distribution to find the p-value.

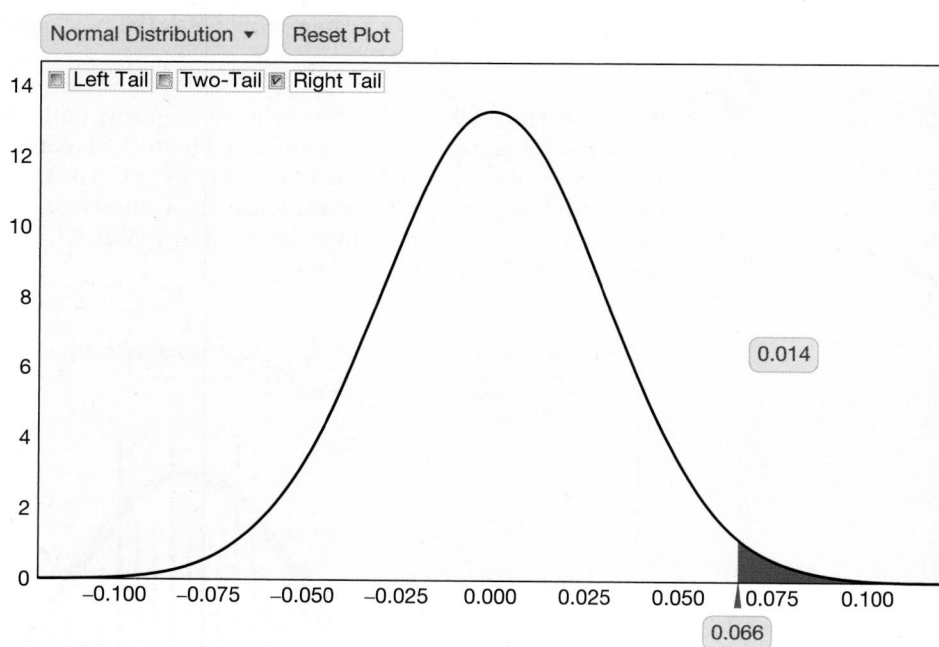

Figure 5.7 *P-value from a N(0, 0.03) distribution*

P-value from a Normal Distribution

The normal distribution that best approximates a bell-shaped randomization distribution has mean equal to the null value of the parameter, with standard deviation equal to the standard error:

$$N(\text{null parameter}, SE).$$

When the randomization distribution is shaped like a normal distribution, a p-value can be found as the proportion of this normal distribution beyond the observed sample statistic in the direction of the alternative (or twice the smaller tail for two-tailed tests).

P-value from a Standard Normal Distribution

We have seen how to use the original sample statistic compared to the normal distribution on the original data scale to find a p-value. Another option is to standardize the sample statistic to give a *standardized test statistic* which we then compare to a standard normal distribution. Recall that we convert to a standard normal distribution using a *z*-score. We convert from the original sample statistic on a N(null parameter, SE) scale to a standardized statistic on a $N(0, 1)$ scale by subtracting the null parameter value and dividing by the standard error. The result is often called a *z-statistic*, and is then used with a standard normal distribution to find the p-value.

P-value from a Standard Normal Distribution

When the distribution of the statistic under H_0 is normal, we compute a standardized **test statistic** using

$$z = \frac{\text{Sample Statistic} - \text{Null Parameter}}{SE}$$

The p-value for the test is the proportion of a standard normal distribution beyond this standardized test statistic, depending on the direction of the alternative hypothesis.

Example 5.7

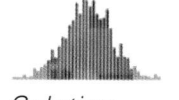

Find the standardized test statistic and compare to the standard normal distribution to find the p-value for the test described in Example 5.3.

Solution

Recall that the original sample statistic for this test is $\hat{p}_o - \hat{p}_c = 0.066$. The null hypothesis is $H_0 : p_o = p_c$ so the null parameter value is $p_o - p_c = 0$. The standard error for $\hat{p}_o - \hat{p}_c$ obtained from the randomization distribution in Figure 5.6 is 0.03. Thus, we compute the standardized test statistic as

$$z = \frac{\text{Sample Statistic} - \text{Null Parameter}}{SE} = \frac{0.066 - 0}{0.03} = 2.2.$$

We find the p-value as the proportion of the standard normal distribution beyond 2.2, as shown in Figure 5.8, yielding a p-value of 0.014.

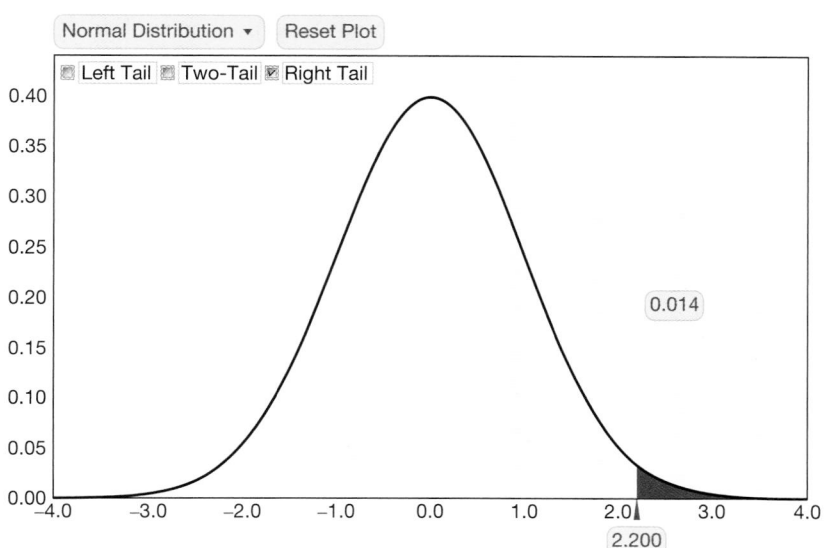

Figure 5.8 *P-value from a standard normal distribution*

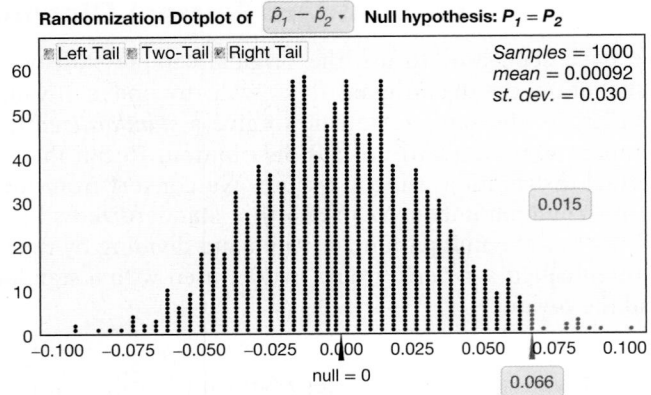

(a) Randomization Distribution

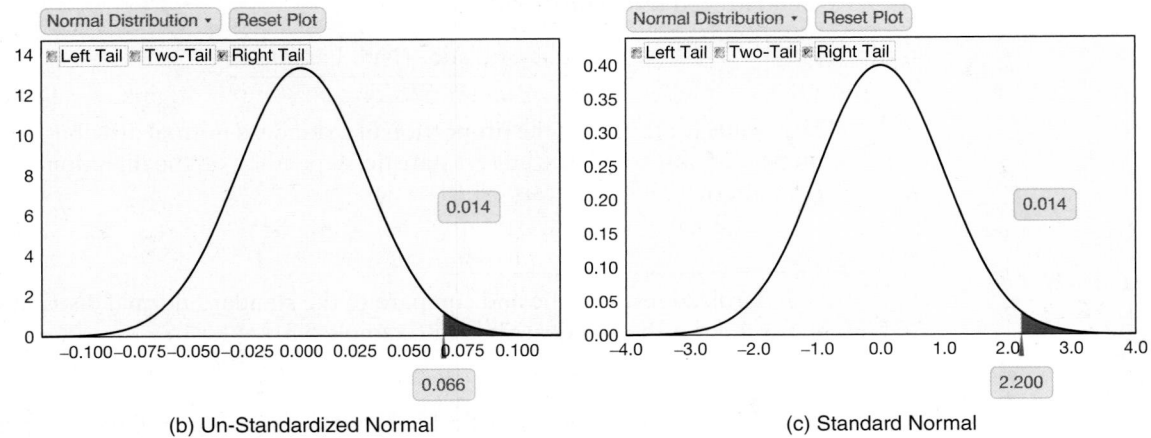

(b) Un-Standardized Normal   (c) Standard Normal

Figure 5.9 *Three ways of computing a p-value*

Note that the picture and p-value using the standard normal in Figure 5.8 are identical to the picture and p-value using the normal distribution on the original scale shown in Figure 5.7, except for the numeric scale on the horizontal axis. If the randomization distribution is shaped like a normal distribution, the picture and p-values resulting from the randomization distribution, the un-standardized normal, and the standard normal should all be similar, as shown in Figure 5.9.

Example 5.8

Testing Average Body Temperature

In the examples that follow Data 4.8 on page 328, we use the data in **BodyTemp50** to test whether average body temperature is different from 98.6°F. The hypotheses are

$$H_0: \mu = 98.6$$
$$H_a: \mu \neq 98.6$$

The sample of body temperatures for $n = 50$ subjects has mean $\bar{x} = 98.26$ and standard deviation $s = 0.765$. Figure 4.32 shows a randomization distribution, with a standard error of 0.1066 for these randomization means. Compute a z-statistic and use the standard normal distribution to find the p-value for this test and interpret the result.

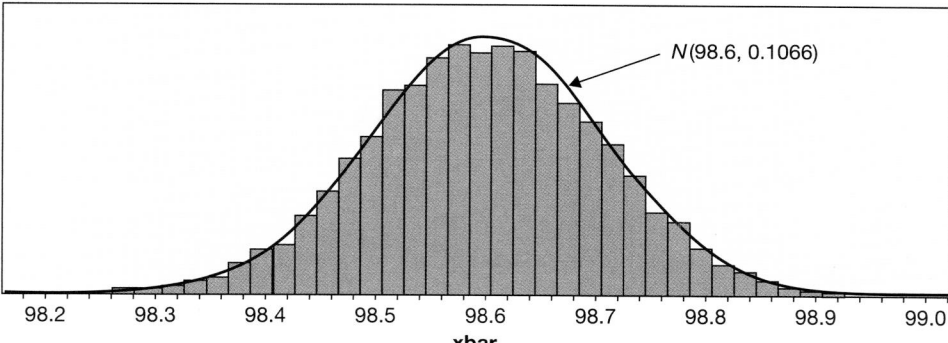

Figure 5.10
Randomization distribution of body temperature means with a normal distribution

Solution

The histogram of the randomization means in Figure 5.10 shows that an overlayed normal curve is an appropriate model. Based on the null hypothesis, we use 98.6 as the mean for this normal distribution and the standard error, $SE = 0.1066$, from the randomization means as the standard deviation.

To find a p-value, we need to measure how unusual the original $\bar{x} = 98.26$ is within this $N(98.6, 0.1066)$ distribution. Converting to a standard normal, we find

$$z = \frac{\text{Sample statistic} - \text{Null parameter}}{SE} = \frac{98.26 - 98.6}{0.1066} = -3.19$$

We use technology to find the area in a standard normal curve below -3.19 and then, since the alternative hypothesis, $H_a : \mu \neq 98.6$, is two-tailed, we double the result to account for the two tails. Figure 5.11 shows that the area below -3.19 is about 0.0007, so we have

$$\text{p-value} = 2 \cdot 0.0007 = 0.0014$$

As always, we interpret this p-value in terms of the question that motivated the test in the first place. The very small p-value (0.0014) indicates strong evidence against $H_0 : \mu = 98.6$ and we conclude that the average body temperature is probably less than 98.6°F.

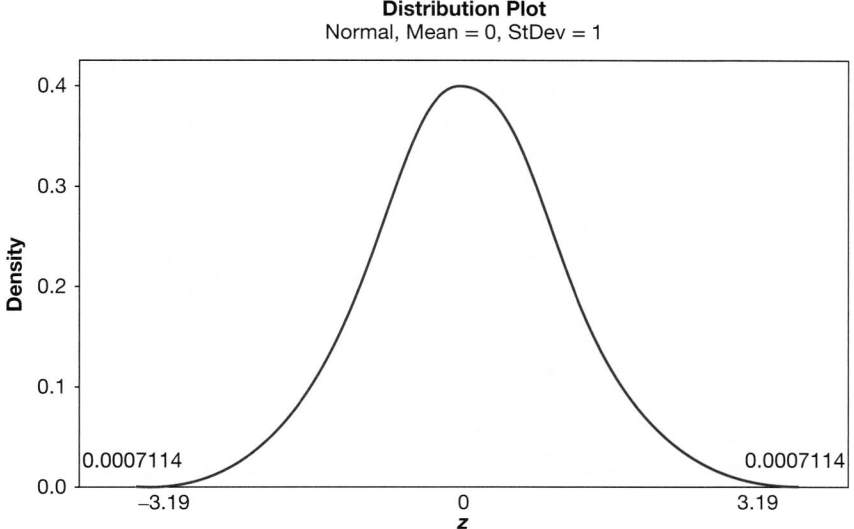

Figure 5.11 *P-value based on the normal distribution for the test of body temperature*

The small region beyond -3.19 in the standard normal curve in Figure 5.11 corresponds to the small region below 98.26 in the histogram of Figure 5.10. In fact, for the randomization distribution in Figure 4.32 on page 330, the p-value obtained by counting the outcomes more extreme than $\bar{x} = 98.26$ was 0.0016, so the normal distribution results are quite consistent with those obtained earlier directly from the randomization distribution.

Example 5.9

Home Field Advantage in American Football

In the 2015 National Football League (NFL) season, there were 263 games played on the home field of one of the teams, and the home team won 142 of those games.[7] If we use this as a sample of all NFL games, do the data provide evidence that the home team wins more than half the time in the NFL? Assume that a randomization distribution for the test is normally distributed with a standard error of $SE = 0.031$. Use a 5% significance level.

Solution

We are testing to see if there is evidence that the proportion, p, of all NFL games won by the home team is greater than 0.5. The relevant hypotheses are

$$H_0: p = 0.5$$
$$H_a: p > 0.5$$

The sample statistic of interest is $\hat{p} = 142/263 = 0.540$. We compute the standardized test statistic using the fact that the null parameter from the null hypothesis is 0.5, and the standard error is 0.031. The test statistic is

$$z = \frac{\text{Sample Statistic} - \text{Null Parameter}}{SE} = \frac{0.540 - 0.5}{0.031} = 1.290$$

This is a right-tailed test, so we find the proportion of a standard normal distribution larger than 1.290. The area in this upper tail is 0.099, so we have

$$\text{p-value} = 0.099$$

Since the p-value is larger than the 5% significance level, we do not find enough evidence to conclude that the home team wins more than half the games in the National Football League. Note that a home field advantage might still exist, we just don't see significant enough results in this sample to draw that conclusion.

One advantage of calculating a standardized test statistic is that the standard normal distribution shown in Figure 5.8 and Figure 5.11 (ignoring the shaded p-values) is always exactly the same. About 95% of the distribution falls between −2 and 2, and values rarely fall below −3 or above 3. Because we have this consistent shape, we can start to recognize extremity from the standardized test statistic alone, before we compute a formal p-value. In Example 5.8, the test statistic is −3.19, more than three standard deviations below the mean, so we should not be surprised by the small p-value. The test statistic of 1.29 in Example 5.9 is only 1.29 standard deviations above the mean, so we should not be surprised at the relatively large p-value.

In Examples 5.8 and 5.9, we could have conducted the test using another normal distribution rather than the standard normal. We chose to use a standard normal distribution here to help build intuition about test statistics as z-scores and to lead in to the methods we use in Chapter 6.

---

[7] *https://www.teamrankings.com/nfl/trends/win_trends/?sc=is_home.*

SECTION LEARNING GOALS

You should now have the understanding and skills to:

▶ • Recognize the shape of a normal distribution, and how the mean and standard deviation relate to the center and spread of a normal distribution

▶ • Find an area in a normal distribution

▶ • Compute a standardized test statistic

▶ • Compute a p-value using a normal distribution

Exercises for Section **5.1**

SKILL BUILDER 1

In Exercises 5.1 to 5.6, find the given area in a standard normal distribution.

5.1 The area in the right tail more extreme than $z = 2.20$

5.2 The area in the right tail more extreme than $z = 0.80$

5.3 The area in the right tail more extreme than $z = -1.25$

5.4 The area in the right tail more extreme than $z = 3.0$

5.5 The area in the left tail more extreme than $z = -1.75$

5.6 The area in the left tail more extreme than $z = -2.60$

SKILL BUILDER 2

Exercises 5.7 to 5.12 include a set of hypotheses, some information from one or more samples, and a standard error from a randomization distribution. Find the value of the standardized z-test statistic in each situation.

5.7 Test $H_0: \mu = 80$ vs $H_a: \mu > 80$ when the sample has $n = 20$, $\bar{x} = 82.4$, and $s = 3.5$, with $SE = 0.8$.

5.8 Test $H_0: p = 0.25$ vs $H_a: p < 0.25$ when the sample has $n = 800$ and $\hat{p} = 0.235$, with $SE = 0.018$.

5.9 Test $H_0: p = 0.5$ vs $H_a: p \neq 0.5$ when the sample has $n = 50$ and $\hat{p} = 0.41$, with $SE = 0.07$.

5.10 Test $H_0: \mu = 10$ vs $H_a: \mu \neq 10$ when the sample has $n = 75$, $\bar{x} = 11.3$, and $s = 0.85$, with $SE = 0.10$.

5.11 Test $H_0: p_1 = p_2$ vs $H_a: p_1 < p_2$ when the samples have $n_1 = 150$ with $\hat{p}_1 = 0.18$ and $n_2 = 100$ with $\hat{p}_2 = 0.23$. The standard error of $\hat{p}_1 - \hat{p}_2$ from the randomization distribution is 0.05.

5.12 Test $H_0: \mu_1 = \mu_2$ vs $H_a: \mu_1 > \mu_2$ when the samples have $n_1 = n_2 = 50$, $\bar{x}_1 = 35.4$, $\bar{x}_2 = 33.1$, $s_1 = 1.28$, and $s_2 = 1.17$. The standard error of $\bar{x}_1 - \bar{x}_2$ from the randomization distribution is 0.25.

SKILL BUILDER 3

In Exercises 5.13 and 5.14, find the p-value based on a standard normal distribution for each of the following standardized test statistics.

5.13 (a) $z = 0.84$ for a right-tail test for a difference in two proportions

(b) $z = -2.38$ for a left-tail test for a difference in two means

(c) $z = 2.25$ for a two-tailed test for a proportion

5.14 (a) $z = -1.08$ for a left-tail test for a mean

(b) $z = 4.12$ for a right-tail test for a proportion

(c) $z = -1.58$ for a two-tailed test for a difference in means

SKILL BUILDER 4

In Exercises 5.15 to 5.17, find the specified areas for a normal distribution.

5.15 For a $N(60, 10)$ distribution

(a) The area to the right of 65

(b) The area to the left of 48

5.16 For a $N(15, 5)$ distribution

(a) The area to the right of 28

(b) The area to the left of 12

5.17 For a $N(160, 25)$ distribution

(a) The area to the right of 140

(b) The area to the left of 200

MORE BENEFITS OF EATING ORGANIC

Using Data 5.1 on page 375, we find a significant difference in the proportion of fruit flies surviving after 13 days between those eating organic potatoes and those eating conventional (not organic) potatoes. Exercises 5.18 to 5.21 ask you to conduct a hypothesis test using additional data from this study.[8] In every case, we are testing

$$H_0: p_o = p_c$$
$$H_a: p_o > p_c$$

where p_o and p_c represent the proportion of fruit flies alive at the end of the given time frame of those eating organic food and those eating conventional food, respectively. In each exercise, show all remaining details of the test, including standardized test statistic, p-value from a standard normal distribution, generic conclusion, and an informative conclusion in context. Use a 5% significance level.

5.18 Effect of Organic Soybeans after 5 Days After 5 days, the proportion of fruit flies eating organic soybeans still alive is 0.90, while the proportion still alive eating conventional soybeans is 0.84. The standard error for the difference in proportions is 0.021.

5.19 Effect of Organic Bananas after 25 Days After 25 days, the proportion of fruit flies eating organic bananas still alive is 0.42, while the proportion still alive eating conventional bananas is 0.40. The standard error for the difference in proportions is 0.031.

5.20 Effect of Organic Potatoes after 11 Days After 11 days, the proportion of fruit flies eating organic potatoes still alive is 0.68, while the proportion still alive eating conventional potatoes is 0.66. The standard error for the difference in proportions is 0.030.

5.21 Effect of Organic Soybeans after 8 Days After 8 days, the proportion of fruit flies eating organic soybeans still alive is 0.79, while the proportion still alive eating conventional soybeans is 0.32. The standard error for the difference in proportions is 0.031.

5.22 How Do You Get Your News? A study by the Pew Research Center[9] reports that in 2010, for the first time, more adults aged 18 to 29 got their news from the Internet than from television. In a random sample of 1500 adults of all ages in the US, 66% said television was one of their main sources of news. Does this provide evidence that more than 65% of all adults in the US used television as one of their main sources for news in 2010? A randomization

distribution for this test shows $SE = 0.013$. Find a standardized test statistic and compare to the standard normal to find the p-value. Show all details of the test.

5.23 To Study Effectively, Test Yourself! Cognitive science consistently shows that one of the most effective studying tools is to self-test. A recent study[10] reinforced this finding. In the study, 118 college students studied 48 pairs of Swahili and English words. All students had an initial study time and then three blocks of practice time. During the practice time, half the students studied the words by reading them side by side, while the other half gave themselves quizzes in which they were shown one word and had to recall its partner. Students were randomly assigned to the two groups, and total practice time was the same for both groups. On the final test one week later, the proportion of items correctly recalled was 15% for the reading-study group and 42% for the self-quiz group. The standard error for the difference in proportions is about 0.07. Test whether giving self-quizzes is more effective and show all details of the test. The sample size is large enough to use the normal distribution.

5.24 Penalty Shots in World Cup Soccer A study[11] of 138 penalty shots in World Cup Finals games between 1982 and 1994 found that the goalkeeper correctly guessed the direction of the kick only 41% of the time. The article notes that this is "slightly worse than random chance." We use these data as a sample of all World Cup penalty shots ever. Test at a 5% significance level to see whether there is evidence that the percent guessed correctly is less than 50%. The sample size is large enough to use the normal distribution. The standard error from a randomization distribution under the null hypothesis is $SE = 0.043$.

5.25 Dogs Ignore an Unreliable Person A study[12] investigated whether dogs change their behavior depending on whether a person displays reliable or unreliable behavior. Dogs were shown two containers, one empty and one containing a dog biscuit. An experimenter pointed to one of the two containers. If the experimenter pointed to the container with

[8]Proportions approximated from information given in the paper.

[9]Pew Research Center, "Internet Gains on Television as Public's Main News Source," January 4, 2011.

[10]Pyc, M. and Rawson, K., "Why testing improves memory: Mediator effectiveness hypothesis," *Science*, October 15, 2010; 330:335.

[11]St. John, A., "Physics of a World Cup Penalty-Kick Shootout—2010 World Cup Penalty Kicks," *Popular Mechanics*, June 14, 2010.

[12]Takaoka, A., Maeda, T., Hori. Y., and Fujita, K., "Do dogs follow behavioral cues from an unreliable human?," *Animal Cognition* (2015), 18: 475–483.

the treat on the first trial, 16 of 26 dogs followed the experimenter's cue on the second trial. However, if the experimenter misled the dog on the first trial, only 7 of 26 dogs followed the cue on the second trial. Test to see if the proportion following the cue is different depending on whether the person exhibited reliable or unreliable behavior. The standard error for the difference in proportions is 0.138. Use a 5% significance level and show all details of the test.

5.26 Exercise and Gender The dataset **Exercise-Hours** contains information on the amount of exercise (hours per week) for a sample of statistics students. The mean amount of exercise was 9.4 hours for the 30 female students in the sample and 12.4 hours for the 20 male students. A randomization distribution of differences in means based on these data, under a null hypothesis of no difference in mean exercise time between females and males, is centered near zero and reasonably normally distributed. The standard error for the difference in means, as estimated from the randomization distribution, is $SE = 2.38$. Use this information to test, at a 5% level, whether the data show that the mean exercise time for female statistics students is less than the mean exercise time of male statistics students.

5.27 Smile Leniency Data 4.5 on page 293 describes an experiment to study the effects of smiling on leniency in judging students accused of cheating. Exercise B.53 on page 361 shows a dotplot, reproduced in Figure 5.12, of a randomization distribution of differences in sample means. The relevant hypotheses are $H_0 : \mu_s = \mu_n$ vs $H_a : \mu_s > \mu_n$, where μ_s and μ_n are the mean leniency scores for smiling and neutral expressions, respectively. This distribution is reasonably bell-shaped and we estimate the standard error of the differences in means under the null hypothesis to be about 0.393. For the actual

sample in **Smiles**, the original difference in the sample means is $D = \bar{x}_s - \bar{x}_n = 4.91 - 4.12 = 0.79$. Use a normal distribution to find and interpret a p-value for this test.

5.28 How Often Do You Use Cash? In a survey[13] of 1000 American adults conducted in April 2012, 43% reported having gone through an entire week without paying for anything in cash. Test to see if this sample provides evidence that the proportion of all American adults going a week without paying cash is greater than 40%. Use the fact that a randomization distribution is approximately normally distributed with a standard error of $SE = 0.016$. Show all details of the test and use a 5% significance level.

5.29 Incentives to Exercise: How Well Do They Work? A study[14] was designed to see what type of incentive might be most effective in encouraging people to exercise. In the study, 281 overweight or obese people were assigned the goal to walk 7000 steps a day, and their activity was tracked for 100 days. The participants were randomly assigned to one of four groups, with different incentives for each group. In this problem, we look at the overall success rate. For each participant, we record the number of days that the participant met the goal. For all 281 participants, the average number of days meeting the goal is 36.5. The standard error for this estimate is 1.80. Test to see if this provides evidence that the mean number of days meeting the goal, for people in a 100-day program to encourage exercise, is greater than 35. Show all details of the test.

[13]"43% Have Gone Through a Week Without Paying Cash," *Rasmussen Reports*, April 11, 2011.
[14]Patel, M.S., et al., "Framing Financial Incentives to Increase Physical Activity Among Overweight and Obese Adults: A Randomized, Controlled Trial," *Annals of Internal Medicine*, 2016; 164(6):385–94.

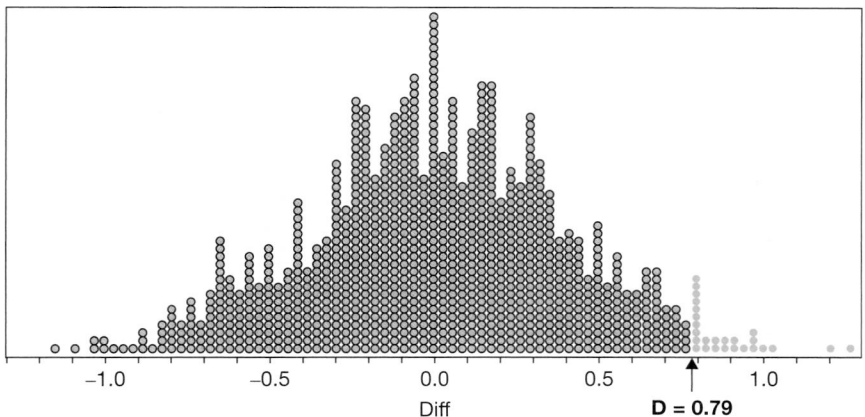

Figure 5.12 *Randomization distribution for 1000 samples testing $H_0 : \mu_s = \mu_n$ using Smiles data*

5.30 Incentives to Exercise: What Works Best? In the study described in Exercise 5.29, the goal for each of the overweight participants in the study was to walk 7000 steps a day. The study lasted 100 days and the number of days that each participant met the goal was recorded. The participants were randomly assigned to one of four different incentive groups: for each day they met the goal, participants in the first group got only an acknowledgement, participants in the second group got entered into a lottery, and participants in the third group received cash (about $1.50 per day). In the fourth group, participants received all the money up front and *lost* money (about $1.50 per day) if they didn't meet the goal. The success rate was almost identical in the first three groups (in other words, giving cash did not work much better than just saying congratulations) and the mean number of days meeting the goal for these participants was 33.7. For the participants who would lose money, however, the mean number of days meeting the goal was 45.0. (People really hate to lose money!) Test to see if this provides evidence of a difference in means between those losing money and those with other types of incentives, using the fact that the standard error for the difference in means is 4.14.

5.31 Incentives for Quitting Smoking: Group or Individual? In a smoking cessation program, over 2000 smokers who were trying to quit were randomly assigned to either a group program or an individual program. After six months in the program, 148 of the 1080 in the group program were successfully abstaining from smoking, while 120 of the 990 in the individual program were successful.[15] We wish to test to see if this data provide evidence of a difference in the proportion able to quit smoking between smokers in a group program and smokers in an individual program.

(a) State the null and alternative hypotheses, and give the notation and value of the sample statistic.

(b) Use a randomization distribution and the observed sample statistic to find the p-value.

(c) Give the mean and standard error of the normal distribution that most closely matches the randomization distribution, and then use this normal distribution with the observed sample statistic to find the p-value.

(d) Use the standard error found from the randomization distribution in part (b) to find the standardized test statistic, and then use that test statistic to find the p-value using a standard normal distribution.

(e) Compare the p-values from parts (b), (c), and (d). Use any of these p-values to give the conclusion of the test.

5.32 Incentives for Quitting Smoking: Do They Work? Exercise 5.31 describes a study examining incentives to quit smoking. With no incentives, the proportion of smokers trying to quit who are still abstaining six months later is about 0.06. Participants in the study were randomly assigned to one of four different incentives, and the proportion successful was measured six months later. Of the 498 participants in the group with the least success, 47 were still abstaining from smoking six months later. We wish to test to see if this provides evidence that even the smallest incentive works better than the proportion of 0.06 with no incentive at all.

(a) State the null and alternative hypotheses, and give the notation and value of the sample statistic.

(b) Use a randomization distribution and the observed sample statistic to find the p-value.

(c) Give the mean and standard error of the normal distribution that most closely matches the randomization distribution, and then use this normal distribution with the observed sample statistic to find the p-value.

(d) Use the standard error found from the randomization distribution in part (b) to find the standardized test statistic, and then use that test statistic to find the p-value using a standard normal distribution.

(e) Compare the p-values from parts (b), (c), and (d). Use any of these p-values to give the conclusion of the test.

5.33 Lucky numbers If people choose lottery numbers at random, the last digit should be equally likely to be any of the ten digits from 0 to 9. Let p measure the proportion of choices that end with the digit 7. If choices are random, we would expect $p = 0.10$, but if people have a special preference for numbers ending in 7 the proportion will be greater than 0.10. Suppose that we test this by asking a random sample of 20 people to give a three-digit lottery number and find that four of the numbers have 7 as the last digit. Figure 5.13 shows a randomization distribution of proportions for 5000 simulated samples under the null hypothesis $H_0: p = 0.10$.

[15]Halpern, S.D., and French, B., et al., "Randomized Trial of Four Financial-Incentive Programs for Smoking Cessation," *The New England Journal of Medicine*, 2015; 372:2108–17, May 13, 2015.

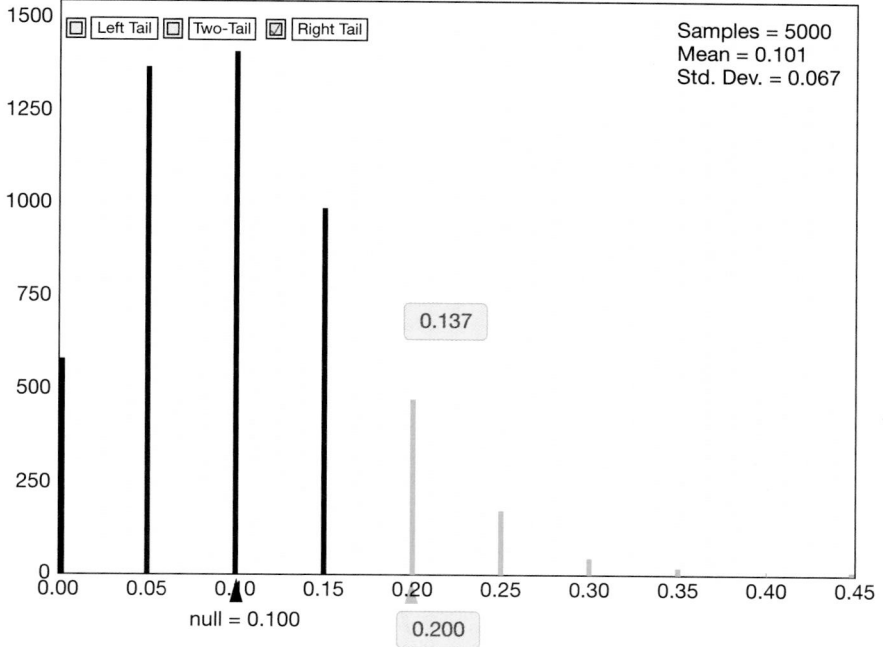

Figure 5.13 *Randomization distribution for proportions in Exercise 5.33 when p = 0.10 and n = 20*

(a) Use the sample proportion $\hat{p} = 0.20$ and a standard error estimated from the randomization distribution to compute a standardized test statistic.

(b) Use the normal distribution to find a p-value for an upper tail alternative based on the test statistic found in part (a).

(c) Compare the p-value obtained from the normal distribution in part (b) to the p-value shown for the randomization distribution. Explain why there might be a discrepancy between these two values.

5.2 CONFIDENCE INTERVALS USING NORMAL DISTRIBUTIONS

In Section 5.1, we see that many bootstrap and randomization distributions are normally distributed, and we see how to use the normal distribution to find p-values in a hypothesis test. In this section, we see how to use the normal distribution to find a confidence interval by approximating a bell-shaped bootstrap distribution with a normal distribution. To find a 95% bootstrap confidence interval in Chapter 3, we find an interval that contains the middle 95% of bootstrap statistics. The equivalent idea on a normal curve is to find the interval that captures the middle 95% of area in the normal distribution.

As in Section 5.1, there are many different methods that can be used to find the endpoints of this interval. Again, we generally rely on technology to handle the computational details. Some technology options, like *StatKey*, allow us to directly specify 95% for the middle range, while others have us find the interval endpoints based on the tail proportions (a 95% interval would leave 2.5% in each tail). This is a good time to get comfortable with the method that you will use.

Example 5.10

Find an interval that contains the middle 95% of area in a normal distribution with mean 40 and standard deviation 5.

Solution

We see in Figure 5.14 that the interval containing the middle 95% of area in this distribution is 30.201 to 49.799. Notice that this interval is approximately two standard deviations away from the mean on each side, as we expect with the 95% Rule from Chapter 2.

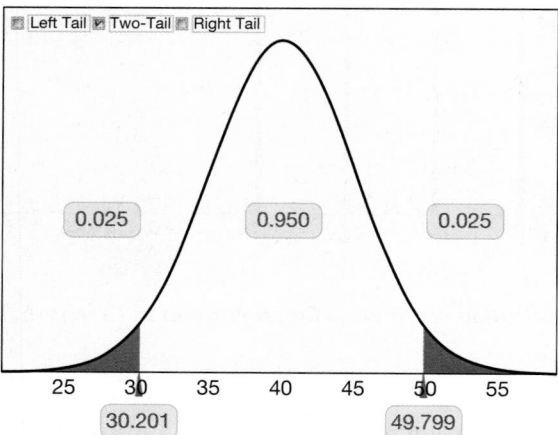

Figure 5.14 *Finding the middle 95% in the normal distribution N(40, 5)*

Example 5.11

Find an interval in the standard normal distribution that contains the middle 99% of area.

Solution

Recall that the standard normal distribution has mean 0 and standard deviation 1. We see in Figure 5.15 that the interval that contains the middle 99% of area in this distribution is −2.575 to 2.575. Other technologies might give the slightly different values ±2.576.

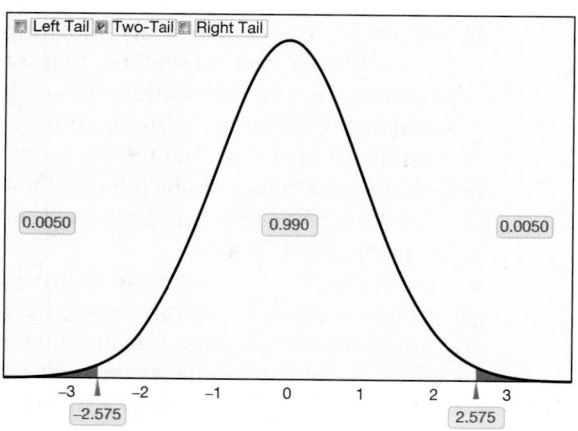

Figure 5.15 *Finding the middle 99% in the standard normal distribution*

Ilona Baha / Shutterstock

What percentage of American adults have used online dating?

DATA 5.2 **Online Dating**

A survey conducted in July 2015 asked a random sample of $n = 2001$ American adults whether they had ever used online dating (either an online dating site or a dating app on their cell phone).[16] Overall, 15% said that they had used online dating, showing an increase over the percentage in a similar survey conducted two years earlier. (Two groups in particular showed a change over this two-year period: Usage by 18- to 24-year-olds nearly tripled while usage by 55- to 64-year-olds more than doubled.) ■

We are interested in using the 15% found in the survey to estimate the proportion of all American adults to use online dating. First, we return to the bootstrap confidence intervals from Chapter 3, to connect the methods of Chapter 3 to the use of the normal distribution.

Review: Confidence Intervals Using a Bootstrap Distribution

Example 5.12

Use a bootstrap distribution to find a 90% confidence interval for the true proportion of all American adults who have used online dating.

Solution

We repeatedly resample, with replacement, from the observed sample values to get bootstrap samples of size $n = 2001$. For each bootstrap sample, we calculate the bootstrap statistic proportion of adults who have used either an online dating site or a mobile dating app. We repeat this process many times to form a bootstrap distribution, as shown in Figure 5.16. We then find a 90% confidence interval as the interval containing the middle 90% of bootstrap statistics, shown in Figure 5.16, yielding a confidence interval of (0.137, 0.163). We are 90% confident that between 13.7% and 16.3% of American adults have used online dating.

[16]Smith, A., (2016). "15% of American Adults Have Used Online Dating Sites or Mobile Dating Apps," *Pew Research Center*, 2/11/16, *http://www.pewinternet.org/2016/02/11/15-percent-of-american-adults-have-used-online-dating-sites-or-mobile-dating-apps/*.

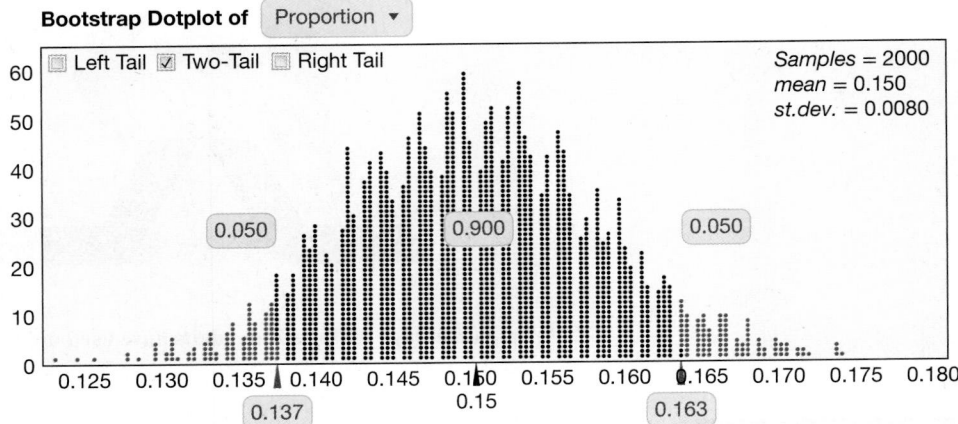

Figure 5.16 *Bootstrap distribution and 90% confidence interval*

Confidence Intervals Based on a Normal Distribution

Note that the bootstrap distribution in Figure 5.16 is bell-shaped, so we can use a normal distribution to approximate it.

Example 5.13

Which normal distribution best approximates the bootstrap distribution shown in Figure 5.16? Give the mean and standard deviation.

Solution

Bootstrap distributions are centered at the original sample statistic, in this case $\hat{p} = 0.15$.[17] The standard deviation of the bootstrap distribution is equal to the standard error, 0.008. With a mean of 0.15 and a standard deviation of 0.008, the $N(0.15, 0.008)$ distribution best approximates this bootstrap distribution. This normal curve is shown as the blue curve over the bootstrap distribution in Figure 5.17.

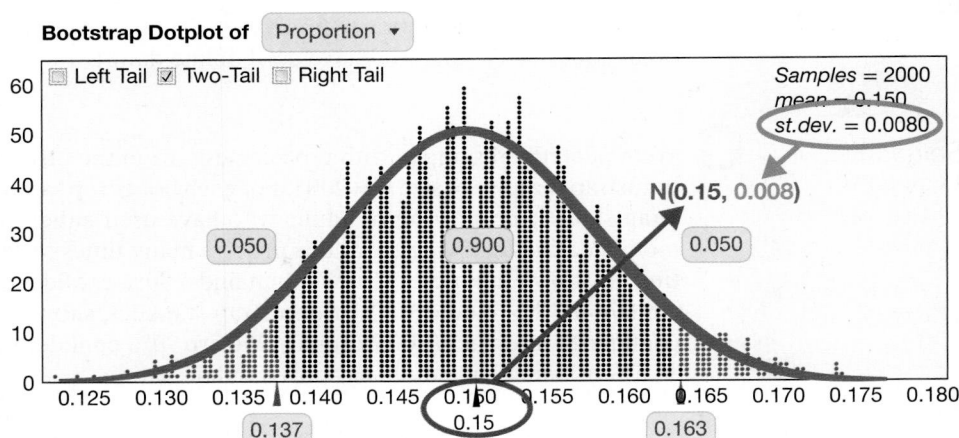

Figure 5.17 *Normal approximation to a bootstrap distribution*

As with randomization distributions, once we have the normal distribution, we can work with that directly to find the confidence interval. We find the confidence interval in exactly the same way we did when using percentiles from the bootstrap

[17]The mean for a particular set of statistics in a bootstrap distribution might vary a little from the original sample statistic, but we still use the original statistic as the mean for the normal approximation to the bootstrap distribution.

distribution: we find the interval that contains the middle percentage of the distribution that matches the desired confidence level.

Confidence Intervals Using Normal Distributions

If a bootstrap distribution is bell-shaped, a $P\%$ confidence interval can be found as the interval containing the middle $P\%$ of the normal distribution with mean equal to the observed sample statistic and standard deviation equal to the standard error of the statistic:

$$N(\text{sample statistic}, SE)$$

Example 5.14

Online Dating with a Normal Distribution

Use the normal distribution that approximates the bootstrap distribution in Example 5.13 to find a 90% confidence interval for the proportion of American adults who have used online dating.

Solution

We find the interval that contains the middle 90% of the $N(0.15, 0.008)$ distribution, as shown in Figure 5.18. This yields the confidence interval $(0.137, 0.163)$, exactly the same as that found from the bootstrap distribution. The interpretation of this confidence interval remains the same: We are 90% confident that between 13.7% and 16.3% of American adults have used an online dating site or mobile dating app.

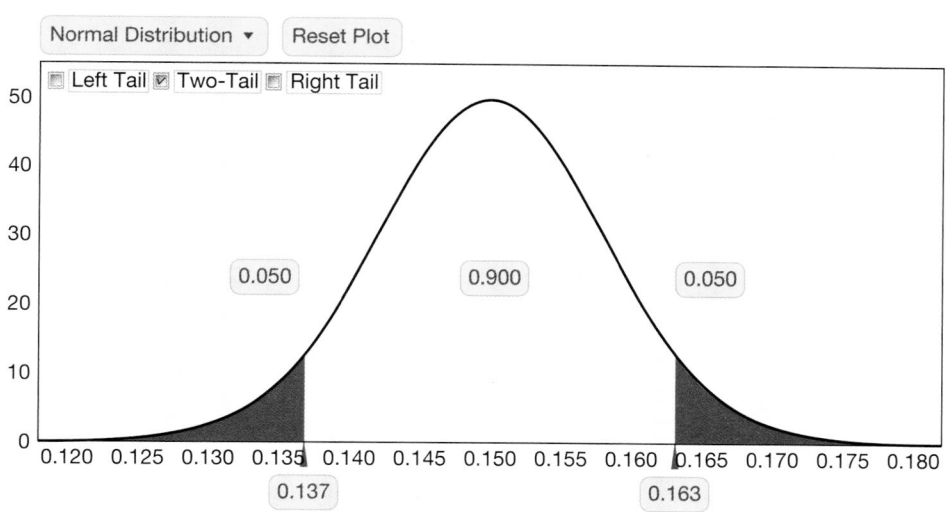

Figure 5.18 *Confidence interval from a N(0.15, 0.008) distribution*

Confidence Intervals Using a Standard Normal Distribution

As with hypothesis testing, it is often convenient to work with only one normal distribution: the standard normal distribution with mean 0 and standard deviation 1, $N(0, 1)$. Thus far we have connected back to the percentile method of Chapter 3, but what about the standard error method? Recall from Section 3.2 on page 235 that we first found a rough 95% confidence interval using

$$\text{Sample Statistic} \pm 2 \cdot SE$$

where SE is the standard error in a bootstrap distribution. Where does this 2 come from? You guessed it: The standard normal!

The "2" in the formula *statistic* $\pm 2 \cdot SE$ comes from the fact that (approximately) 95% of the standard normal distribution falls between -2 and 2. To be more precise, if we ask for the middle 95% of a standard normal distribution, we find the endpoints are ± 1.96 (as we see in Figure 5.22). Thus, we should actually go 1.96 standard errors in either direction to get the interval (but 2 SE is a reasonable quick approximation). We can use the same idea for other levels of confidence. For example, Figure 5.15 shows that the middle 99% of a standard normal distribution lies between -2.575 and 2.575. Thus we can get 99% confidence by going 2.575 standard errors on either side of the original sample statistic, *sample statistic* $\pm 2.575 \cdot SE$. This idea generalizes to any level of confidence.

Confidence Intervals Using a Standard Normal Distribution

If the distribution for a statistic follows the shape of a normal distribution with standard error SE, we find a confidence interval for the parameter using

$$\text{Sample Statistic} \pm z^* \cdot SE$$

where z^* is chosen so that the proportion between $-z^*$ and $+z^*$ in the standard normal distribution is the desired level of confidence.

Example 5.15

Online Dating with the Standard Normal

Use the standard normal distribution to find a 90% confidence interval for the proportion of American adults who have used online dating.

Solution

The observed sample statistic is $\hat{p} = 0.15$, and the standard error is 0.008 (as found from the bootstrap distribution shown in Figure 5.16). In Figure 5.19 we see that z^* for a 90% confidence interval is 1.645. Thus we can find a 90% confidence interval for the proportion of adults using online dating with

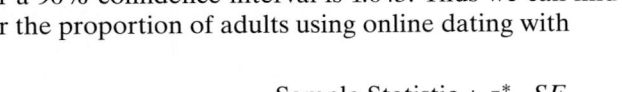

$$\text{Sample Statistic} \pm z^* \cdot SE$$
$$0.15 \pm 1.645 \cdot 0.008$$
$$0.137 \text{ to } 0.163$$

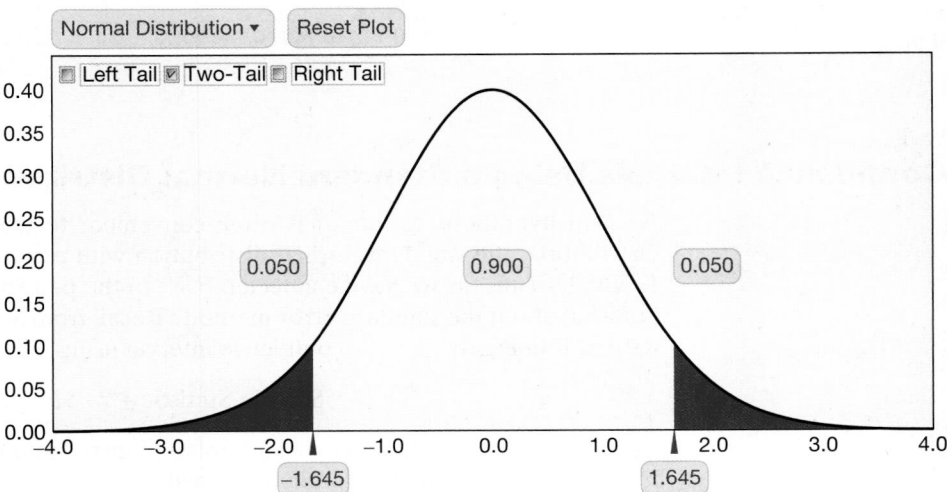

Figure 5.19 *Finding z^* for a 90% confidence interval*

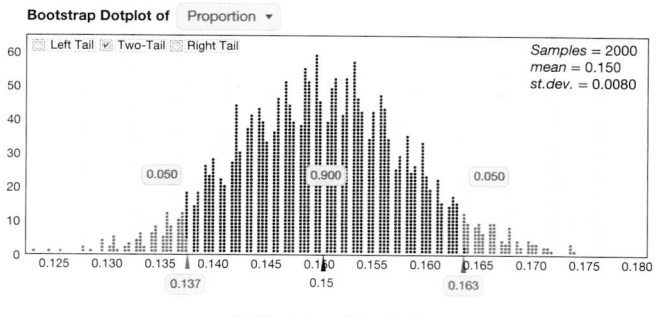

(a) Bootstrap Distribution

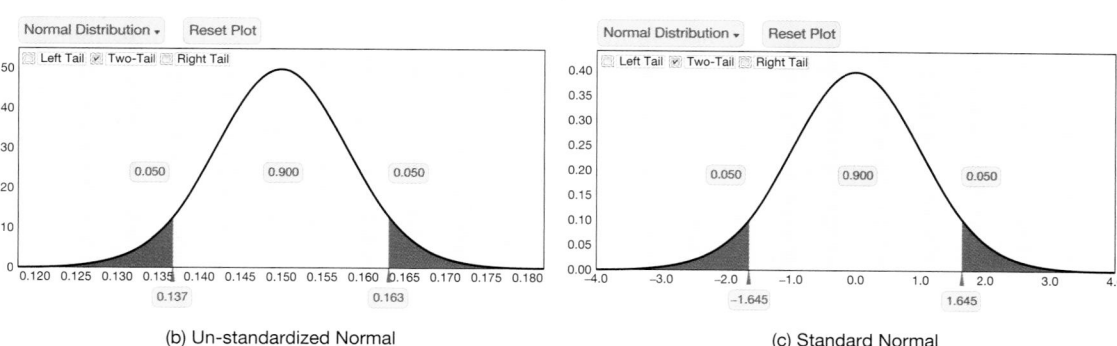

(b) Un-standardized Normal

(c) Standard Normal

Figure 5.20 *Three ways of generating a confidence interval*

This is exactly the same confidence interval found using a bootstrap distribution in Example 5.12 and using the un-standardized normal distribution in Example 5.14. These three ways of finding a confidence interval are shown in Figure 5.20. In all three figures, we are finding the interval corresponding to the middle 90% of the distribution, and only the distribution differs. The un-standardized normal replaces the bootstrap distribution with a smooth curve, and the standard normal just adjusts the scale to have mean 0 and standard deviation 1.

You can also think of the formula

$$\text{Sample Statistic} \pm z^* \cdot SE$$

as the reverse of the standardization process achieved through z-scores. Recall that a value X from a $N(\mu, \sigma)$ can be converted to a value Z on the standard normal scale by

$$Z = \frac{X - \mu}{\sigma}.$$

Doing a little algebra to rewrite this equation to find X in terms of Z, we can reverse the standardization with

$$X = \mu + Z \cdot \sigma.$$

When finding confidence intervals directly on the original data scale, we use $N(Sample\ Statistic, SE)$ with μ equal to the sample statistic (mean of the bootstrap distribution) and σ equal to the standard error. Therefore, we can find the confidence interval $(-z^*, z^*)$ on the standard normal scale, and convert back to the original data scale with $(Sample\ Statistic - z^* \cdot SE, Sample\ Statistic + z^* \cdot SE)$.

Example 5.16

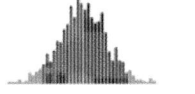

Confidence Intervals for Atlanta Commute Time

Figure 5.21 reproduces the bootstrap distribution of mean commute times in Atlanta that appears in Figure 3.15 on page 231. In that example, the original sample of

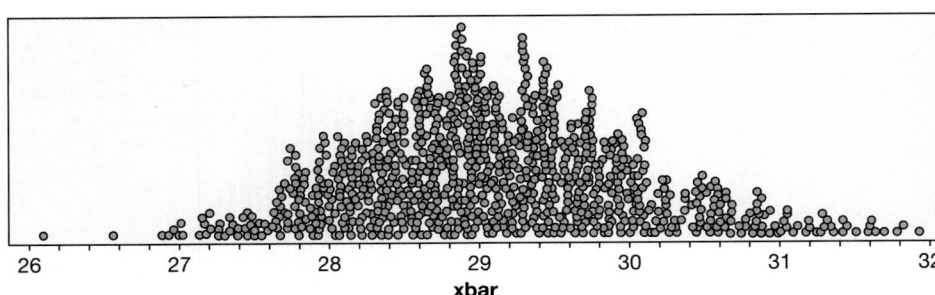

Figure 5.21 *Means for 1000 bootstrap samples from CommuteAtlanta*

$n = 500$ Atlanta commute times has a mean of $\bar{x} = 29.11$ minutes. The standard error from the bootstrap distribution is 0.915 minutes. Use this information to find 95%, 99%, and 90% confidence intervals for the mean Atlanta commute time.

Solution To find a 95% confidence interval, we find the endpoints on a standard normal distribution[18] that give 95% area in the middle. See Figure 5.22 which shows that $z^* = 1.96$.

The resulting 95% confidence interval for the mean is

$$\text{Sample Statistic} \pm z^* \cdot SE$$
$$\bar{x} \pm z^* \cdot SE$$
$$29.11 \pm 1.96(0.915)$$
$$29.11 \pm 1.79$$
$$27.32 \text{ to } 30.90$$

As usual we interpret this interval to say that we are 95% sure that the mean commute time for all commuters in Atlanta is between 27.32 and 30.90 minutes.

For a 99% confidence interval, Figure 5.15 gives $z^* = 2.575$ and the interval is

$$29.11 \pm 2.575(0.915) = 29.11 \pm 2.36 = (26.75, 31.47)$$

For a 90% confidence interval, Figure 5.19 gives $z^* = 1.645$ and the interval is

$$29.11 \pm 1.645(0.915) = 29.11 \pm 1.51 = (27.60, 30.62)$$

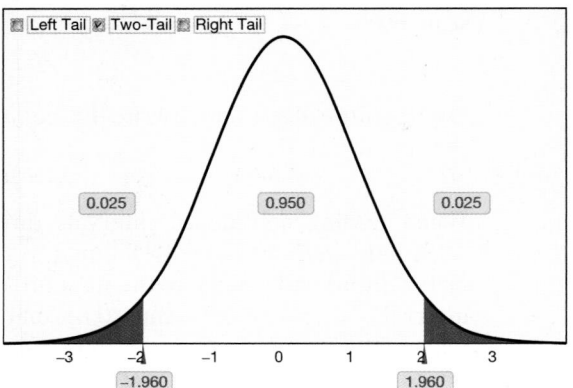

Figure 5.22 *Standard normal percentiles for a 95% confidence interval*

[18]In the next chapter, we use the t-distribution for means and the standard normal for proportions. However, for sample sizes this large the t-distribution and the standard normal are practically indistinguishable, and yield the same z^* values.

We see in Example 5.16 that for a larger level of confidence, the normal z^* gets larger and the interval widens to have a better chance of capturing the true mean commute time.

We can compare the confidence intervals computed in Example 5.16 to the ones based on the percentiles computed directly from the 1000 bootstrap samples shown in Figure 3.23 on page 243. These were (26.98, 31.63) and (27.20, 30.71) for 99% and 90%, respectively. While not exactly the same, the intervals are consistent with each other. Remember that even the percentile method would yield different intervals for a different 1000 bootstrap samples and a new sample of another 500 Atlanta drivers would also produce a slightly different interval.

One of the advantages of using the standard normal percentile for finding a confidence interval is that the z^* values become familiar for common confidence levels. Table 5.1 gives some of these common values.

Table 5.1 *Normal percentiles for common confidence levels*

Confidence level	80%	90%	95%	98%	99%
z^*	1.282	1.645	1.960	2.326	2.576

Example 5.17

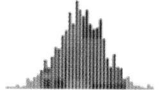

Change in Online Dating Use from 2013 to 2015

In our 2015 study, 15% of 2001 randomly sampled American adults had used online dating. From a 2013 random sample of 2,252 American adults, 11% had used online dating (again, meaning either an online dating site or a dating app on their cell phone).[19] A bootstrap distribution finds the standard error for the difference in proportions to be $SE = 0.01$. Find and interpret a 99% confidence interval for the difference in proportions of American adults using online dating between these two years.

Solution

To construct the confidence interval, we use

$$\text{Sample Statistic} \pm z^* \cdot SE$$

The sample statistic is $\hat{p}_{2015} - \hat{p}_{2013} = 0.15 - 0.11 = 0.04$ and the standard error is $SE = 0.01$. For a 99% confidence interval, we find the value of z^* using technology (or by looking at Table 5.1.) We see that $z^* = 2.576$. The 99% confidence interval is

$$(\hat{p}_{2015} - \hat{p}_{2013}) \pm z^* \cdot SE$$
$$0.04 \pm 2.576(0.01)$$
$$0.04 \pm 0.02576$$
$$0.014 \text{ to } 0.066.$$

We are 99% confident that the proportion of American adults using online dating was between 0.014 and 0.066 higher in 2015 than in 2013. Because this interval does not contain 0, we have significant evidence of an increase over time.

[19] Smith, A., and Duggan, M. (2013). "Online Dating & Relationships," *Pew Research Center*, 10/21/13, *http://www.pewinternet.org/2013/10/21/online-dating-relationships/*.

Looking Ahead

We now have two very easy formulas that apply when a randomization or bootstrap distribution for a sample statistic is approximately normally distributed:

General Formulas when using a Standard Normal Distribution

- **Confidence Intervals:**

$$\text{Sample Statistic} \pm z^* \cdot SE$$

where z^* is a standard normal endpoint based on the desired confidence level.

- **Hypothesis Tests:** We compute a standardized test statistic using

$$z = \frac{\text{Sample Statistic} - \text{Null Parameter}}{SE}$$

and find a p-value using this test statistic on a standard normal distribution.

In both cases, we need to know the standard error, SE. While computers have made it relatively easy to generate thousands of simulations to find SE from a randomization or bootstrap distribution, wouldn't it be nice if we had simple formulas we could use to compute SE directly? If this sounds nice to you, keep reading! The next chapter gives "shortcut" formulas to estimate the standard error for common parameters. As we work through the many shortcut formulas for the standard error, keep the big picture in mind. In every case, we return to the two general formulas displayed above.

SECTION LEARNING GOALS

You should now have the understanding and skills to:

- Find endpoints in a normal distribution
- Recognize how the normal distribution can be used to approximate a bootstrap distribution
- Compute a confidence interval using a normal distribution

Exercises for Section **5.2**

SKILL BUILDER 1

In Exercises 5.34 and 5.35, find the z^* values based on a standard normal distribution for each of the following.

5.34 (a) An 80% confidence interval for a proportion.

(b) An 84% confidence interval for a slope.

(c) A 92% confidence interval for a standard deviation.

5.35 (a) An 86% confidence interval for a correlation.

(b) A 94% confidence interval for a difference in proportions.

(c) A 96% confidence interval for a proportion.

SKILL BUILDER 2

In Exercises 5.36 to 5.41, find the indicated confidence interval. Assume the standard error comes from a bootstrap distribution that is approximately normally distributed.

5.36 A 95% confidence interval for a proportion p if the sample has $n = 100$ with $\hat{p} = 0.43$, and the standard error is $SE = 0.05$.

5.37 A 95% confidence interval for a mean μ if the sample has $n = 50$ with $\bar{x} = 72$ and $s = 12$, and the standard error is $SE = 1.70$.

5.38 A 90% confidence interval for a mean μ if the sample has $n = 30$ with $\bar{x} = 23.1$ and $s = 5.7$, and the standard error is $SE = 1.04$.

5.39 A 99% confidence interval for a proportion p if the sample has $n = 200$ with $\hat{p} = 0.78$, and the standard error is $SE = 0.03$.

5.40 A 95% confidence interval for a difference in proportions $p_1 - p_2$ if the samples have $n_1 = 50$ with $\hat{p}_1 = 0.68$ and $n_2 = 80$ with $\hat{p}_2 = 0.61$, and the standard error is $SE = 0.085$.

5.41 A 95% confidence interval for a difference in means $\mu_1 - \mu_2$ if the samples have $n_1 = 100$ with $\bar{x}_1 = 256$ and $s_1 = 51$ and $n_2 = 120$ with $\bar{x}_2 = 242$ and $s_2 = 47$, and the standard error is $SE = 6.70$.

5.42 What Is Your Sleep Position? In a study conducted in the United Kingdom about sleeping positions, 1000 adults in the UK were asked their starting position when they fall asleep at night. The most common answer was the fetal position (on the side, with legs pulled up), with 41% of the participants saying they start in this position. Use a normal distribution to find and interpret a 95% confidence interval for the proportion of all UK adults who start sleep in this position. Use the fact that the standard error of the estimate is 0.016.

5.43 Hearing Loss in Teenagers A recent study[20] found that, of the 1771 participants aged 12 to 19 in the National Health and Nutrition Examination Survey, 19.5% had some hearing loss (defined as a loss of 15 decibels in at least one ear). This is a dramatic increase from a decade ago. The sample size is large enough to use the normal distribution, and a bootstrap distribution shows that the standard error

for the proportion is $SE = 0.009$. Find and interpret a 90% confidence interval for the proportion of teenagers with some hearing loss.

5.44 Where Is the Best Seat on the Plane? A survey of 1000 air travelers[21] found that 60% prefer a window seat. The sample size is large enough to use the normal distribution, and a bootstrap distribution shows that the standard error is $SE = 0.015$. Use a normal distribution to find and interpret a 99% confidence interval for the proportion of air travelers who prefer a window seat.

5.45 Average Age for ICU Patients The **ICU-Admissions** dataset includes a variable indicating the age of the patient. Find and interpret a 95% confidence interval for mean age of ICU patients using the facts that, in the sample, the mean is 57.55 years and the standard error for such means is $SE = 1.42$. The sample size of 200 is large enough to use a normal distribution.

5.46 Smoke-Free Legislation and Asthma Hospital admissions for asthma in children younger than 15 years was studied[22] in Scotland both before and after comprehensive smoke-free legislation was passed in March 2006. Monthly records were kept of the annualized percent change in asthma admissions. For the sample studied, before the legislation, admissions for asthma were increasing at a mean rate of 5.2% per year. The standard error for this estimate is 0.7% per year. After the legislation, admissions were decreasing at a mean rate of 18.2% per year, with a standard error for this mean of 1.79%. In both cases, the sample size is large enough to use a normal distribution.

(a) Find and interpret a 95% confidence interval for the mean annual percent rate of change in childhood asthma hospital admissions in Scotland before the smoke-free legislation.

(b) Find a 95% confidence interval for the same quantity after the legislation.

(c) Is this an experiment or an observational study?

(d) The evidence is quite compelling. Can we conclude cause and effect?

5.47 How Much More Effective Is It to Test Yourself in Studying? In Exercise 5.23, we see that students who study by giving themselves quizzes recall a greater proportion of words than students who

[20]Rabin, R., "Childhood: Hearing Loss Grows Among Teenagers," www.nytimes.com, August 23, 2010.

[21]Willingham, A., "And the best seat on a plane is... 6A!," HLNtv.com, April 25, 2012.
[22]Mackay, D., et. al., "Smoke-free Legislation and Hospitalizations for Childhood Asthma," *The New England Journal of Medicine*, September 16, 2010; 363(12):1139–45.

study by reading. In Exercise 5.23 we see that there is an effect, but often the question of interest is not "Is there an effect?" but instead "How big is the effect?" To address this second question, use the fact that $\hat{p}_Q = 0.42$ and $\hat{p}_R = 0.15$ to find and interpret a 99% confidence interval for the difference in proportions $p_Q - p_R$, where p_Q represents the proportion of items correctly recalled by all students who study using a self-quiz method and p_R represents the proportion of items correctly recalled by all students who study using a reading-only approach. Assume that the standard error for a bootstrap distribution of such differences is about 0.07.

5.48 Incentives to Exercise: How Much More Effective Is It to Lose Money? In Exercise 5.30, we see that overweight participants who *lose* money when they don't meet a specific exercise goal meet the goal more often, on average, than those who *win* money when they meet the goal, even if the final result is the same financially. In particular, participants who lost money met the goal for an average of 45.0 days (out of 100) while those winning money or receiving other incentives met the goal for an average of 33.7 days. In Exercise 5.30 we see that the incentive does make a difference. In this exercise, we ask how big the effect is between the two types of incentives. Find and interpret a 95% confidence interval for the difference in mean number of days meeting the goal, between people who lose money when they don't meet the goal and those who win money or receive other similar incentives when they do meet the goal. The standard error for the difference in means from a bootstrap distribution is 4.14.

MORE ON THE ONLINE DATING SURVEY
Exercises 5.49 to 5.52 refer to Data 5.2 on page 389, which describes a survey conducted in July 2015 asking a random sample of American adults whether they had ever used online dating (either an online dating site or a dating app on their cell phone).

5.49 18- to 24-year-olds The survey included 194 young adults (ages 18 to 24) and 53 of them said that they had used online dating. If we use this sample to estimate the proportion of all young adults to use online dating, the standard error is 0.032. Find a 95% confidence interval for the proportion of all US adults ages 18 to 24 to use online dating.

5.50 55- to 64-year-olds The survey included 411 adults between the ages of 55 and 64, and 50 of them said that they had used online dating. If we use this sample to estimate the proportion of all American adults ages 55 to 64 to use online dating, the standard error is 0.016. Find a 95% confidence interval

for the proportion of all US adults ages 55 to 64 to use online dating.

5.51 College-Educated or Not? The survey also asked participants for their level of education, and we wish to estimate the difference in the proportion to use online dating between those with a college degree and those with a high school degree or less. The results are shown in the two-way table in Table 5.2.

(a) What proportion of college graduates answered yes when asked if they had ever used online dating? What proportion of those with a high school degree or less said yes? What is the sample difference in proportions?

(b) Use the fact that the standard error for the estimate is 0.019 to find a 99% confidence interval for the population difference in proportions.

(c) Is it plausible that there is no difference between college graduates and high school graduates or less in how likely they are to use online dating? Use the confidence interval from part (b) to answer, and explain your reasoning.

Table 5.2 *Have you ever used an online dating site or a dating app?*

	College	High School	Total
Yes	157	70	227
No	666	565	1231
Total	823	635	1458

5.52 Comparing Males to Females In the survey, 17% of the men said they had used online dating, while 14% of the women said they had.

(a) Find a 99% confidence interval for the difference in the proportion saying they used online dating, between men and women. The standard error of the estimate is 0.016.

(b) Is it plausible that there is no difference between men and women in how likely they are to use online dating? Use the confidence interval from part (a) to answer and explain your reasoning.

5.53 Prices of Used Mustangs Data 3.4 on page 246 describes a sample of $n = 25$ Mustang cars being offered for sale on the Internet. Use the data in **MustangPrice** to construct a 95% confidence interval to estimate the mean *Price* (in $1000s) for the population of all such Mustangs. Find the 95% confidence interval two ways:

(a) Using percentiles of a bootstrap distribution

(b) Using a normal distribution with *SE* estimated from a bootstrap distribution

Compare your answers.

5.54 Predicting Price of Used Mustangs from Miles Driven Data 3.4 on page 246 describes a sample of $n = 25$ Mustang cars being offered for sale on the Internet. The data are stored in **MustangPrice**, and we want to predict the *Price* of each car (in $1000s) based on the *Miles* it has been driven (also in 1000s).

(a) Find the slope of the regression line for predicting *Price* based on *Miles* for these data.

(b) Estimate the standard error of the slope using a bootstrap distribution and use it and the normal distribution to find a 98% confidence interval for the slope of this relationship in the population of all Mustangs for sale on the Internet.

5.55 Correlation between Time and Distance in Commuting In Exercise B.62 on page 363, we find an interval estimate for the correlation between *Distance* (in miles) and *Time* (in minutes) for Atlanta commuters, based on the sample of size $n = 500$ in **CommuteAtlanta**. The correlation in the original sample is $r = 0.807$.

(a) Use technology and a bootstrap distribution to estimate the standard error of sample correlations between *Distance* and *Time* for samples of 500 Atlanta commutes.

(b) Assuming that the bootstrap correlations can be modeled with a normal distribution, use the results of (a) to find and interpret a 90% confidence interval for the correlation between distance and time of Atlanta commutes.

5.56 Cell Phone Apps In 2010, some researchers with the Pew Internet & American Life project interviewed a random sample of adults about their cell phone usage.[23] One of the questions asked was whether the respondent had ever downloaded an application or "app" to their cell phone. The sample proportion who had, based on 1917 respondents who had cell phones, was $\hat{p} = 0.29$.

(a) Describe how you could construct a bootstrap distribution to estimate the standard error of the proportion in this situation.

(b) One such distribution, based on proportions from 5000 bootstrap samples, is shown in Figure 5.23. The standard deviation of these proportions is 0.0102. Use this information to find a 99% confidence interval for the proportion of cell phone users (in 2010) who have downloaded at least one app to their phone.

5.57 Malevolent Uniforms in Football Figure 5.24 shows a bootstrap distribution of correlations between penalty yards and uniform malevolence using the data on 28 NFL teams in **Malevolent UniformsNFL**. We see from the percentiles of the bootstrap distribution that a 99% confidence interval for the correlation is -0.224 to 0.788. The correlation between the two variables for the original sample is $r = 0.43$.

[23]http://pewinternet.org/Shared-Content/Data-Sets/2010/May-2010–Cell-Phones.aspx.

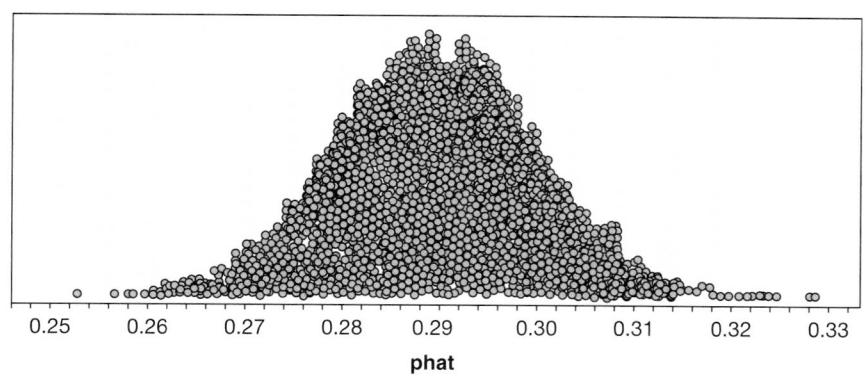

Figure 5.23 *Bootstrap distribution for the proportion of cell phone users who have downloaded an app*

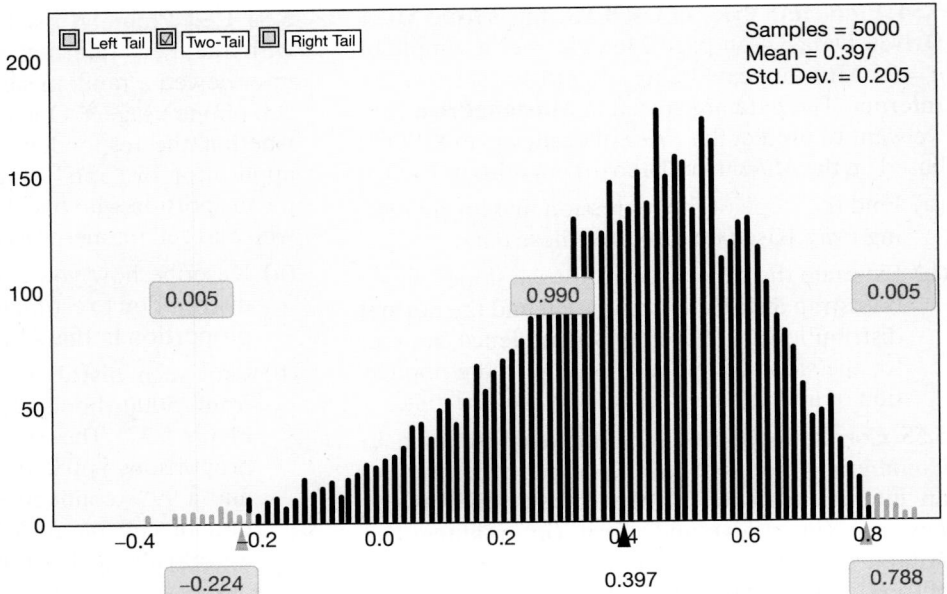

Figure 5.24 *Bootstrap correlations for uniform malevolence vs penalty yards*

(a) Use the original sample correlation and the standard deviation of the bootstrap distribution shown in Figure 5.24 to compute a 99% confidence interval for the correlation using z^* from a normal distribution.

(b) Why is the normal-based interval somewhat different from the percentile interval? (*Hint:* Look at the shape of the bootstrap distribution.)

CHAPTER 6

Inference for Means and Proportions

"To exploit the data flood, America will need many more [data analysts]…The story is similar in fields as varied as science and sports, advertising and public health—a drift toward data-driven discovery and decision-making."

Steve Lohr*

This chapter has many short sections that can be combined and covered in almost any order.

*"The Age of Big Data," *The New York Times*, February 12, 2012, p. SR1.
Top Left: ©Stephen Strathdee/iStockphoto, Top Right: ©slobo/iStockphoto, Bottom Right: Photo by Mike Stobe/Getty Images, Inc.

Questions and Issues

Here are some of the questions and issues we will discuss in this chapter:

- What percent of people would prefer a self-driving car?
- Can fast food disrupt hormones?
- Is taking notes longhand more effective than taking notes on a laptop?
- Do science faculty have a gender bias?
- Do babies prefer speaking or humming? Speaking or singing?
- Why do fingers wrinkle in water?
- Could maternal use of antidepressants lead to autism?
- What percent of US adults believe in ghosts?
- Are polyester clothes polluting our shorelines?
- If you are mean to a bird, will it remember you?
- Can babies tell if you can be trusted?
- Do men behave differently around ovulating women?
- If your college roommate brings a videogame to campus, are your grades negatively affected?
- Does diet cola leach calcium out of your system?
- If you just *think* you should be losing weight, are you more likely to lose weight?
- How strong are the benefits of exercise in helping us be more resilient to stress?
- If we give away the ending of a story, will readers like the story more or less?

6.1-D DISTRIBUTION OF A PROPORTION

For categorical data, the parameter of interest is often a population proportion p. The Central Limit Theorem (CLT) in Section 5.1 says that sample statistics often follow a normal distribution if the sample size is large. This is true for the distribution of sample proportions, $\hat{p}$, which will be centered at the population proportion p. Besides the center, the other important quantity we need to describe a normal distribution is the standard deviation, the standard error (SE) of the sample proportions. As we have seen, one way to estimate this SE is to use a bootstrap or randomization distribution. Statistical theory gives us an alternate method for finding the standard error. We use this in the box below to obtain a version of the Central Limit Theorem that describes the distribution of sample proportions.

> **Distribution of a Sample Proportion**
>
> When selecting random samples of size n from a population with proportion p, the distribution of the sample proportions is centered at the population proportion p, has standard error given by
>
> $$SE = \sqrt{\frac{p(1-p)}{n}}$$
>
> and is reasonably normally distributed if $np \geq 10$ and $n(1-p) \geq 10$.

Using the Formula for Standard Error

In Section 3.1, we see that the proportion of US adults who are college graduates is $p = 0.275$, and we create a sampling distribution to examine sample proportions using sample size $n = 200$. From the simulated sampling distribution, we see that the standard error is $SE = 0.032$. We can use the formula, with $p = 0.275$ and $n = 200$, to verify that the formula gives the same value as the 0.032 arrived at using simulations:

$$SE = \sqrt{\frac{p(1-p)}{n}} = \sqrt{\frac{0.275(1 - 0.275)}{200}} = 0.032.$$

Example 6.1

Use the formula to calculate the standard error in each case:

(a) $p = 0.25$ and $n = 50$

(b) $p = 0.25$ and $n = 200$

Solution

(a) The standard error for a sample of size 50 is $SE = \sqrt{\frac{0.25(1-0.25)}{50}} = 0.0612$.

(b) When the sample size is increased to $n = 200$, the standard error drops to $SE = \sqrt{\frac{0.25(1-0.25)}{200}} = 0.0306$.

In Example 6.1, we see that multiplying the sample size by 4 ($n = 50$ to $n = 200$) cuts the standard error of the sample proportions in half. This is due to the square root of n in the denominator of the formula for SE. In Example 3.9 on page 204 we see that the variability of the sampling distribution for a proportion decreases with increasing sample size — the formula for SE quantifies this relationship.

Justifying the Sample Size Conditions

The distributions of sample proportions are usually well-modeled with a normal distribution, but there are situations where a normal distribution is not appropriate. Looking at the dotplots in Figure 6.1 we see that the normal shape starts to break down when the population proportion p starts to get close to 0 or 1. In Figure 6.2 we see that the sample size also plays a role. For a small proportion, such as $p = 0.10$, normality is a problem for $n = 10$ or $n = 25$ but looks fine for $n = 200$.

As a general rule, the sample size is large enough for the distribution to stay away from 0 if $np \geq 10$. At the other end, we can avoid problems at 1 if $n(1 - p) \geq 10$. This says that the sample size is large enough that we can expect to see at least ten "yes" values and at least ten "no" values in the sample. The distribution of the sample proportions will be reasonably normally distributed if $np \geq 10$ and $n(1 - p) \geq 10$.

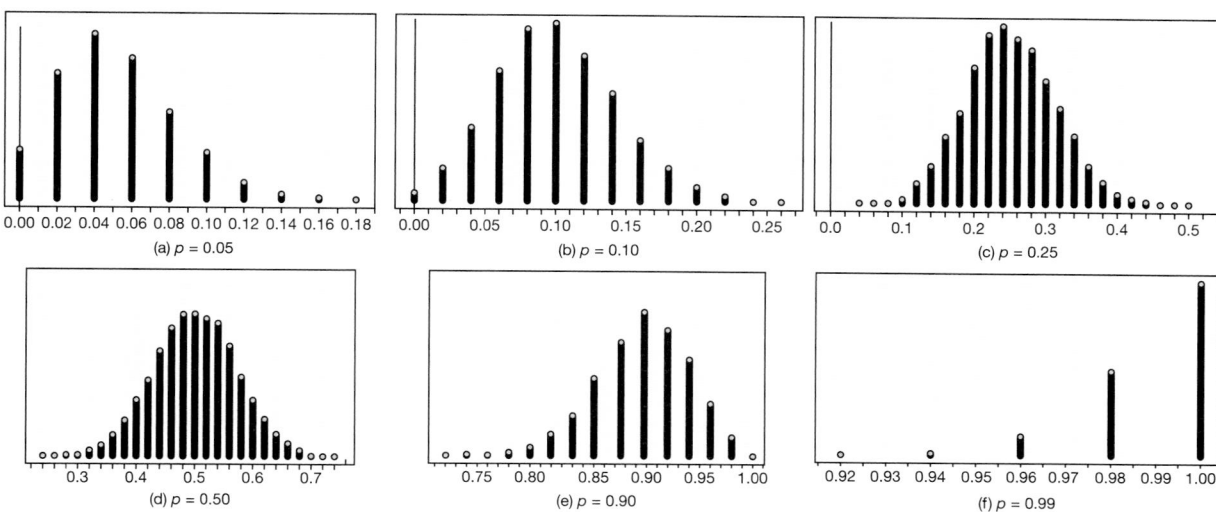

Figure 6.1 *Distributions of sample proportions when n = 50*

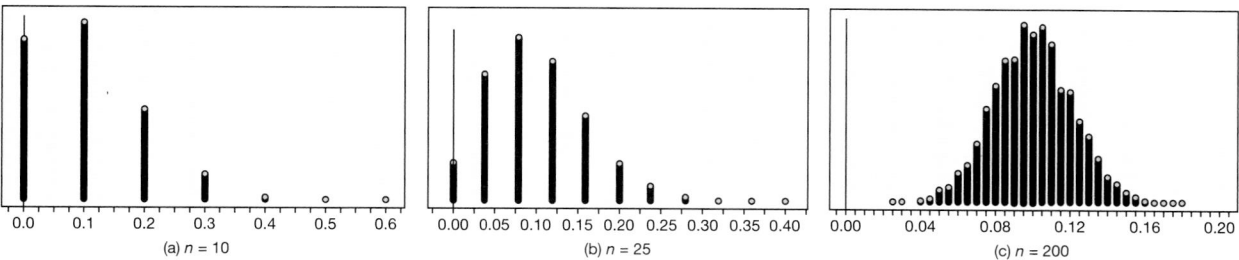

Figure 6.2 *Distributions of sample proportions when p = 0.10*

Example 6.2

Ontime Arrivals

The Bureau of Transportation Statistics[1] tells us that 80% of flights for a particular airline arrive on time (defined as within 15 minutes of the scheduled arrival time). We examine a random sample of 400 flights for this airline and compute the proportion of the sample flights that arrive on time. Compute the standard error for this statistic and verify that the conditions to use a normal distribution apply.

---

[1] For example, see *http://www.transtats.bts.gov/HomeDrillChart.asp* for data collected by the Bureau of Transportation Statistics.

Solution   The mean of the distribution of sample ontime proportions is 0.80, the proportion for the entire population. The standard error for samples of size 400 is

$$SE = \sqrt{\frac{0.8(1 - 0.8)}{400}} = 0.02$$

For $n = 400$ and $p = 0.8$ we have

$$np = 400 \cdot 0.8 = 320 \quad \text{and} \quad n(1 - p) = 400 \cdot (1 - 0.8) = 80$$

Both of these values are well above 10, so we can safely use a normal distribution to model the sample ontime proportions.

We now know how to compute the *SE* for a distribution of sample proportions and how to determine whether this distribution will be approximately normally distributed. In the next two sections we use these tools as an alternate way to create confidence intervals and conduct hypothesis tests for a proportion.

SECTION LEARNING GOALS

You should now have the understanding and skills to:

- • Use a formula to find the standard error for a distribution of sample proportions
- • Identify when a normal distribution is an appropriate model for a distribution of sample proportions

---

Exercises for Section 6.1-D

SKILL BUILDER 1

In Exercises 6.1 to 6.6, if random samples of the given size are drawn from a population with the given proportion, find the standard error of the distribution of sample proportions.

6.1 Samples of size 50 from a population with proportion 0.25

6.2 Samples of size 1000 from a population with proportion 0.70

6.3 Samples of size 60 from a population with proportion 0.90

6.4 Samples of size 30 from a population with proportion 0.27

6.5 Samples of size 300 from a population with proportion 0.08

6.6 Samples of size 100 from a population with proportion 0.41

IMPACT OF SAMPLE SIZE ON ACCURACY

In Exercises 6.7 and 6.8, compute the standard error for sample proportions from a population with the given proportion using three different sample sizes. What effect does increasing the sample size have on the standard error? Using this information about the effect on the standard error, discuss the effect of increasing the sample size on the accuracy of using a sample proportion to estimate a population proportion.

6.7 A population with proportion $p = 0.4$ for sample sizes of $n = 30$, $n = 200$, and $n = 1000$.

6.8 A population with proportion $p = 0.75$ for sample sizes of $n = 40$, $n = 300$, and $n = 1000$.

IS A NORMAL DISTRIBUTION APPROPRIATE?

In Exercises 6.9 and 6.10, indicate whether the Central Limit Theorem applies so that the sample proportions follow a normal distribution.

6.9 In each case below, is the sample size large enough so that the sample proportions follow a normal distribution?

(a) $n = 500$ and $p = 0.1$

(b) $n = 25$ and $p = 0.5$

(c) $n = 30$ and $p = 0.2$

(d) $n = 100$ and $p = 0.92$

6.10 In each case below, is the sample size large enough so that the sample proportions follow a normal distribution?

(a) $n = 80$ and $p = 0.1$

(b) $n = 25$ and $p = 0.8$

(c) $n = 50$ and $p = 0.4$

(d) $n = 200$ and $p = 0.7$

6.1-CI CONFIDENCE INTERVAL FOR A PROPORTION

In Section 5.2 we see that when a distribution for a sample statistic is normally distributed, a confidence interval can be formed using

$$\text{Sample Statistic} \pm z^* \cdot SE$$

where z^* is an appropriate percentile from a standard normal distribution and SE is the standard error of the sample statistic.

In Section 6.1-D we see that, for a large sample, the distribution of sample proportions is reasonably normal with standard error given by

$$SE = \sqrt{\frac{p(1-p)}{n}}$$

where n is the sample size and p is the proportion in the population.

We are almost in a position to combine these facts to produce a formula for computing a confidence interval for a proportion. One small, but very important, detail remains. In order to compute the standard error, SE, for the sample proportions we need to know the proportion, p, for the population—but that's exactly the quantity we are trying to estimate with the confidence interval! Fortunately, there's an easy fix to this predicament. We use the sample proportion, $\hat{p}$, in place of the population proportion, p, when estimating the standard error for a confidence interval. As long as the sample is large enough for the Central Limit Theorem to apply, a normal distribution with this approximated SE is still a good model for the sample proportions.

> **Confidence Interval for a Proportion**
>
> The sample proportion based on a random sample of size n has
>
> $$\text{Sample statistic} = \hat{p} \quad \text{and} \quad SE = \sqrt{\frac{\hat{p}(1-\hat{p})}{n}}$$
>
> If z^* is a standard normal endpoint to give the desired level of confidence, and if the sample size is large enough so that $n\hat{p} \geq 10$ and $n(1 - \hat{p}) \geq 10$, the confidence interval for a population proportion p is
>
> $$\text{Sample statistic} \pm z^* \cdot SE$$
>
> which, in this case, corresponds to
>
> $$\hat{p} \pm z^* \cdot \sqrt{\frac{\hat{p}(1-\hat{p})}{n}}$$

Example 6.3

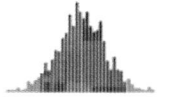

In Example 3.19 on page 232 we describe a sample of 100 mixed nuts that contains 52 peanuts. Verify that the sample is large enough for the Central Limit Theorem to apply and use the formula to find a 95% confidence interval for the proportion of peanuts in this brand of mixed nuts. Interpret the confidence interval in context. Compare the result to the interval 0.420 to 0.620 obtained using a bootstrap distribution in Example 3.22 on page 235.

Solution   The sample has 52 peanuts and 48 other nuts. Since both of these counts are bigger than 10, we can use the normal approximation. The sample proportion is $\hat{p} = 52/100 = 0.52$ and, for a 95% confidence interval, the standard normal $z^* = 1.96$. Using the formula we have

$$\text{Statistic } \pm z^* \cdot SE$$
$$\hat{p} \pm z^* \sqrt{\frac{\hat{p}(1-\hat{p})}{n}}$$
$$0.52 \pm 1.96 \sqrt{\frac{0.52(1-0.52)}{100}}$$
$$0.52 \pm 0.098$$
$$0.422 \text{ to } 0.618$$

Based on this sample, we are 95% sure that between 42.2% and 61.8% of the mixed nuts from this company are peanuts. In Example 3.22 we found a confidence interval of (0.420, 0.620) based on the SE from a bootstrap distribution, which agrees nicely with the result from the formula.

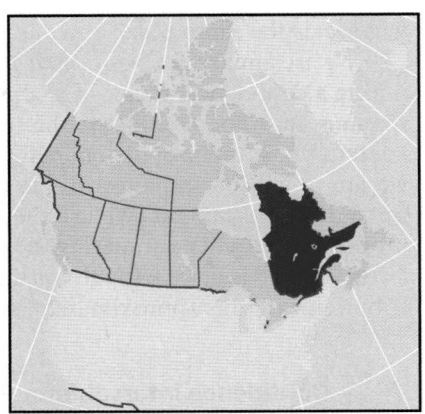

Antenna Audio, Inc/GettyImages, Inc.

What proportion of Quebecers want to secede from Canada?

Example 6.4

Quebec Sovereignty

Quebec is a large province in eastern Canada, and is the only Canadian province with a predominantly French-speaking population. Historically there has been debate over whether Quebec should secede from Canada and establish itself as an independent, sovereign nation. In a survey of 800 Quebec residents, 28% thought that Quebec should separate from Canada. In the same survey, 82% agreed that Quebec society is distinct from the rest of Canada.[2]

(a) Find a 95% confidence interval for the proportion of Quebecers who would like Quebec to separate from Canada.

(b) Find a 95% confidence interval for the proportion of Quebecers who think Quebec society is distinct from the rest of Canada.

[2]"Separation from Canada Unlikely for a Majority of Quebecers," Angus Reid, June 9, 2009.

Solution

The sample size is clearly large enough to use the formula based on the normal approximation.

(a) The proportion in the sample who think Quebec should separate is $\hat{p} = 0.28$ and $z^* = 1.96$ for 95% confidence, so we have

$$0.28 \pm 1.96\sqrt{\frac{0.28(1-0.28)}{800}} = 0.28 \pm 0.031 = (0.249, 0.311)$$

We are 95% sure that between 24.9% and 31.1% of Quebecers would like Quebec to separate from Canada.

(b) The proportion in the sample who think Quebec society is distinct from the rest of Canada is $\hat{p} = 0.82$ so the 95% confidence interval is

$$0.82 \pm 1.96\sqrt{\frac{0.82(1-0.82)}{800}} = 0.82 \pm 0.027 = (0.793, 0.847)$$

We are 95% sure that between 79.3% and 84.7% of Quebecers think Quebec society is distinct.

Examples 6.3 and 6.4 illustrate how the margin of error (ME) in a confidence interval for a proportion depends on both the sample size and the sample proportion.

Example 6.3	$n = 100$	$\hat{p} = 0.52$	$ME = 0.098$
Example 6.4(a)	$n = 800$	$\hat{p} = 0.28$	$ME = 0.031$
Example 6.4(b)	$n = 800$	$\hat{p} = 0.82$	$ME = 0.027$

We know that the margin of error decreases for larger sample sizes. We also see that, even for the same sample size, a sample proportion closer to 0 or 1 gives a smaller margin of error than $\hat{p}$ closer to 0.5. In fact, the largest margin of error (and thus the widest interval) occurs when $\hat{p}$ is 0.5.

Determining Sample Size for Estimating a Proportion

A common question when designing a study is "How large a sample should we collect?" When estimating a proportion with a confidence interval, the answer to this question depends on three related questions:

- How accurate do we want the estimate to be? In other words, what margin of error, ME, do we want?
- How much confidence do we want to have in the interval?
- What sort of proportion do we expect to see?

From the formula for the confidence interval we see that the margin of error is computed with

$$ME = z^*\sqrt{\frac{\hat{p}(1-\hat{p})}{n}}$$

Suppose that, *before getting a sample*, we decide we want the confidence interval for a proportion to have some predetermined margin of error. By choosing a large enough n, we can get the margin of error as small as we'd like. To determine how large a sample size is needed, we could solve this equation for n to find a sample size that gives the specified margin of error with a given level of confidence. With a bit of algebra this gives us

$$n = \left(\frac{z^*}{ME}\right)^2 \hat{p}(1-\hat{p})$$

Unfortunately, we haven't even taken a sample yet so we don't have a value to use for $\hat{p}$. In practice, we address this problem in one of two ways:

- Make a reasonable guess for $\hat{p}$. (We'll refer to this guess as $\tilde{p}$).
- If we are not willing or able to make a reasonable guess, we use $\tilde{p} = 0.5$.

In some cases we might have past experience with a proportion or conduct a small pilot study to get an initial estimate for $\hat{p}$ to use in estimating the required sample size. Remember that the margin of error is largest when $\hat{p} = 0.5$, so if we use that value to estimate a sample size, the resulting interval will have a margin of error within the bound we set or slightly smaller if $\hat{p}$ is farther away from 0.5.

> **Determination of Sample Size to Estimate a Proportion**
>
> If we want to estimate a population proportion to within a desired margin of error, ME, with a given level of confidence, we should select a sample of size
>
> $$n = \left(\frac{z^*}{ME}\right)^2 \tilde{p}(1 - \tilde{p})$$
>
> where we use $\tilde{p} = 0.5$ or, if available, some other estimate of p.

Example 6.5

Quebec Sovereignty (continued)

In Example 6.4 we analyzed a poll of 800 Quebecers, in which 28% thought that Quebec should separate from Canada. About how many Quebecers should we randomly sample to estimate the proportion of residents who think the province should separate to within ±1% with 99% confidence?

Solution

Based on the previous poll, our best guess at the proportion is $\tilde{p} = 0.28$, so we use this to estimate the sample size needed. The margin of error desired is $ME = 0.01$, and $z^* = 2.576$ for 99% confidence, which gives

$$n = \left(\frac{2.576}{0.01}\right)^2 0.28(1 - 0.28) = 13377.72$$

By convention we round up any fractional parts of the sample, so we would require a (rather large) sample size of 13,378 Quebecers to achieve the desired accuracy.

In certain situations, there is a simpler formula for determining sample size. In the special case where we want 95% confidence, so $z^* \approx 2$, and use the conservative estimate of $\tilde{p} = 0.5$, the formula simplifies to

$$n \approx \frac{1}{(ME)^2}.$$

Example 6.6

Political Approval

Polling agencies often ask voters whether they approve or disapprove of the job a politician is doing. Suppose that pollsters want to estimate the proportion who approve of the job done by a particular politician to within ±3% with 95% confidence. How large a sample should they take?

Solution   The desired margin of error is $ME = 0.03$. If we don't assume anything about what the approval rating might be, we use 0.5 for the proportion, and use the approximate formula to find

$$n \approx \frac{1}{(ME)^2} = \frac{1}{(0.03)^2} = 1111.11$$

A random sample of 1112 voters is enough to estimate the approval rating of a politician to within 0.03 with 95% confidence.

_____

In practice, polling organizations often include multiple questions in a survey, so assuming the worst case proportion of $\hat{p} = 0.5$ when choosing a sample size for the whole survey is a prudent decision. The next time you see poll results reported in a news story, check whether the report includes a sample size (often a bit more than 1000) and a general margin of error (often 3%).

SECTION LEARNING GOALS

You should now have the understanding and skills to:

- Use a formula for *SE* and a normal distribution, when appropriate, to compute a confidence interval for a population proportion

- Determine a sample size needed to estimate a proportion within a specified margin of error at a given level of confidence

Exercises for Section **6.1-CI**

SKILL BUILDER 1
In Exercises 6.11 to 6.14, use the normal distribution to find a confidence interval for a proportion p given the relevant sample results. Give the best point estimate for p, the margin of error, and the confidence interval. Assume the results come from a random sample.

6.11 A 95% confidence interval for p given that $\hat{p} = 0.38$ and $n = 500$

6.12 A 90% confidence interval for p given that $\hat{p} = 0.85$ and $n = 120$

6.13 A 99% confidence interval for the proportion who will answer "Yes" to a question, given that 62 answered yes in a random sample of 90 people

6.14 A 95% confidence interval for the proportion of the population in Category A given that 23% of a sample of 400 are in Category A

SKILL BUILDER 2
In Exercises 6.15 to 6.18, what sample size is needed to give the desired margin of error in estimating a population proportion with the indicated level of confidence?

6.15 A margin of error within ±5% with 95% confidence.

6.16 A margin of error within ±1% with 99% confidence.

6.17 A margin of error within ±3% with 90% confidence. We estimate that the population proportion is about 0.3.

6.18 A margin of error within ±2% with 95% confidence. An initial small sample has $\hat{p} = 0.78$.

6.19 How Many Household Cats Hunt Birds? "Domestic cats kill many more wild birds in the United States than scientists thought," states a recent article.[3] Researchers used a sample of $n = 140$ households in the US with cats to estimate that 35% of household cats in the US hunt outdoors.

(a) Find and interpret a 95% confidence interval for the proportion of household cats in the US that hunt outdoors.

[3]Milius, S., "Cats kill more than one billion birds each year," *Science News*, 183(4),February 23, 2013, revised March 8, 2014. Data approximated from information give in the article.

(b) Is it plausible that the proportion of household cats in the US hunting outdoors is 0.45? Is it plausible that it is 0.30?

6.20 Have You Made a Friend Online? A survey[4] of 1060 randomly selected US teens ages 13 to 17 found that 605 of them say they have made a new friend online.

(a) Find and interpret a 90% confidence interval for the proportion, p, of all US teens who have made a new friend online.

(b) Give the best estimate for p and give the margin of error for the estimate.

(c) Use the interval to determine whether we can be 90% confident that more than half of US teens have made a new friend online.

6.21 What Percent of US Adults Say They Never Exercise? In a survey of 1000 US adults, twenty percent say they never exercise. This is the highest level seen in five years.[5] Find and interpret a 99% confidence interval for the proportion of US adults who say they never exercise. What is the margin of error, with 99% confidence?

6.22 Fourth Down Conversions in American Football In analyzing data from over 700 games in the National Football League, economist David Romer identified 1068 fourth-down situations in which, based on his analysis, the right call would have been to go for it and not to punt. Nonetheless, in 959 of those situations, the teams punted. Find and interpret a 95% confidence interval for the proportion of times NFL teams punt on a fourth down when, statistically speaking, they shouldn't be punting.[6] Assume the sample is reasonably representative of all such fourth down situations in the NFL.

6.23 One True Love? Data 2.1 on page 48 deals with a survey that asked whether people agree or disagree with the statement "There is only one true love for each person." The survey results in Table 2.1 show that 735 of the 2625 respondents agreed, 1812 disagreed, and 78 answered "don't know."

(a) Find a 90% confidence interval for the proportion of people who *disagree* with the statement.

(b) Find a 90% confidence interval for the proportion of people who "don't know."

(c) Which estimate has the larger margin of error?

6.24 Social Networking Sites In a survey of 2255 randomly selected US adults (age 18 or older), 1787 of them use the Internet regularly. Of the Internet users, 1054 use a social networking site.[7] Find and interpret a 95% confidence interval for each of the following proportions:

(a) Proportion of US adults who use the Internet regularly.

(b) Proportion of US adult Internet users who use a social networking site.

(c) Proportion of all US adults who use a social networking site. Use the confidence interval to estimate whether it is plausible that 50% of all US adults use a social networking site.

6.25 What Do People Do on Facebook? In the survey of 2255 US adults described in Exercise 6.24, we also learn that 970 of the respondents use the social networking site Facebook. Of the 970 Facebook users, the survey shows that on an average day:

- 15% update their status
- 22% comment on another's post or status
- 20% comment on another user's photo
- 26% "like" another user's content
- 10% send another user a private message

(a) For each of the bulleted activities, find a 95% confidence interval for the proportion of Facebook users engaging in that activity on an average day.

(b) Is it plausible that the proportion commenting on another's post or status is the same as the proportion updating their status? Justify your answer.

6.26 Democrat or Republican? 2011 A survey of 15,000 American adults in 2011 found that 35.3% identify as Democrats and 34.0% identify as Republicans, with the rest identifying as independent or other.[8] If we want 95% confidence, what is the margin of error in the estimate for the proportion of Democrats? For the proportion of Republicans? Do you feel comfortable concluding that in 2011 more American adults self-identified as Democrats than self-identified as Republicans? Explain.

6.27 Democrat or Republican? 2015 Four years after the survey described in Exercise 6.26, in 2015, a different survey of 12,000 American adults found

[4]Lenhart, A., "Teens, Technology, and Friendships," Pew Research Center, *pewresearch.org*, August 6, 2015.
[5]"75% say exercise is important in daily life," Rasmussen Reports, March 26, 2011.
[6]Moskowitz, T. and Wertheim, J., *Scorecasting*, Crown Archetype, New York, 2011, p. 39.
[7]Hampton, K., Goulet, L., Rainie, L., and Purcell, K., "Social Networking Sites and Our Lives," Pew Research Center, June 16, 2011.
[8]"Partisan Trends," Rasmussen Reports, April 1, 2011.

that 29.0% identified as Democrats and 26.0% identified as Republicans, with the rest identifying as independent or other.[9] Answer the same questions about these survey results as in Exercise 6.26: If we want 95% confidence, what is the margin of error in the estimate for the proportion of Democrats? For the proportion of Republicans? Do you feel comfortable concluding that in 2015 more American adults self-identified as Democrats than self-identified as Republicans? Explain.

STANDARD ERROR FROM A FORMULA AND A BOOTSTRAP DISTRIBUTION
In Exercises 6.28 to 6.31, use *StatKey* or other technology to generate a bootstrap distribution of sample proportions and find the standard error for that distribution. Compare the result to the standard error given by the Central Limit Theorem, using the sample proportion as an estimate of the population proportion p.

6.28 Proportion of peanuts in mixed nuts, with $n = 100$ and $\hat{p} = 0.52$

6.29 Proportion of home team wins in soccer, with $n = 120$ and $\hat{p} = 0.583$

6.30 Proportion of lie detector trials in which the technology misses a lie, with $n = 48$ and $\hat{p} = 0.354$

6.31 Proportion of survey respondents who say exercise is important, with $n = 1000$ and $\hat{p} = 0.753$

COMPARING NORMAL AND BOOTSTRAP CONFIDENCE INTERVALS
In Exercises 6.32 and 6.33, find a 95% confidence interval for the proportion two ways: using *StatKey* or other technology and percentiles from a bootstrap distribution, and using the normal distribution and the formula for standard error. Compare the results.

6.32 Proportion of home team wins in soccer, using $\hat{p} = 0.583$ with $n = 120$

6.33 Proportion of Reese's Pieces that are orange, using $\hat{p} = 0.48$ with $n = 150$

WHAT INFLUENCES THE SAMPLE SIZE?
In Exercises 6.34 to 6.36, we examine the effect of different inputs on determining the sample size needed to obtain a specific margin of error when finding a confidence interval for a proportion.

6.34 Find the sample size needed to give, with 95% confidence, a margin of error within ±6% when estimating a proportion. Within ±4%. Within ±1%.

(Assume no prior knowledge about the population proportion p.) Comment on the relationship between the sample size and the desired margin of error.

6.35 Find the sample size needed to give a margin of error to estimate a proportion within ±3% with 99% confidence. With 95% confidence. With 90% confidence. (Assume no prior knowledge about the population proportion p.) Comment on the relationship between the sample size and the confidence level desired.

6.36 Find the sample size needed to give, with 95% confidence, a margin of error within ±3% when estimating a proportion. First, find the sample size needed if we have no prior knowledge about the population proportion p. Then find the sample size needed if we have reason to believe that $p \approx 0.7$. Finally, find the sample size needed if we assume $p \approx 0.9$. Comment on the relationship between the sample size and estimates of p.

6.37 Does the Public Support Sin Taxes? A survey of 1000 adults in the US conducted in March 2011 asked "Do you favor or oppose 'sin taxes' on soda and junk food?" The proportion in favor of taxing these foods was 32%.[10]

(a) Find a 95% confidence interval for the proportion of US adults favoring taxes on soda and junk food.

(b) What is the margin of error?

(c) If we want a margin of error of only 1% (with 95% confidence), what sample size is needed?

6.38 What Proportion Favor a Gun Control Law? A survey is planned to estimate the proportion of voters who support a proposed gun control law. The estimate should be within a margin of error of ±2% with 95% confidence, and we do not have any prior knowledge about the proportion who might support the law. How many people need to be included in the sample?

6.39 Advertising a Sunscreen Pill An advertising firm plans to have a sample of individuals view a commercial on a "sunscreen pill" that one can swallow to provide mild SPF protection throughout the day. After viewing the commercial, each individual will be asked if he/she would consider buying the product. How many individuals should the firm sample to estimate the proportion who would consider buying the product to within a ±4% margin of error with 98% confidence?

[9]*http://www.gallup.com/poll/188096/democratic-republican-identification-near-historical-lows.aspx.*

[10]"32% favor 'sin taxes' on soda, junk food," Rasmussen Reports, April 1, 2011.

USING THE FORMULA $n=1/(ME)^2$ TO DETERMINE SAMPLE SIZE

When we want 95% confidence and use the conservative estimate of $p = 0.5$, we can use the simple formula $n = 1/(ME)^2$ to estimate the sample size needed for a given margin of error ME. In Exercises 6.40 to 6.43, use this formula to determine the sample size needed for the given margin of error.

6.40 A margin of error of 0.01.

6.41 A margin of error of 0.02.

6.42 A margin of error of 0.04.

6.43 A margin of error of 0.05.

6.44 Gender of Atlanta Commuters One of the variables in the dataset **CommuteAtlanta**, introduced in Data 3.3 on page 229, gives the sex of each commuter in the sample. Use technology and the dataset to construct and interpret a 95% confidence interval for the proportion of Atlanta commuters who are male.

6.45 Survival of ICU Patients The dataset **ICUAdmissions**, introduced in Data 2.3 on page 69, includes information on 200 patients admitted to an Intensive Care Unit. One of the variables, *Status*, indicates whether each patient lived (indicated with a 0) or died (indicated with a 1). Use technology and the dataset to construct and interpret a 95% confidence interval for the proportion of ICU patients who live.

6.1-HT HYPOTHESIS TEST FOR A PROPORTION

In Section 5.1 we see that, when a randomization distribution is normal, we can compute a p-value using a standard normal distribution and a standardized test statistic of the form

$$z = \frac{\text{Sample Statistic} - \text{Null Parameter}}{SE}$$

The sample statistic is computed from the sample data and the null parameter is specified by the null hypothesis, H_0.

When testing a hypothesis about a population proportion, the null hypothesis is typically $H_0 : p = p_0$ where p_0 is some specific value of the proportion. Thus the null parameter is p_0 and the sample statistic is the proportion from a sample, $\hat{p}$. We have

$$z = \frac{\hat{p} - p_0}{SE}$$

We can calculate the standard error using the formula given in Section 6.1-D. Remember that, in conducting a hypothesis test, we want to see if $\hat{p}$ is in an unusual place of a distribution we would expect to see *when H_0 is true*. Since we assume H_0 is true, we use p_0 in place of p for computing SE. Using the hypothesized null proportion, p_0, when computing the standard error for a test, we have

$$SE = \sqrt{\frac{p_0(1 - p_0)}{n}}$$

Hypothesis Test for a Proportion

To test $H_0 : p = p_0$ vs $H_a : p \neq p_0$ (or a one-tail alternative), we use the standardized test statistic

$$z = \frac{\text{Statistic} - \text{Null Value}}{SE} = \frac{\hat{p} - p_0}{\sqrt{\frac{p_0(1-p_0)}{n}}}$$

where $\hat{p}$ is the proportion in a random sample of size n. Provided the sample size is reasonably large (so that $np_0 \geq 10$ and $n(1 - p_0) \geq 10$), the p-value of the test is computed using the standard normal distribution.

We find the p-value for the test as the area beyond z in one (or both) tail(s) of the standard normal distribution, depending on the direction of the alternative hypothesis.

Example 6.7

Would You Prefer a Self-Driving Car?

In a Nielsen survey[11] conducted in October 2015, a random sample of $n = 340$ US middle school students were asked whether they would prefer a self-driving car or a car that they control themselves. The majority (59%) said that they would prefer a car that they control themselves, while 41% would prefer a self-driving car. Test, using a 5% significance level, to see if this provides evidence that more than 50% of US middle school students prefer a car that they control themselves.

Solution

The null and alternative hypotheses are

$$H_0 : \quad p = 0.50$$
$$H_a : \quad p > 0.50$$

where p is the proportion of all US middle school students choosing a car they control.

The proportion from the null hypothesis is $p_0 = 0.50$ and the sample size is $n = 340$, so the sample size is definitely large enough for the CLT to apply. We may use the test based on a normal distribution. The proportion from the sample is $\hat{p} = 0.59$ and the standardized test statistic is

$$z = \frac{\text{Statistic} - \text{Null Value}}{SE} = \frac{\hat{p} - p_0}{\sqrt{\frac{p_0(1-p_0)}{n}}} = \frac{0.59 - 0.50}{\sqrt{\frac{0.50(1-0.50)}{340}}} = 3.319$$

Since this is a right-tail test, $H_a : p > 0.50$, the p-value is the proportion of the standard normal distribution in the tail to the right of 3.319. Using technology we find this is 0.00045, so we have

$$\text{p-value} = 0.00045$$

The p-value is very small, so we reject H_0 and have strong evidence that the proportion of US middle school students opting for a car they control themselves is greater than 0.50.

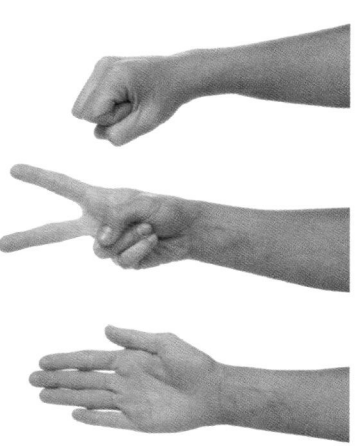

© Nickilford/iStockphoto

Rock-Paper-Scissors

[11]"To Drive, or Not to Drive: The Youth Perspective on Self-Driving Cars," *www.neilsen.com*, Media and Entertainment, March 21, 2016.

DATA 6.1	**Rock-Paper-Scissors**

Rock-Paper-Scissors, also called Roshambo, is a popular two-player game often used to quickly determine a winner and loser. In the game, each player puts out a fist (rock), a flat hand (paper), or a hand with two fingers extended (scissors). In the game, rock beats scissors which beats paper which beats rock. The question is: Are the three options selected equally often? Knowing the relative frequencies with which the options are selected would give a player a significant advantage. A study[12] observed 119 people playing Rock-Paper-Scissors. Their choices for the first turn are shown in Table 6.1. ■

Table 6.1 *Frequencies for first turn in Rock-Paper-Scissors*

Option Selected	Frequency
Rock	66
Paper	39
Scissors	14
Total	119

Example 6.8

Do the data in Table 6.1 provide evidence that the proportion of times players start with "rock" is different from the 1/3 we would expect if the players are choosing randomly?

Solution

We test $H_0 : p = 1/3$ vs $H_a : p \neq 1/3$, where p is the proportion of times a player uses rock on the first turn of a Rock-Paper-Scissors game. For the sample of $n = 119$ players we have $\hat{p} = 66/119 = 0.555$. We compute a standardized test statistic:

$$z = \frac{\text{Statistic} - \text{Null Value}}{SE} = \frac{\hat{p} - p_0}{\sqrt{\frac{p_0(1-p_0)}{n}}} = \frac{0.555 - 1/3}{\sqrt{\frac{1/3(1-1/3)}{119}}} = 5.13$$

Checking that $119 \cdot 1/3 = 39.7$ and $119 \cdot (1 - 1/3) = 79.3$ are both bigger than 10, we can use the standard normal distribution to find the p-value. However, a test statistic of 5.13 is so far out in the tail of a standard normal distribution that we don't really need to use technology to recognize that the p-value is extremely small, even after doubling to account for the two-tailed alternative. Since the p-value is essentially zero, this gives very strong evidence that rock is used more often than random chance would suggest on the first turn of a Rock-Paper-Scissors game.

Example 6.9

Sample Size Matters!

Suppose that a test for extrasensory perception (ESP) is designed where a person tries to guess the suit (spades, diamonds, hearts, clubs) of a randomly selected playing card. In the absence of ESP, the proportion guessed correctly should be $p = 0.25$, so a natural test to consider is $H_0 : p = 0.25$ vs $H_a : p > 0.25$. Complete the test for each possible set of sample results below, and discuss the effect of sample size on the results.

(a) The person guesses 29 correctly out of 100 trials.

(b) The person guesses 290 correctly out of 1000 trials.

[12]Eyler, D., Shalla, Z., Doumaux, A., and McDevitt, T., "Winning at Rock-Paper-Scissors," *The College Mathematics Journal*, March 2009; 40(2):125–128.

Solution   (a) The sample statistic is $\hat{p} = 0.29$. The standardized test statistic is

$$z = \frac{\text{Statistic} - \text{Null Value}}{SE} = \frac{\hat{p} - p_0}{\sqrt{\frac{p_0(1-p_0)}{n}}} = \frac{0.29 - 0.25}{\sqrt{\frac{0.25(0.75)}{100}}} = 0.924$$

This is a right-tail test, and the p-value from a normal distribution is 0.178. This is not a small p-value and we do not reject H_0. We do not have evidence that the person has any special ESP abilities.

(b) The sample statistic in this case is also $\hat{p} = 0.29$. The standardized test statistic is

$$z = \frac{\text{Statistic} - \text{Null Value}}{SE} = \frac{\hat{p} - p_0}{\sqrt{\frac{p_0(1-p_0)}{n}}} = \frac{0.29 - 0.25}{\sqrt{\frac{0.25(0.75)}{1000}}} = 2.921$$

This is a right-tail test, and the p-value from a normal distribution is 0.0017. This is a small p-value and we reject H_0. We have evidence that this person has some ESP abilities! As we saw in Chapter 4, the larger sample size greatly increases our ability to find evidence when there really is an effect.

SECTION LEARNING GOALS

You should now have the understanding and skills to:

- Use a formula for *SE* and a normal distribution, when appropriate, to test a hypothesis about a population proportion

Exercises for Section 6.1-HT

SKILL BUILDER 1

In Exercises 6.46 to 6.51, determine whether it is appropriate to use the normal distribution to estimate the p-value. If it is appropriate, use the normal distribution and the given sample results to complete the test of the given hypotheses. Assume the results come from a random sample and use a 5% significance level.

6.46 Test $H_0 : p = 0.5$ vs $H_a : p > 0.5$ using the sample results $\hat{p} = 0.57$ with $n = 40$

6.47 Test $H_0 : p = 0.3$ vs $H_a : p < 0.3$ using the sample results $\hat{p} = 0.21$ with $n = 200$

6.48 Test $H_0 : p = 0.25$ vs $H_a : p < 0.25$ using the sample results $\hat{p} = 0.16$ with $n = 100$

6.49 Test $H_0 : p = 0.8$ vs $H_a : p > 0.8$ using the sample results $\hat{p} = 0.88$ with $n = 50$

6.50 Test $H_0 : p = 0.75$ vs $H_a : p \neq 0.75$ using the sample results $\hat{p} = 0.69$ with $n = 120$

6.51 Test $H_0 : p = 0.2$ vs $H_a : p \neq 0.2$ using the sample results $\hat{p} = 0.26$ with $n = 1000$

6.52 Antibiotics in Infancy Exercise 2.19 describes a Canadian longitudinal study that examines whether giving antibiotics in infancy increases the likelihood that the child will be overweight later in life. The study included 616 children and found that 438 of the children had received antibiotics during the first year of life. Test to see if this provides evidence that more than 70% of Canadian children receive antibiotics during the first year of life. Show all details of the hypothesis test, including hypotheses, the standardized test statistic, the p-value, the generic conclusion using a 5% significance level, and a conclusion in context.

6.53 Left-Handed Lawyers Approximately 10% of Americans are left-handed (we will treat this as a known population parameter). A study on the relationship between handedness and profession found that in a random sample of 105 lawyers, 16 of them were left-handed.[13] Test the hypothesis that the

[13]Schachter, S. and Ransil, B., "Handedness Distributions in Nine Professional Groups," *Perceptual and Motor Skills*, 1996; 82: 51–63.

proportion of left-handed lawyers differs from the proportion of left-handed Americans.

(a) Clearly state the null and alternative hypotheses.

(b) Calculate the test statistic and p-value.

(c) What do we conclude at the 5% significance level? At the 10% significance level?

6.54 Do You Believe in Ghosts? A telephone survey of 1000 randomly selected US adults found that 31% of them say they believe in ghosts.[14] Does this provide evidence that more than 1 in 4 US adults believe in ghosts? Clearly show all details of the test.

6.55 Home Field Advantage in Baseball 2009 There were 2430 Major League Baseball (MLB) games played in 2009, and the home team won the game in 54.9% of the games.[15] If we consider the games played in 2009 as a sample of all MLB games, test to see if there is evidence, at the 1% level, that the home team wins more than half the games. Show all details of the test.

6.56 Home Field Advantage in Baseball 2014 There were 2428 Major League Baseball (MLB) games played in 2014, and the home team won 1288 of those games.[16] If we consider the games played in 2014 as a sample of all MLB games, test to see if there is evidence, at the 1% level, that the home team wins more than half the games. Show all details of the test.

6.57 Do You Know Your Neighbors? A survey of 2255 randomly selected US adults found that 51% said they know all or most of their neighbors.[17] Does this provide evidence that more than half of US adults know most or all of their neighbors?

6.58 Is B a Good Choice on a Multiple-Choice Exam? Multiple-choice questions on Advanced Placement exams have five options: A, B, C, D, and E. A random sample of the correct choice on 400 multiple-choice questions on a variety of AP exams[18] shows that B was the most common correct choice, with 90 of the 400 questions having B as the answer. Does this provide evidence that B is more likely to be the correct choice than would be expected if all five options were equally likely? Show all details of the test. The data are available in **APMultipleChoice**.

6.59 Euchre In Exercise 4.130 on page 313, we introduce a series of Euchre games played between two teams: Team A and Team B. After 40 games, Team A has won 16 times and Team B has won 24 times. Can we conclude that one team is better than the other? Clearly state the null and alternative hypotheses, calculate the test statistic and p-value, and interpret the result.

6.60 Do Babies Understand Probability? Can babies reason probabilistically? A study[19] investigates this by showing ten- to twelve-month-old infants two jars of lollipop-shaped objects colored pink or black. Each infant first crawled or walked to whichever color they wanted, determining their "preferred" color. They were then given the choice between two jars that had the same number of preferred objects, but that differed in their *probability* of getting the preferred color; each jar had 12 in the preferred color and either 4 or 36 in the other color. Babies choosing randomly or based on the absolute number of their preferred color would choose equally between the two jars, while babies understanding probability would more often choose the jar with the higher proportion of their preferred color. Of the 24 infants studied, 18 chose the jar with the higher proportion of their preferred color. Are infants more likely to choose the jar with the higher proportion of their preferred color?

(a) State the null and alternative hypotheses.

(b) Give the relevant sample statistic, using correct notation.

(c) Which of the following should be used to calculate a p-value for this dataset? A randomization test, a test using the normal distribution, or either one? Why?

(d) Find a p-value using a method appropriate for this data situation.

(e) Make a conclusion in context, using $\alpha = 0.05$.

6.61 Percent of Smokers The data in **NutritionStudy**, introduced in Exercise A.28 on page 174, include information on nutrition and health habits of a sample of 315 people. One of the variables is *Smoke*, indicating whether a person smokes or not (yes or no). Use technology to test whether the data provide evidence that the proportion of smokers is different from 20%.

[14]"31% Believe in Ghosts," Rasmussen Reports, October 30, 2011.

[15]*http://www.baseballprospectus.com/article.php?articleid=9854*, accessed June 2011.

[16]*http://www.betfirm.com/mlb-home-field-advantage/*. Accessed April 2016.

[17]Hampton, K., Goulet, L., Rainie, L. and Purcell, K., "Social Networking Sites and Our Lives," Pew Research Center, *pewresearch.org*, June 16, 2011.

[18]*http://apcentral.collegeboard.com*.

[19]Denison, S., and Xu, F. (2014). "The origins of probabilistic inference in human infants," *Cognition*, 130:335–347.

6.62 Regular Vitamin Use The data in **Nutrition-Study**, introduced in Exercise A.28 on page 174, include information on nutrition and health habits of a sample of 315 people. One of the variables is *VitaminUse*, indicating whether a person takes a multivitamin pill regularly or occasionally or not at all. Use technology to test whether the data provide evidence that the proportion taking a vitamin pill regularly is different from 35%.

6.2-D DISTRIBUTION OF A MEAN

For quantitative data, the parameter of interest is often the population mean, μ. The Central Limit Theorem (CLT) in Section 5.1 says that sample statistics often follow a normal distribution if the sample size is large. This is true for the distribution of sample means, $\bar{x}$, which will be centered at the population mean, μ. Besides the center, the other important quantity we need to describe a normal distribution is the standard deviation, the standard error (*SE*) of the sample means. As we have seen, one way to estimate this *SE* is to use a bootstrap or randomization distribution. Statistical theory gives us an alternate method for finding the standard error. When choosing random samples of size n from a population with mean μ and standard deviation σ, the standard error of the sample means is

$$SE = \frac{\sigma}{\sqrt{n}}$$

Thus we can predict the center and spread of the distribution of sample means. The amazing feature of the Central Limit Theorem is that we can also tell something about the shape. Even if the distribution of the population is heavily skewed or has big outliers, the distribution of the sample means will tend to follow a normal distribution as long as the sample size is large enough. As a general rule, the approximation works well when the sample size is at least 30.

> **Central Limit Theorem for Sample Means**
>
> If the sample size n is large, the distribution of sample means from a population with mean μ and standard deviation σ is approximately normally distributed with mean μ and standard deviation $\sigma/\sqrt{n}$.

Unfortunately, we find two difficulties when trying to use this result in practice. First, we generally don't know the standard deviation σ for the population of interest, we usually have only the information in a sample from the population. This is relatively easy to deal with, since we can just use the standard deviation from the sample, s, when estimating the standard error of the sample means.

$$SE = \frac{s}{\sqrt{n}}$$

The second difficulty is that when we use the SE based on $s/\sqrt{n}$ to standardize the sample mean, the distribution is no longer a standard normal. We'll introduce a new distribution to take care of this difficulty after the next example.

Example 6.10

In each case below, use the formula to compute the standard error and compare it to that obtained earlier using a bootstrap or randomization distribution.

(a) A sample of $n = 500$ Atlanta commute times shows $\bar{x} = 29.11$ minutes with $s = 20.7$. The standard error from simulation was 0.93.

(b) A sample of $n = 50$ body temperatures has mean $\bar{x} = 98.26$ with $s = 0.765$. The standard error from simulation was 0.11.

Solution (a) We have

$$SE = \frac{s}{\sqrt{n}} = \frac{20.7}{\sqrt{500}} = 0.93.$$

(b) We have

$$SE = \frac{s}{\sqrt{n}} = \frac{0.765}{\sqrt{50}} = 0.11.$$

The values of SE obtained with the formula agree nicely with the standard error estimates that we found with the simulated sample means.

The t-Distribution

So we have a formula to estimate the standard error of sample means, but if we standardize $\bar{x}$ using $s/\sqrt{n}$ in the denominator we no longer get a standard normal distribution. Fortunately, a man named William Sealy Gosset[20] worked out the distribution for this quantity in the early 1900s—under an assumption that the underlying population follows a normal distribution. The result is known as a *t-distribution* and it turns out to work well even if the underlying population is not perfectly normal.

A key fact about the t-distribution is that it depends on the size of the sample. We would expect a larger sample to tend to give a better estimate of the true standard deviation σ and thus the standardization based on $s/\sqrt{n}$ should behave more like a standard normal. This is exactly what happens. The sample size is reflected in a parameter called the *degrees of freedom* for the t-distribution. When working with $\bar{x}$ for a sample of size n and $SE = s/\sqrt{n}$, we use a t-distribution with $n - 1$ degrees of freedom (df).

> **The Distribution of Sample Means Using the Sample Standard Deviation**
>
> When choosing random samples of size n from a population with mean μ, the distribution of the sample means is centered at the population mean, μ, and has standard error estimated by
>
> $$SE = \frac{s}{\sqrt{n}}$$
>
> where s is the standard deviation of a sample.
>
> The standardized sample means approximately follow a **t-distribution** with $n - 1$ degrees of freedom (df).
>
> For small sample sizes ($n < 30$), the t-distribution is only a good approximation if the underlying population has a distribution that is approximately normal.

Figure 6.3 shows plots of t-distributions with 5 and 15 degrees of freedom along with a standard normal distribution. Note that the t-distribution is very similar to the classic bell-shaped pattern of the normal curve, only with slightly thicker

[20]Gosset used his statistical knowledge to become Head Brewmaster at Guinness Brewery.

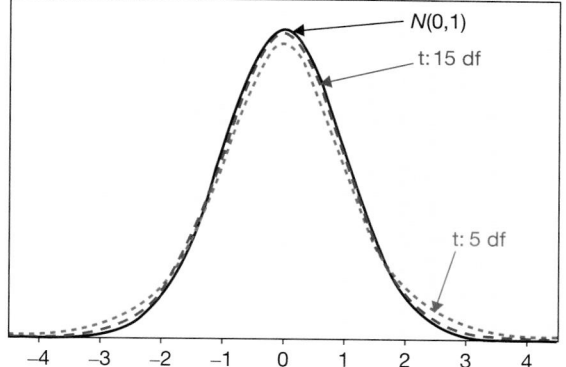

Figure 6.3 *Comparing t-distributions with 5 and 15 df to the standard normal distribution*

tails—especially for small degrees of freedom. Even for 15 degrees of freedom, the t-distribution and standard normal are virtually indistinguishable. The larger the sample size, n, the closer the t-distribution is to the normal distribution.

We see that the t-distribution and the standard normal are very similar, so we weren't far off when we used the standard normal values z in Chapter 5 for inference about means. From now on, though, we will use the slightly more accurate t-distribution with $n - 1$ df, rather than the standard normal, when doing inference for a population mean based on a sample mean $\bar{x}$ and sample standard deviation, s. Fortunately, when using technology to find endpoints or probabilities for a t-distribution, the process is usually very similar to what we have already seen for the normal distribution. For larger samples, the results will be very close to what we get from a standard normal, but we will still use the t-distribution for consistency. For smaller samples, the t-distribution gives an extra measure of safety when using the sample standard deviation s in place of the population standard deviation σ.

Conditions for the t-Distribution

For small samples, the use of the t-distribution requires that the population distribution be approximately normal. If the sample size is small, we need to check that the data are relatively symmetric and have no huge outliers that might indicate a departure from normality in the population. We don't insist on perfect symmetry or an exact bell-shape in the data in order to use the t-distribution. The normality condition is most critical for small sample sizes, since for larger sample sizes the CLT for means kicks in. Unfortunately, it is more difficult to judge whether a sample looks "normal" when the sample size is small. In practice, we avoid using the t-distribution if the sample is small (say less than 20) and the data contain clear outliers or skewness. For more moderate sized samples (20 to 50) we worry if there are very extreme outliers or heavy skewness. When in doubt, we can always go back to the ideas of Chapters 3 and 4 and directly simulate a bootstrap or randomization distribution.

 If the sample size is small and the data are heavily skewed or contain extreme outliers, the t-distribution should not be used.

Example 6.11

Dotplots of three different samples are shown in Figure 6.4. In each case, indicate whether or not it is appropriate to use the t-distribution. If it is appropriate, give the degrees of freedom for the t-distribution and give the estimated standard error.

(a) A sample with $n = 50$, $\bar{x} = 8.0$, and $s = 10.5$, shown in Figure 6.4(a)

(b) A sample with $n = 8$, $\bar{x} = 4.9$, and $s = 1.25$, shown in Figure 6.4(b)

(c) A sample with $n = 10$, $\bar{x} = 12.6$, and $s = 4.8$, shown in Figure 6.4(c)

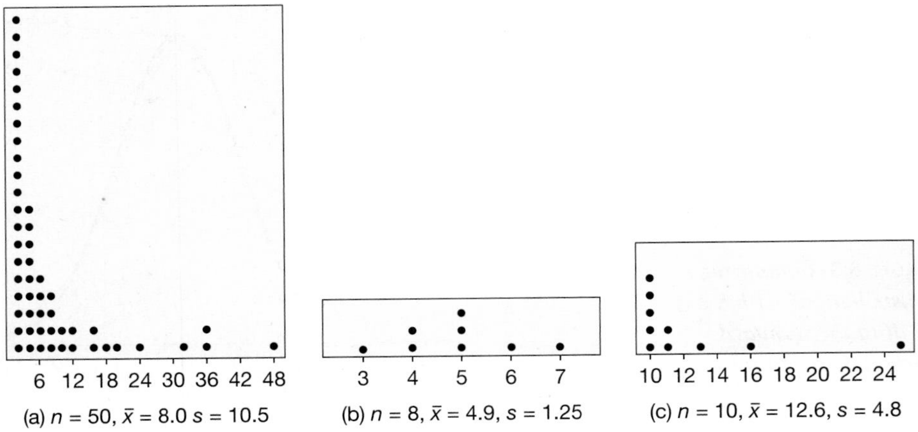

Figure 6.4 *In which cases might we have concerns about using the t-distribution?*

(a) $n = 50, \bar{x} = 8.0\ s = 10.5$ (b) $n = 8, \bar{x} = 4.9, s = 1.25$ (c) $n = 10, \bar{x} = 12.6, s = 4.8$

Solution

The t-distribution is appropriate if the sample size is large ($n \geq 30$) or if the underlying distribution appears to be relatively normal. We have concerns about the t-distribution only for small sample sizes and heavy skewness or outliers.

(a) Since the sample size is large ($n = 50$), the t-distribution is appropriate. For the degrees of freedom df and estimated standard error SE, we have

$$df = n - 1 = 50 - 1 = 49 \quad \text{and} \quad SE = \frac{s}{\sqrt{n}} = \frac{10.5}{\sqrt{50}} = 1.485$$

(b) The sample size is small in this case ($n = 8$) but the sample is not heavily skewed or with outliers, so a condition of normality is reasonable. The t-distribution is appropriate. For the degrees of freedom df and estimated standard error SE, we have

$$df = n - 1 = 8 - 1 = 7 \quad \text{and} \quad SE = \frac{s}{\sqrt{n}} = \frac{1.25}{\sqrt{8}} = 0.442$$

(c) In this case, the sample size is small ($n = 10$) and the data are heavily skewed and have an obvious outlier. It is not appropriate to use the t-distribution with this sample. If we want to do inference using this sample, we might try simulation methods, such as using a bootstrap or randomization distribution.

Using the t-Distribution

As with the normal distribution, the method used to find areas and endpoints in a t-distribution will vary depending on the type of technology being used.

Example 6.12

We select a sample of size 16 from a population that is reasonably normally distributed and use a t-statistic for inference about the sample mean.

(a) Find endpoints in a t-distribution with 0.025 beyond them in each tail.

(b) Find the proportion in a t-distribution above 1.5.

Solution

For a sample size of $n = 16$ we need to use a t-distribution with 15 degrees of freedom.

(a) Figure 6.5(a) shows that the points with 2.5% of the t-distribution in each tail are at -2.131 and $+2.131$.

(b) Figure 6.5(b) shows that about 0.077 of a t-distribution with 15 df will lie above 1.5.

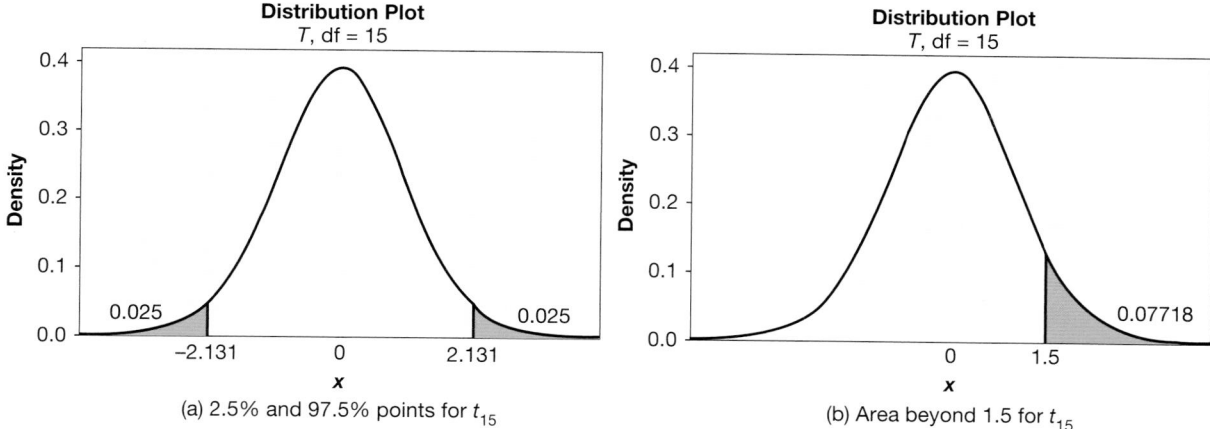

Figure 6.5 *Calculations for t-distribution with 15 df*

In the next two sections, we explore the use of the t-distribution in more detail, first for computing a confidence interval for a population mean and then for computing a p-value to test a hypothesis about a mean.

SECTION LEARNING GOALS

You should now have the understanding and skills to:

▶ • Use the formula to find the standard error for a distribution of sample means

▶ • Recognize when a t-distribution is appropriate for inference about a sample mean

▶ • Find endpoints and proportions for a t-distribution

Exercises for Section 6.2-D

SKILL BUILDER 1

In Exercises 6.63 to 6.66, if random samples of the given size are drawn from a population with the given mean and standard deviation, find the standard error of the distribution of sample means.

6.63 Samples of size 1000 from a population with mean 28 and standard deviation 5

6.64 Samples of size 10 from a population with mean 6 and standard deviation 2

6.65 Samples of size 40 from a population with mean 250 and standard deviation 80

6.66 Samples of size 75 from a population with mean 60 and standard deviation 32

SKILL BUILDER 2

Use a t-distribution to answer the questions in Exercises 6.67 to 6.74. Assume the sample is a random sample from a distribution that is reasonably normally distributed and we are doing inference for a sample mean.

6.67 Find endpoints of a t-distribution with 5% beyond them in each tail if the sample has size $n = 10$.

6.68 Find endpoints of a t-distribution with 1% beyond them in each tail if the sample has size $n = 18$.

6.69 Find endpoints of a t-distribution with 0.025 beyond them in each tail if the sample has size $n = 25$.

6.70 Find endpoints of a t-distribution with 0.005 beyond them in each tail if the sample has size $n = 40$.

6.71 Find the area in a t-distribution above 2.3 if the sample has size $n = 6$.

6.72 Find the area in a t-distribution above 1.5 if the sample has size $n = 8$.

6.73 Find the area in a t-distribution below -1.0 if the sample has size $n = 20$.

6.74 Find the area in a t-distribution below -3.2 if the sample has size $n = 50$.

6.75 Impact of Sample Size on Accuracy Compute the standard error for sample means from a population with mean $\mu = 100$ and standard deviation $\sigma = 25$ for sample sizes of $n = 30, n = 200$, and $n = 1000$. What effect does increasing the sample size have on the standard error? Using this information about the effect on the standard error, discuss the effect of increasing the sample size on the accuracy of using a sample mean to estimate a population mean.

6.76 Impact of the Population Standard Deviation on SE Compute the standard error for sample means from populations all with mean $\mu = 100$ and with standard deviations $\sigma = 5$, $\sigma = 25$, and $\sigma = 75$ using a sample size of $n = 100$. Discuss the effect of the population standard deviation on the standard error of the sample means.

IS A T-DISTRIBUTION APPROPRIATE?

In Exercises 6.77 to 6.80, we give summary statistics and a dotplot for a sample. In each case, indicate whether or not it is appropriate to use the t-distribution. If it is appropriate, give the degrees of freedom for the t-distribution and give the estimated standard error.

6.77 A sample with $n = 12, \bar{x} = 7.6$, and $s = 1.6$

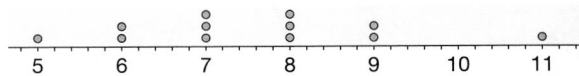

6.78 A sample with $n = 75, \bar{x} = 18.92$, and $s = 10.1$

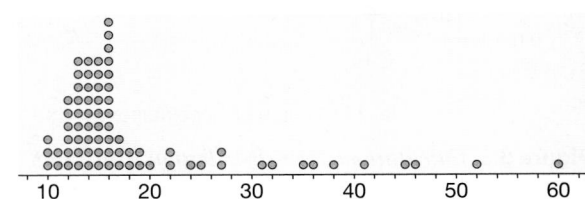

6.79 A sample with $n = 18, \bar{x} = 87.9$, and $s = 10.6$

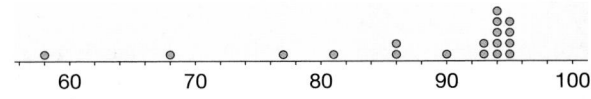

6.80 A sample with $n = 10, \bar{x} = 508.5$, and $s = 21.5$

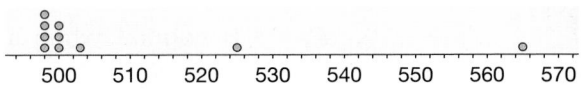

6.2-CI CONFIDENCE INTERVAL FOR A MEAN

Confidence Interval for a Mean Using the t-Distribution

In Section 5.2 we see that when a distribution is normally distributed, a confidence interval can be formed using

$$\text{Sample Statistic} \pm z^* \cdot SE$$

where z^* is an appropriate percentile from a standard normal distribution and SE is the standard error.

In Section 6.2-D we see that we can estimate the standard error for a sample mean using

$$SE = \frac{s}{\sqrt{n}}$$

where n is the sample size and s is the sample standard deviation. However, when we use s rather than the (unknown) population standard deviation σ in computing the SE, the standardized statistic follows a t-distribution with $n - 1$ degrees of freedom, rather than a standard normal (provided the underlying population is reasonably normal).

We combine these facts to produce an easy formula for a confidence interval for a mean.

Confidence Interval for a Mean

A sample mean based on a random sample of size n has

$$\text{Sample statistic} = \bar{x} \quad \text{and} \quad SE = \frac{s}{\sqrt{n}}$$

where $\bar{x}$ and s are the sample mean and standard deviation, respectively. If t^* is an endpoint chosen from a t-distribution with $n - 1$ df to give the desired level of confidence, and if the distribution of the population is approximately normal or the sample size is large ($n \geq 30$), the confidence interval for the population mean, μ, is

$$\text{Sample statistic} \pm t^* \cdot SE$$

which, in this case, corresponds to

$$\bar{x} \pm t^* \cdot \frac{s}{\sqrt{n}}$$

© Stephen Strathdee/iStockphoto

How long does it take to fly from Boston to San Francisco?

DATA 6.2

Boston/San Francisco Flight Times

United Airlines Flight 433 is a nonstop flight from Boston's Logan Airport to San Francisco International Airport. During 2016 it was scheduled to leave each day around 6:00 am (Eastern time) and arrive around 10:00 am (Pacific time). Due to the three hour difference between the time zones, the flight is expected to take about 7 hours (420 minutes) including time spent taxiing on runways and waiting to take off. An important factor in scheduling such flights is the actual airborne flying time from takeoff to touchdown. The data[21] in **Flight433** contain the airborne time (in minutes) for January 2016. ■

[21] Data collected from the Bureau of Transportation Statistics website, *http://www.bts.gov/xml/ ontimesummarystatistics/src/dstat/OntimeSummaryAirtime.xml.*

Example 6.13

Solution ▶

Use the data on airborne times (in minutes) for **Flight433** to find a 95% confidence interval for the mean flight time on this route from Boston to San Francisco.

Although the sample size of $n = 31$ is large enough for the CLT for means to apply, before using the t-distribution we should check a plot of the data for signs of extreme outliers or heavy skewness. The dotplot in Figure 6.6 shows no reasons for serious concerns.

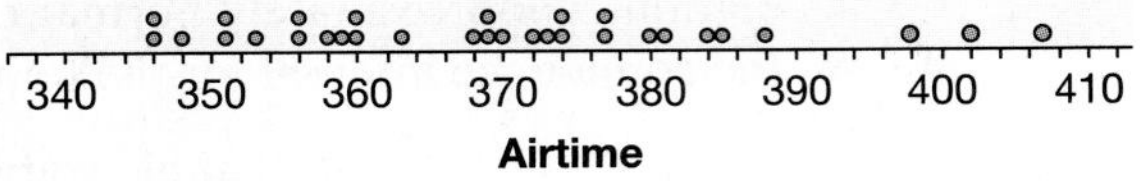

Figure 6.6 *Airborne flight times (in minutes) for Flight 433 from Boston to San Francisco*

```
          o o   o o  o o          o o  o oo  o  o                    o   o   o
    o o   o o   o  o oooo  o      oooo ooo   o  oo  oo  o           o   o   o
  ┼──────┼──────┼──────┼──────┼──────┼──────┼──────┼──────┼
 340    350    360    370    380    390    400    410
                              Airtime
```

Using technology and the data in **Flight433** we find the mean airborne time for the sample of $n = 31$ instances of Flight 433 to be $\bar{x} = 369.52$ minutes and the standard deviation of flight times to be $s = 16.16$ minutes. To find a confidence interval for the population mean, we use a t-distribution with $31 - 1 = 30$ degrees of freedom. For a 95% confidence interval, as in Figure 6.7, we find the endpoints with 2.5% of the distribution in each tail, $t^* = 2.042$.

We put all the pieces together to construct the confidence interval:

$$\text{Statistic} \ \pm \ t^* \cdot SE$$
$$\bar{x} \ \pm \ t^* \frac{s}{\sqrt{n}}$$
$$369.52 \ \pm \ 2.042 \left(\frac{16.16}{\sqrt{31}} \right)$$
$$369.52 \ \pm \ 5.927$$
$$363.593 \ \text{to} \ 375.447$$

Based on these data, we are 95% sure that the mean airborne time for Flight 433 from Boston to San Francisco is between 363.593 and 375.447 minutes.

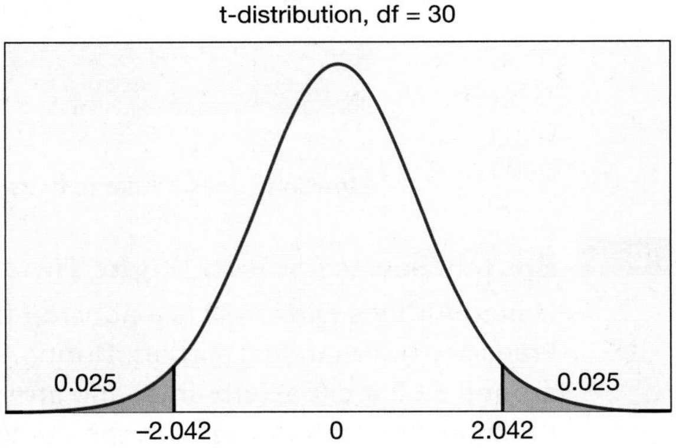

t-distribution, df = 30

Figure 6.7 *Endpoints for a 95% CI using a t-distribution with 30 df*

0.025   0.025

−2.042   0   2.042

Note that the $t^* = 2.042$ value in the previous example is slightly larger than the standard normal $z^* = 1.96$ for a 95% confidence interval. This helps account for the uncertainty in estimating the population standard deviation, σ, using the standard deviation from the sample, s. As the degrees of freedom increase, the t^* values will get closer and closer to the corresponding z^* values.

Example 6.14

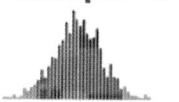

In Example 3.24 on page 243 we compute 99% and 90% confidence intervals using percentiles from a bootstrap distribution for the mean commute time in Atlanta, based on the data in **CommuteAtlanta**. For that sample of $n = 500$ commutes the mean time is $\bar{x} = 29.11$ minutes with $s = 20.7$ minutes. Use the summary statistics to redo both confidence intervals.

Solution

Although the underlying distribution of commute times is somewhat skewed (see Figure 3.13 on page 229), the sample size of 500 is large enough for us to avoid worrying much about the population distribution. For the 99% confidence interval, we use technology to find the points that leave 0.5% in each tail of a t-distribution with 499 degrees of freedom, $t^* = 2.586$. We compute the 99% confidence interval with

$$\text{Statistic} \pm t^* \cdot SE = 29.11 \pm 2.586 \left(\frac{20.7}{\sqrt{500}} \right) = 29.11 \pm 2.39 = (26.72, 31.50)$$

We are 99% sure that the mean commute time for all Atlanta commuters is between 26.72 and 31.50 minutes.

For the 90% interval, the endpoint for the t-distribution is $t^* = 1.648$, so the interval is

$$\text{Statistic} \pm t^* \cdot SE = 29.11 \pm 1.648 \left(\frac{20.7}{\sqrt{500}} \right) = 29.11 \pm 1.53 = (27.58, 30.64)$$

We are 90% sure that the mean commute time for all Atlanta commuters is between 27.58 and 30.64 minutes. As expected, the width goes down as we require less confidence (from 99% to 90%).

In Chapter 3, we used the percentiles of a bootstrap distribution and obtained results of $(26.98, 31.63)$ and $(27.70, 30.71)$ for the 99% and 90% confidence intervals, respectively. These match closely those found in Example 6.14. The advantage of the t-interval is that it is most commonly used in practice when estimating a mean. The advantage of the bootstrap methods is that they remain valid in some cases where conditions, such as normality for a small sample, might be questionable. Also, the bootstrap methods are flexible enough to use for almost all parameters, while the t-distribution only works for certain parameters such as the mean.

Example 6.15

Manhattan Apartments

What is the average rental price of a one-bedroom apartment in Manhattan? We go on Craigslist and record the monthly rent prices for a sample of 20 listed one bedroom apartments in Manhattan. These data[22] are stored in **ManhattanApartments** and are displayed in Figure 6.8. Give a 95% confidence interval for the average monthly rent for a one-bedroom apartment in Manhattan.

Figure 6.8 *Monthly rent for 20 one-bedroom apartments in Manhattan*

[22]Data were obtained from *newyork.craigslist.org*.

Solution This is a small sample size, $n = 20$, so we check the sample data for normality. Figure 6.8 shows a few high outliers, indicating a lack of normality, so we should not use the t-distribution in this case. Instead, we return to the methods of Chapter 3 and use technology to create a bootstrap distribution. The standard error of the bootstrap distribution is \$302, and the bootstrap distribution is approximately normal so we create the interval by

$$\text{Sample Mean} \pm 2 \cdot SE = 3157 \pm 2 \cdot 302 = (2553, 3761)$$

We are 95% confident that the average monthly rent for a one-bedroom apartment in Manhattan is between \$2553 and \$3761. If you are ever looking to rent an apartment, you can collect your own data to estimate the average price in your city.

Determining Sample Size for Estimating a Mean

A common question when designing a study is "How large a sample should we collect?" When estimating a mean with a confidence interval, the answer to this question depends on three related questions:

- How accurate do we want the estimate to be? In other words, what margin of error, *ME*, do we want?
- How much confidence do we want to have in the interval?
- How much variability is there in the population?

From the formula for the confidence interval for a mean we see that the margin of error is computed with

$$ME = t^* \frac{s}{\sqrt{n}}$$

Suppose that, *before getting a sample*, we decide we want the confidence interval for a mean to have some pre-determined margin of error. All else being equal, as *n* gets larger, the margin of error, *ME*, gets smaller. With a bit of algebra, we can solve the equation for the margin of error to get a direct expression to compute *n* for any desired margin of error and level of confidence:

$$n = \left(\frac{t^* \cdot s}{ME} \right)^2$$

However, there are two problems with using this formula in practice to determine a sample size. First, the degrees of freedom for t^* depends on the choice of *n*. Second, the standard deviation, *s*, is computed from the sample—but we haven't even collected a sample yet!

We address the first of these issues by using the standard normal value, z^*, in place of t^*. We know that, as sample size increases, the t^* values get closer and closer to z^*. Our goal is to get a rough estimate for the sample size we need to get the desired margin of error. Unless that indicated sample size is quite small, we don't lose much by using z^* in place of t^*.

The more serious concern is what to do about *s*, since we haven't even taken a sample yet. The solution is to make a reasonable guess about the standard deviation, σ, in the population. We'll call that guess $\tilde{\sigma}$. To do this we might:

- Use the standard deviation from a previous study or a sample from a similar population as $\tilde{\sigma}$.
- Take a small pre-sample and use its standard deviation for $\tilde{\sigma}$.
- Estimate the range (*max − min*) for the population and set $\tilde{\sigma} \approx Range/4$. This assumes that most values tend to be within about two standard deviations on either side of the mean.
- Make a reasonable guess for $\tilde{\sigma}$.

When in doubt, use an estimate or guess on the high side for $\tilde{\sigma}$. If our estimate of $\tilde{\sigma}$ is a bit high, we might end up taking a larger sample than is needed, but the end result would be a margin of error that is smaller (and thus better) than we expected.

Determination of Sample Size to Estimate a Mean

If we want to estimate a population mean to within a desired margin of error, *ME*, with a given level of confidence, we should select a sample of size

$$n = \left(\frac{z^* \cdot \tilde{\sigma}}{ME} \right)^2$$

where $\tilde{\sigma}$ is an estimate for the standard deviation in the population.

Example 6.16

In Example 6.13 on page 426, we consider flying times for a sample of size $n = 31$ Boston to San Francisco flights. The 95% confidence interval for the mean airborne time in that example is 369.52 ± 5.93. Suppose that the schedulers for United Airlines want to get a more accurate estimate, to within just two minutes of the actual mean airborne time, still with 95% confidence. How large a sample of flights would they need to collect to accomplish this?

Solution

The desired margin of error is $ME = 2$ minutes and for 95% confidence the standard normal value is $z^* = 1.96$. We use the standard deviation in the sample from Example 6.13 to estimate $\tilde{\sigma} = 16.16$ minutes. To compute the sample size, we use

$$n = \left(\frac{z^* \cdot \tilde{\sigma}}{ME} \right)^2 = \left(\frac{1.96 \cdot 16.16}{2} \right)^2 = 250.8$$

By convention we round up any factional parts of a sample, so to estimate the mean airborne time of Flight 433 to within 2 minutes with 95% confidence, we should use a random sample of about 251 flights.

SECTION LEARNING GOALS

You should now have the understanding and skills to:

▸ • Use a formula for *SE* and a t-distribution, when appropriate, to compute a confidence interval for a population mean

▸ • Determine a sample size needed to estimate a mean to within a specified margin of error at a given level of confidence

Exercises for Section **6.2-CI**

SKILL BUILDER 1
In Exercises 6.81 to 6.86, use the t-distribution to find a confidence interval for a mean μ given the relevant sample results. Give the best point estimate for μ, the margin of error, and the confidence interval. Assume the results come from a random sample from a population that is approximately normally distributed.

6.81 A 95% confidence interval for μ using the sample results $\bar{x} = 12.7$, $s = 5.6$, and $n = 30$

6.82 A 95% confidence interval for μ using the sample results $\bar{x} = 84.6$, $s = 7.8$, and $n = 42$

6.83 A 90% confidence interval for μ using the sample results $\bar{x} = 3.1$, $s = 0.4$, and $n = 100$

6.84 A 90% confidence interval for μ using the sample results $\bar{x} = 137.0$, $s = 53.9$, and $n = 50$

6.85 A 99% confidence interval for μ using the sample results $\bar{x} = 46.1$, $s = 12.5$, and $n = 10$

6.86 A 99% confidence interval for μ using the sample results $\bar{x} = 88.3$, $s = 32.1$, and $n = 15$

SKILL BUILDER 2
In Exercises 6.87 to 6.90, what sample size is needed to give the desired margin of error in estimating a population mean with the indicated level of confidence?

6.87 A margin of error within ±5 with 95% confidence, assuming a previous sample had $s = 18$

6.88 A margin of error within ±1 with 99% confidence, assuming a sample from a similar population had $s = 3.4$

6.89 A margin of error within ±0.5 with 90% confidence, if we make a reasonable estimate that $\sigma = 25$

6.90 A margin of error within ±12 with 95% confidence, assuming we estimate that $\sigma \approx 125$

6.91 How Many Birds Do Domestic Cats Kill? Exercise 6.19 discusses the headline "Domestic cats kill many more wild birds in the United States than scientists thought," and estimates the proportion of domestic cats that hunt outside. A separate study[23] used KittyCams to record all activity of $n = 55$ domestic cats that hunt outdoors. The video footage showed that the mean number of kills per week for these cats was 2.4 with a standard deviation of 1.51. Find and interpret a 99% confidence interval for the mean number of kills per week by US household cats that hunt outdoors.

6.92 How Much TV Do College Students Watch? In the dataset **StudentSurvey**, 361 students recorded the number of hours of television they watched per week. The average is $\bar{x} = 6.504$ hours with a standard deviation of 5.584. Find a 99% confidence interval for μ and interpret the interval in context. In particular, be sure to indicate the population involved.

6.93 How Many Close Confidants Do People Have? In a recent study,[24] 2006 randomly selected US adults (age 18 or older) were asked to give the number of people in the last six months "with whom you discussed matters that are important to you." The average number of close confidants was 2.2, with a standard deviation of 1.4.

(a) Find the margin of error for this estimate if we want 99% confidence.

(b) Find and interpret a 99% confidence interval for average number of close confidants.

6.94 How Big Are Gribbles? Gribbles are small, pale white, marine worms that bore through wood. While sometimes considered a pest since they can wreck wooden docks and piers, they are now being studied to determine whether the enzyme they secrete will allow us to turn inedible wood and plant waste into biofuel.[25] A sample of 50 gribbles finds an average length of 3.1 mm with a standard deviation of 0.72. Give a best estimate for the length of gribbles, a margin of error for this estimate (with 95% confidence), and a 95% confidence interval. Interpret the confidence interval in context. What do we have to assume about the sample in order to have confidence in our estimate?

6.95 Is Your Stomach Controlling Your Personality? Scientists estimate that there are 10 times more bacterial cells in your body than your own body's cells, and new studies on bacteria in the gut indicate that your gut microbes might be influencing you more than you realize, having positive or negative effects on health, development, and possibly even personality and behavior. A recent study[26] found that the average number of unique genes in gut bacteria, for a sample of 99 healthy European individuals, was 564 million, with a standard deviation of 122 million. Use the t-distribution to find and interpret a 95% confidence interval for the mean number of unique genes in gut bacteria for European individuals.

6.96 Dim Light at Night Makes Fat Mice Data 4.1 on page 258 introduces a study in which mice that

[23]Loyd KAT, et al., "Quantifying free-roaming domestic cat predation using animal-borne video cameras," *Biological Conservation*, 160(2013), 183–189.

[24]Hampton, K., Goulet, L., Rainie, L., and Purcell, K., "Social Networking Sites and Our Lives," Pew Research Center, *pewresearch.org*, June 16, 2011.

[25]Sanderson, K., "A Chewy Problem," *Nature*, 23 June 2011, p. S12.

[26]Qin, J., et al., "A human gut microbial gene catalogue established by metagenomic sequencing," *Nature*, 4 March 2010; 464: 59–65.

had a dim light on at night (rather than complete darkness) ate most of their calories when they should have been resting. These mice gained a significant amount of weight, despite eating the same number of calories as mice kept in total darkness. The time of eating seemed to have a significant effect. We look here at the effect after 8 weeks. There were 10 mice in the group with dim light at night and they gained an average of 7.9 g with a standard deviation of 3.0. We see in Figure 6.9 that the data are not heavily skewed and do not have extreme outliers. Use the t-distribution to find and interpret a 90% confidence interval for weight gain. As always, define the parameter being estimated.

Figure 6.9 *Body mass gain (in grams) for mice with a night light*

6.97 Bright Light at Night Makes Even Fatter Mice Data 4.1 on page 258 introduces a study in which mice that had a light on at night (rather than complete darkness) ate most of their calories when they should have been resting. These mice gained a significant amount of weight, despite eating the same number of calories as mice kept in total darkness. The time of eating seemed to have a significant effect. Exercise 6.96 examines the mice with dim light at night. A second group of mice had bright light on all the time (day and night). There were nine mice in the group with bright light at night and, after 8 weeks, they gained an average of 11.0 g with a standard deviation of 2.6. The data are shown in Figure 6.10. Is it appropriate to use a t-distribution in this situation? Why or why not? If not, how else might we construct a confidence interval for mean weight gain of mice with a bright light on all the time?

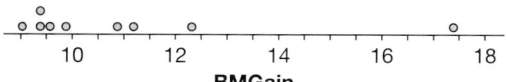

Figure 6.10 *Body mass gain (in grams) for mice with a bright night light*

6.98 United Flights in December In Exercise 4.115 we look at a random sample of 1,000 United flights in the month of December comparing the actual arrival time to the scheduled arrival time. Computer output of the descriptive statistics for the difference

in actual and expected arrival time of these 1,000 flights are shown below.

N	Mean	StDev	SE Mean	Min	Q1	Median	Q3	Max
1000	9.76	41.0	1.30	−45	−12	0	17	458

(a) What is the sample mean difference in actual and expected arrival times? What is the standard deviation of the differences?

(b) Based on the computer output, is it safe to assume this sample is normally distributed? Why or why not?

(c) Can we proceed using the t distribution to build a confidence interval for this mean? Why or why not?

(d) Regardless of your answer to part (c), use the summary statistics to compute a 95% confidence interval for the average difference in actual and scheduled arrival times on United flights in December.

(e) Interpret the confidence interval you found in part (d) in context.

STANDARD ERROR FROM A FORMULA AND A BOOTSTRAP DISTRIBUTION
In Exercises 6.99 to 6.102, use *StatKey* or other technology to generate a bootstrap distribution of sample means and find the standard error for that distribution. Compare the result to the standard error given by the Central Limit Theorem, using the sample standard deviation as an estimate of the population standard deviation.

6.99 Mean number of penalty minutes for NHL players using the data in **OttawaSenators** with $n = 24$, $\bar{x} = 34.13$, and $s = 27.26$

6.100 Mean commute time in Atlanta, in minutes, using the data in **CommuteAtlanta** with $n = 500$, $\bar{x} = 29.11$, and $s = 20.72$

6.101 Mean price of used Mustang cars online (in $1000s) using the data in **MustangPrice** with $n = 25$, $\bar{x} = 15.98$, and $s = 11.11$

6.102 Mean body temperature, in °F, using the data in **BodyTemp50** with $n = 50$, $\bar{x} = 98.26$, and $s = 0.765$

COMPARING NORMAL AND BOOTSTRAP CONFIDENCE INTERVALS
In Exercises 6.103 and 6.104, find a 95% confidence interval for the mean two ways: using *StatKey* or other technology and percentiles from a bootstrap distribution, and using the t-distribution and the formula for standard error. Compare the results.

6.103 Mean distance of a commute for a worker in Atlanta, using data in **CommuteAtlanta** with $\bar{x} =$ 18.156 miles, $s = 13.798$, and $n = 500$

6.104 Mean price of a used Mustang car online, in $1000s, using data in **MustangPrice** with $\bar{x} = 15.98$, $s = 11.11$, and $n = 25$

6.105 How Much Fat Do US Adults Consume? Using the dataset **NutritionStudy**, we calculate that the average number of grams of fat consumed in a day for the sample of $n = 315$ US adults in the study is $\bar{x} = 77.03$ grams with $s = 33.83$ grams.

(a) Find and interpret a 95% confidence interval for the average number of fat grams consumed per day by US adults.

(b) What is the margin of error?

(c) If we want a margin of error of only ± 1, what sample size is needed?

6.106 Estimating Number of Close Confidants More Accurately In Exercise 6.93 on page 430, we see that the average number of close confidants in a random sample of 2006 US adults is 2.2 with a standard deviation of 1.4. If we want to estimate the number of close confidants with a margin of error within ± 0.05 and with 99% confidence, how large a sample is needed?

6.107 Plastic Microfiber Pollution from Clothes Plastic microparticles are contaminating the world's shorelines (see Exercise 6.108), and much of this pollution appears to come from fibers from washing polyester clothes.[27] The worst offender appears to be fleece, and a recent study found that the mean number of polyester fibers discharged into wastewater from washing fleece was 290 fibers per liter of wastewater, with a standard deviation of 87.6 and a sample size of 120.

(a) Find and interpret a 99% confidence interval for the mean number of polyester microfibers per liter of wastewater when washing fleece.

(b) What is the margin of error?

(c) If we want a margin of error of only ± 5 with 99% confidence, what sample size is needed?

6.108 Plastic Microfiber Pollution on Shorelines In Exercise 6.107, we see that plastic microparticles are contaminating the world's shorelines and that

[27]Browne, M., et al., "Accumulation of Microplastic on Shorelines Worldwide: Sources and Sinks," *Environmental Science and Technology*, 2011;45:9175–79. Data are approximated from information given.

much of the pollution appears to come from fibers from washing polyester clothes. The same study referenced in Exercise 6.107 also took samples from ocean beaches. Five samples were taken from each of 18 different shorelines worldwide, for a total of 90 samples of size 250 mL. The mean number of plastic microparticles found per 250 mL of sediment was 18.3 with a standard deviation of 8.2.

(a) Find and interpret a 99% confidence interval for the mean number of polyester microfibers per 250 mL of beach sediment.

(b) What is the margin of error?

(c) If we want a margin of error of only ± 1 with 99% confidence, what sample size is needed?

WHAT INFLUENCES THE SAMPLE SIZE NEEDED?
In Exercises 6.109 to 6.111, we examine the effect of different inputs on determining the sample size needed.

6.109 Find the sample size needed to give, with 95% confidence, a margin of error within ± 10. Within ± 5. Within ± 1. Assume that we use $\tilde{\sigma} = 30$ as our estimate of the standard deviation in each case. Comment on the relationship between the sample size and the margin of error.

6.110 Find the sample size needed to give a margin of error within ± 3 with 99% confidence. With 95% confidence. With 90% confidence. Assume that we use $\tilde{\sigma} = 30$ as our estimate of the standard deviation in each case. Comment on the relationship between the sample size and the confidence level desired.

6.111 Find the sample size needed to give, with 95% confidence, a margin of error within ± 3, if the estimated standard deviation is $\tilde{\sigma} = 100$. If the estimated standard deviation is $\tilde{\sigma} = 50$. If the estimated standard deviation is $\tilde{\sigma} = 10$. Comment on how the variability in the population influences the sample size needed to reach a desired level of accuracy.

6.112 How Big Is the Tip Percentage at a Restaurant? Use technology and the **RestaurantTips** dataset to find a 95% confidence interval for the mean tip percentage (*PctTip*) at the restaurant. Interpret the answer in context.

6.113 How Many Grams of Fiber Do People Get in a Day? Use technology and the **NutritionStudy** dataset to find a 95% confidence interval for the mean number of grams of fiber (*Fiber*) people eat in a day. Interpret the answer in context.

6.2-HT HYPOTHESIS TEST FOR A MEAN

In Section 5.1 we see that, when a randomization distribution is normal, we can compute a p-value using a standard normal curve and a standardized test statistic of the form

$$z = \frac{\text{Sample Statistic} - \text{Null Parameter}}{SE}$$

The sample statistic is computed from the sample data and the null parameter is specified by the null hypothesis, H_0.

When testing a hypothesis about a population mean, the null hypothesis is typically $H_0 : \mu = \mu_0$, where μ_0 is some specific value of the mean. Thus the null parameter is μ_0 and the sample statistic is the mean from a sample, $\bar{x}$:

$$z = \frac{\bar{x} - \mu_0}{SE}$$

As we see in Section 6.2-D, we can estimate the standard error of $\bar{x}$ with $SE = s/\sqrt{n}$, where s is the standard deviation of the sample. However, with this sample estimate for SE, the distribution of the standardized test statistic follows a t-distribution with $n-1$ degrees of freedom rather than a standard normal:

$$t = \frac{\bar{x} - \mu_0}{s/\sqrt{n}}$$

This requires that the underlying population be reasonably normally distributed, although that condition is less critical as the sample size gets larger. In cases where the t-distribution applies, we find a p-value for a given sample by computing this *t-statistic* and using technology to find the proportion beyond it in the tail(s) of the t-distribution. As before, the alternative hypothesis will determine which tail(s) we use when computing the p-value. Because we are using the t-distribution, a hypothesis test conducted this way is often called a *t-test*.

T-Test for a Mean

To test $H_0 : \mu = \mu_0$ vs $H_a : \mu \neq \mu_0$ (or a one-tail alternative) use the t-statistic

$$t = \frac{\text{Statistic} - \text{Null value}}{SE} = \frac{\bar{x} - \mu_0}{s/\sqrt{n}}$$

where $\bar{x}$ is the mean and s is the standard deviation in a random sample of size n. Provided the underlying population is reasonably normal (or the sample size is large), the p-value of the test is computed using the appropriate tail(s) of a t-distribution with $n-1$ degrees of freedom.

Example 6.17

In Data 4.8 on page 328 we consider some data collected to see if the mean body temperature for humans differs from 98.6°F. In that sample of 50 healthy subjects, the mean body temperature is $\bar{x} = 98.26$°F with standard deviation $s = 0.765$. Test whether there is evidence that the mean body temperature is different from 98.6°F.

Solution The relevant hypotheses are $H_0 : \mu = 98.6$ vs $H_a : \mu \neq 98.6$ where μ is the mean body temperature for all healthy humans. Figure 6.11 shows a dotplot of this sample of body temperatures. The plot raises no concerns about a lack of normality in the population and the sample size ($n = 50$) is quite large, so a t-distribution is appropriate for this test.

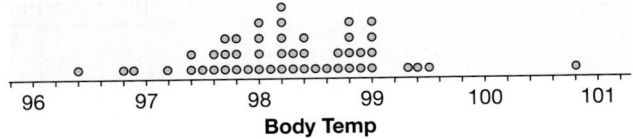

Figure 6.11 *Sample of body temperatures for 50 people*

The t-statistic is computed as

$$t = \frac{\text{Statistic} - \text{Null value}}{SE} = \frac{\bar{x} - \mu_0}{s/\sqrt{n}} = \frac{98.26 - 98.6}{0.765/\sqrt{50}} = -3.14$$

To find the p-value, we use a t-distribution with $50 - 1 = 49$ degrees of freedom and find the proportion that is below -3.14. We see in Figure 6.12 that this is about 0.0014. Since the alternative hypothesis is two-tailed, we double that to find

$$\text{p-value} = 2(0.0014) = 0.0028.$$

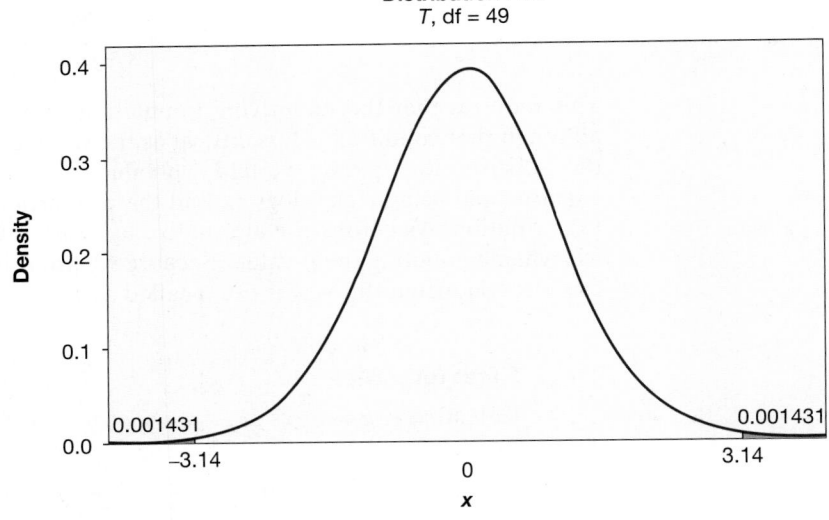

Figure 6.12 *Using a t-distribution with 49 df to find a p-value for t = -3.14*

Based on the small p-value of 0.0028 we have strong evidence to reject the null hypothesis and conclude that the mean body temperature for healthy humans is different from 98.6°F.

We have already found p-values for the body temperature data in two previous examples. From the randomization distribution in Figure 4.32 on page 330 we estimate a two-tailed p-value of 0.0016. In Example 5.8 on page 380 we used a normal distribution after estimating the standard error from the randomization distribution to get a p-value of 0.0014. While the p-value from the t-test is slightly larger than these values, all three p-values indicate that a sample mean as small as 98.26°F would be very unusual to see in a sample of size 50 if the real mean body temperature is 98.6°F. Remember that p-values based on a randomization distribution will also vary as different randomizations are used. The t-test eliminates the need to produce thousands of randomization means. However, if the condition of normality is in doubt, for example with a small sample that is skewed or has significant outliers, we can always return to the randomization procedure as a safe way to assess the strength of evidence.

Example 6.18

The **FloridaLakes** dataset includes information on alkalinity values for 53 Florida lakes. Figure 2.10 on page 72 shows a histogram of the alkalinity values. The mean of the values is $\bar{x} = 37.5$ mg/L with standard deviation $s = 38.20$. Test to see if this sample provides evidence that the average alkalinity of all Florida lakes is greater than 35 mg/L.

Solution

We see in Figure 2.10 that the alkalinity values in this sample are very skewed and don't seem to follow a normal distribution at all. Nonetheless, the sample size of $n = 53$ is large enough that the CLT for means is relevant so a t-test is probably valid. The hypotheses for the test are

$$H_0: \quad \mu = 35$$
$$H_a: \quad \mu > 35$$

where μ represents the mean alkalinity level in all Florida lakes. The t-statistic is

$$t = \frac{\text{Statistic} - \text{Null value}}{SE} = \frac{\bar{x} - \mu_0}{s/\sqrt{n}} = \frac{37.5 - 35}{38.20/\sqrt{53}} = 0.48$$

The proportion beyond 0.48 of a t-distribution with $df = 52$ is 0.317. This is a one-tailed test, so we have

$$\text{p-value} = 0.317$$

The p-value is quite large, so we find no convincing evidence that the average alkalinity levels in all Florida lakes is greater than 35 mg/L.

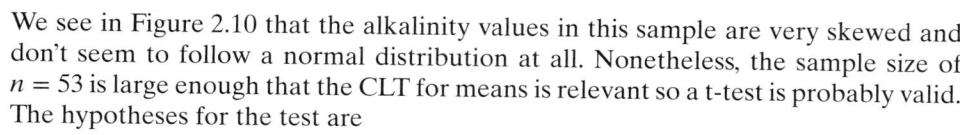

SECTION LEARNING GOALS

You should now have the understanding and skills to:

• Use a formula for *SE* and a t-distribution, when appropriate, to test a hypothesis about a population mean

Exercises for Section **6.2-HT**

SKILL BUILDER 1

In Exercises 6.114 to 6.119, use the t-distribution and the sample results to complete the test of the hypotheses. Use a 5% significance level. Assume the results come from a random sample, and if the sample size is small, assume the underlying distribution is relatively normal.

6.114 Test $H_0: \mu = 15$ vs $H_a: \mu > 15$ using the sample results $\bar{x} = 17.2$, $s = 6.4$, with $n = 40$.

6.115 Test $H_0: \mu = 100$ vs $H_a: \mu < 100$ using the sample results $\bar{x} = 91.7$, $s = 12.5$, with $n = 30$.

6.116 Test $H_0: \mu = 120$ vs $H_a: \mu < 120$ using the sample results $\bar{x} = 112.3$, $s = 18.4$, with $n = 100$.

6.117 Test $H_0: \mu = 10$ vs $H_a: \mu > 10$ using the sample results $\bar{x} = 13.2$, $s = 8.7$, with $n = 12$.

6.118 Test $H_0: \mu = 4$ vs $H_a: \mu \neq 4$ using the sample results $\bar{x} = 4.8$, $s = 2.3$, with $n = 15$.

6.119 Test $H_0: \mu = 500$ vs $H_a: \mu \neq 500$ using the sample results $\bar{x} = 432$, $s = 118$, with $n = 75$.

6.120 How Many Social Ties Do You Have? Most US adults have social ties with a large number of people, including friends, family, co-workers, and other acquaintances. It is nearly impossible for most people to reliably list all the people they know, but using a mathematical model, social analysts estimate that, on average, a US adult has social ties with 634 people.[28] A survey of 1700 randomly selected US adults who are cell phone users finds that the average number of social ties for the cell phone users in the sample was 664 with a standard deviation of 778. Does the sample provide evidence that the average number of social ties for a cell phone user is significantly different from 634, the hypothesized number for all US adults? Define any parameters used and show all details of the test.

6.121 The Autistic Brain Autistic children often have a small head circumference at birth, followed by a sudden and excessive increase in head circumference during the first year of life. A recent study[29] examined the brain tissue in autopsies of seven autistic male children between the ages of 2 and 16. The mean number of neurons in the prefrontal cortex in non-autistic male children of the same age is about 1.15 billion. The prefrontal cortex is the part of the brain most disrupted in autism, as it deals with language and social communication. In the sample of seven autistic children, the mean number of neurons in the prefrontal cortex was 1.94 billion with a standard deviation of 0.50 billion. The values in the sample are not heavily skewed. Use the t-distribution to test whether this sample provides evidence that autistic male children have more neurons (on average) in the prefrontal cortex than non-autistic children. (This study indicates that the causes of autism may be present before birth.)

6.122 Be Nice to Pigeons, As They Remember Your Face In a study[30] conducted in Paris, France, equal amounts of pigeon feed were spread on the ground in two adjacent locations. A person was present in both sites, with one acting hostile and running at the birds to scare them away and the other acting neutral and just observing. The two people were randomly exchanged between the two sites throughout and the birds quickly learned to avoid the hostile

person's site and to eat at the site of the neutral person. At the end of the training session, both people behaved neutrally but the birds continued to remember which one was hostile. In the most interesting part of the experiment, when the two people exchanged coats (orange worn by the hostile one and yellow by the neutral one throughout training), the pigeons were not fooled and continued to recognize and avoid the hostile person. The quantity measured is difference in number of pigeons at the neutral site minus the hostile site. With $n = 32$ measurements, the mean difference in number of pigeons is 3.9 with a standard deviation of 6.8. Test to see if this provides evidence that the mean difference is greater than zero, meaning the pigeons can recognize faces (and hold a grudge!)

6.123 Getting Enough Sleep? It is generally recommended that adults sleep at least 8 hours each night. One of the authors recently asked some of her students (undergraduate and graduate students at Harvard) how many hours each had slept the previous night, curious as to whether her students are getting enough sleep. The data are displayed in Figure 6.13. The 12 students sampled averaged 6.2 hours of sleep with a standard deviation of 1.70 hours. Assuming this sample is representative of all her students, and assuming students need at least 8 hours of sleep a night, does this provide evidence that, on average, her students are not getting enough sleep?

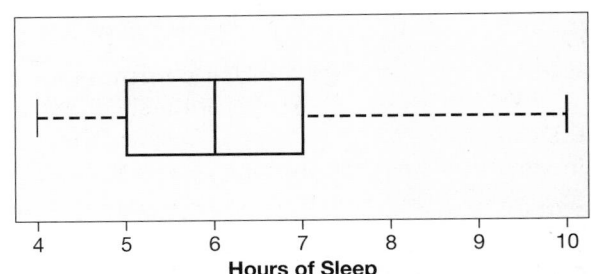

Figure 6.13 *Hours of sleep for Harvard statistics students*

6.124 Football Air Pressure During the National Football League's 2014 AFC championship game, officials measured the air pressure on 11 of the game footballs being used by the New England Patriots. They found that the balls had an average air pressure of 11.1 psi, with a standard deviation of 0.40 psi.

(a) Assuming this is a representative sample of all footballs used by the Patriots in the 2014 season, perform the appropriate test to determine if the average air pressure in footballs used by the

[28]Hampton, K., Goulet, L., Rainie, L., and Purcell, K., "Social Networking Sites and Our Lives," Pew Research Center, *pewresearch.org*, June 16, 2011.
[29]Adapted from Courchesne, E., et al., "Neuron Number and Size in Prefrontal Cortex of Children with Autism," *Journal of the American Medical Association*, November 2011;306(18):2001–2010.
[30]Belguermi, A., "Pigeons discriminate between human feeders," *Animal Cognition*, 2011;14:909–14.

Patriots was significantly less than the allowable limit of 12.5 psi. There is no extreme skewness or outliers in the data, so it is appropriate to use the t-distribution.

(b) Is it fair to assume that this sample is representative of all footballs used by the Patriots during the 2014 season?

6.125 Homes for Sale The dataset **HomesForSale** has data on houses available for sale in three Mid-Atlantic states (NY, NJ, and PA). Table 6.2 shows the mean and standard deviation from the three Mid-Atlantic states, in thousands of dollars. Use this table, the knowledge that within the US the average house sells for about 265 thousand dollars,[31] and a 5% significance level to answer the following questions.

Table 6.2 *Mean housing prices for New York, New Jersey, and Pennsylvania*

State	*n*	Mean	Std. Dev.
New York	30	565.6	697.6
New Jersey	30	388.5	224.7
Pennsylvania	30	249.6	179.3

(a) Is the average cost of a house in New York significantly greater than the US average?

(b) Is the average cost of a house in New Jersey significantly greater than the US average?

(c) Is the average cost of a house in Pennsylvania significantly greater than the US average?

(d) Which state shows the most evidence that the state average is greater than the US average?

6.126 Homes for Sale, Canton We are interested in whether or not the average cost of a house in Canton, NY (the hometown of the Lock family) is significantly different from the national average of $265,000. Table 6.3 and the histogram in Figure 6.14 show the cost (in thousands of dollars) of a sample of 10 houses for sale in Canton. These prices are stored in **HomesForSaleCanton**. Do the appropriate test to determine if the sample provides evidence that the average price of a house in Canton is different from the national average.

Table 6.3 *Price ($1000s) of houses for sale in Canton, NY*

Canton	169	299	325	75	89	105	59	110	168	69

[31]According to the US Census in April 2011.

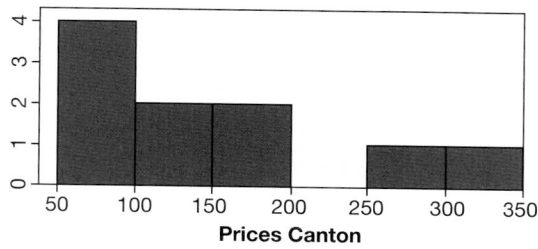

Figure 6.14 *Price ($1000s) of houses for sale in Canton, NY*

6.127 Team Batting Average in Baseball The dataset **BaseballHits** gives 2014 season statistics for all Major League Baseball teams. We treat this as a sample of all MLB teams in all years. Computer output of descriptive statistics for the variable giving the batting average is shown:

```
Descriptive Statistics: BattingAvg
Variable      N     Mean   SE Mean    StDev
BattingAvg   30   0.25110  0.00200  0.01096

  Minimum      Q1     Median      Q3   Maximum
  0.22600   0.24350   0.25300  0.25675  0.27700
```

(a) How many teams are included in the dataset? What is the mean batting average? What is the standard deviation?

(b) Use the descriptive statistics above to conduct a hypothesis test to determine whether there is evidence that average team batting average is different from 0.260. Show all details of the test.

(c) Compare the test statistic and p-value you found in part (b) to the computer output below for the same data:

```
One-Sample T: BattingAvg
Test of mu = 0.26 vs not = 0.26
Variable        N        Mean       StDev
BattingAvg     30      0.25110     0.01096

SE Mean      95% CI              T       P
0.00200   (0.24701, 0.25519)  −4.45   0.000
```

6.128 Are Florida Lakes Acidic or Alkaline? The pH of a liquid is a measure of its acidity or alkalinity. Pure water has a pH of 7, which is neutral. Solutions with a pH less than 7 are acidic while solutions with a pH greater than 7 are basic or alkaline. The dataset **FloridaLakes** gives information, including pH values, for a sample of lakes in Florida. Computer output of descriptive statistics for the pH variable is shown:

```
Descriptive Statistics: pH
Variable    N    N*     Mean    SE Mean    StDev
pH         53     0    6.591     0.177     1.288

  Minimum      Q1     Median      Q3   Maximum
    3.600   5.800      6.800    7.450     9.100
```

(a) How many lakes are included in the dataset? What is the mean pH value? What is the standard deviation?

(b) Use the descriptive statistics above to conduct a hypothesis test to determine whether there is evidence that average pH in Florida lakes is different from the neutral value of 7. Show all details of the test and use a 5% significance level. If there is evidence that it is not neutral, does the mean appear to be more acidic or more alkaline?

(c) Compare the test statistic and p-value you found in part (b) to the computer output below for the same data:

```
One-Sample T: pH
Test of mu = 7 vs not = 7
Variable    N     Mean    StDev   SE Mean
pH          53    6.591   1.288   0.177

   95% CI          T       P
(6.235, 6.946)   -2.31   0.025
```

6.129 Mercury Content in Fish The US Food and Drug Administration has a limit for mercury content in fish of 1.0 ppm (parts per million), while in Canada the limit is 0.5 ppm. Use the variable *Avg-Mercury* in the **FloridaLakes** dataset to test whether there is evidence that average mercury level of fish (large-mouth bass) in Florida lakes is:

(a) Less than 1.0 ppm

(b) Less than 0.5 ppm

6.130 Number of Fouls in a Season by NBA Players The variable *Fouls* in the dataset **NBAPlayers2015** shows the total number of fouls during the 2014–2015 season for all players in the NBA (National Basketball Association) who played at least 24 minutes per game that season. We use this group as a sample of all NBA players in all seasons who play regularly. Use this information to test whether there is evidence that NBA players who play regularly have a mean number of fouls in a season less than 160 (or roughly 2 fouls per game).

6.3-D DISTRIBUTION OF A DIFFERENCE IN PROPORTIONS

In this section we consider the distribution of the difference in proportions between samples taken from two distinct groups. The parameter of interest is $p_1 - p_2$, where p_1 and p_2 represent the proportions in each of the two groups. Note that we are *not* dealing with two different proportions computed for the same group, such as the difference in proportion of voters who choose Candidate A compared to those who choose Candidate B. Even if the data all come from a single sample, we need to identify two groups within the sample and compare the proportions between those two groups.

Example 6.19

One True Love Revisited

In Data 2.1 on page 48 we consider a study that asks whether or not people agree with the statement "There is only one true love for each person." The results for 2625 respondents are broken down by gender: 372 of 1213 males agree and 363 of 1412 females agree. Use this information to estimate the difference in proportions of males and females who agree with the statement.

Solution

We are comparing the proportion who agree with the statement about one true love between two distinct groups in the sample: females and males. The relevant sample proportions are $\hat{p}_f = 363/1412 = 0.257$ and $\hat{p}_m = 372/1213 = 0.307$. To estimate the difference in proportions, $p_f - p_m$, in the population, we use the difference in the sample proportions:

$$\hat{p}_f - \hat{p}_m = 0.257 - 0.307 = -0.050$$

Note that we could have just as easily estimated $p_m - p_f$ with $\hat{p}_m - \hat{p}_f = +0.050$. The interpretation is still the same; we estimate a difference of about 0.05 with males being somewhat more likely to agree.

As always, the key question now is how accurately does the estimate in the sample reflect the true difference in proportions for the population. One way to address this is to use bootstrapping to simulate the difference in proportions. The bootstrap distribution shown in Figure 6.15 is centered around the difference in the original sample (-0.050) and follows the usual bell-shaped pattern. From the bootstrap distribution, we estimate the standard error of the difference in proportions to be about 0.0178.

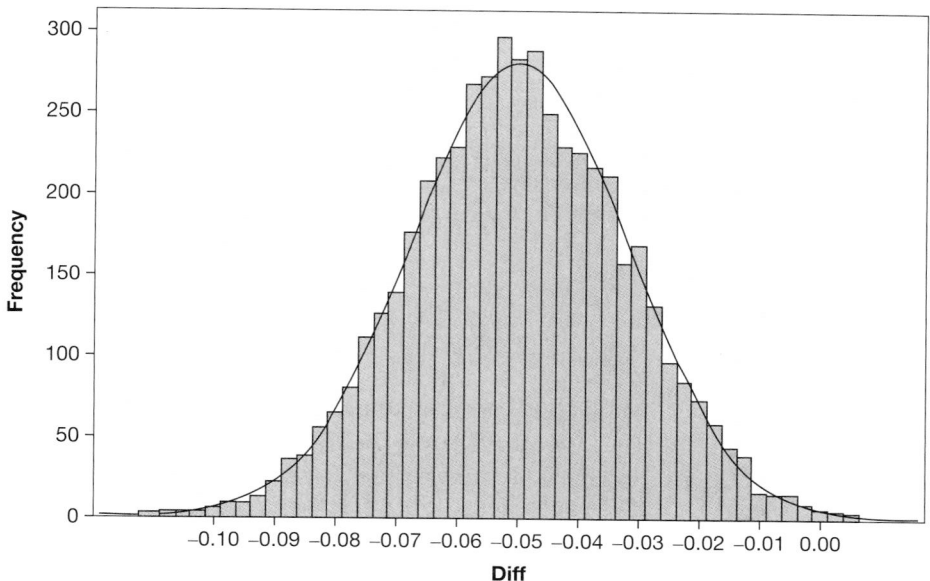

Figure 6.15 *Distribution of $\hat{p}_f - \hat{p}_m$ for 5000 simulations*

In Section 6.1-D we see that the distribution of $\hat{p}$ for a single sample is centered at the population proportion p, has spread given by $SE = \sqrt{\frac{p(1-p)}{n}}$, and approaches a normal curve as the sample size gets large. We extend these ideas to get a similar Central Limit Theorem for differences in proportions for two samples.

Distribution for a Difference in Two Sample Proportions

When choosing random samples of size n_1 and n_2 from populations with proportions p_1 and p_2, respectively, the distribution of the differences in the sample proportions, $\hat{p}_1 - \hat{p}_2$, is centered at the difference in population proportions, $p_1 - p_2$, has standard error given by

$$SE = \sqrt{\frac{p_1(1-p_1)}{n_1} + \frac{p_2(1-p_2)}{n_2}}$$

and is reasonably normally distributed if $n_1 p_1 \geq 10$ and $n_1(1-p_1) \geq 10$ and $n_2 p_2 \geq 10$ and $n_2(1-p_2) \geq 10$.

In finding the standard error for $\hat{p}_1 - \hat{p}_2$, you may be tempted to subtract $\frac{p_2(1-p_2)}{n_2}$ from $\frac{p_1(1-p_1)}{n_1}$ within the square root rather than add those two terms. It's important that the variability of the difference depends on *adding* the variability generated from each of the two samples.

Example 6.20

Figure 6.15 shows the distribution of the difference in sample proportions for samples simulated from the original "one true love" data where the proportions are 0.257 (for females) and 0.307 (for males). Use these sample proportions as approximations of the population proportions and use the sample sizes (1412 females and 1213 males) to compute the standard error and check the conditions for a normal distribution.

Solution   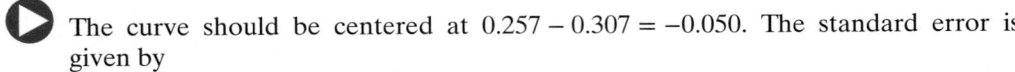 The curve should be centered at $0.257 - 0.307 = -0.050$. The standard error is given by

$$SE = \sqrt{\frac{0.257(1 - 0.257)}{1412} + \frac{0.307(1 - 0.307)}{1213}} = 0.0176$$

Both sample sizes are quite large (well more than 10 people agreeing and disagreeing in each group) so the normal distribution is a reasonable model. This normal curve is plotted in Figure 6.15 and agrees nicely with the simulated results.

SECTION LEARNING GOALS

You should now have the understanding and skills to:

- Recognize when a question calls for comparing proportions for two different groups (as opposed to two proportions within the same group)
- Use a formula to find the standard error for a distribution of differences in sample proportions for two groups
- Recognize when a normal distribution is an appropriate model for a distribution of the differences in two sample proportions

---

Exercises for Section **6.3-D**

SKILL BUILDER 1

In Exercises 6.131 to 6.136, if random samples of the given sizes are drawn from populations with the given proportions:

(a) Find the standard error of the distribution of differences in sample proportions, $\hat{p}_A - \hat{p}_B$.

(b) Determine whether the sample sizes are large enough for the Central Limit Theorem to apply.

6.131 Samples of size 50 from population A with proportion 0.70 and samples of size 75 from population B with proportion 0.60

6.132 Samples of size 300 from population A with proportion 0.15 and samples of size 300 from population B with proportion 0.20

6.133 Samples of size 100 from population A with proportion 0.20 and samples of size 50 from population B with proportion 0.30

6.134 Samples of size 80 from population A with proportion 0.40 and samples of size 60 from population B with proportion 0.10

6.135 Samples of size 40 from population A with proportion 0.30 and samples of size 30 from population B with proportion 0.24

6.136 Samples of size 500 from population A with proportion 0.58 and samples of size 200 from population B with proportion 0.49

TWO GROUPS OR ONE?

In Exercises 6.137 and 6.138, situations comparing two proportions are described. In each case, determine whether the situation involves comparing proportions for two groups or comparing two proportions from the same group. State whether the methods of this section apply to the difference in proportions.

6.137 (a) Compare the proportion of students who use a Windows-based PC to the proportion who use a Mac.

(b) Compare the proportion of students who study abroad between those attending public universities and those at private universities.

(c) Compare the proportion of in-state students at a university to the proportion from outside the state.

(d) Compare the proportion of in-state students who get financial aid to the proportion of out-of-state students who get financial aid.

6.138 (a) In a taste test, compare the proportion of tasters who prefer one brand of cola to the proportion who prefer the other brand.

(b) Compare the proportion of males who voted in the last election to the proportion of females who voted in the last election.

(c) Compare the graduation rate (proportion to graduate) of students on an athletic scholarship to the graduation rate of students who are not on an athletic scholarship.

(d) Compare the proportion of voters who vote in favor of a school budget to the proportion who vote against the budget.

6.3-CI CONFIDENCE INTERVAL FOR A DIFFERENCE IN PROPORTIONS

In Section 5.2 we see that when a distribution of a sample statistic is normally distributed, a confidence interval can be formed using

$$\text{Sample Statistic} \pm z^* \cdot SE$$

where z^* is an appropriate percentile from a standard normal distribution.

In Section 6.3-D we see that we can estimate a difference in proportions, $p_1 - p_2$, for two groups using the difference in the sample proportions from those groups, $\hat{p}_1 - \hat{p}_2$. The standard error found in Section 6.3-D for a difference in proportions uses the population proportions p_1 and p_2. Just as we did for the standard error of a single proportion, we substitute the sample proportions to estimate the standard error for the difference.

Confidence Interval for a Difference in Proportions

The difference in sample proportions based on random samples of size n_1 and n_2, respectively, has

$$\text{Sample statistic} = \hat{p}_1 - \hat{p}_2 \quad \text{and} \quad SE = \sqrt{\frac{\hat{p}_1(1 - \hat{p}_1)}{n_1} + \frac{\hat{p}_2(1 - \hat{p}_2)}{n_2}}$$

If z^* is a standard normal endpoint to give the desired level of confidence, and if the sample sizes are large enough so that $n_1 p_1 \geq 10$ and $n_1(1 - p_1) \geq 10$ and $n_2 p_2 \geq 10$ and $n_2(1 - p_2) \geq 10$, the confidence interval for a difference in population proportions $p_1 - p_2$ is

$$\text{Sample statistic} \pm z^* \cdot SE$$

which, in this case, corresponds to

$$(\hat{p}_1 - \hat{p}_2) \pm z^* \cdot \sqrt{\frac{\hat{p}_1(1 - \hat{p}_1)}{n_1} + \frac{\hat{p}_2(1 - \hat{p}_2)}{n_2}}$$

© slobo/iStockphoto

Crows hold a grudge

DATA 6.3

Crows Never Forget a Face

Biologists studying crows will capture a crow, tag it, and release it. These crows seem to remember the scientists who caught them and will scold them later. A study[32] to examine this effect had several scientists wear a caveman mask while they trapped and tagged seven crows. A control group did not tag any crows and wore a different mask. The two masks did not elicit different reactions from the crows before the tagging. Volunteers then strolled around town wearing one or the other of the two masks. Sure enough, the tagged crows scolded the caveman mask significantly more often. What is really interesting, however, is that even more than two years later and even at sites over a kilometer from the original tagging, crows that did not get tagged and even crows that were not born yet at the time of the tagging continued to scold the caveman mask more than the other mask. The crows had apparently communicated to other crows the fact that the caveman mask was dangerous. It appears that crows hold a grudge for a long time! As one volunteer put it after wearing the caveman mask on a stroll, the reaction to the mask was "quite spectacular. The birds were really raucous, screaming persistently, and it was clear they weren't upset about something in general. They were upset with me." The crows scolded a person wearing a caveman mask in 158 out of 444 encounters with crows, whereas crows scolded a person in a neutral mask in 109 out of 922 encounters. ■

Example 6.21

Use the information in Data 6.3 to find and interpret a 90% confidence interval for the difference in the proportion of crow scoldings between volunteers wearing the caveman mask and those wearing the neutral mask.

---

[32]Cornell, H., Marzluff, J., and Pecoraro, S., "Social learning spreads knowledge about dangerous humans among American crows," *Proceedings of the Royal Society, Biological Sciences*, February 2012; 279(1728): 499–508.

Solution

First, we compute the proportion of scoldings for each of the two groups, using $\hat{p}_c$ for the proportion when volunteers are wearing the caveman mask and $\hat{p}_n$ for the proportion when volunteers are wearing the neutral mask:

$$\hat{p}_c = \frac{158}{444} = 0.356 \qquad \text{and} \qquad \hat{p}_n = \frac{109}{922} = 0.118$$

The estimated difference in proportions is $\hat{p}_c - \hat{p}_n = 0.356 - 0.118 = 0.238$.

The sample sizes are both quite large, well more than 10 scolding and not scolding for each type of mask, so we model the difference in proportions with a normal distribution. For 90% confidence the standard normal endpoint is $z^* = 1.645$. This gives

$$\text{Statistic} \pm z^* \cdot SE$$

$$(\hat{p}_1 - \hat{p}_2) \pm z^* \cdot \sqrt{\frac{\hat{p}_1(1 - \hat{p}_1)}{n_1} + \frac{\hat{p}_2(1 - \hat{p}_2)}{n_2}}$$

$$(0.356 - 0.118) \pm 1.645 \cdot \sqrt{\frac{0.356(1 - 0.356)}{444} + \frac{0.118(1 - 0.118)}{922}}$$

$$0.238 \pm 0.041$$

$$0.197 \text{ to } 0.279$$

We are 90% sure that the proportion of crows that will scold is between 0.197 and 0.279 higher if the volunteer is wearing the caveman mask than if he or she is wearing the neutral mask.

Note that we could easily have switched the order in the previous example and estimated the difference in proportions with $\hat{p}_n - \hat{p}_c = 0.118 - 0.356 = -0.238$. This would only change the signs in the confidence interval and lead to the same interpretation.

Note also that the interpretation includes some direction (caveman mask tends to be more likely to elicit scolding) rather than a less informative statement such as "We are 90% sure that the difference in proportion of crow scoldings between the caveman mask and the neutral mask is between 0.197 and 0.279." In fact, since the interval includes only positive values (and not zero), we can be relatively sure (at least to a 10% significance level) that a hypothesis of no difference in the two proportions would be rejected.

SECTION LEARNING GOALS

You should now have the understanding and skills to:

- Use a formula for *SE* and a normal distribution, when appropriate, to compute a confidence interval for a difference in proportions between two groups

Exercises for Section 6.3-CI

SKILL BUILDER 1

In Exercises 6.139 to 6.142, use the normal distribution to find a confidence interval for a difference in proportions $p_1 - p_2$ given the relevant sample results. Give the best estimate for $p_1 - p_2$, the margin of error, and the confidence interval. Assume the results come from random samples.

6.139 A 95% confidence interval for $p_1 - p_2$ given that $\hat{p}_1 = 0.72$ with $n_1 = 500$ and $\hat{p}_2 = 0.68$ with $n_2 = 300$.

6.140 A 90% confidence interval for $p_1 - p_2$ given that $\hat{p}_1 = 0.20$ with $n_1 = 50$ and $\hat{p}_2 = 0.32$ with $n_2 = 100$.

6.141 A 99% confidence interval for $p_1 - p_2$ given counts of 114 yes out of 150 sampled for Group 1 and 135 yes out of 150 sampled for Group 2.

6.142 A 95% confidence interval for $p_1 - p_2$ given counts of 240 yes out of 500 sampled for Group 1 and 450 yes out of 1000 sampled for Group 2.

6.143 Who Is More Trusting: Internet Users or Non-users? In a randomly selected sample of 2237 US adults, 1754 identified themselves as people who use the Internet regularly while the other 483 indicated that they do not use the Internet regularly. In addition to Internet use, participants were asked if they agree with the statement "most people can be trusted." The results show that 807 of the Internet users agree with this statement, while 130 of the non-users agree.[33] Find and clearly interpret a 90% confidence interval for the difference in the two proportions.

6.144 Gender and Gun Control A survey reported in *Time* magazine included the question "Do you favor a federal law requiring a 15 day waiting period to purchase a gun?" Results from a random sample of US citizens showed that 318 of the 520 men who were surveyed supported this proposed law while 379 of the 460 women sampled said "yes". Use this information to find and interpret a 90% confidence interval for the difference in the proportions of men and women who agree with this proposed law.

[33]Hampton, K., Goulet, L., Rainie, L., and Purcell, K., "Social Networking Sites and Our Lives," Pew Research Center, *pewresearch.org*, June 16, 2011.

6.145 Are Errors Less Likely with Electronic Prescriptions? Errors in medical prescriptions occur, and a study[34] examined whether electronic prescribing may help reduce errors. Two groups of doctors used written prescriptions and had similar error rates before the study. One group switched to e-prescriptions while the other continued to use written prescriptions, and error rates were measured one year later. The results are given in Table 6.4. Find and interpret a 95% confidence interval for the difference in proportion of errors between the two groups. Is it plausible that there is no difference?

Table 6.4 *Are prescription error rates different?*

	Error	No Error	Total
Electronic	254	3594	3848
Written	1478	2370	3848

6.146 Public Libraries and Gender A survey[35] asked a random sample of $n = 2752$ US adults whether they had visited a public library in the last 12 months. The results for males and females are shown in Table 6.5.

(a) Find $\hat{p}_f$ and $\hat{p}_m$, the sample proportions who have visited a public library in the last 12 months, for females and males respectively. What is the difference in sample proportions $\hat{p}_f - \hat{p}_m$?

(b) Find a 95% confidence interval for the difference in population proportions $p_f - p_m$.

Table 6.5 *Have you visited a public library in the last year?*

	Yes	No	Total
Females	726	697	1423
Males	505	824	1329
Total	1231	1521	2752

[34]Kaushal, R., et al., "Electronic Prescribing Improves Medication Safety in Community-Based Office Practices," *Journal of General Internal Medicine*, June 2010; 25(6): 530–536.

[35]Rainie, L., "Libraries and Learning," *www.pewinternet.org/2016/04/07/libraries-and-learning/*, Pew Research Center, April 7, 2016.

(c) Can we conclude from the confidence interval that there is a difference in proportions? If so, are males or females more likely to visit the public library?

6.147 Public Libraries and Children Exercise 6.146 introduces a survey that asked a random sample of $n = 2752$ US adults whether they had visited a public library in the last 12 months. We see in that data that more females than males visit the library, but there is likely a confounding variable. Table 6.6 shows the survey results broken down by whether or not the participant is a parent of a minor. Find and interpret a 90% confidence interval for $p_C - p_N$, the difference in proportion visiting the library between those with children in the house and those without.

Table 6.6 *Have you visited a public library in the last year?*

	Yes	No	Total
Children in house	421	411	832
No children in house	810	1110	1920
Total	1231	1521	2752

6.148 Metal Tags on Penguins and Survival Data 1.3 on page 10 discusses a study designed to test whether applying metal tags is detrimental to penguins. One variable examined is the survival rate 10 years after tagging. The scientists observed that 10 of the 50 metal tagged penguins survived, compared to 18 of the 50 electronic tagged penguins. Construct a 90% confidence interval for the difference in proportion surviving between the metal and electronic tagged penguins ($p_M - p_E$). Interpret the result.

6.149 Metal Tags on Penguins and Breeding Success Data 1.3 on page 10 discusses a study designed to test whether applying metal tags is detrimental to penguins. Exercise 6.148 investigates the survival rate of the penguins. The scientists also studied the breeding success of the metal- and electronic-tagged penguins. Metal-tagged penguins successfully produced offspring in 32% of the 122 total breeding seasons, while the electronic-tagged penguins succeeded in 44% of the 160 total breeding seasons. Construct a 95% confidence interval for the difference in proportion successfully producing offspring ($p_M - p_E$). Interpret the result.

STANDARD ERROR FROM A FORMULA AND A BOOTSTRAP DISTRIBUTION
In Exercises 6.150 and 6.151, use *StatKey* or other technology to generate a bootstrap distribution of sample differences in proportions and find the standard error for that distribution. Compare the result to the value obtained using the formula for the standard error of a difference in proportions from this section.

6.150 Sample A has a count of 30 successes with $n = 100$ and Sample B has a count of 50 successes with $n = 250$.

6.151 Sample A has a count of 90 successes with $n = 120$ and Sample B has a count of 180 successes with $n = 300$.

COMPARING NORMAL AND BOOTSTRAP CONFIDENCE INTERVALS
In Exercises 6.152 and 6.153, find a 95% confidence interval for the difference in proportions two ways: using *StatKey* or other technology and percentiles from a bootstrap distribution, and using the normal distribution and the formula for standard error. Compare the results.

6.152 Difference in proportion who use text messaging, using $\hat{p}_t = 0.87$ with $n = 800$ for teens and $\hat{p}_a = 0.72$ with $n = 2252$ for adults.

6.153 Difference in proportion who favor a gun control proposal, using $\hat{p}_f = 0.82$ for 379 out of 460 females and $\hat{p}_m = 0.61$ for 318 out of 520 for males. (We found a 90% confidence interval for this difference in Exercise 6.144.)

6.154 Survival in the ICU and Infection In the dataset **ICUAdmissions**, the variable *Status* indicates whether the ICU (Intensive Care Unit) patient lived (0) or died (1), while the variable *Infection* indicates whether the patient had an infection (1 for yes, 0 for no) at the time of admission to the ICU. Use technology to find a 95% confidence interval for the difference in the proportion who die between those with an infection and those without.

6.155 Survival in the ICU and Gender The dataset **ICUAdmissions** contains information on patients admitted to an Intensive Care Unit. The variable *Status* indicates whether the patient lived (0) or died (1), while the variable *Sex* indicates whether the patient is male (0) or female (1). Use technology to find a 95% confidence interval for the difference in the proportion who die between males and females.

6.3-HT HYPOTHESIS TEST FOR A DIFFERENCE IN PROPORTIONS

In Section 5.1 we see that, when a randomization distribution is normal, we can compute a p-value using a standard normal curve and a standardized test statistic of the form

$$z = \frac{\text{Sample Statistic} - \text{Null Parameter}}{SE}$$

The sample statistic is computed from the sample data and the null parameter is specified by the null hypothesis, H_0.

When comparing proportions between two groups, the null hypothesis is typically $H_0 : p_1 = p_2$ or, equivalently, $H_0 : p_1 - p_2 = 0$. Thus the "null parameter" is often equal to zero and we use the difference in proportions for two samples, $\hat{p}_1 - \hat{p}_2$, as the "sample statistic":

$$z = \frac{(\hat{p}_1 - \hat{p}_2) - 0}{SE}$$

Once again, we are left with the problem of how to estimate the standard error, SE. From the CLT for a difference in proportions in Section 6.3-D we know that it has the form

$$SE = \sqrt{\frac{p_1(1-p_1)}{n_1} + \frac{p_2(1-p_2)}{n_2}}$$

But what do we use for p_1 and p_2? Recall in Section 6.1-HT that we solved this problem for a single proportion by substituting the proportion specified by the null hypothesis, p_0. In this case, however, the null hypothesis only specifies that the two proportions are equal, but not the value to which they are equal. We can't just substitute $\hat{p}_1$ and $\hat{p}_2$ from the sample, because the null hypothesis says the proportions must be equal.

If the null hypothesis is really true, then the best way to estimate the common proportion is to combine the two samples into one big sample and find its proportion. That is precisely what we do to compute what is known as a *pooled proportion*, denoted $\hat{p}$. We combine the two groups into one big combined group with sample size $n_1 + n_2$ and find the sample proportion $\hat{p}$ for that large group.

Golden Balls: split or steal?

DATA 6.4

Split or Steal?

A popular British TV show called *Golden Balls* features a final round where two contestants each make a decision to either split or steal the final jackpot. If both choose "split," they share the prize, but if one chooses "split" and the other picks "steal," the whole prize goes to the player who steals. If both choose "steal," they both win nothing.

Some researchers[36] collected data from 287 episodes, each with two participants, to give 574 "split" or "steal" decisions. Some results are displayed in Table 6.7 broken down by the age of the participant. ■

Table 6.7 *Split/Steal choice by age group*

Age Group	Split	Steal	Total
Under 40	187	195	382
Over 40	116	76	192
Total	303	271	574

Example 6.22

We use the data in Table 6.7 to test if there is a significant difference in the proportions who choose "split" between younger and older players. Specify the hypotheses and compute the sample proportion within each group as well as the pooled proportion.

Solution

If we let p_1 and p_2 represent the proportions who choose "split" among under and over 40-year-old players, respectively, the relevant hypotheses are $H_0 : p_1 = p_2$ vs $H_a : p_1 \neq p_2$. The sample proportions within each group are

$$\text{Under 40:} \quad \hat{p}_1 = \frac{187}{382} = 0.490 \qquad \text{Over 40:} \quad \hat{p}_2 = \frac{116}{192} = 0.604$$

Under the null hypothesis that the two proportions are the same, we estimate the pooled proportion by combining the two groups into one combined group, in which 303 opted to split out of a total of 574 players. The pooled proportion is

$$\hat{p} = \frac{187 + 116}{382 + 192} = \frac{303}{574} = 0.528$$

Note that the value of the pooled proportion is between the two sample proportions in each group—this is always the case. To be consistent with the null hypothesis of equal proportions, we use the pooled proportion for both samples when computing the standard error for $\hat{p}_1 - \hat{p}_2$.

$$SE = \sqrt{\frac{\hat{p}(1 - \hat{p})}{n_1} + \frac{\hat{p}(1 - \hat{p})}{n_2}}$$

Pay attention to this difference in the standard error for a difference in proportions! For a confidence interval, we use the two sample proportions. For a hypothesis test, in which we use the assumption that the null hypothesis of equal proportions is true, we use a pooled sample proportion in place of both the individual sample proportions.

---

[36]Van den Assem, M., Van Dolder, D., and Thaler, R., "Split or Steal? Cooperative Behavior When the Stakes Are Large," available at SSRN, *http://ssrn.com/abstract=1592456*, February 19, 2011.

Test for Difference in Proportions

To test $H_0 : p_1 = p_2$ vs $H_a : p_1 \neq p_2$ (or a one-tail alternative) based on samples of size n_1 and n_2 from the two groups, the standardized test statistic is

$$z = \frac{\text{Statistic} - \text{Null value}}{SE} = \frac{(\hat{p}_1 - \hat{p}_2) - 0}{\sqrt{\dfrac{\hat{p}(1-\hat{p})}{n_1} + \dfrac{\hat{p}(1-\hat{p})}{n_2}}}$$

where $\hat{p}_1$ and $\hat{p}_2$ are the proportions in the two samples and $\hat{p}$ is the pooled proportion obtained by combining the two samples.

If both samples are sufficiently large (at least 10 successes and failures in each group), the p-value of the test statistic is computed using the standard normal distribution.

Example 6.23

Split or Steal?

Complete the details for testing whether the information in Data 6.4 provide sufficient evidence to suggest that the proportion of participants who choose "split" is different between younger and older players.

Solution

Using the hypotheses and sample proportions from Example 6.22 we compute the standardized test statistic

$$z = \frac{\text{Statistic} - \text{Null value}}{SE} = \frac{(0.490 - 0.604) - 0}{\sqrt{\dfrac{0.528(1-0.528)}{382} + \dfrac{0.528(1-0.528)}{192}}} = \frac{-0.114}{0.04416} = -2.58$$

For the two-tailed alternative, the p-value is twice the proportion in the standard normal tail beyond $z = -2.58$, so we have

$$\text{p-value} = 2(0.0049) = 0.0098.$$

This is a small p-value, indicating that there is little chance that we would see this big a difference in the sample proportions if age really didn't matter. This gives strong evidence that the proportion of younger contestants who choose to cooperate and split the jackpot is lower than the proportion of older contestants willing to cooperate.

SECTION LEARNING GOALS

You should now have the understanding and skills to:

• Use a formula for *SE* and a normal distribution, when appropriate, to test a hypothesis about a difference in proportions between two groups

Exercises for Section **6.3-HT**

SKILL BUILDER 1

In Exercises 6.156 to 6.161:

(a) Find the relevant sample proportions in each group and the pooled proportion.

(b) Complete the hypothesis test using the normal distribution and show all details.

6.156 Test whether there is a difference between two groups in the proportion who voted, if 45 out of a random sample of 70 in Group 1 voted and 56 out of a random sample of 100 in Group 2 voted.

6.157 Test whether patients getting Treatment A are more likely to survive, if 63 out of 82 getting Treatment A survive and 31 out of 67 getting Treatment B survive.

6.158 Test whether people with a specific genetic marker are more likely to have suffered from clinical depression than people without the genetic marker, using the information that 38% of the 42 people in a sample with the genetic marker have had clinical depression while 12% of the 758 people in the sample without the genetic marker have had clinical depression.

6.159 Test whether males are less likely than females to support a ballot initiative, if 24% of a random sample of 50 males plan to vote yes on the initiative and 32% of a random sample of 50 females plan to vote yes.

6.160 Table 6.8 gives flight arrival numbers from a random sample of flights for two airlines. Test whether there is a difference between the two airlines in the percent of flights that arrive late.

Table 6.8 *Arrival times for two airlines*

	Early	On-time	Late	Total
Airline A	133	416	151	700
Airline B	58	355	87	500
Total	191	771	238	1200

6.161 Table 6.9 shows data on whether or not a treatment relieved pain for patients. Test whether the treatment is significantly better than a placebo at relieving pain. The patients were randomly allocated to the two groups and the experiment was double-blind.

Table 6.9 *Is the treatment significantly better than the placebo?*

	Treatment	Placebo	Total
Relieved pain	36	21	57
Did not relieve pain	39	54	93
Total	75	75	150

6.162 Babies Learn Early Who They Can Trust A new study[37] indicates that babies may choose not to learn from someone they don't trust. A group of 60 babies, aged 13 to 16 months, were randomly divided into two groups. Each baby watched an adult express great excitement while looking into a box. The babies were then shown the box and it either had a toy in it (the adult could be trusted) or it was empty (the adult was not reliable). The same adult then turned on a push-on light with her forehead, and the number of babies who imitated the adult's behavior by doing the same thing was counted. The results are in Table 6.10. Test at a 5% level to see if there is evidence that babies are more likely to imitate those they consider reliable.

Table 6.10 *Babies imitate those they trust*

	Imitated	Did not imitate
Reliable	18	12
Unreliable	10	20

6.163 Do Ovulating Women Affect Men's Speech? Studies suggest that when young men interact with a woman who is in the fertile period of her menstrual cycle, they pick up subconsciously on subtle changes in her skin tone, voice, and scent. A study introduced in Exercise B.23 suggests that men may even change their speech patterns around ovulating women. The men were randomly divided into two groups with one group paired with a woman in the fertile phase of her cycle and the other group with a woman in a different stage of her cycle. The same

[37]Wood, J., "Babies Learn Early Who They Can Trust," *PsychCentral, http://psychcentral.com/news/2011/12/07/babies-learn-early-who-they-can-trust/32278.html.*

women were used in the two different stages. For the men paired with a less fertile woman, 38 of the 61 men copied their partner's sentence construction in a task to describe an object. For the men paired with a woman at peak fertility, 30 of the 62 men copied their partner's sentence construction. The experimenters hypothesized that men might be less likely to copy their partner during peak fertility in a (subconscious) attempt to attract more attention to themselves. Use the normal distribution to test at a 5% level whether the proportion of men copying sentence structure is less when the woman is at peak fertility.

6.164 Quebec vs Texas Secession In Example 6.4 on page 408 we analyzed a poll of 800 Quebecers, in which 28% thought that the province of Quebec should separate from Canada. Another poll of 500 Texans found that 18% thought that the state of Texas should separate from the United States.[38]

(a) In the sample of 800 people, about how many Quebecers thought Quebec should separate from Canada? In the sample of 500, how many Texans thought Texas should separate from the US?

(b) In these two samples, what is the pooled proportion of Texans and Quebecers who want to separate?

(c) Can we conclude that the two population proportions differ? Use a two-tailed test and interpret the result.

6.165 Physician's Health Study In the Physician's Health Study, introduced in Data 1.6 on page 37, 22,071 male physicians participated in a study to determine whether taking a daily low-dose aspirin reduced the risk of heart attacks. The men were randomly assigned to two groups and the study was double-blind. After five years, 104 of the 11,037 men taking a daily low-dose aspirin had had a heart attack while 189 of the 11,034 men taking a placebo had had a heart attack.[39] Does taking a daily low-dose aspirin reduce the risk of heart attacks? Conduct the test, and, in addition, explain why we can infer a causal relationship from the results.

MORE BENEFITS OF EATING ORGANIC
Using Data 5.1 on page 375, we find a significant difference in the proportion of fruit flies surviving after

13 days between those eating organic potatoes and those eating conventional (not organic) potatoes. Exercises 6.166 to 6.169 ask you to conduct a hypothesis test using additional data from this study.[40] In every case, we are testing

$$H_0 : \quad p_o = p_c$$
$$H_a : \quad p_o > p_c$$

where p_o and p_c represent the proportion of fruit flies alive at the end of the given time frame of those eating organic food and those eating conventional food, respectively. Also, in every case, we have $n_1 = n_2 = 500$. Show all remaining details in the test, using a 5% significance level.

6.166 Effect of Organic Raisins after 15 Days After 15 days, 320 of the 500 fruit flies eating organic raisins are still alive, while 300 of the 500 eating conventional raisins are still alive.

6.167 Effect of Organic Bananas after 15 Days After 15 days, 345 of the 500 fruit flies eating organic bananas are still alive, while 320 of the 500 eating conventional bananas are still alive.

6.168 Effect of Organic Raisins after 20 Days After 20 days, 275 of the 500 fruit flies eating organic raisins are still alive, while 170 of the 500 eating conventional raisins are still alive.

6.169 Effect of Organic Potatoes after 20 Days After 20 days, 250 of the 500 fruit flies eating organic potatoes are still alive, while 130 of the 500 eating conventional potatoes are still alive.

6.170 Autism and Maternal Antidepressant Use A recent study[41] compared 298 children with Autism Spectrum Disorder to 1507 randomly selected control children without the disorder. Of the children with autism, 20 of the mothers had used antidepressant drugs during the year before pregnancy or the first trimester of pregnancy. Of the control children, 50 of the mothers had used the drugs.

(a) Is there a significant association between prenatal exposure to antidepressant medicine and the risk of autism? Test whether the results are significant at the 5% level.

(b) Can we conclude that prenatal exposure to antidepressant medicine increases the risk of autism in the child? Why or why not?

[38]"In Texas, 31% Say State Has Right to Secede From U.S., But 75% Opt To Stay," Rasmussen Reports, April 17, 2009.
[39]"Final report on the aspirin component of the ongoing Physicians' Health Study. Steering Committee of the Physicians' Health Study Research Group," *New England Journal of Medicine,* 1989;321(3):129–35.
[40]Proportions approximated from information given in the paper.
[41]Croen, L., Grether, J., Yoshida, C., Odouli, R., and Hendrick, V., "Antidepressant Use During Pregnancy and Childhood Autism Spectrum Disorders," *Archives of General Psychiatry,* 2011; 68(11):1104–12.

(c) The article describing the study contains the sentence "No increase in risk was found for mothers with a history of mental health treatment in the absence of prenatal exposure to selective serotonin reuptake inhibitors [antidepressants]." Why did the researchers conduct this extra analysis?

6.171 Electrical Stimulation for Fresh Insight? In Exercise 4.111 on page 301, we used the data in Table 6.11 to conduct a randomization test to see if there was evidence that electrical stimulation of the brain helped people solve a problem that requires fresh insight. Explain why it would not be appropriate to conduct this test using the normal distribution and the formulas in this section.

Table 6.11 *Does electrical brain stimulation bring fresh insight to a problem?*

Treatment	Solved	Not Solved
Sham	4	16
Electrical	12	8

6.172 Green Tea and Prostate Cancer A preliminary study suggests a benefit from green tea for those at risk of prostate cancer.[42] The study involved 60 men with PIN lesions, some of which turn into prostate cancer. Half the men, randomly determined, were given 600 mg a day of a green tea extract while the other half were given a placebo. The study was double-blind, and the results after one year are shown in Table 6.12. Does the sample provide evidence that taking green tea extract reduces the risk of developing prostate cancer?

Table 6.12 *Does green tea extract reduce the risk of prostate cancer?*

Treatment	Cancer	No Cancer
Green tea	1	29
Placebo	9	21

6.173 Can Malaria Parasites Control Mosquito Behavior? Are malaria parasites able to control mosquito behavior to their advantage? A study[43] investigated this question by taking mosquitos and giving them the opportunity to have their first "blood meal" from a mouse. The mosquitoes were randomized to either eat from a mouse infected with malaria or an uninfected mouse. At several time points after this, mosquitoes were put into a cage with a human and it was recorded whether or not each mosquito approached the human (presumably to bite, although mosquitoes were caught before biting). Once infected, the malaria parasites in the mosquitoes go through two stages: the Oocyst stage in which the mosquito has been infected but is not yet infectious to others and then the Sporozoite stage in which the mosquito is infectious to others. Malaria parasites would benefit if mosquitoes sought risky blood meals (such as biting a human) *less* often in the Oocyst stage (because mosquitos are often killed while attempting a blood meal) and *more* often in the Sporozoite stage after becoming infectious (because this is one of the primary ways in which malaria is transmitted). Does exposing mosquitoes to malaria actually impact their behavior in this way?

(a) In the Oocyst stage (after eating from mouse but before becoming infectious), 20 out of 113 mosquitoes in the group exposed to malaria approached the human and 36 out of 117 mosquitoes in the group not exposed to malaria approached the human. Calculate the z-statistic.

(b) Calculate the p-value for testing whether this provides evidence that the proportion of mosquitoes in the Oocyst stage approaching the human is lower in the group exposed to malaria.

(c) In the Sporozoite stage (after becoming infectious), 37 out of 149 mosquitoes in the group exposed to malaria approached the human and 14 out of 144 mosquitoes in the group not exposed to malaria approached the human. Calculate the z-statistic.

(d) Calculate the p-value for testing whether this provides evidence that the proportion of mosquitoes in the Sporozoite stage approaching the human is higher in the group exposed to malaria.

(e) Based on your p-values, make conclusions about what you have learned about mosquito behavior, stage of infection, and exposure to malaria or not.

(f) Can we conclude that being exposed to malaria (as opposed to not being exposed to malaria) *causes* these behavior changes in mosquitoes? Why or why not?

[42] Schardt, D., "What's all the fuss about green tea?" *Nutrition Action Health Letter*, Center for Science in the Public Interest, May 2011, p. 10.

[43] Cator, L.J., George, J., Blanford, S., Murdock, C.C., Baker, T.C., Read, A.F., Thomas, M.B., (2013). 'Manipulation' without the parasite: altered feeding behaviour of mosquitoes is not dependent on infection with malaria parasites. *Proc R Soc B*, 280: 20130711. Data from: 'Manipulation' without the parasite: altered feeding behaviour of mosquitoes is not dependent on infection with malaria parasites. Dryad Digital Repository. *http://dx.doi.org/10.5061/dryad.j4n89.*

HORMONE REPLACEMENT THERAPY

Exercises 6.174 through 6.177 refer to a study on hormone replacement therapy. Until 2002, hormone replacement therapy (HRT), taking hormones to replace those the body no longer makes after menopause, was commonly prescribed to postmenopausal women. However, in 2002 the results of a large clinical trial[44] were published, causing most doctors to stop prescribing it and most women to stop using it, impacting the health of millions of women around the world. In the experiment, 8506 women were randomized to take HRT and 8102 were randomized to take a placebo. Table 6.13 shows the observed counts for several conditions over the five years of the study. (*Note:* The planned duration was 8.5 years. If Exercises 6.174 through 6.177 are done correctly, you will notice that several of the p-values are just below 0.05. The study was terminated as soon as HRT was shown to significantly increase risk [using a significance level of $\alpha = 0.05$], because at that point it was unethical to continue forcing women to take HRT).

6.174 Does HRT influence the chance of a woman getting cardiovascular disease?

6.175 Does HRT influence the chance of a woman getting invasive breast cancer?

6.176 Does HRT influence the chance of a woman getting cancer of any kind?

6.177 Does HRT influence the chance of a woman having a fracture?

6.178 Infections in the ICU and Gender In the dataset **ICUAdmissions**, the variable *Infection* indicates whether the ICU (Intensive Care Unit) patient had an infection (1) or not (0) and the variable *Sex* gives the gender of the patient (0 for males and 1 for females.) Use technology to test at a 5% level whether there is a difference between males and females in the proportion of ICU patients with an infection.

6.179 Surgery in the ICU and Gender In the dataset **ICUAdmissions**, the variable *Service* indicates whether the ICU (Intensive Care Unit) patient had surgery (1) or other medical treatment (0) and the variable *Sex* gives the gender of the patient (0 for males and 1 for females.) Use technology to test at a 5% level whether there is a difference between males and females in the proportion of ICU patients who have surgery.

Table 6.13 *Counts for several conditions within the HRT group and the placebo group*

Condition	HRT Group	Placebo Group
Cardiovascular Disease	164	122
Invasive Breast Cancer	166	124
Cancer (all)	502	458
Fractures	650	788

[44]Rossouw, J., et al., "Risks and benefits of estrogen plus progestin in healthy postmenopausal women: principal results from the women's health initiative randomized controlled trial," *Journal of the American Medical Association*, 2002; 288(3):321–333.

6.4-D DISTRIBUTION OF A DIFFERENCE IN MEANS

In this section we consider the distribution of the differences in means between samples taken from two distinct groups. Those groups might be two different populations, two subsets within a single sample identified by a categorical variable, or different treatments in an experiment. The parameter of interest is $\mu_1 - \mu_2$, where μ_1 and μ_2 represent the "true" means in each of the two groups.

In Section 6.2-D we saw that the distribution of $\bar{x}$ for a single sample is centered at the population mean μ, has spread given by $SE = \sigma/\sqrt{n}$, and approaches a normal curve as the sample size gets large. Notice that we can also write the standard error as $SE = \sqrt{\sigma^2/n}$. This leads us to the standard error for differences in means:

$$SE = \sqrt{\frac{\sigma_1^2}{n_1} + \frac{\sigma_2^2}{n_2}}$$

The t-Distribution

Recall from the work with a single mean in Section 6.2-D that we need to make a small adjustment when working with means, since we almost certainly do not know the population standard deviations. Fortunately, we can apply a similar remedy in the case of differences in two sample means, namely to substitute the sample standard deviations when computing the standard error and then use the t-distribution instead of the normal distribution when finding endpoints for confidence intervals or p-values for tests. Also, as with a single mean, we need to check each sample for heavy skewness or extreme outliers that might indicate serious departures from normality, especially when either sample size is small.

However, we have one additional difficulty in the two-sample case: What should we use for the degrees of freedom? Recall that for a single sample we use $n - 1$ degrees of freedom, but now we have two (possibly different) sample sizes. One solution is a complicated formula, called Satterwaithe's approximation, for estimating the degrees of freedom, which is used in many statistical software packages. As a conservative approach, in this text we will use the smaller of the two degrees of freedom, either $n_1 - 1$ or $n_2 - 1$.

The Distribution of Differences in Sample Means

When choosing random samples of size n_1 and n_2 from populations with means μ_1 and μ_2, respectively, the distribution of the differences in the two sample means, $\bar{x}_1 - \bar{x}_2$, is centered at the difference in population means, $\mu_1 - \mu_2$, and has standard error estimated by

$$SE = \sqrt{\frac{s_1^2}{n_1} + \frac{s_2^2}{n_2}}$$

The standardized differences in sample means follow a t-distribution with degrees of freedom approximately equal to the smaller of $n_1 - 1$ and $n_2 - 1$.

For small sample sizes ($n_1 < 30$ or $n_2 < 30$), the t-distribution is only a good approximation if the underlying population has a distribution that is approximately normal.

 In finding the standard error for the difference $\bar{x}_1 - \bar{x}_2$, you may be tempted to subtract s_2^2/n_2 from s_1^2/n_1 within the square root rather than add those two terms. It's important that the variability of the difference depends on *adding* the variability generated from each of the two samples.

Using the Formula for Standard Error

Example 6.24

Figure 6.16 shows two different simulation distributions for differences in means. In each case, use the formula to compute the standard error, and compare the result to that arrived at by simulation methods.

(a) In Example 3.25, we consider mean number of hours per week spent exercising, between males and females, from the **ExerciseHours** dataset. The mean for the 20 males is $\bar{x}_M = 12.4$ hours with standard deviation $s_M = 8.80$. The mean for the 30 females is $\bar{x}_F = 9.4$ hours with standard deviation $s_F = 7.41$. From the bootstrap distribution of differences in sample means in Figure 6.16(a), we have $SE = 2.34$.

(b) In Data 4.5, we look at an experiment to compare the leniency scores assigned to students charged with an infraction, where the students charged had either a smiling or a neutral expression. The data in **Smiles** show that the 34 scores given

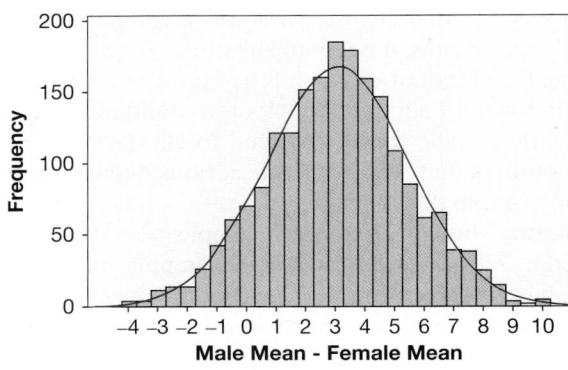

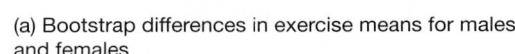

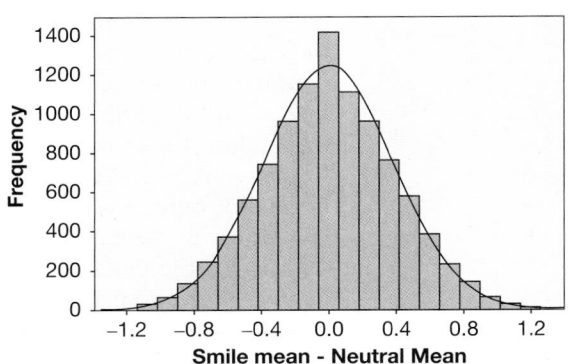

(a) Bootstrap differences in exercise means for males and females

(b) Randomization differences in leniency means when $\mu_s = \mu_c$

Figure 6.16 *Simulation distributions for differences in two sample means*

to smiling faces had a mean leniency score of 4.91 with a standard deviation of 1.68, while the 34 neutral expressions had a mean score of 4.12 and standard deviation of 1.52. From the randomization differences in Figure 6.16(b) we find $SE = 0.40$.

Solution (a) For the bootstrap differences in mean exercise times the standard error should be

$$SE = \sqrt{\frac{8.80^2}{20} + \frac{7.41^2}{30}} = 2.39$$

(b) For the randomization differences in leniency scores the standard error should be

$$SE = \sqrt{\frac{1.68^2}{34} + \frac{1.52^2}{34}} = 0.39$$

These are very similar to the standard errors arrived at using simulation methods.

Example 6.25

Is it appropriate to use a t-distribution for conducting inference on the data on male and female exercise hours described in Example 6.24(a)? If it is appropriate, what degrees of freedom should we use?

Solution The sample size for males is $n_M = 20$ and the sample size for females is $n_F = 30$. Since the sample size for males is less than 30, we check the distribution. A dotplot of the number of hours spent exercising for the 20 males is shown in Figure 6.17. It is not extremely skewed and does not have extreme outliers, so a t-distribution is appropriate.

The degrees of freedom is the smaller of $n_M - 1$ or $n_F - 1$. Since n_M is less than n_F, we have $df = n_M - 1 = 20 - 1 = 19$.

Figure 6.17 *Male exercise times*

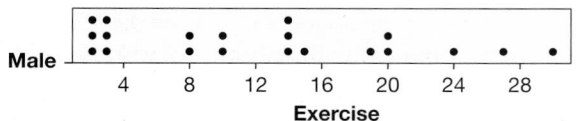

In the next two sections, we explore the use of the t-distribution in more detail—first for computing a confidence interval for a difference in two means and then for computing a p-value to test a hypothesis that two means are equal.

SECTION LEARNING GOALS

You should now have the understanding and skills to:

▶ • Use a formula to find the standard error for a distribution of differences in sample means for two groups

▶ • Recognize when a t-distribution is an appropriate model for a distribution of the standardized difference in two sample means

Exercises for Section 6.4-D

SKILL BUILDER 1

In Exercises 6.180 to 6.183, random samples of the given sizes are drawn from populations with the given means and standard deviations. For each scenario, use the formula to find the standard error of the distribution of differences in sample means, $\bar{x}_1 - \bar{x}_2$.

6.180 Samples of size 100 from Population 1 with mean 87 and standard deviation 12 and samples of size 80 from Population 2 with mean 81 and standard deviation 15

6.181 Samples of size 25 from Population 1 with mean 6.2 and standard deviation 3.7 and samples of size 40 from Population 2 with mean 8.1 and standard deviation 7.6

6.182 Samples of size 50 from Population 1 with mean 3.2 and standard deviation 1.7 and samples of size 50 from Population 2 with mean 2.8 and standard deviation 1.3

6.183 Samples of size 300 from Population 1 with mean 75 and standard deviation 18 and samples of

size 500 from Population 2 with mean 83 and standard deviation 22

SKILL BUILDER 2

Use a t-distribution to answer the questions in Exercises 6.184 to 6.187. Assume the samples are random samples from distributions that are reasonably normally distributed, and that a t-statistic will be used for inference about the difference in sample means. State the degrees of freedom used.

6.184 Find the endpoints of the t-distribution with 2.5% beyond them in each tail if the samples have sizes $n_1 = 15$ and $n_2 = 25$.

6.185 Find the endpoints of the t-distribution with 5% beyond them in each tail if the samples have sizes $n_1 = 8$ and $n_2 = 10$.

6.186 Find the proportion in a t-distribution less than -1.4 if the samples have sizes $n_1 = 30$ and $n_2 = 40$.

6.187 Find the proportion in a t-distribution above 2.1 if the samples have sizes $n_1 = 12$ and $n_2 = 12$.

6.4-CI CONFIDENCE INTERVAL FOR A DIFFERENCE IN MEANS

In Section 5.2 we see that when a distribution of a statistic is normally distributed, a confidence interval can be formed using

$$\text{Sample Statistic} \pm z^* \cdot SE$$

where z^* is an appropriate percentile from a standard normal distribution and SE is the standard error of the statistic.

In Section 6.4-D we see that we can estimate the standard error for a difference in sample means using

$$SE = \sqrt{\frac{s_1^2}{n_1} + \frac{s_2^2}{n_2}}$$

where n_1 and n_2 are the sample sizes and s_1 and s_2 are the standard deviations from the respective samples. However, when we use the sample standard deviations rather than the (unknown) population standard deviations in computing the SE, the standardized statistic follows a t-distribution rather than a standard normal (provided the underlying populations are reasonably normal).

We combine these facts to produce a formula for a confidence interval for a difference in means.

> **Confidence Interval for a Difference in Means**
>
> A difference in sample means based on random samples of sizes n_1 and n_2 has
>
> $$\text{Sample statistic} = \bar{x}_1 - \bar{x}_2 \quad \text{and} \quad SE = \sqrt{\frac{s_1^2}{n_1} + \frac{s_2^2}{n_2}}$$
>
> where $\bar{x}_1$ and $\bar{x}_2$ are the means and s_1 and s_2 are the standard deviations for the respective samples. If t^* is an endpoint chosen from a t-distribution with df equal to the smaller of $n_1 - 1$ or $n_2 - 1$ to give the desired level of confidence, and if the distribution of the populations are approximately normal or the sample sizes are large ($n_1 \geq 30$ and $n_2 \geq 30$), the confidence interval for the difference in population means, $\mu_1 - \mu_2$, is
>
> $$\text{Sample statistic} \pm t^* \cdot SE$$
>
> which, in this case, corresponds to
>
> $$(\bar{x}_1 - \bar{x}_2) \pm t^* \cdot \sqrt{\frac{s_1^2}{n_1} + \frac{s_2^2}{n_2}}$$

Example 6.26

Atlanta vs St. Louis Commute Times

In Data 3.3 on page 229, we introduce a sample of 500 commuting times for people who work in Atlanta, and in Exercise 3.132 we introduce a sample of 500 commuting times for people who work in St. Louis. The data are in **CommuteAtlanta** and **CommuteStLouis**. The summary statistics for these two samples are shown below and Figure 6.18 displays boxplots of the commute times from each city.

Group	n	Mean	Std. Dev.
Atlanta	500	29.11	20.72
St. Louis	500	21.97	14.23

Use these data to compute a 90% confidence interval for the difference in mean commute time between Atlanta and St. Louis.

Solution   From the boxplots we see that both samples are right skewed and have numerous outliers. If these were smaller samples, we would be hesitant to model the difference in means with a t-distribution. However, with these large samples ($n_1 = n_2 = 500$) we can go ahead and use the t-distribution to find the interval.

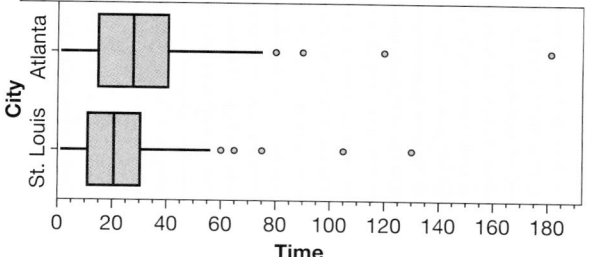

Figure 6.18 *Commute times in Atlanta and St. Louis*

Each sample has $500 - 1 = 499$ degrees of freedom, so we find the t^* value with an area of 0.05 in the tail beyond it in a t-distribution with 499 degrees of freedom. This value is $t^* = 1.648$ and is very close to the standard normal percentile of $z^* = 1.645$.

Substituting into the formula for a confidence interval for a difference in means we get

$$(\bar{x}_1 - \bar{x}_2) \quad \pm \quad t^* \sqrt{\frac{s_1^2}{n_1} + \frac{s_2^2}{n_2}}$$

$$(29.11 - 21.97) \quad \pm \quad 1.648 \sqrt{\frac{20.72^2}{500} + \frac{14.23^2}{500}}$$

$$7.14 \quad \pm \quad 1.85$$

$$5.29 \quad \text{to} \quad 8.99$$

Based on these results, we are 90% sure that the mean commute time in Atlanta is between 5.29 and 8.99 minutes more than the mean commute time in St. Louis.

Note that we could easily have switched the order in the previous example and estimated the difference in means with $\bar{x}_{stl} - \bar{x}_{atl} = 21.97 - 29.11 = -7.14$. This would only change the signs in the confidence interval and lead to the same interpretation.

Note also that the interpretation includes some direction (commute times tend to be longer in Atlanta than St. Louis) rather than a less informative statement such as "We are 90% sure that the difference in mean commute time between Atlanta and St. Louis is between 5.29 and 8.99 minutes." In fact, since the interval includes only positive values (and not zero), we can be relatively sure (at least to a 10% significance level) that a hypothesis of no difference in the two means would be rejected.

In Exercise 3.132 on page 254 we used the bootstrap distribution in Figure 3.34 to estimate this 90% confidence interval for the difference in mean commute time between the two cities. The percentiles of that distribution give an interval from 5.21 to 8.95 minutes, which is very similar to the result from the t-interval in Example 6.26.

SECTION LEARNING GOALS

You should now have the understanding and skills to:

- Use a formula for *SE* and a t-distribution, when appropriate, to compute a confidence interval for the difference in means between two groups

Exercises for Section 6.4-CI

SKILL BUILDER 1

In Exercises 6.188 to 6.191, use the t-distribution to find a confidence interval for a difference in means $\mu_1 - \mu_2$ given the relevant sample results. Give the best estimate for $\mu_1 - \mu_2$, the margin of error, and the confidence interval. Assume the results come from random samples from populations that are approximately normally distributed.

6.188 A 95% confidence interval for $\mu_1 - \mu_2$ using the sample results $\bar{x}_1 = 75.2$, $s_1 = 10.7$, $n_1 = 30$ and $\bar{x}_2 = 69.0$, $s_2 = 8.3$, $n_2 = 20$.

6.189 A 90% confidence interval for $\mu_1 - \mu_2$ using the sample results $\bar{x}_1 = 10.1$, $s_1 = 2.3$, $n_1 = 50$ and $\bar{x}_2 = 12.4$, $s_2 = 5.7$, $n_2 = 50$.

6.190 A 99% confidence interval for $\mu_1 - \mu_2$ using the sample results $\bar{x}_1 = 501$, $s_1 = 115$, $n_1 = 400$ and $\bar{x}_2 = 469$, $s_2 = 96$, $n_2 = 200$.

6.191 A 95% confidence interval for $\mu_1 - \mu_2$ using the sample results $\bar{x}_1 = 5.2$, $s_1 = 2.7$, $n_1 = 10$ and $\bar{x}_2 = 4.9$, $s_2 = 2.8$, $n_2 = 8$.

IS FAST FOOD MESSING WITH YOUR HORMONES?

Exercises 6.192 and 6.193 examine the results of a study[45] investigating whether fast food consumption increases one's concentration of phthalates, an ingredient in plastics that has been linked to multiple health problems including hormone disruption. The study included 8,877 people who recorded all the food they ate over a 24-hour period and then provided a urine sample. Two specific phthalate byproducts were measured (in ng/mL) in the urine: DEHP and DiNP. Find and interpret a 95% confidence interval for the difference, $\mu_F - \mu_N$, in mean concentration between people who have eaten fast food in the last 24 hours and those who haven't.

6.192 The mean concentration of DEHP in the 3095 participants who had eaten fast food was $\bar{x}_F = 83.6$ with $s_F = 194.7$ while the mean for the 5782 participants who had not eaten fast food was $\bar{x}_N = 59.1$ with $s_N = 152.1$.

6.193 The mean concentration of DiNP in the 3095 participants who had eaten fast food was $\bar{x}_F = 10.1$ with $s_F = 38.9$ while the mean for the 5782 participants who had not eaten fast food was $\bar{x}_N = 7.0$ with $s_N = 22.8$.

6.194 Does Red Increase Men's Attraction to Women? Exercise 1.99 on page 44 described a study[46] which examines the impact of the color red on how attractive men perceive women to be. In the study, men were randomly divided into two groups and were asked to rate the attractiveness of women on a scale of 1 (not at all attractive) to 9 (extremely attractive). Men in one group were shown pictures of women on a white background while the men in the other group were shown the same pictures of women on a red background. The results are shown in Table 6.14 and the data for both groups are reasonably symmetric with no outliers. To determine the possible effect size of the red background over the white, find and interpret a 90% confidence interval for the difference in mean attractiveness rating.

Table 6.14 *Does red increase men's attraction to women?*

Color	n	$\bar{x}$	s
Red	15	7.2	0.6
White	12	6.1	0.4

6.195 Dark Chocolate for Good Health A study[47] examines chocolate's effects on blood vessel function in healthy people. In the randomized, double-blind, placebo-controlled study, 11 people received 46 grams (1.6 ounces) of dark chocolate (which is naturally flavonoid-rich) every day for two weeks, while a control group of 10 people received a placebo consisting of dark chocolate with low flavonoid content. Participants had their vascular health measured (by means of flow-mediated dilation) before and after the two-week study. The increase over the two-week period was measured, with larger numbers indicating greater vascular health. For the group getting the good dark chocolate, the mean increase was 1.3 with a standard deviation of 2.32, while the control group had a mean change of −0.96 with a standard deviation of 1.58.

[45]Zota, A.R., Phillips, C.A., and Mitro, S.D., "Recent Fast Food Consumption and Bisphenol A and Phthalates Exposure among the U.S. Population in NHANES, 2003–2010," *Environmental Health Perspectives*, April 13, 2016.

[46]Data approximated from information given in Elliot, A. and Niesta, D., "Romantic Red: Red Enhances Men's Attraction to Women," *Journal of Personality and Social Psychology*, 2008; 95(5):1150–64.

[47]Engler, M., et. al., "Flavonoid-rich dark chocolate improves endothelial function and increases plasma epicatechin concentrations in healthy adults," *Journal of the American College of Nutrition*, 2004 Jun; 23(3):197–204.

(a) Explain what "randomized, double-blind, placebo-controlled study" means.

(b) Find and interpret a 95% confidence interval for the difference in means between the two groups. Be sure to clearly define the parameters you are estimating. You may assume that neither sample shows significant departures from normality.

(c) Is it plausible that there is "no difference" between the two kinds of chocolate? Justify your answer using the confidence interval found in part (b).

6.196 Close Confidants and Social Networking Sites Exercise 6.93 introduces a study[48] in which 2006 randomly selected US adults (age 18 or older) were asked to give the number of people in the last six months "with whom you discussed matters that are important to you." The average number of close confidants for the full sample was 2.2. In addition, the study asked participants whether or not they had a profile on a social networking site. For the 947 participants using a social networking site, the average number of close confidants was 2.5 with a standard deviation of 1.4, and for the other 1059 participants who do not use a social networking site, the average was 1.9 with a standard deviation of 1.3. Find and interpret a 90% confidence interval for the difference in means between the two groups.

6.197 Effect of Splitting the Bill Exercise 2.153 on page 105 describes a study to compare the cost of restaurant meals when people pay individually versus splitting the bill as a group. In the experiment half of the people were told they would each be responsible for individual meal costs and the other half were told the cost would be split equally among the six people at the table. The 24 people paying individually had a mean cost of 37.29 Israeli shekels with a standard deviation of 12.54, while the 24 people splitting the bill had a higher mean cost

of 50.92 Israeli shekels with a standard deviation of 14.33. The raw data can be found in **SplitBill** and both distributions are reasonably bell-shaped. Use this information to find and interpret a 95% confidence interval for the difference in mean meal cost between these two situations.

IMPACT OF COLLEGE ROOMMATES ON GRADES
In Exercises 6.198 to 6.202, we investigate answers to the questions: How much of an effect does your roommate have on your grades? In particular, does it matter whether your roommate brings a videogame to college? Exercise B.5 on page 350 introduces a study involving $n = 210$ first-year students who were randomly assigned a roommate. Table 6.15 gives summary statistics on grade point average (GPA) for the first semester depending on whether the student and/or the roommate brought a videogame to campus.

6.198 Considering only students who do *not* bring a videogame to campus, find and interpret a 95% confidence interval for the difference in mean GPA between students whose roommate does not bring a videogame and those whose roommate does bring a videogame. Comment on the effect on these students of having a roommate bring a videogame.

6.199 Considering only students who *do* bring a videogame to campus, find and interpret a 95% confidence interval for the difference in mean GPA between students whose roommate does not bring a videogame and those whose roommate does bring a videogame. Comment on the effect on these students of having a roommate bring a videogame.

6.200 Considering only students whose roommate does *not* bring a videogame to campus, find and interpret a 95% confidence interval for the difference in mean GPA between students who bring a videogame and those who do not bring a videogame. Comment on the effect on these students of bringing a videogame.

6.201 Considering only students whose roommate *does* bring a videogame to campus, find and interpret a 95% confidence interval for the difference

[48]Hampton, K., Goulet, L., Rainie, L., and Purcell, K., "Social Networking Sites and Our Lives," Pew Research Center, *pewresearch.org*, June 16, 2011.

Table 6.15 *Videogames and GPA*

Student brought videogame	Roommate brought videogame	Sample Size	Mean GPA	Std. Dev.
No	No	88	3.128	0.590
Yes	No	44	3.039	0.689
No	Yes	38	2.932	0.699
Yes	Yes	40	2.754	0.639

in mean GPA between students who bring a videogame and those who do not bring a videogame. Comment on the effect on these students of bringing a videogame.

6.202 We consider the effect of neither student bringing videogames compared to both students bringing them, still using the data in Table 6.15.

(a) Find and interpret a 95% confidence interval for the difference in means between students in rooms in which neither the student nor the roommate brings a videogame and students in rooms in which both the student and the roommate bring a videogame. Comment on the effect of videogames on GPA.

(b) Can we conclude that bringing videogames to campus reduces GPA? Why or why not?

STANDARD ERROR FROM A FORMULA AND A BOOTSTRAP DISTRIBUTION
In Exercises 6.203 and 6.204, use *StatKey* or other technology to generate a bootstrap distribution of sample differences in means and find the standard error for that distribution. Compare the result to the standard error given by the Central Limit Theorem, using the sample standard deviations as estimates of the population standard deviations.

6.203 Difference in mean commuting time (in minutes) between commuters in Atlanta and commuters in St. Louis, using $n_1 = 500$, $\bar{x}_1 = 29.11$, and $s_1 = 20.72$ for Atlanta and $n_2 = 500$, $\bar{x}_2 = 21.97$, and $s_2 = 14.23$ for St. Louis.

6.204 Difference in mean commuting distance (in miles) between commuters in Atlanta and commuters in St. Louis, using $n_1 = 500$, $\bar{x}_1 = 18.16$, and $s_1 = 13.80$ for Atlanta and $n_2 = 500$, $\bar{x}_2 = 14.16$, and $s_2 = 10.75$ for St. Louis.

6.205 Who Exercises More: Males or Females? The dataset **StudentSurvey** has information from males and females on the number of hours spent exercising in a typical week. Computer output of descriptive statistics for the number of hours spent exercising, broken down by gender, is given:

```
Descriptive Statistics: Exercise
Variable   Gender    N    Mean    StDev
Exercise   F        168   8.110   5.199
           M        193   9.876   6.069

Minimum    Q1    Median    Q3    Maximum
 0.000   4.000    7.000  12.000  27.000
 0.000   5.000   10.000  14.000  40.000
```

(a) How many females are in the dataset? How many males?

(b) In the sample, which group exercises more, on average? By how much?

(c) Use the summary statistics to compute a 95% confidence interval for the difference in mean number of hours spent exercising. Be sure to define any parameters you are estimating.

(d) Compare the answer from part (c) to the confidence interval given in the following computer output for the same data:

```
Two-sample T for Exercise
Gender   N   Mean  StDev  SE Mean
F       168  8.11   5.20    0.40
M       193  9.88   6.07    0.44

Difference = mu (F) – mu (M)
Estimate for difference: –1.766
95% CI for difference: (–2.932, –0.599)
```

(e) Interpret the confidence interval in context.

6.206 Who Watches More TV: Males or Females? The dataset **StudentSurvey** has information from males and females on the number of hours spent watching television in a typical week. Computer output of descriptive statistics for the number of hours spent watching TV, broken down by gender, is given:

```
Descriptive Statistics: TV
Variable   Gender    N    Mean    StDev
TV         F        169   5.237   4.100
           M        192   7.620   6.427

Minimum    Q1   Median     Q3    Maximum
 0.000   2.500   4.000    6.000  20.000
 0.000   3.000   5.000   10.000  40.000
```

(a) In the sample, which group watches more TV, on average? By how much?

(b) Use the summary statistics to compute a 99% confidence interval for the difference in mean number of hours spent watching TV. Be sure to define any parameters you are estimating.

(c) Compare the answer from part (c) to the confidence interval given in the following computer output for the same data:

```
Two-sample T for TV
Gender   N   Mean  StDev  SE Mean
F       169  5.24   4.10    0.32
M       192  7.62   6.43    0.46

Difference = mu (F) – mu (M)
Estimate for difference: –2.383
99% CI for difference: (–3.836, –0.930)
```

(d) Interpret the confidence interval in context.

6.207 Who Eats More Fiber: Males or Females?
Use technology and the **NutritionStudy** dataset to find a 95% confidence interval for the difference in number of grams of fiber (*Fiber*) eaten in a day between males and females. Interpret the answer in context. Is "No difference" between males and females a plausible option for the population difference in mean number of grams of fiber eaten?

6.208 Systolic Blood Pressure and Survival Status
Use technology and the **ICUAdmissions** dataset to find a 95% confidence interval for the difference in systolic blood pressure (*Systolic*) upon admission to the Intensive Care Unit at the hospital based on survival of the patient (*Status* with 0 indicating the patient lived and 1 indicating the patient died.) Interpret the answer in context. Is "No difference" between those who lived and died a plausible option for the difference in mean systolic blood pressure? Which group had higher systolic blood pressures on arrival?

6.4-HT HYPOTHESIS TEST FOR A DIFFERENCE IN MEANS

In Section 5.1 we see that, when a randomization distribution is normal, we can compute a p-value using a standard normal curve and a standardized test statistic of the form

$$z = \frac{\text{Sample Statistic} - \text{Null Parameter}}{SE}$$

When comparing means between two groups, the null hypothesis is typically $H_0 : \mu_1 = \mu_2$ or, equivalently, $H_0 : \mu_1 - \mu_2 = 0$. Thus the "Null parameter" is usually equal to zero and we use the difference in means for two samples, $\bar{x}_1 - \bar{x}_2$, as the "Sample statistic".

As we see in Section 6.4-D, we can estimate the standard error of $\bar{x}_1 - \bar{x}_2$ with

$$SE = \sqrt{\frac{s_1^2}{n_1} + \frac{s_2^2}{n_2}}$$

where s_1 and s_2 are the standard deviations in the two samples.[49] However, when we use the sample standard deviations in estimating SE, we need to switch to a t-distribution rather than the standard normal when finding a p-value. This requires either that the underlying populations are reasonably normal or that the sample sizes are large.

Two-Sample t-Test for a Difference in Means

To test $H_0 : \mu_1 = \mu_2$ vs $H_a : \mu_1 \neq \mu_2$ (or a one-tail alternative) based on samples of sizes n_1 and n_2 from the two groups, we use the two-sample t-statistic

$$t = \frac{\text{Statistic} - \text{Null value}}{SE} = \frac{(\bar{x}_1 - \bar{x}_2) - 0}{\sqrt{\frac{s_1^2}{n_1} + \frac{s_2^2}{n_2}}}$$

where $\bar{x}_1$ and $\bar{x}_2$ are the means and s_1 and s_2 are the standard deviations for the respective samples.

If the underlying populations are reasonably normal or the sample sizes are large, we use a t-distribution to find the p-value for this statistic. For degrees of freedom we can either use the smaller of $n_1 - 1$ or $n_2 - 1$, or technology to get a more precise approximation.

---

[49]Some textbooks use a *pooled standard deviation* when standard deviations are approximately equal. This practice offers almost no advantage, however, and is not included in this text.

Example 6.27

Smiles and Leniency

In Data 4.5 on page 293, we look at an experiment to compare the leniency scores assigned to students charged with a disciplinary infraction in which subjects are shown a picture of the alleged wrongdoer randomly selected to show either a smiling or a neutral pose. Summary statistics from the data in **Smiles** are given in Table 6.16.

(a) Construct and interpret a graph to verify that the t-distribution is appropriate for comparing these means.

(b) Use the t-distribution to test whether the mean leniency score for smiling students is higher than the mean score for students with a neutral expression.

Table 6.16 *Summary statistics for leniency scores*

Group	n	Mean	Std. Dev.
Smile	34	4.91	1.68
Neutral	34	4.12	1.52

Solution

(a) Figure 6.19 shows boxplots of the sample leniency scores for the neutral and smiling groups. Both plots are relatively symmetric and have no strong outliers so we don't see strong evidence that the distributions are not normal. Also, the sample sizes of 34 for each group are not very small, so a t-distribution is reasonable to model the standardized distribution of $\bar{x}_s - \bar{x}_n$.

(b) The hypotheses are $H_0 : \mu_s = \mu_n$ vs $H_a : \mu_s > \mu_n$, where μ_s and μ_n are the means, respectively, for leniency scores assigned to smiling and neutral expressions. Based on the summary statistics, we compute the t-statistic

$$t = \frac{\text{Statistic} - \text{Null value}}{SE} = \frac{(\bar{x}_s - \bar{x}_n) - 0}{\sqrt{\frac{s_s^2}{n_s} + \frac{s_n^2}{n_n}}} = \frac{(4.91 - 4.12) - 0}{\sqrt{\frac{1.68^2}{34} + \frac{1.52^2}{34}}} = \frac{0.79}{0.389} = 2.03$$

To find the p-value we use the upper tail of a t-distribution with $34 - 1 = 33$ degrees of freedom. Technology shows this area to give a p-value of 0.025. This gives fairly strong evidence that the mean leniency score for smiling expressions is higher than the mean leniency score for neutral expressions.

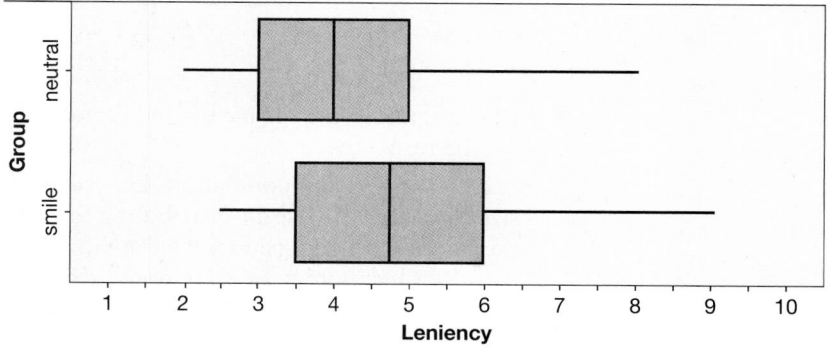

Figure 6.19 *Leniency scores for neutral and smiling faces*

We often use technology to handle the details of a two-sample t-test such as in Example 6.27. Here is some typical output for such a test with the data in **Smiles**:

```
Two-sample T for Leniency
Group     N    Mean   StDev   SE Mean
neutral   34   4.12   1.52    0.26
smile     34   4.91   1.68    0.29
```

Difference = mu (neutral) − mu (smile)
Estimate for difference: −0.794
95% upper bound for difference: −0.145
T-Test of difference = 0 (vs <): T-Value = −2.04 P-Value = 0.023 DF = 65

We see several differences between the computer output and the calculations in Example 6.27. First, the stat package uses the difference in the other direction, $\bar{x}_n - \bar{x}_s$, so the test becomes lower rather than upper tailed. The small difference in the magnitude of the t-statistic (T-value = −2.04 in the bottom line of the output) is due to rounding, but the larger degrees of freedom (65 rather than 33) is because the technology uses a more complicated formula for approximating the degrees of freedom. This gives a slightly smaller p-value (0.023 in the output) than our conservative use of $34 - 1 = 33$ degrees of freedom. Nevertheless, the basic results and interpretation of the test are the same.

Figure 4.23 on page 294 shows a randomization distribution for 1000 differences in means for the **Smiles** data. In that figure there are 23 values that are at or beyond the sample difference of $\bar{x}_s - \bar{x}_n = 0.79$. This gives an estimated p-value of $23/1000 = 0.023$, which agrees nicely with the results of Example 6.27 and the computer output. While the formulas in this section allow us to calculate a test statistic using summary statistics, remember that the randomization procedures of Chapter 4 still apply, even in situations where the conditions for the t-distribution might be in question.

SECTION LEARNING GOALS

You should now have the understanding and skills to:

- Use a formula for *SE* and a t-distribution, when appropriate, to test a hypothesis about a difference in means between two groups

Exercises for Section **6.4-HT**

SKILL BUILDER 1

In Exercises 6.209 to 6.212, use the t-distribution and the given sample results to complete the test of the given hypotheses. Assume the results come from random samples, and if the sample sizes are small, assume the underlying distributions are relatively normal.

6.209 Test $H_0 : \mu_1 = \mu_2$ vs $H_a : \mu_1 > \mu_2$ using the sample results $\bar{x}_1 = 56$, $s_1 = 8.2$ with $n_1 = 30$ and $\bar{x}_2 = 51$, $s_2 = 6.9$ with $n_2 = 40$.

6.210 Test $H_0 : \mu_1 = \mu_2$ vs $H_a : \mu_1 \neq \mu_2$ using the sample results $\bar{x}_1 = 15.3$, $s_1 = 11.6$ with $n_1 = 100$ and $\bar{x}_2 = 18.4$, $s_2 = 14.3$ with $n_2 = 80$.

6.211 Test $H_0 : \mu_A = \mu_B$ vs $H_a : \mu_A \neq \mu_B$ using the fact that Group A has 8 cases with a mean of 125 and a standard deviation of 18 while Group B has 15 cases with a mean of 118 and a standard deviation of 14.

6.212 Test $H_0 : \mu_T = \mu_C$ vs $H_a : \mu_T < \mu_C$ using the fact that the treatment group (*T*) has a sample mean of 8.6 with a standard deviation of 4.1 while the control group (C) has a sample mean of 11.2 with a standard deviation of 3.4. Both groups have 25 cases.

6.213 Take Your Notes Longhand! A study[50] randomly assigned students to take notes either longhand or using a laptop. The resulting scores of the students on a test of the material are given in Table 6.17. Does the data provide evidence that it is more effective to take notes longhand rather than on a laptop? Show all details of the test.

Table 6.17 *Summary statistics for test scores after taking notes longhand or on a laptop*

Note-taking	n	$\bar{x}$	s
Longhand	38	25.6	10.8
Laptop	40	18.3	9.0

6.214 Statistical Inference in Babies Is statistical inference intuitive to babies? In other words, are babies able to generalize from sample to population? In this study,[51] 8-month-old infants watched someone draw a sample of five balls from an opaque box. Each sample consisted of four balls of one color (red or white) and one ball of the other color. After observing the sample, the side of the box was lifted so the infants could see all of the balls inside (the population). Some boxes had an "expected" population, with balls in the same color proportions as the sample, while other boxes had an "unexpected" population, with balls in the opposite color proportion from the sample. Babies looked at the unexpected populations for an average of 9.9 seconds (sd = 4.5 seconds) and the expected populations for an average of 7.5 seconds (sd = 4.2 seconds). The sample size in each group was 20, and you may assume the data in each group are reasonably normally distributed. Is this convincing evidence that babies look longer at the unexpected population, suggesting that they make inferences about the population from the sample?

(a) State the null and alternative hypotheses.

(b) Calculate the relevant sample statistic.

(c) Calculate the t-statistic.

(d) Find the p-value.

(e) Make a generic conclusion about the null hypothesis, using $\alpha = 0.10$.

(f) Make a conclusion in context.

6.215 Gender Bias In a study[52] examining gender bias, a nationwide sample of 127 science professors evaluated the application materials of an undergraduate student who had ostensibly applied for a laboratory manager position. All participants received the same materials, which were randomly assigned either the name of a male ($n_m = 63$) or the name of a female ($n_f = 64$). Participants believed that they were giving feedback to the applicant, including what salary could be expected. The average salary recommended for the male applicant was $30,238 with a standard deviation of $5152 while the average salary recommended for the (identical) female applicant was $26,508 with a standard deviation of $7348. Does this provide evidence of a gender bias, in which applicants with male names are given higher recommended salaries than applicants with female names? Show all details of the test.

6.216 Is Gender Bias Influenced by Faculty Gender? Exercise 6.215 describes a study in which science faculty members are asked to recommend a salary for a lab manager applicant. All the faculty members received the same application, with half randomly given a male name and half randomly given a female name. In Exercise 6.215, we see that the applications with female names received a significantly lower recommended salary. Does gender of the evaluator make a difference? In particular, considering only the 64 applications with female names, is the mean recommended salary different depending on the gender of the evaluating faculty member? The 32 male faculty gave a mean starting salary of $27,111 with a standard deviation of $6948 while the 32 female faculty gave a mean starting salary of $25,000 with a standard deviation of $7966. Show all details of the test.

6.217 Football Air Pressure During the NFL's 2014 AFC championship game, officials measured the air pressure on game balls following a tip that one team's balls were under-inflated. In exercise 6.124 we found that the 11 balls measured for the New England Patriots had a mean psi of 11.10 (well below the legal limit) and a standard deviation of 0.40. Patriot supporters could argue that the

[50]Mueller, P.A., and Oppenheimer, D.M., "The Pen is Mightier Than the Keyboard: Advantages of Longhand Over Laptop Note Taking," *Psychological Science*, May 22, 2014.

[51]Data approximated from Xu, F., and Garcia, V. (2008). "Intuitive statistics by 8-month-old infants," *Proceedings on the National Academy of Sciences*, 105(13):5012–15, doi: 10.1073/pnas.0704450105.

[52]Moss-Racusin, C.A., et al., "Science faculty's subtle gender biases favor male students," *Proceedings of the National Academy of Sciences*, 109(41), October 9, 2012, 16764–479.

under-inflated balls were due to the elements and other outside effects. To test this the officials also measured 4 balls from the opposing team (Indianapolis Colts) to be used in comparison and found a mean psi of 12.63, with a standard deviation of 0.12. There is no significant skewness or outliers in the data. Use the t-distribution to determine if the average air pressure in the New England Patriot's balls was significantly less than the average air pressure in the Indianapolis Colt's balls.

6.218 Mind-Set Matters In 2007 a Harvard psychologist set out to test her theory that "Mind-Set Matters."[53] She recruited 75 female maids[54] working in different hotels to participate in her study, and informed 41 maids (randomly chosen) that the work they do satisfies the Surgeon General's recommendations for an active lifestyle (which is true), giving the maids examples on how their work qualifies as good exercise. The other 34 maids were told nothing. After four weeks, the exercise habits of the two groups had not changed, but the informed group had lost an average of 1.79 lbs ($s = 2.88$) and the uninformed group had lost an average of 0.2 lbs ($s = 2.32$). The data are stored in **MindsetMatters**. Based on this study, does "Mind-Set Matter"? In other words, for maids, does simply thinking they are exercising more actually cause them to lose more weight? Show all details of the test.

6.219 Exercise and Stress Many studies have shown that people who engage in any exercise have improved mental health over those that never exercise. In particular, even a small amount of exercise seems to confer some resilience to stress. Most of these studies, by necessity, have been observational studies. A recent experiment with mice[55] moves us one step closer to determining a causal association. In the study, mice were randomly assigned to either an enriched environment (EE) where there was an exercise wheel available or a standard environment (SE) with no exercise options. After three weeks in the specified environment, for five minutes a day for two weeks, the mice were each exposed to a "mouse bully"—a mouse

that was very strong, aggressive, and territorial. At the end of the two weeks, the mice in the SE group exhibited maladaptive, depressive-like, and anxiety-like behavior across a wide spectrum of activities. This was not true of the mice in the EE group; they behaved similarly to mice that had never had the stress-inducing bully experience. In particular, one measure of mouse anxiety is amount of time hiding in a dark compartment, with mice that are more anxious spending more time in darkness. The amount of time spent in darkness during one trial is recorded for all the mice and is shown in Table 6.18 and available in **StressedMice**. Test to see if mice that have spent time in an enriched environment with options for exercise spend significantly less time in darkness after a stress-inducing experience.

Table 6.18 *Do mice from an enriched environment spend less time in darkness?*

Environment	Time in Darkness (seconds)						
Enriched	359	280	138	227	203	184	231
Standard	394	477	439	428	391	488	454

6.220 Diet Cola and Calcium Exercise B.3 on page 349 introduces a study examining the effect of diet cola consumption on calcium levels in women. A sample of 16 healthy women aged 18 to 40 were randomly assigned to drink 24 ounces of either diet cola or water. Their urine was collected for three hours after ingestion of the beverage and calcium excretion (in mg) was measured. The summary statistics for diet cola are $\bar{x}_C = 56.0$ with $s_C = 4.93$ and $n_C = 8$ and the summary statistics for water are $\bar{x}_W = 49.1$ with $s_W = 3.64$ and $n_W = 8$. Figure 6.20 shows dotplots of the data values. Test whether there is evidence that diet cola leaches calcium out of the system, which would increase the amount of calcium in the urine for diet cola drinkers. In Exercise B.3, we used a randomization distribution to conduct this test. Use a t-distribution here, after first checking that the conditions are met and explaining your reasoning. The data are stored in **ColaCalcium**.

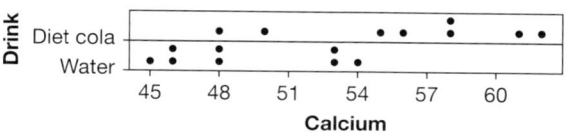

Figure 6.20 *Do diet cola drinkers excrete more calcium?*

[53]Crum, A. and Langer, E., "Mind-Set Matters: Exercise and the Placebo Effect," *Psychological Science*, 2007; 18:165–171.
[54]Maids with missing values for weight change have been removed.
[55]Data approximated from summary statistics in: Lehmann, M. and Herkenham, M., "Environmental Enrichment Confers Stress Resiliency to Social Defeat through an Infralimbic Cortex-Dependent Neuroanatomical Pathway," *The Journal of Neuroscience*, April 20, 2011; 31(16):6159–73.

6.221 Drink Tea for a Stronger Immune System Drinking tea appears to offer a strong boost to the immune system. In a study introduced in Exercise 3.91 on page 239, we see that production of interferon gamma, a molecule that fights bacteria, viruses, and tumors, appears to be enhanced in tea drinkers. In the study, eleven healthy non-tea-drinking individuals were asked to drink five or six cups of tea a day, while ten healthy non-tea- and non-coffee-drinkers were asked to drink the same amount of coffee, which has caffeine but not the L-theanine that is in tea. The groups were randomly assigned. After two weeks, blood samples were exposed to an antigen and production of interferon gamma was measured. The results are shown in Table 6.19 and are available in **ImmuneTea**. The question of interest is whether the data provide evidence that production is enhanced in tea drinkers.

Table 6.19 *Immune system response in tea and coffee drinkers*

Tea	5	11	13	18	20	47
	48	52	55	56	58	
Coffee	0	0	3	11	15	16
	21	21	38	52		

(a) Is this an experiment or an observational study?

(b) What are the null and alternative hypotheses?

(c) Find a standardized test statistic and use the t-distribution to find the p-value and make a conclusion.

(d) Always plot your data! Look at a graph of the data. Does it appear to satisfy a normality condition?

(e) A randomization test might be a more appropriate test to use in this case. Construct a randomization distribution for this test and use it to find a p-value and make a conclusion.

(f) What conclusion can we draw?

6.222 Metal Tags on Penguins and Length of Foraging Trips Data 1.3 on page 10 discusses a study designed to test whether applying metal tags is detrimental to a penguin, as opposed to applying an electronic tag. One variable examined is the length of foraging trips. Longer foraging trips can jeopardize both breeding success and survival of chicks waiting for food. Mean length of 344 foraging trips for penguins with a metal tag was 12.70 days with a standard

deviation of 3.71 days. For those with an electronic tag, the mean was 11.60 days with standard deviation of 4.53 days over 512 trips. Do these data provide evidence that mean foraging trips are longer for penguins with a metal tag? Show all details of the test.

6.223 Metal Tags on Penguins and Arrival Dates Data 1.3 on page 10 discusses a study designed to test whether applying a metal tag is detrimental to a penguin, as opposed to applying an electronic tag. One variable examined is the date penguins arrive at the breeding site, with later arrivals hurting breeding success. Arrival date is measured as the number of days after November 1st. Mean arrival date for the 167 times metal-tagged penguins arrived was December 7th (37 days after November 1st) with a standard deviation of 38.77 days, while mean arrival date for the 189 times electronic-tagged penguins arrived at the breeding site was November 21st (21 days after November 1st) with a standard deviation of 27.50. Do these data provide evidence that metal tagged penguins have a later mean arrival time? Show all details of the test.

6.224 Split the Bill? Exercise 2.153 on page 105 describes a study to compare the cost of restaurant meals when people pay individually versus splitting the bill as a group. In the experiment half of the people were told they would each be responsible for individual meal costs and the other half were told the cost would be split equally among the six people at the table. The data in **SplitBill** includes the cost of what each person ordered (in Israeli shekels) and the payment method (*Individual* or *Split*). Some summary statistics are provided in Table 6.20 and both distributions are reasonably bell-shaped. Use this information to test (at a 5% level) if there is evidence that the mean cost is higher when people split the bill. You may have done this test using randomizations in Exercise 4.118 on page 302.

Table 6.20 *Cost of meals by payment type*

Payment	Sample size	Mean	Std. dev.
Individual	24	37.29	12.54
Split	24	50.92	14.33

6.225 Restaurant Bill by Gender In the study described in Exercise 6.224 the diners were also chosen so that half the people at each table were female and half were male. Thus we can also test for a difference in mean meal cost between females ($n_f = 24$, $\bar{x}_f = 44.46$, $s_f = 15.48$) and males ($n_m = 24$, $\bar{x}_m = 43.75$, $s_m = 14.81$). Show all details for doing this test.

6.226 What Gives a Small P-value? In each case below, two sets of data are given for a two-tail difference in means test. In each case, which version gives a *smaller* p-value relative to the other?

(a) Both options have the same standard deviations and same sample sizes but:

Option 1 has:	$\bar{x}_1 = 25$	$\bar{x}_2 = 23$
Option 2 has:	$\bar{x}_1 = 25$	$\bar{x}_2 = 11$

(b) Both options have the same means ($\bar{x}_1 = 22$, $\bar{x}_2 = 17$) and same sample sizes but:

Option 1 has:	$s_1 = 15$	$s_2 = 14$
Option 2 has:	$s_1 = 3$	$s_2 = 4$

(c) Both options have the same means ($\bar{x}_1 = 22$, $\bar{x}_2 = 17$) and same standard deviations but:

Option 1 has:	$n_1 = 800$	$n_2 = 1000$
Option 2 has:	$n_1 = 25$	$n_2 = 30$

6.227 Quiz Timing A young statistics professor decided to give a quiz in class every week. He was not sure if the quiz should occur at the beginning of class when the students are fresh or at the end of class when they've gotten warmed up with some statistical thinking. Since he was teaching two sections of the same course that performed equally well on past quizzes, he decided to do an experiment. He randomly chose the first class to take the quiz during the second half of the class period (Late) and the other class took the same quiz at the beginning of their hour (Early). He put all of the grades into a data table and ran an analysis to give the results shown below. Use the information from the computer output to give the details of a test to see whether the mean grade depends on the timing of the quiz. (You should not do any computations. State the hypotheses based on the output, read the p-value off the output, and state the conclusion in context.)

```
Two-Sample T-Test and CI
Sample    N    Mean    StDev    SE Mean
Late      32   22.56   5.13     0.91
Early     30   19.73   6.61     1.2

  Difference = mu (Late) − mu (Early)
  Estimate for difference: 2.83
```

95% CI for difference: (−0.20, 5.86)
T-Test of difference = 0 (vs not =): T-Value = 1.87
P-Value = 0.066 DF = 54

6.228 Survival Status and Heart Rate in the ICU The dataset **ICUAdmissions** contains information on a sample of 200 patients being admitted to the Intensive Care Unit (ICU) at a hospital. One of the variables is *HeartRate* and another is *Status* which indicates whether the patient lived (*Status* = 0) or died (*Status* = 1). Use the computer output to give the details of a test to determine whether mean heart rate is different between patients who lived and died. (You should not do any computations. State the hypotheses based on the output, read the p-value off the output, and state the conclusion in context.)

```
Two-sample T for HeartRate
Status   N    Mean    StDev   SE Mean
0        160  98.5    27.0    2.1
1        40   100.6   26.5    4.2

Difference = mu (0) − mu (1)
Estimate for difference: −2.13
95% CI for difference: (−11.53, 7.28)

T-Test of difference = 0 (vs not =): T-Value = −0.45
P-Value = 0.653  DF = 60
```

6.229 Who Exercises More: Males or Females? Use technology and the **StudentSurvey** dataset to test whether the data provide evidence that there is a difference in the mean number of hours of exercise per week between males and females. What are the hypotheses? What is the p-value given by the software? State the conclusion in context.

6.230 Who Watches More TV: Males or Females? Use technology and the **StudentSurvey** dataset to test whether the data provide evidence that there is a difference in the mean number of hours spent watching television per week between males and females. What are the hypotheses? What is the p-value given by the software? State the conclusion in context.

6.231 Gender and Atlanta Commutes The data in **CommuteAtlanta** (see Data 3.3 on page 229 for more description) contain a variable that identifies the sex of each commuter in the Atlanta sample. Test at a 5% level whether there is a difference in mean commute time between female and male commuters in Atlanta.

6.5 PAIRED DIFFERENCE IN MEANS

In Section 6.4 we consider inference for a difference in means when the data consist of two separate samples. However, in situations such as the matched pairs experiments described in Section 1.3, the data being compared consist of pairs of data values. Paired data may consist of two measurements on each unit, such as the same unit being measured under two different conditions, or measurements on a pair of units that go together, such as twin studies. In this section we see how to handle inferences for a difference in means when the data are paired in some way.

Photo by MikeStobe/Getty Images, Inc.

Olympic swimmer Michael Phelps in a wetsuit

DATA 6.5	**Wetsuits and Swimming Speed**

The 2008 Olympics were full of controversy about new swimsuits possibly providing unfair advantages to swimmers, leading to new international rules that came into effect January 1, 2010, regarding swimsuit coverage and material. Can a certain swimsuit really make a swimmer faster? A study[56] tested whether wearing wetsuits influences swimming velocity. Twelve competitive swimmers and triathletes swam 1500 m at maximum speed twice each, once wearing a wetsuit and once wearing a regular bathing suit. The order of the trials was randomized. Each time, the maximum velocity in meters/sec of the swimmer (among other quantities) was recorded. These data are available in **Wetsuits** and shown in Table 6.21. ∎

[56]de Lucas, R., Balildan, P., Neiva, C., Greco, C., and Denadai, B., "The effects of wetsuits on physiological and biomechanical indices during swimming," *Journal of Science and Medicine in Sport*, 2000; 3(1):1–8.

Table 6.21 *Maximum velocity swimming with and without a wetsuit*

Swimmer	1	2	3	4	5	6	7	8	9	10	11	12	$\overline{x}$	s
Wetsuit	1.57	1.47	1.42	1.35	1.22	1.75	1.64	1.57	1.56	1.53	1.49	1.51	1.507	0.136
No Wetsuit	1.49	1.37	1.35	1.27	1.12	1.64	1.59	1.52	1.50	1.45	1.44	1.41	1.429	0.141

Example 6.28

Using Data 6.5 and the methods of Section 6.4-HT, test whether the average maximum velocity for competitive swimmers differs when wearing wetsuits vs not wearing wetsuits.

Solution

The relevant parameters are μ_w and μ_{nw}, the average maximum velocities for swimmers wearing wetsuits and swimmers not wearing wetsuits, respectively. We wish to test $H_0 : \mu_w = \mu_{nw}$ vs $H_a : \mu_w \neq \mu_{nw}$. The relevant sample statistics are $\overline{x}_w = 1.507$, $s_w = 0.136$, $n_w = 12$ for the swimmers in wetsuits and $\overline{x}_{nw} = 1.429$, $s_{nw} = 0.141$, $n_{nw} = 12$ for the swimmers without wetsuits. Figure 6.21 does not show extreme skewness or outliers, so we can proceed with the t-distribution. We compute the t-statistic using the formula from Section 6.4-HT:

$$t = \frac{\overline{x}_w - \overline{x}_{nw}}{\sqrt{\frac{s_w^2}{n_w} + \frac{s_{nw}^2}{n_{nw}}}} = \frac{1.507 - 1.429}{\sqrt{\frac{0.136^2}{12} + \frac{0.141^2}{12}}} = 1.38$$

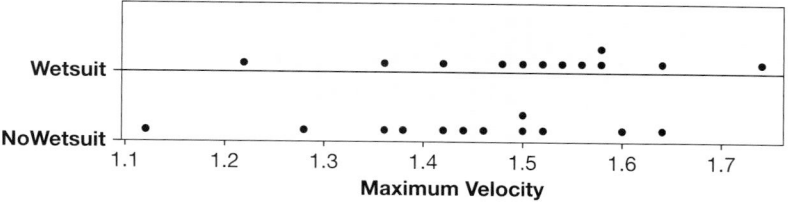

Figure 6.21 *Dotplot of maximum velocity swimming with and without a wetsuit*

We compare this to a t-distribution with 11 df and find the area in the upper tail to be 0.0975. Since this is a two-tailed test, the p-value is 2(0.0975) = 0.195. This study does not provide convincing evidence that swimming speeds are affected by wearing a wetsuit.

Using this naive method, we do not find any association between wearing a wetsuit and swimming speeds. However, taking a closer look at the data in Table 6.21, we see that every single swimmer swam faster wearing the wetsuit! Surely this must provide conclusive evidence that swimmers are faster wearing a wetsuit. What went wrong?

We failed to take into account the paired structure of the data. The formula for the standard error for a difference in means given in Section 6.4-D applies only to data from two different groups, not to paired data. When used on paired data it will often give a standard error that is much too high.

The wetsuit study was conducted on males and females, swimmers and triathletes. Not surprisingly, there is a great deal of variability in the maximum velocities of the swimmers! Because of all this variability, it is difficult to tell whether the difference in mean swim speed observed in the sample (1.507 − 1.429 = 0.078) represents a real difference or is just due to random chance. Rather than comparing females in the wetsuit group to males in the non-wetsuit group, and triathletes in the non-wetsuit group to swimmers in the wetsuit group, we would really like to compare each swimmer's wetsuit and non-wetsuit values directly! The secret is to take the *difference* for each pair of data values. These differences are displayed in the bottom row of Table 6.22.

Table 6.22 *Maximum velocity swimming with and without a wetsuit, including differences*

Swimmer	1	2	3	4	5	6	7	8	9	10	11	12	$\bar{x}$	s
Wetsuit	1.57	1.47	1.42	1.35	1.22	1.75	1.64	1.57	1.56	1.53	1.49	1.51	1.507	0.136
No Wetsuit	1.49	1.37	1.35	1.27	1.12	1.64	1.59	1.52	1.50	1.45	1.44	1.41	1.429	0.141
Difference, d	0.08	0.10	0.07	0.08	0.10	0.11	0.05	0.05	0.06	0.08	0.05	0.10	0.078	0.022

The key to analyzing paired data is to work with these differences rather than the original data. This helps us to eliminate the variability across different units (different swimmers) and instead focus on what we really care about—the difference between the values with and without a wetsuit. We denote the differences with the letter d.

Note that the mean of the sample of differences is equal to the difference of the sample means: $\bar{x}_d = \bar{x}_w - \bar{x}_{nw}$. (For the swimming data, we see that $0.078 = 1.507 - 1.429$.) To make inferences about the difference in means, $\mu_w - \mu_{nw}$, we can equivalently make inferences about the mean difference, μ_d.

The paired differences are a single sample of values for the differences. Thus we estimate μ_d with the sample mean of these differences, $\bar{x}_d$. As with any mean for a single sample, we estimate its standard error by dividing the standard deviation of the sample values by the square root of the sample size. Here the sample size is the number of data pairs (which is also the number of computed differences), n_d, and the relevant standard deviation is the standard deviation of the differences, s_d.

Fortunately, we have already discussed inference for a mean based on a single sample in Section 6.2. The procedures are the same for paired data, once we convert the data to a single sample of differences.

Inference for a Difference in Means with Paired Data

To estimate the difference in means based on paired data, we first subtract to compute the difference for each data pair and compute the mean $\bar{x}_d$, the standard deviation s_d, and the sample size n_d for the sample differences.

Provided the differences are reasonably normally distributed (or the sample size is large), a confidence interval for the difference in means is given by

$$\text{Statistic} \pm t^* \cdot SE = \bar{x}_d \pm t^* \frac{s_d}{\sqrt{n_d}}$$

where t^* is a percentile from a t-distribution with $n_d - 1$ degrees of freedom.

To test $H_0 : \mu_d = 0$ vs $H_a : \mu_d \neq 0$ (or a one-tail alternative) we use the t-test statistic

$$t = \frac{\text{Statistic} - \text{Null value}}{SE} = \frac{\bar{x}_d - 0}{s_d/\sqrt{n_d}}$$

If the differences are reasonably normally distributed (or the sample size is large), we use a t-distribution with $n_d - 1$ degrees of freedom to compute a p-value for this test statistic.

Example 6.29

Wetsuits and Swimming Speed

Using Data 6.5, test whether the average maximum velocity for competitive swimmers differs when wearing wetsuits vs not wearing wetsuits. (Note that this is exactly the same as Example 6.28, except that we now know how to utilize the paired structure.)

Solution

Because the sample size is fairly small, we should check to see whether there are problems with normality in the sample of differences. Figure 6.22 shows a dotplot of the differences.

Figure 6.22 *Dotplot of differences in maximum velocity with and without a wetsuit*

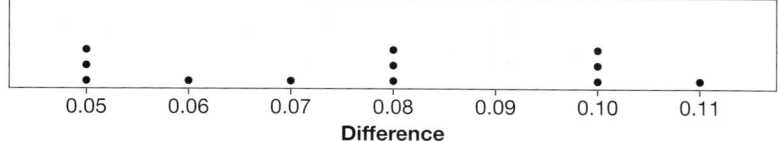

Although it is difficult to see a bell-shape for samples this small, the differences look relatively symmetric and have no clear outliers, so we don't detect any serious problems with normality and can proceed with the t-distribution. Using the differences in Table 6.22, we see that $\bar{x}_d = 0.078$ with $s_d = 0.022$ and $n_d = 12$. We compute the t-statistic as follows:

$$t = \frac{\text{Statistic} - \text{Null value}}{SE} = \frac{\bar{x}_d - 0}{s_d/\sqrt{n_d}} = \frac{0.078}{0.022/\sqrt{12}} = 12.3$$

This is a huge t-statistic, and corresponds to a p-value of essentially zero. (Anytime the t-statistic is bigger than 5, there is really no point in looking up the p-value—it will be very small!) These data provide very convincing evidence that swimmers are faster on average when wearing wetsuits. Because this was a randomized experiment, we can conclude that wetsuits cause swimmers to swim faster.

Notice the difference in conclusions between Examples 6.28 and 6.29. The same data were analyzed in each case, but the conclusions reached were drastically different. It is very important to think about how the data were collected before proceeding with the analysis! Although we can find the mean of the differences, $\bar{x}_d$, from the difference in the individual sample means, $\bar{x}_w - \bar{x}_{nw}$, there is no way to compute the standard deviation of the differences, s_d, from the standard deviations of the individual samples. To compute the standard deviation properly when the data are paired, we must find the individual differences for all pairs.

Example 6.30

Example 6.29 verifies that wetsuits make swimmers faster, but how much faster? Compute a 95% confidence interval for the difference in average maximum velocity for swimmers wearing wetsuits minus swimmers not wearing wetsuits.

Solution

We have already verified from Figure 6.22 that it is appropriate to use the t-distribution. With 11 degrees of freedom for 95% confidence, we have $t^* = 2.20$. We compute a confidence interval as follows:

$$\bar{x}_d \pm t^* \frac{s_d}{\sqrt{n_d}} = 0.078 \pm 2.20 \left(\frac{0.022}{\sqrt{12}}\right) = 0.078 \pm 0.014 = (0.064, 0.092)$$

We are 95% confident that for competitive swimmers and triathletes, wetsuits increase maximum swimming velocity by an average of between 0.064 and 0.092 meters per second.

_____

If this had not been paired data but simply two groups of 12 swimmers each, the 95% confidence interval would be

$$0.078 \pm 2.20(0.057) = 0.078 \pm 0.125 = (-0.047, 0.203)$$

The margin of error is larger, and the interval is wider than in Example 6.30. By using paired data and analyzing it appropriately, we can get a much more accurate and reliable estimate of the difference in means.

Remember that paired data does not always mean that all the cases do both treatments. Data is paired if there is a valid reason for matching each case in one group to a case in the other group. For example, in investigating a new method to teach reading in first grade, we might match each student in the experimental class with another student outside that class who is similar in terms of IQ, family income, parent education level, and so on. Some famous examples of paired data involve identical twins, in which one twin receives the treatment and the other serves as a control.

Identical Twins!! Cal and Axel Lock Morgan

Example 6.31

_____

Genes or Choices: Body Fat and Brain Volume

A study[57] examined 10 male identical twins in their mid-30s, in which one twin engaged in regular physical activity and the other was more sedentary. The journal article reports: "According to pairwise analysis, the active twins had lower body fat percentage ($P = 0.029$)" and also "brain grey matter volumes were larger...in active twins compared to those in inactive co-twins, with a statistical threshold of $P < 0.001$." Since identical twins share the same genetic make-up, what does this study tell us about the effect of individual choices on body fat percentage and brain volume?

[57] Rottensteiner M, et al., "Physical activity, fitness, glucose homeostatis, and brain morphology in twins," *Medicine & Science in Sports & Exercise*, 47(3), March 2015, 509–518.

Solution ▶ Both p-values show significance at the 5% level, so we have evidence that, regardless of genetics, choosing to engage in regular physical activity influences body fat percentage (which is not too surprising) and also influences, even more strongly, brain volume (which might be more surprising.) Be careful, though, not to assume causation, since this is an observational study and not an experiment.

SECTION LEARNING GOALS

You should now have the understanding and skills to:

▶ • Distinguish between separate samples and paired data when comparing two means

▶ • Identify the advantages of using paired data when estimating a difference in means

▶ • Use a t-distribution, when appropriate, to compute a confidence interval for a difference in means based on paired data

▶ • Use a t-distribution, when appropriate, to test a hypothesis about a difference in means based on paired data

Exercises for Section **6.5**

SKILL BUILDER 1

In Exercises 6.232 to 6.235, use a t-distribution to find a confidence interval for the difference in means $\mu_1 - \mu_2$ using the relevant sample results from paired data. Give the best estimate for $\mu_1 - \mu_2$, the margin of error, and the confidence interval. Assume the results come from random samples from populations that are approximately normally distributed, and that differences are computed using $d = x_1 - x_2$.

6.232 A 95% confidence interval for $\mu_1 - \mu_2$ using the paired difference sample results $\bar{x}_d = 3.7$, $s_d = 2.1$, $n_d = 30$

6.233 A 90% confidence interval for $\mu_1 - \mu_2$ using the paired difference sample results $\bar{x}_d = 556.9$, $s_d = 143.6$, $n_d = 100$

6.234 A 99% confidence interval for $\mu_1 - \mu_2$ using the paired data in the following table:

Case	1	2	3	4	5
Treatment 1	22	28	31	25	28
Treatment 2	18	30	25	21	21

6.235 A 95% confidence interval for $\mu_1 - \mu_2$ using the paired data in the following table:

Case	Situation 1	Situation 2
1	77	85
2	81	84
3	94	91
4	62	78
5	70	77
6	71	61
7	85	88
8	90	91

SKILL BUILDER 2

In Exercises 6.236 to 6.239, use a t-distribution and the given matched pair sample results to complete the test of the given hypotheses. Assume the results come from random samples, and if the sample sizes are small, assume the underlying distribution of the differences is relatively normal. Assume that differences are computed using $d = x_1 - x_2$.

6.236 Test $H_0 : \mu_1 = \mu_2$ vs $H_a : \mu_1 \neq \mu_2$ using the paired difference sample results $\bar{x}_d = 15.7$, $s_d = 12.2$, $n_d = 25$.

6.237 Test $H_0 : \mu_1 = \mu_2$ vs $H_a : \mu_1 \neq \mu_2$ using the paired difference sample results $\bar{x}_d = -2.6$, $s_d = 4.1$, $n_d = 18$.

6.238 Test $H_0 : \mu_1 = \mu_2$ vs $H_a : \mu_1 < \mu_2$ using the paired data in the following table:

| Treatment 1 | 16 | 12 | 18 | 21 | 15 | 11 | 14 | 22 |
| Treatment 2 | 18 | 20 | 25 | 21 | 19 | 8 | 15 | 20 |

6.239 Test $H_0 : \mu_1 = \mu_2$ vs $H_a : \mu_1 > \mu_2$ using the paired data in the following table:

| Situation 1 | 125 | 156 | 132 | 175 | 153 | 148 | 180 | 135 | 168 | 157 |
| Situation 2 | 120 | 145 | 142 | 150 | 160 | 148 | 160 | 142 | 162 | 150 |

SKILL BUILDER 3

In Exercises 6.240 to 6.245, a data collection method is described to investigate a difference in means. In each case, determine which data analysis method is more appropriate: paired data difference in means or difference in means with two separate groups.

6.240 To study the effect of sitting with a laptop computer on one's lap on scrotal temperature, 29 men have their scrotal temperature tested before and then after sitting with a laptop for one hour.

6.241 To study the effect of women's tears on men, levels of testosterone are measured in 50 men after they sniff women's tears and after they sniff a salt solution. The order of the two treatments was randomized and the study was double-blind.

6.242 In another study to investigate the effect of women's tears on men, 16 men watch an erotic movie and then half sniff women's tears and half sniff a salt solution while brain activity is monitored.

6.243 To measure the effectiveness of a new teaching method for math in elementary school, each student in a class getting the new instructional method is matched with a student in a separate class on IQ, family income, math ability level the previous year, reading level, and all demographic characteristics. At the end of the year, math ability levels are measured again.

6.244 In a study to determine whether the color red increases how attractive men find women, one group of men rate the attractiveness of a woman after seeing her picture on a red background and another group of men rate the same woman after seeing her picture on a white background.

6.245 A study investigating the effect of exercise on brain activity recruits sets of identical twins in middle age, in which one twin is randomly assigned to engage in regular exercise and the other doesn't exercise.

6.246 Drink Tea for a Stronger Immune System We saw in Exercise 6.221 on page 466 that drinking tea appears to offer a strong boost to the immune system. In a study extending the results of the study described in that exercise,[58] blood samples were taken on five participants before and after one week of drinking about five cups of tea a day (the participants did not drink tea before the study started). The before and after blood samples were exposed to *e. coli* bacteria, and production of interferon gamma, a molecule that fights bacteria, viruses, and tumors, was measured. Mean production went from 155 pg/mL before tea drinking to 448 pg/mL after tea drinking. The mean difference for the five subjects is 293 pg/mL with a standard deviation in the differences of 242. The paper implies that the use of the t-distribution is appropriate.

(a) Why is it appropriate to use paired data in this analysis?

(b) Find and interpret a 90% confidence interval for the mean increase in production of interferon gamma after drinking tea for one week.

6.247 Testing the Effects of Tea on the Immune System Exercise 6.246 describes a study to examine the effects of tea on the immune system. Use the information there to test whether mean production of interferon gamma as a response to bacteria is significantly higher after drinking tea than before drinking tea. Use a 5% significance level.

6.248 Pheromones in Female Tears? On page 11 in Section 1.1, we describe studies to investigate whether there is evidence of pheromones (subconscious chemical signals) in female tears that affect sexual arousal in men. In one of the studies,[59] 50 men had a pad attached to the upper lip that contained either female tears or a salt solution dripped down the same female's face. Each subject participated twice, on consecutive days, once with tears and once with saline, randomized for order, and double-blind. Testosterone levels were measured before sniffing and after sniffing on both days. While normal testosterone levels vary significantly between different men, average levels for the group were the same before sniffing on both days and after sniffing the salt solution (about 155 pg/mL) but were reduced after sniffing the tears

[58] Adapted from Kamath, A., et al., "Antigens in tea-beverage prime human $V\gamma2V\delta2$ T cells *in vitro* and *in vivo* for memory and non-memory antibacterial cytokine responses," *Proceedings of the National Academy of Sciences*, 2003; 100(10):6009–6014.

[59] Data approximated from Gelstein, S., et al., "Human Tears Contain a Chemosignal," *Science*, January 14, 2011; 331(6014): 226–230.

Table 6.23 *Quiz and lecture pulse rates for 10 students*

Student	1	2	3	4	5	6	7	8	9	10	Mean	Std. Dev.
Quiz	75	52	52	80	56	90	76	71	70	66	68.8	12.5
Lecture	73	53	47	88	55	70	61	75	61	78	66.1	12.8

(about 133 pg/mL). The mean difference in testosterone levels after sniffing the tears was 21.7 with standard deviation 46.5.

(a) Why did the investigators choose a matched-pairs design for this experiment?

(b) Test to see if testosterone levels are significantly reduced after sniffing tears?

(c) Can we conclude that sniffing female tears reduces testosterone levels (which is a significant indicator of sexual arousal in men)?

6.249 Measuring the Effect of Female Tears Exercise 6.248 describes a study to measure, in a double-blind randomized experiment, the effect of female tears on male testosterone. Use the information there to find a 99% confidence interval for the effect size (the amount of reduction in testosterone levels after sniffing female tears.)

6.250 Quiz vs Lecture Pulse Rates Do you think your pulse rate is higher when you are taking a quiz than when you are sitting in a lecture? The data in Table 6.23 show pulse rates collected from 10 students in a class lecture and then from the same students during a quiz. The data are stored in **QuizPulse10**. Construct a 95% confidence interval for the difference in mean pulse rate between students in a class lecture and taking a quiz.

6.251 Testing for a Difference in Pulse Rates Exercise 6.250 describes pulse rates collected from 10 students, once during a quiz and once during a lecture. The data are given in Table 6.23 and stored in **QuizPulse10**. We might expect mean pulse rates to increase under the stress of a quiz. Use the information in Exercise 6.250 and the data in Table 6.23 to test whether the data provide sufficient evidence to support this claim.

6.252 Testing Whether Story Spoilers Spoil Stories A story spoiler gives away the ending early. Does having a story spoiled in this way diminish suspense

and hurt enjoyment? A study[60] investigated this question. For 12 different short stories, the study's authors created a second version in which a spoiler paragraph at the beginning discussed the story and revealed the outcome. Each version of the 12 stories was read by at least 30 people and rated on a 1 to 10 scale to create an overall rating for the story, with higher ratings indicating greater enjoyment of the story. The ratings are given in Table 6.24 and stored in **StorySpoilers**. Stories 1 to 4 were ironic twist stories, stories 5 to 8 were mysteries, and stories 9 to 12 were literary stories. Test to see if there is a difference in mean overall enjoyment rating based on whether or not there is a spoiler.

6.253 Measuring the Effect of Story Spoilers Exercise 6.252 describes a study investigating whether giving away the ending of the story makes the story more or less enjoyable to read. The data are given in Table 6.24 and stored in **StorySpoilers**.

(a) Explain why it is appropriate and helpful to use the matched pairs nature of these data in the analysis.

(b) Find and interpret a 95% confidence interval for the difference in mean enjoyment rating between stories with a spoiler and stories without.

6.254 Do Babies Prefer Speech? Psychologists in Montreal and Toronto conducted a study to determine if babies show any preference for speech over general noise.[61] Fifty infants between the ages of 4-13 months were exposed to both happy-sounding infant speech and a hummed lullaby by the same woman. Interest in each sound was measured by the

[60]Leavitt, J., and Christenfeld, N., "Story Spoilers Don't Spoil Stories," *Psychological Science*, published OnlineFirst, August 12, 2011.

[61]Corbeil, M., Trehub, S. E., and Peretz, I., "Speech vs. singing; infants choose happier sounds," *Frontiers in Psychology*, June 25, 2013.

Table 6.24 *Enjoyment ratings for stories with and without spoilers*

Story	1	2	3	4	5	6	7	8	9	10	11	12
With spoiler	4.7	5.1	7.9	7.0	7.1	7.2	7.1	7.2	4.8	5.2	4.6	6.7
Original	3.8	4.9	7.4	7.1	6.2	6.1	6.7	7.0	4.3	5.0	4.1	6.1

Table 6.25 *Are grades higher on the second quiz?*

First Quiz	72	95	56	87	80	98	74	85	77	62
Second Quiz	78	96	72	89	80	95	86	87	82	75

amount of time the baby looked at the woman while she made noise. The mean difference in looking time was 27.79 more seconds when she was speaking, with a standard deviation of 63.18 seconds. Perform the appropriate test to determine if this is sufficient evidence to conclude that babies prefer actual speaking to humming.

6.255 Do Babies Prefer Speaking or Singing? As part of the same study described in Exercise 6.254, the researchers also were interested in whether babies preferred singing or speech. Forty-eight of the original fifty infants were exposed to both singing and speech by the same woman. Interest was again measured by the amount of time the baby looked at the woman while she made noise. In this case the mean time while speaking was 66.97 with a standard deviation of 43.42, and the mean for singing was 56.58 with a standard deviation of 31.57 seconds. The mean of the differences was 10.39 more seconds for the speaking treatment with a standard deviation of 55.37 seconds. Perform the appropriate test to determine if this is sufficient evidence to conclude that babies have a preference (either way) between speaking and singing.

6.256 Do Hands Adapt to Water? Researchers in the UK designed a study to determine if skin wrinkled from submersion in water performed better at handling wet objects.[62] They gathered 20 participants and had each move a set of wet objects and a set of dry objects before and after submerging their hands in water for 30 minutes (order of trials was randomized). The response is the time (seconds) it

[62]Kareklas, K., Nettle, D., and Smulders, T.V., "2013 Water-induced finger wrinkles improve handling of wet objects." *Biol. Lett* vol 9 (2013).

took to move the specific set of objects with wrinkled hands minus the time with unwrinkled hands. The mean difference for moving dry objects was 0.85 seconds with a standard deviation of 11.5 seconds. The mean difference for moving wet objects was −15.1 seconds with a standard deviation of 13.4 seconds.

(a) Perform the appropriate test to determine if the wrinkled hands were significantly faster than unwrinkled hands at moving dry objects.

(b) Perform the appropriate test to determine if the wrinkled hands were significantly faster than unwrinkled hands at moving wet objects.

6.257 Are Grades Significantly Higher on the Second Quiz? Table 6.25 gives a sample of grades on the first two quizzes in an introductory statistics course. We are interested in testing whether the mean grade on the second quiz is significantly higher than the mean grade on the first quiz.

(a) Complete the test if we assume that the grades from the first quiz come from a random sample of 10 students in the course and the grades on the second quiz come from a different separate random sample of 10 students in the course. Clearly state the conclusion.

(b) Now conduct the test if we assume that the grades recorded for the first quiz and the second quiz are from the same 10 students in the same order. (So the first student got a 72 on the first quiz and a 78 on the second quiz.)

(c) Why are the results so different? Which is a better way to collect the data to answer the question of whether grades are higher on the second quiz?

Summary of Inference for Means and Proportions using Distributions

In Unit C, we discuss the process of using the normal and t-distributions, together with formulas for standard errors, to make inferences about means and proportions. The Central Limit Theorem tells us that, when the sample size is large enough, sample means and sample proportions are approximately normally distributed and centered at the value of the corresponding population parameter.

The general formulas we obtain are:

Confidence Interval: Sample Statistic $\pm (t^* \text{ or } z^*) \cdot SE$

Hypothesis Test: $\text{Test Statistic} = \dfrac{\text{Sample Statistic} - \text{Null Parameter}}{SE}$

When deciding which procedure to apply to answer a given question, we need to consider at least three aspects of the question:

- Is the question about a quantitative variable (*mean*) or categorical data (*proportion*)?
- Are we considering a *single sample*, comparing *two samples*, or using *paired data*?
- Are we interested in estimating the size of a parameter or effect (*confidence interval*) or checking if a difference or effect exists (*hypothesis test*)?

When doing inferences for proportions, we approximate the distribution of sample proportions with a normal distribution, provided the sample size is large enough. "Large" usually means some version of $np \geq 10$ and $n(1 - p) \geq 10$.

When doing inference for means and using the standard deviation from a sample, we use a t-distribution, provided the condition of normality is reasonably met. This condition is more critical for smaller sample sizes (less than 30) and can be relaxed for larger samples.

Once we have decided on the method to use, the tables on the next page give a summary of the key formulas, distributions, and conditions for applying each technique.

Summary of Key Formulas and Facts

	Distribution	Conditions	Standard Error
Proportion	Normal	$np \geq 10$ and $n(1-p) \geq 10$	$\sqrt{\dfrac{p(1-p)}{n}}$
Mean	$t, df = n-1$	$n \geq 30$ or reasonably normal	$\dfrac{s}{\sqrt{n}}$
Diff. in Proportions	Normal	$n_1 p_1 \geq 10$, $n_1(1-p_1) \geq 10$, and $n_2 p_2 \geq 10$, $n_2(1-p_2) \geq 10$	$\sqrt{\dfrac{p_1(1-p_1)}{n_1} + \dfrac{p_2(1-p_2)}{n_2}}$
Diff. in Means	$t, df =$ the smaller of $n_1 - 1$ and $n_2 - 1$	$n_1 \geq 30$ or reasonably normal, and $n_2 \geq 30$ or reasonably normal	$\sqrt{\dfrac{s_1^2}{n_1} + \dfrac{s_2^2}{n_2}}$
Paired Diff. in Means	$t, df = n_d - 1$	$n_d \geq 30$ or reasonably normal	$\dfrac{s_d}{\sqrt{n_d}}$

	Confidence Interval	Test Statistic
General	Sample Statistic $\pm\, z^* \cdot SE$	$\dfrac{\text{Sample Statistic} - \text{Null Parameter}}{SE}$
Proportion	$\hat{p} \pm z^* \cdot \sqrt{\dfrac{\hat{p}(1-\hat{p})}{n}}$	$\dfrac{\hat{p} - p_0}{\sqrt{\frac{p_0(1-p_0)}{n}}}$
Mean	$\bar{x} \pm t^* \cdot \dfrac{s}{\sqrt{n}}$	$\dfrac{\bar{x} - \mu_0}{s/\sqrt{n}}$
Difference in Proportions	$(\hat{p}_1 - \hat{p}_2) \pm z^* \cdot \sqrt{\dfrac{\hat{p}_1(1-\hat{p}_1)}{n_1} + \dfrac{\hat{p}_2(1-\hat{p}_2)}{n_2}}$	$\dfrac{(\hat{p}_1 - \hat{p}_2) - 0}{\sqrt{\frac{\hat{p}(1-\hat{p})}{n_1} + \frac{\hat{p}(1-\hat{p})}{n_2}}}$
Difference in Means	$(\bar{x}_1 - \bar{x}_2) \pm t^* \cdot \sqrt{\dfrac{s_1^2}{n_1} + \dfrac{s_2^2}{n_2}}$	$\dfrac{(\bar{x}_1 - \bar{x}_2) - 0}{\sqrt{\frac{s_1^2}{n_1} + \frac{s_2^2}{n_2}}}$
Paired Diff. in Means	$\bar{x}_d \pm t^* \cdot \dfrac{s_d}{\sqrt{n_d}}$	$\dfrac{\bar{x}_d - 0}{s_d/\sqrt{n_d}}$

Case Study: Golden State Warriors Free Throws

The Golden State Warriors set a National Basketball Association (NBA) record in the 2015–2016 season by winning 73 games and losing only 9 games during the regular season. The data described below come from the Warriors' games during their record breaking season.

**Golden State Warriors' Stephen Curry
attempts a free throw**

DATA C.1   **Golden State Warriors 2015–2016 Regular Season**

The data in **GSWarriors** include game results for all 82 games of the 2015–16 regular season.[64] Variables include information such as the number of points scored, shots attempted, steals, rebounds, turnovers, and fouls for both the Warriors and their opponent in each game. There is also a *Location* variable that codes whether each game was played on the Golden State Warriors' home court in Oakland or on the road in their opponent's city.   ■

This case study focuses on the topic of free throws. These are shots awarded to a team for certain infractions (fouls) made by its opponent (hence they are also called foul shots). Free throws are taken at a set distance (15 feet) from the basket with no opponent allowed to defend the shot. This produces a consistent environment for comparing shooting effectiveness. The variables of interest in the **GSWarriors** data are free throws attempted by the Warriors each game (FTA), free throws successfully made (FT), and similar quantities for its opponents ($OppFTA$ and $OppFT$).

[64] Data for the 2015–2016 Golden State Warriors' games *http://www.basketball-reference.com/teams/GSW/2016/gamelog/.*

Here are some of the questions that fans might have about Golden State Warriors free throws:

- What's an average number of free throws for the Warriors to attempt during a game?
- Do the Warriors make more free throws (on average) during games at home than on the road?
- What proportion of free throw attempts do the Warrior players make?
- Players in the NBA as a whole make about 75.6% of their free throws. Is the proportion made by the Warriors different from this?
- Is the mean number of free throw attempts awarded to the Warriors during their games different from the mean number attempted by their opponents?
- Is the proportion of free throws made by the Warriors different between games they play at home and those they play on the road?
- Over the past 10 years, NBA teams have averaged close to 25 free throw attempts per game. Treating this as the population mean, is the mean number of free throw attempts by the Warriors much different?
- How many more (or fewer) free throw attempts do the Warriors take (on average) for home games compared to road games?
- How much better (or worse) are the Warriors at making free throw attempts compared to their opponents?
- How many more (or fewer) free throws do the Warriors make (on average per game) than their opponents?

We address some of these questions in the next few examples and leave the rest for you to try in the exercises. For purposes of this case study, we will regard the 82 games played by the Golden State Warriors in the 2015–2016 regular season as a sample of games they might have played against other NBA opponents in that or future seasons. For some questions the sampling unit is games (for example in determining the mean number of free throws that are attempted in a game). For other questions the sampling unit is each free throw attempt, in which case we have a total sample of 1790 attempts by Warrior players over the season. Before we start any analysis we'll first consider whether conditions are likely to be met.

Example C.1

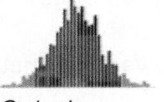

Solution

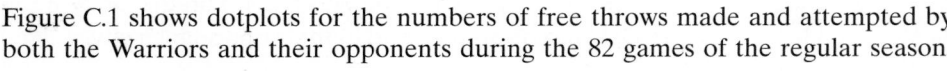

Are normality conditions reasonable for the distributions of numbers of free throws attempted or made per game? Check both the Warriors and their opponents.

Figure C.1 shows dotplots for the numbers of free throws made and attempted by both the Warriors and their opponents during the 82 games of the regular season.

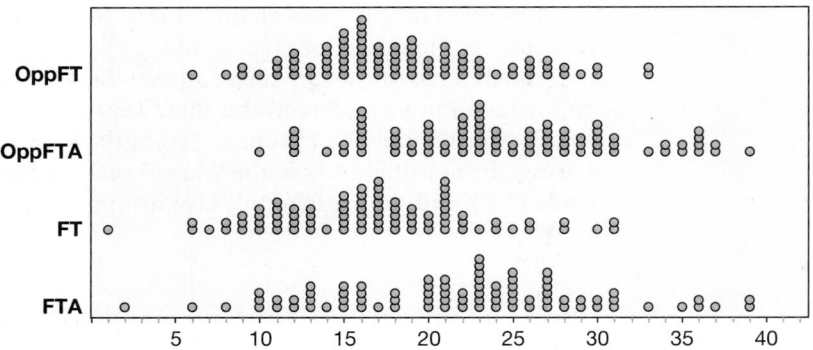

Figure C.1 *Free throws attempted and made by the Golden State Warriors and opponents*

They each are relatively symmetric and bell-shaped with no strong outliers. Also a sample size of 82 (for the full season) or 41 (for home or away games) should be ample for the Central Limit Theorem to apply when doing inference for means.

Example C.2

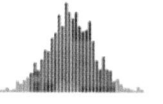

What's an average number of free throws for the Warriors to attempt during a game?

Solution

This is a question about estimating a single mean, μ = mean number of free throw attempts per game. We can use the sample data in the *FTA* variable of **GSWarriors** to estimate the mean and then find a 90% confidence interval for the population mean.

The mean of the *FTA* variable over all 82 games is $\bar{x} = 21.83$ attempts with a standard deviation of $s = 7.73$ attempts. We find a confidence interval for the mean based on a *t*-distribution with $82 - 1 = 81$ degrees of freedom. For 90% confidence this gives a value of $t^* = 1.664$. The 90% confidence interval is

$$\bar{x} \pm t^* \cdot \frac{s}{\sqrt{n}} = 21.83 \pm 1.664 \cdot \frac{7.73}{\sqrt{82}} = 21.83 \pm 1.42 = (20.41, 23.25)$$

Based on this sample of games, we are 90% sure that the Golden State Warriors average somewhere between 20.41 and 23.25 free throw attempts per game.

Example C.3

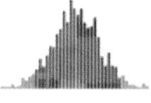

How accurately do the Warrior players make the free throws they attempt?

Solution

This is a question about estimating a single proportion, p = the proportion of free throw attempts that the Warrior players successfully make. In addition to estimating this proportion, we'll construct a 99% confidence interval for the proportion.

By summing the values in *FT* and summing the values in *FTA* for all 82 games we see that the Warriors made 1366 out of 1790 free throw attempts over the entire season. This gives a sample proportion of $\hat{p} = 1366/1790 = 0.763$ or 76.3%. The sample size is very large so we use the normal distribution to find the 99% confidence interval, using $z^* = 2.576$. The confidence interval is

$$\hat{p} \pm z^* \cdot \sqrt{\frac{\hat{p}(1 - \hat{p})}{n}} = 0.763 \pm 2.576 \cdot \sqrt{\frac{0.763(1 - 0.763)}{1790}}$$
$$= 0.763 \pm 0.026 = (0.737, \; 0.789)$$

Based on this sample of free throws, we are 99% sure that the Golden State Warriors make between 73.7% and 78.9% of their free throws.

Some basketball fans suspect that referees are more likely to call fouls on players for the visiting team, possibly in response to reactions from the home crowd. It might also be true that visiting players actually commit more fouls, or that visiting players are less effective at shooting when on the road. In either case, we might expect to see a team average more free throws made at home games than on the road. Let's see if there really is much difference.

Example C.4

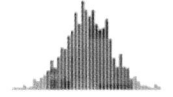

Do the Warriors make more free throws (on average) during games at home than on the road?

Solution

This is a question about comparing two means, μ_H and μ_A, the respective mean number of free throws made at home and away. It suggests an upper tail test of $H_0 : \mu_H = \mu_A$ vs $H_a : \mu_H > \mu_A$ to see if we have evidence that the average number of free throws made by the Warriors when playing games at home is more than the mean number of free throws they make for road games.

Using technology we find the mean and standard deviation for number of free throws made (*FT*) at home and away:

Location	Sample Size	Mean	Std. Dev.
Home	41	17.07	6.10
Away	41	16.24	5.99

Both sample sizes are more than 30 and we see no big outliers in the *FTA* values, so we use a two-sample *t*-test:

$$t = \frac{\bar{x}_H - \bar{x}_A}{\sqrt{\frac{s_H^2}{n_H} + \frac{s_A^2}{n_A}}} = \frac{17.07 - 16.24}{\sqrt{\frac{6.10^2}{41} + \frac{5.99^2}{41}}} = 0.62$$

We compare this to the upper tail of a *t*-distribution with $41 - 1 = 40$ degrees of freedom to get a p-value of 0.269. This is not less than a significance level of $\alpha = 0.05$, so we do not reject H_0 and do not have enough evidence to conclude that the Warriors average more free throws made at home games than at road games.

Example C.5

Is the mean number of free throw attempts awarded to the Warriors during their games different from the mean number attempted by their opponents?

Solution

The **GSWarriors** dataset also has the number of free throws attempted by the Warriors' opponent in each game (*OppFTA*). This question asks for a test of a difference in mean free throw attempts per game between the Warriors and their opponents. Now we have paired data, values for each variable (*FTA* and *OppFTA*) in each of the 82 games. Pairing the data by game makes sense, since we might expect that referees call some games more tightly and others have fewer fouls (for both teams). The hypotheses are $H_0 : \mu_d = 0$ vs $H_a : \mu_d \neq 0$, where μ_d is the mean difference in number of free throw attempts (per game) between the Warriors and their opponents.

Since this is paired data, we first find the difference (*FTA* − *OppFTA*) for each game. Figure C.2 shows a dotplot of these differences which appears relatively symmetric with no big outliers.

The mean for the differences is $\bar{x}_d = -2.72$ and the standard deviation of the differences is $s_d = 10.02$. Treating the differences as a single sample, the t-statistic is

$$t = \frac{\bar{x}_d - 0}{s_d / \sqrt{n_d}} = \frac{-2.72}{10.02 / \sqrt{82}} = -2.46$$

Figure C.2 *Difference in Warriors and opponent free throw attempts*

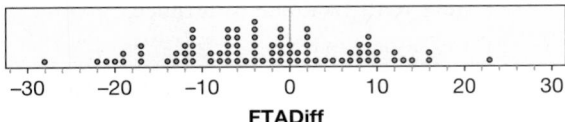

FTADiff

We compare this to a t-distribution with $82 - 1 = 81$ degrees of freedom, doubling the proportion in the tail to get a p-value $= 2(0.008) = 0.016$. This is a small p-value, so we have strong evidence that (on average) the Warriors attempt fewer free throws in games than their opponents.

Example C.6

How much better (or worse) are the Warriors at making their free throw attempts compared to their opponents?

Solution

This question is asking about estimating a difference in proportions, $p_W - p_O$, where p_W and p_O are the proportions of free throw attempts that the Warriors and their opponents make. In Example C.3 we see that the Warriors made 1366 of their 1790 attempts, so $\hat{p}_W = 0.763$. Summing the *OppFT* and *OppFTA* columns in **GSWarriors**, we find that their opponents made 1523 out of 2013 free throw attempts, so $\hat{p}_O = 1523/2013 = 0.757$. The estimated difference in proportions is

$$\hat{p}_W - \hat{p}_O = 0.763 - 0.757 = 0.006$$

We should also construct a confidence interval for this difference in proportions; we'll choose 95% confidence. The sample sizes are large so we use a normal distribution with $z^* = 1.96$. The confidence interval is

$$(\hat{p}_W - \hat{p}_O) \pm z^* \cdot \sqrt{\frac{\hat{p}_W(1 - \hat{p}_W)}{n_W} + \frac{\hat{p}_O(1 - \hat{p}_O)}{n_O}}$$

$$(0.763 - 0.757) \pm 1.96 \cdot \sqrt{\frac{0.763(1 - 0.763)}{1790} + \frac{0.757(1 - 0.757)}{2013}}$$

$$0.006 \pm 0.027$$

$$-0.021 \text{ to } 0.033$$

Based on these results, we are 95% sure that the Warriors successfully make somewhere between 2.1% fewer to 3.3% more of their free throw attempts than their opponents.

While the Warriors tend to get fewer free throw attempts (on average) than their opponents, a key question is how much does that hurt them on the scoreboard?

Example C.7

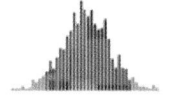

How does the average number of free throws made (per game) by the Warriors compare to their opponents?

Solution

This question is similar to Example C.5, only now we are interested in estimating the magnitude of the difference in mean number of free throws made (rather than testing the difference in mean number of free throws attempted). Again, we have paired data, this time finding the difference in free throws made for each game, $FT - OppFT$. The mean of these differences is $\bar{x}_d = -1.91$ and the standard deviation is $s_d = 8.45$. As with the differences in free throws attempted, the distribution is relatively symmetric with no big outliers. We find a 95% confidence interval for the difference using a t-distribution with 81 degrees of freedom, so $t^* = 1.990$. The confidence interval is

$$\bar{x}_d \pm t^* \cdot \frac{s_d}{\sqrt{n_d}} = -1.91 \pm 1.990 \cdot \frac{8.45}{\sqrt{82}} = -1.91 \pm 1.86 = (-3.77, -0.05)$$

Based on these results, we are 95% sure that the Warriors average somewhere between 3.77 and 0.05 fewer points from free throws made per game than their opponents—yet they still managed to win 73 games and only lose 9 in the regular season!

From the data in **GSWarriors** we can also see that the Warriors outscored their opponents in those games by an average score of 114.89 to 104.13 points, a victory margin of 10.76 points on average. The rest of their game was strong enough to make up for the small deficit in points earned with free throws.

SECTION LEARNING GOALS

You should now have the understanding and skills to:

▶ • Demonstrate an understanding of the complete process of using the normal or *t*-distribution in inference for means and proportions

▶ • Identify which type of inference for means or proportions is appropriate in a given situation

▶ • Put all the pieces together to answer more involved questions using real data

---

Exercises for UNIT C: Essential Synthesis

IDENTIFYING THE METHOD OF ANALYSIS
In Exercises C.1 to C.8, identify the method of analysis needed to answer the question. Indicate whether we should conduct a hypothesis test or find a confidence interval and also indicate whether the analysis will be done on a proportion, a mean, a difference in proportions, a difference in means, or a matched pairs difference in means.

C.1 Use data collected at a retail store to estimate the average amount of money people spend in the store.

C.2 Use results collected at a supermarket to see whether there is a difference in the average amount of time customers have to wait in line between two different check-out cashiers.

C.3 Use data from an experiment on mice to see if there is evidence that mice fed a high-sugar diet are more likely to be classified as insulin-resistant than mice fed a normal diet.

C.4 Use data collected at an online shopping site to estimate the proportion of people visiting the site who make a purchase.

C.5 Use data collected from a sample of applicants at a college admissions office to measure how large the difference is in the average size of the financial aid package between early decision applicants and regular decision applicants.

C.6 Use data from a study done at a college fitness center in which muscle mass of participants was measured before and after a 6-week program working with resistance bands to estimate the mean increase in muscle mass.

C.7 Use a sample of students at a large university to determine whether the proportion of students at the university who are left-handed is different from the national US proportion of 10%.

C.8 Use results from a survey to estimate the difference in the proportion of males and females who say they are trying to lose weight.

USING INTUITION ABOUT TEST STATISTICS
In Exercises C.9 to C.14, a standardized test statistic is given for a hypothesis test involving proportions (using the standard normal distribution) or means (using the *t*-distribution and assuming a relatively large sample size.) *Without using any technology or tables*, in each case:

(a) Is the p-value likely to be relatively *large* or relatively *small*?

(b) Is the conclusion of the test likely to be *Reject H_0* or *Do not reject H_0*?

C.9 $z = 5.6$
C.10 $z = 8.3$
C.11 $z = 0.54$
C.12 $t = 12.2$
C.13 $t = 7.1$
C.14 $t = 0.83$

RESTAURANT TIPS

Exercises C.15 to C.19 refer to the dataset **RestaurantTips**. The data were introduced in Data 2.12 on page 123, and include information from a sample of 157 restaurant bills collected at the *First Crush* bistro.

C.15 Table C.1 shows a two-way table for Servers A, B, and C and for whether a credit/debit card or cash was used for payment (yes for a credit or debit card, no for cash). Do the data in the table provide evidence that Server B is responsible for more than 1/3 of the bills at this restaurant?

Table C.1 *Two-way table of server and method of payment*

	A	B	C
Yes	21	15	15
No	39	50	17

C.16 Use the information in Table C.1 to compute and interpret a 95% confidence interval for the proportion of bills paid with a credit card.

C.17 Use the information in Table C.1 to determine whether the sample provides evidence of a difference between Servers B and C in the proportion of bills paid with cash.

C.18 Table C.2 gives summary statistics for the tip percentage based on whether or not a credit card was used. In the sample, which method of payment has a larger average tip percent? Which method has more variability? Is there evidence of a difference in mean tip percentage depending on the method of payment?

Table C.2 *Summary statistics for tip percent by method of payment*

Credit?	Sample Size	Mean	Std. Dev.
Yes	51	17.10	2.47
No	106	16.39	5.05

C.19 Table C.3 gives summary statistics for the size of the bill based on whether or not a credit card was used. In the sample, which method of payment was used for larger bills? Which method has more variability in the size of the bill? Is there evidence of a difference in the mean size of the bill depending on the method of payment?

Table C.3 *Summary statistics for size of bill by method of payment*

Credit?	Sample Size	Mean	Std. Dev.
Yes	51	29.4	14.5
No	106	19.5	9.4

C.20 Posture and Pain Research shows that people adopting a dominant pose have reduced levels of stress and feel more powerful than those adopting a submissive pose. Furthermore, it is known that if people feel more control over a situation, they have a higher tolerance for pain. Putting these ideas together, a recent study[65] describes three experiments investigating how posture might influence the perception of pain.

(a) In the first experiment, 89 participants were told that they were in a study to examine the health benefits of doing yoga poses at work. All participants had their pain threshold measured both before and after holding a yoga pose for 20 seconds. The pain threshold was measured by inflating a blood pressure cuff until participants said stop: The threshold was measured in mmHg and the difference in before and after thresholds was recorded. Participants were randomly divided into two groups: One group ($n = 45$) was randomly assigned to strike a dominant pose (moving limbs away from the body) while the other group ($n = 44$) was assigned to strike a submissive pose (curling the torso inward). The mean change in pain threshold for the group striking a dominant pose was 14.3 with a standard deviation of 39.8, while the mean change in pain threshold for the group striking a submissive pose was −6.1 with a standard deviation of 40.4. Does the experiment provide evidence that a dominant pose increases one's mean tolerance of pain more than a submissive pose?

(b) Prior research has shown that a person will assume a pose complementary to the pose of a peer or colleague: assuming a more submissive pose if the peer has a dominant pose and vice versa. In the second experiment, 30 participants were told they were participating in a study on relaxation methods and randomly divided into

[65]Data approximated from Bohns, V., and Wiltermuth, S., "It hurts when I do this (or you do that): Posture and pain tolerance," *Journal of Experimental Social Psychology*, 2012; 48(1): 341–345.

two groups of size 15. Each participant took turns describing nature photographs with a peer who was part of the study and was secretly told to strike either a dominant or submissive posture during the interactions. Pain thresholds were measured in the same way as in the first experiment. Mean difference in pain threshold was −13.8 with a standard deviation of 27.1 for the group with a dominant peer and 4.2 with a standard deviation of 22.9 for the group with a submissive peer. Does the experiment provide evidence that mean pain tolerance is higher if one's interaction partner is submissive? The data do not have any significant outliers.

(c) As part of the experiment described in part (b), participants were also given a handgrip strength test both before and after the interaction with the peer, and the difference in handgrip strength was measured in newtons. Mean change in handgrip strength for those with a dominant interaction partner is −45.3 newtons with a standard deviation of 46.5 while for those with a submissive partner mean change was −6.8 with a standard deviation of 31.0. The data do not have any very large outliers. Find a 90% confidence interval for the difference in means and interpret the result. Based on the confidence interval, do you believe that there is a significant difference in mean change in handgrip strength between those with a submissive partner and those with a dominant partner?

(d) Since reducing the perception of pain is a goal in health care, what are the implications of these studies for health care professionals?

C.21 Mental Imaging and Muscle Fatigue Studies suggest that when people mentally rehearse a physical action, they engage similar neural and cognitive operations as when they actually perform the action. Because of this, mental imaging can be a valuable training tool. A new study[66] explores how actual muscle fatigue affects mental imaging. In the study, participants were asked to either perform actual arm pointing motions or to mentally imagine equivalent arm pointing motions. Participants then developed muscle fatigue by holding a heavy weight out horizontally as long as they could. After becoming

fatigued, they were asked to repeat the previous mental or actual motions. Eight participants were assigned to each group, and the time in seconds to complete the motions is given in Table C.4 and stored in **MentalMuscle**. Use a 5% significance level for all tests.

(a) Test to see whether there is a significant difference in mean times between mentally imaging doing the actions and actually doing the actions before any muscle fatigue (pre-fatigue).

(b) Test to see whether people who actually perform the motions are slower, on average, post-fatigue than pre-fatigue.

(c) Test to see whether people who mentally perform the motions are faster, on average, at mentally imaging the actions post-fatigue than pre-fatigue.

(d) Test to see whether there is a significant difference in mean times between mentally imaging doing the actions and actually doing the actions, after experiencing muscle fatigue (post-fatigue).

(e) Write a short paragraph summarizing the results of the experiment.

C.22 Results from the Student Survey Data 1.1 on page 4 describes a dataset giving results of a student survey. We use the students who filled out the survey as a sample from the population of all students at that university. Answer the following questions **using the computer output which follows the exercise**. Justify your answers to the questions using specific values from the computer output. In particular, for all tests, give the null and alternative hypotheses, the p-value from the computer output, and the conclusion in context.

(a) Nationally, about 20% of people smoke. What percent of students in the sample smoke? Is the percent of all students at this university who smoke different from the national percentage?

(b) Is the average math SAT score of students at this university greater than 600?

[66]Data approximated from summary statistics in Demougeot, L., and Papaxanthis, C., "Muscle Fatigue Affects Mental Simulation of Action," *The Journal of Neuroscience*, 2011;31(29): 10712–20.

Table C.4 *Mental and actual times for physical activity after muscle fatigue*

Mental pre-fatigue	5.9	9.9	8.1	7.2	6.6	7.4	6.9	6.7
Mental post-fatigue	7.4	6.0	6.6	6.1	5.5	7.2	5.4	4.6
Actual pre-fatigue	7.3	7.8	6.8	7.1	6.2	7.2	8.4	6.5
Actual post-fatigue	9.8	7.8	7.5	7.6	6.4	7.8	8.1	9.3

(c) One of the variables in the dataset is *HigherSAT*, which indicates whether the math or the verbal SAT score was higher for each student. What is the proportion of females in the sample with a higher verbal SAT score? What is the proportion of males for whom the verbal score is higher? Is the proportion for whom verbal is higher different between males and females for all students at this university? In addition to the test, state and interpret a 95% confidence interval for the gender effect: the difference between the proportion of females with a higher verbal score minus the proportion of males with a higher verbal score.

(d) Who has a higher average pulse rate in the sample: smokers or non-smokers? Is there evidence of a difference in mean pulse rate between smokers and non-smokers for all students at this university?

(e) Who has a higher mean GPA (grade point average) in the sample: smokers or non-smokers? Is there evidence of a difference in mean GPA between smokers and non-smokers for all students at this university? Are the results significant at a 10% level? At a 5% level? At a 1% level?

Computer Output for Exercise C.22

Test and CI for One Proportion: Smoke
Test of p = 0.2 vs p not = 0.2
Event = Yes

Variable	X	N	Sample p	95% CI	P-Value
Smoke	43	362	0.118785	(0.085456, 0.152113)	0.000

One-Sample T: MathSAT
Test of mu = 600 vs > 600

Variable	N	Mean	StDev	SE Mean	95% Lower Bound	T	P
MathSAT	362	609.44	68.49	3.60	603.50	2.62	0.005

Test and CI for Two Proportions: HigherSAT, Gender
Event = Verbal

Gender	X	N	Sample p
F	84	165	0.509091
M	66	190	0.347368

Difference = p (F) − p (M)
Estimate for difference: 0.161722
95% CI for difference: (0.0597323, 0.263713)
Test for difference = 0 (vs not = 0): Z = 3.11 P-Value = 0.002

Two-Sample T-Test and CI: Pulse, Smoke
Two-sample T for Pulse

Smoke	N	Mean	StDev	SE Mean
No	319	69.3	12.3	0.69
Yes	43	71.8	11.7	1.8

Difference = mu (No) −mu (Yes)
Estimate for difference: −2.54
95% CI for difference: (−6.37, 1.28)
T-Test of difference = 0 (vs not =): T-Value = −1.33 P-Value = 0.188 DF = 55

Two-Sample T-Test and CI: GPA, Smoke
Two-sample T for GPA

Smoke	N	Mean	StDev	SE Mean
No	302	3.173	0.399	0.023
Yes	43	3.054	0.379	0.058

Difference = mu (No) −mu (Yes)
Estimate for difference: 0.1188
95% CI for difference: (−0.0059, 0.2435)
T-Test of difference = 0 (vs not =): T-Value = 1.91 P-Value = 0.061 DF = 56

MORE GOLDEN STATE WARRIORS FREE THROWS

The data in **GSWarriors** contain information from 82 regular season games played by the Golden State Warriors basketball team. Exercises C.23 to C.26 involve some of the other questions raised in the Case Study introduced in this section. For each question, decide what inference technique is appropriate for addressing it and use the data in **GSWarriors** to carry out the procedure and reach a conclusion. Use 95% confidence for any intervals and $\alpha = 0.05$ for any hypothesis tests.

C.23 Is there evidence that the mean number of free throw attempts per game by the Warriors is different from the mean for all NBA teams? Assume that the mean number of free throws attempted by teams in all NBA games is 25.0 (based on a very large number of games over the past 10 years).

C.24 Is the proportion of free throws successfully made by the Warrior players different from the overall proportion for all NBA players? Assume that the population proportion of free throws made by all NBA players is about 0.756 (based on many free throw attempts over a ten year period).

C.25 Is the proportion of free throws made by the Warriors better at games they play at home versus those they play on the road? In Example C.4 on page 481 we did not find enough evidence to conclude that the mean number of free throws made by the Warriors in home games is larger than away games. Can we detect a home advantage in the proportion of free throws made being higher than at away games? Here is a table showing Golden State free throws made and attempted in both locations.

Location	Made	Attempts
Home	700	894
Away	666	896
Total	1366	1790

C.26 How many more (or fewer) free throw attempts do the Warriors tend to get (on average) at home games compared to their road games? Find a 95% confidence interval for the difference in mean free throw attempts.

C.27 Patients Admitted to an Intensive Care Unit Data 2.3 on page 69 describes a dataset about patients being admitted to an Intensive Care Unit at a large hospital. We use the patients for whom information is available as a sample from the population of all patients admitted to the ICU at this hospital.

Use technology and the data stored in the file **ICUAdmissions** to answer the following questions. Justify your answers to the questions using specific values from the statistical software package that you use. In particular, for all tests, give the null and alternative hypotheses, the p-value from the computer output, and the conclusion in context.

(a) The average heart rate for healthy adults is 72 beats per minute. What is the average heart rate (*HeartRate*) for the sample of patients admitted to the ICU? Does it give evidence that the average heart rate of patients admitted to this ICU is different than 72?

(b) What proportion of patients in the sample died? Survival is coded as 0 = lived and 1 = died in the *Status* variable. Find and interpret a 95% confidence interval for the proportion of patients admitted to this ICU that die.

(c) Were more males or females admitted to this ICU in this sample? Gender is coded as 0 for male and 1 for female in the *Sex* variable. Is there evidence that the genders of patients are not equally split between males and females among all ICU patients at this hospital?

(d) Does the average age of patients admitted to this ICU differ between males and females?

(e) Does the proportion of patients who die differ between males and females?

C.28 Effect of Diet on Nutrient Levels Data 2.11 on page 114 describes a dataset that gives nutrient levels in people's blood as well as information about their eating habits. We use the people for whom information is available as a sample from the population of all people. Use technology and the data stored in the file **NutritionStudy** to answer the following questions. Justify your answers to the questions using specific values from the statistical software package that you use. In particular, for all tests, give the null and alternative hypotheses, the p-value from the computer output, and the conclusion in context.

(a) Find and interpret a 95% confidence interval for the percent of people that smoke.

(b) Find and interpret a 99% confidence interval for the average number of grams of fiber per day that people eat. Give the best estimate, the margin of error, and the confidence interval.

(c) Find and interpret a 90% confidence interval for the average number of grams of fat per day that people eat.

(d) Is there evidence of a difference in the percent of current smokers by gender?

(e) Is there evidence of a difference in the mean cholesterol level of males and females?

(f) Is there evidence of a difference in mean level of beta carotene in the blood (*BetaPlasma*) between smokers and non-smokers?

Review Exercises for UNIT C

FINDING AREAS IN A STANDARD NORMAL DISTRIBUTION In Exercises C.29 and C.30, find the specified areas for a standard normal distribution.

C.29 (a) The area below $z = -2.10$

(b) The area above $z = 1.25$

C.30 (a) The area below $z = 1.68$

(b) The area above $z = 2.60$

FINDING ENDPOINTS ON A STANDARD NORMAL DISTRIBUTION In Exercises C.31 and C.32, find endpoint(s) on a standard normal distribution with the given property.

C.31 (a) The area to the left of the endpoint is about 0.60.

(b) The area to the left of the endpoint is about 0.02.

C.32 (a) The area to the left of the endpoint is about 0.25.

(b) The area to the right of the endpoint is about 0.08.

USING A *t*-DISTRIBUTION Use a *t*-distribution to answer the questions in Exercises C.33 to C.36. Assume the sample is a random sample from a distribution that is reasonably normally distributed and we are doing inference for a sample mean.

C.33 Find endpoints of a *t*-distribution with 5% beyond them in each tail if the sample has size $n = 25$.

C.34 Find endpoints of a *t*-distribution with 1% beyond them in each tail if the sample has size $n = 12$.

C.35 Find the area in a *t*-distribution to the right of 2.75 if the sample has size $n = 10$.

C.36 Find the area in a *t*-distribution to the left of -1.50 if the sample has size $n = 24$.

C.37 Grams of Fiber per Day The **NutritionStudy** dataset includes a variable indicating the number of grams of fiber consumed per day by the participants. In the sample, the mean is 12.79 grams and the standard error based on a simulation distribution is $SE = 0.30$. The sample size of $n = 315$ is large enough to use a normal distribution.

(a) Find and interpret a 95% confidence interval for mean number of grams of fiber per day.

(b) Use a normal distribution to test whether there is evidence that mean number of grams of fiber is greater than 12. Give all details of the test.

C.38 Online Browsing on a Phone A recent study[67] shows that 17% of a random sample of 1954 cell phone owners do most of their online browsing on their phone. The standard error for the proportion is 0.0085. The sample size is large enough to use a normal distribution.

(a) Find and interpret a 90% confidence interval for the proportion of cell phone owners who do most of their online browsing on their phone.

(b) Use a normal distribution to test whether there is evidence that the proportion is greater than 0.15. Give all details of the test.

C.39 Do Kids Spend Too Much Time on Electronic Devices? In a nationwide poll of 1000 randomly sampled adults, 83% said they think children spend too much time on their computers and other electronic devices (but 37% say time spent on a computer is better than time spent in front of a TV).[68] Find and interpret a 95% confidence interval for the proportion of adults who believe children spend too much time on electronic devices. What is the margin of error for this result? Is it plausible that the proportion of all adults who feel this way is less than 80%? Is it plausible that the proportion is greater than 85%?

C.40 Home Field Advantage in Baseball There were 2430 Major League Baseball (MLB) games played in 2009, and the home team won in 54.9% of the games.[69] If we consider the games played in 2009 as a sample of all MLB games, find and interpret a 90% confidence interval for the proportion of games the home team wins in Major League Baseball.

C.41 Size of the Tip in a Restaurant The dataset **RestaurantTips** has information from *First Crush* bistro in northern New York state. Computer output

[67]Smith, A., "Cell Internet Use 2012," Pew Research Center, pewresearch.org, June 26, 2012.

[68]"83% Say Kids Spend Too Much Time On Electronic Devices," Rasmussen Reports, July 6, 2011.

[69]http://www.baseballprospectus.com/article.php?articleid=9854, accessed June 2011.

of descriptive statistics for the variable giving the size of the tip is shown:

Descriptive Statistics: Tip

Variable	N	N*	Mean	SE Mean	StDev
Tip	157	0	3.849	0.193	2.421

Minimum	Q1	Median	Q3	Maximum
0.250	2.050	3.350	5.000	15.000

(a) How many tips are included in the dataset? What is the mean? What is the standard deviation?

(b) Compute the standard error for the mean using the formula $SE = s/\sqrt{n}$. Compare the result to the value given under "SE Mean" in the computer output.

(c) Use the summary statistics to compute a 95% confidence interval for the average tip given at this restaurant.

(d) Compare the answer in part (c) to the confidence interval given in the following computer output for the same data:

One-Sample T: Tip

Variable	N	Mean	StDev	SE Mean	95% CI
Tip	157	3.849	2.421	0.193	(3.468, 4.231)

(e) Interpret the confidence interval in context.

C.42 Number of Walks for a Baseball Team in a Season The dataset **BaseballHits** gives 2014 season statistics for all Major League Baseball (MLB) teams. We treat this as a sample of all MLB teams in all years. Computer output of descriptive statistics for the variable giving the number of *Walks* is shown:

Descriptive Statistics: Walks

Variable	N	N*	Mean	SE Mean	StDev
Walks	30	0	467.33	9.81	53.75

Minimum	Q1	Median	Q3	Maximum
380.0	417.0	469.5	516.3	586.0

(a) How many teams are included in the dataset? What is the mean number of walks? What is the standard deviation?

(b) Compute the standard error for the mean using the formula $SE = s/\sqrt{n}$. Compare the result to the value given under "SE Mean" in the computer output.

(c) Use the summary statistics to compute a 95% confidence interval for the mean number of walks per team in a season.

(d) Compare the answer from part (c) to the confidence interval given in the following computer output for the same data:

One-Sample T: Walks

Variable	N	Mean	StDev	SE Mean	95% CI
Walks	30	467.33	53.75	9.81	(447.26, 487.41)

(e) Interpret the confidence interval in context.

C.43 The Chips Ahoy! Challenge In the mid-1990s a Nabisco marketing campaign claimed that there were at least 1000 chips in every bag of Chips Ahoy! cookies. A group of Air Force cadets collected a sample of 42 bags of Chips Ahoy! cookies, bought from locations all across the country, to verify this claim.[70] The cookies were dissolved in water and the number of chips (any piece of chocolate) in each bag were hand counted by the cadets. The average number of chips per bag was 1261.6, with standard deviation 117.6 chips.

(a) Why were the cookies bought from locations all over the country?

(b) Test whether the average number of chips per bag is greater than 1000. Show all details.

(c) Does part (b) confirm Nabisco's claim that every bag has at least 1000 chips? Why or why not?

C.44 Has Support for Capital Punishment Changed over Time? The General Social Survey (GSS) has been collecting demographic, behavioral, and attitudinal information since 1972 to monitor changes within the US and to compare the US to other nations.[71] Support for capital punishment (the death penalty) in the US is shown in 1974 and in 2006 in the two-way table in Table C.5. Find a 95% confidence interval for the change in the proportion supporting capital punishment between 1974 and 2006. Is it plausible that the proportion supporting capital punishment has not changed?

Table C.5 *Has public support for capital punishment changed?*

Year	Favor	Oppose	Total
1974	937	473	1410
2006	1945	870	2815

C.45 Who Is More Trusting: Internet Users or Non-users? In a randomly selected sample of 2237 US adults, 1754 identified themselves as people who use the Internet regularly while the other 483 indicated that they do not. In addition to Internet use, participants were asked if they agree with the statement "most people can be trusted." The results show that 807 of the Internet users agree with this statement, while 130 of the non-users agree.[72]

[70]Warner, B., and Rutledge, J., "Checking the Chips Ahoy! Guarantee," *Chance*, 1999; 12(1):10–14.

[71]General Social Survey website, http://www3.norc.org/GSS+Website.

[72]Hampton, K., Goulet, L., Rainie, L., and Purcell, K., "Social Networking Sites and Our Lives," Pew Research Center, pewresearch.org, June 16, 2011.

(a) Which group is more trusting in the sample (in the sense of having a larger percentage who agree): Internet users or people who don't use the Internet?

(b) Can we generalize the result from the sample? In other words, does the sample provide evidence that the level of trust is different between the two groups in the broader population?

(c) Can we conclude that Internet use causes people to be more trusting?

(d) Studies show that formal education makes people more trusting and also more likely to use the Internet. Might this be a confounding factor in this case?

HOMES FOR SALE

Exercises C.46 to C.49 refer to the dataset **Homes-ForSale**, which has data on houses available for sale in three Mid-Atlantic states (NY, NJ, and PA) as well as California (CA). Table C.6 has summary statistics for each of the four states, with prices given in thousands of dollars. (Since $n = 30$, we ask you to use the t-distribution here despite the fact that the data are quite skewed. In practice, we might have enough concern about the skewness to choose to use bootstrap methods instead.)

Table C.6 *Mean housing prices for four states*

State	n	Mean	Std. Dev.
New York	30	565.6	697.6
New Jersey	30	388.5	224.7
Pennsylvania	30	249.6	179.3
California	30	715.1	1112.3

C.46 Find and interpret a 95% confidence interval for the mean price of a home in California.

C.47 Find and interpret a 90% confidence interval for the difference in mean housing price between California and New York.

C.48 Find and interpret a 99% confidence interval for the difference in mean housing price between New Jersey and Pennsylvania.

C.49 Find and interpret a 95% confidence interval for the difference in mean housing price between New York and New Jersey.

C.50 Handedness and Earnings Do left-handed or right-handed people make more money? One study[73] recorded the hourly earnings for a random sample of 2295 American men, of whom 2027 were right-handed and 268 were left-handed. The right-handed men earned an average of $13.10 per hour, while the left-handed men earned an average of $13.40 per hour. The sample standard deviation for both left-handed and right-handed workers was about $7.90. Test the hypothesis that the average earnings for left-handed and right-handed men are the same. Be sure to state the null and alternative hypotheses, find the test statistic and p-value, and interpret the conclusion.

C.51 Comparing Weight Loss Methods Researchers randomly assigned 107 young overweight women to cut 25% of their calories in one of two ways: The continuous group ate about 1500 calories a day every day while the intermittent group ate about 500 calories a day for two days a week and their typical diets the rest of the week. (Interestingly, the women who cut calories on two days a week did not overeat on the other days.)[74] The summary statistics for weight loss (in pounds) after 6 months are shown in Table C.7. Test to see if there is any difference in mean weight loss results between the two methods. Show all details of the test.

Table C.7 *Number of pounds lost under two different calorie restriction methods*

Method	n	$\bar{x}$	s
Continuous	54	14.1	13.2
Intermittent	53	12.2	10.6

C.52 Radiation from Cell Phones and Brain Activity Exercise B.58 on page 362 introduces a matched pairs study in which 47 participants had cell phones put on their ears and then had their brain glucose metabolism (a measure of brain activity) measured under two conditions: with one cell phone turned on for 50 minutes (the "on" condition) and with both cell phones off (the "off" condition). Brain glucose metabolism is measured in μmol/100 g per minute, and the differences of the metabolism rate in the on condition minus the metabolism rate in the off condition were computed for all participants. The mean of the differences was 2.4 with a standard deviation of 6.3. Find and interpret a 95% confidence interval for the effect size of the cell phone waves on mean brain metabolism rate.

---

[73]Ruebeck, C., et al., "Handedness and Earnings," *Laterality*, 2007;12(2):101–20.

[74]Harvie, M., et al., "The Effects of Intermittent or Continuous Energy Restriction on Weight Loss and Metabolic Disease Risk Markers: A Randomised Trial in Young Overweight Women," *International Journal of Obesity* (London), 2011; 35(5):714–27.

C.53 Testing the Effects of Cell Phones on Brain Activity Exercise C.52 describes a matched pairs study examining the effect of cell phones on brain glucose metabolism. Use the information there to test to see if there is evidence that mean glucose metabolism is higher when a cell phone is nearby. Show all details of the test.

C.54 Worldwide, Store Brands Are Gaining on Name Brands In a Nielsen global online survey of about 27,000 people from 53 different countries, 61% of consumers indicated that they purchased more store brands during the economic downturn and 91% indicated that they will continue to purchase the same number of store brands when the economy improves. The results were remarkably consistent across all regions of the world.[75] The survey was conducted during September 2010 and had quotas for age and gender for each country, and the results are weighted to be representative of consumers with Internet access.

(a) What is the sample? What is an appropriate population?

(b) Find and interpret a 99% confidence interval for the proportion of consumers who purchased more store brands during the economic downturn.

(c) Find and interpret a 99% confidence interval for the proportion of consumers who will continue to purchase the same number of store brands.

C.55 Congressional Approval Rating In April 2012, the Gallup Poll reported that in a random sample of 1016 US adults, only 17% approve of the way Congress is handling its job.[76]

(a) Use the poll results to estimate the proportion of all US adults who approve of the way Congress is doing its job. What is the margin of error, with 99% confidence, for this estimate?

(b) If the Gallup Poll wants the estimate to be accurate to within ±1%, with 99% confidence, how large a sample must they use?

C.56 Which Award? In Example 2.6 on page 53 we consider a survey of students who were asked to choose an award they would like to win from among an Academy Award, Nobel Prize, and Olympic gold medal. If the awards are equally popular we would expect about 1/3 to choose each type of award. In one sample of 169 female students (see data in **StudentSurvey**) we find that 73 chose the Olympic

gold medal. Does this provide sufficient evidence to conclude that the proportion of female students choosing the Olympic gold medal is not 1/3?

HORMONE REPLACEMENT THERAPY

Exercises C.57 through C.60 refer to a study on hormone replacement therapy. Until 2002, hormone replacement therapy (HRT), taking hormones to replace those the body no longer makes after menopause, was commonly prescribed to postmenopausal women. However, in 2002 the results of a large clinical trial[77] were published, causing most doctors to stop prescribing it and most women to stop using it, impacting the health of millions of women around the world. In the experiment, 8506 women were randomized to take HRT and 8102 were randomized to take a placebo. Table C.8 shows the observed counts for several conditions over the five years of the study. (*Note:* The planned duration was 8.5 years. If Exercises C.57 through C.60 are done correctly, you will notice that several of the intervals just barely exclude zero. The study was terminated as soon as some of the intervals included only positive values, because at that point it was unethical to continue forcing women to take HRT.)

Table C.8 *Counts for several conditions within the HRT group and the placebo group*

Condition	HRT Group	Placebo Group
Cardiovascular Disease	164	122
Invasive Breast Cancer	166	124
Cancer (all)	502	458
Fractures	650	788

C.57 Find a 95% confidence interval for the difference in proportions of women who get cardiovascular disease taking HRT vs taking a placebo.

C.58 Find a 95% confidence interval for the difference in proportions of women who get invasive breast cancer taking HRT vs taking a placebo.

C.59 Find a 95% confidence interval for the difference in proportions of women who get any type of cancer taking HRT vs taking a placebo.

C.60 Find a 95% confidence interval for the difference in proportions of women who fracture a bone taking HRT vs taking a placebo.

C.61 Number of Bedrooms in Houses in New York and New Jersey The dataset **HomesForSale** has data

[75]"The Rise of the Value-Conscious Shopper," A Nielsen Global Private Label Report, www.nielsen.com, March 2011.

[76]http://www.gallup.com/poll/153968/Congressional-Approval-Recovers-Slightly.aspx, accessed April 2012.

[77]Rossouw, J., et al., "Risks and benefits of estrogen plus progestin in healthy postmenopausal women: principal results from the women's health initiative randomized controlled trial," *Journal of the American Medical Association*, 2002; 288(3):321–33.

on houses available for sale in three Mid-Atlantic states (NY, NJ, and PA). For this exercise we are specifically interested in homes for sale in New York and New Jersey. We have information on 30 homes from each state and observe the proportion of homes with more than three bedrooms. We find that 26.7% of homes in NY ($\hat{p}_{NY}$) and 63.3% of homes in NJ ($\hat{p}_{NJ}$) have more then three bedrooms.

(a) Is the normal distribution appropriate to model this difference?

(b) Test for a difference in proportion of homes with more than three bedrooms between the two states and interpret the result.

C.62 THC vs Prochloroperazine An article in the *New York Times* on January 17, 1980 reported on the results of an experiment that compared an existing treatment drug (prochloroperazine) with using THC (the active ingredient in marijuana) for combating nausea in patients undergoing chemotherapy for cancer. Patients being treated in a cancer clinic were divided at random into two groups which were then assigned to one of the two drugs (so they did a randomized, double-blind, comparative experiment). Table C.9 shows how many patients in each group found the treatment to be effective or not effective.

(a) Use these results to test whether the proportion of patients helped by THC is significantly higher (no pun intended) than the proportion helped by prochloroperazine. Use a 1% significance level since we would require very strong evidence to switch to THC in this case.

(b) Why is it important that these data come from a well-designed experiment?

Table C.9 *Effectiveness of anti-nausea treatments*

Treatment	Sample Size	Effective	Not Effective
THC	79	36	43
Prochloroperazine	78	16	62

C.63 Quality Control Susan is in charge of quality control at a small fruit juice bottling plant. Each bottle produced is supposed to contain exactly 12 fluid ounces (fl oz) of juice. Susan decides to test this by randomly sampling 30 filled bottles and carefully measuring the amount of juice inside each. She will recalibrate the machinery if the average amount of juice per bottle differs from 12 fl oz at the 1% significance level. The sample of 30 bottles has an average of 11.92 fl oz per bottle and a standard deviation of 0.26 fl oz. Should Susan recalibrate the machinery?

C.64 College Football Playoffs As of 2011, Division I college football in the US did not have a tournament-style playoff to pick a national champion. In a random survey conducted by Quinnipiac University at that time,[78] people who identified themselves as college football fans were asked if they favor a playoff system, similar to college basketball, to determine a national champion. Of those sampled, 63% said they would favor such a system, and the margin of error is ±3.1%. Assuming the margin of error corresponds to a 95% confidence interval, about how many college football fans were sampled?

C.65 Light at Night and Weight Gain A study described in Data 4.1 found that mice exposed to light at night gained substantially more weight than mice who had complete darkness at night. How large is the effect of light on weight gain? In the study, 27 mice were randomly divided into two groups. In this exercise, we include mice with any light at night (dim or bright) and we examine the effect after 8 weeks. The 8 mice with darkness at night gained an average of 5.9 grams in body mass, with a standard deviation of 1.9 grams. The 19 mice with light at night gained an average of 9.4 grams with a standard deviation of 3.2 grams. We see in Figure C.3 that there is no extreme skewness or extreme outliers, so it is appropriate to use a *t*-distribution. Find and interpret a 99% confidence interval for the difference in mean weight gain.

[78]"Use Playoff System to Pick College Football Champ, American Fans Tell Quinnipiac University National Poll," Quinnipiac University Polling Institute, December 29, 2009.

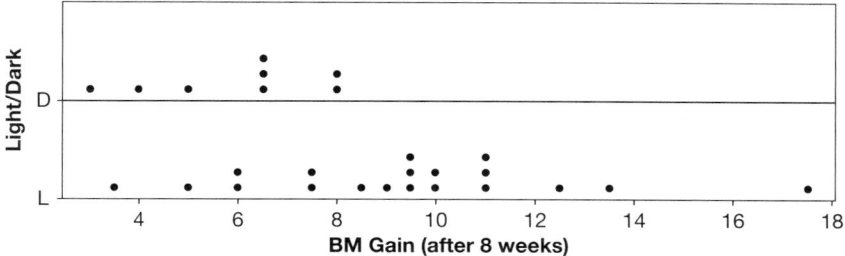

Figure C.3 *Does light at night affect body mass gain?*

C.66 Home Field Advantage in American Football How big is the home field advantage in the National Football League (NFL)? To investigate this question, we select a sample of 80 games from the 2011 regular season[79] and find the home team scored an average of 25.16 points with standard deviation 10.14 points. In a separate sample of 80 different games, the away team scored an average of 21.75 points with a standard deviation of 10.33 points. Use this summary information to estimate the mean home field advantage and find a 90% confidence interval for the mean home field advantage, $\mu_H - \mu_A$, in points scored.

C.67 Home Field Advantage in American Football: Paired Data How big is the home field advantage in the National Football League (NFL)? In Exercise C.66, we examine a difference in means between home and away teams using two separate samples of 80 games from each group. However, many factors impact individual games, such as weather conditions and the scoring of the opponent. It makes more sense to investigate this question using a matched pairs design, using scores for home and away teams matched for the same game. The data in **NFLScores2011** include the points scored by the home and away team in 256 regular season games in 2011. We will treat these games as a sample of all NFL games. Estimate average home field scoring advantage and find a 90% confidence interval for the mean difference.

C.68 NHL Hockey Penalties In the 2010–11 National Hockey League (NHL) regular season, the number of penalty minutes per game for each of the 30 teams ranged from a low of 8.8 for the Florida Panthers to a high of 18.0 for the most penalized New York Islanders. All 30 data values are given in Table C.10 and are also available in the dataset **HockeyPenalties**.

(a) Find the mean and the standard deviation of penalty minutes per game.

(b) Use the data in Table C.10 from the 2010–11 season as a sample for all NHL teams in all years,

and use the t-distribution to find a 95% confidence interval for the average number of penalty minutes per game by team.

(c) Discuss why it may or may not be appropriate to generalize this sample to the population described in part (b).

C.69 Gender and Award Preference In Example 2.6 on page 53 we consider data from a sample of statistics students that is stored in **StudentSurvey**. One of the survey questions asked which award students would most like to win from among an Academy Award, Nobel Prize, and Olympic gold medal. Among the 193 male students who responded, 109 chose the Olympic gold medal, while 73 of the 169 females also picked Olympic gold. Use this information to find a 90% confidence interval for the difference between the proportions of male and female statistics students who choose Olympic gold.

C.70 Stomach Bacteria and Irritable Bowel Syndrome Studies are finding that bacteria in the stomach are essential for healthy functioning of the human body. One study[80] compared the number of unique bacterial genes in stomachs of healthy patients and those of patients with irritable bowel syndrome (IBS). For healthy patients, we have $\bar{x} = 564$ million with $s = 122$ million and $n = 99$. For those with IBS, we have $\bar{x} = 425$ million with $s = 127$ million and $n = 25$. Both distributions appear to be approximately normally distributed. Test to see if people with IBS have, on average, significantly fewer unique bacterial genes in their stomachs. Show all details, including giving the degrees of freedom used.

C.71 Spring Break Effect? A statistics professor was handing out midterm grade slips on a Friday which happened to be the day before the school's Spring break. He noticed that there were an unusually large number of students missing from class that day. So he collected the leftover grade slips and created the data in Table C.11 that summarized the midterm grades (out of a possible 100) for students that attended and missed class.

[79]NFL scores found at http://www.pro-football-reference.com/years/2011/games.htm. Home scores from weeks 2, 4, 12, 14, and 16. Away scores from weeks 1, 3, 13, 15, and 17.

[80]Qin, J., et. al., "A Human Gut Microbial Gene Catalogue Established by Metagenomic Sequencing," *Nature*, March 4, 2010; 464: 59–65.

Table C.10 *Penalty minutes per game for 30 NHL teams*

8.8	8.9	9.3	9.3	9.6	10.2	10.4	10.6	11.0	11.1
11.2	11.2	11.4	11.4	11.6	11.8	12.0	12.7	13.4	13.4
13.5	13.5	13.6	13.9	13.9	14.0	14.5	14.9	16.8	18.0

Table C.11 *Summary statistics for midterm grades*

	n	Mean	Std. Dev.
In class	15	80.9	11.07
Missed class	9	68.2	9.26

(a) The professor had reason to suspect, before even looking at the data, that, in general, students who missed class would tend to have lower mean midterm grades. Write down the null and alternative hypotheses that he should use to test this suspicion using this class as a sample.

(b) Carry out the test in part (a). You may assume that the data for both groups are reasonably symmetric and have no strong outliers.

(c) Can we conclude on the basis of this test that skipping class on the day before break tends to hurt students' grades?

(d) There was one student who had stopped coming to class after the first week and thus had mostly zero grades and an extremely low midterm average. The instructor did not include that student when computing the statistics for Table C.11. Was that a good decision? Explain.

C.72 Mobile Phones in India India has over 600 million mobile phone subscribers. The largest company providing mobile phone service is Bharti Airtel, which has 30% of the market share.[81] If random samples of 500 mobile phone subscribers in India are selected and we compute the proportion using service from Bharti Airtel, find the mean and the standard error of the sample proportions.

C.73 What Percent of Houses Are Owned vs Rented? The 2010 US Census[82] reports that, of all the nation's occupied housing units, 65.1% are owned by the occupants and 34.9% are rented. If we take random samples of 50 occupied housing units and compute the sample proportion that are owned for each sample, what will be the mean and standard deviation of the distribution of sample proportions?

C.74 Percent of Free Throws Made Usually, in sports, we expect top athletes to get better over time. We expect future athletes to run faster, jump higher, throw farther. One thing has remained remarkably constant, however. The percent of free throws made by basketball players has stayed almost exactly the same for 50 years.[83] For college basketball players,

the percent is about 69%, while for players in the NBA (National Basketball Association) it is about 75%. (The percent in each group is also very similar between male and female basketball players.) In each case below, find the mean and standard deviation of the distribution of sample proportions of free throws made if we take random samples of the given size.

(a) Samples of 100 free throw shots in college basketball

(b) Samples of 1000 free throw shots in college basketball

(c) Samples of 100 free throw shots in the NBA

(d) Samples of 1000 free throw shots in the NBA

C.75 How Old Is the US Population? From the US Census,[84] we learn that the average age of all US residents is 36.78 years with a standard deviation of 22.58 years. Find the mean and standard deviation of the distribution of sample means for age if we take random samples of US residents of size:

(a) $n = 10$

(b) $n = 100$

(c) $n = 1000$

C.76 Time to Finish the Boston Marathon The Boston Marathon is the world's oldest annual marathon, held every year since 1897. In 2011, 23,879 runners finished the race, with a mean time for all runners of 3:49:54 (about 230 minutes) with standard deviation 0:37:56 (about 38 minutes).[85] Find the mean and standard deviation (in minutes) of the distribution of sample means if we take random samples of Boston marathon finishers of size:

(a) $n = 10$

(b) $n = 100$

(c) $n = 1000$

C.77 Does Austria or Denmark Have a Greater Percentage of Elderly? We see in the **AllCountries** dataset that the percent of the population that is elderly (over 65 years old) is 18.4 in Austria and 17.9 in Denmark. Suppose we take random samples of size 200 from each of these countries and compute the difference in sample proportions $\hat{p}_A - \hat{p}_D$, where $\hat{p}_A$ represents the proportion of the sample that is elderly in Austria and $\hat{p}_D$ represents the proportion of the sample that is elderly in Denmark. Find the mean and standard deviation of the differences in sample proportions.

[81]"Happy customers, no profit," *The Economist*, June 18, 2011.
[82]www.census.gov.
[83]Branch, J., "For Free Throws, 50 Years of Practice is No Help," *New York Times*, March 3, 2009.
[84]www.census.gov.
[85]Boston Marathon Race Results 2011, http://www.marathon guide.com/results/browse.cfm?MIDD=15110418.

C.78 Does Australia or New Zealand Have a Greater Percentage of Elderly? We see in the **All-Countries** dataset that the percent of the population that is over 65 is 14.3 in Australia and 14.0 in New Zealand. Suppose we take random samples of size 500 from Australia and size 300 from New Zealand, and compute the difference in sample proportions $\hat{p}_A - \hat{p}_{NZ}$, where $\hat{p}_A$ represents the sample proportion of elderly in Australia and $\hat{p}_{NZ}$ represents the sample proportion of elderly in New Zealand. Find the mean and standard deviation of the differences in sample proportions.

C.79 Male-Female Ratios Of the 50 states in the United States, Alaska has the largest percentage of males and Rhode Island has the smallest percentage of males. (Interestingly, Alaska is the largest state and Rhode Island is the smallest.) According to the 2010 US Census, the population of Alaska is 52.0% male and the population of Rhode Island is 48.3% male. If we randomly sample 300 people from Alaska and 300 people from Rhode Island, what is the approximate distribution of $\hat{p}_a - \hat{p}_{ri}$, where $\hat{p}_a$ is the proportion of males in the Alaskan sample and $\hat{p}_{ri}$ is the proportion of males in the Rhode Island sample?

C.80 How Likely Is a Female President of the US? A 2010 headline stated that "73% say Woman President Likely in Next 10 Years." The report gives the results of a survey of 1000 randomly selected likely voters in the US.[86] Find and interpret a 95% confidence interval for the proportion of likely voters in the US in 2010 who thought a woman president is likely in the next 10 years.

C.81 Can Rats Feel Empathy? Can rats feel empathy toward fellow rats? In a recent study,[87] some rats were first habituated to two chambers: a witness chamber adjacent to a shock chamber. The experimental rats ($n = 15$) were then given electric shocks through the floor of the shock chamber, while the control rats ($n = 11$) received no shocks but had all else the same. Twenty-four hours after the shocks were administered, each rat was put in the witness room and observed another rat getting shocked in the shock chamber. When rats get shocked, they freeze. The response variable as a measure of empathy on the part of the witness rats was the percent of time the witness rats spent in "freeze" mode when watching other rats get shocked. The experiment was double-blind. For the experimental rats who had previously received shocks, the mean percent

time spent in freeze mode was 36.6 with a standard deviation of 21.3. For the control rats who had never been shocked, the mean time in freeze mode was 1.2 with a standard deviation of 2.3. Test to see whether the time spent in freeze mode is significantly higher for the rats with prior shock experience. Show all details of the test. You may assume that the data on time spent in freeze mode have no large outliers.

C.82 Laptop Computers and Sperm Count Studies have shown that heating the scrotum by just 1°C can reduce sperm count and sperm quality, so men concerned about fertility are cautioned to avoid too much time in the hot tub or sauna. Exercise 2.113 on page 90 introduces a study suggesting that men also keep their laptop computers off their laps. The study measured scrotal temperature in 29 healthy male volunteers as they sat with legs together and a laptop computer on their lap. Temperature increase in the left scrotum over a 60-minute session is given as 2.31 ± 0.96 and a note tells us that "Temperatures are given as °C; values are shown as mean $\pm$ SD." Test to see if we can conclude that the average temperature increase for a man with a laptop computer on his lap for an hour is above the danger threshold of 1°C.

C.83 NFL Overtime In Exercise B.4 on page 349 we look at some data on the results of overtime games in the National Football League (NFL) from 1974 through the 2009 season. The question of interest is how much advantage (if any) is given to the team that wins the coin flip at the start of the sudden death overtime period. Assume that the overtime games played during this period can be viewed as a sample of all possible NFL overtime games.

(a) The winner of the coin flip has gone on to win 240 of the 428 games where a winner is determined in overtime. Does this provide sufficient evidence to conclude that the team winning the coin flip has an advantage in overtime games?

(b) The NFL changed a rule before the 1994 season (moving the kickoff line back 5 yards) that might affect this analysis. For 188 games (again ignoring ties) from 1974 to 1993 the winner of the coin flip won 94 times and lost 94 times. In 240 games played between 1994 and 2009 (after the rule change) the winner of the coin flip won 146 games and lost 94. Discuss any statistical evidence for a difference in the advantage (if any exists at all) for the team winning the coin flip under the new and old rules.

C.84 Is There a Genetic Marker for Dyslexia? Exercise 2.23 on page 59 describes a study finding

---

[86]*Rasmussen Reports*, June 27, 2010.
[87]Atsak, P., et al., "Experience Modulates Vicarious Freezing in Rats: A Model for Empathy," *PLoS ONE*, 2001; 6(7): e21855.

that a gene disruption may be related to an increased risk of developing dyslexia. Researchers studied the gene in 109 people diagnosed with dyslexia and in a control group of 195 others who had no learning disorder. The disruption occurred in 10 of those with dyslexia and in 5 of those in the control group. Are the conditions met to use the normal distribution to estimate the size of the difference in the proportion of those with the gene disruption between those who have dyslexia and those who don't? Use an appropriate method to estimate the size of this difference, with 95% confidence.

DO LIE DETECTORS WORK? Exercises C.85 to C.88 refer to the data in Table C.12. These data, introduced in Exercise A.53 on page 180, involve participants who read either deceptive material or truthful material while hooked to a lie detector. The two-way table indicates whether the participants were lying or telling the truth and also whether the lie detector indicated that they were lying or not.

Table C.12 *How accurate are lie detectors?*

	Detector Says Lying	Detector Says Not	Total
Person lying	31	17	48
Person not lying	27	21	48

C.85 Find and interpret a 90% confidence interval for the proportion of times a lie detector accurately detects a lying person.

C.86 Test to see if there is evidence that the lie detector says a person is lying more than 50% of the time, regardless of what the person reads.

C.87 Test to see if there is a difference in the proportion the lie detector says is lying depending on whether the person is lying or telling the truth.

C.88 Find and interpret a 95% confidence interval for the difference in the proportion the lie detector says is lying between those lying and those telling the truth.

C.89 Does the US Government Provide Enough Support for Returning Troops? A survey conducted of 1502 randomly selected US adults found that 931 of them believed the government does not provide enough support for soldiers returning from Iraq or Afghanistan.[88] Use this information to construct a 99% confidence interval. Clearly define the parameter you are estimating.

[88]"Four Years After Walter Reed, Government Still Faulted for Troop Support," Pew Research Center, pewresearch.org, June 29, 2011.

C.90 Arsenic in Toenails Arsenic is toxic to humans, and people can be exposed to it through contaminated drinking water, food, dust, and soil. Exercises 2.61 and A.63 on pages 74 and 181, respectively, describe an interesting new way to measure a person's level of arsenic poisoning: by examining toenail clippings. Two samples of arsenic in toenails are given below: one from near an arsenic mine in Great Britain and one from New Hampshire, US. In each case, find a 95% confidence interval for the level of arsenic in toenails in the relevant population. First, determine whether it appears to be appropriate to use a *t*-distribution to construct the confidence interval. If a *t*-distribution is appropriate, use it. If not, use a bootstrap method. Note that the units for measuring arsenic (mg/kg or ppm) are not the same for the two studies.

(a) Levels of arsenic, measured in mg/kg, are given below for 8 people living near a former arsenic mine in Great Britain:

0.8 1.9 2.7 3.4 3.9 7.1 11.9 26.0

(b) Levels of arsenic, measured in ppm, are given for 19 individuals with private wells in New Hampshire in Table C.13 (and stored in **ToenailArsenic**).

Table C.13 *Arsenic concentration in toenail clippings in New Hampshire*

0.119	0.118	0.099	0.118	0.275	0.358	0.080
0.158	0.310	0.105	0.073	0.832	0.517	0.851
0.269	0.433	0.141	0.135	0.175		

C.91 Smoking and Pregnancy Rate Exercise A.51 on page 179 introduces a study investigating whether smoking might negatively effect a person's ability to become pregnant. The study collected data on 678 women who had gone off birth control with the intention of becoming pregnant. Smokers were defined as those who smoked at least one cigarette a day prior to pregnancy. We are interested in the pregnancy rate during the first cycle off birth control. The results are summarized in Table C.14.

Table C.14 *Smoking and Pregnancy Rate*

	Smoker	Non-smoker	Total
Pregnant	38	206	244
Not pregnant	97	337	434
Total	135	543	678

(a) Find a 95% confidence interval for the difference in proportion of women who get pregnant between smokers and non-smokers. From the confidence interval, can we conclude that one group has a significantly higher pregnancy success rate than the other? Explain.

(b) Conduct a hypothesis test to determine if the pregnancy success rates of smokers and non-smokers are significantly different. Does the result agree with the conclusion in part (a)?

(c) Can we conclude that smoking causes women to have less success when trying to become pregnant during the first cycle? Explain.

C.92 Does Red Increase Men's Attraction to Women? Exercise 1.99 on page 44 describes a recent study which examines the impact of the color red on how attractive men perceive women to be. (We examine a confidence interval involving these data in Exercise 6.194 on page 458.) In the study, men were randomly divided into two groups and were asked to rate the attractiveness of women on a scale of 1 (not at all attractive) to 9 (extremely attractive). One group of men were shown pictures of women on a white background and the other group were shown the same pictures of women on a red background. The results are shown in Table C.15. Test to see if men rate women as significantly more attractive (on average) when a red background is used rather than a white background. Show all details and clearly state your conclusion.

Table C.15 *Does red increase men's attraction to women?*

Color	n	$\bar{x}$	s
Red	15	7.2	0.6
White	12	6.1	0.4

C.93 Close Confidants and Social Networking Sites Exercise 6.93 on page 430 introduced a study in which 2006 randomly selected US adults (age 18 or older) were asked to give the number of people in the last six months "with whom you discussed matters that are important to you." The average number of close confidants for the full sample is 2.2. In addition, the study asked participants whether or not they had a profile on a social networking site. For the 947 participants using a social networking site, the average number of close confidants is 2.5 with a

standard deviation of 1.4, and for the other 1059 participants who do not use a social networking site, the average is 1.9 with a standard deviation of 1.3. (We examine a confidence interval involving these data in Exercise 6.196 on page 459.)

(a) What is the sample? What is the intended population?

(b) Is this an experiment or an observational study? Can we make causal conclusions from these data?

(c) Do the sample data provide evidence that those who use social networking sites tend to have more close confidants on average?

(d) Describe a possible confounding variable that might be influencing the association in part (c).

C.94 Near Death Experiences Exercise A.49 on page 178 describes a study of the prevalence of near-death experiences. A near-death experience includes the sensation of seeing a bright light or feeling separated from one's body or sensing time speeding up or slowing down, and sometimes is experienced by people who have a brush with death. Researchers interviewed 1595 people admitted to a hospital cardiac care unit during a recent 30-month period. Patients were classified as cardiac arrest patients (in which the heart briefly stops after beating unusually quickly) or patients suffering other serious heart problems (such as heart attacks). The study found that 27 individuals reported having had a near-death experience, including 11 of the 116 cardiac arrest patients.

(a) What proportion of cardiac arrest patients in the sample reported near-death experiences? What proportion of other heart patients reported them?

(b) Test, at a 5% level, to see if cardiac arrest patients are more likely to have a near-death experience than other heart patients.

C.95 Time Spent Exercising, between Males and Females In the **StudentSurvey** data, there are 36 seniors: 26 males and 10 females. Table C.16 gives the number of hours per week that each said he or she spent exercising. Find a 95% confidence interval for the difference in mean time spent exercising between male and female seniors at this university.

Table C.16 *Number of hours spent exercising a week*

Females	4	2	5	6	12	15	10	5	0	5		
Males	10	10	6	5	7	8	4	12	12	4	15	10	5
	5	2	2	7	3	5	15	6	6	5	0	8	5

C.96 Fighting Insomnia Exercise A.71 on page 183 introduces a study investigating the effectiveness of behavioral changes and prescription medication in helping older people find improvement in fighting insomnia. The results are summarized in Table C.17.

(a) Some of the counts in the cells are very small, definitely too small to use a normal-based test. When this is the case, we often collapse some of the cells together. Combine cells from Table C.17 to fill in the cells in Table C.18 with appropriate values.

(b) Although, when completed, one of the cells in Table C.18 has only a count of 9, this is close to 10 and is only one cell, so we proceed with a normal-based test. Test to see if training helps people improve in the fight against insomnia.

Table C.17 *Treating insomnia*

Improvement	Medication	Training	Both	Neither	Total
Much	5	7	10	0	22
Some	4	5	6	1	16
None	8	6	3	17	34
Total	17	18	19	18	72

Table C.18 *Combine cells to create this table*

Any Improvement	Training	No Training	Total
Yes			
No			
Total			

C.97 Heart Rates The typical resting pulse rate in adults is 60 to 80 beats per minute. For the 200 Intensive Care Unit patients in the dataset **ICUAdmissions**, the average pulse rate in the sample is $\bar{x} = 98.9$ bpm with $s = 26.8$ bpm. Test to see if this provides evidence that the average pulse rate of ICU patients is greater than 80.

C.98 Normal Body Temperature It is commonly believed that normal human body temperature is 98.6°F (or 37°C). In fact, "normal" temperature can vary from person to person, and for a given person it can vary over the course of a day. Table C.19 gives a set of temperature readings of a healthy woman taken over a two-day period. Test to see if the mean body temperature for this person is different from 98.6°F.

C.99 Prostate Cancer and a Drug for Baldness Exercise A.69 on page 183 introduces a study in which the drug finasteride, marketed as Propecia to help protect against male pattern baldness, is investigated for its protective effects against prostate cancer. Men were randomly assigned to receive either a daily finasteride pill or a placebo, and the study was double-blind. At the end of the seven-year study, prostate cancer was found in 804 of 4368 men taking finasteride and in 1145 of 4692 men taking a placebo. Test to see if men taking finasteride are less likely to get prostate cancer than men taking a placebo.

C.100 Dark Chocolate for Good Health A recent study examines chocolate's effects on blood vessel function in healthy people. In the randomized, double-blind, placebo-controlled study, 11 people received 46 grams (1.6 ounces) of flavonoid-rich dark chocolate (which is normal dark chocolate) every day for two weeks, while a control group of 10 people were given dark chocolate with reduced flavonoid content. Participants had their vascular health measured (by means of flow-mediated dilation) before and after the two-week study. The increase in flow-mediated dilation over the two week period was measured, with larger numbers indicating greater vascular health. For the group getting the flavonoid-rich dark chocolate, the mean increase was 1.3 with a standard deviation of 2.32, while the control group had a mean change of −0.96 with a standard deviation of 1.58. (We examine a confidence interval involving these data in Exercise 6.195 on page 458.)

(a) Why were participants in the control group given dark chocolate that had reduced flavonoids? Why weren't they given nothing at all?

(b) Do the results of the study provide evidence that vascular health (as measured by flow-mediated dilation) is greater in those eating dark chocolate daily? Are the results significant at the 5% level? At the 1% level?

(c) Can we conclude that eating dark chocolate improves vascular health? Why or why not?

C.101 How Old Are Scout Honeybees? Honeybee colonies have specific "scout" honeybees, that are entrusted with finding new sites for hives. The mean

Table C.19 *Body temperature during the day*

97.6	98.0	98.8	98.9	98.7	98.1	97.7	98.1	98.9	98.9	98.8	98.3

age of honeybees in a hive is about 12 days.[89] In a sample of 50 scout honeybees, the average age is 29.1 days with a standard deviation of 5.6. Does this provide evidence that scout bees are older, on average, than expected by random chance if they were selected from the hive as a whole? Show all details of the test.

C.102 Tribulus Tribulus is a food supplement that some athletes use to enhance performance. A study[90] on the effects of this supplement involved randomly assigning 20 athletes to take the supplement for a 20-day period and comparing various characteristics to 12 similar athletes who were not given Tribulus. One of the measurements was anaerobic alactic muscular power (AAMP) where the Tribulus group showed a mean performance of 1305.6 with a standard deviation of 177.3 while the control group had a mean AAMP of 1255.9 with a standard deviation of 66.8.

(a) Test whether the data provide evidence that mean AAMP is higher for athletes using Tribulus compared to those not using the supplement.

(b) The authors of the study also report that the mean AAMP of the 20 subjects in the experimental group *before* they started the Tribulus supplements was 1215.5 with a standard deviation of 146.6. Explain why this is *not* sufficient information to let us test for a difference in means before and after using the supplement.

C.103 Homes for Sale The dataset **HomesForSale** has data on houses available for sale in three Mid-Atlantic states (NY, NJ, and PA) and California (CA).[91] We are interested in the proportion of Mid-Atlantic houses that are larger than the mean US household size, which is approximately 2700 square feet. In the sample of 90 Mid-Atlantic homes for sale in **HomesForSale**, we see that 16 are larger then 2700 sq ft.

(a) What is the sample proportion? Estimate the standard error of the sample proportion.

(b) Is the sample size large enough for the Central Limit Theorem to apply?

(c) Find and interpret a 95% confidence interval for the proportion of homes larger than the national mean in the Mid-Atlantic states.

[89]Values approximated from information available in Seeley, T., *Honeybee Democracy*, Princeton University Press, Princeton, NJ, 2010, p. 96.

[90]Milasius, K., Peciukoniene, M., Dadeliene, R., and Skernevicius, J., "Efficacy of the Tribulus Food Supplement Used By Athletes," *Acta Medica Lituanica*, 2010; 17(1–2):65–70.

[91]Data collected from www.zillow.com.

C.104 Homes for Sale in Mid-Atlantic States and California The dataset **HomesForSale** has data on houses available for sale in three Mid-Atlantic states (NY, NJ, and PA) as well as California (CA). In Exercise C.103 we see that in the sample of 90 Mid-Atlantic homes for sale, 16 are larger than the national average. In the sample of 30 California homes, 7 are larger than the national average.

(a) Is the normal distribution appropriate to model the proportion larger than the national average for Mid-Atlantic houses? California houses?

(b) Do we have sufficient evidence at a 5% level that the proportion of Mid-Atlantic homes larger than the national average is less then 25%? Perform an appropriate test.

(c) Do we have sufficient evidence at a 5% level that the proportion of California homes larger than the national average is less than 25%? Perform an appropriate test.

C.105 How Big Is the Difference in Homes For Sale? The dataset **HomesForSale** has data on houses available for sale in three Mid-Atlantic states (NY, NJ, and PA) as well as California (CA). In Exercise C.104 we looked at the proportion of homes for sale that were larger than the national average of 2700 sq ft. We found 16 large houses out of 90 in the Mid-Atlantic states and 7 large houses out of 30 in California.

(a) Find a point estimate for the difference in proportions, $p_M - p_C$.

(b) Find a 90% confidence interval for the difference in proportions, $p_M - p_C$.

(c) Based on the interval in part (b), is there evidence of a difference in the proportion of large houses between the two locations?

C.106 Infection in Dialysis Patients Exercise A.77 on page 184 discusses a study showing the recurrence time to infection at the point of insertion of the catheter for kidney patients using portable dialysis equipment. There are 38 patients, and the mean time to infection is $\bar{x} = 111.7$ days with $s = 144.0$.

(a) Find a 99% confidence interval for the mean time to infection for these patients. Give the best estimate, the margin of error, and give and interpret the confidence interval.

(b) Is it reasonable to find a patient with a time to infection of 24 days? How about 152 days?

(c) Is it reasonable to find the mean time to infection in the population is 24 days? How about 152 days?

Table C.20 *Systolic blood pressure of ICU patients*

Teens	100	100	104	104	112	130	130	136	140	140	142
	146	156									
Eighties	80	100	100	110	110	122	130	135	136	138	140
	141	162	190	190							

C.107 Age and Blood Pressure Table C.20 gives systolic blood pressure readings (first introduced in Exercise A.75 on page 184) for Intensive Care Unit patients in their teens and those in their eighties.

(a) Find and interpret a 95% confidence interval for the mean systolic blood pressure for each group. Which has a larger margin of error? What aspect of the data is the cause of that larger margin of error?

(b) Test to see if there is a difference in systolic blood pressure readings between ICU patients in their teens and those in their eighties.

C.108 Light at Night and Weight Gain A study described in Data 4.1 found that mice exposed to light at night gained substantially more weight than mice who had complete darkness at night. In the study, 27 mice were randomly divided into two groups. (In this exercise, we combine the dim light and bright light groups and consider results after 8 weeks.) After 8 weeks, the 8 mice with darkness at night gained an average of 5.9 grams in body mass, with a standard deviation of 1.9 g. The 19 mice with light at night gained an average of 9.4 g with a standard deviation of 3.2 g. Is there evidence that mice with light at night gain significantly more weight (while eating the same number of calories) than mice with darkness at night? Justify your answer by showing all details of the test. (We examine a confidence interval involving these data in Exercise C.65 on page 493, and the figure in that exercise indicates that it is appropriate to use a t-distribution for this test.)

C.109 CAOS Exam The Comprehensive Assessment of Outcomes in Statistics (CAOS) exam[92] is an online multiple-choice test on concepts covered in a typical introductory statistics course. Students can take a pretest version before instruction and then a posttest version after instruction. Scores on the pretest and posttest for a random sample of $n = 10$

[92]http://app.gen.umn.edu/artist/caos.html.

students with one instructor are shown in Table C.21 and stored in **CAOSExam**. Use this information to compute and interpret a 95% confidence interval for the improvement in mean CAOS scores between the two exams for this instructor's students.

C.110 CAOS Comparisons An article[93] by the developers of the CAOS exam described in Exercise C.109 gives benchmark data based on a very large number of students taking the CAOS pretest and posttest. The mean score on the CAOS pretest was 44.9 and the mean on the CAOS postest was 54.0 for an average improvement of 9.1 points. We treat these values as population means for all students taking the CAOS exams.

(a) Using the data in Table C.21 and **CAOSExam** for a sample from one instructor, is there enough evidence to conclude at the 5% level that this instructor's students have a mean score on the posttest that is higher than 54.0?

(b) Can we conclude that this instructor was starting with stronger students? Test if the mean score for this instructor's students on the CAOS pretest is higher than the benchmark mean of 44.9 points, using a 5% significance level.

(c) Can we conclude, at the 5% level, that the mean *improvement* from pretest to posttest for students with this instructor is higher than the national norm of 9.1 points?

C.111 Gender and Commuting Time–St. Louis Some computer output is shown on the next page from an analysis to compare mean commute time between males and females using data from St. Louis commuters in **CommuteStLouis**. Write a paragraph interpreting what this output shows about the relationship (if any) between commuting time and gender in St. Louis.

[93]DeLMas, R., Garfield, J., Ooms, A., and Chance, B., "Assessing Students' Conceptual Understanding After a First Statistics Course," *Statistics Education Research Journal*, 2007; 6(2): 28–58.

Table C.21 *CAOS pretest and posttest scores*

Student	A	B	C	D	E	F	G	H	I	J
Pretest	42.5	40	47.5	65	60	47.5	42.5	37.5	42.5	37.5
Posttest	60	45	55	80	65	72.5	57.5	55	57.5	55

```
Two-sample T for Time
Sex        N    Mean   StDev   SE Mean
F        240    21.6    13.9      0.90
M        260    22.3    14.6      0.90
```

Difference = mu (F) − mu (M)
Estimate for difference: −0.70
95% CI for difference: (−3.20, 1.80)
T-Test of difference = 0 (vs not =): T-Value = −0.55 P-Value = 0.585 DF = 497

C.112 Gender and Commuting Time–Atlanta Exercise C.111 gives computer output comparing mean commute time between males and females in the city of St. Louis. Use the data in **Commute-Atlanta** to see if a similar relationship (or lack of relationship) holds in Atlanta. Include both a confidence interval and test for the difference in mean commute time by gender in Atlanta.

C.113 Commuting by Bicycle: Which Type of Bike Is Best? Dr. Jeremy Groves, a British anaesthetist, often uses a bicycle for his 27-mile round-trip commute to work. He bought an expensive, lightweight, carbon bike but also had an older, heavier, steel bike—so he decided to do an experiment.[94] On each day he biked to work he flipped a coin to determine which bike he would ride. He used a bicycle computer to accurately record the commute time each day as well as his maximum and average speed for the day. His data for 56 days are stored in **BikeCommute**. The type of bicycle (carbon or steel) is in the *Bike* variable and his time (in minutes) is stored in *Minutes*. Do the data provide evidence that mean commute time differs between the two types of bikes?

C.114 Marriage Ages Exercise B.6 on page 350 introduces the **MarriageAges** dataset that contains the ages for husbands and wives from a sample of 105 marriage licenses in St. Lawrence County, New York. In that exercise we use bootstrap and randomization methods to compare ages between husbands and wives. Repeat the analyses now using technology and the methods of this unit.

(a) For couples marrying in St. Lawrence County, does the sample provide evidence that, on average, husbands are older than wives?

(b) Is the proportion of couples for which the husband is older greater than 50%?

(c) For any significant results in parts (a) and (b), construct and interpret a 95% confidence interval for the parameter of interest.

C.115 Marriage Age Intervals Refer to Exercise C.114 for a description of the data in **MarriageAges** for a sample of 105 newly married couples.

(a) Use technology to find the mean age for the wives in this sample and construct a 95% confidence interval for the mean age at marriage for wives in the population.

(b) Repeat part (a) for the husbands' ages.

(c) Based on the confidence intervals in parts (a) and (b), can we predict what a hypothesis test for a difference in mean marriage age between the husband and wife in a couple might conclude, based on this dataset? If so, explain what that decision would be. If not, explain why not.

C.116 Better Traffic Flow Exercise 2.155 on page 105 describes a study conducted by engineers in Dresden, Germany looking at ways to improve traffic flow by enabling traffic lights to communicate with each other in real time. They simulated buses moving along a street and recorded the delay time (in seconds) for both the current fixed timed system and a flexible interacting system of lights. They repeated the simulation in each case for a total of 24 situations. The data in **TrafficFlow** show the total delay time (in minutes) for both the *Timed* and the *Flexible* simulations for each run, as well as a column showing the *Difference* in the time for each pair, with *Difference = Timed − Flexible* in each case. Use statistical software to find a 95% confidence interval for the mean difference in delay time between the two systems and to conduct a hypothesis test to see if there is a difference in delay times between the two systems. Interpret the confidence interval and include all details of the test. Which method has the least average delay time for traffic?

C.117 Hockey Malevolence Data 4.2 on page 263 describes a study of a possible relationship between the perceived malevolence of a team's uniform and penalties called against the team. In Exercise 4.179 on page 337 we consider a randomization distribution to test for a positive correlation for National Hockey League teams using the *NHLMalevolence* and *ZPenMin* data in **MalevolentUniformsNHL**.

[94]"Bicycle weight and commuting time: randomised trial," *British Medical Journal*, 2010; 341: c6801. Thanks to Dr. Groves for providing his data.

Repeat this test, using the fact that the randomization distribution is reasonably normal to find and interpret a p-value.

C.118 More Hockey Malevolence Refer to the randomization test described in Exercise C.117 for the correlation between uniform malevolence and penalty minutes for NHL teams. Suppose that a student constructs randomization samples by scrambling the *NHLMalevolence* variable in **Malevolent-UniformsNHL** and computing the correlation with *ZPenMin*. The student repeats this process 1000 times to generate a randomization distribution and finds the standard deviation of the randomization correlations to be 0.22. Since the distribution is reasonably normal and the correlation between these variables in the original sample is $r = 0.521$, the student computes a 90% confidence interval for the correlation using $0.521 \pm 1.645 \cdot 0.22 = (0.159, 0.883)$.

(a) Identify a clear error in the process that the student has used to construct a confidence interval for the correlation in this situation.

(b) Suggest a more appropriate method for estimating the standard error to find a confidence interval for the correlation between uniform malevolence and standardized penalty minutes for NHL teams, based on the sample data in **MalevolentUniformsNHL**.

(c) Carry out the procedure you describe in part (b) to obtain an estimate of the standard error. Assuming that a normal distribution is appropriate, use the standard error to find (and interpret) a 90% confidence interval for the correlation in this context.

(d) Does it look like a normal distribution is appropriate in this situation? Explain why or why not.

C.119 Penny Spinning Suppose that you hold a coin vertically on edge, flick one side to make it spin, then see if it settles on heads or tails. Is this a fair process? That is, will half of spins tend to fall heads and the other half on tails? To investigate this question, a student spun the same US penny 200 times and recorded 84 heads and 116 tails. She uses these results to test $H_0 : p = 0.5$ vs $H_a : p \neq 0.5$, where p is the proportion of penny spins that land heads. The file **RandomP50N200** contains counts and proportions of heads for 5000 simulated samples of size $n = 200$ when $p = 0.50$.

(a) Use the randomization proportions, the variable *Phat* in **RandomP50N200**, to estimate the standard error for the sample proportion, $\hat{p}$, in this situation.

(b) Use the standard error from part (a) to compute a standardized test statistic for testing $H_0 : p = 0.5$ based on her original sample of 200 spins.

(c) Under this null hypothesis, what should be the mean of the distribution of *counts* of number of heads in 200 spins?

(d) The counts of heads for each of the 5000 randomization samples are in the variable *Count*. Use these to estimate the standard deviation for the counts (number of heads in 200 spins) under this null hypothesis.

(e) Use the mean from part (c) and standard error from part (d) to compute a standardized test statistic based on the count of 84 heads in the original sample of 200 spins. How does this test statistic compare to the one based on the sample proportions in part (b)?

(f) Use either test statistic and the fact that both randomization distributions are relatively normal to compute and interpret a p-value for this test.

Inference for Multiple Parameters

"The biggest leaps forward in the next several decades—in business, science, and society at large—will come from insights gleaned through perpetual, real-time analysis of data"

–IBM*

UNIT OUTLINE

7 Chi-Square Tests for Categorical Variables
8 ANOVA to Compare Means
9 Inference for Regression
10 Multiple Regression
Essential Synthesis

In this unit, we consider statistical inference for situations with multiple parameters: testing categorical variables with more than two categories, comparing means between more than two groups, making inferences using the slope and intercept of a regression model, and building regression models with more than one explanatory variable.

*Advertisement, May 2010, seen in *The Week*, May 14, 2010, p. 12.

CHAPTER 7

Chi-Square Tests for Categorical Variables

"The marshalling of data to test presumptions and locate paths to success is transforming almost every aspect of human life."

–Larry Summers*

*"What you (really) need to know," *The New York Times*, January 22, 2012.
Top left: PNC/Getty Images, Inc., Top right: ©97/iStockphoto, Bottom right: Sports Illustrated/Getty Images, Inc.

Questions and Issues

Here are some of the questions and issues we will discuss in this chapter:

- What is the most common sleep position?
- How common is binge drinking among college students?
- Which is most important to middle school students: good grades, athletic ability, or popularity?
- How often do people "like" something on Facebook?
- Are left-handed people more likely to choose certain occupations?
- Can people delay death to make it to a special occasion?
- Are Canadian youth hockey players born early in the year more likely to become professional hockey players than those born late in the year?
- Can people tell bottled water and tap water apart in a blind taste test?
- In Rock-Paper-Scissors, are the three options chosen equally often?
- Are hospitals deadlier in July?
- Which James Bond actor is the favorite?
- Do males or females update their status more frequently on Facebook?
- Is there a "sprinting gene"?
- Are children who are the youngest in their class more likely to be diagnosed with ADHD?

7.1 TESTING GOODNESS-OF-FIT FOR A SINGLE CATEGORICAL VARIABLE

In Chapter 6, we discuss tests for proportions that deal with a single category of a categorical variable. In this chapter, we develop methods for testing hypotheses that involve two or more categories. To do this, we use *chi-square tests*. The data for chi-square tests are frequency counts for the different categories, or cells, of one or more categorical variables. In this section, we see how to use a chi-square test to test an assumption about a single categorical variable. In the next section, we see how to use a chi-square test to test for an association between two categorical variables.

Section 6.1-HT discusses a test for a proportion that deals with a single category for a categorical variable. If a categorical variable has more than two categories and we want to test a hypothesis about the proportions across all of the categories, we use a chi-square *goodness-of-fit* test.

1 Ⓐ Ⓑ Ⓒ Ⓓ Ⓔ	26 Ⓐ Ⓑ Ⓒ Ⓓ Ⓔ	51 Ⓐ Ⓑ Ⓒ Ⓓ Ⓔ
2 Ⓐ Ⓑ Ⓒ Ⓓ Ⓔ	27 Ⓐ Ⓑ Ⓒ Ⓓ Ⓔ	52 Ⓐ Ⓑ Ⓒ Ⓓ Ⓔ
3 Ⓐ Ⓑ Ⓒ Ⓓ Ⓔ	28 Ⓐ Ⓑ Ⓒ Ⓓ Ⓔ	53 Ⓐ Ⓑ Ⓒ Ⓓ Ⓔ
4 Ⓐ Ⓑ Ⓒ Ⓓ Ⓔ	29 Ⓐ Ⓑ Ⓒ Ⓓ Ⓔ	54 Ⓐ Ⓑ Ⓒ Ⓓ Ⓔ
5 Ⓐ Ⓑ Ⓒ Ⓓ Ⓔ	30 Ⓐ Ⓑ Ⓒ Ⓓ Ⓔ	55 Ⓐ Ⓑ Ⓒ Ⓓ Ⓔ
6 Ⓐ Ⓑ Ⓒ Ⓓ Ⓔ	31 Ⓐ Ⓑ Ⓒ Ⓓ Ⓔ	56 Ⓐ Ⓑ Ⓒ Ⓓ Ⓔ

Which letter is the best to guess?

DATA 7.1 **Multiple Choice Answers**

Many standardized tests use questions with a multiple choice format. For example, the *Advanced Placement* (AP) exams for various subjects frequently use multiple choice questions for a portion of the exam, with five options labeled A, B, C, D, and E. Are certain choices more or less likely to occur than others as the correct answer? Table 7.1 shows the frequencies of the letters for correct answers in a sample of 400 multiple choice questions selected from previously released exams.[1] The data are also stored in **APMultipleChoice**. ■

Table 7.1 *Choice for correct response of 400 AP multiple choice questions*

Answer	A	B	C	D	E
Frequency	85	90	79	78	68

Example 7.1

Use proper notation in answering each of the following questions.

(a) Find the relative frequency (proportion) of answers for each letter in Table 7.1.

(b) If the letter choices are really made at random, what should be the proportion of *all* AP multiple choice questions that have each letter as the correct answer?

---

[1] Data obtained from released exams at *http://apcentral.collegeboard.com*.

Solution

(a) The data in Table 7.1 are from a sample of exam questions so we use the notation for sample proportions when giving the relative frequencies. For example, for letter A we have

$$\hat{p}_a = \frac{85}{400} = 0.2125$$

The other four proportions are

$$\hat{p}_b = \frac{90}{400} = 0.225 \quad \hat{p}_c = \frac{79}{400} = 0.1975 \quad \hat{p}_d = \frac{78}{400} = 0.195 \quad \hat{p}_e = \frac{68}{400} = 0.170$$

(b) Because we are interested in the proportions for the population of all AP multiple choice questions, we use the notation p for a parameter. If each of the five letters are equally often the correct choice, the proportion for each is 1/5 or 0.20. We have

$$p_a = p_b = p_c = p_d = p_e = 0.20$$

We could test the proportion in any one of the categories, for example $H_0 : p_a = 0.2$ vs $H_a : p_a \neq 0.2$, with the techniques of Chapter 4 or Section 6.1-HT. However, we prefer to assess the evidence for/against "equally likely" using all five categories at once.

Null and Alternative Hypotheses

The answer to part (b) of Example 7.1 is precisely the sort of hypothesis we would like to test for a categorical variable. The null hypothesis specifies proportions for each of the groups defined by the variable's categories. The alternative hypothesis is that at least one of those proportions is wrong. For the five multiple choice answers we have

$$H_0 : \quad p_a = p_b = p_c = p_d = p_e = 0.2$$
$$H_a : \quad \text{Some } p_i \neq 0.2$$

Note that the alternative doesn't specify which specific group has a proportion different from 0.2, just that at least one of the null proportions (denoted by a generic "p_i") is different from what the null hypothesis claims. Also, as the next example illustrates, we don't have to have all proportions equal to each other in the null hypothesis. We can test all sorts of distributions of proportions for the groups.

PNC/Getty Images, Inc.

Are the members of this jury representative?

DATA 7.2	**Alameda County Jury Pools**

Pools of prospective jurors are supposed to be drawn at random from eligible adults in a community. The American Civil Liberties Union (ACLU) conducted an analysis[2] of the jury pools for a sample of 10 trials in Alameda County, California. The racial makeup of those jury pools, a total of 1453 individuals, is shown in Table 7.2. To see if these data are consistent with community percentages, we also have census data on the percentage breakdown by race of eligible jurors for the entire county. These percentages are also shown in Table 7.2. ■

Table 7.2 *Racial makeup of Alameda County juries and community*

Race	White	Black	Hispanic	Asian	Other
Number in jury pools	780	117	114	384	58
Census percentage	54%	18%	12%	15%	1%

Example 7.2

Write down the null and alternative hypotheses for testing if the racial distribution of Alameda County jury pools differs significantly from the census proportions of the community.

Solution If we let p_w, p_b, p_h, p_a, and p_o represent the proportions of Whites, Blacks, Hispanics, Asians and Others, respectively, in Alameda jury pools, the hypotheses of interest are

$$H_0: \quad p_w = 0.54, \quad p_b = 0.18, \quad p_h = 0.12, \quad p_a = 0.15, \quad p_o = 0.01$$
$$H_a: \quad \text{Some } p_i \text{ is not as specified in } H_0$$

Notice that in both examples the alternative hypothesis is that things are not as we expect and that something interesting is going on.

Expected Counts

The null hypothesis in both of these examples is more complicated than those in earlier chapters. Neither reduces to a claim about the value of a single parameter (such as $H_0 : p = 0.2$) or even about two parameters (such as $H_0 : p_1 = p_2$). For this reason we need a more complicated test statistic than the form (*sample statistic − null parameter*)/SE that is common in Chapters 5 and 6. We begin by finding the *expected* frequency counts in each category if the null hypothesis is true.

Example 7.3

In Data 7.1 about Advanced Placement exams, if we have a sample of size $n = 400$ multiple choice questions and assume the null hypothesis is exactly true (that the proportion for each letter is equal to 0.2), what frequency counts do we expect to see for each letter?

Solution If the data fit the null hypothesis perfectly, we would expect to see $400(0.2) = 80$ counts in each cell, as in Table 7.3.

[2]*http://www.aclunc.org/docs/racial_justice/racial_and_ethnic_disparities_in_alameda_county _jury_pools.pdf.*

Table 7.3 *Expected frequencies for n = 400 under*
$H_0 : p_a = p_b = p_c = p_d = p_e = 0.2$

Answer	A	B	C	D	E
Expected count	80	80	80	80	80

In general, for a specific sample size, we compute the *expected count* for each cell in a table by multiplying the sample size by the proportion specified in the null hypothesis:

$$Expected\ count = n \cdot p_i$$

Example 7.4

The data on potential jurors in Table 7.2 is based on a sample containing 1453 individuals. Use the null hypothesis in Example 7.2 to find expected counts for each of the racial groups.

Solution   To find the expected count for each cell we multiply the sample size ($n = 1453$) by the proportion given in the null hypothesis. For example, the expected number of Whites is

$$Expected\ count = n \cdot p_w = 1453(0.54) = 784.6$$

The expected count for each racial group is shown in Table 7.4.

Table 7.4 *Expected juror counts based on null census proportions*

Race	White	Black	Hispanic	Asian	Other
Expected count	784.6	261.5	174.4	218.0	14.5

Of course, in practice we rarely see the actual counts observed in a real sample exactly match the expected counts, even if the null hypothesis is true. The observed counts tend to vary from sample to sample. Once again, we come to the key question of a test of significance: "Are the *observed* counts in the original sample farther from the *expected* counts than we would reasonably tend to see by random chance alone (assuming the actual population proportions are as given in the null hypothesis)?"

Chi-square Statistic

We need a statistic to compare the observed counts from a sample to the expected counts from a null hypothesis and we would like it to combine the information from all cells of the table. One common way of doing this is with a *chi-square statistic*.

Chi-square Statistic

The **chi-square statistic**, denoted with the Greek χ^2, is found by comparing the **observed counts** from a sample with **expected counts** derived from a null hypothesis. The formula for computing the statistic is

$$\chi^2 = \sum \frac{(Observed - Expected)^2}{Expected}$$

where the sum is over all cells of the table.

Example 7.5

Use the information from Tables 7.1 and 7.3 to find the value of the chi-square statistic for the sample of answers from multiple choice questions on AP exams.

Solution The summation is a bit tedious but straightforward:

$$\chi^2 = \frac{(85-80)^2}{80} + \frac{(90-80)^2}{80} + \frac{(79-80)^2}{80} + \frac{(78-80)^2}{80} + \frac{(68-80)^2}{80}$$
$$= 0.3125 + 1.25 + 0.0125 + 0.05 + 1.8$$
$$= 3.425$$

The square term in the formula for the chi-square statistic prevents large positive deviations from canceling large negative deviations when they are summed. Dividing each square by the expected count for that cell is a way of standardizing each term, since an *Observed − Expected* difference of 20 might be a large discrepancy if the expected count is 30, but would be pretty good agreement if the expected count is 3000. Note that large values of the chi-square statistic correspond to samples that do not agree with the null hypothesis.

So now we have a test statistic, $\chi^2 = 3.425$, that measures how close the observed counts for the sample of 400 multiple choice answers are to the expected counts under a null hypothesis of equal proportions. Is that an especially large value? Bigger than we would usually see by chance alone? Fortunately, we can apply randomization methods to address these questions.

Randomization Test for Goodness-of-Fit

Recall from Chapter 4 that we can obtain a p-value for almost any statistic by simulating new randomization samples that are consistent with a null hypothesis, constructing a randomization distribution of the statistics for those samples, and seeing where the value from the original sample lies in that distribution. Let's try that now for the data on AP multiple choice answers.

The null hypothesis, $H_0 : p_a = p_b = p_c = p_d = p_e = 0.2$, states that answers should be chosen at random from among A, B, C, D, and E for each of the 400 questions in a sample. With technology we can randomly sample 400 values with replacement from among those five letters, which is equivalent to sampling from a population with all letters equally likely. For each randomization sample we count how many times each of the five letters appears. Table 7.5 shows the results for one such sample.

The value of the chi-square statistic for this randomization sample is

$$\chi^2 = \frac{(77-80)^2}{80} + \frac{(86-80)^2}{80} + \frac{(73-80)^2}{80} + \frac{(77-80)^2}{80} + \frac{(87-80)^2}{80} = 1.90$$

Figure 7.1 shows a randomization distribution with the chi-square statistics for 1000 such simulated samples. The value from the original sample, $\chi^2 = 3.425$, lies somewhere in the middle of this distribution. In fact, 493 of these 1000 samples, which were simulated with random choices of the letters for each question, produced a chi-square statistic that is bigger (farther from the expected counts) than the original sample. This gives a p-value of 0.493, which is not less than any

Table 7.5 *Observed counts for one randomization sample with n = 400*

Answer	A	B	C	D	E
Frequency	77	86	73	77	87

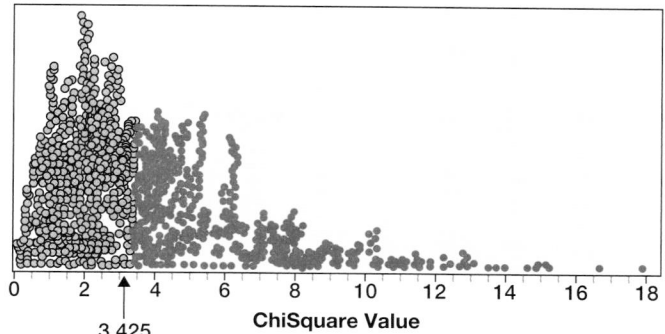

Figure 7.1 *Chi-square statistics for 1000 samples simulated with* $H_0 : p_a = p_b = p_c = p_d = p_e = 0.2$

reasonable significance level. We do not have sufficient evidence to reject a null hypothesis that the proportion for each possible answer is 0.2 for the population of all multiple choice questions on AP exams.

Are we sure that the letters used for the correct responses really are equally likely on AP exams? No. Remember that a lack of evidence to refute a null hypothesis, even a p-value as large as 0.493, should not be misinterpreted as strong evidence to accept H_0. Perhaps there are some small differences that this sample was not large enough to detect, or we happened to pick a sample that was more "random" than the rest of the population.

Chi-square Distribution

In Chapter 5 we see that many distributions of sample statistics can be approximated with normal distributions. However, a quick glance at Figure 7.1 shows that a normal distribution is not an appropriate model for the simulated chi-square statistics. The distribution is clearly skewed to the right and can never have values below zero.

Fortunately, the shape of the distribution of the statistics in Figure 7.1 is quite predictable. We use a new distribution, called a *chi-square* distribution, as a model for this shape. Similar to a t-distribution, the chi-square distribution has a degrees of freedom parameter that is determined by the number of categories (cells) in the table. In general, for a goodness-of-fit test based on a table with k cells, we use a chi-square distribution with $k - 1$ degrees of freedom.

For the five cells in Table 7.1 we use a chi-square distribution with four degrees of freedom. Because we know the total sample size, if we "free"-ly choose any four of the sample counts in Table 7.1 or 7.5, the fifth count is completely determined by the sample size. That is why we have just four degrees of freedom. Figure 7.2 shows a histogram for the 1000 randomization chi-square statistics from Figure 7.1

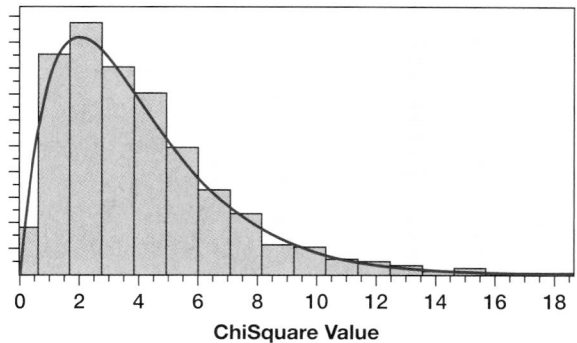

Figure 7.2 *Chi-square distribution with four degrees of freedom fits a histogram of simulated statistics*

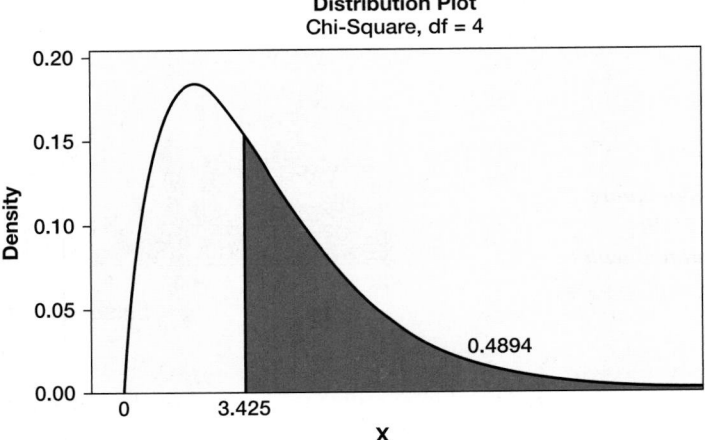

Figure 7.3 *P-value for* $\chi^2 = 3.425$ *from a chi-square distribution with df = 4*

with the density curve for a chi-square distribution with four degrees of freedom. This provides a good model for the distribution of these statistics under the null hypothesis.

We can compute a p-value for a sample statistic, such as $\chi^2 = 3.425$, as the area under a chi-square distribution. Figure 7.3 shows this area using four degrees of freedom. Compare this plot to Figure 7.1. The resulting p-value (0.4894) is quite close to the empirical value (0.493) that we obtained from the randomization distribution. Note that in both cases we use the upper tail (above the observed chi-square statistic) since those values represent samples that are as far (or farther) away from the null expected counts.

As with the Central Limit Theorems of Chapters 5 and 6, the fit of a chi-square distribution tends to get better for larger sample sizes. As a general rule, if the expected count in each cell is at least five, the chi-square distribution should be a good approximation.

Chi-square Goodness-of-Fit Test

To test a hypothesis about the proportions of a categorical variable, based on a table of observed counts in k cells:

H_0 : Specifies proportions, p_i, for each cell

H_a : At least one p_i is not as specified

- Compute the expected count for each cell using $n \cdot p_i$, where n is the sample size and p_i is given in the null hypothesis.
- Compute the value of the chi-square statistic,

$$\chi^2 = \sum \frac{(Observed - Expected)^2}{Expected}$$

- Find the p-value for χ^2 using the upper tail of a chi-square distribution with $k - 1$ degrees of freedom.

The chi-square distribution is appropriate if the sample size is large enough that each of the expected counts is at least 5.

Be sure that all possible categories of a categorical variable are listed when doing a chi-square goodness-of-fit test. The probabilities given in the null hypothesis for all the categories should add up to one, and the sum of the expected counts will be the sample size, which is also the sum of the observed counts.

Example 7.6

Alameda County Jurors

Use the data in Table 7.2 to test whether the racial proportions of jury pools tend to differ from the racial make-up in the Alameda County community.

Solution

 The hypotheses are

$$H_0: \quad p_w = 0.54, \quad p_b = 0.18, \quad p_h = 0.12, \quad p_a = 0.15, \quad p_o = 0.01$$
$$H_a: \quad \text{Some } p_i \text{ is not as specified}$$

Table 7.6 shows the observed counts from the sample together with the expected counts (in parentheses) that we found for a sample of size $n = 1453$ in Example 7.4 on page 511.

We calculate the chi-square statistic using the observed and expected counts:

$$\chi^2 = \frac{(780 - 784.6)^2}{784.6} + \frac{(117 - 261.5)^2}{261.5} + \frac{(114 - 174.4)^2}{174.4}$$
$$+ \frac{(384 - 218.0)^2}{218.0} + \frac{(58 - 14.5)^2}{14.5}$$
$$= 0.03 + 79.8 + 20.9 + 126.4 + 130.5$$
$$= 357.6$$

Checking a chi-square distribution with four degrees of freedom ($df = \#cells - 1$) we see that $\chi^2 = 357.6$ is extremely far in the tail, giving a p-value that is essentially zero. The data provide very strong evidence that the proportions of racial groups in Alameda County jury pools are quite different from the racial makeup of the community.

Table 7.6 *Observed (expected) juror counts in Alameda County*

Race	White	Black	Hispanic	Asian	Other
	780 (784.6)	117 (261.5)	114 (174.4)	384 (218.0)	58 (14.5)

The chi-square statistic of Example 7.6 provides strong evidence against the null hypothesis, but that doesn't tell us which groups might be over- or under-represented. One way to address this is to compare the observed and expected counts and look at the contribution to the chi-square statistic from each cell. Although the agreement is very close for Whites, the other four groups show very large discrepancies. It would appear that Blacks and Hispanics are under-represented in Alameda County jury pools while more Asians and Others are present than we would expect by their share of the population.

In practice we generally use technology to automate the calculations for a goodness-of-fit test. Some typical output (with more decimal places shown) for the test on Alameda County jurors in Example 7.6 is shown below.

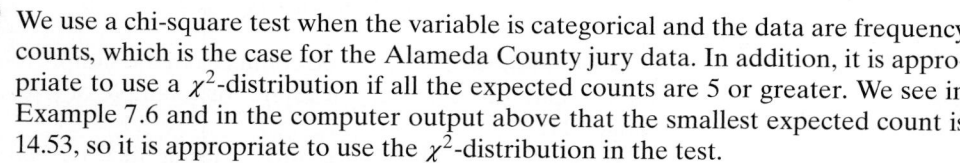

Category	Observed	Test Proportion	Expected	Contribution to Chi-Sq
White	780	0.54	784.62	0.027
Black	117	0.18	261.54	79.880
Hispanic	114	0.12	174.36	20.895
Asian	384	0.15	217.95	126.509
Other	58	0.01	14.53	130.051

N	DF	Chi-Sq	P-Value
1453	4	357.362	0.000

Example 7.7

We complete a chi-square test for testing jury pools in Alameda County in Example 7.6. Explain why a chi-square test is appropriate in that situation.

Solution We use a chi-square test when the variable is categorical and the data are frequency counts, which is the case for the Alameda County jury data. In addition, it is appropriate to use a χ^2-distribution if all the expected counts are 5 or greater. We see in Example 7.6 and in the computer output above that the smallest expected count is 14.53, so it is appropriate to use the χ^2-distribution in the test.

Goodness-of-Fit for Two Categories

Example 7.8

Penny Spins

Exercise C.119 on page 503 describes a sample a student collected by spinning a penny on edge to see if it would land *Heads* or *Tails*. Her data showed 84 heads and 116 tails in 200 spins. Test to see if this provides evidence that spinning a coin is biased away from a 50–50 distribution of heads and tails. Do this test two ways:

(a) Using a normal-based test for a proportion

(b) Using a chi-square goodness-of-fit test

Solution (a) If we let p denote the proportion of all penny spins that land heads, the hypotheses are $H_0 : p = 0.5$ vs $H_a : p \neq 0.5$. Based on seeing 84 heads in 200 sample spins the sample proportion is $\hat{p} = 84/200 = 0.42$. The standardized test statistic is

$$z = \frac{\hat{p} - p_0}{\sqrt{\frac{p_0(1-p_0)}{n}}} = \frac{0.42 - 0.50}{\sqrt{\frac{0.50(1-0.50)}{200}}} = -2.263$$

Comparing to a standard normal distribution, the p-value for this two-tailed test is $2(0.0118) = 0.0236$. At a 5% significance level, this is a small p-value and provides evidence that the proportion of heads when spinning a penny is different from 0.5.

(b) We can think of this as a goodness-of-fit test for a variable with just two categories: heads and tails. The null hypothesis is $H_0 : p_h = p_t = 0.5$ and the alternative is $H_a : p_h \neq 0.5$ or $p_t \neq 0.5$. For 200 spins the expected counts should be $200(0.5) = 100$ in both cells. The observed data and expected counts (in parentheses) are shown in Table 7.7.

We calculate the chi-square statistic.

$$\chi^2 = \frac{(84 - 100)^2}{100} + \frac{(116 - 100)^2}{100} = 2.56 + 2.56 = 5.12$$

Table 7.7 *Observed (expected) counts for 200 penny spins*

Heads	Tails
84 (100)	116 (100)

Comparing this to a chi-square distribution with just one $(2 - 1 = 1)$ degree of freedom, we see that the area beyond $\chi^2 = 5.12$ gives a p-value of 0.0236. At a 5% significance level, this is a small p-value and provides evidence that the proportions of heads and tails when spinning a penny are not both 0.5.

Note that the p-values in both parts of Example 7.8 are the same. This is not a coincidence. The two methods, a two-tailed test for a proportion based on the normal distribution and a chi-square test for two categories, are equivalent. In fact, you can check that the χ^2-test statistic for the goodness-of-fit test (5.12) is just the square of the z-statistic (-2.263). That's one reason for having "square" in the name chi-square. If you are mathematically inclined, you might try to show algebraically (using the formulas for both test statistics) that this always happens. Furthermore, in Exercise C.119 you are asked to do this test using a randomization distribution. The p-value by that method turns out to be 0.024. This is consistent with both the normal and chi-square tests. Remember that the randomization procedure is still a viable option in all situations, including those (like small sample sizes) where the distribution-based tests might not be appropriate.

SECTION LEARNING GOALS

You should now have the understanding and skills to:

▶ • Test a hypothesis about a categorical variable using a chi-square goodness-of-fit test

▶ • Recognize when a chi-square distribution is appropriate for testing a categorical variable

Exercises for Section **7.1**

SKILL BUILDER 1

In Exercises 7.1 to 7.4, find the expected counts in each category using the given sample size and null hypothesis.

7.1 $H_0 : p_1 = p_2 = p_3 = p_4 = 0.25;$ $n = 500$

7.2 H_0: All three categories A, B, C are equally likely; $n = 1200$

7.3 $H_0 : p_A = 0.50, p_B = 0.25, p_C = 0.25;$ $n = 200$

7.4 $H_0 : p_1 = 0.7, p_2 = 0.1, p_3 = 0.1, p_4 = 0.1;$ $n = 400$

SKILL BUILDER 2

In Exercises 7.5 to 7.8, the categories of a categorical variable are given along with the observed counts from a sample. The expected counts from a null hypothesis are given in parentheses. Compute the χ^2-test statistic, and use the χ^2-distribution to find the p-value of the test.

7.5

Category	A	B	C
Observed (Expected)	35 (40)	32 (40)	53 (40)

7.6

Category	A	B	C
Observed (Expected)	61 (50)	35 (50)	54 (50)

7.7

Category	A	B	C	D
Observed (Expected)	132 (160)	181 (160)	45 (40)	42 (40)

7.8

Category	A	B	C	D
Observed (Expected)	38 (30)	55 (60)	79 (90)	128 (120)

SKILL BUILDER 3

Exercises 7.9 to 7.12 give a null hypothesis for a goodness-of-fit test and a frequency table from a sample. For each table, find:

(a) The expected count for the category labeled B.

(b) The contribution to the sum of the chi-square statistic for the category labeled B.

(c) The degrees of freedom for the chi-square distribution for that table.

7.9 $H_0 : p_a = p_b = p_c = p_d = 0.25$
$H_a :$ Some $p_i \neq 0.25$

A	B	C	D	Total
40	36	49	35	160

7.10 $H_0 : p_a = p_b = p_c = p_d = 0.25$
$H_a :$ Some $p_i \neq 0.25$

A	B	C	D	Total
120	148	105	127	500

7.11 $H_0 : p_a = 0.1, p_b = 0.35, p_c = 0.2,$
$\quad p_d = 0.05, p_e = 0.1, p_f = 0.2$
$H_a :$ Some p_i is wrong

A	B	C	D	E	F
210	732	396	125	213	324

7.12 $H_0 : p_a = 0.2, p_b = 0.80$
$H_a :$ Some p_i is wrong

A	B
132	468

7.13 Favorite Skittles Flavor Skittles are a popular fruity candy with five different flavors (colored green, orange, purple, red, and yellow). A sample of 66 people[3] recorded their favorite flavor and the results are shown in Table 7.8. Perform a chi-square test, as indicated in the steps below, to see whether or not the flavors are equally popular.

(a) State the null and alternative hypotheses.

(b) If every flavor of Skittles were equally popular, how many people (in a sample of 66) would we expect to choose each?

(c) How many degrees of freedom do we have?

(d) Calculate the chi-square test statistic.

(e) What is the conclusion about the popularity of the Skittles flavors?

Table 7.8 *Skittles popularity*

Green	Orange	Purple	Red	Yellow
18	9	15	13	11

7.14 Rock-Paper-Scissors In Data 6.1 on page 416 we see a table, reproduced in Table 7.9, that shows the choices made by 119 players on the first turn of a Rock-Paper-Scissors game. Recall that rock beats scissors which beats paper which beats rock. A player gains an advantage in playing this game if there is evidence that the choices made on the first turn are not equally distributed among the three options. Use a goodness-of-fit test to see it there is evidence that any of the proportions are different from 1/3.

Table 7.9 *Frequencies for first turn in Rock-Paper-Scissors*

Option Selected	Frequency
Rock	66
Paper	39
Scissors	14
Total	119

7.15 Age Distribution of Users of Social Networking Sites Between 2008 and 2011, the age distribution of users of social networking sites such as Facebook changed dramatically. In 2008, 70% of users were 35 years old or younger. In 2011, the age distribution was much more spread out.

---
[3]*http://www.quibblo.com.*

Table 7.10 shows the age distribution of 975 users of social networking sites from a survey reported in June 2011.[4]

(a) Test an assumption that users are equally likely to be in each of the five age groups listed. Show all details of the test.

(b) Which age group contributes the largest amount to the sum for the χ^2 test statistic? For this age group, is the observed count smaller or larger than the expected count?

Table 7.10 *Age distribution of users of social networking sites*

Age	18–22	23–35	36–49	50–65	65+
Frequency	156	312	253	195	59

7.16 What is Your Sleep Position? How do you position yourself when you are going to sleep? A website[5] tells us that 41% of us start in the fetal position, another 28% start on our side with legs straight, 13% start on their back, and 7% on their stomach. The remaining 11% have no standard starting sleep position. If a random sample of 1000 people produces the frequencies in Table 7.11, should you doubt the proportions given in the article on the website? Show all details of the test, and use a 5% significance level.

Table 7.11 *Starting sleep positions*

Sleep position	Frequency
Fetal	391
Side, legs straight	257
Back	156
Stomach	89
None	107
Total	1000

ADHD? OR JUST YOUNGEST IN THE CLASS?

A study[6] indicates that the youngest children in a school grade are more likely to be diagnosed with attention-deficit/hyperactivity disorder (ADHD) than their older peers in the same grade. The study involved 937,943 children between 6 and 12 years old in British Columbia, Canada. The cutoff date for entering school in any year in British Columbia is December 31st, so in any given class, those born late in the year are almost a year younger than those born early in the year. Is it possible that the younger students are being over-diagnosed with ADHD? Exercises 7.17 and 7.18 examine this question.

7.17 Boys: ADHD or Just Young? Table 7.12 shows the number of boys diagnosed with ADHD based on the quarter of the year in which they were born, as well as the proportion of all boys born during that quarter.

(a) What is the total number of boys diagnosed with ADHD in the sample?

(b) For the null hypothesis, use the overall proportion of births in a quarter to give the null proportion for that quarter. Compute the expected number of ADHD diagnoses for each quarter under this hypothesis.

(c) Compute the χ^2-statistic.

(d) Give the degrees of freedom and find the p-value.

(e) State the conclusion of the test. For which group of children does ADHD appear to be diagnosed more frequently than we would expect? Less frequently? Write a sentence explaining what this means about ADHD and relative age in school.

Table 7.12 *ADHD diagnoses and birth date for boys*

Birth Date	ADHD Diagnoses	Proportion of Births
Jan–Mar	6880	0.244
Apr–Jun	7982	0.258
Jul–Sep	9161	0.257
Oct–Dec	8945	0.241

7.18 Girls: ADHD or Just Young? Exercise 7.17 examines a relationship between relative age in a class and likelihood of ADHD diagnosis for boys in British Columbia. Girls are less likely overall to be diagnosed with ADHD but does the same relationship exist with relative age in school? Table 7.13 shows the number of girls diagnosed with ADHD based on the quarter of the year in which they were

[4]Hampton, K., Goulet, L., Rainie, L., and Purcell, K., "Social networking sites and our lives," Pew Research Center, *pewresearch.org*, June 16, 2011.

[5]Foltz-Gray, D. "The Best and Worst Sleeping Positions," *http://spryliving.com/articles/best-and-worst-sleeping-positions/*, January 20, 2015.

[6]Morrow, R., et al., "Influence of relative age on diagnosis and treatment of attention-deficit/hyperactivity disorder in children," *Canadian Medical Association Journal*, April 17, 2012; 184(7): 755–762.

born, as well as the proportion of girls born during that quarter. Answer the same questions as in Exercise 7.17, using the data for girls instead of boys.

Table 7.13 *ADHD diagnoses and birth date for girls*

Birth Date	ADHD Diagnoses	Proportion of Births
Jan–Mar	1960	0.243
Apr–Jun	2358	0.258
Jul–Sep	2859	0.257
Oct–Dec	2904	0.242

7.19 Birth Date and Canadian Ice Hockey In his book *Outliers: The Story of Success* (2008), Malcolm Gladwell speculates that Canadian ice hockey players who are born early in the year have an advantage. This is because the birthdate cutoff for different levels of youth hockey leagues in Canada is January 1st, so youth hockey players who are born in January and February are slightly older than teammates born later in the year. Does this slight age advantage in the beginning lead to success later on? A 2010 study[7] examined the birthdate distribution of players in the Ontario Hockey League (OHL), a high-level and selective Canadian hockey league (ages 15–20), for the 2008–2009 season. The number of OHL players born during the 1st quarter (Jan–Mar), 2nd quarter (Apr–Jun), 3rd quarter (Jul–Sep), and 4th quarter (Oct–Dec) of the year is shown in Table 7.14. The overall percentage of live births in Canada (year 1989) are also provided for each quarter. Is this evidence that the birthdate distribution for OHL players differs significantly from the national proportions? State the null and alternative hypotheses, calculate the chi-square statistic, find the p-value, and state the conclusion in context.

Table 7.14 *Birthdates nationally in Canada and for elite hockey players*

	Qtr 1	Qtr 2	Qtr 3	Qtr 4
OHL players	147	110	52	50
% of Canadian births	23.7%	25.9%	25.9%	24.5%

7.20 Birthdate and Australian Football Most Australian youth-sports leagues separate athletes by birthdate, and the cutoff date is January 1st. Thus,

those children born in January and February have some physical advantages in youth sports over those born in November and December. A recent study[8] suggests that those physical advantages enjoyed early in life impact the likelihood of a child becoming a professional athlete. Table 7.15 gives the number of Australian-born 2009 Australian Football League players born in different months of the year, as well as the proportion of births expected if the birthdates of the athletes matched the distribution of all births nationally. Which parts of the year have a higher than expected number of AFL athletes? Which have a lower than expected number? Is there evidence that the distribution of birthdates of AFL athletes is not the same as the distribution of birthdates nationally? If there is evidence of a difference between the actual and expected counts, which categories are contributing the most to the sum for the chi-square test statistic?

Table 7.15 *Birthdates nationally in Australia and for Australian football players*

Months	Proportion Nationally	Actual for AFL Players
Jan–Mar	0.248	196
Apr–June	0.251	162
Jul–Sep	0.254	137
Oct–Dec	0.247	122

7.21 Are Hospitals Deadlier in July? Most medical school graduates in the US enter their residency programs at teaching hospitals in July. A recent study suggests that a spike in deaths due to medication errors coincides with this influx of new practitioners.[9] The study indicates that the number of deaths is significantly higher than expected in July.

(a) What type of statistical analysis was probably done to arrive at this conclusion?

(b) Is the χ^2 statistic likely to be relatively large or relatively small?

(c) Is the p-value likely to be relatively large or relatively small?

(d) What does the relevant categorical variable record?

---

[7]Nolan, J. and Howell, G., "Hockey success and birth date: The relative age effect revisited," *International Review of Sociology of Sport*, 2010; 45(4): 507–512.

[8]Biderman, D., "Born Late Year? Choose Another Sport," *The Wall Street Journal*, March 21, 2010.

[9]Young, J., et al., "July Effect: Impact of the Academic Year-End Changeover on Patient Outcomes. A Systematic Review," *Annals of Internal Medicine*, 2011; 155(5): 309–315.

(e) What cell contributes the most to the χ^2 statistic?

(f) In the cell referred to in part (e), which is higher: the observed count or the expected count?

7.22 Can People Delay Death? A study indicates that elderly people are able to postpone death for a short time to reach an important occasion. The researchers[10] studied deaths from natural causes among 1200 elderly people of Chinese descent in California during six months before and after the Harbor Moon Festival. Thirty-three deaths occurred in the week before the Chinese festival, compared with an estimated 50.82 deaths expected in that period. In the week following the festival, 70 deaths occurred, compared with an estimated 52. "The numbers are so significant that it would be unlikely to occur by chance," said one of the researchers.

(a) Given the information in the problem, is the χ^2 statistic likely to be relatively large or relatively small?

(b) Is the p-value likely to be relatively large or relatively small?

(c) In the week before the festival, which is higher: the observed count or the expected count? What does this tell us about the ability of elderly people to delay death?

(d) What is the contribution to the χ^2-statistic for the week before the festival?

(e) In the week after the festival, which is higher: the observed count or the expected count? What does this tell us about the ability of elderly people to delay death?

(f) What is the contribution to the χ^2-statistic for the week after the festival?

(g) The researchers tell us that in a control group of elderly people in California who are not of Chinese descent, the same effect was not seen. Why did the researchers also include a control group?

7.23 Favorite James Bond Actor? Movies based on Ian Fleming's novels starring British secret agent James Bond have become one of the longest running film series to date. As of 2016, six different actors have portrayed the secret agent. Which actor is the best James Bond? A sample of responses[11] to this question is shown in Table 7.16.

(a) Does the sample provide evidence of a significant difference in popularity among the six actors, at a 5% significance level?

(b) Repeat the test from part (a) if we ignore the results for George Lazenby, who only appeared in one Bond film. Do we find evidence of a significant difference in popularity among the remaining five actors?

(c) The message from Chapter 1 still holds true: Pay attention to where the data come from! These data come from a poll held on a James Bond fan site. Can we generalize the results of this poll to the movie-watching population?

Table 7.16 *Favorite James Bond actor*

Actor	Frequency
Sean Connery	98
George Lazenby	5
Roger Moore	23
Timothy Dalton	9
Pierce Brosnan	25
Daniel Craig	51

7.24 Genetic Variation in Fast-Twitch Muscles Chi-square tests are common in genetics. A gene called *ACTN3* encodes a protein which functions in fast-twitch muscles. People have different variants of this gene, classified as *RR*, *RX*, or *XX*. Computer output is shown for testing whether the proportions in these categories are 0.25, 0.50, and 0.25, respectively. The observed counts come from a study[12] conducted in Australia. (We examine the connection of these variations to fast-twitch muscles in Exercises 7.52 to 7.54 in the next section.)

Chi-Square Goodness-of-Fit Test for Observed Counts

Category	Observed	Test Proportion	Expected	Contribution to Chi-Sq
RR	130	0.25	109	4.04587
RX	226	0.50	218	0.29358
XX	80	0.25	109	7.71560

N	DF	Chi-Sq	P-Value
436	2	12.0550	0.002

(a) What is the sample size?

(b) What is the observed number of people with the variant *RR* for this gene? What is the expected number of people in this group under H_0?

[10]Phillips, D. and Smith, D., "Postponement of Death Until Symbolically Meaningful Occasions," *Journal of the American Medical Association*, 1990; 263(14): 1947–1951.

[11]*http://www.jamesbondwiki.com/page/Poll+Results*.

[12]Yang, N., et al., "ACTN3 genotype is associated with human elite athletic performance," *American Journal of Human Genetics*, September 2003; 73: 627–631.

(c) Which variant contributes the most to the chi-square statistic? For this variant, is the observed value greater than or less than expected?

(d) What are the degrees of freedom for the test?

(e) What is the p-value? Give the conclusion of the test in context.

7.25 Examining Genetic Alleles in Fast-Twitch Muscles Exercise 7.24 discusses a study investigating the *ACTN*3 genotypes *RR*, *RX*, and *XX*. The same study also examines the *ACTN*3 genetic alleles *R* and *X*, also associated with fast-twitch muscles. Of the 436 people in this sample, 244 were classified *R* and 192 were classified *X*. Does the sample provide evidence that the two options are not equally likely?

(a) Conduct the test using a chi-square goodness-of-fit test. Include all details of the test.

(b) Conduct the test using a test for a proportion, using $H_0 : p = 0.5$ where p represents the proportion of the population classified *R*. Include all details of the test.

(c) Compare the p-values and conclusions of the two methods.

7.26 Benford's Law Frank Benford, a physicist working in the 1930s, discovered an interesting fact about some sets of numbers. While you might expect the first digits of numbers such as street addresses or checkbook entries to be randomly distributed, Benford showed that in many cases the distribution of leading digits is not random, but rather tends to have more ones, with decreasing frequencies as the digits get larger.[13] Table 7.17 shows the proportions of leading digits for data that satisfy Benford's law.

Professor Rick Cleary of Bentley University has given several public lectures about Benford's law. As part of his presentation, he rips out pages of a telephone book and asks audience members to select entries at random and record the first digit of the street address. Counts for the leading digits of 1188 such addresses are shown in Table 7.18 and

stored in a variable called *Address* in the dataset **Benford**. Test if these counts are inconsistent with the probabilities given by Benford's law.

7.27 Auditing a Company with Benford's Law Refer to the discussion of Benford's law in Exercise 7.26. While this may seem like a curious oddity, researchers have developed some important applications for these proportions. One involves auditing company records to look for instances of fraud or other financial malfeasance. In many cases accounting records tend to follow Benford's law and significant departures can help auditors discover patterns that should be examined more closely. For example, if a company's policy requires co-signatures for expenses over \$10,000 and auditors find an unusually high number of claims starting with the digit "9," they might be suspicious and examine those claims more closely.

Two of Professor Cleary's students obtained data for 7273 invoices at a company. The observed counts for the leading digits of the invoice amounts are shown in Table 7.19 and stored in the *Invoices* variable of the **Benford** data file. Test if these counts are inconsistent with the probabilities given by Benford's law.

Table 7.19 *Leading digits for 7273 invoices*

Leading digit	1	2	3	4	5	6	7	8	9
Observed count	2225	1214	881	639	655	532	433	362	332

7.28 Craps! The game of craps is a gambling game where players place wagers on the sum from the roll of two six-sided dice.[14] One author rolled a pair of dice 180 times and recorded the results in Table 7.20. He was attempting to make some numbers appear more often than they would by random chance. Use the 180 rolls to test whether he can defeat random chance, and should thus head

---

[13] According to Benford's law, the proportion of leading digits that are d is $log_{10}(1 + 1/d)$.

[14] For complete rules see *http://www.crapsrules.org*.

Table 7.17 *Proportions for leading digits under Benford's law*

Leading digit	1	2	3	4	5	6	7	8	9
Proportion	0.301	0.176	0.125	0.097	0.079	0.067	0.058	0.051	0.046

Table 7.18 *Leading digits for 1188 addresses*

Leading digit	1	2	3	4	5	6	7	8	9
Observed count	345	197	170	126	101	72	69	51	57

to a craps table! (*Hint:* You will need to calculate or find the null proportions for each sum.)

Table 7.20 *180 craps throws*

2	3	4	5	6	7	8	9	10	11	12
5	11	16	13	26	34	19	20	16	13	7

7.29 Random Digits in Students' Random Numbers? How well can people generate random numbers? A sample of students were asked to write down a "random" four-digit number. Responses from 150 students are stored in the file **Digits**. The data file has separate variables (*RND1*, *RND2*, *RND3*, and *RND4*) containing the digits in each of the four positions.

(a) If the numbers are randomly generated, we would expect the last digit to have an equal chance of being any of the 10 digits. Test H_0 :

$p_0 = p_1 = p_2 = \cdots = p_9 = 0.10$ using technology and the data in *RND4*.

(b) Since students were asked to produce four-digit numbers, there are only nine possibilities for the first digit (zero is excluded). Use technology to test whether there is evidence in the values of *RND1* that the first digits are not being chosen by students at random.

7.30 Random Digits in Social Security Numbers? Refer to the data in **Digits** that are described in Exercise 7.29. The 150 students were also asked to give the last two digits of their nine-digit social security number. Those digits are stored in *SSN8* and *SSN9* in the same file. Does the government do a better job at assigning numbers at random? Pick either of the two columns (*SSN8* or *SSN9*) and use technology to test whether there is evidence that the digits are not equally likely (H_a : Some $p_i \neq 0.10$).

7.2 TESTING FOR AN ASSOCIATION BETWEEN TWO CATEGORICAL VARIABLES

In Section 2.1 we consider two-way tables as a way to investigate a relationship between two categorical variables. While we may get some feel for a possible relationship by just looking at a two-way table, we need a mechanism to formally test whether an association is really present or whether the apparent pattern might just be due to random chance.

Example 7.9

One True Love? — Revisited

Data 2.1 on page 48 describes a study in which people are asked how they feel about a statement that "there is only one true love for each person." The results, broken down by gender, are reproduced in Table 7.21. What proportion of the sample agree? Disagree? Don't know? What proportion of the males agree with the statement? What proportion of the females agree?

Solution

We divide each of the row totals in Table 7.21 by the sample size ($n = 2625$) to find the proportion of people in the sample with each attitude. This means $735/2625 = 0.28$ or 28% agree, $1812/2625 = 0.69$ or 69% disagree, and $78/2625 = 0.03$ or 3% don't know. If we look at just the 1213 males the proportion who agree is $372/1213 = 0.307$ or 30.7%, while the 1412 females have $363/1412 = 0.257$ or 25.7% agreeing.

Table 7.21 *Two-way table: Is there one true love for each person?*

↓Attitude/Gender→	Male	Female	Total
Agree	372	363	735
Disagree	807	1005	1812
Don't know	34	44	78
Total	1213	1412	2625

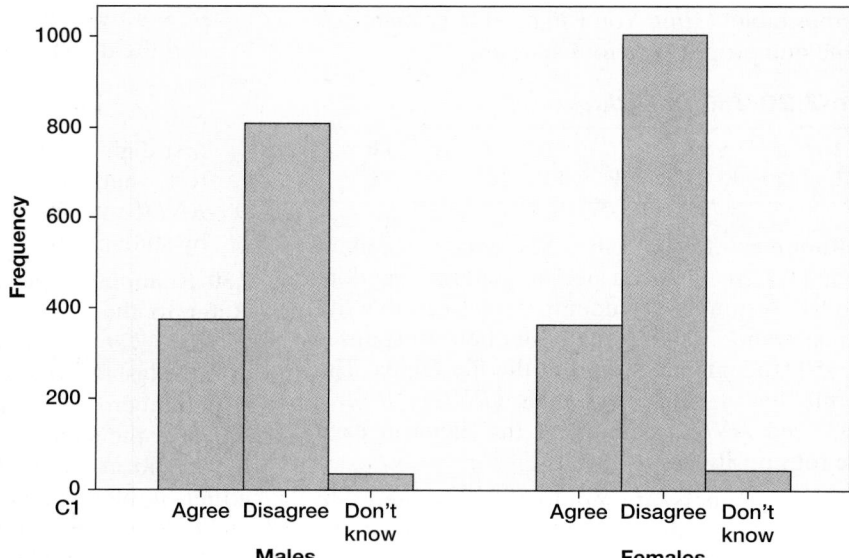

Figure 7.4 *Are these distributions significantly different?*

Is the difference in this sample between the proportion of males who agree (0.307) and the proportion of females who agree (0.257) significant enough to conclude there is a difference in attitude on this subject between males and females in the entire population? While we could use the techniques of Section 6.3-HT to formally test for a difference in these two proportions, we would be ignoring the other two groups (Disagree and Don't know). Just as in Section 7.1 where we want to test all cells of a one-way table simultaneously, we would like to be able to assess a possible association between gender and attitude toward one true love using *all* of the information in Table 7.21, not just the "Agree" row.

The distributions of sample responses for males and females are shown in the comparative bar charts in Figure 7.4. While the distributions are not identical, are the differences likely to be due to random chance or are they significant enough that we can generalize to the entire population? We use a chi-square test to answer this question.

Null and Alternative Hypotheses

Up to this point, most of the hypotheses we have tested make specific claims about one or more parameters of a population. The hypotheses we use when testing association between variables in a two-way table are a bit more general and tend to be expressed in words rather than through parameters. For example, to test for an association between *Attitude* and *Gender* with the one true love data we use

$H_0:$ *Attitude* is not associated with *Gender*

$H_a:$ *Attitude* is associated with *Gender*

As usual, the null hypothesis reflects the belief that nothing significant or interesting is going on. The alternative says that there is some association between the two variables. Remember that we only discard the null and go with the alternative if there is substantial evidence of an association.

Another way to think of an association is that the distribution of one variable (such as *Attitude*) is different for different values of the other variable (such as *Gender*). In this example, an association means that the percent who agree, disagree, and don't know is different between males and females. Of course, the *sample* distributions will always tend to differ by some amount. We need to assess whether the differences are large enough to signal that the *population* distributions are actually different.

Expected Counts for a Two-Way Table

In Section 7.1, we see how to use a chi-square test to compare *observed* to *expected* counts for a single categorical variable. The chi-square test for an association between two categorical variables takes the same approach. How might we compute the expected counts?

The expected counts are computed assuming the null hypothesis is true. Since the null hypothesis states that there is no association, we compute the expected counts to reflect identical percent distributions for males and females that are equal to the overall percent distribution. Furthermore, we keep the row and column totals the same, as in Table 7.22.

Table 7.22 *Expected counts should match overall distribution*

↓Attitude/Gender→	Male	Female	Total	Relative Frequency
Agree			735	0.28
Disagree			1812	0.69
Don't know			78	0.03
Total	1213	1412	2625	1.0

Since 28% of the people in the entire sample agree, we expect 28% of the males and 28% of the females to agree, when the null hypothesis is strictly followed. This gives expected counts of $0.28(1213) = 339.6$ for the first row of the males and $0.28(1412) = 395.4$ for the first row of the females. Similarly, we use 69% to find the expected counts for the Disagree row and 3% for the Don't know row.

Looking again at the computation of the expected count for males who agree, we have

$$339.6 = 0.28 \cdot 1213 = \frac{735}{2625} \cdot 1213 = \frac{735 \cdot 1213}{2625}$$

The computation $(735 \cdot 1213)/2625$ is an alternate way to compute the expected count that reduces round-off error and also shows that the computation is completely symmetric: It doesn't matter which variable is in the rows and which is in the columns.

In general, we find the expected counts for the cells of a two-way table using

$$\text{Expected count} = \frac{\text{Row total} \cdot \text{Column total}}{\text{Sample size}}$$

For the first cell in Table 7.23, this calculation for the expected count for males who agree is

$$\frac{\text{Agree row total} \cdot \text{Male column total}}{n} = \frac{735 \cdot 1213}{2625} = 339.6$$

Computing all the expected counts this way, we arrive at the table of expected counts shown in Table 7.23.

Table 7.23 *Expected counts when Attitude and Gender are not related*

↓Attitude/Gender→	Male	Female	Total
Agree	339.6	395.4	735
Disagree	837.3	974.7	1812
Don't know	36.0	42.0	78
Total	1213	1412	2625

Note that (up to round-off differences) the data in Table 7.23 has the same row and column totals as the original data. However, with the expected counts, the distribution (as proportions) is the same in each column (0.28, 0.69, 0.03) so that the null hypothesis is exactly true.[15]

Chi-square Test for Association

Now that we can compute expected counts to produce a table that exactly matches the null hypothesis of no relationship, the remaining task is to assess how far away our actual sample is from this ideal table. As with the goodness-of-fit test in the previous section, we use a chi-square statistic:

$$\chi^2 = \sum \frac{(Observed - Expected)^2}{Expected}$$

where the sum now is over all of the cells in the two-way table.

For the one true love survey data in Table 7.21, we have 6 cells and the calculation of the chi-square statistic is:

$$\chi^2 = \frac{(372 - 339.6)^2}{339.6} + \frac{(363 - 395.4)^2}{395.4} + \frac{(807 - 837.3)^2}{837.3} + \frac{(1005 - 974.7)^2}{974.7}$$

$$+ \frac{(34 - 36.0)^2}{36.0} + \frac{(44 - 42.0)^2}{42.0}$$

$$= 3.09 + 2.65 + 1.10 + 0.94 + 0.11 + 0.10$$

$$= 7.99$$

Just one more detail before we can find a p-value and finish the test — how many degrees of freedom for the chi-square distribution? We say that a two-way table with 3 categories for rows and 2 categories for columns, such as Table 7.21, has just 2 degrees of freedom because once two cells are "freely" established, the rest of the cell values are automatically determined based on the row and column totals. In general, if a two-way table has r rows and c columns, the degrees of freedom is $(r - 1) \cdot (c - 1)$. Think of covering up the last row and column (not including totals) and counting how many cells are left. That gives the degrees of freedom for the table.

We are finally in a position to complete the test for association between *Attitude* on the one true love issue and *Gender*. We find that the p-value for the statistic, $\chi^2 = 7.99$, using the upper tail of a chi-square distribution with 2 degrees of freedom, is 0.018. This is a small p-value, less than 5%, so we have fairly strong evidence that attitudes on the existence of one true love are associated with gender.

[15]The expected counts in Table 7.23 also have exactly the same distribution across each row, 0.462 male and 0.538 female.

Note that the conclusion of the hypothesis test doesn't tell us how the two variables are associated, only that the relationship in the data is more extreme than we would reasonably expect to see by random chance alone. By comparing observed and expected counts, we see that males tend to be more likely to agree than the expected count would indicate, females are more likely to disagree than the expected count would indicate, and both genders choose don't know at about the rate we would expect.

Chi-square Test for Association

To test for an association between two categorical variables, based on a two-way table that has r rows as categories for variable A and c columns as categories for variable B:

Set up hypotheses:

$$H_0 : \text{Variable A is not associated with variable B}$$
$$H_a : \text{Variable A is associated with variable B}$$

Compute the expected count for each cell in the table using

$$\text{Expected count} = \frac{\text{Row total} \cdot \text{Column total}}{\text{Sample size}}$$

Compute the value for a chi-square statistic using

$$\chi^2 = \sum \frac{(Observed - Expected)^2}{Expected}$$

Find a p-value using the upper tail of a chi-square distribution with $(r - 1)(c - 1)$ degrees of freedom.

The chi-square distribution is appropriate if the expected count is at least five in each cell.

A two-way table in a chi-square test is sometimes referred to as an $r \times c$ *contingency table*, where r denotes the number of rows and c is the number of columns, and the test is also sometimes referred to as a χ^2 *test for independence*.

DATA 7.3

Water Taste Preferences

Some students at Longwood University in Virginia were interested in people's taste preferences among various brands of bottled water.[16] They collected data from community members, including a double-blind taste test where participants ranked water from four sources (tap water, and bottled water from Aquafina, Fiji, and Sam's Choice, presented in a random order). Some of the data from this study are stored in **WaterTaste** and Table 7.24 shows the top choices of 100 participants as well as their usual water source (bottled or tap/filtered). ■

---

[16]Lunsford, M. and Fink, A., "Water Taste Data," *Journal of Statistics Education*, 2010; 18(1).

© 97/iStockphoto

Which type of water is preferred?

Table 7.24 *Usual water source and preferred brand for taste*

↓Usual Source/Top Choice→	Tap	Aquafina	Fiji	Sam's Choice	Total
Bottled	4	14	15	8	41
Tap/Filtered	6	11	26	16	59
Total	10	25	41	24	100

Example 7.10

Based on the information in this 2×4 table, is there evidence that the top choices for taste preference are associated with whether or not people usually drink bottled water?

Solution The hypotheses for this chi-square test are:

H_0 :   Choice for best tasting brand is not related to the usual water source

H_a :   Choice for best tasting brand is related to the usual water source

We see that

$$\text{Expected count for (Bottled, Tap) cell} = \frac{41 \cdot 10}{100} = 4.1$$

Computing the rest of the expected counts similarly, we obtain the values shown in parentheses in Table 7.25. The observed count for the (Bottled, Tap) cell is 4—there are four people who usually drink bottled water that actually prefer tap water! We see that one expected count is less than 5, but because it is just one cell and only slightly less than 5, we proceed but with some caution.

Table 7.25 *Observed (expected) counts for water taste test*

↓Usual Source/Top Choice→	Tap	Aquafina	Fiji	Sam's Choice	Total
Bottled	4 (4.1)	14 (10.3)	15 (16.8)	8 (9.8)	41
Tap/Filtered	6 (5.9)	11 (14.8)	26 (24.2)	16 (14.2)	59
Total	10	25	41	24	100

The value of the chi-square statistic for this table is

$$\chi^2 = \frac{(4-4.1)^2}{4.1} + \frac{(14-10.3)^2}{10.3} + \frac{(15-16.8)^2}{16.8} + \frac{(8-9.8)^2}{9.8}$$
$$+ \frac{(6-5.9)^2}{5.9} + \frac{(11-14.8)^2}{14.8} + \frac{(26-24.2)^2}{24.2} + \frac{(16-14.2)^2}{14.2}$$
$$= 0.002 + 1.329 + 0.193 + 0.331 + 0.002 + 0.976 + 0.134 + 0.228$$
$$= 3.195$$

The degrees of freedom for the 2×4 table are $(2-1) \cdot (4-1) = 1 \cdot 3 = 3$. The upper tail of a chi-square distribution with 3 degrees of freedom gives a p-value of 0.363, which is not small at all. See Figure 7.5. Therefore the table provides no significant evidence that the brand preferred by taste is related to whether or not people usually drink bottled water.

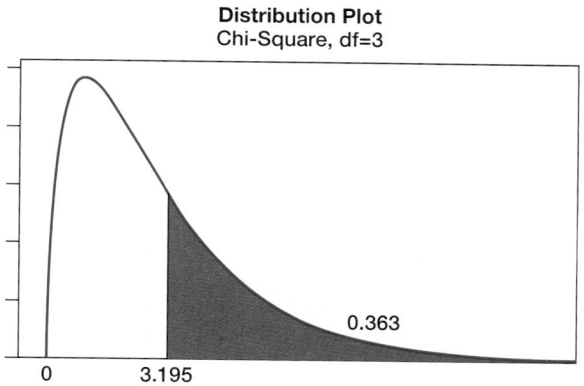

Figure 7.5 *P-value from a chi-square distribution with df = 3*

Special Case for a 2×2 Table

Recall from Section 7.1 that a test for a single proportion can be viewed as a special case of a goodness-of-fit test when the variable has just two categories. As we see in the next example, a similar relationship exists between a z-test to compare two proportions and a chi-square test for a 2×2 table.

Example 7.11

Split or Steal? — Revisited

Data 6.4 on page 447 describes decisions made on the *Golden Balls* game show to split a prize or attempt to steal it from another contestant. The decisions made by a sample of 574 contestants are shown in Table 7.26 along with information on

Table 7.26 *Split/steal choice by age group*

	Split	Steal	Total
Younger	187	195	382
Older	116	76	192
Total	303	271	574

age (Younger = under 40 years old, Older = 40 or older). Use a chi-square test to determine if the table provides evidence that split/steal decisions are related to age.

Solution 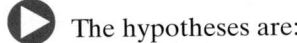 The hypotheses are:

$$H_0 : \quad \text{Split/steal decision is not related to age}$$

$$H_a : \quad \text{Split/steal decision is related to age}$$

Table 7.27 shows the expected counts (in parentheses) along with the observed counts. For example, to get the expected count in the first (Younger, Split) cell we find $(382 \cdot 303)/574 = 201.6$.

Table 7.27 *Observed (expected) counts for split/steal decision by age*

	Split	Steal	Total
Younger	187 (201.6)	195 (180.4)	382
Older	116 (101.4)	76 (90.6)	192
Total	303	271	574

The chi-square statistic for this table is

$$\chi^2 = \frac{(187 - 201.6)^2}{201.6} + \frac{(195 - 180.4)^2}{180.4} + \frac{(116 - 101.4)^2}{101.4} + \frac{(76 - 90.6)^2}{90.6}$$

$$= 1.06 + 1.18 + 2.10 + 2.35$$

$$= 6.69$$

We find the p-value as the area in the tail above $\chi^2 = 6.69$ for a chi-square distribution with $df = 1$ since $(2 - 1) \cdot (2 - 1) = 1 \cdot 1 = 1$. This gives a p-value of 0.0097, which is quite small, providing substantial evidence that the split/steal decision is related to age.

If we look back at Example 6.23 on page 448, we see the same data analyzed with a z-test to compare the proportion of "Younger" who split (0.490 in the sample) to the proportion who split in the "Older" group (0.604). The standardized test statistic for the difference in proportions is $z = -2.60$ and produces a two-tailed p-value from the normal distribution of 0.0094. Note that the p-value matches the p-value from the chi-square test for the 2×2 table (up to round-off error) and you may check that the chi-square statistic is the square of the normal z-statistic. This is always the case when comparing a chi-square test for a 2×2 table with the corresponding two-tailed test for a difference in two proportions.

SECTION LEARNING GOALS

You should now have the understanding and skills to:

- Test for an association between two categorical variables based on data in a two-way table

- Recognize when a chi-square distribution is appropriate for testing with a two-way table

Exercises for Section **7.2**

SKILL BUILDER 1

Exercises 7.31 to 7.34 give a two-way table and specify a particular cell for that table. In each case find the expected count for that cell and the contribution to the chi-square statistic for that cell.

7.31 (Group 3, Yes) cell

	Yes	No	Total
Group 1	56	44	100
Group 2	132	68	200
Group 3	72	28	100
Total	260	140	400

7.32 (B, E) cell

	D	E	F	G	Total
A	39	34	43	34	150
B	78	89	70	93	330
C	23	37	27	33	120
Total	140	160	140	160	600

7.33 (Control, Disagree) cell

	Strongly Agree	Agree	Neutral	Disagree	Strongly Disagree
Control	40	50	5	15	10
Treatment	60	45	10	5	0

7.34 (Group 2, No)

	Yes	No
Group 1	720	280
Group 2	1180	320

SKILL BUILDER 2

Exercises 7.35 to 7.38 refer to the tables in Skill Builder 1. In each case, give the degrees of freedom for the chi-square test based on that two-way table.

7.35 Two-way table in Exercise 7.31.

7.36 Two-way table in Exercise 7.32.

7.37 Two-way table in Exercise 7.33.

7.38 Two-way table in Exercise 7.34.

7.39 Gender and Award Preference Example 2.6 on page 53 contains a two-way table showing preferences for an award (Academy Award, Nobel Prize, Olympic gold medal) by gender for the students sampled in **StudentSurvey**. The data are reproduced in Table 7.28. Test whether the data indicate there is some association between gender and preferred award.

Table 7.28 *Two-way table of gender and preferred award*

	Academy	Nobel	Olympic	Total
Female	20	76	73	169
Male	11	73	109	193
Total	31	149	182	362

7.40 One True Love by Educational Level In Data 2.1 on page 48, we introduce a study in which people were asked whether they agreed or disagreed with the statement that there is only one true love for each person. Table 7.29 gives a two-way table showing the answers to this question as well as the education level of the respondents. A person's education is categorized as HS (high school degree or less), Some (some college), or College (college graduate or higher). Is the level of a person's education related to how the person feels about one true love? If there is a significant association between these two variables, describe how they are related.

Table 7.29 *Educational level and belief in one true love*

	HS	Some	College	Total
Agree	363	176	196	735
Disagree	557	466	789	1812
Don't know	20	26	32	78
Total	940	668	1017	2625

7.41 Metal Tags on Penguins In Exercise 6.148 on page 445 we perform a test for the difference in the proportion of penguins who survive over a ten-year period, between penguins tagged with metal tags and those tagged with electronic tags. We are interested in testing whether the type of tag has an effect on penguin survival rate, this time using a chi-square test. In the study, 10 of the 50 metal-tagged penguins survived while 18 of the 50 electronic-tagged penguins survived.

(a) Create a two-way table from the information given.

(b) State the null and alternative hypotheses.

(c) Give a table with the expected counts for each of the four categories.

(d) Calculate the chi-square test statistic.

(e) Determine the p-value and state the conclusion using a 5% significance level.

7.42 Treatment for Cocaine Addiction Cocaine addiction is very hard to break. Even among addicts trying hard to break the addiction, relapse is common. (A relapse is when a person trying to break out of the addiction fails and uses cocaine again.) Data 4.7 on page 323 introduces a study investigating the effectiveness of two drugs, desipramine and lithium, in the treatment of cocaine addiction. The subjects in the six-week study were cocaine addicts seeking treatment. The 72 subjects were randomly assigned to one of three groups (desipramine, lithium, or a placebo, with 24 subjects in each group) and the study was double-blind. In Example 4.34 we test lithium vs placebo, and in Exercise 4.181 we test desipramine vs placebo. Now we are able to consider all three groups together and test whether relapse rate differs by drug. Ten of the subjects taking desipramine relapsed, 18 of those taking lithium relapsed, and 20 of those taking the placebo relapsed.

(a) Create a two-way table of the data.

(b) Find the expected counts. Is it appropriate to analyze the data with a chi-square test?

(c) If it is appropriate to use a chi-square test, complete the test. Include hypotheses, and give the chi-square statistic, the p-value, and an informative conclusion.

(d) If the results are significant, which drug is most effective? Can we conclude that the choice of treatment drug causes a change in the likelihood of a relapse?

7.43 Painkillers and Miscarriage Exercise A.50 on page 179 describes a study examining the link between miscarriage and the use of painkillers during pregnancy. Scientists interviewed 1009 women soon after they got positive results from pregnancy tests about their use of painkillers around the time of conception or in the early weeks of pregnancy. The researchers then recorded which of the pregnancies were successfully carried to term. The results are in Table 7.30. (NSAIDs refer to a class of painkillers that includes aspirin and ibuprofen.)

Table 7.30 Does the use of painkillers increase the risk of miscarriage?

	Miscarriage	No miscarriage	Total
NSAIDs	18	57	75
Acetaminophen	24	148	172
No painkiller	103	659	762
Total	145	864	1009

Does there appear to be an association between having a miscarriage and the use of painkillers? If so, describe the relationship. If there is an association, can we conclude that the use of painkillers increases the chance of having a miscarriage?

7.44 Binge Drinking The American College Health Association—National College Health Assessment survey,[17] introduced on page 60, was administered at 44 colleges and universities in Fall 2011 with more than 27,000 students participating in the survey. Students in the ACHA-NCHA survey were asked "Within the last two weeks, how many times have you had five or more drinks of alcohol at a sitting?" The results are given in Table 7.31. Is there a significant difference in drinking habits depending on gender? Show all details of the test. If there is an association, use the observed and expected counts to give an informative conclusion in context.

Table 7.31 In the last two weeks, how many times have you had five or more drinks of alcohol?

	Male	Female	Total
0	5,402	13,310	18,712
1–2	2,147	3,678	5,825
3–4	912	966	1,878
5+	495	358	853
Total	8,956	18,312	27,268

7.45 Which Is More Important: Grades, Sports, or Popularity? 478 middle school (grades 4 to 6) students from three school districts in Michigan were asked whether good grades, athletic ability, or popularity was most important to them.[18] The results are shown below, broken down by gender:

	Grades	Sports	Popular
Boy	117	60	50
Girl	130	30	91

(a) Do these data provide evidence that grades, sports, and popularity are not equally valued among middle school students in these school districts? State the null and alternative hypotheses, calculate a test statistic, find a p-value, and answer the question.

[17]www.acha-ncha.org/docs/ACHA-NCHA-II_ReferenceGroup_DataReport_Fall2011.pdf.
[18]Chase, M. and Dummer, G., "The Role of Sports as a Social Determinant for Children," *Research Quarterly for Exercise and Sport*, 1992; 63: 418–424.

(b) Do middle school boys and girls have different priorities regarding grades, sports, and popularity? State the null and alternative hypotheses, calculate a test statistic, find a p-value, and answer the question.

7.46 Favorite Skittles Flavor? Exercise 7.13 on page 518 discusses a sample of people choosing their favorite Skittles flavor by color (green, orange, purple, red, or yellow). A separate poll sampled 91 people, again asking them their favorite Skittles flavor, but rather than by color they asked by the actual flavor (lime, orange, grape, strawberry, and lemon, respectively).[19] Table 7.32 shows the results from both polls. Does the way people choose their favorite Skittles type, by color or flavor, appear to be related to which type is chosen?

(a) State the null and alternative hypotheses.

(b) Give a table with the expected counts for each of the 10 cells.

(c) Are the expected counts large enough for a chi-square test?

(d) How many degrees of freedom do we have for this test?

(e) Calculate the chi-square test statistic.

(f) Determine the p-value. Do we find evidence that method of choice affects which is chosen?

Table 7.32 Skittles popularity

	Green (Lime)	Orange	Purple (Grape)	Red (Strawberry)	Yellow (Lemon)
Color	18	9	15	13	11
Flavor	13	16	19	34	9

7.47 Handedness and Occupation Is the career someone chooses associated with being left- or right-handed? In one study[20] a sample of Americans from a variety of professions were asked if they consider themselves left-handed, right-handed, or ambidextrous (equally skilled with the left and right hand). The results for five professions are shown in Table 7.33.

(a) In this sample, what profession had the greatest proportion of left-handed people? What profession had the greatest proportion of right-handed people?

(b) Test for an association between handedness and career for these five professions. State the null and alternative hypotheses, calculate the test statistic, and find the p-value.

(c) What do you conclude at the 5% significance level? What do you conclude at the 1% significance level?

Table 7.33 Handedness vs profession

	Right	Left	Ambidextrous	Total
Psychiatrist	101	10	7	118
Architect	115	26	7	148
Orthopedic surgeon	121	5	6	132
Lawyer	83	16	6	105
Dentist	116	10	6	132
Total	536	67	32	635

7.48 Age Distribution of Social Networking Site Users The Pew Research Center conducted a survey of randomly sampled American adults in 2008 and in 2010, asking them about their use of social networking sites such as Facebook.[21] Table 7.34 shows age groups of social networking site users in 2008 and in 2010. Did the age distribution change significantly in the two-year time span? Show all details of the test.

Table 7.34 Age distribution of social network site users

↓Age/Year→	2008	2010	Total
18–22	138	152	290
23–35	197	303	500
36–49	108	246	354
50+	52	246	298
Total	495	947	1442

7.49 Age and Frequency of Status Updates on Facebook Exercise 7.48 introduces a study about users of social networking sites such as Facebook. Table 7.35 shows the self-reported frequency of status updates on Facebook by age groups.

(a) Based on the totals, if age and frequency of status updates are really unrelated, how many of the 156 users who are 18 to 22 years olds should we expect to update their status every day?

(b) Since there are 20 cells in this table, we'll save some time and tell you that the chi-square statistic for this table is 210.9. What should we conclude about a relationship (if any) between age and frequency of status updates?

---

[19]http://www.deviantart.com.

[20]Schachter, S. and Ransil, B., "Handedness Distributions in Nine Professional Groups," *Perceptual and Motor Skills*, 1996; 82: 51–63.

[21]Hampton, K., Goulet, L., Rainie, L., and Purcell, K., "Social networking sites and our lives," Pew Research Center, *pewresearch.org*, June 16, 2011.

Table 7.35 *Age and frequency of status updates on Facebook*

↓Status/Age→	18–22	23–35	36–49	50+	Total
Every day	47	59	23	7	136
3–5 days/week	33	47	30	7	117
1–2 days/week	32	69	35	25	161
Every few weeks	23	65	47	34	169
Less often	21	74	99	170	364
Total	156	314	234	243	947

7.50 Gender and Frequency of Status Updates on Facebook Exercise 7.48 introduces a study about users of social networking sites such as Facebook. Table 7.36 shows the self-reported frequency of status updates on Facebook by gender. Are frequency of status updates and gender related? Show all details of the test.

Table 7.36 *Gender and frequency of status updates on Facebook*

↓Status/Gender→	Male	Female	Total
Every day	42	88	130
3–5 days/week	46	59	105
1–2 days/week	70	79	149
Every few weeks	77	79	156
Less often	151	186	337
Total	386	491	877

7.51 Gender and Frequency of "Liking" Content on Facebook Exercise 7.48 introduces a study about users of social networking sites such as Facebook. Table 7.37 shows the frequency of users "liking" content on Facebook, with the data shown by gender. Does the frequency of "liking" depend on the gender of the user? Show all details of the test.

Table 7.37 *Gender and frequency of "liking" content on Facebook*

↓Liking/Gender→	Male	Female	Total
Every day	77	142	219
3–5 days/week	39	54	93
1–2 days/week	62	69	131
Every few weeks	42	44	86
Less often	166	182	348
Total	386	491	877

GENETICS AND FAST-TWITCH MUSCLES
Exercises 7.52 to 7.54 investigate the gene *ACTN3*, which encodes a protein that functions in fast-twitch muscles. People can be classified according to which genotype of this gene they have, *RR*, *RX*, or *XX*, and also according to which genetic allele they have, *R* or *X*. The study[22] described here, and introduced in Exercises 7.24 and 7.25 on page 521, examines the association between this gene and different subgroups of the population. All participants in the study live in Australia. The earlier exercises only included the control group.

7.52 Testing Genotypes for Fast-Twitch Muscles The study on genetics and fast-twitch muscles includes a sample of elite sprinters, a sample of elite endurance athletes, and a control group of non-athletes. Is there an association between genotype classification (*RR*, *RX*, or *XX*) and group (sprinter, endurance, control group)? Computer output is shown for this chi-square test. In each cell, the top number is the observed count, the middle number is the expected count, and the bottom number is the contribution to the chi-square statistic.

	RR	RX	XX	Total
Control	130	226	80	436
	143.76	214.15	78.09	
	1.316	0.655	0.047	
Sprint	53	48	6	107
	35.28	52.56	19.16	
	8.901	0.395	9.043	
Endurance	60	88	46	194
	63.96	95.29	34.75	
	0.246	0.558	3.645	
Total	243	362	132	737

Chi-Sq = 24.805, DF = 4, P-Value = 0.000

(a) What is the expected count for endurance athletes with the *XX* genotype? For this cell, what is the contribution to the chi-square statistic? Verify both values by computing them yourself.

(b) What are the degrees of freedom for the test? Verify this value by computing it yourself.

(c) What is the chi-square test statistic? What is the p-value? What is the conclusion of the test?

(d) Which cell contributes the most to the chi-square statistic? For this cell, is the observed count greater than or less than the expected count?

(e) Which genotype is most over-represented in sprinters? Which genotype is most over-represented in endurance athletes?

[22] Yang, N., et al., "ACTN3 genotype is associated with human elite athletic performance," *American Journal of Human Genetics*, 2003; 73: 627–631.

7.53 Testing Genetic Alleles for Fast-Twitch Muscles The study on genetics and fast-twitch muscles includes a sample of elite sprinters, a sample of elite endurance athletes, and a control group of non-athletes. Is there an association between genetic allele classification (R or X) and group (sprinter, endurance, control)? Computer output is shown for this chi-square test. In each cell, the top number is the observed count, the middle number is the expected count, and the bottom number is the contribution to the chi-square statistic.

	R	X	Total
Control	244	192	436
	251.42	184.58	
	0.219	0.299	
Sprint	77	30	107
	61.70	45.30	
	3.792	5.166	
Endurance	104	90	194
	111.87	82.13	
	0.554	0.755	
Total	425	312	737

Chi-Sq = 10.785, DF = 2, P-Value = 0.005

(a) How many endurance athletes were included in the study?

(b) What is the expected count for sprinters with the R allele? For this cell, what is the contribution to the chi-square statistic? Verify both values by computing them yourself.

(c) What are the degrees of freedom for the test? Verify this value by computing it yourself.

(d) What is the chi-square test statistic? What is the p-value? What is the conclusion of the test?

(e) Which cell contributes the most to the chi-square statistic? For this cell, is the observed count greater than or less than the expected count?

(f) Which allele is most over-represented in sprinters? Which allele is most over-represented in endurance athletes?

7.54 Gender and ACTN3 Genotype We see in the previous two exercises that sprinters are more likely to have allele R and genotype RR versions of the *ACTN3* gene, which makes these versions associated with fast-twitch muscles. Is there an association between genotype and gender? Computer output is shown for this chi-square test, using the control group in the study. In each cell, the top number is the

observed count, the middle number is the expected count, and the bottom number is the contribution to the chi-square statistic. What is the p-value? What is the conclusion of the test? Is gender associated with the likelihood of having a "sprinting gene"?

	RR	RX	XX	Total
Male	40	73	21	134
	40.26	69.20	24.54	
	0.002	0.208	0.509	
Female	88	147	57	292
	87.74	150.80	53.46	
	0.001	0.096	0.234	
Total	128	220	78	426

Chi-Sq = 1.050, DF = 2, P-Value = 0.592

7.55 Another Test for Cocaine Addiction Exercise 7.42 on page 532 describes an experiment on helping cocaine addicts break the cocaine addiction, in which cocaine addicts were randomized to take desipramine, lithium, or a placebo. The results (relapse or no relapse after six weeks) are summarized in Table 7.38.

(a) In Exercise 7.42, we calculate a χ^2 statistic of 10.5 and use a χ^2 distribution to calculate a p-value of 0.005 using these data, but we also could have used a randomization distribution. How would you use cards to generate a randomization sample? What would you write on the cards, how many cards would there be of each type, and what would you do with the cards?

(b) If you generated 1000 randomization samples according to your procedure from part (a) and calculated the χ^2 statistic for each, approximately how many of these statistics do you expect would be greater than or equal to the χ^2 statistic of 10.5 found using the original sample?

Table 7.38 *Breaking the cocaine addiction*

	Relapse	No Relapse	Total
Desipramine	10	14	24
Lithium	18	6	24
Placebo	20	4	24
Total	48	24	72

7.56 Who Is More Likely to Take Vitamins: Males or Females? Data 2.11 on page 114 introduces the dataset **NutritionStudy** which contains, among other things, information about vitamin use and gender of the participants. Is there a significant association between these two variables? Use a statistical software package and the variables *VitaminUse* and *Gender* to conduct a chi-square analysis and clearly give the results.

7.57 Who Is More Likely to Smoke: Males or Females? Data 2.11 on page 114 introduces the dataset **NutritionStudy** which contains, among other things, information about smoking history and gender of the participants. Is there a significant association between these two variables? Use a statistical software package and the variables *PriorSmoke* and *Gender* to conduct a chi-square analysis and clearly give the results. The variable *PriorSmoke* is coded as 1 = never smoked, 2 = prior smoker, and 3 = current smoker.

CHAPTER 8

ANOVA to Compare Means

"When moral posturing is replaced by an honest assessment of the data, the result is often a new, surprising insight."

−Steven Levitt and Stephen Dubner*

Freakonomics, Harper Collins, New York, 2009, p. 11.
Top left: Nick Koudis/Getty Images, Inc., Top right: Ryan McVay/Getty Images, Inc., Bottom right: ©John Wiley & Sons. Photo by Vincent LaRussa

Questions and Issues

Here are some of the questions and issues we will discuss in this chapter:

- What type of sandwich do ants prefer?

- Does hearing a mother's voice reduce stress levels? Does it matter if the voice is on the phone rather than in person? Does it matter if the contact is through texting?

- Does a person's ability to solve a problem depend on the color in which the problem is written?

- We've seen that having a light on at night increases weight gain. Why?

- Does regular exercise help give people (or rats) resilience to stress?

- Does adopting a dominant posture enable one to deal with pain more easily?

- What types of incentives are most effective at encouraging sedentary people to exercise?

- Does synchronized movement (such as dancing or marching) increase feelings of closeness?

- Does drawing an image help a person to remember it?

- Is aggressive treatment or mild treatment better at minimizing drug-resistant pathogens?

- What type of courses have the highest textbook costs?

- If a man uses a laptop computer on his lap, how much does groin temperature increase?

8.1 ANALYSIS OF VARIANCE

In the previous chapter, we use chi-square tests to extend tests for a proportion or a difference in proportions. In this chapter, we investigate a method to extend a difference in means test. This procedure allows us to analyze a relationship between a quantitative variable and a categorical variable.

In Section 6.4-HT we use a t-test to compare the means of a quantitative variable between two groups. What if we want to compare more than two groups? The goal of this chapter is to develop a method to test for a difference in means among several groups. The procedure is quite different from tests introduced earlier in this text. The basic idea is to compare measures of variability, both between the groups and within each group, as a way to assess how different the groups really are. Using variability to test for a difference in means may seem strange to you at first, but this is a general statistical approach that can be applied in many important settings. We call such tests *analysis of variance* or just ANOVA for short.

Nick Koudis/Getty Images,Inc.

DATA 8.1 **Sandwich Ants**

As young students in Australia, Dominic Kelly and his friends enjoyed watching ants gather on pieces of sandwiches. Later, as a university student, Dominic decided to study this with a more formal experiment. He chose three types of sandwich fillings to compare: vegemite, peanut butter, and ham & pickles. To conduct the experiment he randomly chose a sandwich, broke off a piece, and left it on the ground near an ant hill. After several minutes he placed a jar over the sandwich bit and counted the number of ants. He repeated the process, allowing time for ants to return to the hill after each trial, until he had eight samples for each of the three sandwich fillings. The data (number of ants) are shown in Table 8.1 and stored in the file **SandwichAnts**[1] along with some additional information about the sandwiches. ■

[1]Mackisack, M., "Favourite Experiments: An Addendum to What Is the Use of Experiments Conducted by Statistics Students?" *Journal of Statistics Education*, 1994, *http://www.amstat.org/publications /jse/v2n1/mackisack.supp.html.*

Table 8.1 *Number of ants visiting sandwiches*

Vegemite	18	29	42	42	31	21	38	25
Peanut Butter	43	59	22	25	36	47	19	21
Ham & Pickles	44	34	36	49	54	65	59	53

Null and Alternative Hypotheses

As usual, we have two competing hypotheses. The null hypothesis is that all three types of sandwiches are equally liked by the ants and attract the same mean number of ants. If we let μ_1, μ_2, and μ_3 represent the mean number of ants on the respective sandwich types, we have

$$H_0 : \mu_1 = \mu_2 = \mu_3$$

The alternative hypothesis is that the means are *not* all the same. This doesn't imply that all three types of sandwiches have different means, just that the mean for at least one of the fillings is different from the mean for another filling. To state this we use

$$H_a : \text{At least one } \mu_i \neq \mu_j$$

The alternative hypothesis again indicates that something is going on: that there is an association between the categorical variable (which defines the groups) and the quantitative variable for which we compute means. Note that the alternative hypothesis does not say which two sandwich fillings have different means or give a direction as we might see in a one-tailed test. The ANOVA procedure for testing a difference in means is only designed to determine whether we have enough evidence to conclude that a difference exists somewhere among the groups. We leave to Section 8.2 the question of determining which groups might be different.

We examine summary statistics and a graph to help us compare the groups numerically and visually.

Example 8.1

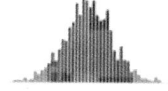

Use the data in **SandwichAnts** or Table 8.1 to find the sample mean and standard deviation of the number of ants for each sandwich filling. Also, produce a plot to compare the samples.

Solution

The means and standard deviations are shown in Table 8.2. We use side-by-side dotplots, as in Figure 8.1, to compare the numbers of ants for each sandwich. The graph also shows the mean number of ants (vertical bar) for each filling.

Table 8.2 *Group means and standard deviations for ant counts*

Filling	Sample size	Mean	Standard deviation
Vegemite	$n_1 = 8$	$\bar{x}_1 = 30.75$	$s_1 = 9.25$
Peanut Butter	$n_2 = 8$	$\bar{x}_2 = 34.0$	$s_2 = 14.63$
Ham & Pickles	$n_3 = 8$	$\bar{x}_3 = 49.25$	$s_3 = 10.79$

We see in Table 8.2 that the sample means are different for each sandwich type and that the mean number of ants for ham & pickles is quite a bit larger than the means for the other two fillings. But is that difference statistically significant? Remember, if we want to measure evidence against the null hypothesis, we need to think about what sort of data we might see if samples were collected when the null hypothesis is true. In this case the null hypothesis says that all the fillings attract

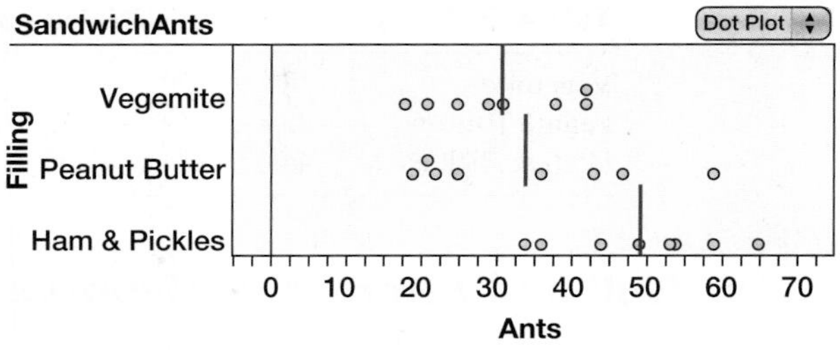

Figure 8.1 *Dotplots (with means) comparing number of ants on three types of sandwiches*

the same mean number of ants. If the null hypothesis is true, we can combine the samples to give one big sample of 24 sandwich pieces. The mean of all 24 values in the sample is $\bar{x} = 38.0$ ants. We can also find the standard deviation of all 24 sandwich bits, $s = 13.95$, which summarizes the deviations of all the ant counts from the overall mean.

In this example, we could also find the overall mean by simply averaging the three sandwich means. However, this only works when the sample sizes are the same for each group. In general, the overall mean and standard deviation should be computed from all the data values in the combined sample.

Why Analyze *Variability* to Test for a Difference in *Means*?

If the null hypothesis is true, the samples are generated from populations with the same mean. If we select samples of size eight from a population where the mean is 38, how likely is it to see sample means as different as 30.75, 34.0, and 49.25? We know that variability in sample means depends not only on the sample size but also on the variability in the population. Furthermore, while we can easily find the difference between any pair of group means, we want a single measure that reflects how far apart the means are for all groups. We address both of these issues by measuring different aspects of the variability in the data.

Example 8.2

Figure 8.2 shows boxplots (with symbols showing the means) for hypothetical data comparing three groups. The group means in Datasets A and B are the same, but the boxes show different spread. Datasets A and C have the same spread for the boxes but different group means. Discuss which of these graphs appear to give strong visual evidence for a difference in the group means.

Solution The boxplots for Dataset A show the weakest evidence for a difference in means between the three groups. There is quite a bit of overlap between the boxes and they could easily be three samples taken from the same population. Datasets B and C both show strong evidence for a difference in group means, especially since all the data in the Group 1 sample is less than every data point in Group 2 for both datasets.

The important point illustrated in Example 8.2 is that an assessment of the difference in means between several groups depends on two kinds of variability: how different the means are from each other AND the amount of variability in the samples. Just knowing the sample means for the groups is not enough. If the values within each sample are very close to each other, we can detect a small difference in means as significant (as in Dataset B). If there is more variability in the samples, we need the group means to be farther apart (as in Dataset C).

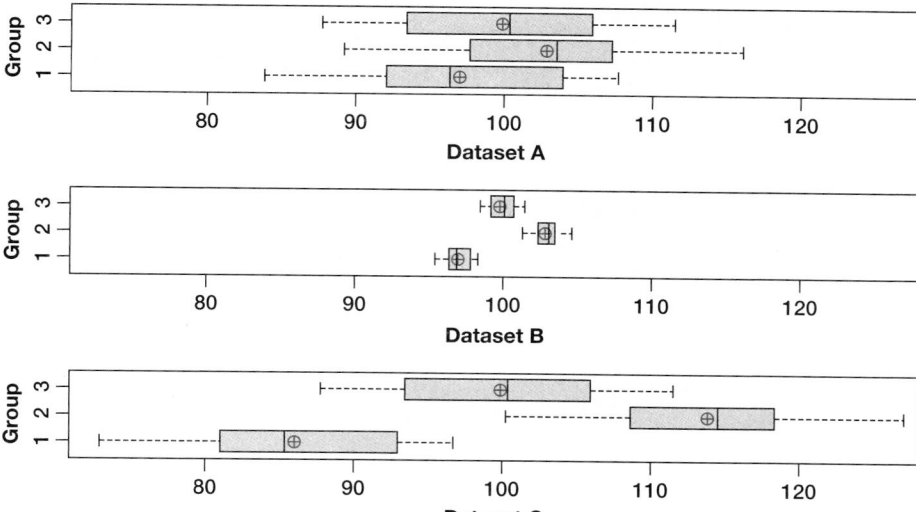

Figure 8.2 *Boxplots comparing samples from three groups for Example 8.2*

Partitioning Variability

The basic idea of ANOVA is to split the *total variability* in data into two (or more) distinct pieces. When comparing means, one of these pieces reflects the variability *between* the groups. If the group means are very different from each other, this portion of the variability will tend to be large. The other piece measures the variability *within* the samples themselves. In the sandwich example, this reflects the fact that the number of ants varies, even for different pieces of the same sandwich filling. Here is a diagram to illustrate this partition.

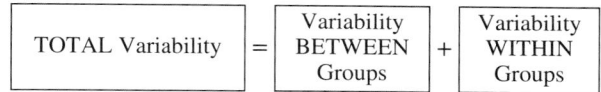

How do we go about actually measuring each of these pieces of variability? As with the sample standard deviation in Chapter 2, we use sums of squared deviations. The calculations to compute these sums of squares (abbreviated SS) are given at the end of this section, but they are tedious to apply by hand and we generally rely on technology to perform the calculations.

TOTAL Variability is denoted *SSTotal* for total sum of squares. It is a measure of the variability of all the data values from the overall combined mean.

Variability BETWEEN Groups is denoted *SSG* for sum of squares for groups. It is a measure of how far apart the group means are. This is the variability we can explain by the fact that there are different groups.

Variability WITHIN Groups is denoted *SSE* for sum of squares for error. It is a measure of how much variability there is within each group. This is the variability that we can't explain by the different groups and for that reason is referred to as the "error" variability.

If all of the group means happen to be exactly the same, the variability *between* groups (*SSG*) would be zero. If the number of ants were always identical on pieces with the same filling, the variability *within* the groups (*SSE*) would be zero. In practice, we rarely see either of these two extremes.

The ANOVA rule for partitioning variability means that

$$SSTotal = SSG + SSE$$

Example 8.3

Typical computer output for testing whether there is a difference in mean number of ants for the three types of sandwich fillings is shown. Identify the three measures of variability and verify that $SSTotal = SSG + SSE$.

One-way ANOVA: Ants versus Filling

Source	DF	SS	MS	F	P
Filling	2	1561	781	5.63	0.011
Error	21	2913	139		
Total	23	4474			

Solution

The partition of variability into sums of squares is shown in the column labeled "SS." Total variability is *SSTotal* and we see that it is 4474. Variability within groups is *SSE* and we see in the "Error" row that it is 2913. The "groups" in this example are the three types of fillings, so the variability between groups is *SSG* and we see in the row labeled "Filling" that it is 1561. Since $1561 + 2913 = 4474$, we see that $SSG + SSE = SSTotal$ as required.

The computer output in Example 8.3 is called an *ANOVA table* and shows more information than just the various sums of squares. This additional information allows us to find the test statistic.

The F-Statistic

Remember that our goal is to test whether the data provide evidence of a difference in means among the groups. The variability between the groups (*SSG*) is a good measure of how much the group means vary, but we need to balance that against the background variation within the groups (*SSE*). Those two pieces of the total variability are not directly comparable, since, in the sandwich ant data, *SSG* measures variability between 3 means while *SSE* measures variability using all 24 data values. To put them on a comparable scale, we use *degrees of freedom*.

You have already seen the idea of degrees of freedom, for example in Section 6.2-D where the t-statistic has $n - 1$ degrees of freedom. In fact, back in Section 2.3, when we introduce the sample standard deviation, we see an $n - 1$ term in the denominator that is this same degrees of freedom. The degrees of freedom for the total row in an ANOVA table is this same $n - 1$. What about the degrees of freedom for groups? If *SSG* is based on k groups, then it has $k - 1$ degrees of freedom. Sum of squared errors (*SSE*) loses one degree of freedom for each group mean, so if there are k groups, the degrees of freedom for *SSE* is $n - k$. Degrees of freedom add up in the same way that sums of squares do. In summary, if we have k groups with a total of n data values, we have

$$\begin{array}{ccccc} (df \text{ for groups}) & + & (df \text{ for error}) & = & \text{Total } df \\ (k - 1) & + & (n - k) & = & n - 1 \end{array}$$

To put the sums of squares on a comparable scale, we divide each sum of squares by its degrees of freedom. We call the result a *mean square*:

$$\text{Mean Square} = \frac{\text{Sum of Squared Deviations}}{\text{Degrees of Freedom}}$$

We compute the mean square for groups (*MSG*) by dividing *SSG* by *df* for groups, and we compute mean square for error (*MSE*) by dividing *SSE* by *df* for error:

$$MSG = \frac{SSG}{k-1} \quad \text{(Groups)} \qquad MSE = \frac{SSE}{n-k} \quad \text{(Error)}$$

We can now define a statistic for testing for a difference among several means. If the null hypothesis (no difference) is true, the two mean squares, *MSG* and *MSE*, should be roughly the same size. If the alternative hypothesis is true and the population means really differ, we expect *MSG* to be larger relative to *MSE* (since the sample means will tend to be more different relative to internal variation). To compare these two variability estimates, we look at their ratio in what is known as an *F-statistic*:

$$F = \frac{MSG}{MSE}$$

Example 8.4

Calculate the degrees of freedom, the mean squares, and the F-statistic for the sandwich ants data. Verify your results using the computer output in Example 8.3.

Solution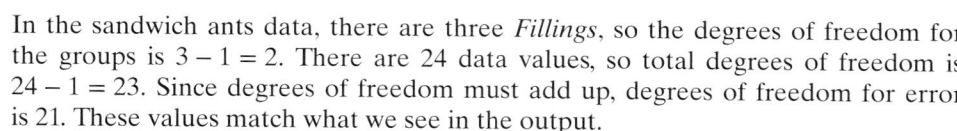

In the sandwich ants data, there are three *Fillings*, so the degrees of freedom for the groups is $3 - 1 = 2$. There are 24 data values, so total degrees of freedom is $24 - 1 = 23$. Since degrees of freedom must add up, degrees of freedom for error is 21. These values match what we see in the output.

To find the mean squares, we divide the sums of squares by degrees of freedom. From Example 8.3 we see that $SSG = 1561$ and $SSE = 2913$. We have

$$MSG = \frac{SSG}{df} = \frac{1561}{2} = 780.5 \quad \text{and} \quad MSE = \frac{SSE}{df} = \frac{2913}{21} = 138.7$$

Finally, the F-statistic is

$$F = \frac{MSG}{MSE} = \frac{780.5}{138.7} = 5.63$$

All these values match what we see in the output in Example 8.3.

From the F-statistic in Example 8.4, we see that the mean square reflecting the spread of the sample group means (*MSG*) is more than five times bigger than we would expect under the null hypothesis, based on the variability within the groups (as reflected in *MSE*). How much larger than the *MSE* does *MSG* need to be for us to conclude that there is some difference in the population means? Although under certain conditions we can find a theoretical model for the F-statistics, we can always use the randomization techniques of Chapter 4 to investigate this sort of question.

Randomization Distribution of F-statistics

If the null hypothesis is true (mean ant counts really don't differ among the three sandwich fillings), any of the 24 values in the **SandwichAnts** data could just as easily be associated with any of the three fillings. To create randomization samples under this null hypothesis, we randomly scramble the filling labels and assign them to the 24 ant counts so that each filling is used eight times. This replicates the way randomization was used to randomly pick which sandwich piece is put out on each trial of the experiment in the original data collection.

For each of the randomization samples we compute the F-statistic (good thing this part is automated!). The ANOVA output for one such randomization sample, with $F = 0.98$, is shown:

Level	N	Mean	StDev
Vegemite	8	40.63	14.39
Peanut Butter	8	41.00	14.43
Ham & Pickles	8	32.38	13.02

Source	DF	SS	MS	F	P
ScrambleFilling	2	380	190	0.98	0.394
Error	21	4094	195		
Total	23	4474			

To see if the original F-value (5.63) is unusual, we repeat the random reallocation process 1000 times to obtain a randomization distribution of F-statistics (when H_0 is true.) One such distribution is shown in Figure 8.3.

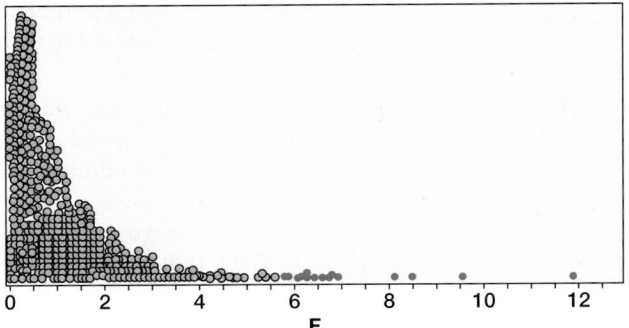

Figure 8.3
Randomization distribution of F-statistics for ant counts

To find a p-value from the randomization distribution we count the proportion of randomization samples that give F-statistics bigger than 5.63. We use the upper tail since large values of the F-statistic occur when *MSG* is large and the group means are more different. In the distribution of Figure 8.3, only 16 of the 1000 randomization statistics are bigger than 5.63, giving an estimated p-value of 0.016. This is close to the p-value of 0.011 that appears in the ANOVA output in Example 8.3. With such a small p-value, we have convincing evidence that there is a difference in mean ant counts among the three types of sandwich filling.

The F-distribution

We can use an *F-distribution* to find the p-value when the following two conditions are met:

Normal distribution: The data from each of the populations should follow a normal distribution, although this is more critical for small sample sizes. In practice, we watch out for clear skewness or extreme outliers if the sample size is small. From Figure 8.1 we don't see any concerns about normality in the sandwich ants data.

Equal Variance: The variability should be roughly the same in each of the groups. This can be a problem, since in some situations the variability tends to be larger for larger means. As a rough rule, we only worry if the sample standard deviation for one group is more than twice the standard deviation for another group. Looking back at Table 8.2 on page 541 we see that the largest standard deviation for those fillings ($s_2 = 14.63$ for Peanut Butter) is not more than twice the smallest standard deviation ($s_1 = 9.25$ for Vegemite).

When these two conditions are reasonably met, the distribution of the F-statistic when the null hypothesis is true follows an *F-distribution*. Since an F-distribution arises as a ratio of two mean squares, it has two values for degrees of freedom: one for the numerator and one for the denominator. In the sandwich ants example, we use an F-distribution with 2 numerator and 21 denominator degrees of freedom (denoted $F_{2,21}$). Figure 8.4 shows a scaled histogram of the randomization F-statistics along with the graph for an $F_{2,21}$ distribution.

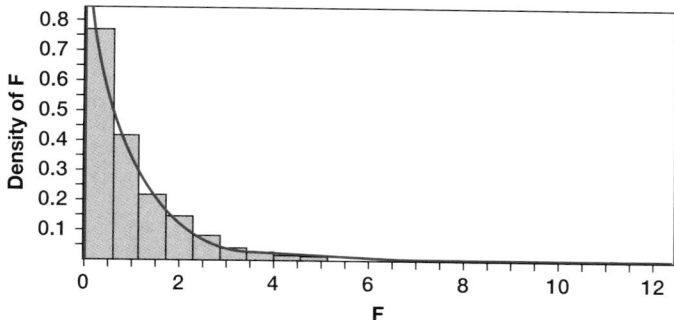

Figure 8.4
Randomization F-statistics and F-distribution with 2 and 21 degrees of freedom

The p-value for the test of means of sandwich ants is the area in the upper tail of this F-distribution beyond the F-statistic of $F = 5.63$. Using technology, we see that the p-value is 0.011, matching the ANOVA output in Example 8.3. There is evidence that ants do not prefer sandwich fillings equally.

ANOVA to Compare Means

To test for a difference in means among k groups:

$$H_0: \quad \mu_1 = \mu_2 = \cdots = \mu_k$$
$$H_a: \quad \text{At least one } \mu_i \neq \mu_j$$

We partition the variability to construct an **ANOVA table**:

Source	df	Sum of Sq.	Mean Square	F-statistic	p-value
Groups	$k-1$	SSG	$MSG = \dfrac{SSG}{k-1}$	$F = \dfrac{MSG}{MSE}$	Upper tail $F_{k-1,n-k}$
Error	$n-k$	SSE	$MSE = \dfrac{SSE}{n-k}$		
Total	$n-1$	$SSTotal$			

Conditions to use an F-distribution:

- Sample sizes are large (each $n_i \geq 30$) or data are relatively normally distributed.
- Variability is similar in all groups.

More Examples of ANOVA

Data 8.1 deals with data from an experiment where the sample sizes are equal in each group. (These are called *balanced* samples.) In the next example we use data from an observational study where the sample sizes differ between the groups.

Example 8.5

Pulse Rates and Award Preference

Data 1.1 on page 4 describes a sample survey collected from students in several statistics classes. One of the variables records which *Award* students would prefer to win from among an Academy Award, Nobel Prize, and Olympic gold medal. The students also measured their pulse rates (in beats per minute). Could pulse rates be related to award preference? Use ANOVA to test whether there is a difference in mean pulse rate between students in the three award categories. Be sure to check that the conditions are reasonably satisfied.

Solution

The computer output below shows the sample size, mean, and standard deviation for the pulse rates in each award group and overall. Figure 8.5 shows boxplots comparing the distribution of the three samples.

Variable	Award	Count	Mean	StDev
Pulse	Academy	31	70.52	12.36
	Nobel	149	72.21	13.09
	Olympic	182	67.253	10.971

Variable	Count	Mean	StDev
Pulse	362	69.575	12.205

The boxplots are relatively symmetric, with a few outliers in both tails for the *Nobel* and *Olympic* groups, but those sample sizes are large ($n_2 = 149$, $n_3 = 182$) so we don't need to be concerned with the normality condition. The standard deviations for the groups are similar; none is close to being twice another. The conditions for applying ANOVA to compare the means look reasonable.

To test $H_0 : \mu_1 = \mu_2 = \mu_3$, where each μ_i denotes the mean pulse rate for one of the award groups, vs $H_a :$ Some $\mu_i \neq \mu_j$, we obtain the ANOVA output shown below:

Source	DF	SS	MS	F	P
Award	2	2047	1024	7.10	0.001
Error	359	51729	144		
Total	361	53776			

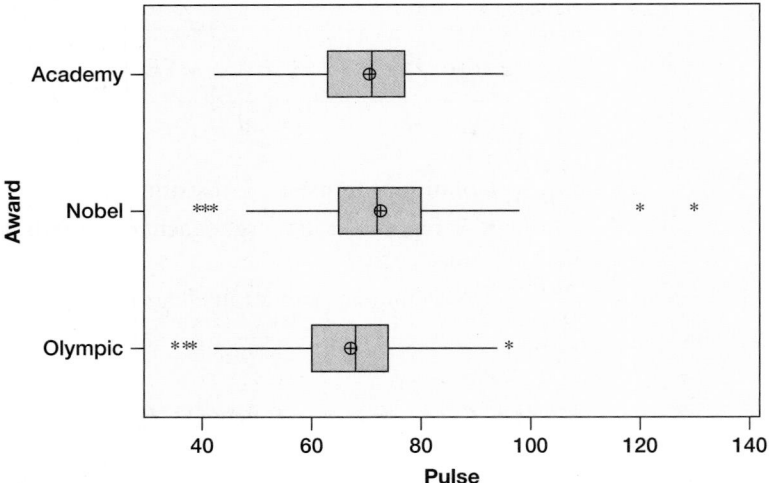

Figure 8.5 *Pulse rates within each Award category*

The F-statistic (7.10) gives a p-value of 0.001 when compared to an F-distribution with 2 numerator and 359 denominator degrees of freedom. This is a very small p-value, so we have strong evidence that the average pulse rates differ depending on the award students prefer.

But which groups are different? For example, is the mean pulse rate for students who prefer an Olympic gold medal different from the mean for those who prefer an Academy Award? Remember that a significant result in the ANOVA only signals that a difference exists—it does not tell us which specific groups differ. We look at techniques for answering that question in the next section.

The data in **StudentSurvey** are from an observational study, so we can't conclude that a desire for a particular award tends to *cause* pulse rates to increase or decrease. Can you think of a possible confounding variable that might help explain the association between award preference and pulse rates? (*Hint:* Try Exercise 8.37.)

Another Look at Variability

The key to using ANOVA to test for a difference in means is the comparison of the variability *between* the sample means (*SSG* and *MSG*) to the variability *within* the samples (*SSE* and *MSE*). In Figure 8.2 on page 543 we see boxplots for three datasets that compare hypothetical data for three groups. Recall that Datasets A and B have the same group means, while A and C have the same spread in the boxplots. Some ANOVA output for each of those datasets is shown below. Pay attention to how the group means and variability affect the values in the ANOVA table and the conclusions about the significance of the differences in means for each dataset.

Dataset A	Df	Sum Sq	Mean Sq	F value	Pr(>F)
Group	2	216	108	1.6875	0.2006
Residuals	33	2112	64		
Total	35	2328			

Dataset B	Df	Sum Sq	Mean Sq	F value	Pr(>F)
Group	2	216	108	108	0.000
Residuals	33	33	1		
Total	35	249			

Dataset C	Df	Sum Sq	Mean Sq	F value	Pr(>F)
Group	2	4704	2352	36.75	0.000
Residuals	33	2112	64		
Total	35	6816			

For Dataset A, we see from the ANOVA table (p-value = 0.2006) that those sample means are not considered significantly different. It is not unusual to see three samples of size $n_i = 12$, drawn from the same population, differ by the amounts shown in the top graph of Figure 8.2.

Dataset B shows the same sample means as Dataset A, but the standard deviation of each sample has been reduced (as seen in the much narrower boxes of the boxplots). The sum of squares and mean square for "Group" is identical to those values in the ANOVA for Dataset A, since the group means are the same in both datasets. However, the smaller standard deviation within the groups of Dataset B reduces both *SSE* and *MSE*. When we divide to find the F-statistic ($F = 108$) we see that it is now very significant (p-value = 0.000). The sample means shown in the middle dataset of Figure 8.2 are considered significantly different, even though the means themselves are identical to Dataset A.

When we get to the ANOVA for Dataset C, we see that the *SSE* and *MSE* values (labeled as "Residuals" in this output) are identical to those in the ANOVA for Dataset A. This makes sense since the boxplots in Dataset C have the same spread within the groups as those in Dataset A. In this case, it's the *SSG* and *MSG* values that change, getting much larger since the means are much farther apart in Dataset C. With a bigger difference among the sample means, the ANOVA table for Dataset C has a small p-value (0.000), meaning we are unlikely to see sample means this far apart if the samples are drawn from the same population. We find strong evidence for a difference in means in Dataset C.

As we see when comparing these ANOVA tables and the graphs in Example 8.2, the important point when using ANOVA to test means is that we assess the differences between the group means against the background variability within the groups. The same set of means can give significant results when there is less variability in the samples, but not be detected as different when the samples themselves are widely scattered.

ANOVA Calculations

While we generally rely on technology to do the nitty-gritty calculations for ANOVA, seeing how the various types of variability are computed can help you distinguish and keep straight their roles. As the notation "SS" suggests, each variability is measured as a sum of squared deviations. The deviations used in each case reflect the type of variability being measured. The mean for all the data values is often referred to as the *grand mean*.

TOTAL Variability: (deviations of the *data* from the *grand mean*)
Under the null hypothesis of equal means, we use the grand mean ($\bar{x}$) to estimate the common population mean. The total variability is just the sum of squared deviations of each of the data points from this grand mean:

$$\text{Total variability} = SSTotal = \sum (x - \bar{x})^2$$

Variability BETWEEN Groups: (deviations of the *group means* from the *grand mean*)
To compare the group means, we square the deviations between the group means and the grand mean. This sum has just one term for each group and we multiply the squared deviation by the sample size in the group when computing the sum:

$$\text{Variability Between Groups} = SSG = \sum n_i(\bar{x}_i - \bar{x})^2$$
$$= n_1(\bar{x}_1 - \bar{x})^2 + \cdots + n_k(\bar{x}_k - \bar{x})^2$$

Variability WITHIN Groups: (deviations of the *data* from their *group mean*)
To measure variability within the groups, we use squared deviations between each data value and the mean for its group, $\bar{x}_i$ (rather than the grand mean). If we think of the group mean as predicting values for the group (like the number of ants for a particular sandwich filling), the deviations from that mean are often called "errors" or "residuals." That's why we use the notation *SSE* for the sum of the squares of these errors to measure the variability within the groups:

$$\text{Variability Within Groups} = SSE = \sum (x - \bar{x}_i)^2$$

Note that the sums for *SSTotal* and *SSE* use the individual data values, so they have *n* terms in each sum, where *n* is the overall sample size. The calculation of *SSG* between the groups is quicker since the number of terms in that sum is just k, the number of groups.

Recall that

$$SSTotal = SSG + SSE$$

so if we know any two of the three sums of squares we can easily find the last one.

Example 8.6

Find each of the sums of squares terms for the sandwich ants data in Table 8.1.

Solution The grand mean is $\bar{x} = 38$ and we sum the deviations for all 24 ant counts to get

$$SSTotal = (18 - 38)^2 + (29 - 38)^2 + \cdots + (59 - 38)^2 + (53 - 38)^2 = 4474$$

To find the variability between the three sandwich types, we need only three terms in the sum

$$SSG = 8(30.75 - 38)^2 + 8(34.0 - 38)^2 + 8(49.25 - 38)^2 = 1561$$

We could subtract to find $SSE = SSTotal - SSG = 4474 - 1561 = 2913$ or do the sum comparing all 24 ant counts to their sandwich mean:

$$SSE = (18 - 30.75)^2 + (29 - 30.75)^2 + \cdots + (59 - 49.25)^2 + (53 - 49.25)^2 = 2913$$

The computer output showing the ANOVA table in Example 8.3 on page 544 confirms these sum of square values.

The most tedious parts of the calculations in Example 8.6 are finding $SSTotal$ and SSE since each is a sum of 24 terms. We can save some time if we know the standard deviation of the overall sample and the standard deviation of the sample in each group. Recall the formula for computing a standard deviation is

$$s = \sqrt{\frac{\sum (x - \bar{x})^2}{n - 1}} = \sqrt{\frac{SSTotal}{n - 1}}$$

With a little algebra, we can turn this into an easy formula for computing the total sum of squares for ANOVA:

$$SSTotal = (n - 1)s^2$$

where s is the standard deviation obtained from the combined sample and n is the overall sample size.

In a similar way, we can find the sum of squared deviations from the group mean within any group using $(n_i - 1)s_i^2$ where n_i is the sample size and s_i is the standard deviation from the sample in the ith group. Summing these for each of the groups, we get a shortcut for finding the sum of squared errors within the groups:

$$SSE = \sum (n_i - 1)s_i^2 = (n_1 - 1)s_1^2 + (n_2 - 1)s_2^2 + \cdots + (n_k - 1)s_k^2$$

Adding up the degrees of freedom for each of these terms, $(n_1 - 1) + (n_2 - 1) + \cdots + (n_k - 1) = n - k$, is another way to see how we get the degrees of freedom for the error in ANOVA.

We can relatively easily find each of the sum of squared variability terms for an ANOVA using basic summary statistics (sample size, mean, and standard deviation) from the groups and overall sample:

Calculating Sum of Squares from Summary Statistics

$$\text{Between Groups} = SSG = \sum n_i(\bar{x}_i - \bar{x})^2 = n_1(\bar{x}_1 - \bar{x})^2 + \cdots + n_k(\bar{x}_k - \bar{x})^2$$

$$\text{Within Groups} = SSE = \sum (n_i - 1)s_i^2 = (n_1 - 1)s_1^2 + \cdots + (n_k - 1)s_k^2$$

$$\text{Total} = SSTotal = (n - 1)s^2$$

If we have any two of these quantities, we can always find the third using

$$SSTotal = SSG + SSE$$

SECTION LEARNING GOALS

You should now have the understanding and skills to:

- Use ANOVA to test for a difference in means among several groups
- Explain how variation between groups and variation within groups are relevant for testing a difference in means between multiple groups

Exercises for Section 8.1

SKILL BUILDER 1

In Exercises 8.1 to 8.6, two sets of sample data, A and B, are given. Without doing any calculations, indicate in which set of sample data, A or B, there is likely to be stronger evidence of a difference in the two population means. Give a brief reason, comparing means and variability, for your answer.

8.1

Dataset A		Dataset B	
Group 1	**Group 2**	**Group 1**	**Group 2**
12	25	15	20
20	18	14	21
8	15	16	19
21	28	15	19
14	14	15	21
$\bar{x}_1 = 15$	$\bar{x}_2 = 20$	$\bar{x}_1 = 15$	$\bar{x}_2 = 20$

8.2

Dataset A		Dataset B	
Group 1	**Group 2**	**Group 1**	**Group 2**
13	18	13	48
14	19	14	49
15	20	15	50
16	21	16	51
17	22	17	52
$\bar{x}_1 = 15$	$\bar{x}_2 = 20$	$\bar{x}_1 = 15$	$\bar{x}_2 = 50$

8.3

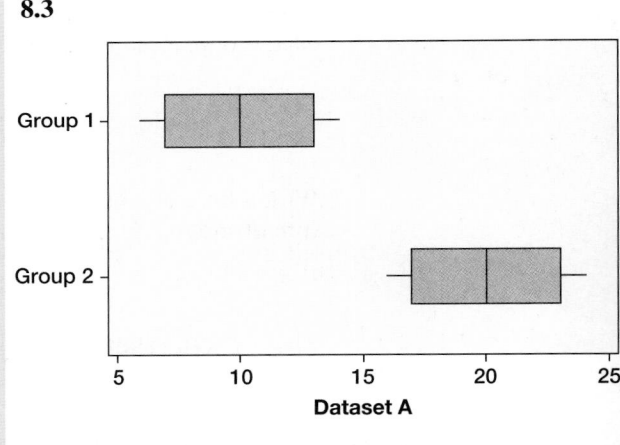

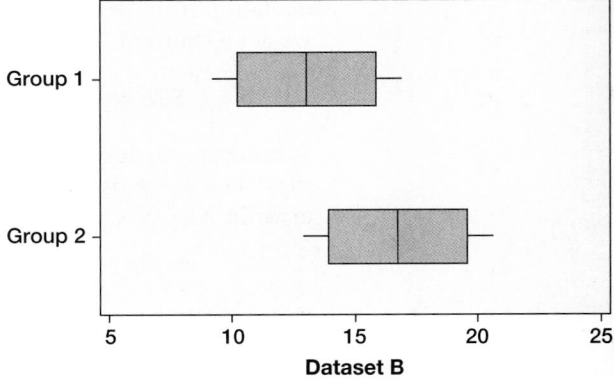

8.4

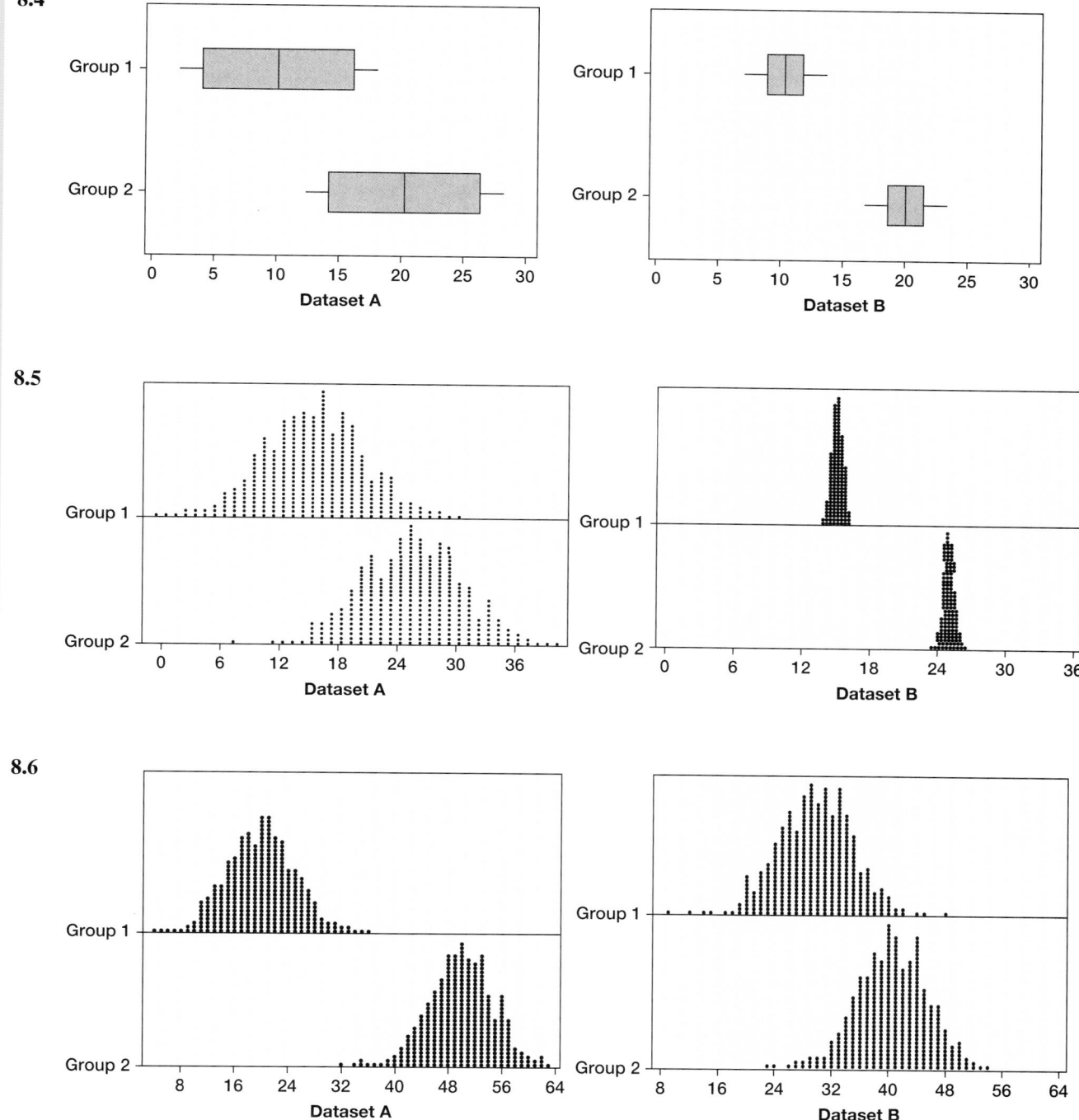

8.5

8.6

SKILL BUILDER 2

In Exercises 8.7 to 8.10, we give sample sizes for the groups in a dataset and an outline of an analysis of variance table with some information on the sums of squares. Fill in the missing parts of the table. What is the value of the F-test statistic?

8.7   Three groups with $n_1 = 5$, $n_2 = 5$, and $n_3 = 5$. ANOVA table includes:

Source	df	SS	MS	F-statistic
Groups		120		
Error		282		
Total		402		

8.8 Four groups with $n_1 = 10, n_2 = 10, n_3 = 10,$ and $n_4 = 10$. ANOVA table includes:

Source	df	SS	MS	F-statistic
Groups		960		
Error		5760		
Total		6720		

8.9 Three groups with $n_1 = 10, n_2 = 8,$ and $n_3 = 11$. ANOVA table includes:

Source	df	SS	MS	F-statistic
Groups		80		
Error				
Total		1380		

8.10 Four groups with $n_1 = 5$, $n_2 = 8$, $n_3 = 7$, and $n_4 = 5$. ANOVA table includes:

Source	df	SS	MS	F-statistic
Groups				
Error		800		
Total		1400		

SKILL BUILDER 3

In Exercises 8.11 to 8.14, some computer output for an analysis of variance test to compare means is given.

(a) How many groups are there?

(b) State the null and alternative hypotheses.

(c) What is the p-value?

(d) Give the conclusion of the test, using a 5% significance level.

8.11

Source	DF	SS	MS	F
Groups	3	360.0	120.0	1.60
Error	16	1200.0	75.0	
Total	19	1560.0		

8.12

Source	DF	SS	MS	F
Groups	4	1200.0	300.0	5.71
Error	35	1837.5	52.5	
Total	39	3037.5		

8.13

Source	DF	SS	MS	F
Groups	2	540.0	270.0	8.60
Error	27	847.8	31.4	
Total	29	1387.8		

8.14

Source	DF	SS	MS	F
Groups	3	450.0	150.0	0.75
Error	16	3200.0	200.0	
Total	19	3650.0		

8.15 Stress Levels and a Mother's Voice A recent study[2] examines the impact of a mother's voice on stress levels in young girls. The study included 68 girls ages 7 to 12 who reported good relationships with their mothers. Each girl gave a speech and then solved mental arithmetic problems in front of strangers. Cortisol levels in saliva were measured for all girls and were high, indicating that the girls felt a high level of stress from these activities. (Cortisol is a stress hormone and higher levels indicate greater stress.) After the stress-inducing activities, the girls were randomly divided into four equal-sized groups: one group talked to their mothers in person, one group talked to their mothers on the phone, one group sent and received text messages with their mothers, and one group had no contact with their mothers. Cortisol levels were measured before and after the interaction with mothers and the change in the cortisol level was recorded for each girl.

(a) What are the two main variables in this study? Identify each as categorical or quantitative.

(b) Is this an experiment or an observational study?

(c) The researchers are testing to see if there is a difference in the change in cortisol level depending on the type of interaction with mom. What are the null and alternative hypotheses? Define any parameters used.

(d) What are the total degrees of freedom? The *df* for groups? The *df* for error?

(e) The results of the study show that hearing mother's voice was important in reducing stress levels. Girls who talk to their mother in person or on the phone show decreases in cortisol significantly greater, at the 5% level, than girls who text with their mothers or have no contact with their mothers. There was not a difference between in person and on the phone and there was not a difference between texting and no contact. Was the p-value of the original ANOVA test above or below 0.05?

8.16 Does Synchronized Dancing Boost Feelings of Closeness? Exercise 3.96 introduces a study designed to examine the effect of doing synchronized movements (such as marching in step or doing synchronized dance steps) and the effect of exertion on many different variables, including how close participants feel to others in their group. In the study, high school students in Brazil were randomly

[2]Seltzer, L., Prososki, A., Ziegler, T., and Pollak, S., "Instant messages vs. speech: Hormones and why we still need to hear each other," *Evolution and Human Behavior*, 2012; 33(1): 42–45.

assigned to an exercise with either high synchronization (HS) or low synchronization (LS) and also either to high exertion (HE) or low exertion (LE). Thus, there are four groups: HS+HE, HS+LE, LS+HE, and LS+LE. Closeness is measured on a 7-point Likert scale (1=least close to 7=most close), and the response variable is the change in how close participants feel to those in their group using the rating after the exercise minus the rating before the exercise. The data are stored in **Synchronized-Movement** and output for an ANOVA test is shown below, along with some summary statistics.

Analysis of Variance

Source	DF	SS	MS	F-Value	P-Value
Group	3	27.04	9.012	2.77	0.042
Error	256	831.52	3.248		
Total	259	858.55			

Group	N	Mean	StDev
HS+HE	72	0.319	1.852
HS+LE	64	0.328	1.861
LS+HE	66	0.379	1.838
LS+LE	58	−0.431	1.623

(a) In both groups with high synchronization (HS), does mean closeness rating go up or down after the synchronized exercise?

(b) In the groups with low synchronization (LS), does mean closeness rating go up or down if the group engages in high exertion (HE) exercise? How about if the group engages in low exertion (LE) exercise?

(c) How many students were included in the analysis?

(d) At a 5% level, what is the conclusion of the test? If there is a difference in means, indicate where the difference is likely to be.

(e) At a 1% level, what is the conclusion of the test?

8.17 The Color Red and Performance Color affects us in many ways. For example, Exercise C.92 on page 498 describes an experiment showing that the color red appears to enhance men's attraction to women. Previous studies have also shown that athletes competing against an opponent wearing red perform worse, and students exposed to red before a test perform worse.[3] Another study[4] states that "red is hypothesized to impair performance on achievement tasks, because red is associated with the danger of failure." In the study, US college students were asked to solve 15 moderately difficult, five-letter, single-solution anagrams during a 5-minute period. Information about the study was given to participants in either red, green, or black ink just before they were given the anagrams. Participants were randomly assigned to a color group and did not know the purpose of the experiment, and all those coming in contact with the participants were blind to color group. The red group contained 19 participants and they correctly solved an average of 4.4 anagrams. The 27 participants in the green group correctly solved an average of 5.7 anagrams and the 25 participants in the black group correctly solved an average of 5.9 anagrams. Work through the details below to test if performance is different based on prior exposure to different colors.

(a) State the hypotheses.

(b) Use the fact that sum of squares for color groups is 27.7 and the total sum of squares is 84.7 to complete an ANOVA table and find the F-statistic.

(c) Use the F-distribution to find the p-value.

(d) Clearly state the conclusion of the test.

8.18 Laptop Computers and Sperm Count Studies have shown that heating the scrotum by just 1°C can reduce sperm count and sperm quality, with long-term consequences. Exercise 2.113 on page 90 introduces a study indicating that males sitting with a laptop on their laps have increased scrotal temperatures. Does a lap pad help reduce the temperature increase? Does sitting with legs apart help? The study investigated all three of these conditions: legs together and a laptop computer on the lap, legs apart and a laptop computer on the lap, and legs together with a lap pad under the laptop computer. Scrotal temperature increase over a 60-minute session was measured in °C, and the summary statistics are given in Table 8.3.

Table 8.3 *Scrotal temperature increase in °C with a laptop computer on lap*

Condition	n	Mean	Std.Dev.
Legs together	29	2.31	0.96
Lap pad	29	2.18	0.69
Legs apart	29	1.41	0.66

[3]"Color Red Increases the Speed and Strength of Reactions," *Science Daily*, *sciencedaily.com*, June 2, 2011.

[4]Elliot, A., et al., "Color and Psychological Functioning: The Effect of Red on Performance Attainment," *Journal of Experimental Psychology: General*, 2007; 136(1): 154–168. Data approximated from summary statistics.

(a) Which condition has the largest mean temperature increase? Which has the smallest?

(b) Do the data appear to satisfy the condition that the standard deviations are roughly the same? (The data satisfy the normality condition.)

(c) Use the fact that sum of squares for groups is 13.7 and error sum of squares is 53.2 to test whether there is a difference in mean temperature increase between the three conditions. Show all details of the test, including an analysis of variance table.

EXERCISE AND STRESS

Exercise 6.219 on page 465 introduces a study showing that exercise appears to offer some resiliency against stress. In the study, mice were randomly assigned to live in an enriched environment (EE), a standard environment (SE), or an impoverished environment (IE) for several weeks. Only the enriched environment provided opportunities for exercise. Half the mice then remained in their home cage (HC) as control groups while half were subjected to stress (SD) by being placed repeatedly with a very aggressive mouse. All the mice in SD exhibited acute signs of stress during these brief exposures to "mouse bullies." The researchers were interested in how resilient the mice were in recovering from the stress after the mouse bullying stopped. Exercises 8.19 to 8.21 discuss the results of their work.[5] There were eight mice in each of the six groups, and conditions for the ANOVA tests are met.

8.19 Time Hiding in Darkness One measure of mouse anxiety is amount of time hiding in a dark compartment, with mice that are more anxious spending more time in darkness. The amount of time (in seconds) spent in darkness during one trial is recorded for all the mice and the mean results are shown in Table 8.4.

(a) In this sample, do the control groups (HC) spend less time in darkness on average than the stressed groups (SD)? Which of the stressed groups spends the least amount of time, on average, in darkness?

(b) The sum of squares for groups is $SSG = 481,776$ and for error is $SSE = 177,835$. Complete a test to determine if there is a difference in mean time spent in darkness between the six groups.

[5]Lehmann, M. and Herkenham, M., "Environmental Enrichment Confers Stress Resiliency to Social Defeat through an Infralimbic Cortex-Dependent Neuroanatomical Pathway," *Journal of Neuroscience*, 2011; 31(16): 6159–6173. Data approximated from summary statistics.

Table 8.4 *Mean time (sec) in darkness*

IE:HC	SE:HC	EE:HC	IE:SD	SE:SD	EE:SD
192	196	205	392	438	231

8.20 Time Immobile One measure of mouse anxiety is amount of time spent immobile; mice tend to freeze when they are scared. The amount of time (in seconds) spent immobile during one trial is recorded for all the mice and the mean results are shown in Table 8.5.

(a) In this sample, do the control groups (HC) spend less time immobile on average than the stressed groups (SD)? Which of the stressed groups spends the least amount of time, on average, immobile?

(b) The sum of squares for groups is $SSG = 188,464$ and for error is $SSE = 197,562$. Complete a test to determine if there is a difference in mean time immobile between the six groups.

Table 8.5 *Mean time (sec) immobile*

IE:HC	SE:HC	EE:HC	IE:SD	SE:SD	EE:SD
50	47	52	191	188	96

8.21 Immunological Effects In addition to the behavioral effects of stress, the researchers studied several immunological effects of stress. One measure studied is stress-induced decline in FosB-positive cells in the FosB/ΔFosA expression. This portion of the study only included seven mice in each of the six groups, and lower levels indicate more stress. The mean levels of FosB+ cells for each combination of environment and stress are shown in Table 8.6.

(a) In each of the three environments (IE, SE, and EE), which sample group (HC or SD) has a lower average level of FosB+ cells? Does this match what we would expect? Within each of the no-stress and stress groups (HC and SD) separately, which environment has the highest average level of FosB+ cells?

(b) The sum of squares for groups is $SSG = 118,286$ and for error is $SSE = 75,074$. Complete a test to determine if there is a difference in mean FosB+ levels between the groups.

Table 8.6 *FosB+ cells*

IE:HC	SE:HC	EE:HC	IE:SD	SE:SD	EE:SD
86	129	178	21	68	152

8.22 Incentives to Exercise Exercise 5.29 on page 385 describes a study designed to see what type of incentive might be most effective in encouraging people to exercise. In the study, 281 overweight or obese people were assigned the goal to walk 7000 steps a day, and their activity was tracked for 100 days. The response variable is the number of days (out of 100) that each participant met the goal. The participants were randomly assigned to one of four different incentive groups: for each day they met the goal, participants in the first group got only praise, participants in the second group got entered into a lottery, and participants in the third group received cash (about $1.50 per day). In the fourth group, participants received all the money up front and *lost* money (about $1.50 per day) if they didn't meet the goal. (The overall financial effect for participants in the third and fourth conditions is identical, but the psychological effect between winning money and losing money is potentially quite different.) The summary statistics[6] for the four conditions and overall are shown in Table 8.7.

(a) In the sample, which incentive had the most success in helping participants meet the goal? Which incentive had the least success?

(b) Do the conditions for using the F-distribution appear to be met?

(c) Test to see if the data provide evidence of a difference in mean success rates depending on the incentive used. Show all details of the test, including showing the ANOVA table.

Table 8.7 *Number of days meeting an exercise goal*

Condition	n	Mean	St.Dev.
Praise	70	30.0	32.0
Lottery	70	35.0	29.9
Get money	70	36.0	29.4
Lose money	71	45.0	30.1
Overall	281	36.5	30.6865

8.23 Posture and Pain Research shows that people adopting a dominant pose have reduced levels of stress and feel more powerful than those adopting a submissive pose. Furthermore, it is known that if people feel more control over a situation, they have a higher tolerance for pain. Putting these

ideas together, a study,[7] introduced in Exercise C.20 on page 485, investigates how posture might influence the perception of pain. In the experiment, participants were told that they were participating in a study to examine the health benefits of doing yoga poses at work. All participants had their pain threshold measured both before and after holding a yoga pose for 20 seconds. The pain threshold was measured by inflating a blood pressure cuff until participants said stop: the threshold was measured in mmHg and the difference in before and after thresholds was recorded for each participant. The participants were randomly divided into three groups: one group was randomly assigned to strike a dominant pose (moving limbs away from the body), another group was assigned to strike a submissive pose (curling the torso inward), and a control group struck a neutral pose. Summary statistics are shown in Table 8.8. Do the data provide evidence of a difference in mean pain tolerance based on the type of pose? Show all details of the test.

Table 8.8 *Difference in pain threshold (mmHg)*

Pose	Sample Size	Mean	Std.Dev.
Dominant	30	14.3	34.8
Neutral	29	−4.4	31.9
Submissive	30	−6.1	35.4
Overall	89	1.33	35.0

8.24 Sandwich Ants and Bread Data 8.1 on page 540 describes an experiment to study how different sandwich fillings might affect the mean number of ants attracted to pieces of a sandwich. The students running this experiment also varied the type of bread for the sandwiches, randomizing between four types: Multigrain, Rye, Wholemeal, and White. The ant counts in 6 trials and summary statistics for each type of bread and the 24 trials as a whole are given in Table 8.9 and stored in **SandwichAnts**.

(a) Show how to use the summary information to compute the three sums of squares needed for using ANOVA to test for a difference in mean number of ants among these four types of bread.

(b) Use the sums of squares from part (a) to construct the ANOVA table and complete the details for this test. Be sure to give a conclusion in the context of this data situation.

---

[6]Summary statistics are estimated from information given in the paper.

[7]Bohns, V. and Wiltermuth, S., "It hurts when I do this (or you do that): Posture and pain tolerance," *Journal of Experimental Social Psychology*, available online May 26, 2011. Data approximated from information in the article.

Table 8.9 *Number of ants by type of bread*

Bread	Ants						Mean	Std.Dev.
Multigrain	42	22	36	38	19	59	36.00	14.52
Rye	18	43	44	31	36	54	37.67	12.40
Wholemeal	29	59	34	21	47	65	42.50	17.41
White	42	25	49	25	21	53	35.83	13.86
						Total	38.00	13.95

DOES DRAWING AN IMAGE HELP MEMORY?
Exercises 8.25 to 8.27 describe three different experiments[8] investigating the effect of drawing on memory. In each experiment, participants were randomly divided into three groups and shown a list of 10 words to memorize. The response variable is the number of words participants are able to recall. For each experiment, construct an ANOVA table to assess the difference in mean word recall between the three groups and answer the following questions.

(a) Which method had the highest mean recall in the sample? Which had the lowest?

(b) What is the F-statistic?

(c) What is the p-value?

(d) At a 5% level, what is the conclusion of the test?

(e) According to this experiment, does it matter what you do if you want to memorize a list of words, and, if it matters, what should you do?

8.25 Participants were instructed to either draw an image for each word, write a list of attributes for each word, or write the word. The summary statistics for number of words recalled (out of 10) are shown in Table 8.10.

Table 8.10 *What helps memory recall? Experiment 1*

Group	n	Mean	St.Dev.
Draw	16	4.8	1.3
List attributes	16	3.4	1.6
Write	16	3.2	1.2
Overall	48	3.8	1.527

8.26 Participants were instructed to either draw an image for each word, visualize an image for each

[8]Wammes, J.D., Meade, M.E., and Fernandes, M.A., "The drawing effect: Evidence for reliable and robust memory benefits in free recall," *The Quarterly Journal of Experimental Psychology*, 69:9, 1752-1776, DOI: 10.1080/17470218.2015.1094494a, February 16, 2016. Summary statistics are approximated from information in the article.

word, or write the word. The summary statistics for number of words recalled (out of 10) are shown in Table 8.11.

Table 8.11 *What helps memory recall? Experiment 2*

Group	n	Mean	St.Dev.
Draw	9	5.1	1.1
Visualize	9	3.7	1.7
Write	9	3.2	1.3
Overall	27	4.0	1.566

8.27 Participants were instructed to either draw an image for each word, view an image provided for each word, or write the word. The summary statistics for number of words recalled (out of 10) are shown in Table 8.12.

Table 8.12 *What helps memory recall? Experiment 3*

Group	n	Mean	St.Dev.
Draw	12	4.4	1.2
View	12	3.4	1.5
Write	12	3.0	1.1
Overall	36	3.6	1.377

LIGHT AT NIGHT MAKES FAT MICE
Studies have shown that exposure to light at night is harmful to human health. Data 4.1 on page 258 introduces a study in mice showing that dim light at night has an effect on weight gain after just three weeks. In the full study, mice were randomly assigned to live in one of three light conditions: LD had a standard light/dark cycle, LL had bright light all the time, and DM had dim light when there normally would have been darkness. Exercises 8.28 to 8.34 analyze additional results of this study, examining results after 4 weeks.

8.28 Checking Conditions for Body Mass Gain The mice in the study had body mass measured throughout the study. Computer output showing body mass gain (in grams) after 4 weeks for each of the three light conditions is shown, and a dotplot of the data is given in Figure 8.6.

Level	N	Mean	StDev
DM	10	7.859	3.009
LD	8	5.926	1.899
LL	9	11.010	2.624

(a) In the sample, which group of mice gained the most, on average, over the four weeks? Which gained the least?

(b) Do the data appear to meet the requirement of having standard deviations that are not dramatically different?

(c) The sample sizes are small, so we check that the data are relatively normally distributed. We see in Figure 8.6 that we have no concerns about the DM and LD samples. However, there is an outlier for the LL sample, at 17.4 grams. We proceed as long as the z-score for this value is within ± 3. Find the z-score. Is it appropriate to proceed with ANOVA?

(d) What are the cases in this analysis? What are the relevant variables? Are the variables categorical or quantitative?

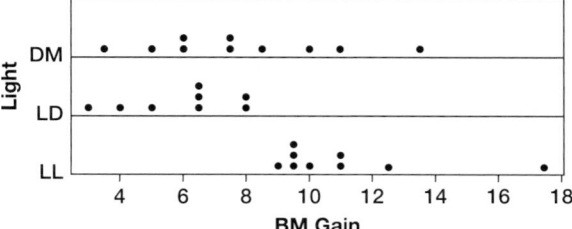

Figure 8.6 *Body mass gain under three light conditions*

8.29 Body Mass Gain The mice in the study had body mass measured throughout the study. Computer output showing an analysis of variance table to test for a difference in mean body mass gain (in grams) after four weeks between mice in the three different light conditions is shown. We see in Exercise 8.28 that the conditions for ANOVA are met, and we also find the summary statistics for each experimental group there.

One-way ANOVA: BM Gain versus Light

Source	DF	SS	MS	F	P
Light	2	113.08	56.54	8.38	0.002
Error	24	161.84	6.74		
Total	26	274.92			

(a) State the null and alternative hypotheses.

(b) What is the F-statistic? What is the p-value? What is the conclusion of the test?

(c) Does there appear to be an association between the two variables (body mass gain and light condition)? If so, discuss the nature of that relationship. Under what light condition do mice appear to gain the most weight?

(d) Can we conclude that there is a cause-and-effect relationship between the variables? Why or why not?

8.30 Activity Levels Perhaps the mice with light at night in Exercise 8.28 gain more weight because they are exercising less. The conditions for an ANOVA test are met and computer output is shown for testing the average activity level for each of the three light conditions. Is there a significant difference in mean activity level? State the null and alternative hypotheses, give the F-statistic and the p-value, and clearly state the conclusion of the test.

Level	N	Mean	StDev
DM	10	2503	1999
LD	8	2433	2266
LL	9	2862	2418

One-way ANOVA: Activity versus Light

Source	DF	SS	MS	F	P
Light	2	935954	467977	0.09	0.910
Error	24	118718447	4946602		
Total	26	119654401			

8.31 Food Consumption Perhaps the mice with light at night in Exercise 8.28 are gaining more weight because they are eating more. Computer output is shown for average food consumption (in grams) during week 4 of the study for each of the three light conditions.

Level	N	Mean	StDev
DM	10	4.1241	0.6938
LD	8	4.3275	0.4337
LL	9	4.5149	1.3149

(a) Is it appropriate to conduct an ANOVA test with these data? Why or why not?

(b) A randomization test is conducted using these data and the randomization distribution is shown in Figure 8.7. The randomization test gives a p-value of 0.652. Clearly state the conclusion of the test in the context of this data situation.

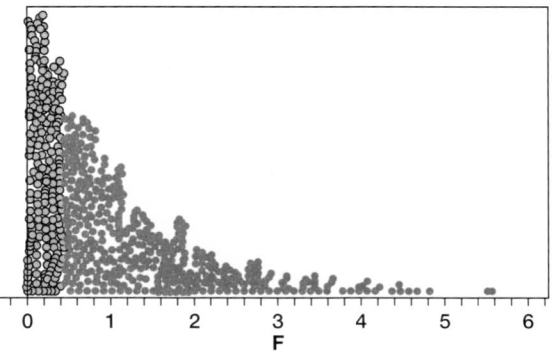

Figure 8.7 *Randomization test for consumption by light condition*

8.32 Stress Levels In addition to monitoring weight gain, food consumed, and activity level, the study measured stress levels in the mice by measuring corticosterone levels in the blood (higher levels indicate more stress). Conditions for ANOVA are met and computer output for corticosterone levels for each of the three light conditions is shown.

Level	N	Mean	StDev
DM	10	73.40	67.49
LD	8	70.02	54.15
LL	9	50.83	42.22

One-way ANOVA: Corticosterone versus Light

Source	DF	SS	MS	F	P
Light	2	2713	1357	0.43	0.656
Error	24	75782	3158		
Total	26	78495			

(a) What is the conclusion of the analysis of variance test?

(b) One group of mice in the sample appears to have very different corticosterone levels than the other two. Which group is different? What aspect of the data explains why the ANOVA test does not find this difference significant? How is this aspect reflected in both the summary statistics and the ANOVA table?

8.33 When Calories Are Consumed Researchers hypothesized that the increased weight gain seen in mice with light at night might be caused by *when* the mice are eating. (As we have seen in the previous exercises, it is not caused by changes in amount of food consumed or activity level.) Perhaps mice with light at night eat a greater percentage of their food during the day, when they normally should be sleeping. Conditions for ANOVA are met and computer output for the percentage of food consumed during the day for each of the three light conditions is shown.

Level	N	Mean	StDev
DM	10	55.516	10.881
LD	8	36.028	8.403
LL	9	76.573	9.646

One-way ANOVA: DayPct versus Light

Source	DF	SS	MS	F	P
Light	2	6987.0	3493.5	36.39	0.000
Error	24	2304.3	96.0		
Total	26	9291.2			

(a) For mice in this sample on a standard light/dark cycle, what is the average percent of food consumed during the day? What percent is consumed at night? What about mice that had dim light at night?

(b) Is there evidence that light at night influences when food is consumed by mice? Justify your answer with a p-value. Can we conclude that there is a cause-and-effect relationship?

8.34 Glucose Tolerance We have seen that light at night increases weight gain in mice and increases the percent of calories consumed when mice are normally sleeping. What effect does light at night have on glucose tolerance? After four weeks in the experimental light conditions, mice were given a glucose tolerance test (GTT). Glucose levels were measured 15 minutes and 120 minutes after an injection of glucose. In healthy mice, glucose levels are high at the 15-minute mark and then return to normal by the 120-minute mark. If a mouse is glucose intolerant, levels tend to stay high much longer. Computer output is shown giving the summary statistics for both measurements under each of the three light conditions.

Descriptive Statistics: GTT-15

Variable	Light	N	Mean	StDev
GTT-15	DM	10	338.8	80.6
	LD	8	318.7	106.5
	LL	9	373.6	59.1

Descriptive Statistics: GTT-120

Variable	Light	N	Mean	StDev
GTT-120	DM	10	258.7	113.0
	LD	8	173.5	41.9
	LL	9	321.4	109.0

(a) Why is it more appropriate to use a randomization test to compare means for the GTT-120 data?

(b) Describe how we might use the 27 data values in GTT-120 to create one randomization sample.

(c) Using a randomization test in both cases, we obtain a p-value of 0.402 for the GTT-15 data and a p-value of 0.015 for the GTT-120 data. Clearly state the results of the tests, using a 5% significance level. Does light at night appear to affect glucose intolerance?

8.35 Football and Brain Size Exercise 2.143 on page 102 describes a study examining a possible relationship of football playing and concussions on hippocampus volume, in μL, in the brain. The study included three groups: controls who had never played football (*Control*), football players with no history of concussions (*FBNoConcuss*), and football players with a history of concussions (*FBConcuss*). The data is available in **FootballBrain**, and the side-by-side boxplots shown in Exercise 2.143 indicate that the conditions for using the F-distribution appear to be met.

(a) Use technology to find the summary statistics for each of the three groups. Which group has the largest mean hippocampus volume? Which group has the smallest?

(b) Use technology to construct an ANOVA table. What is the F-statistic? What is the p-value?

(c) What is the conclusion of the test?

8.36 Football and Brain Size: Randomization Test Exercise 8.35 describes a study examining hippocampus volume in the brain between controls who have never played football, football players with no history of concussion, and football players with a history of concussion. The data are available in **FootballBrain**. Use *StatKey* or other technology to conduct an ANOVA randomization test to determine whether there is a significant difference in mean hippocampus volume between the three categories.

(a) What is the F-statistic for the observed data?

(b) Generate one randomization sample. What is the F-statistic for that sample?

(c) Generate a full randomization distribution and find the p-value. (If you have also completed Exercise 8.35, compare this result with the p-value obtained using a theoretical F-distribution.)

(d) What is the conclusion of the test?

8.37 Exercise and Award Preference In Example 8.5 on page 548 we see a comparison of mean pulse rates between students who prefer each of three different awards (Academy Award, Nobel Prize, Olympic gold medal). The ANOVA test shows that there appears to be a difference in mean pulse rates among those three groups. Can you guess why award preference might be associated with pulse rates? One possibility is exercise. Perhaps students who prefer an Olympic medal are more likely to be athletes who exercise more frequently, stay in shape, and thus have lower pulse rates. Use technology and the data in **StudentSurvey** that includes a variable measuring the typical hours of exercise per week for each student to see if there is a difference in mean exercise amounts depending on award preference. Be sure to check that the conditions for ANOVA are reasonable in this situation.

8.38 Fish Ventilation Most fish use gills for respiration in water and researchers can observe how fast a fish's gill cover beats to study ventilation, much like we might observe breathing rate for a person. Professor Brad Baldwin is interested in how water chemistry might affect gill beat rates. In one experiment he randomly assigned fish to tanks with

different levels of calcium. One tank was low in calcium (0.71 mg/L), the second tank had a medium amount (5.24 mg/L), and the third tank had water with a high calcium level (18.24 mg/L). His research team counted gill rates (beats per minute) for samples of 30 fish in each tank. The results[9] are stored in **FishGills3**. (*Note:* You may also see a file called **FishGills12** which is a more extensive experiment with 12 tanks.)

(a) Use technology to check that the conditions for an ANOVA model are reasonable for these data. Include a plot that compares the gill rates for the three calcium conditions.

(b) If the conditions are met, use technology to find the ANOVA table and complete the test. If the conditions are not reasonable, use a randomization test (scrambling the calcium levels) to complete the test.

8.39 Drug Resistant Pathogens Drug resistant pathogens pose one of the major public health challenges of this century. Conventional wisdom is that aggressive treatment should be used to kill pathogens rapidly, before they have a chance to acquire resistance. However, if drug pathogens are already present, this strategy may actually backfire by intensifying the natural selection placed on drug-resistant pathogens. A study[10] took mice and infected them with a mixture of drug-resistant and drug-susceptible malaria parasites, then randomized mice to one of four different treatments of an antimalarial drug (with 18 mice in each group): untreated (no drug), light (4 mg/kg for 1 day), moderate (8 mg/kg for 1 day), or aggressive (8 mg/kg for 5 or 7 days).[11] Data are available in **DrugResistance**.[12]

[9]Thanks to Professor Baldwin and his team for supplying the data.
[10]Huijben, S., et al., "Aggressive Chemotherapy and the Selection of Drug Resistant Pathogens," *PLOS Pathogens*, 9:9, e1003578, Sep 2013, doi:10.1371/journal.ppat.1003578.
[11]This study includes two different experiments, but the experiments only differ by the initial ratio of resistant to susceptible parasites ($10^1 : 10^6$ in Experiment 1 and $10^1 : 10^9$ in Experiment 2 and the duration (5 or 7 days) for the aggressive treatment), with 18 mice in each group. Here we combine the results from the two experiments, but if you are interested to see if there is a difference between the two experiments, the dataset has a variable indicating which experiment each mouse was part of, and you are welcome to analyze them separately on your own.
[12]Huijben, S., et al., (2013). Data from: Aggressive chemotherapy and the selection of drug resistant pathogens. Dryad Digital Repository. *http://dx.doi.org/10.5061/dryad.09qc0.*

(a) For each mouse, response variables included measures of drug resistance, such as

- Density of drug-resistant parasites (*ResistanceDensity* per µl)
- Number of days (out of the 50-day experiment) each mouse had infectious drug-resistant parasites present (*DaysInfectious*)

and measures of health such as

- Body mass (*Weight* in grams)
- Red blood cell density (*RBC* in million/µl)

The different response variables are displayed in Figure 8.8 by treatment group, with measures of drug resistance in the top row and measures of health outcomes in the bottom row. Higher values correspond to more drug resistance (which is bad) in the drug-resistance response variables and to healthier outcomes (which are good) in the health response variables. Does there appear to be an association between treatment level and the response variables measuring drug resistance? If so, describe the association.

Does there appear to be an association between treatment level and the response variables measuring health? If so, describe the association.

(b) Because we are interested in dose here, exclude the untreated category, and only compare the different actual treatments (Light, Moderate, and Aggressive). Conduct the ANOVA test and give the p-value for each of the four response variables by treatment level. Which variables are significantly associated with treatment level? Which variables are not significantly associated with treatment level?

(c) What have you learned about treatment level and drug resistance? How does this compare with the conventional wisdom that aggressive treatments are more effective at preventing drug resistance?

(d) For the response variables measuring drug resistance, do the conditions for ANOVA appear to be satisfied? If not, which condition is most obviously violated? For the response variables measuring health, do the conditions for

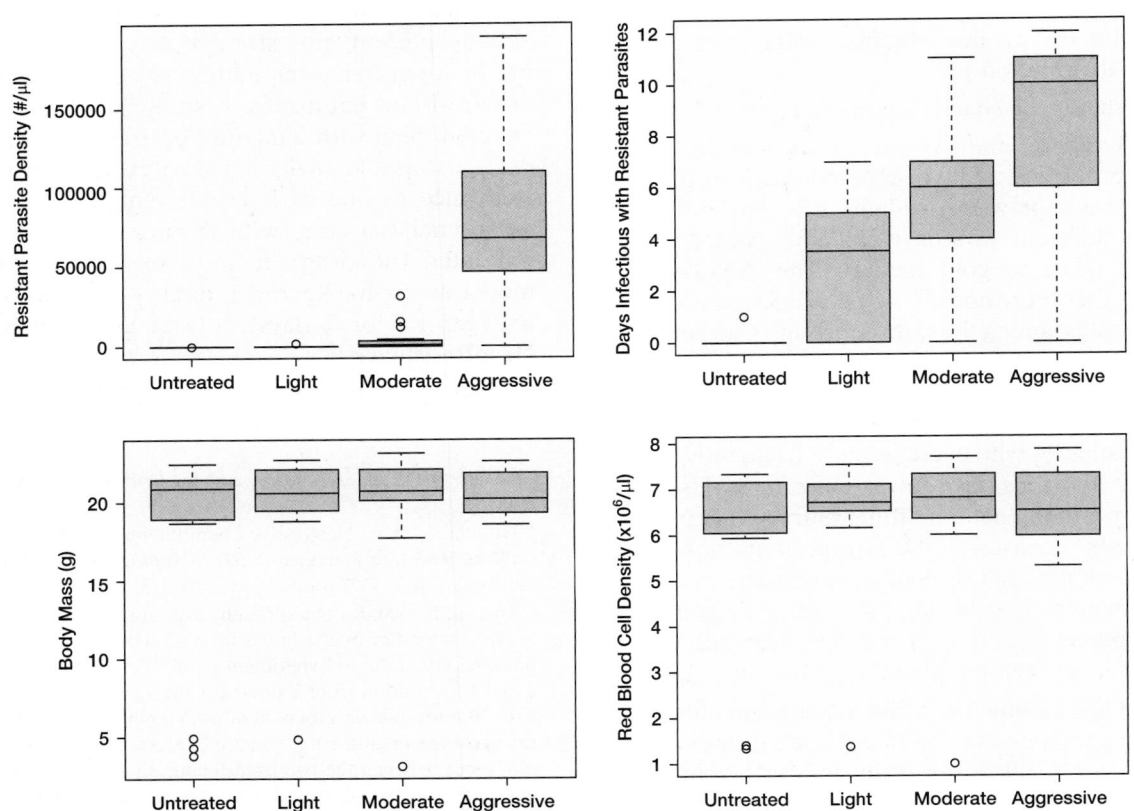

Figure 8.8 *Measures of drug resistance (top row) and health outcomes (bottom row)*

ANOVA appear to be satisfied? If not, which condition is most obviously violated? If the conditions do appear to be violated, keep in mind that the p-values calculated in part (b) will not be entirely accurate.

8.40 Drug Resistance Outliers Exercise 8.39 introduces a study of drug resistant parasites in mice. Figure 8.8 shows boxplots of measures of drug resistance and health for mice getting three different doses of antibacterial drug and one untreated group getting no drug. The plot (bottom right) showing red blood cell density (*RBC*) as the response variable on the vertical axis raises concerns about the normality condition for ANOVA due to a few very low outliers. Eliminate those extreme cases from the data in **DrugResistance** as well as the untreated group (as in Exercise 8.39), and rerun the ANOVA to look for differences in mean red blood cell density between the three groups that got different doses of the treatment drug. Do the findings change much when the outliers are eliminated?

8.2 PAIRWISE COMPARISONS AND INFERENCE AFTER ANOVA

In Section 8.1 we see how to use ANOVA to test for a difference in means among several groups. However, that test only tells us when differences exist, not which specific groups differ. The goal of this section is to adapt the inference procedures of Chapter 6 to use the results of the ANOVA analysis. This allows us to find a confidence interval for the mean in any group, find a confidence interval for a difference in means between two groups, and test when that difference is significant.

Using ANOVA for Inferences about Group Means

In Chapter 6 we use formulas such as those below for doing inference (either confidence intervals or tests) for a single mean and differences in two means:

$$\bar{x} \pm t^* \cdot \frac{s}{\sqrt{n}} \qquad (\bar{x}_1 - \bar{x}_2) \pm t^* \sqrt{\frac{s_1^2}{n_1} + \frac{s_2^2}{n_2}} \qquad t = \frac{\bar{x}_1 - \bar{x}_2}{\sqrt{\frac{s_1^2}{n_1} + \frac{s_2^2}{n_2}}}$$

If we have found an ANOVA table based on samples from several groups, we make a couple of small adjustments to these computations:

- Estimate any standard deviation with $\sqrt{MSE}$ from the ANOVA table.
- Use the error degrees of freedom, $n - k$, for any t-distributions.

For example, to find a confidence interval for the mean of the *i*th group, we use

$$\bar{x}_i \pm t^* \frac{\sqrt{MSE}}{\sqrt{n_i}}$$

Since one of the conditions for the ANOVA is that the standard deviation is the same in each group, using $\sqrt{MSE}$ gives an estimate that is based on all of the samples, rather than just one. That is why we use the *MSE* degrees of freedom, rather than $n_i - 1$. We often call $\sqrt{MSE}$ the *pooled standard deviation*.

Example 8.7

Sandwich Ants

The ANOVA table on page 544 for assessing a difference in average ant counts on three types of sandwiches is reproduced below:

One-way ANOVA: Ants versus Filling

Source	DF	SS	MS	F	P
Filling	2	1561	780.5	5.63	0.011
Error	21	2913	138.7		
Total	23	4474			

The mean number of ants for the sample of eight pieces of peanut butter sandwiches is 34.0. Use this and the ANOVA results to find a 95% confidence interval for the mean number of ants attracted to a peanut butter sandwich.

Solution

From the ANOVA table we find $MSE = 138.7$ with 21 degrees of freedom. For a 95% confidence interval we find percentiles from a t-distribution with 21 degrees of freedom as $t^* = 2.080$. The confidence interval is

$$34.0 \pm 2.080 \frac{\sqrt{138.7}}{\sqrt{8}} = 34.0 \pm 8.66 = (25.34, 42.66)$$

We are 95% sure that the mean number of ants attracted to a peanut butter sandwich is between 25.3 and 42.7.

The ANOVA output in Minitab includes a crude graphical representation of the confidence intervals for the means of each of the groups. Check that the interval shown for the peanut butter filling is consistent with the calculation from Example 8.7. The value labeled "Pooled StDev" is $\sqrt{MSE}$, the estimate of the common standard deviation within the groups.

```
                              Individual 95% CIs For Mean Based on Pooled StDev
Level           N   Mean   StDev  ──+────+────+────+──
Ham & Pickles   8  49.25   10.79            (──-*───)
Peanut Butter   8  34.00   14.63       (──-*───)
Vegemite        8  30.75    9.25  (───*──-)
                                  ──+────+────+────+──
                                   30   40   50   60
Pooled StDev = 11.78
```

We could have found a confidence interval for the peanut butter mean using just the mean ($\overline{x}_2 = 34.0$) and standard deviation ($s_2 = 14.63$) from its sample and a t-distribution with 7 degrees of freedom:

$$34.0 \pm 2.365 \frac{14.63}{\sqrt{8}} = 34.0 \pm 12.23 = (21.77, 46.23)$$

This gives a wider interval since s_2 is larger than $\sqrt{MSE}$ and there is more uncertainty in the estimate of the standard deviation based on just eight values, so the t^* value is larger.

For comparing two means, the usual standard error for $\overline{x}_i - \overline{x}_j$ is $SE = \sqrt{\dfrac{s_i^2}{n_i} + \dfrac{s_j^2}{n_j}}$.

If we substitute MSE for s_i^2 and s_j^2, this simplifies to

$$SE \quad (\text{of } \overline{x}_i - \overline{x}_j) = \sqrt{MSE\left(\frac{1}{n_i} + \frac{1}{n_j}\right)} \qquad \text{(after ANOVA)}$$

We use this in both a confidence interval for $\mu_i - \mu_j$ and a test statistic to compare a pair of means.

Inference for Means after ANOVA

After doing an ANOVA for a difference in means among k groups, based on samples of size $n_1 + n_2 + \cdots + n_k = n$:

Confidence interval for μ_i: $\qquad \bar{x}_i \pm t^* \dfrac{\sqrt{MSE}}{\sqrt{n_i}}$

Confidence interval for $\mu_i - \mu_j$: $\qquad (\bar{x}_i - \bar{x}_j) \pm t^* \sqrt{MSE \left(\dfrac{1}{n_i} + \dfrac{1}{n_j} \right)}$

If the ANOVA indicates that there are differences among the means:

Pairwise test of μ_i vs μ_j: $\qquad t = \dfrac{\bar{x}_i - \bar{x}_j}{\sqrt{MSE \left(\dfrac{1}{n_i} + \dfrac{1}{n_j} \right)}}$

where MSE is the mean square error from the ANOVA table and the t-distributions use $n - k$ degrees of freedom.

Example 8.8

Use the **SandwichAnts** data and ANOVA to find a 95% confidence interval for the difference in average ant counts between vegemite and ham & pickles sandwiches.

Solution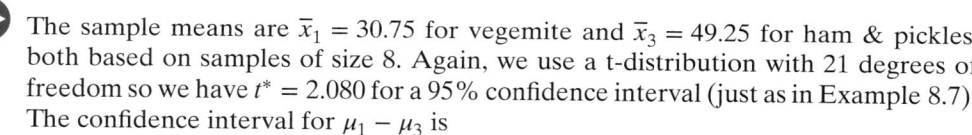

The sample means are $\bar{x}_1 = 30.75$ for vegemite and $\bar{x}_3 = 49.25$ for ham & pickles, both based on samples of size 8. Again, we use a t-distribution with 21 degrees of freedom so we have $t^* = 2.080$ for a 95% confidence interval (just as in Example 8.7). The confidence interval for $\mu_1 - \mu_3$ is

$$(30.75 - 49.25) \pm 2.080 \sqrt{138.7 \left(\frac{1}{8} + \frac{1}{8} \right)} = -18.50 \pm 12.25 = (-30.75, -6.25)$$

We are 95% sure that the mean number of ants for vegemite sandwiches is somewhere between 30.75 and 6.25 *less* than the mean number of ants for ham & pickle sandwiches.

Note that the confidence interval for $\mu_1 - \mu_3$ in the previous example includes only negative differences (and not zero). This implies evidence that the two population means differ with vegemite having a smaller mean than ham & pickles. This result is not so surprising since the ANOVA indicates that at least two of the groups have different means and the sample means are farthest apart for vegemite and ham & pickles. What about vegemite versus peanut butter?

Example 8.9

Based on the ANOVA results, test at a 5% level whether the data provide evidence of a difference in mean number of ants between vegemite and peanut butter sandwiches.

Solution

The relevant hypotheses are $H_0 : \mu_1 = \mu_2$ vs $H_a : \mu_1 \neq \mu_2$. We compare the sample means, $\bar{x}_1 = 30.75$ and $\bar{x}_2 = 34.0$, and standardize using the SE for a difference after ANOVA:

$$t = \frac{30.75 - 34.0}{\sqrt{138.7 \left(\frac{1}{8} + \frac{1}{8} \right)}} = \frac{-3.25}{5.89} = -0.55$$

We find the p-value using a t-distribution with 21 (error) degrees of freedom, doubling the area below $t = -0.55$ to get p-value $= 2(0.2941) = 0.5882$. This is a large p-value so we do not have evidence of a difference in mean number of ants between vegemite and peanut butter sandwiches.

We leave it to Exercise 8.53 to compare the ant attractiveness of the peanut butter and ham & pickle fillings. Note that with three different groups we have three possible pairs of means (vegemite vs peanut butter, vegemite vs ham & pickles, and peanut butter vs ham & pickles) to compare when the ANOVA indicates there is a difference in the means. The number of pairwise comparisons can be much larger when there are more groups. We consider some methods for handling multiple comparisons in the next example.

Lots of Pairwise Comparisons

For cases with a larger number of groups, the number of possible pairs to compare can grow quite quickly ($k = 4 \implies 6$ pairs, $k = 5 \implies 10$ pairs, etc.). This raises two issues: the need to automate such comparisons and a concern about *multiplicity* where the chance of making a Type I error increases as we do more and more tests.

©John Wiley & Sons. Photo by Vincent LaRussa.

How much do these textbooks cost?

DATA 8.2

Textbook Costs

Textbook costs can have a substantial impact on a student's budget. Do costs tend to differ depending on academic area? To investigate this question we selected a random sample of 10 introductory courses at one college within each of four broad areas (Arts, Humanities, Natural Science, and Social Science). For each course we used the college bookstore's website to determine the number of required books and total cost of the books (assuming students purchase new copies), rounding to the nearest dollar. The data for these 40 courses are stored in **TextbookCosts**. ∎

Example 8.10

Check that the conditions for running an ANOVA are reasonable to compare mean textbook costs between courses from different academic fields. Use technology to compute the ANOVA table and explain what it tells us about whether the differences in the sample means are significant.

Solution

Figure 8.9 shows side-by-side boxplots of the textbook costs for the courses, compared between the academic fields. All four samples are relatively symmetric, have no outliers, and appear to have about the same variability, so the conditions are met. We use ANOVA to test for a difference in mean textbook costs among the four fields. The hypotheses are $H_0 : \mu_1 = \mu_2 = \mu_3 = \mu_4$ vs H_a : At least one $\mu_i \neq \mu_j$, where we number the fields so that $1 =$ Arts, $2 =$ Humanities, $3 =$ Natural Science, and $4 =$ Social Science. Some ANOVA output for comparing costs between these fields is shown.

Source	DF	SS	MS	F	P
Field	3	30848	10283	4.05	0.014
Error	36	91294	2536		
Total	39	122142			

Level	N	Mean	StDev
Arts	10	94.60	44.95
Humanities	10	120.30	58.15
NaturalScience	10	170.80	48.49
SocialScience	10	118.30	48.90

Checking the standard deviations of text costs for each academic area, we see that they are roughly the same, giving further evidence that the equal-variance condition is reasonable. The p-value $= 0.014$ from the ANOVA is fairly small, so we have good evidence that there is probably some difference in mean textbook costs depending on the academic field.

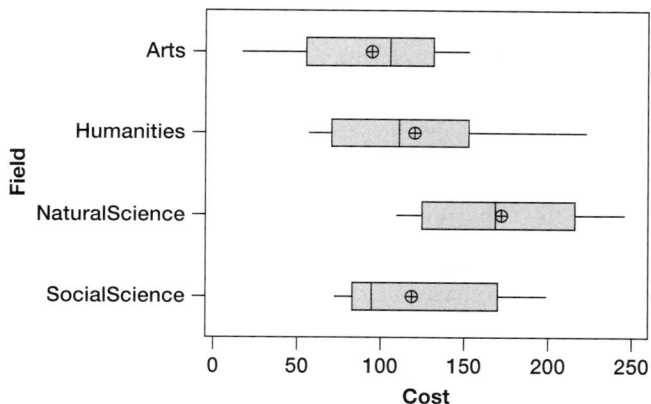

Figure 8.9 *Textbook costs between courses in different academic areas*

Since the ANOVA shows there are differences among the means, we again need to ask the question "Which fields differ from others in their mean textbook costs?" The way we handle automation of pairwise comparisons depends on the software we use. Some software has an option to produce confidence intervals for all pairwise differences. Applied to the ANOVA in Example 8.10 we get the following computer output:

Field	N	Mean	Grouping
NaturalScience	10	170.80	A
Humanities	10	120.30	B
SocialScience	10	118.30	B
Arts	10	94.60	B

Means that do not share a letter are significantly different.

CI for Difference			Lower	Center	Upper
Humanities	-	Arts	−19.97	25.70	71.37
NaturalScience	-	Arts	30.53	76.20	121.87
SocialScience	-	Arts	−21.97	23.70	69.37
NaturalScience	-	Humanities	4.83	50.50	96.17
SocialScience	-	Humanities	−47.67	−2.00	43.67
SocialScience	-	NaturalScience	−98.17	−52.50	-6.83

Remember that a 95% confidence interval for $\mu_i - \mu_j$ contains zero exactly when a 5% test of $H_0 : \mu_i = \mu_j$ vs $H_a : \mu_i \neq \mu_j$ does not have enough evidence to reject H_0. In the bottom half of the computer output, we see intervals for the six possible comparisons of these four fields. Three of the intervals include zero (have one negative and one positive endpoint) so we do not have evidence of a significant difference in mean textbook costs between courses in Arts, Humanities, or Social Sciences. The other three intervals, pairing each of these three fields with Natural Science, fail to contain zero. In each case, the Natural Science mean is enough larger than the mean cost for courses in the other field that a zero difference is outside of the plausible range of values.

Thus the data provide evidence that the mean cost for textbooks in Natural Science courses at this college is different from (and larger than) the mean costs in Arts, Humanities, and Social Sciences. We find no evidence of a difference in mean textbook costs among Arts, Humanities, and Social Science courses. This conclusion is illustrated near the top of the computer output where the letter "A" sets Natural Science off by itself and the letter "B" lumps the other three fields together as not significantly different.

The Problem of Multiplicity

One concern when doing lots of pairwise comparisons after an ANOVA (or anytime we do multiple significance tests) is the issue of *multiplicity*. As we saw in Section 4.5 on page 282, if each test has a 5% chance of making a Type 1 error (in pairwise comparisons that is finding the two groups have different means, when really they are the same), the overall error rate for lots of tests can be quite a bit higher. Even if the null hypothesis (no difference) is true, we would expect about 1 in 20 tests (at a 5% level) to reject H_0. It is important to *only* conduct the pairwise comparisons described in this section if an overall analysis of variance test shows significance.

There are several ways to deal with multiplicity, especially when doing pairwise comparisons, which are beyond the scope of this book. You may find some of these options, with names like Tukey's HSD, Fisher's LSD (see Exercise 8.60), Student Newman-Kuels, or Bonferroni's adjustment, in your statistical software or a later course in statistics.

SECTION LEARNING GOALS

You should now have the understanding and skills to:

▷ • Create confidence intervals for single means and differences of means after doing an ANOVA for means

▷ • Test specific pairs of means after ANOVA indicates a difference in means, recognizing the problem of multiplicity

Exercises for Section **8.2**

SKILL BUILDER 1

Exercises 8.41 to 8.45 refer to the data with analysis shown in the following computer output:

Level	N	Mean	StDev
A	5	10.200	2.864
B	5	16.800	2.168
C	5	10.800	2.387

Source	DF	SS	MS	F	P
Groups	2	133.20	66.60	10.74	0.002
Error	12	74.40	6.20		
Total	14	207.60			

8.41 Is there sufficient evidence of a difference in the population means of the three groups? Justify your answer using specific value(s) from the output.

8.42 What is the pooled standard deviation? What degrees of freedom are used in doing inferences for these means and differences in means?

8.43 Find a 95% confidence interval for the mean of population A.

8.44 Find a 90% confidence interval for the difference in the means of populations B and C.

8.45 Test for a difference in population means between groups A and C. Show all details of the test.

SKILL BUILDER 2

Exercises 8.46 to 8.52 refer to the data with analysis shown in the following computer output:

Level	N	Mean	StDev
A	6	86.833	5.231
B	6	76.167	6.555
C	6	80.000	9.230
D	6	69.333	6.154

Source	DF	SS	MS	F	P
Groups	3	962.8	320.9	6.64	0.003
Error	20	967.0	48.3		
Total	23	1929.8			

8.46 Is there evidence for a difference in the population means of the four groups? Justify your answer using specific value(s) from the output.

8.47 What is the pooled standard deviation? What degrees of freedom are used in doing inferences for these means and differences in means?

8.48 Find a 99% confidence interval for the mean of population A. Is 90 a plausible value for the population mean of group A?

8.49 Find a 95% confidence interval for the difference in the means of populations C and D.

8.50 Test for a difference in population means between groups A and D. Show all details of the test.

8.51 Test for a difference in population means between groups A and B. Show all details of the test.

8.52 Test for a difference in population means between groups B and D. Show all details of the test.

8.53 Peanut Butter vs Ham & Pickles The ANOVA table in Example 8.3 on page 544 for the **SandwichAnts** data indicates that there is a difference in mean number of ants among the three types of sandwich fillings. In Examples 8.8 and 8.9 we find that the difference is significant between vegemite and ham & pickles, but not between vegemite and peanut butter. What about peanut better vs ham & pickles? Test whether the difference in mean ant counts is significant (at a 5% level) between those two fillings, using the information from the ANOVA.

8.54 Pulse Rate and Award Preference In Example 8.5 on page 548 we find evidence from the ANOVA of a difference in mean pulse rate among students depending on their award preference. The

ANOVA table and summary statistics for pulse rates in each group are shown below.

Source	DF	SS	MS	F	P
Award	2	2047	1024	7.10	0.001
Error	359	51729	144		
Total	361	53776			

Level	N	Mean	StDev
Academy	31	70.52	12.36
Nobel	149	72.21	13.09
Olympic	182	67.25	10.97

Use this information and/or the data in **StudentSurvey** to compare mean pulse rates, based on the ANOVA, between each of three possible pairs of groups:

(a) Academy Award vs Nobel Prize.

(b) Academy Award vs Olympic gold medal.

(c) Nobel Prize vs Olympic gold medal.

LIGHT AT NIGHT MAKES FAT MICE, CONTINUED

Data 4.1 introduces a study in mice showing that even low-level light at night can interfere with normal eating and sleeping cycles. In the full study, mice were randomly assigned to live in one of three light conditions: LD had a standard light/dark cycle, LL had bright light all the time, and DM had dim light when there normally would have been darkness. Exercises 8.28 to 8.34 in Section 8.1 show that the groups had significantly different weight gain and time of calorie consumption. In Exercises 8.55 and 8.56, we revisit these significant differences.

8.55 Body Mass Gain Computer output showing body mass gain (in grams) for the mice after four weeks in each of the three light conditions is shown, along with the relevant ANOVA output. Which light conditions give significantly different mean body mass gain?

Level	N	Mean	StDev
DM	10	7.859	3.009
LD	9	5.987	1.786
LL	9	11.010	2.624

One-way ANOVA: BM4Gain versus Light

Source	DF	SS	MS	F	P
Light	2	116.18	58.09	8.96	0.001
Error	25	162.10	6.48		
Total	27	278.28			

8.56 When Calories Are Consumed Researchers hypothesized that the increased weight gain seen in mice with light at night might be caused by *when* the mice are eating. Computer output for the percentage of food consumed during the day (when mice would normally be sleeping) for each of the three light conditions is shown, along with the relevant ANOVA output. Which light conditions give significantly different mean percentage of calories consumed during the day?

Level	N	Mean	StDev
DM	10	55.516	10.881
LD	9	36.485	7.978
LL	9	76.573	9.646

One-way ANOVA: Day/night consumption versus Light

Source	DF	SS	MS	F	P
Light	2	7238.4	3619.2	39.01	0.000
Error	25	2319.3	92.8		
Total	27	9557.7			

8.57 More on Exercise and Stress Exercise 6.219 on page 465 introduces a study showing that exercise appears to offer some resiliency against stress, and Exercise 8.19 on page 556 follows up on this introduction. In the study, mice were randomly assigned to live in an enriched environment (EE), a standard environment (SE), or an impoverished environment (IE) for several weeks. Only the enriched environment provided opportunities for exercise. Half the mice then remained in their home cage (HC) as control groups while half were subjected to stress (SD). The researchers were interested in how resilient the mice were in recovering from the stress. One measure of mouse anxiety is amount of time hiding in a dark compartment, with mice who are more anxious spending more time in darkness. The amount of time (in seconds) spent in darkness during one trial is recorded for all the mice and the means and the results of the ANOVA analysis are shown. There are eight mice in each of the six groups.

Group:	IE:HC	SE:HC	EE:HC	IE:SD	SE:SD	EE:SD
Mean:	192	196	205	392	438	231

Source	DF	SS	MS	F	P
Light	5	481776	96355.2	39.0	0.000
Error	42	177835	2469.9		
Total	47	659611			

(a) Is there a difference between the groups in the amount of time spent in darkness? Between which two groups are we most likely to find a difference in mean time spent in darkness? Between which two groups are we least likely to find a difference?

(b) By looking at the six means, where do you think the differences are likely to lie?

(c) Test to see if there is a difference in mean time spent in darkness between the IE:HC group and the EE:SD group (that is, impoverished but not stressed vs enriched but stressed).

8.58 Drug Resistance and Dosing Exercise 8.39 on page 561 explores the topic of drug dosing and drug resistance by randomizing mice to four different drug treatment levels: untreated (no drug), light (4 mg/kg for 1 day), moderate (8 mg/kg for 1 day), or aggressive (8 mg/kg for 5 or 7 days). Exercise 8.39 found that, contrary to conventional wisdom, higher doses can actually promote drug resistance, rather than prevent it. Here, we further tease apart two different aspects of drug dosing: duration (how many days the drug is given for) and amount per day. Recall that four different response variables were measured; two measuring drug resistance (density of resistant parasites and number of days infectious with resistant parasites) and two measuring health (body mass and red blood cell density). In Exercise 8.39 we don't find any significant differences in the health responses (*Weight* and *RBC*) so we concentrate on the drug resistance measures (*ResistanceDensity* and *DaysInfectious*) in this exercise. The data are available in **DrugResistance** and we are not including the untreated group.

(a) Investigate *duration* by comparing the moderate treatment with the aggressive treatment (both of which gave the same amount of drug per day, but for differing number of days). Which of the two resistance response variables (*ResistanceDensity* and *DaysInfectious*) have means significantly different between these two treatment groups? For significant differences, indicate which group has the higher mean.

(b) Investigate *amount per day* by comparing the light treatment with the moderate treatment (both of which lasted only 1 day, but at differing amounts). Which of the two resistance response variables have means significantly different between these two treatment groups? For significant differences, indicate which group has the higher mean.

(c) Does duration or amount seem to be more influential (at least within the context of this study)? Why?

8.59 Effects of Synchronization and Exertion on Closeness Exercise 8.16 on page 554 looks at possible differences in ratings of closeness to a group after doing a physical activity that involves either high or low levels of synchronization (HS or LS) and high or low levels of exertion (HE or LE). Students were randomly assigned to one of four groups with different combinations of these variables, and the change in their ratings of closeness to their group (on a 1 to 7 scale) were recorded. The data are stored in **SynchronizedMovement** and the means for each treatment group are given below, along with an ANOVA table that indicates a significant difference in the means at a 5% level.

Group	N	Mean	StDev
HS+HE	72	0.319	1.852
HS+LE	64	0.328	1.861
LS+HE	66	0.379	1.838
LS+LE	58	−0.431	1.623

Analysis of Variance

Source	DF	SS	MS	F-Value	P-Value
Group	3	27.04	9.012	2.77	0.042
Error	256	831.52	3.248		
Total	259	858.55			

The first three means look very similar, but the LS+LE group looks quite a bit different from the others. Is that a significant difference? Test this by comparing the mean difference in change in closeness ratings between the synchronized, high exertion activity group (HS+HE) and the non-synchronized, low exertion activity group (LS+LE).

8.60 Fisher's LSD One way to "automate" pairwise comparisons that works particularly well when the sample sizes are balanced is to compute a single value that can serve as a threshold for when a pair of sample means are far enough apart to suggest that the population means differ between those two groups. One such value is called Fisher's *Least Significant Difference* or *LSD* for short.

$$LSD = t^* \sqrt{MSE\left(\frac{1}{n_i} + \frac{1}{n_j}\right)}$$

You may recognize this as the margin of error for a confidence interval for a difference in two means after doing an ANOVA. That is exactly how we compute it. Recall that the test for a pair of means will show a significant difference exactly when the confidence interval fails to include zero. The confidence level should be matched to the significance level of the test (for example, a 95% confidence interval corresponds to a 5% significance level). If the difference in two group means (in absolute value) is smaller than the LSD margin of error, the confidence interval will have one positive and one negative endpoint. Otherwise, the interval will stay either all positive or all negative and we conclude the two means differ:

Reject H_0 and conclude
the two means differ $\iff |\bar{x}_i - \bar{x}_j| > LSD$

Compute *LSD* using a 5% significance level for the ANOVA data comparing textbook costs in Example 8.10 on page 567. Use the value to determine which academic fields appear to show evidence of a difference in mean textbook costs.

8.61 LSD for Exercise and Stress Use Fisher's LSD, as described in Exercise 8.60, to discuss differences in mean time mice spend in darkness for the six combinations of environment and stress that produce the output in Exercise 8.57.

8.62 Fish Ventilation In Exercise 8.38 on page 561 we consider an ANOVA to test for difference in mean gill beat rates for fish in water with three different levels of calcium. The data are stored in **FishGills3**. If the ANOVA table indicates that the mean gill rates differ due to the calcium levels, determine which levels lead to different means. If the ANOVA shows no significant difference, find a confidence interval for the mean gill rate at each level of calcium.

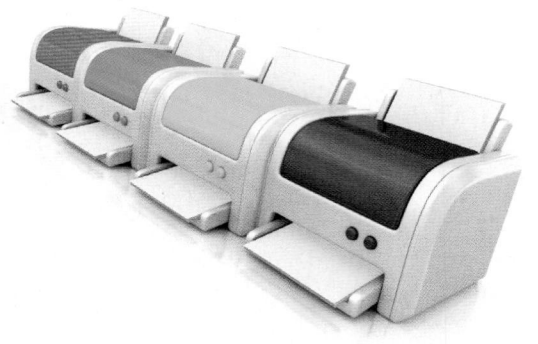

CHAPTER 9

Inference for Regression

"All models are wrong, but some are useful."

–George E. P. Box*

*Box, G. and Draper, N., *Empirical Model-Building and Response Surfaces*, John Wiley and Sons, New York, 1987, p. 424.
Top left: ©pagadesign/iStockphoto, Top right: ©Jennifer Fluharty/iStockphoto, Bottom right: ©NetPhotos/Alamy Limited

Questions and Issues

Here are some of the questions and issues we will discuss in this chapter:

- Is there an association between the number of Facebook friends a person has and the social perception areas of the brain?

- Does the price of an inkjet printer depend on how fast it prints?

- Does the tip percentage in a restaurant depend on the size of the bill?

- How well do SAT scores predict college grade point averages?

- Does when food is eaten affect weight gain?

- Is offense or defense more important in the NBA?

- Is the percent of a country's expenditure on health care associated with life expectancy in that country?

- How do sugar, sodium, and fiber contribute to the number of calories in breakfast cereal?

- What is the relationship between the amount of time a person spends exercising and the amount of time a person spends watching television?

- How well does a president's approval rating predict his or her re-election chances?

- What is the average tip left on a restaurant bill of $30?

- How well do first round scores in the Master's golf tournament do at predicting final scores?

- How well does success in the pre-season predict success in the regular season, in the National Football League?

- Is the honeybee population shrinking?

9.1 INFERENCE FOR SLOPE AND CORRELATION

In Sections 2.5 and 2.6 we introduce summary statistics for correlation and linear regression as ways to describe the relationship between two quantitative variables. In Chapters 3 and 4 we see examples for doing inference for these quantities using bootstrap distributions and randomization tests. In this chapter we develop methods similar to those in Chapters 5 and 6 for applying standard distributions to help with inferences for quantitative vs quantitative relationships.

Simple Linear Model

For a simple linear model we have a quantitative response variable (Y) and a quantitative explanatory variable (X). We assume the values of Y tend to increase or decrease in a regular (linear) way as X increases. This does not mean an exact relationship with all points falling perfectly on a line. A statistical model generally consists of two parts: one specifying the main trend of the relationship and the second allowing for individual deviations from that trend. For a simple linear model, a line (specified with a slope and an intercept) shows the general trend of the data, and individual points tend to be scattered above and below the line.

In Section 2.6, we use the notation $\widehat{Y} = a + bX$ for the least squares line for a sample. Here we switch to the following notation for the least squares line for a sample:

$$\widehat{Y} = b_0 + b_1 X$$

We use the following notation to express a simple linear model for a population:

$$Y = \beta_0 + \beta_1 X + \epsilon$$

The *linear* part of the model ($\beta_0 + \beta_1 X$) reflects the underlying pattern for how the average Y behaves depending on X. We use Greek letters for the intercept and slope[1] in the model since they represent parameters for the entire population. The *error* term in the model (denoted by ϵ) allows for individual points to vary above or below the line.

In practice, just as we rarely know the mean, μ, or proportion, p, for an entire population, we can only estimate the population slope and intercept using the data in a sample. Once we have estimated the line, we can also estimate the error term for any point as the distance away from the fitted line.

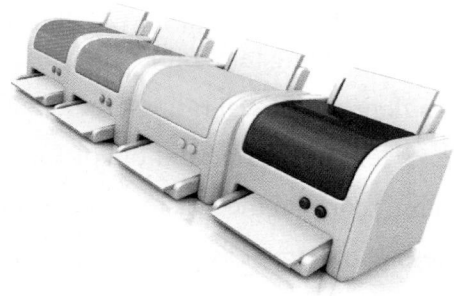

What factors influence the price of these printers?

---

[1] We use subscripts for β_0 and β_1 to make it easy to consider additional explanatory terms in a linear model.

DATA 9.1

Inkjet Printers

Suppose we are interested in purchasing a multifunction inkjet printer. How are performance factors related to the price of the printer? To investigate this question we checked reviews at PCMag.com for a sample of 20 all-in-one printers.[2] The data stored in **InkjetPrinters** include:

PPM	printing rate (pages per minute) for a set of print jobs
PhotoTime	average time (in seconds) to print 4 × 6 color photos
CostBW	average cost per page (in cents) for printing in black & white
CostColor	average cost per page (in cents) for printing in color
Price	typical retail price (in dollars) at the time of the review

Table 9.1 *Printing rate and price for 20 inkjet printers*

PPM	3.9	2.9	2.7	2.9	2.4	4.1	3.4	2.8	3.0	3.2
Price	300	199	79	129	70	348	299	248	150	150

PPM	2.7	2.7	2.2	2.5	2.7	1.7	2.8	1.8	1.8	4.1
Price	87	100	99	189	99	60	199	149	79	199

Example 9.1

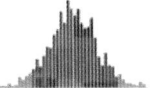

Use technology and data in **InkjetPrinters** (also given in Table 9.1) to estimate and display the least squares line for predicting *Price* based on *PPM*.

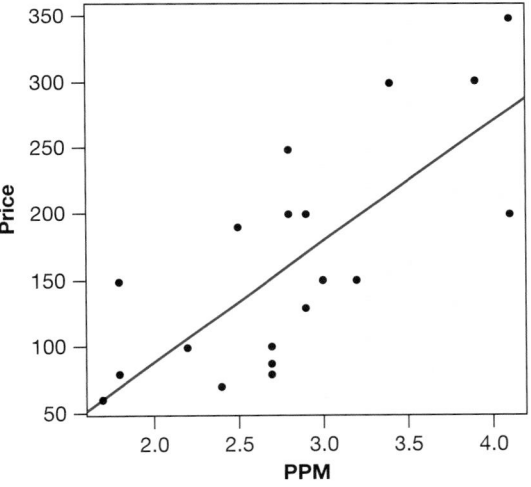

Figure 9.1 *Plot of Price vs PPM with least squares line for 20 inkjet printers*

Solution

Here is some computer output for finding the least squares line using the data on 20 printers in **InkjetPrinters**:

Coefficients:	Estimate	Std.Error	t value	Pr(> |t|)
(Intercept)	−94.22	56.40	−1.671	0.112086
PPM	90.88	19.49	4.663	0.000193

S = 58.5457   R-Sq = 54.7%   R-Sq(adj) = 52.2%

We see that the intercept is $b_0 = -94.22$ and the slope is $b_1 = 90.88$ to produce the least squares prediction equation

$$\widehat{Price} = -94.22 + 90.88 \cdot PPM$$

[2]Reviews found at *http://www.pcmag.com/reviews/printers, August 2011.*

Figure 9.1 shows a scatterplot of *Price* vs *PPM* with this least squares line summarizing the trend in the data. The plot shows a generally increasing trend with faster printers (higher *PPM*) tending to cost more.

Inference for Slope

In Example B.4 on page 344 we use a bootstrap distribution to find a confidence interval for the slope in a regression model to predict *Tip* based on a restaurant *Bill*, using the data in **RestaurantTips**. In Example B.6 on page 346 we create a randomization distribution to test if the slope between the percentage tip (*PctTip*) and *Bill* is different from zero. We can apply these same ideas to construct a bootstrap distribution of slopes for the inkjet printer model or a randomization distribution to test if the slope between *Price* and *PPM* is positive. Figure 9.2 shows 1000 simulated slopes in each of these situations.

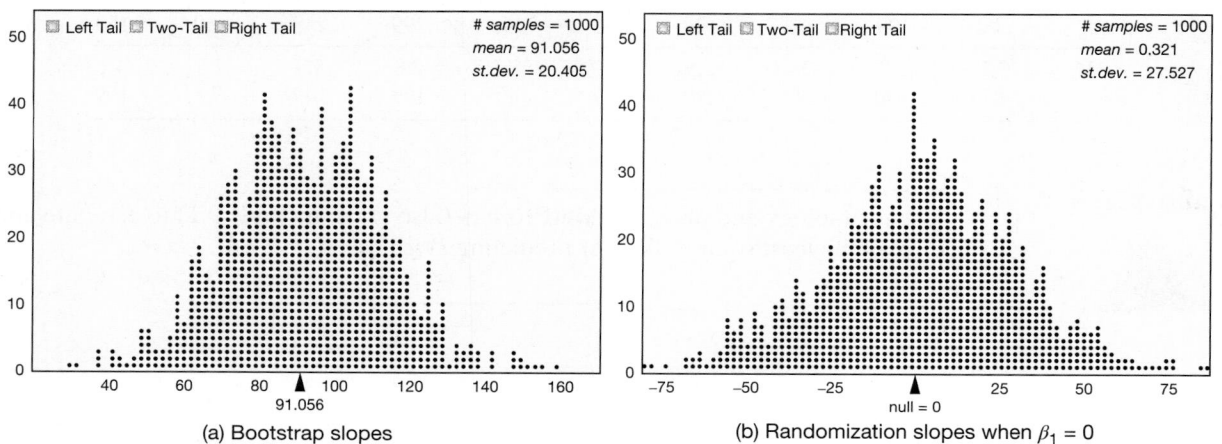

(a) Bootstrap slopes   (b) Randomization slopes when $\beta_1 = 0$

Figure 9.2 *Bootstrap and randomization distributions for Price vs PPM with InkjetPrinters.*

Once again, we see familiar shapes that are predictable in advance when a linear model is appropriate for summarizing the relationship between two quantitative variables. The bootstrap slopes for *Price* vs *PPM* are bell-shaped, are centered near 90.88 (the slope in the original sample), and have a standard error of about $SE = 20.4$. The randomization slopes are also bell-shaped and centered at zero (as expected for a null hypothesis that $\beta_1 = 0$). We see that none of the randomization slopes are as extreme as the slope $b_1 = 90.88$ that was observed in the original sample. This gives very strong evidence for a positive association, indicating that the slope when using *PPM* to predict *Price* for all inkjet printers is greater than zero. As we saw in Chapters 5 and 6, the key to doing inference with a theoretical distribution such as the normal or t (rather than a bootstrap or randomization distribution) is being able to estimate the standard error (*SE*) of the statistic of interest. Fortunately, most statistical software packages provide an estimate for the standard error of the slope when estimating a regression model.

Looking back at the computer output for predicting *Price* based on *PPM* in Example 9.1 we see the standard error for the slope (coefficient of *PPM*) is 19.49, just a bit smaller than the standard error from the bootstrap slopes in Figure 9.2(a). While the formula for computing the standard error of the slope is a bit complicated (see the computational notes at the end of Section 9.2), we can use the standard error from the regression output to form a confidence interval or a standardized test statistic. Given the bell-shape of the bootstrap and randomization distributions for slope in Figure 9.2, it shouldn't surprise you to learn that the appropriate reference distribution is a t-distribution, although in this case we use $n - 2$ degrees of freedom since we are estimating two parameters in this model. The general formulas for confidence intervals and test statistics that we used repeatedly in Chapters 5 and 6 apply here also.

Inference for the Slope in a Simple Linear Model

When the conditions for a simple linear model are reasonably met, we find:

(a) A confidence interval for the population slope using

$$\text{Sample statistic} \pm t^* \cdot SE = b_1 \pm t^* \cdot SE$$

(b) A test statistic for $H_0 : \beta_1 = 0$ using

$$t = \frac{\text{Sample statistic} - \text{Null parameter}}{SE} = \frac{b_1 - 0}{SE} = \frac{b_1}{SE}$$

where b_1 is the slope for the least squares line for the sample and SE is the standard error of the slope (both obtained with technology.)

The appropriate reference distribution is a t-distribution with $n - 2$ degrees of freedom.

Example 9.2

Use information in the computer output to find a 95% confidence interval for the population slope to predict *Price* based on *PPM*. Also test (at a 5% level) whether we have evidence that the printing speed (*PPM*) is an effective linear predictor of the price of such printers.

Solution

Here, again, is the output for fitting a least squares line to predict *Price* using *PPM* for the 20 printers in **InkjetPrinters**:

Coefficients:	Estimate	Std.Error	t value	Pr(> |t|)
(Intercept)	−94.22	56.40	−1.671	0.112086
PPM	90.88	19.49	4.663	0.000193

S = 58.5457   R-Sq = 54.7%   R-Sq(adj) = 52.2%

The estimated slope is $b_1 = 90.88$ and the standard error is $SE = 19.49$. For 95% confidence we use a t-distribution with $20 - 2 = 18$ degrees of freedom to find $t^* = 2.10$. The confidence interval for the slope is

$$b_1 \pm t^* \cdot SE$$
$$90.88 \pm 2.10(19.49)$$
$$90.88 \pm 40.93$$
$$49.95 \text{ to } 131.81$$

Based on these data we are 95% sure that the slope (increase in price for every extra page per minute in printing speed) is somewhere between $49.95 and $131.81.

We know that *PPM* has some relationship as a predictor of *Price* if the population slope β_1 is not zero. To test $H_0 : \beta_1 = 0$ vs $H_a : \beta_1 \neq 0$ using the sample data, we compute a standardized test statistic with

$$t = \frac{\text{Statistic} - \text{Null value}}{SE} = \frac{b_1 - 0}{SE} = \frac{90.88}{19.49} = 4.66$$

The p-value is twice the area above 4.66 in a t-distribution with 18 degrees of freedom, but this is a large t-statistic so we see that p-value ≈ 0. This gives strong evidence that the printing speed (*PPM*) is an effective predictor of the price for inkjet printers. Note that the values of the test statistic (4.663) and the p-value (0.000193) can also be found directly in the computer output for this model.

Example 9.3

Predicting Tips Based on Restaurant Bills

Some computer output for fitting a least squares line to predict the size of a restaurant tip using the amount of the bill as a predictor is given below. The data, a sample of $n = 157$ restaurant bills at the First Crush bistro, are stored in **RestaurantTips**.

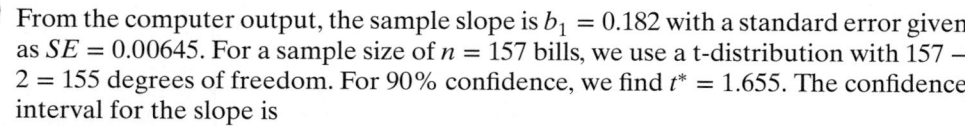

Coefficients:	Estimate	Std. Error	t value	Pr(> |t|)
(Intercept)	−0.292267	0.166160	−1.759	0.0806
Bill	0.182215	0.006451	28.247	<2e-16

Use information from the computer output to find and interpret a 90% confidence interval for the slope of this regression model.

Solution

From the computer output, the sample slope is $b_1 = 0.182$ with a standard error given as $SE = 0.00645$. For a sample size of $n = 157$ bills, we use a t-distribution with $157 - 2 = 155$ degrees of freedom. For 90% confidence, we find $t^* = 1.655$. The confidence interval for the slope is

$$b_1 \pm t^* \cdot SE = 0.182 \pm 1.655(0.00645) = 0.182 \pm 0.011 = (0.171, 0.193)$$

Based on these data we are 90% sure that the increase in the tip for each extra dollar of a bill is somewhere between $0.171 and $0.193.

Example 9.4

Predicting Percent Tip Based on Restaurant Bills

Use the data in **RestaurantTips** to test whether the amount of the bill is an effective predictor of the size of the tip *as a percentage* of the bill (*PctTip*) for customers at the First Crush bistro.

Solution

We are testing $H_0 : \beta_1 = 0$ vs $H_a : \beta_1 \neq 0$, where β_1 is the slope for predicting the percentage tip based on the amount of the bill for all First Crush customers. Here is some computer output for fitting this model:

Coefficients:	Estimate	Std. Error	t value	Pr(> |t|)
(Intercept)	15.50965	0.73956	20.97	<2e-16
Bill	0.04881	0.02871	1.70	0.0911

We see that the sample slope is $b_1 = 0.0488$ with standard error $SE = 0.0287$. We see in the output that the t-statistic for testing the slope is $t = 1.70$, which gives a (two-tail) p-value of 0.0911. This gives some (significant at just a 10% level) but not very convincing evidence that the percent tip is related to the size of the bill.

Note that the computer output for a regression model also gives a standard error and t-test for the intercept in the model, but we rarely need to worry about inference for the intercept. We are generally more concerned with the effectiveness of the predictor and, in many cases, the intercept is not directly interpretable (such as the cost for a printer that prints zero pages per minute!)

t-Test for Correlation

In some situations we may be interested in testing a linear association between two quantitative variables when we don't have a specific predictor/response relationship. In Section 2.5 we introduce the correlation as a measure of the strength of a linear association between two quantitative variables. Recall that a correlation of zero indicates no linear relationship while a positive or negative correlation indicates some linear relationship. We can use the correlation in a sample as a way to test whether the population correlation ρ differs from zero. The standard error for correlation r using a sample of size n is $SE = \sqrt{(1 - r^2)/(n - 2)}$.

---

t-Test for correlation

To test $H_0 : \rho = 0$ vs $H_a : \rho \neq 0$ (or a one-tailed alternative) we use a standardized test statistic

$$t = \frac{\text{Sample statistic} - \text{Null parameter}}{SE} = \frac{r - 0}{\sqrt{\frac{1-r^2}{n-2}}} = \frac{r\sqrt{n - 2}}{\sqrt{1 - r^2}}$$

where r is the correlation for a sample of size n. To find a p-value we use a t-distribution with $n - 2$ degrees of freedom.

Example 9.5

The correlation between printing rate (*PPM*) and cost per page for printing in black & white (*CostBW*) for the 20 inkjet printers in **InkjetPrinters** is $r = -0.636$. Does this provide sufficient evidence to conclude there is a negative association between printing speed and cost?

Solution   We test $H_0 : \rho = 0$ vs $H_a : \rho < 0$, where ρ is the correlation between *CostBW* and *PPM* for all inkjet printers on the market. The relevant test statistic is

$$t = \frac{r\sqrt{n - 2}}{\sqrt{1 - r^2}} = \frac{-0.636\sqrt{20 - 2}}{\sqrt{1 - (-0.636)^2}} = -3.50$$

We find the p-value using the lower tail (below -3.50) of a t-distribution with 18 degrees of freedom. This gives p-value = 0.0013, which is quite small. There is strong evidence of a negative association between printing speed and ink costs per page. Faster printers actually tend to cost less for the ink on each page.

  Note that the strength of evidence when testing a correlation depends on the sample size as well as the magnitude of the sample correlation. The same value of r might be quite insignificant for a small sample but strong evidence of an association for a much larger sample.

Example 9.6

Find the correlation between *Bill* and *PctTip* for the **RestaurantTip** data and use it to test (at a 5% level) whether the correlation for the population differs from zero.

Solution   Using technology we find that the correlation between the size of the bill and percentage tip for the sample of 157 restaurant tips is $r = 0.135$. To test $H_0 : \rho = 0$ vs $H_a : \rho \neq 0$ we use the t-statistic

$$t = \frac{0.135\sqrt{157 - 2}}{\sqrt{1 - 0.135^2}} = 1.70$$

which gives a two-tailed p-value of 0.0911 (based on a t-distribution with 155 degrees of freedom). This p-value is not less than 5% so we do not have strong evidence of a linear association between the size of the bill and percent of the tip.

If you look back at the test for the slope of the model to predict *PctTip* using *Bill* in Example 9.4 on page 580 you should notice an interesting fact. The t-statistic ($t = 1.70$) and p-value (0.0911) for that test are exactly the same as in the test for correlation in Example 9.6. This is not an accident! It turns out that the formula for computing the t-statistic for a slope always gives an identical result to the t-test for correlation.

This means we can use these two tests (for a slope and for a correlation) interchangeably. The t-test for a slope is commonly found in regression output when the slope and intercept are estimated. The t-test for correlation requires only knowing r and n and is often easy to compute when summary statistics are given in an article or report.

Coefficient of Determination: R-squared

Another common connection between a sample correlation and a regression line comes from computing r^2, which is known as the *coefficient of determination*. Since $-1 \leq r \leq 1$ is always true for a correlation r, we know that r^2 is always between zero and one. Amazingly, it turns out that this value gives us the proportion of the total variability in the response variable (Y) that is explained by the explanatory variable (X). When we interpret it this way, we usually denote it as R^2 and state it as a percentage.

Example 9.7

Find and interpret the value of R^2 for the relationship between inkjet *Price* and print speed *PPM*.

Solution ▶ Using technology we find the correlation between *PPM* and *Price* is $r = 0.7397$, which gives $R^2 = (0.7397)^2 = 0.547$. This means that 54.7% of the variability in prices of the inkjet printers in this sample is explained by their print speed.

We see "R-Sq=54.7%" in the computer output for this model in Example 9.1 on page 577. This is a common value to find in regression output and is more often referred to as "R-squared" rather than the more cumbersome "coefficient of determination." The exact notion of a percentage of variability explained by a model is a bit vague at this point, but we examine this in more detail when we consider ANOVA for regression in Section 9.2.

Checking Conditions for a Simple Linear Model

Recall that a t-test for a mean is only valid if certain conditions on the underlying distribution (such as normality) are met. Similarly, when doing inference for a simple linear model, there are conditions on the model that help ensure that a t-distribution is reasonable to use when doing inference for the slope.

Simple Linear Model

A simple linear model for a response variable Y based on a predictor X has the form

$$Y = \beta_0 + \beta_1 X + \epsilon$$

where the random errors are independent values from a $N(0, \sigma_\epsilon)$ distribution.

We can think of this as a distribution of Y values for each different value of the predictor X, where the means increase (or decrease) in a regular way along the line and the errors cause individual points to scatter above and below the line with some fixed variability (denoted in the model by σ_ϵ). This is illustrated in Figure 9.3. Remember that this is a model for the *population*. A sample from this population is drawn from these normal distributions to give the data we see in a scatterplot.

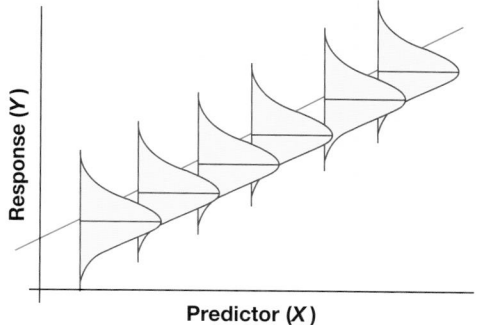

Figure 9.3 *Simple linear model with normal distributions for every predictor value*

How do we use the data in a sample to check if the conditions for a simple linear model are reasonable? Visual checks with graphs are generally the most useful tools. For now, we assess the appropriateness of the simple linear model for a particular dataset relatively informally by looking at a scatterplot with the regression line drawn on it.[3] Ideally, we like to see a consistent band of data stretching relatively symmetrically on either side of the line as in Figure 9.4. Try to visualize how data in

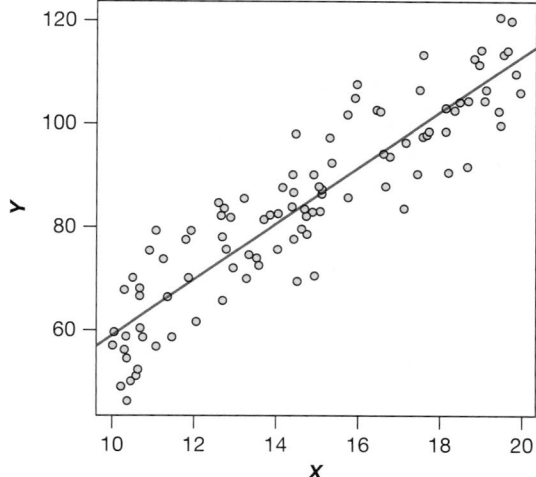

Figure 9.4 *Scatterplot of data from a simple linear model*

[3]We consider additional plots for assessing the conditions of the regression model in Section 10.2.

the scatterplot in Figure 9.4 could arise by sampling from a population as described in the model of Figure 9.3.

What Can Go Wrong?

As when we assess normality for doing a t-test for a mean, be on the lookout for signs of *obvious* departures from the ideal scatterplot for a regression model. Several of these are illustrated in Figure 9.5. Try not to be too picky; don't worry about small departures from an ideal pattern. Pay attention (and view the results of inference with some skepticism) only when we see a consistent departure from the expected pattern.

Watch out for:

- *Departures from linearity.* Figure 9.5(a) shows some obvious curvature with a trend that is clearly increasing, but not in a linear way.

- *Consistently changing variability.* Figure 9.5(b) shows data where the variability above and below the line clearly increases in a "fanning" pattern as the values get larger. We would prefer to see roughly parallel bands above and below the line.

- *Outliers and influential points.* Figure 9.5(c) shows four points that clearly depart from the pattern exhibited by the rest of the data. These large discrepancies produce outliers in the distribution of errors that indicate a lack of normality. Also, the two very large predictor values ($x = 100$ and $x = 150$) could strongly influence the location of the least squares line.

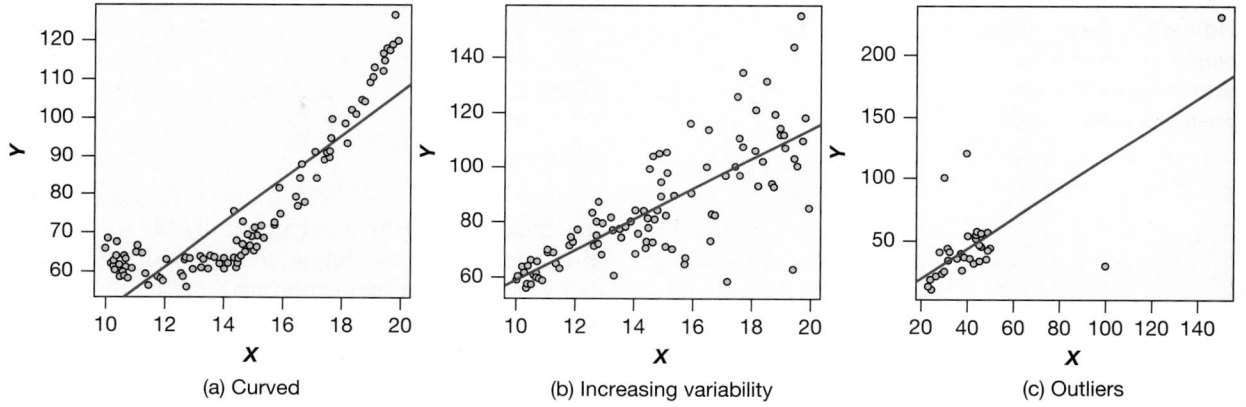

Figure 9.5 *Scatterplots for least squares fits with problems*

Example 9.8

Check Conditions for the Inkjet Printer Model

Using the scatterplot with regression line in Figure 9.1 on page 577, comment on the appropriateness of the simple linear model to predict inkjet printer prices using *PPM* printing rates with the data in **InkjetPrinters**.

Solution The plot in Figure 9.1 shows a general increasing trend, no obvious curvature or big outliers, and a relatively equal scatter of points above and below the line. Nothing in this plot raises serious concerns about the simple linear model conditions. While it is hard to make definitive assessments of the simple linear model conditions based on just 20 data points, we don't see any reason for strong concerns about the conditions for these data.

Example 9.9

Solution

Check Conditions for the Restaurant Tip Models

Produce graphs and comment on the appropriateness of the regression models to predict *Tip* and *PctTip* using the amount of the *Bill* for the data in **RestaurantTips**.

The scatterplots with regression lines for the two models are shown in Figure 9.6.

Other than a few somewhat unusually large tips, the pattern in the scatterplot for *Tip* vs *Bill* looks pretty good. There is no sign of curvature and the points are fairly equally distributed on either side of the line with a consistent (and relatively small) amount of variability. Given the large sample size ($n = 157$), we have no serious concerns about using a t-distribution for inference for the slope in this model. The plot for *PctTip* vs *Bill* shows more variability around the line, but still a relatively linear pattern with reasonably consistent variability. The only mild concern would be the three large outliers where generous customers tipped more than 30%. Note that even a relatively poor fit, such as *PctTip* vs *Bill*, can still follow a simple linear model, just one with a slope near zero and fairly large variability in the errors.

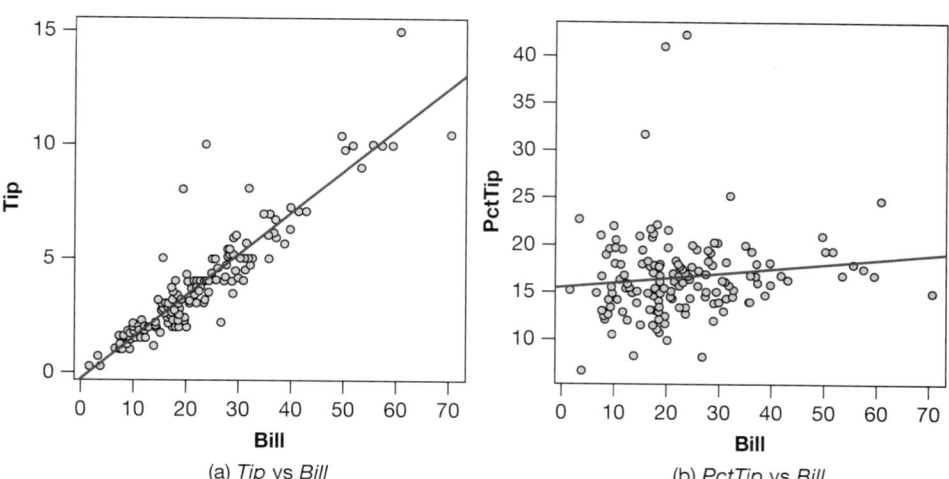

Figure 9.6 *Regressions to predict Tip and PctTip based on restaurant Bill*

(a) *Tip* vs *Bill*

(b) *PctTip* vs *Bill*

SECTION LEARNING GOALS

You should now have the understanding and skills to:

- Use computer output to make predictions and interpret coefficients using a fitted simple linear model
- Construct a confidence interval for the slope in a regression model
- Test a hypothesis about the slope of a regression model
- Test for evidence of a linear association between two quantitative variables using a sample correlation
- Find and interpret the value of R^2 for a regression model
- Check a scatterplot for obvious departures from a simple linear model

Exercises for Section **9.1**

SKILL BUILDER 1

In Exercises 9.1 to 9.4, use the computer output (from different computer packages) to estimate the intercept β_0, the slope β_1, and to give the equation for the least squares line for the sample. Assume the response variable is Y in each case.

9.1 The regression equation is Y = 29.3 + 4.30 X

Predictor	Coef	SE Coef	T	P
Constant	29.266	6.324	4.63	0.000
X	4.2969	0.6473	6.64	0.000

9.2 The regression equation is Y = 808 − 3.66 A

Predictor	Coef	SE Coef	T	P
Constant	807.79	87.78	9.20	0.000
A	−3.659	1.199	−3.05	0.006

9.3

Coefficients:	Estimate	Std.Error	t value	Pr(> |t|)
(Intercept)	77.44	14.43	5.37	0.000
Score	−15.904	5.721	−2.78	0.012

9.4

Coefficients:	Estimate	Std.Error	t value	Pr(> |t|)
(Intercept)	7.277	1.167	6.24	0.000
Dose	−0.3560	0.2007	−1.77	0.087

SKILL BUILDER 2

Exercises 9.5 to 9.8 show some computer output for fitting simple linear models. State the value of the sample slope for each model and give the null and alternative hypotheses for testing if the slope in the population is different from zero. Identify the p-value and use it (and a 5% significance level) to make a clear conclusion about the effectiveness of the model.

9.5 The regression equation is Y = 89.4 - 8.20 X

Predictor	Coef	SE Coef	T	P
Constant	89.406	4.535	19.71	0.000
X	−8.1952	0.9563	−8.57	0.000

9.6 The regression equation is Y = 82.3 - 0.0241 X

Predictor	Coef	SE Coef	T	P
Constant	82.29	11.80	6.97	0.000
X	−0.02413	0.02018	−1.20	0.245

9.7

Coefficients:	Estimate	Std.Error	t value	Pr(> |t|)
(Intercept)	7.277	1.167	6.24	0.000
Dose	−0.3560	0.2007	−1.77	0.087

9.8

Coefficients:	Estimate	Std.Error	t value	Pr(> |t|)
(Intercept)	807.79	87.78	9.30	0.000
A	−3.659	1.199	−3.05	0.006

SKILL BUILDER 3

In Exercises 9.9 and 9.10, find and interpret a 95% confidence interval for the slope of the model indicated.

9.9 The model given by the output in Exercise 9.5, with $n = 24$.

9.10 The model given by the output in Exercise 9.7, with $n = 30$.

SKILL BUILDER 4

In Exercises 9.11 to 9.14, test the correlation, as indicated. Show all details of the test.

9.11 Test for a positive correlation; $r = 0.35$; $n = 30$.

9.12 Test for evidence of a linear association; $r = 0.28$; $n = 10$.

9.13 Test for evidence of a linear association; $r = 0.28$; $n = 100$.

9.14 Test for a negative correlation; $r = -0.41$; $n = 18$.

9.15 Student Survey: Correlation Matrix A correlation matrix allows us to see lots of correlations at once, between many pairs of variables. A correlation matrix for several variables (*Exercise*, *TV*, *Height*, *Weight*, and *GPA*) in the **StudentSurvey** dataset is given. For any pair of variables (indicated by the row and the column), we are given two values: the correlation as the top number and the p-value for a two-tail test of the correlation right beneath it.

Correlations: Exercise, TV, Height, Weight, GPA

	Exercise	TV	Height	Weight
TV	0.010			
	0.852			
Height	0.118	0.181		
	0.026	0.001		
Weight	0.118	0.165	0.619	
	0.026	0.002	0.000	
GPA	−0.159	−0.129	−0.116	−0.217
	0.003	0.017	0.033	0.000

Cell Contents: Pearson correlation
             P-Value

(a) Which two variables are most strongly positively correlated? What is the correlation? What is the p-value? What does a positive correlation mean in this situation?

(b) Which two variables are most strongly negatively correlated? What is the correlation? What is the p-value? What does a negative correlation mean in this situation?

(c) At a 5% significance level, list any pairs of variables for which there is *not* convincing evidence of a linear association.

9.16 NBA Players: Correlation Matrix The dataset **NBAPlayers2015** is introduced on page 91 and contains information on many variables for players in the NBA (National Basketball Association) during the 2014–2015 season. The dataset includes information for all players who averaged more than

24 minutes per game ($n = 182$) and 25 variables, including *Age*, *Points* (number of points for the season per game), *FTPct* (free throw shooting percentage), *Rebounds* (number of rebounds for the season), and *Steals* (number of steals for the season). A correlation matrix for these five variables is shown. A correlation matrix allows us to see lots of correlations at once, between many pairs of variables. For any pair of variables (indicated by the row and the column), we are given two values: the correlation as the top number and the p-value for a two-tail test of the correlation right beneath it.

Correlations: Age, Points, FTPct, Rebounds, Steals

	Age	Points	FTPct	Rebounds
Points	−0.019			
	0.800			
FTPct	0.147	0.288		
	0.047	0.000		
Rebounds	−0.028	0.332	−0.408	
	0.712	0.000	0.000	
Steals	−0.071	0.557	0.008	0.096
	0.021	0.000	0.918	0.198

Cell Contents: Pearson correlation
P-Value

(a) Which two variables are most strongly positively correlated? What is the correlation? What is the p-value? What does a positive correlation mean in this situation?

(b) Which two variables are most strongly negatively correlated? What is the correlation? What is the p-value? What does a negative correlation mean in this situation?

(c) At a 5% significance level, list any pairs of variables for which there is *not* convincing evidence of a linear association.

9.17 Verbal SAT as a Predictor of GPA A scatterplot with regression line is shown in Figure 9.7 for a regression model using Verbal SAT score, *VerbalSAT*, to predict grade point average in college, *GPA*, using the data in **StudentSurvey**. We also show computer output below of the regression analysis.

The regression equation is GPA = 2.03 + 0.00189 VerbalSAT

Predictor	Coef	SE Coef	T	P
Constant	2.0336	0.1621	12.54	0.000
VerbalSAT	0.0018929	0.0002709	6.99	0.000

S = 0.373214   R-Sq = 12.5%   R-Sq(adj) = 12.2%

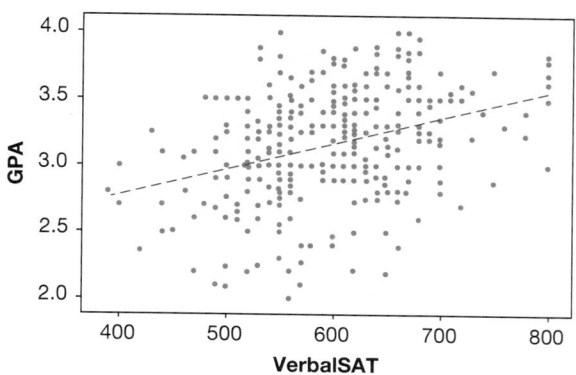

Figure 9.7 *Using Verbal SAT score to predict grade point average*

(a) Use the scatterplot to determine whether we should have any significant concerns about the conditions being met for using a linear model with these data.

(b) Use the fitted model to predict the GPA of a person with a score on the Verbal SAT exam of 650.

(c) What is the estimated slope in this regression model? Interpret the slope in context.

(d) What is the test statistic for a test of the slope? What is the p-value? What is the conclusion of the test, in context?

(e) What is R^2? Interpret it in context.

9.18 Does When Food Is Eaten Affect Weight Gain? Data 4.1 on page 258 introduces a study that examines the effect of light at night on weight gain in mice. In the full study of 27 mice over a four-week period, the mice who had a light on at night gained significantly more weight than the mice with darkness at night, despite eating the same number of calories and exercising the same amount. Researchers noticed that the mice with light at night ate a greater percentage of their calories during the day (when mice are supposed to be sleeping). The computer output shown below allows us to examine the relationship between percent of calories eaten during the day, *DayPct*, and body mass gain in grams, *BMGain*. A scatterplot with regression line is shown in Figure 9.8.

Pearson correlation of BMGain and DayPct = 0.740
P-Value = 0.000

The regression equation is
BMGain = 1.11 + 0.127 DayPct

Predictor	Coef	SE Coef	T	P
Constant	1.113	1.382	0.81	0.428
DayPct	0.12727	0.02315	5.50	0.000

S = 2.23097   R-Sq = 54.7%   R-Sq(adj) = 52.9%

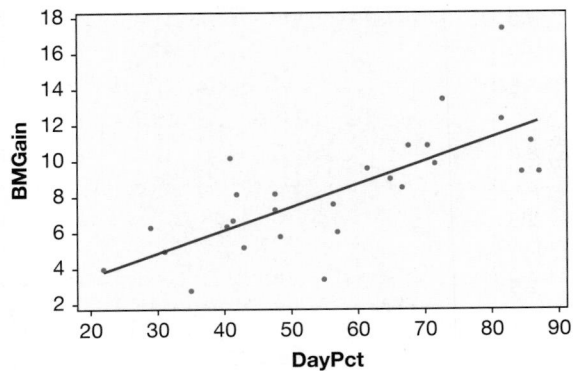

Figure 9.8 *Does when food is eaten affect weight gain?*

(a) Use the scatterplot to determine whether we should have any strong concerns about the conditions being met for using a linear model with these data.

(b) What is the correlation between these two variables? What is the p-value from a test of the correlation? What is the conclusion of the test, in context?

(c) What is the least squares line to predict body mass gain from percent daytime consumption? What gain is predicted for a mouse that eats 50% of its calories during the day ($DayPct = 50$)?

(d) What is the estimated slope for this regression model? Interpret the slope in context.

(e) What is the p-value for a test of the slope? What is the conclusion of the test, in context?

(f) What is the relationship between the p-value of the correlation test and the p-value of the slope test?

(g) What is R^2 for this linear model? Interpret it in context.

(h) Verify that the correlation squared gives the coefficient of determination R^2.

9.19 Social Networks and Brain Structure A recent study in Great Britain[4] examines the relationship between the number of friends an individual has on Facebook and grey matter density in the areas of the brain associated with social perception and associative memory. The data are available in the dataset **FacebookFriends** and the relevant variables are *GMdensity* (normalized z-scores of grey matter density in the relevant regions) and *FBfriends*

[4]Kanai, R., Bahrami, B., Roylance, R., and Rees, G., "Online social network size is reflected in human brain structure," *Proceedings of the Royal Society*, 2012; 279(1732): 1327–1334. Data approximated from information in the article.

(the number of friends on Facebook). The study included 40 students at City University London. A scatterplot of the data is shown in Figure 9.9 and computer output for both correlation and regression is shown below.

Pearson correlation of GMdensity and FBfriends = 0.436
P-Value = 0.005
– – – – – – – –
The regression equation is FBfriends = 367 + 82.4 GMdensity

Predictor	Coef	SE Coef	T	P
Constant	366.64	26.35	13.92	0.000
GMdensity	82.45	27.58	2.99	0.005

S = 165.716   R-Sq = 19.0%   R-Sq(adj) = 16.9%

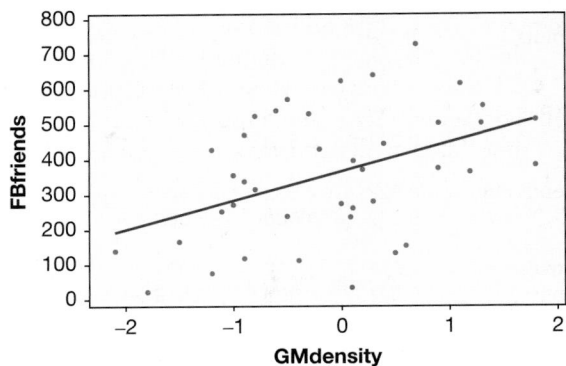

Figure 9.9 *Does brain density influence number of Facebook friends?*

(a) Use the scatterplot to determine whether any of the study participants had grey matter density scores more than two standard deviations from the mean. (*Hint:* The grey matter density scores used in the scatterplot are z-scores!) If so, in each case, indicate if the grey matter density score is above or below the mean and estimate the number of Facebook friends for the individual.

(b) Use the scatterplot to determine whether we should have any significant concerns about the conditions being met for using a linear model with these data.

(c) What is the correlation between these two variables? What is the p-value from a test of the correlation? What is the conclusion of the test, in context?

(d) What is the least squares line to predict the number of Facebook friends based on the normalized grey matter density score? What number of Facebook friends is predicted for a person whose normalized score is 0? Whose normalized score is +1? Whose normalized score is −1?

(e) What is the p-value for a test of the slope? Compare it to the p-value for the test of correlation.

(f) What is R^2 for this linear model? Interpret it in context.

9.20 Inference on the Slope of Facebook Friends and the Brain In Exercise 9.19, we give computer output for a regression line to predict the number of Facebook friends a student will have, based on a normalized score of the grey matter density in the areas of the brain associated with social perception and associative memory. Data for the sample of $n = 40$ students are stored in **FacebookFriends**.

(a) What is the slope in this regression analysis? What is the standard error for the slope?

(b) Use the information from part (a) to calculate the test statistic to test the slope to determine whether *GMdensity* is an effective predictor of *FBfriends*. Give the hypotheses for the test, find the p-value, and make a conclusion. Show your work. Verify the values of the test statistic and the p-value using the computer output in Exercise 9.19.

(c) Use the information from part (a) to find and interpret a 95% confidence interval for the slope.

9.21 Using pH in Lakes as a Predictor of Mercury in Fish The **FloridaLakes** dataset, introduced in Data 2.4, includes data on 53 lakes in Florida. Two of the variables recorded are *pH* (acidity of the lake water) and *AvgMercury* (average mercury level for a sample of fish from each lake). We wish to use the pH of the lake water (which is easy to measure) to predict average mercury levels in fish, which is harder to measure. A scatterplot of the data is shown in Figure 2.49(a) on page 109 and we see that the conditions for fitting a linear model are reasonably met. Computer output for the regression analysis is shown below.

The regression equation is AvgMercury = 1.53 - 0.152 pH

Predictor	Coef	SE Coef	T	P
Constant	1.5309	0.2035	7.52	0.000
pH	−0.15230	0.03031	−5.02	0.000

S = 0.281645   R-Sq = 33.1%   R-Sq(adj) = 31.8%

(a) Use the fitted model to predict the average mercury level in fish for a lake with a pH of 6.0.

(b) What is the slope in the model? Interpret the slope in context.

(c) What is the test statistic for a test of the slope? What is the p-value? What is the conclusion of the test, in context?

(d) Compute and interpret a 95% confidence interval for the slope.

(e) What is R^2? Interpret it in context.

9.22 Alkalinity in Lakes as a Predictor of Mercury in Fish The **FloridaLakes** dataset, introduced in Data 2.4, includes data on 53 lakes in Florida. Figure 9.10 shows a scatterplot of *Alkalinity* (concentration of calcium carbonate in mg/L) and *AvgMercury* (average mercury level for a sample of fish from each lake). Explain using the conditions for a linear model why we might hesitate to fit a linear model to these data to use *Alkalinity* to predict average mercury levels in fish.

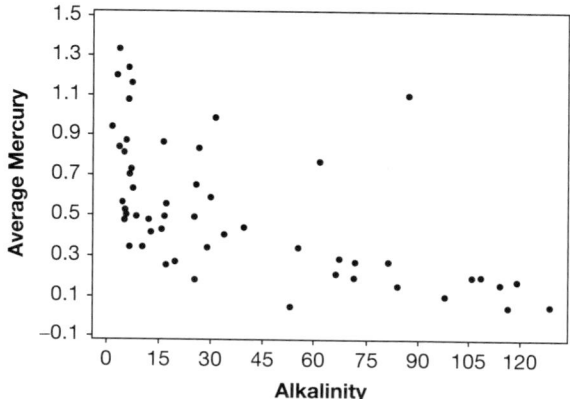

Figure 9.10 *Are the conditions met for fitting a linear model?*

9.23 Rain and Hantavirus in Mice Hantavirus is carried by wild rodents and causes severe lung disease in humans. A study[5] on the California Channel Islands found that increased prevalence of the virus was linked with greater precipitation. The study adds "Precipitation accounted for 79% of the variation in prevalence."

(a) What notation or terminology do we use for the value 79% in this context?

(b) What is the response variable? What is the explanatory variable?

(c) What is the correlation between the two variables?

9.24 NFL Pre-Season Teams in the National Football League (NFL) in the US play four pre-season games each year before the regular season starts. Do teams that do well in the pre-season tend to also do well in the regular season? We are interested in whether there is a positive linear association between the number of wins in the pre-season and

[5]"More Rain, More Virus," *Nature*, April 28, 2011, p. 392.

the number of wins in the regular season for teams in the NFL.

(a) What are the null and alternative hypotheses for this test?

(b) The correlation between these two variables for the 32 NFL teams over the 10 year period from 2005 to 2014 is 0.067. Use this sample (with $n = 320$) to calculate the appropriate test statistic and determine the p-value for the test.

(c) State the conclusion in context, using a 5% significance level.

(d) When an NFL team goes undefeated in the pre-season, should the fans expect lots of wins in the regular season?

9.25 Is the Honeybee Population Shrinking? The **Honeybee** dataset introduced in Exercise 2.218 on page 135 shows an estimated number of honeybee colonies in the United States for the years 1995 through 2012 (18 years). The correlation between year and number of colonies from these data is $r = -0.41$.

(a) Treating these as a sample of years do we have significant evidence that the number of honeybee colonies is linearly related to year? Give the t-statistic and the p-value, as well as a conclusion in context.

(b) What percent of the variability in number of honeybee colonies can be explained by year in these data?

9.26 Homes for Sale The dataset **HomesForSaleCA** contains a random sample of 30 houses for sale in California. We are interested in whether there is a positive association between the number of bathrooms and number of bedrooms in each house.

(a) What are the null and alternative hypotheses for testing the correlation?

(b) Find the correlation in the sample.

(c) Calculate (or use technology to find) the appropriate test statistic, and determine the p-value.

(d) State the conclusion in context.

9.27 Life Expectancy A random sample of 50 countries is stored in the dataset **SampCountries**. Two variables in the dataset are life expectancy (*LifeExpectancy*) and percentage of government expenditure spent on health care (*Health*) for each country. We are interested in whether or not the percent spent on health care can be used to effectively predict life expectancy.

(a) What are the cases in this model?

(b) Create a scatterplot with regression line and use it to determine whether we should have any

serious concerns about the conditions being met for using a linear model with these data.

(c) Run the simple linear regression, and report and interpret the slope.

(d) Find and interpret a 95% confidence interval for the slope.

(e) Is the percentage of government expenditure on health care a significant predictor of life expectancy?

(f) The population slope (for all countries) is 0.467. Is this captured in your 95% CI from part (d)?

(g) Find and interpret R^2 for this linear model.

9.28 NBA: Offense or Defense? A common (and hotly debated) saying among sports fans is "Defense wins championships." Is offensive scoring ability or defensive stinginess a better indicator of a team's success? To investigate this question we'll use data from the 2015–2016 National Basketball Association (NBA) regular season. The data[6] stored in **NBAStandings2016** include each team's record (wins, losses, and winning percentage) along with the average number of points the team scored per game (*PtsFor*) and average number of points scored against them (*PtsAgainst*).

(a) Examine scatterplots for predicting *WinPct* using *PtsFor* and predicting *WinPct* using *PtsAgainst*. In each case, discuss whether conditions for fitting a linear model appear to be met.

(b) Fit a model to predict winning percentage (*WinPct*) using offensive ability (*PtsFor*). Write down the prediction equation and comment on whether *PtsFor* is an effective predictor.

(c) Repeat the process of part (b) using *PtsAgainst* as the predictor.

(d) Compare and interpret R^2 for both models.

(e) The Golden State Warriors set an NBA record by winning 73 games in the regular season and only losing 9 (*WinPct* = 0.890). They scored an average of 114.9 points per game while giving up an average of 104.1 points against. Find the predicted winning percentage for the Warriors using each of the models in (b) and (c).

(f) Overall, does one of the predictors, *PtsFor* or *PtsAgainst*, appear to be more effective at explaining winning percentages for NBA teams? Give some justification for your answer.

9.29 Birth Rate and Life Expectancy Use the dataset **AllCountries** to examine the correlation between birth rate and life expectancy across countries of the world.

(a) Plot the data. Do birth rate and life expectancy appear to be linearly associated?

(b) From this dataset, can we conclude that the population correlation between birth rate and life expectancy is different from zero?

(c) Explain why inference is not necessary to answer part (b).

(d) For every percent increase in birth rate, how much does the predicted life expectancy of a country change?

(e) From this dataset, can we conclude that lowering the birth rate of a country will increase its life expectancy? Why or why not?

9.2 ANOVA FOR REGRESSION

In the previous section we see that the square of the correlation between two quantitative variables, R^2, can be interpreted as the amount of variability in one of the variables that is "explained" by the other variable. How do we go about actually measuring the amount of explained variability? The approach we use falls under the general heading of *analysis of variance*, or ANOVA. If you've already looked at Chapter 8, you have seen how the variability in one variable can be broken down to test for a difference in means among several groups. The approach in this section for regression is similar, with some changes in the computational details.

Partitioning Variability

The general form of a simple linear model for a single quantitative predictor is

$$Y = \beta_0 + \beta_1 X + \epsilon$$

or

$$\text{Response} = \text{Regression line} + \text{Error}$$

To assess how well the model does at explaining the response, we split the total variability in the response into two pieces: one that represents the variability explained by the model (the least squares line) and another that measures the variability that is left unexplained in the errors.

$$\text{TOTAL Variability in the Response} = \text{Variability Explained by the MODEL} + \text{Unexplained Variability in the ERROR}$$

Each of these portions of variability is measured with a sum of squared deviations:

$$SSTotal = SSModel + SSE$$

where

$$\text{Sum of squares explained by the model} = SSModel = \sum (\hat{y} - \bar{y})^2$$
$$\text{Sum of squared errors} = SSE = \sum (y - \hat{y})^2$$
$$\text{Total sum of squares} = SSTotal = \sum (y - \bar{y})^2$$

You might recognize SSE as the sum of squared residuals that is minimized when estimating the least squares line. This represents the errors that still occur when predicting with the fitted model. The $SSTotal$ is the sum of squared deviations from the mean of the responses that is used in computing the standard deviation of the sample responses. Think of $\bar{y}$ as a very crude model that uses a single value to predict all cases with no information from the predictor. The difference between these two, $SSModel = SSTotal - SSE$, is the amount of the original variability in Y that is successfully explained by the model.

Fortunately, statistical software usually displays the three sums of squares as part of the standard regression output, as we see in the next example.

Nutrition Facts		
Serving Size 1.0 cup		
Amount Per Serving		
Calories 108 Calories from Fat 11		
		% Daily Value *
Total Fat 1g		2%
Saturated Fat 0g		1%
Polyunsaturated Fat 0g		
Monounsaturated Fat 0g		
Cholesterol 0mg		0%
Sodium 201mg		8%
Total Carbohydrate 24g		8%
Dietary Fiber 3g		11%
Sugars 6g		
Protein 2g		

© Jennifer Fluharty/iStockphoto

What's in your cereal?

DATA 9.2 **Breakfast Cereals**

Labels on many food products contain a wealth of nutritional information. The data in **Cereal** include the number of calories as well as the grams of fat, carbohydrates, fiber, sugars, and protein and milligrams of sodium in each cup of a sample of 30 breakfast cereals from three different manufacturers (General Mills, Kellogg's, and Quaker).[7] ■

Example 9.10

Some regression output for a model to predict *Calories* in cereals based on the amount of *Sugars* is shown below. In the output, find the sum of squared deviations, "SS," explained by the regression model and the SS due to the error, and verify that they add up to the total sum of squares.

Predictor	Coef	SE Coef	T	P
Constant	88.92	10.81	8.22	0.000
Sugars	4.3103	0.9269	4.65	0.000

S = 26.6147   R-Sq = 43.6%   R-Sq(adj) = 41.6%

Analysis of Variance

Source	DF	SS	MS	F	P
Regression	1	15317	15317	21.62	0.000
Residual Error	28	19834	708		
Total	29	35150			

Solution

The sums of squares are found in the section of the output labeled "Analysis of Variance." This shows that

Sum of squares explained by the regression model is $SSModel = 15{,}317$

Sum of squared errors is $SSE = 19{,}834$

Total sum of squares is $SSTotal = 35{,}150$

---

[7]Cereal data obtained from nutrition labels at *http://www.nutritionresource.com/foodcomp2.cfm?id=0800.*

and we verify that (up to round-off)

$$SSModel + SSE = 15,317 + 19,834 = 35,150 = SSTotal$$

The analysis of variance portion of the regression output also has a value labeled "P" which, as you might suspect, is the p-value for a test. If $SSModel$ is large relative to SSE, the model has done a good job of explaining the variability in the response. In fact, if all of the data lie exactly on the least squares line, we would have $SSE = 0$ and all of the variability would be explained by the model. On the other hand, if $SSModel = 0$, every prediction is the same as the mean response and we get no useful information from the predictor. Of course, in practice, we rarely see either of these two extremes. The key question then becomes: Is the amount of variability explained by the model more than we would expect to see by random chance alone, when compared to the variability in the error term? That is the question that we test with ANOVA.

F-Statistic

To compare the variability explained by the model to the unexplained variability of the error we first need to adjust each by an appropriate degrees of freedom. These are shown in the ANOVA section of the output of Example 9.10 as 1 degree of freedom for the model and 28 degrees of freedom for the error. For the simple linear model the degrees of freedom for the model is always 1 (just one predictor) and $n - 2$ for the error, adding up to $n - 1$ degrees of freedom for the total.

We divide each sum of squares by its degrees of freedom to get a *mean square* for each source of variability:

$$\text{Mean square for the model} = MSModel = \frac{SSModel}{1}$$

$$\text{Mean square error} = MSE = \frac{SSE}{n - 2}$$

When the predictor is ineffective, these mean squares should be roughly the same magnitude. However, when the predictor is useful, the mean square model will tend to be large relative to the mean square error. To compare the two mean squares, we use their ratio to obtain the F-statistic:

$$F = \frac{MSModel}{MSE}$$

The formal hypotheses being tested are:

H_0 : The model is ineffective   (or, equivalently, $\beta_1 = 0$)

H_a : The model is effective   (or, equivalently, $\beta_1 \neq 0$)

How do we know when the F-statistic is large enough to provide evidence of something effective in the model? The appropriate reference distribution[8] for finding a p-value is the upper tail of an F-distribution with 1 degree of freedom for the numerator and $n - 2$ degrees of freedom for the denominator.

Example 9.11

Verify the calculations for the F-statistic and p-value for the ANOVA output in Example 9.10 on page 592.

---

[8]See more details about the F-distribution in Section 8.1 and Exercise 10.37.

Solution From the ANOVA portion of the regression output we see

$$MSModel = \frac{15{,}317}{1} = 15{,}317 \qquad MSE = \frac{19{,}834}{28} = 708.4$$

and the F-statistic is then found as

$$F = \frac{MSModel}{MSE} = \frac{15{,}317}{708.4} = 21.62$$

Using technology we find the area beyond $F = 21.62$ for an F-distribution with 1 and 28 degrees of freedom gives a p-value $= 0.00007$. This provides very strong evidence that there is some relationship between amount of sugar and calories in cereals and that *Sugars* is an effective predictor for *Calories*.

We summarize the details for using ANOVA to assess the effectiveness of a simple linear model in the box below.

ANOVA to Test a Simple Linear Model

To test for the effectiveness of a regression model, $Y = \beta_0 + \beta_1 X + \epsilon$:

H_0 : The model is ineffective   Equivalently, $H_0 : \beta_1 = 0$
H_a : The model is effective $H_a : \beta_1 \neq 0$

We partition the variability to construct an **ANOVA table** for regression:

Source	df	Sum of Sq.	Mean Square	F-statistic	p-value
Model	1	SSModel	$\dfrac{SSModel}{1}$	$F = \dfrac{MSModel}{MSE}$	$F_{1,n-2}$
Error	$n-2$	SSE	$\dfrac{SSE}{n-2}$		
Total	$n-1$	SSTotal			

Example 9.12

Some output for using the amount of *Sodium* to predict *Calories* for the data in **Cereal** is shown in Figure 9.11. Use it to test the effectiveness of the *Sodium* predictor two ways:

(a) Using a t-test for the slope

(b) Using an F-test based on the ANOVA

Solution (a) The information for a t-test of $H_0 : \beta_1 = 0$ vs $H_a : \beta_1 \neq 0$ is given in the line near the top of the output labeled *Sodium*. We see that the t-statistic is given as $t = 1.686$ and the p-value $= 0.1028$. Since this p-value is not small (even at a 10% significance level), we do not have sufficient evidence to conclude that the amount of sodium is an effective predictor of the number of calories in breakfast cereals.

(b) Using the ANOVA table to test whether the model is effective, we see that the F-statistic is $F = 2.844$ and the corresponding p-value (using 1 and 28 df)

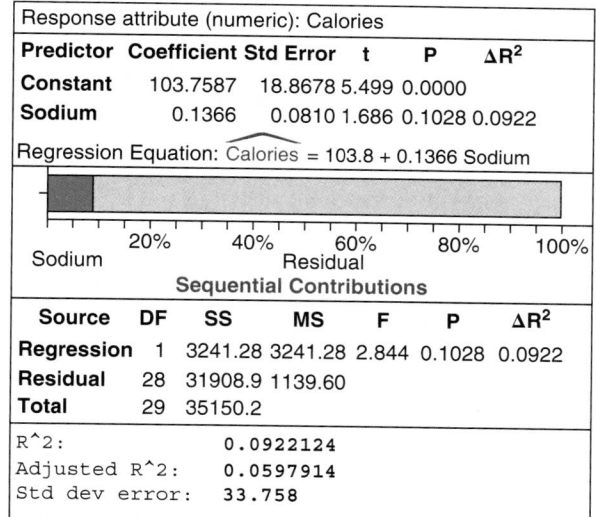

Figure 9.11 *Regression output for predicting Calories based on Sodium*

is 0.1028. The conclusion is the same as the t-test: We lack convincing evidence to show that *Sodium* is an effective predictor of *Calories* in breakfast cereals.

You have probably noticed that the p-values for both tests in Example 9.12 are identical. As with the t-tests for slope and correlation, this is no accident. For a single predictor model, a two-tail t-test for the slope and the ANOVA F-test will always give equivalent results[9] and identical p-values. Why do we need yet another test? One reason, shown in Section 10.1, is that the ANOVA generalizes nicely to test models with more than one predictor. Also, as the following examples illustrate, we can obtain additional information for assessing the regression model from the output shown in an ANOVA table.

Coefficient of Determination: R-squared

We started this section with a question about how to interpret R^2 as the amount of variability in the response variable that is explained by the model. We can now answer that question using values from the ANOVA table for the regression model.

> ### Coefficient of Determination, R^2
>
> For any regression model, the **coefficient of determination** is
>
> $$R^2 = \frac{SSModel}{SSTotal}$$
>
> and is interpreted as the percent of variability in the sample response values that is explained by the regression model.

Example 9.13

Use the information in the regression ANOVA output in Example 9.10 and Figure 9.11 to compute (and interpret) the R^2 values for using *Sugars* and *Sodium*, respectively, as predictors of *Calories* in breakfast cereals.

---

[9]In fact, you can check that the ANOVA F-statistic is always (up to round-off) the square of the t-statistic!

Solution Using the output for the model using *Sugars* in Example 9.10 on page 592, we see that "R-Sq = 43.6%." We can also use SS values from the ANOVA table to compute R^2 ourselves:

$$R^2 = \frac{SSModel}{SSTotal} = \frac{15,317}{35,150} = 0.436$$

The amount of sugars explains 43.6% of the variability in calories for these 30 breakfast cereals.

Using the output for the model using *Sodium* in Figure 9.11 on page 595, we see that "R^2 : 0.0922124", or we can use the SS values in the ANOVA table to compute it ourselves:

$$R^2 = \frac{SSModel}{SSTotal} = \frac{3241}{35,150} = 0.092$$

The amount of sodium explains 9.2% of the variability in calories for these 30 breakfast cereals.

Note that the denominator for computing R^2 is the same for both models in Example 9.13. The total amount of variability in the *Calories* response variable ($SSTotal = 35,150$) is the same regardless of what model we use to try to explain it. We also see that *Sugars* ($R^2 = 43.6\%$) is much more effective at explaining that variability than is *Sodium* ($R^2 = 9.2\%$). We often use R^2 in this way to quickly compare the effectiveness of competing models.

Computational Details

In the rest of this section we consider some additional quantities that appear in typical computer output for a regression model. In many situations we only need to know where to find the information in the output and how to interpret or use it. We include some extra computational details below for those who want to see a bit more about where the numbers come from.

Standard Deviation of the Error Term

For a simple linear model $Y = \beta_0 + \beta_1 X + \epsilon$, we assume that the errors are distributed as $\epsilon \sim N(0, \sigma_\epsilon)$. The *standard deviation of the errors*, denoted by σ_ϵ, is an important quantity to estimate because it measures how much individual data points tend to deviate above and below the regression line. A small σ_ϵ indicates that the points hug the line closely and we should expect fairly accurate predictions, while a large σ_ϵ suggests that, even if we estimate the line perfectly, we can expect individual values to deviate from it by substantial amounts. Fortunately, most regression computer output includes an estimate for the standard deviation of the error term, which we denote by s_ϵ.

Example 9.14

The standard deviation of the error term is given as S=26.6 in the computer output of the model to predict cereal calories based on sugars in Example 9.10 on page 592. Use the fact that about 95% of values from a normal distribution fall within two standard deviations of the center to interpret what the value $s_\epsilon = 26.6$ says about calories for cereals.

Solution For any particular amount of sugar, the calories should be distributed above and below the regression line with standard deviation estimated to be $s_\epsilon = 26.6$. Since $2s_\epsilon = 2(26.6) = 53.2$, we expect most (about 95%) of the calorie counts for cereals to be within about 53 calories of the regression line.

We can compute the estimate of the standard deviation of the error term for a least squares line more directly using the sum of squared errors from the ANOVA table.

Standard Deviation of the Error, s_ϵ

For a simple linear model, we estimate the standard deviation of the error term with

$$s_\epsilon = \sqrt{\frac{\sum(y - \hat{y})^2}{n - 2}} = \sqrt{\frac{SSE}{n - 2}} = \sqrt{MSE}$$

where SSE and MSE are obtained from the ANOVA table.

The computation of s_ϵ might remind you of the standard deviation of a sample defined on page 79. Note that in the simple linear model setting we divide by $n - 2$ (rather than $n - 1$) since we lose one degree of freedom for each of the two estimated parameters (intercept and slope). In most situations we use technology to handle the details of this computation.

Example 9.15

Use the information in the ANOVA table in Figure 9.11 (predicting *Calories* based on *Sodium*) to compute the estimate of the standard deviation of the error term for that model. Confirm your calculation by finding the estimate in the output.

Solution   From the ANOVA table in Figure 9.11 we see that $SSE = 31,908.9$ and the df for the error is $30 - 2 = 28$. The estimated standard deviation for the error term is

$$s_\epsilon = \sqrt{\frac{SSE}{n - 2}} = \sqrt{\frac{31908.9}{28}} = \sqrt{1139.6} = 33.76$$

This value is labeled in the output as `Std dev error: 33.758`. Note also that the ANOVA table shows $MSE = 1139.6$, which is s_ϵ^2 or the variance of the error term.

Standard Error of the Slope

Although we generally rely on technology to find the standard error for the slope, we can also find this quantity directly from other summary statistics. Here is one such formula:

$$SE = \frac{s_\epsilon}{s_x\sqrt{n - 1}}$$

where s_ϵ is the standard deviation of the error term (as found in the regression output or computed above) and s_x is the standard deviation for the sample values of the predictor.

Example 9.16

Use the formula for SE of the slope to verify the standard error value for the coefficient of *Sodium* in the computer output of Figure 9.11. Some summary statistics for the two variables (*Calories* and *Sodium*) are given below. Check that the computed standard error matches (up to round-off) the SE for the slope given in the output.

Variable	N	Mean	StDev
Sodium	30	220.2	77.4
Calories	30	133.8	34.8

Solution In the computer output (or Example 9.15) we see that the standard deviation of the error term in this model is estimated to be $s_\epsilon = 33.76$ and the summary statistics show that the standard deviation of the predictor (*Sodium*) is 77.4. Putting these together with the sample size ($n = 30$) gives the following calculation:

$$SE = \frac{s_\epsilon}{s_x\sqrt{n-1}} = \frac{33.76}{77.4\sqrt{30-1}} = 0.081$$

This matches the value of the standard error for the coefficient of *Sodium* shown in the output for this model in Figure 9.11.

 It is very easy to confuse the various types of standard deviation that occur in the regression setting. We have the *standard error of the slope* (*SE*), standard deviation of the *error* (s_ϵ), standard deviation of the *predictor* (s_x), and standard deviation of the *response* (s_y). Take care to pay attention to which type of standard deviation is called for in a particular setting.

SECTION LEARNING GOALS

You should now have the understanding and skills to:

• Use ANOVA to test the effectiveness of a simple linear model

• Find and interpret the value of R^2 for a simple linear model using values in the ANOVA table

• Find the standard deviation of the error term for a simple linear model

• Find the standard error of the slope for a simple linear model

Exercises for Section 9.2

SKILL BUILDER 1
In Exercises 9.30 to 9.33, we show an ANOVA table for regression. State the hypotheses of the test, give the F-statistic and the p-value, and state the conclusion of the test.

9.30 Analysis of Variance

Source	DF	SS	MS	F	P
Regression	1	303.7	303.7	1.75	0.187
Residual Error	174	30146.8	173.3		
Total	175	30450.5			

9.31 Analysis of Variance

Source	DF	SS	MS	F	P
Regression	1	3396.8	3396.8	21.85	0.000
Residual Error	174	27053.7	155.5		
Total	175	30450.5			

9.32 Response: Y

	Df	Sum Sq	Mean Sq	F value	Pr(>F)	
ModelA	1	352.97	352.97	11.01	0.001	**
Residuals	359	11511.22	32.06			
Total	360	11864.20				

9.33 Response: Y

	Df	Sum Sq	Mean Sq	F value	Pr(>F)
ModelB	1	10.380	10.380	2.18	0.141
Residuals	342	1630.570	4.768		
Total	343	1640.951			

SKILL BUILDER 2
In Exercises 9.34 to 9.37, we refer back to the ANOVA tables for regression given in Exercises 9.30 to 9.33. Use the information in the table to give the sample size and to calculate R^2.

9.34 The ANOVA table in Exercise 9.30.

9.35 The ANOVA table in Exercise 9.31.

9.36 The ANOVA table in Exercise 9.32.

9.37 The ANOVA table in Exercise 9.33.

SKILL BUILDER 3
In Exercises 9.38 to 9.41, we give some information about sums of squares and sample size for a linear

model. Use this information to fill in all values in an analysis of variance for regression table as shown.

Source	df	SS	MS	F-statistic	p-value
Model					
Error					
Total					

9.38 $SSModel = 250$ with $SSTotal = 3000$ and a sample size of $n = 100$.

9.39 $SSModel = 800$ with $SSTotal = 5820$ and a sample size of $n = 40$.

9.40 $SSModel = 8.5$ with $SSError = 247.2$ and a sample size of $n = 25$.

9.41 $SSError = 15,571$ with $SSTotal = 23,693$ and a sample size of $n = 500$.

9.42 Social Networks and Brain Structure Exercise 9.19 on page 588 introduces a study examining the relationship between the number of friends an individual has on Facebook and grey matter density in the areas of the brain associated with social perception and associative memory. The data are available in the dataset **FacebookFriends** and the relevant variables are *GMdensity* (normalized z-scores of grey matter density in the brain) and *FBfriends* (the number of friends on Facebook). The study included 40 students at City University London. Computer output for ANOVA for regression to predict the number of Facebook friends from the normalized brain density score is shown below.

The regression equation is FBfriends = 367 + 82.4 GMdensity
Analysis of Variance

Source	DF	SS	MS	F	P
Regression	1	245400	245400	8.94	0.005
Residual Error	38	1043545	27462		
Total	39	1288946			

Is the linear model effective at predicting the number of Facebook friends? Give the F-statistic from the ANOVA table, the p-value, and state the conclusion in context. (We see in Exercise 9.19 that the conditions are met for fitting a linear model in this situation.)

9.43 Fiber in Cereal In Data 9.2 on page 592, we introduce the dataset **Cereal**, which has nutrition information on 30 breakfast cereals. Computer output is shown for a linear model to predict *Calories* in one cup of cereal based on the number of grams of *Fiber*. Is the linear model effective at predicting the number of calories in a cup of cereal? Give the F-statistic from the ANOVA table, the p-value, and state the conclusion in context.

The regression equation is Calories = 119 + 8.48 Fiber
Analysis of Variance

Source	DF	SS	MS	F	P
Regression	1	7376.1	7376.1	7.44	0.011
Residual Error	28	27774.1	991.9		
Total	29	35150.2			

9.44 Predicting Prices of Printers Data 9.1 on page 577 introduces the dataset **InkjetPrinters**, which includes information on all-in-one printers. Two of the variables are *Price* (the price of the printer in dollars) and *CostColor* (average cost per page in cents for printing in color). Computer output for predicting the price from the cost of printing in color is shown:

The regression equation is Price = 378 - 18.6 CostColor
Analysis of Variance

Source	DF	SS	MS	F	P
Regression	1	57604	57604	13.19	0.002
Residual Error	18	78633	4369		
Total	19	136237			

(a) What is the predicted price of a printer that costs 10 cents a page for color printing?

(b) According to the model, does it tend to cost more or less (per page) to do color printing on a cheaper printer?

(c) Use the information in the ANOVA table to determine the number of printers included in the dataset.

(d) Use the information in the ANOVA table to compute and interpret R^2.

(e) Is the linear model effective at predicting the price of a printer? Use information from the computer output and state the conclusion in context.

9.45 ANOVA for Verbal SAT as a Predictor of GPA How well does a student's Verbal SAT score (on an 800-point scale) predict future college grade point average (on a four-point scale)? Computer output for this regression analysis is shown, using the data in **StudentSurvey**:

The regression equation is GPA = 2.03 + 0.00189 VerbalSAT
Analysis of Variance

Source	DF	SS	MS	F	P
Regression	1	6.8029	6.8029	48.84	0.000
Residual Error	343	47.7760	0.1393		
Total	344	54.5788			

(a) What is the predicted grade point average of a student who receives a 550 on the Verbal SAT exam?

(b) Use the information in the ANOVA table to determine the number of students included in the dataset.

(c) Use the information in the ANOVA table to compute and interpret R^2.

(d) Is the linear model effective at predicting grade point average? Use information from the computer output and state the conclusion in context.

9.46 Football and Cognition Exercise 2.143 on page 102 introduces a study examining years playing football, brain size, and percentile score on a cognitive skills test. We show computer output below for a model to predict *Cognition* score based on *Years* playing football. (The scatterplot given in Exercise 2.143 allows us to proceed without serious concerns about the conditions.)

```
– – – – – – – – – – – –
Pearson correlation of Years and Cognition = -0.366
P-Value = 0.015
– – – – – – – – – – – –
Regression Equation
Cognition = 102.3 -3.34 Years
```

Coefficients

Term	Coef	SE Coef	T-Value	P-Value
Constant	102.3	15.6	6.56	0.000
Years	-3.34	1.31	-2.55	0.015

S	R-sq	R-sq(adj)	R-sq(pred)
25.4993	13.39%	11.33%	5.75%

Analysis of Variance

Source	DF	Adj SS	Adj MS	F-Value	P-Value
Regression	1	4223	4223.2	6.50	0.015
Error	42	27309	650.2		
Total	43	31532			

```
– – – – – – – – – – – –
```

(a) What is the correlation between these two variables? What is the p-value for testing the correlation?

(b) What is the slope of the regression line to predict cognition score based on years playing football? What is the t-statistic for testing the slope? What is the p-value for the test?

(c) The ANOVA table is given for testing the effectiveness of this model. What is the F-statistic for the test? What is the p-value?

(d) What do you notice about the three p-values for the three tests in parts (a), (b), and (c)?

(e) In every case, at a 5% level, what is the conclusion of the test in terms of football and cognition?

9.47 Mating Activity of Water Striders In Exercise A.97 on page 189, we introduce a study about mating activity of water striders. The dataset is available as **WaterStriders** and includes the variables *FemalesHiding*, which gives the proportion of time the female water striders were in hiding, and *MatingActivity*, which is a measure of mean mating activity with higher numbers meaning more mating. The study included 10 groups of water striders. (The study also included an examination of the effect of hyper-aggressive males and concludes that if a male wants mating success, he should not hang out with hyper-aggressive males.) Computer output for a model to predict mating activity based on the proportion of time females are in hiding is shown below, and a scatterplot of the data with the least squares line is shown in Figure 9.12.

```
The regression equation is
MatingActivity = 0.480 - 0.323 FemalesHiding
```

Predictor	Coef	SE Coef	T	P
Constant	0.48014	0.04213	11.40	0.000
FemalesHiding	-0.3232	0.1260	-2.56	0.033

S = 0.101312   R-Sq = 45.1%   R-Sq(adj) = 38.3%

Analysis of Variance

Source	DF	SS	MS	F	P
Regression	1	0.06749	0.06749	6.58	0.033
Residual Error	8	0.08211	0.01026		
Total	9	0.14960			

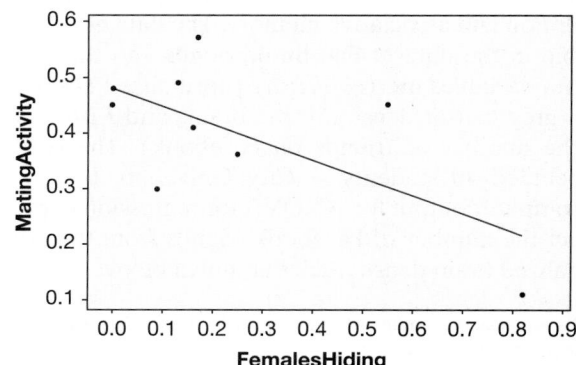

Figure 9.12 *When females hide, mating is hard*

(a) While it is hard to tell with only $n = 10$ data points, determine whether we should have any serious concerns about the conditions for fitting a linear model to these data.

(b) Write down the equation of the least squares line and use it to predict the mating activity of water striders in a group in which females spend 50% of the time in hiding (*FemalesHiding* = 0.50).

(c) Give the hypotheses, t-statistic, p-value, and conclusion of the t-test of the slope to determine whether time in hiding is an effective predictor of mating activity.

(d) Give the hypotheses, F-statistic, p-value, and conclusion of the ANOVA test to determine whether the regression model is effective at predicting mating activity.

(e) How do the two p-values from parts (c) and (d) compare?

(f) Interpret R^2 for this model.

9.48 Points and Penalty Minutes in Hockey The dataset **OttawaSenators** contains information on the number of points and the number of penalty minutes for 24 Ottawa Senators NHL hockey players. Computer output is shown for predicting the number of points from the number of penalty minutes:

The regression equation is Points = 29.53 − 0.113 PenMins

```
Predictor      Coef   SE Coef      T      P
Constant      29.53      7.06   4.18  0.000
PenMins      −0.113     0.163  −0.70  0.494

S = 21.2985    R-Sq = 2.15%    R-Sq(adj) = 0.00%

Analysis of Variance
Source          DF       SS      MS     F      P
Regression       1    219.5   219.5  0.48  0.494
Residual Error  22   9979.8   453.6
Total           23  10199.3
```

(a) Write down the equation of the least squares line and use it to predict the number of points for a player with 20 penalty minutes and for a player with 150 penalty minutes.

(b) Interpret the slope of the regression equation in context.

(c) Give the hypotheses, t-statistic, p-value, and conclusion of the t-test of the slope to determine whether penalty minutes is an effective predictor of number of points.

(d) Give the hypotheses, F-statistic, p-value, and conclusion of the ANOVA test to determine whether the regression model is effective at predicting number of points.

(e) How do the two p-values from parts (c) and (d) compare?

(f) Interpret R^2 for this model.

9.49 More Computation on Points and Penalty Minutes in Hockey Exercise 9.48 gives output for a regression model to predict number of points for a hockey player based on the number of penalty minutes for the hockey player. Use this output, together with any helpful summary statistics from Table 9.2, to show how to calculate the regression quantities given in parts (a) and (b) below. Verify your results by finding the equivalent results in the output.

(a) The standard deviation of the error term, s_ϵ.

(b) The standard error of the slope, SE.

Table 9.2 *Points and penalty minutes for hockey players*

	Sample Size	Mean	Std.Dev.
Points	24	25.67	21.06
PenMin	24	34.13	27.26

9.50 Computations Based on ANOVA for Predicting Mercury in Fish In Exercise 9.21, we see that the conditions are met for using the pH of a lake in Florida to predict the mercury level of fish in the lake. The data are given in **FloridaLakes**. Computer output is shown for the linear model with several values missing:

The regression equation is AvgMercury = 1.53 − 0.152 pH

```
Predictor      Coef   SE Coef      T      P
Constant     1.5309    0.2035   7.52  0.000
pH          −0.15230  **(c)**  −5.02  0.000

S = **(b)**    R-Sq = **(a)**

Analysis of Variance
Source          DF      SS      MS      F      P
Regression       1  2.0024  2.0024  25.24  0.000
Residual Error  51  4.0455  0.0793
Total           52  6.0479
```

(a) Use the information in the ANOVA table to compute and interpret the value of R^2.

(b) Show how to estimate the standard deviation of the error term, s_ϵ.

(c) Use the result from part (b) and the summary statistics below to compute the standard error of the slope, SE, for this model:

```
Variable     N   Mean  StDev  Minimum  Maximum
pH          53  6.591  1.288    3.600    9.100
AvgMercury  53 0.5272 0.3410   0.0400   1.3300
```

9.51 More Computation on Fiber in Cereal Exercise 9.43 gives output for a regression model to predict calories in a serving of breakfast cereal based on the number of grams of fiber in the serving. Use this output, together with any helpful summary statistics from Table 9.3, to show how to calculate the following regression quantities.

(a) The standard deviation of the error term, s_ϵ.

(b) The standard error of the slope, SE.

Table 9.3 *Calories and fiber in cereal*

	Sample Size	Mean	Std.Dev.
Calories	30	133.83	34.812
Fiber	30	1.797	1.880

9.52 More Computation on Predicting GPA
Exercise 9.45 gives output for a regression model
to predict grade point average in college based on
the score on the Verbal SAT exam. Use this output, together with any helpful summary statistics
from Table 9.4, to calculate the following regression
quantities:

(a) The standard deviation of the error term, s_ϵ.

(b) The standard error of the slope, SE.

Table 9.4 *Grade point average and Verbal
SAT score*

	Sample Size	Mean	Std.Dev.
GPA	345	3.1579	0.3983
VerbalSAT	345	594.0	74.29

9.53 Time Spent Exercising or Watching TV? Consider a simple linear model for the number of hours
of exercise students get (per week) based on the
number of hours spent watching TV. Use the data
in **ExerciseHours** to fit this model. Test the effectiveness of the *TV* predictor three different ways
as requested below, giving hypotheses, test statistic,
p-value, and a conclusion for each test.

(a) Use a t-test for the coefficient of *TV*.

(b) Use an ANOVA to test the effectiveness of the
model.

(c) Use a t-test for the correlation between *TV* and
Exercise.

(d) Compare the results for these three tests.

9.54 Life Expectancy and Health Expenditures In
Exercise 9.27 we see that the conditions are met
for fitting a linear model to predict life expectancy
(*LifeExpectancy*) from the percentage of government expenditure spent on health care (*Health*)
using the data in **SampCountries**. Use technology
to examine this relationship further, as requested
below.

(a) Find the correlation between the two variables
and give the p-value for a test of the correlation.

(b) Find the regression line and give the t-statistic
and p-value for testing the slope of the regression line.

(c) Find the F-statistic and the p-value from an
ANOVA test for the effectiveness of the model.

(d) Comment on the effectiveness of this model.

9.55 Homes for Sale in California The dataset
HomesForSaleCA contains a random sample of 30

houses for sale in California. We are interested in
whether we can use number of bathrooms *Baths*
to predict number of bedrooms *Beds* in houses in
California. Use technology to answer the following
questions:

(a) What is the fitted regression equation? Use the
regression equation to predict the number of
bedrooms in a house with three bathrooms.

(b) Give the t-statistic and the p-value for the t-test
for slope in the regression equation. State the
conclusion of the test.

(c) Give the F-statistic and the p-value from an
ANOVA for regression for this model. State the
conclusion of the test.

(d) Give and interpret R^2 for this model.

**9.56 Which Variable Is Best in Homes for Sale
in California** Consider the data described in
Exercise 9.55 on homes for sale in California and
suppose that we are interested in predicting the *Size*
(in thousands of square feet) for such homes.

(a) What is the total variability in the sizes of the
30 homes in this sample? (*Hint:* Try a regression ANOVA with any of the other variables as a
predictor.)

(b) Which other variable in the **HomesForSaleCA**
dataset explains the greatest amount of the
total variability in home sizes? Explain how you
decide on the variable.

(c) How much of the total variability in home sizes
is explained by the "best" variable identified in
part (b)? Give the answer both as a raw number
and as a percentage.

(d) Which of the variables in the dataset is the *weakest* predictor of home sizes? How much of the
variability does it explain?

(e) Is the weakest predictor identified in part (d)
still an effective predictor of home sizes?
Include some justification for your answer.

**9.57 Explore a Relationship between Two Quan-
titative Variables** Select any two quantitative
variables in any dataset used thus far in the text,
avoiding those analyzed so far in Chapter 9.

(a) Identify the dataset and variables you are using.
Indicate which variable you will use as the
response variable and which as the explanatory
variable.

(b) Create a scatterplot and describe it. Are the
conditions for fitting a linear model reasonably
met?

(c) Find the correlation between the two variables and give the p-value for a test of the correlation.

(d) Find the regression line and give the t-statistic and p-value for testing the slope of the regression line.

(e) Find the F-statistic and the p-value from an ANOVA test for the effectiveness of the model.

(f) Find and interpret R^2.

(g) Comment on the effectiveness of this model.

9.3 CONFIDENCE AND PREDICTION INTERVALS

One of the common purposes for fitting a simple linear model is to make predictions about the value of the response variable for a given value of the explanatory variable. For example:

- What size tip should a waitress at the First Crush bistro expect when she hands a table a bill for $30?
- How much should we expect to pay for an inkjet printer that prints about 3.0 pages per minute?
- If a cereal has 10 grams of sugar, how many calories should we expect?

Interpreting Confidence and Prediction Intervals

Once we fit the linear model, we can easily use the least squares line to make a prediction given a specific value of the explanatory variable, but how accurate is that prediction? We know that there is some uncertainty in the estimated coefficients of the model and we can also expect additional random error when trying to predict an individual point. As with estimates for other population parameters, we often want to produce an interval estimate that has some predetermined chance of capturing the quantity of interest. In a regression setting we have two common types of intervals for the response variable.

> **Regression Intervals for a Response Variable**
>
> For a specific value, x^*, of the explanatory variable:
>
> A **confidence interval for the mean response** is an interval which has a given chance of capturing the mean response for all elements of the population where the predictor value is x^*.
>
> A **prediction interval for an individual response** is an interval which has a given chance of capturing the response value for a specific case in the population where the predictor value is x^*.
>
> The conditions for these intervals are the same as we use for the simple linear model.

Although the predicted value, $\hat{y} = b_0 + b_1 \cdot x^*$, is the same for both types of intervals, they have quite different purposes and interpretations. The confidence interval for a mean response is much like other confidence intervals we have encountered for parameters such as the mean or proportion in a population. The only difference is that we are limiting the "population" to only cases that have the specific value of the predictor.

On the other hand, the prediction interval is trying to capture most of the response values in the population for that particular value of the predictor. This usually requires a much wider interval, since it's trying to contain most of the population, rather than just the mean.

Example 9.17

Using the data in **RestaurantTips** we find a least squares line for predicting the size of the tip based on the amount of the bill to be

$$\widehat{Tip} = -0.292 + 0.182 \cdot Bill$$

For a bill of $30, the predicted tip is $\widehat{Tip} = -0.292 + 0.182 \cdot 30 = 5.17$. Using software we obtain the two intervals below using this model, 95% confidence, and a bill of $30. Write a sentence interpreting each interval in the context of this problem.

(a) Confidence interval for mean tip = (4.99, 5.35)

(b) Prediction interval for tip = (3.23, 7.12)

Solution

(a) From the confidence interval, we are 95% sure that the mean tip amount for all bills of $30 at this restaurant is somewhere between $4.99 and $5.35.

(b) From the prediction interval, we are 95% sure that the tip for a specific bill of $30 at this restaurant will be between $3.23 and $7.12. This is the more appropriate interval for answering the question posed at the start of this section.

Note that the predicted tip, $\widehat{Tip} = 5.17$ when the bill is $30, lies at the center of both of the intervals in Example 9.17. The confidence interval for the mean tip has a much smaller margin of error ($0.18) than the prediction interval for individual tips (margin of error = $1.95).

In practice, we generally rely on statistical software to handle the details of computing either type of interval in a regression setting. However, we need to take care that we choose the proper type of interval to answer a specific question.

Example 9.18

Suppose we find an inkjet printer with a printing speed of 3.0 pages per minute for sale at a price of $129. Is this an unusually good deal? Use a regression model and the data in **InkjetPrinters** to address this question.

Solution

Using technology we fit this model and obtain both types of regression intervals for the printer prices when $PPM = 3.0$.

The regression equation is Price = −94.2 + 90.9 PPM

Predictor	Coef	SE Coef	T	P
Constant	−94.22	56.40	−1.67	0.112
PPM	90.88	19.49	4.66	0.000

S = 58.5457   R-Sq = 54.7%   R-Sq(adj) = 52.2%

Predicted Values for New Observations

New Obs	Fit	SE Fit	95% CI	95% PI	PPM
1	178.4	13.6	(149.9, 206.9)	(52.1, 304.7)	3.00

The question asks about the price for a specific printer, so we should use the prediction interval (labeled 95% PI). This shows that roughly 95% of printers that print 3.0 pages per minute are priced somewhere between $52 and $305. The $129 price in the example is on the low side of this interval but well within these bounds, so is not an unusually low price for a printer with this printing speed. (The sale price of $129 is, however, below the mean price for a printer with this printing rate, based on the 95% confidence interval for the mean of $149.9 to $206.9.)

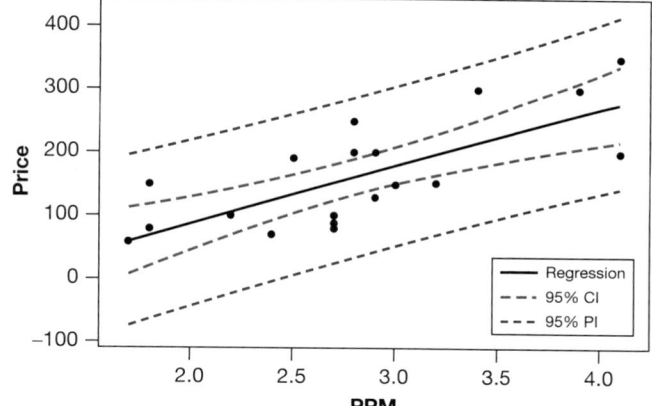

Figure 9.13 *Confidence and prediction intervals for inkjet printer prices*

Figure 9.13 shows a scatterplot of *Price* vs *PPM* with the least squares line and bounds for both the confidence interval for the mean price and prediction interval for individual prices computed for every value of the predictor. For example, you can roughly locate the intervals given in Example 9.18 above *PPM* = 3.0 in the plot.

Note that most of the data values lie outside of the 95% CI bounds. That is not unusual. The 95% confidence intervals for the mean responses are trying to capture the "true" line for the population, not the individual cases. The 95% PI bounds easily capture all 20 of these printer prices. While this is not always the case, we should not expect to see many (perhaps about 5%) of the data cases outside of the prediction bounds.

We also see that the confidence bands (and to a lesser extent the prediction intervals) are narrower near the middle of the plot and wider at the extremes. This is also typical of regression intervals since we have more uncertainty when predicting the response for more unusual predictor values. Visualize a small change in the slope of the line that still goes through the middle of the plot. This gives more substantial changes to the predictions at either extreme.

Computational Details

Although we usually rely on technology to compute confidence and prediction intervals for a regression response, the formulas for the margin of error can be instructive and are given in the box that follows.

Note that the formulas for computing both types of intervals are very similar. The only difference is an extra addition of "1" within the square root for the prediction interval. After multiplying by t^* and s_ϵ, this accounts for the extra variability in individual cases caused by the application of the error term, ϵ, in the simple linear model. Even if we could estimate the regression line perfectly, we would still have uncertainty in predicting individual values due to variability in the error term.

We also see in the $(x^* - \bar{x})^2$ term how the margin of error increases (and the intervals become wider) as the predicted value moves farther from the center. This phenomenon is visible in the confidence bands of Figure 9.13.

Formulas for Confidence and Prediction Intervals in Regression

For a specific value, x^*, of a predictor, the predicted response using a least squares line is

$$\hat{y} = b_0 + b_1 \cdot x^*$$

A confidence interval for the mean response when the predictor is x^* is

$$\hat{y} \pm t^* s_\epsilon \sqrt{\frac{1}{n} + \frac{(x^* - \overline{x})^2}{(n-1)s_x^2}}$$

and the prediction interval for an individual response when the predictor is x^* is

$$\hat{y} \pm t^* s_\epsilon \sqrt{1 + \frac{1}{n} + \frac{(x^* - \overline{x})^2}{(n-1)s_x^2}}$$

where $\overline{x}$ and s_x are the mean and standard deviation, respectively, of the predictor values in the sample. The value s_ϵ is the standard deviation of the error term and t^* comes from a t-distribution with $n-2$ degrees of freedom.

Example 9.19

Use the formulas and data in **InkjetPrinters** to verify the confidence interval for mean price and prediction interval for individual price that are shown for $PPM = 3.0$ in the computer output of Example 9.18.

Solution ▶

From the computer output we find the fitted least squares line and substitute $PPM = 3.0$ to get a predicted price:

$$\widehat{Price} = -94.22 + 90.88 \cdot 3.0 = 178.42$$

We also need the mean of the printer PPM values, $\overline{x} = 2.815$, and standard deviation, $s_x = 0.689$, which we find from the original data. The regression output shows that the standard deviation of the error term is $s_\epsilon = 58.55$. Finally, for a t-distribution with $20 - 2 = 18$ degrees of freedom and 95% confidence, we have $t^* = 2.101$.

To compute the confidence interval for mean price when $PPM = 3.0$, we use

$$178.42 \pm 2.101 \cdot 58.55 \sqrt{\frac{1}{20} + \frac{(3.0 - 2.815)^2}{(20-1)0.689^2}} = 178.42 \pm 28.53 = (149.89, 206.95)$$

To compute the prediction interval for individual price when $PPM = 3.0$, we use

$$178.42 \pm 2.101 \cdot 58.55 \sqrt{1 + \frac{1}{20} + \frac{(3.0 - 2.815)^2}{(20-1)0.689^2}} = 178.42 \pm 126.28 = (52.14, 304.70)$$

The predicted price and both intervals match the computer output.

SECTION LEARNING GOALS

You should now have the understanding and skills to:

▶ • Find and interpret a confidence interval for the mean response for a specific value of the predictor in a regression model

▶ • Find and interpret a prediction interval for the individual responses for a specific value of the predictor in a regression model

Exercises for Section **9.3**

SKILL BUILDER 1

In Exercises 9.58 to 9.61, two intervals are given, A and B, for the same value of the explanatory variable. In each case:

(a) Which interval is the confidence interval for the mean response? Which interval is the prediction interval for the response?

(b) What is the predicted value of the response variable for this value of the explanatory variable?

9.58 A: 10 to 14;   B: 4 to 20.

9.59 A: 94 to 106;   B: 75 to 125.

9.60 A: 2.9 to 7.1;   B: 4.7 to 5.3.

9.61 A: 16.8 to 23.2;   B: 19.2 to 20.8.

WHEN CALORIES ARE CONSUMED AND WEIGHT GAIN IN MICE In Exercise 9.18 on page 587, we look at a model to predict weight gain (in grams) in mice based on the percent of calories the mice eat during the day (when mice should be sleeping instead of eating). In Exercises 9.62 and 9.63, we give computer output with two regression intervals and information about the percent of calories eaten during the day. Interpret each of the intervals in the context of this data situation.

(a) The 95% confidence interval for the mean response

(b) The 95% prediction interval for the response

9.62 The intervals given are for mice that eat 50% of their calories during the day:

DayPct	Fit	SE Fit	95% CI	95% PI
50.0	7.476	0.457	(6.535, 8.417)	(2.786, 12.166)

9.63 The intervals given are for mice that eat 10% of their calories during the day:

DayPct	Fit	SE Fit	95% CI	95% PI
10.0	2.385	1.164	(−0.013, 4.783)	(−2.797, 7.568)

FIBER IN CEREALS AS A PREDICTOR OF CALORIES In Example 9.10 on page 592, we look at a model to predict the number of calories in a cup of breakfast cereal using the number of grams of sugars. In Exercises 9.64 and 9.65, we give computer output with two regression intervals and information about a specific amount of sugar. Interpret each of the intervals in the context of this data situation.

(a) The 95% confidence interval for the mean response

(b) The 95% prediction interval for the response

9.64 The intervals given are for cereals with 10 grams of sugars:

Sugars	Fit	SE Fit	95% CI	95% PI
10	132.02	4.87	(122.04, 142.01)	(76.60, 187.45)

9.65 The intervals given are for cereals with 16 grams of sugars:

Sugars	Fit	SE Fit	95% CI	95% PI
16	157.88	7.10	(143.35, 172.42)	(101.46, 214.31)

FOOTBALL AND COGNITION

Exercises 9.66 and 9.67 refer to the regression line (given in Exercise 9.46):

$$\widehat{Cognition} = 102.3 - 3.34 \cdot Years$$

using years playing football to predict the score on a cognition test. In each exercise,

(a) Find the predicted cognition score for that case.

(b) Two intervals are shown: one is a 95% confidence interval for the mean response and the other is a 95% prediction interval for the response. Which is which?

9.66 A person who has played football for 8 years. Interval I: (22.7, 128.5)   Interval II: (63.4, 87.8)

9.67 A person who has played football for 12 years. Interval I: (10.2, 114.3)   Interval II: (54.4, 70.1)

9.68 Housing Prices People in real estate are interested in predicting the price of a house by the square footage, and predictions will vary based on geographic area. We look at predicting prices (in $1000s) of houses in New York state based on the size (in thousands of square feet). A random sample of 30 houses for sale in New York state is given in the dataset **HomesForSaleNY**. Use technology and this dataset to answer the following questions:

(a) Is square footage an effective predictor of price for houses in New York?

(b) Find a point estimate for the price of a 2000-square-foot New York home.

(c) Find and interpret a 90% confidence interval for the average price of all 2000-square-foot New York homes.

(d) Find and interpret a 90% prediction interval for the price of a specific 2000-square-foot New York home.

9.69 Life Expectancy In Exercise 9.27 on page 607, we consider a regression equation to predict life expectancy from percent of government expenditure on health care, using data for a sample of 50 countries in **SampCountries**. Using technology and this dataset, find and interpret a 95% prediction interval for each of the following situations:

(a) A country which puts only 3% of its expenditure into health care.

(b) A country which puts 10% of its expenditure into health care.

(c) A country which puts 50% of its expenditure into health care.

(d) Calculate the widths of the intervals from (a), (b), and (c). What do you notice about these widths? (Note that for this sample, government expenditures on health care go from a minimum of 4.0% to a maximum of 20.89%, with a mean of 12.31%.)

9.70 Predicting GPA from Verbal SAT score In Exercise 9.17 on page 587, we use the information in **StudentSurvey** to fit a linear model to use Verbal SAT score to predict a student's grade point average in college. The regression equation is
$\widehat{GPA} = 2.03 + 0.00189 \cdot VerbalSAT$.

(a) What GPA does the model predict for a student who gets a 500 on the Verbal SAT exam? What GPA is predicted for a student who gets a 700?

(b) Use technology and the **StudentSurvey** dataset to find and interpret:

  i. A 95% confidence interval for the mean GPA of students who get a 500 Verbal SAT score.

  ii. A 95% prediction interval for the GPA of students who get a 500 Verbal SAT score.

  iii. A 95% confidence interval for the mean GPA of students who get a 700 Verbal SAT score.

  iv. A 95% prediction interval for the GPA of students who get a 700 Verbal SAT score.

9.71 Predicting Re-Election Margin Data 2.9 on page 106 introduces data on the approval rating of an incumbent US president and the margin of victory or defeat in the subsequent election (where negative numbers indicate the margin by which the incumbent president lost the re-election campaign). The data are reproduced in Table 9.5 and are available in **ElectionMargin**.

Computer output for summary statistics for the two variables and for a regression model to predict the margin of victory or defeat from the approval rating is shown:

```
Variable    N     Mean    StDev
Approval    12    52.92   11.04
Margin      12    7.62    10.72
```

The regression equation is Margin = −36.76 + 0.839 Approval

```
Term        Coef    SE Coef   T-Value   P-Value
Constant    −36.76  8.34      −4.41     0.001
Approval    0.839   0.155     5.43      0.000
```

S = 5.66054   R-Sq = 74.64%   R-Sq(adj) = 72.10%

Analysis of Variance

```
Source          DF    SS       MS        F       P
Regression      1     943.0    943.04    29.43   0.000
Residual Error  10    320.4    32.04
Total           11    1263.5
```

Use values from this output to calculate and interpret the following. Show your work.

(a) A 95% confidence interval for the mean margin of victory for all presidents with an approval rating of 50%.

(b) A 95% prediction interval for the margin of victory for a president with an approval rating of 50%.

(c) A 95% confidence interval for the mean margin of victory if we have no information about the approval rating. (*Hint:* This is just an ordinary confidence interval for a mean based only on the single sample of *Margin* values.)

Table 9.5 *Presidential approval rating and margin of victory or defeat*

Approval	62	50	70	67	57	48	31	57	39	55	49	50
Margin	10.0	4.5	15.4	22.6	23.2	−2.1	−9.7	18.2	−5.5	8.5	2.4	3.9

9.72 Golf Scores In a professional golf tournament the players participate in four rounds of golf and the player with the lowest score after all four rounds is the champion. How well does a player's performance in the first round of the tournament predict the final score? Table 9.6 shows the first round score and final score for a random sample of 20 golfers who made the cut in a recent Masters tournament. The data are also stored in **MastersGolf**.

Computer output for a regression model to predict the final score from the first-round score is shown. Use values from this output to calculate and interpret the following. Show your work.

(a) Find a 95% interval to predict the average final score of all golfers who shoot a 0 on the first round at the Masters.

(b) Find a 95% interval to predict the final score of a golfer who shoots a −5 in the first round at the Masters.

(c) Find a 95% interval to predict the average final score of all golfers who shoot a +3 in the first round at the Masters.

Variable	N	Mean	StDev
First	20	−0.550	3.154
Final	20	−0.65	5.82

————————

The regression equation is Final = 0.162 + 1.48 First

Predictor	Coef	SE Coef	T	P
Constant	0.1617	0.8173	0.20	0.845
First	1.4758	0.2618	5.64	0.000

S = 3.59805   R-Sq = 63.8%   R-Sq(adj) = 61.8%

Analysis of Variance

Source	DF	SS	MS	F	P
Regression	1	411.52	411.52	31.79	0.000
Residual Error	18	233.03	12.95		
Total	19	644.55			

Table 9.6 *Golf scores after the first and final rounds of the Masters*

First	−4	−4	−1	−5	−4	−4	0	−1	−3	−1	0	1	−1	0	−1	4	3	4	−1	7
Final	−12	−8	−7	−7	−5	−5	−3	−2	−2	−1	−1	−1	0	3	4	4	6	6	8	10

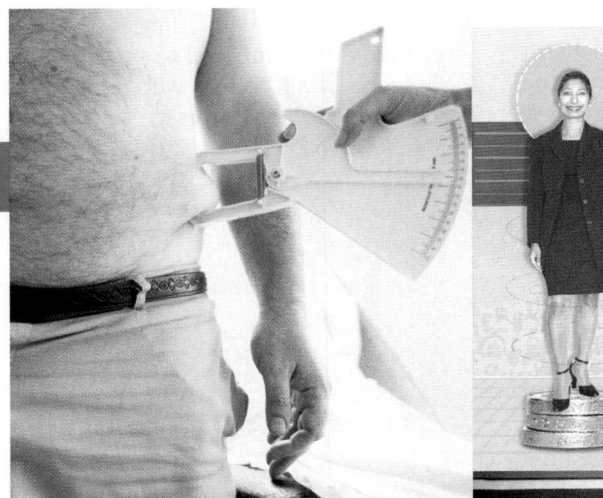

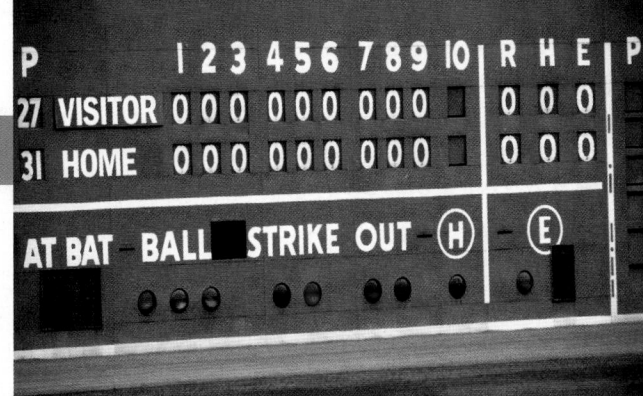

CHAPTER 10

Multiple Regression

"Here's a sector of the economy that's growing: data mining. Businesses accumulate much information about you and the world around you… More companies are trying to use that information, and that is fueling demand for people who can make sense of the data… [There is a] recruitment war for math talent. … Everybody's looking for these people."

–Steve Inskeep, Yuki Noguchi, and D.J. Patil*

*"The Search for Analysts to Make Sense of Big Data," NPR's Morning Edition, November 30, 2011.
Top left: © Science Photo Library/Alamy Limited, Top right: Matt Herring/Getty Images, Inc., Bottom right: © TriggerPhoto/iStockphoto

Questions and Issues

Here are some of the questions and issues we will discuss in this chapter:

- How effective is it to use a person's body measurements to predict the percent of body fat?

- Which variables are most important in predicting the price of a house: square footage, number of bedrooms, or number of bathrooms?

- What variables factor into the life expectancy of a country?

- Is age or miles driven more important in the price of a used car?

- Is there gender discrimination in salaries for college professors?

- How well do hourly exam grades in statistics predict the grade on the final exam?

- What variables are best at predicting profitability of movies?

- What variables are most important in predicting the length of time a baseball game lasts?

- Are carbon or steel bikes faster?

10.1 MULTIPLE PREDICTORS

A simple linear model allows us to predict values for a quantitative variable based on a single quantitative predictor. In many data situations we may have lots of variables that could serve as potential predictors. For example:

- Response: Printer *Price* (**InkjetPrinters**)
  Potential predictors: print speed (*PPM*), photo print time (*PhotoTime*), black & white ink cost (*CostBW*), color ink cost (*CostColor*)
- Response: *Tip* amount (**RestaurantTips**)
  Potential predictors: size of bill (*Bill*), number in the party (*Guests*), pay with a credit card? (*Credit* = 1 for yes, *Credit* = 0 for no)
- Response: Breakfast cereal *Calories* (**Cereal**)
  Potential predictors: amounts of *Fiber, Fat, Sodium, Sugars, Carbs, Protein*
- Response: NBA team winning percentage, *WinPct* (**NBAStandings2016**)
  Potential predictors: points scored per game (*PtsFor*), points allowed per game (*PtsAgainst*)
- Response: Home *Price* (**HomesForSale**)
  Potential predictors: square feet (*Size*), number of bedrooms (*Beds*), number of bathrooms (*Baths*)
- Response: Average household income for states *HouseholdIncome* (**USStates**)
  Potential predictors: Percentage of residents who have a college degree (*College*), smoke (*Smokers*), drink heavily (*HeavyDrinkers*), are non-white (*NonWhite*), are obese (*Obese*)

Multiple Regression Model

Why should we be limited to just one predictor at a time? Might we get a better model if we include information from several explanatory variables in the same model? For this reason, we extend the simple linear model from Section 9.1 to a *multiple regression* model that allows more than one predictor.

> **Multiple Regression Model**
>
> Given a response variable Y and k explanatory variables $X_1, X_2, \ldots X_k$, a **multiple regression model** has the form
>
> $$Y = \beta_0 + \beta_1 X_1 + \beta_2 X_2 + \cdots + \beta_k X_k + \epsilon$$
>
> where ϵ is normally distributed $N(0, \sigma_\epsilon)$ and independent.

This model allows more than one predictor to contribute to the linear part of the model and has the same conditions about the random errors as the simple linear model. With multiple predictors we lose the nice graphical representation of the model as a line on a scatterplot, but statistical software makes it easy to estimate the coefficients from sample data.

The prediction equation has the form

$$\hat{Y} = b_0 + b_1 X_1 + b_2 X_2 + \cdots + b_k X_k$$

where the coefficients are chosen to minimize the sum of squared residuals between the predicted $\hat{Y}$ values and the actual Y responses in the sample.

Example 10.1

Consider a multiple regression model to predict the prices of inkjet printers based on the printing speed (*PPM*) and cost for black & white printing (*CostBW*). Use technology and the data in **InkjetPrinters** to estimate the coefficients of the model and write down the prediction equation. What price does this fitted model predict for a printer that prints at 3.0 pages per minute and costs 3.7 cents per page for black & white printing?

Solution

Here is computer output for fitting the model $Price = \beta_0 + \beta_1 PPM + \beta_2 CostBW + \epsilon$:

```
The regression equation is
Price = 89.2 + 58.1 PPM - 21.1 CostBW

Predictor      Coef    SE Coef       T       P
Constant      89.20      95.74    0.93   0.365
PPM           58.10      22.79    2.55   0.021
CostBW      -21.125       9.341  -2.26   0.037

S = 52.8190    R-Sq = 65.2%    R-Sq(adj) = 61.1%
```

The prediction equation is $\widehat{Price} = 89.20 + 58.10 \cdot PPM - 21.125 \cdot CostBW$. For a printer that prints 3.0 pages per minute with a black & white cost of 3.7 cents per page, we have

$$\widehat{Price} = 89.20 + 58.10(3.0) - 21.125(3.7) = 185.34$$

The predicted price for this printer is $185.34.

The Kodak ESP Office 2170 All-in-One Printer is a case in the **InkjetPrinters** file that has $PPM = 3.0$, $CostBW = 3.7$, and its actual price is $150. The residual for this printer is $150 - 185.34 = -35.34$. In the fitted model using just *PPM* (see Example 9.1 on page 577) we have $\widehat{Price} = -94.22 + 90.88 \cdot PPM$ and the predicted price when $PPM = 3.0$ is $178.42 with a residual of -28.42. Notice that the intercept and coefficient of *PPM* both change when *CostBW* is added to the model. Also, the prediction for this printer is better when *CostBW* is not included in the model. Does this mean the simpler model is better?

When comparing models we should consider the residuals for all of the data cases. One way to do this is to look at the estimated standard deviation of the error term. In the output for the multiple regression model in Example 10.1 we see $s_\epsilon = 52.82$, while the output for the simple linear model on page 577 shows $s_\epsilon = 58.55$. The size of the typical error in predicting these inkjet prices is smaller when we include both *PPM* and *CostBW* as predictors in the model.

Testing Individual Terms in a Model

Once we have more than one predictor in a multiple regression model, the question naturally arises as to which predictors are actually useful to include in the model. We can test the individual variables in a multiple regression model by seeing if their coefficients are significantly different from zero. This is analogous to the t-test for the slope that we have already seen for a simple linear model. However, the presence of additional predictors can make the tests more challenging to interpret.

T-tests for Coefficients in a Multiple Regression Model

To test the effectiveness of any predictor, say X_i, in a multiple regression model we consider $H_0 : \beta_i = 0$ vs $H_a : \beta_i \neq 0$. We usually use computer software to find the t-statistic and p-value.

Alternately, use the test statistic

$$t = \frac{b_i}{SE_{b_i}}$$

where the estimated coefficient, b_i, and its standard error, SE_{b_i}, are given in the computer output. We find a p-value using a t-distribution with $n - k - 1$ degrees of freedom, where n is the sample size and k is the number of predictors in the model.

If we reject the null hypothesis, finding evidence that the coefficient is different from zero, we see that the predictor is an effective contributor to this model.

Example 10.2

Use the output in the solution of Example 10.1 to judge the effectiveness of each of the predictors (*PPM* and *CostBW*) in the multiple regression model to predict inkjet prices.

Solution

To test *PPM* we use $H_0 : \beta_1 = 0$ vs $H_a : \beta_1 \neq 0$. From the regression output we see $t = 58.10/22.79 = 2.55$, which gives a two-tailed p-value of 0.021 for a t-distribution with $20 - 2 - 1 = 17$ degrees of freedom. This is a fairly small p-value (less than 5%) which means we have evidence that the coefficient of *PPM* in the population differs from zero and it is a useful predictor of *Price* in this model.

To test *CostBW* we use $H_0 : \beta_2 = 0$ vs $H_a : \beta_2 \neq 0$. From the regression output we see $t = -21.125/9.341 = -2.26$ which gives a two-tailed p-value of 0.037. This is also less than 5%, so it would appear that *CostBW* is also useful in this model to predict printer *Price*.

Note that we can also find the t-statistic (T) and the p-value (P) for both tests directly from the output.

DATA 10.1 **Body Fat**

The percentage of a person's weight that is made up of body fat is often used as an indicator of health and fitness. However, accurate methods of measuring percent body fat are difficult to implement. One method involves immersing the body in water to estimate its density and then applying a formula to estimate percent body fat. An alternative is to develop a model for percent body fat that is based on body characteristics such as height and weight that are easy to measure. The dataset **BodyFat** contains such measurements for a sample of 100 men.[1] For each subject we have the percent body fat (*Bodyfat*) measured by the water immersion method, *Age*, *Weight* (in pounds), *Height* (in inches), and circumference (in cm) measurements for the *Neck*, *Chest*, *Abdomen*, *Ankle*, *Biceps*, and *Wrist*. ∎

[1] A sample taken from data provided by Johnson, R., "Fitting Percentage of Body Fat to Simple Body Measurements," *Journal of Statistics Education*, 1996, *http://www.amstat.org/publications/jse/v4n1/datasets.johnson.html*.

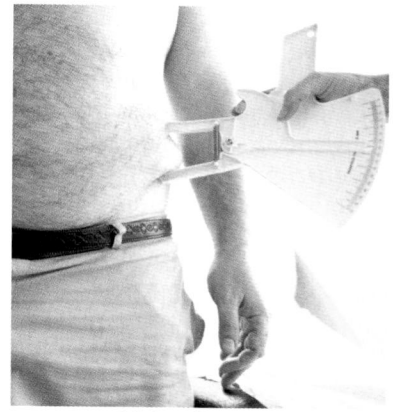

What percent body fat?

Example 10.3

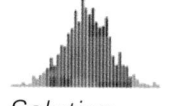

Solution

Fit a model to predict *Bodyfat* using *Height* and *Weight*. Comment on whether either of the predictors appears to be important in the model.

Here is some output for fitting $Bodyfat = \beta_0 + \beta_1 Weight + \beta_2 Height + \epsilon$:

	Estimate	Std. Error	t value	Pr(> |t|)	
(Intercept)	71.48247	16.20086	4.412	2.65e-05	***
Weight	0.23156	0.02382	9.721	5.36e-16	***
Height	−1.33568	0.25891	−5.159	1.32e-06	***

The prediction equation is

$$\widehat{Bodyfat} = 71.48 + 0.232 Weight - 1.336 Height$$

The p-values for the t-tests for the coefficients of *Weight* and *Height* are both very close[2] to zero so we have strong evidence that both terms are important in this model.

Example 10.4

Solution

Add *Abdomen* as a third predictor to the model in Example 10.3 and repeat the assessment of the effectiveness of each predictor.

Here is some output for fitting
$Bodyfat = \beta_0 + \beta_1 Weight + \beta_2 Height + \beta_3 Abdomen + \epsilon$:

	Estimate	Std. Error	t value	Pr(> |t|)	
(Intercept)	−56.1329	18.1372	−3.095	0.002580	**
Weight	−0.1756	0.0472	−3.720	0.000335	***
Height	0.1018	0.2444	0.417	0.677750	
Abdomen	1.0747	0.1158	9.279	5.27e-15	***

The prediction equation is now

$$\widehat{Bodyfat} = -56.13 - 0.1756 Weight + 0.1018 Height + 1.0747 Abdomen$$

The coefficient of *Abdomen* is very significant ($t = 9.279$, p-value ≈ 0). *Weight* is also an effective predictor ($t = -3.720$, p-value $= 0.000335$), but the *Height* coefficient has a large p-value (0.678), indicating that *Height* is not an effective predictor in this model.

[2]The notation 1.32e-06 means $1.32 \times 10^{-6} = 0.00000132$.

You probably find the results of Exercises 10.3 and 10.4 a bit surprising. Why is *Height* considered a strong predictor in the first model but not effective at all in the second? The key lies in understanding that the individual t-tests for a multiple regression model assess the importance of each predictor *after the other predictors are in the model.* When both *Weight* and *Abdomen* are in the model, we really don't need *Height* anymore—most of its information about *Bodyfat* is already supplied by the other two predictors.

Even more surprising, while the coefficient of *Weight* is significant in the three-predictor model, its sign is negative, the opposite of its sign in the two-predictor model. Does it really make sense that people who weigh more should be predicted to have less percentage body fat? Just looking at the two variables together, the correlation between *Bodyfat* and *Weight* is positive ($r = +0.6$) and quite significant. However, when *Abdomen* is also in the model the coefficient of *Weight* is negative and quite strong. Think for a moment about two men with the same abdomen circumference, but one weighs much more than the other (possibly because he is taller or more muscular). Which would you expect to have the higher percentage of body fat? For a fixed abdomen size, more weight is actually an indicator of less body fat as a percentage of weight.

In Section 2.6, we interpret the coefficient in a simple linear model as the predicted change in the response variable given a one unit increase in the predictor. The same interpretation of coefficients applies here in a multiple regression model, with the added condition that *the values of all other variables stay the same.*

Example 10.5

Interpret the coefficient of *Abdomen* in context for the model in Example 10.4.

Solution

The coefficient of *Abdomen* is 1.0747. If a person's weight and height stayed exactly the same and the abdomen circumference increased by 1 cm, we expect the percent body fat to increase by 1.0747. (Alternately, if two people have the same weight and height and one of them has an abdomen 1 cm larger, the one with the larger abdomen is predicted to have a percent body fat 1.0747 higher.)

Interpreting the individual t-tests in a multiple regression model can be quite tricky, especially when the predictors are related to each other. For an introductory course, you should just be aware that the individual tests are assessing the contribution of a predictor to that particular model and avoid making more general statements about relationships with the response variable.

ANOVA for a Multiple Regression Model

The individual t-tests tell us something about the effectiveness of individual predictors in a model, but we also need a way to assess how they do as a group. This is not a big issue for a simple linear model, since the effectiveness of the model depends only on the effectiveness of the single predictor. For a model with multiple predictors we need to measure how well the linear combination of the predictors does at explaining the structure of the response variable. This brings us back to partitioning variability to construct an ANOVA table as we did in Section 9.2.

The general form of a multiple regression model is

$$Y = \beta_0 + \beta_1 X_1 + \beta_2 X_2 + \cdots + \beta_k X_k + \epsilon$$

or

Response = Linear combination of predictors + Error

We split the total variability in the response into two pieces: one that represents the variability explained by the model and another that measures the variability that is left unexplained in the errors:

TOTAL Variability in the Response	=	Variability Explained by the MODEL	+	Unexplained Variability in the ERROR

We generally rely on statistical software to manage the details of computing the sums of squared deviations to measure each of these amounts of variability. However, the formulas are identical to the simple linear case (page 591) and we note that they can be applied in *any* situation where we have an estimated model that yields predicted values ($\hat{y}$) for each data case:

$$SSTotal = SSModel + SSE$$

where

$$SSModel = \sum (\hat{y} - \bar{y})^2$$
$$SSE = \sum (y - \hat{y})^2$$
$$SSTotal = \sum (y - \bar{y})^2$$

Example 10.6

Comparing Regression ANOVA Tables

The ANOVA tables for two regression models for printer prices are shown below. Model A uses just a single predictor (*PPM*) as in Example 9.1 on page 577. Model B adds *CostBW* as a second predictor to the model (as in Example 10.1). Discuss how the ANOVA table changes as we add the new *CostBW* predictor.

Model A: $Price = \beta_0 + \beta_1 PPM + \epsilon$

Analysis of Variance

Source	DF	SS	MS	F	P
Regression	1	74540	74540	21.75	0.000
Residual Error	18	61697	3428		
Total	19	136237			

Model B: $Price = \beta_0 + \beta_1 PPM + \beta_2 CostBW + \epsilon$

Analysis of Variance

Source	DF	SS	MS	F	P
Regression	2	88809	44405	15.92	0.000
Residual Error	17	47427	2790		
Total	19	136237			

Solution

- The sum of squares explained by the regression model (*SSModel*) increases from 74,540 to 88,809 when we add *CostBW*. Adding a new predictor will never explain *less* variability. The individual t-test (as described in Example 10.2) helps determine whether the new variability explained is more than we would expect by random chance alone.

- The sum of squares for the error term (*SSE*) decreases from 61,697 to 47,427. Again, this is expected since the total variability ($SSTotal = 136,237$) is the same for *any* model to predict these printer prices. Adding a new predictor can only improve the overall accuracy for predicting the data cases that are used to fit the model.

- The degrees of freedom for the regression model increase from 1 to 2. In general, for a multiple regression model, this degrees of freedom will be the number of predictors in the model.
- The degrees of freedom for the error decrease from 18 to 17. Again, this is not surprising since the total degrees of freedom remain at 19. In general, we lose one degree of freedom for each parameter in the model plus the constant term, so $20 - 2 - 1 = 17$.
- The rest of the ANOVA table changes as we divide the new sums of squares by the new df to get the mean squares and then the F-statistic. Although the p-values look the same, the first is based on an $F_{1,18}$ distribution and the second uses $F_{2,17}$.

In general, the ANOVA table for a k-predictor multiple regression uses k degrees of freedom for the model and $n - k - 1$ degrees of freedom for the error. The other important change concerns the hypotheses being tested. The ANOVA for regression is testing the whole model, all k predictors as a group. The formal hypotheses are:

$H_0 : \beta_1 = \beta_2 = \cdots = \beta_k = 0$
    (Model is ineffective and all predictors could be dropped)
$H_a :$ At least one $\beta_i \neq 0$ (At least one predictor in the model is effective)

Note that the constant term, β_0, is not included in the regression ANOVA null hypothesis. We are only looking for evidence that at least one of the predictors is more helpful than random chance (and its coefficient is different from zero). When we see a small p-value (such as both ANOVA tables in Example 10.6) we conclude that something in the model is effective for predicting the response, but we need to consider the individual t-tests to judge which predictors are or are not useful.

ANOVA to Test a Regression Model

To test for the overall effectiveness of a regression model, $Y = \beta_0 + \beta_1 X_1 + \beta_2 X_2 + \cdots + \beta_k X_k + \epsilon$:

$H_0 : \beta_1 = \beta_2 = \cdots = \beta_k = 0$ (The model is ineffective)
$H_a :$ At least one $\beta_i \neq 0$
    (At least one predictor in the model is effective)

We partition the variability to construct an **ANOVA table** for regression, and usually use computer software to give the ANOVA table.

Source	df	Sum of Sq.	Mean Square	F-statistic	p-value
Model	k	SSModel	$\dfrac{SSModel}{k}$	$F = \dfrac{MSModel}{MSE}$	$F_{k,n-k-1}$
Error	$n-k-1$	SSE	$\dfrac{SSE}{n-k-1}$		
Total	$n-1$	SSTotal			

Example 10.7

Solution

Suppose that we use the time to print a color picture (*PhotoTime*) and cost per page for color ink (*CostColor*) as two predictors of inkjet printer prices. Use technology to find and interpret an ANOVA table for testing the effectiveness of this model.

Here is some computer output for fitting
$Price = \beta_0 + \beta_1 PhotoTime + \beta_2 CostColor + \epsilon$:

Predictor	Coef	SE Coef	T	P
Constant	371.89	66.89	5.56	0.000
PhotoTime	0.1038	0.3663	0.28	0.780
CostColor	−18.732	5.282	−3.55	0.002

S = 67.8509   R-Sq = 42.6%   R-Sq(adj) = 35.8%

Analysis of Variance

Source	DF	SS	MS	F	P
Regression	2	57973	28987	6.30	0.009
Residual Error	17	78264	4604		
Total	19	136237			

The p-value in the ANOVA is 0.009 which is quite small, giving evidence that at least one term in this model is effective for helping to explain printer prices. Looking at the individual t-tests it appears that *CostColor* is an effective predictor in the model (p-value = 0.002), but *PhotoTime* is not very helpful (p-value = 0.780).

Coefficient of Determination: R-squared

On page 595 of Section 9.2 we see that the portion of total variability in the response variable that is successfully explained by the model is known as the coefficient of determination, or R^2. With multiple predictors we cannot get this value by squaring the correlation (as we did in Section 9.1) with any of the individual predictors, but we can use the information from the ANOVA table to obtain R^2 (as we did in Section 9.2 for simple regression).[3]

Coefficient of Determination, R^2

For any regression model, the **coefficient of determination** is

$$R^2 = \frac{SSModel}{SSTotal}$$

and is interpreted as the percent of variability in the response values in the sample that is explained by the fitted regression model.

Example 10.8

In Examples 10.6 and 10.7 we see ANOVA tables for testing two-predictor models for printer prices: one using *PPM* and *CostBW*, the other using *PhotoTime* and *CostColor*. Use the information from each ANOVA table to compute and interpret the value of R^2 for each model. What do the results tell us about the relative effectiveness of the two models?

---

[3]If we let R be the correlation between actual and predicted response values for the sample, its square is the percentage of variability explained.

Solution   The total variability, *SSTotal* = 136,237, is the same for both models, so we find R^2 in each case by dividing the sum of squares explained by the model (*SSModel*) by this quantity:

$$R^2 = \frac{88,809}{136,237} = 0.652 \quad (PPM \text{ and } CostBW)$$

$$R^2 = \frac{57,973}{136,237} = 0.426 \quad (Photo\,Time \text{ and } CostColor)$$

We see that *PPM* and *CostBW* together explain 65.2% of the variability in the prices for these 20 printers, while *PhotoTime* and *CostColor* together explain only 42.6% of this variability. Although both of these models are judged as "effective" based on their ANOVA tests, we would tend to prefer the one based on *PPM* and *CostBW* that explains a larger portion of the variability in inkjet printer prices.

Would some other combination of predictors in the **InkjetPrinters** dataset give an even more effective model for predicting prices? We consider the question of choosing an effective set of predictors in Section 10.3. Before doing so, we examine the conditions for regression models (and how to check them) in more detail in Section 10.2.

SECTION LEARNING GOALS

You should now have the understanding and skills to:

- Use computer output to make predictions and interpret coefficients using a multiple regression model
- Test the effectiveness of individual terms in a multiple regression model
- Use ANOVA to test the overall effectiveness of a multiple regression model
- Find and interpret the value of R^2 for a multiple regression model

Exercises for Section **10.1**

SKILL BUILDER 1

Exercises 10.1 to 10.11 refer to the multiple regression output shown:

The regression equation is
Y = 43.4 - 6.82 X1 + 1.70 X2 + 1.70 X3 + 0.442 X4

Predictor	Coef	SE Coef	T	P
Constant	43.43	18.76	2.31	0.060
X1	−6.820	1.059	−6.44	0.001
X2	1.704	1.189	1.43	0.202
X3	1.7009	0.6014	2.83	0.030
X4	0.4417	0.1466	3.01	0.024

S = 2.05347   R-Sq = 99.8%   R-Sq(adj) = 99.6%

Analysis of Variance

Source	DF	SS	MS	F	P
Regression	4	10974.7	2743.7	650.66	0.000
Residual Error	6	25.3	4.2		
Total	10	11000.0			

10.1   What are the explanatory variables? What is the response variable?

10.2   One case in the sample has $Y = 30$, $X1 = 8$, $X2 = 6$, $X3 = 4$, and $X4 = 50$. What is the predicted response for this case? What is the residual?

10.3   One case in the sample has $Y = 60$, $X1 = 5$, $X2 = 7$, $X3 = 5$, and $X4 = 75$. What is the predicted response for this case? What is the residual?

10.4   What is the coefficient of $X2$ in the model? What is the p-value for testing this coefficient?

10.5   What is the coefficient of $X1$ in the model? What is the p-value for testing this coefficient?

10.6   Which of the variables are significant at the 5% level?

10.7 Which of the variables are significant at the 1% level?

10.8 Which variable is most significant in this model?

10.9 Which variable is least significant in this model?

10.10 Is the model effective, according to the ANOVA test? Justify your answer.

10.11 State and interpret R^2 for this model.

SKILL BUILDER 2

Exercises 10.12 to 10.22 refer to the multiple regression output shown:

The regression equation is
$Y = -61 + 4.71\ X1 - 0.25\ X2 + 6.46\ X3 + 1.50\ X4 - 1.32\ X5$

Predictor	Coef	SE Coef	T	P
Constant	−60.7	105.7	−0.57	0.575
X1	4.715	2.235	2.11	0.053
X2	−0.253	1.178	−0.21	0.833
X3	6.459	2.426	2.66	0.019
X4	1.5011	0.4931	3.04	0.009
X5	−1.3151	0.8933	−1.47	0.163

S = 32.4047   R-Sq = 77.9%   R-Sq(adj) = 70.0%

Analysis of Variance

Source	DF	SS	MS	F	P
Regression	5	51799	10360	9.87	0.000
Residual Error	14	14701	1050		
Total	19	66500			

10.12 What are the explanatory variables? What is the response variable?

10.13 One case in the sample has $Y = 20$, $X1 = 15$, $X2 = 40$, $X3 = 10$, $X4 = 50$, and $X5 = 95$. What is the predicted response for this case? What is the residual?

10.14 One case in the sample has $Y = 50$, $X1 = 19$, $X2 = 56$, $X3 = 12$, $X4 = 85$, and $X5 = 106$. What is the predicted response for this case? What is the residual?

10.15 What is the coefficient of $X1$ in the model? What is the p-value for testing this coefficient?

10.16 What is the coefficient of $X5$ in the model? What is the p-value for testing this coefficient?

10.17 Which of the variables are significant at the 5% level?

10.18 Which of the variables are significant at the 1% level?

10.19 Which variable is most significant in this model?

10.20 Which variable is least significant in this model?

10.21 Is the model effective, according to the ANOVA test? Justify your answer.

10.22 State and interpret R^2 for this model.

10.23 Predicting Prices of New Homes Here is some output for fitting a model to predict the price of a home (in $1000s) using size (in square feet, *SizeSqFt*, different units than the variable *Size* in **HomesForSale**), number of bedrooms, and number of bathrooms. (The data are based indirectly on information in the **HomesForSale** dataset.)

The regression equation is
Price = - 217 + 0.331 SizeSqFt - 135 Beds + 200 Baths

Predictor	Coef	SE Coef	T	P
Constant	−217.0	145.9	−1.49	0.140
SizeSqFt	0.33058	0.07262	4.55	0.000
Beds	−134.52	57.03	−2.36	0.020
Baths	200.03	78.94	2.53	0.013

S = 507.706   R-Sq = 46.7%   R-Sq(adj) = 45.3%

Analysis of Variance

Source	DF	SS	MS	F	P
Regression	3	26203954	8734651	33.89	0.000
Residual Error	116	29900797	257765		
Total	119	56104751			

(a) What is the predicted price for a 2500 square foot, four bedroom home with 2.5 baths?

(b) Which predictor has the largest coefficient (in magnitude)?

(c) Which predictor appears to be the most important in this model?

(d) Which predictors are significant at the 5% level?

(e) Interpret the coefficient of *SizeSqFt* in context.

(f) Interpret what the ANOVA output says about the effectiveness of this model.

(g) Interpret R^2 for this model.

10.24 Predicting Calories Consumed Using the data in **NutritionStudy**, we show computer output for a model to predict calories consumed in a day based on fat grams consumed in a day, cholesterol consumed in mg per day, and age in years:

The regression equation is
Calories = 513 + 16.3 Fat + 0.421 Cholesterol - 1.42 Age

Predictor	Coef	SE Coef	T	P
Constant	512.95	86.51	5.93	0.000
Fat	16.2645	0.7925	20.52	0.000
Cholesterol	0.4208	0.2015	2.09	0.038
Age	-1.420	1.304	−1.09	0.277

S = 331.904   R-Sq = 76.4%   R-Sq(adj) = 76.2%

Analysis of Variance

Source	DF	SS	MS	F	P
Regression	3	111082085	37027362	336.12	0.000
Residual Error	311	34259921	110161		
Total	314	145342006			

(a) What daily calorie consumption does the model predict for a 35-year-old person who eats 40 grams of fat in a day and 180 mg of cholesterol?

(b) In this model, which variable is least significant? Which is most significant?

(c) Which predictors are significant at a 5% level?

(d) Interpret the coefficient of *Fat* in context.

(e) Interpret the coefficient of *Age* in context.

(f) Interpret what the ANOVA output says about the effectiveness of this model.

(g) Interpret R^2 for this model.

10.25 Predicting Life Expectancy In Exercises 9.27 and 9.69 we attempt to predict a country's life expectancy based on the percent of government expenditure on health care, using a sample of fifty countries in the dataset **SampCountries**. We now add to the model the variables population (in millions), percentage with Internet, and birth rate (births per 1000). (Note that the *Internet* value is missing for South Sudan, so the model is fit with the other 49 countries.) Table 10.1 shows some computer output from fitting this model.

(a) Which variables are significant predictors of life expectancy in this model, at a 5% level? Which is the most significant?

(b) Predict the life expectancy of a country that spends 20% of government expenditures on health care, has a population of 2,500,000, for which 75% of people have access to the Internet, and the birth rate is 30 births per 1000.

(c) How does predicted life expectancy change if we find out the Internet rate is actually higher?

Table 10.1 *Multiple regression output*

	Estimate	**Std. Error**	**t-value**	*Pr(>|t|)*
(Intercept)	76.24	4.83	15.77	0.0000
Health	0.131	0.222	0.59	0.5557
Population	−0.0003	0.0128	−0.03	0.9794
Internet	0.1118	0.0438	2.55	0.0142
BirthRate	−0.594	0.122	−4.86	0.00002

10.26 Binary Categorical Variables: Weight Based on Height and Gender Categorical variables with only two categories (such as male/female or yes/no) can be used in a multiple regression model if we code the answers with numbers. In Chapter 9, we look at a simple linear model to predict *Weight* based on *Height*. What role does gender play? If a male and a female are the same height (say 5'7"), do we predict the same weight for both of them? Is gender

a significant factor in predicting weight? We can answer these questions by using a multiple regression model to predict weight based on height and gender. Using 1 for females and 0 for males in a new variable called *GenderCode* in the dataset **StudentSurvey**, we obtain the following output:

The regression equation is
Weight = - 23.9 + 2.86 Height-25.5 GenderCode

Predictor	Coef	SE Coef	T	P
Constant	−23.92	27.36	−0.87	0.383
Height	2.8589	0.3855	7.42	0.000
GenderCode	−25.470	3.138	−8.12	0.000

S = 22.8603   R-Sq = 48.2%   R-Sq(adj) = 47.9%

(a) What weight does the model predict for a male who is 5'7" (67 inches)? For a female who is 5'7"?

(b) Which predictors are significant at a 5% level?

(c) Interpret the coefficient of *Height* in context.

(d) Interpret the coefficient of *GenderCode* in context. (Pay attention to how the variable is coded.)

(e) What is R^2 for this model? Interpret it in context.

10.27 Binary Categorical Variables: Predicting Cognitive Score Categorical variables with only two categories (such as male/female or yes/no) can be used in a multiple regression model if we code the answers with numbers. Exercise 2.143 on page 102 introduces a study examining years playing football, brain size, and percentile score on a cognitive skills test. We show computer output below for a model to predict *Cognition* score based on *Years* playing football and a categorical variable *Concussion*. The variable *Concussion* is coded 1 if the player has ever been diagnosed with a concussion and is coded 0 if he has not been diagnosed with a concussion.

Regression Equation
Cognition = 100.6 – 3.07 Years – 2.70 Concussion

Coefficients

Term	Coef	SE Coef	T-Value	P-Value
Constant	100.6	16.9	5.97	0.000
Years	−3.07	1.62	−1.90	0.064
Concussion	−2.70	9.49	−0.29	0.777

S = 25.7829 R-sq = 13.56% R-sq(adj) = 9.35%

Analysis of Variance

Source	DF	SS	MS	F-Value	P-Value
Regression	2	4277.3	2138.63	3.22	0.050
Error	41	27255.0	664.76		
Total	43	31532.2			

(a) One of the participants in the study played football for 9 years, had never been diagnosed with a concussion, and scored a 74 on the cognitive skills test. What is his predicted cognition score? What is the residual for this prediction?

(b) Another one of the participants played football for 7 years, had been diagnosed with a concussion, and scored a 42 on the cognitive skills test. What is his predicted cognition score? What is the residual for this prediction?

(c) What is the coefficient of *Years* in this model? Interpret it in context.

(d) What is the coefficient of *Concussion* in this model? Interpret it in context. (Pay attention to how the variable is coded.)

(e) At a 10% level, is the overall model effective at predicting cognition scores? What value in the computer output are you basing your answer on?

(f) There are two variables in this model. How many of them are significant at the 10% level? How many are significant at the 5% level?

(g) Which of the two variables is most significant in the model?

(h) How many football players were included in the analysis?

(i) What is R^2? Interpret it in context.

PRICE OF RACE HORSES
For Exercises 10.28 to 10.31, use information in the ANOVA table below, which comes from fitting a multiple regression model to predict the prices for horses (in $1000s).

Source	DF	SS	MS	F	P
Regression	3	4327.7	1442.6	10.94	0.000
Residual Error	43	5671.4	131.9		
Total	46	9999.1			

10.28 How many predictors are in the model?

10.29 How many horses are in the sample?

10.30 Find and interpret (as best you can with the given context) the value of R^2.

10.31 Is this an effective model for predicting horse prices? Write down the relevant hypotheses as well as a conclusion based on the ANOVA table.

10.32 Hantavirus in Mice In Exercise 9.23 on page 589, we discuss a study[4] conducted on the California Channel Islands investigating the prevalence of hantavirus in mice. This virus can cause severe lung disease in humans. The article states: "Precipitation accounted for 79% of the variation in

prevalence. Adding in island area upped this to 93%, and including predator richness took the total to 98%."

(a) Give the correct notation or terminology for the quantity the scientists are comparing in the quotation.

(b) Based on the information given, do you expect the ANOVA p-value for the model with all three predictors to be relatively large or relatively small? Explain.

10.33 Housing Prices in New York In Exercise 9.68 we look at predicting the price (in $1000s) of New York homes based on the size (in thousands of square feet), using the data in **HomesForSaleNY**. Two other variables in the dataset are the number of bedrooms and the number of bathrooms. Use technology to create a multiple regression model to predict price based on all three variables: size, number of bedrooms, and number of bathrooms. (In Exercise 10.23, we investigate a similar model using homes from more states.)

(a) Which predictors are significant at a 5% level? Which variable is the most significant?

(b) Interpret the two coefficients for *Beds* and *Baths*. Do they both make sense?

(c) What price does the model predict for a 1500 square foot (*Size* = 1.5) New York home with 3 bedrooms and 2 bathrooms?

10.34 Predicting Body Mass Gain in Mice Use technology and the data in **LightatNight4Weeks** to predict body mass gain in mice, *BMGain*, over a four-week experiment based on stress levels measured in *Corticosterone*, percent of calories eaten during the day (most mice in the wild eat all calories at night) *DayPct*, average daily consumption of food in grams *Consumption*, and activity level *Activity*.

(a) Interpret the coefficient of *DayPct* in context.

(b) Interpret the coefficient of *Consumption* in context.

(c) Which variable is most significant in this model?

(d) Which variable is least significant in this model?

(e) According to the ANOVA table, is the model effective at predicting body mass gain of mice in this situation?

(f) Interpret R^2 for this model.

10.35 NBA Winning Percentage In Exercise 9.28 on page 590 we consider simple linear models to predict winning percentages for NBA teams based

[4]"More Rain, More Virus," *Nature*, April 28, 2011, p. 392.

on either their offensive ability (*PtsFor* = average points scored per game) or defensive ability (*PtsAgainst* = average points allowed per game). With multiple regression we can include both factors in the same model. Use the data from the 2015–2016 regular season in **NBAStandings2016** to fit a two-predictor model for *WinPct* based on *PtsFor* and *PtsAgainst*.

(a) Write down the fitted prediction equation.

(b) The Golden State Warriors set an NBA record by winning 73 games while losing only 9 for a 0.890 winning percentage. They scored 114.9 points per game (tops in the league), while allowing 104.1 points against per game. Find the predicted winning percentage for the Warriors using this model and compute the residual.

(c) Comment on the effectiveness of each predictor in this model.

(d) As single predictors, *PtsFor* and *PtsAgainst* are very similar in effectiveness for predicting *WinPct*. Do we do much better by including both predictors? Choose some measure (such as s_ϵ, or R^2) to compare the effectiveness of a simple linear model based on either *PtsFor* or *PtsAgainst* to this two-predictor model.

10.36 Prices of Mustang Cars Data 3.4 on page 246 describes a sample of $n = 25$ Mustang cars being offered for sale on the Internet. We would like to predict the *Price* of used Mustangs (in $1000s) and the possible explanatory variables in **MustangPrice** are the *Age* in years and *Miles* driven (in 1000s).

(a) Fit a simple linear model for *Price* based on *Age*. Does *Age* appear to be an effective predictor of *Price*? Justify your answer.

(b) Fit a multiple regression model for *Price* based on *Age* and *Miles*. Is *Age* an effective predictor in this model? Justify your answer.

(c) Can you think of an explanation for the change from (a) to (b)?

10.37 Comparing Models for NBA Winning Percentage In Exercise 9.28 on page 590 we consider separate simple linear models to predict NBA winning percentages using *PtsFor* and *PtsAgainst*. In Exercise 10.35 we combine these to form a multiple regression model. The data is in **NBAStandings2016**.

(a) Compare the percentages of variability in winning percentages that are explained by these three models (*PtsFor* alone, *PtsAgainst* alone, and the two together in a multiple regression).

(b) Create a new predictor, *Diff* = *PtsFor* − *PtsAgainst*, to measure the average margin of victory (or defeat) for each team. Use it as a single predictor in a simple linear model for *WinPct*. Include a scatterplot with the regression line and comment on how this model compares to the three in part (a).

10.38 Randomization Test for a Multiple Regression Model When deriving the F-statistic on page 593 we include a note that the use of the F-distribution can be simulated with a randomization procedure. That is the purpose of this exercise. Consider the model in Example 10.7 that uses *PhotoTime* and *CostColor* to predict inkjet printer prices. The F-statistic in that ANOVA table is $F = 6.30$. Suppose that we want to estimate how unusual that is, when H_0 is true and there is no relationship between either predictor and *Price*. To simulate this, randomly scramble the 20 printer prices in **InkjetPrinters** and assign them to the various combinations of *PhotoTime* and *CostColor*. For each such randomization, estimate the fitted model and compute the F-statistic. (Technology is a must here!) Repeat many times to obtain a randomization distribution and see how far the original $F = 6.30$ is in the tail. Compare the results to what you get with the F-distribution with 2 and 17 degrees of freedom.

10.2 CHECKING CONDITIONS FOR A REGRESSION MODEL

For a given set of k predictors, the multiple regression model is

$$Y = \beta_0 + \beta_1 X_1 + \beta_2 X_2 + \cdots + \beta_k X_k + \epsilon$$

where ϵ is normally distributed $N(0, \sigma_\epsilon)$ and independent.

As with the simple linear model, when $k = 1$, most of the conditions for the model are reflected in the error term. We expect the deviations of actual response values from the regression model to have:

• A mean of zero

• The same variability at different values of the explanatory variable(s)

• A normal distribution

These conditions are reflected in Figure 9.3 on page 583 which shows the simple linear model as a string of normal distributions running along the population regression line. For single predictors we can often check the conditions (as we did, somewhat informally, in Section 9.1) by looking at a scatterplot of the data with the least squares line drawn on it. This is often sufficient to reveal problems such as a curved rather than linear pattern, variability that consistently increases or decreases over time, or large outliers that might affect the fit or indicate points that are poorly predicted.

What about checking conditions for a multiple regression model? Although we could produce a scatterplot of a response Y versus any single explanatory variable X_i, we could not easily visualize the multiple regression fit on that plot. We need a different method for checking the conditions on the error term when we have more than one predictor.

Residuals vs Fitted Values Plot

The multiple regression model says that the errors from the regression equation should follow a normal distribution with mean zero and constant standard deviation, σ_ϵ. We estimate these errors with the residuals, $y_i - \hat{y}_i$, which compare the actual values for the response variable to the predicted values from the fitted model. Fortunately, the residuals for least squares regression models always have mean zero, so the main things we need to look for are changing variance, lack of normality, or departures from the linear pattern. We generally accomplish this with *residual plots* that give graphical displays of the residuals to check model conditions. One common method is to produce a scatterplot of the residuals vs the predicted values.

Residuals vs Fits Plot

To assess possible departures from the linear model and/or consistently changing variance we use a **residuals vs fits plot** with the residuals on the vertical axis and the predicted values on the horizontal axis.

The ideal pattern is a horizontal band of residuals scattered on either side of a horizontal line at zero.

Watch out for

- Curved patterns or other obvious trends that indicate problems with the linearity of the model
- Fanning patterns that show the variability consistently increasing or decreasing
- Outliers that are unusually far above or below the zero line.

Example 10.9

Figure 10.1(a) shows a scatterplot with regression line using data from a simple linear model in Figure 9.4 on page 583. Figure 10.1(b) shows the residuals vs fits plot for the same data. Both plots show a random, parallel scatter above and below the line that we expect to see for a linear model. Although the scales are different, points that are close to the regression line in Figure 10.1(a) correspond to points near the zero line in Figure 10.1(b).

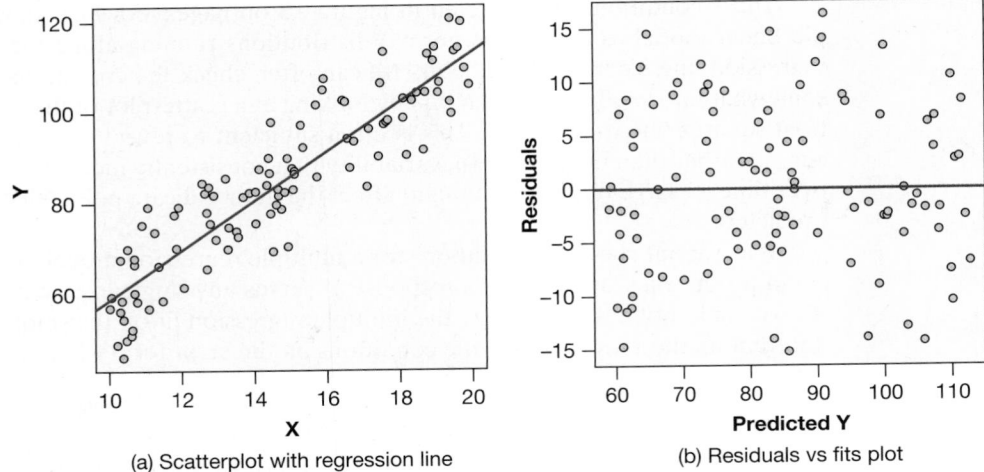

(a) Scatterplot with regression line   (b) Residuals vs fits plot

Figure 10.1 *Plots from the same simple linear model*

Example 10.10

Recognizing Problems

Figure 10.2 shows three scatterplots from Figure 9.5 on page 584 that illustrate problems with the regression model assumptions. The corresponding residuals vs fits plots for each dataset are shown below them. Pay attention to how the problems (curvature, increasing variance, and outliers) translate from the original scatterplot to the residual plot.

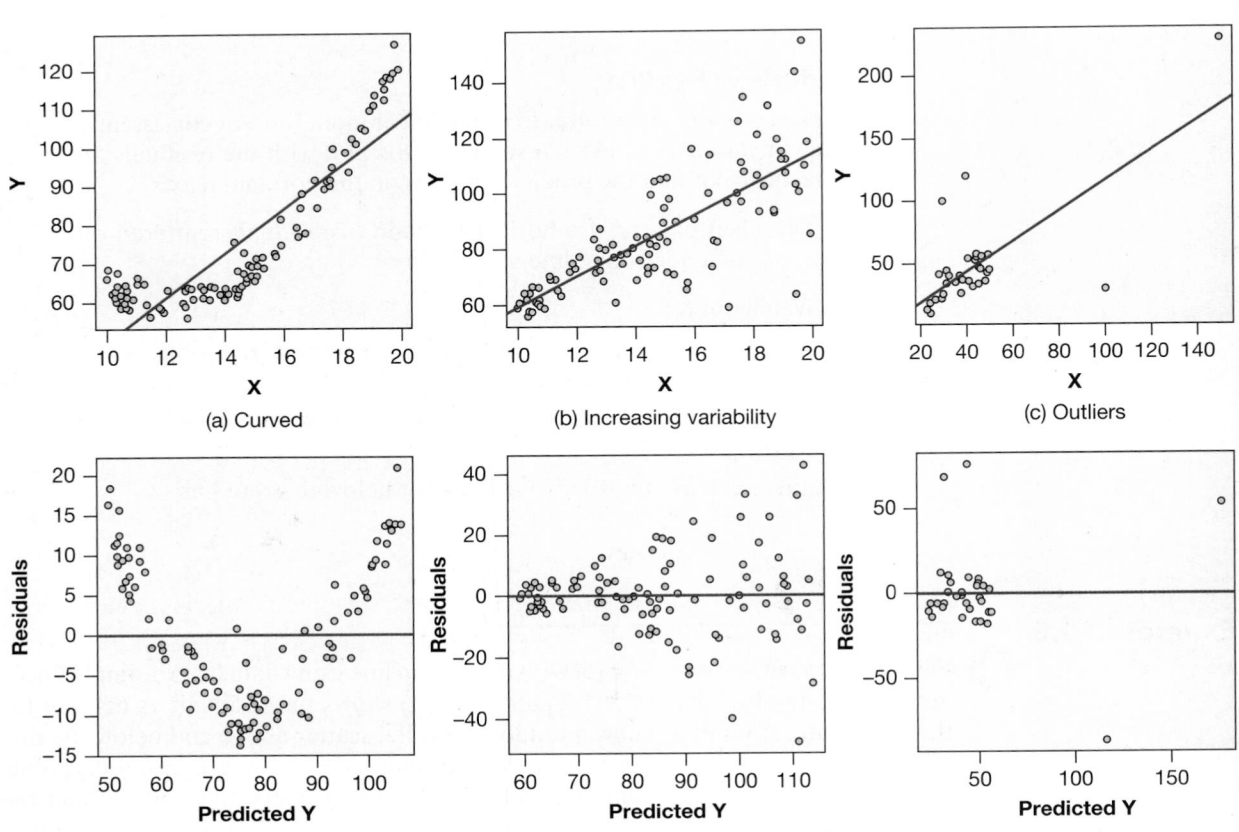

(a) Curved   (b) Increasing variability   (c) Outliers

Figure 10.2 *Residual plots for least squares fits with problems*

Histogram/Dotplot/Boxplot of Residuals

A dotplot, histogram, or boxplot of the residuals from a model is a good way to check the normality condition and look for outliers. As with any plot of a single quantitative variable, we look for reasonable symmetry.

Example 10.11

Figure 10.3 shows three histograms of the same three datasets as in Figure 10.2. Comment on the appropriateness of a normality condition in each case.

Solution

(a) The histogram of the residuals for the curved data in Figure 10.3(a) shows some mild skewness with a peak to the left of zero and very little "tail" in that direction. However, the sample size is fairly large (most of the histogram bars show a frequency more than 10) so we shouldn't be too concerned.

(b) The distribution of residuals for the middle model shows no concerns with normality. The graph is symmetric, bell-shaped, and centered around zero.

(c) The outliers in Figure 10.3(c) make it difficult to tell more about the distribution, but this is not what we expect to see in a normal distribution.

Note that the normality condition applies to the *residuals* of the model. There is no specific restriction on the distribution of either the explanatory or response variables. For example, we might have values of the predictor in the sample that are highly skewed with lots of small values that produce a similar skewness in the responses. As long as the errors are relatively bell-shaped, we can still use inference based on the t and F distributions in assessing the regression model.

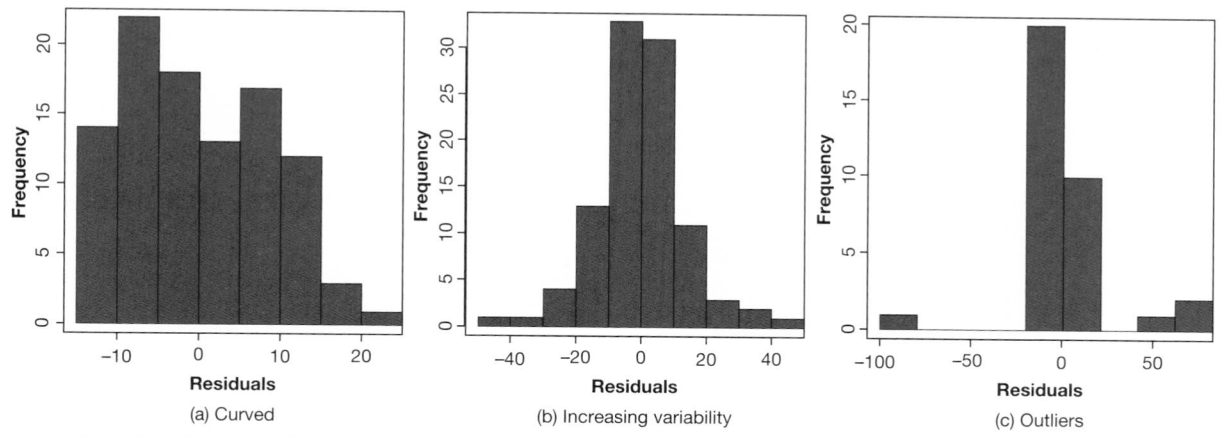

Figure 10.3 *Residual plots for least squares fits with problems*

Example 10.12

Checking Conditions for an Inkjet Printer Model

Produce graphs and comment on the appropriateness of the simple linear model conditions for the regression model to predict inkjet printer prices using *PPM* printing rates with the data in **InkjetPrinters**.

Solution

We have already seen the scatterplot with regression line in Figure 9.1 on page 577. It shows a general increasing trend, no obvious curvature or big outliers, and a relatively equal scatter of points above and below the line. Nothing in this plot raises serious concerns about the simple linear model conditions.

A dotplot of the 20 residuals is shown in Figure 10.4. It looks reasonably symmetric and has no huge outliers in either tail so we don't need to worry about significant departures from normality of the residuals.

A residuals vs fits plot for this model is displayed in Figure 10.5. It shows a reasonable horizontal band of points on either side of the zero line.

While it is hard to make definitive assessments of the simple linear model conditions based on just 20 data points, we don't see any reason for strong concerns about any of the conditions for these data.

Figure 10.4 *Residuals from regression to predict inkjet printer price based on PPM*

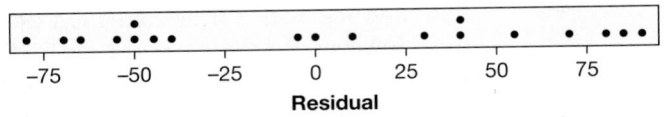

Figure 10.5 *Residuals vs Fits for predicting printer price based on PPM*

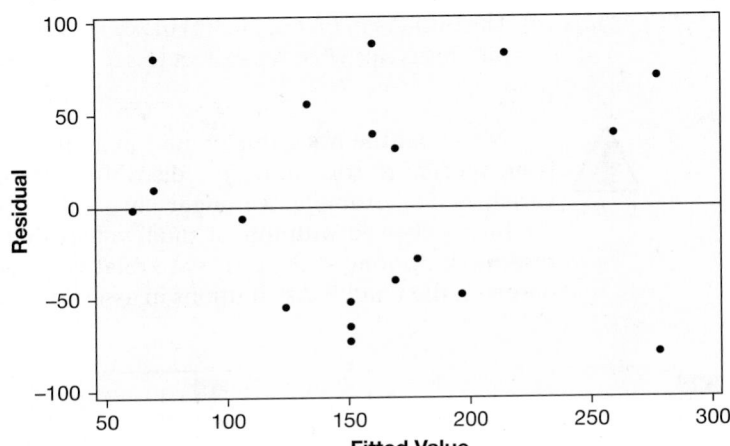

Checking Conditions for a Multiple Regression Model

The conditions for a multiple regression model are basically the same as for a simple linear model in that the errors should be normally distributed with zero mean and a constant variance for any combination of the predictors. We can use most of the same graphical tools to assess the residuals: A residuals vs fits plot and some sort of histogram or dotplot of the residuals to assess normality and look for skewness/outliers.

Example 10.13

Checking Conditions for a Bodyfat Model

Since the output in Example 10.4 on page 615 indicates that *Height* isn't an especially effective predictor in that model, run a new model with just *Weight* and *Abdomen* as predictors for *Bodyfat*. Use residual plots to assess the conditions for the multiple regression model.

Solution

Here is some output for fitting $Bodyfat = \beta_0 + \beta_1 Weight + \beta_2 Abdomen + \epsilon$:

	Estimate	Std. Error	t value	Pr(> |t|)
(Intercept)	−48.77854	4.18098	−11.667	< 2e-16 ***
Weight	−0.16082	0.03101	−5.185	1.18e-06 ***
Abdomen	1.04408	0.08918	11.707	< 2e-16 ***

Both predictors have very small p-values and appear to be important in this model. Figure 10.6 shows a histogram of the residuals and a plot of residuals vs fitted values for this model. The distribution of the residuals in the histogram is symmetric and bell-shaped so the normality condition is quite reasonable. The residuals vs fits plot shows an even scatter on either side of the zero line with no unusual patterns so constant variability is also reasonable.

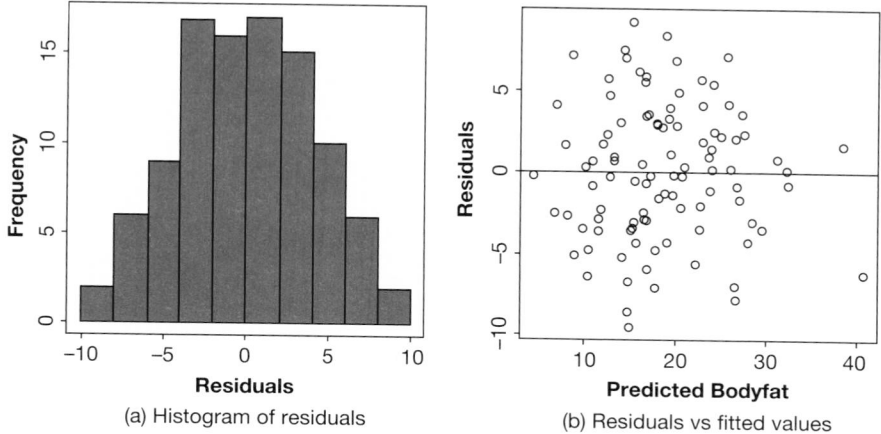

Figure 10.6 *Residual plot for predicting Bodyfat with Weight and Abdomen*

(a) Histogram of residuals

(b) Residuals vs fitted values

 Take care to recognize which type of plot is helpful to check the different conditions. A residual vs fit plot can detect lack of linearity or increasing variability, but is not so helpful at assessing normality. A histogram or dotplot of residuals can detect departures from normality or show outliers, but won't tell us if there are nonlinear patterns in the data.

What about the condition of independence of the residuals? This means that the fact that one value in the sample tends to lie above or below the regression fit does not affect where the next point lies. This is tougher to check visually, but we can generally rely on the method of randomization used in collecting the data (a random sample for observational data or random assignment of treatments for an experiment) to satisfy this condition.

What do we do if we have serious concerns about any of the regression conditions?

• Use a bootstrap or randomization procedure that is not so dependent on specific conditions to perform the inference.

• Consider deleting one or two particularly troublesome data points—especially if they might represent errors in the data—but take care not to blindly exclude any extreme point. Those are often the most interesting parts of an analysis.

• Although it is beyond the scope of this course, many statisticians use transformations of the data (such as a square, square root, or logarithm) to help meet the regression conditions.

SECTION LEARNING GOALS

You should now have the understanding and skills to:

▶ • Use a residual vs fits plot to check for linearity and consistent variability of a regression model

▶ • Use a histogram, dotplot, or boxplot to check for normality and outliers in the distribution of residuals for a regression model

Exercises for Section **10.2**

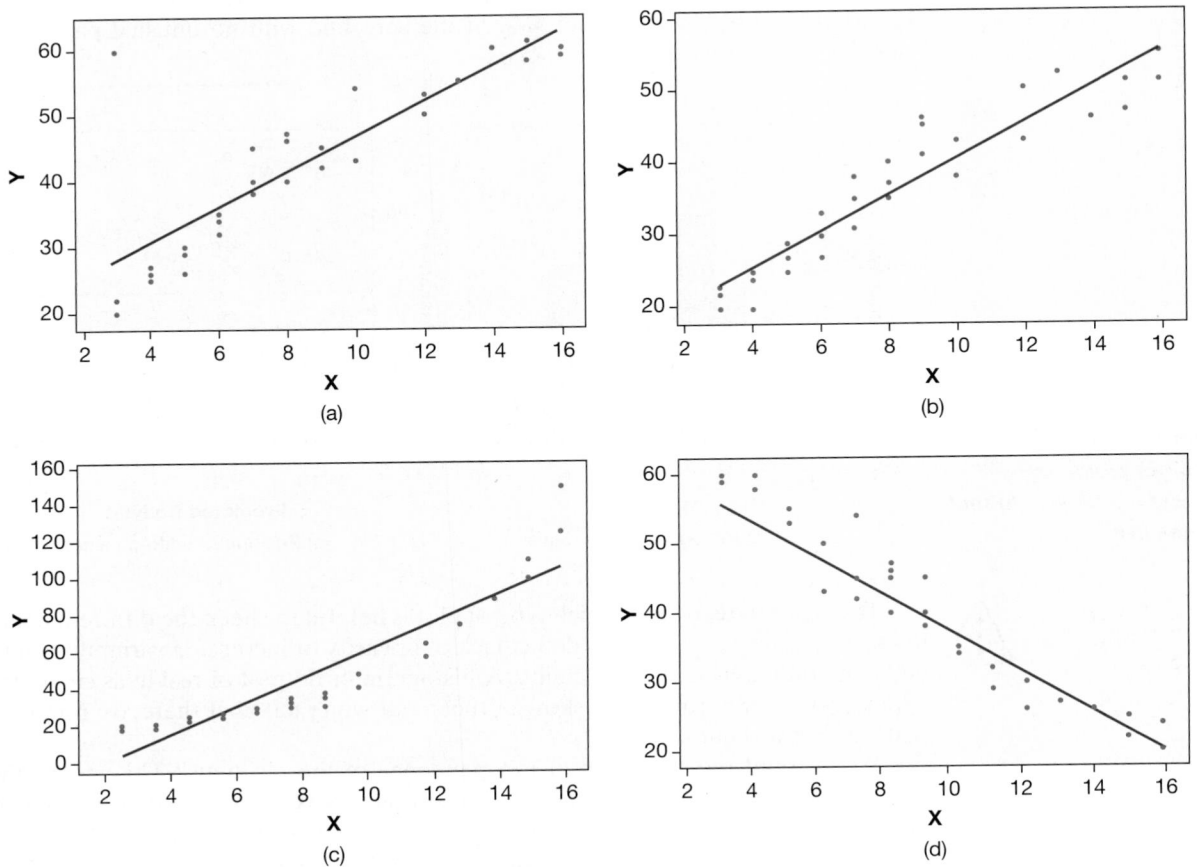

Figure 10.7 *Match Exercises 10.39 to 10.42 with these scatterplots*

SKILL BUILDER 1
Exercises 10.39 to 10.42 give scatterplots of residuals against predicted values. Match each with one of the scatterplots shown in Figure 10.7.

10.39

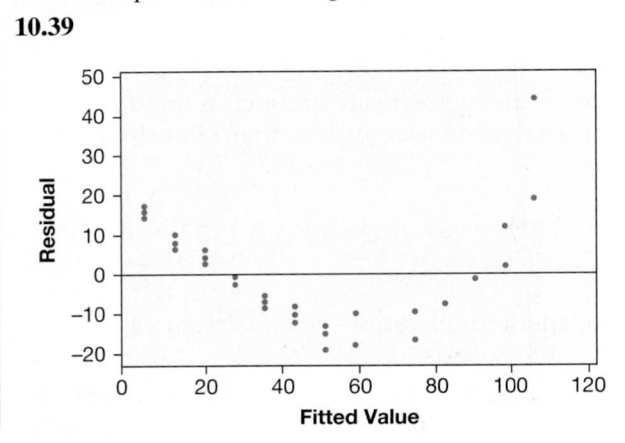

10.40

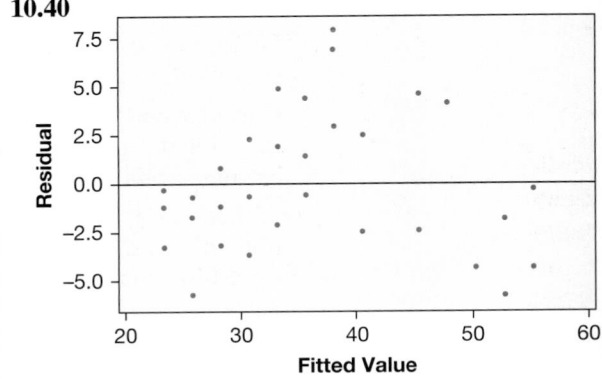

10.41

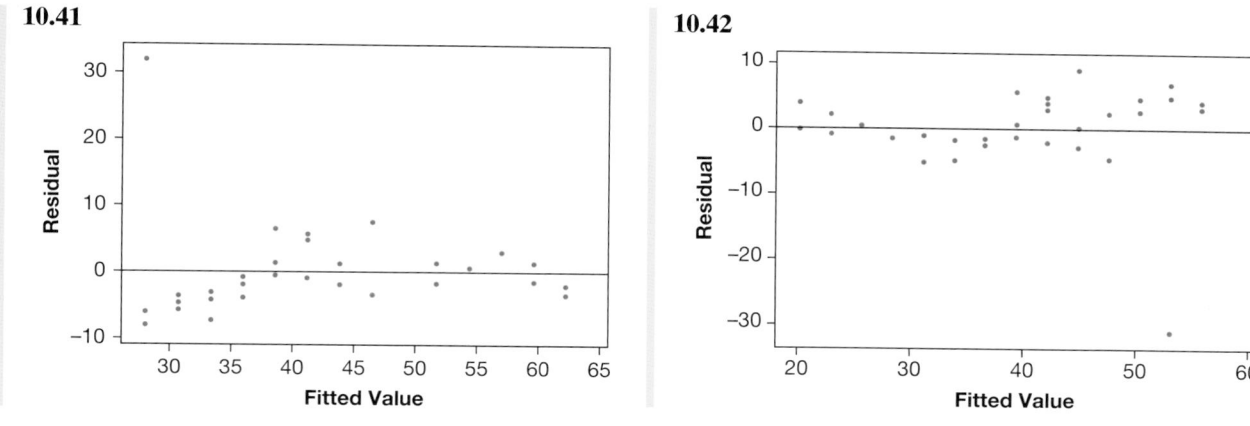

10.42

CHECK CONDITIONS

In Exercises 10.43 and 10.44, three graphs are shown for a linear model: the scatterplot with least squares line, a histogram of the residuals, and a scatterplot of residuals against predicted values. Determine whether the conditions are met and explain your reasoning.

10.43

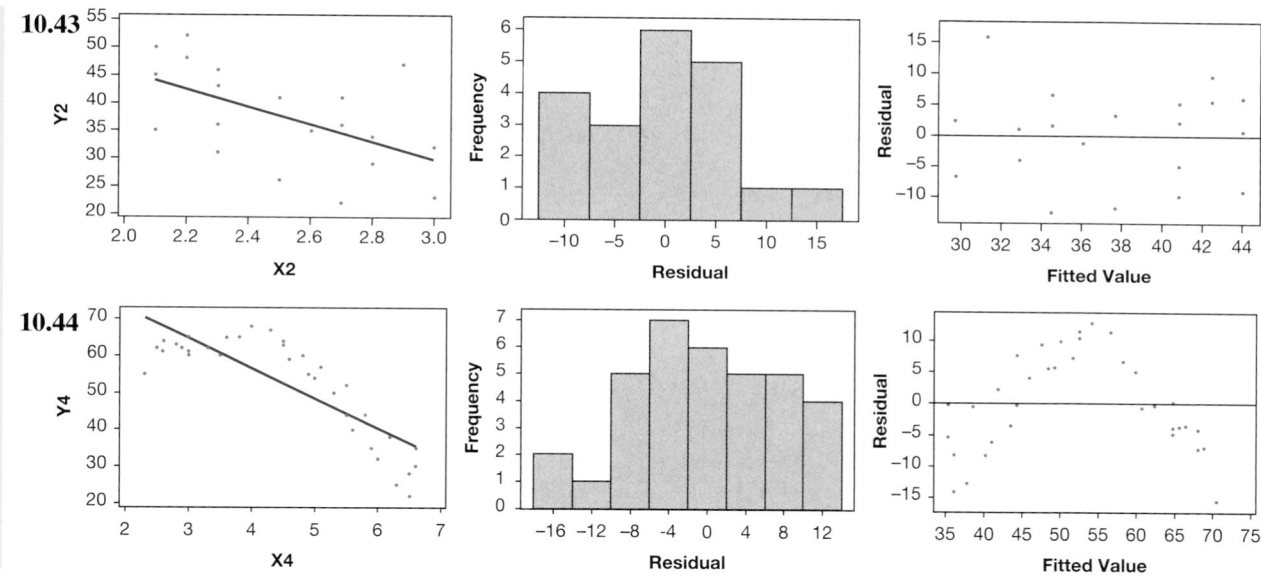

10.44

10.45 Height and Weight Using the data in **StudentSurvey**, we see that the regression line to predict *Weight* from *Height* is $\widehat{Weight} = -170 + 4.82 Height$. Figure 10.8 shows three graphs for this linear model: the scatterplot with least squares line, a histogram of the residuals, and a scatterplot of residuals against predicted values.

(a) One of the students in the dataset has a height of 63 inches and a weight of 200 pounds. Put an arrow showing the dot representing this person on the scatterplot with least squares line (or a rough sketch of the plot).

(b) Calculate the predicted value and the residual for the person described in part (a).

(c) Put an arrow showing where the person from part (a) is represented in the histogram of residuals. Also, put an arrow showing where the person from part (a) is represented in the scatterplot of residuals against predicted values.

(d) Determine whether the conditions are met for inference on this regression model.

10.46 More on Height and Weight As we see in Exercise 10.45, or by using the data in **StudentSurvey**, the regression line to predict *Weight* from

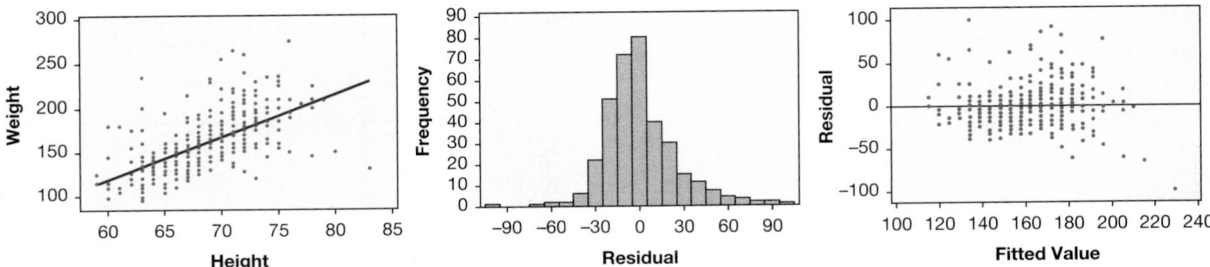

Figure 10.8 *Plots to assess Height predicting Weight*

Height is $\widehat{Weight} = -170 + 4.82 Height$. Figure 10.8 shows three graphs for this linear model: the scatterplot with least squares line, a histogram of the residuals, and a scatterplot of residuals against predicted values.

(a) One of the students in the dataset has a height of 73 inches and a weight of 120 pounds. Put an arrow showing the dot representing this person on the scatterplot with least squares line (or a rough sketch of the plot).

(b) Calculate the predicted value and the residual for the person described in part (a).

(c) Put an arrow showing where the person from part (a) is represented in the histogram of residuals. Also, put an arrow showing where the person from part (a) is represented in the scatterplot of residuals against predicted values.

(d) Use these plots to assess the conditions for inference on this regression model.

10.47 Exercise and Pulse Rate Use the data in **StudentSurvey** to assess the conditions for doing inference on a regression model to predict a person's pulse rate, *Pulse*, from the number of hours a week spent exercising, *Exercise*. Explain your reasoning, using each of the three relevant graphs.

10.48 Grams of Fat and Number of Calories Use the data in **NutritionStudy** to assess the conditions for doing inference on a regression model to predict a person's daily calories, *Calories*, from the daily grams of fat, *Fat*. Explain your reasoning, using each of the three relevant graphs.

10.49 Grams of Fat and Cholesterol Level Use the data in **NutritionStudy** to assess the conditions for doing inference on a regression model to predict a person's cholesterol level, *Cholesterol*, from the daily grams of fat, *Fat*. Explain your reasoning, using each of the three relevant graphs.

10.50 Restaurant Bill and Tip Use the data in **RestaurantTips** to assess the conditions for doing inference on a regression line to predict the size of a customer's tip, *Tip*, from the size of the bill, *Bill*. Explain your reasoning, using each of the three relevant graphs.

10.51 Predicting Atlanta Commute Time The data in **CommuteAtlanta** show information on both the commute distance (in miles) and time (in minutes) for a sample of 500 Atlanta commuters. Suppose that we want to build a model for predicting the commute time based on the distance.

(a) Fit the simple linear model, $Time = \beta_0 + \beta_1 Distance + \epsilon$, for the sample of Atlanta commuters and write down the prediction equation.

(b) What time (in minutes) does the fitted model predict for a 20-mile commute?

(c) Produce a scatterplot of the relationship between *Time* and *Distance* and comment on any interesting patterns in the plot.

(d) Produce a dotplot or histogram to show the distribution of the residuals for this model. Comment on whether the normality condition is reasonable.

(e) Produce a plot of the residuals vs the fitted values. Comment on what this plot says about the simple linear model conditions in this situation.

10.52 Predicting St. Louis Commute Time Refer to Exercise 10.51. The file **CommuteStLouis** contains similar information for a sample of 500 commuters in St. Louis. Answer the same questions as Exercise 10.51 using the St. Louis data. Are the results in St. Louis much different from Atlanta?

10.53 How Accurate Are Our Estimates in Predicting Atlanta Commute Time? In Exercise 10.51 we consider a simple linear model to predict *Time* in minutes for Atlanta commuters based on *Distance* in miles using the data in **CommuteAtlanta**. For a 20-mile commute the predicted time is 31.34 minutes. Here is some output containing intervals for this prediction.

NewObs	Fit	SE Fit	95% CI	95% PI
1	31.343	0.553	(30.257, 32.430)	(7.235, 55.452)

(a) Interpret the "95% CI" in the context of this data situation.

(b) In Exercise 10.51 we find that the residuals for this model are skewed to the right with some large positive outliers. This might cause some problems with a prediction interval that tries to capture this variability. Explain why the 95% prediction interval in the output is not very realistic. (*Hint:* The speed limit on most Atlanta freeways is 55 mph.)

10.54 Checking Conditions for Predicting Housing Prices In Exercise 10.33 on page 623, we use the data in **HomesForSaleNY** to predict prices for houses based on size, number of bedrooms, and number of bathrooms. Use technology to find the residuals for fitting that model and construct appropriate residual plots to assess whether the conditions for a multiple regression model are reasonable.

10.55 Checking Conditions for Predicting Body Mass Gain in Mice In Exercise 10.34 on page 623, we use the data in **LightatNight4Weeks** to predict body mass gain in mice (*BMGain*) over a four-week experiment based on stress levels measured in *Corticosterone*, percent of calories eaten during the day (most mice in the wild eat all calories at night) *DayPct*, average daily consumption of food in grams *Consumption*, and activity level *Activity*. Use technology to find the residuals for that model and construct appropriate residual plots to assess whether the conditions for a multiple regression model are reasonable.

10.56 Checking Conditions for Predicting NBA Winning Percentage In Exercise 10.35 on page 623, we use the data in **NBAStandings2016** to predict NBA winning percentage based on *PtsFor* and *PtsAgainst*. Use technology to find the residuals for fitting that model and construct appropriate residual plots to assess whether the conditions for a multiple regression model are reasonable.

10.57 Checking Conditions for Predicting Mustang Prices In Exercise 10.36 on page 624, we use the data in **MustangPrice** to predict the *Price* of used Mustang cars based on the *Age* in years and number of *Miles* driven. Use technology to find the residuals for fitting that model and construct appropriate residual plots to assess whether the conditions for a multiple regression model are reasonable.

10.3 USING MULTIPLE REGRESSION

For most of this book, data analysis has been restricted to just one or two variables at a time. Multiple regression, allowing for the inclusion of many variables, opens up a whole new world of possibilities! The goal of this section is to delve a little deeper into multiple regression, and to provide a taste of the type of data analysis you are now capable of.

Choosing a Model

When several explanatory variables are available, how do we decide which combination of variables form the best model? If the goal is to use the model for prediction, we want our model to include all explanatory variables that are helpful for predicting the response, but to not include superfluous variables. Choosing a final regression model from several potential predictors is somewhat of an art that often requires a good deal of experience. While the intricacies of model selection are beyond the scope of this course (time to take Stat2?) you can still experiment with competing models and become aware of some criteria available for choosing a model.

Example 10.14

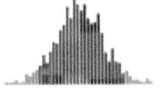

More Predictors for Body Fat

In Section 10.1 we use *Weight, Height,* and *Abdomen* circumference to predict percent body fat, *Bodyfat*. We also have data in **Bodyfat** on *Age* and *Wrist* circumference for the 100 men in this sample. What combination of these predictors will do the best job at predicting percent body fat? As an initial step, run a multiple regression with all five possible explanatory variables included. Determine if the overall model is effective and identify the least effective of the explanatory variables. Should this variable be dropped from the model?

Solution

Here is some computer output for fitting a model for *Bodyfat* based on these five predictors:

Coefficients:

	Estimate	Std. Error	t value	Pr(> |t|)	
(Intercept)	−24.94157	20.77414	−1.201	0.2329	
Weight	−0.08434	0.05891	−1.432	0.1555	
Height	0.05177	0.23849	0.217	0.8286	
Abdomen	0.96762	0.13040	7.421	5.15e-11	***
Age	0.07740	0.04868	1.590	0.1152	
Wrist	−2.05797	0.72893	−2.823	0.0058	**

Residual standard error: 4.074 on 94 degrees of freedom
Multiple R-squared: 0.7542, Adjusted R-squared: 0.7411
F-statistic: 57.67 on 5 and 94 DF, p-value: < 2.2e-16

With all of these predictors, we explain 75.4% of the variability in *Bodyfat* and the p-value for the F-statistic is essentially zero, giving strong evidence that some part of the model is effective for predicting body fat. Looking at the individual t-tests, only the coefficients corresponding to *Abdomen* and *Wrist* are significant. The p-value for *Height* is very large (0.8286), indicating that *Height* is not contributing significantly to the model.

Based on the result of Example 10.14, we drop *Height* from the model and obtain the following output:

Coefficients:

	Estimate	Std. Error	t value	Pr(> |t|)	
(Intercept)	−21.06107	10.52814	−2.000	0.04831	*
Weight	−0.07608	0.04474	−1.700	0.09231	
Abdomen	0.95069	0.10399	9.142	1.13e-14	***
Age	0.07854	0.04815	1.631	0.10620	
Wrist	−2.06898	0.72350	−2.860	0.00521	**

Residual standard error: 4.054 on 95 degrees of freedom
Multiple R-squared: 0.754, Adjusted R-squared: 0.7437
F-statistic: 72.81 on 4 and 95 DF, p-value: < 2.2e-16

First, notice that R^2 has only decreased by 0.0002, from 0.7542 to 0.7540. Removing a predictor, even a predictor entirely unrelated to the response variable, can never explain *more* variability, so rather than looking for an increase in R^2 we check that R^2 has not decreased by any substantial amount. Here R^2 barely changes, so the new model can explain essentially as much variability in body fat without the unnecessary variable *Height*.

Beyond R^2, we can look at several other numbers to compare models. We could compare ANOVA p-values and choose in favor of the model with the smaller p-value, but, as is often the case, here both p-values are too small to observe a meaningful difference. The residual standard error has decreased from 4.074 to 4.054, indicating that the model without *Height* yields somewhat more accurate predictions. The F-statistic has increased from 57.67 to 72.81, more evidence in favor of the model without *Height*. Lastly, a number called "Adjusted R^2" has increased from 0.7411 to 0.7437. *Adjusted* R^2 is essentially R^2, but with a small penalty for including extra variables that are not helping the model. Because of this penalty, removing a superfluous variable can cause *Adjusted R^2* to increase, and the fact that it increases here is further evidence that removing *Height* is a good idea.

Criteria for Choosing a Model

There are several numbers which can help in choosing a model:

- Individual t-test p-values
- R^2
- Residual standard error
- Overall model p-value
- F-statistic from ANOVA
- Adjusted R^2

We want the individual t-test p-values, overall p-value, and residual standard error to be *low*, and R^2, adjusted R^2, and the F-statistic to be *high*.

In practice, we may not find a single model that is best for all of these criteria and we need to use some judgement to balance between them.

Can we improve the body fat model further by removing *Age*, which has the next highest p-value of 0.1062? Let's find out!

Example 10.15

Multiple regression output using *Weight*, *Abdomen* and *Wrist* to predict *Bodyfat* (without *Age*) is given below. Compare this to the previous output to determine whether the model with or without *Age* is better.

Coefficients:

	Estimate	Std. Error	t value	Pr(> |t|)
(Intercept)	−28.75313	9.49382	−3.029	0.003156 **
Weight	−0.12360	0.03425	−3.609	0.000491 ***
Abdomen	1.04495	0.08720	11.983	< 2e-16 ***
Wrist	−1.46586	0.62722	−2.337	0.021513 *

Residual standard error: 4.089 on 96 degrees of freedom
Multiple R-squared: 0.7471, Adjusted R-squared: 0.7392
F-statistic: 94.56 on 3 and 96 DF, p-value: < 2.2e-16

Solution

Some statisticians would say that the model without *Age* is superior, because the p-value for *Age* is insignificant in the earlier model, because removing *Age* only caused R^2 to decrease from 0.754 to 0.747, and because the F-statistic increased from 72.81 to 94.56. Other statisticians would say that the model with *Age* included is superior, because its residual standard error is lower (4.054 to 4.089) and because its Adjusted R^2 is higher in that model (0.7437 to 0.7392).

The above solution is somewhat unsatisfying. Which answer is correct? This is where statistics starts to be just as much an art as a science. Often in model selection there is no "right" answer, and even experienced statisticians will disagree as to which is the best model. You can make your own decision about whether *Age* should be left in the model. This is part of the fun of statistics!

Based on the output in Example 10.15, we would probably not consider removing any of the remaining variables from the model. *Weight*, *Abdomen*, and *Wrist* all yield low individual p-values, indicating that all three of these variables contribute significantly to the model.

Notice that we consider removing variables one at a time, rather than removing all insignificant variables immediately from the original model. This is because coefficients and significance of variables change, depending on what else is included in the model. For example, *Weight* was insignificant (p-value = 0.1555) in the original model including *Height* and *Age*, but became quite significant (p-value = 0.00889) once *Height* and *Age* were removed. It's good practice to consider the removal of one variable at a time, and then reassess the importance of the other variables based on the new model.

The coefficients and p-values of remaining variables change when we remove a variable from a multiple regression model. For this reason, we usually remove one variable at a time and reassess the model at each stage.

Categorical Variables

Multiple regression is powerful because it can include not only quantitative variables as predictors, but, when used correctly, categorical explanatory variables as well. For the model

$$Y = \beta_0 + \beta_1 X_1 + \cdots + \beta_k X_k + \epsilon$$

to make sense, the values of all explanatory variables need to be numbers. We can include a binary categorical variable simply by coding its two categories with 0 and 1.

Matt Herring/Getty Images, Inc.

DATA 10.2 **Gender Discrimination among College Teachers?**

SalaryGender contains data collected in 2010 on a random sample[5] of 100 postsecondary (college) teachers including *Gender* (coded 0 = female and 1 = male), yearly *Salary* (in thousands of dollars), *Age*, and whether or not the teacher has a *PhD* (coded 0 = no and 1 = yes). We are interested in whether the data provide evidence for discrimination based on gender among college teachers. ∎

---

[5] A random sample taken from the 2010 American Community Survey (ACS) 1-year Public Use Microdata Sample (PUMS), *http://www.census.gov/acs/www/data_documentation/public_use_microdata_sample/*.

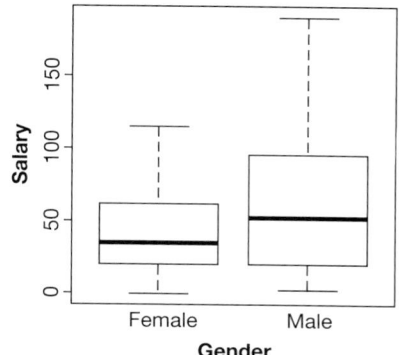

Figure 10.9 *Yearly salary (in thousands of dollars) of college teachers by gender*

Side-by-side boxplots of salary by gender are shown in Figure 10.9. It appears that the males are making more than the females; is this difference significant? The sample difference in means is

$$\bar{x}_m - \bar{x}_f = 63.418 - 41.631 = 21.787$$

On average, the males in the sample make $21,787 more per year than the females. A randomization test or t-test for a difference in means yields a two-tail p-value around 0.0096, indicating that male college teachers have significantly different (and higher) mean salaries than female college teachers.

We can also do this test using regression. (In fact, you may notice that much of what we've covered in earlier chapters of this book can also be accomplished using regression.)

Example 10.16

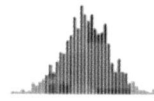

Run a regression model using *Salary* as the response variable and the 0/1 coded *Gender* as the explanatory variable. Comment on how the estimated coefficients relate to the salary means.

Solution

Here is some output for fitting the model $Salary = \beta_0 + \beta_1 Gender + \epsilon$ using the data in **SalaryGender**:

```
Coefficients:
             Estimate   Std. Error   t value   Pr(> |t|)
(Intercept)    41.631        5.796     7.183   1.34e-10 ***
Gender         21.787        8.197     2.658   0.00918  **
```

Residual standard error: 40.98 on 98 degrees of freedom
Multiple R-squared: 0.06724, Adjusted R-squared: 0.05772
F-statistic: 7.065 on 1 and 98 DF, p-value: 0.009181

We notice that the intercept, 41.631, matches the mean salary for the females in the sample. The coefficient of *Gender*, 21.787, is exactly the same as the sample difference in means $(\bar{x}_m - \bar{x}_f)$, and the p-value is quite close to the one we achieved using a test for difference in means.

This is not a coincidence! To interpret this more fully, we need to think a little harder about what it actually means to code a categorical variable to use in a regression model.

From the regression output in Example 10.16, we get the prediction equation

$$\widehat{Salary} = 41.631 + 21.787 \times Gender$$

We find the predicted salary for males using *Gender* = 1:

$$\widehat{Salary} = 41.631 + 21.787 \times 1 = 63.418$$

and the predicted salary for females using *Gender* = 0:

$$\widehat{Salary} = 41.631 + 21.787 \times 0 = 41.631$$

In this sample, the males make an average of $63,418, and the females make an average of $41,631 a year. These results based on the fitted regression model match the means for the females and males in the sample and we see that the slope of the regression measures the difference in those means.

Example 10.17

Salary and PhD

Output from a model using *PhD* (1 = teacher has a PhD, 0 = teacher does not have a PhD) as an explanatory variable for the response variable *Salary* is shown below. Use the regression output to answer the questions that follow.

Coefficients:

	Estimate	Std. Error	t value	Pr(> |t|)
(Intercept)	33.863	4.518	7.496	2.97e-11 ***
PhD	47.850	7.234	6.614	1.98e-09 ***

Residual standard error: 35.28 on 98 degrees of freedom
Multiple R-squared: 0.3086, Adjusted R-squared: 0.3016
F-statistic: 43.75 on 1 and 98 DF, p-value: 1.979e-09

About how big is the difference in mean salary between the two groups in this sample? Is there evidence that college teachers with a PhD earn significantly more, on average, than those without a PhD? Give a 95% confidence interval for the size of the difference in the mean salaries of those with and without PhD's among all college teachers from which this sample was drawn.

Solution

The estimated coefficient of *PhD* for this model is 47.85, which indicates that college teachers with PhD's average about $47,850 more in salary than those without PhD's. The very small p-value of 1.98×10^{-9} for testing this coefficient in the model gives strong evidence that a difference this large would be very surprising to see by random chance alone. Since this slope estimates the difference in means, a 95% confidence interval for the slope can be interpreted as a 95% confidence interval for the difference in the two means. Using a t-distribution with $100 - 2 = 98$ degrees of freedom and the standard error for the slope in the regression output we have

$$b_1 \pm t^* \cdot SE = 47.85 \pm 1.98(7.234) = 47.85 \pm 14.32 = (33.53, 62.17)$$

We are 95% sure that college teachers with a PhD average between $33,530 and $62,170 more dollars a year than college teachers without a PhD.

Categorical variables with more than two levels can be included similarly, with multiple 0-1 variables for multiple levels, although the details are beyond the scope of this course.

We have seen how to use binary categorical variables (with only two possible categories) in a multiple regression model. In general, it is *not* appropriate to simply assign numbers to the categories of a categorical variable with more than two variables to use in a model.

Accounting for Confounding Variables

We have learned that among college teachers, males get paid significantly more than females. However, is this alone evidence of gender discrimination? There may be a confounding variable, or another variable that is associated with both gender and salary, that can explain the salary difference by gender. For example, we've just learned in Example 10.17 that college teachers with a PhD earn significantly more, on average, than those without a PhD. Also, in this sample 48% of the males have a PhD, while only 30% of the females have a PhD. Could this explain the difference in salary due to gender? How do we account for this?

Until now in the course, the only way we've had to deal with confounding variables is to conduct a randomized experiment. With gender and salary, a randomized experiment would be extremely difficult—how would you randomize a teacher to be female? While a randomized experiment is the only way to truly eliminate all confounding variables, multiple regression provides a powerful way to *account for* confounding variables by including them as additional explanatory variables in the model.

To test whether mean salaries are significantly higher for male college teachers, even after accounting for the variable *PhD*, we simply include both *Gender* and *PhD* as explanatory variables in a multiple regression model. This output is below:

Coefficients:

	Estimate	Std. Error	t value	Pr(> |t|)
(Intercept)	28.050	5.383	5.211	1.06e-06 ***
Gender	13.638	7.083	1.926	0.0571
PhD	45.270	7.261	6.235	1.18e-08 ***

Residual standard error: 34.81 on 97 degrees of freedom
Multiple R-squared: 0.3341, Adjusted R-squared: 0.3204
F-statistic: 24.33 on 2 and 97 DF, p-value: 2.725e-09

The fitted model is

$$\widehat{Salary} = 28.05 + 13.64 \cdot Gender + 45.27 \cdot PhD$$

After accounting for whether or not the teachers have a PhD, *Gender* becomes only marginally significant (p-value = 0.0571 is not significant at a 5% level), and the estimated difference in mean salary due to gender drops from $21,878 without accounting for *PhD* to $13,638 after accounting for *PhD*.

Example 10.18

Accounting for both PhD and Age

PhD is not the only potential confounding variable. The sample mean age for males in **SalaryGender** is $\bar{x}_M = 49.32$, and only $\bar{x}_F = 44.44$ for females. Also, regressing *Salary* on *Age*, we find that *Age* and *Salary* are significantly associated, with predicted salary increasing by $1,319 for every year increase in *Age*. We account for *Age* by including it in the model as an additional explanatory variable, with the relevant output given below:

Coefficients:

	Estimate	Std. Error	t value	Pr(> |t|)
(Intercept)	−6.9549	10.8364	−0.642	0.52253
Gender	11.0944	6.7070	1.654	0.10136
PhD	36.4305	7.2534	5.023	2.35e-06 ***
Age	0.8474	0.2318	3.655	0.00042 ***

Residual standard error: 32.78 on 96 degrees of freedom
Multiple R-squared: 0.4154, Adjusted R-squared: 0.3972
F-statistic: 22.74 on 3 and 96 DF, p-value: 3.308e-11

Is gender a significant predictor of salary, after accounting for the age of the teachers and for whether or not they have a PhD?

Solution No. After accounting for *Age* and *PhD*, *Gender* is no longer a significant predictor of *Salary* (p-value = 0.101). In other words, based solely on this dataset, we do not have significant evidence of gender discrimination in salaries for college teachers.

It may be tempting to conclude that all differences in salary due to gender can be explained by the fact that male college teachers tend to be older and more likely to have a PhD, and therefore that gender discrimination does not exist among salaries of college teachers. However, remember that lack of significance does NOT mean the null hypothesis is true! In fact, this is only a subset of a much larger dataset, and running the same model on the larger dataset yields a significant p-value for *Gender*, even after accounting for *PhD* and *Age*.

It also may be tempting to make causal conclusions once we have accounted for confounding variables, but multiple regression only allows us to account for confounding variables *which we have data on*. The only way to really make causal conclusions is to eliminate all possible confounding variables, which can only be done with a randomized experiment.

Association between Explanatory Variables

The ability to include multiple explanatory variables opens up endless possibilities for modeling. However, multiple explanatory variables can also make interpreting models much more complicated, particularly when explanatory variables are associated with each other.

DATA 10.3

Exam Grades

It's that time of year; your first class in statistics is ending, and the final exam is probably on your mind. Wouldn't it be nice to be able to predict your final exam score? **StatGrades** contains data on exam scores for 50 students who have completed a course using this textbook.[6] The dataset contains scores on *Exam1* (Chapters 1 to 4), *Exam2* (Chapters 5 to 8), and the *Final* exam (entire book). ∎

Example 10.19

How well do scores on the first two exams predict scores on the final exam?

Solution We fit a multiple regression model with both exam scores as explanatory variables and final exam score as the response:

Coefficients:

	Estimate	Std. Error	t value	Pr(> |t|)
(Intercept)	30.8952	7.9973	3.863	0.000342 ***
Exam1	0.4468	0.1606	2.783	0.007733 **
Exam2	0.2212	0.1760	1.257	0.215086

Residual standard error: 6.377 on 47 degrees of freedom
Multiple R-squared: 0.5251, Adjusted R-squared: 0.5049
F-statistic: 25.98 on 2 and 47 DF, p-value: 2.515e-08

---

[6]These students were randomly selected from all students who took a course using this textbook and were taught by a particular member of the Lock family.

The first two exam scores explain 52.5% of the variability in final exam scores. We expect the actual final exam score to be within approximately two residual standard errors, or $2 \times 6.38 = 12.76$, of the predicted final exam score.

You may have noticed that *Exam2*, with a p-value of 0.215, is not a significant predictor in this model. The final exam covers material from the entire book, including the material from Chapters 5 to 8 that is tested on *Exam2*, so intuitively it seems as if a student's score on *Exam2* should help predict that student's score on the final. In fact, running a simple linear regression with *Exam2* as the only explanatory variable, we see that *Exam2* is a very significant predictor of *Final* exam score:

Coefficients:

	Estimate	Std. Error	t value	Pr(> |t|)
(Intercept)	32.8699	8.5071	3.864	0.000334 ***
Exam2	0.6332	0.1017	6.227	1.13e-07 ***

Residual standard error: 6.81 on 48 degrees of freedom
Multiple R-squared: 0.4468, Adjusted R-squared: 0.4353
F-statistic: 38.77 on 1 and 48 DF, p-value: 1.129e-07

The key is that *Exam2* is very helpful for predicting *Final* exam score *on it's own*, but if *Exam1* is already in the model, also knowing how the student did on *Exam2* becomes unnecessary and redundant. This is because *Exam1* and *Exam2* are very highly associated ($r = 0.84$), so if *Exam1* is known, also knowing *Exam2* does not contribute that much additional information. The relationships between exam scores are shown in Figure 10.10.

 Particularly when explanatory variables are associated with each other, it is important to remember that estimates and significance of coefficients in multiple regression *depend on the other explanatory variables included in the model*.

Of course, we don't recommend using these models to predict your own final exam score. These are scores of students of one professor so they cannot be generalized to other statistics classes and other exams. Always remember to think about how the data were collected and how this affects the scope of inference!

This section presents several examples of how multiple regression is used. There are entire statistics courses devoted solely to multiple regression, and there is much more to discover on the topic. If this has begun to whet your appetite, we encourage you to take Stat2!

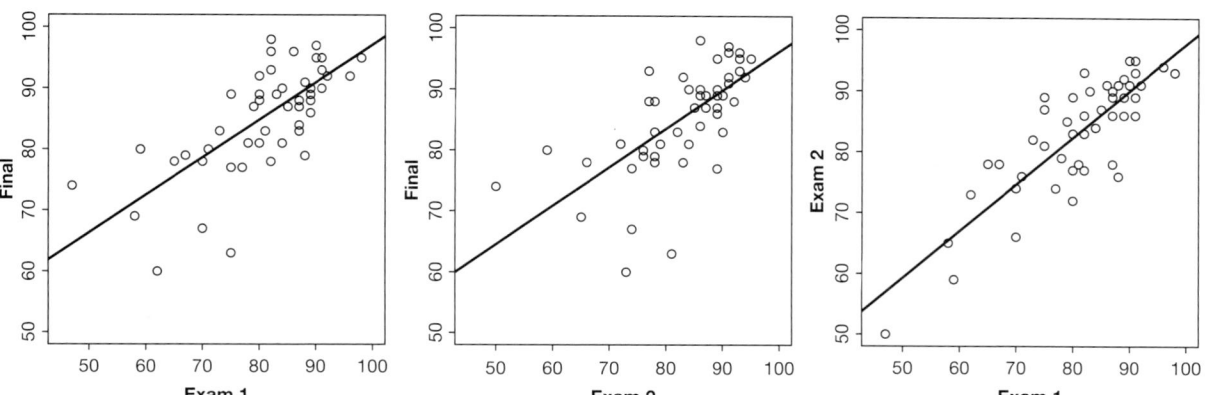

Figure 10.10 *Relationships between Exam1, Exam2, and Final exam scores.*

Exercises for Section 10.3

SKILL BUILDER 1

In Exercises 10.58 and 10.59, use the given output to answer questions about how a model might change.

10.58 Use the multiple regression output shown to answer the following questions.

The regression equation is $Y = 9.78 + 0.244\ X1 + 0.065\ X2 - 0.219\ X3$

Predictor	Coef	SE Coef	T	P
Constant	9.781	4.047	2.42	0.025
X1	0.2440	0.1777	1.37	0.184
X2	0.0653	0.1771	0.37	0.716
X3	−0.2186	0.1706	−1.28	0.214

$S = 4.93734$ R-Sq = 15.0%   R-Sq(adj) = 2.9%

Analysis of Variance

Source	DF	SS	MS	F	P
Regression	3	90.32	30.11	1.23	0.322
Residual Error	21	511.92	24.38		
Total	24	602.24			

(a) Which variable might we try eliminating first to possibly improve this model?

(b) What is R^2 for this model? Do we expect R^2 to *increase*, *decrease*, or *remain the same* if we eliminate the variable chosen in part (a)? What type of change in R^2 would indicate that removing the variable chosen in part (a) was a good idea? A bad idea?

(c) What is the p-value for ANOVA for the original 3-predictor model? Is the p-value most likely to *increase*, *decrease*, or *remain the same* if we eliminate the variable chosen in part (a)? What type of change in the p-value for ANOVA would indicate that removing the variable chosen in part (a) was a good idea? A bad idea?

(d) What is the F-statistic from ANOVA for this model? Is this F-statistic most likely to *increase*, *decrease*, or *remain the same* if we eliminate an insignificant variable? What do we hope is true about any change in this F-statistic when we eliminate such a variable?

10.59 Use the multiple regression output shown to answer the following questions.

The regression equation is $Y = 15.1 + 0.135\ X1 - 0.696\ X2 + 0.025\ X3$

Predictor	Coef	SE Coef	T	P
Constant	15.069	5.821	2.59	0.020
X1	0.1353	0.2354	0.57	0.573
X2	−0.6962	0.3029	−2.30	0.035
X3	0.0253	0.1920	0.13	0.897

$S = 4.92431$ R-Sq = 41.7%   R-Sq(adj) = 30.7%

Analysis of Variance

Source	DF	SS	MS	F	P
Regression	3	277.02	92.34	3.81	0.031
Residual Error	16	387.98	24.25		
Total	19	665.00			

(a) Which variable might we try eliminating first to possibly improve this model?

(b) What is R^2 for this model? Do we expect R^2 to *increase*, *decrease*, or *remain the same* if we eliminate the variable chosen in part (a)? What type of change in R^2 would indicate that removing the variable chosen in part (a) was a good idea? A bad idea?

(c) What is the p-value for ANOVA for the original 3-predictor model? Is the p-value most likely to *increase*, *decrease*, or *remain the same* if we eliminate the variable chosen in part (a)? What type of change in the p-value for ANOVA would indicate that removing the variable chosen in part (a) was a good idea? A bad idea?

(d) What is the F-statistic from ANOVA for this model? Is this F-statistic most likely to *increase*, *decrease*, or *remain the same* if we eliminate an insignificant variable? What type of change in the F-statistic would indicate that removing the variable chosen in part (a) was a good idea?

10.60 Predicting Profitability of Hollywood Movies The dataset **HollywoodMovies** includes information on movies that came out of Hollywood between 2007 and 2013. We want to build a model to predict *OpeningWeekend*, which is the gross income (in millions) from the movie's opening weekend. Start with a model including the following five explanatory variables:

RottenTomatoes	Meta rating of critical reviews, from the Rotten Tomatoes website
AudienceScore	Average audience score, from the Rotten Tomatoes website
TheatersOpenWeek	Number of theaters showing the movie on opening weekend
Budget	Production budget (in millions)
Year	Year movie was released

Eliminate variables (and justify your decisions) and comment on how the model changes. Decide which model you believe is best using only these variables or a subset of them. (Note: Several of the variables have missing values, so the movies used to fit each model might change slightly depending on which predictors are in the model.) Give the model you believe is best and explain why and how you chose it as the best model.

10.61 Predicting Mercury Levels in Fish The dataset **FloridaLakes** includes information on lake water in Florida. We want to build a model to predict *AvgMercury*, which is the average mercury level of fish in the lake. Start with a model including the following four explanatory variables: *Alkalinity*, *pH*, *Calcium*, and *Chlorophyll*. Eliminate variables (and justify your decisions) and comment on how the model changes. Decide which model you believe is best using only these variables or a subset of them. Give the model you believe is best and explain why and how you chose it as the best model.

10.62 Predicting Blood Levels of Beta-Carotene We wish to find a model to predict levels of beta-carotene in the blood, which is the variable

BetaPlasma in the dataset **NutritionStudy**, using the following variables as potential predictors: *Age*, *Fat*, *Fiber*, *Alcohol*, and *BetaDiet*. The last is the amount of beta-carotene consumed by a person.

(a) Use technology to find the correlation between each of the predictors and the response variable *BetaPlasma*. Identify the predictors that appear to be potentially useful based on these correlations.

(b) Try different models and combinations of predictors to help explain the beta-carotene plasma levels. Try to get a good R^2 and a good ANOVA p-value, but also have significant predictors. Decide on a final model and briefly indicate why you chose it.

10.63 Predicting Length of Games in Baseball Baseball is played at a fairly leisurely pace — in fact, sometimes too slow for some sports fans. What contributes to the length of a major league baseball game? The file **BaseballTimes** contains information from a sample of 30 games to help build a model for the time of a game (in minutes). Potential predictors include:

Runs	Total runs scored by both teams
Margin	Difference between the winner's and loser's scores
Hits	Total base hits for both teams
Errors	Total number of errors charged to both teams
Pitchers	Total number of pitchers used by both teams
Walks	Total number of walks issued by pitchers from both teams

(a) Use technology to find the correlation between each of the predictors and the response variable *Time*. Identify the predictors that appear to be potentially useful based on these correlations.

(b) Try different models and combinations of predictors to help explain the game times. Try to get a good R^2 and a good ANOVA p-value, but also have significant predictors. Decide on a final model and briefly indicate why you chose it.

10.64 Life Expectancy and Electricity Use Use the data in **AllCountries** to answer the following questions.

(a) Is electricity use a significant single predictor of life expectancy?

(b) Explain why *GDP* (per-capita Gross Domestic Product) is a potential confounding variable in the relationship between *Electricity* and *LifeExpectancy*.

(c) Is electricity use a significant predictor of life expectancy, even after accounting for *GDP*?

10.65 Life Expectancy and Cell Phones Use the data in **AllCountries** to answer the following questions.

(a) Is the number of mobile subscriptions per 100 people, *Cell*, a significant single predictor of life expectancy?

(b) Explain why *GDP* (per-capita Gross Domestic Product) is a potential confounding variable in the relationship between *Cell* and *Life-Expectancy*.

(c) Is *Cell* a significant predictor of life expectancy, even after accounting for *GDP*?

10.66 Predicting Prices of New Homes In Exercise 10.23 on page 621 we fit a model predicting the price of a home (in $1000s), using size (in square feet), number of bedrooms, and number of bathrooms, based on data in **HomesForSale**. Output for fitting a slightly revised model is shown below where the *Size* variable is measured in 1000s of square feet (rather than just the raw square footage in *SizeSqFt* in the previous model):

```
Coefficients:
              Estimate   Std. Error   t value   Pr(> |t|)
(Intercept)   -217.00      145.90      -1.487    0.1396
Size           330.58       72.62       4.552    1.32e-05 ***
Beds          -134.52       57.03      -2.359    0.0200 *
Baths          200.03       78.94       2.534    0.0126 *

Residual standard error: 507.7 on 116 degrees of freedom
Multiple R-squared: 0.4671, Adjusted R-squared: 0.4533
F-statistic: 33.89 on 3 and 116 DF, p-value: 8.383e-16
```

(a) Compare this output to the regression output in Exercise 10.23 and comment on how the coefficient, standard error, and t-statistic for *Size* change when we code the variable in 1000s of square feet rather than square feet as in *SizeSqFt*.

(b) Interpret the coefficient for *Beds*, the number of bedrooms in this fitted model.

(c) An architect (who has not taken statistics) sees this output and decides to build houses with fewer bedrooms so they will sell for more money. As someone who has taken statistics, help him to correctly interpret this output.

ARE CARBON OR STEEL BIKES FASTER?
Exercises 10.67 through 10.70 refer to data introduced in Exercise C.113 on page 502 from an experiment in which Dr. Jeremy Groves flipped a coin each day to randomly decide whether to ride his 20.9 lb (9.5 kg) carbon bike or his 29.75 lb (13.5 kg) steel bike for his 27-mile round-trip commute. His data for 56 days are stored in **BikeCommute**. The type of bicycle (carbon or steel) is the *Bike* variable and his time (in minutes) is stored in *Minutes*. We've created a new variable *BikeSteel*, which is 1 if the ride is on a steel bike and 0 if the ride is on a carbon bike.

10.67 Minutes vs Bike Type Output regressing *Minutes* on *BikeSteel* is shown below.

```
Response: Minutes
Coefficients:
              Estimate   Std. Error   t value   Pr(> |t|)
(Intercept)   108.342       1.087      99.624    <2e-16 ***
BikeSteel      -0.553       1.486      -0.372    0.711

Residual standard error: 5.545 on 54 degrees of freedom
Multiple R-squared: 0.002558, Adjusted R-squared: -0.01591
F-statistic: 0.1385 on 1 and 54 DF, p-value: 0.7112
```

(a) What is Dr. Grove's predicted commute time if riding the steel bike?

(b) What is Dr. Grove's predicted commute time if riding the carbon bike?

(c) Based on this experiment, is there a significant difference between the commute time for the carbon bike and steel bike?

10.68 Distance vs Bike Type The commute is about 27 miles round trip, but actual biking distances, *Distance*, ranged from 25.86 to 27.52 miles. Output regressing *Distance* on *BikeSteel* is given below. Is the predicted distance higher for the carbon or the steel bike? By how much? Is this difference significant?

```
Response: Distance
Coefficients:
              Estimate   Std. Error   t value   Pr(> |t|)
(Intercept)   27.37446     0.04164    657.363    < 2e-16 ***
BikeSteel     -0.39613     0.05689     -6.962    4.74e-09 ***

Residual standard error: 0.2123 on 54 degrees of freedom
Multiple R-squared: 0.473, Adjusted R-squared: 0.4633
F-statistic: 48.48 on 1 and 54 DF, p-value: 4.741e-09
```

10.69 Minutes vs Distance and Bike Type *Distance* is associated with both the type of bike and commute time, so if we are really interested in which type of bike is faster, we should account for the confounding variable *Distance*. Output regressing *Minutes* on both *BikeSteel* and *Distance* (measured in miles) is shown below.

```
Response: Minutes
Coefficients:
              Estimate   Std. Error   t value   Pr(> |t|)
(Intercept)   -176.620     90.065      -1.961    0.05514
BikeSteel        3.571      1.895       1.884    0.06500
Distance        10.410      3.290       3.164    0.00258 **

Residual standard error: 5.133 on 53 degrees of freedom
Multiple R-squared: 0.161, Adjusted R-squared: 0.1294
F-statistic: 5.087 on 2 and 53 DF, p-value: 0.00953
```

(a) Interpret the coefficient of *BikeSteel*.

(b) Interpret the coefficient of *Distance*.

(c) What is the predicted commute time for a 27-mile commute on the steel bike? On the carbon bike?

10.70 Predicting Average Bike Speed In Exercise 10.67, regressing *Minutes* on *BikeSteel*, the coefficient for *BikeSteel* is negative. In Exercise 10.69, regressing *Minutes* on *BikeSteel* and *Distance*, the coefficient for *BikeSteel* is positive.

(a) A biker interested in whether carbon or steel bikes are faster is not sure what to make of these seemingly contradictory results. Explain to her why the coefficient can be negative in one model and positive in the other. (*Hint:* In Exercise 10.68 we see regressing *Distance* on *BikeSteel* yields a negative coefficient for *BikeSteel*.)

(b) If we were to regress the variable *AvgSpeed* = *Distance*/*Minutes* on *BikeSteel*, would the coefficient for *BikeSteel* be negative or positive? Explain. (You should not need technology to answer this question.)

Summary of Inference for Multiple Parameters

In Unit D, we discuss the methods for inference with multiple parameters; either pertaining to multiple categories in a categorical variable (chi-square tests and analysis of variance) or multiple parameters in a regression model.

Chi-square tests and analysis of variance extend the hypothesis tests of Chapters 4 and 6 to allow for categorical variables with multiple categories. Chi-square tests allow for testing one or two categorical variables with multiple categories, and analysis of variance allows for testing one quantitative and one categorical variable with multiple categories.

Regression involves building models to predict a quantitative response variable based on one quantitative explanatory variable (simple regression), or multiple explanatory variables (multiple regression). Most of the inferential methods presented in this book pertain to one or two variables, but multiple regression provides a way of incorporating more than two variables.

As with the rest of the book, the appropriate method of analysis is determined by the type of variables, categorical or quantitative, although now we also have to determine whether the categorical variables have two categories or more than two categories. The appropriate methods of inference, based on the number of variables, whether the variable(s) are categorical or quantitative, and whether the categorical variable(s) have two or more categories, are summarized in Table D.1.

Table D.1 *Guide to choosing the appropriate method based on the variables and number of categories*

Variables	Number of Categories	Appropriate Inference
One Categorical	Two Categories	Single Proportion or Chi-Square Goodness of Fit
	More Categories	Chi-Square Goodness of Fit
One Quantitative	—	Single Mean
Two Categorical	Two Categories	Difference in Proportions or Chi-Square Test for Association
	More Categories	Chi-Square Test for Association
One Categorical, One Quantitative	Two Categories	Difference in Means or Analysis of Variance
	More Categories	Analysis of Variance
Two Quantitative	—	Correlation, Simple Regression
Quantitative Response, Multiple Explanatory	—	Multiple Regression
Categorical Response, Multiple Explanatory	—	Take STAT2!

Hypothesis Testing

Although the methods of this unit are quite different, they all include hypothesis testing. In every instance of hypothesis testing:

- We have a null and an alternative hypothesis, a test statistic, a p-value, and a conclusion to reject or not reject the null hypothesis.

- The null hypothesis is usually that nothing interesting is going on (no association, proportions are as specified, variable not useful in model), whereas the alternative is that there is something interesting going on.
- A smaller p-value provides more evidence to support the alternative hypothesis, often that there is a relationship between variables.
- The fundamental question in every case: Do the sample data provide enough evidence against the null hypothesis to rule out random chance as an explanation for the data (if the null hypothesis is true)?

Although we encounter new distributions and more sophisticated computations, the basic ideas all build on the fundamental concepts we introduce in Chapter 4 and have carried throughout the text.

Confidence Intervals

The methods in Unit D tend to be more focused on testing hypotheses (is there a relationship between these variables?) rather than estimating with confidence intervals (how big is the effect?). In many of the settings, we use intervals *after* running an initial test to see if a relationship exists or *after* fitting a model. For example:

- After an ANOVA for difference in means, find a confidence interval for the difference in means between a specific pair of groups.
- After finding an effective simple linear regression model, find a prediction interval for the response for a specific value of the explanatory variable.
- After fitting a regression model, find a confidence interval for the coefficient (slope) for a predictor.

For each of these intervals, we use the familiar formula

$$\text{Sample Statistic} \pm t^* \cdot SE$$

where we use technology or a formula to find the appropriate standard error.

Chi-Square Tests: Tests for Categorical Variables

Chi-square tests are used for testing hypotheses about one or two categorical variables and are appropriate when the data can be summarized by counts in a table. The variables can have multiple categories. The type of chi-square test depends on whether there are one or two categorical variables:

- One Categorical Variable: Chi-Square Goodness-of-Fit Test
- Two Categorical Variables: Chi-Square Test for Association

Chi-square tests compare observed counts to expected counts (if the null hypothesis were true). If the observed counts are farther away from the expected counts than can be explained just by random chance, we have evidence against the null hypothesis and in favor of the alternative. The distance between observed and expected counts is quantified with the χ^2-statistic, which is compared to a χ^2-distribution to calculate the p-value. The details are laid out below:

1. State hypotheses

- For one categorical variable:
  - Null hypothesis: The proportions match some pre-assumed set of proportions.
  - Alternative hypothesis: At least one category has a proportion different from the null values.

- For two categorical variables:
  - Null hypothesis: There is no association between the variables (distribution of proportions for one variable is the same within each category of the second variable).
  - Alternative hypothesis: There is an association between the variables.

2. Summarize the data in a table with observed counts

3. Calculate the expected counts for each cell (as if the null hypothesis were true)

- For one categorical variable:
  Expected count for a cell $= n \cdot p_i$, where p_i is given in H_0.
- For two categorical variables:
  $$\text{Expected count for a cell} = \frac{\text{Row total} \cdot \text{Column total}}{\text{Total sample size}}.$$

4. Compute the χ^2-statistic:

$$\chi^2 = \sum \frac{(Observed - Expected)^2}{Expected}$$

5. Find the p-value as the upper tail in a χ^2-distribution

- For one categorical variable: $df = k - 1$, where k is the number of categories in the variable.
- For two categorical variables: $df = (r - 1) \cdot (c - 1)$, where r is the number of rows (categories in one variable) and c is the number of columns (categories in the other).

6. Make a conclusion

- If the results are significant, we have evidence in favor of the alternative hypothesis. A more informative conclusion can be given by comparing the relative sizes of observed and expected counts of individual cells, and the relative contribution of cells to the chi-square statistic.

With only two categories the chi-square goodness-of-fit test is equivalent to a test for a single proportion, and the chi-square test for association is equivalent to a test for a difference in two proportions.

Analysis of Variance: Test for a Difference in Means

Analysis of variance is used to test for an association between one quantitative variable and one categorical variable or, equivalently, to test for a difference in means across categories of a categorical variable. The categorical variable can have multiple categories. This method is appropriate when the summary statistics include sample means calculated within groups.

Analysis of variance compares variability within groups to variability between groups. If the ratio of variability between groups to variability within groups is higher than we would expect just by random chance, we have evidence of a difference in means. This ratio is called the F-statistic, which we compare to an F-distribution to find the p-value. The details are laid out below.

1. State hypotheses

- Null hypothesis: $\mu_1 = \mu_2 = \cdots = \mu_k$ (no difference in means by category).
- Alternative hypothesis: Some $\mu_i \neq \mu_j$ (difference in means between categories).

2. Compute the F-statistic using an ANOVA table:

Source	df	Sum of Sq.	Mean Square	F-statistic	p-value
Groups	$k-1$	SSG	$MSG = \dfrac{SSG}{k-1}$	$F = \dfrac{MSG}{MSE}$	Upper tail $F_{k-1, n-k}$
Error	$n-k$	SSE	$MSE = \dfrac{SSE}{n-k}$		
Total	$n-1$	SSTotal			

The sums of squares $SSTotal = SSG + SSE$ are obtained by technology or formula.

3. Find the p-value as the upper tail in an F-distribution

- Use df for Groups and df for Error from the ANOVA table.

4. Make a conclusion

- If the results are significant, we have evidence of an association between the variables (and a difference in means between the groups defined by the categorical variable). A more informative conclusion can be given if desired by using the methods of pairwise comparison presented in Section 8.2.

  If the categorical variable has only two categories, analysis of variance is equivalent to a test for a difference in means between two groups.

Inference after ANOVA: Confidence Intervals or Pairwise Tests

- Use t-distribution with Error df and $\sqrt{MSE}$ from ANOVA to estimate variability. Use technology or see formulas on page 565.

Regression

Regression is used to predict a quantitative response variable based on one or more explanatory variables, and to model relationships between explanatory variable(s) and a quantitative response variable. In order to use regression, all variables need to be measured on the same set of cases.

The simple linear regression model (one quantitative explanatory variable) is introduced in Section 2.6, and Chapter 9 extends this analysis to include inference. Multiple regression extends simple linear regression to include multiple explanatory variables.

Some important aspects of regression are summarized below:

- R^2 gives the percent of variability in the response variable that is explained by the explanatory variable(s) in the model.
- Test for Correlation (only for simple regression)

  –Null hypothesis: There is no linear relationship between the variables ($\rho = 0$)

  –Test statistic: $t = \dfrac{r\sqrt{n-2}}{\sqrt{1-r^2}}$

  –Distribution: t-distribution with $df = n - 2$
- Test for Slope (in simple regression this is equivalent to a test for correlation)

  –Null hypothesis: The variable is not significant in the model ($\beta_i = 0$)

  –Test statistic: $t = \dfrac{b_i - 0}{SE}$, where SE is the standard error of the slope

  –Distribution: t-distribution with $df = n - k - 1$, where k is the number of explanatory variables

- Analysis of Variance for Regression (check overall model fit)
    - Null hypothesis: The model is not effective at predicting the response variable
    - Test statistic: F-statistic from an ANOVA table (see details on page 594)
    - Distribution: Upper tail in an F-distribution with *df* for Model and *df* for Error
- Conditions for Regression
    - In a scatterplot (in simple regression) or residuals vs fits plot (in multiple regression), watch out for curvature (or any non-linear trend), increasing or decreasing variability, or outliers.
    - In a histogram or dotplot of the residuals, watch out for obvious skewness or outliers.
- Regression intervals (use technology or formulas on page 605)
    - Confidence interval for the mean response at some specific explanatory value
    - Prediction interval for an individual response at some specific explanatory value
- Variables in Multiple Regression
    - The coefficient and significance of each explanatory variable depend on the other explanatory variables included in the model.
    - More variables are not always better; consider pruning insignificant variables from the model.

© Klubovy/iStockphoto

Is this an early morning class?

Case Study: Sleep, Circadian Preference, and Grade Point Average

In Data A.1 on page 163, we introduce a study examining the relationship between class start times, sleep, circadian preference, alcohol use, academic performance, and other variables in college students. The data were obtained from a sample of students who did skills tests to measure cognitive function, completed a survey that asked many questions about attitudes and habits, and kept a sleep diary to record time and quality of sleep over a two-week period. Some data from this study, including 27 different variables, are available in **SleepStudy**. When we introduced this study in Unit A, we had only finished Chapters 1 and 2 and could only describe the data. We return to some of the same questions now, when we have many more powerful statistical tools at our disposal.

Example D.1

Are You a Lark or an Owl?

Past studies[7] indicate that about 10% of us are morning people (Larks) while 20% are evening people (Owls) and the rest aren't specifically classified as either. Studies also indicate that this circadian preference may not be settled until 22 years of age or later. Table D.2 shows the number in each category for the 253 college students in the study described in Data A.1. Is there evidence that the owl/lark preferences for college students differ from the claimed proportions?

Table D.2 *Circadian preference: Are you a lark or an owl?*

Type	Frequency
Lark	41
Neither	163
Owl	49
Total	253

Solution

We are comparing frequency counts from a sample to some preconceived proportions, so we use a chi-square goodness-of-fit test for this analysis. The hypotheses are

$$H_0 : p_L = 0.1, \ p_N = 0.7, \ p_O = 0.2$$
$$H_a : \text{Some } p_i \text{ is wrong}$$

where p_L, p_N, p_O represent the proportion in each category of Lark, Neither, Owl, respectively, for the population of college students represented by this sample. We find the expected counts using $n \cdot p_i$, so we have

$$E_L = 253(0.1) = 25.3 \qquad E_N = 253(0.7) = 177.1 \qquad E_O = 253(0.2) = 50.6$$

All of the expected counts are well above 5 so we proceed with a chi-square test. We find the chi-square statistic using

$$\chi^2 = \frac{(41 - 25.3)^2}{25.3} + \frac{(163 - 177.1)^2}{177.1} + \frac{(49 - 50.6)^2}{50.6}$$
$$= 9.743 + 1.123 + 0.051 = 10.917$$

We use the upper tail of a chi-square distribution to find the p-value, and there are three categories so $df = 3 - 1 = 2$. The area above $\chi^2 = 10.917$ gives a p-value of 0.004. This is a small p-value and we find evidence that the proportions are significantly different than those expected. The largest contribution comes from the "Lark" cell and we see (surprisingly, for college students!) that there are more Larks than expected.

Example D.2

Stress and Circadian Preference

Are stress levels of college students affected by circadian preference? Test to see if there is evidence of a relationship between the variables *LarkOwl* and *Stress*. Stress level is identified as either normal or high and circadian preference is identified as lark, owl, or neither. Table D.3 shows the counts in each cell.

[7] *http://www.nasw.org/users/llamberg/larkowl.htm.*

Table D.3 *Stress levels and circadian preference*

	Lark	Neither	Owl	Total
Normal	31	125	41	197
High	10	38	8	56
Total	41	163	49	253

Solution

The data are frequency counts for two categorical variables, so we use a chi-square test of association for the analysis. The null hypothesis is that there is no relationship between *Stress* and *LarkOwl* status and the alternative hypothesis is that there is a relationship. We find the expected count in each cell, and then the contribution to the chi-square statistic, using

$$\text{Expected count} = \frac{\text{Row total} \cdot \text{Column total}}{\text{Total sample size}}$$

$$\text{Contribution} = \frac{(Observed - Expected)^2}{Expected}$$

The computer output below shows, for each cell, the observed count from the table with the expected count below it and the contribution to the chi-square statistic below that:

	Lark	Neither	Owl
Normal	31	125	41
	31.92	126.92	38.15
	0.027	0.029	0.212
High	10	38	8
	9.08	36.08	10.85
	0.094	0.102	0.747

We add up all the contributions to obtain the χ^2-statistic:

$$\chi^2 = 0.027 + 0.029 + 0.212 + 0.094 + 0.102 + 0.747 = 1.211$$

We use a chi-square distribution to find the p-value, with $df = (r-1) \cdot (c-1) = 1 \cdot 2 = 2$. We find a p-value of 0.546. This is a large p-value and does not offer sufficient evidence that circadian preference and stress levels are related.

Example D.3

Are Cognitive Skills and Alcohol Use Related?

One of the variables in **SleepStudy** is *CognitionZscore* which assigns each person a z-score based on results on several cognitive skills tests, with higher scores indicating stronger cognitive ability. Another is *AlcoholUse* which shows self-reported levels of alcohol use in one of four categories: Abstain, Light, Moderate, or Heavy. Is there a relationship between cognitive skills and alcohol use? Summary statistics are given in Table D.4 and side-by-side boxplots are shown in Figure D.1.

Solution

Notice that the light drinkers have the highest mean cognitive score while the heavy drinkers have the lowest. Is the difference between these four alcohol use groups significant? We are investigating a relationship between a quantitative variable

Table D.4 *Cognitive skills and alcohol use*

	Sample Size	Mean	Std.Dev.
Abstain	34	0.0688	0.7157
Light	83	0.1302	0.7482
Moderate	120	−0.0785	0.6714
Heavy	16	−0.2338	0.6469
Overall	253	0.000	0.7068

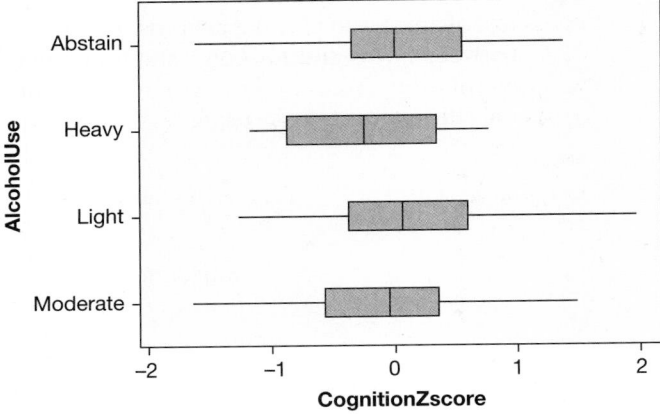

Figure D.1 *Is there a difference in cognitive skills based on alcohol use?*

(*CognitionZscore*) and a categorical variable (*AlcoholUse*), so the appropriate test is analysis of variance for a difference in means. The relevant hypotheses are:

H_0 : Mean cognition score is the same for each alcohol use group

H_a : Mean cognition score differs among some alcohol use groups

We see in Table D.4 and Figure D.1 that the normality and equal variance conditions appear to be met, so we proceed to construct the ANOVA table.

The Groups df is $k - 1 = 4 - 1 = 3$ in this case, while the Total df is $n - 1 = 253 - 1 = 252$ and the Error df is $n - k = 253 - 4 = 249$. We find the sum of squares using the formulas or using technology. Using the formulas (page 552), we have

$$SSG = 34(0.069 - 0)^2 + 83(0.130 - 0)^2 + 120(0.079 - 0)^2$$
$$+ 16(-0.234 - 0)^2 = 3.2$$
$$SSError = 33(0.7157^2) + 82(0.7482^2) + 119(0.6714^2) + 15(0.6469^2) = 122.7$$
$$SSTotal = 252(0.7068)^2 = 125.9$$

Completing the ANOVA table, we arrive at the results shown in the following computer output:

```
Source        DF        SS        MS       F       P
AlcoholUse     3      3.183     1.061    2.15    0.094
Error        249    122.718     0.493
Total        252    125.901
```

The p-value is found using the F-distribution with 3 numerator df and 249 denominator df. The p-value of 0.094 is not significant at a 5% level, so we do not find evidence of a difference in mean cognition score based on alcohol use. At a 10% level, however, we do find this evidence. As always, we should be careful not to infer cause and effect (in either direction) since these data come from an observational study.

Example D.4

Comparing Cognitive Skill between Light and Heavy Drinkers

Based on the ANOVA results of Example D.3, find and interpret a 95% confidence interval for the difference in mean cognitive skill z-scores between students who classify themselves as light and heavy alcohol users.

Solution

From the ANOVA output we find $MSE = 0.493$ with 249 degrees of freedom. For 95% confidence and this many degrees of freedom we find $t^* = 1.97$. Using the means and sample sizes for the light and heavy categories of Table D.4 we compute the confidence interval for the difference in means as

$$(\bar{x}_L - \bar{x}_H) \pm t^* \cdot \sqrt{MSE \left(\frac{1}{n_L} + \frac{1}{n_H} \right)}$$

$$(0.1302 - (-0.2338)) \pm 1.97 \cdot \sqrt{0.493 \left(\frac{1}{83} + \frac{1}{16} \right)}$$

$$0.3640 \pm 0.3780$$

$$-0.014 \text{ to } 0.742$$

We are 95% sure that students who classify themselves as light alcohol users have an average cognitive z-score that is somewhere between 0.014 points lower and 0.742 points higher than students classified as heavy alcohol users. Note that this interval includes zero (no difference at a 5% level) as we expect given the result of the ANOVA test in Example D.3.

Example D.5

Early Classes and Grade Point Average

The correlation between number of early classes (starting at or before 8:30 am) per week and grade point average is $r = 0.101$ with $n = 253$. Explain what a positive correlation means in this situation, and test whether this sample correlation provides evidence of an association between these two variables.

Solution

A positive correlation implies that grades tend to be higher for those taking more early classes. To test for an association, we test $H_0 : \rho = 0$ vs $H_a : \rho \neq 0$. The t-statistic is

$$t = \frac{r\sqrt{n-2}}{\sqrt{1-r^2}} = \frac{0.101\sqrt{251}}{\sqrt{1-(0.101^2)}} = 1.608$$

Using a t-distribution with $n - 2 = 253 - 2 = 251$ degrees of freedom, and using a two-tail test, we find a p-value of $2(0.0545) = 0.109$. We do not find sufficient evidence to show an association between the number of early classes and grade point average.

Example D.6

Sleep Quality and DASScore

Students were rated on sleep quality and the results are in the quantitative variable *PoorSleepQuality*, with higher values indicating poorer sleep quality. Students were also rated on Depression, Anxiety, and Stress scales, with the DAS score (*DASScore*) giving a composite of the three scores, with higher values indicting more depression, anxiety, and/or stress. How well does the DAS score predict sleep quality?

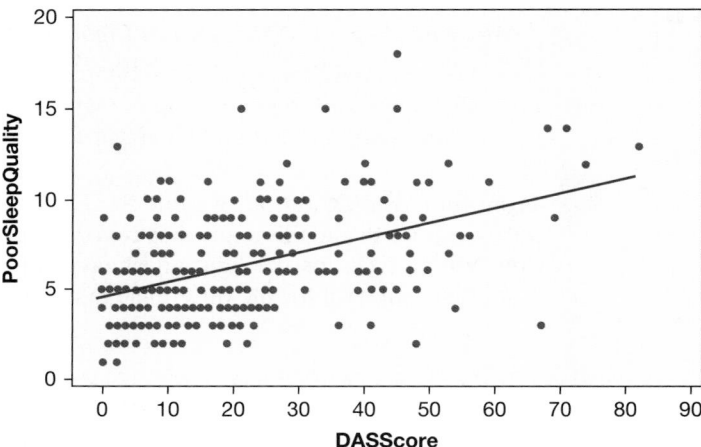

Figure D.2 *Sleep quality and depression scores*

A scatterplot of the data is in Figure D.2 and computer output for the regression analysis is given below:

```
The regression equation is
PoorSleepQuality = 4.64 + 0.0806 DASScore

Predictor              Coef      SE Coef         T        P
Constant             4.6418       0.2574     18.04    0.000
DASScore           0.080594     0.009912      8.13    0.000

S = 2.60279     R-Sq = 20.8%     R-Sq(adj) = 20.5%

Analysis of Variance
Source              DF          SS        MS        F        P
Regression           1      447.90    447.90    66.12    0.000
Residual Error     251     1700.40      6.77
Total              252     2148.30
```

(a) Do the conditions for fitting a linear model appear to be met?

(b) Interpret R^2 in this context.

(c) Identify and interpret the t-test of the slope.

(d) What conclusion can we draw based on the ANOVA table?

Solution

(a) Judging from the scatterplot of the data, the conditions for a linear model appear to be met. There is no obvious curvature and the data appear to be scattered in roughly parallel bands above and below the least squares line.

(b) We see that $R^2 = 20.8\%$, so 20.8% of the variability in sleep quality can be explained by students' DAS scores.

(c) The slope is 0.0806 and we see from the large t-statistic of 8.13 and small p-value of 0.000 that DAS score is an effective predictor of sleep quality.

(d) From the ANOVA test, we see that the F-statistic is 66.12 and the p-value is 0.000. This simple linear regression model based on DAS score is effective at predicting sleep quality.

Example D.7

Predicting Sleep Quality for a Specific Student

Suppose that one of the students at this college has a fairly high DAS score of 40. Predict the sleep quality for this student and find an interval that will be 95% sure to contain the actual value for her *PoorSleepQuality*.

Solution   Using the regression equation from Example D.6 we find that the predicted *PoorSleepQuality* score for a student with *DASScore* = 40 is

$$\widehat{PoorSleepQuality} = 4.642 + 0.0806(40) = 7.87$$

We use technology to request regression intervals when *DASScore* = 40 to produce the output below:

New Obs	Fit	SE Fit	95% CI	95% PI
1	7.866	0.257	(7.360, 8.371)	(2.715, 13.017)

Since we need an interval to contain the sleep quality for a particular student, we use the prediction interval. We are 95% sure that a student with DAS score of 40 will have a poor sleep quality score somewhere between 2.7 and 13.0.

Example D.8

Predicting Grade Point Average with Multiple Predictors

We create a multiple regression model to predict grade point average (*GPA*) from the number of early classes, the number of classes missed, the quality of sleep, a happiness score (with higher values indicating greater happiness), the number of alcoholic drinks per week, and the average number of hours of sleep per night. Graphs of residuals raise no serious concerns about performing the analysis, and computer output is shown:

The regression equation is
GPA = 3.77 + 0.0111 NumEarlyClass − 0.0198 ClassesMissed
    − 0.00424 PoorSleepQuality − 0.00244 Happiness − 0.0254 Drinks
    − 0.0339 AverageSleep

Predictor	Coef	SE Coef	T	P
Constant	3.7699	0.2825	13.34	0.000
NumEarlyClass	0.01110	0.01634	0.68	0.497
ClassesMissed	−0.019795	0.007907	−2.50	0.013
PoorSleepQuality	−0.004241	0.009540	−0.44	0.657
Happiness	−0.002439	0.004774	−0.51	0.610
Drinks	−0.025418	0.006040	−4.21	0.000
AverageSleep	−0.03388	0.02623	−1.29	0.198

S = 0.386280   R-Sq = 10.9%   R-Sq(adj) = 8.7%

Analysis of Variance

Source	DF	SS	MS	F	P
Regression	6	4.4822	0.7470	5.01	0.000
Residual Error	246	36.7062	0.1492		
Total	252	41.1884			

(a) Interpret (in context) the signs of the coefficients for *NumEarlyClasses* and *ClassesMissed*.

(b) Which of the six explanatory variables is most significant in the model? Is the coefficient of this variable positive or negative? Interpret the sign in context.

(c) Which variables are significant at a 5% level?

(d) Use output in the ANOVA table to determine whether the overall model is effective at predicting GPA.

(e) Interpret R^2.

(f) How might we try to improve the model?

Solution   (a) The coefficient of *NumEarlyClasses* is positive, which means that, given the other variables in the model, GPA tends to go up as the number of early classes goes up. (This makes sense, since studies show that more motivated students tend to be more likely to take early classes.) The coefficient of *ClassesMissed* is negative, which means that, given the other variables in the model, GPA tends to go down as the number of classes missed goes up. (This also makes sense, since more motivated students are less likely to miss class.)

(b) The variable *Drinks* is most significant in the model. The coefficient is negative which means that, given the other variables in the model, GPA tends to be lower for students who have more alcoholic drinks. This variable is very significant in the model, with a p-value of 0.000.

(c) Only two variables are significant at a 5% level: *Drinks* and *ClassesMissed*. Both have negative coefficients.

(d) The p-value from the ANOVA table is 0.000 so the model as a whole is effective at predicting grade point average.

(e) We see that 10.9% of the variability in grade point averages can be explained by these six explanatory variables.

(f) Several of the explanatory variables (such as *PoorSleepQuality* and *Happiness*) are very insignificant in the model. It makes sense to try eliminating one of them from the model and running the regression again. For example, using *ClassesMissed*, *Drinks*, and *AverageSleep* will still give an R^2 of 10.5%. Are there other explanatory variables that might do better? Try it and see!

There are many interesting variables in this dataset and much more analysis that can be conducted. Make up some of your own questions using these variables. Then use technology and the dataset to see what you can discover!

SECTION LEARNING GOALS

You should now have the understanding and skills to:

- Identify which type of inference for multiple categories or multiple variables is appropriate in a given situation

- Put all the pieces together to answer more involved questions using real data for which chi-square, ANOVA, or regression analysis is appropriate

---

Exercises for UNIT D: Essential Synthesis

RESTAURANT TIPS

In Data 2.12 on page 123, we introduce the dataset **RestaurantTips** containing information on the tipping patterns of patrons of the *First Crush* bistro in northern New York state. The data from 157 bills include the amount of the bill, size of the tip, percentage tip, number of customers in the group, whether or not a credit card was used, day of the week, and a coded identity of the server. The first four variables are quantitative and the last three are categorical. In Exercises D.1 to D.10, we use the **RestaurantTips** dataset to analyze relationships between these variables.

D.1   Do the Servers Serve Equal Numbers of Tables? The data come from three different servers, coded as A, B, and C to preserve anonymity. The number of bills for each server is shown in Table D.5. Do the servers serve equal numbers of tables?

Table D.5 *Number of bills by server*

Server	A	B	C	Total
Number of Bills	60	65	32	157

D.2   Are Bills Evenly Distributed between the Days of the Week? The number of bills for each day of the week is shown in Table D.6. Does this provide evidence that some days of the week are more popular (have more bills) than others?

Table D.6 *Number of bills by day*

Day	Mon	Tues	Wed	Thurs	Fri	Total
Number of Bills	20	13	62	36	26	157

D.3   Are Credit Cards used Equally Often among the Three Servers? The data come from three different servers, coded as A, B, and C, and we also have information on whether or not a credit (or debit) card was used to pay the bill rather than cash. The frequency counts are shown in the two-way table in Table D.7. At a 5% significance level, is there an association between who the server is and whether the bill is paid in cash or with a credit/debit card?

Table D.7 *Cash or credit card by server*

	A	B	C	Total
Cash	39	50	17	106
Card	21	15	15	51
Total	60	65	32	157

D.4   Are Credit Cards used Equally Often between the Days of the Week? The data come from all five week days and we also have information on whether or not a credit (or debit) card was used to pay the bill rather than cash. The frequency counts are shown in the two-way table in Table D.8. Find the chi-square statistic for this two-way table. Are the conditions met to use the chi-square distribution? Why or why not? Conduct the appropriate test to determine whether there is evidence of an association between the day of the week and whether the bill is paid in cash or with a credit/debit card.

Table D.8 *Cash or credit card by day*

	Mon	Tues	Wed	Thurs	Fri	Total
Cash	14	5	41	24	22	106
Card	6	8	21	12	4	51
Total	20	13	62	36	26	157

D.5   Does Average Tip Percentage Vary by Server? Most restaurant patrons leave a tip between 15% and 20% of the bill, and some restaurant customers determine the percent tip to leave based on the quality of service. Summary statistics for mean tip percentage left between the three servers, coded A, B, and C, are given in Table D.9. Is there a difference in mean tip percentage between the three servers? Be sure to check conditions of the test, and conduct the appropriate test.

Table D.9 *Percent tip by server*

Server	Sample Size	Mean	Std.Dev.
A	60	17.543	5.504
B	65	16.017	3.485
C	32	16.109	3.376
Overall	157	16.619	4.386

D.6   Does Size of the Bill Vary by Server? Are some servers given the big spenders (or large groups) while others tend to those having only a cup of coffee or a glass of wine? Is there a difference in the mean size of the bill between the three different servers? Summary statistics for mean bill size between the three servers, coded A, B, and C, are given in Table D.10. Is there a difference in mean bill size between the three servers? Be sure to check conditions of the test, and conduct the appropriate test.

Table D.10 *Size of the bill by server*

Server	Sample Size	Mean	Std.Dev.
A	60	22.76	12.71
B	65	21.14	10.19
C	32	25.92	14.35
Overall	157	22.729	12.157

D.7 **Tip by Number of Guests: Correlation** Do larger parties tend to leave a larger tip? Figure D.3 shows a scatterplot between the size of the tip and the number of guests.

(a) Does there appear to be an association in the data between number of guests and tip size? If so, is it positive or negative? Are there any outliers?

(b) The correlation between the two variables is $r = 0.504$ with $n = 157$. Test to see if this shows a significant positive correlation.

(c) If the correlation is significant, does that imply that more guests *cause* the tip to go up? If not, what is an obvious confounding variable?

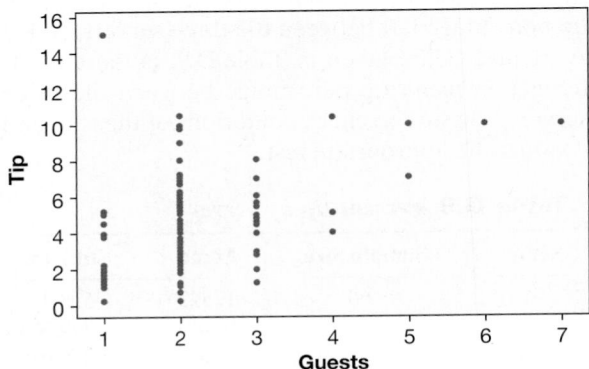

Figure D.3 *Size of the tip by number of guests*

D.8 **Tip by Number of Guests: Regression** How much larger do tips get with larger parties? Figure D.3 shows a scatterplot between the size of the tip and the number of guests. Output regressing *Tip* on *Guests* is shown below:

Coefficients:

	Estimate	Std. Error	t value	Pr(> |t|)	
(Intercept)	1.1068	0.4130	2.680	0.00816	**
Guests	1.3087	0.1802	7.264	1.72e-11	***

Residual standard error: 2.098 on 155 degrees of freedom
Multiple R-squared: 0.254, Adjusted R-squared: 0.2492
F-statistic: 52.77 on 1 and 155 DF, p-value: 1.715e-11

(a) Interpret the value of the coefficient for *Guests* in context.

(b) A server begins to wait on a table with three guests. What is her predicted tip?

(c) The server calculates a 95% confidence interval for the average tip for a table of three guests to be (4.57, 5.49), and a 95% prediction interval to be (0.86, 9.20), but is unsure what these intervals

tell about how much she should expect to get as a tip. Help her interpret both intervals.

D.9 **ANOVA for Regression to Predict Tip from Bill** We have seen in earlier exercises that the conditions are met for using a regression line to predict the *Tip* from the size of the *Bill*. Some regression output is shown for fitting this linear model:

The regression equation is
Tip = −0.292 + 0.182 Bill

S = 0.979523   R-Sq = 83.7%   R-Sq(adj) = 83.6%

Analysis of Variance

Source	DF	SS	MS	F	P
Regression	1	765.53	765.53	797.87	0.000
Residual Error	155	148.72	0.96		
Total	156	914.25			

(a) Find the value for R^2 in the output and interpret it in context.

(b) Find the F-statistic and p-value in the regression ANOVA. What is the conclusion of this test?

D.10 **Predicting Tip from Bill and Number of Guests** In Exercise D.9, we use the size of the bill to predict the tip. In this exercise, we use both the size of the bill and the number of guests to predict the size of the tip. Some regression output is shown for this analysis:

The regression equation is
Tip = −0.252 + 0.184 Bill − 0.036 Guests

Predictor	Coef	SE Coef	T	P
Constant	−0.2524	0.2019	−1.25	0.213
Bill	0.183751	0.007815	23.51	0.000
Guests	−0.0357	0.1019	−0.35	0.727

S = 0.982307   R-Sq = 83.7%   R-Sq(adj) = 83.5%

Analysis of Variance

Source	DF	SS	MS	F	P
Regression	2	765.65	382.82	396.74	0.000
Residual Error	154	148.60	0.96		
Total	156	914.25			

(a) What is the regression equation? What tip is predicted for three guests with a $30 bill?

(b) There are two explanatory variables in the model. Interpret the coefficient of each.

(c) Give the p-value for testing the slope of each explanatory variable, and indicate whether each is significant in the model.

(d) Give the value of R^2 and interpret it in context.

(e) Find the F-statistic and p-value in the regression ANOVA. What are the hypotheses and conclusion of this test?

Review Exercises for **UNIT D**

USING THEORETICAL DISTRIBUTIONS

In Exercises D.11 to D.18, a test statistic for one of the tests in this unit is given, along with information about sample size(s) or degrees of freedom. Give the p-value and indicate whether the results are significant at the 5% level.

D.11 An upper-tail test for correlation with t-statistic $= 1.36$ and df $= 15$

D.12 An analysis of variance test with F-statistic $= 7.42$ and df-numerator $= 3$ and df-denominator $= 56$

D.13 A chi-square goodness-of-fit test with χ^2-statistic $= 4.18$ and 5 groups

D.14 A two-tailed test for slope in a one-predictor regression model with t-statistic $= 2.89$ and $n = 30$

D.15 ANOVA for difference in means for 100 people separated into six groups with F-statistic $= 2.51$

D.16 A chi-square test for a 2×4 table with χ^2-statistic $= 6.83$

D.17 A two-tailed test for the coefficient of the first predictor in a three-predictor regression model with t-statistic $= 1.83$ and $n = 26$

D.18 A lower-tail test for correlation with t-statistic $= -4.51$ and $n = 81$

WHICH TEST IS APPROPRIATE?

In this unit, we have covered six specific tests, listed below. For each situation given in Exercises D.19 to D.27, identify which of these tests is most appropriate. If multiple tests are appropriate, list them all.

- Chi-square goodness-of-fit test
- Chi-square test for association
- Analysis of variance for difference in means
- Test for correlation
- Test for a slope/coefficient in a regression model
- Analysis of variance for regression

D.19 Three different drugs are being tested on patients who have leukemia and the response variable is white blood cell count.

D.20 Three different drugs are being tested on patients who are HIV-positive and the response variable is whether or not the person develops AIDS.

D.21 Data are collected from 50 different towns on number of wood-burning houses and number of people with asthma, and the study is investigating

whether there is a linear relationship between the two.

D.22 The admissions office at a university uses data from high school transcripts such as number of honors courses, number of AP courses, grade in 11th grade English, and grade in 9th grade math to develop a model to predict success in college as measured by grade point average. They wish to test the effectiveness of this model.

D.23 A polling agency working in a large city knows (from census data) the distribution of all city residents by race. They select a sample of 2000 residents and would like to check that the distribution of racial groups within their sample is not significantly different from the proportions in the city as a whole.

D.24 A breakfast cereal company wants to know how useful the height of the display for that brand is in the store, in a model predicting sales of the boxes of cereal based on height of the display, price of the cereal, width of the aisle in the store, and amount spent on advertising in that community.

D.25 A hockey team wants to determine how effective a model is to predict winning percentage based on power-play percentage, penalty-kill percentage, number of checks per game, face-off win percentage, and number of penalties per game.

D.26 A test is being conducted to see if the average time it takes for a case to go to trial differs between counties in a state. Seven counties will be included and the data will include a random sample of 25 cases from each county.

D.27 A test is being conducted to see if the proportion of cases that get settled out of court is different between the different counties in a state. Seven counties will be included and the data will include a random sample of 60 cases from each county.

D.28 Ford Car Sales Assume you are working as a statistician for the automotive company Ford. Ford's three most popular cars are the Escape (SUV), the Focus (midsize sedan), and the Fusion (hybrid sedan). Your boss is putting together next year's production numbers and asks you to determine if any of the three models are selling better or worse than the other two. You have the sales numbers from one month of sales: the Escape sold 22,274, the Focus 21,385, and the Fusion sold 20,808.[8] Perform

---

[8]Sales from *The Wall Street Journal* for June 2011.

the necessary test to determine if sales are significantly different and report the conclusion(s) of interest to your boss.

D.29 Higher SAT Score? One of the variables in our **StudentSurvey** dataset is a categorical variable indicating in which SAT subject the student scored higher (Math, Verbal, or the same). The results are in Table D.11. Perform the appropriate test to determine if the overall number of intro stat students who score higher on the math section is different from the number who score higher on the verbal. (*Hint:* Ignore ties.)

Table D.11 *Higher SAT subject*

Math	Verbal	Same
205	150	7

D.30 Rain in California Amy is interested in moving to California but isn't certain which city she'd prefer. One variable of interest to her is the proportion of rainy days in each city. She took a random sample of days for each of the four biggest cities in California (Los Angeles, San Francisco, San Diego, and San Jose) and recorded the results in Table D.12.[9] If a day contained any precipitation it is considered a rainy day.

(a) Is the chi-square distribution appropriate for testing this two-way table?

(b) Perform the appropriate test to determine if the proportion of rainy days is different among these four cities.

(c) State your conclusion(s). If there is a significant difference, which city would you advise Amy to move to if she does not like rainy days?

Table D.12 *Count of rainy days in four California cities*

	LA	SF	SD	SJ	Total
Rain	4	6	22	3	35
No Rain	21	17	63	20	121
Total	25	23	85	23	156

D.31 Rain in San Diego In Exercise D.30 we observe that within the random sample of 85 days the city of San Diego has rainy weather just over a quarter of the time. Amy is now curious as to whether there is a rainy season, or if the rainy days are dispersed evenly throughout the year. She

separates the 85 days into the four seasons, Spring (Mar–May), Summer (Jun–Aug), Fall (Sep–Nov), and Winter (Dec–Feb) and observes rainy day counts as seen in Table D.13.

(a) Is the chi-square distribution appropriate for testing this two-way table?

(b) Perform the appropriate test to determine if the proportion of rainy days is different among the four seasons.

(c) State your conclusion(s). If there is a significant difference, when is the rainy season?

Table D.13 *Count of rainy days in four San Diego seasons*

	Spring	Summer	Fall	Winter
Rain	5	0	6	11
No Rain	16	22	14	11

D.32 Homes for Sale: Chi-Square Test Throughout Unit C we examine various relationships between the variables regarding houses for sale in the dataset **HomesForSale**. One topic of repeated interest is differences between states. We previously only had the capability to test for a difference between two states at a time, but we can now test for relationships across all four states at once. Table D.14 shows the number of big (greater than the national average of 2700 sqft) and not big houses for each of the four states. Is location associated with the proportion of big houses? Determine whether conditions are met for a chi-square test and explain your reasoning. Perform the appropriate test to see if the proportion of big houses is related to location (states).

Table D.14 *Houses for sale in four states*

	California	New Jersey	New York	Pennsylvania	Total
Big	7	6	7	3	23
Not	23	24	23	27	97
Total	30	30	30	30	120

D.33 Homes for Sale: ANOVA In Exercise C.105 on page 500, we look at differences in average housing price between two states at a time with data from **HomesForSale**. We can now set up an ANOVA to test and see if there is a difference between all four states at once. We have a total of $n = 120$ homes from the $k = 4$ states.

(a) What are the null and alternative hypotheses?

(b) What is the degrees of freedom for Groups?

[9]Sample collected from *http://www.wunderground.com*.

(c) What is the degrees of freedom for Error?

(d) Without looking at the data, which (groups or error) would you guess has a greater sum of squares?

D.34 Checking Conditions for Homes for Sale In Exercise D.33 we outline and discuss an ANOVA approach to test for a difference in average housing price between all four states using the dataset **HomesForSale**. Figure D.4 shows one of the samples, specifically for New York, and Table D.15 shows the mean and standard deviation for each of the four states. Are the conditions for ANOVA met? If not, what is the problem or problems?

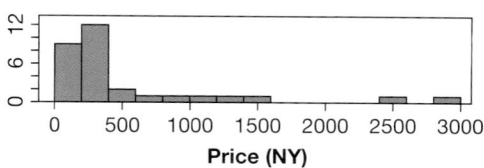

Figure D.4 *Histogram of sales for New York*

Table D.15 *Mean housing prices for four states*

State	n	Mean	Std.Dev.
New York	30	565.6	697.6
New Jersey	30	388.5	224.7
Pennsylvania	30	249.6	179.3
California	30	715.1	1112.3

D.35 Fiber and Brand in Breakfast Cereal Data 9.2 on page 592 introduces the dataset **Cereal**, which includes information on the number of grams of fiber in a serving for 30 different breakfast cereals. The cereals come from three different companies: General Mills, Kellogg's, and Quaker. Use the fact that *SSGroups* is 4.96 and *SSTotal* is 102.47 to conduct an analysis of variance test to determine whether there is a difference in mean number of grams of fiber per serving between the three companies. The conditions for ANOVA are reasonably met.

D.36 Height and Voice Data were collected on the heights of singers[10] and are summarized and displayed below. Does average height differ by voice?

---

[10]Chambers, J., Cleveland, W., Kleiner, B., and Tukey P., *Graphical Methods for Data Analysis*, Wadsworth International Group, Belmont, CA, 1983, p. 350.

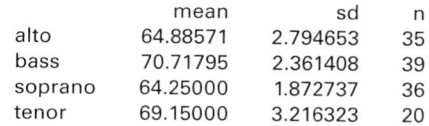

	mean	sd	n
alto	64.88571	2.794653	35
bass	70.71795	2.361408	39
soprano	64.25000	1.872737	36
tenor	69.15000	3.216323	20

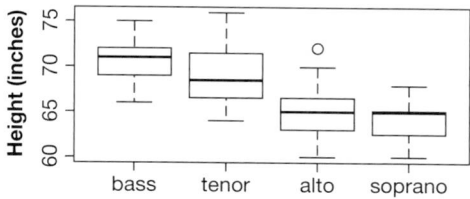

(a) State the null and alternative hypotheses.

(b) Are the conditions for using the F-distribution satisfied? Why or why not?

(c) Complete the ANOVA table given below, and make a conclusion in context.

Source	df	Sum of Sq.	Mean Square	F-statistic	p-value
Groups		1058.5			
Error		796.7			
Total					

D.37 Effect of Color on Performance: In Exercise 8.17 on page 555, we discuss a study investigating the effect of ink color on performance in an anagram test. Three different colors were used with a total of 71 participants. The red group contained 19 participants and they correctly solved an average of 4.4 anagrams. The 27 participants in the green group correctly solved an average of 5.7 anagrams and the 25 participants in the black group correctly solved an average of 5.9 anagrams. From the analysis of variance in Exercise 8.17, we see that there is a significant difference between the groups and that the mean square error from the ANOVA table is 0.84.

(a) Find and interpret a 95% confidence interval for the mean number of anagrams we expect people to solve when the ink is red.

(b) Find and interpret a 95% confidence interval for the difference in mean number of anagrams we expect people to solve between when the ink is green and when it is red.

(c) Test whether there is a significant difference in means between when the ink is red and when it is black.

D.38 Heat from Laptop Computers and Sperm Count In Exercise 8.18 on page 555, we conduct an ANOVA test to see if mean scrotal temperature increase is different between three different conditions. In each condition, males sit with a laptop computer on the lap. In one condition, they sit with legs together, in another with legs apart, and in a third with legs together but a lap pad under the laptop. The ANOVA test found a significant difference in temperature increase between the groups. The summary statistics are shown in Table D.16 and the mean square error from the ANOVA table is 0.63. Conduct tests of all three pairwise comparisons and summarize the findings.

Table D.16 *Scrotal temperature increase in °C with a laptop computer on lap*

Condition	n	Mean	Std.Dev.
Legs together	29	2.31	0.96
Lap pad	29	2.18	0.69
Legs apart	29	1.41	0.66

D.39 Cognition Skills Test and Grade Point Average How closely related are results on short cognitive skills tests and a grade point average over several years in college? In Data A.1 on page 163, we introduce the data in **SleepStudy**. Two of the variables in that study are *CognitionZscore*, which is a standardized z-score derived from multiple tests of cognitive skills such as recalling a list of words, and *GPA*, grade point average on a 4-point scale. The sample correlation between these two variables is $r = 0.267$ with $n = 253$. Test to see if there is evidence to show a positive association between these two variables in the population.

D.40 Happiness and Hours of Sleep Does a good night's sleep make you happier? In Data A.1 on page 163, we introduce the data in **SleepStudy**. Two of the variables in that study are *Happiness*, scores on a standard happiness scale with higher numbers indicating greater happiness, and *AverageSleep*, average number of hours slept in a night. The sample correlation between these two variables is $r = 0.104$ with $n = 253$. Test to see if there is evidence to find a positive association between these two variables in the population.

D.41 Depression and Missed Classes Is depression a possible factor in students missing classes? In Data A.1 on page 163, we introduce the data in **SleepStudy**. Two of the variables in that study are *DepressionScore*, scores on a standard depression scale with higher numbers indicating greater depression, and *ClassesMissed*, the number of classes missed during the semester. Computer output is

shown for predicting the number of classes missed based on the depression score.

Coefficients:

	Estimate	Std. Error	t value	Pr(> |t|)	
(Intercept)	1.77712	0.26714	6.652	1.79e-10	***
DepressionScore	0.08312	0.03368	2.468	0.0142	*

Residual standard error: 3.208 on 251 degrees of freedom
Multiple R-squared: 0.0237, Adjusted R-squared: 0.01981
F-statistic: 6.092 on 1 and 251 DF, p-value: 0.01424

(a) Interpret the slope of the regression line in context.

(b) Identify the t-statistic and the p-value for testing the slope. What is the conclusion, at a 5% level?

(c) Interpret R^2 in context.

(d) Identify the F-statistic and p-value from the ANOVA for regression. What is the conclusion of that test?

D.42 Alcoholic Drinks and Missed Classes Is drinking alcohol a possible factor in students missing classes? In Data A.1 on page 163, we introduce the data in **SleepStudy**. Two of the variables in that study are *Drinks*, the number of alcoholic drinks in a week, and *ClassesMissed*, the number of classes missed during the semester. Computer output is shown for predicting the number of classes missed based on the number of drinks.

Coefficients:

	Estimate	Std. Error	t value	Pr(> |t|)
(Intercept)	1.86444	0.34395	5.421	1.39e-07 ***
Drinks	0.06196	0.04979	1.244	0.215

Residual standard error: 3.237 on 251 degrees of freedom
Multiple R-squared: 0.006131, Adjusted R-squared: 0.002172
F-statistic: 1.548 on 1 and 251 DF, p-value: 0.2145

(a) Interpret the slope of the regression line in context.

(b) Identify the t-statistic and the p-value for testing the slope. What is the conclusion, at a 5% level?

(c) Interpret R^2 in context.

(d) Identify the F-statistic and p-value from the ANOVA for regression. What is the conclusion of that test?

D.43 Checking Conditions for Depression and Missing Classes In Exercise D.41, we consider a regression model to use a student's depression score to predict the number of classes missed in a semester. Here we check the conditions for using that regression model. Three graphs for this model are shown in Figure D.5: the scatterplot with regression line, a histogram of the residuals, and a scatterplot of residuals against predicted values. Discuss whether the conditions are met. Be sure to comment on all three graphs.

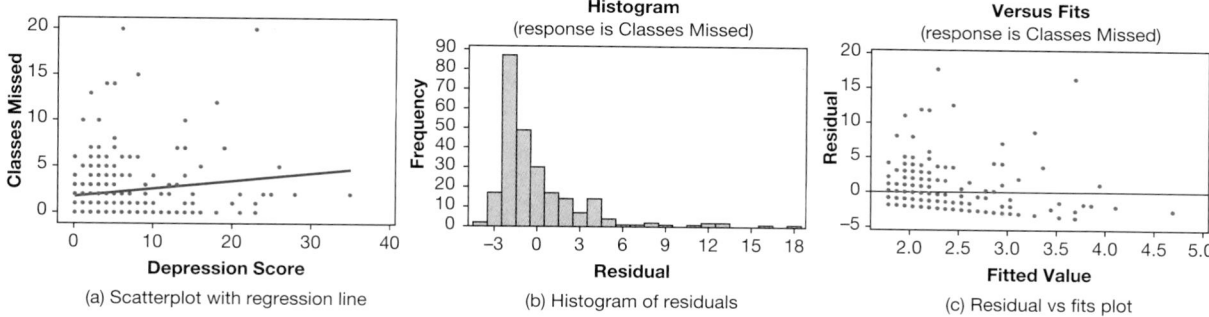

Figure D.5 *Checking conditions for predicting classes missed by depression scores*

D.44 Checking Conditions for Alcoholic Drinks and Missing Classes In Exercise D.42, we consider a regression model to use the number of alcoholic drinks a student has in a week to predict the number of classes missed in a semester. Here we check the conditions for using that regression model. Three graphs for this model are shown in Figure D.6: the scatterplot with regression line, a histogram of the residuals, and a scatterplot of residuals against predicted values. Discuss whether the conditions are met. Be sure to comment on all three graphs.

PREDICTING POINTS SCORED BY A BASKETBALL PLAYER

Using the data in **NBAPlayers2015**, we can create a regression model to predict points in a season for an NBA basketball player based on the number of free throws made. For our sample data, the number of free throws made in a season ranges from 14 to 715, while the number of points ranges from 175 to 2217. In Exercises D.45 and D.46, use the given information and computer output to give the values for each interval and interpret it in context:

(a) The 95% confidence interval for the mean response

(b) The 95% prediction interval for the response

D.45 The intervals given are for a player who makes 100 free throws in a season:

FTMade	Fit	SE Fit	95% CI	95% PI
100	726.0	17.15	(692.2, 759.8)	(339.3, 1112.7)

D.46 The intervals given are for a player who makes 400 free throws in a season:

FTMade	Fit	SE Fit	95% CI	95% PI
400	1616.5	37.02	(1543.4, 1689.5)	(1224.4, 2008.5)

D.47 Predicting Grade Point Average The computer output below shows a multiple regression model to predict grade point average (*GPA*) using six variables from the dataset **SleepStudy**. *Gender* is coded 0 for females and 1 for males; *ClassYear* is coded 1 for first year, 2 for sophomore, 3 for junior, and 4 for senior; *ClassesMissed* is number of classes missed during the semester; *CognitionZscore* is a normalized *z*-score of results from cognitive skills tests; *DASScore* is a combined measure of depression, anxiety, and stress with higher numbers indicating more depression, anxiety, or stress; *Drinks* is the number of alcoholic drinks consumed in a week.

The regression equation is
GPA = 3.49 − 0.0971 Gender − 0.0558 ClassYear
      − 0.0146 ClassesMissed + 0.118 CognitionZscore
      + 0.00284 DASScore − 0.0163 Drinks

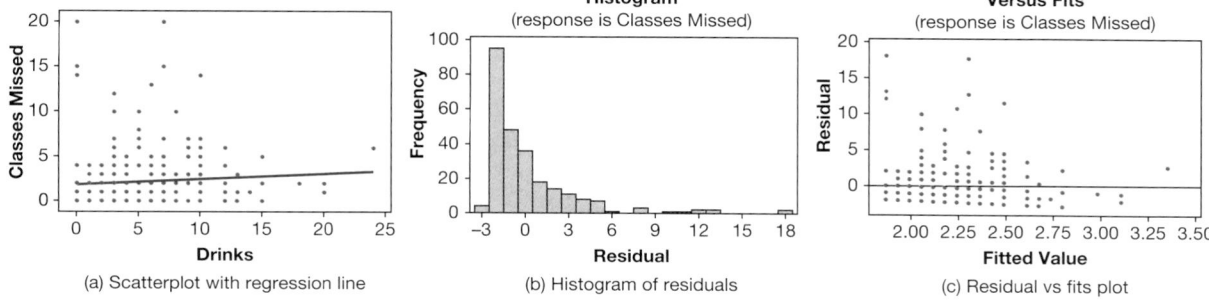

Figure D.6 *Checking conditions for predicting classes missed by alcoholic drinks*

Predictor	Coef	SE Coef	T	P
Constant	3.48759	0.07497	46.52	0.000
Gender	−0.09714	0.05326	−1.82	0.069
ClassYear	−0.05583	0.02284	−2.44	0.015
ClassesMissed	−0.014613	0.007467	−1.96	0.051
CognitionZscore	0.11837	0.03421	3.46	0.001
DASScore	0.002844	0.001441	1.97	0.049
Drinks	−0.016336	0.006241	−2.62	0.009

S = 0.369552   R-Sq = 18.4%   R-Sq(adj) = 16.4%

Analysis of Variance

Source	DF	SS	MS	F	P
Regression	6	7.5925	1.2654	9.27	0.000
Residual Error	246	33.5958	0.1366		
Total	252	41.1884			

(a) Interpret the coefficients of *Gender*, *ClassYear*, and *ClassesMissed* in context. Be sure to pay attention to how the first two variables are coded.

(b) Use the p-value from the ANOVA test to determine whether the model is effective.

(c) Interpret R^2 in context.

(d) Which explanatory variable is most significant in the model? Which is least significant?

(e) Which variables are significant at a 5% level?

D.48 Predicting Percent Body Fat Data 10.1 on page 614 introduces the dataset **BodyFat**. Computer output is shown for using this sample to create a multiple regression model to predict percent body fat using the other nine variables.

The regression equation is
Bodyfat = − 23.7 + 0.0838 Age − 0.0833 Weight + 0.036 Height
+ 0.001 Neck − 0.139 Chest + 1.03 Abdomen + 0.226 Ankle
+ 0.148 Biceps − 2.20 Wrist

Predictor	Coef	SE Coef	T	P
Constant	−23.66	29.46	−0.80	0.424
Age	0.08378	0.05066	1.65	0.102
Weight	−0.08332	0.08471	−0.98	0.328
Height	0.0359	0.2658	0.14	0.893
Neck	0.0011	0.3801	0.00	0.998
Chest	−0.1387	0.1609	−0.86	0.391
Abdomen	1.0327	0.1459	7.08	0.000
Ankle	0.2259	0.5417	0.42	0.678
Biceps	0.1483	0.2295	0.65	0.520
Wrist	−2.2034	0.8129	−2.71	0.008

S = 4.13552   R-Sq = 75.7%   R-Sq(adj) = 73.3%

Analysis of Variance

Source	DF	SS	MS	F	P
Regression	9	4807.36	534.15	31.23	0.000
Residual Error	90	1539.23	17.10		
Total	99	6346.59			

(a) Interpret the coefficients of *Age* and *Abdomen* in context. *Age* is measured in years and *Abdomen* is abdomen circumference in centimeters.

(b) Use the p-value from the ANOVA test to determine whether the model is effective.

(c) Interpret R^2 in context.

(d) Which explanatory variable is most significant in the model? Which is least significant?

(e) Which variables are significant at a 5% level?

D.49 Exercise, Gender, and GPA The dataset **StudentSurvey**, introduced on page 4, contains information on hours of *Exercise* per week and *GPA*. Here we use a slightly modified version called **GPAGender** which eliminates missing values (leaving $n = 343$ students) and codes the gender with 1 for males and 0 for females in a new *GenderCode* variable. This allows us to use information on gender in a regression model.

(a) Test for an association between *Exercise* and *GPA* using the data in **GPAGender**. Give the p-value and make a conclusion in context.

(b) Tests for difference in means reveal that gender is significantly associated with both *GPA* and *Exercise* (males have lower GPAs and exercise more on average), so gender may be a confounding variable in the association between *Exercise* and *GPA*. Use multiple regression to determine whether *Exercise* is a significant predictor of *GPA*, even after accounting for gender as coded in *GenderCode*.

D.50 Number of Piercings, SAT Score, and GPA The dataset **GPAGender**, described in Exercise D.49, contains a subset of the **StudentSurvey** data which also has information on total *SAT* scores, *GPA*, and number of *Piercings* for those $n = 343$ students.

(a) Test for an association between number of *Piercings* and *GPA* using the data in **GPAGender**. Give the p-value and make a conclusion in context.

(b) Use multiple regression to test for an association between number of *Piercings* and *GPA*, after accounting for *SAT* score.

D.51 Finding a Model for Percent Body Fat Exercise D.48 shows a multiple regression model to predict percent body fat using the nine other variables in the dataset **BodyFat**. Try to improve on this model, using a subset of these predictors to balance the desire to use important individual predictors with explaining a significant portion of the variability in body fat. Describe the process you use to obtain your model and discuss the merits of your final choice. (*Note:* There are a number of reasonable final models.)

D.52 Finding a Model for Happiness The **Sleep-Study** dataset introduced in Data A.1 on page 163 contains many variables measuring different characteristics of the students in the study. One of these is a *Happiness* score that includes values from a minimum of 0 (very unhappy) to a maximum of 35 (extremely happy!). Find a reasonable regression model to predict *Happiness* based on some of the other variables available in the **SleepStudy** dataset. Describe the process you use to obtain your model and discuss the merits of your final choice. (*Note:* There are a number of reasonable choices for final models.)

We began this journey of investigating the power of statistics, way back in Section 1.1, by classifying variables as either *categorical* or *quantitative*. This way of organizing our thinking remains effective as we briefly summarize the methods of description and inference discussed on the journey. Before we examine individual variables or relationships between variables, however, we revisit again the important messages of Chapter 1.

Summary: When Is Statistical Inference Appropriate?

Statistical inference is appropriate when we wish to use data from a sample to *estimate* one or more parameters for a population or to *test* a hypothesis about one or more parameters for a population.

Before conducting any inference procedures, we should always stop and think about the way the data were collected. *Statistical inference is only appropriate if we believe the sample data are representative of the population about which we hope to make inferences.* The best way to ensure that a sample avoids sampling bias and is representative of a population is to collect a random sample from the population.

In examining relationships between variables, be wary of confounding variables and remember that *the only way to infer a causal association between variables statistically is through data obtained from a randomized experiment.*

We have seen many methods of analysis in this text. The best analysis there is, however, will not make up for data that is collected in a biased manner. It is important to reiterate that appropriate data collection is at least as important as appropriate data analysis.

Summary: The Key Ideas of Inference

As we have seen, data from just a small subset of a population, if collected well, can be used to give very accurate estimates and make very specific conclusions about the entire population. We use an understanding of the variability of sample statistics to estimate how far the true population parameter may be from the sample statistic. This distance is the margin of error. The sample statistic, together with a margin of error, give us a confidence interval for a population parameter.

How do we find evidence for a claim about a population? In order to conclude that a result holds for an entire population, we need the evidence from the sample to be quite conclusive. This means the evidence has to be strong *against* the "status quo" (which we call the null hypothesis) and in support of the claim we are testing. We determine the strength of this evidence by asking: "How likely is it that sample results this extreme would happen just by random chance if the null hypothesis were true?" This is what motivates the important idea of a *p-value*. If sample results so extreme are very *unlikely* to have happened just by random chance, then the p-value is small and we have strong evidence against the null and in favor of our (alternative hypothesis) claim.

These key ideas, introduced early in Chapters 3 and 4, have provided the framework for almost everything we have done since then, and they form the foundation for most of statistical inference.

Summary: Investigating Variables and Relationships between Variables

We organize the summary by quantitative and categorical variables, and look first at individual variables and then at relationships between variables. In every case, we discuss descriptive statistics (graphs and/or summary statistics) and methods of statistical inference. We hope this brief summary is useful, although be sure to recognize that it leaves out many details. In particular, when using the theoretical distributions to conduct statistical inference, we need to make sure that the relevant conditions are met. See the individual unit summaries for more details.

Analyzing a Single Quantitative Variable
- Descriptive statistics
  - Graphical display: dotplot, histogram, boxplot
  - Summary statistics: mean, standard deviation, five number summary
- Statistical inference
  - Estimating with a confidence interval for a mean (using bootstrap or t-distribution)
  - Testing a hypothesis about a population mean (using randomization or t-distribution)

Analyzing a Single Categorical Variable
- Descriptive statistics
  - Graphical display: bar chart, pie chart
  - Summary statistics: frequency, relative frequency, proportion
- Statistical inference if the categorical variable has one particular category of interest
  - Estimating with a confidence interval for a proportion (using bootstrap or normal distribution)
  - Testing a hypothesis about a population proportion (using randomization or normal distribution)
- Statistical inference if the categorical variable has more than two categories of interest
  - Testing a hypothesis about population proportions using chi-square goodness-of-fit test

Analyzing a Relationship between One Categorical Variable and One Quantitative Variable
- Descriptive statistics
  - Graphical display: side-by-side boxplots, dotplots, or histograms
  - Summary statistics: statistics for the quantitative variable within each category, difference in means
- Statistical inference if the categorical variable has two relevant categories
  - Estimating a confidence interval for difference in means (using bootstrap or t-distribution)
  - Testing a hypothesis about a difference in population means (using randomization or t-distribution)

- Statistical inference if the categorical variable has more than two categories
  - Testing a hypothesis about several population means using ANOVA for difference in means
  - Intervals and tests for means and differences in pairs of means after ANOVA

Analyzing a Relationship between Two Categorical Variables

- Descriptive statistics
  - Graphical display: segmented or side-by-side bar chart
  - Summary statistics: two-way table, row proportions, column proportions, difference in proportions
- Statistical inference if both categorical variables have two relevant categories
  - Estimating a confidence interval for a difference in proportions (using bootstrap or normal distribution)
  - Testing a hypothesis about a difference in proportions (using randomization or normal distribution)
- Statistical inference if either categorical variable has more than two categories
  - Testing a hypothesis about a relationship using chi-square test for association

Analyzing a Relationship between Two Quantitative Variables

- Descriptive statistics
  - Graphical display: scatterplot
  - Summary statistics: correlation, regression line
- Statistical inference
  - Estimation
    * Estimating a confidence interval for the correlation (using bootstrap or t-distribution)
    * Estimating a confidence interval for the slope of the regression line (using bootstrap or t-distribution)
    * Estimating the response variable at one value of the explanatory variable (using confidence interval for mean response or prediction interval for individual responses)
  - Testing
    * Testing a hypothesis about the correlation (using randomization or t-distribution)
    * Testing a hypothesis about the slope for a regression line (using randomization or t-distribution)
    * Testing the effectiveness of the regression model using analysis of variance for regression

Analyzing Relationships between More Than Two Variables

- Multiple regression!

As we said at the start of this book, "We are being inundated with data… The people who are able to analyze this information are going to have great jobs and are going to be valuable in virtually every field. One of the wonderful things about statistics is that it is relevant in so many areas. Whatever your focus and your future career plans, it is likely that you will need statistical knowledge to make smart decisions in your field and in everyday life." We finish with an example of such an application in the final case study. We hope this journey we have taken together has helped you understand the power of data!

Case Study: Speed Dating

© Spencer Grant/Alamy Limited

Speed Dating

<table>
<tr><td>

DATA E.1

</td><td>

Speed Dating

Between 2002 and 2004, a series of speed dating experiments were conducted at Columbia University.[15,16] Participants were students at Columbia's graduate and professional schools, recruited by mass email, posted fliers, and fliers handed out by research assistants. Each participant attended one speed dating session, in which they met with each participant of the opposite sex for four minutes. Order and session assignments were randomly determined. After each four-minute "speed date," participants filled out a form rating their date on a scale of 1 to 10 on the attributes *Attractive*, *Sincere*, *Intelligent*, *Fun*, *Ambitious*, and *SharedInterests*. They also made a *Decision* indicating whether they would like to see the date partner again, rated how much they *Like* the partner overall, and answered how likely they think it is that the partner will say yes to them, *PartnerYes*. The data are stored in **SpeedDating**. Each row is a date, and each variable name is followed by either an *M*, indicating male answers, or an *F*, indicating female answers. In other words, each case includes answers from both parties involved in the date. To avoid dependencies in the data, we only look at data from the first dating round of each session, giving $n = 276$ dates.[17] We also have data on *Age* and *Race*. ■

</td></tr>
</table>

These data can help us gain insights into the mysterious world of dating! How likely is it that a pair will both like each other enough to want a second date? Are people more likely to want a second date with someone they think is interested in them? With someone who actually is interested in them? Do opinions of dates differ by race/ethnicity? How much is romantic interest reciprocated? Can we predict how much a male will like a female based on how much the female likes the male? Which

[15]Fisman, R., Iyengar, S., Kamenica, E., and Simonson, I., "Gender differences in mate selection: Evidence from a speed dating experiment," *Quarterly Journal of Economics*, 2006; 121(2): 673–697.
[16]Gelman, A. and Hill, J., *Data Analysis using Regression and Multilevel/Hierarchical Models*, Cambridge University Press, New York, 2007.
[17]Many of the techniques we have used in this book assume independence of the cases. If we were to include multiple dates for each individual, this assumption would be violated.

attributes are most helpful for predicting how much someone likes his or her date overall? Does the answer to this question differ by gender? We'll explore all of these questions in this section.

Example E.1

Thinking critically about the data

The following are questions you should *always* ask yourself before jumping into data analysis.

(a) What are the cases?

(b) What are the variables, and is each categorical or quantitative?

(c) What is the sample? To what population would you like to make inferences? What population is more realistic?

(d) What are some ways in which your sample may differ from your ideal population? From your more realistic population?

(e) Do these data allow for possible conclusions about causality? Why or why not?

Solution   (a) The cases are first dates during the speed dating experiment.

(b) Categorical variables are whether a person wants to see his or her partner again (*DecisionM, DecisionF*) and race (*RaceM, RaceF*). Quantitative variables are how much a person likes the partner (*LikeM, LikeF*), how likely they think it is that the partner will say yes to them (*PartnerYesM, PartnerYesF*), age (*AgeM, AgeF*), and the 1-10 ratings for each of the six attributes.

(c) The sample is the 276 first-round speed dates on which we have data. The ideal population would be all first speed dates, or even all first dates. A more realistic population may be all first-round heterosexual speed dates between graduate or professional students at prestigious urban American universities.

(d) Regular first dates allow much more time to get to know the other person than speed dates. Speed dates between graduate and professional students at Columbia may involve different types of people than typical speed dates. Some students may participate in this speed dating in the context of the research experiment, but would not participate otherwise. Because this wasn't a random sample from the population, there may be other forms of sampling bias that we are not aware of.

(e) No. This was an observational study. Although the pairings were randomized, none of the variables were, and for causal conclusions the explanatory variable must be randomly determined.

Example E.2

How likely is a match?

A match is declared if both the male and the female want a second date. Of the 276 speed dates, 63 resulted in matches. Create a 95% confidence interval for the population proportion of matches using:

(a) Bootstrapping and the percentile method

(b) Bootstrapping and $Statistic \pm 2 \times SE$

(c) The normal distribution and the relevant formula for the standard error

Solution   (a) Using *StatKey* or other technology, we create the bootstrap distribution for a single proportion, shown in Figure E.1. Keeping only the middle 95% of bootstrap statistics, our 95% confidence interval for the proportion of matches is $(0.181, 0.279)$.

Bootstrap Dotplot of [Proportion ▾]

☐ Left Tail ☑ Two Tail ☐ Right Tail

samples = 1000
mean = 0.227
st. dev. = 0.025

0.025   0.95   0.025

0.150   0.175   0.200   0.225   0.250   0.275   0.300   0.325
0.227

0.181   0.279

Figure E.1 *Bootstrap distribution for proportion of matches*

(b) We calculate the sample statistic to be $\hat{p} = 63/276 = 0.228$. We find the standard error as the standard deviation of the bootstrap distribution, $SE = 0.025$. Therefore our 95% confidence interval for the proportion of matches is

$$Statistic \pm 2 \times SE = 0.228 \pm 2 \times 0.025 = (0.178, 0.278)$$

(c) The formula for the standard error for a sample proportion is

$$SE = \sqrt{\frac{\hat{p}(1-\hat{p})}{n}} = \sqrt{\frac{0.228(1-0.228)}{276}} = 0.025$$

A 95% confidence interval for the proportion of matches using $z^* = 1.96$ from a normal distribution is

$$\hat{p} \pm z^* \cdot SE = 0.228 \pm 1.96 \times 0.025 = (0.179, 0.277)$$

Notice that however you decide to create a confidence interval, you get approximately the same answer. In all cases, the generated confidence interval is interpreted the same: We are 95% confident that the proportion of matches among first-round speed dates between graduate or professional students at prestigious urban American universities is between 0.18 and 0.28.

Example E.3

Do people tend to want a second date with people they think *will also want a second date?*

Each person was asked "How probable do you think it is that this person will say 'yes' for you? (0 = not probable, 10 = extremely probable)," with answers stored in *PartnerYesM* and *PartnerYesF*.[18] Is the perception of the likelihood of a return yes from a partner higher for males who want a second date than for males who don't? (When dating, should you be obviously "into" your dating partner?)

[18]The sample sizes in this example are slightly lower than in other examples because of some missing values for *PartnerYesM* and *PartnerYesF*.

(a) Create the relevant visualization for this relationship.

(b) Calculate the relevant summary statistic(s).

(c) State the relevant hypotheses, use a randomization test to generate a p-value, and make a conclusion in context.

(d) Repeat (b) and (c), but for females rather than males, and using a theoretical distribution and formula rather than a randomization test.

Solution ▶ (a) We are comparing one "yes/no" categorical variable, *DecisionM*, with one quantitative assessment of the partner's inclination, *PartnerYesM*, and so visualize with side-by-side boxplots, as shown in Figure E.2. When comparing the boxplots, we see that the *PartnerYesM* values tend to be higher for those where the male says "yes".

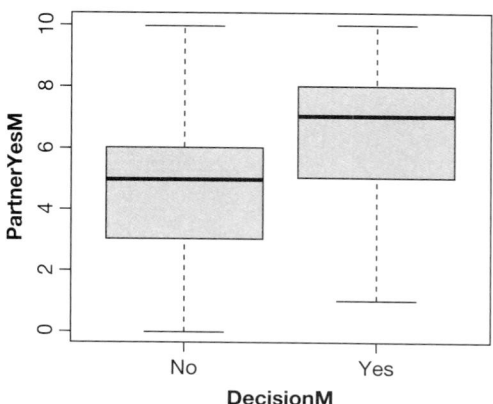

Figure E.2 *Perceived chance of a return yes (PartnerYesM) by male decision (DecisionM)*

(b) The relevant statistics are the mean *PartnerYesM* score for two groups, the males who said "yes" and those who said "no," and the difference between those two means. The sample mean *PartnerYesM* score among males who said yes to their dates is $\bar{x}_{yes} = 6.5$, and among males who said no to their dates is $\bar{x}_{no} = 4.9$, so the sample difference in means is $\bar{x}_{yes} - \bar{x}_{no} = 1.6$.

(c) The null and alternative hypotheses are

$$H_0 : \mu_{yes} = \mu_{no}$$
$$H_a : \mu_{yes} > \mu_{no}$$

Using *StatKey* or other technology we create a randomization distribution such as the one shown in Figure E.3. In those 1000 randomization samples there were no statistics even close to as extreme as the observed sample statistic of $\bar{x}_{yes} - \bar{x}_{no} = 1.6$, so the p-value ≈ 0. There is very strong evidence that males are more optimistic about getting a return "yes" from females they say "yes" to than those they choose not to date again. Females, perhaps playing hard to get may not be the best strategy to get a return date, at least in speed dating.

(d) The relevant statistics are the mean *PartnerYesF* score for two groups, the females who said "yes" and those who said "no," and the difference between those two means. The sample mean *PartnerYesF* score among females who said yes to their dates is $\bar{x}_{yes} = 6.68$, and among females who said no to their dates is $\bar{x}_{no} = 5.11$, so the sample difference in means is $\bar{x}_{yes} - \bar{x}_{no} = 1.57$.

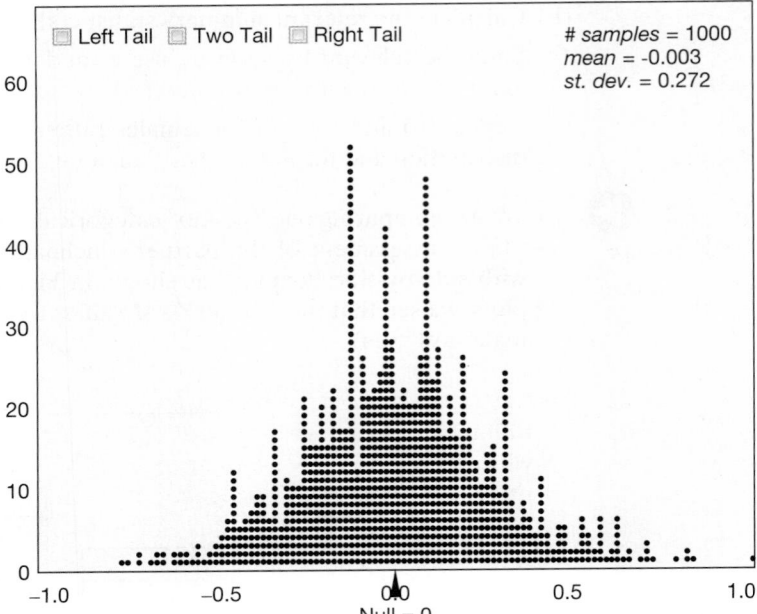

Figure E.3
Randomization distribution

The hypotheses are the same as (c), although pertaining to females instead of males. The t-statistic is

$$ t = \frac{Statistic - Null}{SE} = \frac{(\overline{x}_{yes} - \overline{x}_{no}) - 0}{\sqrt{\frac{s_{yes}^2}{n_{yes}} + \frac{s_{no}^2}{n_{no}}}} = \frac{6.68 - 5.11}{\sqrt{\frac{1.92^2}{126} + \frac{2.24^2}{146}}} = \frac{1.57}{0.25} = 6.28 $$

We find the area above $t = 6.28$ in a t-distribution with 125 df to be approximately 0, so the p-value ≈ 0. There is very strong evidence that females are more optimistic about getting a return "yes" from males they say "yes" to than those they choose not to date again. Males, perhaps playing hard to get may not be the best strategy to get a return date, at least in speed dating.

Example E.4

Do people tend to want a second date with people who actually *also want a second date?*

In Example E.3 we see that males and females say yes more often to dates they *think* will say yes back. Do they also say yes more often to dates that *actually* say yes back? Table E.1 provides counts on the number of females and males who say yes or no to each other.

(a) Visualize the relationship between whether or not the male says yes and whether or not the female says yes, and comment on what you see.

Table E.1 *Male and female decisions on whether they want a second date*

Second Date?	Female Yes	Female No
Male Yes	63	83
Male No	64	66

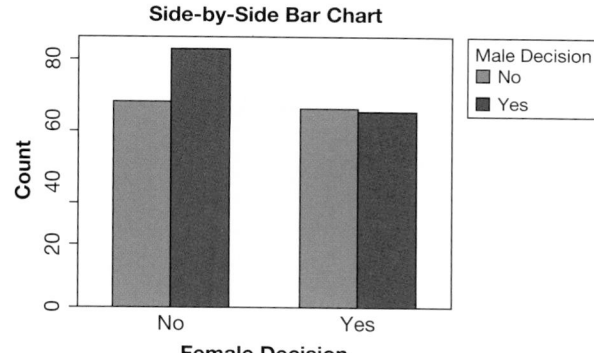

Figure E.4 *Male yes/no answers by female yes/no answers*

(b) Calculate the sample difference in proportions: proportion of male yes answers to females who say yes back minus proportion of male yes answers to females who say no back. (Note that there are several other interesting differences in proportions you could also calculate from this table.)

(c) Perform a hypothesis test to see if the proportion of male yes answers is significantly different between answers to females who say yes and answers to females who say no.

Solution

(a) This is a relationship between two categorical variables, which we can visualize with a side-by-side bar chart, shown in Figure E.4. In this sample males say yes about half the time to girls who say yes back, and say yes more than half the time to girls who say no. This is quite interesting! Although in Example E.3 we found males say yes more often to girls they *think* will say yes back, here we see that they say yes more often to girls who *actually* say no back (at least in this sample).

(b) The sample difference in proportions is

$$\frac{63}{63 + 64} - \frac{83}{83 + 66} = 0.496 - 0.557 = -0.061$$

(c) Let p_{Fyes} and p_{Fno} denote the proportion of male yes answers to females who say yes and no back, respectively. Our hypotheses are

$$H_0 : p_{Fyes} = p_{Fno}$$
$$H_a : p_{Fyes} \neq p_{Fno}$$

We can use *StatKey* or other technology to create a randomization distribution, shown in Figure E.5. The proportion of randomization statistics less than or equal to the observed sample difference in proportions of -0.061 is 0.155, which we double because of the two-sided alternative for a p-value of $0.155 \cdot 2 = 0.31$.

Because the counts are large enough, we could also use formulas and the normal distribution to obtain a p-value. The pooled proportion of male yes answers is

$$\hat{p} = \frac{63 + 83}{63 + 64 + 83 + 66} = 0.529$$

so the standard error is

$$SE = \sqrt{\frac{\hat{p}(1 - \hat{p})}{n_{Fyes}} + \frac{\hat{p}(1 - \hat{p})}{n_{Fno}}} = \sqrt{\frac{0.529(1 - 0.529)}{127} + \frac{0.529(1 - 0.529)}{149}} = 0.0603$$

The z-statistic is then

$$z = \frac{Statistic - Null}{SE} = \frac{-0.061 - 0}{0.0603} = -1.01$$

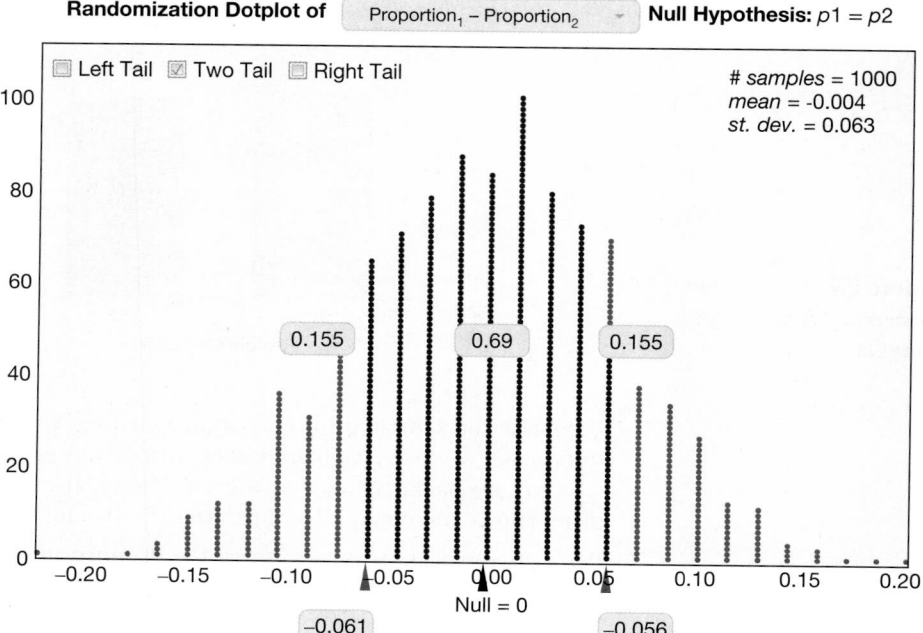

Figure E.5
Randomization distribution

The area below $z = -1.01$ in a normal distribution is 0.156, which we double because of the two-sided alternative for a p-value of $2 \cdot 0.156 = 0.312$. This is very similar to the p-value obtained by the randomization test.

The p-value is not small, so we do not have sufficient evidence against the null hypothesis. From these data we cannot determine whether males are more or less likely to say yes to females who say yes back.

Example E.5

Do opinions of partners differ by race?

Participants rated how much they like their partner overall, on a scale of 1 to 10 (1 = don't like at all, 10 = like a lot), in the variables *LikeM* and *LikeF*. The sample mean, sample standard deviation, and sample size[19] are given in Tables E.2 and E.3 by the race of the rater, along with the proportion of "yes" decisions.

(a) Does how much males like their partners differ by race/ethnicity of the rater? Use technology and the data in **SpeedDating** to construct the appropriate visualization, conduct the appropriate test, and make a conclusion in context.

Table E.2 *Male responses to females, by race of the male raters*

MALES	Asian	Black	Caucasian	Latino	Other	Overall
$\bar{x}$	6.61	6.44	6.66	6.94	6.95	6.68
s	1.85	1.74	1.78	1.68	1.77	1.78
$\hat{p}$	0.59	0.56	0.50	0.59	0.57	0.53
n	64	9	161	17	21	272

[19]The sample size after excluding cases with missing values for *Like* or *Race*.

Table E.3 *Female responses to males, by race of the female raters*

FEMALES	Asian	Black	Caucasian	Latino	Other	Overall
$\bar{x}$	6.47	7.36	6.27	6.02	6.6	6.37
s	1.66	1.28	1.77	1.96	1.88	1.76
$\hat{p}$	0.50	0.57	0.42	0.48	0.53	0.46
n	70	14	146	23	15	268

(b) Does the proportion of females who say yes to their partners differ by race/ethnicity of the female? Construct the appropriate visualization, conduct the appropriate test, and make a conclusion in context.

Solution ▶ (a) This is exploring the relationship between a quantitative variable, *LikeM*, and a categorical variable, *RaceM*, which we visualize with side-by-side boxplots, as shown in Figure E.6. We see that the boxplots all have the same median (7) and are mostly symmetric with only a few mild outliers.

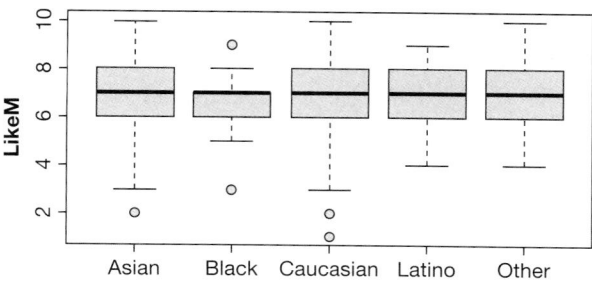

Figure E.6 *Male Like ratings of females, by race of the male*

Because race has more than two categories, we use analysis of variance for difference in means to conduct our hypothesis test. The sample standard deviations are approximately equal, and the distributions appear to be roughly symmetric, so we can proceed with the F-distribution. Computer output is given below:

	Df	Sum Sq	Mean Sq	F value	Pr(>F)
RaceM	4	3.6	0.893	0.278	0.892
Residuals	267	856.2	3.207		

The p-value of 0.892 indicates that the differences between races are not significant, so we cannot determine from these data whether mean *LikeM* differs by race/ethnicity of the rater.

(b) This is exploring the relationship between two categorical variables, *DecisionF* and *RaceF*, which we can visualize with side-by-side bar charts, as shown in Figure E.7.

Because we are interested in a relationship between two categorical variables, one of which has multiple categories, we use a chi-square test for association. We first create the relevant two-way table, Table E.4, based on the information in Table E.3 (using $\hat{p} \cdot n$ to find the number of yes answers for each group) or using technology and the data in **SpeedDating**.

To perform the chi-square test, we could calculate the expected count for each cell using $\frac{(Row\ total)(Column\ total)}{Total}$, and then calculate the χ^2 statistic using

$$\chi^2 = \sum \frac{(Observed - Expected)^2}{Expected}$$

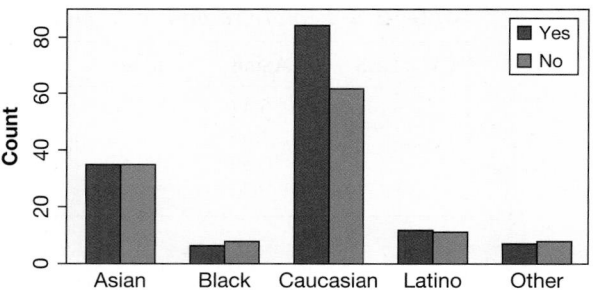

Figure E.7 *Proportion of females who want a second date, by race of the female*

Table E.4 *Counts for females who want or do not want a second date, by race*

FEMALES	Asian	Black	Caucasian	Latino	Other	Total
Yes	35	8	62	11	8	124
No	35	6	84	12	7	144
Total	70	14	146	23	15	268

and compare this to a χ^2-distribution with $(2-1) \cdot (5-1) = 4$ degrees of freedom to find the p-value. Instead, we simply use a computer to calculate this χ^2-statistic and p-value for us:

Pearson's Chi-squared test

```
data:  table(DecisionF, RaceF)
X-squared = 2.2308, df = 4, p-value = 0.6934
```

The p-value of 0.6934 does not provide enough evidence to reject the null. We cannot determine whether the proportions of females who say yes to their dates differ by race/ethnicity of the females.

Example E.6

Predicting a partner's opinion

How well can you predict a male's overall rating of a female, *LikeM*, based on the female's overall rating of the male, *LikeF*?

(a) Use technology to fit the appropriate model, and comment on pieces of the output relevant to answering the question.

(b) Suppose you are a female participating in speed dating and you really like one male, whom you gave a 10 (out of 10) for *LikeF*. Naturally, you want to predict how much this male likes you! Use the model from part (a) to give a prediction, and also generate and interpret a prediction interval.

Solution

(a) We fit the simple linear regression model $\widehat{LikeM} = \beta_0 + \beta_1 \cdot LikeF + \epsilon$. Some computer output is given below:

```
Coefficients:
             Estimate   Std. Error   t value   Pr(> |t|)
(Intercept)  5.61417    0.40305      13.93     < 2e-16 ***
LikeF        0.16781    0.06103      2.75      0.00637 **
```

Residual standard error: 1.763 on 269 degrees of freedom
  (5 observations deleted due to missingness)
Multiple R-squared: 0.02734,   Adjusted R-squared: 0.02373
F-statistic: 7.562 on 1 and 269 DF, p-value: 0.006366

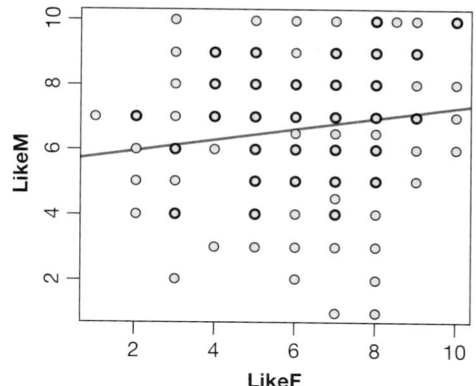

Figure E.8 *Scatterplot of female rating of male against male rating of female for each date*

The scatterplot with this regression line overlaid is shown in Figure E.8. It's a litle difficult to read the scatterplot since many of the integer ratings overlap at the same point.

The low p-value of 0.006 (either from the test for slope or ANOVA) indicates a significant association between *LikeF* and *LikeM*. Knowing how much a female likes a male helps to predict how much the male will like the female. However, the low R^2 value of 2.7% indicates that only 2.7% of the variability in *LikeM* can be explained by *LikeF*. Unfortunately, this is what makes romance so hard! There are many factors which contribute to whether a male likes a female, and sadly how much the female likes him back is only a small part of the equation (at least initially).

(b) The predicted value for *LikeM* when $LikeF = 10$ is $\widehat{LikeM} = 5.61 + 0.168 \cdot 10 = 7.29$. (The average for all males is 6.68, so this is slightly above average.) Using technology, we find the 95% prediction interval is $(3.8, 10.8)$. In fact, the upper bound doesn't even make sense (the max score is 10). This interval is very wide and so not very informative, but it tells us that we are 95% confident that the male will give you a score between 3.8 and 10. However, looking at the scatterplot, we see that in our sample of males to which females gave a 10 (the right-most column of data points), no scores below 6 were given. Remember that a prediction interval is based on the assumption of equal variability, although in this case the variability seems to be greater for less extreme values of *LikeF*, which would give an interval that is slightly too wide for extreme values of *LikeF* such as 10.

Example E.7

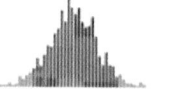

Which attributes are most important for romantic interest?

Each participant rated each partner on a scale of 1 to 10 on the attributes *Attractive*, *Sincere*, *Intelligent*, *Fun*, *Ambitious*, and *SharedInterests*. Which of these attributes are most helpful for predicting how much each person likes his or her partner overall? Answer this question separately for males and females, and compare the results.

Solution

 We have multiple explanatory variables and one response variable, so we fit a multiple regression model. First, some output from the model for males (males rating females) is given below:

Response: LikeM
Coefficients:

	Estimate	Std. Error	t value	Pr(> |t|)	
(Intercept)	−0.175578	0.430784	−0.408	0.683965	
AttractiveM	0.551738	0.054138	10.191	< 2e-16	***
SincereM	0.188545	0.058899	3.201	0.001563	**
IntelligentM	−0.006483	0.075537	−0.086	0.931678	
FunM	0.174910	0.055365	3.159	0.001795	**
AmbitiousM	−0.055786	0.055317	−1.008	0.314292	
SharedInterestsM	0.151640	0.039938	3.797	0.000188	***

Residual standard error: 1.075 on 229 degrees of freedom
 (40 observations deleted due to missingness)
Multiple R-squared: 0.6563,   Adjusted R-squared: 0.6473
F-statistic: 72.89 on 6 and 229 DF, p-value: < 2.2e-16

The attractiveness rating is by far the most significant in the model. Besides attractiveness, how sincere and fun the partner is perceived to be, as well as the extent to which interests are shared, all seem to be helpful in this model for predicting how much a male likes a female overall.

Similar output, but this time for females rating males, is given below:

Response: LikeF
Coefficients:

	Estimate	Std. Error	t value	Pr(> |t|)	
(Intercept)	0.06755	0.44106	0.153	0.8784	
AttractiveF	0.28156	0.04476	6.290	1.58e-09	***
SincereF	0.08820	0.05122	1.722	0.0864	
IntelligentF	0.16064	0.06617	2.428	0.0160	*
FunF	0.24523	0.05009	4.896	1.84e-06	***
AmbitiousF	−0.01628	0.04910	−0.332	0.7405	
SharedInterestsF	0.20264	0.03907	5.186	4.71e-07	***

Residual standard error: 1.095 on 230 degrees of freedom
 (39 observations deleted due to missingness)
Multiple R-squared: 0.6223,   Adjusted R-squared: 0.6125
F-statistic: 63.16 on 6 and 230 DF, p-value: < 2.2e-16

As with males, attractiveness rating is the most significant explanatory variable. For females, both shared interests and fun are also extremely significant. Intelligence is significant with a p-value of 0.016 (this was nowhere near significant for males), and sincerity is moderately significant with a p-value of 0.086.

Attractiveness, fun, and shared interests are very helpful in predicting romantic interest for both genders. For males rating females, sincerity is also significant, while for females rating males, intelligence is significant. Keep in mind that just because variables are insignificant in the multiple regression model does not mean they are not important predictors individually. In fact, if you were to do a test for correlation between *Ambitious* and *Like* for either males or females (try it!), you would find that both correlations are positive and extremely significant.

It's important to keep in mind the limitations with these data—always think about how the data were collected when making conclusions! These are speed dates, and while it is easy to judge someone's physical attractiveness in four minutes, it is much more difficult to judge attributes such as ambition, sincerity, intelligence, etc.

Recognize that these results can only be generalized to speed dating, not dating in general!

Also, if you have learned nothing else from this course, we hope you have learned that observational data cannot be used to draw conclusions about causality! Just because coefficients for *Attractive*, *Sincere*, and *SharedInterests* are positive and significant in both models does not necessarily imply that becoming more attractive, sincere, or having interests in common with the person you are speed dating will raise the *Like* score. Perhaps the relationship actually works in reverse; maybe if someone really likes a partner (for whatever reason), he or she tends to see that person as more attractive, sincere, etc.

There are many other interesting questions to be asked and answered from this **SpeedDating** data, and you now have the knowledge and tools to answer them on your own! Do women prefer older men? Do men prefer younger women? Are people of the same race more likely to result in a match? Are males or females more selective? How are the attributes correlated with each other? For example, is attractiveness positively or negatively correlated with intelligence? Is this relationship significant? You can make up your own questions and answer them. At this stage, having completed an entire course in statistics, all you need is appropriately collected data in order to answer many questions you may have, whether they pertain to dating, your academic field of interest, your health, your future job, or almost anything else!

SECTION LEARNING GOALS

You should now have the understanding and skills to:

▶ • Unlock the power of data!

Exercises for the Big Picture: Essential Synthesis

AMERICAN COMMUNITY SURVEY

The American Community Survey,[20] administered by the US Census Bureau, is given every year to a random sample of about 3.5 million households (about 3% of all US households). It has been crucial for government and policy decisions, and helps to determine how over 400 billion dollars of government funds are distributed each year. Unfortunately, the House voted in May 2012 to eliminate this valuable source of information.[21] Data on a random sample of 1% of all US residents are made public (after ensuring anonymity), and we have selected a random sub-sample of $n = 1000$ from the 2010 data, stored in **ACS**.[22] Exercises E.1 to E.10 pertain to this dataset.

E.1 Random Sample Daniel Webster, a first-term Republican congressman from Florida, sponsored the relevant legislation to eliminate the American Community Survey. Part of his reasoning was that "…this is not a scientific survey. It's a random survey." As you know, and as was pointed out by many, including this quote in the *New York Times*,[23] "the randomness of the survey is precisely what makes the survey scientific." Write a short

---

[20]http://www.census.gov/acs.

[21]For more information, see "The Beginning of the End of the Census?" by C. Rampell, http://www.nytimes.com/2012/05/20/sunday-review/the-debate-over-the-american-community-survey.html?_r=4&emc=eta1.

[22]We have selected a small subset of cases and variables to work with. The full public dataset can be downloaded at http://www.census.gov/acs/www/data_documentation/pums_data/, and the full list of variables are at http://www.census.gov/acs/www/Downloads/data_documentation/pums/DataDict/PUMSDataDict10.pdf.

[23]See, for example, Rampell, C., "The Beginning of the End of the Census," http://www.nytimes.com/2012/05/20/sunday-review/the-debate-over-the-american-community-survey.html?_r=4&emc=eta1, May 19, 2012.

letter to Congressman Daniel Webster explaining this concept to him.

E.2 Margin of Error The US Census Bureau provides a document[24] to assist people with statistical inference using the data from the American Community Survey. Below is an excerpt from this document. Use the information given to fill in the value that goes in the two blanks.

> All ACS estimates from tables on AFF include either the _?_ percent margin of error or _?_ percent confidence bounds. The margin of error is the maximum difference between the estimate and the upper and lower confidence bounds. Most tables on AFF containing 2005 or later ACS data display the margin of error. Use the margin of error to calculate the standard error (dropping the +/− from the displayed value first) as: Standard Error = Margin of Error / Z where Z = 1.645 for 2006 ACS data and recent years.

E.3 What Percentage of Americans have Health Insurance? In our sample of 1000 people, 861 have health insurance and 139 do not.

(a) Use bootstrapping to generate a 90% confidence interval for the proportion of US residents who do not have health insurance.

(b) Use the appropriate formula and distribution to calculate a 90% confidence interval for the proportion of US residents who do not have health insurance. How does this answer compare to your answer from part (a)?

(c) Interpret your answer to either (a) or (b) in context.

(d) What is the sample statistic and corresponding margin of error in part (b)?

(e) We can also use the website *http://factfinder2.census.gov* to find the sample statistic and corresponding margin of error based on the *entire* American Community Survey sample. There we find $\hat{p} = 0.155$ with a margin of error of 0.001 (for 90% confidence). Why is this margin of error so different from the one you computed in (d)?

(f) Use the information given in (e) to generate and interpret a 90% confidence interval for the proportion of US residents who do not have health insurance, based on the entire American Community Survey sample.

[24]http://www.census.gov/acs/www/Downloads/data_documentation/Statistical_Testing/2010StatisticalTesting1year.pdf.

E.4 What Proportion of US Adults are Married? The *Married* variable in **ACS** codes whether each respondent is married (1) or not (0).

(a) In the **ACS** dataset, there are 825 people who are 15 years of age or older, and $\hat{p} = 0.53$ of these people are married. Generate and interpret a 90% confidence interval for the proportion of US adults aged 15 and older who are married.

(b) While the American Community Survey surveys a sample of US residents every year, the decennial census surveys the entire US population every 10 years. The American Community Survey has been conducted in some form yearly since 1850, and the decennial census has been conducted in years ending in "0" since 1790, meaning that there have been 17 years of overlap. Suppose we were to use the ACS from each of these 17 years to create a 90% confidence interval for the proportion of people aged 15 and older in the US who were married (as of that year). If there were no sampling bias and everyone responded to the census, approximately how many of these intervals do you think would have contained the true proportion of married adults as obtained by the census?

(c) Based on the 2000 census, 54.5% of people aged 15 and older in the US were married in 2000. Do the data from our **ACS** sub-sample provide significant evidence that the proportion of people aged 15 and older in the US who were married in 2010 is different than in 2000?

E.5 Income and Gender The dataset **EmployedACS** contains just the subset of people in the dataset **ACS** who were employed ($n = 431$). *Income* gives each person's wages or salary income in the past 12 months (in thousands of dollars), and *Sex* is coded with 1 for males, 0 for females.

(a) Produce a plot of the distribution of yearly income for US residents and comment on the distribution.

(b) Give relevant summary statistics for the yearly income for US residents.

(c) Produce a plot showing the relationship between *Income* and *Sex* and comment on what you see.

(d) Summarize the relationship between *Income* and *Sex* with appropriate summary statistics.

(e) Does the **EmployedACS** dataset provide significant evidence that average yearly income differs for employed males and females in the US? State hypotheses, use a randomization test to calculate a p-value, and make a conclusion in context.

E.6 Working Hours and Gender The dataset **EmployedACS** contains just the subset of people in the dataset **ACS** who were employed ($n = 431$), and includes *HoursWk*, the usual number of hours worked per week in the past 12 months.

(a) Produce a plot to examine the relationship between *HoursWk* and *Sex* (coded as 1 = male, 0 = female). Comment on what the plot shows.

(b) Summarize the relationship between *HoursWk* and *Sex* with appropriate summary statistics.

(c) Does the **EmployedACS** dataset provide significant evidence that employed males work more hours per week, on average, than employed females in the US? State hypotheses, use the appropriate formula and distribution to calculate a p-value, and make a conclusion in context.

E.7 Health Insurance and Race Of the $n = 1000$ people in **ACS**, 761 are White, 106 are Black, 70 are Asian, and 63 are other races. The sample proportion of people who have health insurance for each racial group are $\hat{p}_{white} = 0.880$, $\hat{p}_{black} = 0.811$, $\hat{p}_{asian} = 0.843$, and $\hat{p}_{other} = 0.730$.

(a) Create the two-way table for counts of health insurance status by race.

(b) Produce a graph to visualize the relationship between health insurance status and race.

(c) Is there a significant association between whether or not a person has health insurance and race? State hypotheses, calculate a p-value, and make a conclusion in context.

E.8 Age and Race

(a) Use the ACS dataset to create a plot for the overall age distribution in the US, and comment on the distribution.

(b) Give and interpret a 95% confidence interval for the average age of a US resident.

(c) Create a plot for the age distribution in the US by racial group, and comment on what you see.

(d) Does average age differ by racial group in the US? State hypotheses, calculate a p-value, and make a conclusion in context. Some relevant sample statistics are given below:

	mean	sd	n
asian	41.17143	20.45991	70
black	33.00000	23.53437	106
other	31.31746	19.99260	63
white	41.67937	23.06364	761

(e) After doing the analysis in (d), give a 95% confidence interval for the average age of an Asian US resident. How does this compare to your answer for all US residents from (b)?

E.9 Income and Hours Worked Answer the following questions using **EmployedACS**, a subset of ACS that only includes people who are employed.

(a) Construct a graph to visualize the relationship between *HoursWk* and *Income*. Comment on what you see.

(b) Is there a significant positive association between hours worked per week and yearly income?

(c) Fit a regression model regressing *Income* on *HoursWk*. What is the fitted prediction equation?

(d) What is the predicted salary for someone who typically works a 40-hour work week?

(e) How much of the variability in yearly income is explained by the number of hours worked per week?

(f) Are the conditions met for performing inference based on this regression model?

E.10 Income, Hours per Week, and Gender In Exercise E.5 we find that males have a higher average yearly income than females, in Exercise E.6 we find that males work more hours per week than females, on average, and in Exercise E.9 we find that people who work more hours per week make a higher yearly income. Using the **EmployedACS** dataset:

(a) Explain why *HoursWk* is a confounding variable in the relationship between *Income* and *Sex*.

(b) Use multiple regression to see if *Sex* is a significant predictor of *Income*, even after accounting for the number of hours worked per week.

(c) What is the predicted yearly income for a male who works 40 hours per week? For a female who works 40 hours per week? What is the difference?

WHICH METHOD?
We have learned a lot of methods in this course, some of which are given below. In Exercises E.11 to E.27, state which statistical method would be most appropriate for the given question or situation.

Interval for a Proportion	Test for a Proportion	Chi-Square Goodness of Fit
Interval for a Mean	Test for a Mean	Chi-Square Test for Association
Interval for Difference in Proportions	Test for Difference in Proportions	ANOVA for Difference in Means
Interval for Difference in Means	Test for Difference in Means	Simple Linear Regression
Interval for Correlation	Test for Correlation	Multiple Regression

E.11 Anthropologists have found two burial mounds in the same region. They know that several different tribes lived in the region and that the tribes have been classified according to different lengths of skulls. They measure the skulls found in each burial mound and wish to determine if the two mounds were made by different tribes.

E.12 The Hawaiian Planters Association is developing three new strains of pineapple (call them A, B, and C) to yield pulp with higher sugar content. Twenty plants of each variety (60 plants in all) are randomly distributed into a two acre field. After harvesting, the resulting pineapples are measured for sugar content and the yields are recorded for each strain. Are there significant differences in average sugar content between the three strains?

E.13 Researchers were commissioned by the Violence In Children's Television Investigative Monitors (VICTIM) to study the frequency of depictions of violent acts in Saturday morning TV fare. They selected a random sample of 40 shows which aired during this time period over a 12-week period. Suppose that 28 of the 40 shows in the sample were judged to contain scenes depicting overtly violent acts. How should they use this information to make a statement about the population of all Saturday morning TV shows?

E.14 The Career Planning Office is interested in seniors' plans and how they might relate to their majors. A large number of students are surveyed and classified according to their major (Natural Science, Social Science, Humanities) and future plans (Graduate School, Job, Undecided). Are the type of major and future plans related?

E.15 Every week during the Vietnam War, a body count (number of enemy killed) was reported by each army unit. The last digits of these numbers should be fairly random. However, suspicions arose that the counts might have been fabricated. To test this, a large random sample of body count figures was examined and the frequency with which the last digit was a 0 or a 5 was recorded. Psychologists have shown that people making up their own "random"

numbers will use these digits less often than random chance would suggest (i.e. 103 sounds like a more "real" count than 100). If the data were authentic counts, the proportion of numbers ending in 0 or 5 should be about 0.20. Do these data show evidence of fabrication?

E.16 In one of his adventures, Sherlock Holmes found footprints made by the criminal at the scene of a crime and measured the distance between them. After sampling many people, measuring their height and length of stride, he confidently announced that he could predict the height of the suspect. How?

E.17 Do people drive less (fewer miles) when gas prices are higher?

E.18 How many times a day do humans urinate, on average?

E.19 Is there an association between whether or not a person is smiling and whether or not the sun is shining?

E.20 Does average number of ounces of alcohol consumed each week, per person, differ by class year (First Year, Sophomore, Junior, Senior) among college students?

E.21 What percentage of Americans support same-sex marriage?

E.22 Is percentage of the national budget spent on health care associated with life expectancy for countries?

E.23 Common wisdom says that every dog year corresponds to 7 human years, so the human equivalent maturity level of a dog can be found by multiplying a dog's age by 7. If the human maturity age of many dogs were measured by dog experts, along with their true age, how would the number 7 be determined?

E.24 A student wants to predict his score on the MCAT exam (which was revised to include statistical concepts in 2015) and wants to use all information available to him. He has anonymous data on MCAT exam scores from previous students at his school, as well as data on each student's GPA, number of science/math courses, and whether or not the person graduated with honors.

E.25 People were recruited for a study on weight loss, in which participants were randomly assigned to one of four groups. Group 1 was given exercise instructions but no dietary instructions, group 2 was given dietary instructions but no exercise instructions, group 3 was given both exercise and dietary instructions, and group 4 was given neither exercise nor dietary instructions. The researchers have many questions, but one goal is to estimate how much more weight was lost on average by people who were given exercise instructions, as opposed to those who weren't.

E.26 Ultimate Frisbee games often begin by flipping two discs (tossing them as you would when flipping a coin), while someone calls "Same" (both face up or both face down) or "Different" (one face up, one face down). Whichever team wins gets to decide whether to start on offense or defense, or which side to start on (which may matter if it is windy). It would be advantageous to know if one choice is more likely, so a team decides to spend a practice repeating this flipping of the discs many times, each time recording same or different. Once the data are collected, they want to know whether they have learned anything significant.

E.27 One of the authors of this book used to be a professional figure skater. For her project when she took introductory statistics (from another author of this book), she was interested in which of two jumps she landed more consistently. She did 50 double loops (landed 47 successfully) and 50 double flips (landed 48 successfully), and wanted to determine whether this was enough evidence to conclude that she had a higher success rate with one jump than the other.

Probability Basics

"Probability theory is nothing but common sense reduced to calculation."

–Pierre-Simon Laplace*

The material in this chapter is independent of the other chapters and can be covered at any point in a course.

* *Théorie Analytique des Probabilités*, Ve. Courcier, Paris, 1814.
Top left: © Design Pics Inc./Alamy Limited, Top right: © Photo by Frederick Breedon/Getty Images Inc.
Bottom right: © Ash Waechter/iStockphoto

Questions and Issues

Here are some of the questions and issues we will discuss in this chapter:

- If a driver is involved in an accident, what is the chance he or she is under 25 years old?

- How does an insurance company estimate average losses to help price its policies?

- What percent of players in the Hockey Hall of Fame are Canadian?

- What percent of Americans are color-blind?

- If a man in the US lives until his 60th birthday, what is the probability that he lives to be at least 90 years old?

- Are changes in the stock market independent from day to day?

- If a person has a positive tuberculosis test, what is the probability that the person actually has tuberculosis?

- How do filters identify spam emails and text messages?

- What is the probability that Stephen Curry of the Golden State Warriors makes two free throws in a row?

- What percent of housing units in the US have only one person living there?

P.1 PROBABILITY RULES

A process is *random* if its outcome is uncertain. For example, when we flip a coin, it is not known beforehand whether the coin will land heads or tails. Nevertheless, if we flip a coin over and over and over again, the distribution is predictable: We can expect that it will land heads about 50% of the time and tails about 50% of the time. Given some random process, the *probability* of an event (e.g., a coin flip comes up heads, a cancer patient survives more than five years, a poker hand contains all hearts) is the long run frequency or proportion of times the event would occur if the process were repeated many, many times.

> **Probability**
>
> The probability of an event is the long run frequency or proportion of times the event occurs.
>
> Probabilities are always between 0 and 1.

Throughout this book we often take advantage of computer technology to estimate probabilities by simulating many occurrences of a random process and counting how often the event occurs. In some situations we can determine a probability theoretically. In this chapter we examine some theoretical methods to compute probabilities for single events or combinations of events.

Equally Likely Outcomes

The easiest case to handle is when the process consists of selecting one outcome, at random, from a fixed set of equally likely possible outcomes. For example, a fair coin flip represents an equally likely choice between two outcomes, heads and tails. Rolling a standard six-sided die gives a random selection from the outcomes 1, 2, 3, 4, 5, and 6. Putting the names of all students in your class in a hat, then drawing one out to select a student to solve a homework problem is a random process that fits this "equally likely outcomes" description.

For any single outcome, the probability of it occurring in such a random process is just $1/n$, where n is the number of possible outcomes. In many situations we are interested in combinations of several outcomes, for example, the result of a die roll is an even number or the gender of the student chosen to solve a problem is female. We call such a combination of one or more outcomes an *event*. When the outcomes are equally likely, we can easily find the probability of any event as a proportion:

> **Probabilities When Outcomes Are Equally Likely**
>
> When outcomes are equally likely, we have
>
> $$\text{Probability of an event} = P(event) = \frac{\text{Number of outcomes in the event}}{\text{Total number of outcomes}}$$

We use the notation $P(\ )$ as a shorthand for the "probability of" whatever event is described inside the parentheses. For example, when flipping a fair coin, $P(Heads) = 1/2$.

Example P.1

Pick a Card

Suppose that we start with a standard deck of 52 playing cards. Each card shows one of four suits (spades, hearts, diamonds, or clubs) and one of 13 denominations (ace, 2, 3, 4, 5, 6, 7, 8, 9, 10, jack, queen, king). We shuffle the cards and draw one at random. What is the probability that the card:

(a) Is a heart?

(b) Is a jack?

(c) Is the jack of hearts?

(d) Is a face card (jack, queen, or king) of any suit?

Solution   Since the deck contains 52 cards and we are selecting one at random, we just need to count how many cards satisfy each of these events and divide by 52 to get the probability.

(a) There are 13 hearts in the deck (one of each denomination) so $P(\text{Heart}) = 13/52 = 1/4$.

(b) There are 4 jacks in the deck (one of each suit) so $P(\text{Jack}) = 4/52 = 1/13$.

(c) There is only 1 jack of hearts in the deck, so $P(\text{Jack of Hearts}) = 1/52$.

(d) There are 12 face cards in the deck (4 jacks, 4 queens, and 4 kings) so $P(\text{Face Card}) = 12/52$.

In other random processes we have outcomes that are not equally likely. For example, if we toss a tack onto a flat surface, it might land with the point up or with the point down, but these outcomes may not be equally likely. We could estimate $P(\text{Point Up})$ by tossing a tack many times, but it would be difficult to derive this probability theoretically (without a deep understanding of the physics of tack tossing). Similarly, we might be interested in the probability that a basketball player makes a free throw, a stock goes up in price, or a surgical procedure relieves a symptom. In each case we could observe many trials and estimate the probability of the event occurring.

In other situations we may talk about the probability of an event, even though repeated trials are not feasible. You might wonder, what's the probability that I get an A in this course? Or what's the probability that it rains tomorrow? You might have a reasonable estimate for the chance of getting an A in a course, even if it's impractical to take the course many times to see how often you get an A. We often refer to such probabilities as *subjective* or *personal* probabilities.

Combinations of Events

Outcomes in events can be combined in various ways to form new events. For example, we might be interested in the probability that two events occur simultaneously, at least one of the two events occurs, an event does not occur, or one of the events occurs if the other one occurs. The next example illustrates several of these possibilities.

Example P.2

Gender and the US Senate

The United States Senate consists of two senators for each of the 50 states. Table P.1 shows a breakdown of the party affiliation (by caucus) and gender of the 100 senators who were in office on January 1, 2016.

If we were to select one senator at random from the 2016 US Senate, what is the probability that senator is:

(a) A woman?

(b) A Republican?

(c) A Republican woman?

(d) A woman if we know the senator is a Republican?

(e) A Republican if we know the senator is a woman?

(f) Either a woman or a Republican (or both)?

Table P.1 *Gender and party caucus in the 2016 US Senate*

	Democrat	Republican	Total
Female	14	6	20
Male	32	48	80
Total	46	54	100

Solution

Since we are choosing at random, we can find each of these probabilities by computing a proportion.

(a) There are 20 women out of the 100 senators, so the probability of selecting a woman is 20/100 = 0.20.

(b) There are 54 Republicans out of the 100 senators, so the probability of selecting a Republican is 54/100 = 0.54.

(c) There are only 6 senators who are both female and Republican, so the probability of selecting a Republican woman is 6/100 = 0.06.

(d) If we know the senator is a Republican, there are 54 possible choices. Of these 54, we see that 6 are women, so the probability of being a woman if the senator is a Republican is 6/54 = 0.11.

(e) If we know the senator is a woman, there are only 20 possible choices. Of these 20, we see that 6 are Republicans, so the probability of being a Republican if the senator is a woman is 6/20 = 0.30.

(f) There are 14 + 6 + 48 = 68 senators who are either women or Republicans, so the probability of selecting a woman or a Republican is 68/100 = 0.68.

The questions and answers such as those in Example P.2 can be difficult to discuss in ordinary prose. Phrases such as "Republican woman," "Republican or woman," and "Republican if she is a woman" are easily confused with each other. To help us express relationships involving probabilities, we often assign a letter to denote the event of interest. For example, we might let F and M denote the gender of the randomly chosen senator and use D or R to denote the party affiliation. Based on the results of Example P.2, we have $P(F) = 0.20$ and $P(R) = 0.54$. More

Table P.2 *Common combinations of events*

Notation	Meaning	Terminology
A **and** B	The event must satisfy *both* conditions.	*joint* or *intersection*
A **or** B	The event can satisfy *either* of the two conditions (or both).	*union*
not A	The event A does not happen.	*complement*
B **if** A	The event B happens if A also happens.	*conditional*

complicated expressions typically involve one of four basic operations shown in Table P.2.

Example P.3

Let F or M denote the gender of the randomly chosen senator and label the party caucus with R or D. Write each of the events described in parts (c) to (f) of Example P.2 as probability expressions.

Solution

(c) A Republican woman $\Rightarrow P(R \text{ and } F)$.

(d) A woman if the senator is a Republican $\Rightarrow P(F \text{ if } R)$.

(e) A Republican if the senator is a woman $\Rightarrow P(R \text{ if } F)$.

(f) Either a woman or a Republican $\Rightarrow P(F \text{ or } R)$.

The various methods for combining events are shown schematically in Figure P.1, where the desired probability is the shaded area as a fraction of the box with a heavy border. For each of these combinations, we have rules to help us compute the probability based on other probabilities.

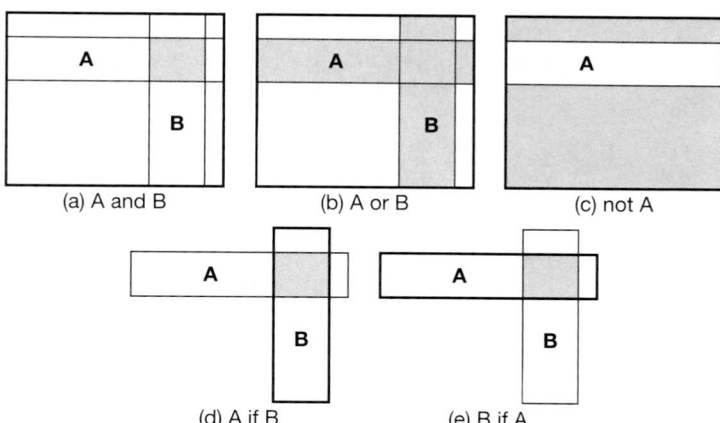

Figure P.1 *Common combinations of events*

Complement Rule

If an event A occurs on 1/3 of all trials, $P(A) = 1/3$, then it does not occur on the other 2/3 of the trials. This reasoning leads to the obvious rule for the *complement* of an event:

$$P(\text{not } A) = 1 - P(A)$$

Example P.4

If we draw a single card at random from a standard deck, what's the probability it's not a face card?

Solution In Example P.1 we see that $P(\text{Face Card}) = 12/52$, so $P(\text{not Face Card}) = 1 - 12/52 = 40/52$. We could also have counted how many non-face cards are in the deck and divided by 52 to get this probability directly, but sometimes it is much easier or more convenient to count how often something doesn't happen than to count how often it happens.

Additive Rule

In part (f) of Example P.2 we see that the probability a randomly chosen senator is either a woman or a Republican, $P(F \text{ or } R)$, is 0.68. Note that this is *not* just the sum of $P(F) = 0.20$ and $P(R) = 0.54$. The reason is that the six female senators who are also Republicans are double counted if we compute $P(F) + P(R) = 0.20 + 0.54 = 0.74$. To adjust the sum to avoid the double count, we can subtract the overlap. This gives the *additive rule* for the probability that an event A or an event B occurs:

$$P(A \text{ or } B) = P(A) + P(B) - P(A \text{ and } B)$$

For a senator being a woman or a Republican this means

$$P(F \text{ or } R) = P(F) + P(R) - P(F \text{ and } R)$$
$$= 0.20 + 0.54 + 0.06$$
$$= 0.68$$

Example P.5

ICU Admissions

Suppose that 35% of all patients admitted to a hospital's intensive care unit have high blood pressure, 42% have some sort of infection, and 12% have both problems.[1] Find the probability that a randomly chosen patient in this ICU has either high blood pressure or an infection.

Solution Let HBP = high blood pressure and INF = infection. From the given information we have $P(HBP) = 0.35$, $P(INF) = 0.42$, and $P(HBP \text{ and } INF) = 0.12$. We need to find the chance of high blood pressure or infection, $P(HBP \text{ or } INF)$. Applying the additive rule gives

$$P(HBP \text{ or } INF) = P(HBP) + P(INF) - P(HBP \text{ and } INF)$$
$$= 0.35 + 0.42 - 0.12$$
$$= 0.65$$

Special Case: Disjoint Events

If two events have no outcomes in common, we say they are *disjoint*. For example, it's impossible for a standard playing card to be both a jack and a seven. Therefore the events "jack" and "seven" are disjoint, as are "spade" and "heart," while "ace" and "spade" are not disjoint events. If events A and B have no outcomes in common, $P(A \text{ and } B)$ must be zero—they both can't happen at the same time. This leads to a special case of the additive rule:

$$P(A \text{ or } B) = P(A) + P(B) \qquad \text{whenever A and B are disjoint}$$

---

[1] Probabilities estimated from the data in **ICUAdmissions**.

Conditional Probability

Part (d) of Example P.2 on page 692 asks for the probability of an event (senator is female) when we assume that some other event (senator is a Republican) has also occurred. This is known as a *conditional* probability. Note that in computing this probability we divide the count of female Republican senators (6) by the number of Republican senators (54), not the total number of senators. This same reasoning applies if we have probabilities for the events rather than counts, that is, we divide the probability of both occurring, $P(R \text{ and } F)$, by the probability of the condition, $P(R)$. In general, to find the *conditional* probability of an event B occurring if A occurs, we use

$$P(B \text{ if } A) = \frac{P(A \text{ and } B)}{P(A)}$$

For the US senators this means the probability that a Republican senator is a woman is

$$P(F \text{ if } R) = \frac{P(R \text{ and } F)}{P(R)} = \frac{0.06}{0.54} = 0.11$$

Be careful to distinguish properly between the event you want the probability of and the event that determines the condition. For example, part (e) of Example P.2 asks for the probability that a female senator is Republican. This would be

$$P(R \text{ if } F) = \frac{P(F \text{ and } R)}{P(F)} = \frac{0.06}{0.20} = 0.30$$

Example P.6

Refer to the probabilities of high blood pressure given in Example P.5. Find the probability that a patient admitted to the ICU:

(a) With high blood pressure also has an infection.

(b) Has high blood pressure if an infection is present.

Solution

From information in Example P.5 we have $P(HBP) = 0.35$, $P(INF) = 0.42$, and $P(HBP \text{ and } INF) = 0.12$.

(a) The conditional probability of infection given that the patient has high blood pressure is

$$P(INF \text{ if } HBP) = \frac{P(HBP \text{ and } INF)}{P(HBP)} = \frac{0.12}{0.35} = 0.343$$

(b) The conditional probability of high blood pressure given that the patient has an infection is

$$P(HBP \text{ if } INF) = \frac{P(HBP \text{ and } INF)}{P(INF)} = \frac{0.12}{0.42} = 0.286$$

Multiplicative Rule

In some situations it is more natural or convenient to start with information on conditional probabilities and use it to find the joint probability of two events. To do so we merely rearrange the conditional probability rule to get the *multiplicative* rule:

$$P(B \text{ if } A) = \frac{P(A \text{ and } B)}{P(A)} \qquad \Rightarrow \qquad P(A \text{ and } B) = P(A) \cdot P(B \text{ if } A)$$

Thus to find the probability that both A and B occur, we find the probability of A occurring and multiply by the chance that B occurs if A does.

Example P.7

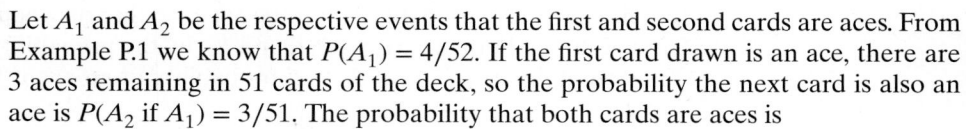

Suppose that we draw two cards (without replacement) from a standard 52 card deck. What is the probability that both the first and second cards are aces?

Solution

Let A_1 and A_2 be the respective events that the first and second cards are aces. From Example P.1 we know that $P(A_1) = 4/52$. If the first card drawn is an ace, there are 3 aces remaining in 51 cards of the deck, so the probability the next card is also an ace is $P(A_2$ if $A_1) = 3/51$. The probability that both cards are aces is

$$P(A_1 \text{ and } A_2) = P(A_1) \cdot P(A_2 \text{ if } A_1) = \frac{4}{52} \cdot \frac{3}{51} = 0.0045$$

Example P.8

Suppose that 37% of the students in an introductory statistics course are athletes on one of the school's sports teams. When asked to pick which award they would prefer to win, 5% of the athletes chose an Academy Award, 22% picked a Nobel Prize, and 73% wanted an Olympic Gold. Among non-athletes, the percentages were 15% for the Academy Award, 54% for the Nobel Prize, and 31% for the Olympic Gold. If we pick a student at random from this class, what is the probability that we get:

(a) An athlete who wants an Olympic Gold

(b) A non-athlete who wants an Academy Award

Solution

We know that $P(Athlete) = 0.37$ and $P(Non\text{-}athlete) = 1 - 0.37 = 0.63$. The other information gives conditional probabilities of the various awards (AA, NP, OG), given the status of athlete or non-athlete:

(a) $P(OG$ and $Athlete) = P(Athlete) \cdot P(OG$ if $Athlete) = 0.37 \cdot 0.73 = 0.27$.

(b) $P(AA$ and $Non\text{-}athlete) = P(Non\text{-}athlete) \cdot P(AA$ if $Non\text{-}athlete) = 0.63 \cdot 0.15 = 0.095$.

Special Case: Independent Events

Notice in Example P.8 that the probabilities for the awards *depend* on the athletic status of the students. In some special circumstances, as the next example illustrates, the conditional probabilities don't depend on the condition.

Example P.9

Suppose that we draw a single card from a standard deck and are told the card is a queen. What is the probability that the card is also a heart?

Solution

We need to compute

$$P(Heart \text{ if } Queen) = \frac{P(Heart \text{ and } Queen)}{P(Queen)} = \frac{1/52}{4/52} = \frac{1}{4}$$

Notice that the answer $P(Heart$ if $Queen) = 1/4$ is the same as $P(Heart)$ by itself without the condition. In fact, if we were told any other denomination (say a "seven" or an "ace") for the selected card, the probability it is a heart remains 1/4. When this happens, we say the events "Heart" and "Queen" are *independent.* The formal definition is

Events A and B are independent whenever $P(B \text{ if } A) = P(B)$

In many circumstances we can infer independence from the way the outcomes are determined. For example, if we select a card from the deck, record its denomination, put it back in the deck, shuffle, and select a card at random, the denomination of the second card is independent of the first. Thus if we choose two cards with replacement, the chance both are aces is

$$P(A_1 \text{ and } A_2) = P(A_1) \cdot P(A_2 \text{ if } A_1) = P(A_1) \cdot P(A_2) = \frac{1}{4} \cdot \frac{1}{4} = \frac{1}{16}$$

This shows a special form of the multiplicative rule.

$$P(A \text{ and } B) = P(A) \cdot P(B) \qquad \text{whenever A and B are independent}$$

Example P.10

Traffic Lights

Suppose that a set of three traffic lights along one section of road operate independently (i.e., no communication or special timing between the lights). Since this is a fairly main road, the lights are green with a probability of 0.7 and red with probability 0.3. As you go through this stretch of road, find the probability that:

(a) All three lights are green.

(b) The first two lights are green but the third is red.

(c) At least one of the lights is red.

Solution Since the lights operate independently, we can apply the special case of the multiplicative rule to find the probability of any sequence of red and green lights by multiplying the probabilities for the individual lights:

(a) $P(G_1 \text{ and } G_2 \text{ and } G_3) = P(G_1)P(G_2)P(G_3) = 0.7 \cdot 0.7 \cdot 0.7 = 0.343.$

(b) $P(G_1 \text{ and } G_2 \text{ and } R_3) = P(G_1)P(G_2)P(R_3) = 0.7 \cdot 0.7 \cdot 0.3 = 0.147.$

(c) Note that "at least one red" is the same as "not all green," so by the complement rule and the result of part (a), we have $P(\text{At least one red}) = 1 - P(\text{All three green}) = 1 - 0.343 = 0.657.$

 It is easy to confuse the concepts of "disjoint" events and "independent" events, but they are very different. If A and B are disjoint, when A occurs we know that B can't possibly also occur on the same trial. However, if A and B are independent, when A occurs it doesn't change at all the probability that B occurs on the same trial. Thus, independent events need common outcomes and disjoint events can't have any.

Example P.11

In Exercise P.5, we learn that for one Intensive Care Unit, 35% of admitted patients have high blood pressure (HPB), 42% have an infection (INF), and 12% have both problems. Are the two events HPB and INF:

(a) Disjoint?

(b) Independent?

Solution (a) Events A and B are disjoint if they cannot happen at the same time, which means $P(A \text{ and } B) = 0$. In this case, HPB and INF can occur simultaneously, and in fact these events do occur together in 12% of patients. We have $P(HPB \text{ and } INF) = 0.12$. These events are not disjoint.

(b) Events A and B are independent if knowing one gives us no information about the other. In symbols, this means $P(A \text{ and } B) = P(A) \cdot P(B)$. In this case, we see that $P(HPB) \cdot P(INF) = 0.35 \cdot 0.42 = 0.147$. Since this does not equal $P(HPB \text{ and } INF) = 0.12$, the events are not independent. (We could also have used the conditional probabilities of Example P.6 to answer this question.)

Basic Probability Rules: Summary

If A and B represent any two events:

Complement: $P(\text{not } A) = 1 - P(A)$

Additive: $P(A \text{ or } B) = P(A) + P(B) - P(A \text{ and } B)$

Multiplicative: $P(A \text{ and } B) = P(A) \cdot P(B \text{ if } A)$

Conditional: $P(B \text{ if } A) = \dfrac{P(A \text{ and } B)}{P(A)}$

Special cases:

If A and B are disjoint: $P(A \text{ or } B) = P(A) + P(B)$

If A and B are independent: $P(A \text{ and } B) = P(A) \cdot P(B)$

SECTION LEARNING GOALS

You should now have the understanding and skills to:

- Compute the probability of events if outcomes are equally likely
- Identify when a probability question is asking for A and B, A or B, not A, or A if B
- Use the complement, additive, multiplicative, and conditional rules to compute probabilities of events
- Recognize when two events are disjoint and when two events are independent

---

Exercises for Section **P.1**

SKILL BUILDER 1

In Exercises P.1 to P.7, use the information that, for events A and B, we have $P(A) = 0.4$, $P(B) = 0.3$, and $P(A \text{ and } B) = 0.1$.

P.1   Find $P(\text{not } A)$.

P.2   Find $P(\text{not } B)$.

P.3   Find $P(A \text{ or } B)$.

P.4   Find $P(A \text{ if } B)$.

P.5   Find $P(B \text{ if } A)$.

P.6   Are events A and B disjoint?

P.7   Are events A and B independent?

SKILL BUILDER 2

In Exercises P.8 to P.14, use the information that, for events A and B, we have $P(A) = 0.8$, $P(B) = 0.4$, and $P(A \text{ and } B) = 0.25$.

P.8   Find $P(\text{not } A)$.

P.9 Find $P(\text{not } B)$.

P.10 Find $P(A \text{ or } B)$.

P.11 Find $P(A \text{ if } B)$.

P.12 Find $P(B \text{ if } A)$.

P.13 Are events A and B disjoint?

P.14 Are events A and B independent?

SKILL BUILDER 3
In Exercises P.15 to P.18, use the fact that we have independent events A and B with $P(A) = 0.7$ and $P(B) = 0.6$.

P.15 Find $P(A \text{ if } B)$.

P.16 Find $P(B \text{ if } A)$.

P.17 Find $P(A \text{ and } B)$.

P.18 Find $P(A \text{ or } B)$.

SKILL BUILDER 4
Table P.3 gives probabilities for various combinations of events A, B, and their complements. Use the information from this table in Exercises P.19 to P.26.

P.19 Find $P(A)$.

P.20 Find $P(\text{not } B)$.

P.21 Find $P(A \text{ and } B)$.

P.22 Find $P(A \text{ or } B)$.

P.23 Find $P(A \text{ if } B)$.

P.24 Find $P(B \text{ if } A)$.

P.25 Are events A and B disjoint?

P.26 Are events A and B independent?

Table P.3 *Probability of being in each cell of a two-way table*

	A	not A
B	0.2	0.4
not B	0.1	0.3

DISJOINT, INDEPENDENT, AND COMPLEMENT
For Exercises P.27 to P.30, state whether the two events (A and B) described are **disjoint**, **independent**, and/or **complements**. (It is possible that the two events fall into more than one of the three categories, or none of them.)

P.27 Draw three skittles (possible colors: yellow, green, red, purple, and orange) from a bag. Let A be the event that all three skittles are green and B be the event that at least one skittle is red.

P.28 South Africa plays Australia for the championship in the Rugby World Cup. Let A be the event

that Australia wins and B be the event that South Africa wins. (The game cannot end in a tie.)

P.29 South Africa plays Australia for the championship in the Rugby World Cup. At the same time, Poland plays Russia for the World Team Chess Championship. Let A be the event that Australia wins their rugby match and B be the event that Poland wins their chess match.

P.30 Roll two (six-sided) dice. Let A be the event that the first die is a 3 and B be the event that the sum of the two dice is 8.

P.31 Explain What Is Wrong Each of the following statements demonstrate a common misuse of probability. Explain what is wrong with each statement:

(a) Approximately 10% of adults are left-handed. So, if we take a simple random sample of 10 adults, 1 of them will be left-handed.

(b) A pitch in baseball can be called a ball or a strike or can be hit by the batter. As there are three possible outcomes, the probability of each is 1/3.

(c) The probability that a die lands with a 1 face up is 1/6. So, since rolls of the die are independent, the probability that two consecutive rolls land with a 1 face up is $1/6 + 1/6 = 1/3$.

(d) The probability of surviving a heart attack is 2.35.

P.32 Studio and Genre of Movies About 26% of movies coming out of Hollywood are comedies, Warner Bros has been the lead studio for about 13% of recent movies, and about 3% of recent movies are comedies from Warner Bros.[2] Let C denote the event a movie is a comedy and W denote the event a movie is produced by Warner Bros.

(a) Write probability expressions for each of the three facts given in the first sentence of the exercise.

(b) What is the probability that a movie is either a comedy or produced by Warner Bros?

(c) What is the probability that a Warner Bros movie is a comedy?

(d) What is the probability that a comedy has Warner Bros as its producer?

(e) What is the probability that a movie coming out of Hollywood is not a comedy?

(f) In terms of movies, what would it mean to say that C and W are disjoint events? Are they disjoint events?

[2]Probabilities based on data from **HollywoodMovies**.

(g) In terms of movies, what would it mean to say that C and W are independent events? Are they independent events?

P.33 Rock and Roll Hall of Fame From its founding through 2015, the Rock and Roll Hall of Fame has inducted 303 groups or individuals.[3] Table P.4 shows how many of the inductees have been female or have included female members and also shows how many of the inductees have been performers. (The full dataset is available in **RockandRoll**.) Letting F represent the event of having female members (or being a female) and MP represent the event of being a (music) performer, write each of the following questions as a probability expression and find the probability.

What is the probability that an inductee chosen at random:

(a) Is a performer?

(b) Does not have any female members?

(c) Has female members if it is a performer?

(d) Is not a performer if it has no female members?

(e) Is a performer with no female members?

(f) Is either not a performer or has female members?

Table P.4 *Members of the Rock and Roll Hall of Fame*

	Female members	No female members	Total
Performer	38	168	206
Not performer	9	88	97
Total	47	256	303

P.34 Hockey Hall of Fame From its founding through 2016, the Hockey Hall of Fame has inducted 268 players.[4] Table P.5 shows the number of players by place of birth and by position played. If a player is chosen at random from all player inductees into the Hockey Hall of Fame, let *C* represent the event of being born in Canada and *D* represent the event of being a defenseman. Write each of the following questions as a probability expression and find the probability.

[3]Rock and Roll Hall of Fame website: *http://www.rockhall.com/inductees.*
[4]Hockey Hall of Fame website: *http://www.hhof.com.*

Table P.5 *Hockey Hall of Fame by place of birth and position*

	Offense	Defense	Goal	Total
Canada	126	74	33	233
USA	8	5	1	14
Europe	10	5	3	18
Other	2	1	0	3
Total	146	85	37	268

(a) What is the probability that an inductee chosen at random is Canadian?

(b) What is the probability that an inductee chosen at random is not a defenseman?

(c) What is the probability that a player chosen at random is a defenseman born in Canada?

(d) What is the probability that a player chosen at random is either born in Canada or a defenseman?

(e) What is the probability that a Canadian inductee plays defense?

(f) What is the probability that an inductee who plays defense is Canadian?

P.35 Peanut M & Ms In a bag of peanut M & M's, there are 80 M & Ms, with 11 red ones, 12 orange ones, 20 blue ones, 11 green ones, 18 yellow ones, and 8 brown ones. They are mixed up so that each candy piece is equally likely to be selected if we pick one.

(a) If we select one at random, what is the probability that it is red?

(b) If we select one at random, what is the probability that it is not blue?

(c) If we select one at random, what is the probability that it is red or orange?

(d) If we select one at random, then put it back, mix them up well (so the selections are independent) and select another one, what is the probability that both the first and second ones are blue?

(e) If we select one, keep it, and then select a second one, what is the probability that the first one is red and the second one is green?

P.36 More Peanut M & Ms As in Exercise P.35, we have a bag of peanut M & M's with 80 M & Ms in it, and there are 11 red ones, 12 orange ones, 20 blue ones, 11 green ones, 18 yellow ones, and 8 brown ones. They are mixed up so that each is equally likely to be selected if we pick one.

(a) If we select one at random, what is the probability that it is yellow?

(b) If we select one at random, what is the probability that it is not brown?

(c) If we select one at random, what is the probability that it is blue or green?

(d) If we select one at random, then put it back, mix them up well (so the selections are independent) and select another one, what is the probability that both the first and second ones are red?

(e) If we select one, keep it, and then select a second one, what is the probability that the first one is yellow and the second one is blue?

P.37 Free Throws During the 2015–16 NBA season, Stephen Curry of the Golden State Warriors had a free throw shooting percentage of 0.908. Assume that the probability Stephen Curry makes any given free throw is fixed at 0.908, and that free throws are independent.

(a) If Stephen Curry shoots two free throws, what is the probability that he makes both of them?

(b) If Stephen Curry shoots two free throws, what is the probability that he misses both of them?

(c) If Stephen Curry shoots two free throws, what is the probability that he makes exactly one of them?

P.38 Color Blindness in Men and Women The most common form of color blindness is an inability to distinguish red from green. However, this particular form of color blindness is much more common in men than in women (this is because the genes corresponding to the red and green receptors are located on the X-chromosome). Approximately 7% of American men and 0.4% of American women are red-green color-blind.[5]

(a) If an American male is selected at random, what is the probability that he is red-green color-blind?

(b) If an American female is selected at random, what is the probability that she is NOT red-green color-blind?

(c) If one man and one woman are selected at random, what is the probability that neither are red-green color-blind?

(d) If one man and one woman are selected at random, what is the probability that at least one of them are red-green color-blind?

P.39 More Color Blindness in Men and Women Approximately 7% of men and 0.4% of women are red-green color-blind (as in Exercise P.38). Assume that a statistics class has 15 men and 25 women.

(a) What is the probability that nobody in the class is red-green color-blind?

(b) What is the probability that at least one person in the class is red-green color-blind?

(c) If a student from the class is selected at random, what is the probability that he or she will be red-green color-blind?

P.40 Probabilities of Death The US Social Security Administration collects information on the life expectancy and death rates of the population. Table P.6 gives the number of US men out of 100,000 born alive who will survive to a given age, based on 2011 mortality rates.[6]
For example, 50,344 of 100,000 US males live to their 80th birthday.

(a) What is the probability that a man lives to age 60?

(b) What is the probability that a man dies before age 70?

(c) What is the probability that a man dies at age 90 (after his 90th and before his 91st birthday)?

(d) If a man lives until his 90th birthday, what is the probability that he will die at the age of 90?

(e) If a man lives until his 80th birthday, what is the probability that he will die at the age of 90?

(f) What is the probability that a man dies between the ages of 60 and 89?

(g) If a man lives until his 60th birthday, what is the probability that he lives to be at least 90 years old?

Table P.6 *Life Table for US males, 2011*

Age	60	70	80	90	91
Number of lives	85,995	73,548	50,344	17,429	14,493

P.41 Is the Stock Market Independent? The Standard and Poor 500 (S&P 500) is a weighted average of the stocks for 500 large companies in the United States. It is commonly used as a measure of the overall performance of the US stock market. Between January 1, 2009 and January 1, 2012, the S&P 500 increased for 423 of the 756 days that the stock market was open. We will investigate whether changes

[5]Montgomery, G., "Color Blindness: More Prevalent Among Males," in *Seeing, Hearing, and Smelling the World, http://www. hhmi.org/senses/b130.html*, accessed April 27, 2012.

[6]Period Life Table 2011, *http://www.ssa.gov/oact/STATS/table4c6 .html*, accessed May 30, 2016.

to the S&P 500 are independent from day to day. This is important, because if changes are not independent, we should be able to use the performance on the current day to help predict performance on the next day.

(a) What is the probability that the S&P 500 increased on a randomly selected market day between January 1, 2009 and January 1, 2012?

(b) If we assume that daily changes to the S&P 500 are independent, what is the probability that the S&P 500 increases for two consecutive days? What is the probability that the S&P 500 increases on a day, given that it increased the day before?

(c) Between January 1, 2009 and January 1, 2012 the S&P 500 increased on two consecutive market days 234 times out of a possible 755. Based on this information, what is the probability that the S&P 500 increases for two consecutive days? What is the probability that the S&P 500 increases on a day, given that it increased the day before?

(d) Compare your answers to part (b) and part (c). Do you think that this analysis proves that daily changes to the S&P 500 are not independent?

P.42 Pancakes A friend makes three pancakes for breakfast. One of the pancakes is burned on both sides, one is burned on only one side, and the other is not burned on either side. You are served one of the pancakes at random, and the side facing you is burned. What is the probability that the other side is burned? (*Hint:* Use conditional probability.)

P.2 TREE DIAGRAMS AND BAYES' RULE

Example P.12

Automobile Accident Rates

According to data from the US Census Bureau's 2012 Statistical Abstract, the probability a young person (under the age of 25) is involved as a driver in an automobile accident during a given year is about 0.16. For a driver whose age is in the middle (25 to 54 years old), the probability drops to 0.08 and for older drivers (55 and older) the rate is about 0.04.[7] The US Census also tells us that about 13.2% of all licensed drivers are young, 35.6% are between 25 and 54, and the remaining 51.2% are older. What is the overall probability that a licensed driver has an accident during the year?

Solution If we let Y, M, and O represent the age groups (young, middle, and old) and A be the event that a driver has an accident, the given information states that

$$P(A \text{ if } Y) = 0.16 \quad P(A \text{ if } M) = 0.08 \quad P(A \text{ if } O) = 0.04$$
$$P(Y) = 0.132 \quad\quad P(M) = 0.356 \quad\quad P(O) = 0.512$$

We need to find $P(A)$.

Using the multiplicative rule we can find the probability of having an accident and being in each of the respective age groups:

$$P(A \text{ and } Y) = P(Y) \cdot P(A \text{ if } Y) = 0.132 \cdot 0.16 = 0.0211$$
$$P(A \text{ and } M) = P(M) \cdot P(A \text{ if } M) = 0.356 \cdot 0.08 = 0.0285$$
$$P(A \text{ and } O) = P(O) \cdot P(A \text{ if } O) = 0.512 \cdot 0.04 = 0.0205$$

Since the three age groups are disjoint and cover all possible ages, we get the overall probability of an accident, $P(A)$, by adding these three results:

$$P(A) = P(A \text{ and } Y) + P(A \text{ and } M) + P(A \text{ and } O)$$
$$= 0.0211 + 0.0285 + 0.0205$$
$$= 0.0701$$

The probability a licensed driver has an accident during the year is about 0.0701, or about a 7% chance.

---

[7]Based on accident rates given in the US Census Bureau's 2012 Statistical Abstract, Table 1114, downloaded at *http://www.census.gov/compendia/statab/cats/transportation/motor_vehicle_accidents_and_fatalities.html*.

Total Probability

The last calculation of Example P.12 involves finding the probability of an event (A) by adding the probabilities that it occurs along with each of a set of disjoint events (Y, M, and O). This is an example of the *total probability* rule. As long as the disjoint events include all the possible outcomes in the event of interest we can find its probability this way. In particular, we can always use an event, A, and its complement, *not* A, as a pair of disjoint events to help find the probability of some other event. Depending on the types of probabilities we have, this can be done with joint probabilities (*and*), or the multiplicative rule and conditional probabilities (*if*). These options are summarized in the formulas below.

> ### Total Probability Rule
>
> For any two events A and B,
>
> $$P(B) = P(A \text{ and } B) + P(\text{ not } A \text{ and } B)$$
> $$= P(A)P(B \text{ if } A) + P(\text{not } A)P(B \text{ if not } A)$$
>
> More generally if A, B, and C are disjoint events which contain all of the outcomes of another event D, then
>
> $$P(D) = P(A \text{ and } D) + P(B \text{ and } D) + P(C \text{ and } D)$$
> $$= P(A)P(D \text{ if } A) + P(B)P(D \text{ if } B) + P(C)P(D \text{ if } C)$$
>
> We can extend this in equivalent ways to more than three events.

Example P.13

Olympic Gold

Refer to Example P.8 on page 696 where we see that 73% of athletes in a statistics course would prefer to win an Olympic Gold medal (over an Academy Award or Nobel Prize) while 31% of non-athletes make that choice. If 37% of the students in the class are athletes, what is the probability that a student chosen at random from this class would pick the Olympic Gold?

Solution

If we let A denote the event "athlete" and OG denote "Olympic Gold," the given probabilities are P(A) = 0.37, P(OG if A) = 0.73, and P(OG if not A) = 0.31. Applying the total probability rule, we have

$$P(OG) = P(A)P(OG \text{ if } A) + P(\text{not } A)P(OG \text{ if not } A)$$
$$= 0.37(0.73) + (1 - 0.37)(0.31)$$
$$= 0.2701 + 0.1953 = 0.4654$$

Overall, about 46.5% of students in the class pick the Olympic Gold.

Tree Diagrams

When the given information is in terms of probabilities for one type of event and conditional probabilities for another, such as in Examples P.12 and P.13, we can often organize the calculation of the multiplicative rule with a tree diagram such as those shown in Figure P.2. The initial set of "branches" show the probabilities from one set of events and the second set of branches show conditional probabilities (with the initial branch as the condition). Multiplying along any set of branches uses the multiplicative rule to find the joint probability for that pair of events.

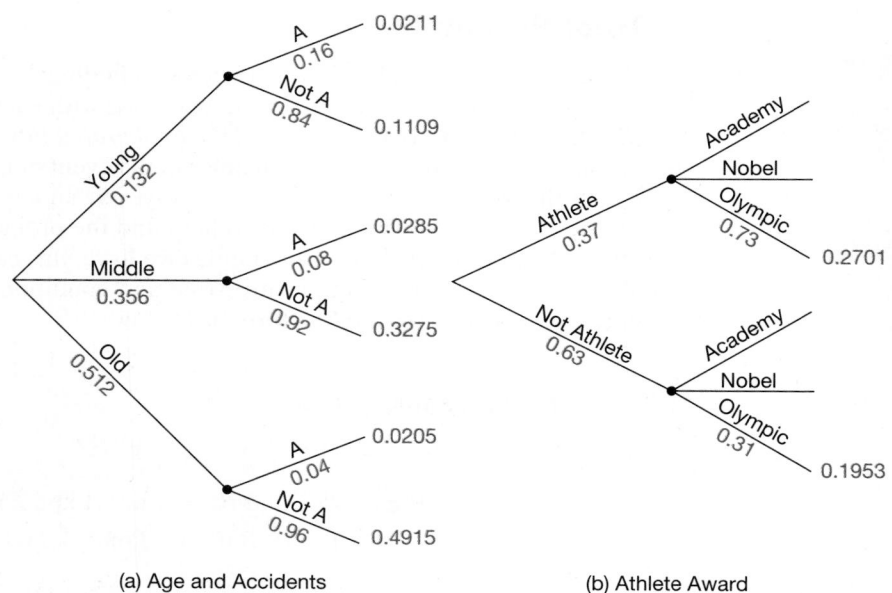

Figure P.2 *Tree diagrams for Examples P.12 to P.14*

(a) Age and Accidents   (b) Athlete Award

Example P.14

An insurance company gets a report that one of its drivers was in an accident. Using the probabilities given in Example P.12 and calculations shown in Figure P.2(a), what is the probability that the driver is under 25 years old?

Solution

We need to find the probability that a driver is young if we know he or she has had an accident, that is, $P(Y \text{ if } A)$. By the conditional probability rule this is

$$P(Y \text{ if } A) = \frac{P(Y \text{ and } A)}{P(A)}$$

From the work in Example P.12 and the tree diagram in Figure P.2 we know that $P(Y \text{ and } A) = 0.0211$ and $P(A) = 0.0211 + 0.0285 + 0.0205 = 0.0701$, so

$$P(Y \text{ if } A) = \frac{P(Y \text{ and } A)}{P(A)} = \frac{0.0211}{0.0701} = 0.301$$

The probability is 0.301. If the insurance company hears about an accident, there is about a 30% chance that the driver was young.

A question such as the one posed in Example P.14 asks for a *posterior* probability, since we are given some information (the driver had an accident) and asked to go back and revise our estimate of some initial probability (the driver is young). Note that the information that the driver had an accident makes it more likely that the driver is under 25 (30% compared to the original 13.2%), since younger drivers are more prone to accidents. We can do a similar calculation for the other two age groups:

$$P(M \text{ if } A) = \frac{P(M \text{ and } A)}{P(A)} = \frac{0.0285}{0.0701} = 0.41$$

$$P(O \text{ if } A) = \frac{P(O \text{ and } A)}{P(A)} = \frac{0.0205}{0.0701} = 0.29$$

When restricted to drivers having an accident, the age distribution is 30% young, 41% middle, and 29% older.

Bayes' Rule

In Examples P.12 and P.14 we know something about the probabilities for one type of event (the driver age groups) and conditional probabilities for a different event (having an accident) given each of those groups. When the latter event occurs, the question of interest is often the chance it is associated with one of the initial events. Put another way, if we know $P(A$ if $B)$, what can we say about $P(B$ if $A)$?

Although the calculations are fairly intuitive from a tree diagram, we can also use formulas to compute the conditional probability directly. This method is known as Bayes' rule.[8] We give several equivalent formulas below for the case of two events, but they can easily be generalized to more complicated situations. The form we choose depends on the nature of the given information. If the formulas look a bit intimidating, remember that we can also (or instead) use a tree diagram (as we did in Example P.14).

Bayes' Rule

If A and D are any two events,

$$P(A \text{ if } D) = \frac{P(A \text{ and } D)}{P(D)}$$

$$= \frac{P(A)P(D \text{ if } A)}{P(D)}$$

$$= \frac{P(A \text{ and } D)}{P(A \text{ and } D) + P(\text{ not } A \text{ and } D)}$$

$$= \frac{P(A)P(D \text{ if } A)}{P(A)P(D \text{ if } A) + P(\text{ not } A)P(D \text{ if not } A)}$$

The last two versions can easily be generalized using the total probability rule to handle more than two events.

Example P.15

Tuberculosis Tests

One of the common tests for tuberculosis (TB) is a skin test where a substance is injected into a subject's arm and we see if a rash develops in a few days. The test is relatively accurate, but in a few cases a rash might be detected even when the subject does not have TB (a false positive) or the rash may not be seen even when the subject has TB (a false negative). Assume that the probability of the first event, $P(\text{Rash if not TB})$, is about 5% and the chance of the other, $P(\text{not Rash if TB})$, is about 1%. Suppose also that only about 4 in 10,000 people have TB. An applicant for a teaching position is required to get a TB test and the test comes back positive (shows a rash). What is the probability that the applicant really has TB?

[8] Reverend Thomas Bayes, a Presbyterian minister in England who also studied mathematics, was credited, after his death in 1761, with the discovery of this method for finding "inverse" probabilities.

Solution Our initial information about an applicant is one of two possibilities, $P(\text{TB}) = 0.0004$ and $P(\text{not TB}) = 1 - 0.0004 = 0.9996$. How does this change when we include the additional information of a positive TB test? Applying Bayes' rule we have

$$P(\text{TB if Rash}) = \frac{P(\text{TB})P(\text{Rash if TB})}{P(\text{TB})P(\text{Rash if TB}) + P(\text{not TB})P(\text{Rash if not TB})}$$

$$= \frac{0.0004(0.99)}{0.0004(0.99) + 0.9996(0.05)}$$

$$= \frac{0.000396}{0.050376} = 0.00786$$

Although the TB test came back positive for the applicant, the probability that the person actually has tuberculosis is only 0.00786.

You may be surprised that the probability found in Example P.15 is so small, when it seems as if the test is pretty accurate. Note, however, that the initial probability of having TB (0.0004) is so small that, when picking a person at random, it is far more likely that person does not have TB and thus an error to the test would be a false positive. Figure P.3 shows a tree diagram attacking this problem, in which case

$$P(\text{TB if Rash}) = \frac{P(\text{TB and Rash})}{P(\text{Rash})} = \frac{0.000396}{0.050376} = 0.00786$$

Note that we get the same answer whether we use a visual display like the tree diagram or the mathematical formulation of Bayes' rule.

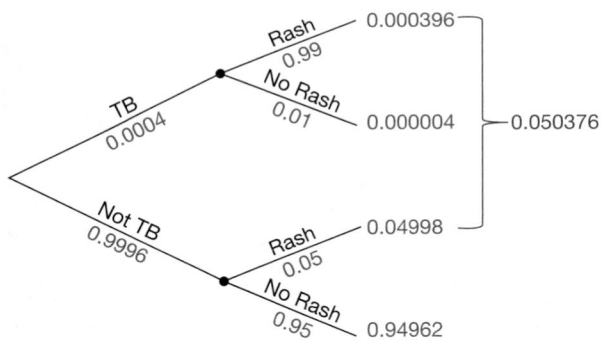

Figure P.3 *Tree diagram for TB tests*

Exercises for Section **P.2**

SKILL BUILDER 1: INCOMPLETE TREE DIAGRAMS

In Exercises P.43 to P.46, complete each tree diagram by filling in the missing entries (marked with a "?").

P.43

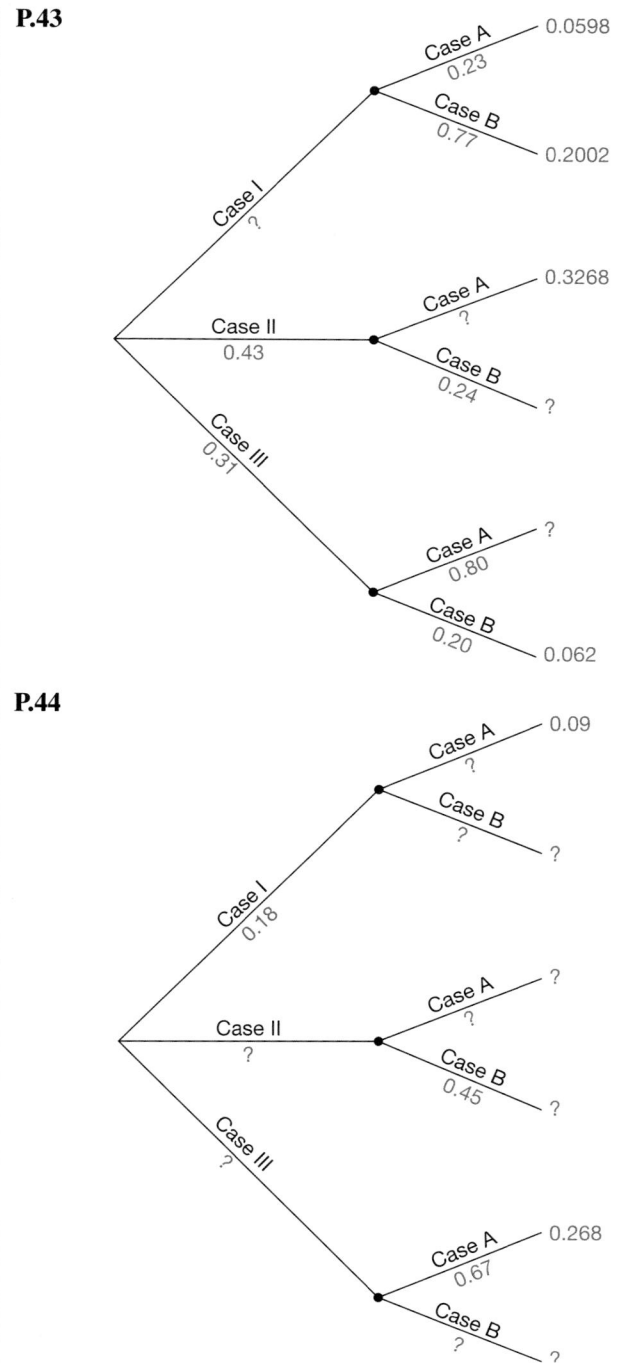

P.44

P.45

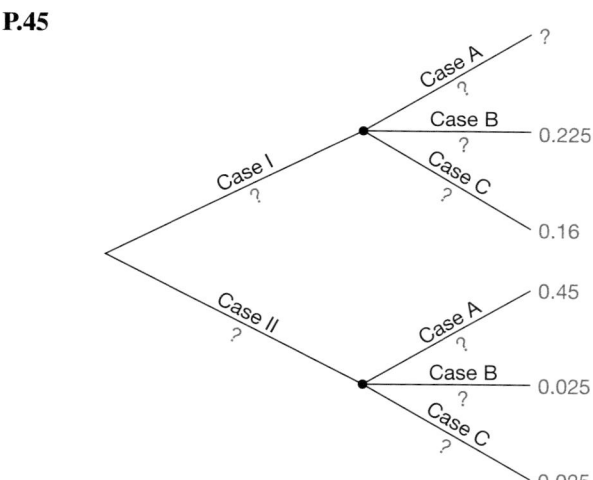

P.46

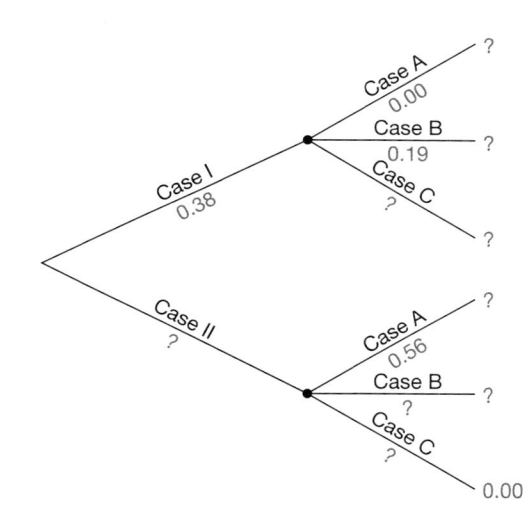

SKILL BUILDER 2: FINDING PROBABILITIES USING TREE DIAGRAMS

In Exercises P.47 to P.54, find the requested probabilities using the tree diagram in Figure P.4.

P.47 $P(B \text{ and } R)$

P.48 $P(A \text{ and } S)$

P.49 $P(R \text{ if } A)$

P.50 $P(S \text{ if } B)$

P.51 $P(R)$

P.52 $P(S)$

P.53 $P(A \text{ if } S)$

P.54 $P(B \text{ if } R)$

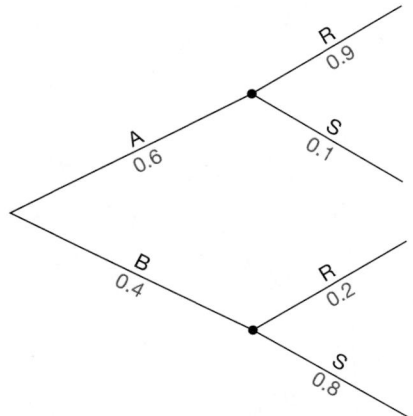

Figure P.4 *Tree diagram for Exercises P.47 to P.54*

P.55 Housing Units in the US According to the 2010 US Census, 65% of housing units in the US are owner-occupied while the other 35% are renter-occupied.[9] Table P.7 shows the probabilities of the number of occupants in a housing unit under each of the two conditions. Create a tree diagram using this information and use it to answer the following questions:

(a) What is the probability that a US housing unit is rented with exactly two occupants?

(b) What is the probability that a US housing unit has three or more occupants?

(c) What is the probability that a unit with one occupant is rented?

Table P.7 *Conditional probabilities of number of occupants in US housing units*

Condition	1	2	3 or more
Owner-occupied	0.217	0.363	0.420
Renter-occupied	0.362	0.261	0.377

P.56 Restless Leg Syndrome and Fibromyalgia People with restless leg syndrome have a strong urge to move their legs to stop uncomfortable sensations. People with fibromyalgia suffer pain and tenderness in joints throughout the body. A recent study indicates that people with fibromyalgia are much more likely to have restless leg syndrome than people without the disease.[10] The study indicates that, for people with fibromyalgia, the probability is 0.33 of having restless leg syndrome, while for people without fibromyalgia, the probability is 0.03. About 2% of the population has fibromyalgia. Create a tree diagram from this information and use it to find the probability that a person with restless leg syndrome has fibromyalgia.

P.57 Mammograms and Breast Cancer The mammogram is helpful for detecting breast cancer in its early stages. However, it is an imperfect diagnostic tool. According to one study,[11] 86.6 of every 1000 women between the ages of 50 and 59 that do not have cancer are wrongly diagnosed (a "false positive"), while 1.1 of every 1000 women between the ages of 50 and 59 that do have cancer are not diagnosed (a "false negative"). One in 38 women between 50 and 59 will develop breast cancer. If a woman between the ages of 50 and 59 has a positive mammogram, what is the probability that she will have breast cancer?

P.58 What's the Pitch? Slippery Elum is a baseball pitcher who uses three pitches, 60% fastballs, 25% curveballs, and the rest spitballs. Slippery is pretty accurate with his fastball (about 70% are strikes), less accurate with his curveball (50% strikes), and very wild with his spitball (only 30% strikes). Slippery ends one game with a strike on the last pitch he throws. What is the probability that pitch was a curveball?

IDENTIFYING SPAM TEXT MESSAGES
Bayes' rule can be used to identify and filter spam emails and text messages. Exercises P.59 to P.62 refer to a large collection of real SMS text messages from participating cellphone users.[12] In this collection, 747 of the 5574 total messages (13.40%) are identified as spam.

P.59 The word "free" is contained in 4.75% of all messages, and 3.57% of all messages both contain the word "free" and are marked as spam.

(a) What is the probability that a message contains the word "free", given that it is spam?

(b) What is the probability that a message is spam, given that it contains the word "free"?

[9]*http://www.census.gov.*

[10]"Fibromyalgia, Restless Legs Linked in New Study," *http://www.healthcentral.com/chronic-pain/news-540062-98 .html*, 2010.

[11]Nelson, H., et al., "Screening for Breast Cancer: Systematic Evidence Review Update for the U.S. Preventive Services Task Force," Evidence Review Update No. 74, AHRQ Publication No. 10-05142-EF-1, Rockville, MD, Agency for Healthcare Research and Quality, 2009.

[12]Almeida, T., Hidalgo, J., and Yamakami, A., "Contributions to the Study of SMS Spam Filtering: New Collection and Results," *Proceedings of the 2011 ACM Symposium on Document Engineering (DOCENG'11)*, Association for Computing Machinery, Mountain View, CA, 2011.

P.60 The word "text" (or "txt") is contained in 7.01% of all messages, and in 38.55% of all spam messages. What is the probability that a message is spam, given that it contains the word "text" (or "txt")?

P.61 Of all spam messages, 17.00% contain both the word "free" and the word "text" (or "txt"). For example, "Congrats!! You are selected to receive a free camera phone, txt ******* to claim your prize."

Of all non-spam messages, 0.06% contain both the word "free" and the word "text" (or "txt"). Given that a message contains both the word "free" and the word "text" (or "txt"), what is the probability that it is spam?

P.62 Given that a message contains the word "free" but does NOT contain the word "text" (or "txt"), what is the probability that it is spam? (*Hint:* Use the information in Exercises P.59 to P.61.)

P.3 RANDOM VARIABLES AND PROBABILITY FUNCTIONS

Random Variables

A *random variable* denotes a numeric quantity that changes from trial to trial in a random process. We often use a capital letter, like X or Y, to denote a random variable. Here are some examples:

X = number showing when a six-sided die is rolled

Y = sum of the numbers shown on two dice rolls

Z = number of girls among three children in a family

W = weight change after six months in an exercise program

T = time needed to read this section of the text

C = cost to repair damage to a car after an accident

We would like answers to probability questions about events determined by a random variable. For example, what is the probability that the sum of two dice rolls is eight ($Y = 8$), a family has three children who are all girls ($Z = 3$), or the cost to repair a car is more than \$2000 ($C > 2000$)?

Discrete vs Continuous

We say a random variable is *discrete* if it has a finite set of possible values. The result of a die roll {1, 2, 3, 4, 5, or 6}, the sum of two dice {2, 3, ..., or 12}, and the number of girls among three children{0, 1, 2, or 3} are all discrete random variables.

A variable that can take any value within some interval is called *continuous*. The amount of weight change and time needed to read the text are examples of continuous random variables.

In some cases a variable might technically be discrete (like the cost in dollars to repair a car), but there are so many possible values that we may decide to treat it as if it were continuous. In other situations, we might compress/round a continuous variable to a relatively few discrete values.

The distinction between discrete and continuous random variables is important for determining how we express their probabilities. For continuous random variables we use a density curve, as described in Section P.5, where the probability of being in some region is found as the area under the density curve.

For the rest of this section we assume that a random variable is discrete with a relatively small set of possible values. In that case we determine probabilities by specifying a probability for each possible value of the random variable.

Probability Functions

For a discrete random variable, a *probability function* gives the probability for each of its possible values. We often use notation such as $p(2)$ as shorthand to denote the probability of the event, $P(X = 2)$. In some cases (see Exercises P.93 and P.94), the probability function may be given as a mathematical expression, but often we simply give a table of the probabilities for each possible value.

> **Probability Function for a Discrete Random Variable**
>
> A probability function assigns a probability, between 0 and 1, to every value of a discrete random variable. The sum of all of these probabilities must be one, i.e. $\sum p(x) = 1$.

Example P.16

Roll a Die

Suppose that we roll a fair six-sided die and let the random variable X be the value showing at the top of the die. Find the probability function for X.

Solution   There are six possible values, $\{1, 2, 3, 4, 5, 6\}$, that are all equally likely, so the probability of each is $1/6$. We can express this as a table:

x	1	2	3	4	5	6
$p(x)$	$\dfrac{1}{6}$	$\dfrac{1}{6}$	$\dfrac{1}{6}$	$\dfrac{1}{6}$	$\dfrac{1}{6}$	$\dfrac{1}{6}$

or simply write $p(x) = 1/6$ for $x = 1, 2, \dots, 6$.

Example P.17

Sum of Two Dice

Suppose that we roll two six-sided dice and let a random variable X measure the sum of the two rolls. There are 6 possible outcomes for each die, so $6 \times 6 = 36$ possible pairs of rolls. The possible sums are values from 2 to 12, but they are not all equally likely. For example, there is only one pair, $6 + 6$, that gives a sum of 12, but three ways, $4 + 6$, $5 + 5$, and $6 + 4$, to get a sum of 10. The probability function for X is shown in Table P.8.

Use the probability function to find:

(a) $P(X = 7 \text{ or } X = 11)$

(b) $P(X > 8)$

Solution   (a) The events $X = 7$ and $X = 11$ are disjoint so we find the probability that one or the other occurs by adding the individual probabilities:

$$P(X = 7 \text{ or } X = 11) = p(7) + p(11) = \frac{6}{36} + \frac{2}{36} = \frac{8}{36} = 0.222$$

Table P.8 *Probability function for sum of two dice*

x	2	3	4	5	6	7	8	9	10	11	12
$p(x)$	$\dfrac{1}{36}$	$\dfrac{2}{36}$	$\dfrac{3}{36}$	$\dfrac{4}{36}$	$\dfrac{5}{36}$	$\dfrac{6}{36}$	$\dfrac{5}{36}$	$\dfrac{4}{36}$	$\dfrac{3}{36}$	$\dfrac{2}{36}$	$\dfrac{1}{36}$

(b) To find the probability that the sum is greater than 8, we add the individual probabilities from the probability function for the values of X that satisfy this condition:

$$P(X > 8) = p(9) + p(10) + p(11) + p(12) = \frac{4}{36} + \frac{3}{36} + \frac{2}{36} + \frac{1}{36} = \frac{10}{36} = 0.278$$

Mean of a Random Variable

Example P.18

Raffle Winnings

A charitable organization is running a raffle as a fundraiser. They offer a grand prize of $500, two second prizes of $100, and ten third prizes of $20 each. They plan to sell 1000 tickets at $2 per ticket. What is the *average* amount of money won with each ticket in the lottery?

Solution The total amount of prize money is $500 \cdot 1 + \$100 \cdot 2 + \$20 \cdot 10 = \$900$. Since there are 1000 tickets sold, the average amount won per ticket is $\$900/1000 = \$0.90$, or about 90 cents.

We can formalize the process of Example P.18 to find the mean of any random variable if we know its probability function. Letting X represent the amount won with a raffle ticket, the probability function is shown in Table P.9.

Table P.9 *Probability function for raffle winnings*

x	500	100	20	0
$p(x)$	$\dfrac{1}{1000}$	$\dfrac{2}{1000}$	$\dfrac{10}{1000}$	$\dfrac{987}{1000}$

The process of calculating the total winnings and dividing by the number of tickets sold is equivalent to multiplying each value of the random variable by its corresponding probability and adding the results:

$$500 \cdot \frac{1}{1000} + 100 \cdot \frac{2}{1000} + 20 \cdot \frac{10}{100} + 0 \cdot \frac{987}{1000} = 0.90$$

We call this the *mean* or *expected value* of the random variable X. Since this represents the average value over the "population" of all tickets, we use the notation $\mu = 0.90$ to represent this mean.

In general, we find the mean for a random variable from its probability function by multiplying each of the possible values by the probability of getting that value and summing the results.

Mean of a Random Variable

For a random variable X with probability function $p(x)$, the mean, μ, is

$$\mu = \sum x \cdot p(x)$$

Example P.19

Find the mean of the sum of two dice rolls using the probability function given in Example P.17 on page 710.

Solution

We multiply each of the possible values from the sum of two dice rolls by its corresponding probability given in Table P.8, and add up the results:

$$\mu = 2 \cdot \frac{1}{36} + 3 \cdot \frac{2}{36} + 4 \cdot \frac{3}{36} + 5 \cdot \frac{4}{36} + 6 \cdot \frac{5}{36} + 7 \cdot \frac{6}{36}$$
$$+ \; 8 \cdot \frac{5}{36} + 9 \cdot \frac{4}{36} + 10 \cdot \frac{3}{36} + 11 \cdot \frac{2}{36} + 12 \cdot \frac{1}{36} = 7.0$$

The average sum on a roll of two dice is 7. This is not very surprising based on the symmetry of this variable.

Example P.20

Actuarial Analysis

An actuary is a person who assesses various forms of risk. For example, suppose that past data indicate that the holder of an automobile insurance policy has a 5% chance of an accident causing $1000 of damage, 2% chance of $5000 damage, 1% chance of totaling the car ($25,000), and a 92% chance of making it through the year with no accidents.[13] If the insurance company charges $1200 for such a policy, are they likely to make or lose money?

Solution

The mean of the damages according to the given probabilities is

$$\mu = \$1000 \cdot 0.05 + \$5000 \cdot 0.02 + \$25{,}000 \cdot 0.01 + \$0 \cdot 0.92 = \$400$$

So, on average, the insurance company will make about $800 per policy under these circumstances, if they charge a $1200 premium.

Standard Deviation of a Random Variable

In Section 2.3 on page 78 we introduce the notion of standard deviation as a way to measure the variability in a sample:

$$s = \sqrt{\frac{\sum (x - \overline{x})^2}{n - 1}}$$

We apply similar reasoning to measure the standard deviation in a population that is defined by a random variable with probability function $p(x)$. To do this, we find the average squared deviation from the mean, μ, and then take a square root.

Standard Deviation of a Random Variable

For a random variable X with probability function $p(x)$ and mean μ, the *variance*, σ^2, is

$$\sigma^2 = \sum (x - \mu)^2 \cdot p(x)$$

and the *standard deviation* is $\sigma = \sqrt{\sigma^2}$.

---

[13] In reality, actuaries use much more extensive data than the few values shown here.

Example P.21

Find the standard deviation of the random variable X = the sum of two dice rolls.

Solution

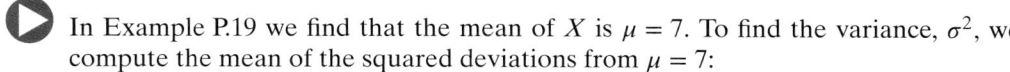

In Example P.19 we find that the mean of X is $\mu = 7$. To find the variance, σ^2, we compute the mean of the squared deviations from $\mu = 7$:

$$\sigma^2 = (2-7)^2 \frac{1}{36} + (3-7)^2 \frac{2}{36} + (4-7)^2 \frac{3}{36} + \cdots + (11-7)^2 \frac{2}{36} + (12-7)^2 \frac{1}{36}$$

$$= 25 \cdot \frac{1}{36} + 16 \cdot \frac{2}{36} + 9 \cdot \frac{3}{36} + \cdots + 16 \cdot \frac{2}{36} + 25 \cdot \frac{1}{36}$$

$$= 5.8333$$

The standard deviation of X is

$$\sigma = \sqrt{5.833} = 2.42$$

The value $\sigma = 2.42$ is the population standard deviation for the sums of all possible throws of two dice. If you were to roll a sample pair of dice many times, record the sums, and compute the sample standard deviation, s, you should get a number fairly close to 2.42. Try it!

Example P.22

Find the standard deviation of the damage amounts for the auto insurance situation described in Example P.20.

Solution

Given the mean damage, $\mu = 400$, the variance is

$$\sigma^2 = (1000-400)^2 \cdot 0.05 + (5000-400)^2 \cdot 0.02$$
$$+ (25{,}000-400)^2 \cdot 0.01 + (0-400)^2 \cdot 0.92$$
$$= 6{,}640{,}000$$

The standard deviation of damage amounts is $\sigma = \sqrt{6{,}640{,}000} = \$2576.82$.

Note that we need to be sure to include the deviation from \$0 (no accidents) as part of this calculation.

SECTION LEARNING GOALS

You should now have the understanding and skills to:

- Compute probabilities for a random variable using a probability function
- Compute the mean of a random variable from a probability function
- Compute the variance and standard deviation of a random variable from a probability function

Exercises for Section P.3

SKILL BUILDER 1: DISCRETE OR CONTINUOUS?

In Exercises P.63 to P.67, state whether the process described is a discrete random variable, is a continuous random variable, or is not a random variable.

P.63 Draw 10 cards from a deck and count the number of hearts.

P.64 Draw 10 cards from a deck and find the proportion that are hearts.

P.65 Deal cards one at a time from a deck. Keep going until you deal an ace. Stop and count the total number of cards dealt.

P.66 Draw one M&M from a bag. Observe whether it is blue, green, brown, orange, red, or yellow.

P.67 Observe the average weight, in pounds, of everything you catch during a day of fishing.

SKILL BUILDER 2: A PROBABILITY FUNCTION

Exercises P.68 to P.73 refer to the probability function given in Table P.10 for a random variable X that takes on the values 1, 2, 3, and 4.

P.68 Verify that the values given in Table P.10 meet the conditions for being a probability function. Justify your answer.

P.69 Find $P(X = 3 \text{ or } X = 4)$.

P.70 Find $P(X > 1)$.

P.71 Find $P(X < 3)$.

P.72 Find $P(X \text{ is an odd number})$.

P.73 Find $P(X \text{ is an even number})$.

Table P.10 *Probability function for Exercises P.68 to P.73*

x	1	2	3	4
$p(x)$	0.4	0.3	0.2	0.1

SKILL BUILDER 3: MORE PROBABILITY FUNCTIONS

In Exercises P.74 to P.77, fill in the ? to make $p(x)$ a probability function. If not possible, say so.

P.74

x	1	2	3	4
$p(x)$	0.1	0.1	0.2	?

P.75

x	10	20	30	40
$p(x)$	0.2	0.2	?	0.2

P.76

x	1	2	3
$p(x)$	0.5	0.6	?

P.77

x	1	2	3	4	5
$p(x)$	0.3	?	0.3	0.3	0.3

SKILL BUILDER 4: MEAN AND STANDARD DEVIATION

In Exercises P.78 to P.81, use the probability function given in the table to calculate:

(a) The mean of the random variable

(b) The standard deviation of the random variable

P.78

x	1	2	3
$p(x)$	0.2	0.3	0.5

P.79

x	10	20	30
$p(x)$	0.7	0.2	0.1

P.80

x	20	30	40	50
$p(x)$	0.6	0.2	0.1	0.1

P.81

x	10	12	14	16
$p(x)$	0.25	0.25	0.25	0.25

P.82 Owner-Occupied Household Size Table P.11 gives the probability function for the random variable[14] giving the household size for an owner-occupied housing unit in the US.[15]

(a) Verify that the sum of the probabilities is 1 (up to round-off error).

(b) What is the probability that a unit has only one or two people in it?

(c) What is the probability that a unit has five or more people in it?

---

[14]The largest category is actually "7 or more" but we have cut it off at 7 to make it a random variable. Can you explain why this was necessary?

[15]*http://www.census.gov.*

Table P.11 *Household size in owner-occupied units*

x	1	2	3	4	5	6	7
$p(x)$	0.217	0.363	0.165	0.145	0.067	0.026	0.018

(d) What is the probability that more than one person lives in a US owner-occupied housing unit?

P.83 Renter-Occupied Household Size Table P.12 gives the probability function for the random variable giving the household size for a renter-occupied housing unit in the US.

(a) Verify that the sum of the probabilities is 1 (up to round-off error.)

(b) What is the probability that a unit has only one or two people in it?

(c) What is the probability that a unit has five or more people in it?

(d) What is the probability that more than one person lives in a US renter-occupied housing unit?

Table P.12 *Household size in renter-occupied units*

x	1	2	3	4	5	6	7
$p(x)$	0.362	0.261	0.153	0.114	0.061	0.027	0.022

P.84 Average Household Size for Owner-Occupied Units Table P.11 in Exercise P.82 gives the probability function for the random variable giving the household size for an owner-occupied housing unit in the US.

(a) Find the mean household size.

(b) Find the standard deviation for household size.

P.85 Average Household Size for Renter-Occupied Units Table P.12 in Exercise P.83 gives the probability function for the random variable giving the household size for a renter-occupied housing unit in the US.

(a) Find the mean household size.

(b) Find the standard deviation for household size.

P.86 Fruit Fly Lifetimes Suppose that the probability function in Table P.13 reflects the possible lifetimes (in months after emergence) for fruit flies.

(a) What proportion of fruit flies die in their second month?

(b) What is the probability that a fruit fly lives more than four months?

Table P.13 *Fruit fly lifetimes (in months)*

x	1	2	3	4	5	6
$p(x)$	0.30	?	0.20	0.15	0.10	0.05

(c) What is the mean lifetime for a fruit fly?

(d) What is the standard deviation of fruit fly lifetimes?

P.87 Used Car Sales A used car dealership uses past data to estimate the probability distribution for the number of cars they sell in a day, X. The probability distribution of X is given in Table P.14.

(a) What is $P(X = 4)$?

(b) What is the probability that the dealership sells less than two cars during a day?

(c) What is the expected number (mean) of cars sold in a day?

(d) What is the standard deviation of the number of cars sold in a day?

Table P.14 *Cars sold in a day*

x	0	1	2	3	4
$p(x)$	0.29	0.3	0.2	0.17	?

P.88 More Fruit Fly Lifetimes Refer to Table P.13 in Exercise P.86 that gives probabilities for fruit fly lifetimes.

(a) If we know a fruit fly died before the end of its second month, what is the probability it died in its first month?

(b) If a fruit fly makes it past its second month, what is the probability it will live more than four months?

P.89 More Used Car Sales Refer to Table P.14 in Exercise P.87 that gives probabilities for the number of used cars a dealer sells in a day. What is the probability that the dealership sells no cars on three consecutive days? (Assume daily sales are independent.)

P.90 Stephen Curry's Free Throws As we see in Exercise P.37 on page 701, during the 2015–16 NBA season, Stephen Curry of the Golden State Warriors had a free throw shooting percentage of 0.908. Assume that the probability Stephen Curry makes any given free throw is fixed at 0.908, and that free throws are independent. Let X be the number of free throws Stephen Curry makes in two attempts.

(a) What is the probability distribution of X?

(b) What is the mean of X?

P.91 Life Insurance A non-profit organization plans to offer a life insurance service. Participants agree to a five-year contract in which they pay the organization a yearly fee. The fee does not change over the course of the contract. If the policy holder dies during the five-year period, the organization will pay $100,000 to her family and there will be no more yearly fee. The probabilities of death at ages 60, 61, 62, 63, and 64 for a US woman on her 60th birthday are given in Table P.15.[16] For example, on her 60th birthday a woman will have a 0.76% chance of dying at the age of 62.

(a) Let X be the organization's total profit, in dollars, five years after selling a contract to a woman on her 60th birthday. Write the probability distribution of X, where the values of X are given in terms of the yearly fee c.

(b) Write the mean of X in terms of c.

(c) What yearly fee should the organization charge 60-year-old women if they hope to break even? (The organization can expect to break even if they have a mean profit of $0.)

Table P.15 *Probabilities of death for a US woman on her 60th birthday*

Age	60	61	62	63	64
Probability	0.00648	0.00700	0.00760	0.00829	0.00908

P.92 Internet Pricing An Internet Service Provider (ISP) offers its customers three options:

- **Basic:** Standard internet for everyday needs, at $23.95 per month.
- **Premium:** Fast internet speeds for streaming video and downloading music, at $29.95 per month.
- **Ultra:** Super-fast internet speeds for online gaming, at $39.95 per month.

Ultra is the company's least popular option; they have twice as many Premium customers, and three times as many Basic customers:

[16]Period Life Table 2011, *http://www.ssa.gov/oact/STATS/table 4c6.html*, accessed June 8, 2016.

(a) Let X be the monthly fee paid by a randomly selected customer. Give the probability distribution of X.

(b) What is the mean of X? (This is the company's average monthly revenue per customer.)

(c) What is the standard deviation of X?

P.93 Benford's Law Frank Benford, a physicist working in the 1930s, discovered an interesting fact about some sets of numbers. While you might expect the first digits of numbers such as street addresses or checkbook entries to be randomly distributed (each with probability 1/9), Benford showed that in many cases the distribution of leading digits is not random, but rather tends to have more ones, with decreasing frequencies as the digits get larger. If a random variable X records the first digit in a street address, Benford's law says the probability function for X is

$$P(X = k) = \log_{10}(1 + 1/k)$$

(a) According to Benford's law, what is the probability that a leading digit of a street address is 1? What is the probability for 9?

(b) Using this probability function, what proportion of street addresses begin with a digit greater than 2?

P.94 Getting to the Finish In a certain board game participants roll a standard six-sided die and need to hit a particular value to get to the finish line exactly. For example, if Carol is three spots from the finish, only a roll of 3 will let her win; anything else and she must wait another turn to roll again. The chance of getting the number she wants on any roll is $p = 1/6$ and the rolls are independent of each other. We let a random variable X count the number of turns until a player gets the number needed to win. The possible values of X are 1, 2, 3, ... and the probability function for any particular count is given by the formula

$$P(X = k) = p(1 - p)^{k-1}$$

(a) Find the probability a player finishes on the third turn.

(b) Find the probability a player takes more than three turns to finish.

P.4 BINOMIAL PROBABILITIES

It is not always necessary to start from basic principles when computing probabilities for a random variable. Sometimes the probability function is already well known. For example, a *normal* distribution (see Section P.5) can be used to find probabilities in many applications that require a continuous random variable. In this section we describe a *binomial* probability function, which can be used to find probabilities for an important class of discrete random variables.

Conditions for a Binomial Random Variable

A binomial random variable counts the number of times that something occurs in a fixed number of independent trials. What that "something" is, and what each "trial" represents, depends on the context. For example, the number of times that a coin lands heads in a series of 10 tosses is a binomial random variable. On each trial the outcome that is counted by the random variable (the coin lands heads) is often called a *success*, while anything else is often called a *failure*. This does not necessarily mean that one is good and the other is bad; we just use these terms to distinguish between the two outcomes. The conditions that define a binomial random variable are given in detail below.

> ### Conditions for a Binomial Random Variable
>
> For a process to give a binomial random variable we need the following characteristics:
> - A number of trials, n, that is fixed in advance
> - A probability of success, p, that does not change from trial to trial
> - Independence of the outcomes from trial to trial
>
> A **binomial random variable** counts the number of successes in the n independent trials.

Consider the coin-tossing example above. The number of trials is fixed ($n = 10$ tosses), the trials are independent (the outcome of one toss will not influence another toss), and the probability of success is fixed at $p = P(\text{Head}) = 1/2$ (assuming a fair coin). This satisfies the conditions for a binomial random variable.[17]

Example P.23

In each of the following cases, state whether or not the process describes a binomial random variable. If it is binomial, give the values of n and p.

(a) Count the number of times a soccer player scores in five penalty shots against the same goalkeeper. Each shot has a 1/3 probability of scoring.

(b) Count the number of times a coin lands heads before it lands tails.

(c) Draw 10 cards from the top of a deck and record the number of cards that are aces.

(d) Conduct a simple random sample of 500 registered voters, and record whether each voter is Republican, Democrat, or Independent.

(e) Conduct a simple random sample of 500 registered voters, and count the number that are Republican.

(f) Randomly select one registered voter from each of the 50 US states, and count the number that are Republican.

Solution (a) Binomial (assuming shots are independent), with $n = 5$ and $p = 1/3$.

(b) Not binomial, since the number of trials n is not fixed.

---

[17]We sometimes use $B(n, p)$ as shorthand to denote a binomial, so the number of heads in 10 flips is $B(10, 1/2)$.

(c) Not binomial, since the trials are not independent (if the first card is an ace, it is less likely that the next card is an ace because there are less of them remaining in the deck).

(d) Not binomial, because it is not clear what defines a success.

(e) Binomial, with $n = 500$ and p is the population proportion of registered voters who are Republican.

(f) Not binomial, because the probability of selecting a Republican can vary from state to state.

Each trial for a binomial random variable must result in an outcome we call a success or a failure. In Example P.23(e), a success means a selected voter is Republican and a failure means a selected voter is not Republican (remember, a success does not imply "good" or "bad"!) However, each trial need not have only two possible outcomes. In Example P.23(e), a selected voter could be Republican, Democrat, or Independent.

When sampling from a finite population, the outcomes are not completely independent. This is illustrated in Example P.23(b), where the population is a deck of 52 cards. However, if the population is much larger than the sample size (as a rule of thumb, 10 times bigger), then the outcomes are very close to being independent and the binomial distribution is appropriate. This is the case in Example P.23(e).

Calculating Binomial Probabilities

Example P.24

Roulette

A European roulette wheel contains 37 colored pockets. One of the pockets is colored green, 18 are colored red, and 18 are colored black. A small ball spins around the inside of the wheel before eventually falling into one of the colored pockets. Each pocket has an equal probability, and gamblers often bet on which color pocket the ball will fall into.

A gambler decides to place four bets at the roulette wheel, with all four bets on black. Let X be the number of times the participant bets correctly. What is the probability function of X?

Solution With each bet, the participant is interested in whether the ball lands on black (a success: S) or not on black (a failure: F). The probability of success is $P(S) = \frac{18}{37}$ and the probability of failure is $P(F) = \frac{19}{37}$. As each bet is independent, we can multiply their probabilities. So, the probability of success on all four bets is

$$P(X = 4) = P(SSSS) = P(S)P(S)P(S)P(S) = \left(\frac{18}{37}\right)^4 = 0.0560$$

and the probability of failure on all four bets is

$$P(X = 0) = P(FFFF) = P(F)P(F)P(F)P(F) = \left(\frac{19}{37}\right)^4 = 0.0695$$

There are four possible outcomes in which the participant wins just one bet: $\{SFFF, FSFF, FFSF, FFFS\}$. These outcomes are disjoint and each has probability $\frac{18}{37} \times \left(\frac{19}{37}\right)^3$, so

$$P(X = 1) = 4 \times \frac{18}{37} \times \left(\frac{19}{37}\right)^3 = 0.2635$$

Similarly, there are four possible outcomes in which the participant wins three bets $\{SSSF, SSFS, SFSS, FSSS\}$, each with probability $\left(\frac{18}{37}\right)^3 \times \frac{19}{37}$, so

$$P(X = 3) = 4 \times \left(\frac{18}{37}\right)^3 \times \frac{19}{37} = 0.2365$$

There are six possible outcomes in which the participant wins two bets $\{SSFF, SFSF,$ $SFFS, FSSF, FSFS, FFSS\}$, each with probability $\left(\frac{18}{37}\right)^2 \times \left(\frac{19}{37}\right)^2$, so

$$P(X = 2) = 6 \times \left(\frac{18}{37}\right)^2 \times \left(\frac{19}{37}\right)^2 = 0.3745$$

Note that the number of wins (black) in the previous example is a binomial random variable with $n = 4$ and $p = 18/37$. Calculating these probabilities can be time consuming, and are even more complicated for larger n. Luckily, there is a shortcut to calculate binomial probabilities. Before we get there, though, we need to introduce some notation to ease the computation:

- *Factorials:* The factorial of a number, $n!$, is the product of all positive integers less than or equal to n:

$$n! = n \times (n - 1) \times (n - 2) \times \cdots \times 1$$

For example, $5! = 5 \times 4 \times 3 \times 2 \times 1 = 120$. By convention, $0! = 1$.
- *Binomial Coefficients:* The binomial coefficient $\binom{n}{k}$, read "n choose k," is given by

$$\binom{n}{k} = \frac{n!}{k!(n - k)!}$$

The binomial coefficient gives the number of possible ways to arrange k successes in n trials. In Example P.24 we see there are six possible ways for the participant to win 2 bets in 4 trials, and a quick calculation shows

$$\binom{4}{2} = \frac{4!}{2!(4 - 2)!} = \frac{4 \times 3 \times 2 \times 1}{(2 \times 1)(2 \times 1)} = 6$$

 Although they look similar, the binomial coefficient $\binom{n}{k}$ is not the same as the fraction $\left(\frac{n}{k}\right)$.

We are now ready for the formula to compute binomial probabilities.

Binomial Probabilities

If a random variable X is binomial with n trials and probability of success p, the probability of getting exactly k successes is

$$P(X = k) = \binom{n}{k} p^k (1 - p)^{n-k}$$

for $k = 0, 1, \ldots, n$.

Example P.25

Roulette (continued)

In Example P.24, let X represent the number of times a gambler wins with four consecutive bets on black. Use the binomial formula to find the probability distribution of X.

Solution

Standard calculations give $\binom{4}{0} = 1$, $\binom{4}{1} = 4$, $\binom{4}{2} = 6$, $\binom{4}{3} = 4$, and $\binom{4}{4} = 1$. Using the binomial formula,

$$P(X = 0) = 1 \times \left(\frac{18}{37}\right)^0 \times \left(\frac{19}{37}\right)^4 = 0.0695$$

$$P(X = 1) = 4 \times \left(\frac{18}{37}\right)^1 \times \left(\frac{19}{37}\right)^3 = 0.2635$$

$$P(X = 2) = 6 \times \left(\frac{18}{37}\right)^2 \times \left(\frac{19}{37}\right)^2 = 0.3745$$

$$P(X = 3) = 4 \times \left(\frac{18}{37}\right)^3 \times \left(\frac{19}{37}\right)^1 = 0.2365$$

$$P(X = 4) = 1 \times \left(\frac{18}{37}\right)^4 \times \left(\frac{19}{37}\right)^0 = 0.0560$$

These agree with our raw calculations in Example P.24.

For the binomial coefficients it is always the case that $\binom{n}{k} = \binom{n}{n-k}$. However, unless $p = 1/2$, it is not the case that $P(X = k) = P(X = n - k)$.

Example P.26

Norwegian Coffee Consumption

Norwegians drink the most coffee in the world (it must be the cold winters). In one survey[18] of 389,624 Norwegians in their early 40s, more than half claimed to drink five or more cups of coffee per day! Furthermore, 89.4% drink at least one cup of coffee per day. Assume that the overall proportion of Norwegian adults who drink at least five cups of coffee per day is 50%, and the proportion of all Norwegian adults who drink at least one cup of coffee per day is 89.4%.

(a) In a random sample of 10 Norwegian adults, what is the probability that exactly 6 drink at least five cups of coffee per day?

(b) In a random sample of 10 Norwegian adults, what is the probability that exactly 6 drink at least one cup of coffee per day?

(c) In a random sample of 50 Norwegian adults, what is the probability that more than 45 will drink at least one cup of coffee per day?

Solution

(a) Using the formula for binomial probabilities with $p = 0.5$ and $n = 10$,

$$P(X = 6) = \binom{10}{6}0.5^6(1 - 0.5)^4 = 210 \cdot 0.5^6 0.5^4 = 0.205$$

[18]Tverdal, A., Hjellvik, V., and Selmer, R., "Coffee intake and oral-oesophageal cancer: Follow-up of 389,624 Norwegian men and women 40–45 years," *British Journal of Cancer*, 2011; 105:157–161.

(b) Using the formula for binomial probabilities with $p = 0.894$ and $n = 10$,

$$P(X = 6) = \binom{10}{6} \times 0.894^6 \times (1 - 0.894)^4 = 210 \cdot 0.894^6 0.106^4 = 0.014$$

(c) For "more than 45" we need to find probabilities for $X = 46, \ldots, 50$. Using the formula for binomial probabilities with $p = 0.894$ and $n = 50$,

$$P(X = 46) = \binom{50}{46} \times 0.894^{46} \times (1 - 0.894)^4 = 0.1679$$

Similar calculations show that $P(X = 47) = 0.1205$, $P(X = 48) = 0.0635$, $P(X = 49) = 0.0219$, and $P(X = 50) = 0.0037$. So,

$$P(X > 45) = 0.1679 + 0.1205 + 0.0635 + 0.0219 + 0.0037 = 0.3775$$

We can also use technology to further automate the computation of binomial probabilities.

Mean and Standard Deviation of a Binomial Random Variable

Example P.27

Roulette (continued)

In Example P.24, X represents the number of times a gambler wins with four consecutive bets on black. Find the mean μ and standard deviation σ of X.

Solution Here is the probability function found in Example P.24:

x	0	1	2	3	4
$p(x)$	0.0695	0.2635	0.3745	0.2365	0.0560

Using the formulas from Section P.3,

$$\mu = \sum x \cdot p(x)$$
$$= 0 \cdot 0.0695 + 1 \cdot 0.2635 + 2 \cdot 0.3745 + 3 \cdot 0.2365 + 4 \cdot 0.0560$$
$$= 1.946$$

and

$$\sigma^2 = \sum (x - \mu)^2 \cdot p(x)$$
$$= (0 - 1.946)^2 \cdot 0.0695 + (1 - 1.946)^2 \cdot 0.2635 + \cdots + (4 - 1.946)^2 \cdot 0.0560$$
$$= 0.99908$$

so $\sigma = \sqrt{0.99908} = 0.99954$.

Just as recognizing that a random variable is binomial can make it easier to compute probabilities, it can also make it easier to compute the mean and standard deviation. We do not need to go through the tedious calculations in Example P.27. Shortcuts for the mean and standard deviation of a binomial random variable are given below.

> **Mean and Standard Deviation of a Binomial Random Variable**
>
> If a random variable X is binomial with n trials and probability of success p, then its mean μ and standard deviation σ are given by
>
> $$\mu = np \quad \text{and} \quad \sigma = \sqrt{np(1-p)}$$

Example P.28

Roulette (continued)

Find the mean μ and standard deviation σ of X from Example P.27, using the short-cuts for a binomial random variable.

Solution

In this case $n = 4$ and $p = 18/37$, so

$$\mu = 4 \cdot \frac{18}{37} = 1.946$$

and

$$\sigma = \sqrt{4 \cdot \frac{18}{37} \cdot \left(1 - \frac{18}{37}\right)} = 0.9996$$

which agree with our answers in Example P.27.

Example P.29

Norwegian Coffee Drinkers

In Example P.26 we see that the proportion of Norwegians in their early 40s who drink at least one cup of coffee per day is about $p = 0.894$. Suppose that we take random samples of 50 Norwegians from this age group. Find the mean and standard deviation for the number of regular (at least a cup per day) coffee drinkers in such samples. Would you be surprised to find fewer than 35 coffee drinkers in such a sample?

Solution

The mean and standard deviation for the number of regular coffee drinkers in samples of size 50 when $p = 0.894$ are

$$\mu = np = 50 \cdot 0.894 = 44.7 \quad \text{and} \quad \sigma = \sqrt{50 \cdot 0.894 \cdot (1 - 0.894)} = 2.18$$

We see that 35 is almost 4.5 standard deviations below $\mu = 44.7$, so it would be very surprising for a random sample of 50 Norwegians in this age group to contain fewer than 35 regular coffee drinkers.

> **SECTION LEARNING GOALS**
>
> *You should now have the understanding and skills to:*
>
> - Identify when a discrete random variable is binomial
> - Compute probabilities for a binomial random variable
> - Compute the mean and standard deviation for a binomial random variable

Exercises for Section **P.4**

SKILL BUILDER 1: BINOMIAL OR NOT?
In Exercises P.95 to P.99, determine whether the process describes a binomial random variable. If it is binomial, give values for n and p. If it is not binomial, state why not.

P.95   Count the number of sixes in 10 dice rolls.

P.96   Roll a die until you get 5 sixes and count the number of rolls required.

P.97   Sample 50 students who have taken Intro Stats and record the final grade in the course for each.

P.98   Suppose 30% of students at a large university take Intro Stats. Randomly sample 75 students from this university and count the number who have taken Intro Stats.

P.99   Worldwide, the proportion of babies who are boys is about 0.51. We randomly sample 100 babies born and count the number of boys.

SKILL BUILDER 2: FACTORIALS AND BINOMIAL COEFFICIENTS
In Exercises P.100 to P.107, calculate the requested quantity.

P.100 4!

P.101 7!

P.102 8!

P.103 6!

P.104 $\binom{8}{3}$

P.105 $\binom{5}{2}$

P.106 $\binom{10}{8}$

P.107 $\binom{6}{5}$

SKILL BUILDER 3: COMPUTING BINOMIAL PROBABILITIES
In Exercises P.108 to P.111, calculate the requested binomial probability.

P.108 Find $P(X = 2)$ if X is a binomial random variable with $n = 6$ and $p = 0.3$.

P.109 Find $P(X = 7)$ if X is a binomial random variable with $n = 8$ and $p = 0.9$.

P.110 Find $P(X = 3)$ if X is a binomial random variable with $n = 10$ and $p = 0.4$.

P.111 Find $P(X = 8)$ if X is a binomial random variable with $n = 12$ and $p = 0.75$.

SKILL BUILDER 4: MEAN AND STANDARD DEVIATION OF A BINOMIAL
In Exercises P.112 to P.115, calculate the mean and standard deviation of the binomial random variable.

P.112 A binomial random variable with $n = 6$ and $p = 0.4$

P.113 A binomial random variable with $n = 10$ and $p = 0.8$

P.114 A binomial random variable with $n = 30$ and $p = 0.5$

P.115 A binomial random variable with $n = 800$ and $p = 0.25$

P.116   Boys or Girls? Worldwide, the proportion of babies who are boys is about 0.51. A couple hopes to have three children and we assume that the sex of each child is independent of the others. Let the random variable X represent the number of girls in the three children, so X might be 0, 1, 2, or 3. Give the probability function for each value of X.

P.117   Class Year Suppose that undergraduate students at a university are equally divided between the four class years (first-year, sophomore, junior, senior) so that the probability of a randomly chosen student being in any one of the years is 0.25. If we randomly select four students, give the probability function for each value of the random variable $X =$ the number of seniors in the four students.

P.118   College Graduates From the 2010 US Census, we learn that 27.5% of US adults have graduated from college. If we take a random sample of 12 US adults, what is the probability that exactly 6 of them are college graduates?

P.119   Senior Citizens In the 2010 US Census, we learn that 13% of all people in the US are 65 years old or older. If we take a random sample of 10 people, what is the probability that 3 of them are 65 or older? That 4 of them are 65 or older?

P.120   Owner-Occupied Housing Units In the 2010 US Census, we learn that 65% of all housing units are owner-occupied while the rest are rented. If we take a random sample of 20 housing units, find the probability that:

(a) Exactly 15 of them are owner-occupied

(b) 18 or more of them are owner-occupied

P.121 Mean and Standard Deviation of Boys or Girls In Exercise P.116, we discuss the random variable counting the number of girls in three babies, given that the proportion of babies who are girls is about 0.49. Find the mean and standard deviation of this random variable.

P.122 Mean and Standard Deviation of Class Year In Exercise P.117, we discuss the random variable counting the number of seniors in a sample of four undergraduate students at a university, given that the proportion of undergraduate students who are seniors is 0.25. Find the mean and standard deviation of this random variable.

P.123 Mean and Standard Deviation of College Graduates Exercise P.118 describes a random variable that counts the number of college graduates in a sample. Use the information in that exercise to find the mean and standard deviation of this random variable.

P.124 Mean and Standard Deviation of Senior Citizens Exercise P.119 describes a random variable that counts the number of senior citizens in a sample. Use the information in that exercise to find the mean and standard deviation of this random variable.

P.125 Mean and Standard Deviation of Owner-Occupied Housing Units Exercise P.120 describes a random variable that counts the number of owner-occupied units in a sample of housing units. Use the information in that exercise to find the mean and standard deviation of this random variable.

P.126 Stephen Curry's Free Throws As we see in Exercise P.37 on page 701, during the 2015-16 NBA season, Stephen Curry of the Golden State Warriors had a free throw shooting percentage of 0.908. Assume that the probability Stephen Curry makes any given free throw is fixed at 0.908, and that free throws are independent.

(a) If Stephen Curry shoots 8 free throws in a game, what is the probability that he makes at least 7 of them?

(b) If Stephen Curry shoots 80 free throws in the playoffs, what is the probability that he makes at least 70 of them?

(c) If Stephen Curry shoots 8 free throws in a game, what are the mean and standard deviation for the number of free throws he makes during the game?

(d) If Stephen Curry shoots 80 free throws in the playoffs, what are the mean and standard deviation for the number of free throws he makes during the playoffs?

P.127 Airline Overbooking Suppose that past experience shows that about 10% of passengers who are scheduled to take a particular flight fail to show up. For this reason, airlines sometimes overbook flights, selling more tickets than they have seats, with the expectation that they will have some no shows. Suppose an airline uses a small jet with seating for 30 passengers on a regional route and assume that passengers are independent of each other in whether they show up for the flight. Suppose that the airline consistently sells 32 tickets for every one of these flights.

(a) On average, how many passengers will be on each flight?

(b) How often will they have enough seats for all of the passengers who show up for the flight?

P.128 Mean and Standard Deviation of a Proportion To find the proportion of times something occurs, we divide the count (often a binomial random variable) by the number of trials n. Using the formula for the mean and standard deviation of a binomial random variable, derive the mean and standard deviation of a proportion resulting from n trials and probability of success p.

P.5 DENSITY CURVES AND THE NORMAL DISTRIBUTION

In Section P.3 we introduce the idea of *discrete* and *continuous* random variables. In that section we use a probability function to assign probabilities to each value of a discrete random variable. This is not feasible for a continuous random variable where the values can occur anywhere within some range of possible values—far too many to let us give separate probabilities to every possible result. In this section we introduce a method for describing the distribution of continuous random variables and use it to work with one of the most important continuous random variables, the normal distribution.[19]

---

[19]Chapter 5 discusses using a normal distribution to find confidence intervals and p-values. In this section we consider more general uses of the normal distribution to describe any population.

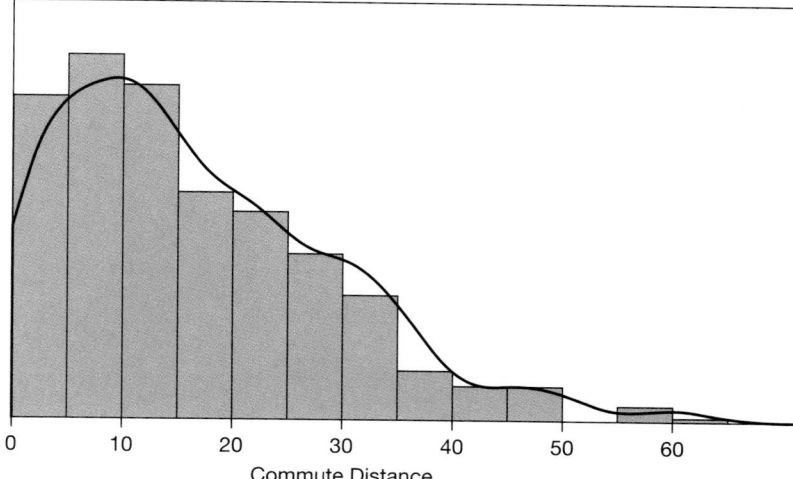

Figure P.5 *Histogram of Atlanta commute distances with density curve*

Density Curves

A theoretical model for the distribution of a continuous random variable is called a *density curve*. A density curve is a curve that reflects the location, spread, and general shape of the distribution over the range of possible values. Figure P.5 shows a histogram representing the distances of commutes for workers in the city of Atlanta.[20] We add a rough density curve to smooth out the sharp edges of the histogram. This density curve is a relatively simple curve that follows the general pattern of the data, thus providing a model for the underlying distribution.

Density Curves

A density curve is scaled to have two important properties:

- The total area under the curve is equal to one, to correspond to 100% of the distribution.
- The area over any interval is the proportion of the distribution in that interval.

The density curve represents the entire distribution and regions under the curve correspond to portions of the distribution.

Example P.30

The density curve in Figure P.6 is a model for the distribution of commute distances in Atlanta, without the underlying histogram of Figure P.5. Three regions of commute distances are given. Match the proportion in each region with one of the following values:

| 0.05 | 0.15 | 0.40 | 0.50 | 0.80 | 1.0 | 30 |

(a) Between 30 and 40 miles (the shaded region)

(b) Less than 30 miles

(c) More than 40 miles

[20]The histogram is based roughly on distances for a sample of Atlanta commutes in **CommuteAtlanta**. We assume that the general shape is representative of the population.

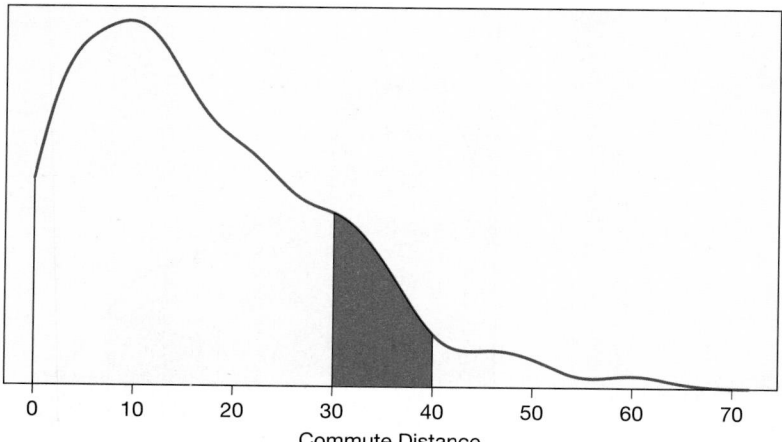

Figure P.6 *Density curve for Atlanta commute distances*

Solution ▶ The proportions correspond to the area under the curve over that region. All three areas are less than the total area that is 1.0, so all three proportions are less than 1.

(a) The shaded area of the region between 30 and 40 miles is well less than half of the total area, so the area is definitely less than 0.5. A guess between 10% and 25% would be reasonable, so we estimate that the proportion is about 0.15.

(b) The area below 30 miles is clearly more than half of the total area. A guess between 70% and 85% would be reasonable, so we match this proportion with 0.80.

(c) There's not much area above 40 miles. Since the three areas need to add up to the total of 100%, a guess of 5% for the tail area to the right of 40 miles would be consistent with the previous estimates of 15% and 80% for the other two regions. We match this proportion with 0.05.

In practice, specifying the exact form of a density curve and finding the exact areas of regions such as those shown in Figure P.6 require tools (such as calculus) that are beyond the scope of this text. However, one general type of density curve occurs in many common applications. In the rest of this section, we focus on working with this important distribution.

Normal Distributions

While a density curve can have almost any shape, a *normal density* curve has the special form of a bell-shaped curve. The actual equation of this curve is fairly complicated, but the general shape is readily recognized.

Normal Density Curve

A **normal distribution** follows a bell-shaped curve. We use the two parameters *mean*, μ, and *standard deviation*, σ, to distinguish one normal curve from another.

For shorthand we often use the notation $N(\mu, \sigma)$ to specify that a distribution is normal (N) with some mean (μ) and standard deviation (σ).

We use the population parameters μ and σ when specifying a normal density. The reason is that the normal curve is a model for the population, even if that

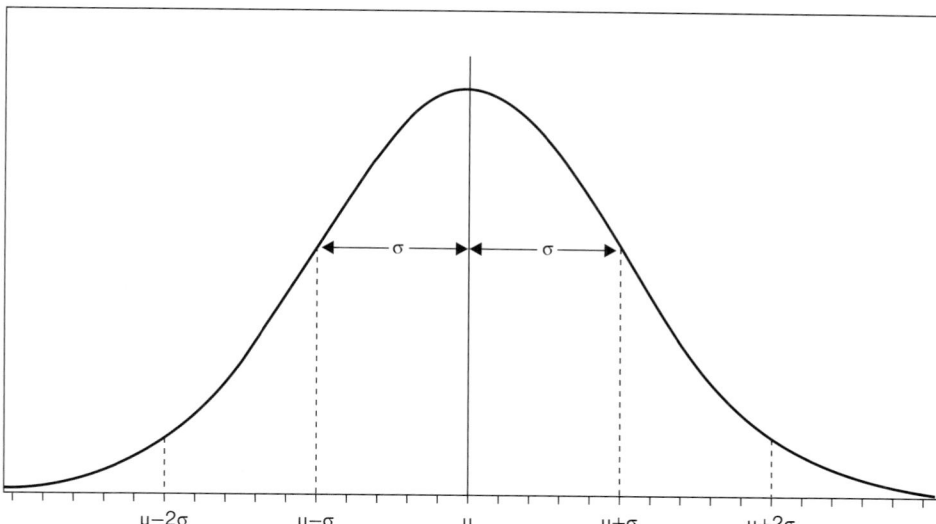

Figure P.7 *Normal density curve*

"population" is a bootstrap or randomization distribution for some sample statistic. In practice, we often use sample values ($\bar{x}$ and s) or a hypothesized value of a parameter to estimate or specify the mean or standard deviation for a normal distribution.

Figure P.7 shows a normal density curve centered at some mean, μ. The standard deviation helps determine the horizontal scale. Recall that roughly 95% of data fall within two standard deviations of the mean. This amount corresponds to the area within $\mu \pm 2\sigma$.

Graph of a Normal Density Curve

The graph of the normal density curve $N(\mu, \sigma)$ is a bell-shaped curve which:

- Is centered at the mean μ

- Has a horizontal scale such that 95% of the area under the curve falls within two standard deviations of the mean (within $\mu \pm 2\sigma$)

Figure P.8 shows how the normal distribution changes as the mean μ is shifted to move the curve horizontally or the standard deviation σ is changed to stretch or shrink the curve. Remember that the area under each of these curves is equal to one.

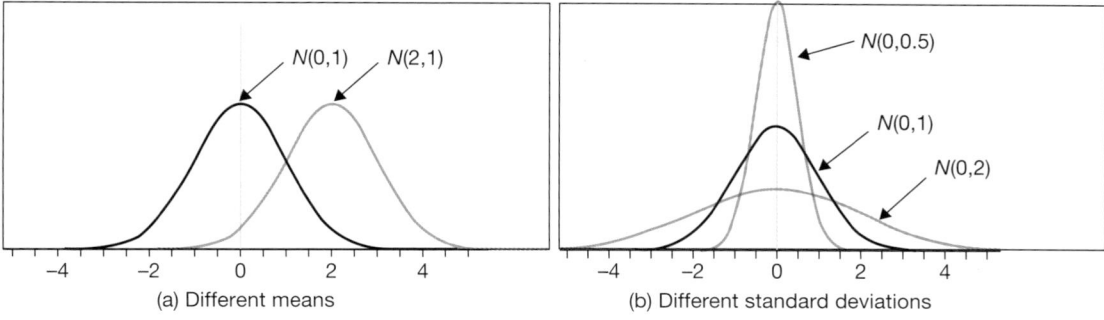

(a) Different means   (b) Different standard deviations

Figure P.8 *Comparing normal curves*

Example P.31

Drawing Normal Curves

Sketch a normal density curve for each of the following situations:

(a) Scores on an exam which have a $N(75, 10)$ distribution.

(b) Heights of college students if the distribution is bell-shaped with a mean of 68 inches and standard deviation of 4 inches.

(c) Grade point average (GPA) of students taking introductory statistics that follow a normal distribution centered at 3.16 with a standard deviation of 0.40. [21]

Solution

(a) The normal density curve $N(75, 10)$ is a bell-shaped curve centered at the mean of 75 and with a scale on the horizontal axis so that 95% of the area under the curve is within the range $\mu \pm 2\sigma = 75 \pm 2(10) = 75 \pm 20$, or between 55 and 95. See Figure P.9(a).

(b) The normal density curve $N(68, 4)$ is a bell-shaped curve centered at the mean of 68 and with a scale so that 95% of the area is within the range $\mu \pm 2\sigma = 68 \pm 2(4) = 68 \pm 8$, or between 60 and 76. See Figure P.9(b).

(c) The normal density curve $N(3.16, 0.40)$ is a bell-shaped curve centered at the mean of 3.16 and with a scale so that 95% of the area is within the range $\mu \pm 2\sigma = 3.16 \pm 2(0.40) = 3.16 \pm 0.80$, or between 2.36 and 3.96. See Figure P.9(c).

Note that the shapes of the three curves in Figure P.9 are the same; only the scaling on the horizontal axis changes in each case. The mean locates the center of the distribution and the standard deviation on either side controls the spread.

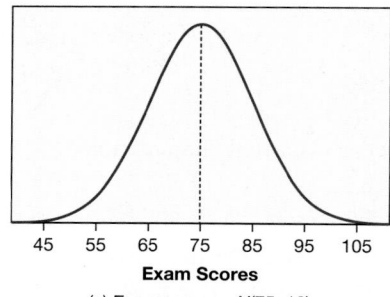

Exam Scores	
(a) Exam scores~ $N(75, 10)$	

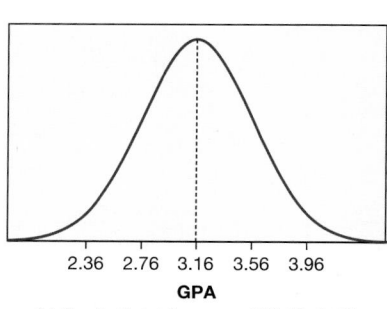

Figure P.9 *Three normal curves*

Finding Normal Probabilities and Percentiles

We find probabilities of intervals using the area under the density curve, but no convenient formulas exist for computing areas with a normal density. [22] For this reason we rely on technology, such as *StatKey*, statistical software, or a calculator, to compute probabilities for normal distributions. In most of these applications we need to specify:

• The mean and standard deviation for the normal distribution

• The endpoint(s) of the interval

• The direction (above, below, or between) the endpoint(s)

[21] Distribution of student heights and GPA approximated from the data in **StudentSurvey**.
[22] Even those of you with a calculus background will find that there is no antiderivative to help find areas under a normal density function.

© Robert Kneschke/iStockphoto

Will scores on this exam follow a normal curve?

Example P.32

Exam Scores

Suppose that scores on an exam follow a normal distribution with mean $\mu = 75$ and standard deviation $\sigma = 10$. What proportion of the scores are above 90 points?

Solution

Figure P.10 shows the results using *StatKey*, a TI-83 calculator, and the R statistical package to find the area above 90 for a $N(75, 10)$ density. The probability of getting an exam grade above 90 is about 0.067.

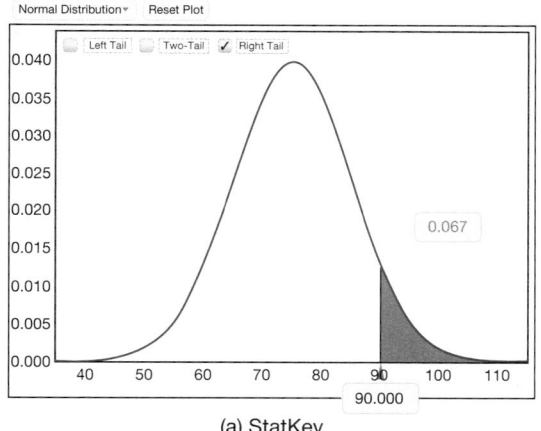

(a) StatKey

TI-83:

```
normalcdf(90,1000,75,10)
        0.0668072287
```

R:

```
> 1-pnorm(90,75,10)
[1] 0.0668072
```

(b) TI-83 calculator and R package

Figure P.10 *Technologies for computing a normal probability*

In addition to finding probabilities of regions as areas under a normal density, we are also often interested in going in the other direction and finding a region with a specific proportion. Again, we generally rely on technology to handle the computational details.

Example P.33

Still assuming that the distribution of scores for an exam is $N(75, 10)$, suppose that an instructor determines that students with scores in the lowest 20% of the distribution need to see a tutor for extra help. What is the cutoff for scores that fall in this category?

Solution ▶

Notice now that the given information is the area under the normal curve (20%) and the goal is to find the endpoint that has that amount of area below it. Figure P.11 shows the results, again using several different technologies. The instructor should ask students with exam grades below 66.6 to sign up for the extra tutoring.

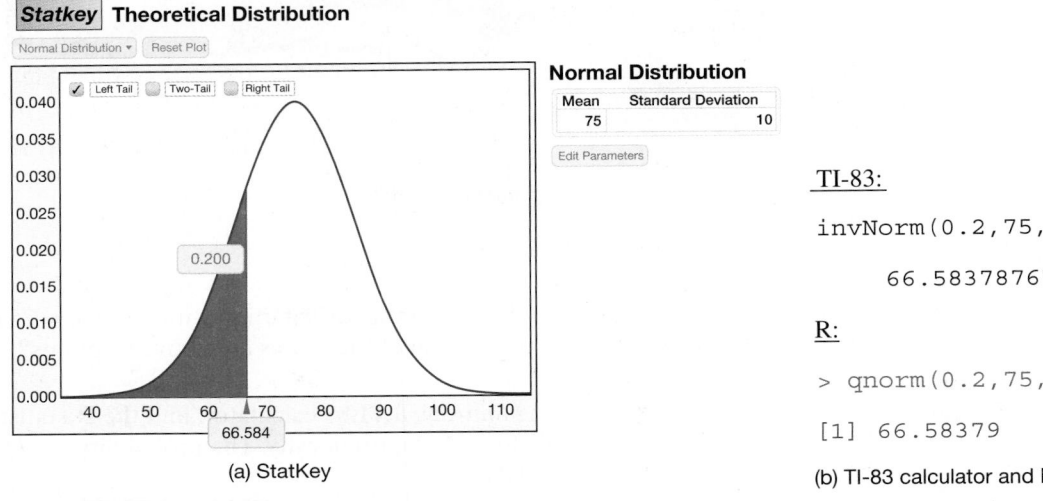

Normal Distribution

Mean	Standard Deviation
75	10

TI-83:

`invNorm(0.2,75,10)`

    `66.58378767`

R:

`> qnorm(0.2,75,10)`

`[1] 66.58379`

(a) StatKey   (b) TI-83 calculator and R package

Figure P.11 *Technologies for finding a normal endpoint*

Standard Normal $N(0, 1)$

Because all the normal distributions look the same except for the horizontal scale, another common way to deal with normal calculations is to convert problems to one specific *standard normal* scale. The standard normal has a mean of 0 and a standard deviation of 1, so $\mu = 0$ and $\sigma = 1$. We often use the letter Z to denote a standard normal, so that $Z \sim N(0, 1)$.

To convert a value from a $N(\mu, \sigma)$ scale to a standard normal scale, we subtract the mean μ to shift the center to zero, then divide the result by the standard deviation σ to stretch (or shrink) the difference to match a standard deviation of one. If X is a value on the $N(\mu, \sigma)$ scale, then $Z = (X - \mu)/\sigma$ is the corresponding point on the $N(0, 1)$ scale.[23] You should recognize this as the *z-score* from page 82, because the standardized value just measures how many standard deviations a value is above or below the mean. The process of converting from the standard normal back to $N(\mu, \sigma)$ just reverses the process of finding a *z*-score. Namely, we multiply the *z*-value by the standard deviation and then add the mean.

---

[23] When technology is not available, a printed table with probabilities for certain standard normal endpoints can be used.

Standard Normal

The **standard normal** distribution has mean zero and standard deviation equal to one, $Z \sim N(0, 1)$.

To convert from any $X \sim N(\mu, \sigma)$ to $Z \sim N(0, 1)$, we standardize values with the z-score:

$$Z = \frac{X - \mu}{\sigma}$$

To convert from $Z \sim N(0, 1)$ to any $X \sim N(\mu, \sigma)$, we reverse the standardization with:

$$X = \mu + Z \cdot \sigma$$

Example P.34

Example P.31(b) on page 728 uses a $N(68, 4)$ distribution to describe the heights (in inches) of college students. Suppose we are interested in finding the proportion of students who are at least six feet (72 inches) tall.

(a) Convert the endpoint (72 inches) to a standard normal scale.

(b) Sketch both distributions, shade the corresponding regions of interest, and compare the probabilities.

Solution (a) We compute the z-score for the original endpoint.

$$z = \frac{X - \mu}{\sigma} = \frac{72 - 68}{4} = 1.0$$

The region above 72 in the $N(68, 4)$ scale translates to a region above 1.0 in the $N(0, 1)$ scale.

(b) The regions in both the original and standardized scales are shown in Figure P.12. Notice that they look identical except for the horizontal scale and show that about 16% of students should be at least 72 inches tall.

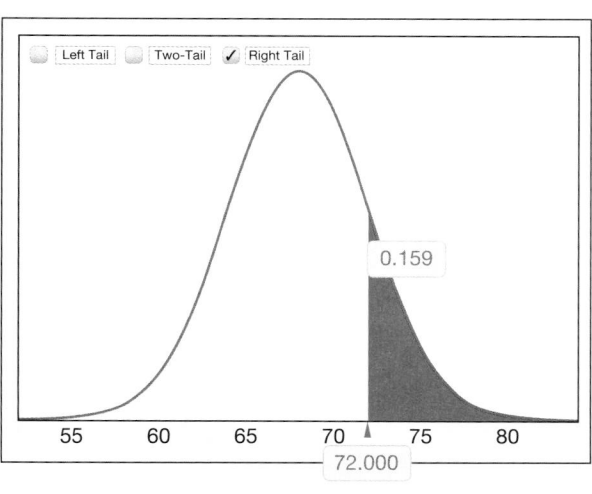

(a) $N(68, 4)$

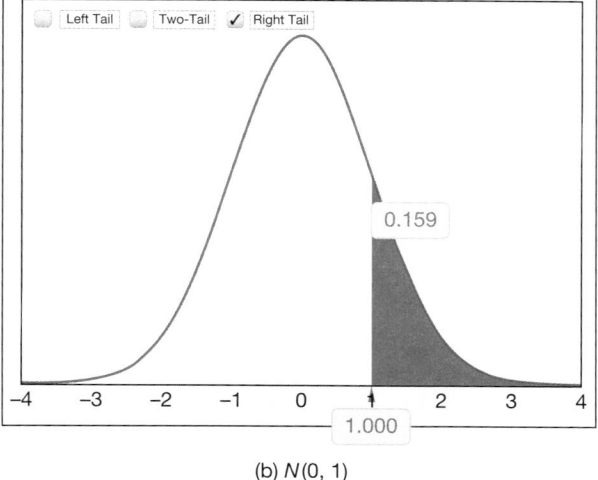

(b) $N(0, 1)$

Figure P.12 *Converting heights above 72 to a standard normal*

What about getting percentiles using a standard normal distribution? For that we must reverse the process. We first find an endpoint (or endpoints) on the standard normal curve that has the desired property, then convert that value to the given normal distribution.

Example P.35

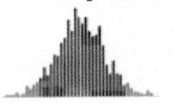

More Exam Scores

Use the standard normal distribution to find a point in a $N(75, 10)$ exam distribution that has 20% of the scores below it.

Solution

This is the same question as in Example P.33 on page 730. The difference now is that we illustrate the use of the standard normal distribution. Using technology, we find that the lowest 20% of a standard normal distribution is all of the values below $z = -0.842$. Thus we need to find the point on a $N(75, 10)$ curve that is -0.842 standard deviations from its mean. The relevant calculation is

$$x = \mu + z \cdot \sigma = 75 - 0.842 \cdot 10 = 66.58$$

This answer, recommending tutors for students with exam grades below 66.6, is consistent with what we found without standardizing in Example P.33.

SECTION LEARNING GOALS

You should now have the understanding and skills to:

- Estimate probabilities as areas under a density function
- Recognize how the mean and standard deviation relate to the center and spread of a normal distribution
- Use technology to compute probabilities of intervals for any normal distribution
- Use technology to find endpoint(s) of intervals with a specified probability for any normal distribution
- Convert in either direction between a general $N(\mu, \sigma)$ distribution and a standard $N(0, 1)$ distribution

---

Exercises for Section P.5

SKILL BUILDER 1

Exercises P.129 to P.131 refer to the density function shown in Figure P.13. In each exercise, use the density function to choose the best estimate for the proportion of the population found in the specified region.

P.129 The percentage of the population that is less than 25 is closest to:

    10%   28%   50%   62%   95%

P.130 The percentage of the population that is more than 30 is closest to:

    4%   27%   50%   73%   95%

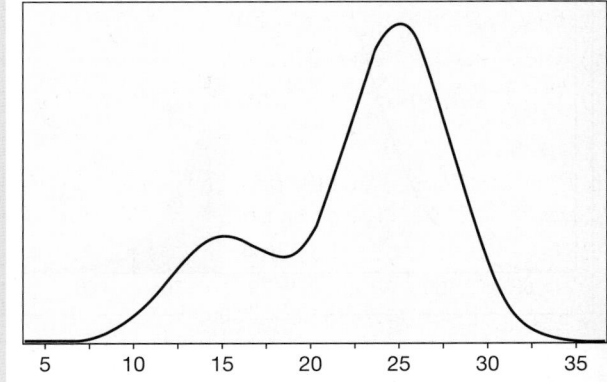

Figure P.13 *Density curve for Exercises P.129 to P.131*

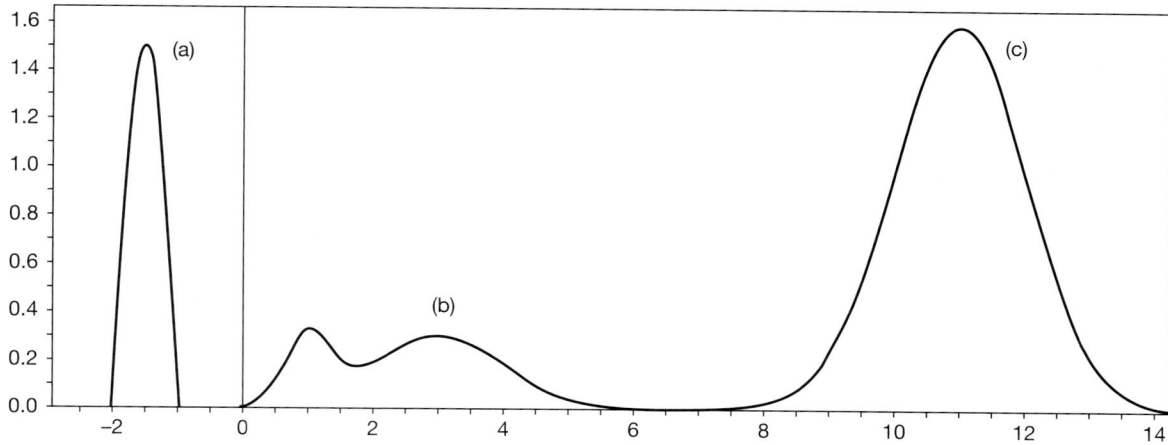

Figure P.14 *Two density curves and one that isn't*

P.131 The percentage of the population that is between 10 and 30 is closest to:

3%   33%   50%   67%   95%

SKILL BUILDER 2

P.132 Two of the curves shown in Figure P.14 are valid density curves and one is not. Identify the one that is not a density. Give a reason for your choice.

SKILL BUILDER 3

In Exercises P.133 to P.136, find the specified areas for a $N(0, 1)$ density.

P.133 (a) The area below $z = 1.04$

(b) The area above $z = -1.5$

(c) The area between $z = 1$ and $z = 2$

P.134 (a) The area below $z = 0.8$

(b) The area above $z = 1.2$

(c) The area between $z = -1.75$ and $z = -1.25$

P.135 (a) The area above $z = -2.10$

(b) The area below $z = -0.5$

(c) The area between $z = -1.5$ and $z = 0.5$

P.136 (a) The area above $z = 1.35$.

(b) The area below $z = -0.8$.

(c) The area between $z = -1.23$ and $z = 0.75$.

SKILL BUILDER 4

In Exercises P.137 to P.140, find endpoint(s) on a $N(0, 1)$ density with the given property.

P.137 (a) The area to the left of the endpoint is about 0.10.

(b) The area to the right of the endpoint is about 0.80.

(c) The area between $\pm z$ is about 0.95.

P.138 (a) The area to the left of the endpoint is about 0.70.

(b) The area to the right of the endpoint is about 0.01.

(c) The area between $\pm z$ is about 0.90.

P.139 (a) The area to the right of the endpoint is about 0.90.

(b) The area to the left of the endpoint is about 0.65.

P.140 (a) The area to the right of the endpoint is about 0.02.

(b) The area to the left of the endpoint is about 0.40.

SKILL BUILDER 5

In Exercises P.141 to P.144, find the specified areas for a normal density.

P.141 (a) The area below 80 on a $N(75, 10)$ distribution

(b) The area above 25 on a $N(20, 6)$ distribution

(c) The area between 11 and 14 on a $N(12.2, 1.6)$ distribution

P.142 (a) The area above 6 on a $N(5, 1.5)$ distribution

(b) The area below 15 on a $N(20, 3)$ distribution

(c) The area between 90 and 100 on a $N(100, 6)$ distribution

P.143 (a) The area above 200 on a $N(120, 40)$ distribution

(b) The area below 49.5 on a $N(50, 0.2)$ distribution

(c) The area between 0.8 and 1.5 on a $N(1, 0.3)$ distribution

P.144 (a) The area below 0.21 on a $N(0.3, 0.04)$ distribution

(b) The area above 472 on a $N(500, 25)$ distribution

(c) The area between 8 and 10 on a $N(15, 6)$ distribution

SKILL BUILDER 6

In Exercises P.145 to P.148, find endpoint(s) on the given normal density curve with the given property.

P.145 (a) The area to the right of the endpoint on a $N(50, 4)$ curve is about 0.01.

(b) The area to the left of the endpoint on a $N(2, 0.05)$ curve is about 0.70.

(c) The symmetric middle area on a $N(100, 20)$ curve is about 0.95.

P.146 (a) The area to the right of the endpoint on a $N(25, 8)$ curve is about 0.25.

(b) The area to the left of the endpoint on a $N(500, 80)$ curve is about 0.02.

(c) The symmetric middle area on a $N(10, 3)$ curve is about 0.95.

P.147 (a) The area to the left of the endpoint on a $N(100, 15)$ curve is about 0.75.

(b) The area to the right of the endpoint on a $N(8, 1)$ curve is about 0.03.

P.148 (a) The area to the left of the endpoint on a $N(5, 2)$ curve is about 0.10.

(b) The area to the right of the endpoint on a $N(500, 25)$ curve is about 0.05.

SKILL BUILDER 7

Exercises P.149 to P.156 ask you to convert an area from one normal distribution to an equivalent area for a different normal distribution. Draw sketches of both normal distributions, find and label the endpoints, and shade the regions on both curves.

P.149 The area below 40 for a $N(48, 5)$ distribution converted to a standard normal distribution

P.150 The upper 30% for a $N(48, 5)$ distribution converted to a standard normal distribution

P.151 The upper 5% for a $N(10, 2)$ distribution converted to a standard normal distribution

P.152 The area above 13.4 for a $N(10, 2)$ distribution converted to a standard normal distribution

P.153 The lower 10% for a standard normal distribution converted to a $N(500, 80)$ distribution

P.154 The area above 2.1 for a standard normal distribution converted to a $N(500, 80)$ distribution

P.155 The area between 1 and 2 for a standard normal distribution converted to a $N(100, 15)$ distribution

P.156 The middle 80% for a standard normal distribution converted to a $N(100, 15)$ distribution

P.157 SAT scores The Scholastic Aptitude Test (SAT) was taken by 1,698,521 college-bound students in the class of 2015.[24] The test has three parts: Critical Reading, Mathematics, and Writing. Scores on all three parts range from 200 to 800. The means and standard deviations for the three tests are shown in Table P.16. Assuming that the Critical Reading scores follow a normal distribution, draw a sketch of the normal distribution and label at least 3 points on the horizontal axis.

Table P.16 *SAT scores for the class of 2015*

	Mean	St. Dev.
Critical Reading	495	116
Mathematics	511	120
Writing	484	115

P.158 Critical Reading on the SAT Exam In Table P.16 with Exercise P.157, we see that scores on the Critical Reading portion of the SAT (Scholastic Aptitude Test) exam are normally distributed with mean 495 and standard deviation 116. Use the normal distribution to answer the following questions:

(a) What is the estimated percentile for a student who scores 700 on Critical Reading?

(b) What is the approximate score for a student who is at the 30th percentile for Critical Reading?

P.159 Writing on the SAT Exam In Table P.16 with Exercise P.157, we see that scores on the Writing portion of the SAT (Scholastic Aptitude Test) exam are normally distributed with mean 484 and standard deviation 115. Use the normal distribution to answer the following questions:

(a) What is the estimated percentile for a student who scores 450 on Writing?

(b) What is the approximate score for a student who is at the 90th percentile for Writing?

P.160 Boys Heights Heights of 10-year-old boys (5th graders) follow an approximate normal distribution with mean $\mu = 55.5$ inches and standard deviation $\sigma = 2.7$ inches.[25]

(a) Draw a sketch of this normal distribution and label at least three points on the horizontal axis.

[24] *https://research.collegeboard.org/programs/sat/data/cb-seniors-2015..*

[25] Centers for Disease Control and Prevention growth chart, *http://www.cdc.gov/growthcharts/html_charts/statage.htm.*

(b) According to this normal distribution, what proportion of 10-year-old boys are between 4 ft 4 in and 5 ft tall (between 52 inches and 60 inches)?

(c) A parent says his 10-year-old son is in the 99th percentile in height. How tall is this boy?

P.161 Heights of Men in the US Heights of adult males in the US are approximately normally distributed with mean 70 inches (5 ft 10 in) and standard deviation 3 inches.

(a) What proportion of US men are between 5 ft 8 in and 6 ft tall (68 and 72 inches, respectively)?

(b) If a man is at the 10th percentile in height, how tall is he?

P.162 What Proportion Have College Degrees? According to the US Census Bureau,[26] about 32.5% of US adults over the age of 25 have a bachelor's level (or higher) college degree. For random samples of $n = 500$ US adults over the age of 25, the sample proportions, $\hat{p}$, with at least a bachelor's degree follow a normal distribution with mean 0.325 and standard deviation 0.021. Draw a sketch of this normal distribution and label at least three points on the horizontal axis.

P.163 Quartiles for GPA In Example P.31 on page 728 we see that the grade point averages (GPA) for students in introductory statistics at one college are modeled with a N(3.16, 0.40) distribution. Find the first and third quartiles of this normal distribution. That is, find a value (Q_1) where about 25% of the GPAs are below it and a value (Q_3) that is larger than about 75% of the GPAs.

P.164 Random Samples of College Degree Proportions In Exercise P.162, we see that the distribution of sample proportions of US adults with a

[26]From the 2015 Current Population Survey at *http://www.census.gov/hhes/socdemo/education/data/cps/*.

college degree for random samples of size $n = 500$ is N(0.325, 0.021). How often will such samples have a proportion, $\hat{p}$, that is more than 0.35?

P.165 Commuting Times in St. Louis A bootstrap distribution of mean commute times (in minutes) based on a sample of 500 St. Louis workers stored in **CommuteStLouis** is shown in Figure P.15. The pattern in this dotplot is reasonably bell-shaped so we use a normal curve to model this distribution of bootstrap means. The mean for this distribution is 21.97 minutes and the standard deviation is 0.65 minutes. Based on this normal distribution, what proportion of bootstrap means should be in each of the following regions?

(a) More than 23 minutes

(b) Less than 20 minutes

(c) Between 21.5 and 22.5 minutes

P.166 Randomization Slopes A randomization distribution is created to test a null hypothesis that the slope of a regression line is zero. The randomization distribution of sample slopes follows a normal distribution, centered at zero, with a standard deviation of 2.5.

(a) Draw a rough sketch of this randomization distribution, including a scale for the horizontal axis.

(b) Under this normal distribution, how likely are we to see a sample slope above 3.0?

(c) Find the location of the 5%-tile in this normal distribution of sample slopes.

P.167 Exam Grades Exam grades across all sections of introductory statistics at a large university are approximately normally distributed with a mean of 72 and a standard deviation of 11. Use the normal distribution to answer the following questions.

(a) What percent of students scored above a 90?

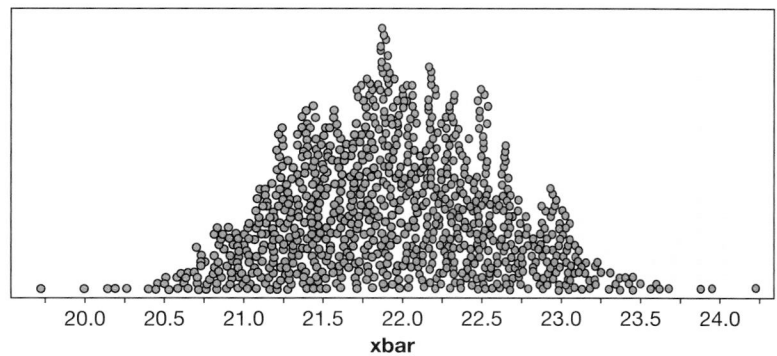

Figure P.15 *Bootstrap means for commute times in CommuteStLouis*

(b) What percent of students scored below a 60?

(c) If the lowest 5% of students will be required to attend peer tutoring sessions, what grade is the cutoff for being required to attend these sessions?

(d) If the highest 10% of students will be given a grade of A, what is the cutoff to get an A?

P.168 Curving Grades on an Exam A statistics instructor designed an exam so that the grades would be roughly normally distributed with mean $\mu = 75$ and standard deviation $\sigma = 10$. Unfortunately, a fire alarm with ten minutes to go in the exam made it difficult for some students to finish. When the instructor graded the exams, he found they were roughly normally distributed, but the mean grade was 62 and the standard deviation was 18. To be fair, he decides to "curve" the scores to match the desired $N(75, 10)$ distribution. To do this, he standardizes the actual scores to z-scores using the $N(62, 18)$ distribution and then "unstandardizes" those z-scores to shift to $N(75, 10)$. What is the new grade assigned for a student whose original score was 47? How about a student who originally scores a 90?

P.169 Empirical Rule for Normal Distributions Pick any positive values for the mean and the standard deviation of a normal distribution. Use your selection of a normal distribution to answer the questions below. The results of parts (a) to (c) form what is often called the *Empirical Rule* for the standard deviation in a normal distribution.

(a) Verify that about 95% of the values fall within two standard deviations of the mean.

(b) What proportion of values fall within one standard deviation of the mean?

(c) What proportion of values fall within three standard deviations of the mean?

(d) Will the answers to (b) and (c) be the same for *any* normal distribution? Explain why or why not.

Chapter Summaries

Guide to choosing the appropriate method based on the variables and number of categories:

Variables	Visualization	Number of Categories	Appropriate Inference
One Categorical	Bar chart, Pie chart	Two Categories	Single Proportion or Chi-Square Goodness of Fit
		More Categories	Chi-Square Goodness of Fit
One Quantitative	Histogram, Dotplot, Boxplot	—	Single Mean
Two Categorical	Side-by-Side or Segmented Bar Chart	Two Categories	Difference in Proportions or Chi-Square Test for Association
		More Categories	Chi-Square Test for Association
One Quantitative One Categorical,	Side-by-Side Plots	Two Categories	Difference in Means or Analysis of Variance
		More Categories	Analysis of Variance
Two Quantitative	Scatterplot	—	Correlation, Simple Regression
Quantitative Response, Multiple Explanatory	—	—	Multiple Regression
Categorical Response, Multiple Explanatory	—	—	Take STAT2!

Chapter 1: Collecting Data

In Chapter 1, we learn about appropriate ways to *collect* data. A dataset consists of values for one or more variables that record or measure information for each of the cases in a sample or population. A variable is generally classified as either *categorical*, if it divides the data cases into groups, or *quantitative*, if it measures some numerical quantity.

What we can infer about a population based on the data in a sample depends on the method of data collection. We try to collect a sample that is representative of the population and that avoids sampling bias. The most effective way to avoid sampling bias is to select a random sample. Also, we try to avoid other possible sources of bias by considering things like the wording of a question. The key is to always think carefully about whether the method used to collect data might introduce any bias.

Data collected to analyze a relationship between variables can come from an observational study or a randomized experiment. In an observational study, we need to be wary of confounding variables. A randomized experiment allows us to avoid confounding variables by actively (and randomly) manipulating the explanatory variables. The handling of different treatment groups in an experiment should be as similar as possible, with the use of blinding and/or a placebo treatment when appropriate.

The only way to infer a causal association between variables statistically is through data obtained from a randomized experiment. One of the most common and serious mistakes in all of statistics comes from a failure to appreciate the importance of this statement.

There are many questions to ask involving how data are collected, but two stand out, both involving randomness. These questions, and their simplified conclusions, are summarized in the diagram below.

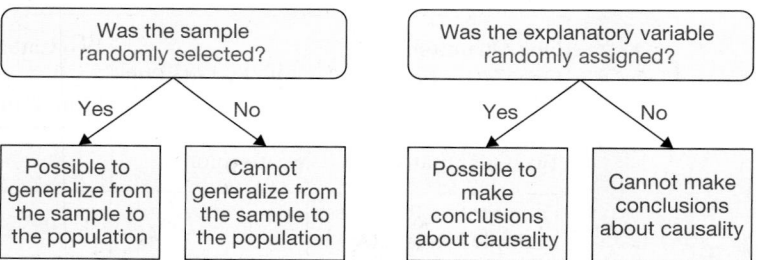

Two fundamental questions about data collection

Chapter 2: Describing Data

In Chapter 2, we learn about methods to *display* and *summarize* data. We use statistical plots to display information and summary statistics to quantify aspects of that information. The type of visualization or statistic we use often depends on the types of variables (quantitative or categorical), as summarized below:

Describing a Single Variable

- Quantitative variable
  - Graphical display: *dotplot, histogram, boxplot*
  - Summary statistics:
    * Center: *mean, median*
    * Other locations: *maximum, minimum, first quartile, third quartile*
    * Spread: *standard deviation, interquartile range, range*
- Categorical variable
  - Graphical display: *bar chart, pie chart*
  - Summary statistics: *frequency, relative frequency, proportion*

Describing a Relationship between Two Variables

- Categorical vs Categorical
  - Graphical display: *segmented or side-by-side bar chart*
  - Summary statistics: *two-way table, row/column proportions, difference in proportions*
- Categorical vs Quantitative
  - Graphical display: *side-by-side boxplots, dotplots, or histograms*
  - Summary statistics: *quantitative statistics* within each category, *difference in means*
- Quantitative vs Quantitative
  - Graphical display: *scatterplot*
  - Summary statistics: *correlation, regression line*

We also discuss, in Section 2.7, a variety of creative and effective ways to display data with additional variables and/or data that include geographic or time variables.

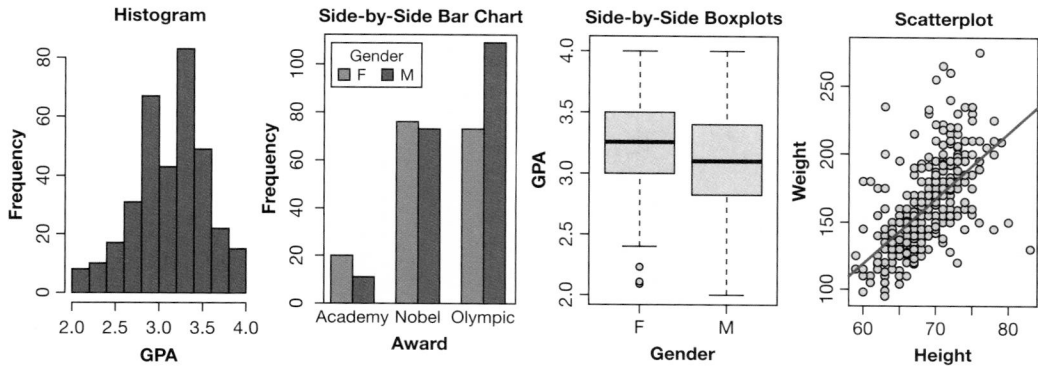

Chapter 3: Confidence Intervals

We estimate a population *parameter* using a sample *statistic*. Since such statistics vary from sample to sample, we need an idea for how far the population parameter could be from the sample statistic, a *margin of error*. An *interval estimate* is a range of plausible values for the population parameter. When we construct this interval using a method that has some predetermined chance of capturing the true parameter, we get a *confidence interval*.

We assess the variability in sample statistics with a *bootstrap distribution*, created using the key idea that if the sample is representative of the population, then the population can be approximated by many copies of the sample. To construct a bootstrap distribution we:

- Generate bootstrap samples with replacement from the original sample, using the same sample size
- Compute the statistic of interest for each of the bootstrap samples
- Collect the statistics from many (usually at least 1000) bootstrap samples into a bootstrap distribution

From a symmetric bootstrap distribution, we have two methods to construct an interval estimate:

Method 1: Estimate *SE*, the standard error of the statistic, as the standard deviation of the bootstrap distribution. The roughly 95% confidence interval for the parameter is then *Sample statistic* $\pm 2 \cdot SE$.

Method 2: Use percentiles of the bootstrap distribution to chop off the tails of the bootstrap distribution and keep a specified percentage (determined by the confidence level) of the values in the middle.

A bootstrap distribution is shown for mean body temperature. The bootstrap distribution is centered around the sample statistic, $\bar{x} = 98.26$, with $SE = 0.109$, so a 95% confidence interval is *Statistic* $\pm 2 \cdot SE = 98.26 \pm 2 \cdot 0.109 = (98.042, 98.478)$. A 95% confidence interval can also be found as the middle 95% of bootstrap statistics, shown in the figure to be $(98.044, 98.476)$. We are 95% confident that mean body temperature is between 98.04°F and 98.48°F.

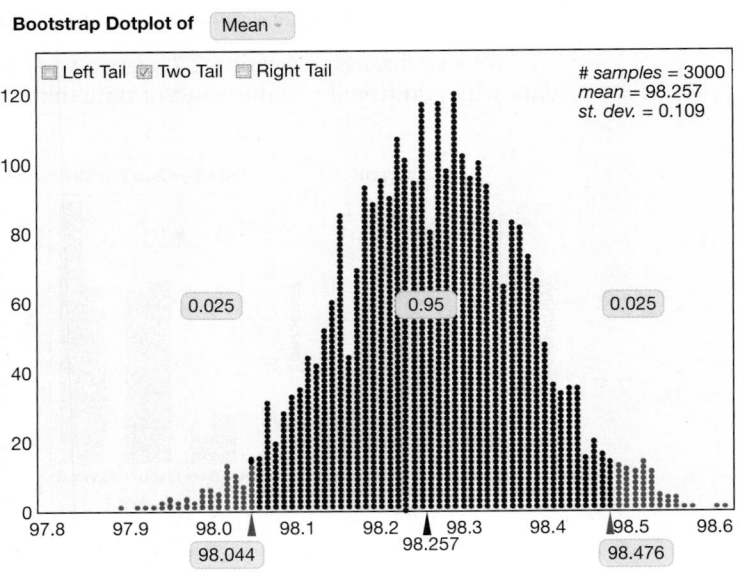

Bootstrap distribution of sample means

Chapter 4: Hypothesis Tests

Hypothesis tests are used to investigate claims about population parameters. We use the question of interest to determine the two competing *hypotheses*: The null hypothesis (H_0) is generally that there is no effect or no difference while the alternative hypothesis (H_a) is the claim for which we seek evidence. We conclude in favor of the alternative hypothesis if the sample supports the alternative hypothesis and provides strong evidence against the null hypothesis. We measure the strength of evidence a sample shows against the null hypothesis with a p-value.

The *p-value* is the probability of obtaining a sample statistic as extreme as (or more extreme than) the observed sample statistic, when the null hypothesis is true.

A small p-value means that the observed sample results would be unlikely to happen, when the null hypothesis is true, just by random chance. When making formal decisions based on the p-value, we use a pre-specified *significance level*, α.

- If p-value $< \alpha$, we reject H_0 and have statistically significant evidence for H_a.
- If p-value $\geq \alpha$, we do not reject H_0, the test is inconclusive, and the results are not statistically significant.

The key idea is: *The smaller the p-value, the stronger the evidence against the null hypothesis and in support of the alternative hypothesis.* Rather than making a formal reject/do not reject decision, we sometimes interpret the p-value as a measure of strength of evidence.

One way to estimate a p-value is to construct a randomization distribution of sample statistics that we might see by random chance, if the null hypothesis were true. The p-value is the proportion of randomization statistics that are as extreme as the observed sample statistic. If the original sample falls out in the tails, then a result that extreme is unlikely to occur if the null hypothesis is true, providing evidence against the null.

A randomization distribution for difference in mean memory recall between sleep and caffeine groups for data in **SleepCaffeine** is shown. Each dot is a difference in means that might be observed just by random assignment to treatment groups, if there were no difference in terms of mean (memory) response. We see that 0.042 of the simulated statistics are as extreme as the observed statistic ($\bar{x}_s - \bar{x}_c = 3$), so the p-value is 0.042. This p-value is less than 0.05, so the results are statistically significant at $\alpha = 0.05$, giving moderately strong evidence that sleeping is better than drinking caffeine for memory.

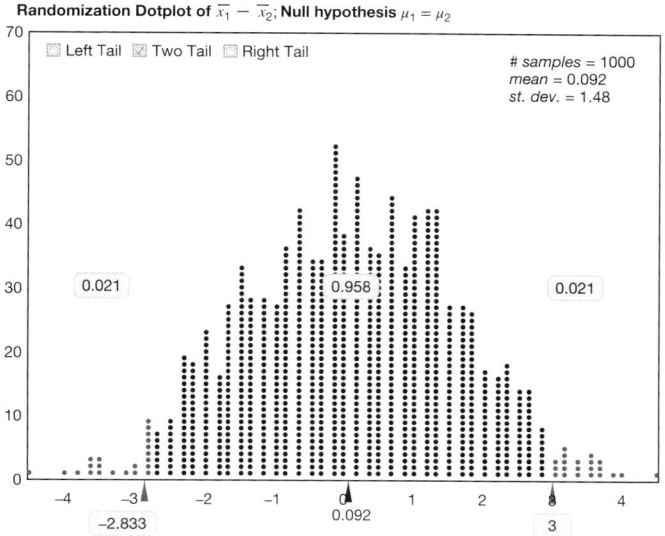

Randomization distribution of differences in means

Chapter 5: Approximating with a Distribution

In Chapter 5 we see that the familiar bell-shape we encounter repeatedly in bootstrap and randomization distributions is predictable, and is called the *normal distribution*. Although a normal distribution can have any mean and standard deviation, $X \sim N(\mu, \sigma)$, we often work with a standard normal, $Z \sim N(0, 1)$, by converting to a z-score:

$$Z = \frac{X - \mu}{\sigma}$$

We generally rely on technology (such as the *StatKey* figures shown below) to compute areas or endpoints for normal distributions.

The Central Limit Theorem tells us that, when the sample size is large enough, sample means, proportions, and other statistics are approximately normally distributed and centered at the value of the corresponding population parameter.

When sample statistics are normally distributed we can utilize the following general formulas:

Confidence Interval :   Sample Statistic $\pm z^* \cdot SE$

Hypothesis Test :   Test Statistic $= \dfrac{\text{Sample Statistic} - \text{Null Parameter}}{SE}$

The z^* in the confidence interval is based on a threshold keeping the desired level of confidence in the middle of a standard normal distribution. The test statistic has a standard normal distribution if the null hypothesis is true, so should be compared to a standard normal distribution to find the p-value. In general we can find the standard error, SE, from a bootstrap or randomization distribution. In the next chapter we see a number of shortcut formulas for estimating SE in common situations and replace the standard normal z with a t-distribution for inference involving means.

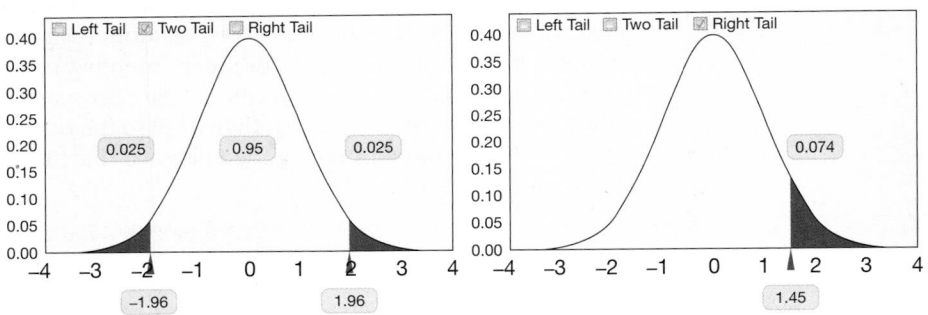

Areas and endpoints for the standard normal

Chapter 6: Inference for Means and Proportions

Under general conditions we can find formulas for the standard errors of sample means, proportions, or their differences. This leads to formulas for computing confidence intervals or test statistics based on normal or t-distributions.

	Distribution	**Conditions**	**Standard Error**
Proportion	Normal	$np \geq 10$ and $n(1-p) \geq 10$	$\sqrt{\dfrac{p(1-p)}{n}}$
Mean	$t, df = n-1$	$n \geq 30$ or reasonably normal	$\dfrac{s}{\sqrt{n}}$
Difference in Proportions	Normal	$n_1 p_1 \geq 10, n_1(1-p_1) \geq 10$, and $n_2 p_2 \geq 10$, $n_2(1-p_2) \geq 10$	$\sqrt{\dfrac{p_1(1-p_1)}{n_1} + \dfrac{p_2(1-p_2)}{n_2}}$
Difference in Means	$t, df =$ the smaller of $n_1 - 1$ and $n_2 - 1$	$n_1 \geq 30$ or reasonably normal, and $n_2 \geq 30$ or reasonably normal	$\sqrt{\dfrac{s_1^2}{n_1} + \dfrac{s_2^2}{n_2}}$
Paired Difference in Means	$t, df = n_d - 1$	$n_d \geq 30$ or reasonably normal	$\dfrac{s_d}{\sqrt{n_d}}$

	Confidence Interval	**Test Statistic**
General	Sample statistic $\pm z^* \cdot SE$	$\dfrac{\text{Sample statistic} - \text{Null parameter}}{SE}$
Proportion	$\hat{p} \pm z^* \cdot \sqrt{\dfrac{\hat{p}(1-\hat{p})}{n}}$	$\dfrac{\hat{p} - p_0}{\sqrt{\frac{p_0(1-p_0)}{n}}}$
Mean	$\bar{x} \pm t^* \cdot s/\sqrt{n}$	$\dfrac{\bar{x} - \mu_0}{s/\sqrt{n}}$
Difference in Proportions	$(\hat{p}_1 - \hat{p}_2) \pm z^* \cdot \sqrt{\dfrac{\hat{p}_1(1-\hat{p}_1)}{n_1} + \dfrac{\hat{p}_2(1-\hat{p}_2)}{n_2}}$	$\dfrac{(\hat{p}_1 - \hat{p}_2) - 0}{\sqrt{\frac{\hat{p}(1-\hat{p})}{n_1} + \frac{\hat{p}(1-\hat{p})}{n_2}}}$
Difference in Means	$(\bar{x}_1 - \bar{x}_2) \pm t^* \cdot \sqrt{\dfrac{s_1^2}{n_1} + \dfrac{s_2^2}{n_2}}$	$\dfrac{(\bar{x}_1 - \bar{x}_2) - 0}{\sqrt{\frac{s_1^2}{n_1} + \frac{s_2^2}{n_2}}}$
Paired Diff. in Means	$\bar{x}_d \pm t^* \cdot \dfrac{s_d}{\sqrt{n_d}}$	$\dfrac{\bar{x}_d - 0}{s_d/\sqrt{n_d}}$

Chapter 7: Chi-Square Tests for Categorical Variables

Chi-square tests are used for testing hypotheses about one or two categorical variables, and are appropriate when the data can be summarized by counts in a table. The variables can have multiple categories. The type of chi-square test depends on whether there are one or two categorical variables:

- One Categorical Variable: Chi-Square Goodness-of-Fit Test
- Two Categorical Variables: Chi-Square Test for Association

Chi-square tests compare observed counts to expected counts (if the null hypothesis were true). If the observed counts are farther away from the expected counts than can be explained just by random chance, we have evidence against the null hypothesis and in favor of the alternative. The distance between observed and expected counts is quantified with the χ^2-statistic, which is compared to a χ^2-distribution to calculate the p-value. The details are laid out below:

1. State hypotheses
   - For one categorical variable:
     - Null hypothesis: The proportions match an assumed set of proportions.
     - Alternative hypothesis: At least one category has a different proportion.
   - For two categorical variables:
     - Null hypothesis: There is no association between the two variables.
     - Alternative hypothesis: There is an association between the two variables.

2. Calculate the expected counts for each cell (as if the null hypothesis were true)
   - For one categorical variable: Expected count for a cell $= n \cdot p_i$, where p_i is given in H_0
   - For two categorical variables:

$$\text{Expected count for a cell } = \frac{\text{Row total} \cdot \text{Column total}}{\text{Total sample size}}.$$

3. Compute the χ^2-statistic:

$$\chi^2 = \sum \frac{(Observed - Expected)^2}{Expected}$$

4. Find the p-value as the upper tail in a χ^2 distribution
   - For one categorical variable: $df = k - 1$, where k is the number of categories.
   - For two categorical variables: $df = (r - 1) \cdot (c - 1)$, where r is the number of categories in one variable and c is the number of categories in the other.

5. Make a conclusion
   - If the results are significant, we have evidence in favor of the alternative hypothesis. A more informative conclusion can be given by comparing the relative sizes of observed and expected counts of individual cells, and the relative contribution of cells to the chi-square statistic.

With only two categories the chi-square goodness-of-fit test is equivalent to a test for a single proportion, and the chi-square test for association is equivalent to a test for a difference in two proportions.

Chapter 8: ANOVA to Compare Means

Analysis of variance is used to test for an association between one quantitative variable and one categorical variable or, equivalently, to test for a difference in means across categories of a categorical variable. The categorical variable can have multiple categories. This method is appropriate when the summary statistics include sample means calculated within groups.

Analysis of variance compares variability within groups to variability between groups. If the ratio of variability between groups to variability within groups is higher than we would expect just by random chance, we have evidence of a difference in means. This ratio is called the F-statistic, which we compare to an F-distribution to find the p-value. The details are laid out below.

1. State hypotheses
   - Null hypothesis: $\mu_1 = \mu_2 = \cdots = \mu_k$ (no difference in means by category).
   - Alternative hypothesis: Some $\mu_i \neq \mu_j$ (difference in means between categories).

2. Compute the F-statistic, using an ANOVA table:

Source	d.f.	SS	MS	F-statistic	p-value
Groups	$k-1$	SSG	$MSG = \dfrac{SSG}{k-1}$	$F = \dfrac{MSG}{MSE}$	Upper tail $F_{k-1,n-k}$
Error	$n-k$	SSE	$MSE = \dfrac{SSE}{n-k}$		
Total	$n-1$	$SSTotal$			

The sums of squares $SSTotal = SSG + SSE$ are obtained by technology or formula.

3. Find the p-value as the upper tail in an F-distribution
   - Use df for Groups and df for Error from the ANOVA table.

4. Make a conclusion
   - If the results are significant, we have evidence of an association between the variables (and a difference in means between the groups defined by the categorical variable). A more informative conclusion can be given if desired by using the methods of pairwise comparison presented in Section 8.2.

   If the categorical variable has only two categories, analysis of variance is equivalent to a test for a difference in means between two groups.

Inference after ANOVA: Confidence Intervals or Pairwise Tests

   - Use a t-distribution with Error df and $\sqrt{MSE}$ from ANOVA to estimate variability. Use technology or see formulas on page 565.

One-way ANOVA: Ants versus Filling

Source	DF	SS	MS	F	P
Filling	2	1561	781	5.63	0.011
Error	21	2913	139		
Total	23	4474			

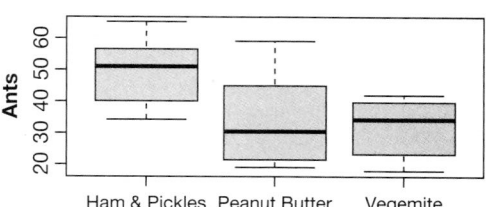

Does mean number of ants differ by sandwich filling?

Chapter 9: Inference for Regression

Simple linear regression predicts a quantitative response variable, Y, based on a quantitative explanatory variable, X. In order to use regression, both variables need to be measured on the same set of cases. The simple linear regression model is introduced in Section 2.6, and Chapter 9 extends this to include inference.

The simple linear regression model is $Y = \beta_0 + \beta_1 \cdot X + \epsilon$. For prediction we use the estimated coefficients: $\hat{Y} = b_0 + b_1 \cdot X$.

There are three different ways to test for an association between two quantitative variables:

- Test for Correlation
  * Null hypothesis: There is no linear relationship ($\rho = 0$).
  * Test statistic: $t = \dfrac{r\sqrt{n-2}}{\sqrt{1-r^2}}$.
  * Distribution: t-distribution with $df = n - 2$.
- Test for Slope
  * Null hypothesis: The variable is not significant in the model ($\beta_1 = 0$).
  * Test statistic $t = \dfrac{b_1 - 0}{SE}$, where SE is the standard error of the slope.
  * Distribution: t-distribution with $df = n - 2$.
- Analysis of Variance for Regression
  * Null hypothesis: The model is not effective at predicting the response.
  * Test statistic: F-statistic from an ANOVA table (see details on page 594).
  * Distribution: Upper tail of F-distribution with df Model and df Error.

A scatterplot should always be checked to make sure the trend is approximately linear, the variability of points around the line is relatively constant for different x values, and there are not major outliers.

R^2 gives the percent of variability in the response variable that is explained by the explanatory variable in the model, and is equivalent to the squared correlation between y and x.

Confidence intervals for the mean response value at a specific x value, or prediction intervals for an individual response value at a specific x value, can be created with technology or the formulas on page 605.

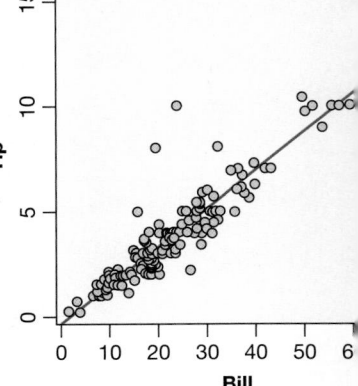

The regression equation is Tip = −0.292 + 0.182 Bill

Predictor	Coef	SE Coef	T	P
Constant	−0.2923	0.1662	−1.76	0.081
Bill	0.182215	0.006451	28.25	0.000

S = 0.979523   R-Sq = 83.7%   R-Sq(adj) = 83.6%

Chapter 10: Multiple Regression

Multiple regression extends simple linear regression to include multiple explanatory variables. It allows us to incorporate multiple variables in a single analysis. Multiple regression is used to predict a quantitative response variable based on multiple explanatory variables, and to model relationships between explanatory variable(s) and a quantitative response variable.

Many concepts introduced for simple linear regression also apply to multiple regression:

- Test for Slope Coefficient
  * Null hypothesis: The variable is not significant in the model ($\beta_i = 0$).
  * Test statistic: $t = \dfrac{b_i - 0}{SE}$, where SE is the standard error of the slope.
  * Distribution: t-distribution with $df = n - k - 1$, where k is the number of explanatory variables.

- Analysis of Variance for Regression (check overall model fit)
  * Null hypothesis: The model is not effective at predicting the response.
  * Test statistic: F-statistic from an ANOVA table (see details on page 618).
  * Distribution: Upper tail of F-distribution with df Model and df Error.

- R^2 gives the percent of variability in the response variable that is explained by the explanatory variables in the model.

Each slope coefficient is interpreted as the amount that the predicted response changes for a unit increase in that explanatory variable, if all the other explanatory variables in the model are held constant.

In simple linear regression we can assess the conditions by looking at a scatterplot. In multiple regression we need to look at a *plot of residuals versus fitted values*. We should watch out for curvature (or any nonlinear trend), increasing or decreasing variability, or outliers. We also watch out for obvious skewness or outliers on a histogram or dotplot of the residuals.

More variables are not always better; consider pruning insignificant variables from the model. There are many ways of deciding between competing models; for details see the box on page 635.

With multiple explanatory variables, it is very important to remember *the coefficient and significance of each explanatory variable depend on the other explanatory variables included in the model.*

Multiple regression output for predicting percent body fat is given below:

Coefficients:

	Estimate	Std. Error	t value	Pr(> |t|)	
(Intercept)	−24.94157	20.77414	−1.201	0.2329	
Weight	−0.08434	0.05891	−1.432	0.1555	
Height	0.05177	0.23849	0.217	0.8286	
Abdomen	0.96762	0.13040	7.421	5.15e-11	***
Age	0.07740	0.04868	1.590	0.1152	
Wrist	−2.05797	0.72893	−2.823	0.0058	**

Residual standard error: 4.074 on 94 degrees of freedom
Multiple R-squared: 0.7542,   Adjusted R-squared: 0.7411
F-statistic: 57.67 on 5 and 94 DF, p-value: < 2.2e-16

Selected Dataset Descriptions

Descriptions of Variables for Selected Larger Datasets

There are more than 100 datasets to accompany this text, all described and available in the web resources. We offer additional descriptions here of a few of the ones with many variables:

AllCountries

BaseballHits

Cars2015

FloridaLakes

GSWarriors

HappyPlanetIndex

HollywoodMovies

ICUAdmissions

MindsetMatters

NBAPlayers2015

NutritionStudy

SleepStudy

SpeedDating

StudentSurvey

SynchronizedMovement

USStates

AllCountries

Information about 215 individual countries as determined by the World Bank between 2012 and 2014.
Source: *www.worldbank.org*

Country	Name of the country
LandArea	Size in 1000 sq. kilometers
Population	Population in millions
Density	Number of people per square kilometer
GDP	Gross Domestic Product (per capita in US$)
Rural	Percentage of population living in rural areas
CO2	CO_2 emissions (metric tons per capita)
PumpPrice	Price for a liter of gasoline (in US$)
Military	Percentage of government expenditures directed toward the military
Health	Percentage of government expenditures directed toward healthcare
ArmedForces	Number of active duty military personnel (in 1000s)
Internet	Percentage of the population with access to the Internet
Cell	Cell phone subscriptions (per 100 people)
HIV	Percentage of the population with HIV
Hunger	Percent of the population considered undernourished
Diabetes	Percent of the population diagnosed with diabetes
BirthRate	Births per 1000 people
DeathRate	Deaths per 1000 people
ElderlyPop	Percentage of the population at least 65 years old
LifeExpectancy	Average life expectancy (years)
FemaleLabor	Percent of females 15–64 in the labor force
Unemployment	Percent of labor force unemployed
EnergyUse	Energy usage (kilotons of oil)
Electricity	Electric power consumption (kWh per capita)
Developed	Categories for kilowatt hours per capita, 1 = under 2500, 2 = 2500 to 5000, 3 = over 5000

BaseballHits

Team level data for 30 major league baseball teams from the 2014 regular season.
Source: *www.baseball-reference.com/leagues/MLB/2014-standard-batting.shtml*

Team	Name of baseball team
League	Either AL (American League) or NL (National League)
Wins	Number of wins for the season
Runs	Number of runs scored
Hits	Number of hits
Doubles	Number of doubles
Triples	Number of triples
HomeRuns	Number of home runs
RBI	Number of runs batted in
StolenBases	Number of stolen bases
CaughtStealing	Number of times caught stealing
Walks	Number of walks
Strikeouts	Number of strikeouts
BattingAvg	Team batting average

Cars2015

Information about new car models in 2015
Source: Consumer Reports 2015 New Car Buying Guide *http://www.magastack.com/ issue/6053-consumer-reports-new-car-buying-guide-february-2015?page=1*

Make	Manufacturer
Model	Car model
Type	Vehicle category (Small, Hatchback, Sedan, Sporty, Wagon, SUV, 7Pass)
LowPrice	Lowest suggested retail price (in $1000)
HighPrice	Highest suggested retail price (in $1000)
Drive	Type of drive (front = FWD, rear = RWD, both = AWD)
CityMPG	City miles per gallon (EPA)
HwyMPG	Highway miles per gallon (EPA)
FuelCap	Fuel capacity (in gallons)
Length	Length (in inches)
Width	Width (in inches)
Wheelbase	Wheelbase (in inches)
Height	Height (in inches)
UTurn	Diameter (in feet) needed for a U-turn
Weight	Curb weight (in pounds)
Acc030	Time (in seconds) to go from 0 to 30 mph
Acc060	Time (in seconds) to go from 0 to 60 mph
QtrMile	Time (in seconds) to go 1/4 mile
PageNum	Page number in the *Consumer Report 2015 New Car Buyers Guide*
Size	Small, Midsized, Large

FloridaLakes

This dataset describes characteristics of water and fish samples from 53 Florida lakes. Some variables (e.g., Alkalinity, pH, and Calcium) reflect the chemistry of the water samples. Mercury levels were recorded for a sample of largemouth bass selected at each lake.
Source: Lange, Royals, and Connor, Transactions of the American Fisheries Society (1993)

ID	An identifying number for each lake
Lake	Name of the lake
Alkalinity	Concentration of calcium carbonate (in mg/L)
pH	Acidity
Calcium	Amount of calcium in water
Chlorophyll	Amount of chlorophyll in water
AvgMercury	Average mercury level for a sample of fish (largemouth bass) from each lake
NumSamples	Number of fish sampled at each lake
MinMercury	Minimum mercury level in a sampled fish
MaxMercury	Maximum mercury level in a sampled fish
ThreeYrStdMercury	Adjusted mercury level to account for the age of the fish
AgeData	Mean age of fish in each sample

GSWarriors

Information from online boxscores for all 82 regular season games played by the Golden State Warriors basketball team during the 2015–2016 regular season.
Source: Data downloaded from *http://www.basketball-reference.com/teams/GSW/2016/gamelog/*

Game	ID number for each game
Date	Date the game was played
Location	Away or Home
Opp	Opponent team
Win	Game result: L or W
Points	Number of points scored
OppPoints	Opponent's points scored
FG	Field goals made
FGA	Field goals attempted
FG3	Three-point field goals made
FG3A	Three-point field goals attempted
FT	Free throws made
FTA	Free throws attempted
Rebounds	Total rebounds
OffReb	Offensive rebounds
Assists	Number of assists
Steals	Number of steals
Blocks	Number of shots blocked
Turnovers	Number of turnovers
Fouls	Number of fouls
OppFG	Opponent's field goals made
OppFGA	Opponent's field goals attempted
OppFG3	Opponent's three-point field goals made
OppFG3A	Opponent's three-point field goals attempted
OppFT	Opponent's free throws made
OppFTA	Opponent's free throws attempted
OppRebounds	Opponent's total rebounds
OppOffReb	Opponent's offensive rebounds
OppAssists	Opponent's assists
OppSteals	Opponent's steals
OppBlocks	Opponent's shots blocked
OppTurnovers	Opponent's turnovers
OppFouls	Opponent's fouls

HappyPlanetIndex

Data for 143 countries from the Happy Planet Index Project, *http://www.happyplanetindex.org*, that works to quantify indicators of happiness, well-being, and ecological footprint at a country level. Region of the world is coded as: 1 = Latin America, 2 = Western nations, 3 = Middle East, 4 = Sub-Saharan Africa, 5 = South Asia, 6 = East Asia, 7 = former Communist countries.
Source: Downloaded from *http://www.happyplanetindex.org/data/*

Country	Name of country
Region	Code for region of the world, with code given in the description above.

Happiness	Score on a 0 to 10 scale for average level of happiness (10 is happiest)
LifeExpectancy	Average life expectancy (in years)
Footprint	Ecological footprint—a measure of the (per capita) ecological impact, with higher numbers indicating greater environmental impact
HLY	Happy Life Years—combines life expectancy with well-being
HPI	Happy Planet Index (0–100 scale)
HPIRank	HPI rank for the country
GDPperCapita	Gross Domestic Product (per capita)
HDI	Human Development Index
Population	Population (in millions)

HollywoodMovies

Information for 970 movies released from Hollywood between 2007 and 2013. Source: McCandless, D., "Most Profitable Hollywood Movies," from "Information is Beautiful," *http://www.informationisbeautiful.net/data/* and *http://bit.ly/hollywood budgets*

Movie	Title of movie
LeadStudio	Studio that released the movie
RottenTomatoes	Rotten Tomatoes rating (reviewers)
AudienceScore	Audience rating (via Rotten Tomatoes)
Story	General theme—one of 21 themes
Genre	One of 14 possible genres
TheatersOpenWeek	Number of screens for opening weekend
OpeningWeekend	Opening weekend gross (in millions)
BOAverageOpenWeek	Average box office income per theater—opening weekend
DomesticGross	Gross income for domestic viewers (in millions)
ForeignGross	Gross income for foreign viewers (in millions)
WorldGross	Gross income for all viewers (in millions)
Budget	Production budget (in millions)
Profitability	WorldGross as a percentage of Budget
OpenProfit	Percentage of budget recovered on opening weekend
Year	Year the movie was released

ICUAdmissions

Data from a sample of 200 patients following admission to an adult intensive care unit (ICU). Source: DASL dataset downloaded from *http://lib.stat.cmu.edu/DASL/Datafiles/ICU.html*

ID	Patient ID number
Status	Patient status: 0 = lived or 1 = died
Age	Patient's age (in years)
Sex	0 = male or 1 = female
Race	Patient's race: 1 = white, 2 = black, or 3 = other
Service	Type of service: 0 = medical or 1 = surgical
Cancer	Is cancer involved? 0 = no or 1 = yes

Renal	Is chronic renal failure involved? 0 = no or 1 = yes
Infection	Is infection involved? 0 = no or 1 = yes
CPR	Patient gets CPR prior to admission? 0 = no or 1 = yes
Systolic	Systolic blood pressure (in mm Hg)
HeartRate	Pulse rate (beats per minute)
Previous	Previous admission to ICU within 6 months? 0 = no or 1 = yes
Type	Admission type: 0 = elective or 1 = emergency
Fracture	Fractured bone involved? 0 = no or 1 = yes
PO2	Partial oxygen level from blood gases under 60? 0 = no or 1 = yes
PH	pH from blood gas under 7.25? 0 = no or 1 = yes
PCO2	Partial carbon dioxide level from blood gas over 45? 0 = no or 1 = yes
Bicarbonate	Bicarbonate from blood gas under 18? 0 = no or 1 = yes
Creatinine	Creatinine from blood gas over 2.0? 0 = no or 1 = yes
Consciousness	Level: 0 = conscious, 1 = deep stupor, or 2 = coma

MindsetMatters

In 2007 a Harvard psychologist recruited 75 female maids working in different hotels to participate in a study. She informed 41 maids (randomly chosen) that the work they do satisfies the Surgeon General's recommendations for an active lifestyle (which is true), giving them examples showing that their work is good exercise. The other 34 maids were told nothing (uninformed). Various characteristics (weight, body mass index,…) were recorded for each subject at the start of the experiment and again four weeks later. Maids with missing values for weight change have been removed.
Source: Crum, A.J. and Langer, E.J. (2007). Mind-Set Matters: Exercise and the Placebo Effect, *Psychological Science*; 18:165–171. Thanks to the authors for supplying the data.

Cond	Treatment condition: 0 = uninformed or 1 = informed
Age	Age (in years)
Wt	Original weight (in pounds)
Wt2	Weight after 4 weeks (in pounds)
BMI	Original body mass index
BMI2	Body mass index after 4 weeks
Fat	Original body fat percentage
Fat2	Body fat percentage after 4 weeks
WHR	Original waist to hip ratio
WHR2	Waist to hip ratio after 4 weeks
Syst	Original systolic blood pressure
Syst2	Systolic blood pressure after 4 weeks
Diast	Original diastolic blood pressure
Diast2	Diastolic blood pressure after 4 weeks

NBAPlayers2015

Data for 182 NBA basketball players from the 2014–2015 regular season. Includes all players who averaged more than 24 minutes per game that season.
Source: *http://www.basketball-reference.com/leagues/NBA_2015_stats.html*

Player	Name of player
Pos	Position (PG = point guard, SG = shooting guard, PF = power forward, SF = small forward, C = center)
Age	Age (in years)
Team	Team name
Games	Games played (out of 82)
Starts	Games started
Mins	Minutes played
MinPerGame	Minutes per game
FGMade	Field goals made
FGAttempt	Field goals attempted
FGPct	Field goal percentage
FG3Made	Three-point field goals made
FG3Attempt	Three-point field goals attempted
FG3Pct	Three-point field goal percentage
FTMade	Free throws made
FTAttempt	Free throws attempted
FTPct	Free throw percentage
OffRebound	Offensive rebounds
DefRebound	Defensive rebounds
Rebounds	Total rebounds
Assists	Number of assists
Steals	Number of steals
Blocks	Number of blocked shots
Turnovers	Number of turnovers
Fouls	Number of personal fouls
Points	Number of points scored

NutritionStudy

Data on 315 patients undergoing elective surgery from a cross-sectional study to investigate the relationship between personal characteristics and dietary factors, and plasma concentrations of retinol, beta-carotene, and other carotenoids. Study subjects were patients who had an elective surgical procedure during a three-year period to biopsy or remove a lesion of the lung, colon, breast, skin, ovary, or uterus that was found to be non-cancerous.

Source: *http://lib.stat.cmu.edu/datasets/Plasma_Retinol* Original source: Nierenberg, D., Stukel, T., Baron, J., Dain, B., and Greenberg, E., "Determinants of plasma levels of beta-carotene and retinol," *American Journal of Epidemiology*, 1989, 130(3):511–521.

ID	ID number for each subject in this sample
Age	Subject's age (in years)
Smoke	Does the subject smoke: Yes or No
Quetelet	Weight/(Height^2)
Vitamin	Vitamin use coded as: 1 = Regularly, 2 = Occasionally, or 3 = No
Calories	Number of calories consumed per day
Fat	Grams of fat consumed per day
Fiber	Grams of fiber consumed per day
Alcohol	Number of alcoholic drinks consumed per week
Cholesterol	Cholesterol consumed (mg per day)
BetaDiet	Dietary beta-carotene consumed (mcg per day)
RetinolDiet	Dietary retinol consumed (mcg per day)

BetaPlasma	Concentration of beta-carotene (ng/ml) in the blood
RetinolPlasma	Concentration of retinol (ng/ml) in the blood
Gender	`Female` or `Male`
VitaminUse	Coded as `No`, `Occasional`, or `Regular`
PriorSmoke	Smoking status coded as: `1` = Never, `2` = Former, or `3` = Current

SleepStudy

The data were obtained from a sample of students who did skills tests to measure cognitive function, completed a survey that asked many questions about attitudes and habits, and kept a sleep diary to record time and quality of sleep over a two-week period.

Source: Onyper, S., Thacher, P., Gilbert, J., and Gradess, S., "Class start times, sleep, and academic performance in college: A path analysis," *Chronobiology International*, April 2012; 29(3):318–335. Thanks to the authors for supplying the data.

Gender	`1` = male, `0` = female
ClassYear	Year in school, `1` = first year,..., `4` = senior
LarkOwl	Early riser or night owl? `Lark`, `Neither`, or `Owl`
NumEarlyClass	Number of classes per week before 9 am
EarlyClass	Indicator for any early classes
GPA	Grade point average (0–4 scale)
ClassesMissed	Number of classes missed in a semester
CognitionZscore	Z-score on a test of cognitive skills
PoorSleepQuality	Measure of sleep quality (higher values are poorer sleep)
DepressionScore	Measure of degree of depression (higher values mean more depression)
AnxietyScore	Measure of amount of anxiety (higher values mean more anxiety)
StressScore	Measure of amount of stress (higher values mean more stress)
DepressionStatus	Coded depression score: `normal`, `moderate`, or `severe`
AnxietyStatus	Coded anxiety score: `normal`, `moderate`, or `severe`
Stress	Coded stress score: `normal` or `high`
DASScore	Combined score for depression, anxiety and stress
Happiness	Measure of degree of happiness
AlcoholUse	Self-reported: `Abstain`, `Light`, `Moderate`, or `Heavy`
Drinks	Number of alcoholic drinks per week
WeekdayBed	Average weekday bedtime (24.0 = midnight)
WeekdayRise	Average weekday rise time (8.0 = 8 am)
WeekdaySleep	Average hours of sleep on weekdays
WeekendBed	Average weekend bedtime (24.0 = midnight)
WeekendRise	Average weekend rise time (8.0 = 8 am)
WeekendSleep	Average hours of sleep on weekend days
AverageSleep	Average hours of sleep for all days
AllNighter	Had an all-nighter this semester? `1` = yes, `0` = no

SpeedDating

Participants were students at Columbia's graduate and professional schools, recruited by mass email, posted fliers, and fliers handed out by research assistants. Each participant attended one speed dating session, in which they met with each participant of the opposite sex for four minutes. Order and session assignments were randomly determined. After each four-minute "speed date," participants filled out a form rating their date on a scale of 1 to 10 on various attributes. Only data from the first date in each session is recorded here, for a total of 276 dates.

Source: Gelman, A. and Hill, J., *Data analysis using regression and multi-level/hierarchical models*, Cambridge University Press: New York, 2007.

DecisionM	Would the male like another date? `1 = yes 0 = no`
DecisionF	Would the female like another date? `1 = yes 0 = no`
LikeM	How much the male likes his partner (1–10 scale)
LikeF	How much the female likes her partner (1–10 scale)
PartnerYesM	Male's estimate of chance the female wants another date (1–10 scale)
PartnerYesF	Female's estimate of chance the male wants another date (1–10 scale)
AgeM	Male's age (in years)
AgeF	Female's age (in years)
RaceM	Male's race: `Asian, Black, Caucasian, Latino,` or `Other`
RaceF	Female's race: `Asian, Black, Caucasian, Latino,` or `Other`
AttractiveM	Male's rating of female's attractiveness (1–10 scale)
AttractiveF	Female's rating of male's attractiveness (1–10 scale)
SincereM	Male's rating of female's sincerity (1–10 scale)
SincereF	Female's rating of male's sincerity (1–10 scale)
IntelligentM	Male's rating of female's intelligence (1–10 scale)
IntelligentF	Female's rating of male's intelligence (1–10 scale)
FunM	Male's rating of female as fun (1–10 scale)
FunF	Female's rating of male as fun (1–10 scale)
AmbitiousM	Male's rating of female's ambition (1–10 scale)
AmbitiousF	Female's rating of male's ambition (1–10 scale)
SharedInterestsM	Male's rating of female's shared interests (1–10 scale)
SharedInterestsF	Female's rating of male's shared interests (1–10 scale)

StudentSurvey

Data on 362 introductory statistics students from an in-class survey given over several years. All values are self-reported.
Source: Authors

Year	Year in school: `FirstYear, Sophomore, Junior,` or `Senior`
Gender	Student's gender: `F` or `M`
Smoke	Smoker? `No` or `Yes`
Award	Preferred award: `Academy, Nobel,` or `Olympic`
HigherSAT	Which SAT is higher? `Math` or `Verbal`
Exercise	Hours of exercise per week
TV	Hours of TV viewing per week
Height	Height (in inches)
Weight	Weight (in pounds)
Siblings	Number of siblings
BirthOrder	Birth order, `1` = oldest, `2` = second oldest, etc.
VerbalSAT	Verbal SAT score
MathSAT	Math SAT score
SAT	Combined Verbal + Math SAT
GPA	College grade point average
Pulse	Pulse rate (beats per minute)
Piercings	Number of body piercings

SynchronizedMovement

A study of 264 high school students in Brazil examined the effects of doing synchronized movements (such as marching in step or doing synchronized dance steps) and the effects of exertion on variables, such as pain tolerance and attitudes toward others. Students were randomly assigned to activities that involved synchronized or non-synchronized movements involving high or low levels of exertion. Pain tolerance was measured with a blood pressure cuff, going to a maximum possible reading of 300 mmHg.
Source: Tarr, B., Launay, J., Cohen, E., and Dunbar, R., "Synchrony and exertion during dance independently raise pain threshold and encourage social bonding," *Biology Letters*, 11(10), October 2015.

Sex	f = Female or m = Male
Group	Type of activity. Coded as
	HS+HE, HS+LE, LS+HE, LS+LE for High/Low Synchronizaton + High/Low Exertion
Synch	Synchronized activity? yes or no
Exertion	Exertion level: high or low
PainToleranceBefore	Measure of pain tolerance (mm HG) before activity
PainTolerance	Measure of pain tolerance (mm Hg) after activity
PainTolDiff	Change in pain tolerance (*PainTolerance* − *PainToleranceBefore*)
MaxPressure	Reached the maximum pressure (300 mm Hg) when testing pain tolerance (after)
CloseBefore	Rating of closeness to the group before activity (1 = least close to 7 = most close)
CloseAfter	Rating of closeness to the group after activity (1 = least close to 7 = most close)
CloseDiff	Change on closeness rating (*CloseAfter* − *CloseBefore*)

USStates

Data for all 50 US states.
Source: Various online sources, mostly at *www.census.gov*

State	Name of state
HouseholdIncome	Mean household income (in dollars)
Region	Area of the country: MW = Midwest, NE = Northeast, S = South, or W = West
Population	Number of residents (in millions)
EighthGradeMath	Average score on NAEP Mathematics test for 8th graders (2013)
HighSchool	Percentage of residents (ages 25–34) who are high school graduates
College	Percentage of residents (ages 25–34) with college degrees
IQ	Mean IQ score of residents
GSP	Gross State Product ($1000s per capita in 2013)
Vegetables	Percentage of residents who eat vegetables at least once per day
Fruit	Percentage of residents who eat fruit at least once per day
Smokers	Percentage of residents who smoke

PhysicalActivity	Percentage of residents who do 150+ minutes of aerobic physical activity per week
Obese	Percentage of residents classified as obese
NonWhite	Percentage of residents who are not white
HeavyDrinkers	Percentage of residents who drink heavily
Electoral	Number of votes in the presidential electoral college
ObamaVote	Proportion of votes for Barack Obama in 2012 US Presidential election
ObamaRomney	Which 2012 Presidential candidate won state? R = Romney or O = Obama
TwoParents	Percentage of children living in two-parent households
StudentSpending	School spending (in $1000 per pupil in 2013)
Insured	Percentage of adults (ages 18–64) with health coverage

UNIT A: Data
CHAPTER 1

Section 1.1 Partial Answers

1.1 (a) The people who are asked
    (b) Support the law or not; Categorical

1.3 (a) The teenagers in the sample
    (b) At least five servings or not; Categorical

1.5 (a) The 10 beams
    (b) Force at which each beam broke; Quantitative

1.7 Explanatory: Years smoking;
    Response: Lung capacity

1.9 Explanatory: Number of drinks;
    Response: Blood alcohol content

1.11 (a) *Year* and *HigherSAT* are categorical;
      Other six are quantitative although *Siblings*
      might be either
    (b) Answers will vary
    (c) Answers will vary

1.13 Mock sex or not, categorical, explanatory;
    Time for mating, quantitative, response

1.15 Gene variant or not, Ethnic group;
    Both categorical

1.17 (a) 10 cases; 2 variables: Population (quantitative);
      Hemisphere (categorical)
    (b) Short answer not appropriate

1.19 Explanatory: enriched or not (categorical);
    Response: offspring time to learn (quantitative)

1.21 (a) 6
    (b) 1 categorical; 5 quantitative
    (c) 6 columns, 859 rows

1.23 (a) The 40 people with insomnia
    (b) Which group (categorical);
      Sleep improvement or not (categorical)
    (c) 2 columns and 40 rows

1.25 Cases: people eligible to vote
    Variables: political party and voted or not

1.27 Answers will vary

Section 1.2 Partial Answers

1.29 Population
1.31 Sample
1.33 Sample: 500 Canadian adults;
    Population: All Canadian adults
1.35 Sample: The 1000 households
    Population: All US households with TV
1.37 (a) The 10 selected Twitter accounts
    (b) All twitter accounts
    (c) The author's followers' twitter accounts

1.39 (a) Girls on the selected basketball teams
    (b) All female high school students
    (c) Female HS students on a basketball team

1.41 Yes
1.43 No
1.45 No
1.47 Biased; wording biases the results
1.49 Not biased
1.51 All parents in Kansas City
1.53 (a) Yes
    (b) Yes
1.55 (a) Cases: 6000 restroom patrons observed;
      3 categorical variables: Wash, Gender, Location
    (b) People not always honest in self-reporting
1.57 No, volunteer sample is biased
1.59 (a) No, non-random selection
    (b) How the questions were worded
1.61 The sample of planes that return is biased
1.63 (a) All US residents
    (b) All US emergency room patients
      i. NHANES   ii. NHAMCS
      iii. NHAMCS iv. NHANES
1.65 Answers will vary

Section 1.3 Partial Answers

1.67 Neither
1.69 Association
1.71 Association and causation
1.73 Population
1.75 Snow
1.77 Gender
1.79 Experiment
1.81 Observational study
1.83 Experiment
1.85 Amount of snow and ice on the roads
1.87 Yes
1.89 (a) Yes
    (b) No
1.91 (a) The 2623 schoolchildren
    (b) Amount of greenery
    (c) Test scores
    (d) Yes
    (e) Observational study
    (f) No!
    (g) Short answer not appropriate
1.93 (a) Explanatory: time spent sitting; Response:
      cancer
    (b) Observational study

(c) No

(d) No

1.95 (a) Observational study

(b) No

(c) No

(d) No

1.97 (a) Explanatory: sleep or not; Response: ability to recognize facial expressions

(b) Randomized experiment; matched pairs

(c) Yes

(d) No

1.99 (a) Experiment

(b) Explanatory: color (categorical); Response: rating (quantitative)

(c) Short answer not appropriate

(d) Yes

1.101 Exp: *payment* (randomized), *sex* (not randomized) Resp: *items*, *cost*

1.103 (a) Randomly divide into two groups

(b) Short answer not appropriate

(c) Short answer not appropriate

1.105 Short answer not appropriate

1.107 Answers will vary

CHAPTER 2

Section 2.1 Partial Answers

2.1 0.4669, or 46.69%

2.3 0.1972, or 19.72%

2.5 $p = 0.124$

2.7 $\hat{p} = 0.571$

2.9 Academy Award: 0.086; Nobel Prize: 0.412; Olympic gold medal: 0.503

2.11 (a) $80/200 = 0.40$

(b) $100/200 = 0.5$

(c) $80/100 = 0.80$

(d) $60/80 = 0.75$

2.13 Population; 66,000 soccer games; Whether the home team wins; Categorical; $p = 0.624$

2.15 (a) Sample: 119 players; Population: All RPS players; Variable: the option played

(b) 0.555 (Rock), 0.328 (Paper), 0.118 (Scissors)

(c) Paper

(d) Scissors

2.17 (a) Short answer not appropriate

(b) HS: 38.6%; Some college: 26.3%; College: 19.3%

As education goes up, percent agreeing goes down

(c) 38.7%

(d) 30.7%

2.19 (a) $438/616 = 0.711$

(b) $181/616 = 0.294$

(c) $144/438 = 0.329$

(d) $37/178 = 0.208$

(e) $\hat{p}_A - \hat{p}_N = 0.329 - 0.208 = 0.121$

(f) $144/181 = 0.796$

2.21 $p_H - p_C = 0.045$

2.23 (a) Observational study; Two variables: Dyslexia or not and Gene break or not

(b) 304 rows; 2 columns

(c) Short answer not appropriate

(d) Dyslexia: 0.092; Control: 0.026

(e) There appears to be an association

(f) No, since not from an experiment

2.25 (a) Experiment

(b) Single-blind

(c) Two variables, both categorical

(d) Short answer not appropriate

(e) 75%

(f) $\hat{p}_E - \hat{p}_S = 0.60 - 0.20 = 0.40$

(g) Yes

2.27 (a) 68.6%

(b) 58.0%

(c) 15.7%

(d) 7.2%

2.29 (a) Females; Graph (a)

(b) Approximately equal; Graph (a)

(c) Males; Graph (b)

(d) Females; Graph (b)

2.31 Proportion who accepted in Reward = 0.899 vs Deposit = 0.139

2.33 (a) Answers will vary

(b) Answers will vary

2.35 (a) FY 94, Soph 195, Jr 35, Sr 36

(b) Soph 54.2%

2.37 Short answer not appropriate

2.39 Short answer not appropriate

2.41 Graph (b)

Section 2.2 Partial Answers

2.43 F

2.45 B, C, E

2.47 E, G: Mean ≈ Median; F: Mean > Median; H: Mean < Median

2.49 Answers will vary

2.51 Answers will vary

2.53 (a) $\bar{x} = 11.2$

(b) $m = 12$

(c) No outliers

2.55 (a) $\bar{x} = 24.5$

(b) $m = 20$

(c) 58 is a likely outlier

2.57 $\bar{x} = 2386$

2.59 $\mu = 41.5$

2.61 (a) Mean
(b) Mean = 7.2 mg/kg; Median = 3.65 mg/kg
2.63 (a) mean
(b) greater than
2.65 (a) Population
(b) Skewed to the right, with one outlier
(c) Rough estimate is about 4 million
(d) Rough estimate is about 6 million
2.67 (a) Skewed to the left
(b) About 74
(c) Less than the median
2.69 (a) $\bar{x}_Y = 56.76$ minutes
(b) $\bar{x}_O = 34.69$ minutes
(c) $\bar{x}_Y - \bar{x}_O = 22.07$ minutes
Mice getting young blood ran 22 minutes more, on average
(d) Experiment
(e) Yes
2.71 $m_H - m_C = -399$
2.73 Answers will vary
2.75 Strongly right skewed
2.77 (a) Married women; Never married women
(b) $m = 2$ for married women;
$m = 0$ for never married women

Section 2.3 Partial Answers
2.79 (a) $\bar{x} = 15.09$; $s = 13.30$
(b) $(1, 4, 10, 25, 42)$
2.81 (a) $\bar{x} = 59.73$; $s = 17.89$
(b) $(25, 43, 64, 75, 80)$
2.83 (a) $\bar{x} = 6.50$; $s = 5.58$
(b) $(0, 3, 5, 9.5, 40)$
2.85 (a) V
(b) III
(c) IV
(d) I
(e) VI
(f) II
2.87 (a) W
(b) X
(c) Y
(d) Z
2.89 About 460; About 540
2.91 $\bar{x} \approx 68$; $s \approx 4.5$
2.93 $(58, 65, 68, 70, 77)$
2.95 Skewed right
2.97 Symmetric
2.99 -0.8, standard deviations below mean
2.101 1.55, standard deviations above the mean
2.103 4 to 16
2.105 900 to 2100
2.107 (a) $\mu = 28.766$; $\sigma = 3.369$
(b) 1.880; -2.216
(c) Between 22.028% and 35.504%

2.109 (a) Positive
(b) 0.25
(c) 12.2
(d) 2.4
2.111 (a) 4, 0, 3, -2, -17
(b) Mean -2.4; StDev 8.5
2.113 (a) About 2.6; About 3.4
(b) 2.0
2.115 Mean: 296.44 billion dollars
StDev: 37.97 billion dollars
Interval: 220.50 to 372.38 billion dollars
2.117 (a) Mean: 38.6; StDev: 39.82
(b) $(0, 14, 23, 48, 200)$
(c) Part (b): Five number summary
(d) Skewed to the right
(e) No, not bell-shaped.
2.119 (a) Critical Reading: 0.896
Math: 0.725; Writing: 0.983
(b) Writing; Math
(c) Writing
2.121 (a) 6217; 772
(b) 1796
(c) 680
2.123 Short answer not appropriate
2.125 Short answer not appropriate
2.127 Mean = 51.708; StDev = 26.821
5 Number: $(0, 28, 52, 75, 99)$
2.129 Short answer not appropriate

Section 2.4 Partial Answers
2.131 (a) S
(b) R
(c) Q
(d) T
2.133 (a) Skewed left
(b) 3 low outliers
(c) About 575 or 580; Answers may vary
2.135 (a) Symmetric
(b) No outliers
(c) About 135
2.137 (a) No outliers
(b) Short answer not appropriate
2.139 (a) Outliers: 42, 95, 96, 99
(b) Short answer not appropriate
2.141 (a) Left-skewed
(b) 73% and 98%
(c) $(15, 73, 92, 98, 100)$
(d) Lower
2.143 (a) Categorical; Quantitative
(b) Control Group; Football with Concussion
(c) Yes; Football with Concussion
(d) About 7000 μL
(e) Yes
(f) No; Not an experiment

2.145 (a) Skewed right
(b) 10,000
(c) About 250
(d) Greater than

2.147 (a) Action; Horror and Drama
(b) Action; Drama
(c) Yes

2.149 (a) South; West
(b) No
(c) Yes

2.151 Vitamin use has little effect on retinol levels

2.153 (a) Five number summaries: Individual (12, 31, 39.5, 45.5, 59) Split (22, 40, 46.5, 61, 81). Mean and standard deviation: Individual (37.29, 12.54) Split (50.92, 14.33)
(b) Side-by-side plot shows costs tend to be higher when split.

2.155 (a) Exp: fixed or flexible (categorical); Resp: delay time (quantitative)
(b) Fixed: $\bar{x}_T = 105$ sec, $s_T = 14.1$ sec; Flexible: $\bar{x}_F = 44$ sec, $s_F = 3.4$ sec
(c) Difference: $\bar{x}_D = 61$ sec, $s_D = 15.2$ sec
(d) Three large outliers

2.157 Answers will vary

2.159 Answers will vary

Section 2.5 Partial Answers

2.161 (c)

2.163 (a)

2.165 (a)

2.167 (b)

2.169 Negative

2.171 Positive

2.173 Positive

2.175 Short answer not appropriate

2.177 $r = -0.932$

2.179 (a) Positive
(b) Positive
(c) Short answer not appropriate
(d) No; Yes

2.181 (a) Positive: BMI, cortisol, depression, heart rate; Negative: weekday sleep hours, activity levels
(b) No

2.183 (a) Pre-season success = more wins in regular season, Pre-season success = fewer wins in regular season
(b) Almost no relationship

2.185 Variables are not quantitative

2.187 (a) Positive
(b) Low happiness, low use of resources
(c) About 2.0
(d) Yes; No; More resources mean greater happiness but only up to a point
(e) High on happiness, low on resource use

(f) Answers will vary
(g) Short answer not appropriate

2.189 (a) Three quantitative variables
(b) Negative; Yes
(c) Positive; No; Yes

2.191 (a) Positive: spend time on both or neither; Negative: spend time on one or the other but not both
(b) Lots of exercise and little TV; Lots of time on both; Very little time doing either; Lots of TV and little exercise
(c) Lots of TV and little exercise; Lots of exercise and little TV
(d) Almost no linear relationship

2.193 (a) Age 20
(b) Early 20s
(c) 0

2.195 (a) Positive
(b) Relatively strong
(c) $r \approx 0.9$
(d) No
(e) 11 mm
(f) Two clumps; one type has smaller petals

2.197 (a) Short answer not appropriate
(b) Positive; Quite strong
(c) DeAndre Jordan, Andre Drummond
(d) 0.800

2.199 Answers will vary

Section 2.6 Partial Answers

2.201 (a) 0.0413, 0.0387
(b) BAC increases about 0.018 per drink
(c) Consuming no drinks, BAC is -0.0127

2.203 (a) 79; 2
(b) Grade goes up 3.8 for one more hour of studying
(c) Expected grade is 41 if there is no studying

2.205 $\hat{Y} = 47.267 + 1.843X$

2.207 $\hat{Y} = 641.62 - 8.42X$

2.209 (a) Duration; Distance
(b) Yes; Positive
(c) $r = 0.994$
(d) $\widehat{Distance} = -399 + 1174 \cdot Duration$
(e) Short answer not appropriate
(f) 775 meters; 3123 meters

2.211 (a) Negative
(b) Slope
(c) Explanatory: elevation; Response: cancer incidence

2.213 (a) 18; 2700; 2400; -300
(b) 12; 3000; 3900; 900
(c) 13; 2900; 2200; -700
Answers may vary slightly

2.215 Short answer not appropriate

2.217 (a) Exp: pre-season wins;
  Resp: regular season wins
  (b) 7.9 wins
  (c) 0.2
  (d) 7.5
  (e) 27.5 wins, extrapolation
2.219 190 lbs, 40% body fat;
  Predicted body fat = 20%; Residual 20
2.221 (a) 13.35%; 22.1%
  (b) Short answer not appropriate
  (c) −8.525%
2.223 (a) Strong linear trend
  (b) $\widehat{Happiness} = -1.09 + 0.103 \cdot LifeExpectancy$
  (c) Short answer not appropriate
2.225 (a) Three variables: Growth, Ratio, Deficit
  (b) $\widehat{Ratio} = 129 - 19.1(Growth)$
  (c) Short answer not appropriate
  (d) 90.8%; 52.6%
  (e) 2.04%
  (f) $\widehat{Deficit} = 2765 - 680(Growth)$
  (g) Short answer not appropriate
  (h) $1.405 trillion; $45 billion
  (i) 2.007%

Section 2.7 Partial Answers

2.227 (a) *Happiness:* Quantitative;
  Footprint: Quantitative;
  Region: Categorical
  (b) 1 (Latin America) and 2 (Western nations);
  2 (Western nations)
  (c) 4 (Sub-Saharan Africa);
  low ecological footprint
  (d) Yes
  (e) No
  (f) Top left
  (g) 4: Increase happiness;
  2: Decrease footprint
2.229 (a) Hippocampus size: Quantitative;
  Years of football: Quantitative;
  Group: Categorical
  (b) Control group all 0 football years
  (c) Negative association
  (d) Football with concussion
  (e) Football with no concussion
2.231 (a) Faster
  (b) DC
  (c) No
2.233 (a) 7
  (b) 2
2.235 (a) Asia
  (b) Africa
  (c) Short answer not appropriate
2.237 (a) 6
  (b) 10–14%; 20–24%; ≥ 30%

2.239 (a) Every state
  (b) 2014
  (c) Mississippi, 15.0%;
  Colorado, 21.3%
2.241 (a) Texas
  (b) Midwest
2.243 (a) East
  (b) Answers will vary
2.245 (a) Two distinct clusters
  (b) Setosa; Virginica
2.247 Sepal length; Green
2.249 (a) Whites
  (b) Hispanics
  (c) Blacks
2.251 (a) Distance higher for carbon
  (b) Not a short answer
  (c) Distance
  (d) Ride carbon; minimize distance
2.253 (a) New England
  (b) South Central
  (c) Mountain; Midwest
  (d) Spaghetti plot, dynamic heat map
2.255 Answers will vary.
2.257 Short answer not appropriate
2.259 (a) Early morning (6 am)
  (b) Summer
2.261 (a) Endangered status of gazelles
  (b) Answers will vary
2.263 Answers will vary
2.265 Answers will vary.
2.267 Answers will vary
2.269 Answers will vary
2.271 Answers will vary

UNIT A: Essential Synthesis Partial Answers

A.1 (a) Yes
  (b) No
A.3 (a) No
  (b) Yes
A.5 (a) No
  (b) No
A.7 (a) One categorical variable
  (b) Bar chart or pie chart
  (c) Frequency or relative frequency table,
  proportion
A.9 (a) One categorical variable and one quantitative
  variable
  (b) Side-by-side boxplots, dotplots,
  or histograms
  (c) Statistics by group or difference
  in means
A.11 (a) Two categorical variables
  (b) Segmented or side-by-side bar charts
  (c) Two-way table or difference in proportions

A.13 (a) One quantitative variable
(b) Histogram, dotplot, or boxplot
(c) Mean, median, standard deviation, range, IQR

A.15 (a) Two quantitative variables
(b) Scatterplot
(c) Correlation or slope from regression

A.17 (a) Experiment
(b) Subjects can see which treatment
(c) Sample is 46 subjects; Population answers may vary
(d) One quantitative, one categorical
(e) Side-by-side boxplots

A.19 (a) The students/computers; 45; Not random
(b) Observational study
(c) Four variables, all quantitative
(d) Histogram, dotplot, or boxplot; Boxplot
(e) Scatterplot; Correlation; Negative
(f) No, not an experiment
(g) Explanatory: time on distracting websites; Response: exam score
(h) Randomized experiment

A.21 (a) Sample: 86 patients; Population: all people with bladder cancer
(b) Two categorical variables
(c) Experiment
(d) Two-way table
(e) Yes, the drug appears to be more effective

A.23 (a) Cases are the bills; $n = 157$
(b) Seven variables;
Bill, Tip, PctTip are quantitative;
Credit, Server, Day are categorical;
Guests could be either
(c) $\bar{x} = 16.62$; $s = 4.39$; (6.7, 14.3, 16.2, 18.2, 42.2);
Values below 8.45 and above 24.05 are outliers
(d) Short answer not appropriate
(e) $\hat{p}_{th} = 0.333$; $\hat{p}_f = 0.154$;
Appears to be an association
(f) Side-by-side graphs; Server A
(g) *Bill; PctTip*; Slightly positive relationship
(h) $r = 0.135$

UNIT A: Review Exercise Partial Answers

A.25 Sample; Answers will vary

A.27 (a) Sample: 48 men; Population: all men
(b) Which group; Protein conversion; Age
(c) Categorical; Quantitative; Quantitative

A.29 (a) The 41 participants
(b) One categorical variable (meditation or not); at least 9 quantitative variables.
(c) Meditation or not
(d) 41 rows and at least 10 columns

A.31 Sample: The fish in that day's catch
Population: All fish in that area

A.33 No, volunteer sample is biased;
Sample: 38,485 people who voted

A.35 (a) Short answer not appropriate
(b) Answers will vary

A.37 Short answer not appropriate

A.39 Snow is associated with colder days;
Shoveling snow leads to back pain

A.41 (a) Students; type of dorm (categorical);
number of hook-ups (quantitative)
(b) Explanatory: type of dorm;
Response: number of hook-ups
(c) Yes
(d) Yes
(e) Observational studies
(f) Students self-select the type of dorm
(g) No!
(h) Assuming causation when he shouldn't be

A.43 (a) Students; 70
(b) Treatment group; Three ratings
(c) Experiment
(d) Short answer not appropriate
(e) Side-by-side boxplots; Scatterplot

A.45 Answers will vary

A.47 (a) Experiment
(b) Short answer not appropriate
(c) $\bar{x}_S - \bar{x}_N = 1.9$
(d) Yes

A.49 Cardiac arrest: 9.5%; Other: 1.1%

A.51 (a) Observational study; No
(b) Women attempting to become pregnant
(c) $\hat{p} = 0.36, \hat{p}_s = 0.28, \hat{p}_{ns} = 0.38$
(d) $\hat{p}_{ns} - \hat{p}_s = 0.10$

A.53 (a) Sample: 48 participants;
Population: All people;
Variable: Whether a person's lie is detected
(b) $\hat{p} = 0.35$
(c) No

A.55 Yes; $\bar{x} \approx 7$; $s \approx 2$.

A.57 (a) There are likely low outliers
(b) Short answer not appropriate
(c) Skewed to the left

A.59 (a) Skewed to right with outliers on right
(b) median ≈ 140; estimates may vary
(c) mean ≈ 190, estimates may vary

A.61 $m = 1$; $\bar{x} = 3.2$

A.63 (a) $\bar{x} = 0.272$; $s = 0.237$
(b) 2.44
(c) (0.073, 0.118, 0.158, 0.358, 0.851)
(d) Range $= 0.778$; $IQR = 0.24$

A.65 (a) $\bar{x} = 45$ for both
(b) Different standard deviation

A.67 (a) 47; 48
(b) 72.6%
(c) 19.2%

(d) 56.25%

(e) Rosiglitazone: 89.4%;
   Placebo: 56.3%

(f) Drug appears effective at reducing blockage

A.69 (a) Experiment

(b) Variables: cancer or not and which drug

(c) Short answer not appropriate

(d) 48.2%

(e) 58.7%

(f) 18.4% and 24.4%; Yes

A.71 (a) 0.306

(b) 0.417

(c) 0.324

(d) 0.286

A.73 (a) Yes

(b) Short answer not appropriate

(c) Yes; No

(d) Short answer not appropriate

A.75 Teens: $(100, 104, 130, 140, 156)$;
   Range $= 56$; IQR $= 31$; $s = 19.57$
   80s: $(80, 110, 135, 141, 190)$;
   Range $= 110$; IQR $= 36$; $s = 31.23$
   Variability is less for the teens

A.77 (a) Four outliers, at $402, 447, 511, 536$

(b) Short answer not appropriate

A.79 (a) Males; Males; Females

(b) Yes; Males much higher calorie consumption

A.81 Blood pressures slightly higher for survivors;
   Descriptions will vary

A.83 (a) Yes, distribution is bell-shaped

(b) 66.38 and 198.18

(c) $186/200 = 93\%$

(d) Yes

A.85 (a) Short answer not appropriate

(b) $r = -0.071$

(c) No

A.87 (a) Short answer not appropriate

(b) $r = -0.189$

(c) Short answer not appropriate

(d) $r = 0.836$

(e) Very substantial effect

A.89 (a) Positive

(b) A: short and light
   B: tall and heavy
   C: short and heavy
   D: tall and thin

A.91 (a) Positive: tall people tend to weigh more;
   Negative: tall people tend to weigh less;
   Expect a positive relationship

(b) Positive; Moderately strong;
   Approximately linear

(c) Very tall thin person!

A.93 (a) Short answer not appropriate

(b) -0.096

(c) 0.562

(d) Yes

A.95 (a) Land area; Percent rural

(b) 0.60

(c) $\widehat{Rural} = 30.35 + 0.075(LandArea)$

(d) No

(e) Uzbekistan (UZB)

(f) 716%; Not reasonable; Extrapolate too far

A.97 (a) No aggressive male: $\bar{x} = 0.044$, $s = 0.062$;
   Aggressive male: $\bar{x} = 0.390$, $s = 0.288$

(b) Short answer not appropriate

(c) $\widehat{MatingActivity} = 0.48 - 0.323 \cdot FemalesHiding$

(d) No hyper-aggressive male: 0.466;
   Hyper-aggressive male: 0.354

(e) Don't hang out with hyper-aggressive males!

A.99 (a) Yes; Positive; No

(b) Income

(c) 67 thousand; 48%; 37%

(d) 71 thousand; 25%; 40%

A.101 Short answer not appropriate

UNIT B: Understanding Inference
CHAPTER 3

Section 3.1 Partial Answers

3.1 Parameter; μ

3.3 Statistic; $\hat{p}$

3.5 Statistic; $\bar{x}$

3.7 $\mu = 401.3$

3.9 $r = 0.037$

3.11 $\rho = -0.147$

3.13 $\mu = 85$; $SE \approx 20$

3.15 $p = 0.80$; $SE \approx 0.03$

3.17 (a) (*i*)

(b) (*i*)

(c) (*ii*)

3.19 (a) (*ii*)

(b) (*ii*)

(c) (*iii*)

3.21 (a) $\mu =$ number apps downloaded for all US
   smartphone users

(b) $\bar{x} = 19.7$

(c) Ask all smartphone users in the US!

3.23 (a) ρ; r; -0.575

(b) Would need to test all 7700 lakes

3.25 (a) Biased: B and C

(b) n = 100: A, n = 500: D

3.27 $\mu_m - \mu_o$; $\bar{x}_m - \bar{x}_o$; best estimate = 39

3.29 (a) Single sample: Boxplot B
   Sampling distribution: Boxplot A

(b) Each value is the budget for one Hollywood
   movie;
   30 values from about 1 to 225 million;
   $\bar{x} \approx 53$ million dollars

(c) Each value is a sample mean;
1000 values from about 30 to 90 million;
$\mu \approx 56$ million dollars

3.31 (a) $\bar{x}$; Values will vary

(b) $\bar{x}$; Values will vary

(c) $\mu = 53.54$

(d) Roughly symmetric and centered at 53.54

3.33 Symmetric and bell-shaped;
Centered at approximately 54;
Standard error is approximately 10.9

3.35 (a) $\mu = 56.1$, $\sigma = 53.8$

(b) Bell-shaped; centered near 56.1; $SE \approx 12$

3.37 (a) $p = 0.155$

(b) Bell-shaped; centered at 0.155

3.39 (a) $SE \approx 0.11$;
Farthest $\hat{p}$ will vary.

(b) $SE \approx 0.08$;
Farthest $\hat{p}$ will vary.

(c) $SE \approx 0.05$;
Farthest $\hat{p}$ will vary.

(d) Increasing n increases accuracy

Section 3.2 Partial Answers

3.41 (a) p

(b) $\hat{p}$

3.43 (a) $\mu_1 - \mu_2$

(b) $\bar{x}_1 - \bar{x}_2$

3.45 22 to 28

3.47 0.57 to 0.67

3.49 (a) Yes

(b) Yes

(c) No

3.51 0.24 to 0.40; p

3.53 0.30 to 0.38; ρ

3.55 1.8 to 4.2; $\mu_1 - \mu_2$

3.57 (a) $\mu_A - \mu_T$

(b) $\bar{x}_A - \bar{x}_T = 0.166$

(c) -0.016 to 0.348

(d) Observational study

3.59 (a) $p_m - p_f$

(b) $\hat{p}_m - \hat{p}_f = 0.417$

(c) 0.109 to 0.725

(d) No

3.61 (a) Statistic; $\hat{p} = 0.30$

(b) Proportion, p, of *all* US young people arrested by 23; $\hat{p} = 0.30$

(c) 0.29 to 0.31

(d) Very unlikely

3.63 Short answer not appropriate

3.65 Point estimate: $\hat{p} = 0.28$; Margin of error: ± 0.018;
95% CI: 0.262 to 0.298

3.67 (a) Short answer not appropriate

(b) Not the same; Game players are faster

(c) Short answer not appropriate

(d) Yes; Similar accuracy is plausible

3.69 (a) No

(b) Yes

(c) Yes

(d) No

3.71 46.2 to 55.8 seconds; No, Yes

3.73 Parameter: $p_1 - p_2$;
Point estimate: $\hat{p}_1 - \hat{p}_2 = 0.44$;
95% CI: 0.16 to 0.72;
Not likely

Section 3.3 Partial Answers

3.75 (a) No

(b) No

(c) Yes

(d) No

(e) Yes

3.77 Point estimate ≈ 0.7; $SE \approx 0.1$;
Interval: 0.5 to 0.9; Parameter: p

3.79 Point estimate ≈ 0.4; $SE \approx 0.05$;
Interval: 0.3 to 0.5; Parameter: ρ

3.81 $SE \approx 0.048$;
Interval: 0.254 to 0.446

3.83 $SE \approx 0.022$;
Interval: 0.236 to 0.324

3.85 (a) $\hat{p} = 0.149$

(b) $SE \approx 0.028$

(c) 0.093 to 0.205

(d) Yes

3.87 (a) Data: Dotplot I; Bootstrap: Dotplot II

(b) $\bar{x} \approx 7600$

(c) $SE \approx 200$

(d) Larger

(e) 7200 to 8000 μL

3.89 Point estimate: $\hat{p}_F - \hat{p}_M = 0.292$;
$SE \approx 0.094$; Interval: 0.104 to 0.480;
Equally compassionate not plausible

3.91 $SE \approx 7.9$; 1.32 to 29.92; Tea

3.93 (a) $\bar{x} = 2356$, $s = 858$

(b) Bell-shaped, mean ≈ 2356, SE ≈ 187

(c) about $(1982, 2730)$

3.95 (a) $\hat{p} = \frac{24}{273} = 0.088$.

(b) $(0.054, 0.122)$

(c) $\hat{p} = \frac{42}{477} = 0.088$.

(d) $(0.064, 0.114)$

(e) Migraine CI is narrower; n is larger

3.97 (a) $\mu_S - \mu_N$

(b) $\bar{x}_S - \bar{x}_N = 0.465$

(c) $SE \approx 0.228$

(d) 0.009 to 0.921

3.99 (a) $\mu_H - \mu_L$

(b) $\bar{x}_H - \bar{x}_L = 12.717$

(c) $SE \approx 9.0$

(d) -5.283 to 30.717

3.101 (a) $p_F - p_M$

(b) $\hat{p}_F - \hat{p}_M = -0.224$

(c) $SE \approx 0.059$

(d) -0.342 to -0.106

3.103 16.9 to 37.7 minutes

Section 3.4 Partial Answers

3.105 (a) 25

(b) 50

(c) 10

(d) 5

3.107 C

3.109 A

3.111 B

3.113 Approximately 0.66 to 0.78;
Answers may vary

3.115 Approximately 0.34 to 0.42;
Answers may vary

3.117 22.8 to 47.7

3.119 Approximately 0.467 to 0.493; Yes

3.121 (a) $(15.8, 32.4)$

(b) Not a short answer

3.123 (a) 0.36

(b) $(0.28, 0.45)$

(c) Not a short answer

3.125 (a) $p_T - p_N$

(b) $\hat{p}_T - \hat{p}_N = 0.366$

(c) Approximately 0.100 to 0.614.

(d) Yes; No

3.127 (a) 90%: A; 99%: B

(b) 90%: \$71 to \$83 in a day;
99%: \$67 to \$87 in a day

3.129 (a) All FA Premier League matches;
Proportion of home team victories

(b) 0.583

(c) 0.508 to 0.650

(d) 0.467 to 0.692

(e) Yes; No

3.131 Around -21.2 to -6.1

3.133 (a) ρ; $r = 0.21$

(b) Select ordered pairs, with replacement

(c) Sample correlation, r

(d) $SE =$ standard deviation of the correlations

(e) -0.07 to 0.49

(f) Zero correlation is plausible

(g) Narrower

3.135 Not appropriate

CHAPTER 4

Section 4.1 Partial Answers

4.1 (a) Sample A

(b) Sample C

4.3 (a) Sample A

(b) Samples B and C

4.5 $H_0 : \mu_A = \mu_B$ vs $H_a : \mu_A \neq \mu_B$

4.7 $H_0 : \mu = 50$ vs $H_a : \mu < 50$

4.9 $H_0 : p_m = p_f$ vs $H_a : p_m > p_f$

4.11 $H_0 : p = 0.20$ vs $H_a : p < 0.20$

4.13 $H_0 : \mu_f = \mu_u$ vs $H_a : \mu_f \neq \mu_u$

4.15 (a) valid

(b) invalid

(c) invalid

(d) invalid

4.17 (a) $H_0 : \mu_b = \mu_w$ vs $H_a : \mu_b > \mu_w$

(b) 17.36 and 18.72; Yes, Yes

(c) 23.60 and 19.17; Yes, Probably not

(d) Drinking beer attracts mosquitoes!

(e) Yes, since it was an experiment

4.19 (a) $H_0 : \mu_e = \mu_s$ vs $H_a : \mu_e < \mu_s$

(b) $H_0 : \mu_e = \mu_s$ vs $H_a : \mu_e > \mu_s$

(c) $H_0 : \rho = 0$ vs $H_a : \rho < 0$

4.21 $H_0 : \mu_m = \mu_f$ vs $H_a : \mu_m > \mu_f$

4.23 $H_0 : \rho = 0$ vs $H_a : \rho > 0$

4.25 $H_0 : \mu = 50$ vs $H_a : \mu > 50$

4.27 (a) $H_0 : \rho = 0$ vs $H_a : \rho \neq 0$

(b) $r = 0.75$

(c) Same (just opposite directions)

4.29 (a) H_0: Muriel's guesses are no better than random;
H_a Muriel's guesses are better than random.

(b) $H_0 : p = 0.5$ vs $H_a : p > 0.5$

4.31 Not a test

4.33 Not a test

4.35 Not a test

4.37 Not a test

4.39 (a) $H_0 : p_c = p_f$ vs $H_a : p_c > p_f$

(b) Answers will vary

(c) Answers will vary

(d) Answers will vary

Section 4.2 Partial Answers

4.41 $\bar{x}$

4.43 $\bar{x}_1 - \bar{x}_2$

4.45 0.5; Left-tail

4.47 0; Two-tail

4.49 0; Right-tail

4.51 (a) (*iii*)

(b) (*i*)

(c) (*ii*)

4.53 (a) 0.30

(b) 0.04

(c) 0.70

4.55 (a) Short answer not appropriate

(b) $D = 2.8$

4.57 (a) Short answer not appropriate

(b) $\bar{x}_1 = 19.0, \bar{x}_2 = 15.4$

4.59 p-value ≈ 0.10

4.61 p-value ≈ 0.01

4.63 p-value ≈ 0.12

4.65 (a) $H_0 : \mu_c = \mu_n$ vs $H_a : \mu_c > \mu_n$

(b) 0.11

(c) 0.03

(d) $\bar{x}_c - \bar{x}_n = 2.4$

4.67 (a) Short answer not appropriate

(b) $\hat{p}_c - \hat{p}_f = 0.2$, p-value = 0.392

$\hat{p}_c - \hat{p}_f = -0.4$, p-value = 0.066

(c) $\hat{p}_c - \hat{p}_f = -0.4$

4.69 (a) $H_0 : p = 0.01$ vs $H_a : p < 0.01$

(b) $\hat{p} = 0.0046$.

(c) p-value = 0.004

4.71 (a) $H_0 : \mu_C = \mu_F$ vs $H_a : \mu_C > \mu_F$

(b) $\bar{x}_C - \bar{x}_F = 7602.6 - 6459.2 = 1143.4$

(c) p-value ≈ 0.000

(d) Very unlikely to be just random chance

4.73 (a) $H_0 : p_M = p_W$ vs $H_a : p_M \neq p_W$

(b) $\hat{p}_M - \hat{p}_W = 0.696 - 0.663 = 0.033$; Men

(c) p-value ≈ 0.10

4.75 (a) $\hat{p}_T = 0.649$; $\hat{p}_N = 0.283$

$\hat{p}_T - \hat{p}_N = 0.366$

(b) $H_0 : p_T = p_N$ vs $H_a : p_T > p_N$

(c) p-value ≈ 0.001

4.77 Short answer not appropriate

4.79 (a) 5; -2; Can't have negative values

(b) 0.008

(c) 0.046; no zeros

Section 4.3 Partial Answers

4.81 0.04

4.83 0.0008

4.85 Reject H_0

4.87 Do not reject H_0

4.89 No; No; No

4.91 Yes; No; No

4.93 The results are significant if the p-value is less than the significance level.

(a) II. 0.0571

(b) I. 0.00008

(c) IV. 0.1753

(d) III. 0.0368

4.95 (a) 0.0001; 0.7

(b) Yes; Randomized experiment

4.97 (a) Short answer now appropriate

(b) Supplier A

(c) Supplier B

4.99 (a) $H_0 : \mu_M = \mu_F$ vs $H_a : \mu_M \neq \mu_F$

(b) Do not reject H_0; No evidence of a difference

4.101 (a) Exp: Antibiotics or not;

Resp: Overweight or not

(b) Observational study

(c) $H_0 : p_A = p_N$ vs $H_a : p_A > p_N$

(d) $\hat{p}_A - \hat{p}_N = 0.142$

(e) Reject H_0; Strong evidence

(f) No

4.103 (a) Evidence that price affects effectiveness

(b) Short answer not appropriate

4.105 (a) $H_0 : p = 0.5$ vs $H_a : p \neq 0.5$

(b) Do not reject H_0; No

(c) Reject H_0; Yes

4.107 (a) Randomized experiment

(b) $H_0 : \mu_{FA} = \mu_{PM}$ vs $H_a : \mu_{FA} \neq \mu_{PM}$

(c) Yes, very strong evidence to reject H_0

(d) Essentially zero

(e) Diabetes is influenced by pollution

4.109 (a) Strong evidence that exercise improves performance

(b) Strong evidence that exercise reduces BMP

(c) Very strong evidence that exercise increases noggin

(d) Noggin level

(e) Exercise has a significant positive effect on brain function in mice

4.111 (a) $H_0 : p_E = p_N$ vs $H_a : p_E > p_N$; Sample statistic is 0.40; p-value ≈ 0.01; People getting electrical stimulation solve the problem more readily

(b) Yes, since results are significant and experiment allows a causal conclusion

4.113 (a) $H_0 : p = 1/3$ and $H_a : p \neq 1/3$

(b) p-value ≈ 0.000

(c) US citizens do *worse* than random guessing

4.115 (a) $H_0 : \mu_D = \mu_U$, $H_a : \mu_D \neq \mu_U$

(b) -2.6 and 9.8

(c) p-value ≈ 0

(d) Reject H_0

4.117 (a) $H_0 : \mu_i = \mu_u$, $H_a : \mu_i > \mu_u$

(b) p-value ≈ 0.0002

(c) Mean twitches is higher for invaded habitat

4.119 (a) No

(b) Eliminate some confounding variables

(c) 0.098

(d) p-value ≈ 0.004

(e) Yes

Section 4.4 Partial Answers

4.121 (a) p-value ≈ 0.16; Not significant

(b) p-value ≈ 0.0002; Significant

Strongest evidence when $n = 500$

4.123 (a) p-value ≈ 0.27; Not significant

(b) p-value ≈ 0.001; Significant

(c) p-value ≈ 0.000; Significant

Strongest evidence when $n_1 = n_2 = 2000$

4.125 3

4.127 40

4.129 (a) Not valid
(b) Valid
(c) Valid

4.131 (a) $H_0 : p_U = p_D, H_a : p_U \neq p_D$
(b) $\hat{p}_D - \hat{p}_U$, -0.093
(c) p-value $\approx 0.$
(d) Reject H_0, proportion late is smaller for Delta
(e) No, p-value ≈ 0.12.

4.133 Small, $\alpha = 0.01$

4.135 Large, $\alpha = 0.10$

4.137 Large, $\alpha = 0.10$

4.139 Short answer not appropriate

4.141 Short answer not appropriate

4.143 Short answer not appropriate

4.145 (a) Reject H_0; No error or Type I error
(b) Do not reject H_0; No error or Type II error
(c) Need to know the actual value of the parameter

4.147 (a) $H_0 : \mu_I = \mu_C$ vs $H_a : \mu_I > \mu_C$
(b) Kindergartners with iPads do better; Yes, statistically significant
(c) Short answer not appropriate

4.149 (a) One or two tests
(b) A Type I error is likely
(c) No, not an experiment

4.151 (a) We should be less confident; The problem of multiple tests
(b) No
(c) Yes

4.153 (a) 0.10
(b) Insufficient evidence
(c) 0.21
(d) Insufficient evidence
(e) Type II error
(f) Type I error

Section 4.5 Partial Answers

4.155 (a) Confidence interval
(b) Hypothesis test
(c) Hypothesis test
(d) Inference is not relevant

4.157 (a) Do not reject H_0, $\alpha = 0.05$
(b) Reject H_0, $\alpha = 0.05$
(c) Do not reject H_0, $\alpha = 0.10$

4.159 (a) Reject H_0; $\alpha = 0.05$; Positive
(b) Reject H_0; $\alpha = 0.10$; Negative
(c) Do not reject H_0, $\alpha = 0.01$

4.161 (a) Do not reject H_0, $\alpha = 0.05$
(b) Reject H_0, $\alpha = 0.05$
(c) Reject H_0, $\alpha = 0.05$

4.163 (a) Reject H_0, $\alpha = 0.10$
(b) Reject H_0, $\alpha = 0.05$
(c) Do not reject H_0, any reasonable α

4.165 (a) Observational study
(b) $\hat{p} = 0.738$

(c) Short answer not appropriate
(d) $H_0 : p = 0.5$ vs $H_a : p > 0.5$
(e) One-tailed test
(f) Reject H_0
(g) Reject H_0, melanomas more likely on left
(h) No, not an experiment.

4.167 (a) Reject H_0; 5% level
(b) Do not reject H_0; 5% level

4.169 (a) 0.508 to 0.658
(b) $H_0 : p = 0.5$ vs $H_a : p \neq 0.5$
(c) Reject H_0 at 10% level
(d) p-value ≈ 0.082
(e) Reject H_0 at 10% level; Yes
(f) Short answer not appropriate
(g) Bootstrap centered at $\hat{p} = 0.583$, randomization centered at $H_0 : p = 0.5$

4.171 (a) $H_0 : \mu = 10$ vs $H_a : \mu \neq 10$; p-value ≈ 0.01; Reject H_0
(b) Not included

4.173 Answers will vary

4.175 p-value ≈ 0.12; Do not reject H_0; No

4.177 (a) $H_0 : \rho = 0, H_a : \rho > 0$
(b) Short answer not appropriate

4.179 $H_0 : \rho = 0$ vs $H_a : \rho > 0$; p-value ≈ 0.02; At 5% level, reject H_0; Positively correlated

4.181 (a) p-value ≈ 0.004; Reject H_0; Strong evidence that desipramine works better
(b) Desipramine, much smaller p-value

4.183 (a) $H_0 : \mu_Q = \mu_L$ vs $H_a : \mu_Q > \mu_L$
(b) $\bar{x}_D = 2.7$
(c) Answers will vary; many possible methods
(d) Answers will vary
(e) Answers will vary

4.185 (a) Inappropriate; doesn't match $H_0 : p = 0.8$
(b) Appropriate

4.187 p-value ≈ 0.20 for each method; Do not reject H_0.

UNIT B: Essential Synthesis Partial Answers

B.1 (a) Reject H_0
(b) Short answer not appropriate
(c) Randomize, placebo, double-blind
(d) Vitamin C reduces mean time to recover

B.3 (a) $H_0 : \mu_{dc} = \mu_w$ vs $H_a : \mu_{dc} > \mu_w$; p-value ≈ 0.005, reject H_0
(b) 95% confidence interval for $\mu_{dc} - \mu_w$ is (2.88, 10.75)

B.5 (a) Roommates are assigned at random
(b) $H_0 : \mu_v = \mu_n$ vs $H_a : \mu_v < \mu_n$
(c) Reject H_0
(d) Negative differences indicate $\mu_v < \mu_n$
(e) Do not reject H_0

(f) Reject H_0

(g) Larger effect on those who bring a videogame themselves

(h) Short answer not appropriate

(i) More videogames associated with lower mean GPA

(j) Answers will vary

B.7 (a) Positive

(b) $r = 0.914$

(c) 0.88 to 0.94

(d) No

UNIT B: Review Exercise Partial Answers

B.9 p; $\hat{p}$; 0.55

B.11 (a) All American adults, p, $\hat{p} = 0.57$

(b) 0.54 to 0.60

B.13 (a) $p = 0.256$

(b) Bell-shaped; centered at 0.256

B.15 (a) 0.05

(b) About 0 to 0.12; About 0.25 to 0.7

(c) $SE \approx 0.02$; $SE \approx 0.005$

(d) Yes; No

B.17 (a) Answers will vary

(b) Answers will vary

(c) $\mu = 25.67$ points

(d) Roughly symmetric and centered at 25.67

B.19 Minimum ≈ 5 to maximum ≈ 50; st.dev. about 8.4. Answers will vary

B.21 0.0004: effect of ringing phone on learning; 0.93: effect of proximity to phone; Strong evidence that ringing phone affects learning

B.23 $H_0 : p_F = p_N$ vs $H_a : p_F < p_N$; p-value ≈ 0.09; No; Yes

B.25 (a) Answers vary, pick 10 values well above 100

(b) Answers vary, pick 10 numbers with $\overline{x} < 100$

(c) Answers vary, pick 10 numbers with $\overline{x}$ just a bit over 100.

B.27 (a) $\overline{x} = 67.59$; $s = 50.02$

(b) Short answer not appropriate

(c) Bell-shaped, centered at 67.59

(d) $45.79 to $89.39

B.29 (a) (15.1,35.5)

(b) $s = 27.3$; No

B.31 (a) 0.075

(b) (−0.006, 0.142)

(c) Answers vary

B.33 (a) One-tailed

(b) $H_0 : p_s = p_{ns}$ vs $H_a : p_s < p_{ns}$

(c) 135 for smoking, 543 for non-smoking

(d) 0.04

B.35 (a) Sample A

(b) Sample B

(c) Sample A

B.37 (a) $p = 0.24$

(b) Bell-shaped; centered at 0.24; $SE \approx 0.068$

B.39 (a) Same

(b) Different

(c) Same

(d) Different

(e) Different

B.41 (a) $H_0 : p = 0.5$ vs $H_a : p > 0.5$

(b) Answers will vary

(c) The proportion of heads is $p = 0.5$

B.43 $p = 0.5$, the proportion for H_0

B.45 (a) American adults; the 7293 contacted

(b) Short answer not appropriate

(c) 0.585 with margin of error 0.012

B.47 (a) $H_0 : \mu = 160$ vs $H_a : \mu < 160$

(b) Short answer not appropriate

B.49 (a) Plausible

(b) Plausible

(c) Plausible

B.51 (a) $H_0 : \rho = 0$ vs $H_a : \rho > 0$

(b) Yes

(c) No

(d) Placebo could give as extreme a correlation

B.53 (a) p-value = 0.027

(b) No change in distribution; p-value = 0.054

B.55 (a) $\hat{p} = 0.57$

(b) 0.43 to 0.71

(c) No

B.57 (a) Reject H_0

(b) Reject H_0

(c) Reject H_0

(d) Do not reject H_0

(e) Body mass gain; Locomotor activity

(f) Answers to (a) and (c) change

(g) Strong; Very strong; Strong; None

(h) Yes, if the 6 lemurs are a random sample

B.59 0.185 to 0.424

B.61 (17.2, 19.2)

B.63 (a) Short answer not appropriate

(b) (0.706, 0.876)

(c) 95%: (0.729, 0.867); 90%: (0.742, 0.859)

(d) Interval gets narrower

B.65 (a) Once

(b) Yes, p-value is very small

(c) No

B.67 (a) $H_0 : \rho = 0$ vs $H_a : \rho < 0$

(b) $r \approx -0.15$

(c) $r \approx -0.50$

B.69 (a) $H_0 : p_1 = p_2$ vs $H_a : p_1 > p_2$

(b) $\hat{p}_1 = 0.519$; $\hat{p}_2 = 0.250$; Yes

(c) Small α, such as 0.01

(d) Pain response same either way; 0

(e) $\hat{p}_1 - \hat{p}_2 = 0.27$

(f) Fairly strong evidence to reject H_0

(g) At 1% level, do not reject H_0

B.71 (a) $H_0 : p_O = 0.5$ vs $H_a : p_O > 0.5$
where p_O is the proportion supporting Obama

(b) Short answer not appropriate

B.73 (a) Short answer not appropriate

(b) Short answer not appropriate

(c) Harmful side effects

(d) Do not reject H_0

(e) Not necessarily, results are inconclusive

B.75 p-value ≈ 0; Reject H_0

B.77 p-value ≈ 0.24; Do not reject H_0

B.79 Short answer not appropriate

B.81 Short answer not appropriate

UNIT C: Inference with Normal and t-Distributions
CHAPTER 5
Section 5.1 Partial Answers

5.1 0.014

5.3 0.894

5.5 0.040

5.7 $z = 3.0$

5.9 $z = -1.29$

5.11 $z = -1.0$

5.13 (a) p-value $= 0.20$

(b) p-value $= 0.0087$

(c) p-value $= 0.024$

5.15 (a) 0.309

(b) 0.115

5.17 (a) 0.788

(b) 0.945

5.19 $z = 0.645$; p-value $= 0.259$
Not a significant difference

5.21 $z = 15.161$; p-value ≈ 0.000
Very significant difference

5.23 $H_0 : p_Q = p_R$ vs $H_a : p_Q > p_R$;
$z = 3.86$; p-value $= 0.000056$; Reject H_0

5.25 $H_0 : p_R = p_U$ vs $H_a : p_R \neq p_U$;
$z = 2.507$; p-value $= 0.012$; Reject H_0

5.27 $z = 2.01$; p-value $= 0.022$; Reject H_0

5.29 $H_0 : \mu = 35$ vs $H_a : \mu > 35$;
$z = 0.833$; p-value $= 0.202$; Do not reject H_0

5.31 (a) $H_0 : p_G = p_I$ vs $H_a : p_G \neq p_I$;
$\hat{p}_G - \hat{p}_I = 0.016$

(b) p-value $= 0.286$

(c) $N(0, 0.015)$, p-value $= 0.286$

(d) $z = 1.067$; p-value $= 0.286$

(e) Do not reject H_0

5.33 (a) $z = 1.49$

(b) p-value $= 0.068$

(c) The randomization distribution is not symmetric

Section 5.2 Partial Answers

5.35 (a) $z^* = 1.476$

(b) $z^* = 1.881$

(c) $z^* = 2.054$

5.37 68.668 to 75.332

5.39 0.703 to 0.857

5.41 0.868 to 27.132

5.43 0.18 to 0.21

5.45 54.8 and 60.3

5.47 0.09 to 0.45 higher using quizzes

5.49 0.210 to 0.336

5.51 (a) 0.191; 0.110; 0.081

(b) 0.032 to 0.130

(c) No

5.53 (a) Around (11.9, 20.4)

(b) Around (11.7, 20.3)

5.55 (a) $SE \approx 0.036$

(b) (0.749, 0.865)

5.57 (a) -0.158 to 0.898

(b) Bootstrap distribution is slightly skewed

CHAPTER 6
Section 6.1-D Partial Answers

6.1 $SE = 0.061$

6.3 $SE = 0.039$

6.5 $SE = 0.016$

6.7 0.089; 0.035; 0.015;
SE goes down; Accuracy is better

6.9 (a) Yes

(b) Yes

(c) No

(d) No

Section 6.1-CI Partial Answers

6.11 $\hat{p} = 0.38$; $ME = 0.043$;
95% CI is 0.337 to 0.423

6.13 $\hat{p} = 0.689$; $ME = 0.126$;
99% CI is 0.563 to 0.815

6.15 $n \geq 385$

6.17 $n \geq 632$

6.19 (a) 0.271 to 0.429

(b) No; Yes

6.21 0.167 to 0.233; $ME = 0.033$

6.23 (a) 0.675 to 0.705

(b) 0.025 to 0.035

(c) Part (a)

6.25 (a) Update status: 0.128 to 0.172;
Comment on another's content: 0.194 to 0.246;
Comment on another's photo: 0.175 to 0.225;
"Like" another's content: 0.232 to 0.288;
Send a private message: 0.081 to 0.119

(b) No; intervals do not overlap

6.27 0.0081; 0.0078; Yes

6.29 About 0.045 both ways; Matches very closely

6.31 About 0.014 both ways; Matches very closely

6.33 0.40 to 0.56; Both give similar results

6.35 $n = 1844$; $n = 1068$; $n = 752$;
For a higher level of confidence, need larger n

6.37 (a) Between 0.291 and 0.349
(b) $ME = \pm 2.9\%$
(c) $n \geq 8360$

6.39 $n \geq 846$

6.41 $n = 2500$

6.43 $n = 400$

6.45 0.745 to 0.855

Section 6.1-HT Partial Answers

6.47 $z = -2.78$; p-value $= 0.003$; Reject H_0

6.49 $z = 1.41$; p-value $= 0.079$; Do not reject H_0

6.51 $z = 4.74$; p-value ≈ 0; Reject H_0

6.53 (a) $H_0 : p_l = 0.10$ vs. $H_a : p_l \neq 0.10$
(b) $z = 1.79$; $p = 0.074$
(c) Reject H_0 at 10%, but not at 5%

6.55 Test statistic $= 4.83$; p-value ≈ 0;
Strong evidence that $p > 0.50$

6.57 $z = 0.95$; p-value $= 0.171$; Do not reject H_0

6.59 $H_0 : p = 0.5$ vs $H_a : p \neq 0.5$;
$z = 1.26$; p-value $= 0.208$; Do not reject H_0

6.61 $z = -2.82$; p-value $= 0.0048$; Reject H_0

Section 6.2-D Partial Answers

6.63 $SE = 0.158$

6.65 $SE = 12.65$

6.67 ± 1.83

6.69 ± 2.06

6.71 0.0349

6.73 0.165

6.75 4.56; 1.77; 0.79;
SE goes down; Accuracy is better

6.77 Appropriate; $df = 11$; $SE = 0.46$

6.79 Not appropriate

Section 6.2-CI Partial Answers

6.81 $\bar{x} = 12.7$; $ME = 2.1$; 10.6 to 14.8

6.83 $\bar{x} = 3.1$; $ME = 0.066$; 3.034 to 3.166

6.85 $\bar{x} = 46.1$; $ME = 12.85$; 33.25 to 58.95

6.87 $n \geq 50$

6.89 $n \geq 6766$

6.91 1.856 to 2.944

6.93 (a) $ME = 0.08$
(b) 2.12 to 2.28

6.95 539.7 to 588.3 million

6.97 Probably not; Skewed and outliers;
Try a bootstrap distribution

6.99 About 5.5 both ways; Matches closely

6.101 About 2.2 both ways; Matches closely

6.103 16.95 to 19.37; Methods give similar results

6.105 (a) 73.27 and 80.79 grams
(b) $ME = \pm 3.76$ grams

6.107 (a) 269 to 311 particles per liter
(b) ± 21.0 particles
(c) $n \geq 2037$

6.109 $n = 35$; $n = 139$; $n = 3458$;
Smaller margin of error requires larger sample size

6.111 $n = 4269$; $n = 1068$; $n = 43$;
High variability requires larger sample size

6.113 12.20 grams to 13.38 grams

Section 6.2-HT Partial Answers

6.115 $t = -3.64$; p-value $= 0.0005$; Reject H_0

6.117 $t = 1.27$; p-value $= 0.115$; Do not reject H_0

6.119 $t = -4.99$; p-value ≈ 0; Reject H_0

6.121 $t = 4.18$; p-value $= 0.003$; Reject H_0

6.123 Yes, $t = -3.67$, p-value $= 0.0018$

6.125 (a) Yes; $t = 2.36$; p-value $= 0.0126$
(b) Yes; $t = 3.01$; p-value $= 0.0027$
(c) No; $t = -0.47$; p-value $= 0.6791$
(d) New Jersey

6.127 (a) $n = 30$; $\bar{x} = 0.2511$; $s = 0.01096$
(b) $t = -4.45$; p-value $= 0.00012$; Reject H_0
(c) The same up to round-off

6.129 (a) $t = -10.09$; p-value ≈ 0; Reject H_0
(b) $t = 0.58$; p-value $= 0.718$; Do not reject H_0

Section 6.3-D Partial Answers

6.131 (a) $SE = 0.086$
(b) Normal curve applies

6.133 (a) $SE = 0.076$
(b) Normal curve applies

6.135 (a) $SE = 0.106$
(b) Normal curve is not appropriate

6.137 (a) One group; No
(b) Two groups; Yes
(c) One group; No
(d) Two groups; Yes

Section 6.3-CI Partial Answers

6.139 $\hat{p}_1 - \hat{p}_2 = 0.04$; $ME = 0.066$;
95% CI for $p_1 - p_2$ is -0.026 to 0.106

6.141 $\hat{p}_1 - \hat{p}_2 = -0.14$; $ME = 0.11$;
99% CI for $p_1 - p_2$ is -0.25 to -0.03

6.143 0.151 to 0.229 higher for Internet users

6.145 0.301 to 0.335 lower with electronic; No

6.147 0.050 to 0.118

6.149 0.23 to 0.01 lower for metal

6.151 About 0.049 both ways; Matches very closely

6.153 0.155 to 0.265; Similar results

6.155 -0.132 to 0.098

Section 6.3-HT Partial Answers

6.157 (a) $\hat{p}_A = 0.768$, $\hat{p}_B = 0.463$, $\hat{p} = 0.631$
(b) $z = 3.84$; p-value ≈ 0; Reject H_0

6.159 (a) $\hat{p}_m = 0.24, \hat{p}_f = 0.32, \hat{p} = 0.28$
(b) $z = -0.89$; p-value $= 0.187$; Do not reject H_0

6.161 (a) $\hat{p}_T = 0.48, \hat{p}_C = 0.28, \hat{p} = 0.38$
(b) $z = 2.52$; p-value $= 0.006$; Reject H_0

6.163 $z = -1.55$; p-value $= 0.0606$;
Do not reject H_0 at 5% level

6.165 $z = -4.99$; p-value ≈ 0; Reject H_0
Experiment $\Longrightarrow$ can infer causation

6.167 $z = 1.675$; p-value $= 0.047$
Significantly higher

6.169 $z = 7.818$; p-value ≈ 0.000
Very significant difference

6.171 Sample size is too small

6.173 (a) -2.30
(b) 0.011
(c) 3.43
(d) 0.0003
(e) Yes

6.175 HRT increases risk; $z = 2.07$; p-value $= 0.038$

6.177 HRT decreases risk; $z = -4.77$, p-value ≈ 0

6.179 $z = 1.07$; p-value $= 0.285$; Do not reject H_0

Section 6.4-D Partial Answers

6.181 $SE = 1.41$

6.183 $SE = 1.431$

6.185 ± 1.89; $df = 7$

6.187 0.0298; $df = 11$

Section 6.4-CI Partial Answers

6.189 $\bar{x}_1 - \bar{x}_2 = -2.3$; $ME = 1.46$;
95% CI is -3.76 to -0.84

6.191 $\bar{x}_1 - \bar{x}_2 = 0.3$; $ME = 3.09$;
95% CI is -2.79 to 3.39

6.193 1.61 to 4.59

6.195 (a) Short answer not appropriate
(b) 0.32 to 4.20
(c) No

6.197 $(-21.67, -5.59)$

6.199 -0.008 to 0.578

6.201 -0.130 to 0.486

6.203 SE$\approx$ 1.12 both ways; Matches closely

6.205 (a) 168; 193
(b) Males; 1.766 hours per week more
(c) -2.93 to -0.60
(d) Same to two decimal places
(e) Short answer not appropriate

6.207 -2.356 to 0.912 grams; Yes

Section 6.4-HT Partial Answers

6.209 $t = 2.70$; p-value $= 0.006$; Reject H_0

6.211 $t = 0.96$; p-value $= 0.37$; Do not reject H_0

6.213 $t = 3.234$; p-value $= 0.0013$;
Reject H_0; Take notes longhand!

6.215 $t = 3.316$; p-value $= 0.0008$; Reject H_0

6.217 $t = -11.36$; p-value $= 0.0073$

6.219 $t = -6.80$; p-value $= 0.00025$; Reject H_0

6.221 (a) Experiment
(b) $H_0 : \mu_T = \mu_C$ vs $H_a : \mu_T > \mu_C$
(c) $t = 2.07$; p-value $= 0.0342$; Reject H_0
(d) Possibly not normal
(e) p-value ≈ 0.029
(f) Drinking tea enhances immune response

6.223 $t = 4.44$; p-value ≈ 0; Reject H_0

6.225 $t = 0.16$, p-value$= 0.874$, Do not reject H_0

6.227 $t = 1.87$, p-value $= 0.066$; Do not reject H_0

6.229 $H_0 : \mu_f = \mu_m$ vs $H_a : \mu_f \neq \mu_m$;
$t = -2.98$; p-value $= 0.003$; Reject H_0

6.231 $t = -2.45$; p-value $= 0.014$; Reject H_0

Section 6.5 Partial Answers

6.233 $\bar{x}_d = 556.9$; $ME = 23.8$;
90% CI is 533.1 to 580.7

6.235 $\bar{x}_d = -3.13$; $ME = 6.46$;
95% CI is -9.59 to 3.33

6.237 $t = -2.69$; p-value $= 0.016$; Reject H_0

6.239 $t = 1.36$; p-value $= 0.103$; Do not reject H_0

6.241 Paired data

6.243 Paired data

6.245 Paired data

6.247 $t = 2.71$; p-value $= 0.027$; Reject H_0

6.249 4.1 to 39.3 pg/ml reduction in testosterone

6.251 $t = 1.52$, p-value $= 0.0814$; Do not reject H_0

6.253 (a) Short answer not appropriate
(b) 0.271 to 0.713 higher with a spoiler

6.255 $t = 1.30$; p-value $= 0.200$;
Do not reject H_0

6.257 (a) $t = -1.09$; p-value $= 0.152$; Do not reject H_0
(b) $t = -2.71$; p-value $= 0.012$; Reject H_0
(c) Short answer not appropriate

UNIT C: Essential Synthesis Partial Answers

C.1 CI for a mean

C.3 Test for difference in proportions

C.5 CI for difference in means

C.7 Test for a proportion

C.9 (a) Small
(b) Reject H_0

C.11 (a) Large
(b) Do not reject H_0

C.13 (a) Small
(b) Reject H_0

C.15 $z = 2.14$; p-value $= 0.0162$; Reject H_0

C.17 $z = 2.39$; p-value $= 0.0168$; Reject H_0

C.19 $t = 4.45$; p-value ≈ 0; Reject H_0

C.21 (a) $t = 0.36$; p-value $= 0.73$; Do not reject H_0
(b) $t = 2.16$; p-value $= 0.0338$; Reject H_0

(c) $t = -2.28$; p-value $= 0.0283$; Reject H_0

(d) $t = -3.83$; p-value $= 0.0064$; Reject H_0

(e) Short answer not appropriate

C.23 Test mean; $t = -3.71$; p-value $= 0.0038$; Reject H_0

C.25 Test difference in proportions; $z = 2.0$; p-value $= 0.023$; Reject H_0

C.27 (a) $\bar{x} = 98.92$; p-value ≈ 0; Reject H_0

(b) $\hat{p} = 0.2$; 0.147 to 0.262

(c) Females; p-value $= 0.001$; Reject H_0

(d) p-value $= 0.184$; Do not reject H_0

(e) p-value $= 0.772$; Do not Reject H_0

UNIT C: Review Exercise Partial Answers

C.29 (a) 0.018

(b) 0.106

C.31 (a) $z = 0.253$

(b) $z = -2.054$

C.33 ± 1.711

C.35 0.011

C.37 (a) 12.20 to 13.38

(b) $H_0 : \mu = 12$ vs $H_a : \mu > 12$; $z = 2.63$; p-value $= 0.004$; Reject H_0

C.39 0.807 to 0.853; 0.023; No; Yes

C.41 (a) $n = 157$; $\bar{x} = 3.849$; $s = 2.421$

(b) $SE = 0.193$; Same as in computer output

(c) \$3.466 to \$4.232

(d) Same up to round off

(e) Short answer not appropriate

C.43 (a) To avoid sampling bias

(b) $t = 14.4$; p-value ≈ 0

(c) No, test is about the *mean*

C.45 (a) Internet users

(b) $z = 7.49$; p-value ≈ 0; Reject H_0

(c) No, not an experiment

(d) Yes

C.47 -257.8 to 556.8

C.49 -96.5 to 450.6

C.51 $t = 0.82$; p-value $= 0.416$; Do not reject H_0

C.53 $t = 2.61$; p-value $= 0.006$; Reject H_0

C.55 (a) $\hat{p} = 0.17$; ME$= 0.03$

(b) $n \geq 9364$

C.57 0.0003 to 0.0081

C.59 -0.0046 to 0.0096

C.61 (a) No, sample size is too small

(b) Randomization p-value ≈ 0.01; Reject H_0

C.63 $t = -1.69$; p-value $= 0.102$; No

C.65 0.02 and 6.98 g more gained in light

C.67 1.70 to 4.84 in favor of home teams

C.69 0.047 to 0.219 higher for males

C.71 (a) $H_0 : \mu_1 = \mu_2$ vs $H_a : \mu_1 > \mu_2$, where μ_1 is "in class" and μ_2 is "missed"

(b) $t = 3.02$; p-value $= 0.008$; Reject H_0

(c) No, not an experiment.

(d) Yes

C.73 Mean $= 0.651$; $SE = 0.067$

C.75 (a) Mean $= 36.78$ yrs; $SE = 7.14$ yrs

(b) Mean $= 36.78$ yrs; $SE = 2.258$ yrs

(c) Mean $= 36.78$ yrs; $SE = 0.714$ yrs

C.77 Mean $= 0.005$; $SE = 0.039$

C.79 $N(0.037, 0.041)$

C.81 $t = 6.39$; p-value ≈ 0; Reject H_0

C.83 (a) $z = 2.52$; p-value $= 0.006$; Reject H_0

(b) $z = -2.23$; p-value $= 0.0258$; Reject H_0

C.85 0.532 to 0.760

C.87 $z = 0.83$; p-value $= 0.406$; Do not Reject H_0

C.89 0.588 to 0.652

C.91 (a) -0.186 to -0.014; Yes

(b) $z = -2.17$; p-value $= 0.030$; Reject H_0

(c) No; observational study

C.93 (a) 2006 people; All US adults

(b) Observational study; No

(c) $t = 9.91$; p-value ≈ 0; Reject H_0

(d) Answers will vary

C.95 -4.11 to 3.29

C.97 $t = 9.97$; p-value ≈ 0; Reject H_0

C.99 $z = -6.95$; p-value ≈ 0; Reject H_0

C.101 $t = 21.6$; p-value ≈ 0; Reject H_0

C.103 (a) $\hat{p} = 0.178$; $SE = 0.04$

(b) Yes

(c) 0.100 to 0.256

C.105 (a) $\hat{p}_M - \hat{p}_C = -0.055$

(b) -0.200 to 0.078

(c) No

C.107 (a) Teens: 114.32 to 137.98; Eighties: 115.01 to 149.53; Eighties; Much greater variability in eighties

(b) $t = -0.63$; p-value $= 0.540$; Do not Reject H_0

C.109 9.44 to 18.56 points

C.111 Short answer not appropriate

C.113 Test $H_0 : \mu_c = \mu_s$ vs $H_a : \mu_c \neq \mu_s$; $t = 0.36$; p-value $= 0.722$, Do not reject H_0

C.115 (a) $\bar{x}_w = 31.8$; CI is 29.8 to 33.9 years old

(b) $\bar{x}_h = 34.7$; CI is 32.3 to 37.0 years old

(c) Short answer not appropriate

C.117 $z \approx 2.35$; p-value ≈ 0.009.

C.119 (a) $SE = 0.0354$

(b) $z = -2.26$

(c) 100

(d) $SE = 7.081$

(e) $z = -2.26$

(f) p-value $= 0.024$

UNIT D: Inference for Multiple Paramaters
CHAPTER 7

Section 7.1 Partial Answers

7.1 125; 125; 125; 125

7.3 100; 50; 50

7.5 $\chi^2 = 6.45$; p-value = 0.0398

7.7 $\chi^2 = 8.38$; p-value = 0.039

7.9 (a) 40

    (b) 0.4

    (c) $df = 3$

7.11 (a) 700

    (b) 1.46

    (c) $df = 5$

7.13 (a) $H_0 : p_g = p_o = p_p = p_r = p_y = 0.2$

       H_a : Some $p_i \neq 0.2$

    (b) 13.2

    (c) 4

    (d) 3.70

    (e) p-value = 0.449; Do not reject H_0

7.15 (a) $\chi^2 = 190.2$; p-value ≈ 0; Reject H_0

    (b) Age 65+; observed < expected

7.17 (a) 32,968

    (b) 8044.2; 8505.7; 8472.8; 7945.3

    (c) $\chi^2 = 382.4$

    (d) $df = 3$; p-value ≈ 0

    (e) Short answer not appropriate

7.19 $\chi^2 = 82.6$; p-value ≈ 0; Reject H_0

7.21 (a) χ^2 goodness-of-fit test

    (b) Large

    (c) Small

    (d) Monthly deaths due to medication errors

    (e) July

    (f) Observed

7.23 (a) Yes, $\chi^2 = 172.0$; p-value ≈ 0

    (b) Yes, $\chi^2 = 120.2$; p-value ≈ 0

    (c) No, not a random sample

7.25 (a) $\chi^2 = 6.20$; p-value = 0.0128; Reject H_0

    (b) $z = 2.490$; p-value = 0.0128; Reject H_0

    (c) Same

7.27 $\chi^2 = 26.66$; p-value = 0.0008; Reject H_0

7.29 (a) $\chi^2 = 17.87$; p-value = 0.037; Reject H_0

    (b) $\chi^2 = 61.68$; p-value ≈ 0; Reject H_0

Section 7.2 Partial Answers

7.31 Expected = 65; Contribution = 0.754

7.33 Expected = 10; Contribution = 2.5

7.35 $df = 2$

7.37 $df = 4$

7.39 $\chi^2 = 8.20$; p-value = 0.017; Reject H_0

7.41 (a) Short answer not appropriate

    (b) Short answer not appropriate

    (c) Short Answer not appropriate

    (d) $\chi^2 = 3.17$

    (e) p-value = 0.075; Do not reject H_0

7.43 $\chi^2 = 6.08$; p-value = 0.0478; Reject H_0;

    Miscarriages higher than expected with NSAIDs;

    Cannot assume causation

7.45 (a) $\chi^2 = 80.53$; p-value ≈ 0; yes

    (b) $\chi^2 = 21.6$; p-value = 0.00002; yes

7.47 (a) Architect; Orthopedic surgeon

    (b) $\chi^2 = 19.0$; $p = 0.015$

    (c) Reject H_0 at 5%, but not at 1%

7.49 (a) Expected = 22.4

    (b) $\chi^2 = 210.9$; p-value ≈ 0; Reject H_0

7.51 $\chi^2 = 10.45$; p-value = 0.034; Reject H_0

7.53 (a) $n = 194$

    (b) 61.70; 3.792

    (c) $df = 2$

    (d) $\chi^2 = 10.785$; p-value = 0.005; Reject H_0

    (e) (Sprint, X); Observed less than expected

    (f) Sprinters: R; Endurance: X

7.55 (a) Short answer not appropriate

    (b) 5

7.57 $\chi^2 = 7.137$; p-value = 0.028; Some evidence of an association

CHAPTER 8

Section 8.1 Partial Answers

8.1 Dataset B; less variability within groups

8.3 Dataset A; means farther apart

8.5 Dataset B; less variability within groups

8.7 $F = 2.55$

8.9 $F = 0.8$

8.11 (a) 4 groups

    (b) $H_0 : \mu_1 = \mu_2 = \mu_3 = \mu_4$ vs H_a : Some $\mu_i \neq \mu_j$

    (c) p-value = 0.229

    (d) Do not reject H_0

8.13 (a) 3 groups

    (b) $H_0 : \mu_1 = \mu_2 = \mu_3$ vs H_a : Some $\mu_i \neq \mu_j$

    (c) p-value = 0.0013

    (d) Reject H_0

8.15 (a) Treatment group (categorical);

       Change in cortisol level (quantitative)

    (b) Experiment

    (c) $H_0 : \mu_1 = \mu_2 = \mu_3 = \mu_4$ vs H_a : Some $\mu_i \neq \mu_j$

    (d) 67; 3; 64

    (e) p-value less than 0.05

8.17 (a) $H_0 : \mu_r = \mu_g = \mu_b$ vs H_a : Some $\mu_i \neq \mu_j$

    (b) $F = 16.5$

    (c) p-value ≈ 0

    (d) Reject H_0

8.19 (a) Yes; EE

    (b) $F = 22.76$; p-value ≈ 0; Reject H_0

8.21 (a) SD lower for all three groups;

       EE highest for both HC and SD

    (b) $F = 11.3$; p-value ≈ 0; Reject H_0

8.23 $F = 3.29$; p-value = 0.042; Reject H_0

8.25 (a) Draw an image; Write the word

    (b) $F = 6.40$

    (c) p-value = 0.0036

    (d) Reject H_0, strong evidence of a difference

    (e) It matters; Draw an image!

8.27 (a) Draw an image; Write the word
(b) $F = 3.83$
(c) p-value = 0.032
(d) Reject H_0; Evidence of a difference
(e) It matters; Draw an image!

8.29 (a) Short answer not appropriate
(b) $F = 8.38$; p-value = 0.002; Reject H_0
(c) Short answer not appropriate
(d) Yes, results from a randomized experiment

8.31 (a) No, standard deviations too different
(b) Do not reject H_0

8.33 (a) LD: 36.0% day, 64.0% night;
DM: 55.5% day, 44.5% night
(b) p-value = 0.000; Reject H_0; Yes

8.35 (a) No football; Football with concussion
(b) F-statistic = 31.47 ; p-value = 0.000
(c) Reject H_0; Strong evidence of a difference

8.37 Conditions are reasonable.
F = 27.86; p-value $\approx$ 0.000; Reject H_0.

8.39 (a) Drug resistance: associated;
Health: not associated
(b) Resistance density: $\approx$ 0;
Days infectious: 0.0002;
Weight: 0.906; RBC: 0.911
(c) Short answer not appropriate
(d) Short answer not appropriate

Section 8.2 Partial Answers

8.41 Yes, p-value = 0.002
8.43 7.77 to 12.63
8.45 $H_0 : \mu_A = \mu_C$ vs $H_a : \mu_A \neq \mu_C$;
$t = -0.38$; p-value = 0.71; Do not reject H_0
8.47 $\sqrt{MSE} = 6.95$; $df = 20$
8.49 2.28 to 19.06
8.51 $t = 2.66$; p-value = 0.015; Reject H_0
8.53 $t = -2.59$; p-value = 0.017; Reject H_0.
8.55 DM vs LD: $t = 1.60$; p-value = 0.122
DM vs LL: $t = -2.69$; p-value = 0.0126
LD vs LL: $t = -4.19$; p-value = 0.0003
8.57 (a) Yes, ANOVA p-value $\approx$ 0;
IE:HC vs SE:SD; IE:HC vs SE:HC
(b) Short answer not appropriate
(c) $t = -1.57$; p-value = 0.124; Do not reject H_0
8.59 $t = 2.36$, p-value = 0.019, Reject H_0
8.61 LSD = 50.1;
IE:HC,SE:HC,EE:HC,EE:SD < IE:SD,SE:SD

CHAPTER 9
Section 9.1 Partial Answers

9.1 $\beta_0 \approx b_0 = 29.3$; $\beta_1 \approx b_1 = 4.30$;
$\widehat{Y} = 29.3 + 4.30 \cdot X$
9.3 $\beta_0 \approx b_0 = 77.44$; $\beta_1 \approx b_1 = -15.904$;
$\widehat{Y} = 77.44 - 15.904 \cdot Score$
9.5 $b_1 = -8.20$; $H_0 : \beta_1 = 0$ vs $H_a : \beta_1 \neq 0$;
p-value = 0.000; Reject H_0

9.7 $b_1 = -0.3560$; $H_0 : \beta_1 = 0$ vs $H_a : \beta_1 \neq 0$;
p-value = 0.087; Do not reject H_0
9.9 -10.18 to -6.22
9.11 $t = 1.98$; p-value = 0.029; Reject H_0
9.13 $t = 2.89$; p-value = 0.0048; Reject H_0
9.15 (a) *Height*, *Weight*; $r = 0.619$; p-value $\approx$ 0;
Taller people tend to weigh more
(b) *GPA*, *Weight*; $r = -0.217$; p-value $\approx$ 0;
Heavier people tend to have lower GPA
(c) *Exercise*, *TV*; $r = 0.010$; p-value = 0.852
9.17 (a) No concerns
(b) $\widehat{GPA} = 3.26$
(c) $b_1 = 0.00189$
(d) $t = 6.99$; p-value $\approx$ 0, Reject H_0
(e) $R^2 = 12.5\%$
9.19 (a) One person; Below the mean;
140 FB friends
(b) No concerns
(c) $r = 0.436$; p-value = 0.005; Reject H_0
(d) $\widehat{FBfriends} = 367 + 82.4 \cdot GMdensity$;
367; 449.4; 284.6
(e) p-value = 0.005; Same
(f) $R^2 = 19.0\%$
9.21 (a) 0.618
(b) $b_1 = -0.152$
(c) $t = -5.02$; p-value $\approx$ 0; Reject H_0
(d) -0.213 to -0.091
(e) $R^2 = 33.1\%$
9.23 (a) R^2
(b) Response = *Prevalence*;
Explanatory = *Precipitation*
(c) $r = 0.889$
9.25 (a) t-statistic = -1.80; p-value = 0.09; No
(b) 16.8%
9.27 (a) Countries
(b) Reasonable, perhaps negative residuals are too large
(c) $b_1 = 0.729$
(d) -0.003 to 1.461
(e) p-value = 0.051; Do not reject H_0
(f) $\beta_1 = 0.467$ is in the CI
(g) $R^2 = 7.71\%$
9.29 (a) Yes
(b) Yes, $\rho = -0.857$
(c) Data on entire population
(d) $\beta_1 = -0.7379$
(e) No, not an experiment

Section 9.2 Partial Answers

9.31 $F = 21.85$; p-value = 0.000;
The model is effective
9.33 $F = 2.18$; p-value = 0.141;
The model is not effective
9.35 $n = 176$; $R^2 = 11.2\%$

9.37 $n = 344$; $R^2 = 0.6\%$

9.39 $F = 6.06$; p-value $= 0.0185$

9.41 $F = 259.76$; p-value ≈ 0

9.43 $F = 7.44$; p-value $= 0.011$;
The model is effective

9.45 (a) $\widehat{GPA} = 3.07$
(b) $n = 345$
(c) $R^2 = 12.5\%$
(d) $F = 48.84$; p-value $= 0.000$;
The model is effective

9.47 (a) One possible extreme point
(b) $\widehat{MatingActiviy} = 0.319$
(c) $t = -2.56$; p-value $= 0.033$; Reject H_0
(d) $F = 6.58$; p-value $= 0.033$; Reject H_0
(e) They are the same
(f) $R^2 = 45.1\%$

9.49 (a) $s_e = 21.3$
(b) $SE = 0.163$

9.51 (a) $s_e = 31.5$
(b) $SE = 3.11$

9.53 (a) $t = 2.04$; p-value $= 0.047$; Reject H_0
(b) $F = 4.15$; p-value $= 0.047$; Reject H_0
(c) $t = 2.04$; p-value $= 0.047$; Reject H_0
(d) p-values and conclusions are the same

9.55 (a) $\widehat{Beds} = 1.367 + 0.7465 Baths$; 3.6
(b) $t = 6.385$; p-value ≈ 0
(c) $F = 40.77$; p-value ≈ 0
(d) $R^2 = 59.3\%$

9.57 Answers will vary

Section 9.3 Partial Answers

9.59 (a) A; B
(b) 100

9.61 (a) B; A
(b) 20

9.63 (a) -0.013 to 4.783
(b) -2.797 to 7.568

9.65 (a) 143.4 to 172.4
(b) 101.5 to 214.3

9.67 (a) 62.22
(b) 95% CI for mean: II; 95% PI: I

9.69 (a) 41.79 to 85.22
(b) 47.92 to 89.29
(c) 63.33 to 132.17
(d) (a) 43.43, (b) 41.37, (c) 68.84

9.71 (a) 1.4 to 9.0
(b) -8.0 to 18.4
(c) 0.8 to 14.4

CHAPTER 10

Section 10.1 Partial Answers

10.1 $X1$, $X2$, $X3$, and $X4$; Y.

10.3 62.85; -2.85

10.5 -6.820; 0.001

10.7 $X1$

10.9 $X2$

10.11 $R^2 = 99.8\%$

10.13 13.85; 6.15

10.15 4.715; 0.053

10.17 $X3$ and $X4$

10.19 $X4$

10.21 Yes, p-value $= 0.000$

10.23 (a) $570{,}500$
(b) $Baths$, $\hat{\beta}_3 = 200$
(c) $SizeSqFt$, $t = 4.55$, p-value $= 0.000$
(d) All three
(e) Short answer not appropriate
(f) Effective; p-value $= 0.000$
(g) $R^2 = 46.7\%$

10.25 (a) Internet, BirthRate; BirthRate
(b) 69.42
(c) Increase

10.27 (a) 72.97; 1.03
(b) 76.41; -34.41
(c) -3.07; Short answer not appropriate
(d) -2.70; Short answer not appropriate
(e) Yes; ANOVA p-value $= 0.050$
(f) One; None
(g) $Years$
(h) $n = 44$
(i) 13.56%; Short answer not appropriate

10.29 47

10.31 p-value $= 0.000$; reject H_0

10.33 (a) All 3; Size
(b) Beds: -254.86; Baths: 228.92
(c) $335{,}605$

10.35 (a) $\widehat{WinPct} = 0.492 + 0.03223 \cdot PtsFor - 0.03215$
$\cdot PtsAgainst$
(b) $\widehat{WinPct} = 0.848$; Residual $= 0.042$
(c) Both p-values ≈ 0, both strong predictors
(d) Two-predictor model much better

10.37 (a) $PtsFor$: $R^2 = 45.87\%$;
$PtsAgainst$: $R^2 = 44.16\%$;
Both: $R^2 = 95.97\%$.
(b) $\widehat{WinPct} = 0.500 + 0.033 Diff$;
$R^2 = 95.97\%$

Section 10.2 Partial Answers

10.39 (c)

10.41 (a)

10.43 Conditions appear to be met

10.45 (a) Short answer not appropriate
(b) 133.66 lbs; 66.34
(c) Short answer not appropriate
(d) Conditions appear to be met

10.47 Conditions are reasonably well met

10.49 Conditions are reasonably well met

10.51 (a) $\widehat{Time} = 7.12 + 1.211 \cdot Distance$
  (b) $\widehat{Time} = 31.34$ minutes
  (c) Short answer not appropriate
  (d) Short answer not appropriate
  (e) Short answer not appropriate

10.53 (a) Short answer not appropriate
  (b) 7.235 implies speed of 166 mph!

10.55 Conditions appear to be met

10.57 Conditions are not well met

Section 10.3 Partial Answers

10.59 (a) $X3$
  (b) $R^2 = 41.7\%$; Decrease; Small decrease; Large decrease
  (c) 0.031; Decrease; Decrease; Increase
  (d) 3.81; Increase; Increase

10.61 Answers will vary

10.63 (a) Runs: 0.504, Margin: −0.116, Hits: 0.349, Errors: −0.040, Pitchers: 0.721, Walks: 0.565
  (b) $\widehat{Time} = 120.55 + 2.3632 \cdot Walks + 5.851 \cdot Pitchers$

10.65 (a) Yes; p-value ≈ 0
  (b) GDP associated with $Cell$ and $LifeExpectancy$
  (c) Yes; p-value ≈ 0

10.67 (a) 107.79
  (b) 108.34
  (c) No; p-value = 0.711

10.69 (a) Short answer not appropriate
  (b) Short answer not appropriate
  (c) Steel: 108.021; Carbon:104.45

UNIT D: Essential Synthesis Partial Answers

D.1 $\chi^2 = 12.09$; p-value = 0.002; Reject H_0

D.3 $\chi^2 = 5.818$; p-value = 0.055; Do not Reject H_0

D.5 Conditions are met;
  $F = 2.19$; p-value = 0.115; Do not Reject H_0

D.7 (a) Positive association; One outlier
  (b) $t = 7.265$; p-value ≈ 0; Reject H_0
  (c) Confounding variable is size of bill

D.9 (a) $R^2 = 83.7\%$
  (b) $F = 797.87$; p-value ≈ 0; Reject H_0

UNIT D: Review Exercise Partial Answers

D.11 p-value = 0.0970; Not significant

D.13 p-value = 0.382; Not significant

D.15 p-value = 0.035; Significant

D.17 p-value = 0.0808; Not significant

D.19 ANOVA for difference in means

D.21 Test correlation, slope, or regression ANOVA

D.23 Chi-square goodness-of-fit test

D.25 ANOVA for regression

D.27 Chi-square test for association

D.29 $\chi^2 = 8.52$; p-value = 0.0035; Reject H_0

D.31 (a) Yes (just barely)

  (b) $\chi^2 = 14.6$; p-value = 0.002; Reject H_0
  (c) Rainy season in winter

D.33 (a) $H_0 : \mu_{CA} = \mu_{NY} = \mu_{NJ} = \mu_{PA}$
      $H_a :$ Some $\mu_i \neq \mu_j$
  (b) Groups df = 3
  (c) Error df = 116
  (d) Sum of squares for error

D.35 $F = 0.69$; p-value = 0.512; Do not Reject H_0

D.37 (a) 3.98 to 4.82
  (b) 0.75 to 1.85
  (c) $t = -5.38$; p-value ≈ 0; Reject H_0

D.39 $t = 4.39$; p-value ≈ 0; Reject H_0

D.41 (a) $b_1 = 0.0831$
  (b) $t = 2.47$; p-value = 0.014; Reject H_0
  (c) $R^2 = 2.4\%$
  (d) $F = 6.09$; p-value = 0.014; Reject H_0

D.43 Conditions not met;
  Outlier positive residuals

D.45 (a) 692.2 to 759.8
  (b) 339.3 to 1112.7

D.47 (a) Short answer not appropriate
  (b) p-value = 0.000; Model is effective
  (c) $R^2 = 18.4\%$
  (d) $CognitionZscore$; $Gender$
  (e) $ClassYear, CognitionZscore, DASScore, Drinks$

D.49 (a) $t = -2.98$; p-value = 0.003; Reject H_0
  (b) $t = -2.54$; p-value = 0.011; Reject H_0

D.51 Answers will vary

Final Essential Synthesis Partial Answers

E.1 Short answer not appropriate

E.3 (a) around 0.121 to 0.157
  (b) 0.121 to 0.157
  (c) Short answer not appropriate
  (d) $\hat{p} = 0.139$; ME = 0.018
  (e) Larger sample size
  (f) 0.154 to 0.156

E.5 (a) Short answer not appropriate
  (b) $\bar{x} = \$41,494$; $M = \$29,000$; $s = \$52,248$
  (c) Short answer not appropriate
  (d) $\bar{x}_M = \$50,971, \bar{x}_F = \$32,158, \bar{x}_M - \bar{x}_F = \$18,813$
  (e) Yes; p-value = 0.0002

E.7 (a) Short answer not appropriate
  (b) Short answer not appropriate
  (c) Yes; $\chi^2 = 13.79$, p-value = 0.0032

E.9 (a) Short answer not appropriate
  (b) Yes; t = 8.48, p-value ≈ 0
  (c) $\widehat{Income} = -18.3468 + 1.5529 \cdot HoursWk$
  (d) $43,769
  (e) $R^2 = 14.35\%$
  (f) No; variability not constant

E.11 Test for difference in means

E.13 Interval for a proportion

E.15 Test for a proportion

E.17 Test for correlation

E.19 Test for diff. in proportions or chi-square test for association

E.21 Interval for a proportion

E.23 Simple linear regression

E.25 Interval for difference in means

E.27 Test for difference in proportions

CHAPTER P

Section P.1 Partial Answers

P.1 0.6

P.3 0.6

P.5 0.25

P.7 No

P.9 0.6

P.11 0.625

P.13 No

P.15 0.7

P.17 0.42

P.19 0.3

P.21 0.2

P.23 0.333

P.25 No

P.27 Disjoint

P.29 Independent

P.31 Short answer not appropriate

P.33 (a) $P(MP) = 0.680$
(b) $P(\text{not } F) = 0.845$
(c) $P(F \text{ if } MP) = 0.184$
(d) $P(\text{not } MP \text{ if not } F) = 0.344$
(e) $P(MP \text{ and not } F) = 0.554$
(f) $P(\text{not } MP \text{ or } F) = 0.446$

P.35 (a) $P(Red) = 0.1375$
(b) $P(\text{not } Blue) = 0.75$
(c) $P(Red \text{ or } Orange) = 0.2875$
(d) $P(Blue_1 \text{ and } Blue_2) = 0.0625$
(e) $P(Red_1 \text{ and } Green_2) = 0.191$

P.37 (a) 0.824
(b) 0.008
(c) 0.167

P.39 (a) 0.305
(b) 0.695
(c) 0.029

P.41 (a) 0.5595
(b) 0.3130; 0.5595
(c) 0.3099; 0.5539
(d) Answers vary

Section P.2 Partial Answers

P.43 $P(\text{I}) = 0.26$; $P(A \text{ if } \text{II}) = 0.76$;
$P(\text{II and B}) = 0.1032$; $P(\text{III and A}) = 0.248$

P.45 $P(\text{I and A}) = 0.115$; $P(\text{I}) = 0.5$; $P(\text{II}) = 0.5$;
$P(A \text{ if } \text{I}) = 0.23$; $P(B \text{ if } \text{I}) = 0.45$;
$P(C \text{ if } \text{I}) = 0.32$; $P(A \text{ if } \text{II}) = 0.9$;
$P(B \text{ if } \text{II}) = 0.05$; $P(C \text{ if } \text{II}) = 0.05$

P.47 $P(\text{B and R}) = 0.08$

P.49 $P(\text{R if A}) = 0.9$

P.51 $P(R) = 0.62$

P.53 $P(\text{A if S}) = 0.158$

P.55 (a) 0.091
(b) 0.405
(c) 0.474

P.57 $P(\text{Cancer if Positive}) = 0.2375$

P.59 (a) $P(\text{Free if Spam}) = 0.266$
(b) $P(\text{Spam if Free}) = 0.752$

P.61 $P(\text{Spam if Free and Text}) = 0.978$

Section P.3 Partial Answers

P.63 Discrete

P.65 Discrete

P.67 Continuous

P.69 0.3

P.71 0.7

P.73 0.4

P.75 0.4

P.77 Not a probability function

P.79 (a) $\mu = 14$
(b) $\sigma = 6.63$

P.81 (a) $\mu = 13$
(b) $\sigma = 2.236$

P.83 (a) Short answer not appropriate
(b) 0.623
(c) 0.110
(d) 0.638

P.85 (a) $\mu = 2.42$
(b) $\sigma = 1.525$

P.87 (a) $P(X = 4) = 0.04$
(b) $P(X < 2) = 0.59$
(c) $\mu = 1.37$ cars
(d) $\sigma = 1.180$ cars

P.89 0.0244

P.91 (a) Short answer not appropriate
(b) $\mu = 4.9296c - \$3845$
(c) $c = \$779.98$

P.93 (a) $p(1) = 0.301$; $p(9) = 0.046$
(b) $P(X > 2) = 0.523$

Section P.4 Partial Answers

P.95 Binomial

P.97 Not binomial

P.99 Binomial

P.101 5040

P.103 720

P.105 10

P.107 6

P.109 0.383

P.111 0.194

P.113 $\mu = 8$; $\sigma = 1.265$

P.115 $\mu = 200$; $\sigma = 12.25$

P.117 $p(0) = 0.316$; $p(1) = 0.422$;
$p(2) = 0.211$; $p(3) = 0.047$; $p(4) = 0.004$

P.119 0.099; 0.026

P.121 $\mu = 1.47$; $\sigma = 0.866$

P.123 $\mu = 3.3$; $\sigma = 1.55$

P.125 $\mu = 13.0$; $\sigma = 2.13$

P.127 (a) $\mu = 28.8$ people

(b) 0.844

Section P.5 Partial Answers

P.129 62%

P.131 95%

P.133 (a) 0.8508

(b) 0.9332

(c) 0.1359

P.135 (a) 0.982

(b) 0.309

(c) 0.625

P.137 (a) -1.282

(b) -0.8416

(c) ± 1.960

P.139 (a) -1.28

(b) 0.385

P.141 (a) 0.691

(b) 0.202

(c) 0.643

P.143 (a) 0.023

(b) 0.006

(c) 0.700

P.145 (a) 59.3

(b) 2.03

(c) 60.8 and 139.2

P.147 (a) 110

(b) 9.88

P.149 $z = -1.6$

P.151 $x = 13.3$, $z = 1.64$

P.153 $z = -1.28$; $x = 397.6$

P.155 $x = 115$ and $x = 130$

P.157 Short answer not appropriate

P.159 (a) 38th percentile

(b) 631

P.161 (a) 0.495

(b) 66.2 inches

P.163 $Q_1 = 2.89$, $Q_3 = 3.43$

P.165 (a) 0.0565

(b) 0.0012

(c) 0.5578

P.167 (a) 0.0509 or 5.09%

(b) 0.138 or 13.8%

(c) Grades below 53.9

(d) Grades above 86.1

P.169 (a) 0.954

(b) 0.683

(c) 0.997

(d) Yes

GENERAL INDEX